McGraw-Hill Encyclopedia of Science and Technology

12 | sab
spin

McGraw-Hill Encyclopedia

McGRAW-HILL BOOK COMPANY INC.

NEW YORK CHICAGO SAN FRANCISCO DALLAS TORONTO LONDON

# of Science and Technology

AN INTERNATIONAL REFERENCE WORK

IN FIFTEEN VOLUMES INCLUDING AN INDEX

VOLUME 12 SAB-SPIN

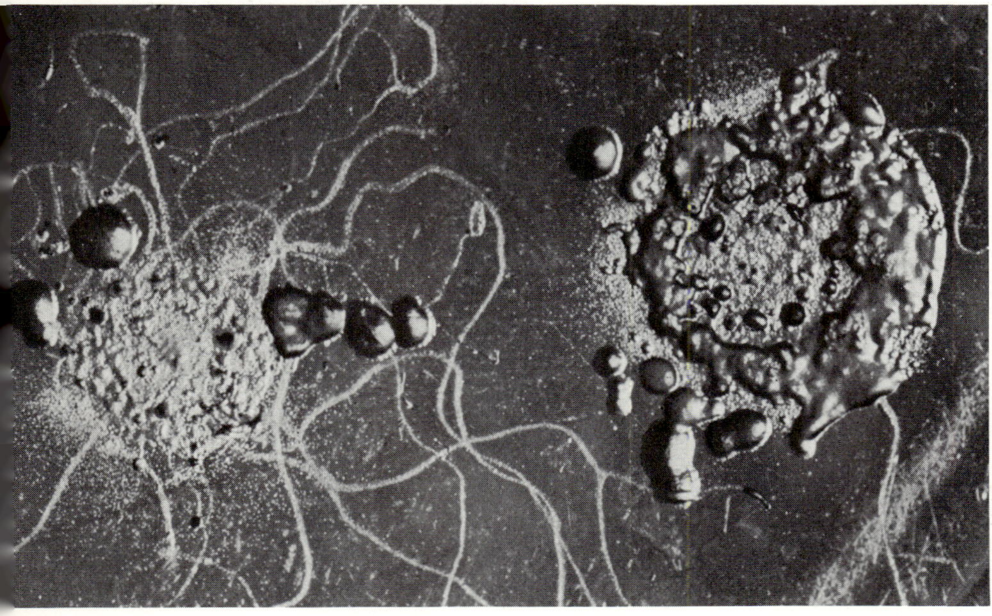

(LEFT) A platinum replica of a noninfectious mutant tuberculosis bacillus under an electron microscope. X6800 *(Battelle Memorial Institute)*. (RIGHT) Bacterial colonies being devoured by slime molds *(photograph by R. Vishniac)*.

Library of Congress Catalog Card Number: 60-11000

Suggestions to the Readers

The basic plan of the Encyclopedia is explained here in order to facilitate its use.

The subject matter of the various disciplines or branches of science and technology is organized systematically: a general article provides a broad survey of the field, and a number of separate articles, alphabetically arranged, cover its main subdivisions and more specific aspects.

Cross references guide the reader from the general articles to the other articles into which the subject is subdivided, and from these to articles on more highly specialized phases of the subject. The cross references—there are about 50,000 of them—are printed in small capital letters so that they can be easily recognized. By means of the cross references a reader may find his way from ELECTRICAL ENGINEERING, through ELECTRONICS and VACUUM TUBE, to ELECTRON MOTION IN VACUUM or ELECTRON EMISSION. Or, following another line of cross references, the reader would be led to ELECTRIC POWER SYSTEMS, TRANSMISSION LINES, ELECTROMAGNETIC WAVE, and so on.

In general, each article begins with a definition of the title that states its scope and coverage. Usually, only the scientific or technological sense is discussed. Most of the articles, after this statement, go on to increasingly complex and detailed considerations. A reader thus needs to proceed only as far as his inclinations and requirements dictate.

The Index, Volume 15, should be consulted to locate the discussion of topics covered in the Encyclopedia but not given in separate entries.

Every phylum, class, and order in the plant and animal kingdoms is allotted a separate article. Many of the more common families, genera, and species are covered either in one of the order articles or in a separate article under its own scientific or common name.

The adjectives electric and electrical are used in the following senses. Electric—containing, producing, arising from, actuated by, or carrying electricity, or capable of doing so; as, for instance, electric generator, electric motor, electric wiring. Electrical—related to, pertaining to, or associated with electricity, but not having its properties or characteristics; as, for example, electrical code, electrical engineering.

Words used as titles are, wherever possible, given in the singular to permit a consistent alphabetic arrangement. Titles are alphabetized by word and not by letter; for example,

Earth sciences
Earth tides
Earthmover
Earthquake

A word used as a noun precedes the same word used adjectivally; thus:

Mercury (element)
Mercury (planet)
Mercury battery

or

Circuit, electronic
Circuit breaker

Hyphened terms are alphabetized as single words; for example,

Animal virus
Animal-feed composition

Most of the longer articles contain bibliographies citing useful sources of further information. For additional bibliographical citations, the reader should refer to related articles (as indicated by the cross references in the article). Bibliographies are placed at the ends of articles or sometimes at the ends of major sections in long articles.

A list of initials and names of the contributors to the Encyclopedia is to be found in Volume 15. This list will permit quick identification of a contributor's initials after an article. Immediately following this list is a second list of encyclopedia contributors with their affiliations and the titles of articles each has written for the Encyclopedia.

McGraw-Hill Encyclopedia of Science and Technology

S *Sable to Spinulosa*

Sable

A North American carnivore, *Martes americana*, of the family Mustelidae. This animal, also called marten, is found in the dense coniferous forest across Canada and Alaska, southward to northern New York and New England, and in the western mountains. The marten is an active, arboreal predator, about 24–27 in. long, yellowish-brown above, slightly paler below, and with a buffy throat patch. The fur is dense, soft, and of high quality. It is rather easily trapped, and is now rare, or extinct, over much of its original range. In spite of their size, martens are agile, successfully pursuing squirrels through the tree tops but also preying upon mice and other animals. *See* CARNIVORA. [J.D.B.]

The sable, or marten, *Martes americana;* length 17 in. (*From E. L. Palmer, Fieldbook of Natural History, McGraw-Hill, 1949*)

Saccharin

An organosulfur compound first prepared by Ira Remsen, and also called *o*-sulfobenzoic imide (II). The material used as a sweetening agent (about 500–700 times as sweet as cane sugar) is the sodium salt (III), which passes largely unchanged through the body and is excreted in the urine. The slightly bitter aftertaste of sodium saccharin is mainly that of impurities from the conventional synthesis, and can reputedly be avoided by new syntheses (starting from anthranilic acid or benzothiophene). Saccharin is used in food preparation for low-caloric diets and in diabetes therapy, where normal sugars cannot be tolerated. Sodium saccharin is also compounded with the more recent sweetening agent, Succaryl. *See* ORGANOSULFUR COMPOUND; SULFAMATE. [N.K.]

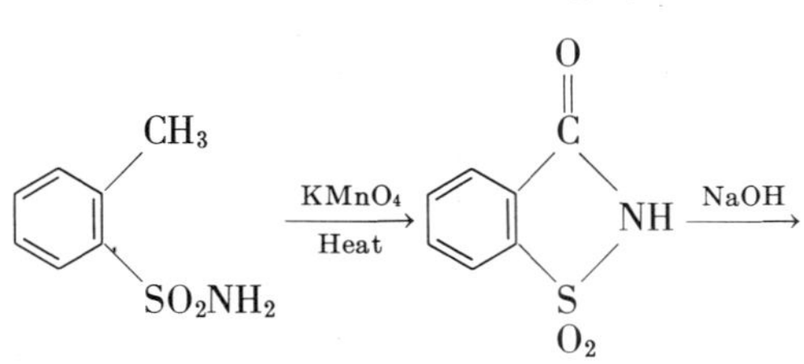

(I) *o*-Toluenesulfonamide (II)

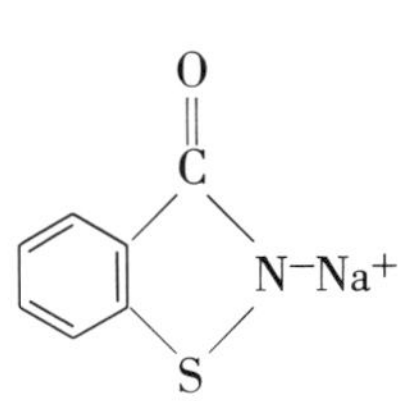

(III)

Saccharomycetales

An order in the subclass Hemi-Ascomycetes, which accommodates the ascosporogenous yeasts. The name Endomycetales is a synonym which should be discontinued in favor of Saccharomycetales for reason of priority. The order has one family, the Saccharomycetaceae. The order and family are characterized by the presence of naked asci in which spores are formed by free cell formation; diploidization immediately precedes ascus formation or occurs directly after or during ascospore germination; a dikaryotic condition and ascogenous hyphae are absent. The family is usually divided into five subfamilies, mainly on the basis of vegetative reproduction. *See* ASCOMYCETES; YEAST.

Eremascoideae. This subfamily contains only the genus *Eremascus*. The mycelia are mostly septate and the asci are spherical with eight oval to round ascospores. Conjugation by specialized hyphae (gametangia) precedes ascus formation.

Endomycetoideae. This subfamily is composed of two genera, *Endomyces* and *Schizosaccharomyces*. *Endomyces* forms a septate mycelium, which splits up into arthrospores. Asci are formed on the hyphae after conjugation and contain four hat-shaped or hemispherical spores. In *Schizosaccharomyces* the mycelium is usually much reduced and the thallus normally consists of arthrospores only. It is a typical fission yeast. Asci contain four or eight spores, depending on the species. Conjugation precedes sporulation.

Saccharomycetoideae. This subfamily comprises the genera *Endomycopsis*, *Saccharomyces*, *Pichia*, *Hansenula*, *Debaryomyces*, *Schwanniomyces*, *Saccharomycopsis*, *Saccharomycodes*, *Hanseniaspora*, and *Nadsonia*. In the genus *Endomycopsis* a septate mycelium is found as well as budding cells. Depending on the species, the spores are hat-shaped, sickle-shaped, round, or oval. Some species are heterothallic.

Saccharomyces species have either simple budding cells or a pseudomycelium and sometimes both. Spores are round, oval, or kidney-shaped. There are 1–4 spores per ascus. The species are haploid or diploid in chromosome number, the metabolism is fermentative and oxidative, but surface pellicles are not formed.

Pichia species may or may not form pseudomycelium or surface pellicles, although the common species isolated from various foods produce both. They are usually oxidative; the spores are hat-shaped or round.

Hansenula is similar to *Pichia* but it is the only genus of the Saccharomycetaceae in which all species utilize nitrate as the sole nitrogen source.

Debaryomyces species usually have spherical budding cells with a haploid chromosome number. Normally a single warty spore is formed per ascus. They are principally oxidative yeasts.

Schwanniomyces has oval budding cells (haploid) and usually forms a single warty spore, containing a ledge and a large lipid globule. The metabolism is mainly fermentative.

Saccharomycopsis has large, elongate, budding cells (diploid) and grows only at 37°C. It has 1–4 oval ascospores per ascus. It has unusual nutrient requirements and is weakly fermentative.

Saccharomycodes has apiculate cells (diploid) and four spherical spores per ascus. It is mainly fermentative.

Hanseniaspora also has apiculate cells (diploid) and forms either four hat-shaped spores or a single round spore depending on the species. This genus is mainly fermentative.

Nadsonia is a haploid apiculate yeast, forming a single spiny spore per ascus. It is oxidative (pellicle formation) or fermentative.

Nematosporoideae. The only genus in this subfamily which has been studied in culture is *Nematospora*. It forms mycelium and budding cells. The asci contain eight spindle-shaped ascospores, each containing a whiplike appendage. The metabolism is fermentative as well as oxidative.

Lipomycetoideae. This subfamily contains the single genus *Lipomyces* characterized by budding cells and a saclike appendage, which develops into an ascus containing up to 16 brownish ascospores. Its metabolism is only oxidative. [H.J.P.]

Bibliography: J. Lodder and N. J. W. Kreger van Rij, *The Yeasts—a Taxonomic Study*, 1952.

Saccopharyngiformes

A small order of teleost fishes, the gulpers, highly modified for life in the deep seas. This order was formerly the Lyomeri. They have been regarded as related to eels but are more likely allied to the Myctophiformes. Their degenerative adaptations include loss of opercular bones, branchiostegal rays, swim bladder, pelvic and caudal fins, scales, and ribs. One genus reportedly lacks the upper jaw. The tremendous mouth is greatly modified structurally, the eyes are tiny and placed far forward, the gill openings are minute, and the pharynx is enormously distensible. The tail is slender and tapering (see illustration). Gulpers are rare oceanic fishes and are classified in three families, three genera, and nine species. One is reported to attain a length of 6 ft. *See* ACTINOPTERYGII. [R.M.B.]

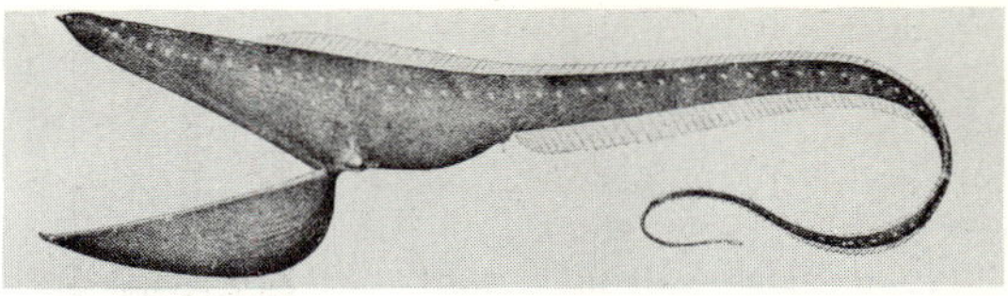

Gulper, Eupharynx bairdi. (After G. B. Goode and T. H. Bean, Oceanic Ichthyology, Mem. Museum Comp. Zool., vol. 22, U.S. Natl. Museum, 1895)

Safe

A protected place. A bank safe is fireproof, strongly built, and equipped with a well-locked door. A furrier's safe provides, in addition, cooling, ventilation, and protection against insects. A safe or vault is a complete structure, the walls, floor, and ceiling being as much a part of the protective features as the more conspicuous door. For utmost protection, a vault is so large and heavy that it cannot be removed. Combination locks eliminate the keyhole as a weakness, and eliminate the key, which could be stolen or duplicated. A time clock built into the locking mechanism unlocks the vault entirely from the inside. The time lock operates only after the elapsed time, set when the vault was closed.

Walk-in vaults are furnished with shelves, locked drawers, and cabinets for systematic and compact storage of valuables. The vault may be air conditioned and may have a telephone extension inside. [F.H.R.]

Safety factor

An empirical number by which the strength of a material is divided to obtain a conservative design stress. A safety factor is used because of uncertainties in operating conditions that may be encountered, nonuniformities in materials, simplifying design assumptions, effects of aging such as corrosion, and strains introduced inadvertently during fabrication and transportation and because of the seriousness of failure. Safety factors vary widely depending on the material, consequences of failure, and operating conditions. For ductile materials, safety factors applied to yield strength are often between 1.5 and 4. For brittle materials that fracture with no prior evidence of incipient failure, factors of 5–8 may be appropriate. [W.J.KR.]

Safety glass

A laminated glass consisting originally (1905) of two sheets of glass glued to a middle sheet of celluloid. By 1950, safety glass was made almost exclusively of two sheets of plate glass laminated with polyvinyl acetate resin, a plastic that makes an extraordinarily tough film on setting. Modern safety glass will not shatter, and thus prevents many deaths and serious injuries. Triplex glass is standard in automobile windshields, and is used in other applications, including locomotives and steamships. In thinner versions, it is found in some goggles and glasses. Multiple-laminated (bullet-proof) glass 1 in. thick or thicker is common in armored cars and in the pressurized cabins of certain aircraft. Heat-treated glass is substituted for the laminated as a safety glass in the side and rear openings of many American automobiles; and wire glass is also occasionally called safety glass, particularly when used to prevent shattering by fire in buildings. *See* GLASS AND GLASS PRODUCTS. [G.CO.]

Safety lamp

A protected-flame lamp for testing the safety of mine atmospheres. It is used particularly to detect methane-air mixtures (firedamp), but it also indicates oxygen-deficient atmospheres.

The forerunners of present-day flame safety lamps resulted from the eighteenth century investigations of three British scientists—Sir Humphry Davy, W. R. Clanny, and George Stephenson—who attempted to provide safe, portable illumination for their country's coal mines. By today's standards these first safety lamps are not considered safe—either for light or for detecting unsafe gaseous atmospheres.

Only two types of flame safety lamps—the Koehler and the Wolf—are used in American mines. Several models of each have been approved as "permissible" by the U.S. Bureau of Mines. The principal elements of a modern safety lamp are a metal base containing fuel, a wick, a globe, double 28-mesh wire gauzes, gaskets, a metal bonnet, an internal igniter, and a magnetic lock to prevent disassembly of lamps except at an authorized place, such as a lamphouse.

Safety lamp: (*a*) Wolf permissible Flame Safety Lamp. (*b*) Koehler permissible Flame Safety Lamp. (*U.S. Bureau of Mines*)

The lamp's safety principle is that the hot products of the combustion of a methane-air mixture burned inside the lamp are cooled below the ignition temperature of the surrounding atmosphere by passing through the gauzes.

Methane-laden atmospheres are tested with normal, luminous flame, approximately ¾–1 in. high, or with nonluminous flame. Flammable methane-air mixtures will burn inside the lamp and elongate a normal flame. By turning the wick down so that only blue (nonluminous) flame shows, a blue cap (cone) will form above the flame when the lamp is inserted into a methane-air mixture. As methane content increases, the flame or the cap will lengthen until the lower explosive limit (5% methane) is reached. Experienced persons can detect as little as 1% methane.

In normal air a luminous flame burns brightly. When an atmosphere is deficient in oxygen, the flame will be dim. The lamp will not burn in a methane-free atmosphere containing less than 16% oxygen. [M.J.A.]

Bibliography: J. W. Paul, L. C. Ilsley, and E. J. Gleim, *Flame Safety Lamps*, U.S. Bureau of Mines Bulletin 227, 1924.

Safety valve

A relief valve set to open at a pressure safely below the bursting pressure of a container, such as a boiler or compressed air receiver. Typically, a disk is held against a seat by a spring; excessive pressure forces the disk open. Construction is such that when the valve opens slightly the opening force builds up to open it fully and to hold the valve open until the pressure drops a predetermined amount, such as 2–4% of the opening pressure. This differential or blowdown pressure and the initial relieving pressure are adjustable. Adjustments must be set by licensed operators, and settings must be tamperproof. The ASME *Boiler Construction Code* gives typical requirements for safety valves. *See* VALVE. [T.B.]

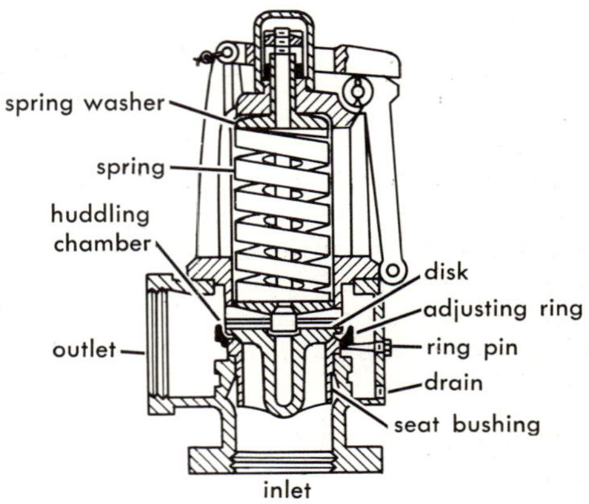

Typical safety valve.

Safflower

This plant, *Carthamnus tinctorius,* is an annual thistlelike herb belonging to the composite family (Compositae). A native of India, the plant has become one of the important crops of the tropics, and

Safflower, *Carthamnus tinctorius.* (USDA)

a new crop of the Great Plains in the United States and Canada. The leaves are sometimes used as salad. The flowers yield two dyes, a red and a yellow, used in the coloring of fabrics; the red is also used in rouge. The seed is processed into a vegetable cooking oil used in cardiac and hypertension diets. *See* CAMPANULALES. [P.D.S.]

Saffron

This plant, *Crocus sativus,* is a member of the iris family (Iridaceae). A native of Greece and Asia

Saffron (*Crocus sativus*).

Minor, it is now cultivated in various parts of Europe, India, and China. This crocus is the source of a potent yellow dye used for coloring foods and medicine. The dye is extracted from the styles and stigmas of the flowers, which appear in autumn. Four thousand flowers are required to produce one ounce of the dye. *See* LILIALES. [P.D.S.]

Sage

This plant, *Salvia officinalis,* is a member of the mint family (Labiatae), the leaves of which yield a spice and an aromatic oil. It is a half-shrub native to the Mediterranean region but is now widely

Sage (*Salvia officinalis*). (USDA)

cultivated. It is much used as a flavoring in stuffing for fowl and in meats, especially sausage. Oil of sage is used in making perfumes. *See* TUBIFLORALES; *see also* SPICE AND FLAVORING. [P.D.S.]

Sagittarius

The Archer, in astronomy, is a zodiacal and summer constellation, the major portion of which lies directly in the Milky Way. Sagittarius is the ninth sign and the southernmost constellation of the Zodiac. In mythology, it is represented by a centaur, Chiron, drawing his bow to release an arrow. Its most prominent feature is a star group commonly called the little Milk Dipper. It is an inverted dipper with four stars to form the bowl and one to form the handle. The Milky Way in Sagittarius is very bright, containing rich star fields and clusters, because its direction lies in the center of the Milky Way stellar system. *See* CONSTELLATION. [C.S.Y.]

Sailplane

An unpowered heavier-than-air vehicle. As compared to a glider, the high gliding ratio and low sinking speed of a sailplane permit advanced soaring flight. It represents a high degree of aerodynamic and structural refinement (Fig. 1). The design of sailplanes has progressed to a point where glide ratio has reached 33:1 for normal airfoils and 42:1 for laminar flow airfoils. Sinking speeds as low as 1.6 ft/sec have been achieved. The useful performance of a sailplane cannot be measured only by its maximum glide and minimum sinking speed because its operation is complicated by natural conditions. Although a light wing loading results in lower speed, and hence lower sinking speed, the necessity for safe structure and reasonable size limits this tendency of design. The low wing loading seriously limits the usable speed range. The modern trend is towards high wing loadings which tend to increase the sinking speed but permit the use of higher aspect ratios. This results in higher glide ratios and greater speed range, but also in large minimum circling diameter, which is a handicap for thermal soaring.

Fig. 1. High-performance sailplane.

Sailplane competition flying requires ability to fly specified courses, such as to specific goals, goal and return, and triangular courses. Under such conditions, with strong winds, a slow sailplane is often handicapped as compared with a high-speed sailplane. Figure 2 shows the effect on performance of varying wing loadings on a hypothetical sailplane. The three sets of curves are for aerodynamically identical vehicles with different wing loadings and show the glide rates in still air and with headwind, and the sinking speed, which is not affected by horizontal winds. Cruising speed is also

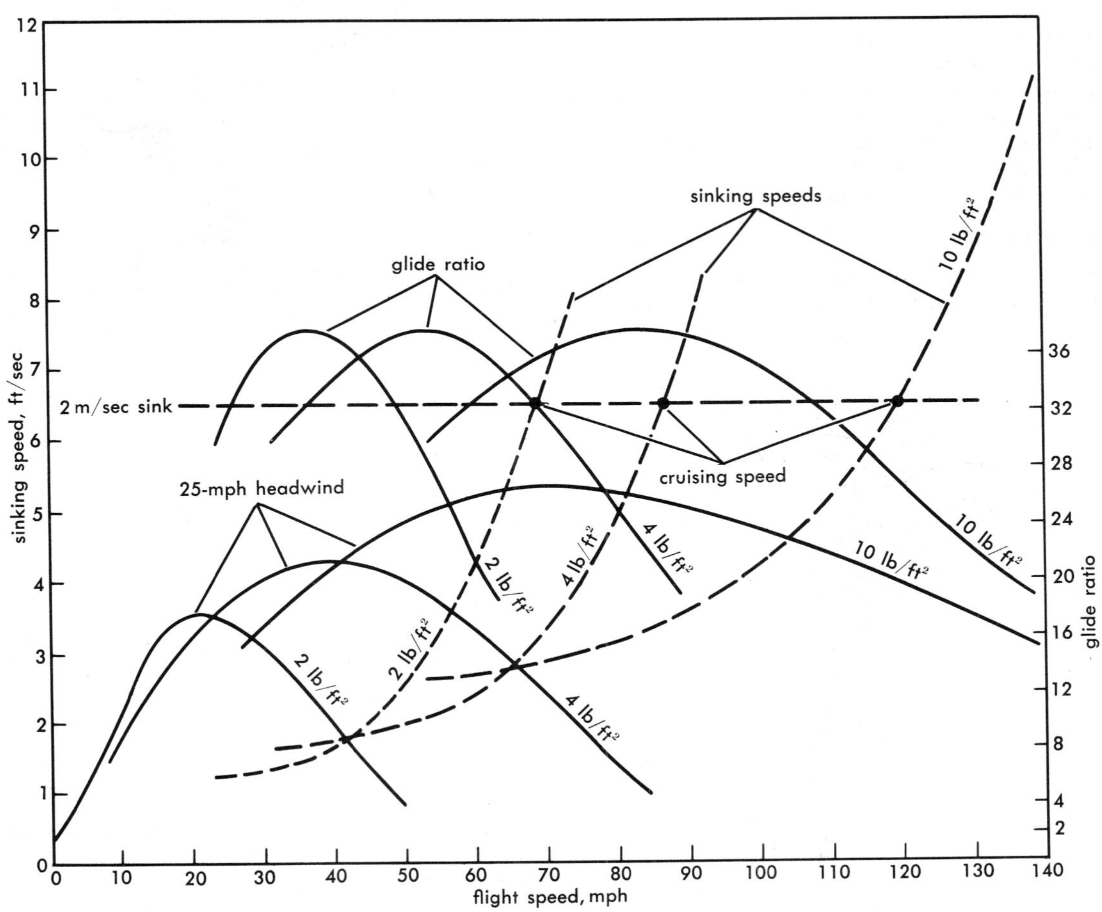

Fig. 2. Performance of three hypothetical sailplanes, identical except for different wing loadings.

shown. This is the speed at which the sailplane has a sinking speed of 2 m/sec. In actual practice, speeds may be higher or lower, depending on the strength and frequency of the thermal conditions. Optimum performance of a sailplane can be achieved only by a highly skilled pilot, because pilot experience, technique, and meterological knowledge are important factors. *See* GLIDER. [E.S.]

Saint Elmo's fire

A type of corona discharge observed on ships under conditions approaching those of an electrical storm. The charge in the atmosphere induces a charge on the masts and other elevated structures. The result of this is a corona discharge which causes a spectacular glow around these points. This effect was accentuated by the contours of early sailing vessels, which usually had several masts and much rigging. The display thus created under darkened conditions was quite striking. The effect was not understood through much of that era and, as a consequence, led to much superstition on the part of sailors. *See* CORONA DISCHARGE. [G.H.MI.]

Salamander

Any member of the amphibian order Caudata, frequently also called Urodela. Salamanders are found throughout most of the temperate parts of the Northern Hemisphere and southward into South America. The United States has representatives of 7 of the 8 families, and 86 species, or about one-third of the known total. Salamanders are most abundant in the eastern part of the United States and along the Pacific Coast. The southern Appalachian region is the major distribution center for salamanders in the United States.

Salamanders appear to have retained the essential primitive traits of the amphibians but are greatly modified from the ancestral forms, especially in the total absence of scales or other exoskeletal structures. They have tails and usually four well-developed legs, but several species have undergone varying degrees of leg reduction. Salamanders are commonly mistaken for lizards, but their slimy, scaleless skin readily distinguishes them from the dry-skinned, scaly lizards.

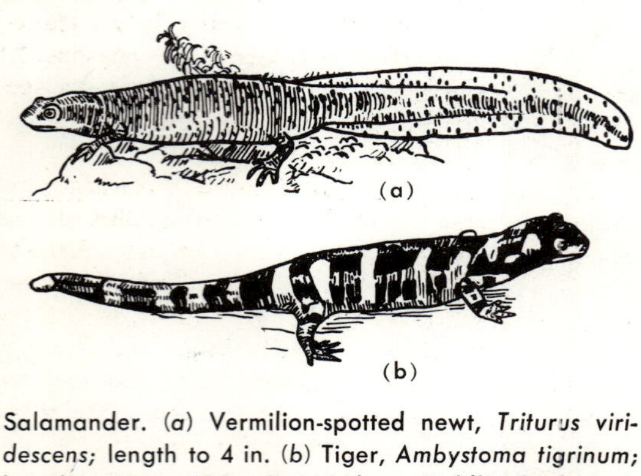

(a)

(b)

Salamander. (*a*) Vermilion-spotted newt, *Triturus viridescens*; length to 4 in. (*b*) Tiger, *Ambystoma tigrinum*; length to 10 in. (*From E. L. Palmer, Fieldbook of Natural History, McGraw-Hill, 1949*)

The young of salamanders have external gills and are essentially like other amphibian tadpoles. A few species are permanent tadpoles and possess both external gills and lungs as adults. Most others leave the water when they become sexually mature, and lose their gills. However, the tiger salamander may or may not transform into a land form, depending upon environmental factors. Some species have lost their lungs and respire entirely through the skin and pharynx.

Salamanders feed primarily upon insects, crustaceans, and earthworms, but will eat whatever animal food they can catch. Individuals of some species are cannibalistic. They are mostly nocturnal. As adults, most salamanders are either terrestrial or aquatic; a very few are arboreal. Several are efficient burrowers. Some of the aquatic species show a tendency toward reduction or loss of the legs.

Certain salamanders, notably some of the newts and *Plethodon* salamanders, are brilliantly colored. The aquatic adult of the common newt of the eastern United States is beautifully marked with black spots, some bordered with brilliant red, over a ground color of olive green. The immature, terrestrial form, commonly called an eft, is orange-red above, yellow to orange below, and marked with similar crimson bordered black spots. Several others are equally brilliant. *See* CAUDATA; HELLBENDER; MUD PUPPY. [J.D.B.]

Salamandroidea

The largest suborder of the Caudata, the salamanders, with some 250 species in 5 families and 49 genera. All living salamanders belong to this group except the members of the suborder Cryptobranchoidea which have external fertilization and lack the fusion of the angular and prearticular bones of the lower jaw seen in higher forms, and the suborder Meantes, neotenic forms lacking hind limbs.

Ambystomatidae. The family Ambystomatidae, sometimes considered as a separate suborder Ambystomoidea, is composed of 5 genera and 30 species limited in distribution to North America. The genus *Ambystoma*, with 21 species, is the largest and most widespread of the family. They range from southern Alaska to Mexico and from the Atlantic to the Pacific. The tiger salamander *Ambystoma tigrinum*, in a variety of subspecific forms, is found from southern Canada to Mexico and over most of the United States. It lives in both arid and humid regions and is the only salamander in much of the region of the Great Plains and the Rocky Mountains. Heavily forested regions of western North America are the home of the Pacific giant salamander, *Dicamptodon*. This largest of terrestrial salamanders may attain a length of 1 ft. All ambystomatids have aquatic larvae, but adults are typically terrestrial, returning to the water only to breed. However, neoteny is of frequent occurrence in the family. The famed axolotl of Mexico, which is neotenic in the natural state but will transform in captivity, is an ambystomatid, *Siredon mexicanum. See* NEOTENY.

Salamandridae. The family Salamandridae of North America, Europe, and Asia includes 40 species in 16 genera. The majority of forms are Eurasian, with only two genera and six species found in North America; *Taricha* on the Pacific Coast and *Diemictylus* in the eastern United States and northeastern Mexico. These genera are endemic to North America, although a more simplified classification places several American and Eurasian genera in a single genus, *Triturus.* The newts, as these forms are called, are often strikingly colored animals with brilliant red or orange underparts and darker brown or green upper surfaces. All newts have an aquatic larval stage, while the adults are variously aquatic or terrestrial, according to species. A complicated life history is shown by the red-spotted newt of eastern North America, in which there is a land-dwelling stage, the red eft, commonly but not always interposed between the aquatic larval and the aquatic adult stages. Although most salamanders are nocturnal, the red eft and adults of the genera *Taricha* and *Diemictylus* are often found walking about in daylight. These animals seem to be almost immune to attack by predators, possibly because of noxious skin secretions.

The salamanders that bear the generic name *Salamandra* are two European species with breeding habits that are peculiar for salamanders. *Salamandra atra,* a high mountain form, retains the eggs and larvae within the body until they emerge as fully transformed young. *Salamandra salamandra,* a species of lower elevations, gives birth to larvae. In both species highly developed gills enable the larvae to absorb oxygen from the maternal oviduct.

Amphiumidae. Two large, eel-like salamanders of the southeastern United States are the only members of the family Amphiumidae. These congo eels of the genus Amphiuma are partly neotenic aquatic animals with very tiny limbs of no use in locomotion.

Plethodontidae. The family Plethodontidae is a large and diverse group of about 175 species and 24 genera characterized by the absence of lungs and the presence of a fine groove from nostril to upper lip. With the exception of two species of *Hydromantes* in Europe, the family is wholly American. Three species of this genus are found in California, but the greatest concentration of genera and species is found in the eastern United States, with a secondary center in Mexico and Central America. The few species found in South America are the world's southernmost.

The lungs are reduced or absent in some salamanders of other families, but this feature is consistent only in the Plethodontidae. It is thought that the absence of lungs serves a hydrostatic function in stream-dwelling salamanders, and for this reason plethodontids are thought to have evolved from ancestors with such habits. However, the

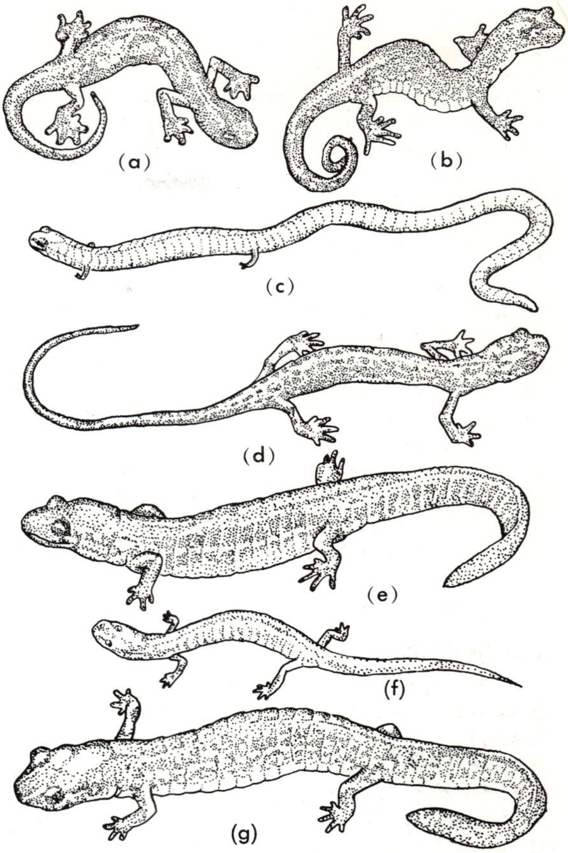

Plethodontid salamanders. (*a*) *Hydromantes italicus,* to 65 mm. (*b*) *Aneides lugubris,* to 100 mm. (*c*) *Batrachoseps attenuatus,* to 62 mm. (*d*) *Eurycea lucifuga,* to 66 mm. (*e*) *Desmognathus quadramaculatus,* to 104 mm. (*f*) *Typhlotriton spelaeus,* to 77 mm. (*g*) *Gyrinophilus porphyriticus,* to 136 mm. (From G. K. Noble, *The Biology of the Amphibia,* Dover, 1954)

present-day members of this family are an ecologically diversified group. A majority of the species are terrestrial and lay eggs that develop directly into salamanders with adult morphology, skipping the aquatic larval stage. Other forms have aquatic larvae and terrestrial or aquatic adults. Many tropical species are arboreal, living in air-plants, bromeliads, that retain moisture in dry seasons. At the other extreme are neotenic, subterranean species that come to light only when a cave is explored or a deep well dug.

The salamanders of the families Proteidae and Necturidae were formerly grouped together in the suborder Proteida, but it is now thought that the similarity between them does not signify especially close relationship. Both *Necturus,* with three species in the eastern United States, and *Proteus,* with a single species in southeastern Europe, are permanently larval, or neotenic, types. *Necturus* lives in streams and lakes, while *Proteus* is a subterranean form found only in underground waters. *See* CAUDATA. [R.G.Z.]

Salicales

An order of the plant subclass Dicotyledoneae, having a single family (Salicaceae). There are 2 genera (*Salix* and *Populus*) with about 340 species of shrubs and trees. The group is very widely distributed with the main center of distribution in the north temperate zone. The plants are dioecious. Flowers are naked and borne in pendant catkins. *Salix*, including the willows (300 species) and *Populus*, including the poplars, aspens, and cottonwoods (40 species), are economically important. Some are timber trees, some are used as ornamentals, and others have tough, pliant twigs used in basketry. *S. alba* yields salicin, a glucoside used in medicine. *See* POPLAR; WILLOW; *see also* DICOTYLEDONEAE; EMBRYOPHYTA; PLANT KINGDOM; TREE. [P.D.S.]

Salicylate

A salt or ester of salicylic acid having the general formula

$$(o)\text{-}C_6H_4(OH)C(=O)\text{—}O\text{—}M \quad \text{or} \quad \text{—}R$$

and formed by replacing the carboxylic hydrogen of the acid by a metal (M) to give a salt or by an organic radical (R) to give an ester. Alkali-metal salts are water soluble; the others, insoluble. Sodium salicylate is used in medicines as an antirheumatic and antiseptic, in the manufacture of dyes, and as a preservative (illegal in foods). Salicylic acid is used in the preparation of aspirin. The methyl ester, the chief component of oil of wintergreen, occurs free and as the glycoside in many plants. This ester is used as a pharmaceutical, flavoring agent, and odorant. The phenyl ester (salol) and others are used medicinally. *See* ASPIRIN. [E.H.H.]

Salientia

One of the three living orders of the class Amphibia which includes the frogs and toads. A frog, a common name that may be used for any salientian, differs most obviously from a salamander in lacking a tail; therefore, frogs are also known as anurans. Usually, the frog has long hind limbs adapted for the hopping locomotion so characteristic of members of the order. There can be no confusion of frogs with the limbless caecilians, members of the third order of amphibians, the Gymnophiona.

Morphology. Frogs are short-bodied animals, usually with long hind limbs, a large mouth, and protruding eyes. The externally visible part of the ear, absent in some forms, is the round, smooth tympanum situated on the side of the head behind the eye. There are five digits on the hind feet and four on the front. Teeth may be present on the upper jaw and the vomerine bones of the roof of the mouth. but are found on the lower jaw of only one species. Often teeth are totally lacking, as in toads

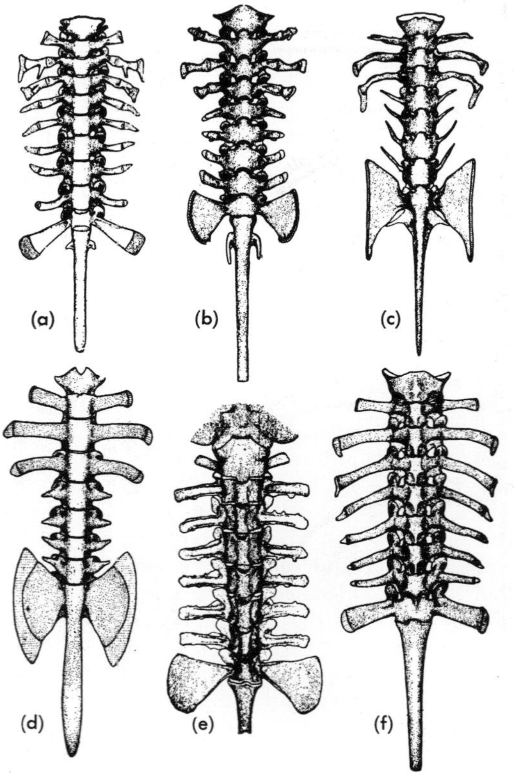

Fig. 1. The principal types of vertebral columns of the Salientia. (*a*) Amphicoelous—*Ascaphus truei.* (*b*) Opisthocoelous—*Alytes obstetricans.* (c) Opisthocoelous with fused coccyx—*Xenopus tropicalis.* (*d*) Anomocoelous—*Scaphiopus couchii.* (e) Procoelous—*Atelopus varius.* (*f*) Diplasiocoelous—*Rana virgatipes.* The vertebral columns are viewed from the ventral aspect. (*From G. K. Noble, The Biology of the Amphibia, Dover, 1954*)

of the genus *Bufo*. The short vertebral column (Fig. 1) consists of from six to ten vertebrae, usually nine, and the elongate coccyx. The sacral vertebra precedes the coccyx and bears more or less enlarged lateral processes with which the pelvic girdle articulates. A characteristic feature of frogs is the fusion of the bones in the lower arm and lower leg, so that a single bone, the radio-ulna in the arm and the tibio-fibula in the leg, occupies the position of two in most other vertebrates.

Taxonomy. There are five suborders of the Salientia: the Amphicoela, Opisthocoela, Anomalocoela, Procoela, and Diplasiocoela. These suborders are distinguished chiefly on the basis of characters of the skeleton and musculature. The phylogeny of the Salientia is illustrated in Fig. 2. Almost 2000 species of frogs are known, so these animals are far more diversified than the salamanders with less than 300 species or caecilians with about 70 species. Only the frozen polar regions and remote oceanic islands are without native frogs, but 80% of the species live in the tropics. The concentration of species in tropical regions is in contrast to

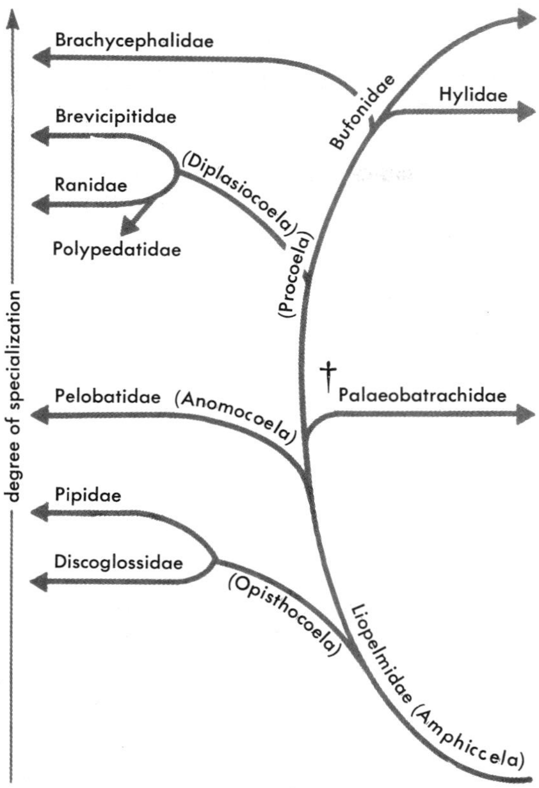

Fig. 2. Diagram illustrating the phylogeny of the Salientia. (*From G. K. Noble, The Biology of the Amphibia, Dover, 1954*)

the distribution of salamanders, most of which are found in more temperate areas.

The one character of frogs that comes to the attention of the most persons, including many who may never see a frog, is the voice. Unlike salamanders, which are mute or nearly so, most frogs have voices and use them in a variety of ways. In the breeding season, great numbers of male frogs may congregate in favorable sites and call, each species giving its own characteristic vocalization. Because no two species breeding at the same time and place have identical calls, it is assumed that the call is important in aiding individuals to find the proper mate. In some species, it appears that the female is active in selecting the mate, and may be responding to the mating call, but the call may not act in exactly the same way in other species. The mating call is given with the mouth closed. Air is shunted back and forth between the lungs and the mouth, so frogs can call even though submerged. Many species possess one or two vocal sacs which are expansible pockets of skin beneath the chin or behind the jaws. These sacs (Fig. 3), which may inflate to a volume as great as that of the frog itself, serve as resonators.

Other noises made by frogs include the so-called fright scream given with the mouth open, and the warning chirp, which evidently serves as a sex recognition signal when one male contacts another.

Reproduction. Breeding and development typically take place in the following manner. The male grasps the female about the body with the forelegs, a procedure called amplexus, and fertilizes the eggs externally as they are extruded. The number of eggs may be quite large (up to 20,000 in the bullfrog or 25,000 in a common toad) or may be as few as one in a frog of the West Indies. The eggs are each surrounded by concentric coats of clear jelly and may be deposited, according to the habit of the species, singly, in groups of various sizes and shapes, or in strings. The larva, called a tadpole, is at first limbless and has external gills and a muscular tail with dorsal and ventral fins. At hatching there is no mouth opening present, but one soon forms that develops a horny beak and several rows of labial teeth not at all like the true teeth of the adult frog. Shortly after the tadpole hatches, the gills become enclosed within chambers and are no longer visible externally. Except for the gradual development of the hind limbs, no additional external changes take place as the tadpole grows until the time for metamorphosis. The anterior limbs, which have been forming hidden in the gill chambers, break through the covering skin as metamorphosis begins. The tail dwindles in size as it is absorbed, while the mouth assumes the shape of that of the adult frog. Many other changes are taking place internally, including shortening of the intestine and adapting it to the carnivorous diet of the adult frog.

The pattern of breeding and development outlined above is widespread among frogs and is undoubtedly the primitive one. However, many modifications of this pattern have evolved. Many species

Fig. 3. A toad (genus *Bufo*) gives its mating call with the vocal sac expanded. (*American Museum of Natural History photograph*)

lay eggs in moist places on land. Each egg is provided with sufficient yolk to allow the embryo to pass an abbreviated larval stage within the egg and emerge as a transformed small frog. The female marsupial frog *Gastrotheca* of South America carries the eggs in a pouch on the back, from which tadpoles or fully-formed frogs emerge, according to the species. In the Surinam toad *Pipa*, also of South America, the eggs undergo development while situated in pits in the back of the mother. The male of Darwin's frog, *Rhinoderma darwini*, another South American species, has the remarkable habit of carrying the eggs and larvae in his vocal sac until metamorphosis occurs. The most highly specialized breeding habit among the Salientia is seen in an African genus, *Nectophrynoides*, which is ovoviviparous; that is, the young develop within the maternal oviduct.

Nutrition. All frogs are carnivorous. The kind of food seems to depend largely upon the size of the frog, and the capacious mouth of a frog permits somewhat astonishing feats of swallowing. A large bullfrog, for example, may snap up low-flying bats, ducklings, snakes, and turtles. Insects and other invertebrates form the bulk of the diet of most frogs. The tongue, moistened by a sticky secretion from the intermaxillary gland in the roof of the mouth, is used to catch smaller prey while larger items of food may bring the front limbs into play. When swallowing, a frog will usually depress the eyeballs into the head to aid in forcing the food down the pharynx. In contrast to transformed frogs, most tadpoles are vegetarian and feed on algae. A few are largely carnivorous or sometimes cannibalistic, and even vegetarian species will scavenge for dead animal matter. Striking feeding specializations occur, such as the funnel mouth of certain tadpoles that skim food from the surface of the water.

Ecology. The habitats of frogs are as various as the places where fresh water accumulates. Lakes and streams are tenanted year-round by many species, and others migrate to these places in the breeding season. Any permanent source of water in the desert is likely to support a population of one or more species, and when rainstorms occur the air around a temporary pool may be filled with mating calls for a few nights while the frogs take advantage of the water for breeding. As often as not, the pool goes dry before the tadpoles metamorphose, and the adult frogs retreat underground to await another rain. Moist tropical regions provide an abundance of habitats little known to temperate regions, such as the air-plants (bromeliads) that hold water and so provide a moist home and breeding site for frogs that may never leave the forest canopy.

Economics. Frogs are used by man in two important ways, as food and as laboratory animals. Many thousands of frog's legs are consumed annually, and the demand in the United States is sufficiently great that the domestic supply is supplemented by imports from Mexico, Cuba, and Japan. Thousands more frogs are used each year as laboratory animals, both as specimens for dissection and study in zoology classes, and as experimental animals for research on a variety of zoological and medical topics. Perhaps a more important service of frogs results from their ecological position as consumers of insects. Indeed, the giant toad *Bufo marinus* of tropical America has been carried to many remote tropical islands to aid in insect control. *See* AMPHIBIA; AMPHICOELA; ANOMOCOELA; CAUDATA; DIPLASIOCOELA; GYMNOPHIONA; OPISTHOCOELA; PROCOELA. [R.G.Z.]

Saline water reclamation

The partial demineralization of sea or brackish water sufficient to make the "fresh-water" product suitable for human or animal consumption, industrial uses, or irrigation. Brackish water is generally regarded as containing at least 1000 parts by weight of dissolved minerals in 1,000,000 parts of water (1000 parts per million, ppm, or 1 gram per liter) but less than sea water, which contains about 35,000 ppm (35 g/liter). Salinity requirements for the fresh-water product depend upon its use. The U.S. Public Health Service recommends that drinking water of good chemical quality should not exceed 500 ppm but permits use of water containing 1000 ppm if necessary. Water containing several thousand ppm is consumed by man in many localities without noticeable ill effects, particularly where perspiration is high. Suitable salinities for irrigation waters depend upon the chemistry of the soil and the mineral requirements of the crop but generally should not exceed about 1200 ppm, particularly if the sodium content is high. Ruminant animals have developed tolerances for salinities up to 12,000 ppm. Industrial water requirements vary greatly, from 1–2 ppm for boiler waters up to 35,000 or more for some flushing and cooling.

Water can be separated from saline solutions by several means. Under a change of phase such as evaporation or freezing, part of the pure water will reach the second phase, but the salt, having a different phase equilibrium, remains in solution. Under the influence of a direct electrical current, salt ions in solution are forced toward the electrodes in electrolysis and electrodialysis. Chemicals may be added to cause exchange of ions in the solution or precipitation of the salts. Chemicals having greater affinity for water than for salt can be used to remove the water from a solution by solvent extraction. In hydration, a hydrocarbon such as propane will, at certain temperatures, combine with the water in a saline solution to form a salt-free hydrate from which the water can then be separated by a small change in temperature or pressure. Under pressure exceeding the osmotic pressure, the fresh water in solution can be forced through an osmotic membrane in reverse osmosis. Other methods which have been studied in various degrees, but so far with little success, include separation by thermal diffusion, adsorption of fresh water on desiccants, and the use of electromagnetic effects, ultra-high-fre-

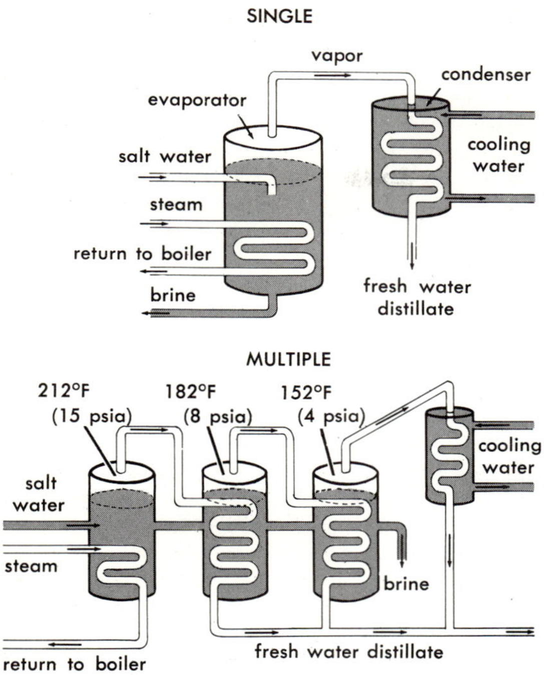

Fig. 1. Single- and multiple-effect distillation.

quency currents, and ultrasonics. The processes by which some plants accumulate certain minerals and reject others and by which animal kidneys separate salt solutions are only now being studied in connection with this problem.

Research and development in saline water reclamation is directed to reduction of the total cost of the product water. Cost includes capital investment, energy, and operation. The absolute minimum theoretical thermodynamic energy required for separating 1000 gal of fresh water from sea water is 2.8 kw-hr. Several types of processes are desirable to meet the requirements and economies of different areas where salt water conversion may be employed.

Multiple-effect distillation. In single-stage distillation (Fig. 1), water is evaporated and condensed once. In multiple distillation, also termed evaporation, several stages, or effects, recapture and reuse the latent heat of vaporization. At a given temperature and pressure, salt water is evaporated in the first effect. The vapor is led to a second effect where additional evaporation occurs at a lower temperature and pressure, using the latent heat of the steam from the first effect, which condenses on the opposite side of a heat-transfer barrier such as a boiler tube. Further evaporation of part of the water occurs in successive effects, each at lower temperature and pressure. The flow of water may be in either direction through the several effects, and either the water or the steam may be within the tubes.

The choice of number of effects, and thus reuse of the heat, depends upon the relative cost of fuel and equipment. Where equipment cost is high and fuel cost low, the number of effects will be smaller than where fuel is expensive and a greater expenditure in equipment (number of effects) is justified for maximum reuse of the heat.

As in all distillation, salt deposits, known as scale, form on the evaporating surfaces at temperatures above about 160–180°F. Scale greatly impedes heat transfer and productivity. Scale-prevention methods are being improved.

Flash distillation. In flash distillation, water at a given pressure and temperature is released into a chamber of slightly lower pressure where it flashes into vapor and is condensed. A vacuum multiple-stage flash-type distilling plant (Fig. 2) is based on the progressive heating of salt water to a temperature of approximately 180°F and the subsequent flashing of a portion in successive chambers, each operating under slightly higher vacuum. The flash vapor from each stage is condensed on tubes containing the incoming cooler sea water and constitutes the fresh-water product.

As with multiple-effect distillation, the flash cycle may be designed to cause water to flow in either direction through the stages. This process has been used extensively in ships, but beginning in 1955

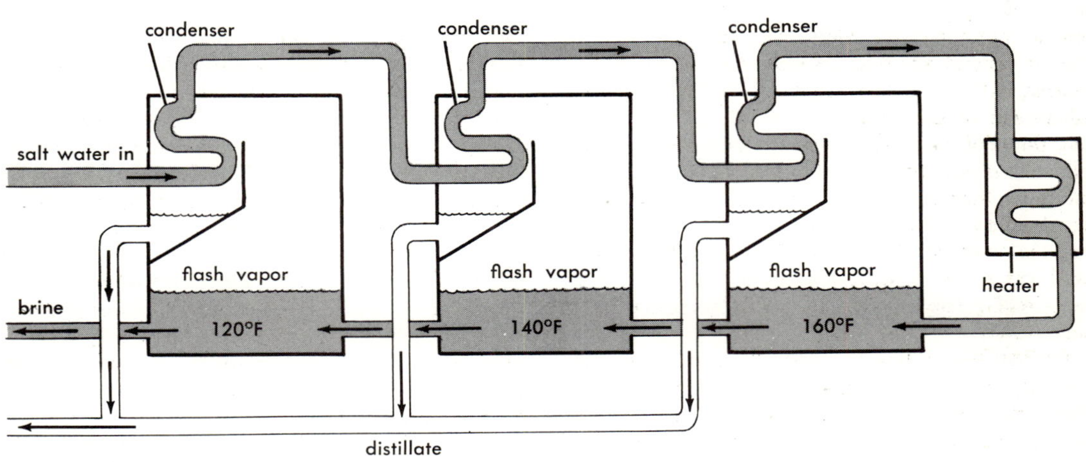

Fig. 2. Flash distillation.

land-based plants were designed, and the process has been improved rapidly to economic feasibility for many purposes.

Vapor compression. Salt water is boiled on one side of a heat-transfer barrier such as a boiler tube (Fig. 3). The vapor is compressed, raising its pressure about 3 psi and temperature about 9°F. The heated steam is returned to the outer side of the boiler tube where its latent heat of condensation evaporates additional water in the tube. The condensed vapor is removed through a heat exchanger as distilled water. In this way, the energy is applied as power to drive the compressor, rather than as heat to boil the water directly. A small quantity of heat is applied to balance radiation losses. Forced circulation of the water and means of causing vapor to condense in droplets have greatly increased the heat-transfer rates.

In the rotary vapor-compression still, salt water is sprayed onto one side of a heated rotating circular plate where it evaporates. The vapor is compressed slightly and, at the resulting higher temperature, is fed to the other side of the plate where it condenses, causing further evaporation on the approach side, as in tubular systems. The rotation causes the salt water to spread into a thin turbulent film and the condensate to be constantly thrown off, both of which promote high heat-transfer rates even when operating at the low pressures and temperatures needed to eliminate scale formation. The early rotary stills developed in 1953 and 1954 utilized a conical shape as shown in Fig. 4. Multiple conical rotors were developed in 1955. Beginning in 1956, attention has been directed to the use of flat rather than conical plates to facilitate the use of more rotors on one shaft, and to the use of the rotation for compression in place of a separate compressor. By using a stack of closely spaced flat rotors, multiple-effect distillation is achieved without compression by evaporating from the top of one plate, condensing on the bottom of the next, and evaporating on its upper surface under reduced pressure.

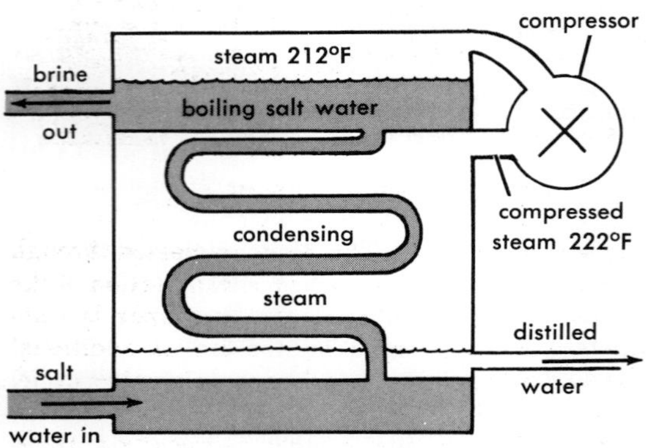

Fig. 3. Vapor-compression distillation.

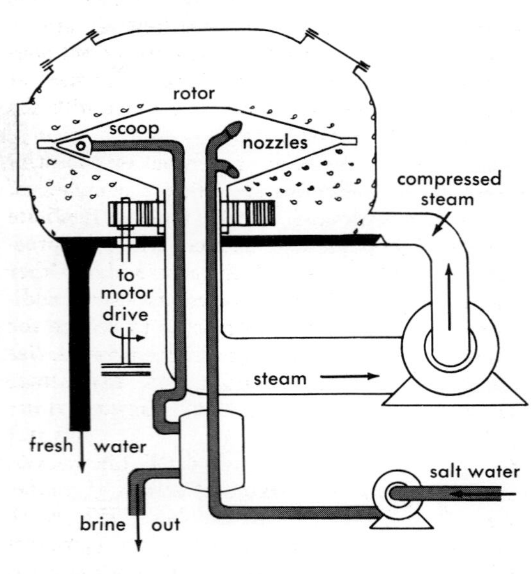

Fig. 4. Single rotary still.

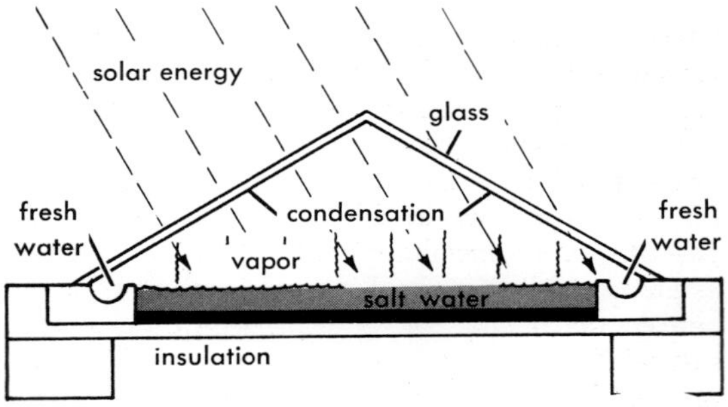

Fig. 5. Simple solar still.

Solar distillation. In solar distillation, the sun's heat is used to evaporate salt water. Solar distillers are classified in three groups: those in which salt water is evaporated directly and condensed in one unit (Fig. 5); those employing focusing devices to acquire higher temperatures for use in various mechanisms; and those in which water is heated in one compartment with evaporation and condensation in another, including multiple-effect mechanisms. Since the magnitude of the incident solar energy is low and cannot be increased by focusing, practical application requires large areas for the collection of this heat energy. Simplified construction methods, including the use of plastic films in place of the glass collectors, are being developed. Several new plastics, including the polyfluorocarbons are relatively inert and resistant to the ultraviolet effects. However, plastics are hydrophobic, and the condensed vapor collects on them in drops and reduces their transparency. The use of wetting agents to prevent this is being developed. *See* DISTILLATION.

Electrodialysis. When an electric current is transmitted through a saline solution, the cations in the solution migrate toward the cathode and the anions toward the anode. Membranes in sheets of cation- or anion-exchange material, a development since 1930, permit the passage of either cations or anions, respectively, in the solution. If a series of alternate cation- and anion-permeable membranes is placed between the electrodes (Fig. 6), the anions pass through the anion-permeable membrane toward the anode but are stopped at the next membrane, which is permeable only to cations. Likewise the cations moving in the opposite direction will pass the cation membrane but not the next. Thus, as anions and cations collect and combine in alternate compartments, the water there is enriched with salt while that in the other compartments is depleted. The salt water flows parallel to the membranes, and the depleted and enriched streams are withdrawn separately as demineralized water and enriched brine. Electrodialysis units are in operation on brackish water at a score of locations throughout the world. The electrical energy required is dependent upon the amount of salt to be removed; this may be contrasted with the various distillation processes, in which the heat energy required is dependent on the amount of fresh water separated from the saline water. Thus, in electrodialysis, greater energy is needed to remove the salt from sea water than from dilute brackish water. *See* DIALYSIS; ION-PERMEABLE MEMBRANE.

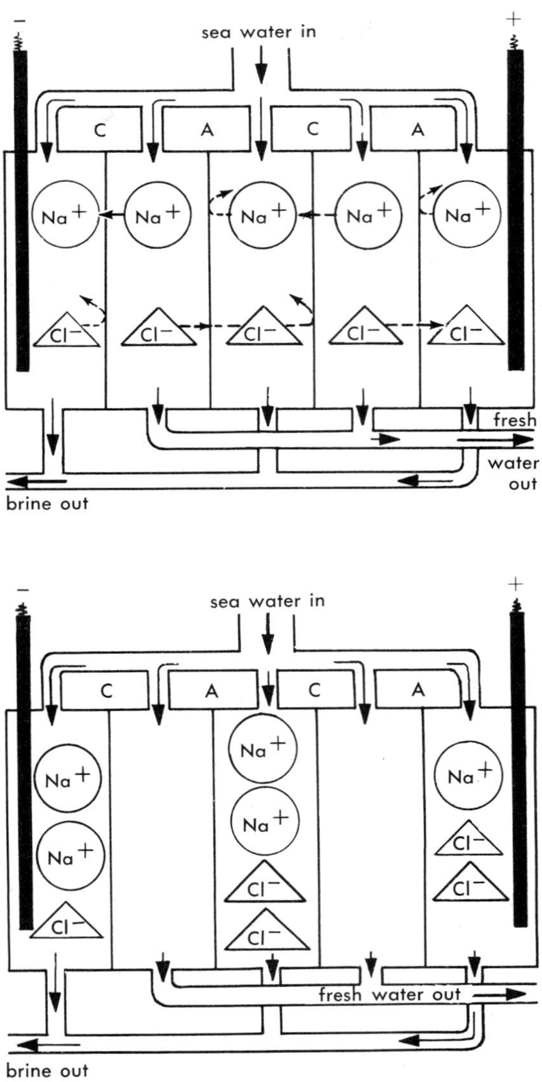

Fig. 6. Electrodialysis.

Freezing methods. Pure ice can be frozen from brine and melted to produce fresh water. Unfortunately, the ice occludes some brine between the crystals. Therefore, two operations are needed: one to form the pure ice and the second to separate the ice from the entrapped brine. One of the more promising techniques (Fig. 7) consists of admitting cold sea water to a chamber under high vacuum where a portion of the water immediately vaporizes. The evaporation process absorbs heat from the remaining salt water, causing a portion of it to freeze to

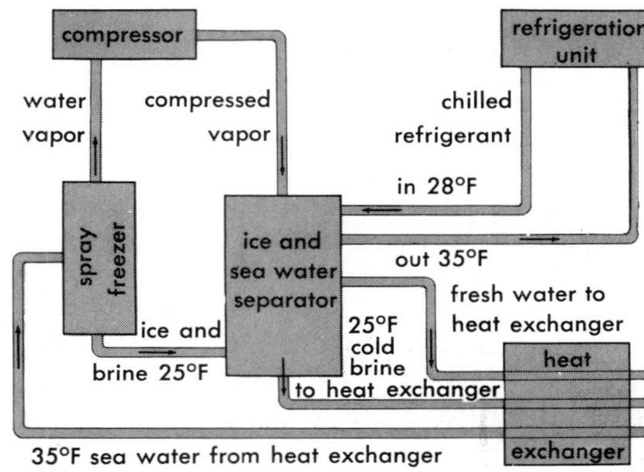

Fig. 7. Freeze-evaporation process.

an ice-brine mixture. This slurry is passed through a separator where it is washed with a portion of the product water. Simultaneously, the vapor is compressed and condensed on the ice as additional fresh water while melting the ice. In another cycle, by using a refrigerant which is immiscible with the water, such as butane or other hydrocarbon, freezing of the water occurs as the refrigerant vaporizes upon passing directly through the solution. This system has the advantage of operation at atmospheric pressure with relatively low vapor volumes and nearly complete recovery of the latent heat of freezing as the pure ice thaws in contact with the compressed refrigerant vapor.

Some research has been carried out on the principle of zone purification in which a chamber of brine is moved through a cold zone to facilitate exclusion of the brine from the ice. While the process is technically successful, the cost is high. Other research has been directed toward the use of additives to modify crystal growth so as to reduce the amount of brine occluded with the ice crystals. *See* CRYSTALLIZATION; HYDROLOGY; ION EXCHANGE; SOLVENT EXTRACTION; WATER CONSERVATION; WATER TREATMENT. [D.S.JE.]

Bibliography: O. L. Chapman, G. W. Lineweaver, and D. S. Jenkins, *Demineralization of Saline Waters*, U.S. Dept. Interior, October, 1952; E. D. Howe, *Sea Water Conversion Program* (progress report for the State of California Legislature), 1956; D. S. Jenkins, Fresh water from salt, *Sci.*

American, vol. 196, no. 3, March, 1957; *Proceedings of the Symposium on Saline Water Conversion 1957*, Nat. Acad, Sci. Natl. Research Council Publ. 568, 1958; U.S. Dept. Interior, Office of Saline Water, Saline Water Conversion Program Research and Development Progress Reports.

Salivary gland virus disease

A viral infection which appears in its most severe, and sometimes fatal, form in infants under 2 years of age. It is also known as cytomegalic inclusion disease. In infants less than 2 years old, clinical cases are characterized by jaundice, enlargement of liver and spleen, and blood and circulatory disturbances.

The virus produces enlargement of cells in the affected organs, or in tissue cultures, as well as large intranuclear inclusion bodies and sometimes intracytoplasmic inclusions. *See* INCLUSION BODIES (VIRUS).

Several animal species (monkeys, mice, guinea pigs) also have salivary gland viruses. The virus of the monkey is closely related to that of man. Subclinical infection is common and the virus is excreted in the urine for months. *See* ANIMAL VIRUS. [J.L.M.]

Bibliography: T. M. Rivers and F. L. Horsfall, Jr. (eds.), *Viral and Rickettsial Infections of Man*, 3d ed., 1959.

Salmon

The family Salmonidae contains the trout, char, and salmon. All salmon are anadromous; that is, they hatch in fresh water, migrate to the sea after a period of growth, and return to spawn in fresh water, usually to the identical spot where they were hatched. The Atlantic salmon, *Salmo salar*, formerly ranged from northwestern Spain throughout the North Atlantic drainage of Europe to the White Sea, westward to Labrador, and south to the Delaware River in the United States. Over much of this range it has now been exterminated; in the United States it is found only in Maine. It was a valuable food fish and is still eagerly sought by anglers. It may spawn several times. There are also landlocked races which never run to the sea.

There are five species of the Pacific salmon found in North America, all of which die soon after spawning. Four of these, as well as two other species, occur in Japanese and Siberian waters. American salmon are still an extremely valuable food resource, producing over 300,000,000 lb annually, worth about $40,000,000. Salmon are now greatly depleted, especially in the southern part of their range. They were at one time abundant as far south as the Sacramento River of California.

The Pacific salmon, *Oncorhynchus tschawytscha*. (*From E. L. Palmer, Fieldbook of Natural History, McGraw-Hill, 1949*)

The largest of the Pacific salmon is *Oncorhynchus tshawytscha*, called king, chinook, or spring salmon, and weighing up to 100 lb. *O. nerka*, the red, sockeye, or blueback salmon, weighs about 7 lb; the coho, or silver salmon, *O. kisutch*, is also a small fish, weighing up to 10 lb. The smallest, with a common maximum of 6 lb, are *O. gorbuscha*, the pink, or humpback salmon, and *O. keta*, the chum, keta, fall, or dog salmon. *See* CLUPEIFORMES. [J.D.B.]

Salmonella

The bacteria of the typhoid-paratyphoid group, a genus of the family of Enterobacteriaceae (*see* ENTEROBACTERIACEAE). The *Salmonellae* are usually motile by peritrichous flagella and lack proteolytic enzymes. Biochemical tests are used as part of the identification process. The tests and their reactions are as follows: methyl red test is positive; Voges-Proskauer and indole tests are negative. Citrate is utilized while lactose is not (*see* IMViC TEST).

Many varieties of *Salmonellae* are distinguished by differences in cultural behavior and antigenic structure. All are pathogenic for either man or animals, or both. The individual types differ considerably in severity of infection caused and in species of animals affected. In man, salmonella infection—collectively called salmonellosis—may give rise to either a predominantly gastrointestinal febrile affection or a septicemic disease (*see* PARATYPHOID FEVER; PARATYPHOID GASTROENTERITIS; TYPHOID FEVER). Occasionally, the dominant manifestations may be localized in the urinary tract, the meninges of the brain, lungs, or bone marrow.

The antigenic structure provides an important tool for the differentiation of all Enterobacteriaceae (*see* ANTIGEN). For this purpose the following antigens are utilized:

1. The somatic, or O, antigens are polysaccharide-lipid-protein complexes extractable from the bacterial body. The polysaccharidic moiety is the distinctive antigenic part. These complexes are toxic and, therefore, sometimes referred to as endotoxins.

2. The flagellar, or H, antigens are of protein nature.

3. Substances from the capsules of slimy envelopes, when present, are of differential value. However, they are often lacking or poorly developed, as for example in most *Salmonellae*. They are present in *S. typhi* as Vi antigen.

Individual entities can be recognized within each genus of Enterobacteriaceae by the stepwise determination of the antigenic structure with specific antisera, which are sera containing specific antibodies. These entities are called serotypes.

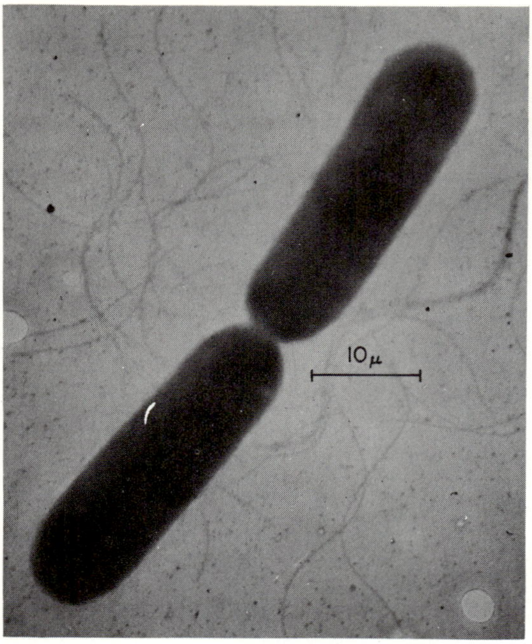

Electron micrograph of *Salmonella typhi* showing rodlike bacterial bodies with inner structure and numerous flagella. (*From A. J. Weil and I. Saphra, Salmonellae and Shigellae, Charles C Thomas, 1953*)

The Kauffmann-White scheme is the internationally recognized code of nomenclature for the serotypes of *Salmonella*. It is essentially a two-step procedure, in which antigens are designated according to an agreed system.

The somatic antigens provide the primary markers for units called groups. Thus all *Salmonellae* with a given somatic antigen belong in the same group. Groups are designated by letters A, B, C, and so on.

Within the group, the H antigens furnish the markers for the individual serotypes, with each type having a name. It often refers to the locality of the first isolation of the species. Approximately 400 serotypes of *Salmonella* can be distinguished by this method. Fortunately, only about a dozen of these serotypes are involved in the great majority of the infections of medical or veterinary importance. Thus for ordinary purposes, a means for diagnosis can be readily provided. For the identification of rare types, a network of international Salmonella centers has been established, to which diagnostic problems can be referred. The table lists serotypes of medical and veterinary importance.

Group	*Serotype*	*Importance*
A	*S. paratyphi A*	Man only; often severe; paratyphoid fever
B	*S. paratyphi B*	Same as *S. paratyphi A*
	S. typhimurium	Man and animals; most common and ubiquitous; synonyms, *S. aertrycke* and *S. breslau*
	S. abortus bovis	Mammals; uterine infection and abortion of cows
C	*S. cholerae suis*	Swine and man; in man, often severe, with high mortality
	S. oranienburg	Man; frequent in human gastroenteritis
	S. montevideo	Same as *S. oranienburg*
	S. newport	Same as *S. oranienburg*
D	*S. typhi*	Man only; typhoid fever
	S. enteritidis	Man and animals
	S. gallinarum	Epidemics in chicken flocks
E	*S. anatum*	Man and animals, particularly ducks

Convalescent hosts may harbor *Salmonellae* in their intestines for long periods of time. The organisms are sometimes also found in the bowels of men or animals without clinical signs or a history of infection. Such people or animals are called carriers (*see* EPIDEMIOLOGY). They are infective to others and are important as links in the chain of infection.

Infection is by the oral route. The vehicles are food or water contaminated by fecal matter, un-

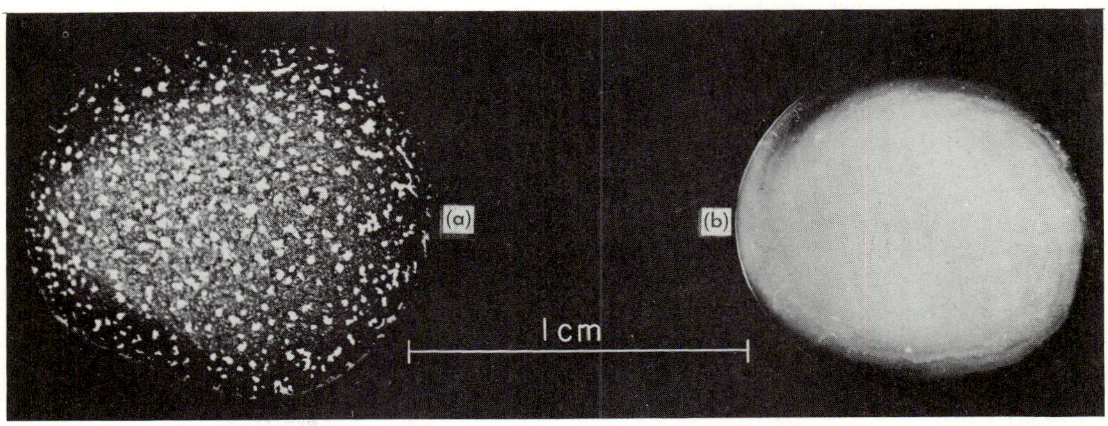

Serological identification of *Salmonellae* and other Enterobacteriaceae. A drop of antiserum is mixed with a suspension of the bacteria in question. (*a*) Positive reaction showing agglutination, or clumping of the bacteria. (*b*) A negative reaction; bacteria stay in suspension. (*A. J. Weil*)

clean hands, or flies. Occasionally, infection is initiated by ingestion of the meat of infected animals.

Effective prevention can be achieved by cleanliness and sanitation, including such measures as identification of carriers, their removal from contact with food, the safeguarding of water, milk, and other food, and proper disposal of fecal matter.

Preventive immunization is practiced in man, against *S. typhi* infection, and in fowl, against *S. gallinarum* infection.

Therapy remains symptomatic; that is, the symptoms are relieved and not the cause of the disease. Sulfonamides and antibiotics, except for chloramphenicol, are not effective for typhoid fever.

Specific diagnosis is made by the identification of *Salmonellae* isolated from stools, blood, or other materials of the sick or carriers, or cultured from inculpated water or food. Indirect diagnosis is made possible by finding antibodies, specific for *Salmonellae*, in the blood serum of man and animals after infection. Since such antibodies persist for a considerable length of time, this may permit retrospective diagnosis of great epidemiological interest. This test method, used for typhoid and paratyphoid fever, is often referred to as Widal's reaction.

The isolation of the enteric pathogens, *Salmonellae* or *Shigellae*, is not difficult when they are present in large numbers. The recovery of organisms becomes a formidable task if only a few pathogens are found among the numerous saprophytic Enterobacteriaceae and other bacteria in a specimen of stool or other material. Most of the saprophytic bacteria produce acid from the sugars lactose and sucrose (saccharose), whereas the pathogens lack this ability. Advantage is taken of this feature when culturing a mixture of pathogenic and nonpathogenic enteric bacteria. Most of the nonpathogens can be recognized by the color change, caused by the acid formation, in and around a colony grown on special media. The media ordinarily contain agar, lactose, and possibly sucrose, as well as a suitable indicator. The colonies without such color changes can then be selected for further investigation by cultural and serological methods.

Certain chemicals promote the growth of *Salmonellae* while restraining that of other enteric bacteria. Among these compounds are bile salts, sodium thiosulfate, and sodium selenite. Thus if feces or other materials suspected of containing *Salmonellae* are seeded in fluid nutrient media containing such compounds, the *Salmonellae* will tend to outgrow the saprophytes. They can then be further checked for pathogenicity by transfer to media of the type used to separate pathogens and nonpathogens. This procedure is often referred to as enrichment. *See* BACTERIOLOGY, MEDICAL; CULTURE, ELECTIVE. [A.J.W.]

Bibliography: P. R. Edwards and W. H. Ewing, *Identification of Enterobacteriaceae*, 1955; A. J. Weil and I. Saphra, *Salmonellae and Shigellae*, 1953.

Salt (chemical)

A compound formed when one or more of the hydrogen atoms of an acid are replaced by one or more cations of a base. The common example is sodium chloride in which the hydrogen ions of hydrochloric acid are replaced by the sodium ions (cations) of sodium hydroxide. There is a great variety of salts because of the large number of acids and bases now known.

Classification. Salts are classified in several ways. One method—normal, acid, and basic salts—depends upon whether all the hydrogen ions of the acid or all the hydroxide ions of the base have been replaced.

Class	*Examples*
Normal salts	$NaCl$, NH_4Cl, Na_2SO_4, Na_2CO_3, Na_3PO_4, $Ca_3(PO_4)_2$
Acid salts	$NaHCO_3$, NaH_2PO_4, Na_2HPO_4, $NaHSO_4$
Basic salts	$Pb(OH)Cl$, $Sn(OH)Cl$

The other method—simple salts, double salts (including alums), and complex salts—depends upon the character of completeness of the ionization.

Class	*Examples*
Simple salts	$NaCl$, $NaHCO_3$, $Pb(OH)Cl$
Double salts	$KCl \cdot MgCl_2$
Alums	$KAl(SO_4)_2$, $NaFe(SO_4)_2$, $NH_4Cr(SO_4)_2$
Complex salts	$K_3Fe(CN)_6$, $Cu(NH_3)_4Cl_2$, $K_2Cr_2O_7$

In general all salts in solution will give ions of each of the metal ions; an exception is the complex type of salt such as $K_3Fe(CN)_6$ and $K_2Cr_2O_7$. In such salts the ionization is entirely as

$$K_3Fe(CN)_6 \rightarrow 3K^+ + Fe(CN)_6^{---}$$

and

$$K_2Cr_2O_7 \rightarrow 2K^+ + Cr_2O_7^{--}$$

No detectable quantities of Fe^{+++} or Cr^{+++} from these salts exist in solution because of the strong bonding of these ions in the complex ions. However, in those complex salts where the bonding is weak, ions of the metal can be detected; for example in Na_2CdCl_4, the cadmium complex ion ionizes appreciably as follows:

$$CdCl_4^{--} \rightarrow Cd^{++} + 4Cl^-$$

The elements with unfilled inner electron shells form complex salts readily. The alum type of salt is a sulfate including the univalent cation of a relatively strong base and a trivalent metal ion such as Al^{+++}, Fe^{+++}, or Cr^{+++}.

Double salts include ions of nearly enough the same size to fit into the same crystal lattice.

Hydrolysis. The solutions of some normal salts are neutral, but those of others are acidic or basic. This results from the reaction of the ions of salt with water. This reaction is called hydrolysis. For example

$$CO_3^{--} + HOH \rightleftharpoons HCO_3^- + OH^-$$
$$CH_3COO^- + HOH \rightleftharpoons CH_3COOH + OH^-$$
$$NH_4^+ + HOH \rightleftharpoons NH_4OH + H^+$$

The resulting solution will be acidic or basic, depending upon whether the hydrolysis produces an excess of hydrogen or hydroxide ion. *See* HYDROLYSIS.

Modern theories of acids and the definition of salts. The development of more general theories of acids and bases in the twentieth century has required a broadening of the concept of salts.

The Brönsted theory of acids lays emphasis on the process of the reaction between acids and bases, and not so much on the product except that the products are other acids and bases. For example

$$\text{Acid}_1 + \text{Base}_2 \rightleftharpoons \text{Acid}_2 + \text{Base}_1$$
$$HCl + H_2O \rightleftharpoons H_3O^+ + Cl^-$$

Since the Brönsted theory extends the proton theory of acids to solvents other than water, the original definition of a salt must be expanded as follows: A salt is an electrovalent compound that contains some cation other than the solvated proton and some anion other than the anion which is the conjugate base of the solvent. In the water system, the salt should not contain the H_3O^+ and OH^- ions alone; in the liquid ammonia system, it should not contain the NH_4^+ and NH_2^- ions alone.

In terms of the Lewis theory of acids and bases, the compound

$$\begin{array}{ccc} & & \cdot\cdot \\ & & :\text{Cl}: \\ & \cdot\cdot & \cdot\cdot \\ :\!\!&\text{Cl}:&\text{Al} \\ & \cdot\cdot & \cdot\cdot \\ & & :\text{Cl}: \\ & & \cdot\cdot \end{array}$$

is an acid, and

$$\begin{array}{ccc} & \text{H} & \\ & \cdot\cdot & \\ \text{H}: & \text{N} & : \\ & \cdot\cdot & \\ & \text{H} & \end{array}$$

is a base. Hence the salt should be

$$\begin{array}{cccc} & \text{Cl} & \text{H} & \\ & \cdot\cdot & \cdot\cdot & \\ \text{Cl}: & \text{Al}: & \text{N}: & \text{H} \\ & \cdot\cdot & \cdot\cdot & \\ & \text{Cl} & \text{H} & \end{array}$$

Here the salt is not limited to replacement of the H^+ with a cation of a base. Rather, it is any aggregate of molecules, atoms, or ions joined together with a coordinate covalent bond. Such compounds correctly can be called salts; however, by common parlance, the term salt usually refers to an electrovalent compound, the classical example of which is sodium chloride. *See* ACID AND BASE; CHEMICAL BINDING. [A.B.G.]

Bibliography: E. Mack, A. B. Garrett, J. F. Haskins, F. H. Verhoek, *Textbook of Chemistry*, 2d ed., 1956.

Salt (food)

The chemical compound sodium chloride. While salt is used extensively in the food industry as a preservative and flavoring, it is also used in the chemical industry to make chlorine and sodium. Historically, salt is one of the oldest materials used in man's food. *See* CHLORINE; SODIUM.

Method of manufacture. Salt was originally made by evaporating sea water (solar salt). This method is still in common usage today; however, impurities in solar salt make it unsatisfactory for most commercial uses and these impurities also lead to clumping. Salt, freshly produced from sea water evaporation ponds, may contain large numbers of halophilic (salt-loving) microorganisms. These occasionally cause spoilage of meat, fish, vegetables, and hides when salt has been used in the preservation process (*see* BRINE, MICROBIOLOGY OF).

Refined salt is obtained from underground mines located in Michigan and Louisiana. Salt is usually handled during the refining processes as brine. These processes are discussed below.

Grainer salt. This type of salt is made by evaporation of brine in long shallow pans, as large as 18 ft wide, 1.5 ft deep, and 150 ft long. The daily capacity of such a grainer may be 80 tons. A scraping conveyor continually removes the crystallizing salt from the bottom of the grainer. The salt is then filtered, dewatered, dried, cooled, and rolled to break clumps. Grainer salt is usually the coarsest in grain and highest in impurities.

Vacuum pan salt. Salt brine is boiled at reduced pressure. A triple-effect evaporator is used; the first stage uses relatively light vacuum but this is increased until in the third it is quite high and the salt solution boils at about 110°F. Production is continuous and the production cycle takes 48 hours. A 20% salt slurry is brought out from the bottom of the third evaporator at the same time fresh brine is admitted; thus impurities are washed from the surface of the outgoing crystals. The salt slurry is filtered, dewatered and high temperature dried at 350°F before screening and packing.

Alberger process. Salt brine is heated to high pressures in heaters and then is passed to a graveller. A graveller is a large cylindrical vessel filled with stones which serves as a deposition site for calcium sulfate. The brine proceeds to flashers where the pressure is gradually reduced to that of the atmosphere, and salt begins to crystallize. The brine and salt mixture is then discharged to a large open pan where the crystallized salt is pumped to a centrifuge for dewatering before being dried in a rotary dryer.

Use in food industry. Large users of salt in the food industry are pickle makers and meat packers. In the pickle industry salt is used as brine to which fresh cucumbers are added. A selective fermentation then proceeds which is governed by salt concentration. In the meat packing industry salt is added to fresh meat as a preservative, as in salt pork, or in combination with nitrates or nitrites to mark the first step in the production of cured meats, such as hams or bacon. *See* FOOD ENGINEERING; FOOD PRESERVATION.

Salt additives. Salt is liable to clumping during periods of high humidity, and preventives to clumping are added to avoid this. Materials used include magnesium carbonate and certain silicates. Iodides are also added to aid in those areas where iodine deficiencies exist. [R.E.M.]

Bibliography: S. C. Prescott and B. E. Proctor, *Food Technology*, 1937.

Salt bridge

A bridge of solution of some salt, usually potassium chloride, which is placed between the two half-cells of a galvanic cell. It is frequently made in tubular form of siphon design. It is used either (1) to reduce to a minimum the potential of the liquid junction between the solutions of the two half-cells or (2) to isolate a solution under study from a reference half-cell and prevent chemical precipitations. When a salt bridge is used to reduce the liquid junction potential, a saturated solution of potassium chloride is chosen, because the ionic mobilities of the potassium and chloride ions are nearly identical. This reduces the liquid-junction potential to a small value. When a salt bridge is used to isolate a solution from a reference half-cell, various salt solutions may be used for the salt bridge, depending on the circumstances. For example, if the potential of silver in a silver nitrate solution were to be compared with a calomel half-cell, the two half-cells could not be in contact, for then the potassium chloride of the calomel half-cell would precipitate the silver nitrate solution surrounding the silver electrode. In this case, then, a salt bridge of potassium nitrate would be placed between the solutions of the two half-cells. *See* ELECTRODE POTENTIAL; ELECTROMOTIVE FORCE (CELLS). [W.J.H.]

Salt dome

An intrusive body of rock salt which has penetrated large thicknesses of overlying sedimentary rock. Salt domes are distinguished from other geological deformations involving salt in being roughly circular in cross section and in having horizontal dimensions of the same order of magnitude or less than their vertical dimensions.

Salt domes are best known along the Gulf Coast of the United States where they are important economically because of their association with oil and sulfur deposits. Most Gulf Coast salt domes have diameters of 1–3 miles and vertical dimensions which vary from 3 to 6 miles. Depths to the tops of salt domes range from less than 100 ft to many thousands of feet.

Formation of domes. The geological process by which domes are formed is not completely understood, but the general mechanism seems fairly clear. The domes occur in areas underlain by evaporite deposits containing large thicknesses of salt (primarily sodium chloride), which flow into the dome from the immediately surrounding area. The salt, being relatively plastic, is highly deformed by the flowage. The flowage apparently results from the fact that the salt is of lower density than the overlying sediments. This density difference is demonstrated by the fact that salt domes produce definite negative gravity anomalies. A model to illustrate this process can be made by filling a glass-sided box nearly full of a high-density viscous liquid, such as boiled-down corn syrup, and then filling the small remainder with a low-density viscous liquid, such as soft asphalt, to represent the salt. When such a model is inverted so that the low-density fluid is on the bottom, the asphalt will flow up through the syrup and take a form which is quite similar to the general form known (by drilling) to be that of many salt domes. Dimensional analyses indicate that the ratios of the physical properties of the fluids in the model and those of the prototype in nature are approximately correct for dynamical similitude. *See* PROSPECTING; PROSPECTING, PETROLEUM.

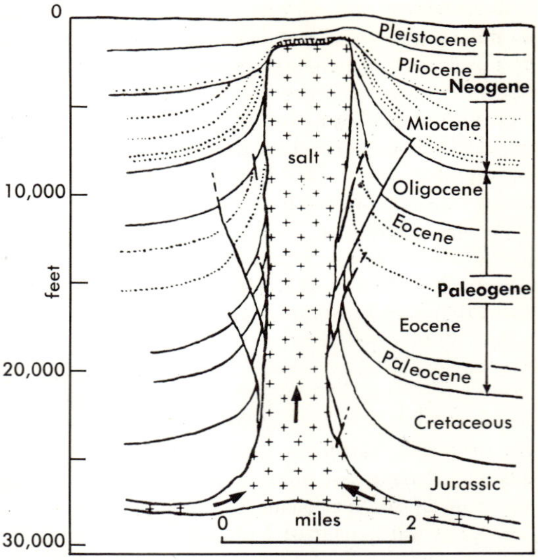

Diagrammatic section of a salt dome in the Gulf region. The drawing represents in composite manner data obtained from several domes, for no wells have been drilled deeply enough to penetrate all the rock divisions shown. The base of the salt plug is hypothetical. (*From R. C. Moore, Introduction to Historical Geology, 2d ed., McGraw-Hill, 1958*)

The model explains in a general way the mechanism by which salt flows into the domes because of the difference in density, but in many cases the movement of salt is significantly influenced by the geological conditions and tectonic forces of the area. It appears that the salt must be covered by a minimum of some 10,000 ft of overburden before flowage into domes takes place. This suggests that the salt may become more plastic under the pressure of a thick overburden, but a clear demonstration of such a property of salt has not been made.

Distribution of domes. The Gulf Coast salt dome belt of Texas and Louisiana, including those domes known to exist as a result of oil exploration offshore, contains approximately 150 known domes. The northeast Texas basin has 20 domes, the interior basin of north Louisiana and southern Mississippi has about 75 domes, and the Rio Grande basin of south Texas has 6 domes. These are the only true salt dome basins in the United States. The Paradox Basin of western Colorado and eastern Utah contains salt structures, but these are elongated anticlines rather than domes.

Salt domes are known in many other parts of the world. They are found in northern Germany, Denmark, Rumania, northeastern Spain, southwestern France, the Persian Gulf and along its northeasterly coast in northern Iran, the Emba basin of Russia just north of the Caspian Sea, and in several smaller basins in northern Pakistan, along the west coast of French Equatorial Africa, and on the Isthmus of Tehauntepec in Mexico.

Cap rock. In many of the domes in southeastern United States and to a certain extent in other parts of the world, particularly in Germany, a cap rock of limestone, $CaCO_3$, anhydrite, $CaSO_4$, or both, occurs on top of the salt. The process by which the cap rock is formed is not completely understood. The most acceptable theory supposes that anhydrite, which is usually a disseminated constituent of the salt, is concentrated as salt is dissolved away by circulating meteoric waters.

Economic importance. Where the salt has intruded petroliferous sediments, the resulting deformation often forms traps in which oil accumulation occurs. Salt dome oil fields of the Gulf Coast are currently producing approximately 300,000,000 barrels per year. Salt dome oil fields are also important in Germany, Rumania, Mexico, and the Emba Basin of Russia. *See* PETROLEUM GEOLOGY.

A relatively small number of Gulf Coast domes are a very important source of sulfur, which occurs in the pores of the limestone cap rock. The sulfur is recovered by the Frasch process in which superheated water is circulated to heat the body of cap rock above the melting point of sulfur, which is pumped out as a liquid (*see* SULFUR). Of the large number of salt domes in the Gulf Coast region of the United States, only about 18 have had sufficient sulfur for economical production. The only other cap rock sulfur production is from the Isthmus of Tehauntepec in Mexico. The Frasch process of sulfur production from salt domes currently recovers about 7,000,000 long tons per year or approximately 40% of the total world's sulfur supply from all sources.

In a few domes salt is mined either by opening shafts into the salt or by drilling wells through which water is circulated to dissolve the salt. Salt production from domes is about 5,000,000 tons per year. *See* DIAPIRIC STRUCTURES. [L.L.N.]

Bibliography: D. C. Barton and G. Sawtelle (eds.), *Gulf Coast Oil Fields*, 1936; R. C. Moore (ed.), *Geology of Salt Dome Oil Fields*, 1926.

Salt gland

A specialized gland located around the eyes and nasal passages in marine turtles, snakes, lizards, and in birds such as the petrels, gulls, and albatrosses, which spend much time at sea. They are compound tubular glands which apparently arise as invaginations of the epithelium of the nasal passage or from the developing conjunctival sac of the eye. In the marine turtle it is an accessory lacrimal gland which opens into the conjunctival sac. In sea-going birds and in marine lizards it opens into the nasal passageway. Salt glands copiously secrete a watery fluid containing a high percentage of salt, higher than the salt content of urine in these species. As a consequence, these animals are able to drink salt-laden sea water without experiencing the dehydration necessary to eliminate the excess salt via the kidney route. *See* GLAND.

[O.E.N.]

Salting-out effect

The decrease in solubility of a species in water resulting from the addition of an electrolyte to the aqueous solution. Gases and organic solids and liquids become less soluble in water in the presence of a salt chemically inert toward them. The solubility of such substances in water depends upon a chemical or physical interaction between solute and water molecules. Ions from a dissolved salt interact through fairly strong electrical forces with water, and become surrounded by a sphere of oriented water molecules. The water held by the salt is then less free to interact with other solute molecules.

Some colloidal suspensions may also be precipitated by the addition of a salt to the suspending medium. *See* ISOELECTRIC POINT; SOLUTION; SOLVENT EXTRACTION. [F.J.J.]

Samarium

Element number 62, samarium, Sm, is a metallic element belonging to the rare-earth group. Its atomic weight is 150.35, and the naturally occurring stable isotopes are Sm^{144} 3.09%, Sm^{147} 14.97%, Sm^{148} 11.24%, Sm^{149} 13.83%, Sm^{150} 7.44%, Sm^{152} 26.72%, Sm^{154} 22.71%. The element was discovered in 1879 by L. de Boisbaudran, and was obtained in the form of very pure compounds

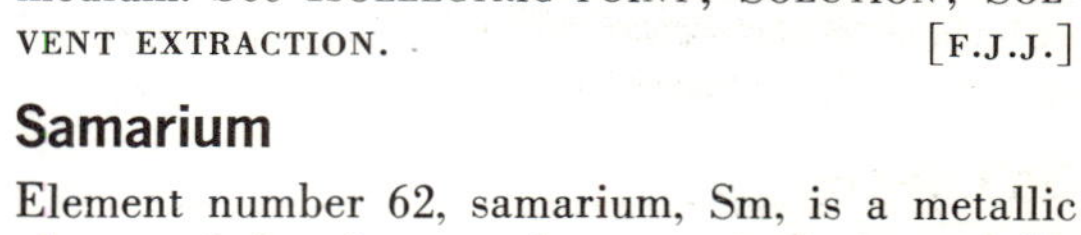

*mass number of most stable known isotope

by E. Demarçay in 1901. The element can be separated from the other rare earths in its trivalent form by ion-exchange methods. The element also exists in the divalent form, but the divalent ion in solution slowly decomposes water. The metal can be extracted from solution into sodium amalgam but cannot be prepared pure in this manner, because samarium distills with the mercury as an intermetallic compound. Reduction of samarium salts with calcium or alkali metals usually gives the divalent salt. The metal, however, can be prepared in the very pure form by mixing the oxide with lanthanum metal or misch metal which has had the other divalent metals removed, and distilling under a vacuum. Samarium metal has an appreciable vapor pressure at the melting point so that it distills away from the mixture and can be condensed in a very pure form, to leave lanthanum oxide in the residue. For a discussion of the properties of the metal, *see* RARE-EARTH ELEMENTS.

Samarium oxide is a pale yellow color, is readily soluble in most acids, and gives topaz-yellow salts in solutions. Samarium has found rather limited use in the ceramic industry, and it is used as a catalyst for certain organic reactions. One of its isotopes has a very high cross section for the capture of neutrons, and therefore there has been some interest in samarium in the atomic industry for use as control rods and nuclear poisons. *See* LANTHANUM. [F.H.SP.]

Sampled-data control system

A form of control system in which the signal appears at one or more points in the system as a sequence of pulses or numbers usually equally spaced in time. The system may be either open-loop or closed-loop, although the latter form is of more significance. *See* CONTROL SYSTEMS.

The operation which converts continuous data into pulse sequence form is referred to as sampling. Sampling is introduced into a system by elements such as scanning radars, digital data links, digital computers, time-shared data links, or mechanical switching devices. These elements all have the property in common that they can receive, transmit, or transduce intermittently, that is, once every sampling instant. For example, a digital computer used as an in-line digital controller will accept a number only once every cycle time or sampling interval. Sampling instants are separated from each other by time intervals whose length depends on the complexity of the computation. *See* COMPUTER CONTROL SYSTEM.

A typical but not necessarily unique structure for a sampled-data control system is shown in the figure. The basic elements in the system are identified. The control error is sampled by means of a sampler, shown schematically as a mechanical switch, which closes momentarily at each sampling instant separated T seconds from the previous instant. The digital controller that follows the first switch accepts a data sample every T seconds. These input data samples are processed mathematically to generate an output number which is applied to the plant or process as a manipulated number sequence. Since physical plants or processes are continuous, the manipulated number sequence must be reconstructed into a continuous signal before being applied to the plant. The data reconstruction device shown in the figure is variously called a data hold, desampling filter, or data extrapolator. The remainder of the control system is conventional. Many other system configurations are possible. For instance, the digital controller may be in the feedback line, the sampling operations may take place at a number of other places in the system, or there may be samplers operating with different sampling intervals.

The process of sampling causes a certain loss of information that is contained in the continuous signal from which the sample sequence is obtained. Signal variations may occur in the continuous signal between sampling instants and may not be detected in the number sequence. If the continuous signal is varying very rapidly, the sampling rate must be fast enough to sense these variations. A practical rule is that the frequency of sampling should be four or five times the highest significant frequency contained in the continuous signal. Because the signal is intermittent at the output of the sampler, a small oscillation known as ripple is induced in the system unless special precautions are taken.

Sampled-data feedback systems have the same properties and limitations of continuous systems but are complicated by the effects of the sampling operations. The analytical study of sampled-data systems is facilitated considerably by the use of a transform calculus analogous to that used in continuous systems. This technique, known as the z transformation, is a modified form of the Laplace transformation used in continuous systems. The main difference is the use of an auxiliary complex variable z which is defined as e^{Ts} rather than the usual Laplace complex frequency variable s. By this use of this auxiliary variable, such concepts as the Nyquist stability criterion, the transfer function, and input-output relationships are found to be very similar to those of continuous systems. A powerful concept in sampled-data systems is the pulse

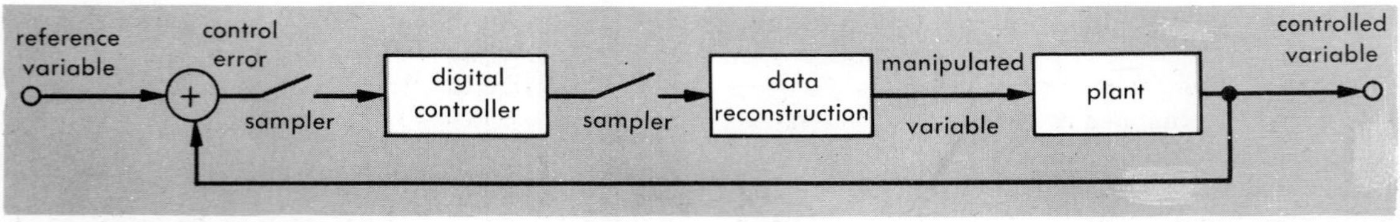

Typical sampled-data control system.

transfer function which relates the output and input pulse sequences in a linear system. Tables of pulse transfer functions and z transforms are available.

An advantage of sampled-data control systems is that a digital controller can be programmed to produce a linear control system that settles completely after a few sampling intervals have elapsed. Such a system responds to an input or disturbance with a transient response that disappears completely after a finite time has elapsed and achieves a neutral steady state. While completely continuous linear systems can settle to a prescribed percentage of the steady state in a given time, their transients do, in fact, have infinite duration. Some of these desirable properties of sampled-data systems can be enhanced by the use of multirate digital controllers, that is, digital controllers which produce a number of solutions for each input number to the controller. [J.R.R.]

Bibliography: J. R. Ragazzini and G. F. Franklin, *Sampled-Data Control Systems*, 1958; J. R. Ragazzini and L. A. Zadeh, The analysis of sampled-data systems, *Trans. A.I.E.E.*, Part II, 71:225–234, 1952; J. G. Truxal, *Automatic Feedback Control System Synthesis*, 1955.

Sampling techniques

In analytical chemistry, the operations required to obtain a laboratory sample from a large quantity of raw material. Most commercial materials are not homogeneous, that is, their compositions vary from portion to portion. If the analysis for a constituent is to be significant, the portion used in the laboratory must have the same composition as that of the original material. The problem of obtaining a representative sample may be more difficult than the problems of analysis.

Sampling of a gas is difficult. Special equipment and procedures are required to obtain a representative, homogeneous portion for analysis. *See* GAS ANALYSIS.

Liquids. Homogeneous liquids in tanks or barrels are sampled by dipping several portions from different locations or by drawing off portions from several containers. Liquids flowing in open rivers or in pipes are sampled by dipping from the open stream or by using a bypass system on a pipe. It is important that sufficient liquid be taken to have a representative sample. The sampling of a multiphase liquid must be done so that the portion removed has the same relative amounts of the different phases as the original material. One way to do this is to use a sampling thief, a long tube with holes which can be opened in the liquid and then closed so that portions representative of different levels are obtained. A system with solid dispersed in a liquid should be agitated before sampling.

Solids. Segregation is a major problem, so usually 0.5–2.0% of the material is taken as the gross sample to be certain of a representative portion.

Metallic materials such as steels, brasses, sheet metals, and wires are sampled by drilling, cutting, milling, or filing if reasonably homogenous, or by taking very fine drillings from several locations if segregation of components is known to exist. After thorough mixing by rolling the drillings, or particles, on paper, the sample is assumed to be representative. This is not always true, however, as different size particles may yield different analyses. If the surface is not the same as the interior, for instance, in a case-hardened steel, care is required to distinguish between the two parts of the sample.

Nonmetallic materials, such as coal, ores, rocks, ore veins, and soils, consist of particles of varying size as well as of varying composition. The gross sample is taken by a sampling thief if the material is in bags or barrels. If the material is shoveled, every nth shovel is set aside. If the material is handled by conveyor, an automatic arrangement such as a slot or divider removes a definite fraction. Several portions of ore veins or soil must be removed and mixed. The gross sample is crushed with mechanical equipment and piled into a flat cone. Two opposite quarters of the cone are removed, mixed by shoveling if a large quantity or by rolling on cloth if a small amount, and ground to a smaller particle size. This process is repeated until only 8–10 lb is left. This portion is ground in a ball-mill to the particle size required for the laboratory sample. Care is necessary to be certain that no impurities are introduced and that no material is lost in the process.

Aliquot portions. Even with care, the final laboratory sample may be somewhat heterogeneous. In this case a portion several times the amount needed for an analysis is dissolved in a suitable solvent, and the sample solution is diluted to a definite volume. A fraction of this solution, usually one-fifth or one-tenth, is taken for analysis. This fraction taken is called an aliquot.

In general, standard sampling procedures have been recommended for most materials of commerce. *See* ANALYTICAL CHEMISTRY. [K.G.S.]

Bibliography: N. H. Furman (ed.), *Scott's Standard Methods of Chemical Analysis*, 5th ed., 1939.

Sand

A loose material consisting of small mineral particles, or rock and mineral particles, distinguishable by the naked eye. Most sands are formed by natural agencies. Many deposits of such sands contain clay and silt in varying amounts; some deposits contain pebbles. The mineral composition of sands varies as does the size of the grains composing them. Sands are widely distributed and have many industrial uses. In 1956 234,570,120 short tons of sand having an average value of $1.05 per ton were sold or used by producers in the United States. All states reported production.

The term sand also is applied, especially commercially, to small mineral or rock particles produced by crushing larger materials; for example, limestone sand made by crushing limestone, slag sand from slag, or sand made by crushing quartzite. Various granular materials, not necessarily of inorganic composition, likewise may be called sand because they consist of sand size particles.

Sand sold or used by producers in 1956*

Use	Tons	Value	Average value
Building	117,149,729	$110,610,696	$0.94
Paving	85,904,199	68,104,395	0.79
Molding	7,961,849	16,639,515	2.09
Glass	6,837,237	19,575,063	2.86
Grinding and polishing	1,668,502	5,250,606	3.15
Engine	1,356,386	1,825,532	1.35
Railroad ballast	917,491	551,718	0.60
Fire or furnace	686,647	1,395,552	2.03
Filter	548,557	848,820	1.55
Other	11,539,523	21,312,882	1.85
Total	234,570,120	$246,114,779	$1.05

SOURCE: U.S. Bureau of Mines, *1956 Minerals Yearbook*, vol. 1, p. 986.

* Data include both commercial and government-and-contractor operations.

Natural sands result primarily from the disintegration of rocks by weathering or erosion. Streams and the waves and currents of lakes and oceans are major agencies eroding rock into sand. The grinding action of glaciers is another important sand-producing agency.

Some sand deposits are formed in place by the weathering of rocks, such as sandstone. Others are the result of the sorting out and concentration of sand from particulate material (composed of particles of various sizes) by the running water of streams or by the waves and currents of lakes and seas. Wind also concentrates sand from certain materials. Deposits of sand accumulate as bars in rivers and streams, in river deltas, as beaches and bars of lakes or seas, and as dunes built up by wind. They may be designated by the name of the place where they are found as river sand, lake sand, or dune sand. More technical terms may be used, as fluvial, lacustrine, and eolian sand, for the same three occurrences.

The mineral quartz is probably the most common major constituent of sands, but locally other materials may predominate as in coral sands, gypsum sands, or black sands (composed of fragments of volcanic rocks). The sands in which quartz is the major component usually contain various amounts of other mineral grains. Coarse-grained sands also may contain small rock fragments. The nonquartz particles are of diverse character and in many cases reflect the mineral composition of the original source materials from which the sands were derived.

Sand grains vary from almost spherical to angular. Industrially, angular sands are sometimes referred to as sharp sands. The degree of rounding of sand grains is in large measure an indication of the amount of wear to which they have been subjected.

There are no generally accepted size limits for sands. Sands used in the construction industry are commonly finer than ¼ in., or will pass a 4 mesh sieve but will be retained on a 200 mesh sieve. Geologists use maximum and minimum limits of 2 mm and 1/16 mm.

Sand has many commercial uses. It is extensively utilized in construction as fine aggregate for concrete, mortars, plasters and for many other purposes. Black sands, such as those in Florida, contain ilmenite ($FeTiO_3$) and rutile (TiO_2) in such quantities that they can be recovered for commercial use. Green sands contain the mineral glauconite and have been employed as fertilizers because of their potash content.

The term silica sand is applied to sands composed almost exclusively of grains of the mineral quartz (SiO_2). There are no exact limits for the silica content of silica sands but they commonly contain more than 95% SiO_2 and some of them more than 99%. Silica sand is used in glass making, as molding sand, refractory sand, filter sand, grinding and polishing sand, and for many other purposes. Silica sands are sometimes referred to as industrial sands.

Natural-bonded molding sand contains sufficient clay and other bonding material so that it can be used for making molds in which metal is cast. Synthetic molding sand consists of silica sand to which is added a controlled amount of fireclay, bentonite, or other bond. *See* GLASS AND GLASS PRODUCTS; QUARTZ; REFRACTORY; SEDIMENTARY ROCKS; SEDIMENTATION (GEOLOGY). [J.E.L.]

Bibliography: R. B. Ladoo and W. M. Myers, *Nonmetallic Minerals*, 2d ed., 1951; F. J. Pettijohn, *Sedimentary Rocks*, 2d ed., 1957; J. R. Thoenen and O. Bowles, *Mineral Facts and Problems*, U.S. Bureau of Mines Bull. 556, 1956.

Sand dollar

A member of the order Clypeasteroidea, class Echinoidea, phylum Echinodermata. There are several species, perhaps the best known being the common sand dollar, *Echinarachnius parma*. This species is common on the Atlantic Coast from New

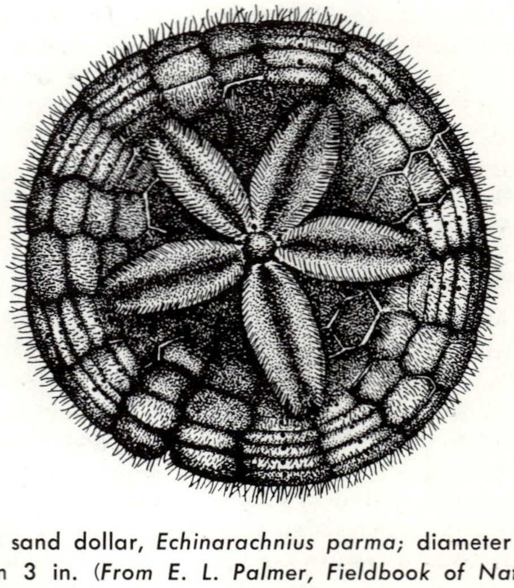

The sand dollar, *Echinarachnius parma;* diameter less than 3 in. (*From E. L. Palmer, Fieldbook of Natural History, McGraw-Hill, 1949*)

Jersey northward, and on the Pacific Coast south of Puget Sound. Sand dollars are basically similar to the related sea urchins, and like the latter, sand dollars have the same fundamental organization as the starfish. The flat discs are circular or oval.

Sand dollars have short tube feet and numerous short spines, about 1/16 in. long, so closely packed that they look and feel like velvet. These animals range in color from gray or brown through purple to black. They may be found in great numbers on quiet water sand flats, and also occur at substantial depths, avoiding surf at all times. The whitened skeletons of sand dollars are among the most common animal remains picked up on exposed beaches. Their general physiology, food, and reproduction are similar to those of sea urchins. In some localities these animals are called sea biscuits or cake urchins. *See* ECHINODERMATA. [J.D.B.]

Sandalwood

This name is applied to any species of the genus *Santalum* of the sandalwood family (Santalaceae). However, the true sandalwood is the hard, close-grained, aromatic heartwood of a parasitic tree,

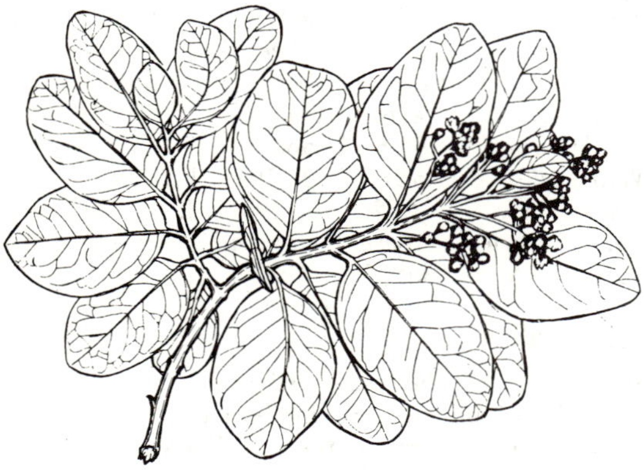

Sandalwood. (*From O. Degener, Ferns and Flowering Plants of Hawaii National Park, 1930*)

S. album, of the Indo-Malayan region. This fragrant wood is used in ornamental carving, cabinet work, and as a source of certain perfumes. The odor of the wood is an insect repellent and on this account the wood is much used in making boxes and chests. The fragrant wood of a number of species in other families bears the same name but none of these is the real sandalwood. *See* SANTALALES. [P.D.S.]

Sandpiper

Any of a large number of shore birds of the family Scolopacidae. This is a cosmopolitan family of 83 species, 32 of which occur in the United States. Several are not called sandpipers, this name being applied to only 13 species in the United States. They are alike in being relatively long-legged wading birds, usually with long bills which are pitted, soft at the tip, and abundantly supplied with nerves. They generally probe in the mud for their food, which is detected by the sensitive bill tip. All

The spotted sandpiper, *Actitis macularia;* length to 8 in. (*From E. L. Palmer, Fieldbook of Natural History, McGraw-Hill, 1949*)

are highly migratory, and most of the United States species winter in South America.

Most sandpipers are streaked brown or gray above and white below, a color pattern seemingly revealing on the bare shores they frequent but demonstrated to be effectively concealing. Sandpipers frequent shores of both inland waters and the sea, although they are more common along American seacoasts than in the interior. They nest singly but frequently gather in flocks during migrations. Formerly considered game birds, their numbers have been greatly depleted by a combination of overshooting and destruction of habitat, both in this country and in their winter homes. Woodcock are still hunted in parts of the United States. Legal protection has helped to restore the numbers of the larger sandpipers. The smaller species have not been depleted as much as the larger ones. *See* CHARADRIIFORMES; SNIPE; WOODCOCK. [J.D.B.]

Sandstone

A detrital sedimentary rock formed by the cementation of individual grains of sand-size particles 2–1/16 mm in diameter. The grains in most sands are commonly composed of the mineral quartz, but many other minerals and even fragments of other rocks may be included. Sandstone may grade into shale which is composed of particles less than 1/16 mm in diameter or conglomerate containing fragments greater than 2 mm in diameter. Sandstones, particularly those with a silica cement, are used for structural purposes. *See* SAND; STONE AND STONE PRODUCTS.

Perhaps more thought has been given to the description and classification of sandstones on a combined descriptive and genetic basis than to that of any other group of sedimentary rocks. This is partly because of their abundance (second only to shale) and because they lend themselves more easily to detailed study, particularly microscopy.

Particle size. The simplest subdivision of the sandstones is on the basis of size of the particles:

very coarse sand, with a range in particle size of 2–1 mm; coarse sand, 1–½ mm; medium sand, ½–¼ mm; fine sand, ¼–⅛ mm; and very fine sand, ⅛–$\frac{1}{16}$ mm. Silt is divided similarly, going from coarse silt, $\frac{1}{16}$–$\frac{1}{32}$ mm, to medium silt, $\frac{1}{32}$–$\frac{1}{64}$ mm, to fine silt, $\frac{1}{64}$–$\frac{1}{128}$ mm, and to very fine silt, $\frac{1}{128}$–$\frac{1}{256}$ mm. This division of sandstones on the basis of particle size is useful in relating the deposit to the strength of the current that transported the sand, for, in general, the stronger the current, the larger the particle size. *See* SEDIMENTATION (GEOLOGY).

Mineralogy. More useful for genetic interpretation than size distributions is the mineralogy of the sandstones. Both composition and stable characteristics of the minerals are important.

Mineral composition. Sandstone particles vary greatly in mineral composition, including such common minerals as quartz, feldspar, and the several clay minerals; a great number of minor minerals, normally found only in small quantities, such as garnet, tourmaline, zircon, rutile, staurolite; and some of the ore minerals, magnetite, pyrite, and chromite. In addition to single mineral grains, rock fragments of many kinds are found in sandstones. The fundamental basis for using mineralogy as a criterion for classification is the chemical and mechanical stability of minerals at earth surface environments.

Stability of minerals. Quartz is stable in surface environments but feldspar generally is not. Some of the rock fragments, such as quartzite, or chert, are stable but fragments of many igneous and metamorphic rocks are not. Thus the presence of only stable minerals and rock fragments in a sandstone implies that any unstable minerals or rocks present in the source area, whose erosion products supply the sediment, have disappeared as a result of chemical activity during weathering and erosion. It could also mean that the source area consisted only of sedimentary rocks with stable minerals. On the other hand, the presence of much feldspar, an unstable mineral, in a sandstone must mean that feldspathic rocks (igneous and metamorphic) were being eroded mechanically at such a rate that chemical weathering did not have the opportunity to destroy them. This interpretation can be utilized further by relating the ratio of mechanical to chemical weathering and topographic relief in the source area. The greater the relief, the more rigorous the mechanical erosion; the less the relief, the greater the effect of chemical erosion. Ultimately, the topographic relief of any area must be a function of tectonic activity, including broad uplifts and mountain-building activity. Therefore, by this chain of reasoning, the mineralogy of the sandstone may reflect source area rock composition, topographic relief, and tectonic state.

Texture. The second fundamental property of sandstones is texture. Any sandstone consists of two textural elements, the larger particles that make up the bulk of the rock or the framework, and the voids between the grains, which may or may not be filled. The framework fraction is described primarily in terms of its mineralogy. The voids may be vacant as in a modern sand, partially filled, or completely filled. The filling of the voids may be an original sedimentation effect or it may be a postdepositional chemical precipitate. The voids in many sandstones are completely filled with a fine-grained clay, referred to as the clay matrix. The sandstones with a high proportion of clay matrix were deposited under sedimentation conditions radically different from those conditions under which the "clean" (nonclayey) sandstones were laid down. It is thought that many of the clay rich sandstones were laid down by turbidity currents. *See* TURBIDITY CURRENT.

Structure. A third property, suggested in the 1950s as being significant in the classification of sandstones, is the kind of sedimentary structure characteristic of the rock type. Thus some sandstones are cross-bedded; others are not. Graded bedding is characteristic of some sands but not of others. It has become apparent that some of these sedimentary structures correlate with mineralogic and textural features and so fit well with other classifications.

Classification. The classification of sandstones that is becoming generally accepted is based on the work of P. D. Krynine, F. J. Pettijohn, and others

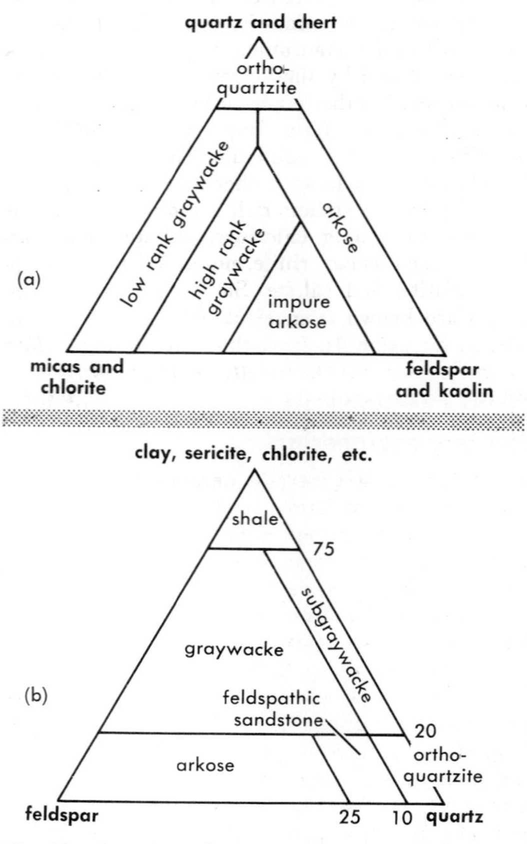

Classification of sandstone. (*a*) P. D. Krynine's classification (*J. Geol., 1948*); (*b*) F. J. Pettijohn's classification (*from unpublished chart, 1944*).

in the 1940s and 1950s. According to this scheme, sandstones can be divided into four general classes, orthoquartzite, arkose, graywacke, subgraywacke. The distinctions between these types are made partially on the basis of mineralogy and partially on the basis of texture (primarily the amount of clay matrix). Each of these types tends to have its distinctive set of characters, including sedimentary structures. They also tend to have preferred associations with other clastic and nonclastic rocks with which they are interbedded. Although there are genetic implications in this classification, any sandstone can be put into its appropriate class by the application of objective, measureable criteria. *See* ARENACEOUS ROCK; ARKOSE; GRAYWACKE; ORTHOQUARTZITE; SEDIMENTARY ROCKS; SUBGRAYWACKE. [R.S.]

Sanidinite

A group of rocks formed by high-grade contact metamorphism (pyrometamorphism) and pneumatolysis of certain sediments, mostly argillites, which have been either trapped by the lavas in a volcanic vent, or thrown out of the vent as ejecta. The most famous sanidinites are those of the Laacher See (Lake) area, West Germany, which have given name to the sanidinite facies of P. Eskola, a facies characterized by extreme high temperature and low pressure. The original material was mostly an alumina-rich schist with quartz, feldspar, kyanite, staurolite, and garnet. This rock has been altered by simple remelting to a glass in some places; in other places it has been completely recrystallized to form hypersthene, cordierite, corundum, and other minerals. The pneumatolytic introduction of soda and other gases has produced large crystals of sodium-rich sanidine and various minerals containing chlorides, sulfates, and carbonates, such as cancrinite, noselite, haüyne, scapolite, apatite, and calcite. Similar mineral assemblages are known from ejecta of many other volcanoes, for example, from those of Vesuvius (Mte Somma). *See* METAMORPHIC ROCKS; METAMORPHISM; PNEUMATOLYSIS. [T.F.W.B.]

Sanitary engineering

A specialty field generally developed in civil engineering but not limited to that branch. The National Research Council defines the sanitary engineer as "a graduate of a full 4-year, or longer, course leading to a Bachelor's, or higher, degree at an educational institution of recognized standing with major study in engineering, who has fitted himself by suitable specialized training, study, and experience (1) to conceive, design, appraise, direct and manage engineering works and projects developed, as a whole or in part, for the protection and promotion of the public health, particularly as it relates to the improvement of man's environment, and (2) to investigate and correct engineering works and other projects that are capable of injury to the public health by being or becoming faulty in conception, design, direction, or management."

Sanitary engineering practice includes surveys, reports, designs, reviews, management, operation and investigation of works or programs for (1) water supply, treatment and distribution; (2) sewage collection, treatment and disposal; (3) control of pollution in surface and underground waters; (4) collection, treatment, and disposal of refuse; (5) sanitary handling of milk and food; (6) housing and institutional sanitation; (7) rodent and insect control; (8) recreational place sanitation; (9) control of atmospheric pollution and air quality in both the general air of communities and in industrial work spaces; (10) control of radiation hazards exposure; (11) and other environmental factors affecting health, comfort, safety, and well being of people.

Sanitary engineers engage in research in engineering sciences and such related sciences as chemistry, physics, and microbiology and apply these in development of works for protection of man and control of his environment. [W.T.I.]

Bibliography: The APHA-ASCE-AWWA-FSWA Joint Committee Report, *Glossary of Water and Sewage Control Engineering*, 1948; APWA Committee on Refuse Collection, *Refuse Collection Practice*, 2d ed., 1958; H. E. Babbitt, *Engineering in Public Health*, 1952; W. D. Claus (ed.), *Radiation Biology and Medicine*, Atoms for Peace Series, 1958; V. M. Ehlers and E. W. Steel, *Municipal and Rural Sanitation*, 5th ed., 1958; G. M. Fair and J. C. Geyer, *Elements of Water Supply and Waste-Water Disposal*, 1958; W. C. L. Hemeon, *Plant and Process Ventilation*, 1955; R. K. Linsley, Jr. and J. B. Franzini, *Elements of Hydraulic Engineering*, 1955; L. C. McCabe (ed.), *Air Pollution*, 1952; K. F. Maxcy (ed.), *Rosenau Preventive Medicine and Public Health*, 8th ed., 1956; F. A. Patty (ed.), *Industrial Hygiene and Toxicology*, vol. 1, 2d ed., 1958.

Santalales

An order of the plant subclass Dicotyledoneae including 4 families with 96 genera and about 1860 species. They have very little economic importance. The sandalwood family (Santalaceae) contains many semiparasites. Sandalwood (*Santalum album*) of Indomalaysia is noted for its sweetly aromatic wood used in cabinetmaking, carving, and perfumes. The mistletoe family (Loranthaceae) consists of semiparasitic herbs or shrubs growing on branches of trees. The true mistletoe (*Viscum album*) is used in Christmas decorations. The malla-tree family (Olacaceae) is a group of tropical shrubs and trees. The balanophora family (Balanophoraceae) is made up of nongreen, total parasites growing on the roots of tropical trees. *See* MISTLETOE; SANDALWOOD; *see also* DICOTYLEDONEAE; EMBRYOPHYTA; PLANT KINGDOM. [P.D.S.]

Sapindales

An order of the plant subclass Dicotyledoneae including 16 families with 323 genera and about 3560 species. There is little agreement concerning the family relationships. Two families, the Limnanthaceae and the Balsaminaceae are herbaceous, but all the others are shrubs and trees. Most of the families are quite small and of no economic importance. The maple family (Aceraceae) produces valuable timber and maple sugar. Other members yielding useful products or used as ornamentals are mango (*Mangifera indica*), cashew, pistachio, box, holly, euonymus, horse-chestnut, and lychee. Poison ivy, poison oak, and poison sumac—members of the sumac family (Anacardiaceae)—cause a dermatitis and therefore are plants of ill repute. Another member of this family, the varnish tree (*Rhus vernicifera*), is the source of lacquer. *See* BUCKEYE; CASHEW; HOLLY; LYCHEE; MANGO; MAPLE; PISTACHIO; POISON IVY; POISON SUMAC; QUEBRACHO; VARNISH TREE; *see also* DICOTYLEDONEAE; EMBRYOPHYTA; PLANT KINGDOM; TREE. [P.D.S.]

Sapphire

The name given to all gem varieties of the mineral corundum, except those that have medium to dark tones of red that characterize ruby. Although the name sapphire is most commonly associated with the blue variety, there are many other colors of gem corundum to which sapphire is applied correctly; these include yellow, brown, green, pink, orange, purple, colorless, and black. The lovely orange variety of gem corundum is known by the exotic name of padparadsha and as orange sapphire. In addition to transparent varieties of sapphire, an equal number of translucent types are fashioned in high-domed cabochons (curved cuttings) to bring out the six-rayed stars for which sapphire is famous. Asterism, the star effect, is the result of reflections from tiny, lustrous, needlelike inclusions of the mineral rutile, plus the domed form of cutting. The minute rutile crystals are oriented in three sets parallel to the base of the corundum crystal, with one set parallel to each of the three pairs of parallel prism faces. *See* CORUNDUM; RUTILE.

The most famous source of blue sapphires is Kashmir, the northernmost state in the Indian Peninsula; however, this deposit appears to have been exhausted. The most important source today is the Mogok area of Burma, which is also the most important source of ruby. Mogok produces a number of other colors, as well as blue and red. Thailand and Ceylon are fairly important sources, the former particularly of inky, dark-blue stones, and the latter of stones too light in tone to achieve maximum value. Australia is the source of very dark-blue transparent sapphires, black-star material, and transparent golden sapphires. Light- to medium-blue sapphires are mined from a basic igneous dike in Montana. Most of the gem-sapphire sources occur either in alluvial deposits or in the type of marble which results from the intrusion of an igneous mass into an impure limestone.

Blue sapphire is most valuable when it is a medium to medium-dark tone of a slightly violet-blue; this is often referred to as a cornflower blue. The Kashmir grade has a slightly "sleepy" appearance, caused by inclusions that reduce transparency somewhat. Blue sapphire is much more valuable than any other color. The best of the transparent stones are slightly more expensive than the finest stars. Sapphire has a hardness of 9, a specific gravity near 4.00, and refractive indices of 1.76–1.77. *See* GEM. [R.T.L.]

Sapropel

A mud, slime, or ooze deposited in more or less open water. Sapropel may vary widely in composition depending upon relative contributions from decomposing substances derived from plants and animals. Hydrogen sulfide, produced during the initial biochemical degradation of these substances, promotes preservation of the more resistant parts of the organisms. Most marine sapropelic deposits contain no appreciable contribution from humic substances of terrestrial origin, but certain carbonaceous marine shales appear to contain some humic matter that was part of the original sapropelic deposit. Metamorphism of a sapropelic deposit leads to such products as torbanite, oil shales, and asphaltites. *See* ASPHALT AND ASPHALTITE; OIL SHALE; TORBANITE. [I.A.B.]

Sapsucker

Any woodpecker of the genus *Sphyrapicus*, represented in the United States by two species, one with several distinct races. Best known is the yellow-bellied sapsucker, *S. varius varius*, of the eastern United States and southern Canada. Sapsuckers are perhaps the only woodpeckers that can be

The yellow-bellied sapsucker, *Sphyrapicus varius*; length to 8⅘ in. (*From E. L. Palmer, Fieldbook of Natural History, McGraw-Hill, 1949*)

accused of damaging trees. This is because of their habit of drilling closely spaced rings of holes around trees, especially maples, to obtain the inner bark, which they eat. Sapsuckers also revisit old drillings to eat the sap and the insects that are trapped in it. *See* PICIFORMES; WOODPECKER.

[J.D.B.]

Sarcodina

A subphylum of the Protozoa in which movement involves protoplasmic flow, sometimes with recognizable pseudopodia. However, certain species develop flagella at a particular stage such as the gametes of certain Foraminiferida. These flagellate stages in the life cycles show dimorphism. Most Sarcodina are floating or creeping; a few are sessile. The pellicle is quite thin, hence, the body is plastic and shows ameboid movement unless restrained by skeletal structures. Sarcodina are represented in fresh, salt, and brackish waters and as ectoparasites and endoparasites of many hosts.

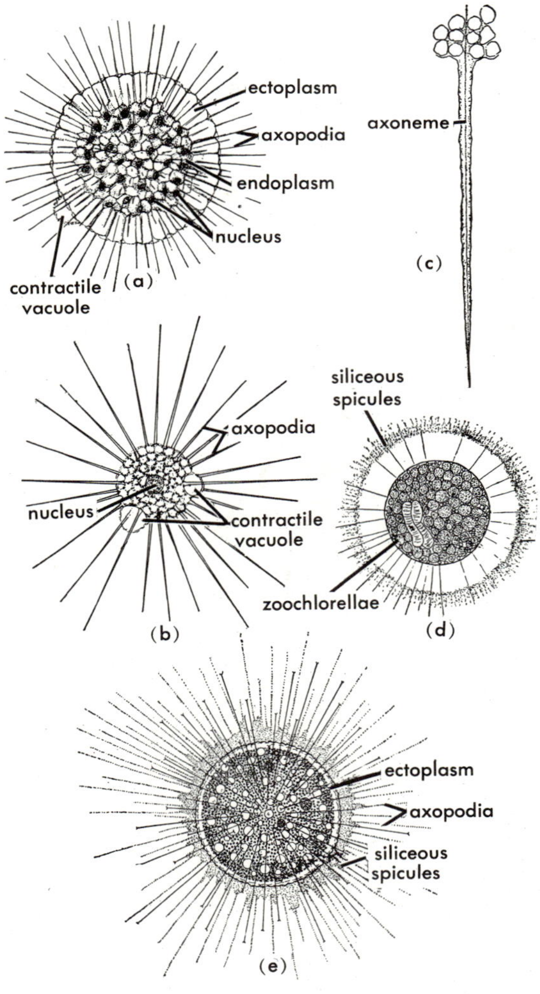

Fig. 1. Heliozoa. (*a*) *Actinosphaerium*, from life. (*b*) *Actinophrys*, from life. (*c*) Axopodium of *Actinosphaerium*, highly magnified. (*d*) *Heterophrys*, from life. (*e*) *Acanthocystis* (*after Leidy*). (*From L. H. Hyman, The Invertebrates, vol. 1, McGraw-Hill, 1940*)

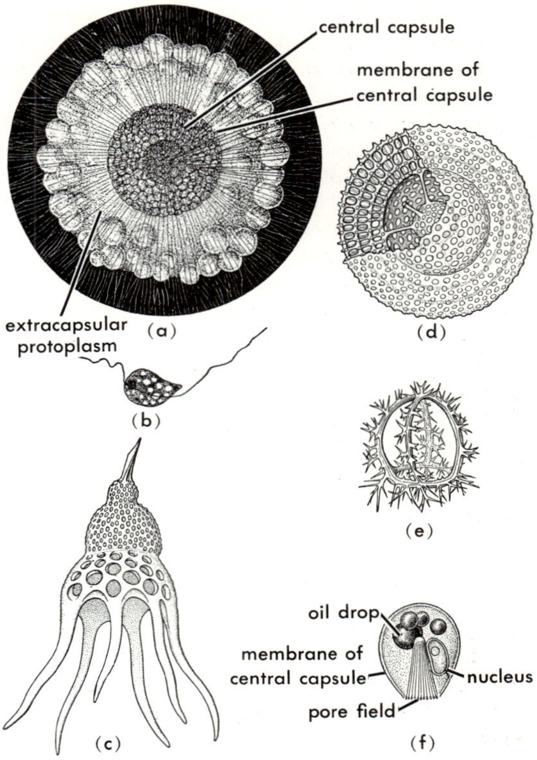

Fig. 2. Radiolaria. (*a*) *Thalassicola*, one of the Peripylina without skeletal elements (*after Huth*). (*b*) Biflagellate gamete (*after Le Calvez*). (*c, d, e*) Skeletons representing certain Monopylina (or Nassellina), Peripylina (or Spumellina), and Monopylina (*after Haeckel*). (*f*) Central capsule, one of Monopylina showing one group of pores (*after Haeckel*). (*From L. H. Hyman, The Invertebrates, vol. 1, McGraw-Hill, 1940*)

On the basis of pseudopodial structure, the Sarcodina have been divided into two classes, the Actinopodea, typically with axopodia, and Rhizopodea, in which the pseudopodia are of types other than axopodia. The Actinopodea include the orders Helioflagellida, Heliozoida, and Radiolarida and the Rhizopodea comprise the orders Proteomyxida, Mycetozoida, Amoebida, Testacida, and Foraminiferida.

Orders of Actinopodea. Helioflagellida, in addition to axopodia, have one or more flagella, either constantly or in one phase of a dimorphic cycle. Some are marine; others occur in fresh water.

Heliozoida (Fig. 1) have thin, radially arranged pseudopodia along which a flow of granules is characteristic. Although axonemes have not been seen in certain genera such as *Choanocystis* and *Hedriocystis*, typical axopodia occur in most Heliozoida. Food is captured when a microorganism adheres to the sticky axopodia. Axonemes may disappear in this region and the prey becomes surrounded by cytoplasm.

Radiolarida (Fig. 2) are pelagic organisms represented by fossils dating from Lower Silurian deposits. Pseudopodia are supposedly axopodia but axonomes are missing in some species. Unlike the Heliozoida, Radiolarida have a central capsule separating the inner from the outer cytoplasm. The capsule is commonly spherical to ovoid and is perforated to permit cytoplasmic continuity. Nuclei, which are intracapsular, vary in number. Usually there are many in Actipylina and one in Monopylina and Tripylina. Skeletal elements may be composed of strontium sulfate in the suborder Actipylina, or of siliceous material as in most of the Radiolarida. In Actipylina the major elements are spines radiating from the center of the body. In siliceous types the skeletal elements, which lie outside the central capsule, show a variety of intricate patterns (Fig. 2) such as perforated shells with single or concentric layers, bivalve shells, helmet-shaped shells, tripods, and other forms.

Orders of Rhizopodea. Skeletal elements of Rhizopodea range from none in amebae to complicated foraminiferan tests (Fig. 3). In certain Testacida, scales or plates, secreted by the organism, are cemented into a solid test as in *Euglypha.* With one reported exception, *Paraquadrula,* in which the plates are of calcium carbonate, the tests of Testacida are mainly siliceous except in a few primitive genera with flexible tests. In *Euglypha,* skeletal plates are formed during growth and stored in the cytoplasm. At fission, one of the daughter organisms receives the stored plates for its new test. In *Difflugia* and related genera, sand grains or other foreign particles are cemented into an arenaceous test.

In certain Foraminiferida, a primitive chitinous test becomes the test of the adult. Usually, however, an initial chitinous test is strengthened during growth by the addition of inorganic salts or foreign particles. In formation of arenaceous tests (Fig. 3*j*), sand grains, sponge spicules, and other materials are cemented onto a primary chitinous test. However, tests of most Foraminiferida are mainly calcareous, rarely siliceous, secretions, the material presumably corresponding to the cements used in arenaceous tests. During growth of some primitive types, the old test is discarded and replaced by a new and larger one. More specialized multilocular types merely add new chambers to the preceding ones as growth continues (Fig. 3). The protoplasm is continuous from one chamber to the next, since the aperture of each chamber remains partly or completely open after a new chamber is added. These openings, or foramina, between chambers suggested the taxonomic name for the group. At maturity, the final aperture may be different from the apertures of the earlier chambers.

The pseudopodia of Foraminiferida are myxopodia (rhizopodia) which form a sticky network, often covering an area many times the diameter of the test. They thus serve as an efficient food trap (Fig. 3*m*, *p*). Once surrounded by pseudopodia, the food undergoes digestion, which may be partial or complete in certain species, outside the test. Myxopodia may, in addition, take part in construction of test and cyst walls and, in locomotion, can pull the body along a surface toward some point of attachment. Some species creep at the rate of several millimeters an hour.

Life cycles include syngamy in certain Heliozoida, Mycetozoida, and Foraminiferida. Evidence is less than conclusive for other groups. A few foraminiferan cycles show dimorphism correlated

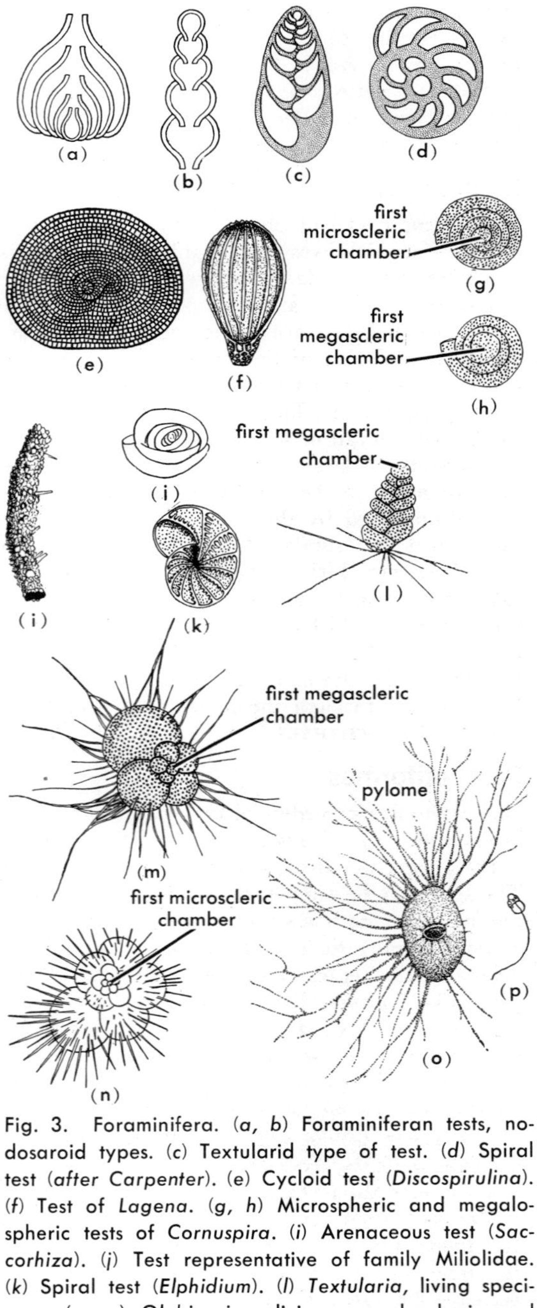

Fig. 3. Foraminifera. (*a, b*) Foraminiferan tests, nodosaroid types. (*c*) Textularid type of test. (*d*) Spiral test (*after Carpenter*). (*e*) Cycloid test (*Discospirulina*). (*f*) Test of *Lagena*. (*g, h*) Microspheric and megalospheric tests of *Cornuspira*. (*i*) Arenaceous test (*Saccorhiza*). (*j*) Test representative of family Miliolidae. (*k*) Spiral test (*Elphidium*). (*l*) *Textularia*, living specimen. (*m, n*) *Globigerina*, living, megalospheric and microspheric. (*o*) *Gromia* (*after Jepps*). (*p*) Supposed flagellated gamete of *Gromia*. (*From L. H. Hyman, The Invertebrates, vol. 1, McGraw-Hill, 1940*)

with an alternation of generations (Fig. 3*n*, *o*) in which there is an asexual microspheric adult, with a small initial chamber and a gamete-producing magalospheric generation, with a large initial chamber. The asexual adult divides into a number of young organisms which build new primary tests. In some species, a cyst wall is laid down before reproduction begins. The gamete-producing organisms produce gametes which are ameboid or flagellated in different species. Syngamy occurs, and each zygote begins construction of a new microspheric test. According to a few reports, the gamete-producing generation is haploid in certain species. Length of the life-cycle ranges from 2–3 weeks to a year or longer in different species. *See* PROTOZOA. [R.P.H.]

Sarcopterygii

The Sarcopterygii or choanate fishes, also known as the Choanichthyes or Amphibioidei, comprise one of the two subclasses of the bony fishes (class Osteichthyes). They are characterized by internal nares, the primitive occurrence of two dorsal fins, an epichordal lobe of the caudal fin, and marked extension of flesh and skeletal supporting elements into the paired fins. Two superorders are included, the Crossopterygii or lobefin fishes and the Dipnoi or lungfishes. These were already distinct when they appeared in the mid-Devonian and most species disappeared by the early Mesozoic, but one crossopterygian survives at middle-depths of the sea off eastern Africa, and three genera of lungfishes persist in fresh-water swamps of the southern continents. Although of minor importance in modern life, the Sarcopterygii are of high evolutionary significance since the terrestrial vertebrates have evolved through them. *See* CROSSOPTERYGII; DIPNOI; OSTEICHTHYES. [R.M.B.]

Sarcoptiformes

One of the five suborders of the order Acari. These are minute (0.3–1.5 mm long) globular mites without stigmata, but there may be a tracheal system. The legs may be simple, or enlarged in the male, and may terminate in suckers, claws or in a modification of both; their coxae form subdermal apodemes on the venter. The chelicerae are normally pincerlike. In normal development, these creatures pass through three stages, larva, first nymph, and second nymph, before becoming adult. Under certain conditions a dispersal form, the hypopus, may develop between the two nymphal stages.

This group can be separated rather easily into the pale, weakly sclerotized Acaridiae and the dark, heavily sclerotized Oribatei. The former group includes such economically significant forms as the free-living grain and cheese mites, the parasitic itch mites, and feather mites, while the latter group includes the oribatid mites, all of which are free-living, though some act as intermediate hosts of tapeworms.

In the Acaridiae are found some of the most serious pests of stored food products, such as *Acarus siro* L. (= *Tyroglyphus farinae* Latr.) and *Tyrophagus castellanii* (Hirst) (Fig. 1). These destroy large quantities of wheat and other grains by their direct action, and also by changing the relative humidity of these products in storage. They also damage stored cheese and dried fruits. The related *Caloglyphus* and *Rhizoglyphus* spp. are responsible for destruction of many varieties of cultivated bulbs, tubers, and corms. Economically, the parasitological significance of the Acaridiae is secondary to their destruction of food. *Glycyphagus domesticus* (de Geer), which lives normally on dried fruits and organic matter, can cause a skin irritation known as grocer's itch. The same mite has been reported as the intermediate host of *Catenotaenia pusilla* (Goeze), a cestode parasite of rodents. The Sarcoptids (Fig. 2) are skin parasites of warm-blooded vertebrates. *Sarcoptes scabiei* and its varieties burrow in the skin of man, cattle, sheep, goats, camels, horses, and dogs under crowded winter conditions and produce an itching that can lead to secondary infections. *Notoedres* spp. can cause severe mange in the head region of the cat and

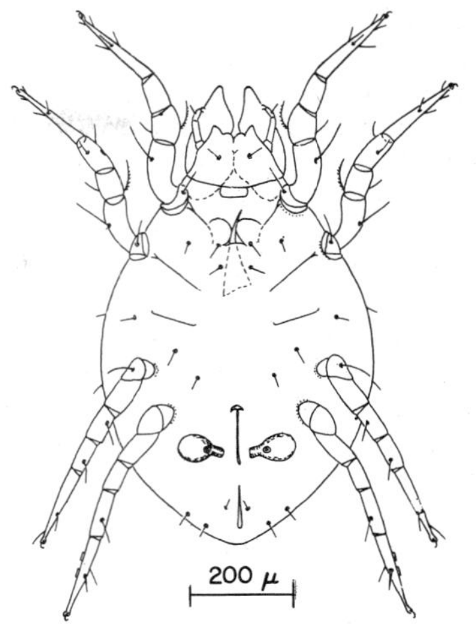

Fig. 1. A sarcoptiform mite pest of stored food. (*The Institute of Acarology*)

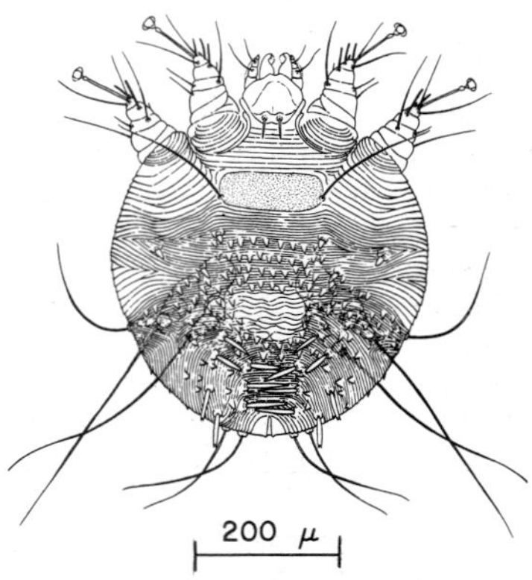

Fig. 2. The human itch mite. (*The Institute of Acarology*)

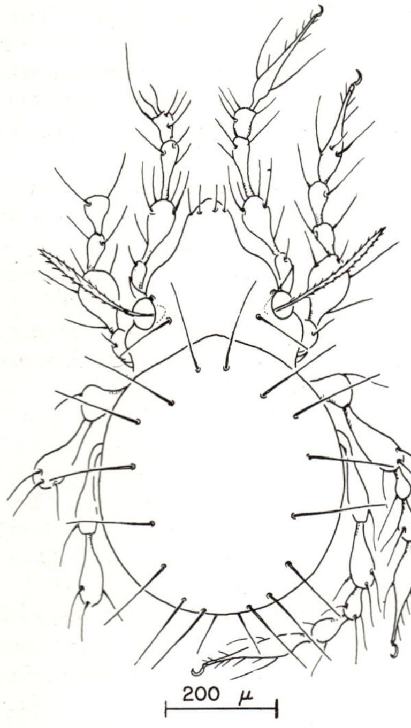

Fig. 3. An oribatid mite. (*The Institute of Acarology*)

dog, and *Knemidokoptes mutans* R. and L. causes the highly contagious "scaly leg" of poultry. A variety of this species can deplume poultry. The psoroptid mites are represented by *Psoroptes equi* and its many varieties that cause sheep-scabs and psoroptic mange in horses and cows. A milder type of mange is caused by the species of *Chorioptes*, and *Otodectes cynotis* Hering is the causal agent of canker of the ears of cats, dogs, foxes, and ferrets.

Until recently the oribatid mites (Fig. 3) were regarded as free-living soil or compost organisms of no economic significance. Recent work, however, has shown that they play an important part in the breakdown of organic matter, and that they act as an intermediate host of tapeworms. To date, 13 species of cyclophyllidean tapeworms of the families Catenotaeneidae and Anoplocephalidae have oribatid mites as intermediate hosts (*see* CYCLOPHYLLIDEA). *Moniezia expansa*, a cosmopolitan parasite of ruminants, has been taken from at least nine different species of oribatids of the families Galumnidae, Oribatulidae, and Pelopidae; and *Bertiella studeri*, a parasite of man, from the Galumnidae and Oribatulidae. One mite, *Scheloribates laevigatus* of the family Oribatulidae, is the vector for *Anoplocephala magna*, *A. perfoliata*, *Bertiella studeri*, *Cittotaenia ctenoides*, *C. denticulata*, *Moniezia benedeni*, *M. expansa*, and *Thysaniezia giardi*. D. M. Allred gives a complete list of oribatid mites and their tapeworms. [H.H.J.N.]

Bibliography: D. M. Allred, Mites as intermediate hosts of tapeworms, *Proc. Utah. Acad. Sci.*, 31:44–51, 1954; E. W. Baker and G. W. Wharton, *Introduction to Acarology*, 1952; H. G. Vitzthum, *Acarina*, in H. G. Bronn (ed.), Klassen und Ordnungen des Tierreichs, vol. 5, pt. 4, 1943.

Sarcosporidia

A large group of microorganisms of uncertain systematic position which are parasitic in skeletal and cardiac muscle of many species of vertebrates. Their host range is extremely wide and includes reptiles, birds, and mammals. Man has occasionally been found to be infected, and the human parasite is often referred to as *Sarcocystis lindemanni*. It is doubtful if Sarcosporidia found in different host species are themselves different species, although this has often been assumed. Some think that only one valid species exists, *Sarcocystis miescheriana* (Kuhn).

Even though sarcosporidiosis is common in nature because these parasites are so widespread, much remains unknown about them. The Sarcosporidia have been variously regarded as an order, or a subclass of Sporozoa, or simply as a group "of undetermined position." Recent work by L. Spindler (not yet confirmed) suggests that they may be fungi. Of considerable interest is the fact that the Sabin-Feldman dye tests for toxoplasma infection are said to give a high proportion of positive results in sarcosporidiosis. A close relationship between the two diseases is perhaps indicated by this and other facts which point in the same direction.

The life cycle of *Sarcocystis* of the pig, as developed by Spindler, involves the production of minute ovoid bodies from the infective spores (Rainey's corpuscles). These form mycelia and hyphae, as does the typical mold *Aspergillus*, which is possibly closely related.

The spores are thought to give rise to new infections when ingested with feces-contaminated food and water. Upon reaching the muscles of the new host, presumably through the blood stream, they form "Miescher's tubes." These structures are often large enough to be visible to the naked eye.

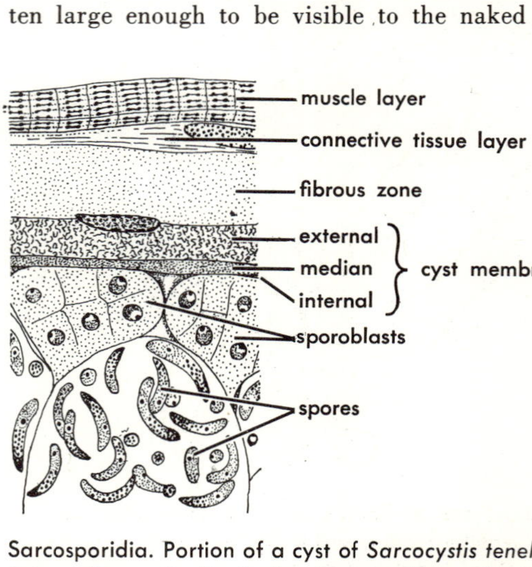

Sarcosporidia. Portion of a cyst of *Sarcocystis tenella* in sheep. (*From R. R. Kudo, Protozoology, Charles C Thomas, 4th ed., 1954*)

Mild infections are probably the rule in nature and seem to do little harm; heavy infections are sometimes serious. *See* Eurotiales. [R.D.M.]

Bibliography: J. W. Scott, The Sarcosporidia: a critical review, *J. Parasitol.*, 16:111–130, 1930; L. A. Spindler and Harry E. Zimmerman, The biological status of *Sarcocystis*, *J. Parasitol.* (suppl.), 31:15, 1945.

Sargasso Sea

A region of the North Atlantic Ocean. The boundaries of the Sargasso Sea are clearly defined in the west and north by the Gulf Stream. In the east it extends to 40°W and in the south to 20°N, though these boundaries are less definite (Fig. 1).

The Sargasso Sea gets its name from the indigenous, yellow-brown, floating seaweed, *Sargassum*, which is found throughout the sea. The floating masses of *Sargassum* contrast vividly with the deep blue of the water. It has been estimated by A. E. Parr that there are 4,000,000–11,000,000 tons of this weed floating in the Sargasso Sea and its environs (Fig. 2).

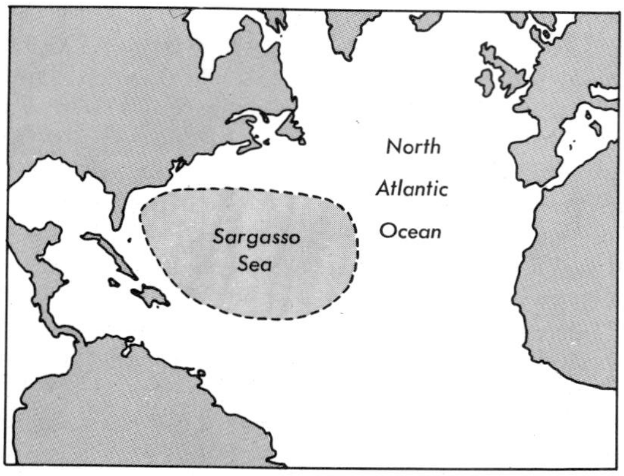

Fig. 1. Sargasso Sea.

Fig. 2. Piece of *Sargassum*, gulf weed, with stemlike stipe, leaflike blades, and berrylike bladders, or floats. (*From H. J. Fuller and O. Tippo, College Botany, rev. ed., Holt, 1954*)

Dynamically, the Sargasso Sea is a high cell (clockwise gyre) and the currents revolve anticyclonically about its center. Much of its deep circulation is involved with the Gulf Stream, to which it contributes a volume of about 45,000,000 m^3/sec between the Straits of Florida and Cape Hatteras. It is not yet clear how this water is recirculated. *See* Ocean currents.

At the surface the contribution is small and sometimes (especially in the summer months) it is reversed. In consequence the upper layers of the Sargasso Sea have a closed circulation. Water is cooled in the northern end of the sea to 18°C and mixes to a depth of about 350 m at the end of each winter. The excess quantities of this 18°C water flow off to the south and a distinct wedge of water with a temperature of 18°C can be found throughout the Sargasso Sea at a depth of 300 m. *See* Atlantic Ocean; Gulf Stream. [L.V.W.]

Bibliography: A. E. Parr, Quantative observations on the pelagic *Sargassum* vegetation of the western North Atlantic, *Bull. Bingham Oceanog. Collection*, 6(7):1–94, 1939; L. V. Worthington, The 18° water in the Sargasso Sea, *Deep-Sea Research*, 5(4):297–305, 1959.

Sarraceniales

A small order of the plant subclass Dicotyledoneae including 3 families with 9 genera and 114 species of interesting insectivorous plants. Insects are captured by the plant and digested by the plant's enzymes. In the Sarraceniaceae, an American family, and in the Nepenthaceae of the Old World tropics, the leaves form pitchers containing water. Insects falling into this liquid are drowned and digested. In the sundew family (Droseraceae) insects are trapped in two ways: in Venus' flytrap (*Dionaea muscipula*), the halves of the leaf fold together capturing the insect between them; in the sundews (*Drosera*), numerous glandular hairs (tentacles) on the leaf secrete a viscous fluid which traps a visiting insect. The tentacles then bend inward about their victim bringing it into contact with the surface of the leaf where it is digested. *See* Insectivorous plants; Pitcher plant; Sundew; Venus' flytrap; *see also* Dicotyledoneae; Embryophyta; Plant kingdom. [P.D.S.]

Sarsaparilla

A flavoring material obtained from the roots of at least four species of the genus *Smilax*. These are tropical American vines with prickly stems. *Smilax medica* of Mexico, *S. officinalis* of Honduras, *S. papyracea* of Brazil, and *S. ornata* of Jamaica, all plants of the dense, moist jungles, are the principal sources of sarsaparilla. The flavoring is used

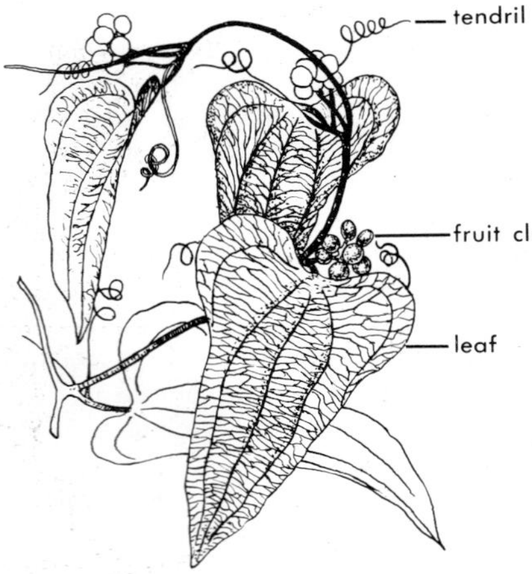

Smilax aristolochiaefolia yielding Mexican sarsaparilla. (*After Bentley and Trimen from E. P. Claus, Gathercoal and Wirth Pharmacognosy, 3d ed., Lea and Febiger, 1956*)

mostly in combination with other aromatics such as wintergreen. *See* LILIALES; SPICE AND FLAVORING. [P.D.S.]

Sassafras

A medium-sized tree, *Sassafras albidum*, of the eastern United States and extending north as far as southern Maine. Sometimes it is only a shrub in the north, but from Pennsylvania southward heights of 90 ft or more with diameters of 4–7 ft have been reported for this plant. Sassafras is said to live from 700 to 1000 years. It can be recognized by the bright green color and aromatic odor of the twigs and leaves. The leaves are simple or mitten-shaped (hence a common name "mitten-tree"), or they may have lobes on both sides of the leaf blade. The wood is soft, brittle, coarse-grained, somewhat aromatic, and has reddish heartwood. *See* STEM (BOTANY); XYLEM. Because it is durable in contact with moisture, it is used for fence posts, rail fences, sills for houses, and in the construction of boats. However, the supply is limited, and little is sold. The aromatic substance, found especially in the roots, is a volatile oil (oil of sassafras, USP) used medicinally as a stimulant, diaphoretic, and also as a flavoring agent. *See* FOREST AND FORESTRY; TREE. [A.H.G.]

Sassafras albidum.

Satellite (astronomy)

A small, solid celestial body circulating around a planet. The Moon is the satellite of Earth. The various planets of the solar system have the following numbers of known satellites: Mercury, Venus, and Pluto, none; Earth, 1; Mars, 2; Neptune, 2; Uranus, 5; Saturn, 9; and Jupiter, 12.

The largest satellites are over 3000 miles in diameter (Ganymede and Callisto of Jupiter, Titan of Saturn), and exceed the diameter of Mercury. The smallest known satellites (Jupiter VII to XII) are probably less than 40 miles in diameter. The mass and surface gravity of nearly all satellites are too small to retain atmospheric gases around them, although the presence of an atmosphere of methane has been detected in the case of Titan.

The mass of a planet is derived most directly from the semimajor axes of the orbits and periods of its satellites.

A number of artificial satellites have been placed in temporary orbits around Earth since the launching of Sputnik I by the Soviet Union on October 4, 1957. *See* SATELLITE, ARTIFICIAL; *see also* JUPITER; MARS; MOON; NEPTUNE; PLANET; SATURN; URANUS. [G.D.V.]

Satellite, artificial

An object placed by man in a semipermanent orbit around a celestial body such as Earth or the Moon (Fig. 1). A satellite revolves around the primary body under the influence of the gravitational field in an orbit which is usually elliptical but may be circular. The required velocity for a close-in satellite of Earth is 18,000 mph. The plane of the orbit will remain relatively fixed in inertial space, precessing at a relatively slow rate if the satellite moves in a nonuniform gravitational field such as that of Earth. This plane will always contain the center of mass of the primary body; hence its inclination to the equatorial plane of the primary body can never be less than the latitude of the launching place (Fig. 2). Because a satellite in orbit is independent of the axial rotation of the primary body, to an observer on the primary body there is an apparent motion of the orbit plane in the direction opposite that rotation.

Orbital energy. Placing an object in orbit about a primary body requires that energy be imparted, first to lift the object against the force of gravity to the desired height, and then to give the object, in the case of a circular orbit, an orbital velocity just great enough to counteract the force of gravity. Artificial satellites of Earth were first put in orbit in October, 1957.

Work done in lifting a satellite to height h above the surface of Earth is

$$w_p = m_s \int_0^h g\, dh$$

where m_s is the mass of the satellite in slugs and g the acceleration of gravity. The value of g varies with height in this fashion:

$$g = g_0 \left(\frac{R_0}{R_0 + h}\right)^2$$

where g_0 is the value of g at Earth's surface (where $h = 0$) corrected for the centrifugal effect of Earth's rotation, about 32.2 ft/sec^2; and R_0 is

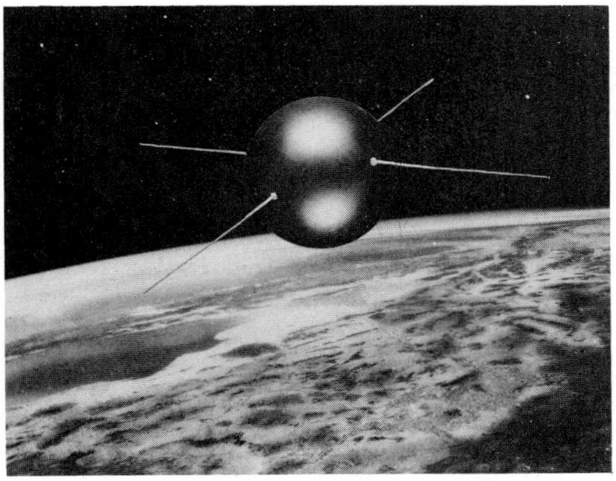

Fig. 1. Spherical artificial satellite with radio telemetering antenna.

Earth's mean equatorial radius, about 2.09×10^7 ft. Hence

$$\begin{aligned} w_p &= m_s g_0 R_0^2 \int_0^h \frac{dh}{(R_0 + h)^2} \\ &= m_s g_0 R_0 \frac{h}{R_0 + h} \\ &= \frac{m_s h}{R_0 + h} \times 6.7 \times 10^8 \text{ slug-feet} \end{aligned}$$

This work equals the kinetic energy the satellite must have, when ejected from the surface of Earth, to reach altitude h, so that

$$w_p = m_s g_0 R_0 \frac{h}{R_0 + h} = \tfrac{1}{2} m_s v_h^2$$

whence the initial velocity in feet per second of ejection must be

$$v_h = 3.66 \times 10^4 \sqrt{\frac{h}{R_0 + h}}$$

For escape from Earth $h = \infty$, and the term in the radical is therefore unity. The resulting value of v_h is known as escape velocity v_e, which is

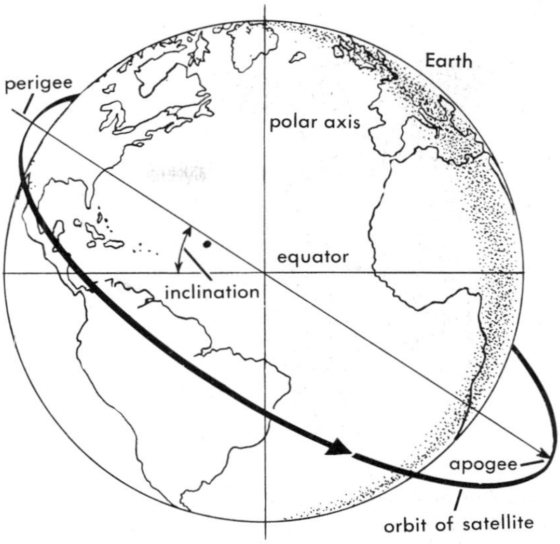

Fig. 2. Typical close-in orbit of an Earth satellite.

$$\begin{aligned} v_e &= 3.66 \times 10^4 \text{ ft/sec} \\ &= 6.93 \text{ mi/sec} \end{aligned}$$

The orbit of one body about another is a conic section; if stable, the orbit is either an ellipse or a circle, with the primary body at a foci or at the center of the circle (Fig. 3). The velocity of the satellite in an elliptical orbit is

$$v^2 = g_0 R_0^2 \left(\frac{2}{R_0 + h} - \frac{1}{a}\right)$$

and its kinetic energy w_k is

$$\tfrac{1}{2} m_s v^2 = m_s g_0 R_0^2 \left(\frac{1}{R_0 + h} - \frac{1}{2a}\right)$$

The total energy of the satellite in orbit, launched from surface R_0 where $h = 0$ is

$$\begin{aligned} T_e &= w_p + w_k \\ &= m_s g_0 R_0^2 \frac{h}{R_0 + h} + m_s g_0 R_0^2 \left(\frac{1}{R_0 + h} - \frac{1}{2a}\right) \\ &= m_s g_0 R_0 \left(1 - \frac{R_0}{2a}\right) \end{aligned}$$

Because all other terms are fixed, the energy in orbit is a function only of the semimajor axis a and is independent of the shape of the orbit. A typical satellite orbit is drawn to scale in Fig. 4.

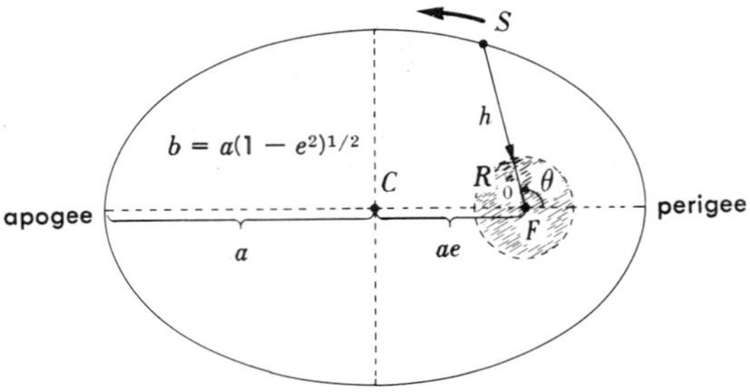

Fig. 3. Principal dimensions of an elliptical orbit.

Fig. 4. Plan view, to scale, of an orbit with 200-mile perigee and 1400-mile apogee. Short spiral at perigee is ascent trajectory.

Orbital velocity. A satellite travels slower as the size of its orbit increases, as shown by the table.

Effect of height on period of circular orbit

Height, miles	Velocity, fps	Period	
		Min	Days
0	25,900	84.5	
100	25,600	88	
400	24,700	98	
1000	23,100	118	
23,000	10,060	1440	1
230,000	3,360		27

In elliptical orbits, the velocity changes around the orbit, being greatest at perigee and least at apogee. The ratio of the two velocities is

$$\frac{v_{\text{per}}}{v_{\text{ap}}} = \frac{1+e}{1-e}$$

where e is eccentricity defined in Fig. 3.

Period P of an Earth satellite is proportional to the 3/2 power of its semimajor axis

$$P = \frac{2\pi a^{3/2}}{g_0 R_0} = 84.5a^{3/2} \text{ min}$$

where a is expressed in units of Earth's radius.

The velocity that must be imparted to a satellite to raise it to the desired orbital height and then to cause it to move forward at such a velocity that it will stay in orbit is in excess of 25,000 fps. A body moving at this velocity in the lower, denser atmosphere would overheat before it could reach the near vacuum of the outer atmosphere. Further, drag forces would be excessively great. For this reason, rockets have proven to be the only feasible means of placing satellites in orbits. The rocket engine can be adjusted so that its impulse is exerted over a long period of time, thereby allowing the vehicle to climb out of the lower atmosphere at safe speeds. In addition, the low acceleration reduces the mechanical load on the delicate instruments in the vehicle.

The terminal velocity of a rocket in vertical flight is reduced by the work done against gravity and by the aerodynamic drag. This terminal velocity is

$$v_r = c \ln \frac{m_0}{m_e} - gt - A$$

where c = exhaust velocity of rocket motor
m_0 = initial weight, rocket and fuel
m_e = final weight, empty rocket
g = acceleration of gravity
t = time
A = velocity loss due to drag

The final velocity is thus determined by the exhaust velocity c and mass ratio m_0/m_e attainable. With chemical fuels and presently available materials it is not possible using a single rocket to achieve sufficiently high values for c and m_0/m_e to give a final velocity sufficient for satellite launching. A technique known as staging, wherein a smaller rocket is carried aloft by a larger booster, must be employed. The empty casing of the booster is discarded after use, the second rocket accelerating only its own weight during burning. Two or more stages may be used. *See* ROCKET STAGING.

A successful technique has been to use the first one or two rockets to lift the vehicle to the desired height and then, after properly orienting the vehicle, to use the last one or two stages to bring the satellite payload up to orbiting velocity (Fig. 5). During the launching it is necessary to control the velocity and angle of projection to close limits. If the velocity is too low, the satellite will fall to earth; if it is too high, the apogee will become large; and if it exceeds the circular velocity by 40%, the satellite will escape from Earth.

If the velocity of projection is parallel to Earth's surface, the optimum orbit will be obtained. If there is an error in angle in the vertical plane, either above or below the horizontal, the orbit will approach more closely to Earth than the point of injection; if the error is too large, the satellite will strike Earth.

Satellite lifetime. The drag on the satellite is proportional to the atmospheric density, which decreases exponentially with height, and to the projected area of the satellite. For a lifetime in orbit of 2 weeks or more, perigee must exceed 110 miles, depending on the apogee attained and the mass-to-area ratio of the satellite.

The achieved or initially predicted lifetimes of early satellites placed in orbit plotted against perigee distance show that the lifetime increases rapidly as the perigee increases (Fig. 6). The Vanguard I satellite, which has a perigee of 406 miles, is expected to stay in orbit for at least 200 years.

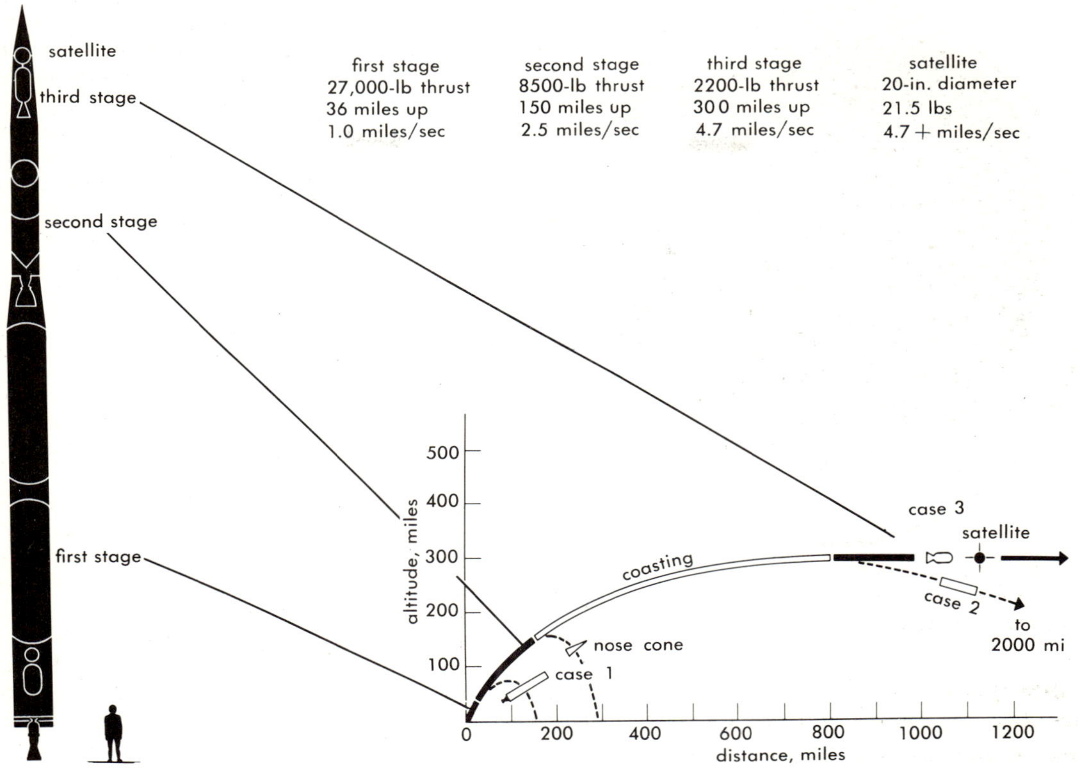

Fig. 5. Sample trajectory used with Vanguard vehicle to launch satellite.

The orbit of an Earth satellite is not perfectly elliptical. As the satellite passes through the rarified air, collisions with air molecules absorb energy from the satellite. This loss in energy causes the semimajor axis to contract and, anomalously, causes the satellite to speed up. Apogee changes more rapidly than does perigee in this decay process. As the end of life approaches, the decay becomes more rapid because the satellite travels more in the denser atmosphere. The satellite spirals into the lower atmosphere and becomes incandescent. By careful measurement of the rate of change of period, and with a knowledge of the geometry of the satellite, the density of the air at great heights has been measured.

Motion of the satellite is further perturbed by the nonspherical shape of Earth. Because of the proximity of the satellite to Earth, the gravitational attraction of the equatorial bulge is a force which has one component toward the center of Earth and the other perpendicular to the plane of the orbit. This results, through gyroscopic action, in a precession of the orbit about Earth's axis. Accurate measurement of the precession, as is possible with a satellite relatively free of atmospheric drag such as Vanguard I, allows a determination of the ratio of the polar to the equatorial axes of Earth.

Tracking techniques. Earth satellites are tracked by both optical and radio techniques. To be visible,

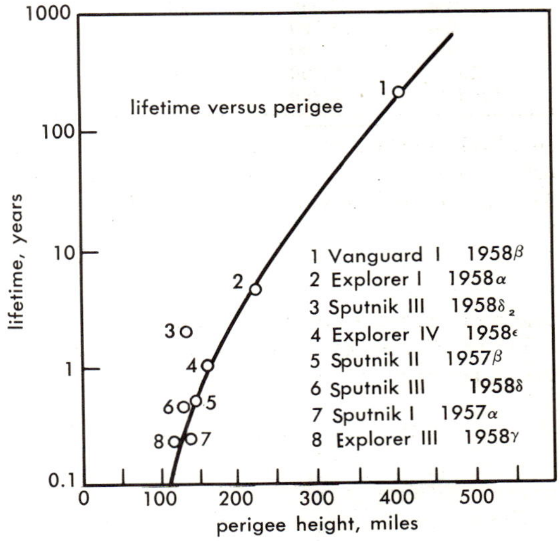

Fig. 6. Satellite lifetime increases with orbital height.

the satellite must be in full sunlight, and the observer must be in relative darkness. At best, most satellites will be no brighter than a first- or second-magnitude star and may be as faint as eighth or ninth magnitude. Precision optical tracking is done with fast cameras similar to meteor cameras, using Schmidt optics and special timing shutters.

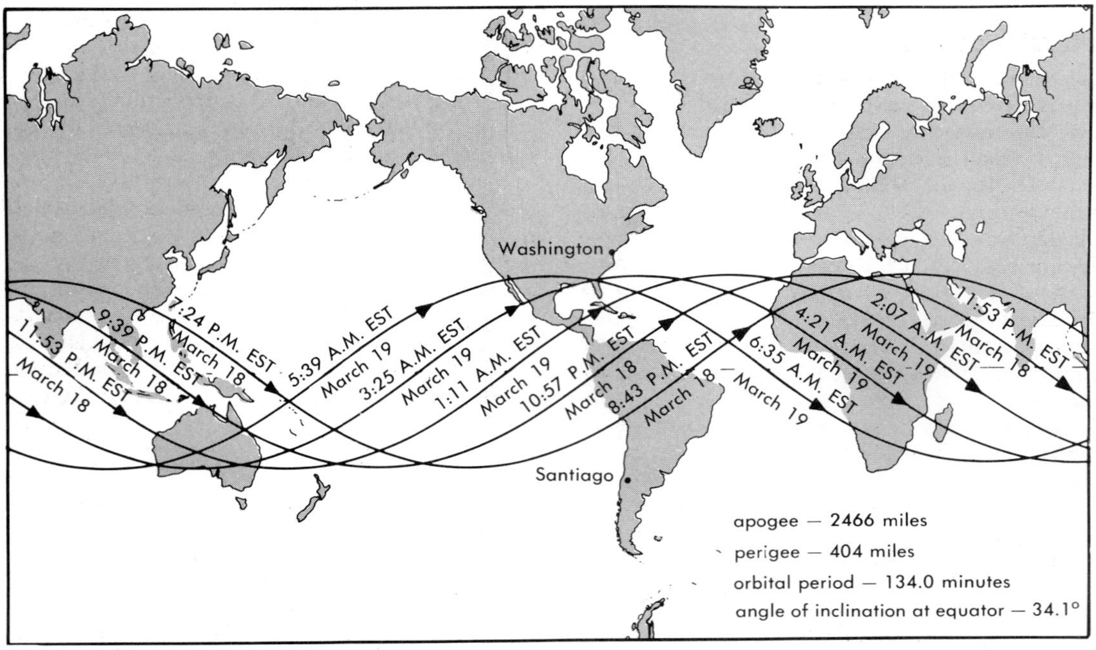

Fig. 7. Orbits of Vanguard satellite, 1958. Times show computed equatorial crossings for five successive orbits shortly after launching on March 17, 1958.

A more certain but somewhat less accurate method of tracking is through the use of radio. Several radio methods are used. In one, radiation from a radio beacon in the satellite is detected and measured by an interferometer-type antenna system (*see* MINITRACK). From the relative phase of the signals on two widely spaced antennas, the direction of the satellite may be obtained.

A second radio method uses the Doppler effect on the incoming beacon signal. As the satellite approaches the observer the apparent frequency of its transmissions falls, goes through a point of inflection where the signal is nearly constant in frequency when the satellite is its nearest to the receiver antenna, then starts to fall again as it moves away. Observations of the time of zero change in frequency and the rate of change in frequency can yield orbital information.

A third radio method is the use of radar. Only extremely powerful radars with large antennas can detect the relatively tiny satellites. The most accurate quantity measured by radar is the distance of the satellite.

Observers in the United States using the Minitrack system have employed large fan-beam antennas located in a string of stations stretching from Washington, D.C., to Santiago, Chile, to construct a radio "fence" along the 75th meridian. By this means, satellites with active beacons are intercepted and measured each time they circle Earth (Fig. 7). In the time the satellite requires to complete one revolution, Earth turns about 25° on its axis, so that the next pass is shifted to the west by that amount.

Satellites are placed in orbit for a purpose—usually to make geophysical or astronomical measurements outside the blanketing atmosphere of Earth. The results of the measurements are telemetered to Earth by impressing the information as a modulation on the tracking beacon signal. Sensitive receivers at stations on Earth record these signals and make them available to the experimenter.

An artificial satellite is a carefully constructed device. It must make maximum use of every ounce of weight, and it must be able to withstand the acceleration of 10–40 *g* during launch as well as high vibrational accelerations due to the noise of the rocket motors. Finally, it must contain its own temperature control. A satellite will absorb heat from the Sun, from Earth, and from the electrical losses in its own circuits. It can dispose of this heat only by radiation. It is therefore necessary to control the visual and infrared emissivity and absorptivity of the surface to keep the temperature within the range tolerated by the experiments and the electronic circuits. For example, satellites launched may have a coating that is transparent to visible light but opaque to infrared rays. In this way much of the Sun's radiation is reflected away, yet the surface remains an excellent heat radiator.

[J.P.H.]

Bibliography: M. Benton, *The Literature of Space Science and Exploration*, U.S. Naval Research Lab., Bibliography no. 13, September, 1958; L. V. Berkner, et al. (eds.), *Manual on Rockets and Satellites*, Ann. Spec. Comm. Intern. Geophys. Year, vol. 6, 1958; W. Ley, *Rockets, Missiles, and Space Travel*, rev. ed., 1957.

Saturable reactor

An iron-core inductor in which the effective inductance is changed by varying the permeability of the core. Saturable-core reactors are used to control large alternating currents where rheostats are impractical. Theater light dimmers often employ saturable reactors.

In the simplified diagram of two types of saturable-core reactors (*a*) shows use of two separate cores, while in (*b*) a 3-legged core is formed by placing two 2-legged cores together. The load winding, connected in series with the load, carries the alternating current and acts as an inductive element. The control winding carries a direct current, of adjustable magnitude, which can saturate the magnetic core. When the core is saturated by the direct current, the effective inductance of the coils in the ac circuit will be small. Little voltage can therefore be induced in the coil to reduce the voltage applied to the load.

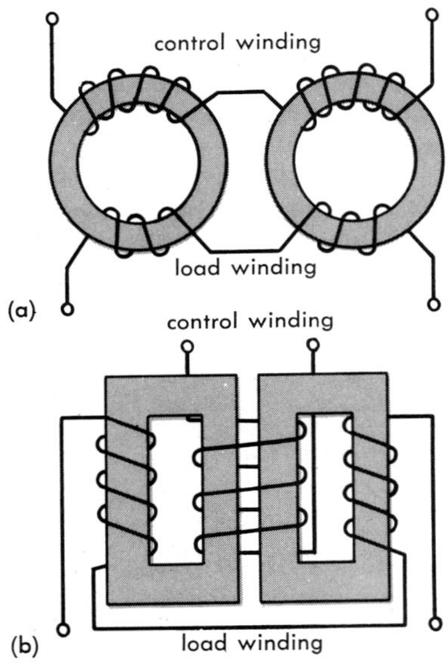

Typical construction of saturable-core reactors. (*a*) Two separate cores. (*b*) Three-legged cores.

If the dc magnetization is reduced to a minimum, the alternating current can induce a voltage in its own windings. The induced voltage opposes the applied voltage and therefore reduces the voltage applied to the load. *See* INDUCTANCE; MAGNETIZATION; PERMEABILITY, MAGNETIC. [B.L.R.; W.S.P.]

Saturation

The condition in which an increase in the independent variable quantity causes substantially no change in the dependent variable. Thus, magnetic saturation represents the maximum magnetization that a given magnetic material can have. Temperature saturation in a thermionic vacuum tube is the condition wherein further increases in cathode temperature cause substantially no further increase in anode current. Similarly, in anode saturation, further increases in anode voltage cause no further increases in anode current because all of the emitted electrons are already being drawn to the anode.

In color television, color saturation is the degree to which a color is mixed with white. Here high saturation means that there is little or no white, as in a deep red color. Low saturation means that there is a great deal of white, as in light pink.

In an induced nuclear reaction, saturation exists when the decay rate of a given radionuclide is equal to its rate of production. In an ionization chamber, saturation exists when the applied voltage is high enough to collect all the ions formed by radiation but not high enough to produce ionization by collision. [J.MR.]

Saturation current

A term having a variety of specific applications but generally meaning the maximum current which can be obtained under certain conditions.

In a simple two-element vacuum tube, it refers to either the space-charge-limited current on one hand or the temperature-limited current on the other. In the first case, further increase in filament temperature produces no significant increase in anode current, whereas in the latter, a further increase in anode-cathode potential difference produces only a relatively small increase in current. *See* VACUUM TUBE.

In a gaseous-discharge device, the saturation current is the maximum current which can be obtained for a given mode of discharge. Attempts to increase the current result in a different type of discharge. Such a case would be the transition from a glow discharge to an arc discharge. *See* ELECTRICAL CONDUCTION IN GASES.

A third case is that of a semiconductor. Here again, the saturation current is that maximum current which just precedes a change in conduction mode. *See* SEMICONDUCTOR. [G.H.MI.]

Saturation of solutions

This situation occurs when a solute gas, liquid, or solid has attained its maximum solubility in a solvent. In such a situation, the two phases are said to be in equilibrium. The attainment of saturation is sometimes a difficult experimental problem, for the rate of attaining equilibrium may be very slow. Repeated stirring or shaking over extended periods of time may be necessary.

If the second phase is removed from the system, the saturated solution may be cooled (or if solubility decreases with increasing temperature, heated) to become a metastable supersaturated solution. Given the proper catalyst (for example, dust, a seed crystal, or charged particles), a supersaturated solution may change spontaneously (and often rapidly) to the thermodynamically stable saturated solution by throwing out (precipitating in the case of a solid) a second phase. The Wilson cloud chamber uses a supersaturated gaseous solution of air and water vapor; the passage of a high-speed charged particle induces condensation of the water vapor to small visible droplets.

At equilibrium, the saturated solution has the same vapor pressure as that of the second phase; the unsaturated solution has a lower vapor pressure; and the supersaturated solution has a higher vapor pressure. *See* EQUILIBRIUM, PHASE; SUPERSATURATION; VAPOR PRESSURE. [R.L.S.]

Saturn

The second largest planet in the solar system and the sixth in the order of distance to the Sun. It is visible to the naked eye as a yellowish first-magnitude star, except during short periods near its conjunctions with the Sun. The outermost planet known until the seventeenth century, it is unique in having a large flat ring surrounding its globe.

Planet and its orbit. The main orbital elements are a semimajor axis or mean distance to Sun of 895×10^6 mi; eccentricity of 0.056, causing the distance to the Sun to vary by 10^8 mi between aphelion and perihelion; sidereal revolution period of 29.458 years; mean orbital velocity of 6.0 mi/sec; inclination of orbital plane to ecliptic of 2°5. *See* PLANET.

The mean apparent diameter of its disk varies between 14″ and 20″ depending on its distance from Earth; the apparent equatorial diameter is about 19″5 at mean opposition. The polar flattening due to the rapid rotation is the largest of all planets; the ellipticity $(r_e - r_p)/r_e = 0.105$. Here r_e is the equatorial radius and r_p is the polar radius. The equatorial diameter is about 75,500 mi and the polar diameter is 67,500 mi. The volume is about 762 (Earth = 1) with an uncertainty of a few per cent. The mass, about 95.3 (Earth = 1) or 1/3500 (Sun = 1), is accurately determined from the motion of its brighter satellites. The mean density is about 0.7 g/cm^3, the lowest mean density of all planets. The corresponding value of the mean gravity at the visible surface is 1.14 (Earth = 1) or about 11.2 m/sec^2; however, because of rapid rotation, the centrifugal force at the equator amounts to 1.76 m/sec^2, reducing the effective gravity to about the same value as on Earth.

Phases. Saturn's phases are always small. The maximum phase angle is about 6° at quadrature and the phase effect is barely detected as a slightly increased darkening of the edge at the terminator. The apparent visual magnitude at mean opposition is +0.8 when the ring is seen edgewise and the corresponding value of the reflectivity (geometric albedo) is about 0.4; the estimated value of the physical albedo is about 0.45. This high value is characteristic of the four major planets and indicates the presence of a dense cloud-laden atmosphere. *See* ALBEDO.

Telescopic appearance. Through an optical telescope Saturn appears as an elliptical disk, darkened near the limb and crossed by a series of bands parallel to the equator. As a rule, however, the bands are weaker than on Jupiter, and frequently only the bright equatorial zone and the

Fig. 1. Telescopic appearance of Saturn.

two dark tropical bands surrounding it are visible (Fig. 1).

The mean rotation period of a few occasional, short-lived spots observed in the equatorial zone is 10h14m; it is 10h38m at intermediate latitudes. The rotation axis is inclined 26°45′ to the perpendicular to the orbital plane.

As on Jupiter the belts vary in strength, width, and occasionally in latitude. They are very faint on photographs in red light, but show up much more strongly in violet and ultraviolet light, by which a dark polar cap is often made visible. Occasional bright spots are observed in the equatorial zone, which diffuse rapidly and spread over most of the zone; their nature is unknown.

Atmosphere. The optical spectrum of Saturn is characterized by strong absorption bands of methane, CH_4, that are stronger than on Jupiter, and by much weaker bands of ammonia, NH_3. The estimated quantities of these gases present in the atmosphere above the cloud level is of the order of 350 m in thickness at standard temperature and pressure (STP) and less than 2.5 m STP, respectively. Theoretical considerations indicate that, as on Jupiter, the main constituents of the atmosphere must be hydrogen and helium. The theoretical temperature, in agreement with radiometric observations, is about 120°K, indicating that most of the ammonia must be condensed and in the form of tiny ice crystals, which probably constitute the visible clouds.

Internal structure. The theoretical models of the internal structure are similar to those for Jupiter. The thermal radiation of Saturn at centimeter wavelengths has been detected, but no intense pulses of radio emission have been observed at longer wavelengths. *See* JUPITER.

Ring system. The most remarkable feature of Saturn's system is the ring or rings surrounding it. The ring system is divided in three main regions designated A, B, and C.

The outer ring A, of moderate brightness, has an outside diameter of 171,000 mi and an inner diameter of 150,000 mi. It is separated from ring B by the dark Cassini division which is 2,500 mi wide. Ring B, which is very bright (occasionally brighter than the equatorial belt on the globe), has an outer diameter of 145,000 mi and an inner diameter of 112,000 mi. It is slightly less bright in its inner re-

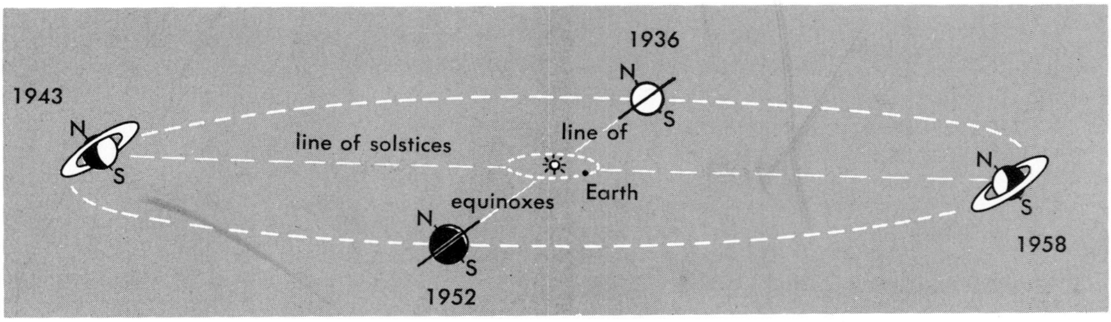

Fig. 2. Different presentations of Saturn's ring system.

gions and may be separated from the innermost ring C by a narrow dark division, perhaps 600 mi wide. Ring C, which is sometimes called the "crape" ring, is much fainter and semitransparent; it appears as a dark band in projection against the disk of the planet and as a faint dusky band against the sky. Its outer diameter is 111,000 mi. Its inner edge is only 7,000 mi above the surface of the planet at the equator. Additional divisions have been noted in the rings by some observers, but their reality remains in doubt; the only definite marking is the Encke division, which marks the limit between the outer darker zone and the inner brighter zone of ring A, but it is more nearly a lane of minimum brightness than a dark division.

Appearance. Depending on the relative positions of Earth and of Saturn in its orbit, the circular ring appears as an ellipse of variable ellipticity which reduces to a thin line when Earth crosses its plane. The maximum opening of the ring takes place when the ring system is tilted 27° to the line of sight (Fig. 2). The slightly variable point of view, depending on the position of Earth in its orbit, causes a slight annual oscillation of the tilt angle. When the tilt is near 0°, Earth may cross the plane of the ring system either once or three times. When Earth is exactly in this plane, the ring becomes invisible for a day or two, indicating that the thickness of the system is very small, perhaps less than 12 mi. For short periods near such times, the Sun and Earth may be on opposite sides of the plane of the ring, which then remains visible, being faintly illuminated by light reflected by the globe of Saturn and by multiple reflections in the ring system (Fig. 3).

Structure. Both theory and observations prove that the ring system is made up of myriads of separate particles which move independently in circular coplanar orbits in the equatorial plane of Saturn. The visibility of the globe of Saturn through ring C, and to a small extent through ring A, the incomplete disappearance of the satellites when in the shadows of these rings, and the visibility of stars shining through them, prove that at least rings A and C are relatively transparent. Photometric observations of the variation of brightness of the ring as a function of phase angle (up to its maximum value of 6°) also show that the particles are rather far apart and occupy perhaps not more than a few per cent of the volume of rings A and B, and a much smaller fraction in ring C.

The discontinuous, meteoric nature of the ring is also demonstrated directly by spectroscopic observations, which show that the inner edge of the ring rotates faster than the outer edge and precisely with the velocities that independent satellites would have at those distances from the planet.

Origin and nature. The origin and nature of Saturn's ring was first determined by E. Roche in 1849 and by J. C. Maxwell in 1859. Roche established that a satellite of a planet of the same density cannot form, because of destructive tidal forces, if it is closer to the planet than 2.44 times the planet's radius. Actually the outer edge of Saturn's ring is at 2.30 radii, inside Roche's limit, whereas the nearest satellite, Mimas, is at 3.11 radii, well outside the limit. Maxwell proved, further, that a ring system of small mass formed of a large number of independent particles is fairly stable against external perturbations such as those caused by the larger satellites. The aggregate mass of the ring system is very small, probably much less than one-quarter of the mass of the Moon. Periodic perturbations by the major satellites are, however, responsible for the main divisions of Saturn's ring in the same way that perturbations by Jupiter cause the Kirkwood gaps in the asteroidal belt (*see* ASTEROID). The radius of Cassini's division between rings A and B corresponds to a revolution period equal, respectively, to ½, ⅓, and ¼ of the revolution periods of the first three satellites; the limit or gap between rings B and C corresponds to a revolution period equal to ⅓ of that of the first satellite and the Encke division in ring A to ⅗ of the same period.

Because of the small but continual interactions between the particles of the ring and their occasional grazing collisions, it is probable that a

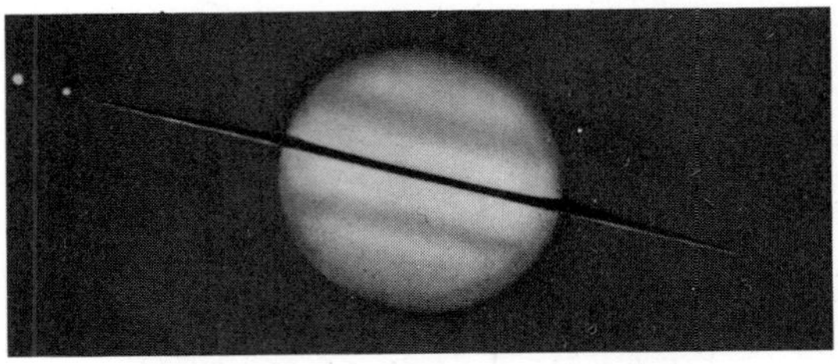

Fig. 3. Saturn's ring system seen edgewise.

grinding process similar to that taking place in the asteroidal belt must have caused the destruction of the larger original bodies and the formation of ever smaller particles. However, except perhaps when a very long time scale is envisaged, Saturn's ring appears to be a stable and practically permanent feature of the system.

Infrared spectra of Saturn's ring indicate that its particles are covered with frost of ordinary water-ice at a low temperature; it is even possible that the particles are themselves entirely made up of water-ice and solid ammonia.

Satellites. Saturn has nine known satellites. The largest and brightest, Titan, was discovered by C. Huygens in 1655 and is visible with small telescopes; the others are much fainter. The outermost satellite was discovered photographically by W. H. Pickering in 1898; a tenth satellite reported by him in 1905 has not been confirmed. Their main elements and orbits are given in the table.

Satellites of Saturn

Satellite	Mean distance from Saturn, 10^3 miles	Sidereal period, days	Diameter, miles	Visual magnitude at mean opposition
I Mimas	115.4	0.942	320	12.1
II Enceladus	149.0	1.370	370	11.7
III Tethys	183.4	1.888	800	10.6
IV Dione	234.5	2.737	800	10.7
V Rhea	327.8	4.518	1100	10.0
VI Titan	760	15.945	3100	8.3
VII Hyperion	922	21.277	250	14
VIII Iapetus	2213	79.331	750	11
IX Phoebe	8043	550.45	190	14.5

The satellite nearest to the planet, Mimas, moves in a period of 22h37m at a distance of only 30,000 mi beyond the outer edge of the ring. The largest satellite, Titan, moves in a period of about 16 days at a mean distance of 760,000 mi from the center of the planet. The outer main satellite, Iapetus, has a period of about 79 days and a mean distance of 2.2×10^6 mi. The very small outermost satellite, Phoebe, moves in a retrograde direction (direction in the opposite sense from that of the 8 inner satellites) at a mean distance of over 8×10^6 mi in a period of 550 days. Phoebe was the first satellite found to have a retrograde motion; its high orbital eccentricity (0.17) and relatively large inclination on Saturn's equatorial plane (5°.3) clearly separate it from the others. *See* RETROGRADE MOTION (ASTRONOMY).

Titan shows a measurable disk in large telescopes; the mean apparent diameter, about 0″.6, corresponds to a linear diameter of approximately 3100 miles, making it larger than Mercury and only slightly smaller than Mars. Its mass can be roughly estimated from the perturbations it exerts on the motions of the other satellites; it is 1–2 times the mass of the Moon, from which follows a density 2–4 times that of water. The escape velocity at the surface of Titan is of the order of 3 km/sec; this is large enough to retain an atmosphere of methane at the low temperature prevailing at this large distance from the Sun. Titan is the only satellite known to have an atmosphere.

All the other satellites are too small to show measurable disks, and their diameters can be only roughly estimated from their apparent brightness and an assumed value of the albedo. The periodic variations of brightness, depending on the positions on the orbits, indicate that all satellites of Saturn always turn the same face toward the planet; that is, as in the case of the Moon and of the main satellites of Jupiter, the periods of rotation are equal to the periods of revolution. Iapetus has the largest variation in apparent brightness, being 5 times brighter at western than at eastern elongation; the albedoes of its opposite hemisphere must therefore differ in the same ratio. The variations are much larger than for any other planet or satellite. The masses of Enceladus and Mimas can be only roughly estimated; these estimates indicate masses of about 1/1000 and 1/2000 respectively of that of the Moon. In conjunction with their observed brightnesses, such masses are possible only if these satellites have a density as low as or lower than that of ice and an albedo as high as that of snow. It is assumed that similar conditions exist for the other major satellites, except Titan. [G.D.V.]

Bibliography: H. N. Russell, R. S. Dugan, and J. Q. Stewart, *Astronomy*, vol. 1, rev. ed., 1945.

Saurischia

An extinct group of dinosaurian reptiles, now placed in the subclass Archosauria (superorder in some classifications) as a separate order (*see* ARCHOSAURIA). For a description of dinosaurian reptiles, *see* DINOSAUR.

Sauropterygia

An order of Mesozoic reptiles (subclass Euryapsida) that are, without exception, adapted to the marine environment. The order includes the closely related nothosaurs and plesiosaurs and the differently specialized placodonts. These reptiles along with the ichthyosaurs played a significant role as predators within the marine animal community of the Mesozoic Era.

Nothosauria. The Nothosauria are the relatively generalized stem group from which the plesiosaurs evolved. With the exception of a single New World species and a doubtful record from Japan the nothosaurs were of essentially European distribution during Middle Triassic time. The nothosaurs are notably diverse in the mode and degree of secondary aquatic modification. The early phase of this process manifests itself (1) by a change in the histological structure of the bones, characterized by a total absence of Haversian systems (*see* SKELETAL SYSTEM) and often a swollen appearance of the bones (pachyostosis); (2) by a notable reduction in size of the lateral girdle bones, scapulae (shoulder) and ilia (pelvis); and

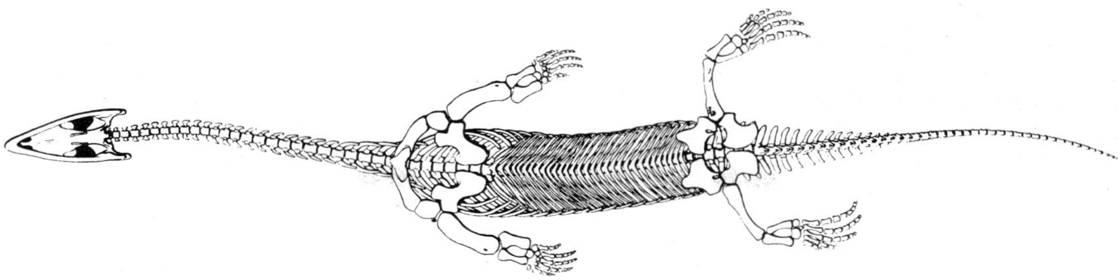

Fig. 1. *Ceresiosaurus calcagnii,* seen from below. A nothosaur with elongated neck. About 3½ ft long. Triassic, Switzerland. (*From B. Peyer, 1944*)

(3) by enlargement of the ventral elements and by the loss of digital claws. The directions of aquatic specialization involve shortening, or more often lengthening, of the neck (Fig. 1); enlargement of the orbits or the temporal fenestrae; reduction, or more commonly increase, in the number of phalanges in manus (hand), pes (foot), or both. A feature of considerable evolutionary significance in the light of plesiosaurian differentiation is the great individual variability in the number of presacral vertebrae (32–42) in *Pachypleurosaurus edwardsi.*

Although most genera of nothosaurs are not in direct line of ancestry of the plesiosaurs, such a relationship has been suggested for *Pistosaurus.*

Plesiosauria. The Plesiosauria are the successful, compact, and highly specialized offshoot of the nothosaurs that attained world-wide distribution. The early steps of aquatic adaptation, initiated by the nothosaurs, have led to extensive anatomical modifications: the region comprising chest and abdomen became short, stout, and inflexible; the ventral bones of shoulder girdle and pelvis increased in area enormously; the limbs, transformed into large flippers, became the principal organs of propulsion.

Two major trends of plesiosaur evolution may be discerned since the Early Jurassic: in the one group there was a tendency toward a shortening of the neck from 27 to 13 vertebrae and an increase in skull size; in the other group the opposite trend led to forms of bizarre body proportions, for example *Elasmosaurus,* with a neck containing 76 vertebrae. The Plesiosaurs were carnivorous animals.

Placodontia. The Placodontia constitute a well-defined group of peculiarly specialized marine reptiles that had a short but interesting history. As the name implies, placodonts are reptiles in which at least part of the dentition consists of flat-crowned teeth adapted to a durophagous diet, probably of hard-shelled mollusks. Within the suborder the modification of the dentition ranges from the primitive reptilian condition of numerous sharply pointed teeth (in *Helveticosaurus*) to one in which only a single, flat tooth remains in each jaw quadrant, as in *Henodus.* An early stage in tooth modification is seen in *Paraplacodus* (Fig. 2). The more advanced placodonts are heavily armored with dermal ossicles. In the earliest occurrence of the group

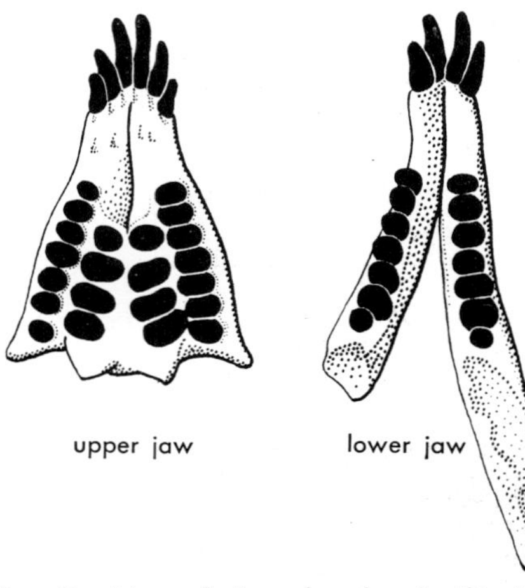

Fig. 2. Dentition of *Paraplacodus broilii,* Triassic, Switzerland. (*After B. Peyer, 1935*)

primitive and notably advanced forms occur together in the same deposit. The group did not survive the Triassic. *See* EURYAPSIDA. [R.Z.]

Bibliography: J. Piveteau (ed.), *Traité de Paléontologie,* vol. 5, 1955; A. S. Romer, *Osteology of the Reptiles,* 1956.

Sawing

The parting of material by using metal disks, blades, bands, or abrasive disks as the cutting tools. Sawing a piece from stock for further machining is called cutoff sawing, while shaping or forming a piece is referred to as contour sawing.

Machine sawing of metal is performed by five types of saws or processes: hack sawing, band sawing, cold sawing, friction sawing, and abrasive sawing.

Hack saws are used principally as cutoff tools. The toothed blade, held in tension, is reciprocated across the workpiece. A vise holds the stock in position. The blade is fed into the work by gravity or by springs. In some cases a mechanical or hydraulic feed is used. Automatic machines, handling bar-length stock, are used for continuous production.

Band saws cut rapidly and are suited for either cutoff or contour sawing. The plane in which the blade operates classifies the machine as being ei-

Vertical saw with power feed table is used to split the part shown. An advantage of sawing is that the slicing action of the tool separates the work into sections rather than reducing the unwanted section to chips. (*The Do All Company*)

ther vertical or horizontal. Band saws are basically a flexible endless band of steel running over pulleys or wheels. The band has teeth on one side and is operated under tension. Guides keep it running true as illustrated. The frame of the horizontal type is pivoted to allow positioning of the workpiece in the vise. Horizontal machines are used for either straight or angular cuts. A table that supports the workpiece and the wide throat between the upright portions of the blade makes the vertical band saw ideal for contour work. Band saws operating at high speed are frequently used as friction saws.

Cold sawing is principally a cutoff operation. The blade is a circular disk with cutting teeth on its periphery. Blades range in size from a few inches to several feet in diameter. The cutting teeth may be cut into the periphery of the disk or they may be inserts of a harder material. The blade moves into the stock with a positive feed. Stock is positioned manually in some cold sawing machines, while other models are equipped for automatic cycle sawing.

Friction sawing is a rapid process used to cut steel as well as certain plastics. This process is not satisfactory for cast iron and nonferrous metals. Cutting is done as the high-speed blade wipes the metal from the kerf after softening it with frictional heat. Circular alloy-steel blades perform cutoff work while frictional band saws do both cutoff and contour sawing. Circular blades are frequently cooled by water or air. Circular blades are advanced into the work, while thick workpieces require power table feed when friction cut on a band saw.

Abrasive sawing is a cutoff process using thin rubber or bakelite bonded abrasive disks. In addition to steel, other materials such as nonferrous metals, ceramics, glass, certain plastics, and hard rubber are cut by this method. Cutting is done by the abrasive action of the grit in the disk.

Abrasive disks are operated either wet or dry. For heavy cutting a cooling agent is generally used. The workpiece is firmly held while the wheel traverses through it. Machines are made in manually operated and automatic models. *See* MACHINING OPERATIONS; *see also* WOODWORKING. [A.T.]

Saw-tooth wave

A waveform that is a continuously increasing function of time for a fixed interval, returns to its initial state during a retrace interval, and then repeats the sequence periodically. The most widely used saw-tooth waveform is ideally a linear function of time during the forward or rising interval and appears as shown in Fig. 1*a*, with the total period T made up of the active forward interval T_f, a retrace interval T_r, and an inactive interval T_i.

Mathematically the saw-tooth wave may be expressed in terms of a Fourier series of harmonically related components with the fundamental having a period equal to the total period T of the saw tooth. If a sufficient number of harmonics having the proper amplitude and phase relationship is included, the mathematical representation will be an accurate approximation of the waveform.

Electronic circuits can generate only an approximation of the idealized waveform which, as a result, tends to have the deviations shown in Fig. 1*b*. Generally the delayed start (with respect to the

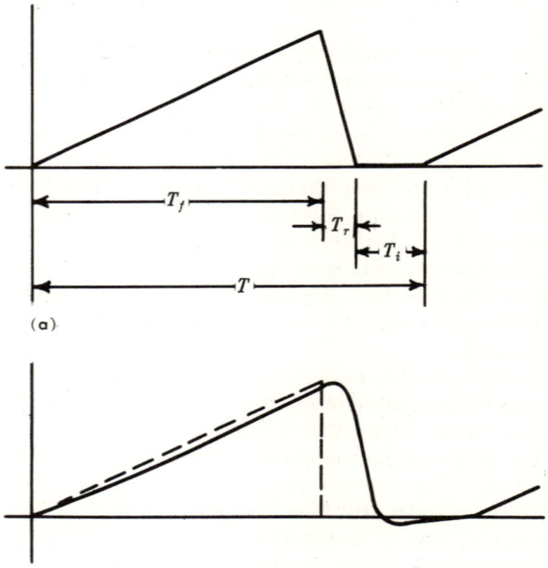

Fig. 1. Saw-tooth wave. (*a*) Ideal linear saw tooth. (*b*) Approximate saw tooth generated by actual circuits.

ideal) and the extended retrace time are caused by high-frequency deficiencies (inadequate generation or transmission of the higher-order harmonics), whereas the inability to maintain a constant slope for large values of t is a low-frequency deficiency in the generation or transmission of components near the fundamental frequency.

Saw-tooth waveforms are used as time-base elements or sweep generators (*see* SWEEP GENERATOR) and in time-delay and -measuring equipment (*see* TIME-DELAY CIRCUITS).

Saw-tooth voltage generation. The complexity of electronic circuitry required to generate linear saw-tooth voltage waves depends upon the accuracy to which such generation is specified. An approximate linear saw tooth may be generated by a dc voltage source, a series RC circuit, and a switch which when closed has a small but finite resistance as shown in Fig. 2.

If the switch S in the illustration is suddenly opened, the voltage at point A will rise from an initial value of $V_B R_s/(R + R_s)$ toward V_B according to the exponential equation

$$v_A = \frac{R_s}{R + R_s} V_B + V_B \left(1 - \frac{R_s}{R + R_s}\right) (1 - \epsilon^{-(t/RC)}) \tag{1}$$

as shown by the dotted curve. Now if the switch is closed after a time T_o, the rise will be interrupted at a value $V_{\max}$ obtained from the solution for v_A in Eq. (1) for $t = T_o$. While the switch is closed during the period T_c the potential will fall in accordance with

$$v_A = \frac{R_s}{R + R_s} V_B + \left(V_{\max} - \frac{R_s}{R + R_s} V_B\right) \epsilon^{-(t/R_sC)} \tag{2}$$

If R_s is very small, the waveform will essentially have recovered to its initial value in a short retrace interval T_r less than T_o. This complete cycle will repeat itself for alternate opening and closing of the switch. If the time T_c during which the switch is closed is less than the time T_r required for complete recovery, the minimum value for the waveform will be higher than $V_B R_s/(R + R_s)$. The maximum and minimum excursions of the waveform can then be found by simultaneous solution of Eqs. (1) and (2) for the appropriate periods, T_o and T_c.

A periodic saw-tooth waveform can be generated by using an astable relaxation oscillator as a switch in the above circuit (*see* RELAXATION OSCILLATOR). One of the simplest of these is the thyratron shown in Fig. 3. The potential $V_{\max}$ is the breakdown voltage for a given grid bias V_G, and the voltage $V_{\min}$ is the minimum maintaining ionization potential. Vacuum-tube relaxation oscillators usually maintain a more stable period than those employing gas tubes and are more widely used. When used as switches in saw-tooth generators, both may be synchronized with external pulses to maintain an accurately controlled period.

If the switch of Fig. 2 is the switch in a synchronous or keyed clamp (*see* CLAMPING CIRCUIT) each interval of the total period of the waveform may be controlled directly from an external source of pulses. A simple triode unidirectional clamp is shown in Fig. 4. During the time T_c that the switch is closed, the grid is held at a slightly positive value by limiting, and the plate resistance R_s is low. During the open time, the grid voltage is sufficiently negative that no plate current will flow and the switch is open.

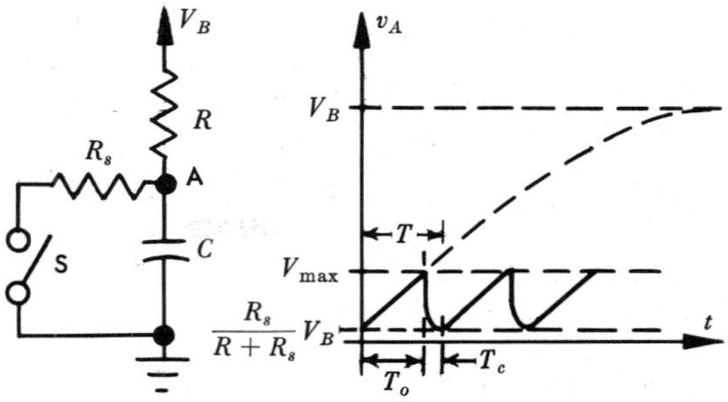

Fig. 2. Elements of saw-tooth sweep circuit.

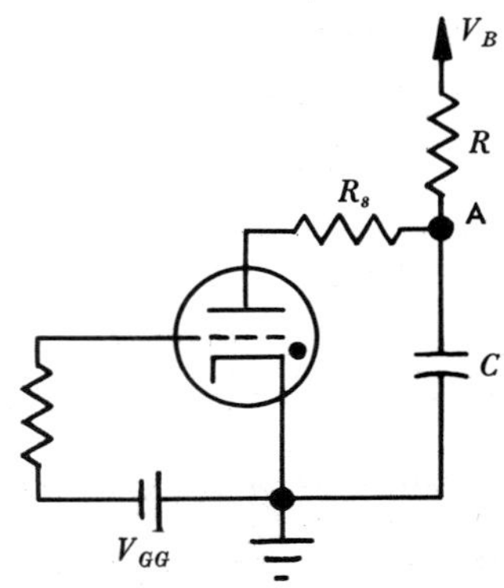

Fig. 3. Thyratron saw-tooth generator.

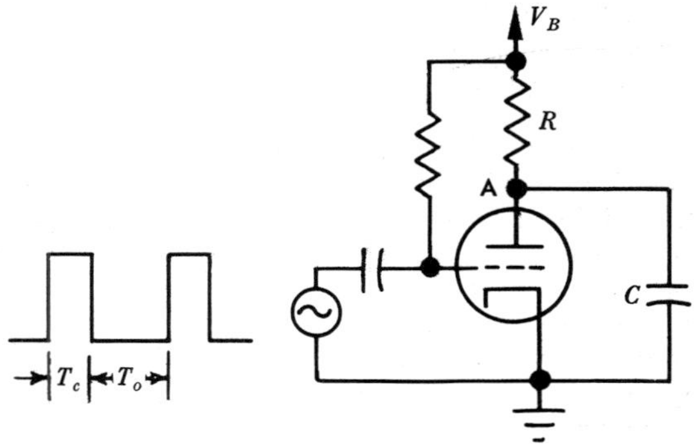

Fig. 4. Pulsed triode saw-tooth generator.

Either a positive-going or a negative-going saw-tooth may be generated by making V_B either positive or negative and replacing the clamp with a bidirectional clamp as shown in Fig. 5. If the potential V_B is variable at a rate slow compared to the periodicity of the waveform, a succession of saw-tooth waveforms of varying amplitude will be generated. One particular application of such a circuit would be to generate one of the components of the rotating radial sweep if V_B were to be made to vary sinusoidally with the desired angular modulation.

Improvement of linearity. A linearly increasing voltage

$$v_c = (I_c/C)t \qquad (3)$$

will appear across the terminals of a capacitor C if a constant current I_c is flowing, since

$$v_c = (1/C) \int I_c \, dt \qquad (4)$$

Therefore, improvement of linearity of the simple RC sweep generators basically depends upon maintaining the current through the capacitor more constant. For a specified saw-tooth amplitude this may be done by increasing the supply voltage V_B. This is a practical solution only within narrow limits. Another method in effect replaces R with an active device, such as the output circuit of a vacuum-tube pentode or transistor, which has a relatively low absolute resistance but an extremely high ac or incremental resistance over the limits of the desired amplitude range. In the circuit of Fig. 6, when point A is not clamped to ground and the base-emitter bias voltage, as determined by the diode operating in the Zener breakdown region, is such that the transistor is in its normal range, the collector current will be relatively independent of collector voltage.

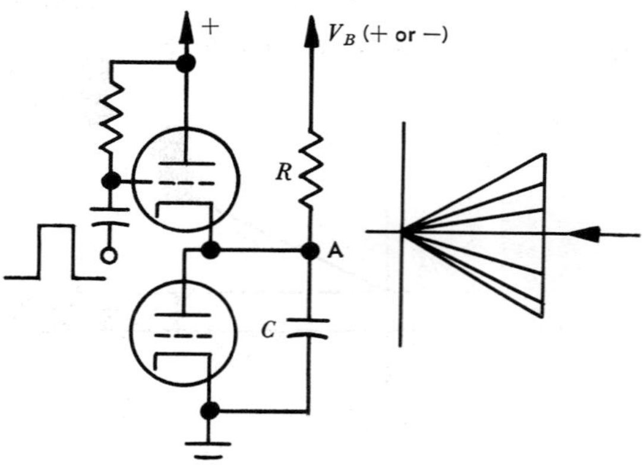

Fig. 5. Sweep generator using bidirectional clamp.

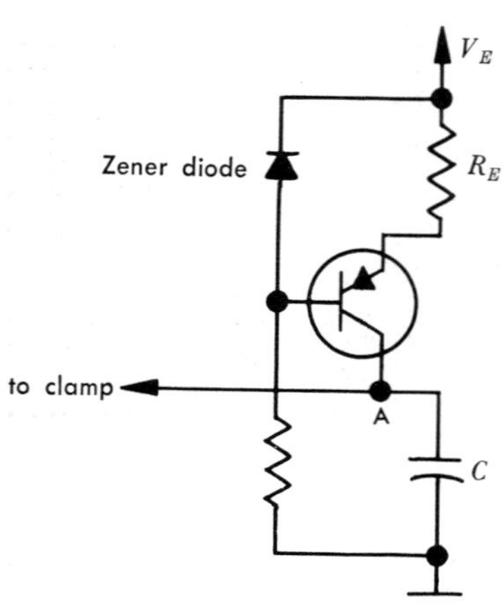

Fig. 6. Constant-current saw-tooth generator.

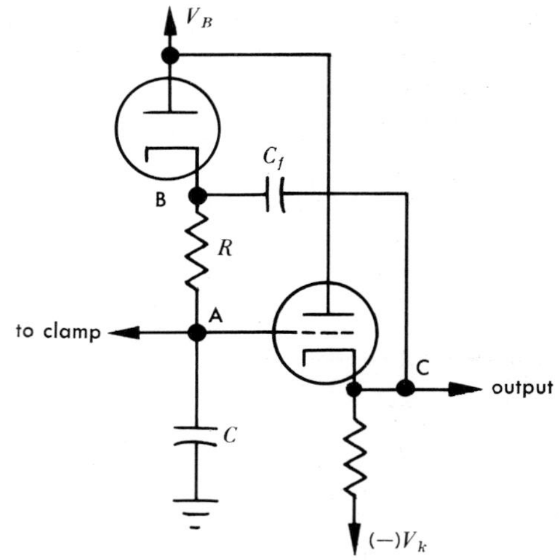

Fig. 7. Bootstrap saw-tooth generator.

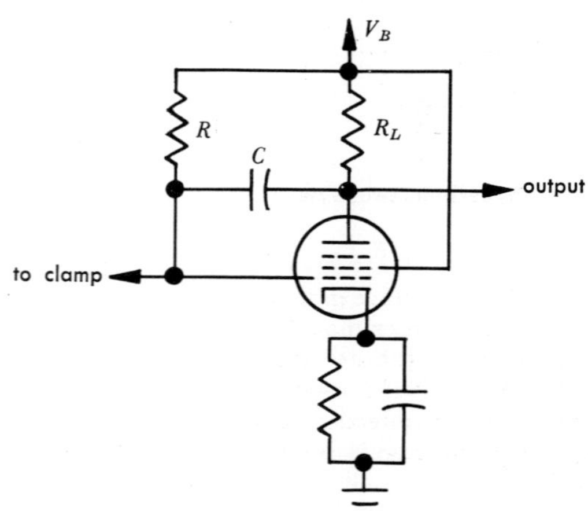

Fig. 8. Single-tube Miller integrator.

The circuit of Fig. 7 is often referred to as the bootstrap saw-tooth generator. When the clamp is closed, point A is at ground potential, point B is at V_B minus the small diode voltage drop, and point C is approximately at the potential of A. When the clamp is opened, capacitor C starts to charge toward V_B through R and the diode resistance. Point C follows closely. If capacitor C_f is very large compared to capacitor C, point B will rise the same amount. This causes the diode to stop conducting. Capacitor C will continue to charge through R and C_f. If sufficiently large C_f functions as a constant-voltage source, and to the extent that it does, and if the gain of the cathode follower is near unity, the charging current will be nearly constant and a nearly linear saw tooth of a magnitude approaching the supply voltage can be generated. To a good approximation, the voltage at point C can be expressed as

$$v_C(t) = AV_B \frac{C_f}{C + C_f(1-A)} \left(1 - \epsilon^{-[C+C_f(1-A)/RCC_f]t}\right) \qquad (5)$$

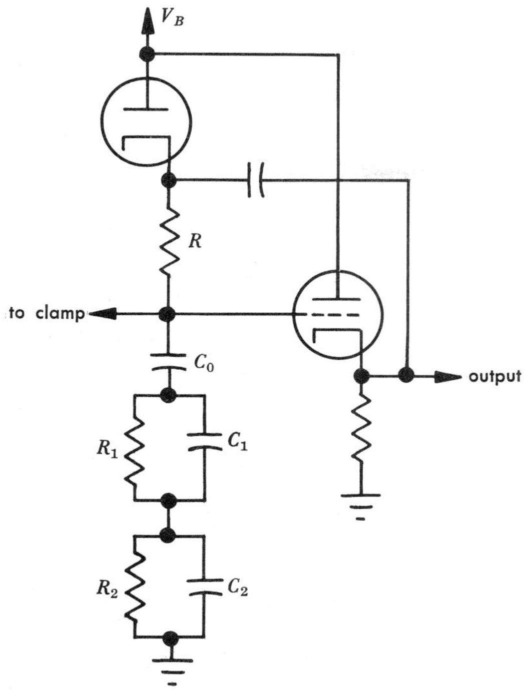

Fig. 9. Hyperbolic sweep generator.

where A is the voltage gain of the cathode follower. If C_f is much larger than C, this is an exponential charging curve with an effective supply voltage of $(A/1 - A)V_B$. If A is near unity, this represents a great increase in effective supply voltage and a corresponding increase in linearity for a required amplitude.

A circuit in which an integrating amplifier is used in addition to the clamp and RC time constant is sometimes referred to as the Miller integrating circuit (see Fig. 8). If the input impedance of the amplifier is high, the output impedance low, and the gain high, the output approximates the integral of the suddenly impressed constant-amplitude supply voltage V_B. The approximate equation of the output waveform is

$$v_o = AV_B(1 - \epsilon^{-(t/ARC)}) \tag{6}$$

where A is the magnitude of the gain of the amplifier. Since A is negative, the result will be a negative-going saw tooth, the first part of an exponential charging toward the effective supply voltage AV_B. Thus the linearity is increased by the same amount that a charging voltage V_B increased by a factor A would increase it. Use of a multistage transistor or vacuum-tube amplifier rather than the single pentode will thus increase the gain and improve the linearity.

Hyperbolic and other waveforms. A waveform other than a linear saw tooth may be generated by using more complicated RC circuits than the simple ones which have been described. For example, the bootstrap generator shown in Fig. 9 generates an approximate hyperbola.

Saw-tooth current generation. Often a saw-tooth current waveform must be applied to a circuit having an inductive component, such as the deflection coil of a magnetically deflected cathode-ray device. If the coil can be represented by an inductance and resistance in series as shown in Fig. 10, the voltage appearing across the terminals of the coil will be

$$v = Ri(t) + L\frac{di(t)}{dt} \tag{7}$$

and if the current $i(t)$ is specified as a linear saw tooth, $i(t) = kt$, then

$$v = Rkt + Lk \tag{8}$$

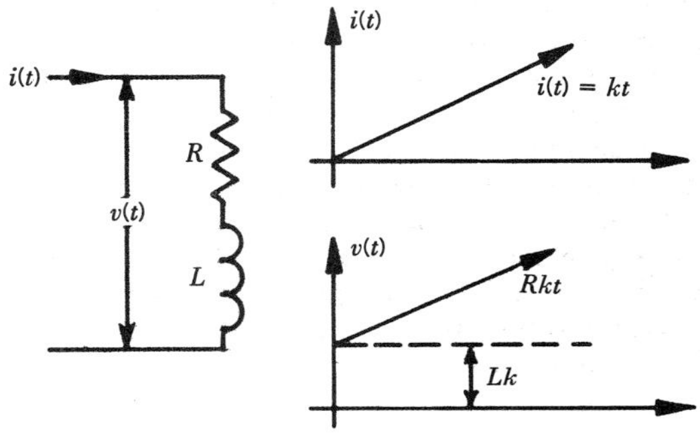

Fig. 10. Linear current in inductive circuit.

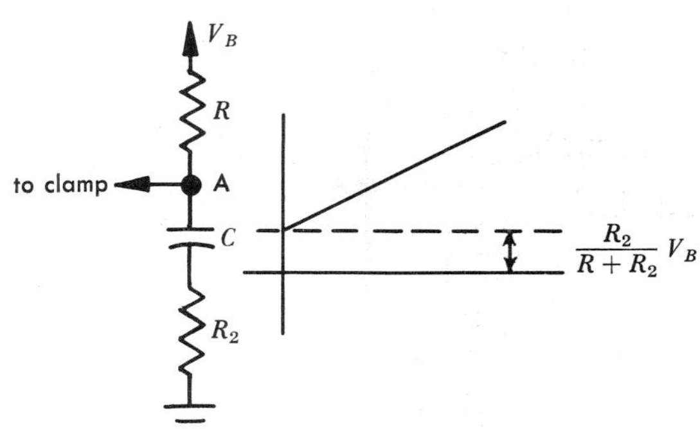

Fig. 11. Trapezoid voltage generator.

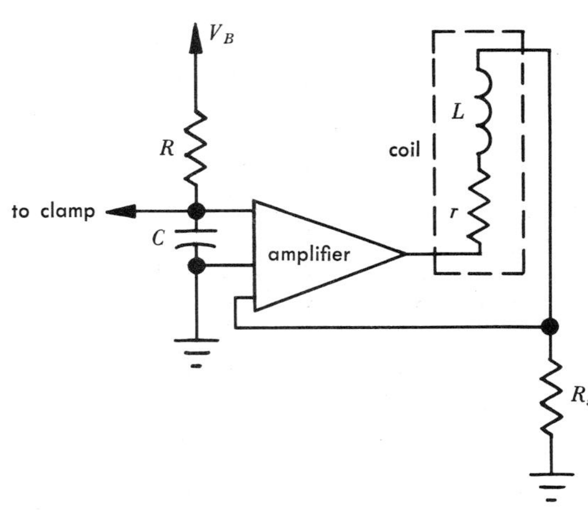

Fig. 12. Current feedback sweep generator.

This voltage is a step added to a linear saw tooth as shown. Such a voltage waveform, required for a linear saw tooth of current, may be generated by using any of the previous circuits with an additional circuit element R_2 as shown in Fig. 11. Initially, since the voltage across C cannot change instantaneously when the clamp is opened, the total voltage V_B will divide between R and R_1, making the potential at point A suddenly rise to $V_B R_2 / (R + R_2)$.

A more common method of generating a current waveform is to generate a voltage waveform of the same form and apply it to a negative feedback amplifier with a large amount of current feedback, which forces the current in the coil to be approximately the form of the generated voltage. Such a system is shown in block form in Fig. 12. Such current feedback makes the output impedance of the amplifier high and thus approximates a current source which is a replica of the applied voltage. An actual circuit is shown in Fig. 13. Here, the feedback from the cathode of the output tube is nearly a true sampling of the coil current.

If dc levels do not have to be preserved, the deflection coils may be transformer-coupled to the output tube as shown in Fig. 14.

High-frequency limitations. Any actual deflection coil can be represented as a series inductance and resistance only at relatively low frequencies.

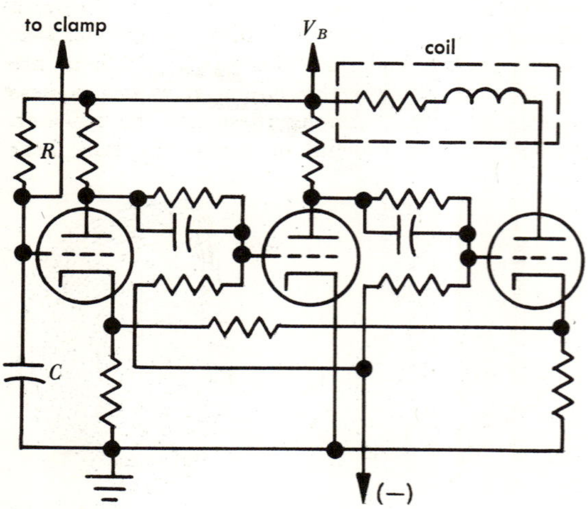

Fig. 13. Feedback sweep generator.

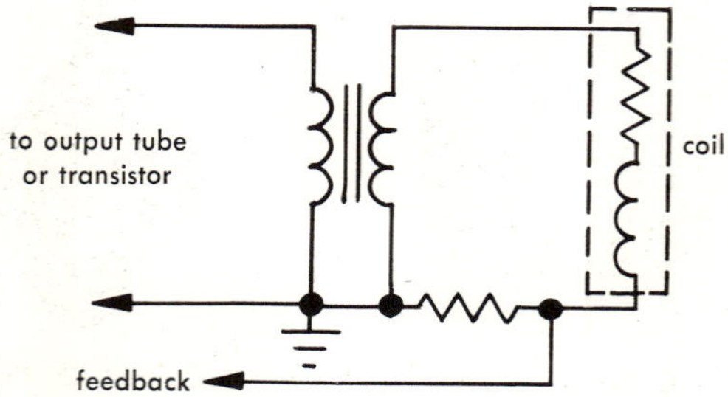

Fig. 14. Transformer-coupled deflection coil.

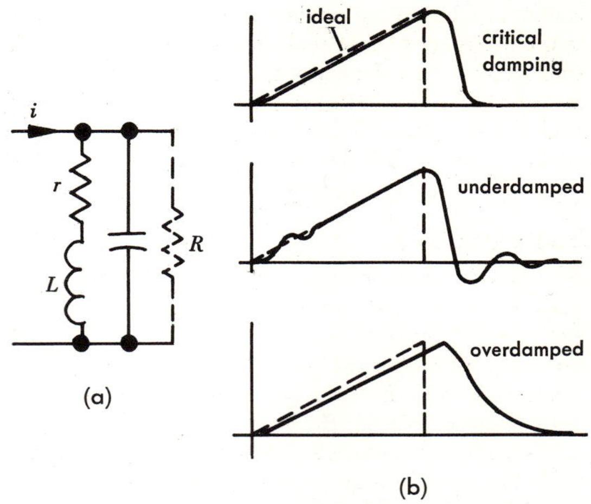

Fig. 15. Effect of shunt capacitance on deflection system. (*a*) Circuit representation. (*b*) Typical waveforms.

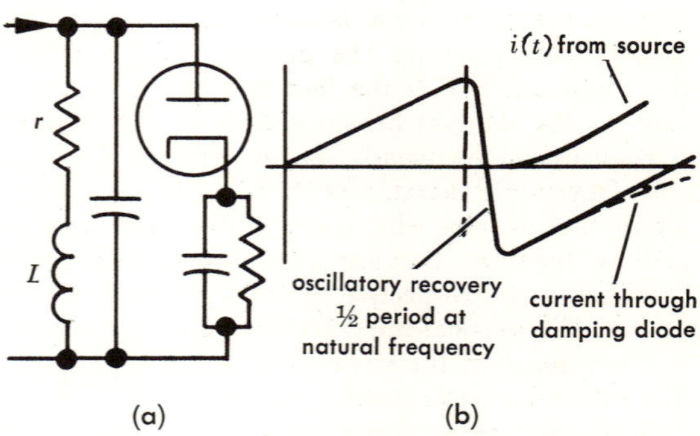

Fig. 16. Deflection system with diode damping. (*a*) Circuit diagram. (*b*) Typical waveform.

It also has distributed capacitance, which can be represented crudely by a shunt capacitance as in Fig. 15. This capacitance accounts principally for the delay in the start of the sweep and for the minimum retrace time T_r necessary for recovery. The best conditions occur when the oscillatory circuit is critically damped by the addition of the shunt resistance R. Typical sweep waveforms for critical damping and departures from it are shown.

The horizontal deflection circuit of the television system represents a special case where the circuit is underdamped, or allowed to be oscillatory, for half a period with the beginning of the sweep waveform controlled by special diode circuits as shown in Fig. 16. For other types of waves, *see* WAVE-SHAPING CIRCUITS. [G.M.G.]

Bibliography: B. Chance et al. (eds.), *Waveforms*, 1949; G. M. Glasford, *Fundamentals of Television Engineering*, 1955; J. Millman and H. Taub, *Pulse and Digital Circuits*, 1956.

Scalar

A term synonymous in mathematics with real in real number or real function. The magnitude of a vector two units in length is the real number or scalar 2. The dot or scalar product of two vectors is the product of three real numbers associated with them and is therefore a scalar. If in the functional relationships $S = S(x,y,z)$, $\mathbf{F} = \mathbf{F}(x,y,z)$, S is a real number while $\mathbf{F}$ is a vector, then $S(x,y,z)$ is a scalar function but $\mathbf{F}(x,y,z)$ is a vector function. *See* CALCULUS OF VECTORS. [H.V.C.]

Scale (music)

A series of notes arranged from low to high by a specified scheme of intervals, suitable for musical purposes. A musical scale is an arrangement of intervals evolved in the making of music. When a given note of a musical scale is repeated, it need not be given exactly the same frequency; the frequency may be modified significantly in accordance with the artistic requirements of the musical context. Thus it is difficult to be very specific about the "specified scheme of intervals."

The number of musical scales and the intervals they contain are myriad, but most of them have in common the interval that is today called the octave. A possible reason for the general acceptance of the octave interval is the fact that it occurs naturally as the interval between the first and second partials of sounds usually considered to be musical (*see* OCTAVE; PARTIAL TONE). Also, this is the interval that results when men and women sing a melody together because the natural difference between their respective voices is roughly an octave. The musical import of two notes an octave apart is so often the same that the two notes are given the same letter name.

A diatonic scale is one in which the octave is divided into intervals of two different sizes, five of one and two of the other. If the size of the smaller interval (often called a half-step) is taken as one unit, the sequence of intervals in the major diatonic scale is 2, 2, 1, 2, 2, 2, 1; in the minor diatonic scale (melodic ascending), the sequence is 2, 1, 2, 2, 2, 2, 1. The relative sizes shown by these numbers are approximate; only in equal temperament are the sizes exactly as indicated. For information on tuning and temperament, *see* MUSICAL ACOUSTICS.

In these days of familiarity with the piano keyboard, a diatonic (heptatonic) scale can well be called a white-key scale because the major scale in the key of C is played on the white keys only. The remaining keys within the octave delineate five intervals that constitute a black-key pentatonic scale.

Numerous mathematical arguments have been propounded to explain the choice of smaller intervals into which the octave can be divided. Also, a case can be made for the evolution of the intervals as a consequence of the instruments used. Many primitive flutes, for example, have only six holes, presumably because the three central fingers of each hand are most expert in covering the holes. If all the holes are covered and then opened one at a time, starting at the end distant from the mouthpiece, seven tones can be had; an eighth tone, obtained by covering all the holes again and blowing still harder, is roughly an octave higher than the first, and a "scale" of seven intervals within an octave is thus generated. It may be added that the holes of primitive instruments are often equally spaced in accordance with a decorative pattern, and their number is not necessarily six. [R.W.Y.]

Scale (zoology)

The body covering of many vertebrates. Scales are of two types, epidermal and dermal, based on their origin and development. Both types are structurally different.

Epidermal scales. This type of scale is usually associated with terrestrial vertebrates, and it is derived specifically from the stratum germinativum. Among the cyclostomes, the tongue and buccal cavity possess teeth which are modified epidermal scales. Fish lack scales of epidermal origin. Among the amphibians, species of the spadefoot toad have a cornified epidermal structure on the hindfoot which is used in digging. It is considered by some to represent an epidermal scale. Epidermal scales are characteristic of reptiles and are well-developed in the class. These are shed periodically by snakes and lizards when they molt, the new set of scales already being formed beneath the old before molting. The rattle of the rattlesnake is a series of specialized scales, as are the scales which cover the carapace and plastron of turtles. The legs and feet of birds are covered by overlapping epidermal scales. However, the most characteristic structure of birds is the feather, which is an epidermal derivative. Feathers are modified reptilian scales. Many mammals possess epidermal scales. Among these are the anteater, armadillo, and rodents of which the tails bear

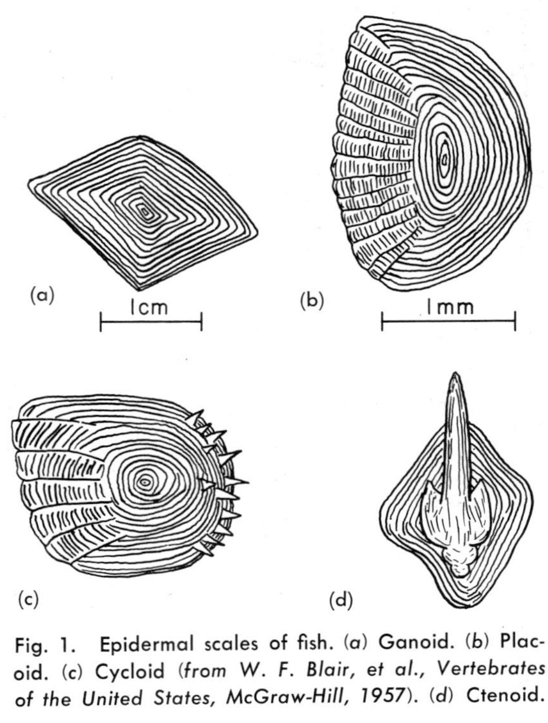

Fig. 1. Epidermal scales of fish. (*a*) Ganoid. (*b*) Placoid. (*c*) Cycloid (*from W. F. Blair, et al., Vertebrates of the United States, McGraw-Hill, 1957*). (*d*) Ctenoid.

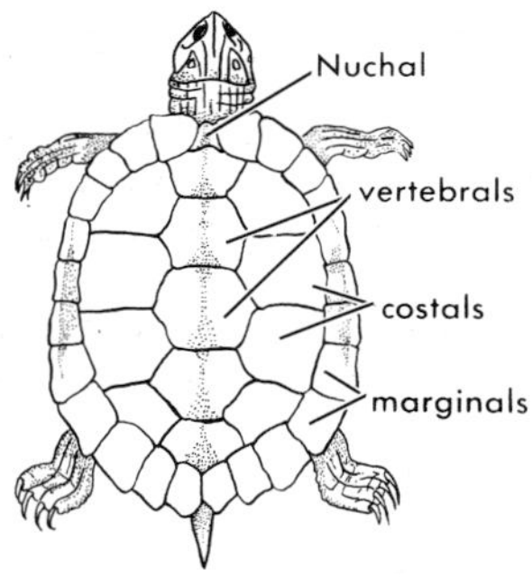

Fig. 2. Dermal elements of a turtle carapace. (*From W. F. Blair, et al., Vertebrates of the United States, McGraw-Hill, 1957*)

scales. Some authorities also consider hair to be modified scales. *See* FEATHER (BIRD); HAIR.

Dermal scales. These structures represent remnants of the dermal skeleton and are most common in the fishes. In living fishes, scales are classified as ganoid, placoid, ctenoid, and cycloid (Fig. 1). Cosmoid scales were characteristic of primitive crossopterygians such as *Osteolepis*. In actinopterygians, the characteristic scale was the ganoid which is present today in the garpikes and the bichir (*Polypterus*). Among the amphibians, some toads have bony plates imbedded in the skin whereas caecilians have scales in dermal pockets. Reptiles, especially the turtles, have a well-developed dermal skeleton, covered with epidermal scales, in the bony plates which comprise the carapace and plastron (Fig. 2). Dermal ribs or gastralia are also present in turtles, as well as membrane bones of the skull in certain lizards and snakes. Most birds and mammals lack dermal skeletal structures except for the membrane bones. [C.B.C.]

Scale insect

Any of several species of the family Coccidae, order Homoptera. The scale insects are highly modified and frequently have lost their wings, antennae, and legs. Usually the males undergo a complete metamorphosis and emerge as minute, two-winged adults without mouthparts, which do not feed. The wingless females have a gradual metamorphosis. All are covered with some type of thickened integument or waxy secretion. Many of them are legless during a large part of their life cycle. Included among the scale insects are some of the most serious plant pests, especially of shade and fruit trees. *See* HOMOPTERA; INSECTA; TREE. [J.D.B.]

Scaling circuit

An electronic circuit that produces one output pulse for a specific number n of input pulses. Such a circuit is referred to as a scale-of-n or an n-counter circuit.

A somewhat specialized form of counter, otherwise known as a frequency divider, employs an astable relaxation oscillator, such as a multivibrator or blocking oscillator. The natural period τ_1 of one state is made slightly greater than the period between n input pulses of an equally spaced pulse train, while the period τ_2 of the other state is less than the period between each pulse. If the input pulses are used as synchronizing pulses superimposed on the input waveform, they cause the relaxation oscillator to recycle in the manner shown by the input waveform of Fig. 1, instead of the natural period shown by the extended dotted lines. *See* MULTIVIBRATOR.

A more widely useful class of circuits functions even when the input pulses have random spacing. The basic unit of such a scale-of-n circuit is often the bistable multivibrator, which is itself a scale-of-two, or binary, circuit. If the square wave from one of the output terminals of the multivibrator is differentiated and the resultant pulses of one polarity used to trigger a second such circuit, the result is a count of 4. A cascaded group of n such circuits produces a count of 2^n. The waveforms of three such cascaded stages are shown in Fig. 2. Feedback may be used in cascaded counter circuits to achieve a count of any number other than 2^n.

Many other forms of counters are possible, some using a different arrangement of scale-of-two circuits known as the ring counter, others using special tubes such as beam-switching tubes, and

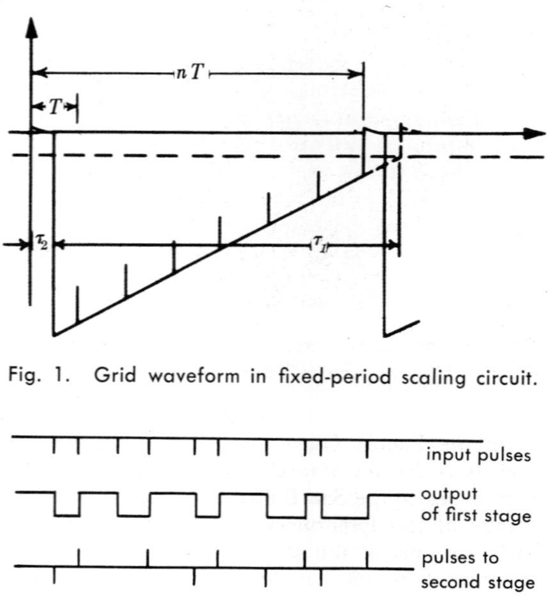

Fig. 1. Grid waveform in fixed-period scaling circuit.

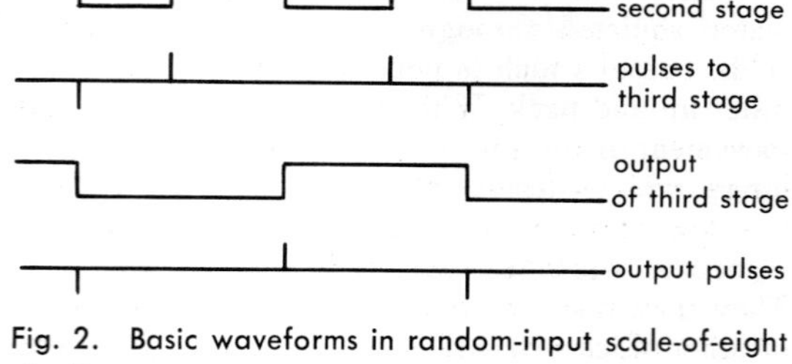

Fig. 2. Basic waveforms in random-input scale-of-eight circuit.

still others using the switching properties of square-loop magnetic cores. *See* COUNTING CIRCUIT.

Scaling circuits are used for direct counting of a series of events and for basic measurements of time and frequency (*see* FREQUENCY COUNTER). The scale-of-two circuit is the basic building block in the binary number system of digital computers (*see* DIGITAL COMPUTER). [G.M.G.]

Bibliography: J. Millman and H. Taub, *Pulse and Digital Circuits*, 1956.

Scallop

Any of several species of the genus *Pecten*, a group of free-swimming, marine bivalve mollusks. The most common commercial species in the eastern United States is *Pecten irradians* which is found along the coast from Nova Scotia to Texas. It varies considerably in color but the shape of the shell is well-known, because this is the species whose shell is the trademark of a major oil company. Although other parts of the pectens are edible, it is only the single, large adductor muscle, which closes the shell, that is marketed as the scallop.

The scallop, *Pecten jacobeus*. (From J. G. Wood, *Popular Natural History*, Porter and Coates, *1885*)

On the Pacific Coast the common commercial species is *Pecten hindsii*, which ranges from the Bering Sea to the San Diego area. There are other species of less importance along both coasts. The basic anatomy of the scallop is similar to that of the mussel except for the presence of only one adductor muscle, rather than two.

Scallops move by the force created by jets of water squirted through openings in the mantle folds, one of which is pointed down and back, the other up and back. With these jets, aided by some movement of the shells, they move slowly forward, hinge side posterior. When startled they reverse the flow of water and clap the shells together, a maneuver which thrusts them quickly backward. They may rest by attaching themselves to plants or other objects by means of a threadlike extension, or byssus. Each mantle half is fringed with sensory tentacles and marked with a row of small blue eyes called ocelli.

Scallops are hermaphroditic, shedding both sperm and eggs into the sea. A short free-swimming stage elapses before the larva assumes the adult form. They grow rapidly, spawn during the second summer, and usually die by the end of their second year. *See* GASTROPODA; MOLLUSCA; MUSSEL. [J.D.B.]

Scandium

Element number 21, scandium, Sc, is one of the transition elements, but it is usually classified with the rare-earth elements, since it has very similar chemical properties. The isotope of mass 45 is the only one occurring in nature.

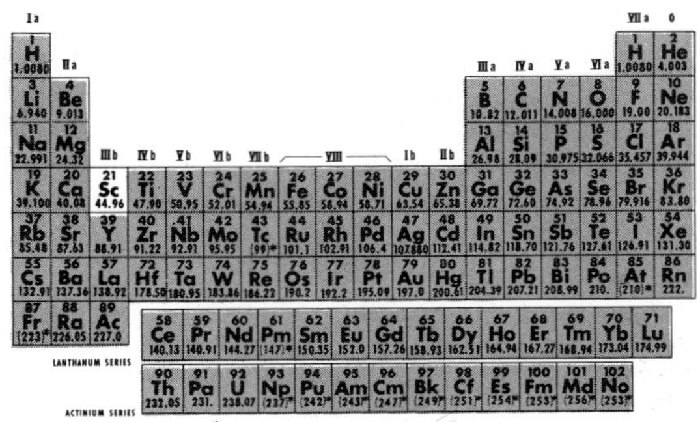

*mass number of most stable known isotope

Its existence was first predicted by Dimitri Mendeleev in 1871, and it was first isolated in 1897. Properties of the element include a specific gravity of 2.5, a melting point of 1200°C, a boiling point of 2400°C, and an atomic weight of 44.96. The metallic element does not yet have an important use. The oxide, Sc_2O_3, is available commercially at about $20 per gram.

Occurrence and distribution. In the years immediately following the discovery of the element, various workers examined many different ores, minerals, and other substances for scandium. In a remarkably large percentage of cases they found the element present but almost invariably in concentrations considerably less than 1%. The only mineral which may be described as a scandium ore is Norwegian thortveitite. An analysis of a specimen showed 34% Sc_2O_3 together with 45% SiO_2, 10% Y_2O_3, 5% Al_2O_3, 3% Fe_2O_3, 1.5% La_2O_3, with lesser amounts of the oxides of copper, magnesium, manganese, and thorium. It also occurs in ores of tin, of tungsten, and of the rare earths.

Compounds. Scandium is the first of the transition family of the periodic table, scandium (atomic number Z, 21) to zinc (Z, 30). Its atom has two electrons in the outermost shell and one electron in the incomplete $3d$ shell. All three electrons participate as valence electrons and the normal oxidation state is 3+. The normal salts may be represented by the formula ScX_3 where X is a halide, nitrate, perchlorate, or other monovalent anion; and by the type Sc_2X_3 for the bivalent anions such as sulfate, oxalate, and carbonate. Scandium forms an aqueous complex ion $[Sc(H_2O)_6]^{3+}$ and consequently forms many basic salts of the inorganic and organic acids. The trivalent cation itself is an acid only slightly weaker than acetic acid in aqueous solution.

$$\begin{array}{ccc} [Sc(H_2O)_6]^{3+} + H_2O \rightleftharpoons & \left[\begin{matrix}OH\\ Sc(H_2O)_5\end{matrix}\right]^{2+} & + H_3O^+ \\ & \upharpoonleft\downharpoonright & \\ & \left[\begin{matrix}OH\\ Sc(H_2O)_5\end{matrix}\right]_2^{4+} & \end{array}$$

The divalent cation resulting from the transfer of one proton to a water molecule forms dimers and also tends to form trimers and higher associates. Since hydroxyl ion and even water can be replaced by practically all the anions, it is not surprising that compounds of the formula

$$Sc(OH)_2Cl\cdot xH_2O,\ Sc(OH)(SO_4)xH_2O,$$
and
$$Sc(OH)(CH_3CO_2)\cdot 7.1H_2O$$

have been reported. In fact, it is difficult to obtain normal salts in the solid state from aqueous solution with the exact stoichiometric ratio of anion to cation.

Scandium also forms a number of organic complexes such as the oxinate, $Sc(C_9H_6ON)_2\cdot C_6H_7ON$, which forms the basis for a quantitative determination of the element, and a number of colored complexes with reagents such as carminic acid, alizarin sulfonic acid, 1,2,5,8-tetrahydroxyanthroquinone, ammonium purpurate, phenylarsonic acid, and 2,5-dihydroxy-1,4-benzoquinone. The last reagent is the basis for a sensitive qualitative test for scandium.

Separation and purification. The element was originally separated as oxide from other elements by precipitation as oxalate, fluoride, or fluosilicate, and the final oxalate precipitate was ignited to the oxide. Other methods involve the extraction of the chloride with ether from aqueous solution in the presence of hydrogen chloride or ammonium thiocyanate. More recently purification by complexing with thenoyltrifluoroacetone has been reported. The use of ion-exchange resins for separation is also possible. *See* RARE-EARTH ELEMENTS; TRANSITION ELEMENTS. [M.KI.]

Scaphopoda

A class of the phylum Mollusca. It is a small but distinct group. The soft body closely fits the external, curved, and tapering, nonchambered, aragonitic shell which is open at both ends. The shell is commonly attenuated posteriorly and superficially resembles an elephant tusk or a carnivorous canine tooth, hence the common name of tusk or tooth shell. It is commonly sculptured with longitudinal or annular ribs, or rarely lacks surface ornamentation.

The class is divided into two families composed of approximately 350 living species. In the family Dentaliidae, the foot is pointed with the epipodial collar interrupted dorsally to give a trifid appearance. Siphinodentaliidae is characterized by a subterminal epipodial ridge which is not slit dorsally and terminates with a crenulated disk.

Scaphopods are largely restricted to subtidal waters, but live in a variety of substrates. The animals burrow into the bottom by means of the foot which is extended through the anterior opening of the tube. The shell is held in an oblique position in the substratum with the anterior portion buried and the posterior opening exposed at the surface of the bottom to provide for the circulation of water. Numerous food-gathering, prehensile filaments extend from the cephalic region through the anterior aperture to remove organic material from the substrate. *See* MOLLUSCA. [W.K.E.]

Scapolite

A complex aluminosilicate of sodium and calcium belonging to the tectosilicate group of silicate minerals. It crystallizes in the tetragonal system, usually in coarse prismatic crystals with pyramidal terminations. There are four directions of cleavage at 45° to each other parallel to two prism forms. The hardness is 5–6 on Mohs scale. The specific gravity varies from 2.65 to 2.74 depending upon the composition. The luster is vitreous; the color is usually white, gray, or pale green and more rarely reddish or blue. *See* SILICATE MINERALS.

Scapolite varies greatly in composition, for the name is used for all intermediate members of a complete solid-solution series. The name wernerite is also given to these intermediate members. The end members of the series are mirialite, $Na_4Al_3Si_9O_{24}Cl$, and meionite, $Ca_4Al_6Si_6O_{24}CO_3$, whose compositions are suggestive of the plagioclase feldspars. There is complete substitution of Ca for Na, and as in the feldspars concomitant substitution of Al for Si. There is also complete substitution of Cl, CO_3, and SO_4 for one another.

Scapolite is a metamorphic mineral found in crystalline limestone as a product of contact metamorphism, and in schists and gneisses. Minerals commonly associated are pyroxenes, amphiboles, garnets, apatite, zircon, and sphene. Transparent yellow crystals from Madagascar and Brazil have been cut as gem stones. Ordinary scapolite has been found at various places in Massachusetts and New York and in Ontario. [C.S.HU.]

Scarlet fever

An acute contagious disease which is the classic example of infection by the microorganism *Streptococcus hemolyticus*. Fever, sore throat, rash, headache, and vomiting follow from 2 to 7 days after contact with a carrier whose illness was probably not scarlet fever. The bright red rash over the body is papular, or rough, with sandpaperlike quality. It blanches on pressure to leave a transient image, as in sunburn. A deeply flushed face with circumoral pallor, or paleness around the mouth, completes the patient's outward appearance. The tongue is aptly described as "strawberry," sometimes "raspberry." There is a fiery redness of the throat and the swollen tonsils are covered with patches of white exudate. The rash fades in a few days to be followed by desquamation, or peeling, while symptoms abate.

Early complications such as visibly swollen neck glands and draining ears mean bacterial spread from the nose and throat. Late complications, such as rheumatic fever and nephritis, or kidney disease, appear during convalescence. Since these are neither from actual infection nor toxemia, that is toxin in blood, a sensitivity to some bacterial substance is likely. In "surgical" scarlet fever the source of the erythrogenic, or rash-producing, toxin is an infected burn or other wound.

Scarlet fever, spontaneously milder for some years, is shortened by prompt penicillin treatment, which prevents both types of complications. Second attacks are rare. The Dick skin test determines susceptibility to scarlet fever. *See* BACTERIOLOGY, MEDICAL; LANCEFIELD DIFFERENTIATION SCHEME; RHEUMATIC FEVER; SKIN TEST; STREPTOCOCCUS.

[P.L.B.]

Scattering (electromagnetic radiation)

The process in which energy is removed from a beam of electromagnetic radiation and emitted without appreciable change in wavelength. This article is restricted to the scattering of visible light; for discussions of scattering of electromagnetic radiation important at other wavelengths, *see* COMPTON EFFECT; RADIO-WAVE PROPAGATION.

In other processes, energy removed from a light beam through interaction with a material is either reemitted with a change in wavelength, or converted to molecular motion and, ultimately, to heat. *See* ABSORPTION (ELECTROMAGNETIC RADIATION); FLUORESCENCE; RAMAN EFFECT.

Important instances of light scattering are Rayleigh scattering by gases, which explains the blue color of the sky; scattering by liquids and by solutions of large molecules; and Tyndall scattering by colloidal particles (*see* TYNDALL EFFECT). Light scattering is an important analytical tool with applications in many fields. For example, it is used to measure the size of particles and also to determine molecular weights. Interstellar dust scatters starlight as it passes through space. Although the scattered light itself is much too faint to be detected, analysis of the transmitted beams yields information on the nature and amount of dust through which the beams have passed.

Theory of scattering. When a beam of light encounters a particle, the electrically charged nuclei and electrons in the material undergo induced vibrations in phase with the incident wave. The oscillating charges act as sources of light which is propagated with the same wavelength as the exciting beam. Since light travels only at right angles to the direction of vibration of the oscillating charges, the horizontally and vertically polarized components of the incident beam produce scattering amplitudes proportional to cos θ (θ being the angle between the incident and scattered rays) and independent of θ, respectively (Fig. 1). The intensity of scattering is proportional to the sum of the squares of the amplitudes, that is, to $(1 + \cos^2 \theta)$ for unpolarized incident light (*see* POLARIZED LIGHT). This analysis holds true only for small isotropic particles; extension to other systems is considered later.

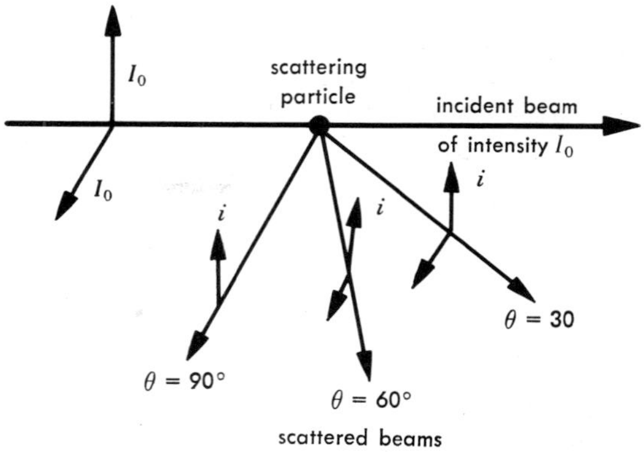

Fig. 1. Relative intensities i of horizontally and vertically polarized light scattered from a small isotropic particle.

Small independent particles. Scattering by small independent particles is called Rayleigh scattering. Lord Rayleigh (1871) showed that the intensity i_θ of light scattered by a gas from an incident beam of intensity I_0 and observed at a distance r is given by

$$\frac{i_\theta r^2}{I_0} = R_\theta = \frac{2\pi^2(n-1)^2}{\lambda^4}\frac{1}{\nu}(1 + \cos^2 \theta)$$

where ν, n, and λ are, respectively, the number of scattering particles per cm^3, the refractive index of the gas, and the wavelength of the light. The quantity R_θ is known as Rayleigh's ratio. This formula allows the calculation, from ν, of either Avogadro's number N or the molecular weight M of the gas, if the other is known.

The dependence of scattered intensity on the inverse fourth power of wavelength accounts for the color of the sky: blue light, of shorter wavelength, is scattered more strongly than light of other colors. The scattering of sunlight by clean air (J. Tyndall, 1868) can be observed directly in the apparatus of Fig. 2.

If the scattering particles are anisotropic (optically unsymmetrical), as is the case even with most gases, the scattered light is partially depolarized, that is, some horizontally polarized light is observed at $\theta = 90°$. The ratio of the intensities of the vertical and horizontal components gives information on the shape and symmetry of the particles (J. Cabannes, 1929).

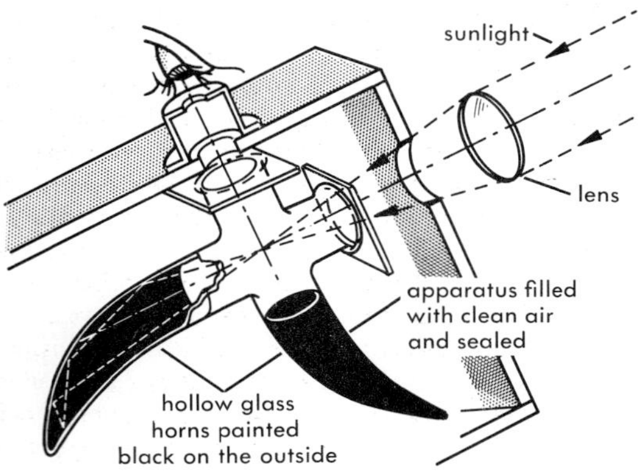

Fig. 2. Laboratory demonstration of the scattering of sunlight by clean air. (*After R. W. Wood, 1934, from J. Strong, Concepts of Classical Optics, W. H. Freeman, 1958*)

Small nonindependent particles. In a gas, each molecule scatters light independently of its neighbors. In condensed phases, however, the positions of neighboring particles are approximately fixed rather than randomly variable. This leads to fixed phase relations and destructive interference of most of the scattered light. The remaining scattering arises from random thermal fluctuations in the density of particles within small elements of the volume (A. Einstein, 1910; M. Smoluchowski, 1912). The magnitude of the fluctuations is derived by comparing the thermal energy kT (k is Boltzmann's constant, T is the absolute temperature) with the work required to cause a change in density through application of an external pressure p. The scattered intensity is proportional to $(kT\kappa/\lambda^4)(pn\ dn/dp)^2$ where κ is the compressibility of the liquid.

Large molecules in solution. In mixtures of liquids and in solutions, irregular changes in density and refractive index also arise from fluctuations in composition. The effect of these fluctuations is calculated (P. Debye, 1944) in terms of concentration changes arising from the osmotic pressure P, that is, the applied pressure that would be required to prevent the flow of the solvent across a perfectly semipermeable membrane. The scattered intensity is proportional to $(ckT/\lambda^4)(n\ dn/dc)^2/(dP/dc)$ where c is the solute concentration.

In application to solutions of polymer molecules two cases must be considered, depending on whether the size of the dissolved molecules is small compared to the wavelength of the light. For small molecules, insertion of the appropriate concentration dependence of the osmotic pressure leads to the relation

$$Kc/R_\theta = 1/M + 2A_2c$$

in which

$$K = (2\pi^2n^2/N\lambda^4)(dn/dc)^2$$

and A_2, called the second virial coefficient, characterizes deviations from ideal solution behavior due to polymer-solvent interactions. It is often convenient to express light scattering in terms of the turbidity τ, the fractional decrease in intensity of the incident beam due to scattering, thus: $\tau = (16\pi/3)R_{90}$, whence

$$Hc/\tau = 1/M + 2A_2c$$

in which

$$H = (32\pi^3n^2/3N\lambda^4)(dn/dc)^2$$

If the solute contains more than one molecular weight species, M is the weight average molecular weight. *See* MOLECULAR WEIGHT; OSMOSIS; POLYMER.

If the scattering particle is larger than about one-tenth the wavelength of the light, it can no longer be considered as a point source. Instead, scattered rays from different parts of the particle undergo destructive interference. This results in the diminution of scattered intensity and the introduction of dissymmetry into its dependence on θ; more light is scattered in the forward than in the backward direction. To a first approximation, suitable for most polymer solutions,

$$Kc/R_\theta = 1/MP(\theta) + 2A_2c$$

where $P(\theta)$ is a particle scattering function. For spheres (Lord Rayleigh, 1911; R. Gans, 1925),

$$P(\theta) = [(3/x^3)(\sin x - x\cos x)]^2$$
$$x = 2\pi(D/\lambda_s)\sin(\theta/2)$$

where D is the diameter of the sphere and $\lambda_s = \lambda/n$ is the wavelength of light in the solution. For random coil polymers (P. Debye, 1947),

$$P(\theta) = (2/x^2)[e^x - (1 - x)]$$
$$x = (8\pi^2/3)(\overline{r^2}/\lambda_s^2)\sin^2(\theta/2)$$

where $\overline{r^2}$ is the mean square distance between the ends of the random coil.

In the usual experimental treatment (B. H. Zimm, 1948), observations are made as a function of both c and θ. The values of M and A_2 are obtained after extrapolation to $\theta = 0$ (where

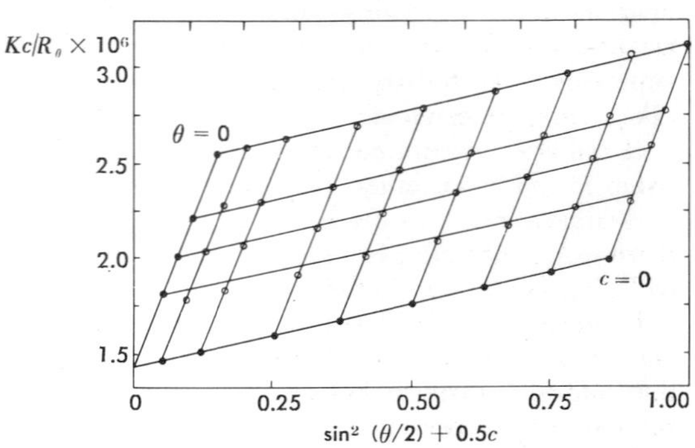

Fig. 3. Treatment of light-scattering data for a polymer solution (polymethyl methacrylate in butanone). Intercept at abscissa = 0 is $1/M$, and A_2 and $\overline{r^2}$ are derived from the slopes of the $\theta = 0$ and $c = 0$ lines, respectively.

$P(\theta) = 1$), and those of D or $\overline{r^2}$ after extrapolation to $c = 0$ (Fig. 3).

Colloidal particles. For scattering by particles larger than macromolecules, the Rayleigh-Gans approximate treatment is no longer adequate. The dependence of scattered intensity on particle size, refractive index, and angle becomes complex; the complete theory is developed only for spheres (G. Mie, 1908). The intensity of Mie scattering may show several maxima and minima as a function of angle (higher-order Tyndall spectra). The dependence on wavelength is to a power less than the inverse fourth; thus the scattered light often appears white instead of blue. Due to the complexity of the Mie equations, it is usual to base quantitative applications on tables of numerical calculations. [F.W.B.]

Bibliography: F. W. Billmeyer, Jr., *Textbook of Polymer Chemistry*, 1957; M. Fishman, *Light Scattering by Colloidal Systems*, 1957; V. K. La Mer and M. Kerker, Light scattered by particles, *Sci. American*, 188(2): 69–76, 1953; K. A. Stacey, *Light Scattering in Physical Chemistry*, 1956; H. C. van de Hulst, *Light Scattering by Small Particles*, 1957.

Scattering experiments, atomic and molecular

Experiments in which an incident particle or system of particles, such as an electron, atom, or molecule, is deflected by collision with an atom or a molecule. Such experiments are useful for many reasons; they provide checks on the theory of scattering and yield information on the nature of atomic and molecular forces. They are also important since the experiments can be designed to simulate conditions in the upper atmosphere, to provide information on electric discharges, and to aid in the theory of stellar absorption and planetary nebulae.

Classification of collisions. An impact between two atomic systems is said to be elastic if it involves no transfer of energy between the internal motions within the two systems and the motion of relative translation. Otherwise it is inelastic or superelastic according as energy is given to, or taken from, internal motion. *See* COLLISION.

If radiation is emitted during the impact the collision is radiative, otherwise it is nonradiative.

Rearrangement collisions are those in which there is a redistribution of particles between the colliding systems after the impact.

In general, in any type of collision, scattering occurs; that is, the direction of relative motion of the colliding systems before and after impact is rotated to a new direction.

Any number of systems may be involved in an impact. Usually collisions between two initial systems are studied. More than two systems may result from the impact, in which case the directions and energies of motion of all the resultants need to be specified.

Collision rates. Specification of collision rates is best done by introducing first the concept of total collision cross section. Consider a beam of electrons passing through a gas. If an electron is regarded as lost from the beam because of a change in direction or energy by a collision with a gas molecule, the beam current will be reduced by a factor $e^{-\alpha x}$ in passing a distance x through the gas. The quantity α can be written as NQ, where N is the number of gas molecules per unit volume and Q is the total effective cross section for collisions between electrons of speed v and the gas molecules.

Effective cross sections for different types of collisions follow by introducing probabilities p_j which give the chance that a collision is of a particular type j. The quantity p_jQ is then the effective cross section Q_j for a collision of this type.

Scattering in a particular type of collision is specified in terms of a differential cross section. If $i_j(\theta, \phi)$ is the chance that, in a collision of type j, the direction of motion of the electron is turned through an angle θ into the solid angle $\sin\theta\, d\theta\, d\phi$, the corresponding differential cross section or scattered intensity is

$$p_jQi_j(\theta,\phi) \sin\theta\, d\theta\, d\phi$$

In many experiments the electrons diffuse as a swarm rather than as a directed beam. If $f(E)dE$ is the number of electrons of the swarm with energy between E and $E + dE$, the number of collisions of type j occurring per second is given by

$$N \int f(E) Q_j(E) v\, dE$$

where v is the speed of an electron of energy E. These definitions and formulas may be extended to cover collisions between more complicated systems.

In general, cross sections defined in this manner have definite values and may be measured with apparatus of sufficient resolving power. For impact between charged particles the total cross section is unbounded, but the scattered intensity remains definite at all angles greater than zero.

Cross-section evaluation. The simplest problem is that of scattering of a beam of structureless particles of mass m and speed v by a center which exerts a force of potential $V(r)$, r being the distance of a particle from the center. Differential cross sections for this case may be calculated without approximation by the methods of quantum theory. If $V(r)$ exceeds the kinetic energy $\frac{1}{2}mv^2$ at distances $r < a$ where $a \gg \hbar/mv$, classical mechanics may be used to calculate the differential cross section for all angles $\theta > \hbar/mva$. If $V(r)$ is much less than $\frac{1}{2}mv^2$ for all r or for $r < a$ where $a \ll \hbar/mv$, a quantum mechanical perturbation treatment, known as Born's first approximation, is valid.

For the special case $V = A/r$, where A is a constant, the classical theory and Born's first approximation both give the exact value for the differential cross sections for all values of θ.

For collisions between systems with internal structure, no exact theoretical calculation of cross sections is possible. If the interaction between the systems is weak, Born's first approximation may be extended to these cases. In general this will be so if the velocity of relative translation of the colliding systems is much greater than the velocities of the internal motions. When this is not satisfied, there is no general method of approximation, but methods applicable to certain types of collision are available.

Electron-atom collisions. These have been studied intensively, both experimentally and theoretically. Different types of collision which may occur include (1) elastic; (2) inelastic, involving excitation of discrete atomic states; (3) ionization; (4) radiative capture to form negative ions; (5) radiative collisions in which the electron is not captured.

Radiative collisions are much less probable than nonradiative collisions, whereas the cross sections for inelastic collisions are comparable with those for elastic collisions at electron energies that are large compared with the minimum energy necessary to produce excitation.

Total cross sections have been measured for low-velocity electrons, using beams as well as diffusing swarms of electrons. Great variability in magnitude and velocity dependence is observed for different atoms (see Fig. 1). In general the observed

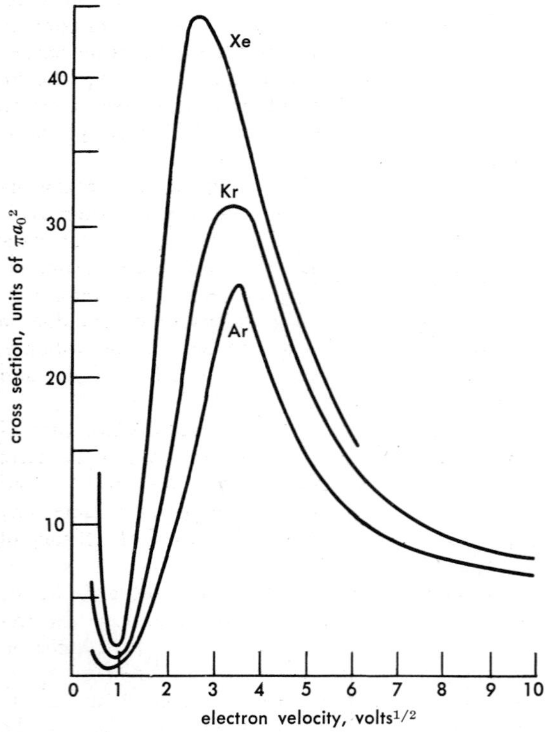

Fig. 1. Observed variation of the total cross sections for collisions of slow electrons in argon, krypton, and xenon; $a_0 = 0.53 \times 10^{-8}$ cm.

differential cross sections for elastic scattering in this velocity range exhibit maxima and minima as functions of the angle of scattering θ, the number increasing with the atomic number and, for a given atom, with electron velocity (see Fig. 2).

These effects are due to diffraction of the electrons by the scattering atoms. For low-velocity electrons elastic scattering is the most important, and the full quantum theory of scattering by the undisturbed field of the atom can be used to calculate the cross sections. In a complete theory, allowance must be made for exchange of electrons between the atom and incident beam and for polarization of the atom during the impact. The latter leads to increased scattering at small angles.

Inelastic nonradiative collisions have been studied experimentally by a variety of methods. Excitation of discrete states has been investigated by optical and electrical techniques. In the optical method one measures the intensity of radiation emitted at a particular wavelength from an electron beam of definite length that is passed through a gas at low pressure. For excitation of metastable states, relative yield at different electron velocities is observed either by use of optical absorption or by observing the electric current emitted from a surface on which the metastable atoms impinge. An important electrical method involves measurement of the fraction of electrons which have lost discrete amounts of energy in diffusing to the walls of a cylinder from an axial source.

Ionizing collisions have been studied by measuring the number of positive ions produced by an electron beam of definite energy passing for a definite distance through the gas at low pressure. The relative probabilities of collisions in which one, two, or more electrons are removed from the atom have been observed in some cases by performing a mass-spectrographic analysis of the product ions.

For all inelastic collisions involving excitation of an optically allowed transition, the variation of cross section with electron velocity has the form illustrated in curve 1 of Fig. 3, although for energies E close to the threshold E_0, some irregularities may occur. The maximum occurs for energies a few times E_0, and at high electron speeds v, the cross section falls off as $v^{-2} \ln \alpha v$ where α is a constant.

Excitation of a state with multiplicity different from that of the ground state can only take place through electron exchange (except for heavy atoms such as mercury for which it is not a good approximation to assign a definite multiplicity to a particular state). The excitation cross sections are only large for electron energies within a few electron volts of the threshold (see Fig. 3, curve 2). Very sharp maxima may occur in this energy region, but detailed information is not yet available except for the excitation of the 2^1S and 2^3S states of helium.

Cross sections for excitation of transitions involving no change of multiplicity but which are optically forbidden (see Fig. 3, curve 3) have the

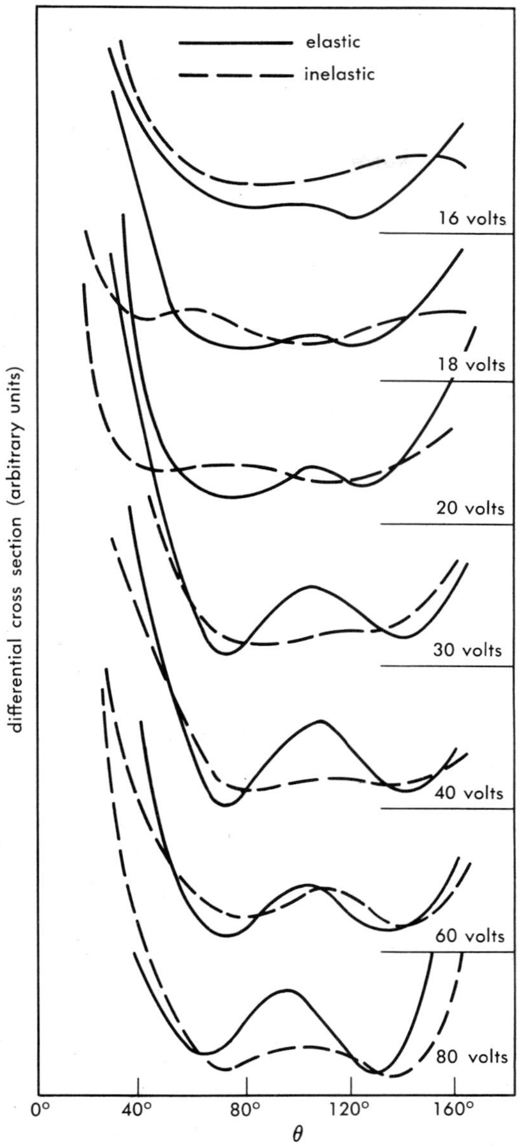

Fig. 2. Observed differential cross sections for elastic and inelastic scattering of electrons in argon for various electron energies. The latter are given as electron volts.

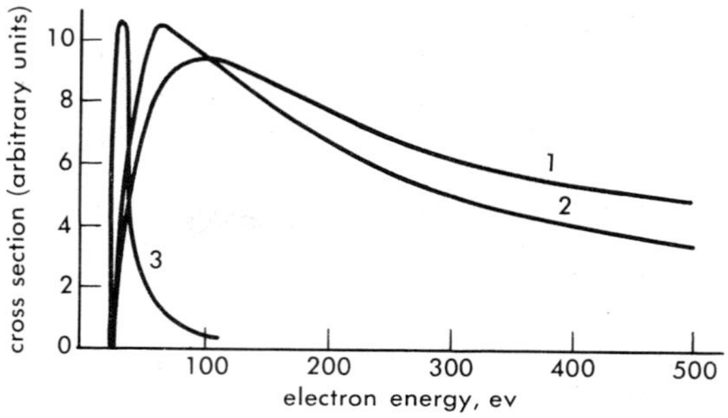

Fig. 3. Observed cross sections for excitation of different states of helium by electron impact: curve 1, 3^1P, an optically allowed transition; curve 2, 4^1D, an optically disallowed transition involving no change of multiplicity; curve 3, 4^3S, an optically disallowed transition with change of multiplicity. The cross sections are given in arbitrary units which are different for each curve—the emphasis is on illustration of the variation of cross sections with electron energy.

same general form as in curve 1 of Fig. 3, but the maximum occurs at a smaller value of E/E_0 and the decrease at high velocity is proportional to v^{-2}.

Born's first approximation gives good results for the cross sections for electron velocities well beyond the maximum. In collisions which do not involve change of multiplicity at lower energies, it gives an overestimate. Improved theory has been applied with some success to calculation of the excitation of 2^1S and 2^3S states of helium and of the $2s$ and $2p$ states of hydrogen.

Differential cross sections for inelastic collisions have not been observed so extensively, but for collisions involving the most probable excitation and also for ionizing collisions, data are available. At not too large electron velocities, maxima and minima are found in the scattered intensity for excitation of a particular state. These resemble corresponding results for elastic scattering and are due to the effect of the atomic field in producing strong distortions of the incident and outgoing electron waves from the plane-wave form.

Polarization of the light emitted by electron impact has been studied for some atoms. This depends on the relative probability of excitation of states with angular momenta of the same magnitude but with different components along the direction of the electron beam. Agreement between theory and experiment in this field is not yet satisfactory.

The cross section for capture of an electron by an atom to give a negative ion is of the order 10^{-21} cm^2 or less. At small electron velocities v, the cross section varies as v^2 for capture into an s state, but is independent of v for capture into a p state. Very few observations of the affinity spectrum due to these capture processes have been made, although some evidence has been reported for oxygen and hydrogen.

Cross sections for the reverse process, that of photodetachment of an electron from a negative ion, have been measured for H^-, O^-, and S^-, and from these the radiative capture cross sections may be obtained, using what is known as the theory of detailed balancing.

Photodetachment is of major importance in the solar atmosphere. The H^- ions determine the frequency distribution of the continuous emission from the sun in the visible region. At longer wavelengths, absorption by free electrons in the neighborhood of H atoms is important. This is the inverse process to electron scattering by hydrogen atoms, in which the electrons emit radiation without being captured.

Electron-molecule collisions. If the electrons collide with a molecule instead of with an atom, some additional possibilities arise. Molecular vibration or rotation may be excited, and dissociation of the molecule into two or more neutral or ionized fragments may occur. The only experimental evidence about electron excitation of vibration and rotation has been obtained from observation of the mean energy of electron swarms diffusing through gases in an electric field. The probability of vibrational excitation is of the order of 1% per collision; that for rotation is much higher for molecules with permanent electric dipole or quadrupole moments. *See* MOLECULAR STRUCTURE AND SPECTRA.

Extensive studies have been made of the relative probabilities of production of different positively charged fragments due to impact of electrons having energies near 100 ev with various molecules. This has been done using mass spectrography and is useful for analysis of industrial gases and vapors, especially hydrocarbons.

Negative ion formation by dissociation attachment may occur in electron impact with a molecule via a process such as

$$AB + e^- \rightarrow A + B^-$$

In O_2 and CO, these cross sections are of the order 10^{-19}–10^{-21} cm^2 in certain narrow electron energy ranges. Capture of very slow electrons without dissociation may occur in polyatomic gases, as in SF_6.

Dissociation of a molecule into positive and negative ions may occur if the electrons are sufficiently energetic.

Collisions of atomic systems. If both the colliding systems are atoms or molecules, elastic, excitation, or ionization processes can occur. The inelastic processes occur only if the energy of relative motion is great enough. Either or both of the colliding systems may be excited.

There are certain additional possibilities which can be called transfer collisions. They include the following:

$$\left.\begin{array}{l} A' + B \rightarrow A + B' \\ A' + B \rightarrow A + B^+ + e^- \end{array}\right\} \text{transfer of excitation}$$

$$\left.\begin{array}{l} A^\pm + B \rightarrow A + B^\pm \\ A^+ + B^- \rightarrow A' + B'' \end{array}\right\} \text{transfer of charge}$$

The last process is often called mutual neutralization.

Elastic scattering. Elastic cross sections for atom-atom impacts may be calculated by treating the interaction between the atoms as static, so that the problem is reducible to that of scattering by a fixed center of force. Except at very small angles of scattering and very low energies of relative motion (much less than the mean value at room temperature) the differential cross section may be calculated from classical mechanics. Very few direct experimental observations have been made, but indirect information is available from measurements of positive ion mobility and the viscosity and diffusion coefficients of gases. For helium, at temperatures below 50°K, quantum effects become appreciable. The main aim of these studies is to derive the interaction energy between the gas atoms.

Transfer collisions. Transfer of charge may be studied by firing an ion beam of homogeneous energy through a gas and measuring the net current of slow positive ions produced for a given path length. Information about excitation transfer is available only from indirect methods involving the quenching of radiation by foreign atoms or other methods.

The magnitude and dependence of the cross section on relative velocity of impact, for a transfer collision, depends very strongly on the amount of energy, ΔE, transferred between translational and internal motion.

Figure 4 illustrates typical observed cross sections for charge transfer collisions between positive ions and neutral atoms as functions of the relative impact velocity, v. In general, the maximum cross section occurs roughly when

$$2\pi a \Delta E/(hv) \cong 1$$

where a is a length of the order of atomic dimensions. When $\Delta E = 0$, as for example in

$$He^+ + He \rightarrow He + He^+$$

the cross section falls steadily as v increases.

In all these cases, the magnitude of the cross section never exceeds gas kinetic values by a large factor. This is because there is no long range interaction between either the initial or the final system. For charge transfer, in which either the initial or the final system is charged, as in

$$A^+ + B^- \rightarrow A' + B''$$

or

$$A^{++} + B \rightarrow A^+ + B^+$$

this is no longer true, and cross sections greatly in excess of gas kinetic values may result. For a given

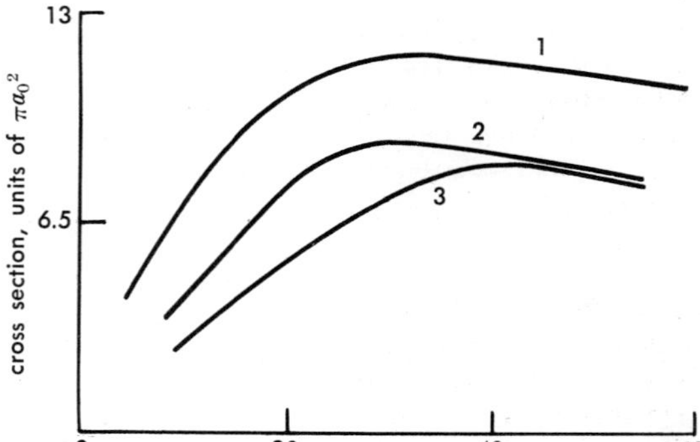

Fig. 4. Some observed cross sections for charge transfer processes:

(1) $O^+ + Kr \rightarrow Kr^+ + O + 0.43$ ev
(2) $Br^+ + Xe \rightarrow Xe^+ + Br - 0.28$ ev
(3) $C^+ + Xe \rightarrow Xe^+ + C - 0.86$ ev

pair of reactants, the minimum cross section need not occur when $\Delta E = 0$ but will usually occur for small values of ΔE.

A similar situation arises with excitation transfer. Cross sections much greater than gas kinetic values may occur for small ΔE only when both transitions which take place in the reactants are allowed.

In general, the cross section for a transfer collision can greatly exceed the gas kinetic value only when there is a long range interaction between either the initial or the final reacting systems.

Excitation and ionization. Provided the velocity v of relative motion is large compared with that, u, of the internal electronic motions concerned, the cross section for excitation of a particular atomic state is nearly the same for impact of singly charged positive ions as for electrons of the same relative velocity v (see Fig. 5). This is true also for ionization.

When $v < u$, the cross section for excitation or ionization by positive ion impact is in general small and decreases rapidly as v decreases. Neutral atoms or molecules may be more effective than ions of the same relative velocity in this range, the de-

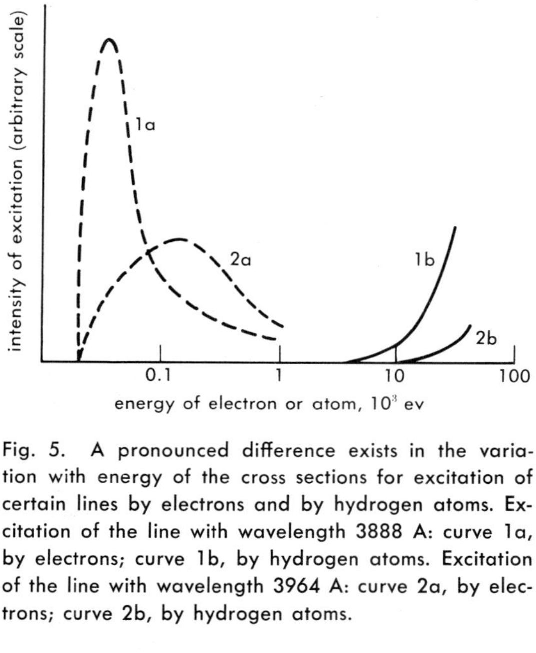

Fig. 5. A pronounced difference exists in the variation with energy of the cross sections for excitation of certain lines by electrons and by hydrogen atoms. Excitation of the line with wavelength 3888 A: curve 1a, by electrons; curve 1b, by hydrogen atoms. Excitation of the line with wavelength 3964 A: curve 2a, by electrons; curve 2b, by hydrogen atoms.

crease of the cross section as v decreases being often more gradual. There may be many exceptions to these general rules, but the subject has not been thoroughly explored. See ELECTRON DIFFRACTION; MOLECULAR BEAMS; QUANTUM THEORY, NONRELATIVISTIC; RAMAN EFFECT; SCATTERING EXPERIMENTS, NUCLEAR. [H.S.W.M.]

Bibliography: S. Fluegge (ed.), *Handbuch der Physik*, vol. 36, 1956; H. S. W. Massey and E. H. S. Burhop, *Electronic and Ionic Impact Phenomena*, 1952; N. F. Mott and H. S. W. Massey, *The Theory of Atomic Collisions*, 2d ed., 1949.

Scattering experiments, nuclear

Experiments in which particles such as electrons, nucleons, α-particles, and mesons are deflected by collisions with atomic nuclei. Much is learned from such experiments about the nature of the scattered particle, the scattering center, and the forces acting between them. Scattering experiments, made possible by the construction of high-energy particle accelerators, are the main source of information regarding the structure of matter.

In the broad sense, any nuclear reaction is an example of scattering. However, this article treats scattering only in a more restricted sense as given in the definition. See NUCLEAR REACTION.

Modern views on nuclear and atomic physics had their beginnings in the experiments on the scattering of α-particles in their passage through matter and their interpretation by Sir Ernest Rutherford. See ATOMIC STRUCTURE AND SPECTRA.

Definitions. The word elastic is used to indicate the absence of energy loss. If particle A collides with particle B of finite mass which is originally at rest, there is a loss in the energy of A even if no energy has been transferred to the internal degrees of freedom of either A or B. Sometimes such a collision is referred to as inelastic in order to distinguish its character from a collision with a particle having an infinite mass or its idealization, a fixed center of force. This terminology is not useful in the present context because in the center-of-mass system of the two particles the sum of kinetic energies after the collision is the same as before (see COLLISION). The distinction between elastic and inelastic scattering is made therefore on the basis of whether there are internal energy changes in the colliding particles. The collision is said to be inelastic even if the energy changes of the two particles compensate so as to leave the sum of the kinetic energies in the center-of-mass system unaltered. The theory of inelastic scattering is connected with nuclear reaction theory because nuclear reactions have a marked influence on the scattering.

It is also useful to distinguish between coherent and incoherent scattering; the distinction is made on the basis of the ability of the scattered wave to interfere with the incident one. Inelastic scattering is always incoherent. There are situations in which elastic scattering is incoherent.

Low energy n-p scattering. Deuterons, the nuclei of heavy hydrogen, have a mass nearly twice that of the neutron. Under the action of γ-rays they dissociate into protons, p, and neutrons, n. It is usually supposed, therefore, that each deuteron is composed of one p and one n, and the first attempts to estimate the magnitude of nuclear forces made use of this assumption together with the measured value of the deuteron binding energy. It appeared reasonable to suppose that the spatial extension of the n-p force, the so-called range of force, is relatively small and this assumption made it possible to estimate the scattering cross section, that is, the probability that a neutron will be scattered by a proton (*see* NEUTRON CROSS SECTION). The measured *n-p* scattering cross section is larger than would be expected if the forces between free neutrons and protons were the same as those in the deuteron. E. Wigner introduced therefore the hypothesis of spin dependence for the *n-p* interaction. This explanation is made possible by the fact that *p* and *n* are known to have a spin of ½; that is, their intrinsic angular momenta are known to be $\frac{1}{2}(h/2\pi) = \hbar/2$, where *h* is Planck's constant. According to quantum mechanics, when two spins s_1, s_2 combine vectorially only the values

$$s = s_1 + s_2, s_1 + s_2 - 1, \ldots, s_1 - s_2$$

are possible for the resultant. In the present case, therefore, the resultant spins are 0 or 1. In the first case one speaks of a singlet, in the second of a triplet. The singlet state behaves much like a round and perfectly smooth object which has the same appearance no matter how it is viewed, corresponding to only one possibility of forming a state with $s = 0$. The state with $s = 1$, on the other hand, can have three distinct spin orientations. Measurement of the projection of *s* on an axis fixed in space can give only the three values (1,0,−1), again in units $\hbar$. When protons with random spin directions collide with neutrons also having random spin directions, the triplet state is formed three times as often as the singlet. Wigner concluded that nuclear forces are spin-dependent, that is, that they depend on the mutual orientation of the spins of the interacting particles.

The neutron-hydrogen scattering experiments on which these conclusions were based, were performed with slow neutrons having an energy equivalent of 1 electron volt (ev) or less. The general quantum-mechanical theory of scattering is much simplified in this case. Because of the small range of nuclear forces, only the collisions with zero orbital angular momentum $L\hbar$ play a role, collisions with higher orbital angular momenta missing the region within which nuclear interactions take place. States with $L = 0$ (called *S* states) have the property of spherical symmetry, and nuclear forces matter in this case only inasmuch as they modify the spherically symmetric part of the wave function. (For definition of the term wave function, *see* QUANTUM THEORY, NONRELATIVISTIC.)

The long wavelength scattering cross section can be described completely by a quantity called the scattering length. For interparticle distances *r* greater than the range of nuclear forces the spherically symmetric part of the wave function has the form

$$C[1 + (a/r)]$$

where *a* and *C* are constants. The constant *a* is called the scattering length. It has the following meaning. Calling the wave function $R(r)$, the product $rR(r)$ when plotted against *r* is represented by a straight line which cuts the axis of *r* at a distance *a* to the left of the origin of coordinates. If the

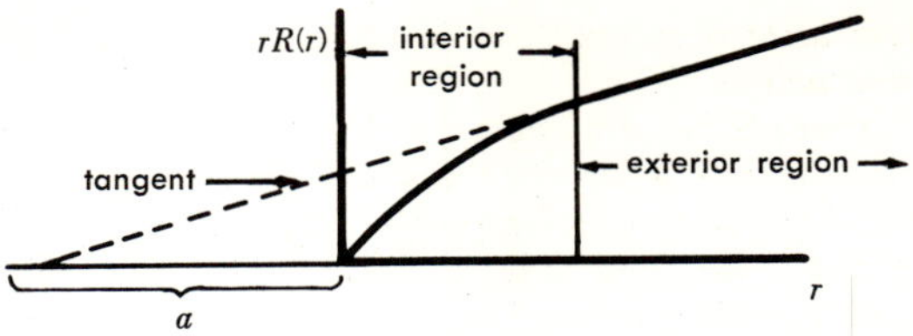

Fig. 1. Scattering length in the case of a virtual level.

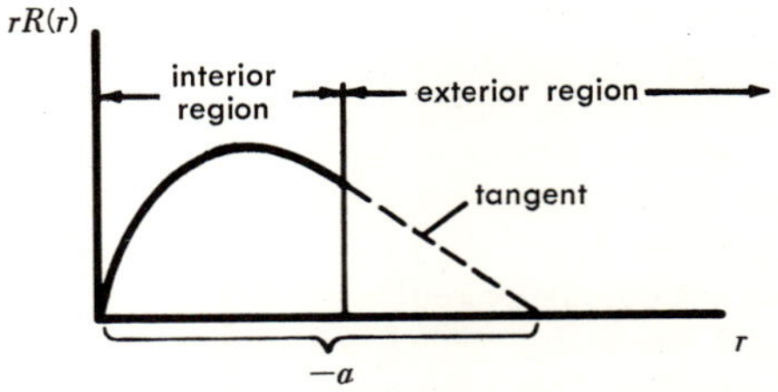

Fig. 2. Scattering length in the case of a real level.

intersection is to the right of the origin, *a* is counted as negative. The two conditions are illustrated in Figs. 1 and 2.

The scattering cross section for a state with well-defined *a* is $4\pi a^2$. Wigner's reasoning determined $4\pi a_0^2$, where a_0 is the *a* for the singlet state. The sign of a_0 was determined from slow neutron scattering and partly from the phenomenon of photomagnetic capture of neutrons by protons. Before the sign was determined one referred to the case $a_0 > 0$ as that of the virtual level and to $a_0 < 0$ as that of the real level in the $^1S_0$ state.

The term potential well is generally used to describe the potential energy for *n-p* scattering because a graph of the potential energy against distance usually shows a characteristic depression similar to the shape of a water well. The potential energy may be expressed as $V(r) = -Av(r/b)$ where *A* and *b* represent constants called depth and range parameters and *v* is a function which determines the shape of the well. The scattering length determines approximately the product Ab^2, for a potential well of assigned shape. The variation of the cross section with energy *E* through the energy range of a few Mev can be used for the determination of *b*. *See* NUCLEAR STRUCTURE.

Low-energy p-p scattering. The *p-p* and *n-n* interactions are closely equal. This equality gives rise to the hypothesis of charge independence of nuclear forces, which supposes that nuclear forces acting in addition to the electrostatic (Coulomb) repulsion in the *p-p*, *p-n*, and *n-n* cases are equal to each other. Because of the limits of experimental accuracy and uncertainties in the theoretical interpretation the hypothesis is not established with perfect accuracy, but it is believed to hold within a few per cent for the depth parameter *A* if the range parameter *b* is specified as the same. Meson theory suggests that nucleon-nucleon interactions are not exactly charge-independent, the underlying law being presumably valid for the meson-nucleon interactions.

There is no proof that the shapes of the *p-p* and *n-p* potential wells are the same, the data up to a few Mev energy determining primarily values of A and b for any assumed reasonable shape but distinguishing poorly between different assumed shapes.

Other n-p and p-p scattering. In the intermediate energy region (10–440 Mev) the analysis of experimental material is less certain than in the low energy region. It has not been found necessary to abandon the hypothesis of charge independence. Some evidence for the existence of strong repulsive regions of the interaction at the smallest distances (hard cores) has been found, but this interpretation is not unique since the nucleons may be merging and producing many mesons when they are close together. It is claimed by some that the data indicate the presence of a strong spin-orbit potential having the form $V(r)$ $(\mathbf{L} \cdot \mathbf{S})$ where $\mathbf{L}$ and $\mathbf{S}$ are respectively the relative orbital angular momentum and the total spin of the particles. Others claim that the data should be interpreted instead in terms of a strong tensor potential, which is similar in mathematical form to the potential between two small magnets. Attempts at interpretation have been aided by measurements of spin polarization obtained by means of double and triple scattering experiments.

Above 440 Mev, meson production is important and the experiments are usually interpreted by means of the so-called optical model, employing complex numbers for the potential. The imaginary part of the number accounts for inelastic processes.

Nucleon scattering. Scattering of nucleons by nuclei at high energies (100–500 Mev) is of interest on account of its connection with nucleon-nucleon scattering. It is possible to calculate approximately the scattering of a nucleon from a nucleus.

Nucleon-nucleus scattering experiments can be accounted for by a potential well model with a complex potential (optical model) making use of a spin-orbit interaction term. The details of angular distributions of the cross section and polarization are reproduced surprisingly well and experiments favor some potential well shapes over others. Wells thus determined are wider than those obtained from electron-nucleus scattering experiments, and similar shapes work in both cases. The potential energy represented by the wells is added to the electrostatic potential energy in the calculations. The latter is approximated by a central potential corresponding to the average distribution of nuclear charge. Many nuclear radii have been determined, and the variation of density of nuclear matter within a nucleus is believed to have been partly determined. The electron scattering experiments are concerned with the charge distribution, and if one supposes that in a nucleus (as is the case outside of it) the neutrons remain neutral, the experiments determine the distribution of proton density. There is evidence from the coherent neutral photopion production that the neutron and proton densities are not very different, and it appears likely therefore that electron scattering gives the density of nuclear matter to a fair approximation. This view has qualitative support in the scattering of high-energy pions ($\sim$1 Bev) which are in agreement with the same shape of potential energy well as used for electron scattering experiments and lead to only a 6% larger radius. A direct comparison of π^+ with π^- scattering indicates a difference of less than 3% in the effective radius for Pb. The 6% increase in radius is perhaps caused by the finite range of nucleon-pion interaction while the neutron and proton density distributions appear to be nearly the same.

Optical model potential fits to nucleon-nucleus data are in general agreement with the data in a wide energy range from several Mev up to about 300 Mev, but the parameters of the potential have to be varied progressively. One of the earlier confirmations of the adequacy of this method has been found in the representation of observed maxima in the scattering of neutrons. The spin-orbit potential found to represent the scattering data at the lower energies has the same value of its ratio to $\frac{dV_{ct}}{r\,dr}$, where V_{ct} is the central potential, as it has in the shell theory of nuclear structure. At higher energies (300 Mev), it has been found that the spin-orbit potential has to be used with a smaller strength than in shell theory. The real part of the central potential decreases with energy and becomes almost zero at 300 Mev. Data at 1 Bev on proton scattering from carbon can be accounted for on the optical model by means of an imaginary central and real spin-orbit potential.

Electron-nucleon scattering. Scattering of electrons by hydrogen gives information regarding the electron-proton interaction. From measurements of the variation of the differential cross section it has been found necessary to postulate that both the proton charge and its intrinsic magnetic moment are distributed through a finite volume. Existing work favors the assumption of similarity of shape of these distributions. The important energies for the detection of these effects are between 200 and 500 Mev, and the anomalous magnetic moment of the proton plays an important part at high energies. The experiments distinguish between different shapes of charge density distributions and are well accounted for by an exponentially decreasing charge density with an rms radius of approximately 0.80×10^{-13} cm. A Gauss error function is excluded as a representation of the charge density. Some other more complicated distributions cannot be excluded, however, and the charge distribution cannot be claimed to be known. Measurements on the scattering of electrons by hydrogen at large angles and high energies are essential for the conclusion that the magnetic moment of the neutron is not a point. If it is assumed that neutron and proton magnetic moments are distributed through nearly the same volumes about each other's centers, a good representation of the scattering measurements is obtained. Analyses of scattering data at various angles and energies indicate that there is no charge density within the volume occupied by the neutron. This result is in agreement with measurements of the neutron-electron interaction made by scattering very slow neutrons from atomic electrons and atomic nuclei. While there is an interaction equivalent to a potential energy of approximately −3900 ev through a distance of e^2/mc^2 amounting to approximately 2.8×10^{-13} cm, it is accounted for qualitatively as a consequence of what E. Schrödinger called the *Zitterbewegung* (tremblatory motion) expected for the neutron. In the interpretation of these experiments there is also no reason for a charge distribution within the neutron. This circumstance is one of the newer puzzles of physics. The charge distribution of the proton is usually explained as originating in a temporary dissociation of a proton into a neutron and a π^+ meson. Similar dissociation of a neutron is expected into a proton and π^-. It is not known whether the charge density thus expected to appear somewhat away from the neutron center is masked by effects of the neutron core which could be anticipated from known views or whether a misconception regarding nucleon structure is at fault.

Meson scattering. Scattering of π mesons (pions) by nuclei plays the same role in nuclear physics as the scattering of light by electrons played in atomic physics. The scattering showed the existence of a resonance at approximately 200 Mev, which fits in with observations on photopion production.

Observations have been made on the scattering of K and τ mesons by nuclei. While in some early work it appeared that the K-p differential cross section showed forward peaking and was supposed to indicate the superposition of a repulsive S-wave interaction onto the Coulomb (electrostatic) wave function, later work does not support this view. This interaction is still believed to be mainly of the S- and perhaps partly P-wave type and to occur in a small volume. The data do not compare in accuracy with nucleon-nucleon experiments, and conclusions regarding the K-nucleon interaction may be premature.

Scattering of μ mesons (muons) and studies of their polarization have been suggested as a possible means of obtaining information concerning nuclei and also regarding the asymmetry of muon decay, the latter being of interest in the theory of the so-called weak interactions. Behavior of atoms containing μ mesons indicates that μ mesons interact with protons and neutrons entirely or almost entirely through electromagnetic forces. No evidence for the existence of another type of μ-p and μ-n interaction had been found through 1959. Studies with μ mesons are thus likely to be similar in content to those performed with electrons. The number of mesons of controlled energy available from modern accelerators is small and such experiments as have been performed have made use of cosmic ray muons. Because of the small number of muons, experimental results have not led to clear-cut conclusions. *See* COMPTON EFFECT; ELECTRON DIFFRACTION; NEUTRON DIFFRACTION; SCATTERING (ELECTROMAGNETIC RADIATION); SCATTERING EXPERIMENTS, ATOMIC AND MOLECULAR; SYMMETRY LAWS (PHYSICS); X-RAY DIFFRACTION. [G.BR.]

Bibliography: H. A. Bethe and P. Morrison, *Elementary Nuclear Theory*, 2d ed., 1956; L. Eisenbud and E. P. Wigner, *Nuclear Structure*, 1958; R. D. Evans, *The Atomic Nucleus*, 1955; H. A. Bethe, F. De Hoffmann, and S. S. Schweber, *Mesons and Fields*, 2 vols., 1955.

Scattering layer

A term in oceanography referring to layers of animals or organisms in the sea which cause sound to scatter and return echoes. Recordings by sonic sounding devices of echoes from sound scatterers indicate that the scattering organisms are arranged in approximately horizontal layers in the water, usually well above the bottom. The layers are found in both shallow and deep water.

Deep scattering layer. In deep water one or more well-defined layers generally are present. They are readily detected by echo-sounding equipment capable of scanning the sound spectrum from 1 to 60 kc/sec, each layer having maximum scattering at different frequencies. Commonly, but not universally, the deep-water layers migrate vertically in apparent response to changes in natural illumination. The most pronounced migration follows a diurnal cycle, the layers rising at night, sometimes to the surface, and descending to greater depth during the day. The common range of daytime depths is 100–400 fathoms. The migration is modified by moonlight at night and has been observed to be modified during the day by heavy local cloud cover, as in a squall. The occurrence of the layers in deep water was first demonstrated by C. Eyring, R. Christiansen, and R. Raitt. *See* ECHO SOUNDER; UNDERWATER SOUND.

Scattering organisms. Sound scatterers have not yet been specifically identified. Shallow-water scatterers have been identified in some instances as fish, in others as plankton. Their occurrence is extremely variable. M. W. Johnson pointed out in 1946 that the layers which migrate diurnally must be animals capable of swimming to change their depth, rather than plant life or some physical boundary such as an abrupt temperature change in the water. In 1953 V. C. Anderson demonstrated that some of the deep-water scatterers have a much smaller acoustical impedance than sea water. This fact fits the suggestion by N. B. Marshall in 1951 that the scatterers may be small fishes with gas-filled swimbladders, many of which are known to be geographically distributed much as the layers are. In 1954 J. B. Hersey and R. H. Backus found that the principal layers in several localities migrate in frequency of peak response while migrating in depth, thus indicating that the majority of scatterers fit Marshall's suggestion.

New techniques, combining acoustic and photographic recording, offer promise of specific identification of scattering organisms. When such identification is made possible, sonic methods should prove even more useful in studying the ecology of bathypelagic animals. *See* DEEP-SEA FAUNA; UNDERWATER PHOTOGRAPHY; UNDERWATER TELEVISION. [J.B.H.]

Bibliography: N. B. Marshall, Bathypelagic fishes as sound scatterers in the ocean, *J. Marine Research*, 10:1–17, 1951; L. A. Walford, The deep-sea layer of life, *Sci. Am.*, 185(2):24–28, 1951.

Scent gland

A specialized skin gland of the tubuloalveolar or acinous variety found in many mammals. These glands produce substances having peculiar odors. In some instances they are large; in others small. Examples of large glands are the civet gland in the civet cat, the musk gland in the musk deer, and the castoreum gland in the beaver. The civet gland is an anal gland, whereas the musk and castoreum are preputial. Examples of small scent glands are the preputial or Tyson's glands in the human male which secrete the smegma, and the vulval glands in the female. The secretions in all of the above glands are sebaceous.

Many other vertebrates have glands whose secretions give off various types of odors. The mucus-secreting skin glands of fishes produce their fishy odor. Amphibia have glands which emit pungent, sweetish, or onionlike odors. Many urodeles have specialized courtship or hedonic glands. These amphibian glands are mucus-secreting. Scent glands are protective devices for many animals; in some species they serve to attract members of the same species or the opposite sex. The femoral glands on the inner aspect of the upper region of the hindlimbs of male lizards are specialized sebaceous glands associated with copulation. They give off a musty odor. The uropygial or preen glands opening upon the upper tail surface of birds produce an odorous, oily material used to waterproof feathers. *See* EPITHELIUM; GLAND; UROPYGIAL GLAND. [O.E.N.]

Scheelite

A mineral consisting of calcium tungstate, $CaWO_4$. Scheelite occurs in colorless to white, tetragonal crystals; it may also be massive and granular. Its fracture is uneven. Its luster is vitreous to adamantine. Scheelite has a hardness of 4.5–5, and a specific gravity of 6.1. Its streak is white. The mineral is transparent and fluoresces bright bluish-white under ultraviolet light.

Scheelite may contain small amounts of molybdenum. It is an important tungsten mineral and occurs principally in contact metamorphosed deposits (known as tactite) associated with garnet, diopside, tremolite, epidote, wollastonite, sphene, molybdenite, and fluorite, with minor amounts of pyrite and chalcopyrite. It also occurs in small amounts in vein deposits. The most important scheelite deposit in the United States is near Mill City, in the Humboldt Range, Nevada. *See* TUNGSTEN. [E.C.T.C.]

Schematic drawing

Concise, graphical symbolism whereby the engineer communicates to others the functional relationship of the parts in a component and, in turn, of the components in a system. The symbols do not

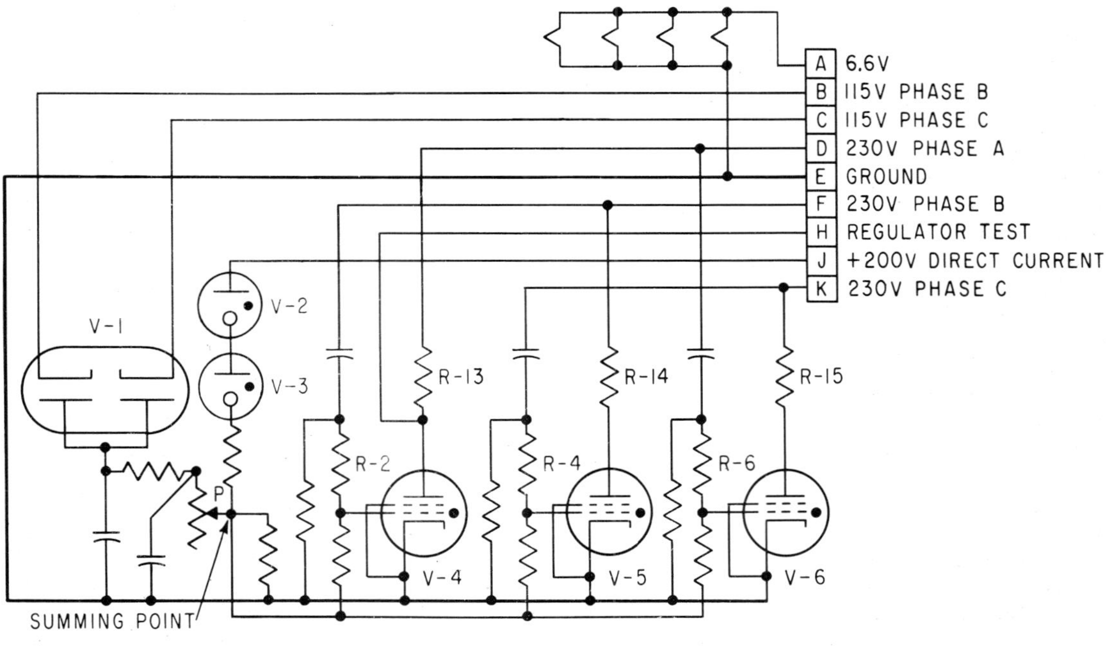

Fig. 1. Electrical schematic of voltage regulator.

attempt to describe in complete detail the characteristics or physical form of the elements, but they do suggest the functional form which the ensemble of elements will take in satisfying the functional requirements of the component. They are different from a block diagram in that schematics describe more specifically the physical process by which the functional specifications of a block diagram are satisfied. Rather than expressing a mathematical relationship between, for example, an input and an output variable as in a block diagram, a schematic illustrates the physical principles and techniques by which the mathematical requirements of the element are realized. For instance, the schematic indicates whether electrical, hydraulic, mechanical or pneumatic techniques are employed, and suitable symbols indicate the appropriate elements, such as batteries, resistors, valves, and motors by which the functional requirements of the component are achieved.

Electrical schematic. An electrical schematic is a functional schematic which defines the interrelationship of the electrical elements in a circuit, equipment, or system. The symbols describing the electrical elements are stylized, simplified, and standardized to the point of universal acceptance (Fig. 1). The simple nature of the element symbol makes possible reporting in a small area the interrelationship of the electrical elements in complex systems. This has the double advantage of economy of space and an increased facility of understanding, because one experienced in the symbolism can easily follow the various functional paths in the electrical schematic. The tracing of a signal path through an electrical schematic is considerably enhanced by the existence of more or less accepted rules with regard to the arrangement of the symbols and of the interconnections between the symbols, all contrived so as to make more lucid the functional interrelationship of the elements.

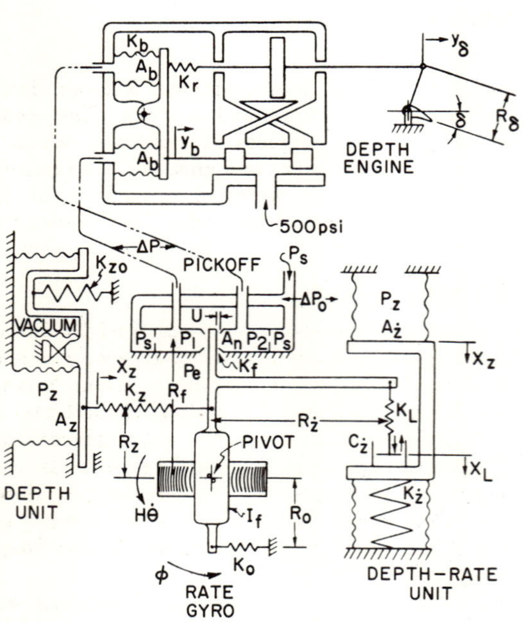

DEFINITIONS OF SUBSCRIPTS

b = DIFFERENTIAL–PRESSURE BELLOWS
e = ENVIRONMENT
f = FLAPPER
L = DEPTH-RATE LINKAGE
n = NOZZLES
o = GROUND, OR REFERENCE
r = RAM FEEDBACK
s = SUPPLY
z = DEPTH UNIT
ż = DEPTH–RATE UNIT
δ = ELEVATOR

Fig. 2. Mechanical schematic of torpedo depth-control mechanism.

Mechanical schematic. A mechanical schematic is also a functional schematic. The graphical description of elements of a mechanical system are more complex and more intimately interrelated than the symbolism of an electrical system and so the graphical characterizations are not nearly as well standardized or simplified (Fig. 2). However, a mechanical schematic illustrates such features as lever-arm arrangements or gear ratios between components, and viscous damping devices. The symbols are arranged in such a manner and with such simplification as to economize on space and to facilitate an understanding of the functional interrelationship of the components in the system. *See* ENGINEERING DRAWING. [R.W.M.]

Schiff base

One of a class of organic compounds represented by the general formula RCH=N—R′, where R and R′ are aliphatic or aromatic hydrocarbon substituents. The term Schiff base usually refers to such a compound in which R and R′ are aromatic, but in a broader sense, it encompasses all substances having this particular structure.

Schiff bases are obtained by the condensation reaction of an aldehyde and a primary amine with concurrent loss of water. The aromatic Schiff bases, and many of the aliphatic-aromatic type, can be prepared simply by mixing and warming equimolar amounts of the aldehyde and the amine. They are stable toward alkali but are easily cleaved by acids to give the parent compounds. Schiff bases also can be prepared from aliphatic aldehydes and aliphatic amines; however, these bases frequently undergo further reaction to give compounds of higher molecular weight.

Schiff bases have been used to characterize aldehydes and primary aromatic amines since they are stable crystalline compounds with definite melting points. Schiff bases are converted to secondary amines by catalytic hydrogenation, and they also have served as intermediates in the preparation of many other organic products. Certain compounds of this structure find use as dyes, as sequestering agents, and as accelerators in the vulcanization of rubber. *See* ALDEHYDE; AMINE; CONDENSATION REACTION. [A.E.B.]

Schist

A large group of rocks which by deformation during regional metamorphism have acquired a schistosity, that is, they show a more or less perfect cleavage along which they easily split up in flaggy slabs. There are orthoschists derived from igneous rocks and paraschists derived from sedimentary rocks. Crystalline schist is a common designation for all kinds of rocks which have recrystallized during regional metamorphism. *See* METAMORPHISM.

Composition. The chemical composition of a schist is similar to that of most silicate rocks, but mineralogically all schists have one feature in common, namely, that one or more of the major mineral constituents are flaky or fibrous (having the crystal structure of phyllosilicates or inosilicates). These minerals are arranged as flakes parallel to the schistosity planes or as subparallel fibers in the schistosity planes, thus emphasizing the schistosity or inducing a linear element in addition to the schistosity. With increase of feldspar and quartz and decrease of schist-forming minerals, the schists pass into the less schistose and more irregular foliated gneisses. *See* GNEISS; SILICATE MINERALS.

Development of schistosity. The metamorphism of the rocks has a direct relation to the orogenic movements in the mountain ranges. At great depths differential movements have taken place by which the constituent minerals of the rocks have acquired a certain spatial arrangement. If this arrangement results in the alignment of flaky, prismatic, or fibrous crystals, a schistose structure is produced. It may be platy or linear. The preferred mineral orientation is either according to the external crystal form (*Formregelung*) or according to the crystal structure (*Gitterregelung*); for explanation *see* METAMORPHIC ROCKS. Rocks that show any kind of preferred mineral orientation are called tectonites by B. Sander. However, not all tectonites are truly schistose. The following types of differential movements will result in preferred mineral orientation.

Laminar gliding. Laminar gliding is similar to the relative displacement of the leaves of a paperbound book which is bent or folded. Laminar gliding is caused by shearing stress. The flaky minerals align on the shear planes, but even the atomic structure of crystals may be shorn and will then slip; that is, a whole layer of ions parallel to so-called glide planes or glide lines in the structure will be displaced relative to the rest of the crystal. Atoms, therefore, move in one preferred direction. resulting in a transport along the greatest stress component. The mechanism is a type of directed self-diffusion and results in creep in this direction. One factor in gliding is therefore slip, another is creep.

Homogeneous irrotational strain. When flattened by homogeneous irrotational strain (an example is the flattening of a ball of dough to a cake), the fibrous and flaky minerals are arranged perpendicularly to the pressure. Flattening is probably of great importance by deformations related to vertical or steeply inclined schistosity planes, whereas laminar gliding is more important in thrusts and other shearing movements in the nappes of the mountain ranges. Usually neither type of movement occurs alone, but always in various, often complicated, combinations.

In mechanically sheared rocks the individual crystals are deformed; this is called fracture cleavage which may pass into flow cleavage. However, recrystallization and blastesis of preferred minerals during the shearing results in undeformed crystals, and the rock then shows crystallization cleavage. The rock deformation may be paracrystalline, postcrystalline, or precrystalline. *See* CLEAVAGE, ROCK; PETROFABRIC ANALYSIS; SLATE. *See also* AMPHIBOLITE; FLASER ROCK; MICA SCHIST; PHYLLITE.

[T.F.W.B.]

Schistosity, rock

The property of certain rocks to split somewhat irregularly into leaflike fragments having approximately parallel, slightly wavy, planar surfaces. Rocks having schistosity are characterized by a roughly parallel arrangement of micaceous and prismatic minerals and by a preferred orientation of the ionic lattices of nonmicaceous, equidimensional minerals, such as quartz and feldspar. Schistosity occurs in gneisses and in schists. The capacity to split along surfaces of schistosity, which is characteristic of schists and gneisses, is apparently connected with the preferred orientation of the mineral constituents of the rocks. *See* PETROFABRIC ANALYSIS.

[E.B.K.]

Schistosity in rock from Massachusetts.

Schistosomiasis

A disease in which man is parasitized by any of three species of blood flukes: *Schistosoma mansoni*, *S. haematobium*, and *S. japonicum*. Adult *S. mansoni* prefer the veins of the hemorrhoidal plexus, *S. haematobium* those of the vesical plexus, and *S. japonicum* those of the small intestine. In contrast to other trematodes, the sexes are separate; that is, male and female reproductive systems occur in individual worms. The disease is also known as bilharziasis. *See* DIGENEA.

The approximate dimensions of the egg and male and female adult in each of the three species are shown in the table.

Distribution. *S. mansoni* is found in Africa, Brazil, Venezuela, Dutch Guiana, Lesser Antilles, and Puerto Rico. *S. haematobium*, originally found in Africa, the Near East, and Mediterranean basin, was introduced into India during World War II. *S. japonicum* is widely spread in Eastern Asia and the Southwest Pacific; however, in Formosa it infects only animals, not man.

Dimensions of human blood fluke

Species	Egg, μ	Adult	
		Length, cm	Diameter, mm
S. mansoni	150 × 65	female 1–1.5	0.25
		male 2	1
S. haematobium	150 × 60	female 1–1.5	0.25
		male 2	1
S. japonicum	80 × 65	female 1–2	0.25
		male 1–2.5	1

Biology. Embryonated eggs passed in feces or urine hatch in fresh water into miracidia. The ciliated miracidium penetrates into specific gastropods or snails. There the miracidium is transformed into a mother sporocyst which gives rise by paedogenesis, or sexual reproduction by larval stages, to several hundred daughter sporocysts (*see* PAEDOGENESIS). These move away to settle elsewhere in the snail and give rise to several thousand cercariae. The larval cycle lasts for about one month. The cercaria emerges from the mollusk, swims in the water for up to 48 hours, and penetrates the skin of the final host upon coming in contact with it. After a phase of growth in the lungs, the schistosomule moves to the intrahepatic veins for further development. Then it moves to its final habitat. *S. mansoni* lives in the inferior mesenteric vein of the lower bowel, *S. haematobium* in the blood vessels of the bladder, and *S. japonicum* in the superior mesenteric vein of the small intestine. Eggs may start to appear within two months after exposure. The adult may live for many years.

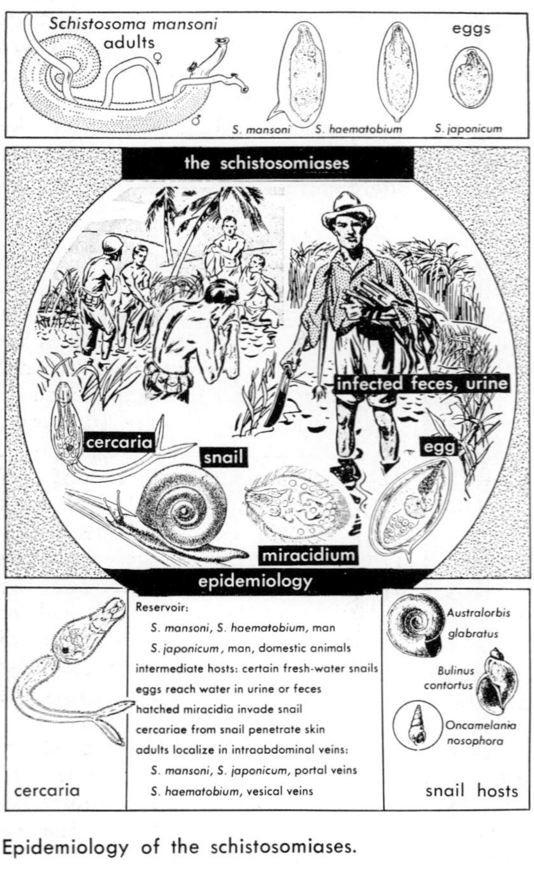

Epidemiology of the schistosomiases.

Epidemiology. As an intermediate host, *S. mansoni* uses planorbid snails of the genus *Taphius*, *S. haematobium* the genera *Bulinus* and *Physopsis*, and *S. japonicum* the operculate amphibian *Onchomelania*. *S. mansoni* is essentially a human parasite. Monkeys may act as natural reservoirs in Africa and a rat (*Rattus rattus frigivorous*) in Brazil. *S. haematobium* has no reservoirs, while *S. japonicum* possesses many. Various animals may be infected in the laboratory with the three species.

Schistosomiasis is an agricultural hazard for all ages in irrigated lands or swamps. Elsewhere fluvial waters are the main source of infection, in which case incidence is marked in individuals less than 15 years old, with boys more often parasitized than girls.

Pathognomy. The pathogenesis of schistosomiasis is presented in the following sections.

Prepatent period. The penetration of the cercariae may or may not produce skin irritation. Heavy initial exposures result in inflammatory disease of the skin (urticaria), toxic symptoms, an increase in the size of the liver and spleen, and eosinophilia before egg laying starts.

Patent period. S. mansoni may lay 350 eggs daily, *S. japonicum* 3000. About 10% reach the feces, the rest remain imprisoned in tissues, particularly liver and intestine or urinary bladder, where they provoke fibrotic reaction. Egg extrusion may result in dysentery or hematuria over a protracted period. Liver and spleen enlarge and the lungs may become involved.

Postpatent period. The fibrotic changes may result eventually in absence of eggs from dejecta. The changes interfere with intestinal, hepatic, and bladder functions with serious consequences. Liver fibrosis leads to splenomegaly and esophageal varices. Death may occur from profuse hematemesis. Lung fibrosis may cause cardiac failure.

Diagnosis. Diagnostic rectal biopsy, complement-fixation, and intradermal tests are used for all stages; fecal examination is positive during the patent period. *See* COMPLEMENT-FIXATION TEST.

Treatment. Medication is difficult and dangerous, particularly in severe infections. Trivalent antimony, either intramuscular (Fuadin) or intravenous (tartar emetic), or lucanthone hydrochloride (Miracil D) by mouth, is used as a chemotherapeutic agent. The last is specific for *S. haematobium*. [J.F.MA.]

Schizocoela

This group includes the animal phyla Bryozoa, Brachiopoda, Phoronida, Sipunculoidea, Echiuroidea, Priapuloidea, Mollusca, Annelida, and Arthropoda. Schizocoelous refers to the manner of formation of the coelom or body cavity during development by its appearance as a space in the embryonic mesoderms. This contrasts with the method of derivation of the coelom in the Enterocoela, where it forms as pouches from the early embryonic gut. *See* ENTEROCOELA. [T.I.S.]

Schizogregarinida

An order of the subclass Gregarinidia in the class Telosporidea. These protozoans are found as parasites of various invertebrates such as insects and annelids. On rare occasions, they have been reported as parasites from tunicates. They are distinguished from the other subclass, the Eugregarinida, in that both sexual and asexual reproduction occurs during the life cycle. Merogony may be intracellular or occur in a body cavity. *Ophryocystis mesnili* (see illustration), which is an unusually small gregarine, occurs in the Malpighian tubules of *Tenebrio molitor*, the meal worm. *See* GREGARINIDIA. [E.R.BE.]

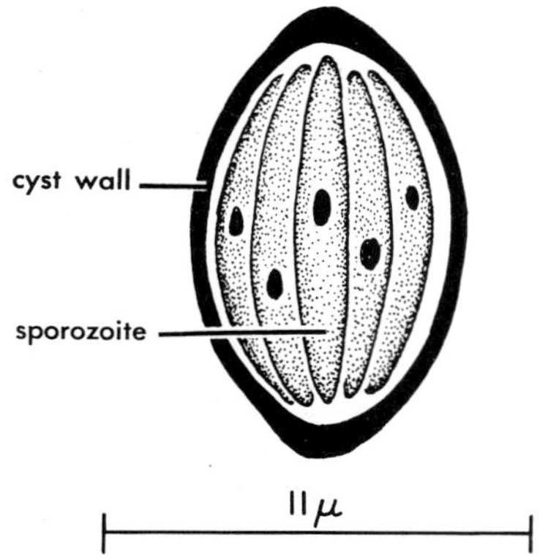

An oocyst of *Ophryocystis mesnili*. (*From T. L. Jahn and F. F. Jahn, How to Know the Protozoa, Wm. C. Brown, 1949*)

Schizomycetes

A class, of the division Protophyta, which contains the bacteria. The name Schizomycetes suggests a fission fungus. Although the true fungus has a true nucleus and belongs to the Thallophyta, it does not have any chlorophyll. Thus the bacteria—the Schizomycetes—because of their asexual reproduction by simple binary fission and their usual lack of chlorophyll, were called the fission fungi. They differ from the true fungi in the absence of a true nucleus, and the absence of chitin from the cell walls.

Modes of reproduction. Binary fission is the mode of reproduction found in all the known bacteria except those of the order Hyphomicrobiales, which produce a tubular outgrowth from the cell and form a new cell at the tip, and of the genus *Blastocaulis* in which terminal buds are produced. In addition to simple binary fission, some bacteria reproduce asexually by the production of conidialike bodies as in the genus *Streptomyces* and related genera. Others also have a cyclic development in which spherical bodies alternate with rod-shaped forms, for example, in the genera *Spirillum* and *Arthrobacter* and in the fruiting Myxobacterales.

A true nucleus is absent but chromatinic material characteristic of nuclei has been shown in many bacteria. It divides when the cell divides. No sexual organs are produced although recent studies have indicated that genetic particles can be transferred from one cell to another. *See* BACTERIAL GENETICS.

Photosynthetic pigments. Photosynthetic pigments do occur in some bacteria, the family Athiorhodaceae, the family Thiorhodaceae, and the genus *Rhodomicrobium*. The chemical composition and the light-absorbing characters are different from those of the plant chlorophylls.

Morphology. Most of the Schizomycetes have a division of the cell into a cell wall and cytoplasm. The cell wall is, in most cells, quite rigid and can be easily separated from the cytoplasm. It gives the cells their characteristic shape—spherical, rod-shaped, or spiral. In others it is barely differentiated from the cytoplasm and is very flexible, as in the orders Myxobacterales and Spirochaetales. Only in the order Mycoplasmatales does a cell wall appear to be completely lacking.

Both unicellular and multicellular organisms occur. The latter are confined mainly to the orders Caryophanales and Beggiatoales. In both, cells are usually longitudinally compressed and are joined for the greater part of their width.

The Schizomycetes are rarely more than 2 microns (μ) wide or more than 10 μ long. A few, like *Beggiatoa*, may be 70–80 μ wide and several hundred microns long.

Cells occur singly, in clusters, in chains, or in branched filaments. They are usually free, but in some, the order Chlamydobacterales and some genera in the order Beggiatoales, they are found in a tubular sheath anchored to a surface by means of a holdfast or simply attached by a stalk as in the genus *Caulobacter* (order Pseudomonadales) and the genus *Blastocaulis* (order Hyphomicrobiales).

Motility. Many Schizomycetes are nonmotile. The motility in others may be due to flagella, to a nonflagellated gliding movement along solid surfaces as in the order Myxobacterales, or to a sinuous flexing movement as in the order Spirochaetales. The latter have one or more very fine fibrils wound around the main body of the cell. The fibrils are responsible for the spiral shape and for the flexing movement. Flagella usually consist of simple fibrils inserted at the poles as in the order Pseudomonadales, or around the cells as in the order Eubacteriales. *See* BACTERIAL MOTILITY.

Metabolism. The photosynthetic bacteria obtain energy from sunlight. The rest obtain their food and energy from simple inorganic substances (autotrophs) or from organic materials (heterotrophs). *See* BACTERIAL METABOLISM.

Taxonomy. The class is divided into 10 orders.

1. Pseudomonadales—small rod-to-spiral-shaped organisms with rigid cell walls and polar flagella and a mode of nutrition usually dependent on free oxygen.
2. Chlamydobacteriales—rod-shaped and spherical organisms occurring in chains within tubular sheaths.
3. Hyphomicrobiales—in which new cells are produced on the ends of fine tubular outgrowths from existing cells.
4. Eubacteriales—simple rod-shaped or spherical cells with rigid cell walls and peritrichous flagella and a mode of nutrition which is usually not dependent on oxygen.
5. Actinomycetales—gram-positive filamentous branching bacteria which reproduce by production of conidialike bodies or by fragmentation of the branching filaments.
6. Caryophanales—multicellular rod-shaped bacteria with peritrichous flagella.
7. Beggiatoales—usually multicellular, rod-shaped bacteria without flagella, which are capable of gliding on solid surfaces.
8. Myxobacterales—slime bacteria which are very flexible, glide on solid surfaces without flagella, and usually produce resting cells in characteristic fruiting bodies.
9. Mycoplasmatales—very minute, highly plastic bacteria which may be filamentous or spherical, the latter usually arising from the filamentous types by fragmentation. The spherical forms are rarely more than 150–200 millimicrons wide.
10. Spirochaetales—very flexible, spirally twisted organisms without a rigid cell wall and without flagella. The mechanism of motion is not clearly understood but in all cases so far examined one or more axial filaments are found wound spirally around the cell, being fixed only at or near the ends of the cell.

See BACTERIA, TAXONOMY OF; see also separate articles on each order. [V.B.D.S.]

Schizophrenia

A group of mental disorders which are characterized by withdrawal from reality and by disturbances in thinking and feeling. It is also called dementia praecox. In many cases it leads to disorganization of the personality, but not necessarily to mental deterioration. One of the outstanding symptoms of schizophrenia is a lack of rapport and serious misjudgment of reality processes. Certain types of schizophrenics are refractory to influence and suggestion; this behavior in its extreme form, is called negativism.

The disorder occurs in at least one person in 200 of the population, particularly in the younger age group. So-called childhood schizophrenia, in many cases, is probably an organic brain disease and is not comparable to the schizophrenias of adults and adolescents. American psychiatrists, particularly of the psychoanalytic school, use the schizophrenia concept more broadly than their European colleagues. Schizophrenia, in the more narrow sense, was divided into the following subtypes by the German psychiatrist, E. Kraepelin: (1) hebephrenia, a form occurring in adolescents and young adults, leading to early deterioration; (2) paranoid schizophrenia, characterized by fantastic hallucinations and paranoid delusions; (3) catatonia, manifesting itself in serious disturbances of volition and movements, with forms ranging from stupor to extreme excitement; and (4) a slow progressive deterioration called dementia simplex, often combined with mental deficiency. These types are rarely pure, and often change from one into another.

The main symptom of schizophrenia is a rather severe dissociation in thinking and feeling. The patients think and speak incoherently and illogically. There is a blocking of the thinking process; ideas become very vague. Emotions become flattened, and also extremely ambivalent or contradictory at the same time.

The etiology of schizophrenia is unknown. Many suspect that it is an organic illness, but so far no conclusive evidence for such an assumption has been produced. Genetic factors are considered of importance; identical twins show a much higher concordance of illness than fraternal twins. E. Kretschmer linked schizophrenia with the occurrence of leptosomic and athletic body types. Psychoanalysts have claimed that schizophrenia is due to a faulty development of the early personality, and recently T. Lidz and others have demonstrated pathological patterns of communication among most members of families with schizophrenic patients, such as severe symbolic distortions and faulty evaluations of self and others.

The course of schizophrenia is often progressive, leading to a terminal state of deterioration; however, there are many cases which do not reach such a terminal state and many even recover spontaneously. The tendency to remission makes any assessment of therapeutic claims very difficult.

Treatment of schizophrenia is empirical. There have been attempts to help the patient overcome his isolation and withdrawn feelings by individual and group psychotherapy. A variety of techniques has been prescribed to achieve this. Such treatment may be carried out in an active hospital setting, although hospitalization for schizophrenia is often not necessary, and at times is not indicated. Organic treatment for schizophrenia includes psychosurgery, insulin coma treatment, and convulsive treatment. All of these methods are definitely in a waning stage. Today, treatment with tranquilizing drugs has become more important, but it does not have a specific effect on the schizophrenic process. The most effective treatment is skillful psychotherapy, possibly in a combination with drug therapy. *See* ABNORMAL BEHAVIOR; PSYCHOPHARMACOLOGIC DRUGS; PSYCHOSIS; PSYCHOSURGERY; PSYCHOTHERAPY. [F.C.R.]

Bibliography: J. R. Ewalt, E. A. Strecker, and F. G. Ebaugh, *Practical Clinical Psychiatry*, 8th ed., 1957.

Schizophyceae

A class, of the division Protophyta, which contains the blue-green algae. According to the taxonomic system used by most botanists, the Schizophyceae are known as Cyanophyta. They reproduce by simple binary fission. They contain chlorophyll a similar to that in the common plants; but instead of being confined to discrete plastids it is distributed throughout the peripheral portion of the cytoplasm of the cell, where it is confined to special units known as chromatophores or grana. The clear central body in the center of the cell represents the nucleus. *See* CHLOROPHYLL; PROTOPHYTA.

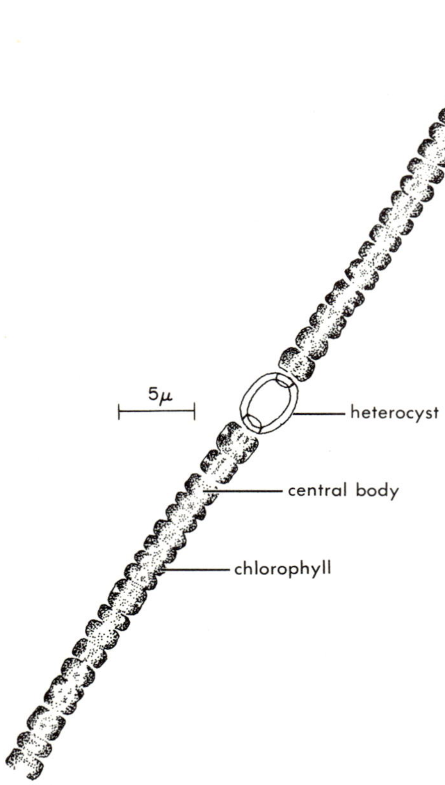

Anabaena, a schizophyte.

In addition to chlorophyll a, the Schizophyceae contain a blue pigment, phycocyanin; reddish carotenes and phycoerythrin; and yellowish myxoxanthin and myxoxanthophyll. The blue and green pigments usually predominate, giving these algae their typical blue-green color. But the proportions in which the pigments are present may vary with the environmental conditions, especially the lighting, and occasionally the red pigments may predominate. The resulting change in the color of the algae, first described by T. W. Engelmann and N. Gaidukov, is believed to have adaptive significance.

The cell wall of the blue-green algae usually contains some cellulose and diaminopimelic acid. There is no true nucleus; but chromatinic material forms the central body and divides with the cell. In the larger chain-forming algae, the central bodies of the successive cells may be connected.

Motile species have no flagella or cilia or other appendages to aid their movement. Movement is thought to be due to the secretion of a gelatinous material which takes up water and swells. The long chain-forming species rotate as they move.

Spherical, rod-shaped, and spiral forms, for example *Spirulina,* have been described. They sometimes occur as single cells, as in *Chroococcus,* or in definite families of cells (not a tissue) joined together by a gelatinous envelope, as in *Aphanothece,* or in long chains as in *Oscillatoria* and *Anabaena.*

The chain-forming species often have different types of cells in the chain—clear heterocysts and opaque akinetes. These are special types of cells formed to protect the algae under adverse conditions. The long chains often separate into smaller units known as hormogonia.

The blue-green algae are common in still waters—both fresh and salt—and on the surface of rocks and in damp soil. *See* CYANOPHYTA. [V.B.D.S.]

Schlieren photography

An optical technique that detects density gradients occurring in a gas flow. The Schlieren system is used particularly in supersonic wind tunnels because it clearly shows the density gradients created by the shock and expansion waves of the airflow around the wind tunnel model. A simple Schlieren system operates as illustrated in the figure. A source of light is shielded so that only a small rectangular slit emits light. A lens is placed at its focal distance from the slit so that the light is bent into a parallel beam. A second lens collects the parallel beam into an image of the slit and forms an inverted image on the screen or photographic plate. If a knife edge is moved into the light stream near the slit image, the image at the screen darkens uniformly. Consider the system just described to be oriented so that the parallel light beam crosses a wind tunnel section. A light ray *a,* bent from the parallel path by density gradients in the test section, cannot be brought to focus at the slit image and is interrupted by the knife edge so that a dark spot occurs at *a′* on the screen. Light ray *b,* deflected the opposite way by a different density gradient, escapes the knife edge and appears as a

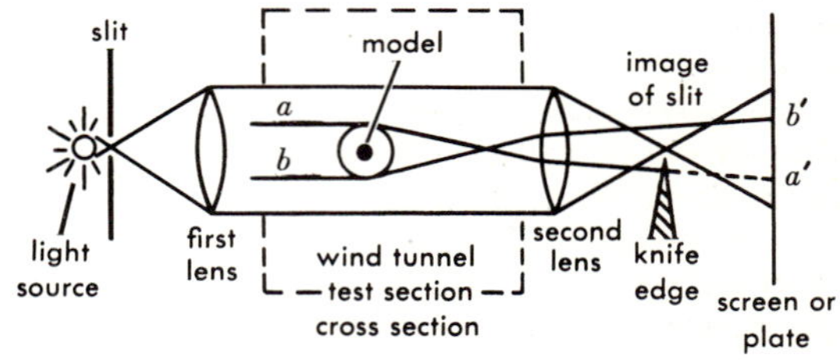

Basic Schlieren optical system.

light spot at b' on the screen. Thus a picture of the density gradients appears on the screen. Sometimes light intensity charts can be correlated with numerical values of the density gradients. However, numerical values of density can be obtained only from Schlieren pictures of airflow about two-dimensional models or about simple axisymmetric models.

For high-speed photographs (3000–4000 frames per second) of unsteady flows, dc light sources are used. Microsecond flash durations from high-intensity gas sources are used in the study of turbulences. To better study specific regions of a flow pattern, the knife edge is rotated to different positions and moved up, down, left, or right. For sharp images at the center of a test section, a multiple-slot or focusing Schlieren system is used. *See* SHOCK-WAVE DISPLAY; WIND TUNNEL INSTRUMENTATION. [D.P.AN.]

Schmidt camera

An optical system consisting of a spherical mirror with a corrector plate near its focus. Cameras of the Schmidt type are of great value in direct astronomical photography and also as parts of spectroscopes.

The corrector plate is a figured plate with one plane surface, its second surface being an almost plane aspheric with a figure such that the aperture aberration of the mirror is balanced as shown in the illustration. The curvature of the plate at the axis is zero. *See* SPHERICAL AND ASPHERIC SURFACES, OPTICAL.

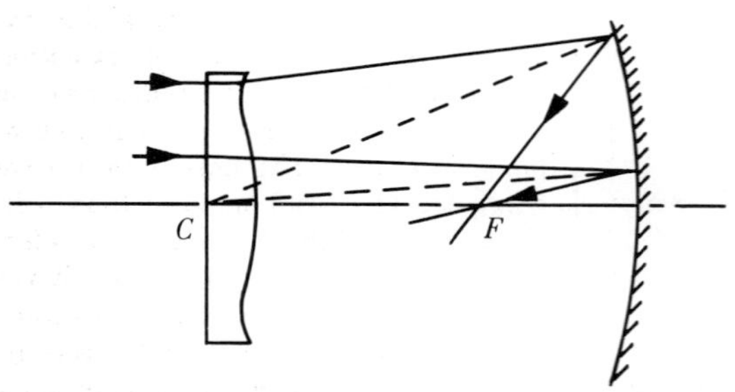

Original Schmidt camera with irregular corrector plate. *C*, center of curvature; *F*, focal point of mirror.

The Schmidt camera is a catadioptric system which is free from aperture aberration, asymmetry, and chromatic aberrations. It can therefore be used at the very large aperture of $f/2.0$ or more. The image, although sharp, has curvature of field, but the field can be flattened by adding a second mirror to produce what could be called a Schmidt-Cassegrain system.

A Newtonian type of mirror with an additional concentric lens in place of the irregular corrector plate was suggested at about the same time (1941) in four different countries. In 1944 a proposal was made to replace the concentric lens by a meniscus lens free from color aberrations.

The Cassegrain type of mirror has also been used in combination with an additional, more or less afocal system, in order to increase the aperture and correct the monochromatic errors of the mirror, without introducing large color aberrations. Additional systems, which are quite complex, have been used.

The Cassegrain type of Schmidt camera has the disadvantage of considerable vignetting, and thus of a very small field. It requires careful baffling besides to prevent undesirable light called flare light from going through the instrument. Frequently the Cassegrain mirrors are oriented 90° with respect to each other by means of a prism or a mirror.

The success of the Schmidt camera, which at first was used mostly for telescopes, has led to numerous designs of catadioptric systems for other purposes.

The application of such systems is very successful in microscope optics, where large aperture and freedom from color aberrations combined with a small field are needed. Most microscopic work in the ultraviolet is now done with catadioptric systems (*see* MICROSCOPE, OPTICAL). Catadioptric systems have also been designed for use as camera lenses, especially when a great focal length is required.

The disadvantage of catadioptric systems is that the mirrors and lenses are in one another's way, thus vignetting the center of the aperture. The diffraction effect connected with this cutting out of the central part of the beam may jeopardize the recognition of very fine detail in the object. *See* ASTRONOMICAL PHOTOGRAPHY; TELESCOPE, ASTRONOMICAL. [M.H.]

Bibliography: H. F. Bennett, U.S. Patent 2,571,-657, 1945; A. Bouwers, *Achievements in Optics*, 2d ed., 1950; C. Carathéodory, *Elementare Theorie des Spiegelteleskops*, 1940; *Handbuch der Wissenschaftlichen und Angewandten Photographie*, 1:191–199, 1955; D. D. Maksutov, *J. Opt. Soc. Am.*, 34(5):270–284, 1944; B. Schmidt, *Mitt. Hamburger Sternwarte*, 1932.

Schottky effect

The enhancement of the thermionic emission of a conductor resulting from an electric field at the conductor surface. Since the thermionic emission current is given by the Richardson formula, an increase in the current at a given temperature implies a reduction in the work function ϕ of the emitter. *See* THERMIONIC EMISSION; WORK FUNCTION (ELECTRONIC).

With reference to Fig. 1 let the vacuum level represent the energy of an electron at rest in free space and let *CD* be the energy of a conduction electron at rest in a metal. If an electron approaches the metal surface from infinity, its potential energy V relative to the vacuum level is given by the well-known image potential $V(x) = -e^2/4x$ where x is the distance from the surface and $e = 1.6 \times 10^{-19}$ coulomb is the electron's charge. The image potential is valid only for $x > x_0$, where x_0 is of the order of the distance between neighboring atoms in the metal; that is, x_0 is a few angstroms. In the absence of an applied field, CAB then represents the potential energy of an electron as a function of x. AB corresponds to the image potential; the exact shape of the curve between C and A is uncertain.

Suppose now a constant field F is applied externally between the surface of the emitting cathode and an anode; this produces a potential energy of an electron of $-eFx$ (line PQ in Fig. 1), and hence the total potential energy of an electron for $x > x_0$ is given by

$$V(x) = -(e^2/4x) - eFx$$

indicated by the dashed line CAQ in Fig. 1. This function has a maximum for

$$x = x_m = (1/2)(e/F)^{1/2}$$

the maximum lies below the vacuum level by an amount

$$\Delta\phi = V(x_m) = -e(eF)^{1/2}$$

where $\Delta\phi$ represents the reduction in the work function of the metal. For $F = 1000$ volt/cm, $x_m \simeq 10^{-5}$ cm and $\Delta\phi \simeq 10^{-2}$ electron volt; the actual change in the work function is thus small. If a field is present, the work function ϕ in the Richardson formula should be replaced by $(\phi - \Delta\phi)$. Hence, the current increases by a factor

$$\exp[(e/kT)(eF)^{1/2}]$$

According to this interpretation a plot of the logarithm of the current versus the square root of the anode voltage should yield a straight line. An example is given in Fig. 2 for tungsten; the deviation from the straight line for low anode voltages is due to space-charge effects (*see* SPACE CHARGE). The straight portion of the line (the Schottky line) confirms the interpretation; the true saturation current for zero field is obtained by extrapolation of the Schottky line as indicated. Detailed studies have shown extremely small periodic deviations with reference to the Schottky line; these deviations are interpreted on the basis of the wave-mechanical theory describing the motion of electrons across the image potential barrier shown in Fig. 1.

[A.J.DE.]

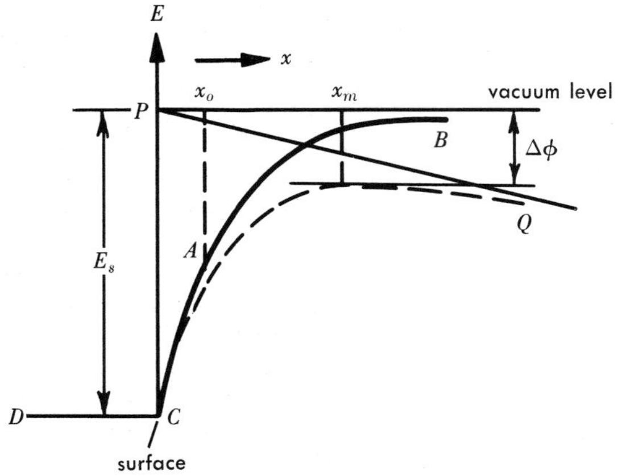

Fig. 1. Illustrating the surface potential barrier and the Schottky effect; the lowering of the work function $\Delta\phi$ is highly exaggerated.

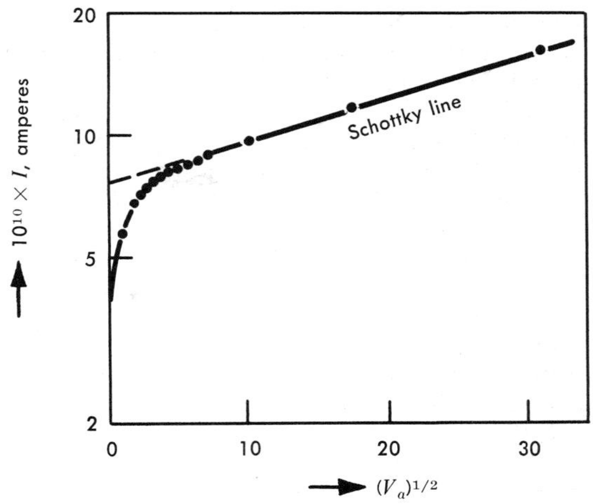

Fig. 2. The logarithm of the thermionic emission current I of tungsten as a function of the square root of the anode voltage V_a. (After W. B. Nottingham, Phys. Rev., 58:927–928, 1940)

Schrödinger's wave equation

The differential equation of quantum mechanics (first proposed by Erwin Schrödinger in 1926) whose solution determines the average result, also termed expectation value, of every conceivable experiment on the physical system under examination. When solved, the Schrödinger equation yields the wave function; from the wave function, expectation values are computed. The designation wave equation comes from its resemblance to those differential equations describing acoustic and electromagnetic waves (*see* WAVE EQUATION; WAVE MOTION). Also, its consequences and mode of derivation are consistent with the tenet that electrons and other particulate constituents of matter have wavelike properties. *See* QUANTUM MECHANICS; QUANTUM THEORY, NONRELATIVISTIC. [E.G.]

Schuler pendulum

Any apparatus which swings, because of gravity, with a natural period of 84 minutes, that is, with the same period as a hypothetical simple pendulum whose length is the Earth's radius. In 1923 Max Schuler showed that such an apparatus has the unique property that the pendulum arm will remain vertical despite any motions of its pivot. It is, therefore, useful as a base for navigational instruments. Schuler also showed how gyroscopes can be used to increase the period of a physical pendulum to the desired 84 min.

Gyrocompasses employ the Schuler principle to avoid errors due to ship accelerations. More recently the principle has become the foundation of the science of inertial navigation.

Basic principle. The principle is most easily explained with the simple pendulum of Fig. 1. The period T for one complete swing of this device is

$$T = 2\pi\sqrt{l/g}$$

in which l is the length of the pendulum and g is the acceleration, due to gravity, of a freely falling object (32.2 ft/sec^2). If length l could be made equal to the Earth's radius R, as in Fig. 2, gravity would hold the bob fixed, and the pendulum arm would stay vertical for all motions of point P. If l equals R, the period becomes 84 min.

While a pendulum of such length cannot be built, the same effect can be achieved using the solid pendulous rod of Fig. 3. Although only a few feet long, this pendulum can be given a period of 84 min by placing the pivot close to the center of mass. (If the pivot were at the mass center, the period would be infinitely long; that is, the rod would remain indefinitely in any position.)

Figure 4*b* demonstrates that a Schuler-period pendulum will also stay locally vertical during accelerations of its pivot, just as the simple pendulum did in Fig. 2. The proof is by comparison. In Fig. 4*a* the pivot is high, and the pendulum swings back from vertical when the pivot is accelerated. In Fig. 4*c* the pivot is at the mass center and the pendulum does not swing at all, so that after traveling some distance around the Earth it is forward of the local vertical. In Fig. 4*b* the pivot is placed at just the right point, so that whatever the acceleration of the pivot, the pendulum swings back just enough to stay locally vertical.

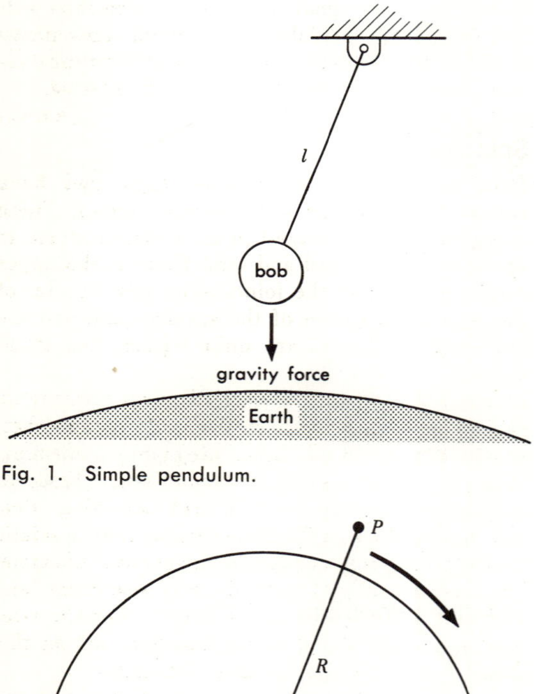

Fig. 1. Simple pendulum.

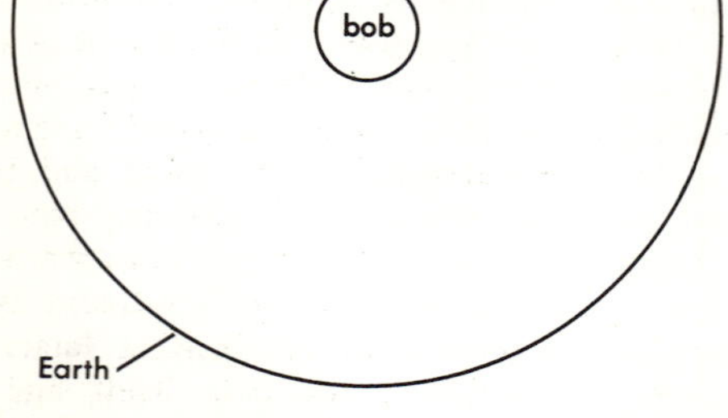

Fig. 2. Earth-radius pendulum.

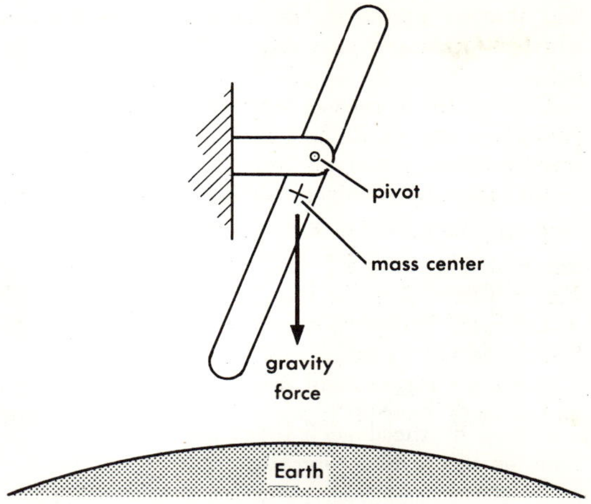

Fig. 3. Pendulous rod.

Unfortunately, for a 2-ft pendulum like that of Fig. 3 to have an 84-min period, the pivot must be placed only about 0.2×10^{-6} in. above the mass center, an unrealizable physical tolerance.

Gyropendulum and gyrocompass. The period of a pendulous mass can be greatly increased by attaching to it a rapidly spinning gyro wheel as, for example, in Fig. 5. Extra freedom, provided by the thrust bearing in Fig. 5, is necessary to permit the gyroscopic action.

The gyrocompass is a special pendulous gyroscope that is sensitive to Earth rotation. If the pendulous mass is m, the horizontal spin axis has a natural period of oscillation, about north, of

$$T = 2\pi\sqrt{\frac{H}{mgl\Omega_e \cos \lambda}}$$

in which H is gyro momentum, λ is latitude, and Ω_e is Earth rate (360°/day). Ship-borne gyrocompasses are rendered insensitive to ship accelerations by making $T = 84$ min. *See* GYROCOMPASS.

Inertial guidance. The stable platform of an inertial guidance system is made to behave like a pendulum by the use of feedback from an accelerometer mounted on the platform. The accelerometer signal controls a restoring torque to the platform.

The principle is demonstrated in Fig. 6. The accelerometer consists of a spring-restrained mass. If the platform tilts, gravity pulls the mass to one side; the amount is measured electrically, integrated, and used to drive the platform motor. (Actual inertial platforms are controlled by a gyro, which is precessed at a rate by the integrator signal.)

The system of Fig. 6 behaves like the physical pendulum of Fig. 1. When either device is tilted from vertical, gravity produces a restoring action so that the device swings back and forth about the vertical. The period of swing is $2\pi\sqrt{l/g}$ for the pendulum of Fig. 1, and $2\pi\sqrt{1/Kg}$ for the platform of Fig. 6, where K is the loop gain from tilt angle to motor rate. The platform is accordingly given an 84-min period by making K equal to $1/R$.

If the platform of Fig. 6 is installed in a vehicle, and is once leveled, it will always remain locally

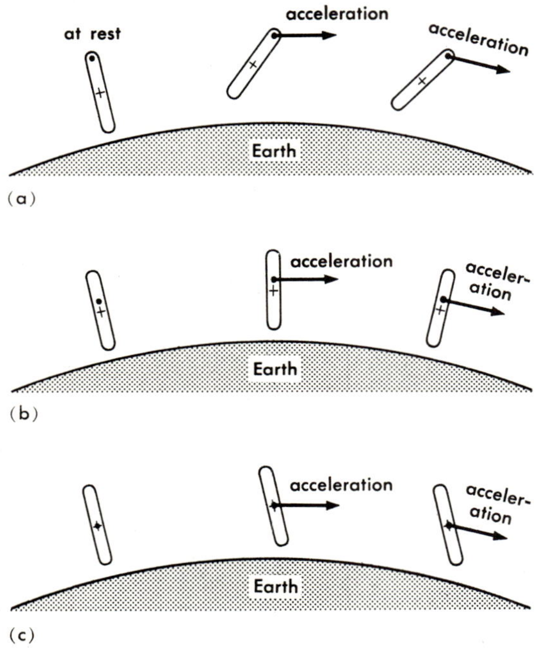

Fig. 4. Behavior of pendulous rod when accelerated about the Earth. (*a*) Short-period pendulous rod. (*b*) Schuler-period pendulous rod. (*c*) Nonpendulous rod.

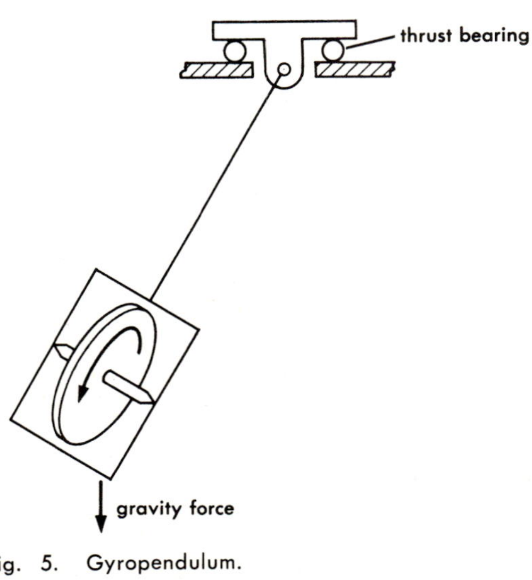

Fig. 5. Gyropendulum.

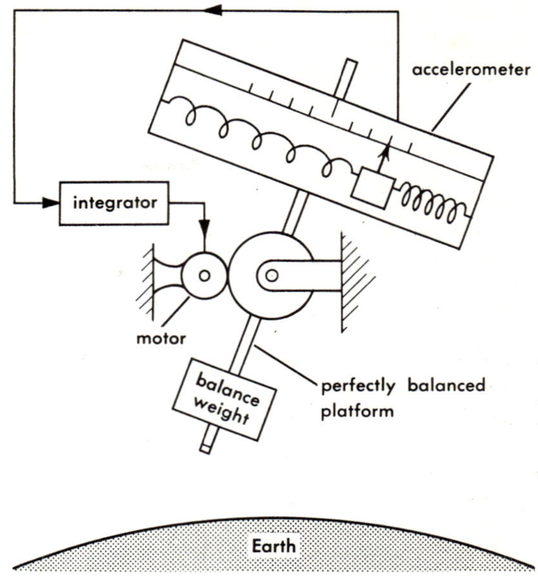

Fig. 6. Principle of the Schuler-tuned stable platform.

level despite motions of the vehicle about the Earth. Then the signal from the accelerometer will represent only vehicle acceleration, uncontaminated by gravity, and can be used to determine vehicle position. *See* INERTIAL GUIDANCE SYSTEM.

[R.H.C.]

Sciatica

Pain in the lower extremities, hips, and back caused by irritation of the sciatic nerves. These paired nerves, the longest in the body, originate in the lower spinal cord and send fibers to the upper thigh muscles and the joints, skin, and muscles of the leg. The location of the specific pain and the causes producing it are quite varied. *See* PAIN, DEEP.

Sciatica may result from mechanical pressure on the nerves or their roots in the cord. Trauma, herniated intervertebral disks, pregnancy, inflammation, or tumors may cause compression. Toxic or metabolic disorders, such as lead poisoning, alcoholism, and vitamin-B deficiency may induce sciatic pain by producing changes in the nerves. Inflammations, both local and systemic, may also cause temporary or permanent nerve injury. Certain viral diseases, syphilis, and local infections act in this manner. *See* SYPHILIS; VIRUS; VITAMIN.

In addition, lesions in the anal region and the prostate may induce sciatica through reflex stimulation. Joint diseases, pelvic strain, and injury most often precipitate an attack. The pain of sciatica may begin suddenly and violently or gradually, as a nagging discomfort. The pain is usually along the leg, with later extension to the thigh and back. Numbness of the outside of the foot may occur.

A complete study of the patient is in order since so many possible causes exist. Treatment is directed toward removal of the inciting factors, if possible, and relief of the pain itself and the cramps of associated muscle spasms. [E.G.ST.]

Science

In common usage the word science is applied to a wide variety of disciplines or intellectual activities which have certain features in common. Application of the term did not begin with any formal definition; rather the various disciplines arose independently, each in response to some particular need. It was then observed that certain of these disciplines had enough traits in common to justify classifying them together as one of the sciences. Usage is not, however, always unanimous as to whether some disciplines should be called sciences or not, and there is often lively controversy as to the propriety of speaking of the social or historical sciences.

Usually a science is characterized by the possibility of making precise statements which are susceptible of some sort of check or proof. This often implies that the situations with which the special science is concerned can be made to recur in order to submit themselves to check, although this is by no means always the case. There are observational sciences such as astronomy or geology in which repetition of a situation at will is intrinsically impossible, and the possible precision is limited to precision of description. There is also usually the implication that the subject matter of the individual science is something in the world of phenomena. Thus it is not usual to speak of the "science" of mathematics or the "science" of logic, even though both these disciplines are capable of the highest precision.

A common method of classifying sciences is to refer to them as either exact sciences or descriptive sciences. Examples of the former are physics and, to a lesser degree, chemistry; and of the latter, taxonomical botany or zoology. The exact sciences are in general characterized by the possibility of exact measurement. Measurement is fundamentally description by the use of numbers. Given the system of measurement and the measuring numbers for any special situation, that situation has been adequately described if it is possible to reconstruct a situation such that measurement on it gives the same numbers. Because mathematics operates to a large extent with numbers, systems subject to exact measurement are also susceptible of mathematical analysis; this susceptibility is one of the most important characteristics of the exact sciences. One of the most important tasks of a descriptive science is to develop a method of description or classification that will permit precision of reference to the subject matter. *See* PHYSICAL SCIENCE; SCIENTIFIC METHODS. [P.W.B.]

Scientific methods

The methods employed in the various sciences are determined both by the general nature of the objective in view and by the nature of the subject matter. A prerequisite to nearly every science is a suitable method of description of its subject matter. The language of such description must be capable of reproducing or recalling the subject matter with precision and uniqueness. If the description is of an object, there should be only one corresponding object, which it should be possible to reproduce or reconstruct from the description; or, given an object, it must be possible to check whether it does or does not satisfy the corresponding description. In many cases, the language of everyday usage does not have the precision to meet these requirements, and many of the sciences have therefore developed their own specialized vocabulary to meet their special needs. This specialized vocabulary may involve the use of both new words (which are often coined for the occasion or derived from Greek or Latin roots) and familiar words in conventionalized and restricted meanings that avoid the ambiguities of common usage and permit a unique designation. The specialized languages of the various sciences also usually permit statements to be made much more concisely than is possible with conventional language.

Classification. To accompany conciseness in description, some method of systematizing or classifying the material to be described is usually adopted. Without such methods of classifying, precise description in a subject dealing with material as complicated as that in any of the biological sciences, for example, would be driven to the inordinate length of treating each individual example as a class in itself. The economy of description that results from such things as, for example, the classification of plants into species on the basis of particular features is obvious. Perhaps classification or some other method of systematization is to be considered the most primitive, ubiquitous, and necessary of the methods of science.

It is an ideal of a good scheme of classification to be applicable without serious revision to new, presently unknown situations as they arise; the undesirability of continually revising the scheme of classification with the discovery of every new fact requires no argument. Of course in a given epoch, while the new fact is still unknown, the presumptive extensibility of the scheme of classification can be judged only on a probability basis, in the light of past experience. The probability can be much enhanced by some theoretical understanding of the matter subject to classification, and it seems to be true that no formal scheme of classification is likely to survive for any length of time before it is made the subject of theoretical speculation. Such theoretical speculation serves the twofold purpose of (1) confirming the method of classification and (2) satisfying the demand of the scientist for understanding. The latter is an almost universal attribute of scientific activity.

If the description that the scientist achieves with the help of his scheme of classification and specialized vocabulary is to be precise and correct, there must be some method, in any individual instance, of checking its truth. The necessity and possibility of checking for correctness or truth is one of the universal attributes of any scientific enterprise and one which, to a large extent, controls the possible methods and also the subject matter of the sciences.

Repetition. One of the most potent methods of checking for correctness or truth is repetition. It is a matter of experience that there are permanent objects and situations which repeat. It is part of the task of a science to formulate the conditions under which a situation repeats. If a scientist can establish the conditions necessary for repetition, he can verify a previous description or observation by finding whether he now gets the same result as before. He thus guards against possible previous mistakes on his part and at the same time increases the presumptive probability that he has correctly stated the conditions necessary for repetition. However, scientists never achieve complete independence between their observation and their formulation of the conditions of observation; always some element of circularity remains. This seems to be an unavoidable accompaniment of all human enterprise.

The need for checking by repetition is so important that it drastically limits the nature of the material deemed suitable, by common consent, for scientific investigation. If a situation cannot be made to repeat, it is commonly regarded as of little or no scientific interest, and none of the usual scientific methods are applicable to it. It is for this reason that so small a part of human artistic or poetical activity, for example, can be subjected to scientific investigation. This is also the primary reason why there is so much unwillingness to grant full scientific status to ESP (extrasensory perception), although many of the detailed tools of the ESP researchers are used in other scientific activities.

The possibility of repetition implies the possibility of control, although such control may be in the realm of the ideal rather than of the actual. Astronomy is regarded as a scientific activity, although man as yet has no control over the behavior of the heavenly bodies. Here the control is an idealized one through man's understanding of mechanics and physics. Neither need all the details of a situation be controlled. In the situations contemplated in wave mechanics, man has no detailed control, but only a statistical control over the average of certain phenomena.

Consensus. Another method of checking or confirming the correctness of an observation or report is agreement between different observers. Here the multiplicity of observations by different persons corresponds to the multiplicity of the repeated occurrence. The matter of consensus among different observers is regarded by many people as so important that it is often incorporated into the definition of science itself, which is sometimes partially defined as the consensus of qualified persons. It must be conceded that, when consensus is attainable, one may have a high degree of confidence that he has not made an error because of some personal idiosyncracy or inadvertence. But logically, it must be recognized that it is possible that everyone may be mistaken, and as a matter of fact there have been instances where public opinion, even in a discipline seemingly so completely objective as mathematics, has been in error for long intervals of time. Furthermore, there have apparently been examples of mass hypnosis. From the point of view of the individual, consensus can be regarded only as a device for increasing the probability of absence of error or as a device to obviate the necessity of detailed confirmation in every individual instance, thus making for economy of effort.

As part of the same picture, acceptance of authority can never be tolerated as a method in any scientific enterprise. No report of experimental observation or theoretical deduction is scientifically acceptable unless made in such terms that it can be repeated and confirmed by any qualified individual.

There would appear to be no method, scientific or other, by which the individual may be relieved of ultimate responsibility or may attain absolute certainty. Neither does there appear to be any method by which, in the last analysis, reference to the individual case can be avoided, a fact which makes illusory the ostensible ideal of science to deal only with the universal, the general, and the repeatable. Complete logical rigor is as unattainable in science as in any other human activity.

Experiment. One of the most potent tools of many of the sciences, both for the discovery of new facts and for more adequate understanding of existing facts, is the experiment. The experimenter artificially varies the conditions under which phenomena occur. In this way, he may greatly increase the frequency with which certain rare but significant conditions occur in nature and thus compress into a practical length of time occurrences which, in the natural course of events, might stretch over many generations, as in the study of genetics by the biologist. Or he may create conditions never observed in nature, as when he applies pressures of tens of thousands of atmospheres to water at ordinary temperatures and thus discovers how matter behaves under completely novel conditions. Used in this way, experiment becomes an enormously potent instrument in acquiring understanding, for man's understanding in a physical situation cannot be regarded as adequate until he can correctly anticipate what will occur under every conceivable range of circumstances, whether imposed naturally or by artifice in the laboratory.

The experimenter often proceeds by isolating the different factors supposed to control a phenomenon and studying the effect of varying these factors independently of each other. His conception of what the significant factors are will often be determined by his theoretical understanding, and often experimentation and theory go hand in hand, one suggesting, modifying, and determining the other. *See* EXPERIMENT.

Cause and effect. In modern scientific activity, the theoretical analysis preceding the isolation of factors to be experimentally varied is, in the vast majority of cases, predicated on the operation of the law of cause and effect. The discovery that this law is operative in almost all experience and the conviction that it is profitable to assume it operates, even where this has not yet been demonstrated, is perhaps the one thing that sets off science since the time of Galileo from earlier scientific activity. Without this law, the development of the method of experiment would hardly have been possible. In spite of this, it does not appear to be justifiable to claim, as many people have, that science is committed to the assumption that there is a law of causality or perhaps more generally, to the assertion that there are regularities which control natural phenomena. It may well be that science is committed to discover what regularities it can, but it is not committed to the thesis that regularities exist. One of the most important tasks of quantum theory is to find how to proceed in the face of the discovery that the regularities in the operation of nature are not as pervasive as, or of the kind that, had been supposed. *See* CAUSALITY.

Discovery of regularities and laws is probably the most important function of experiment, but it is not the only function. Experiment may be used as a simple tool of exploration into new territory, to determine the facts without any expectations of what they will be, allowing one's theoretical understanding to await the accumulation of a sufficient body of facts to make speculation profitable.

Measurement. In many of the sciences, quantitative measurement is employed. In fact, the possibility of measurement affords one of the commonest schemes of classification of the sciences into the exact, or quantitative, sciences and the qualitative sciences. Fundamentally, measurement amounts to description by the use of numbers, but not every use of numbers for the purposes of description is measurement—the individuals on a football team may be referred to by the numbers on their uniforms rather than by their names. The numbers which measure an aspect of a situation or object are obtained by performing certain operations. It is the nature of the operations which determines what is being measured. The number obtained by counting how many times a meterstick can be applied to an object measures the object's length, and the number at which the pointer on a spring balance comes to rest measures the weight of the object in the pan of the balance.

The operations by which the measuring numbers are obtained, involve in the vast majority of cases the use of some sort of instrument, for example, in the case of length a meterstick and in the case of weight a balance. The systematic design and use of instruments is one of the marks of well-developed scientific method. The number which is given by a measuring operation is always subject to some uncertainty or error. The magnitude of such error can be reduced by such devices as repetition or improvement in design of the instrument, but error can never be entirely eradicated.

The numbers obtained by measurement may be subjected to mathematical analysis, and mathematical regularities revealing the operation of various laws of nature often can be discovered and made the basis of theoretical understanding. It is frequently regarded as an ideal of a science that it be capable of mathematical analysis, and the more highly developed the science, the more susceptible it is of such analysis. From this point of view, physics is often regarded as the most highly developed science.

The operations by which the measuring numbers are obtained are not usually of complete generality but, at least in the more highly developed sciences, are subject to an important restriction. For example, the length of an object may be measured either in feet or in yards, and since a yard contains 3 feet, the length in yards will be one-third the length in feet. Systems of measurement which have properties of this sort are subject to limitations such that it is possible to make certain general statements from the mere knowledge that such systems are possible. This gives rise to the method of dimensional analysis, one of the most powerful of scientific methods. By the use of dimensional analysis, it is possible to predict, for example, that the period of a pendulum is proportional to the square root of its length, without the necessity for detailed mathematical analysis or experimental verification. *See* DIMENSIONAL ANALYSIS; *see also* PHYSICAL MEASUREMENT; PROBABILITY; PROBABILITY IN PHYSICS. [P.W.B.]

Bibliography: R. B. Lindsay and H. Margenau, *Foundations of Physics*, 1957.

Scintillation counter

A particle detector which emits light when a charged particle passes through it. The scintillation counter is one of the most versatile of the particle detectors, and is widely used in industry, research, radiation monitoring, and in exploration for those radioactive minerals that emit γ-rays. Many low-level radioactivity measurements are made with scintillation counters. For a discussion of these measurements, *see* LOW-LEVEL COUNTING; *see also* PARTICLE DETECTOR.

The flashes of light produced by the passage of charged particles through certain crystalline materials are visible to the dark-adapted eye as scintillations of the material. Although visual observations were used in the first experiments by Lord Rutherford, modern techniques employ photomultiplier tubes to detect and amplify the scintillations (*see* PHOTOTUBE, MULTIPLIER). The flashes of light result from fluorescent radiation emitted by atoms of the material as they return to their normal energy state after having been ionized or excited by the charged particles passing through the material (*see* FLUORESCENCE).

The first scintillation counters were made of a layer of zinc sulfide coated on glass. In modern counters, a transparent crystalline or plastic material is used as the scintillator. Such a material must be transparent to its own fluorescent radiation. The scintillator is placed in optical contact with the photosensitive surface of the photomultiplier, as shown in the illustration, and is surrounded, except on that surface, by reflecting material. When the particle traverses the scintillator, it leaves a trail of excited atoms which emit the radiation that is detected and amplified in the photomultiplier tube.

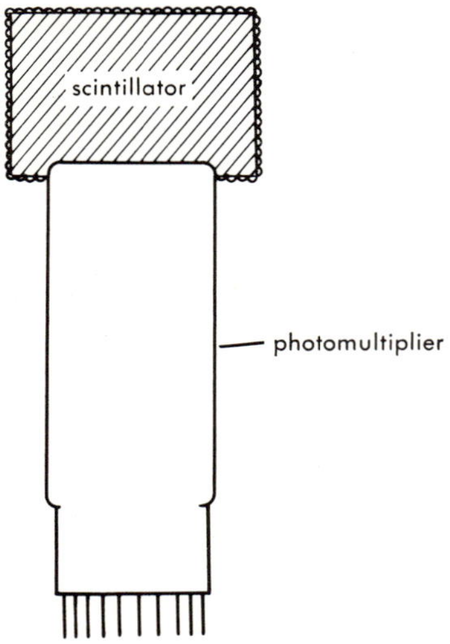

Scintillation counter.

Characteristics. Scintillation counters have several characteristics which make them particularly useful as particle detectors. They are highly efficient and can be made quite thin and small, so as to define the particle position accurately. In combination with more than one photomultiplier, and particularly with liquid as the scintillator, they can be made quite large (*see* SCINTILLATION DETECTOR, LIQUID). The pulse of light emitted is very short (10^{-8} to 10^{-9} sec for some scintillators), and, with proper photomultipliers and amplifying circuits, the speed and resolving time of scintillation counters are better than in any other counter except Cerenkov counters (*see* CERENKOV RADIATION). The maximum counting rate can therefore be very large, and scintillation counters are used near accelerators and other sources of radiation where some detectors, such as Geiger-Müller counters, are useless.

When properly operated, scintillation counters are proportional, that is, the magnitude of the pulse of light is proportional to the energy lost by the particle traversing the scintillator. They can thus give information about the energy of the particle.

Substances used. The materials used in scintillation counters may be classified as inorganic or organic. Inorganic substances include cadmium tungstate, calcium tungstate, and sodium or cesium iodide activated with thallium. These substances have decay times for the pulse of light of about 10^{-6} sec. They are particularly useful for detecting γ-rays because they contain heavy atoms which are efficient in converting γ-rays to electrons. Inorganic crystals are used in scintillation spectrometers, instruments which measure the energy and intensity of γ-rays from radioactive elements. The γ-rays produce electrons in the crystal which are of sufficiently low energy so that the electrons lose all their energy in the crystal, and are absorbed. The energy of the electron is measured by a pulse height analyzer on the output of the photomultiplier, and the γ-ray energies are inferred from the energy distribution of the electrons. For additional information, *see* GAMMA RAYS.

Organic scintillators. These are typically naphthalene, anthracene, transtilbene, and terphenyl. Organic scintillators may also be made by dissolving terphenyl in an organic liquid such as xylene, or in a polymer to make a solid plastic scintillator. The outstanding characteristic of organic scintillators is their very short decay time, from 10^{-8} to 10^{-9} sec, depending on the material. Organic scintillators are therefore used for high counting rates. They are also proportional in their counting action for electrons and protons. They are much less efficient for γ-ray counting than are the inorganic scintillators.

In some cases it is not practical to have the photomultiplier tube close to the scintillator. A "light pipe" is then used to conduct the flashes of light from the scintillator to the photomultiplier. This consists of a piece of glass, plastic, or quartz, cemented to the scintillator on one end and the photomultiplier on the other. Under good conditions, more than half of the light can be transmitted in this way. Systems of lenses and mirrors may also be used.

Coincidence counting. Scintillation counters are often used in coincidence with other particle counters. The coincidence technique eliminates or reduces the problem of local background. Fast organic scintillators are used to measure the time of flight of fast particles. The delay in the pulse from the second scintillation counter is a measure of the time required for the particle to go from one scintillator to the other, and the speed of the particle may thus be measured. Speeds up to 2.9×10^8 m/sec have been measured in this way. Organic scintillators may also be used to measure radioactive decay times as short as 10^{-9} sec. A particle with a short decay time is stopped in the scintillator, giving a pulse of light. If it decays a short time later and emits another charged particle, a second pulse is observed delayed slightly from the first, and measurement of the delay time gives the lifetime of the radioactive substance. [W.B.F.]

Bibliography: S. C. Curran, *Luminescence and the Scintillation Counter*, 1953.

Scintillation detector, liquid

A particle detector in which the sensitive material is a liquid scintillator. For a general discussion of scintillators as particle detectors, *see* SCINTILLATION COUNTER; *see also* PARTICLE DETECTOR.

Liquid scintillation detectors are used when cost is an important factor (they are relatively inexpensive), or when the detector must be made very large, or when the radiation levels to be measured are very low (*see* LOW-LEVEL COUNTING). A typical liquid scintillator consists of 5 g of *p*-terphenyl dissolved in 1 liter of toluene. The liquid organic solvent scintillates satisfactorily, and is inexpensive compared to the pure organic terphenyl crystal. The scintillating material must be placed in a glass or metal container and has the serious disadvantage of being a fire hazard. If a large area is required, the scintillator is placed in a metal drum, the inside surface of which is coated with a white material. Several photomultipliers are mounted in the drum so as to observe the flashes of light emanating from the scintillator.

Liquid scintillators have short decay times, as do other organic scintillators. The size of the electrical pulse resulting from the passage of a given particle is somewhat smaller in a liquid scintillator than in the corresponding pure solid organic scintillator. A large, liquid scintillator was used to verify experimentally the existence of neutrinos. [W.B.F.]

Bibliography: R. K. Swank, Characteristics of scintillators, *Ann. Rev. Nuclear Sci.*, 4:110–140, 1954.

Scitaminales

An order of the plant subclass Monocotyledoneae consisting of four tropical families. Some members of the banana family (Musaceae) are among the world's largest herbs. The ginger family (Zingiberaceae) with 47 genera and 1400 species is the largest. Some of these are cultivated as ornamentals. *Curcuma longa* supplies turmeric, a yellow dye and spice. *Zingiber* has several species whose rhizomes yield the important spice, ginger. *Elettaria* yields the favorite spice of the Orient, cardamon. The canna family (Cannaceae) has numerous ornamentals and the rhizome of *Canna edulis* is edible. The arrowroot family (Marantaceae) furnishes the arrowroot of commerce. *See* ARROWROOT STARCH; BANANA; CARDAMON; GINGER; TURMERIC; *see also* EMBRYOPHYTA; MONOCOTYLEDONEAE; PLANT KINGDOM. [P.D.S.]

Scleractinia

An order of the subclass Zoantharia which comprise the true or stony corals (Fig. 1*a*). These are solitary or colonial anthozoans which attach to a firm substrate. They are profuse in tropical and subtropical waters and contribute to the formation of coral reefs or islands. Some species are free and unattached. *See* CORAL REEF.

Most of the polyp is impregnated with a hard calcareous skeleton secreted from ectodermal calcioblasts. As soon as the planula settles, it secretes a thin skeletal basal plate, from which radiating vertical platelike partitions or septa arise; then characteristic thecal formation follows. The basal parts of the inner septal edges are often fused to form a columella (Fig. 2), which is sometimes ridged by vertical thin plates or pali. The outer

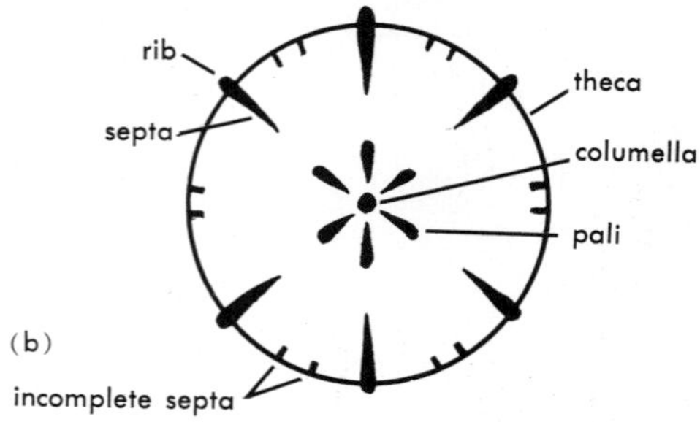

Fig. 1. (*a*) Coral polyps, *Oulangia* sp. (*b*) Diagrammatic figure of cross section of a solitary coral. (*After S. Hickson*)

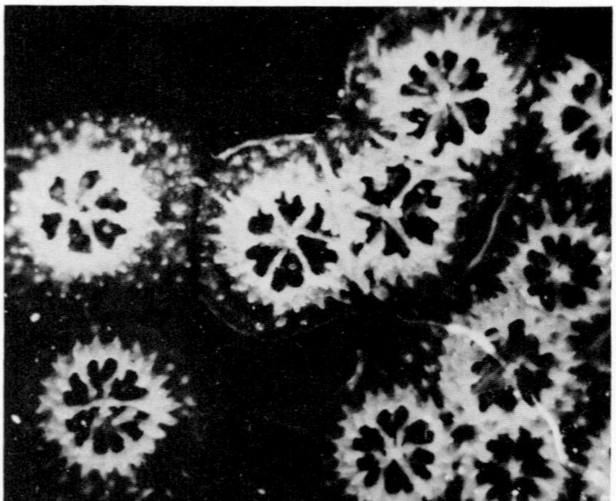

Fig. 2. Skeletons of the aggregated colonies of *Pocillopora damicornis espitosa* formed in 15 days after fixing.

septal edges, projecting beyond the theca, are the costae or ribs. Such is the outline of the corallum. Various other structures develop, such as the endo- or exothecae which comprise the dissepiments. These are inside and outside the wall of the corallum respectively (Fig. 1*b*).

Skeleton. The solitary corals form cylindrical, discoidal, or cuneiform skeletons (Fig. 3*a-e*), whereas colonial skeletons are multifarious. *Stylophora*, *Seriatopora*, *Pocillopora*, and *Acropora* form branching skeletons (Fig. 3*f-h*); while the skeletal systems of *Favia*, *Favites*, *Porites*, *Coeloseris*, and *Goniastrea* are massive (Fig. 3*i-k*). In

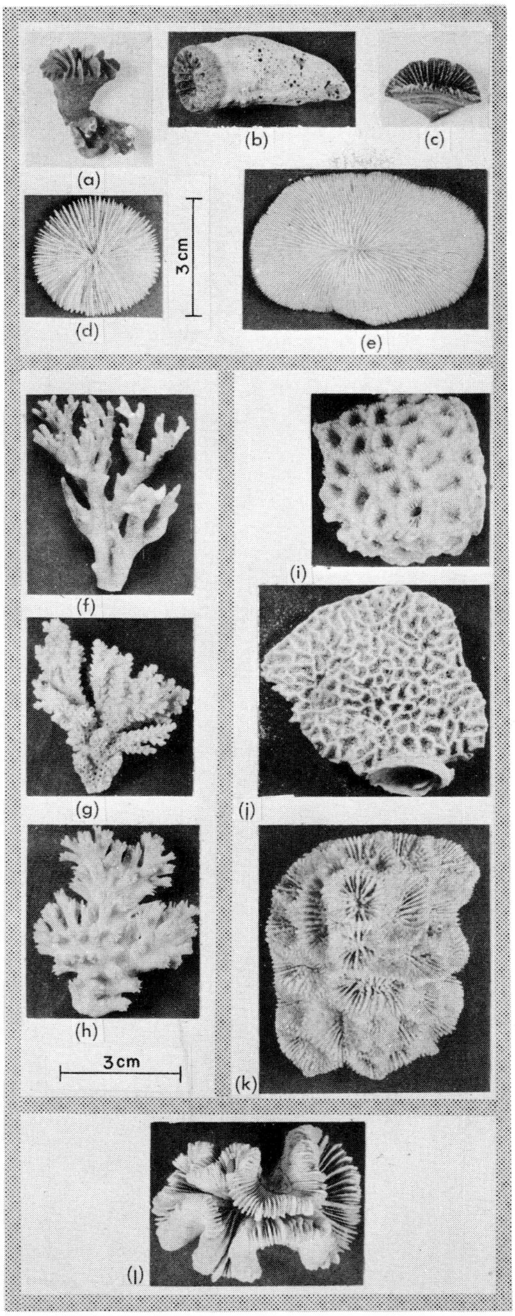

Fig. 3. Coral skeletons. (*a*) *Desmophyllum dianthus.* (*b*) *Flabellum distinctum.* (*c*) *Balanophyllia gigas.* (*d*) *Fungia actiniformis palauensis.* (*e*) *Fungia scutaria.* (*f*) *Pocillopora acuta.* (*g*) *Acropora hyacinthus.* (*h*) *Galaxea fascicularis.* (*i*) *Favia pallida.* (*j*) *Meandra lamellina.* (*k*) *Acanthastrea echinata.* (*l*) *Trachyphyllia amarantum.*

others, such as *Echinopora, Plerogyra, Lobophyllia,* and *Trachychyllia* (Fig. 3*l*), all sorts of shapes are found.

Colony formation. The polyps increase rapidly by intra- or extratentacular budding, and the skeletons of polyps which settle in groups are easily fused (Fig. 2). Consequently, the colonies are readily formed. The pyriform, ciliated planula swims with its aboral extremity, which is composed of an ectodermal sensory layer, directed anteriorly. Planulation occurs periodically in conformity with lunar phases in a good many tropical species. *See* ZOANTHARIA. [K.A.]

Scleractinia fossils

Fossil scleractinian corals (Madreporaria) were the ancestors of the living reef and deep- or cold-water corals and, as such, should provide information explanatory of the present wide diversity of these forms both structurally and ecologically.

Geologic record. The general pattern of deployment in geologic time of the scleractinians is summarized in the accompanying diagram which shows the relation of the five suborders to the pertinent part of the geological time scale. These corals appeared suddenly in rocks of Middle Triassic age in Europe, where they lived on banks or patch reefs in shoal parts of warm seas. Two major lines were then represented: (1) the colonial astrocoenids with small corallites (hence small polyps) and septa composed of a few trabecular elements—features they have always retained; and (2) relatively primitive fungids and favids with larger, solitary or colonial corallites with many trabecular elements in the septa. The relationship of these lines to each other is nearly as obscure as that of the group as a whole is to earlier types of corals. Attempts have been made to explain their origin from their ecological predecessors, the rugose corals of the Paleozoic Era, but to date no truly intermediate types have been discovered. Either the Scleractinia were derived from actiniarianlike polyps previously without skeletons, or they de-

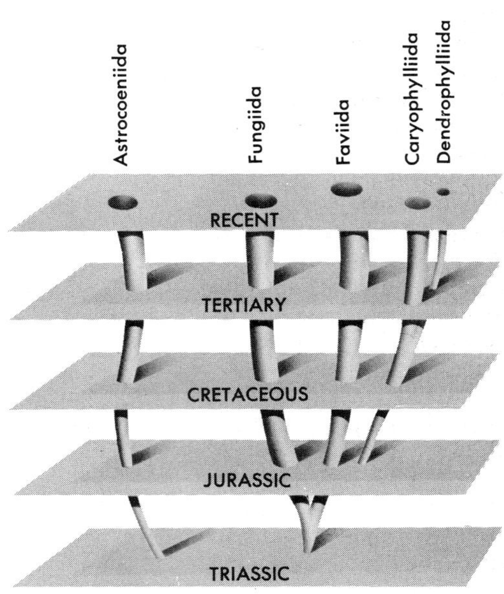

Evolutionary pattern of the Scleractinian suborders. (*After J. W. Wells, 1954*)

veloped from the rugose corals in the time interval between the Late Permian and the Middle Triassic. As yet no fossil corals have been found in rocks of this time. *See* COELENTERATA FOSSILS.

During the Jurassic Period the scleractinians swiftly evolved from the six or seven families of Triassic time into about seventeen. Most of this expansion was in the fungid-favid line in which the corallites became larger, with more complicated and numerous septa. Colonies became larger and larger as new modes of colony formation appeared. At this time, as is characteristic of many groups of organisms, some corals began to become adapted to other environments than the ancestral one in the warm, shallow seas. The caryophyllids, notably, appeared about this time, and some of them became adapted to cooler and often deeper waters outside the tropical zone. Their later offshoot, the dendrophyllids, continued this trend, and today most of the deep-sea corals belong to these two suborders. Most of the astrocoenids, fungids, and favids remained in the tropical waters where they acquired algal symbionts, zooxanthellae, an ecological partnership that made possible not only the growth of huge colonies but also the existence of large numbers of colonies in relatively small areas. As a result of this development, the first true coral reefs appeared in the Middle Jurassic, and by Late Jurassic and Early Cretaceous times reefs grew everywhere in the tropics where conditions of depth and bottom were suitable.

Occurrence. Notable sites of these reefs are in Texas, Mexico, western Europe, across southern Asia, and Japan. This coral reef belt had about the same extent as today but the zone as a whole was nearly 20 degrees farther north. The Late Cretaceous also saw extensive reef building, especially in the Mediterranean region. By the latter part of the Tertiary Period (Cenozoic) the reef zone had shifted southward to about the position it has today and the corals were essentially modernized. The major change since then has been the profuse development of the pocilloporid, acroporid (both of the suborder Astrocoeniida), and poritid (Fungiida) corals. Corals of these groups, with very rapidly growing colonies and with innumerable small polyps, account for more than two-thirds of the present reef coral fauna, both in sheer bulk and number of species. *See* ANTHOZOA; CORAL REEF.

[J.W.WE.]

Sclerenchyma

Single cells or aggregates of cells whose principal function is thought to be mechanical support of plants or plant parts. Sclerenchyma cells have thick secondary walls and may or may not remain alive when mature. They vary greatly in form and are of widespread occurrence in vascular plants. Two general types, sclereids and fibers, are widely recognized, but since these intergrade, the distinction is sometimes arbitrary.

Sclereids. These range from isodiametric to much elongated cells and may be branched. They may occur as isolated cells (idioblasts), groups of cells, or as extensive tissues. Five major kinds of sclereids are described here: brachysclereids, macrosclereids, osteosclereids, astrosclereids, and trichosclereids.

Brachysclereids (stone cells) have the form of the parenchyma cells from which they are derived by secondary sclerosis (hardening). They have conspicuous, frequently ramiform (branched) pits in their thickened walls and are found in the shells of nuts, the pits of stone fruits, the bark of many trees, and in the xylem and pith of some plants. Nests of brachysclereids form the grit in the flesh of pears and quinces (see illustration *a*).

Macrosclereids are rodlike cells formed from the protoderm (embryonic epidermis) of certain seed coats (*b*). Their secondary walls are unevenly thickened.

Osteosclereids, or "bone" cells, are rodlike with swollen ends. They occur in seed coats and in some leaves.

Astrosclereids tend to be radiately branched but are otherwise quite variable (*c*). They occur in the leaves of many plants (*Trochodendron, Pseudotsuga, Mouriria*) and in the petioles of *Camellia*. Astrosclereids sometimes have crystals embedded in their walls (*Castalia*).

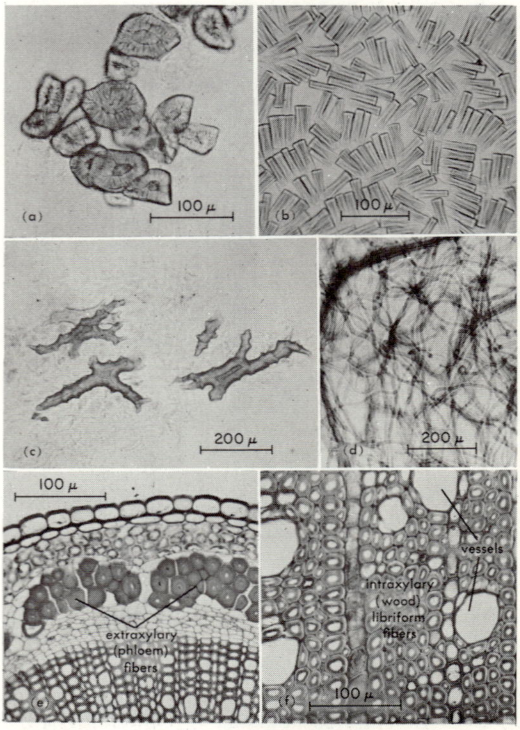

Sclerenchyma. (*a*) Brachysclereids (stone cells) from fruit of pear. (*b*) Macrosclereids (rod cells) from macerated seed coat of bean. (*c*) Branched sclereids from petiole of *Camellia*. (*d*) Fiberlike sclereids in cleared leaf of olive. (*e*) Transection of portion of stem of *Linum* showing extraxylary (phloem) fibers. (*f*) Transection of wood of *Cornus* showing intraxylary (wood) libriform fibers and vessels.

Trichosclereids (*d*) are long and slender, resembling fibers, with which they intergrade (leaves of *Monstera* and olive).

In general, foliar sclereids are idioblastic. They often arise from initials in the ground meristem. Branched forms extend themselves into intercellular spaces and between other cells. Sometimes they are associated with the ends of veinlets.

Fibers. Typically, fibers have tapering ends and are very long in proportion to their width. They usually occur as coherent strands of tissue and are rarely idioblastic. Individual fiber cells vary in length from a few millimeters to more than half a meter (*Boehmeria nivea*). Their thick walls are often sparsely pitted and the pits usually appear simple. Fibers are widely distributed both in the primary and in the secondary body of vascular plants. Two broad types are generally recognized: intraxylary, or wood, fibers occur in the xylem (*e*); extraxylary, or bast, fibers are found in the phloem and cortex of dicotyledons (*f*) and in association with vascular bundles in the stems and leaves of monocotyledons.

Libriform (elongated, thick-walled) wood fibers (*f*), which may constitute as much as 50% of an angiosperm wood, are so called because of their structural resemblance to phloem fibers. They often intergrade with tracheary (water-conducting) elements. Intermediate forms are called fiber-tracheids. Septate wood fibers and fiber-tracheids have transverse partitions which develop after the secondary wall is laid down (*Vitis, Hypericum*). Gelatinous or mucilaginous fibers have hygroscopic cell walls. They are found in many angiosperm woods such as oak (*Quercus*) and black locust (*Robinia*).

The fibers of commerce, excepting cotton and kapok (which consist of unicellular hairs), are bast fibers. The leaf fibers from monocotyledons, such as sisal (*Agave sisalina*), henequen (*Agave fourcroydes*), and abaca (*Musa textilis*), are known as hard fibers. They are lignified and coarse and are used in the manufacture of cordage. The soft, or stem, fibers include jute (*Corchorus*), true hemp (*Cannabis*), ramie (*Boehmeria nivea*), and flax (*Linum usitatissimum*). Flax fibers, the raw material for linen, contain little or no lignin. *See* FIBER CROPS; PHLOEM; XYLEM. [N.H.B.]

Bibliography: *See* PLANT ANATOMY.

Scorpius

The Scorpion, in astronomy, is one of most beautiful and vivid constellations in the sky. Scorpius is the eighth sign of the Zodiac. The constellation resembles a scorpion even to the sting. The bright red star Antares is situated at the heart. Its name (Ant-Ares) means the Rival of Mars, since both the planet and the star are bright and red in color, and the two are often found near each other. Antares is one of the largest stars known, having a diameter over 450 times that of the Sun. As in Sagittarius, the Milky Way in Scorpius is bright and rich in star clouds and clusters. *See* CONSTELLATION. [C.S.Y.]

Scorpion

Any member of the order Scorpionida, class Arachnida, phylum Arthropoda. There are about 650 species, 40 of which occur in the United States. They prefer warm, dry situations, and in the United States are most common in the arid Southwest, although they are found northward to North Dakota.

Scorpions have a sharp poison claw on the end of the abdomen with which they can inflict a painful sting. Two species in the United States are dangerous; at least five species in Mexico are considered capable of inflicting a fatal sting to children. One of the five can kill adults.

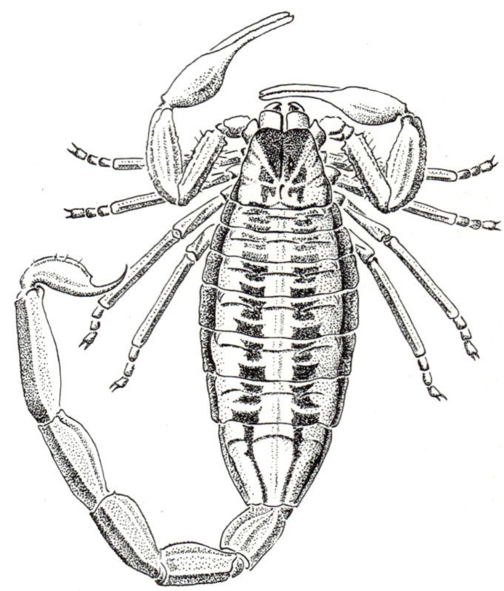

The scorpion, *Centrurus* sp.; length about 3 in. (*From E. L. Palmer, Fieldbook of Natural History, McGraw-Hill, 1949*)

Scorpions have elongated, somewhat flattened bodies, with a cephalothorax and 12 abdominal segments terminating in the poison claw. The abdomen is marked off into a preabdomen, continuous in pattern and size with the cephalothorax, and the slender, highly flexible postabdomen. The pedipalpi are long and powerful. With these appendages they grasp their prey and sting them into submission. They also have four pairs of walking legs. They breathe by spiracles. In general their anatomy is similar to that of other Arachnida.

Scorpions are nocturnal and they feed mainly on insects and spiders. Sexes are separate and fertilization is internal. The young are born alive and are carried about by the mother for several days after birth. *See* SCORPIONIDA. [J.D.B.]

Scorpionida

An order of the Arachnida which have chelate pedipalps and chelicera, a terminal caudal sting, and abdominal pectines (see illustration, *a*). The body is divided into a cephalothorax (prosoma), which is covered by the unsegmented carapace, and a segmented abdomen (opisthosoma). This latter divi-

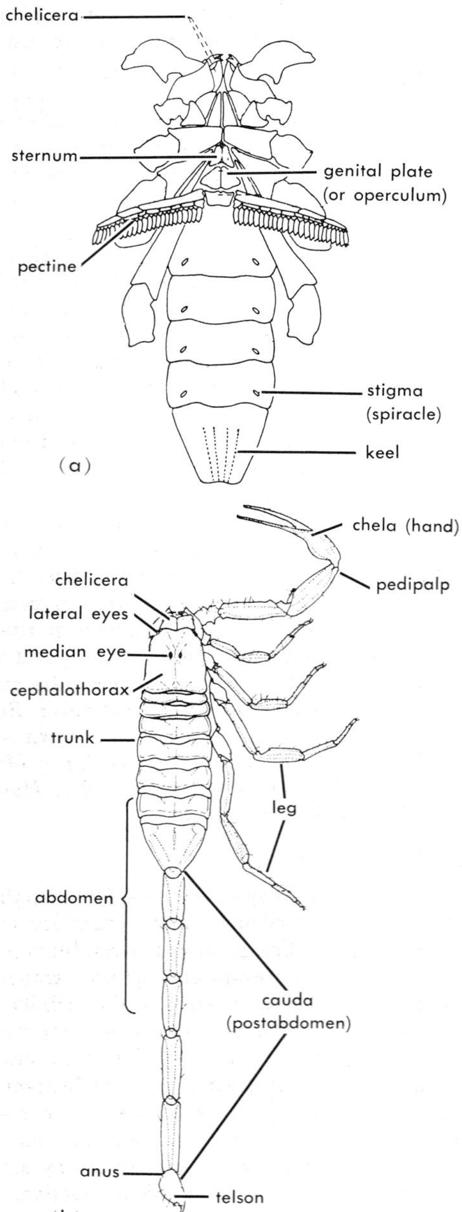

Scorpion. (*a*) Ventral view. (*b*) Dorsal view. (*H. L. Stahnke, Scorpions for laboratory study, Am. Biol. Teacher, vol. 19, no. 5, May, 1957*)

sion is differentiated into an anterior preabdomen (mesosoma) and a postabdomen (metasoma) which, plus the terminal sting, constitutes the "tail."

The cephalothorax bears the chelicerae, the pedipalps, and four pairs of walking legs. The preabdomen contains seven segments, while the postabdomen has five. The carapace bears three groups of small, simple eyes, a median pair and two groups of anterolateral eyes. Each group has two to five ocelli according to the species, while a few species lack lateral eyes. The dorsum of each of the seven preabdominal segments is covered by a distinct tergal plate, and the ventral surface by a similar sternal plate. The genital aperture, covered by the genital operculum, opens ventrally on the first segment. On the second segment is a small, quadrate basal piece, which articulates laterally with the pair of pectines. These toothed structures are tactile organs, used by these nocturnal and relatively blind animals to survey the surface over which they crawl. The lateral areas of sterna III–VI have four pairs of oblique, or oval, slits, which are the apertures, or spiracles, of the internal respiratory organs known as book lungs.

Approximately 650 species are known. They range in the adult form from approximately $\frac{3}{4}$ to 8 in. long. Scorpions are widely distributed throughout the Tropical Zone and the warmer areas of the Temperate Zone. About 40 species are found in the United States, 22 of which are in Arizona. California and Florida also have a rich scorpion fauna. Scorpions are found over three-fourths of the United States; they are absent from the New England states, Iowa, and areas immediately surrounding the Great Lakes. The only known lethal species in the United States are *Centruroides sculpturatus* Ewing and *C. gertschi* Stahnke, found in southern and central Arizona and the adjacent areas of California, New Mexico, and Mexico. The venom of these scorpions has proved fatal to healthy children up to 16 years of age and to adults suffering from hypertension and general debility. The Poisonous Animals Research Laboratory of Arizona State College, Tempe, Arizona, manufactures a scorpion antivenin for free distribution to the medical profession of that state. *See* ARACHNIDA. [H.L.ST.]

Screening

A mechanical method of separating a mixture of solid particles into fractions by size. The mixture to be separated, called the feed, is passed over a screen surface containing openings of definite size. Particles smaller than the openings fall through the screen and are collected as undersize. Particles larger than the openings slide off the screen and are caught as oversize. A single screen separates the feed into only two fractions. Two or more screens may be operated in series to give additional fractions. Screening occasionally is done wet, but most commonly it is done dry.

Industrial screens may be constructed of metal bars, perforated or slotted metal plates, woven wire cloth, or bolting cloth. The openings are usually square, but may be circular or rectangular. In rectangular openings the separation is controlled by the smaller dimension.

When the opening in a screen is larger than 1 in., the actual opening size in inches is specified. When the opening is 1 in. or less, the screen size is designated as the number of openings per linear inch, that is, a 20-mesh screen has 20 openings per inch. The actual size of an opening is less than that corresponding to the mesh number by the thickness of the metal between openings. Mesh sizes range from about 4-in. to 400-mesh, but screens finer than 100- or 150-mesh are seldom used industrially. Particles smaller than this are usually separated by other methods. *See* CLASSIFICATION, MECHANICAL; SEDIMENTATION (INDUSTRIAL).

Testing sieves are used to measure the size distribution of mixtures of particles and to specify the performance of commercial screens and other equipment. They are made of woven wire screening. Mesh and wire size of each screen are standardized. The usual range is from 1- to 200-mesh. In use, a set of standard screens is arranged serially in a stack, with the finest mesh on the bottom and the coarsest on top. An analysis is conducted by placing a weighed sample on the top screen and shaking the stack mechanically for a definite time. Particles passing the bottom screen are caught in a pan. The particles remaining on each screen are removed and weighed, and the weights of the individual increments are converted to percentages of the total sample.

In most screens, the particles are pulled through the openings by gravity only. Large particles are heavy enough to pass through the openings easily, but intermediate and smaller particles are too light to pass through the screen unaided. Most screens are agitated to speed the particles through the openings. Common methods are to revolve a cylindrical screen about an axis slightly inclined to the horizontal, or to shake, gyrate, or vibrate a flat screen.

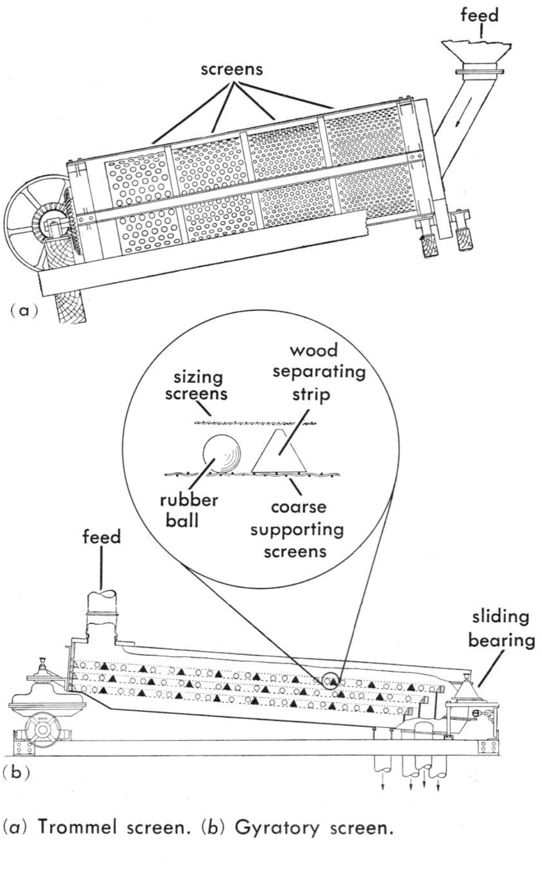

(*a*) Trommel screen. (*b*) Gyratory screen.

Two effects tend to restrict the fall of small particles through a screen: blinding of the screen by wedging of particles into the openings, and sticking of individual particles to each other and to the screen. The motion of the screen reduces blinding, and sometimes positive means are provided to drive wedged particles back through the openings. Sticking is severe if the particles are damp, and may be reduced by drying the feed.

Two common screens are shown in the illustration. The upper one is a revolving screen, called a trommel. This is a combination of four screens in series. This unit gives five products. The lower screen is a horizontal gyrating screen. It uses bouncing balls under the sizing screens to dislodge particles caught in the meshes. The balls are supported on coarse screens which offer no resistance to the materials passing the sizing screens. This screen gives four products.

Ideally, a screen should effect a sharp separation between undersize and oversize, and the largest particle in the undersize should be just smaller than the smallest particles in the oversize. An actual screen does not do this, and appreciable portions of both fractions have the same size range. Even testing screens show this overlap between adjoining fractions. *See* SEPARATION (MECHANICAL); SIZE REDUCTION. [W.L.M.]

Bibliography: A. M. Gaudin, *Principles of Mineral Dressing*, 1939; A. F. Taggart (ed.), *Handbook of Mineral Dressing*, 1945.

Screening smoke

Smoke clouds have many uses in warfare, ranging from screening military operations to marking and signaling purposes. The principal nonmilitary use of smoke has been to prevent atmospheric temperatures from falling below freezing in the vicinity of fruit crops. The obscuring action of screening smokes is associated with the reflection and scattering phenomena associated with fine particle suspensions. The optimum particle diameter for such smokes has been calculated to be about 1 micron.

Military screening smokes are produced by three general means: combustion, chemical reaction, or physical condensation.

Smokes by combustion. The most notable example of a combustion smoke is that generated by white phosphorus, coded "WP" by the United States Army. This material burns spontaneously in air according to the reaction

$$P_4 + 5O_2 \rightarrow 2P_2O_5$$

The phosphorus pentoxide formed is extremely hygroscopic and forms dense clouds of white smoke. The heat of combustion causes the smoke to rise rapidly from the ground in calm weather. This effect, called pillaring, is undesirable, limiting the use of WP as a screen. Because it is easily ignited and has a fairly high burning temperature, white phosphorus is also considered an incendiary.

Munitions for the dissemination of WP smoke all contain an explosive burster which scatters the phosphorus as small burning fragments. These munitions are designed as either shells or grenades.

Smokes by chemical reaction. These smokes are typified by several commonly used military chemicals. Hexachloroethane (HC) smoke is made by the reaction of zinc oxide with hexachloroethane. The reaction is initiated by heat and produces a cloud of hygroscopic zinc chloride particles, which absorb water and form a dense screen. A typical HC smoke mix contains 7% aluminum, 47% zinc oxide, and 46% hexachloroethane. The aluminum content is varied to control the rate of reaction.

HC smoke munitions are not explosive but the reaction temperature is high. The principal military munitions for HC are grenades and smoke pots.

Titanium tetrachloride (FM) and a mixture of sulfur trioxide and chlorosulfonic acid (FS) are examples of chemicals that produce smoke by hydrolysis reactions. These depend on atmospheric moisture. Under extremely dry conditions these materials are not effective.

Titanium tetrachloride is an easily volatilized liquid whose vapors react instantly with water to form a variety of solid, finely divided titanium oxychlorides. Although liquid FM can be stored easily, in the presence of moisture it forms hydrochloric acid, which is highly corrosive. FM can be sprayed from tanks carried by aircraft, but the solid products of hydrolysis have been known to clog the spray orifices. It can also be used as a filling for shells and bombs.

FS mixture was developed in the United States in 1929–1930 to replace the more expensive and less effective FM. The general effect of exposing a spray of sulfur trioxide–chlorosulfonic acid mixture to humid air is to produce a dense screen of fine particles of sulfuric acid. FS is especially adaptable to airplane spray dissemination. It produces no solid reaction products and the high specific gravity of the mixture enables the screen to reach the ground when sprayed at low altitudes. It is, however, highly corrosive as a liquid and as a smoke.

Fig. 1. Smoke screen produced by oil smoke generators.

Fig. 2. Aerial view of smoke screen being laid by airplane.

All chemical smokes, being acidic, are irritating to the respiratory tract.

Condensation smokes. The need for a screening smoke that was nonirritating, inexpensive, and producible in large quantities led in World War II to the development of oil smokes, produced by purely physical means. When high-boiling petroleum oil is vaporized by heat and the vapor cooled rapidly it condenses into numerous fine droplets that form a stable cloud, the life of which depends solely on meteorological conditions. Average field concentrations of oil smoke are harmless by inhalation. Oils suitable for smoke use resemble SAE 10–40 motor oils.

The volume of smoke produced by generating devices is determined by the quantity of heat available for vaporization and by the power available for pumping the oil to the heat source. Present military smoke generators are capable of converting over 50 gal of oil per hour to smoke. Many sources of heat may be utilized. The simultaneous operation of many generators makes it possible to screen several square miles from air observation.

These smokes are also useful for nonmilitary purposes. Farmers have used smoke for many years to prevent the freezing of fruit and other crops. The first smokes were generated by burning wet hay or other moist combustibles in smudge pots. Recent trends have been toward the use of oil smokes. The blanket of smoke reduces the loss of heat by radiation from ground or plant surfaces, and has been quite successful in keeping local temperatures above the freezing point.

Signaling or marking smokes. Smoke is used at times for signaling or marking in military operations. Such smoke signals must be colored to be distinguishable. These are usually burning mixtures containing dyes which vaporize and recondense. Any color can be produced, depending on the thermal stability and other characteristics of the dye. The most satisfactory dyes have been of the azo, anthraquinone, azine, or diphenylmethyl types. *See* AEROSOL; CHEMICAL WARFARE; SMOKE. [S.D.S.]

Screw

A cylindrical body with a helical groove cut into its surface. For practical purposes a screw may be considered to be a wedge wound in the form of a helix so that the input motion is a rotation while the output remains translation. The screw is to the wedge much the same as the wheel and axle is to the lever in that it permits the exertion of force through a greatly increased distance.

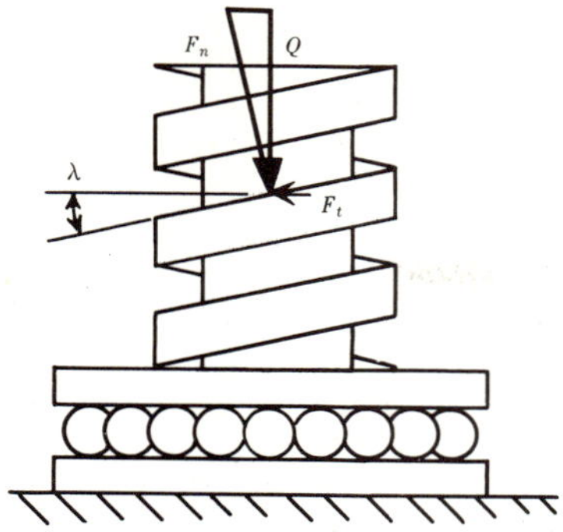

Fig. 1. Screw.

Figure 1 shows a frictionless screw with square threads mounted on a ball thrust bearing being used to raise a load Q. The load and other forces may be considered to be concentrated at an effective radius called the pitch radius. The force diagram is similar to the diagram for a wedge except that wedge angle θ is replaced by screw lead angle λ, which is the angle between the thread of the helix and a plane perpendicular to the axis of rotation. Subscript t is added to indicate that force F is applied tangent to a circle. Therefore,

$$F_t = Q \tan \lambda$$

and, because F_t acts at the pitch radius R, the torque on the screw is

$$T = F_t R = QR \tan \lambda$$

When a practical screw is considered, friction becomes important and the torque on a screw with square threads becomes

$$T = QR \frac{\tan \lambda + \mu}{1 - \mu \tan \lambda}$$

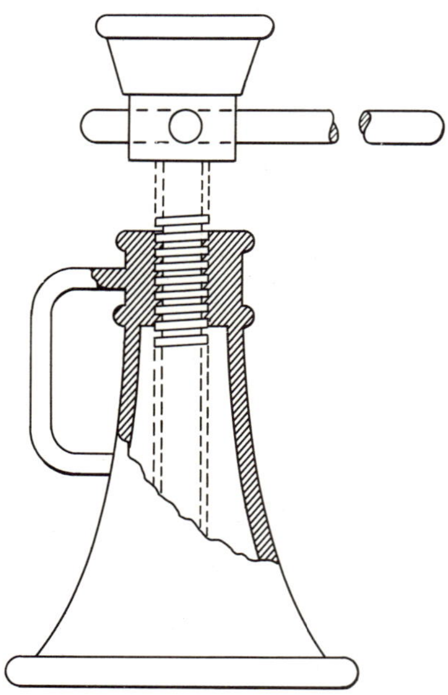

Fig. 2. Screw jack.

and the efficiency of a screw with square threads

$$\eta = \tan \lambda \frac{1 - \mu \tan \lambda}{\tan \lambda + \mu}$$

where μ = the coefficient of friction.

If a screw jack (Fig. 2) is used to raise a heavy object, such as a house or machine, it is normally desirable for the screw to be self-locking, that is, for the screw not to rotate and lower the load when the torque is removed with the load remaining on the jack. For a screw to be self-locking, the coefficient of friction must be greater than the tangent of the lead angle: $\mu > \tan \lambda$.

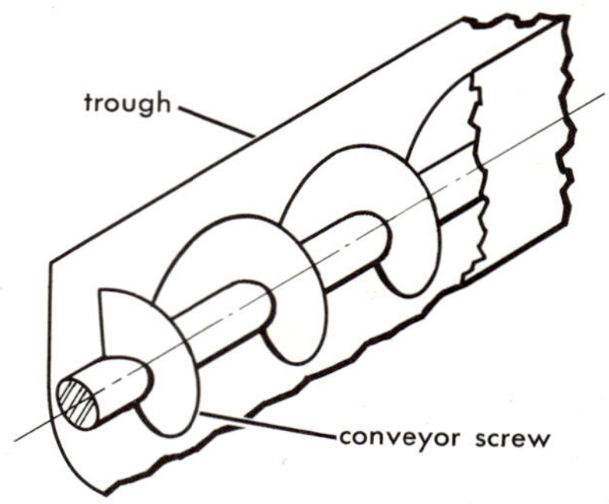

Fig. 3. Screw conveyor.

The screw is by far the most useful form of inclined plane or wedge and finds application in the bolts and nuts used to fasten parts together; in lead and feed screws used to advance cutting tools or parts in machine tools; in screw jacks used to lift such objects as automobiles, houses, and heavy machinery; in screw-type conveyors (Fig. 3) used to move bulk materials; and in propellers for airplanes and ships. *See* SIMPLE MACHINE. [R.M.PH.]

Screw fastener

A threaded machine part used to join parts of a machine or structure. Screw fasteners are used when a connection is required that can be disassembled and reconnected and that must resist tension and shear. A nut and bolt is a common screw fastener. Bolt material is chosen to have an extended stress-strain characteristic free from a pronounced yield point. Nut material is chosen for slight plastic flow.

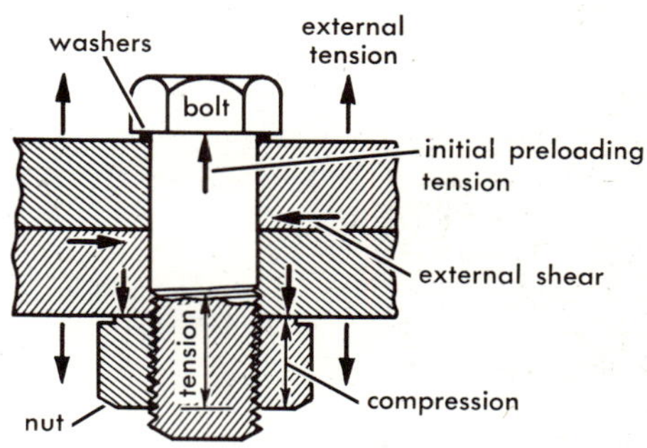

Forces on a bolt fastener under preloading.

The nut is tightened on the bolt to produce a preload tension in the bolt, as illustrated. This preload has several advantageous effects. It places the bolt under sufficient tension so that during vibration the relative stress change is slight with consequent improved fatigue resistance and locking of the nut. Preloading also increases the friction between bearing surfaces of the joined members so that shear loads are carried by the friction forces rather than by the bolt.

The tightened nut is under compression; the bolt is under tension. The deformation that accompanies these forces tends to place the entire preload on the thread nearest the bearing surface. Thus, concentration of loading is counteracted if the nut is slightly plastic so as to set under load. This yielding may be achieved by choice of material or by special shape. If a soft gasket is used in the assembly, preloading is less effective; the bolt may carry the full tension and shear loads. *See* BOLT; JOINT (MECHANICAL); SCREW THREADS.

[F.H.R.]

Bibliography: O. J. Horger (ed.), *Metals Engineering Design*, 1953.

Screw jack

A mechanism for lifting and supporting loads usually of large size. A screw jack mechanism consists of a thrust collar and a nut which rides on a bolt; the threads between the nut and bolt normally have a square shape. A standard form of screw jack has a heavy metal base with a central threaded hole into which fits a bolt capable of rotation under a collar thrusting against the load. Screw jacks are also used for positioning mechanical parts on machine tools in order to carry out manufacturing processes. These jacks are small in size and have standard V threads. Load and stress calculations may be performed on screw jacks in the same manner as in power screws and screw fastenings.

[J.J.R.]

Screw threads

Continuous helical ribs on a cylindrical shank. Screw threads are used principally for fastening, adjusting, and transmitting power. To perform these specific functions, various thread forms have been developed. A thread on the outside of a cylinder or cone is an external (male) thread; a thread on the inside of a member is an internal (female) thread (Fig. 1).

Types of thread. A thread may be either right-hand or left-hand. A right-hand thread on an external member advances into an internal thread when turned clockwise; a left-hand thread advances when turned counterclockwise. If a single helical groove is cut or formed on a cylinder, it is called a single-thread screw. Should the helix angle be increased sufficiently for a second thread to be cut between the grooves of the first thread, a double thread will be formed on the screw. Double, triple, and even quadruple threads are used whenever a

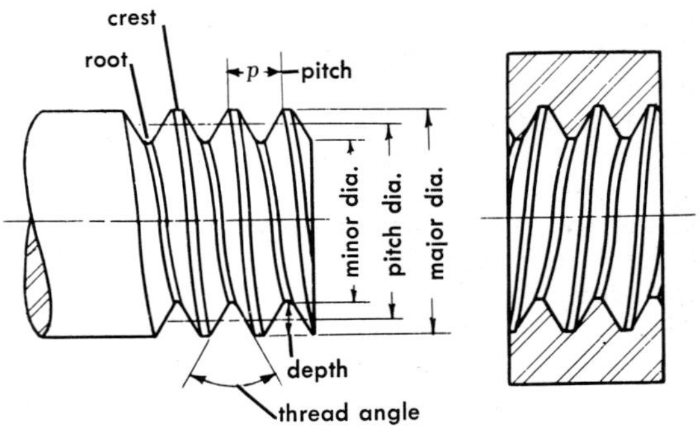

Fig. 1. Screw thread nomenclature.

rapid advance is desired, as on fountain pens and valves. The helices on a double thread start 180° apart while those on a triple begin 120° apart. A multiple thread produces a rapid advance without resort to a coarse thread.

Pitch and major diameter designate a thread. Lead is the distance advanced parallel to the axis when the screw is turned one revolution. For a single thread, lead is equal to the pitch; for a double thread lead is twice the pitch. For a straight thread, the pitch diameter is the diameter of an imaginary coaxial cylinder that would cut the thread forms at a height where the width of the thread and groove would be equal.

Thread forms now in use have been developed to satisfy particular requirements (Fig. 2). Those employed on fasteners and couplings and those used for making adjustments are generally of the modified 60° V type. Where strength is required for the transmission of power and motion, a thread having faces that are more nearly perpendicular to the axis is preferred such as the modified square and the acme. These threads, with their strong thread sections, transmit power nearly parallel to the axis of the screw. The sharp V, formerly found

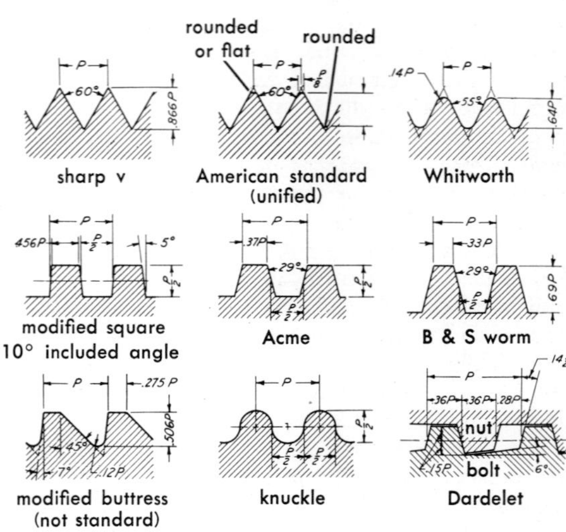

Fig. 2. Screw threads. (*From W. J. Luzadder, Graphics for Engineers, Prentice-Hall, 1957*)

on set screws, is now rarely used because of the difficulty of cutting sharp roots and crests in quantity production.

Standard threads. The unified thread form accompanied by the unified thread standards fulfills the necessary requirements for interchangeability of threaded products between the United States, Great Britain, and Canada. This modified thread form represents two compromises between British Whitworth and American National thread forms. The British accepted the 60° thread angle of the American thread and the Americans accepted a rounded root and crest similar to the Whitworth form. The American National unified form for external threads has a rounded root and may have either a rounded or flat crest. The new thread contours provide advantages over both the earlier forms. The design provides for greater fatigue strength and allows for longer wear of cutting tools. The unified thread is a general purpose thread that is particularly suited for fasteners.

Threads on bolts and screws may be formed by rolling, cutting, or extruding and rolling. Cut threads frequently produce a tighter fit than do rolled threads because of irregularities and burrs. To facilitate assembly, cut threads should have a finished point. Rolled threads ordinarily can be manufactured at less cost but the range of sizes of bolts that can be produced is limited.

The Whitworth 55° thread, similar in many respects to the American Standard form, is the standard V-form used in Great Britain. The unified form, identical with the American Standard, is also standard in Great Britain.

A modified square thread, with an angle of 10° between the sides, is sometimes used for jacks and vises where strength is needed for the transmission of power in the direction of the thread axis.

When originally formulated, the acme thread was intended as an alternate to the square thread and other thread forms that were being used to produce transverse motion on machines. Now used for a variety of purposes, the acme thread is easier to produce than the square thread and its design permits the use of a split nut to provide on-off engagement. A modification of the acme is the 29° stub acme. Because the basic depth of the thread is reduced to 0.30 of the pitch, the thread section is strong and well suited to power applications where space limitations make a shallow thread desirable. A still further modification of the acme is the 60° stub having a 60° angle between the sides of the thread. This thread has a basic depth of 0.433 of the pitch. The Brown and Sharp worm thread, a modified form of the acme thread, is used for transmitting power to a worm wheel.

The buttress or breech-block thread is designed for transmitting an exceptionally high pressure in one direction only. In its original form, the pressure flank was perpendicular to the thread axis and the trailing flank sloped at 45°. To simplify the cutting of the face, modern practice is to give the pressure flank a 7° slope. This form of thread is applicable for assembled tubular members because of the small radial thrust. Buttress threads are used on breech mechanisms of large guns and for airplane propeller hubs.

The knuckle rolled thread is found on sheet metal shells of lamp bases, fuse plugs, and in electric sockets. The thread form, consisting of circular segments forming crests and roots tangent to each other, has been standardized. The thread is usually rolled but it may be molded or cast. The Dardelet thread is self-locking in assembly.

Unified and American screw thread series. Unified and American screw thread standards for screws, bolts, nuts, and other threaded parts consist of six series of threads and a selection of special threads that cover nonstandard combinations of diameter and pitch. Each series of standard threads differs from another by the number of threads specified per inch for a particular diameter. The unified screw threads are limited to two series, coarse (UNC) and fine (UNF). A ¼-in. diameter thread in the UNC series has 20 threads/in., while in the UNF series it has 28 threads/in.

In the unified and American standards, the coarse thread series (UNC and NC) is recommended for general industrial use on threaded machine parts and for screws and bolts. Because a coarse thread has fewer threads per inch than a fine thread of the same diameter, the helix angle is greater and the thread travels farther in one turn. A coarse thread is preferred where rapidity and ease of assembly are desired and where strength and clamping power are not of prime importance.

The fine thread series (UNF and NF) is recommended for general use in the automotive and aircraft fields for threads subject to strong vibration. In general, a fine thread should be used only where close adjustment, extra strength, or increased resistance to loosening is a factor.

The extra-fine thread series (UNEF and NEF) is used principally in aeronautical structures where an extremely shallow thread is needed for thin-walled material and where a maximum practicable number of threads is required for a given length.

The 8-thread series (8N) with 8 threads/in. for all diameters is used on bolts for high-pressure flanges, cylinder-head studs, and other fasteners against pressure where it is required that an initial tension be set up so that the joint will not open when steam or other pressure is applied. This series has come into general use for many types of engineering work. It is sometimes used as a substitute for the coarse thread series for diameters greater than 1 in.

The 12-thread series (12UN or 12N) is a uniform pitch series that is widely used in industry for thin nuts on shafts and sleeves. This series is considered to be a continuation of the fine-thread series for diameters greater than 1½ in.

The 16-thread series (16UN or 16N) is a uniform pitch series for applications that require a very fine thread as in threaded adjusting collars and bearing retaining nuts. This series is used as a continuation of the extra-fine thread series for diameters greater than 2 in.

The manufacturing tolerance and allowance permitted distinguishes one class of thread from another under the unified thread system. Because allowance is an intentional difference between correlated size dimensions of mating threads, and tolerance is the difference between limits of size, a thread class may be considered to control the looseness or tightness between mating threaded parts. The American standard for unified and American screw threads provides classes 1A, 2A, and 3A for external threads, classes 1B, 2B, and 3B for internal threads, and classes 2 and 3 for both internal and external threads. Classes 1A and 1B are for ordnance and other special uses. Classes 2A and 2B are the recognized standards for the bulk of screw thread work and for the normal production of threads on screws, bolts, and nuts. Classes 3A and 3B are for applications where smaller tolerances than those afforded by class 2A and 2B are justified and where closeness of fit between mating threads is important. Class 3A has no allowance.

The screw thread fit needed for a specific application can be obtained by combining suitable classes of thread. For example a class 2A external thread might be used with a class 3B internal thread to meet requirements.

Methods of designating screw threads. Threads are designated bv standardized notes (Fig. 3). Under the unified system, threads are specified by

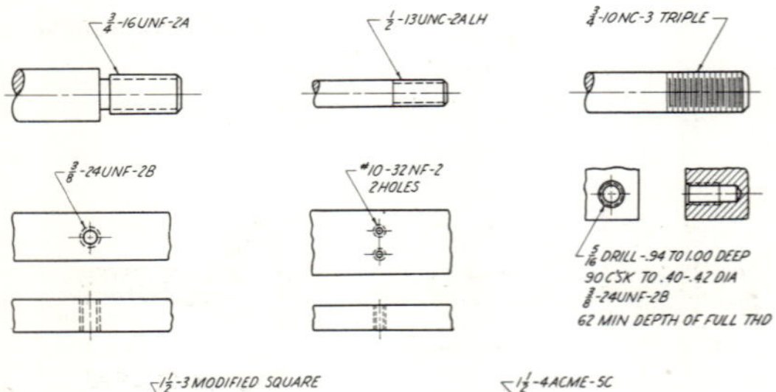

Fig. 3. Thread identification symbols. (*From W. J. Luzadder, Fundamentals of Engineering Drawing, 4th ed., Prentice-Hall, 1959*)

giving in order the nominal diameter, number of threads per inch, the initial letters such as UNC or UNF that identify the series, and class of thread (1A, 2A, and 3A or 1B, 2B, and 3B). Threads are considered to be right-hand and single unless otherwise noted, therefore the specification for a left-hand thread must have the letters LH included. To indicate multiplicity, the word DOUBLE, TRIPLE, and so on must follow the class svmbol.

Threaded fastener uses. Most threaded fasteners are a threaded cylindrical rod with some form of head on one end. There are various types of threaded fasteners available; some, such as bolts, cap screws, and machine screws, have been standardized; others are special designs. The use of removable threaded fasteners is necessary on machines and structures for holding together those parts that must be frequently disassembled and reassembled. Because standard threaded fasteners are mass produced at relatively low cost and are uniform and interchangeable, they are used whenever possible.

Fasteners are identified by names that are descriptive of either their form or application: set screw, shoulder screw, self-tapping screw, thumb screw, and eye-bolt. Of the many forms available, five types meet most requirements for threaded fasteners and are used for the bulk of production work: bolt, stud, cap screw, machine screw, and set screw (Fig. 4). Bolts and screws can be obtained with varied heads and points (Fig. 5).

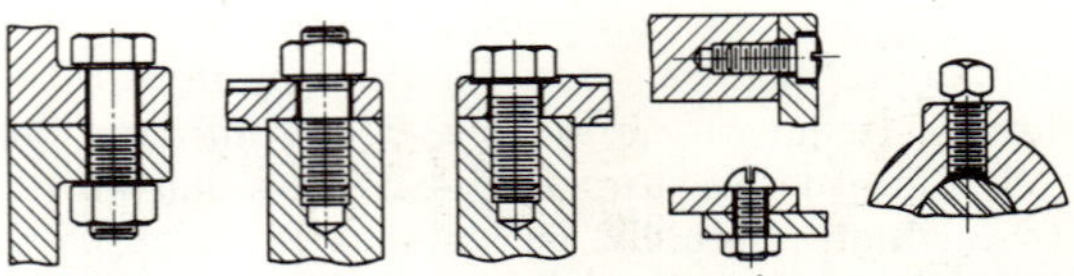

Fig. 4. Common types of fasteners. (*From T. E. French and C. J. Vierck, Graphic Science, McGraw-Hill, 1958*)

A bolt is generally used for drawing two parts together. Having a threaded end and an integral head formed in manufacture, it is passed through aligned clearance holes in the two parts and a nut is applied.

A stud is a rod threaded on both ends. Studs are used for parts that must be removed frequently and for applications where bolts would be impractical. They are first screwed more or less permanently into one part before the removable member with corresponding clearance holes is placed into position. Nuts are used on the projecting ends to draw the parts together.

Cap screws (plated or unplated) are widely used in machine tools and for assembling parts in automotive and aeronautical equipment. They are available in four standard heads: hexagon, flat, round, and fillister. Cap screws may be steel, brass, bronze, or aluminum alloy. Flat- and round-head screws have slotted heads (Fig. 6). Fillister-head screws have either slotted or socket heads. When mating parts are assembled, the cap screws pass through clear holes in one member and screw into threaded holes in the other. The head, an integral part of the fastener along with the thread, holds the parts together. In the automotive industry, the hexagon-head cap screw in combination with a nut is often used as an automotive hexagon-head bolt.

Machine screws, which are similar to cap screws and fulfill the same purpose, are employed principally in the numbered diameter sizes on small

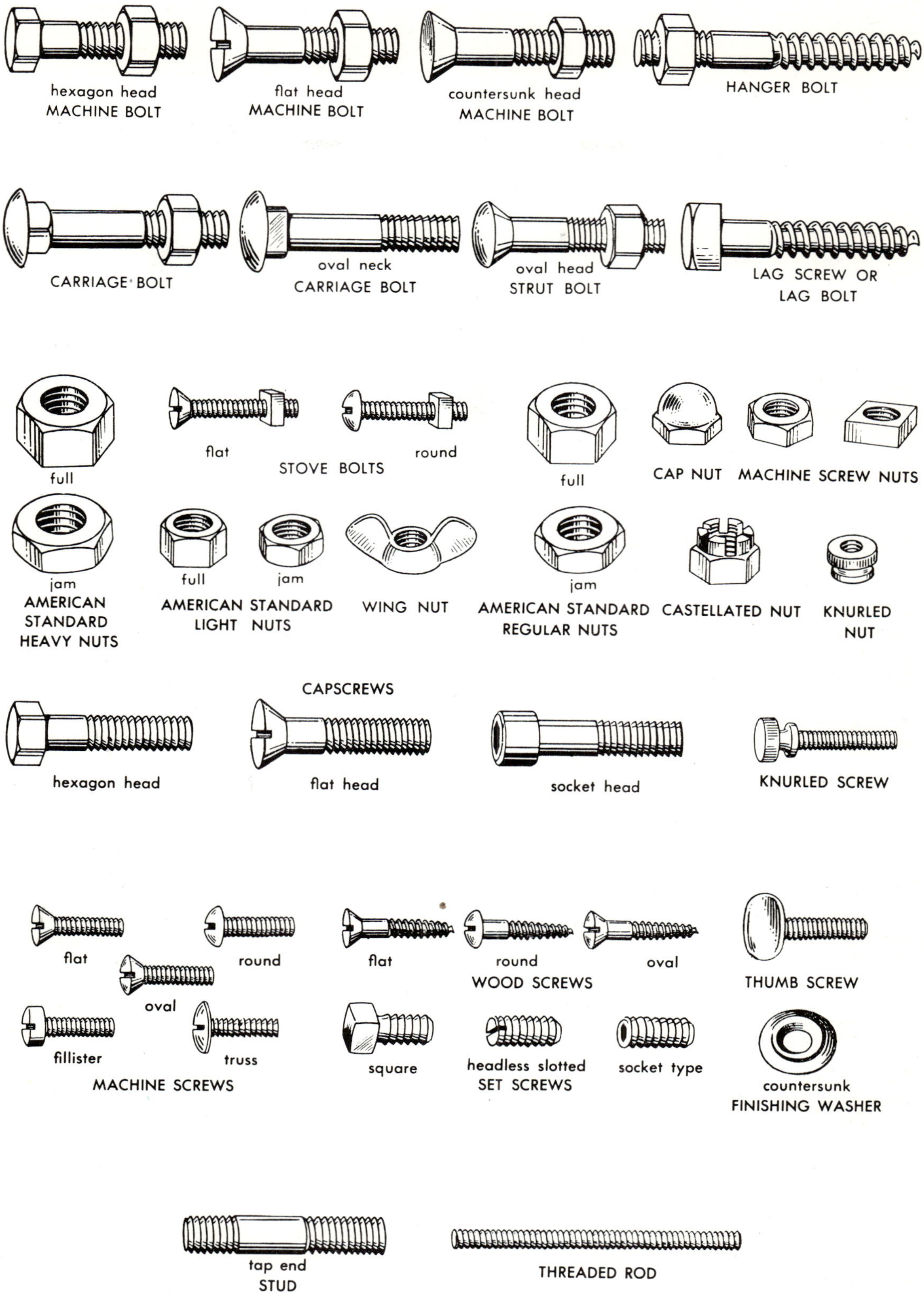

Fig. 5. Threaded fasteners are made in a wide variety of types and used for many different purposes. Some of the most common are shown here. Screws may be slotted or hollow-headed. Hollow-headed screws may have a socket head, Phillips head, or any one of a number of special contours for patented turning systems. (*Reynolds Metals* Co.)

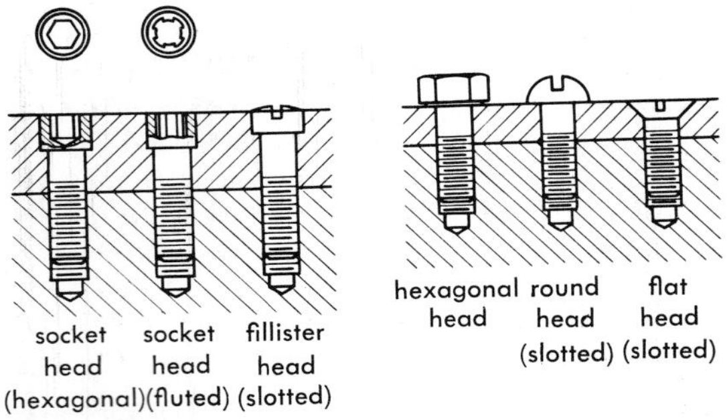

Fig. 6. Cap screws. (*From W. J. Luzadder, Fundamentals of Engineering Drawing, 4th ed., Prentice-Hall, 1959*)

work having thin sections (Fig. 5*d*) or with a machine nut to function as a small bolt.

Set screws made of hardened steel are used to hold parts in a relative position to one another. Normally, their purpose is to prevent rotary motion between two parts, such as would occur in the case of a rotating pulley and shaft combination (Fig. 5*e*). Set screws can be purchased with any one of six types of points in combination with any style of head (Fig. 7). In the application of set screws, a flat surface may be formed on a shaft to provide a seat for a flat point. A cone point fits into a conical spot.

Wood screws, lag screws, and hanger bolts are used in wood (Fig. 4). Wood screws may have either slotted or recessed heads.

Self-tapping screws (Fig. 8) have a specially hardened thread that makes it possible for the screws to form their own internal thread in sheet metal and soft materials when driven into a hole that has been drilled, punched, or punched and reamed. The use of self-tapping screws eliminates costly tapping operations and saves time in assembling parts.

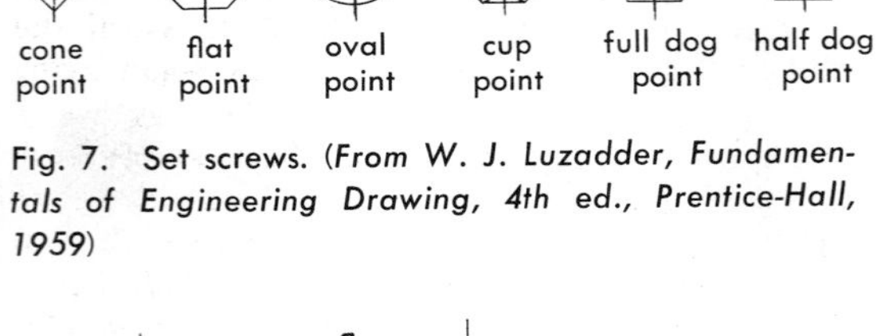

Fig. 7. Set screws. (*From W. J. Luzadder, Fundamentals of Engineering Drawing, 4th ed., Prentice-Hall, 1959*)

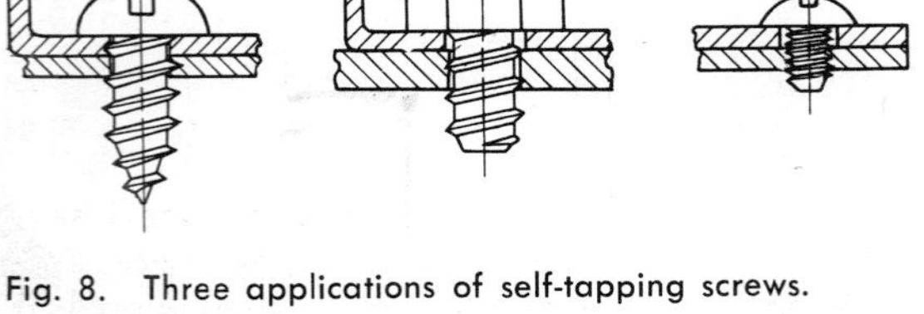

Fig. 8. Three applications of self-tapping screws.

Pipe thread. In the American standard (Briggs) taper pipe thread (Fig. 9), threads are cut (on a cone) to a taper of ¹⁄₁₆ in./in. to ensure a tight joint. Although a normal connection employs a taper external and a taper internal thread, an American standard straight pipe thread having the same pitch angle and depth of thread as the corresponding taper pipe thread is used for pressure-tight joints for couplings, pressure-tight joints for fuel- and oil-line fittings, and for loose-fitting and free-fitting mechanical joints. Assemblies made with taper external threads and straight internal threads are frequently preferred to assemblies employing all taper threads; the assumption is made that relatively soft or ductile metals will adjust to the taper external pipe thread. A modified pipe thread, the American standard Dryseal pipe thread

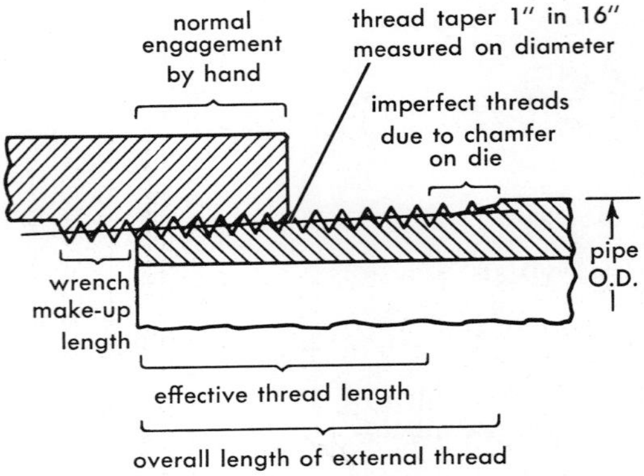

Fig. 9. American standard taper pipe thread.

(taper and internal straight) is used for pressure-tight connections that are to be assembled without lubricant or sealer. Except for a difference in the truncation at the roots and crests, the general form and dimensions of this thread are the same as those of the American standard taper pipe thread. The principal uses for the Dryseal thread are in refrigerant pipes, automotive and aircraft fuel-line fittings, and for gas and chemical shells (ordnance). For railing joints, where a rigid mechanical thread joint is needed, external and internal taper threads are used. Basically the external thread is the same as the American standard taper pipe thread except that the length of the thread is shortened at the end to permit use of the larger diameter portion of the thread. [W.J.L.]

Bibliography: American Standards Association (ASA), *Unified and American Screw Threads Standard*, 3d ed., 1951; ASA, *Acme Screw Threads*, 1952; T. E. French and C. J. Vierck, *Graphic Science*, 1958; W. J. Luzadder, *Graphics for Engineers*, 1957; W. J. Luzadder, *Fundamentals of Engineering Drawing*, 4th ed., 1959; U.S. National Bureau of Standards, *Screw-thread Standards for Federal Services*, Handbook H28, 1945.

SCUBA

Self-Contained-Underwater-Breathing-Apparatus such as the Aqualung (see illustration), extensively used by trained personnel as a tool for direct observation in marine research and underwater engineering. This equipment is designed to deliver a breathable gas mixture at the same pressure as that exerted on the diver by the overlying water column. There is, therefore, no difference in pressure between the interior and exterior parts of his body and he has no feeling of pressure regardless of depth. The gas which he breathes is carried in high pressure cylinders (at starting pressures of 2000–3000 psi). Because the SCUBA diver is free from surface tending, he has much more maneuverability and freedom than the conventional "hard hat" diver.

A diver is shown wearing an Aqualung SCUBA, exposure suit, swim fins, and face mask. Bubbles are from the exhaled air which is allowed to escape into the water in this type of open-circuit equipment. (*Photo by R. F. Dill, U.S. Navy Electronics Laboratory*)

SCUBA can be classified into three types: (1) closed-circuit, (2) semiclosed-circuit, and (3) open-circuit. In the first two, which use pure oxygen or various combinations of oxygen, helium, and nitrogen, exhaled gas is retained and passed through a canister containing a CO_2 absorbent for purification. It is then recirculated to a bag worn by the diver for rebreathing. Additional gas to replace that used by the diver during inhalation is supplied to the bag by various automatic devices from the high pressure cylinders. These two types of equipment are much more efficient than the open-circuit system, which does not take advantage of the unused oxygen in the exhaled gas, the unused oxygen being discharged directly into the water after breathing. The closed- and semiclosed-circuit types are, however, more complicated and can be dangerous if they malfunction or are used by inexperienced divers. The open-circuit system usually has compressed air, which is easy to obtain, as the breathing gas. This type is not as efficient as the other two systems, but for scientific, engineering, and sport diving it has been shown to be much safer and is simpler in construction. The open-circuit SCUBA is preferred by most scientists and engineers who work underwater because of its safety, the ease in learning its use, and its relatively low cost. *See* SKIN DIVING; UNDERWATER PHOTOGRAPHY. [R.F.DI.]

Sculpin

Any of several species of fishes of the family Cottidae, characterized by enlarged, flattened heads, and large pectoral fins. They do not have scales, or the scales are reduced to vestiges. Most of the sculpins are marine, occurring primarily in the colder seas, and most abundantly in the North Pacific. There is one fresh-water genus, *Cottus*. The species of this genus prefer cool to cold water, and in the United States range southward in mountain streams of the Ozarks and the Carolinas. Their dark mottled color, expansive pectoral fins, big mouth, and enlarged head, set off by preopercular spines, give them a grotesque, gnomelike appearance. They are known by many names, miller's thumb, blob, and muddler being among the more common. They are primarily nocturnal and bottom-dwellers, hiding under rocks during the daytime. *See* PERCIFORMES. [J.D.B.]

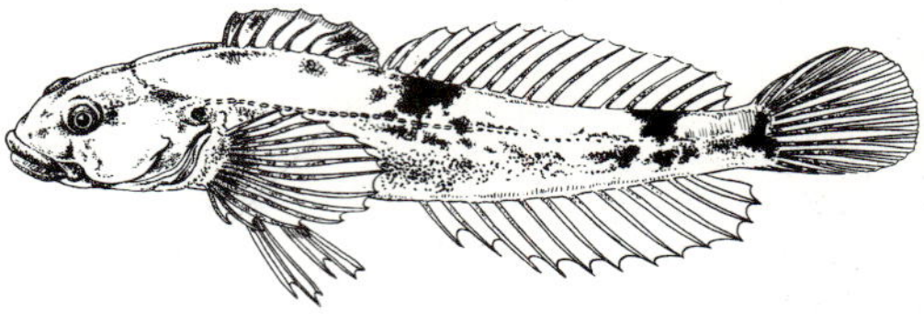

The sculpin, or muddler, *Cottus cognatus*; length to 6 in. (*From E. L. Palmer, Fieldbook of Natural History, McGraw-Hill, 1949*)

Scurvy

An acute or chronic disease due to vitamin-C (ascorbic acid) deficiency. Vitamin C aids in maintaining the proper metabolism of connective tissue, bone, and teeth, as well as the integrity of the walls of small blood vessels. *See* ASCORBIC ACID; SPECIALIZED TISSUE.

Infantile scurvy usually occurs between the sixth and twelfth months of life and is due to a lack of dietary vitamin, especially in artificially fed infants. The characteristic symptoms include irritability, failing appetite, and failure to gain weight. The child cries when moved or handled and may not move a limb. Bony malformations of the ribs and legs follow severe deficiency, often accompanied by a bleeding tendency. Hemorrhage, anemia, and infection are serious complications.

In adults scurvy is rarely due to dietary deficiency in normally fed populations, since 3–12 months of severe deficiency are required to produce symptoms. These include weakness, weight loss, irritability, vague pains, and a progressive bleeding tendency. Stress, infection, and hemorrhage may aggravate the situation. The disease is often related to chronic or severe gastrointestinal disease or to food idiosyncrasies. Any type of healing requires increased vitamin C supply, and infection similarly increases body demand.

Response to treatment is dramatic and effective if permanent bone and tissue damage has not prevailed. *See* VITAMIN. [E.G.ST.]

Scyphozoa

A class of the phylum Coelenterata, containing five living orders—Stauromedusae, Cubomedusae, Coronatae, Semaeostomeae, and Rhizostomeae—and a fossil order, Stromatoporoidea. They are all marine and usually take two forms, the polyp, or scyphopolyp, and the medusa, or scyphomedusa. However, some are polyplike and sessile throughout their lives, while others are always pelagic and lack the sessile polyp stage. Among the Coelenterata, the Scyphozoa are characterized by having well-developed medusae of large size and fairly well-organized polyps of small size. In relation to the other classes in the subphylum Cnidaria, the Scyphozoa stand higher than the Hydrozoa and are more adapted to pelagic life than the Anthozoa.

Morphology. Scyphopolyps are usually small in size. Sometimes they form colonies by asexual budding and strobilation. *Stephanoscyphus*, the polyp of *Nausithoë*, forms large colonies, sometimes attaining 100 mm in height. The scyphopolyp is composed of three parts: oral disk, stem, and pedal disks. The oral disk is crowned by a circlet of tentacles and has a four-sided mouth at its center. The axes running from the four corners of the mouth are called perradii. Halfway between the perradii are the interradii. At each interradius of the polyp there is an infundibulum, a canal which leads down into the stem. The mouth opens into the stomodeum which leads to the stomach. The stem, which gradually narrows, then widens into the pedal disk. A muscular strand runs along each of the four interradial ridges of the stem. The pedal disk has many gland cells that secrete a sticky fluid.

Scyphomedusae are generally large. Sometimes they attain a diameter of 1 m and often weigh 15 kg. The scyphomedusa is composed of an upper part, the umbrella, and a lower part, the oral arms. The convex upper surface of the umbrella is called the exumbrella; the concave inner surface, the subumbrella. The umbrella may have various shapes, being disk-, cone-, or domelike. The margin of the umbrella is divided into many lappets (Fig. 1) in all orders except the Stauromedusae and Cubomedusae. Sensory organs, usually eight in number, are found between the lappets. Tentacles vary from eight to several hundred with the exception of the Rhizostomeae, which have no tentacles. The sensory organs are composed of a statocyst or vesicle containing a round inorganic concretion, and an ocellus, an organ composed of a lens, retina cells, and pigment cells. The tentacles have numerous nematocysts or nettle cells, especially on the abaxial surface. The mouth, situated in the center of the subumbrella, is surrounded by four simple lips in some forms, but generally there are four well-developed oral arms. The oral arms are best developed and most complicated in structure in the Rhizostomeae, in which the central mouth is generally closed. Instead, many small suctorial mouths lead inward from the surface of the oral arms by way of canals. The mouth leads to the stomach, which is large and more or less cross-shaped. There are several rows of gastral filaments in each interradial section of the stomach. The gonads develop from endoderm cells just above the gastral filaments. Medusae are generally dioecious. Muscles are distributed mainly in the subumbrella and the tentacles. In the former, the most important muscles are the ring muscles, which serve to contract the umbrella for locomotion of the medusa; in the latter, there are only longitudinal muscles, which extend and contract the tentacle to catch prey.

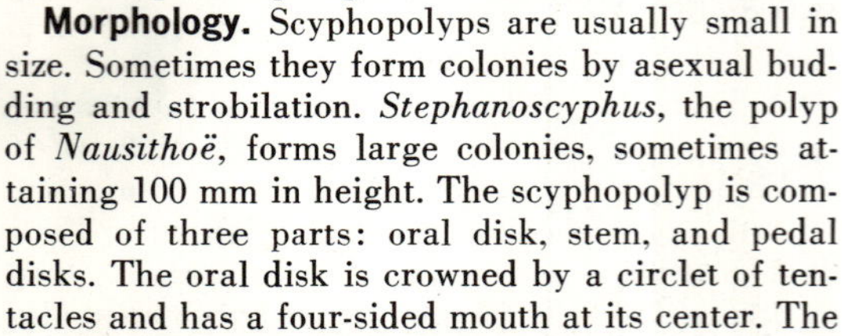

Fig. 1. Schema of scyphomedusa, oral view.

Embryology. The ripe ova, which are liberated from the broken endodermal wall of the ovary, enter the stomach cavity and then are discharged from the mouth into sea water. Fertilization generally occurs in sea water, but sometimes in the stomach cavity. Cleavage is mainly radial, and the endoderm is formed by invagination, or epiboly (*see* CLEAVAGE, EMBRYONIC). The egg develops into a ciliated planula that swims freely. The planula of the Stauromedusae, however, lacks cilia and creeps on the substratum. The planula attaches to seaweed or rocks and soon changes into a scyphopolyp, which grows gradually and increases its tentacles. The scyphopolyp generally produces more scyphopolyps by asexual budding. When fully grown, it metamorphoses into a strobila, which reproduces ephyrae asexually. The ephyra is a young medusa, and it undergoes further metamorphosis. Thus the life history of the scyphozoa shows alternation of generations (*see* METAGENESIS). There are exceptions however, and the scyphopolyps of the Stauromedusae metamorphose directly into combined forms of polyp and medusa, and the planula of the semaeostome genus, *Pelagia*, changes directly into an ephyra.

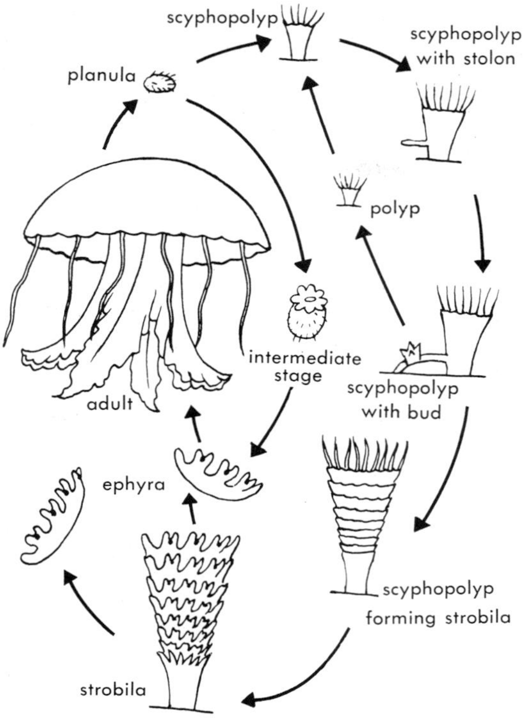

Fig. 2. Life cycle of a scyphomedusa, *Pelagia*.

Physiology. The speed of the contractive movements of the medusa is influenced by temperature. Though some medusae are almost indifferent to light, most are sensitive, especially Cubomedusae. Statocysts enable the medusae to maintain an upright position. Chemical senses enable them to catch food although they have no gustatory or olfactory structures. They are sensitive to several fatty acids and skatol but not to carbohydrates.

Ecology. The scyphomedusae are generally found near the coast. Exceptions are *Pelagia* and other pelagic forms, and the Coronatae, which are abyssal. Most of the Stauromedusae have a circumpolar distribution, while the Cubomedusae and Rhizostomeae are found mostly in warm and tropical seas. Scyphomedusae are carnivorous, except the Rhizostomeae which are plankton-eaters. Food of medusae is mainly fish and crustaceans, but some medusae protect young fish and crustaceans which seek shelter among their oral arms.

Some Semaeostomeae and Cubomedusae are injurious to man because of their nematocysts. On the other hand, some Rhizostomeae are used as food in the Orient. *See* COELENTERATA. [T.U.]

Sea

The term sea has several meanings: (1) the ocean; (2) a major subdivision of an ocean (*see* OCEANS AND SEAS); (3) a lake lacking an outlet to the ocean, therefore usually salty; and (4) ocean waves still under the influence of the wind that produced them, or a single such wave (*see* OCEAN WAVES; SEA STATE). [J.LY.]

Sea cucumber

Any member of the class Holothuroidea, phylum Echinodermata. These are common marine animals; about 500 species are widely distributed in a diversity of habitats.

Although in the same phylum with the starfish, these elongated, saclike, or in some species wormlike, animals bear little resemblance to the starfish. They lack spines and the skeleton is limited to small ossicles buried in the skin. They are usually leathery, but some have thin skins. Typically, the anterior mouth is ringed with tentacles; there is a long, coiled intestine, and a posterior anus. Breathing, in most forms, is by means of a pair of much-branched respiratory trees opening into the cloaca. There are rows of tube-feet along both dorsal and ventral surfaces; the ventral ones are used for locomotion, whereas those on the dorsal surface are used for tactile purposes and also respiration in some species. Those with thin walls lack the respiratory trees and breathe primarily through their skin.

Sea cucumbers are sluggish animals, either lying on or under rocks, or burrowing in the mud. Quite a number are U-shaped. These bury the body in the mud with only mouth and anus at the surface. Some species have the cloacal section extended and may burrow deeply in the mud with only the anal opening above the mud surface.

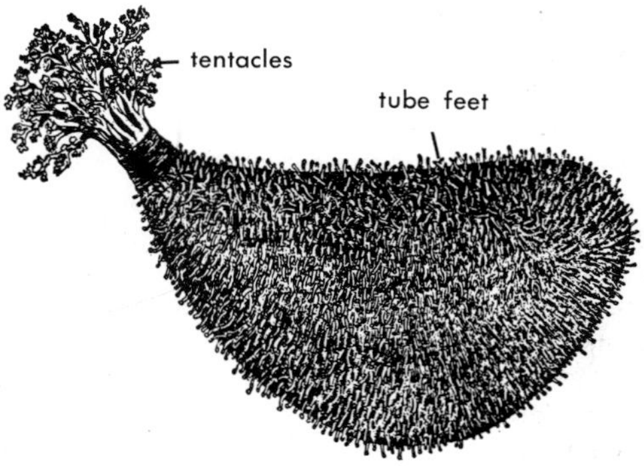

The sea cucumber, *Thyone briareus;* length to 4 in. (*After Coe, 1912, from T. I. Storer and R. L. Usinger, General Zoology, 3d ed., McGraw-Hill, 1957*)

The sexes are separate and most species extrude their eggs into the water to be fertilized. The fertilized egg gives rise, during development, to an auricularia larva. However, a few retain the eggs in or on their body wall.

Food is primarily such detritus or plankton as they can capture in the mud. Many sea cucumbers exhibit a remarkable reaction in which they cast out all or part of their internal organs, which they regenerate in 6–8 weeks. Others may fragment by breaking off sections of the posterior part of the body. In these the anterior portion will regenerate, whereas the posterior segment, or segments, will die. *See* ECHINODERMATA. [J.D.B.]

Sea fan

Any of several species of the order Gorgonacea of the phylum Coelenterata. This is the group known as horn corals.

Sea fans are of no use save for their ornamental appearance. They are highly variable in growth pattern and color. Typically they are flattened and fanlike, and wave gently in the water because of their somewhat flexible skeleton. In this entire order the typical stony skeleton of the corals is replaced by a hornlike substance called gorgonin, which is similar to true horn but lower in sulfur content. The living animals, or polyps, are arranged along the surface of this horny base. Structurally similar to the true corals, the polyps are different in that they have only eight tentacles, each pinnately subdivided, and have a gastrovascular cavity divided by eight septa. They occur mainly in shallow, warm seas. Often sea fans are associated with a variety of corals in the more colorful undersea gardens.

Sea fans are sessile, and only the immature planula larva is able to swim. Reproduction is sexual or asexual. The zygote develops into a planula, and this in turn develops directly into the hydroid form. Medusae are unknown.

Sea whips and sea feathers belong to the same group, differing primarily from the sea fans only in the skeletal pattern. These are also sessile forms.

Sea pens and sea pansies are closely related to sea fans, but are more fleshy and have a central skeletal arrangement more on the order of a central shaft than a fan. They belong to the order Pennatulacea. They can move from place to place, but prefer soft ocean bottoms. Some sea pens are short and stout, while others grow to be 6 ft long, with the polyps bilaterally arranged on either side of a slender stalk. Some sea pens move up and down on the central shaft with the flow of the tide, placing themselves in the best position to feed on passing organisms. Sea pens are frequently highly luminescent, more so than any of the other Coelenterata.

The sea fan, *Gorgonia flabellum;* height to 20 in. (*From P. Martin Duncan, ed., Cassell's Natural History, Cassell*)

The semiprecious red coral is a member of the order Gorgonacea, having a skeleton of red limestone rather than one of gorgonin. *See* COELENTERATA; CORAL; GORGONACEA. [J.D.B.]

Sea horse

Any of five or more species of small fishes of the genus *Hippocampus;* all are marine and are found throughout the warmer seas of the world. One species, *H. hudsonius*, occurs from Florida northward to Nova Scotia.

The sea horse, *Hippocampus hudsonius;* length to 6 in. (*From E. L. Palmer, Fieldbook of Natural History, McGraw-Hill, 1949*)

These strange little fishes, from 2 to 6 in. long, live hidden in seaweed or other vegetation, to which they cling with their unique prehensile tails. They swim in an upright position by undulations of the small dorsal fin. The tubular, suctorial mouth adds to their horselike appearance. They feed almost entirely upon crustaceans. The male has a brood pouch on the abdomen in which the eggs are placed by the female. The male carries the eggs until they hatch. *See* GASTEROSTEIFORMES. [J.D.B.]

Sea ice

Ice in the sea includes sea ice, river ice, and land ice. Land ice is principally icebergs which are prominent in some areas, such as the Ross Sea and Baffin Bay (*see* ICEBERG). River ice is carried into the sea during spring breakup and is important only near river mouths. The greatest part, probably 99% of ice in the sea, is formed by the freezing of sea water and is referred to as sea ice.

Properties of sea ice. The freezing point temperature and the temperature of maximum density of sea water vary with salinity (Fig. 1). When freezing occurs, small flat plates of pure ice freeze out of solution to form a network which entraps

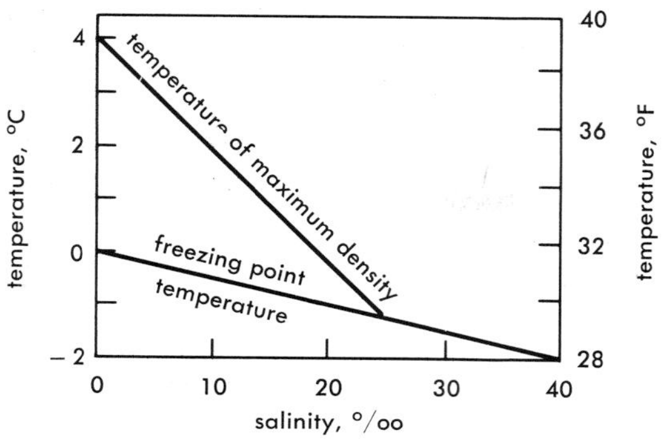

Fig. 1. Sea water. Graph showing change of freezing point and maximum density with varying salinity.

brine in layers of cells. As the temperature decreases more water freezes out of the brine cells, further concentrating the remaining brine so that the freezing point of the brine equals the temperature of the surrounding pure ice structure. At temperatures of −23°C (−9.4°F) or lower some salt in the brine crystallizes out.

The brine cells migrate and change size with change in temperature and pressure. The general downward migration of brine cells through the ice sheet leads to freshening of the top layers to near zero salinity by late summer. During winter, the top surface temperature closely follows the air temperature, whereas the temperature of the underside remains at freezing point corresponding to the salinity of water in contact. Heat flux up through the ice permits freezing at the underside. In summer, freezing can also take place under sea ice in regions where complete melting does not occur. Surface melt water (temperature 0°C) runs down through cracks in the ice to spread out underneath and contact the still cold ice masses and underlying colder sea water. Soft slush ice forms with large cells of entrapped sea water which then solidifies the following winter.

The salinity of recently formed sea ice depends on rate of freezing, thus sea ice formed at −10°C (14°F) has a salinity from 4 to 6 parts per thousand (ppt), whereas that formed at −40°C may have a salinity from 10 to 15 ppt. Sea ice is a

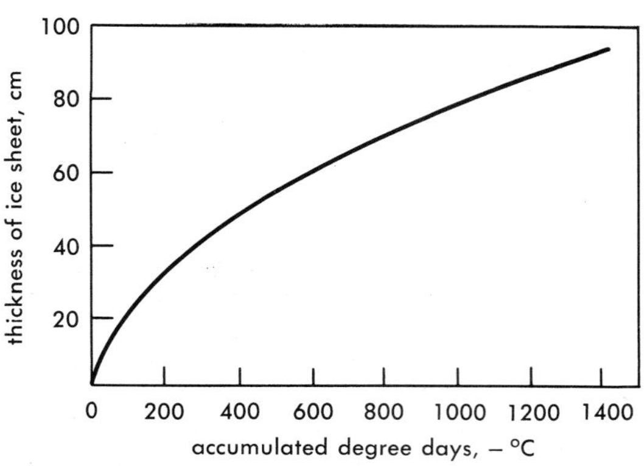

Fig. 2. Growth of undisturbed ice sheet.

poor conductor of heat and the rate of ice formation drops appreciably after 4–6 in. are formed. An undisturbed sheet grows in relation to accumulated degree-days of frost. Figure 2 shows an empirical relation between ice thickness and the sum of the mean diurnal negative air temperature (°C). The thermal conductivity varies greatly with air bubble content, perhaps between 1.5 and 5.0 × 10^{-3} g-cal/(cm)(sec)(°C).

The specific gravity of sea ice varies between 0.85 and 0.95 depending on the amount of entrapped air bubbles. The specific heat varies greatly because changing temperature involves freezing or melting of ice. Near 0°C, amounts that freeze or melt at slight change of temperature are large and "specific heat" is anomalous. At low temperatures, the value approaches that of pure ice; thus, specific heat for 4 ppt saline ice is 4.6 g-cal/(g)(°C) at −2°C and 0.6 at −14°C; for 8 ppt saline ice, 8.8 at −2°C and 0.6 at −14°C.

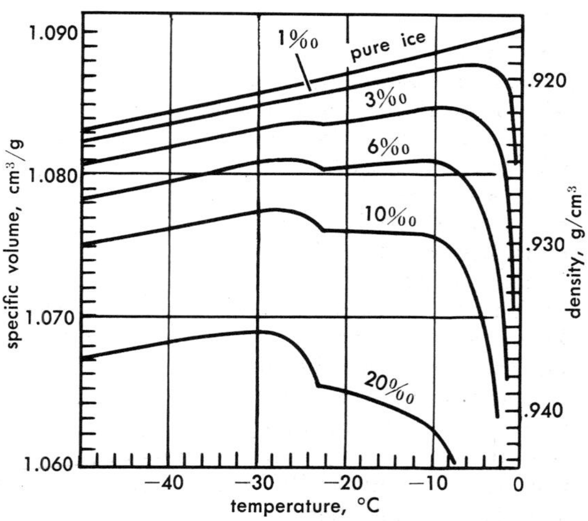

Fig. 3. Specific volume of sea ice for varying salinity and temperature, computed on basis of chemical model. (*By D. L. Anderson, based on data in Arctic Sea Ice, NAS-NRC Publ, 598, 1958*)

Sea ice of high salinity may expand when cooled because further freezing out occurs with attendant increase of specific volume, for example, ice of salinity 8 ppt at −2°C expands at a rate of about 93×10^{-4} per°C decrease in temperature, at −14°C expands 0.1×10^{-4}, but at −20°C contracts 0.4×10^{-4} per °C decrease. Change of specific volume with temperature and salinity is illustrated in Fig. 3.

Sea ice is viscoelastic. Its elasticity varies widely due to brine content, which is very sensitive to temperature and to air bubble content. Young's modulus measured by dynamic methods varies from 5.5×10^{10} dynes/cm² during autumn freezing to 7.3×10^{10} in winter to 3×10^{10} at spring breakup. Static tests give much smaller values, as low as 0.2×10^{10}. The flexural strength varies between 0.5 and 17.3 kg/cm² over salinity range of 7–16 ppt and temperatures −2 to −19°C.

Types and characteristics. The sea ice in any locality is commonly a mixture of recently formed ice, old ice which has survived one or more summers, and possibly old ridges of ice that formed

Fig. 4. Pancake ice with blocks of young ice. (*U.S. Navy Hydrographic Office*)

Fig. 5. Honeycombed structure of an overturned rotten block. (*U.S. Navy Hydrographic Office*)

Fig. 6. Hummocky floes that have weathered. (*U.S. Navy Hydrographic Office*)

Fig. 7. Unweathered pressure ridges. (*U.S. Navy Hydrographic Office*)

against a coast and contain beach material. The various descriptive forms are shown in Figs. 4–7. Except in sheltered bays, sea ice is continually in motion because of wind and current. It is constantly breaking up; floes are driven together, rafting or piling one on another to form pressure ridges which may reach a thickness of 90 ft in the open sea, or if pressed against a beach may reach a thickness of 120 ft or more. Under wind stress there are always leads, or lanes of open water, which soon close while others open somewhere else. [W.LN.]

Bibliography: *A Functional Glossary of Ice Terminology*, U.S. Navy H.O. 609, 1952; *Ice Atlas of the Northern Hemisphere*, U.S. Navy H.O. 550, 1946; *Oceanographic Atlas of the Polar Seas*, Part 1, U.S. Navy H.O. 705, 1958; National Research Council, *National Academy of Sciences Sea Ice Conference*, 1958; *Proceedings of the Conference on Arctic Sea Ice*, Natl. Acad. Sci.–Natl. Res. Council Publ. 598, 1958.

Sea level (datum planes)

Sea level is the elevation of the sea surface measured as the vertical distance between the surface and some fixed point on land—a rocky outcrop on a beach, a mountain peak, or a reference point installed by man. Mean sea level is the average elevation and is frequently used as a reference level in describing the elevation of points on land, or of depths in the sea.

Mean sea level. The surface of the sea is by no means a stationary spheroidal surface (*see* SEA LEVEL FLUCTUATIONS), so that the accurate determination of its average elevation requires a long series of observations. Usually a recording tide gage is installed, and the records are read each hour for several weeks, months, or years. The recommended length of record, to eliminate as nearly as possible all tidal constituents, is 19 years. However, shorter series are frequently used; the value so obtained can often be adjusted to a more representative value of mean sea level by comparison with the records at a nearby gage (where sea level is accurately known) for the same period of time. Such a mean value, computed for observations for a stated period in time, has been adopted for United States surveying and is called sea level datum. *See* GEODESY; SURVEYING.

If hourly values cannot be obtained, selected hourly values throughout each day may be used, with some loss of accuracy. On the other hand, if precise daily means are to be determined, the hourly values must be multiplied by weighting factors determined from the known periods of the tidal constituents.

Other datum planes. Other datum planes are used for special purposes. Half tide level is the average of all the highest and lowest readings during the period (called high and low waters). This was in early use before automatically recording gages were generally distributed, and at many localities only the high and low waters were recorded by an observer. Mean low water and mean high water are the averages of all of the low waters and high waters, respectively; half tide level lies midway between these two means.

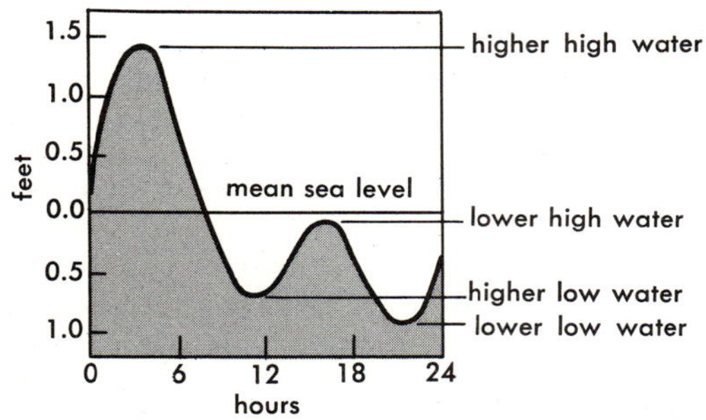

Diagram of tide curve showing diurnal inequality in heights of high waters and low waters.

When there is a large diurnal inequality in the tide (*see* TIDE), mean lower low water is frequently used as the datum on navigational charts. This is the average of only the lower of the two daily low waters, which makes it a safer reference to operators of vessels in tidal waters. Similarly, mean higher high water is the average of the higher of the two high waters each day and is a more realistic estimate of the daily excursion of water up a beach where the highs are quite different in elevation. Other datum planes with features useful for the particular area are in use. H. A. Marmer (1951) has given a detailed treatment of the definitions, determination, and usage of datum planes. [J.G.P.]

Bibliography: G. W. Groves, Numerical filters for discrimination against tidal periodicities, *Trans., Am. Geophys. Union,* 36(6):1073–1084, 1955; H. A. Marmer, *Tidal Datum Planes*, 5th ed., USCGS Spec. Publ. 135, 1951; J. R. Rossiter, Note on methods of determining monthly and annual values of mean water level, *Intern. Hydrograph. Rev.*, May, 1958.

Sea level fluctuations

The sea-level fluctuations first observed by man were surely waves; next he must have noted the rise and fall of the tides. But more concentrated attention (and carefully made records) reveal that the height of sea level is constantly changing and that its average value (mean sea level) will depend critically on the length of the series of observations and the period in time during which they are made. Furthermore, the sea surface is not even level. *See* SEA LEVEL (DATUM PLANES).

Fluctuations in space. Concrete evidence that the sea surface is not perpendicular to a plumb line has been found by precision leveling along the Atlantic Coast of the United States (Fig. 1).

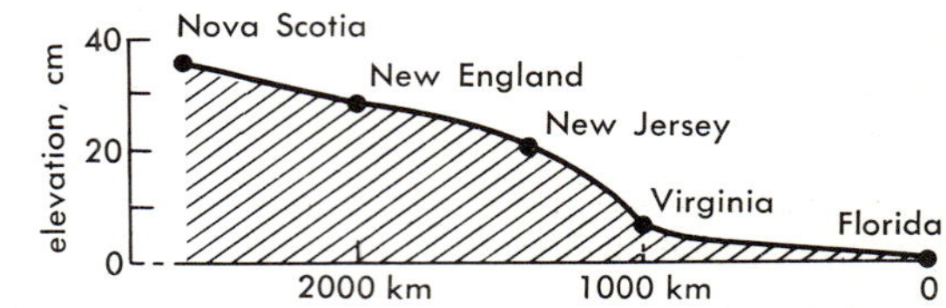

Fig. 1. Elevation of mean sea level from north to south along the eastern coast of North America. Heights refer to the elevation at Florida. (*Based on data from H. U. Sverdrup et al., The Oceans, Prentice-Hall, 1942*)

Effect of Coriolis acceleration. Sea level stands about 30 cm higher along the coast of Maine than it does off Florida, and the steepest slope occurs off Cape Hatteras, North Carolina. This occurs because the water in the sea is always in motion. The motion of any fluid over the surface of the earth is influenced by the rotation of the earth itself; fluids are deflected to the right (looking downstream) in the Northern Hemisphere, to the left in the Southern Hemisphere. This effect results from the Coriolis acceleration. *See* CORIOLIS ACCELERATION AND FORCE.

Offshore from the east coast of the United States the Gulf Stream flows northward, but inshore, on the continental shelf, flow is to the south. The water is then deflected to the west and piles up along the coast. More water piles up along the New England and Middle Atlantic States where southward flow is greatest; less, south of Cape Hatteras where the northward-flowing Gulf Stream lies closer to shore.

This effect is not confined to continental coasts. Each of the major oceans contains at least one current gyre where the water flows more or less continually around in a circle. In the Northern Hemisphere, if the flow is clockwise, water will collect in the center of the gyre until the deflecting force of the earth's rotation is just balanced by the gravitational force urging the water to run downhill out of the center of the gyre. On the other hand, if the flow is counterclockwise, there will be a lowering of sea level in the center of the gyre. *See* OCEAN CURRENTS.

In the North Atlantic, for example, sea level must stand higher near Bermuda, which lies near the center of a clockwise gyre, than it does southeast of Iceland, where the water is revolving in a counterclockwise flow. It is not known exactly how large this difference in elevation is, because the precision leveling network has not yet been extended out over the open ocean. From oceanographic measurements the difference is estimated as between 1 and 2 m.

If the Coriolis effect were the only important factor, then the sea surface could be visualized as a nearly smooth surface approximately perpendicular everywhere to a plumb line, but with rises of a meter or less near Bermuda, off Japan, and also in middle latitudes of the Southern Hemisphere. There would be dips in the elevation of the sea surface off Iceland, in the Gulf of Alaska, and near the Antarctic Continent, which is girdled by a great clockwise flow.

Effect of atmospheric pressure. Sea level also deviates from the mean, however, under the influence of differential atmospheric pressure. If pressure is high over one area of the ocean and low somewhere else, the water will tend to flow toward the low-pressure area. Thus the sea surface behaves as an inverted barometer; it stands high where air pressure is low, and low where air pressure is high. *See* AIR PRESSURE; ATMOSPHERIC HIGH; ATMOSPHERIC LOW.

A difference in air pressure of 1 millibar corresponds very nearly to a difference in sea level of 1 cm. Owing to the average distribution of air pressure, this effect tends to reduce the current-induced differences in elevation just considered. That is, off Iceland the air pressure is low. Near Bermuda and the Azores, air pressure is high. The average difference is about 20 millibars, so that if this effect alone were acting on the sea its surface level would be 20 cm higher off Iceland than near Bermuda and the Azores. The actual difference in level must show the combined effects of the forces related to the relative motions of the earth and the ocean waters (Coriolis effect) and the force resulting from differences in atmospheric pressure from place to place.

Barriers and density differences. A third effect can be observed where parts of the sea surface are separated from each other by a land barrier. Precise leveling across the continent of North America shows that sea level is about 50 cm higher on the Pacific than on the Atlantic side of the continent. Currents and winds probably cause part of this difference in level but another important factor is the difference in the density of the sea water.

Water, including sea water, expands and contracts with changing temperature. In the sea the effect is complicated by the presence of the salt, since a mass of water also takes up more or less room as the relative salt content is changed. Furthermore, the two effects are not completely independent, but as treated here they may be considered so. Thus, when sea water is cold and salty it will take up less space per unit mass and hence stand at a lower level than when it is warm and fresh. *See* SEA WATER.

Let it be assumed that somewhere very deep in the oceans all isobaric surfaces (surfaces of constant pressure) are level. This is equivalent to saying that any two water columns that are of the same area and contain the same mass of water will have their bases on a common level surface. This is very nearly true regardless of where the columns are or what the temperature and salinity distributions are. However, if the water in one column is less dense than in the other, its surface will stand higher.

Fluctuations in time. The height of sea level at a given point changes from second to second, hour to hour, even century to century. The slowest regular changes in sea level that are large enough to be readily observed by eye have periods of approximately 12 and 24 hours. *See* TIDE.

Short-period variations. Less regular than tides are fluctuations where the period is several days to a week or two. These are caused by the moving high- and low-pressure systems in the atmosphere; they are another example of the sea in its role as an inverted barometer. As a low-pressure area moves in over the coast, sea level rises; as the low pressure moves inland and a high pressure overlies the coast, sea level falls again. The area of sea surface that rises and falls is controlled by the size of the meteorological disturbance, and has a diameter, usually, of a few hundred kilometers. The height of the variation in level is dependent on the pressure difference in the atmosphere and may be 10 or 20 cm—sometimes greater.

Occasionally very large and destructive rises in sea level that last for one or several days occur in connection with severe atmospheric storms. It has been estimated that more than three-fourths of the loss of life that has resulted in hurricanes has been caused by the inundations from storm waves, rather than by the direct effects of the high winds. These storm waves are partly induced by the low central pressures in the eye of the storm and are partly the result of water being driven onshore (especially over shoaling bottom areas) by the winds. *See* STORM SURGE.

Annual variations. A more strictly periodic variation, although not nearly so striking, is the fluctuation which has an annual period. Sea level is highest in autumn, lowest in spring, over all middle latitude areas of the sea. This means, of course, that when sea level is high along the coast of the United States, Europe, and Japan (northern autumn, southern spring), it is low along South America, South Africa, and Australia.

The average annual change in level is 20 cm although it varies from almost zero at some equatorial islands to more than 1 m along the north shore of the Bay of Bengal. The oscillation seems to be smaller at mid-ocean near offshore islands (in the open ocean) than at continental coasts, but it has the same phase in either case. That is, in September, sea level is high all across the Atlantic Ocean from the United States to Europe, between latitudes 20–40°N. At the same time it is high in the same general band of latitude across the North Pacific, but low in the middle latitudes of the Southern Hemisphere.

North of 40°N a similar oscillation of about equal amplitude occurs, but not at the same time. That is, the maximum heights are observed in winter and the lowest levels in summer. Data in the Southern Hemisphere are inadequate for determining whether a similar change of time of maximum occurs at high southern latitudes. The variations in the two zones of the Northern Hemisphere are illustrated in Fig. 2.

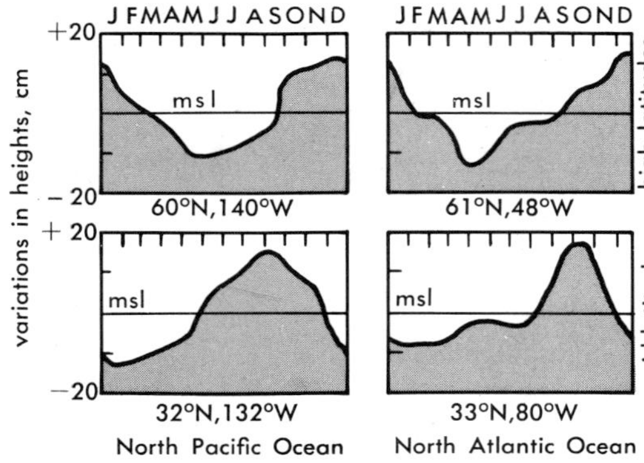

Fig. 2. Seasonal variations in sea level.

There are as yet insufficient data on ocean currents to compute the annual variation in the effect of Coriolis acceleration except at one or two locations. Data from the United States east coast and Bermuda indicate the changing speed of the Gulf Stream is probably closely related to differences in the slope between these two points, but this effect cannot explain why the level should rise at the coast and at Bermuda at the same time.

In most locations annual variations in atmospheric pressure are not large enough to account for the annual change in sea level as an inverted barometer effect. (One must be careful to exclude from this computation annual changes in the total air mass over all the oceans of the world. Sea water is almost incompressible, and only relative changes from one part of the sea to another can cause appreciable changes in sea surface elevation. The known migration of air mass to continents during winter and back over the sea during summer must be excluded.) Even near Japan and Iceland, where pressure changes throughout the year are large, only about one-third of the sea level change can be attributed to the effects of atmospheric pressure. In many parts of the world, indeed, the air pressure changes should cause the sea surface to move up when it is moving down, and vice versa.

In middle latitudes the density of sea water changes measurably throughout the year. At the end of summer the upper 400 m of the sea water have become several degrees warmer than they are at the end of winter. This warmer water stands some 10–20 cm higher than the same mass of cool water does in winter. This effect accounts for about two-thirds of the measured rise in sea level in middle latitudes.

In the more northerly parts of the northern oceans (where the maximum height occurs 3 months later than in middle latitudes), the seasonal changes in ocean temperature are neither large enough nor at the right time to explain the observed variations.

Seasonal changes in salinity usually have a much smaller effect than the changes in temperature. One area where this statement does not hold is in the Bay of Bengal, where the largest seasonal changes in sea level are observed. Here the summer monsoon is accompanied by very heavy rainfall and onshore winds. These effects combine to form a lens of low-salinity water near the coast in summer, which is replaced by cool, high-salinity ocean water in winter. Most of the observed changes in height are due to the low salinity of this freshwater lens, although its high temperature and the onshore winds that prevail at that season are probably also factors.

A fourth possibility must be considered. In discussing salinity above it was tacitly assumed that the total mass of water plus salt in the oceans remains constant throughout the year. Certainly if a particular area is considered this may not be true. After all, changes in salinity are largely brought about by seasonal changes in evaporation or precipitation, such as the removal or addition of fresh water near the surface; this may not be completely compensated by internal flow.

Furthermore, there is some slight evidence that even the total mass of fresh water in the oceans is not constant throughout the year. Both oceanographic evidence and hydrological estimates suggest that there is less water in the oceans at the end of Northern Hemisphere winter in March.

The data indicate that this water is held on the large continents in the form of snow and ground water, and gradually returned to the sea during summer months. A similar effect must occur in the Southern Hemisphere, but the land masses in cool regions of the Southern Hemisphere are not nearly so large and therefore not so effective in storing moisture during the Southern Hemisphere winter season.

The amount of water involved is very great, but spread over all the oceans of the world it would change the average height of sea level by only 1 or 2 cm. Because this change in level is so small, measurement from the sea-level data cannot be considered reliable, but the oceanographic and hydrological estimates agree in both magnitude and phase.

The problem of the causes of the annual variation in sea level cannot be considered solved. As a case in point, if sea level is high in the Northern

Hemisphere while it is low in the Southern, what is the force acting to maintain this slope?

Long-period variations. Still slower changes in sea level have been observed, although these become more difficult to discuss quantitatively because continuous observations are not yet available over all oceans.

Long records from recording gages exist only in Europe, the United States, Japan, and Indonesia. These show that sea level may rise more or less continuously for several years, rising perhaps as much as 1 m, and then suddenly or slowly subside again. Another form of evidence, ancient beach lines, suggests that in earlier centuries sea level was several, perhaps tens of meters, higher than it is today. The causes for these changes cannot be explained until there is more complete information on their geographical extent, when they took place, and how long the changes persisted. For example, slow changes in sea level elevation may have been caused, and may again be caused, by the melting or refreezing of the gigantic Antarctic and Greenland icecaps, which contain enough water to raise the sea level by several tens of meters. *See* COASTAL LANDFORMS; WARPING, EARTH CRUST; *see also* GLACIAL EPOCH; GLACIER; HYDROLOGY.

[J.G.P.]

Sea lion

Either of two large seals belonging to the family Otariidae, the eared seals, both living in the Pacific. The California sea lion, *Zalophus californianus,* occurs off the coast of southern California, as well as in New Zealand, Australia, and Japan. It frequents rocky shores and sandy beaches, where its honking bark is repeated almost continuously. Males are about 8 ft long, the females somewhat smaller. The fur is brown.

The northern, or Steller, sea lion, *Eumetopias jubata,* is yellowish-brown and somewhat larger, the males reaching a length of 10 ft. It does not bark except when disturbed. The northern sea lion ranges from central California and Japan northward to the Bering Strait. Although the fur of sea lions is of little value, they are sometimes hunted for fertilizer and dog food. *See* CARNIVORA; SEAL (ZOOLOGY).

The Steller sea lion, *Eumetopias jubata;* length to 10 ft. (From E. L. Palmer, *Fieldbook of Natural History,* McGraw-Hill, 1949)

[J.D.B.]

Sea squirt

Any member of the class Ascidiacea, subphylum Tunicata, phylum Chordata.

The sea squirts, or ascidians, are a relatively unimportant group of marine animals, but are of considerable interest to zoologists. The young are of some value as fish food, and in a few places the larger species are sometimes used as food for man.

The sea squirt, *Botryllus* sp.; diameter 1 in. (From P. Martin Duncan, ed., *Cassell's Natural History,* Cassell)

Sea squirts are hermaphroditic, but the eggs must be fertilized by sperm from another individual. The egg develops into a small, transparent larva, strikingly similar to a small tadpole in appearance and exhibiting at this stage all the fundamental characteristics of the Chordata. After 4 days of free-swimming life, the larva settles to the bottom, attaches to the substrate, and undergoes a retrograde metamorphosis which results in a sessile animal, without a notochord, and with only a vestige of a central nervous system. The adult is virtually reduced to an enlarged pharynx, with many gill slits. It is supplied with water from an incurrent siphon connected to a short, simple intestine which opens, along with the atrial cavity, into an excurrent siphon. A tough test, or tunic, encloses the entire animal.

Sea squirts live in all the seas of the world at various depths, tending to occupy deeper water in the polar limits and shallow water in the tropics. There are both solitary and colonial species of varying size. *See* CHORDATA; TUNICATA. [J.D.B.]

Sea state

The description of the ocean surface or state of the sea surface with regard to wave action. Wind waves in the sea are of two types: those still growing under the force of the wind are called sea; and those no longer under the influence of the wind that produced them are called swell. Differences between the two types are important in forecasting ocean wave conditions. Properties of sea and swell and their influence upon sea state are described in this article. For a discussion of wave characteristics and mechanics of wave motion *see* WAVE MOTION IN LIQUIDS.

Sea. Those waves which are still growing under the force of the wind have irregular, chaotic, and unpredictable forms (Fig. 1*a*). The unconnected wave crests are only two to three times as long as the distance between crests and commonly appear to be traveling in different directions, varying as much as 20° from the dominant direction. As the waves grow, they form regular series of connected troughs and crests with wave lengths commonly ranging from 12 to 35 times the wave heights. Heights rarely exceed 55 ft. The appearance of the sea surface is described as state of the sea (see Table 1).

The height of a sea is dependent on the strength of the wind, the duration of time the wind has blown, and the fetch or distance of sea surface over which the wind has blown. *See* OCEAN WAVES.

Swell. As sea waves move out of the generating area into a region of weaker winds, a calm, or opposing winds, their height decreases as they advance, their crests become rounded, and their surface is smoothed (Fig. 1*b*). These waves are more regular and more predictable than sea waves and, in a series, tend to show the same form or the same trend in characteristics. Wave lengths generally range from 35 to 200 times the wave heights.

The presence of swell indicates that recently there may have been a strong wind, or even a severe storm, hundreds or thousands of miles away. Along the coast of southern California long-period waves are believed to have traveled distances greater than 5000 miles from generating areas in the South Pacific. Swell can usually be felt by the roll of a ship and, under certain conditions, extremely long and high swells may cause a ship to take solid water over its bow regularly in a glassy

Table 1. Sea height code*

Code	Height, ft	Description of sea surface
0	0	Calm, with mirror-smooth surface
1	1–	Smooth, small wavelets or ripples with appearance of scales but without crests
2	1–3	Slight, short pronounced waves or small rollers; crests have glassy appearance
3	3–5	Moderate, waves or large rollers; scattered whitecaps on wave crests
4	5–8	Rough, waves with frequent whitecaps; chance of some spray
5	8–12	Very rough, waves tend to heap up; continuous whitecapping; foam from whitecaps occasionally blown along by wind
6	12–20	High, waves show visible increase in height, with extensive whitecaps from which foam is blown in dense streaks
7	20–40	Very high, waves heaping up with long frothy crests that are breaking continuously; amount of foam being blown from the crests causes sea surface to take on white appearance and may affect visibility
8	40+	Mountainous, waves so high that ships close by are lost from view in the wave troughs for a time; wind carries off crests of all waves, and sea is entirely covered with dense streaks of foam; air so filled with foam and spray as to affect visibility seriously
9		Confused, waves cross each other from many and unpredictable directions, developing complicated interference pattern that is difficult to describe; applicable to conditions 5–8

* Modified from *Instruction Manual for Oceanographic Observations*, 2d ed., H.O. Publ. 607, 1955.

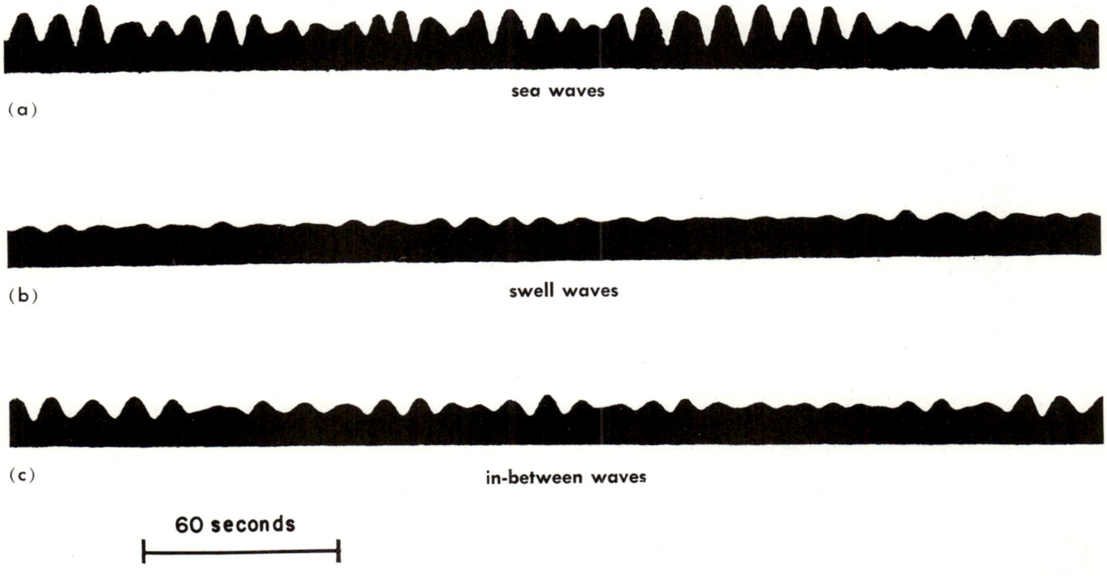

Fig. 1. (a-c) Records of surface waves. (*Adapted from W. J. Pierson, Jr., et al., Observing and Forecasting Ocean Waves, H.O. Publ. 603, U.S. Navy Hydrographic Office, 1955*)

Table 2. Swell-condition code*

Code	Description	Height, ft	Length, ft
0	No swell	0	0
	Low swell	1–6	
1	Short or average		0–600
2	Long		600+
	Moderate swell	6–12	
3	Short		0–300
4	Average		300–600
5	Long		600+
	High swell	12+	
6	Short		0–300
7	Average		300–600
8	Long		600+
9	Confused		

* *Instruction Manual for Oceanographic Observations*, 2d ed., H.O. Publ. 607, 1955.

sea. A descriptive classification of swell waves is given in Table 2. When swell is obscured by sea waves, or the components are so poorly defined that it is impossible to separate them, it is reported as confused.

In-between state. Often both sea waves and swell waves, or two or more systems of swell, are present in the same area (Fig. 1*c*). When waves of one system are superimposed upon those of another, crests may coincide with crests and accentuate wave height, or troughs may coincide with crests and cancel each other to produce flat zones (Fig. 2). This phenomenon is known as wave interference, and the wave forms produced are extremely irregular. Where wave systems cross each other at a considerable angle, the apparently unrelated peaks and hollows are known as a cross sea.

Breaking waves. The action of strong winds (greater than 12 knots) sometimes causes waves in deeper water to steepen too rapidly. As the height-length ratio becomes too large, the water at the crest moves faster than the crest itself and topples forward to form whitecaps.

Breakers. As waves travel over a gradually shoaling bottom, the motion of the water is restricted and the wave train is telescoped together. The wave length decreases and the height first decreases slightly until the water depth is about one-sixth the deep-water wave length and then rapidly increases until the crest curves over and plunges to the water surface below. Swell coming into a beach usually increases in height before breaking, but wind waves are often so steep that there is little if any increase in height before breaking. For this reason, swell that is obscured by wind waves in deeper water often defines the period of the breakers.

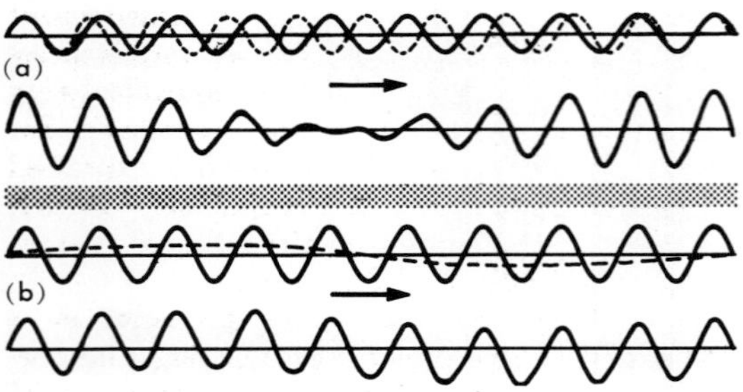

Fig. 2. Wave patterns resulting from interference. (*a*) Interference of waves of equal height and nearly equal length, forming wave groups. (*b*) Interference between short wind waves and long swell. (*From Techniques for Forecasting Wind Waves and Swell, H.O. Publ. 604, U.S. Navy Hydrographic Office, 1951*)

Surf. The zone of breakers, or surf, includes the region of white water between the outermost breaker and the waterline on the beach. If the sea is rough, it may be impossible to differentiate between the surf inshore and the whitecaps in deep water just beyond. *See* SHORE PROCESSES. [N.A.B.]

Bibliography: W. Bascom, Ocean waves, *Sci. American*, 201(2):75–84, 1959; G. E. R. Deacon, Ocean waves, *Endeavour*, 17(67):134–139, 1958.

Sea urchin

A marine invertebrate belonging to the class Echinoidea, phylum Echinodermata. This class includes the sea urchins, heart urchins, and sand dollars. There are about 860 living species.

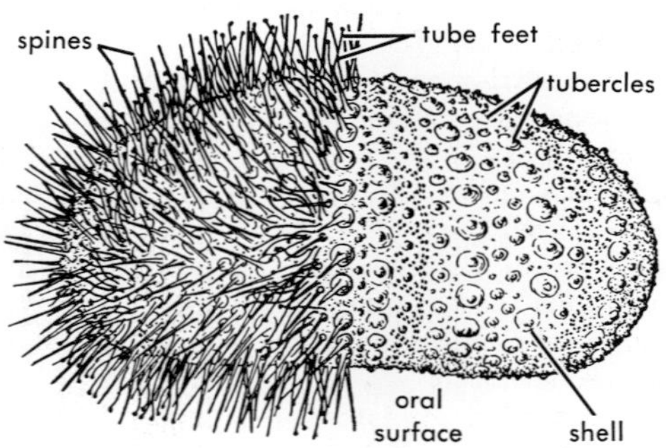

The sea urchin. (*From T. I. Storer and R. L. Usinger, General Zoology, 3d ed., McGraw-Hill, 1957*)

They are of no great economic importance, seldom being utilized for food by other animals, and of no direct use to man. Their eggs and young, however, are frequently used in experimental embryology studies.

The basic anatomy of the sea urchin is similar to that of the starfish, but the arms are lacking and the rounded body is covered by many sharp, movable spines. Some sea urchins have spines which are attached to poison glands and may inflict a painful wound if stepped on. The fundamental five-sided pattern of the phylum is shown in the ambulacra, which are equipped with a double row of tube feet, corresponding to the arms of the starfish. Pedicillariae are scattered over the body and keep it free of foreign matter. The shell is made up of 10 double rows of plates. A five-sided skeletal framework, each side bearing a strong tooth, makes up the food-securing and crushing apparatus surrounding the mouth. It is the skeleton of this mouth apparatus that is often found on the beach and is known as "Aristotle's lantern." Some orders lack the lantern.

Sea urchins are bottom-dwelling animals, living in various habitats from low tide level to a depth of 5000 meters. They feed on organic matter in the mud, dead animal remains, small living organisms, and seaweed. Their reproduction is similar to that of the starfish, and they have the same capacity to regenerate damaged parts. *See* ECHINODERMATA. [J.D.B.]

Sea wall

A structure at the water's edge to resist encroachment of the sea and to retain the natural soil or deposited fill behind the wall. A usual type of sea wall has a sloped or inclined face toward the sea to deflect and dissipate the wave forces. Sometimes it resembles a breakwater. Timber, steel, and concrete are used in a variety of forms. Failures sometimes occur because of scouring and eventual undermining at the outshore face or because of wave overtopping during highest tides. *See* COASTAL ENGINEERING. [E.J.Q.]

Sea water

Water is most often found in nature as sea water (≅98%). The rest is found as ice, water vapor, and as fresh water. Sea water is an aqueous solution of salts of a rather constant composition of elements. Its presence determines our climate and makes life possible on earth. The boundaries of sea water are the boundaries of the oceans, the mediterranean seas, and their embayments. The physical, chemical, biological, and geological events in the hydroplane within these boundaries are the studies which are grouped together and called oceanography. The basic properties of sea water, the distribution of these properties, the interchange of properties between sea and atmosphere or land, the transmission of energy within the sea, and the geochemical laws governing the composition of sea water and sediments are the fundamentals of oceanography.

The discussion of sea water which follows is divided into six sections: (1) physical properties of sea water; (2) interchange of properties between sea and atmosphere; (3) transmission of energy within the sea; (4) composition of sea water; (5) distribution of properties; and (6) sampling and measuring techniques. For further treatment of related aspects of physical character, composition, and constituents *see* HYDROSPHERE, GEOCHEMISTRY OF; MARINE RESOURCES; SEA WATER FERTILITY; UNDERWATER SOUND.

PHYSICAL PROPERTIES OF SEA WATER

Sea water is a rather concentrated solution of salts in water, the most variable part of which is the water. The basic properties of water, such as density, specific volume, compressibility, electric conductivity, sound velocity, viscosity, surface tension, and others, depend upon temperature, salinity, and pressure. The basic properties of pure water are rather abnormal compared with other fluids because of the high dipole moment of the water molecule (*see* WATER). The properties of water are remarkably changed by the salt content. The colligative properties—osmotic pressure, lowering of freezing point, increase in boiling point, and lowering of vapor pressure—depend only on the number of molecules in solution. As the composition of the salts of sea water is nearly constant, the colligative properties depend only on the ratio of dissolved salts to solvent (pure water).

Basic properties. Some of the more important physical properties of sea water and their relation to temperature, salinity, pressure, and other variables are discussed.

Density and specific volume. In oceanographic practice, the specific gravity of sea water (ρ, g/ml) is reported, although it is conventionally described as density. The density depends upon the pressure, temperature, and salinity of the specimen. Pressure is conventionally specified as sea pressure p in oceanographic work; that is, the physical pressure is $p + 1$ atm when the sea pressure is p. Temperature T is given in degrees Celsius and salinity S in parts per thousand by weight (‰). Instead of density in g/ml, the grams excess over 1 kg per liter

$$\sigma = 1000(\rho - 1)$$

is usually reported; the specific volume (ml/g)

$$\alpha = 1/\rho$$

may also be used. Its dependence on S, T, and p is indicated by subscripts, and the specific volume anomaly

$$\delta = \alpha_{S,T,p} - \alpha_{35,0,p}$$

is often tabulated. To a number of significant figures that is entirely adequate for all purposes of oceanography, the specific volume anomaly is identical in either cm^3/g or ml/g, although σ values in g/ml are about 0.03 units higher than those in g/cm^3.

In modern oceanographic practice, the primary observations are S, T, and depth d, from which σ (or δ) is calculated using standard tables and empirical formulas. The salinity is determined to four significant figures (for example, 34.82‰) either by Mohr titration of the halides or by measurement of the electrical conductivity; temperature is measured to 0.01°C, and depth to 1 m (equivalent very nearly to pressure of 0.1 atm or 1 decibar). For computation of the quantity σ, which is expressed to four significant figures (for example $\sigma = 29.15$), the depth d is taken as numerically equal to pressure p in decibars. *See* TITRATION.

This conventional four-figure precision in σ should not be interpreted as indicating that specific gravity values for sea water are known to an accuracy of one part in 10^5 at all depths. Thus, it is very doubtful that the density of sea water at elevated pressures has ever been determined in the laboratory to a precision better than $\Delta\sigma = \pm 0.20$. At atmospheric pressure, the corresponding probable error may be as small as $\Delta\sigma = \pm 0.01$, but the probable error increases rapidly with pressure.

There is some uncertainty as to the need for and justification of the conventional precision of calculation. The geostrophic current theory is often cited as the source of the requirement; however, the known errors inherent in this theory are large, and no final conclusion on the question of need is yet possible. The known errors in the tables and formulae are so great that convention is the only possible justification for the accepted calculation.

Errors also enter these calculations because the chemical composition of sea water is not uniquely determined by salinity (or chlorinity); these errors are all small compared to those mentioned above.

Compressibility: Thermal-saline factors. It has been suggested that the observational data be reduced with calculations based on compressibility and coefficient of thermal expansion. According to the geostrophic theory, the surface currents depend primarily on thermal expansion ($\partial\alpha/\partial T$) and saline contraction ($-\partial\alpha/\partial S$), and are independent of the isothermal compressibility ($-\partial\alpha/\partial p$). Unfortunately, only compressibility has been measured at elevated pressures. The other two coefficients have been measured only at atmospheric pressure and are calculated at elevated pressures from the compressibility measurements. This calculation is subject to large cumulative error, and at $p = 1000$ atm. $\partial\alpha/\partial T$ and $\partial\alpha/\partial p$ are known to little more than order of magnitude.

These two coefficients also determine the stability of the stratification of the oceans. There is an undoubted need for more precise measurements of these quantities at elevated pressures.

Velocity of sound. Because of its importance in echo sounding and echo ranging, the velocity of sound in water (meters per second) has been the subject of numerous investigations. Calculations based on the isothermal compressibility claim an accuracy of about 1.0 m/sec; these have led to various tables, such as that of S. Kuwahara (1939). Laboratory measurements can be made to an apparent accuracy of less than 0.1 m/sec, but as yet, different experimenters obtain values that differ by more than 0.5 m/sec. There is general agreement that Kuwahara's values are systematically low by several meters per second.

Specific heat. The specific heat (constant pressure) of sea water has been measured at atmospheric pressure and 17.5°C, as a function of salinity. The dependence on pressure can be calculated from a thermodynamic formula that involves the thermal expansion coefficient; since present knowledge of this coefficient is scant, the calculation cannot be made reliably at present.

Electrical conductivity. Of all physical properties of sea water, it seems that electrical conductivity can most easily be measured precisely (at atmospheric pressure). For this reason, there is a growing tendency to use conductivity rather than chlorinity to characterize small differences between various samples of sea water. An electrical conductivity salinometer has been designed to measure conductivity and the conductivity method of determining salinity is gaining general acceptance in oceanographic work. [C.E.]

Colligative properties. The colligative property of sea water most readily determined is freezing-point (or melting-point) lowering at 1 atm pressure (Δt). Although it is a rather complicated function of chlorinity or salinity, it is readily obtainable from expressions of the form

$$\Delta t = -kZ$$

where Z is grams of salt per kilogram of solvent water.

The linearity has been shown to hold down to −8°C. The quantity Z can be obtained from chlorinity by

$$Z = \frac{73 + 1811\text{Cl}‰}{999.927 - 1.8110\text{Cl}‰}$$

Three values of k have been reported. J. Lyman and R. H. Fleming (1940) found 0.05241 from measurements made by Hansen (in M. Knudsen, 1903). Y. Miyake (1939) obtained results that yield $k = 0.05474$. A. Assur (1958) calculated the value 0.05411 from measurements by Nelson and T. G. Thompson. It is not clear whether these differences result from variations in experimental technique or represent real differences between sea water from different areas.

Vapor pressure lowering, Δe, was computed by R. Witting (1908), from theoretical considerations and Hansen's freezing points, to be

$$\Delta e/e = 0.000537S‰$$

where e is the vapor pressure of pure water at the temperature under consideration. This expression is equivalent to

$$\Delta e/e = 0.00097\text{Cl}‰$$

However, direct measurements at 25°C by R. A. Robinson (1954) are better satisfied by

$$\Delta e/e = 0.0009206\text{Cl} + .00000236\text{Cl}^2$$

or

$$\Delta e/e = 0.0005072Z + .00000019Z^2$$

From the measurements at higher temperature by A. B. Arons and C. F. Kientzler (1954), it can be computed that the boiling point of sea water at 1 atm pressure is raised about 0.018°C for each unit of chlorinity.

Osmotic pressure at 0°C is readily obtainable from freezing point data by

$$\pi_0 = -12.08\,\Delta t$$

and at any other temperature from

$$\pi_t = \pi_0 \frac{(273 + t)}{273}$$

or it can be obtained from vapor pressure data by

$$\pi_t = -\frac{RT}{V} \ln \frac{e - \Delta e}{e}$$

where R is the gas constant, T absolute temperature, and V the partial molal volume of water, which can be taken as equal to the molar volume of pure water at the temperature in question.

The freezing point of sea water of salinity 35‰ is −1.909°C (Hansen). Its osmolality is 1.909 ÷ 1.86 or 1.026. The gram-ionic concentration at salinity 35‰ is 1.154; thus at this salinity the average activity coefficient of the dissolved salts is 1.026 ÷ 1.154 or 0.89. *See* SEA ICE. [J.LY.]

Bibliography: A. B. Arons and C. F. Kientzler, Vapor pressure of seasalt solutions, *Trans. Am. Geophys. Union,* 355:722–728, 1954; A. Assur, Composition of sea ice and its tensile strength, in *Arctic Sea Ice,* NAS–NRC Publ. 598:106–138, 1958; C. Eckart, The equation of state of water and sea water at low temperatures and pressures, *Am. J. Sci.,* 256:225–240, 1958; J. Lyman and R. H. Fleming, Composition of sea water, *J. Marine Research,* 3:134–146, 1940; *Physical and Chemical Properties of Sea Water,* NAS-NRC Publ. 600, vol. 9, 1959; R. A. Robinson, The vapour pressure and osmotic equivalence of sea water, *J. Mar. Biol. Assoc. United Kingdom,* 33:449–455, 1954; H. U. Sverdrup, M. W. Johnson, and R. H. Fleming, *The Oceans,* 1942.

INTERCHANGE BETWEEN SEA AND ATMOSPHERE

The sea and the atmosphere are fluids in contact with one another, but in different energy states—the liquid and the gaseous. The free surface boundary between them inhibits, but by no means totally prevents, exchange of mass and energy between the two. Almost all interchanges across this boundary occur most effectively when turbulent conditions prevail: a roughened sea surface, large differences in properties between the water and the air, or an unstable air column that facilitates the transport of air volumes from sea surface to high in the atmosphere.

Heat and water vapor. Both heat and water (vapor) tend to migrate across the boundary in the direction from sea to air. Heat is exchanged by three processes: radiation, conduction, and evaporation. The largest net exchange is through evaporation, the process of transferring water from sea to air by vaporization of the water.

Evaporation depends on the difference between the partial pressure of water vapor in the air and the vapor pressure of sea water. Vapor pressure increases with temperature, and partial pressure increases with both temperature and humidity; therefore the difference will be greatest when the sea (always saturated) is warm and the air is cool and dry. In winter, off east coasts of continents, this condition is most ideally met, and very large quantities of water are absorbed by the air. On the average, 100 g water per square centimeter of ocean surface are evaporated per year (W. C. Jacobs, 1951).

Since it takes nearly 600 cal to evaporate 1 g water, the heat lost to each square centimeter of sea surface averages 150 cal/per day. This heat is stored in the atmospheric volume but is not actually transferred to the air parcels until condensation takes place (releasing the latent heat of vaporization); this may occur 1000 miles away and 1 week later.

Radiation of heat from the water surface to the atmosphere and back again are both large—of the order of 800 cal/(cm^2)(day) according to E. R. Anderson (1953). However, the net flux is out from the sea; it averages about 100 cal/day.

Conduction usually plays a much smaller role than either of the above; it may transfer heat in either direction, but usually it contributes a small net transfer from sea to air.

Momentum. Momentum can be exchanged between these two fluids by a process related to evaporation, that is, migration of molecules of air or water across the boundary, carrying their momentum with them. However, in natural conditions the more effective mechanism is the collision of "parcels" of the fluids, as distinct from motions of individual molecules. Also, momentum is usually transferred from air to sea, rather than vice versa. Winds whip up waves; these irregular shapes are more easily attacked by wind action than is the flat sea surface, and both waves and currents are initiated and maintained by the push and stress of the wind on the water surface. *See* OCEAN-METEOROLOGICAL RELATIONS. [J.G.P.]

Projection of droplets. Water, salts, organic materials, and a net electric charge are transferred to the air through the ejection of droplets by bubbles bursting at the sea surface. The exchange of these properties between the sea and the atmosphere is of importance in meteorology and geochemistry. Upon evaporation of the water, the droplet residues are carried great distances by winds. These wind-borne aerosols become nuclei for cloud drop and raindrop formations and probably represent a large part of the cyclic salts of geochemistry.

Air bubbles are forced into surface waters of the sea by wave action, by impinging raindrops, by melting snowflakes, and by other means. The larger bubbles rise to the surface, burst, and eject droplets. Many of the smaller bubbles dissolve before reaching the surface. The photomicrograph (Fig. 1) shows stages in the collapse of a bubble, and the jet and droplet formations which result. The graph (Fig. 2) shows the approximate relationships between the sizes of the bubbles, the sizes of the ejected droplets, and weight of sea salt in these droplets.

The amounts of water which become airborne as droplets near the sea surface are not known. The best estimates which can now be made (from the limited information about the number and size of bubbles in the sea) range from about 2 to 10 g/(m^2)(day) during fresh winds.

The average amounts of sea salt which become airborne at considerable altitudes are shown in Table 1. The total range of observed amounts in individual samples is from about 4×10^{-13}g/m^3

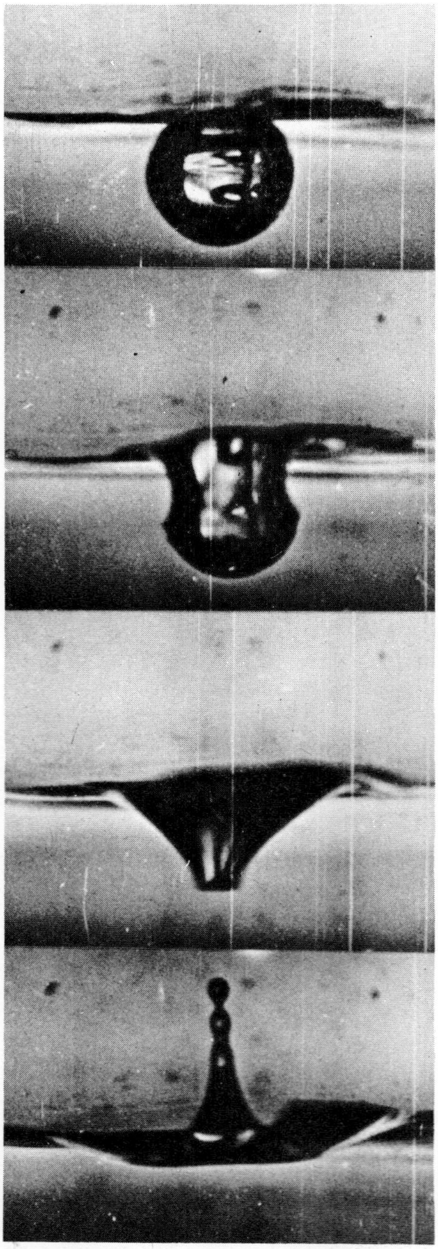

(a)

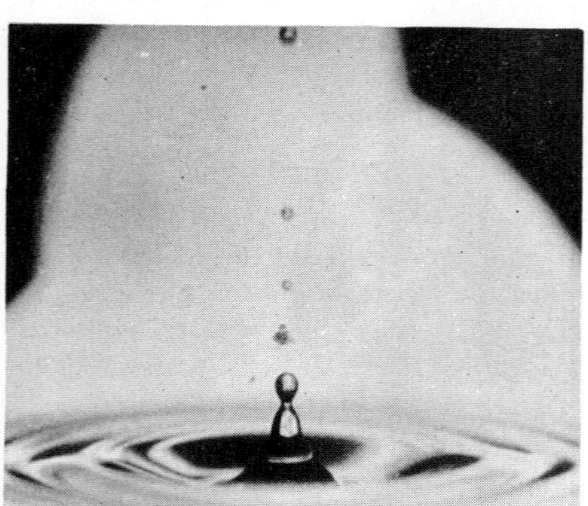

(b)

Fig. 1. Collapse of air bubble and formation of jet and droplets. (*a*) Composite photograph of high-speed motion pictures illustrating stages in the process. Time interval between top and bottom frames about 0.002 sec; bubble diameter 1.7 mm; angle of view horizontal through glass wall. (*b*) Oblique view of jet and droplets from bubble 1.0 mm in diameter. Diameter of smallest droplet 0.09 mm; exposure time 30 μsec.

in a wind of Beaufort force 1 to 10^{-9}g/m^3 in a wind of force 12. *See* WIND.

Parts of marine organisms are seen in droplets ejected from plankton-rich water. The droplets also become coated with organic monolayers when they arise through contaminated surfaces. No information is yet available about the quantities of organic materials which thus become airborne.

Table 1. Airborne sea salts in relation to wind force

Beaufort wind force	Concentrations,* μg/m^3	Flux* mg/(m^2)(day)	Total,† mg/m^2
2–3	2.7	0.42	6.0
4–5	9.9	3.9	11.6
6–7	21.3	24.0	21.8

* At about 500 m
† Integrated through lowest 2000 m

Electrification of droplets from bursting bubbles is predominantly positive and ranges in magnitude from about 100 fundamental units for the smallest particles to several thousands of units for the larger particles. Using reasonable estimates for the num-

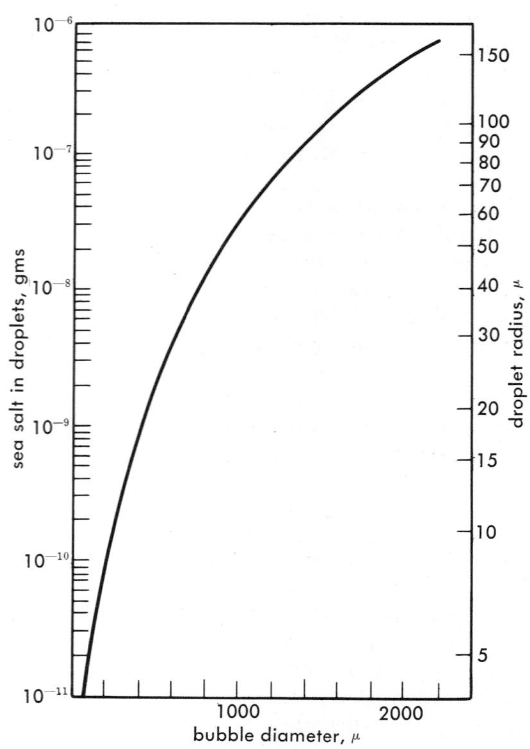

Fig. 2. Approximate relationships between sizes of bursting bubbles and sizes and salt contents of ejected droplets.

bers and sizes of bubbles bursting at the sea surface, a net flux of positive charge from the sea to the atmosphere of about 2000 fundamental units/(cm^2) (sec) is derived. [A.H.W.]

Exchange of gases. Surface water, under ordinary conditions, is saturated with the atmospheric gases—oxygen, nitrogen, carbon dioxide, and the inert gases. Occasionally, however, equilibrium may not have been reached, as in areas of rapid upwelling where deep water comes to the surface, and the water may be undersaturated with oxygen or supersaturated with carbon dioxide. Changes in the atmosphere, the result of its more rapid circulation, may also cause local deviations from saturation.

Water saturated with gases at the surface may eventually sink and travel long distances beneath the surface, during which time chemical processes may either decrease or increase its content of dissolved gas. Oxygen is increased by photosynthesis in the upper levels where there is sufficient light. On the other hand, oxygen is consumed by the processes of respiration and organic decomposition. The most striking feature of the distribution of oxygen in the ocean is the very widespread occurrence of a depth or zone of minimum concentration, which may be only a few hundred feet below the surface in some places or deeper than a mile in others. This phenomenon is the result of the combination of several processes, chemical and physical, which take place at different rates under different conditions. The consumption of oxygen is accompanied by an increase in the proportion of its heavy isotope, O^{18}.

There is little evidence to show that nitrogen is ever utilized as such in the ocean, but there is a possibility that under some conditions it may be produced, presumably as a result of organic decomposition.

The content of carbon dioxide in surface water is variable, depending upon biological conditions determining its production or absorption, and also upon the temperature, which determines its solubility and equilibrium with the combined forms, carbonate and bicarbonate. It often decreases beneath the surface, where photosynthesis is more rapid, and increases greatly in the deep oxygen-minimum zone where oxidation and the consumption of oxygen is accompanied by the production of carbon dioxide.

Consequently, the exchange of carbon dioxide between water and atmosphere may proceed in either direction, depending upon the conditions in the water as well as the content of carbon dioxide in the atmosphere, which is generally less variable. In certain areas of upwelling the gas seems to escape to the atmosphere. Seasonal biological changes, particularly concerned with the growth of plants, can also affect the direction of exchange.

The inert atmospheric gases—neon, argon, and helium—are apparently saturated in the ocean and undergo no changes, except probably those associated with radioactivity. [N.W.R.]

Bibliography: E. R. Anderson, *The Exchange of Energy between a Body of Water and the Atmosphere with Application to Evaporation*, Doctoral dissertation, Scripps Inst. Oceanog., U.C.L.A., 1953; D. C. Blanchard and A. H. Woodcock, Bubble formation and modification in the sea and its meteorological significance, *Tellus*, 9:145–158, 1957; W. C. Jacobs, The energy exchange between sea and atmosphere and some of its consequences, *Bull. Scripps Inst. Oceanog. Univ. Calif.*, 6(2):27–122, 1951; A. H. Woodcock, Salt nuclei in marine air as a function of altitude and wind force, *J. Meteorol.*, 10:362–371, 1953.

TRANSMISSION OF ENERGY

Electromagnetic and acoustic energy from various natural sources permeates the sea, supplying it with heat, supporting its ecology, and providing for sensory perception by its inhabitants; artificial sources afford man the means for underwater communication and detection.

Light. The primary source of energy which heats the ocean and supports its ecology is light from the sun. On a clear day as much as 1 kw of radiant power from the sun and sky may impinge on each square meter of sea surface. Of this power, 4–8% is reflected and the remainder is absorbed within the water as heat or as chemical potential energy due to photosynthesis. The peak of the irradiation is close to the wavelength of greatest transparency for clear sea water, 480 millimicrons (mμ), but nearly half of the radiant power is infrared radiation which water absorbs so strongly that virtually none penetrates more than 1 m beneath the surface. As much as one fifth of the incident power may be ultraviolet (below 400 mμ) and this radiation may penetrate a few tens of meters if little or no "yellow substance" (humic acids and other materials associated with organic decomposition) is present. Only a narrow spectral band of blue-green light, representing less than 10% of the total irradiation, penetrates deeply into the sea. This radiation has been detected by multiplier-phototube photometers at depths of more than 600 m. Visibility, important to predators in the feeding grounds of the sea, is possible chiefly because of this blue-green light.

Irradiance. Irradiance on a flat surface oriented in any manner decreases exponentially with depth, as illustrated by Fig. 3, which depicts experimental values of irradiance on an upward-facing surface. Irradiance on any other surface could be represented by a curve parallel (within 5%) to the one shown; irradiance on downward-facing surfaces is approximately one-fiftieth of the irradiance on upward-facing surfaces at all depths.

Absorption. Light, to be useful for heating or for photosynthesis, must be absorbed. The quantity of radiant power absorbed per unit of volume depends upon the amount of power present and the magnitude of the absorption coefficient; to a useful (5%) approximation power absorbed per unit of volume at any depth can be calculated by multiplying the irradiance at that depth, as in Fig. 3, by the slope

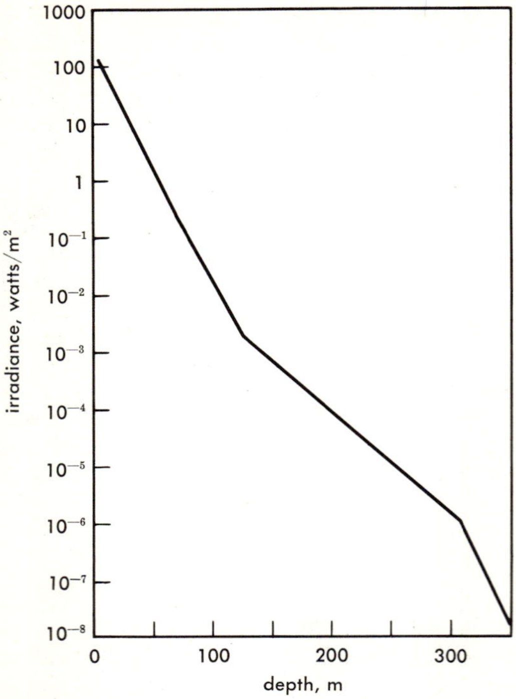

Fig. 3. Experimental values of irradiance on an upward-facing surface.

of the curve expressed in natural log-units per unit of depth, that is, the attenuation coefficient K. Thus in Fig. 3, at a depth of 64 m where the irradiance is 0.5 watt/m^2 and is decreasing with depth at the rate of 0.08 natural log-units/m, approximately $0.5 \times 0.08 = 0.04$ watt of radiant power is absorbed by every cubic meter of sea water.

Visibility. Visibility under water is accomplished by image-forming light (rays) which must pass from the object to the observer without being scattered. The transmission of water for image-forming light is less than for diffused light, since scattering in any direction constitutes a loss of image-forming light, whereas only scattering in rearward directions is a loss for diffused light. Image-forming light is exponentially attenuated with distance but the attenuation coefficient α averages 2.7 times greater than the attenuation coefficient for irradiance K, defined above. Apparent contrast of an underwater object having deep water as a background is exponentially attenuated with distance; the effective attenuation coefficient being $\alpha + K \cos \theta$, where θ is the inclination angle of the path of sight; $\cos \theta = 1$ when the observer looks straight down. See discussion of water color and transparency in section on sampling and measuring techniques. [S.Q.D.]

Compensation intensity and depth. As daylight penetrating into the sea diminishes, the photosynthesis of plants is reduced but respiration remains approximately the same. The light value at which the rates of photosynthesis and respiration are equal is the compensation intensity. The depth at which the compensation intensity is found is the compensation depth. Both of the foregoing have also been termed compensation point, but since ambiguity may occur, it is best to avoid this term. The compensation intensity varies according to the species, the physiological condition of the plants, and other factors, particularly temperature. Lowered temperature depresses respiration more than photosynthesis. The compensation depth depends upon the intensity of the incident radiation, the transparency of the water, and the period considered, since illumination varies with time. Compensation intensities of 10–200 ft-c have been measured for phytoplankton and of 17–45 ft-c for filamentous algae. Compensation depths for 24-hour periods for phytoplankton range from less than 1 m in turbid water to more than 30 m in coastal areas and to 80 m or more in the clearest tropical waters; and for attached plants, to 50 m along the coast and to 160 m in especially clear water, as in the Mediterranean. Generally the compensation depth is found where daylight is reduced to about 1% of its value at the surface, for phytoplankton or about 0.3% for bottom plants. The compensation depth is of particular significance since it marks the lower limit of the photic zone within which green plants can carry on primary production necessary as an energy source for the whole marine ecosystem. [G.L.CA.]

Electromagnetic fields. In sea water, as in any conductor, the electromagnetic behavior is determined by the magnetic permeability μ and the electrical conductivity σ. From these one may find the skin depth δ and the characteristic impedance η by

$$\delta = (\pi f \mu \sigma)^{-1/2} \quad \text{and} \quad \eta = (2\pi f \mu / \sigma)^{1/2}$$

Both δ and η relate to electromagnetic waves of frequency f, for which the wavelength is $2\pi\delta$, the absorption over a path length x reduces the amplitude in the ratio $e^{-x/2\delta}$, and the ratio of electric to magnetic field amplitude in a plane wave is η, the former leading in phase by 45°. *See* ELECTROMAGNETIC RADIATION; SKIN EFFECT; WAVE EQUATION.

For sea water, magnetic permeability μ is nearly the same as for free space and electrical conductivity σ is given to about 1% by

$$\sigma = [4.00 + a(t - 12)]\,[1 + .0269(S - 35)]$$

in which t is temperature in °C, S is salinity in parts per thousand by weight, and a is .10 for $t > 12$ or $a = .092$ for $t < 12$ (see Fig. 4). Using $\sigma = 4.0$ mho/m as a typical value, one obtains

$$\delta = 250 f^{-1/2} \text{ meter} \quad \text{and} \quad \eta = .0014 f^{1/2} \text{ ohm}$$

These formulas are expected to hold for all frequencies below about 900 Mc.

Absorption limits the penetration of a field, either inward from a boundary or outward from an electric or magnetic source, to a small multiple of the skin depth δ. A submerged horizontal dipole source near the surface will, however, have a more extensive field in the air and a shallow layer of

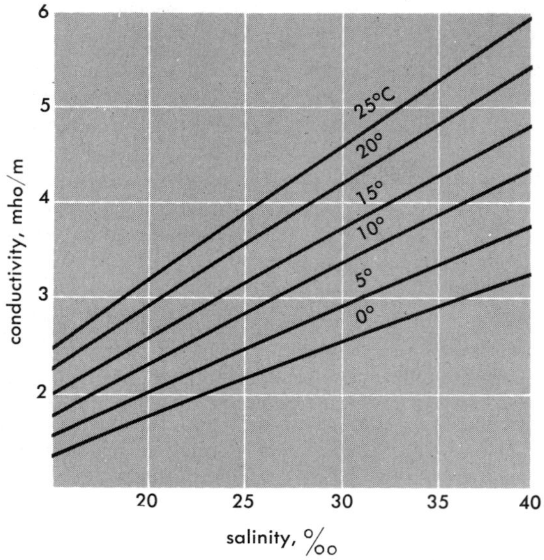

Fig. 4. Conductivity of sea water. (*From B. D. Thomas, T. G. Thompson, and C. L. Utterback, The electrical conductivity of sea water, J. conseil, Conseil permanent intern. exploration mer, 9:28–35, 1934*)

water. At .01 cps δ is 2.5 km; this is a rough upper frequency limit for field fluctuations (such as those of the geomagnetic field) which can penetrate the entire thickness of the ocean layer. At 10 kc δ is 2.5 m. Fields of this and higher frequencies, existing for example in the natural atmospheric noise, can penetrate only a thin surface layer. For radio signals, the sea acts as an excellent ground plane, involving lower losses than transmission over land. [P.R.]

Sound. Sound in sea water travels four or five times its speed in air, or about 1500 m/sec. As sound propagates in the sea, its intensity diminishes inversely as the square of the distance from the source, in the absence of appreciable absorption, refraction, reflection, and scattering. Although losses by absorption are small compared with those that occur when sound of the same frequency travels through air, the increase in absorption with higher frequencies limits the effective range of ultrasonic waves, that is, waves having frequencies above those audible to the human ear. This is an important factor in submarine detection, where ultrasonic frequencies are used because of their desirable directional properties. *See* INVERSE-SQUARE LAW; SONAR.

The velocity of sound in sea water varies with temperature, salinity, and pressure. Hence a beam of ultrasonic waves, when transmitted in a horizontal direction, may be refracted and then reflected one or more times from the surface, ocean bottom, or some layer within the vertical water structure. In this manner several different rays may be received at different intervals from a single source. The direct transmission is limited to specific distances, depending on depth of the bottom, and its theoretical velocity may be computed if the temperature and salinity are known. At greater distances the apparent horizontal velocity is less than the theoretical velocity because of such factors as distance between source and receiver, depth, bottom profile and character, physical properties of the water, and so on.

The vertical velocity is a function of depth (pressure) and the distribution of temperature and salinity. Except in polar latitudes it generally decreases from the surface to some moderate depth (from 500 to 1500 meters) because of decreasing temperature. Below this depth the velocity gradually increases again as the effect of increasing pressure becomes dominant. Because most sonic depth-finding instruments are calibrated for a constant velocity, when a very accurate depth reading is required, it is necessary to correct the readings to true depths. *See* ECHO SOUNDER; SUBMARINE TOPOGRAPHY.

Investigations of sound in the ocean offer many promising areas for further study, particularly studies of underwater noises produced by marine life, the relation of these noises to other sounds in the sea, the seasonal rhythm and geographic variation of the noise makers, and the ecological significance of sound-producing organisms. A variety of problems in underwater acoustics may result from the presence of sound-producing organisms. *See* SCATTERING LAYER; SOUND; UNDERWATER SOUND. [N.A.B.]

Bibliography: G. L. Clarke, *Elements of Ecology*, 1954; R. W. Holmes, Solar radiation, submarine daylight, and photosynthesis, in *Treatise on Marine Ecology and Paleoecology*, Geol. Soc. Am. Mem. 67, vol. 1, 1957; R. H. Lien, Radiation from a horizontal dipole in a semi-infinite dissipative medium, *J. Appl. Phys.* 24(1):1–4, 1953; J. R. Wait, The radiation fields of a horizontal dipole in a semi-infinite dissipative medium, *J. Appl. Phys.*, 24(7): 958–959, 1953.

COMPOSITION OF SEA WATER

Inorganic regulation of composition. The present-day compositions of sea waters (Table 2) are controlled both by the make-up of the ultimate source materials and by the large number of reactions, of chemical and physical natures, occurring in the oceans. For a discussion of the weathered and weathering substances that give rise to the waters of the world, *see* HYDROSPHERE, GEOCHEMISTRY OF. This section considers the nonbiological regulatory mechanisms, most conveniently defined as those reactions occurring in a sterile ocean.

Solubility. Only calcium, among the major cations of sea water, is present in a state of saturation, and its concentration is governed by the solubility of calcium carbonate. The rare gases and dissolved gaseous nitrogen have their marine concentrations determined by the temperature at which their water mass was in contact with the atmosphere and are in states of saturation or very nearly so.

Table 2. Chemical abundances in the marine hydrosphere

Element	Abundance, mg/liter	Element	Abundance, mg/liter
H	108,000	Ag	0.0003
He	0.000005	Cd	0.000055
Li	0.2	In	<0.02
Be		Sn	0.003
B	4.8	Sb	<0.0005
C	28	Te	
N	0.5	I	0.05
O	857,000	Xe	0.0001
F	1.3	Cs	0.0005
Ne	0.0003	Ba	0.0062
Na	10,500	La	0.0003
Mg	1,300	Ce	0.0004
Al	0.01	Pr	
Si	3	Nd	
P	0.07	Pm	
S	900	Sm	
Cl	19,000	Eu	
A	0.6	Gd	
K	380	Tb	
Ca	400	Dy	
Sc	0.00004	Ho	
Ti	0.001	Er	
V	0.002	Tm	
Cr	0.00005	Yb	
Mn	0.002	Lu	
Fe	0.01	Hg	
Co	0.0005	Ta	
Ni	0.0005	W	0.0001
Cu	0.003	Re	
Zn	0.01	Os	
Ga	0.00003	Ir	
Ge	<0.00006	Pt	
As	0.003	Au	0.000004
Se	0.004	Hg	0.00003
Br	65	Tl	<0.00001
Kr	0.0003	Pb	0.003
Rb	0.12	Bi	0.0002
Sr	8	Po	
Y	0.0003	At	
Zr		Rn	9.0×10^{-15}
Nb	0.000005	Fr	
Mo	0.01	Ra	3.0×10^{-11}
Tc		Ac	
Ru		Th	0.0007
Rh		Pa	0.003
Pd		U	

Cation and anion exchange. Cation exchange reactions between positively charged species in sea water and such minerals as the marine clays and zeolites appear to regulate, at least in part, the amounts of sodium, potassium, and magnesium, as well as other members of the alkali and alkaline-earth metals which are not limited by solubility product considerations. High charge and large radius influence favorably the uptake on cation exchange minerals. It appears, for example, that while 65% of the sodium weathered from the continental rocks now resides in the oceans, only 2.5%, 0.15%, and 0.025% of the total amounts of potassium, rubidium, and cesium ions (ions increasingly larger than sodium), have remained there. Further, magnesium and potassium are depleted in the ocean relative to sodium on the basis of igneous rock data (Table 3). The curious fact that magnesium remains in solution to a much higher degree than potassium is as yet not resolved but may be explained by the ability of such ubiquitous clay minerals as the illites to fix potassium into nonexchangeable or difficultly exchangeable sites.

Table 3. Compositions of sea water, igneous rocks, and gases from fumaroles and hot springs

Substance	Sea water, wt %	Igneous rocks, wt %	Gases from fumaroles, steam wells, and geysers, wt %
H_2O	96.5	~1	99.4
Na	1.05	2.8	
K	0.04	2.5	
Mg	0.13	2.0	
Ca	0.04	3.5	
Cl	1.9	0.03	0.12
S	0.09	0.05	0.03

Similarly, anion exchange processes may regulate the composition of some of the negatively charged ions in the oceans. For example, the chlorine-bromine ratio in sea water of 300 is displaced to values around 50 in sediments. Such a result may well arise from the replacement of chlorine by bromine in clays; however, the meager amounts of work in this field preclude any unqualified statements.

Sorption. This category covers some of the surface interactions of the dissolved substances with the solid phases in sea water, such as the hydrated iron oxides; it also covers interactions between dissolved substances and the sediments. For example, the iron oxides scavenge from solution ionic species containing such elements as titanium, zirconium, or vanadium.

Authigenic mineral formation. The formation and alteration of minerals on the sea floor apparently are responsible for controlling the concentrations of a group of metals, all in highly undersaturated states. A suite of such metals, including zinc, manganese, nickel, cobalt, and copper, are accumulated by the ferromanganese minerals in concentrations one or two orders of magnitude greater than the average values for crustal rocks (Table 4). The ferromanganese minerals, which form uniquely on the deep-sea floor, are composed of oxides or iron and manganese and build these trace metals into their layer-lattice structures. They exist in the

Table 4. Observed concentrations of some trace metals in sea water*

Ion	Observed sea-water concentration, moles/liter	Limiting compound	Calculated limiting concentration, moles/liter
Mn^{++}	4×10^{-8}	$MnCO_3$	$\sim 10^{-5}$
Ni^{++}	4×10^{-8}	$Ni(OH)_2$	$\sim 10^{-3}$
Co^{++}	$\sim 10^{-8}$	$CoCO_3$	10^{-6}–10^{-7}
Zn^{++}	2×10^{-7}	$ZnCO_3$	$\sim 10^{-4}$
Cu^{++}	6×10^{-8}	$CuCO_3$	$\sim 10^{-4}$

* Calculated upon the basis of their most insoluble compound.

form of nodular concretions, the so-called manganese nodules, which range in size from millimeters to about a meter, in the form of coatings, and as components of the unconsolidated sediments. It should be pointed out that calcium phosphate minerals, originating from the skeletal remains of fish, amass on the ocean bottom a similar group of metals and may in part account for their low marine concentrations. *See* AUTHIGENIC MINERALS.

Physical processes. Superimposed upon these chemical processes are changes in the chemical make-up of sea water by the melting of ice, evaporation, mixing with runoff waters from the continents, and upwelling of deeper waters. The net effects of the first three processes are changes in the absolute concentrations of all of the elements but with no major changes in the relative amounts of the dissolved species.

Changes with time. Changes in the composition of sea water through geologic times reflect not only differences in the extent and types of weathering processes on the earth's surface but also the relative intensities of the biological and inorganic reactions. The most influential parameter controlling the inorganic processes appears to be the seawater temperature.

Changes in the abyssal temperatures of the oceans from their present values of near 0 to 2.2, 7.0, and 10.4°C in the upper, middle, and lower Tertiary, respectively, have been postulated from studies on the oxygen isotopic composition of the tests of benthic foraminifera. Such temperature increases would of necessity be accompanied by similar ones in the surface and intermediate waters. One obvious effect from the recent cooling of the oceans is an increase in either or both of the calcium and carbonate ions, since the solubility product of calcium carbonate has a negative temperature coefficient. Similarly, the saturated amounts of gases that can dissolve in sea water in equilibrium with the atmosphere increase with decreasing water temperatures. *See* GEOLOGIC THERMOMETRY. [E.D.G.]

Biological regulation of composition. In the open sea all the organic matter is produced by the photosynthesis and growth of unicellular planktonic forms. During this growth all the elements essential for living matter are obtained from the sea water. Some elements are present in great excess, such as the carbon of CO_2, the potassium and the sulfur (as sulfate). Other elements—for example, phosphorus, nitrogen, and silicon—are present in small enough quantities so that plant growth removes virtually all of the supply from the water. During photosynthesis, as these elements are being removed from the water, oxygen is released.

The biochemical cycle. The organic matter formed by photosynthesis and growth of the unicellular plants may be largely eaten by the zooplankton, and these in turn form the food for larger organisms. At each step of the food chain a large proportion of the eaten material is digested and excreted, and this, along with dead organisms, is decomposed by bacterial action. The decomposition process removes oxygen from the water and returns to the water those elements previously absorbed by the phytoplankton.

The distribution of oxygen and essential nutrient elements in the sea is modified by the spatial separation of these biological processes. Photosynthesis is limited to the surface layers of the ocean, generally no more than 100 m or so of depth, but the decomposition of organic material may take place at any depth. Reflecting this separation of processes the concentration of nutrient elements in the surface is low, rises to maximum values at intermediate depths (300–800 m), and decreases slightly to fairly constant values which extend nearly to the bottom. Frequently a slight increase in the concentration of essential elements is observed near the bottom. The oxygen distribution is the opposite of the one just described, with high values at the surface, a minimum value at mid-depth, and intermediate values in the deep water. The oxygen-minimum–nutrient-maximum level in the ocean is the result of two processes working simultaneously. In part it is formed by the decomposition of organic matter sinking from the surface, and in part it results from the fact that this water was originally at the surface in high latitudes where it contained organic matter and subsequently cooled, sank, and spread out over the oceans at the appropriate density levels.

Because of the nearly constant composition of marine organisms, the elements required in the formation of organic material vary in a correlated way. Analyses of marine organisms indicate that in their protoplasm the elements carbon, nitrogen, and phosphorus are present in the ratios of 100:15:1 by atoms. In the production of organic matter these elements are removed from the water in these ratios, and during the decomposition of organic matter they are returned to the water in the same ratios. However, since the decomposition of organic material is not an instantaneous process which releases all elements simultaneously, it is not unusual to find different ratios of concentration of these elements in the sea. Particularly in coastal waters and in confined seas the ratio of concentration of nitrogen to phosphorus, for example, may differ widely from the 15:1 ratio of composition within the organisms.

The biochemical circulation. Unlike the major elements in sea water, the concentrations of these nutrients are widely different in different oceans of the world. Pacific Ocean water contains nearly twice the concentration of nitrogen and phosphorus found at the same depth in the North Atlantic, and intermediate concentrations are found in the South Atlantic and Antarctic Oceans. The lowest concentrations for any extensive body of water are found in the Mediterranean, where they are only about one-third of those found in the North Atlantic Ocean.

These variations can be attributed to the ways in which the water circulates in these oceans, and to the effect of the biological processes on the distribution of elements. The Mediterranean, for example, receives surface water, already low in nutrients, from the North Atlantic and loses water from a greater depth through the Straits of Gibraltar. While the water is in the Mediterranean, the surface layers are further impoverished by growth of phytoplankton, and the organic material formed sinks to the bottom water and is lost in the deeper outflow. A similar process explains the low nutrient concentrations in the North Atlantic, which receives surface water from the South Atlantic and loses an equivalent volume of water from greater depths. *See* OCEAN CURRENTS; SEA WATER FERTILITY. [B.H.K.]

Buffer mechanism. The constituents of sea water include a number of cations, all of which are weak acids, and a smaller number of anions, some of which are strong bases. Thus sea water is always somewhat on the alkaline side of neutrality, ranging in pH roughly between the limits of 7.5 and 8.3. In chemical oceanography, the term alkalinity is used to denote not the concentration of hydroxyl ions, as might be expected, but the concentration of strong bases. Alkalinity can be defined as the number of equivalents of strong acid required to convert stoichiometrically all the strong bases to weak acids. The addition or subtraction of weak acids thus does not affect the alkalinity of sea water, although through the operation of the buffer mechanism it changes the pH.

The principal weak acid in sea water is carbonic, resulting from the hydrolysis of dissolved carbon dioxide. Boric acid is also present in significant amounts. Salts of these two weak acids are the strong bases which make up the alkalinity. Total combined boron, whether as acid or borate, is in virtually constant proportion to chlorinity, the ratio of boron to chlorinity being about 0.00024, which is equivalent to a specific ratio of boric acid to chlorinity of about 0.022 (concentrations in millimoles per liter).

The total alkalinity is more variable in the ocean. Near the surface it can be increased by addition of dissolved carbonates in river discharge or decreased by the precipitation of lime in the formation of coral and shells. In deeper layers it can be increased by the solution of calcareous debris sinking from the surface. F. Koczy (1956) gives a thorough discussion of the variation of specific alkalinity in the oceans. The average ratio of alkalinity to chlorinity is about 0.120 for surface sea water, increasing somewhat with depth (concentrations in milliequivalents per liter).

Even more variable than the alkalinity is the total dissolved carbon dioxide. At the surface, CO_2 moves between the sea and the atmosphere, and in the euphotic zone CO_2 is removed by photosynthesis to be incorporated into organic matter. Below the euphotic zone, CO_2 is regenerated through the biological oxidation of organic matter. Total CO_2 thus typically may vary from 2.0 or 2.1 millimoles/liter at the surface to 2.8 or more at the depth of the oxygen minimum.

Some of this CO_2 is in physical solution and some is undissociated carbonic acid; but most of it is in ionic form, mainly bicarbonate ion with some carbonate. The proportions of the various forms are governed by the dissociation constants of carbonic and boric acids, which vary with temperature and pressure, and by the activity coefficients of the various ions concerned, which vary with temperature, pressure, and salinity. In practice, the activity coefficients and the dissociation constants are combined into apparent dissociation constants, which are tabulated by H. W. Harvey (1957) as functions of temperature and salinity at 1 atmosphere pressure. [J.LY.]

Bibliography: E. D. Goldberg, The processes regulating the composition of sea water, *J. Chem. Educ.*, 35:116–119, 1958; H. W. Harvey, *Chemistry and Fertility of Sea Waters*, 1957; F. Koczy, The specific alkalinity, *Deep-Sea Research*, 3:279–288, 1956; J. Lyman and R. B. Abel, Chemical aspects of physical oceanography, *J. Chem. Educ.*, 35:113–115, 1958; A. C. Redfield, The biological control of chemical factors in the environment, *Am. Scientist*, 46(3):205–221, 1958; F. A. Richards, Some current aspects of chemical oceanography, in L. H. Ahrens et al. (eds.), *Physics and Chemistry of the Earth*, vol. 2, 1957.

DISTRIBUTION OF PROPERTIES

The distribution of physical and chemical properties in the ocean is principally the result of the following: (1) radiation (of heat); (2) exchange with the land (of heat, water, and solids such as salts) and with the atmosphere (of water, salt, and heat dissolved gases); (3) organic processes (photosynthesis, respiration, and decay); and (4) mixing and stirring processes. These processes are largely responsible for the formation of particular water types and ocean water masses.

Horizontal distributions. The general distribution of properties in the oceans shows a marked latitudinal effect which corresponds with radiation income and differences between evaporation and precipitation.

Temperature. Heat is received from the sun at the sea surface where parts of it are reflected and radiated back. Equatorward of 30° latitude the incoming radiation exceeds back radiation and reflection, and poleward it is less. The result is high sea-surface temperature (more than 28°C) in equatorial regions and low sea-surface temperatures (less than 1°) in polar regions.

Salinity. Various dissolved solids have entered the sea from the land and have been so mixed that their relative amounts are everywhere nearly constant, yet the total concentration (salinity) varies considerably. In middle latitudes the evaporation of water exceeds precipitation, and the surface salinity is high; in low and high latitudes precipitation exceeds evaporation, and dilution reduces the surface salinity. Open ocean-surface salinities range from lows of about 32.5‰ in the North Pacific, 34.0 in the Antarctic, 35.0 in the equatorial Atlantic, 34.0 in the equatorial Indian, and 33.5 in the equatorial

Pacific, to highs in the great evaporation centers of the middle latitudes of 35.5 and 36.5 in the North and South Pacific, 37.0 in the North and South Atlantic, and 36.0 in the Indian Ocean.

Dissolved oxygen. Dissolved oxygen is both consumed (respiration and decay) and produced (photosynthesis) in the ocean as well as being exchanged with the atmosphere at the sea surface. Above the thermocline the waters are nearly always near saturation in oxygen content (>7 ml/liter in the cold waters of high latitudes, <5 ml/liter in the warm equatorial waters).

Density. The density of sea water depends upon its temperature, salinity, and depth (pressure) but can vary horizontally only in the presence of currents, and hence its distribution depends closely upon the current structure. In low and middle latitudes the effect of the high temperature exceeds that of the high salinity and the surface waters are lighter than those in high latitudes, with values ranging from less than 1.022 g/ml to more than 1.027. The density at great depth (10,000 m) may exceed 1.065. The greatest vertical gradient is associated with the thermocline and the halocline and is therefore very near to the surface. The heavier surface waters from high latitudes move and mix underneath the lighter water at depths which depend upon their density. The difference in surface density is usually not great, since the high latitude salinity is low, and the waters usually sink only a few hundred meters, forming intermediate water. But in cases where water of high salinity has been carried into high latitudes by the currents and cooled (as in the North Atlantic) or where water of relatively high salinity freezes and gives up part of its water (as on the continental shelf of Antarctica), deep and bottom waters with temperature less than 1°C are formed. *See* HALOCLINE; THERMOCLINE. [J.L.R.]

Vertical distributions. The subsurface distribution of properties is controlled largely by external factors, particularly those which influence surface density, and the type of deep-sea circulation.

Temperature. The thermohaline circulation results in a vertical distribution of temperature such that in low and middle latitudes the deeper waters are colder than the surface waters, and at very high latitudes where surface temperatures are low, the deeper waters are as warm as those at the surface, or warmer. Seasonal cooling in high latitudes may cause the temperature to be at a minimum at some intermediate depth, and the circulation may cause a temperature minimum or maximum at intermediate depths. Over a large part of the Pacific Ocean, where no bottom water is formed, the temperature increases downward from about 4000 m. Since the gradient is not greater than the adiabatic, the water is not unstable.

Density and salinity. Since much of the flow of the ocean is geostrophically balanced, surfaces of constant density slope in various ways with respect to the sea surface and other surfaces of constant pressure, and density varies in the east-west as well as the north-south direction. Mixing and movement of intermediate and deep water along these surfaces cause more complex distributions of other variables. The intermediate waters of the North and South Pacific and of the South Atlantic appear to have salinity minima at intermediate depths in middle and low latitudes, since they originate in the high-latitude regions of low salinity and pass between the high-salinity surface waters of middle latitudes and the bottom waters. In the North Pacific this minimum varies from 33.4–34.1‰, in the South Pacific and South Atlantic from 34.2 to about 34.6‰. In the North Atlantic Ocean there is no intermediate water of low salinity formed, but very saline water (36.5‰) flows in from the Mediterranean at depth and results in a more complicated distribution of salinity than is found in the other oceans. For the temperatures and salinities of the bottom waters, which are more homogeneous than the others, see the later discussion on ocean water masses.

Dissolved oxygen. Where the cold surface waters sink in high latitudes, quantities of oxygen are carried downward. Below the compensation depth consumption gradually reduces the concentration. The bottom waters of the Atlantic Ocean contain from 5–6 ml/liter of dissolved oxygen, the Indian and South Pacific Ocean about 4, and the North Pacific less than 4. Between these bottom waters and those at the surface, which are saturated, smaller values of oxygen are found ranging from less than 0.10 in the eastern tropical Pacific and less than 1.0 in the eastern tropical Atlantic to greater than 4 ml/liter in the South Atlantic.

Nutrients. Other properties such as the nutrients, phosphate and nitrate, have low concentrations in the surface layer, where they are consumed by the plants, and high values at depths, where they are concentrated by the sinking and decay of organisms.

The maximum values of phosphate are found at intermediate depths, usually beneath the layer of minimum oxygen, and vary from less than 1 microgram-atom/liter in the North Atlantic to more than 2 in the Antarctic, Indian, and South Pacific, and more than 3.5 in the North Pacific. Surface values vary from less than 0.1 in the North Atlantic and 0.5 in the North Pacific to more than 1.5 in the Antarctic. Bottom values vary from about 1.0 in the North Atlantic and 2.0 in the Antarctic, South Pacific, and Indian Oceans to 2.5 in the North Pacific.

Nitrate-nitrogen is present in a ratio of about 8:1 by weight to phosphate-phosphorus.

Silicate has no intermediate maximum but increases monotonically toward the bottom, where the values vary from more than 150 microgram-atoms/liter in the North Pacific to less than 40 in the North Atlantic. Surface values are generally less than 10. [J.L.R.]

Stirring and mixing processes. Stirring and mixing are processes of prime importance in determining the distribution of properties and in the formation of water masses in the ocean. Stirring refers here to motions which increase the average magnitude of the gradient of a property throughout a specified region. Mixing is employed in a narrow sense to denote the molecular diffusion or conduction processes by which gradients are decreased. Stirring and subsequent mixing can decrease the gradients in a region much more quickly than molecular processes acting alone. Combined stirring and mixing are often called turbulent diffusion.

Stirring motions involve shearing, or more precisely, deformations of the water. Stirring scales range from those of the permanent currents down nearly to molecular scale. Smaller-scale motions, often highly complex, are the most effective in stirring.

With very important exceptions noted below, the preferred direction of stirring is along surfaces of constant potential density. (Potential density is the density of water when brought adiabatically to atmospheric pressure.) The preference is due to the fact that potential density almost always increases with depth. Under this condition, vertical displacement of a water parcel from a potential density surface is resisted by a stabilizing force proportional to the vertical gradient of potential density.

At the sea surface, powerful mechanisms exist which induce stirring across potential density surfaces. These are the wind stresses and the thermal convections resulting from cooling and evaporation. Near the sea surface, they greatly outweigh the stabilizing forces to produce and maintain a shallow, homogeneous top layer over much of the ocean. Stirring induced at the surface penetrates across the potential density surfaces just below the homogeneous layer, but is damped out as it extends deeper into the thermocline or halocline layer. Thus, at some depth in these layers, the more usual stirring along potential density surfaces becomes dominant. In great depths and in high latitudes, the dominance is less strong since the vertical gradient of potential density is small. Stirring across potential density surfaces may also be brought about by tidal currents, but these are strong only in the shallow, coastal parts of the ocean.

Because of the variety and complexity of stirring motions, no general method of treating them quantitatively has proved really satisfactory. It is often expedient to assume that stirring and mixing follow rules analogous to those for molecular diffusion. Sometimes analysis of the motions in detail is possible, although difficulties in both theory and observation are great. Statistical treatment of the detailed motion seems to provide the most realistic approach. *See* OCEAN CURRENTS. [J.D.CO.]

Ocean water masses. Ocean water masses are extensive bodies of subsurface ocean water characterized by a relatively constant relationship between temperature and salinity or some other conservative dissolved constituent. The concept was developed to permit identification and tracing of such water bodies. The assumption is made that the characteristic properties of the water mass were acquired in a region of origin, usually at the surface, and were subsequently modified by lateral and vertical mixing. The observed characteristics in situ thus depend both on the original properties and on the degree of modification en route to the region where observed.

A water mass is usually defined by means of a characteristic diagram, on which temperature or some other thermodynamic variable is plotted against an expression for the amount of one component of the mixture. A point on such a diagram defines a water type (representing conditions in the region of origin), and the line between points observes, at least approximately, the property of mixtures; that is, on the line connecting the two points, the proportion lying between the point representing one water type and the point for any mixture equals the proportion of the second water type in the mixture. The resulting curve for a vertical water column has been called a characteristic curve, because for a given water mass its shape is invariant regardless of depth. The existence of such a curve implies continuous renewal of water types, since otherwise mixing would lead to homogeneous water, represented by a point on the diagram.

Temperature-salinity relationships. In oceanography the characteristic diagram of temperature against salinity is usually used in studies of water masses, and the resulting temperature-salinity curve (the *T-S* curve) is used to define a water mass (Fig. 5). In drawing such a curve, data from the upper 100 m are usually omitted because of seasonal variation and local modification in the surface layer, so that strictly speaking, a water mass as defined extends only to within 100 m of the sea surface. Although ideally a water mass is defined by a single *T-S* curve, because of random errors in field measurements and perhaps fine structure in

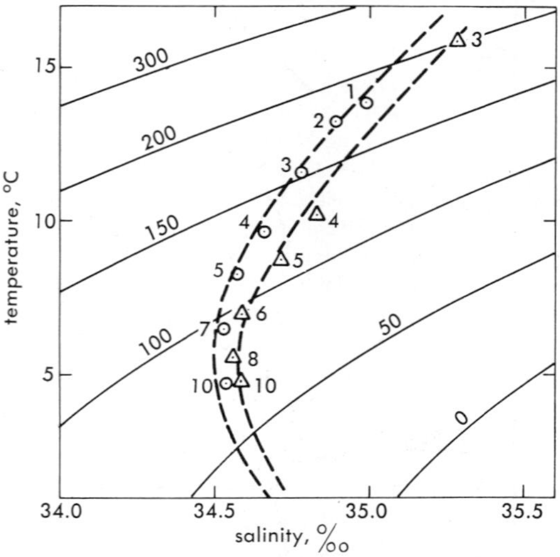

Fig. 5. Temperature-salinity values for *Carnegie* station 40 (circles) and *Dana* station 3756 (triangles); see Fig. 7 for locations. Depths of observations in hectometers. Dashed lines represent definition of Pacific Equatorial water. Solid lines represent specific volume as thermosteric anomaly, in centiliters per ton.

the water mass itself, in practice an envelope of values provides a more useful definition.

On the *T-S* diagram any property which is a function only of temperature and salinity can be represented by the appropriate family of isopleths (such as values at constant pressure of density expressed as σ_t, or thermosteric anomaly, sound speed, saturation concentration of dissolved gases). Therefore, the *T-S* diagram with isopleths can be used to determine values of such temperature-salinity dependent functions. Since the ocean is inherently stable (that is, density increases monotonically with depth), examination of the slope of a *T-S* curve (on which depth is indicated) relative to the isopleths of density permits an estimate of the vertical distribution of stability. The diagram is often useful for the detection of faulty observations and as a guide to interpolation on neighboring stations. When a uniform series of data is available, it can also be used for the quantitative representation of the frequency distribution of water characteristics.

Water-mass types. The most important and best-established water masses (characterized by *T-S* curves in Fig. 6; distribution shown in Fig. 7) occur in the upper 1000 m of the ocean. These are of three general types: (1) polar water, present south of 40°S in all oceans, and north of 40°N in the Pacific; (2) central water, occurring at mid-latitudes over most of the world ocean; and (3) equatorial water, present in the equatorial zones of the Pacific and Indian Oceans.

Polar waters, including the Subarctic, Subantarctic and Antarctic Circumpolar water masses, are formed at the surface in high latitudes, and thus are cold and have relatively low salinity. Subantarctic water is bounded in the south by the well-defined Antarctic Convergence, south of which circumpolar water is found; Subarctic water has no clear-cut northern boundary.

The central water masses appear to sink in the regions of the subtropical convergences (35–40°S and N) where during certain seasons of the year horizontal *T-S* relations at the surface are similar to the vertical distributions characteristic of the various water masses. The great differences in their properties are attributed to differences in the amounts of evaporation and precipitation, heating and cooling, atmospheric and oceanic circulation, and the distributions of land and sea in the source regions.

The widespread and well-defined equatorial waters (Fig. 5) separate the central water masses of the Indian and Pacific Oceans. These equatorial water masses are apparently formed by subsurface mixing at low latitudes, although the place and manner of their formation is not well known.

Intermediate waters underlie the central water masses in all oceans. Antarctic Intermediate Water sinks as a water type along the Antarctic Con-

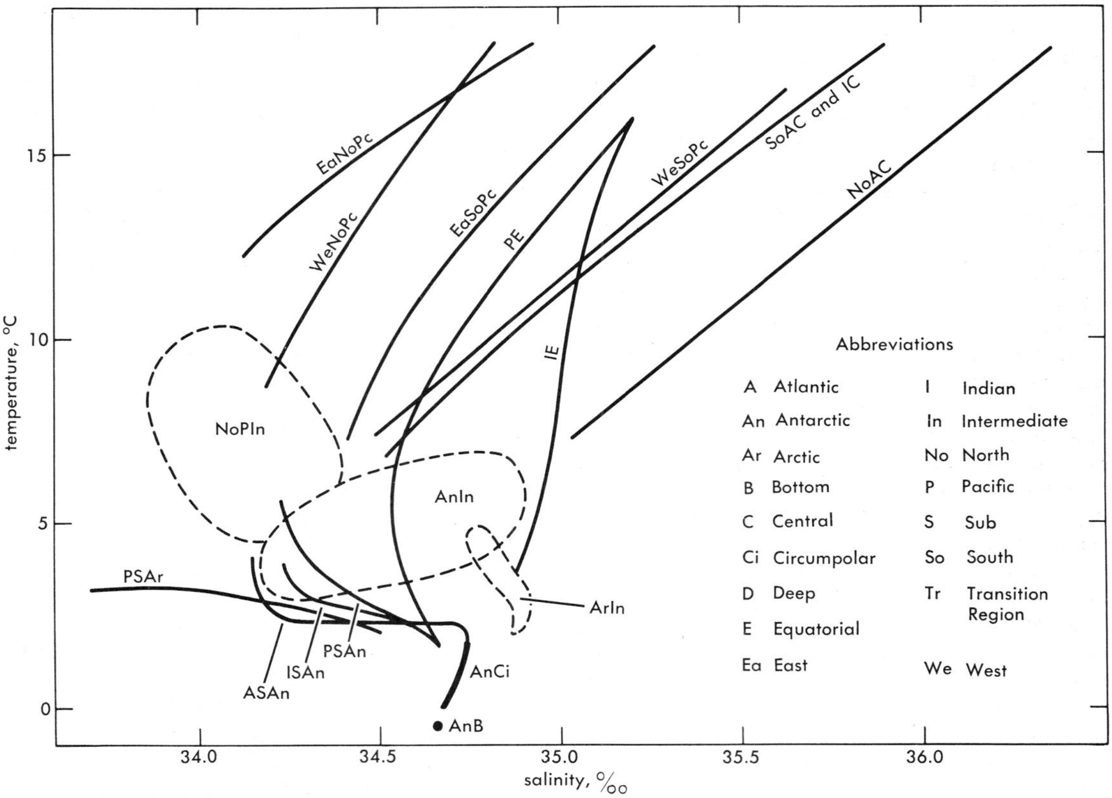

Fig. 6. Temperature-salinity relationships for representative water masses of world ocean. (*Adapted from H. U. Sverdrup, R. H. Fleming, and M. W. Johnson, The Oceans, Prentice-Hall, 1942*)

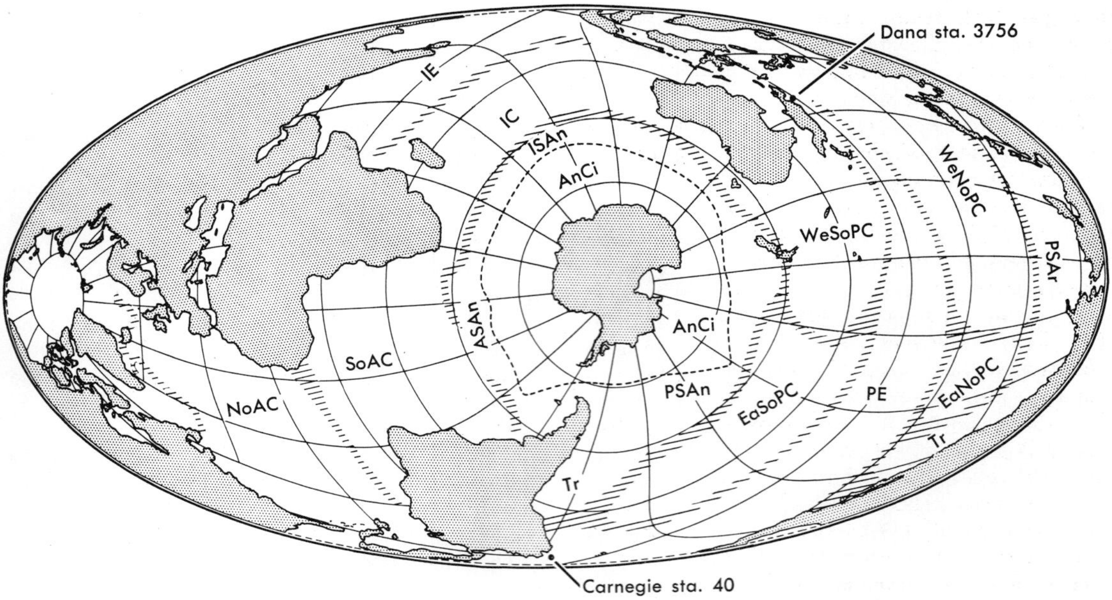

Fig. 7. Distribution of representative water masses of upper 1000 m (symbols as in Fig. 6). Dashed line around Antarctica represents Antarctic Convergence. *(Adapted from H. U. Sverdrup, R. H. Fleming, and M. W. Johnson, The Oceans, Prentice-Hall, 1942)*

vergence; the water mass then formed by subsequent mixing is characterized by a salinity minimum. Arctic Intermediate Water, of little importance in the Atlantic, is widespread in the Pacific and is apparently formed northeast of Japan. Other important intermediate water masses are formed in the Atlantic and Indian Oceans by addition of Mediterranean and Red Sea water, respectively. Deep and bottom waters of the world ocean are formed in high latitudes of the North Atlantic, South Atlantic (Weddell Sea) and Indian Oceans. *See* ANTARCTIC OCEAN; ARCTIC OCEAN; ATLANTIC OCEAN; INDIAN OCEAN; PACIFIC OCEAN. [W.S.W.]

Bibliography: J. D. Cochrane, The frequency distribution of water characteristics in the Pacific Ocean, *Deep-Sea Research*, 5(2):111–127, 1958; C. Eckart, An analysis of the stirring and mixing processes in incompressible fluids, *J. Marine Research*, 7:265, 1948; R. B. Montgomery, Water characteristics of Atlantic Ocean and of world ocean, *Deep-Sea Research*, 5(2):134–148, 1958; M. J. Pollak, Frequency distribution of potential temperatures and salinities in the Indian Ocean, *Deep-Sea Research*, 5(2):128–133, 1958; D. Rochford, Total phosphorus as a means of identifying East Australian water masses, *Deep-Sea Research*, 5(2):89–110, 1958.

SAMPLING AND MEASURING TECHNIQUES

Observations of conditions in the sea generally must be made in situ and the information transmitted back to the observer, or the result must be recorded in situ and retained for reading when withdrawn from the sea. Consequently, the marine scientist is faced with peculiar problems of technique and the use of special equipment to obtain much of his information about sea water. Some of the more commonly used methods and devices for obtaining observations relative to the physical properties of sea water are described here. For information pertaining to the collection of living organisms in the sea, the sampling of the ocean bottom, and the measurement of subsurface currents, *see* MARINE SEDIMENTS; OCEAN CURRENTS.

Temperature-measuring devices. Temperature measurements in the surface layer of the ocean, to depths of 900 ft (275 m), are usually made with the bathythermograph, or BT, a nonelectric device which gives a continuous record of water temperature and depth as it is lowered and raised. The instrument can be operated at frequent intervals while underway and therefore provides a rapid means of obtaining a detailed picture of temperature distribution within the surface layer. *See* BATHYTHERMOGRAPH.

Reversing thermometers. The most reliable and widely used temperature-measuring device is the deep-sea reversing thermometer. This mercury-in-glass thermometer records the temperature at the time it is inverted. It is reliable to about 0.01°C with proper corrections. These thermometers are usually in a pressure-proof glass tube. In this manner the thermometer is protected from the effect of pressure, and the true water temperature is given. These same thermometers are available with a mercury bulb which can be exposed to sea pressure. The compression of the bulb on these unprotected thermometers causes them to read about 1°C high for each 100 m depth. When protected and unprotected reversing thermometers are used in pairs, they give both temperature and depth (thermometric depth). Reversing thermometers usually are attached to a reversing water bottle which collects a water sample when it is inverted.

Electrical thermometers. Since 1950 a rapidly increasing number of electric temperature recorders have been developed around thermistor beads encased in glass. A typical thermistor will change resistance 4% or about 100 ohms/°C and will have a thermal time constant of a few tenths of 1 sec. Electric temperature recorders are usually made to plot temperature against time. They are often used to study microstructure and may have a sensitivity of 0.01–0.001°C. When measuring elements are to be lowered from a ship, the recorder is usually made to plot temperature against depth.

In 1958 a system was developed utilizing electrical thermometers attached to a special cable which could be towed behind a vessel at 500 ft and at speeds of 10 knots. The thermometers are attached to the cable at 25-ft intervals. A facsimile type of recorder draws continuous isotherms on a depth-distance plot. The depth of each degree or tenth-degree isotherm is plotted every 2 sec.

Radiation thermometers have been built and flown from low-flying aircraft. These measure changes in temperature of the water surface to about 0.1°C. Radiotelemetering buoys permit water and air temperatures to be observed on unattended buoys and transmitted to land.

Water samplers. For many chemical and gas analyses it is essential to obtain samples of 100–1000 ml of sea water. These usually are obtained by a series of Nansen bottles attached on 3⁄16-in. hydrographic cable. The bottles are designed to flush continuously when lowered in an upright position. A weight messenger is then sent down the cable. As it strikes the tripping device of the Nansen bottle, the bottle is inverted and the lids are closed (Fig. 8). At the same time another messenger is released to trip the next lower bottle and so on. Usually two reversing thermometers are attached to each bottle. The thermometer and water bottle give accurate results but the method is very time-consuming. To obtain a synoptic picture of temperature and salinity, or density, a number of closely spaced observations must be taken in a relatively short period of time.

In an effort to eliminate contamination from a metallic case, samplers such as the Van Dorn have been made from plastic tubing. Large rubber stoppers on each end are pulled shut by rubber bands. Simple open-tube-type samplers of 1⁄10–10-liter capacity can easily be made.

Measurements such as carbon-14 require samples as large as 200–400 liters. Flushing of such large samplers requires either a large open-ended hose construction or a barrel with two flapper-type ports and water scoops to aid ventilation. After the sample is brought to the surface, some large samplers are retrieved on deck while full. Others are emptied while still in the water by means of a hose.

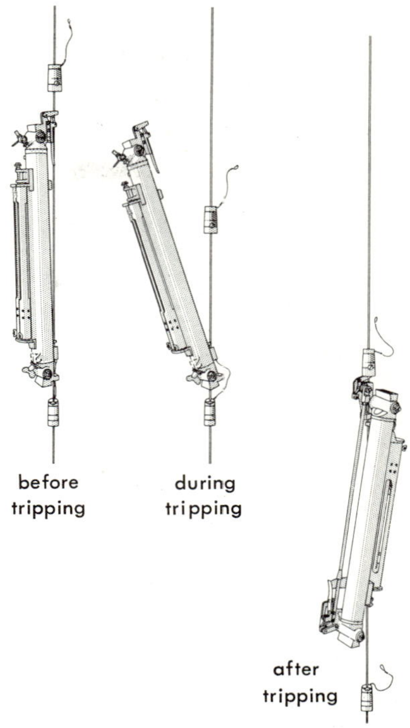

Fig. 8. Nansen bottle in three positions. (*From Instruction Manual for Oceanographic Observations, U.S. Navy Hydrographic Office Publ. 607, 2d ed., 1955*)

Continuous electrical temperature recorders that obtain temperatures with a single probe will almost certainly be used in the future. This will increase the need for a single collecting device that will take many water samples when used at the end of the cable. [A.C.V.]

Serial observations. These are measurements of temperature, salinity, and other properties at a series of depths at some location in the ocean (an oceanographic station), by which the distribution in space and time of these properties (and others computed from them such as density and geopotential) may be described.

Bottles in series. A number of water samplers with thermometers attached are usually lowered on the same cast. As many as 26 samplers have been lowered at once. The number depends on the number of levels to be sampled, the strength of the wire, and the extent of possible damage to the equipment which may result from the roll of the ship or from dragging the bottom.

After the first (deepest) bottle is attached, the wire is paid out and the next bottle and its messenger are attached. When all bottles have been lowered it is necessary to wait for the thermometers to approach equilibrium (about 10 min) before releasing a messenger to trip the cast. If the wire is nearly vertical the messenger falls about 200 m/min. At high wire angles (60° and more have occurred under high wind conditions or strong current shear) the messenger will fall more slowly and may stick. The angle can sometimes be reduced by maneuvering the ship.

After allowing time for the final messenger on the cable to trip the deepest bottle the wire is pulled in and the bottles removed. Water samples are drawn into laboratory bottles and the thermometers are read.

Thermometric depths. When protected and unprotected thermometers are reversed at the same depth, the unprotected will give a higher reading because of the pressure on its bulb. The difference in the two readings depends upon the pressure at reversal, and since this is proportional to depth the "thermometric depth" can be computed. With information from both protected and unprotected thermometers at several of the levels, the shape of the wire can be estimated and the depths of the other samplers computed. Unprotected thermometers are ordinarily used at depths greater than 200 m, since the depth of the upper bottles can be computed from the wire angle and length. Depth computations are estimated to be accurate to ± 5 m in the upper 1000 m and to about 0.5% of wire length below that.

Standard depths. In 1936 the International Association of Physical Oceanography proposed certain standard levels at which observations should be made or values interpolated in reporting. They are (in meters) 0, 10, 20, 30, 50, 75, 100, 150, 200, (250), 300, 400, 500, 600, (700), 800, 1000, 1200, 1500, 2000, 2500, 3000, and by 1000-m intervals at greater depths (depths in parentheses being optional). These values are recommended as a convenient standard of comparison, not as being sufficient for measuring the ocean everywhere. Where the precise level of maxima or minima in the various properties is to be determined, the standard depths may not be adequate, and more depths must be sampled. [J.L.R.]

Analysis of water samples. The development of analytical methods to measure the kind and quantities of dissolved substances in sea water has paralleled advances in analytical chemistry; such methods are usually modifications of techniques and procedures developed for other purposes.

In addition to the usual considerations of accuracy, precision, speed, and cost that control the choice of analytical methods in most applications, methods for sea-water analyses are further restricted by the necessities of performing some analyses on shipboard immediately after the samples are obtained and of storing other samples for analyses that can be performed only in a shore-based laboratory. For example, analyses for biologically active substances, especially those present in trace quantities, are performed on shipboard; analyses for which the highest precision and accuracy are demanded, frequently those which require precision weighing, are performed in shore-based laboratories.

The most complete study now on record of the concentrations of major sea-water constituents—chloride, sodium, magnesium, sulfate, calcium, and potassium—was conducted by W. Dittmar on 77 samples taken during the round-the-world cruise of HMS *Challenger* in 1873–1876. Dittmar's analyses were made by what are now considered classical gravimetric and weight titration procedures: chloride was analyzed by the Volhard method; calcium, by precipitation of the oxalate followed by ignition to and weighing of the oxide; magnesium, by precipitation of magnesium ammonium phosphate followed by ignition to and weighing of pyrophosphate; sulfate, by precipitation and weighing of barium sulfate; potassium, by precipitation of potassium chloroplatinate followed by weighing of metallic platinum after reduction with hydrogen; and sodium, indirectly, as the difference between the measured sum of all cations as sulfates and the sum of the magnesium, calcium, and potassium calculated as sulfates. Despite empirical corrections and analyses by difference in Dittmar's study, more recent analyses have produced only small changes in the mean values of the concentrations of the major constituents.

New analytical methods give results which on the average are not very different from those obtained by classical methods, but the increased accuracy and precision that can be obtained with some of the newer methods permit estimates to be made of relatively small, local changes in the concentration of the major constituents. For example, J. H. Carpenter devised a method for the measurement of calcium having a precision (and probable accuracy) of a few parts in 10,000. The method incorporates separation of calcium from all other cations using ion-exchange chromatography and a complexing elutant, followed by a spectrophotometric weight titration. Analyses of samples taken on the Bahama Banks clearly show the effect of calcium carbonate precipitation, a process that is active in the region.

Many biologically active substances are present in concentrations of micrograms per liter or less. For example, dissolved phosphorus and nitrogen compounds are measured at sea by colorimetric and spectrophotometric methods. Dissolved oxygen, in the parts per million range, is measured by the Winkler titration, a procedure that terminates in an iodimetric titration.

Analytical procedures used on shipboard must be unaffected by ship's motion, both the roll and pitch, caused by waves and winds, and the vibration from engines and generators; and by high humidity and wide variations in temperature. Analytical instruments must be relatively insensitive to poorly regulated power supplies. Lengthy or complicated procedures are not suited for use at sea. The instrumentation of chemical procedures is one of the problems of oceanographic instrumentation.

The measurement of the halide concentration, (predominantly chloride), a property called chlorinity and used in the computation of in situ density, is frequently done with a modification of the classical Mohr titration of halides with silver nitrate using chromate as an indicator. Adaptation to oceanography has produced a highly standardized technique using specially designed pipets and burets. Laboratories throughout the world calibrate the entire chlorinity measurement with a carefully analyzed sea-water standard (Eau de Mer Normale) which is prepared and analyzed in Copenhagen by the International Council for the Exploration of the Sea and distributed at nominal cost to all users.

The practical application of electrical conductivity instruments for the indirect measurement of chlorinity at sea has been demonstrated and precise measurements of the conductivity, temperature, chlorinity relationship have been made. D. W. Jacobson designed an instrument, using electrical conductivity, which provided a record of in situ salinity, temperature, and depth. Various modifications of existing instruments are being tested. Although conductivity has a high temperature coefficient, thereby making it necessary either to measure or control the temperature at the conductivity cell to at least ±0.01°C, the fact that several samples can be processed at one time in the laboratory model makes this a desirable method for the routine measurement of chlorinity.

More recently (1957) an "electrodeless" method of measuring conductivity has been devised at the Chesapeake Bay Institute. Working on an inductance principle—the underwater unit is essentially a tertiary transformer—this method provides the means of in situ measurement and does away with the need for the frequent cell calibration of the conventional conductivity procedures. [D.E.C.]

Water color and transparency. The physical relationships governing the penetration and absorption of light, the color of the water, and the transparency of the sea are of prime importance to physical and biological oceanography. Instruments for measuring the color (transmittance) and transparency of the water are discussed below.

Water color. The color of water in the visual sense is a phenomenon which has both objective and subjective aspects. From an objective point of view the color of water is primarily the result of selective scattering and absorption of visible light by the water itself or by the dissolved or suspended material in the water. The color which is brought about by these basic mechanisms can, however, be drastically altered by the color of the bottom, when visible, by surface films, by the color of the sky or the reflected images of other objects, by the spectral quality of the source of light, and by the optical state of the water surface, as well as by subjective phenomena such as the chromatic adaptation of the observer.

From a physical point of view the color of water is the color of the hydrosol only and can be observed on an overcast day in deep water with the aid of a face mask.

Transmittance. The transmittance of water is closely related to its color. In measuring the transmittance of a path length R of water, consideration must be given to the directional distribution of the light as well as to its spectral composition. For monochromatic light the general equation for transmittance is $T = e^{-CR}$. The attenuation coefficient C varies with the directional distribution of the light, being maximum for collimated light and minimum for completely diffuse light. Modern theory makes use of the attenuation coefficients for both collimated and diffuse light. Their independent measurement is therefore essential. The coefficient α, equal to C for collimated light, is commonly measured with an instrument having a collimated source of light (Fig. 9). The coefficient K, equal to C for diffuse light, is commonly measured under natural illumination by means of a photo detector and a diffusing plate (Fig. 10) having cosine collecting properties. Downwelling irradiance readings H_1 and H_2 are taken at two depths, d_1 and d_2, and the experimental coefficient K is computed from the equation

$$H_1/H_2 = e^{K(d_2-d_1)}$$

Data on the geographic and chronological variations of α and K are still being collected (1958). Some monochromatic data for α are available. Recent monochromatic data (1958) for K are given in Table 5.

Transparency. Measurements of water transparency or clarity are often made with a Secchi disk, an opaque white disk which is held horizontally and lowered into the water until it disappears. The greatest depth at which it can be visually detected is called the Secchi-disk reading and is related in a complex way to the optical properties of the water. Secchi-disk readings depend on the size and reflectance of the disk, the state of the sea surface, the state of the sky, the light adaptation level of the observer, and the technique of observing, as well as many other minor factors.

When the above factors are controlled it can be shown that the apparent contrast of the disk will be

$$C_R = C_0 e^{-(\alpha+K)d}$$

where C_0 is the inherent contrast of the submerged disk. C_0 depends on the submerged reflectance r of the disk and the reflectance of the surrounding water r_b as follows: $C_0 = (r - r_b)/r_b$. Definitions of

Table 5. Spectral values of diffuse attenuation coefficient, K per meter*

Wavelength, mμ	Near-surface values, 28–56 m	Deep-ocean values, 56–159 m
400	0.121	
420	0.139	0.0902
440	0.133	0.0820
460	0.118	0.0749
480	0.115	0.0611
500	0.106	0.0527
520	0.0995	0.0545
540	0.107	0.0806
560	0.105	0.0852
580	0.110	0.0908
600	0.107	0.0933

* Computed from measurements of relative downwelling irradiance obtained by Dr. J. C. Hubbard, Woods Hole Oceanographic Institution, with a submerged monochromater having a bandwidth of approximately 10 mμ. Location of measurements lat. 39°38′N, long. 68°42′W.

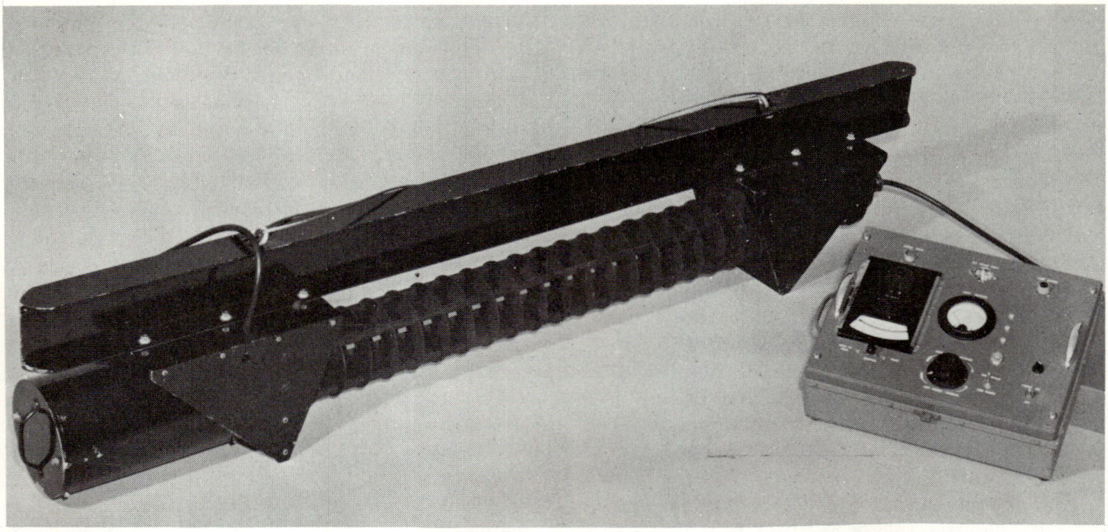

Fig. 9. Hydrophotometer for obtaining the volume attenuation coefficient α. (*Visibility Laboratory, Scripps Institution of Oceanography*)

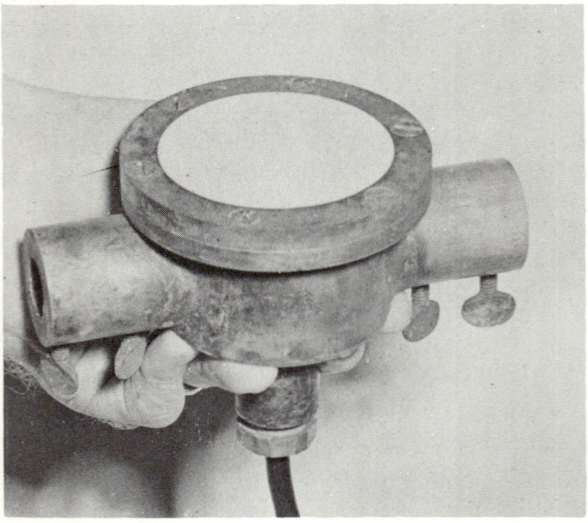

Fig. 10. Irradiance collector for obtaining experimental values of the diffuse attentuation coefficient K. (*Visibility Laboratory, Scripps Institution of Oceanography*)

α and K appear in previous sections; d is the depth of the disk below the surface.

If one uses the utmost precaution and careful technique and does not overlook the effect of human eye capabilities in the final computation, it is possible to obtain the sum $\alpha + K$ from a Secchi-disk reading. It is obvious that Secchi-disk measurements taken from the deck of a ship can be used only to describe the surface strata. [J.E.T.]

Bibliography: H. Barnes, *Apparatus and Methods of Oceanography*, 1959; H. Barnes, *Oceanography and Marine Biology*, 1959; J. H. Carpenter, The determination of calcium in natural waters, *Limnol. Oceanogr.*, 2:271–280, 1957; D. E. Carritt, Analytical chemistry in oceanography, *J. Chem. Educ.*, 35(3):119–122, 1958; J. B. Hersey, Electronics in oceanography, *Advances in Electronics and Electron Physics*, 9:239–295, 1957; E. O. Hulburt, Optics of distilled and natural water, *J. Opt. Soc. Am.*, 35(11):698–705, 1945; *Instruction Manual for Oceanographic Observations*, U.S. Navy H.O. 607, 2d ed., 1955; H. U. Sverdrup, M. W. Johnson, and R. H. Fleming, *The Oceans*, 1942.

Sea water fertility

The fertility of a given area of the sea may be defined in terms of the capacity of the organisms which it contains for the production of organic matter. In practice, measurement of the productive capacity of the plants is sufficient, because their fixation of carbon from carbon dioxide is the primary productive process. The rate of this process can be measured directly, or the amount of organic matter can be estimated in other ways. The estimation of the quantity of living organisms present at any time in a particular place (the standing crop) is difficult and is a measure only of a momentary point of balance between rates of production and destruction. It is not a measure of fertility. *See* BIOLOGICAL PRODUCTIVITY; BIOMASS; FOOD CHAIN; *see also* ECOLOGICAL SYSTEMS, ENERGY IN.

The following list gives estimates of the annual production of biomass in the English Channel. They rest on many assumptions and are approximate only.

Producers	*Biomass, g/m^2 living material*
Phytoplankton	730–910
Zooplankton	275
Pelagic fish	2.9
Demersal fish	1.9
Benthic organisms	55

Coastal and oceanic waters differ considerably in the types of organisms which they support. The chief difference is in the bottom fauna, which is often abundant in shallow water and sparse in the deep oceans where the deeper-living creatures depend on what falls to them from the upper productive layers. *See* DEEP-SEA FAUNA.

Fertility of the oceans. This depends primarily on the production of phytoplankton in the upper, illuminated layers. The plants are mostly diatoms and flagellates which reproduce by division, on the average probably less than once per day. They sink slowly, but are usually eaten by the small zooplankton before reaching the sea bottom. Their chemical composition, unlike that of most land plants, shows a high protein (40–50%) and a high fat (20–27%) content and resembles that of the animals which follow them in the food chain in the sea.

The phytoplankton needs radiant energy, carbon dioxide and water, nitrogen and phosphorus, and a range of other elements of which there seems usually to be no shortage, with the possible exceptions of silicon, iron, and manganese. Since there is always enough carbon dioxide and water, plant growth is normally limited by the supply of radiant energy, nitrogen, and phosphorus. The rate of growth is somewhat temperature-dependent. Daylight is absorbed by sea water and at depths usually less than 100 m is attenuated to an intensity of less than 0.5 cal/(cm^2) (day) which is too low to allow continued growth of the phytoplankton; the latter requires intensities of up to about 50 cal/(cm^2) (day). In latitudes where low solar elevation in winter gives too low an irradiation there is a seasonal variation. When illumination is adequate, plant growth may be limited by the supply of nitrogen, phosphorus, and trace elements. In the surface layers in winter, nitrogen (mostly as nitrate) may be present in concentrations up to 200 μg N/liter, and phosphorus (as phosphate) up to 40 μg P/liter. These concentrations may be reduced almost to zero in summer. A further supply is in circulation in plants and animals. The deep waters of the oceans contain greater quantities of these nutrients, and where water movements bring them to the surface, increased phytoplankton growths occur.

Geographical variations are found, therefore, both in productive capacity and in standing crops. Measured rates of productivity vary up to well over 100 g carbon/(m^2) (year); 1 g carbon may be taken as equivalent to about 16 g living material. The standing crop of plants is in the range 0.1–10 g C/m^2. The productivity of all the oceans has been put at 1.6–15.5 $\times 10^{10}$ tons C/year, compared with 1.9 $\times 10^{10}$ tons C/year for the land areas of the world. [F.A.J.A.]

Productivity and its measurement. The production of organic matter in the sea, as on land, is accomplished by the photosynthetic activity of autotrophic plants. In coastal waters, where sunlight penetrates to the bottom, both rooted plants and benthic algae contribute to this process. In the open sea, organic production is limited to unicellular algae, the phytoplankton, which live suspended in the upper layers of all ocean waters.

Productivity of benthic plant communities may be determined by periodic harvest and measurement of their growth over discrete time intervals. Such direct methods are impossible in the study of the short-lived plankton because of unmeasurable losses from natural death, predation, sedimentation, and advection.

A more satisfactory approach to both benthic and planktonic plant production is through measurement of chemical changes of the water accompanying photosynthesis and growth. These may be followed in situ for periods ranging from 1 day to several weeks, or in vitro by exposing representative samples of the plant population to natural conditions for periods not exceeding 1–2 days.

Photosynthetic activity is indicated by the changes in the water of nitrogen and phosphorus salts, oxygen, and carbon dioxide, and by the degree of acidity (pH). Calculations based on in situ changes of these indicators must allow for gas exchanges across the water surface and the effects of vertical mixing between surface and deep waters. In such calculations the horizontal advection is generally neglected, and complete chemical recycling between sampling periods cannot be accounted for. For the last reason, the method tends to give conservative estimates of productivity.

Experimental in vitro studies include measurement of oxygen production, carbon dioxide assimilation, and pH change. Both in situ and in vitro changes of these properties represent the net effect of photosynthesis and respiration (by both plants and animals), and hence measure net production. Respiration may be measured separately in vitro in dark-bottle experiments. This measurement, when added to the net change observed in transparent bottles, gives a measure of real photosynthesis or gross production. Oxygen-bottle experiments lack the necessary sensitivity for use in the open sea, where plankton are sparse, and have been largely replaced by the extremely sensitive method of measuring CO_2 uptake using C^{14} as a tracer. $C^{14}O_2$ uptake appears to be equivalent to net production (photosynthesis minus respiration) by the plant community.

A third method for estimating productivity is based on the premise that photosynthesis is a function of two independent variables, the chlorophyll content of the plants and the light intensity which they receive. Production may be calculated from simultaneous measurements of these factors in the ocean and from their experimentally derived relationship to photosynthesis.

The few existing measurements of dense benthic plant communities indicate that they may produce as much as 20 g organic matter/(m^2) (day), an amount equivalent to the best agricultural yields on land. Plankton production seldom if ever attains this level, though values half as great are not uncommon. The productivity of shallow, inshore waters is generally higher than that of the open sea, but the seasonal range of most marine areas includes two orders of magnitude. The mean annual rate of production in the oceans as a whole is a matter of some controversy, but probably lies between 100 and 300 g organic matter/m^2 sea surface, which represents an efficiency of utilization of 0.1–0.2% of the incident, visible solar energy.

[J.H.RY.]

Geographic variations in productivity. Strictly speaking, variations in productivity imply variations in the rate of entry of carbon into the organic cycle, or gross photosynthesis. The extent to which this takes place in the sea is determined by the amount of photosynthesizing plant tissue present, the temperature, and the available light energy. The net productivity is the rate of plant growth. This is of greater value as a measurement because it eliminates from the determination the respiratory and excretory losses of the plants and specifies the production of food for the planktonic animals. Variations in net productivity depend on the physiological and oceanographical factors affecting algal growth in the sea, to which the availability of nutrient salts is of prime importance.

The limited penetration of daylight into the sea restricts plant growth to the upper *euphotic* layer where the nutrients can be assimilated. The plants sink to deeper layers where the nutrients are released by decomposition of plant material by microorganisms and returned to the sea. Acting against this downward transport of nutrients is eddy diffusion (produced by turbulence), which brings nutrients up to the surface from the richer deeper waters. This is facilitated where the water is homogeneous and suppressed by stratification. Currents perform the major transport of nutrients, and where surface divergences occur, upwelling currents bring deeper water to the surface (see illustration). In the illustration the north Indian Ocean and the China Sea appear under southwestern monsoon conditions; these currents are reversed, with a shift in regions of divergence, in the northeastern monsoon.

In temperate and higher latitudes there is a pronounced seasonal cycle, with suppressed production in winter because of excessive turbulence and lack of light, a rapid burst of growth in spring, and limitation of this growth by lack of nutrients when the waters stabilize in summer. Frequently in autumn there is a subsidiary flowering before growth is again limited by lack of light.

In tropical and subtropical regions a more permanent stratification limits nutrient supply to the euphotic zone, and there appears to be a low net production rate. Exceptions are found in regions of divergence, principally on the western coasts of the continents and to a lesser extent in the open ocean in the equatorial region, where upwelling of rich deeper waters permits a high productivity, often throughout the year.

The question of relative production in high and low latitudes is still undecided, for the high productivity of the polar seas is of short seasonal duration and the tropics may in fact equal it on an annual basis, although running at a lower instantaneous rate. As yet, measurements of production rates are inconclusive on this point.

Some estimate of the production of higher animals can be obtained from commercial fishing and whaling statistics, and the correlation with net production rate appears to be fairly good. but is modi-

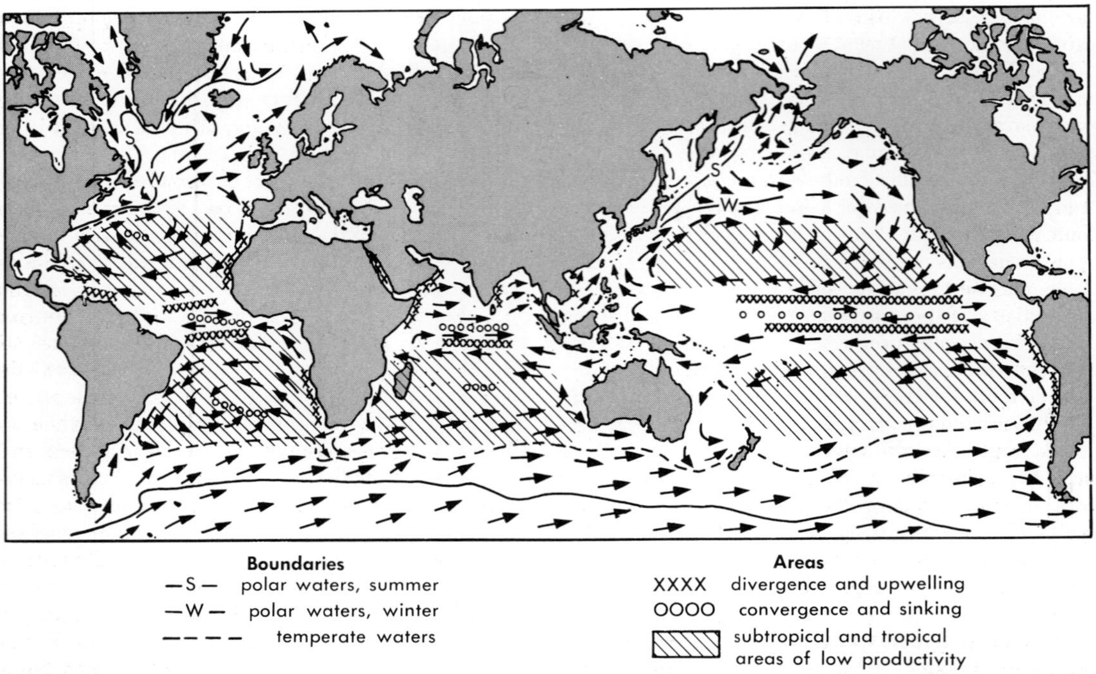

World map (Mercator's projection) showing ocean surface currents, boundaries between areas of high and low productivity, and principal regions of divergence and sinking.

fied by feeding and breeding requirements of the animals in question and by the fact that many higher animals of no commercial interest may be produced in some areas. [R.I.C.]

Size of populations and fluctuations. A population is defined as the smallest collection of organisms of a given species whose numbers are maintained solely by reproductive processes to which every mature member of the population contributes, or is potentially able to contribute. A population usually occupies a definable area, has one or more specific centers of reproduction, and its members usually possess a set of morphological or physiological characteristics by which they may be recognised and distinguished from other populations of the same species; such features, however, are not in themselves sufficient to establish the concept of population.

Methods of measuring the size of marine populations are of three basic kinds. The first is by a census based on samples which together constitute a known fraction either of the whole population, as in the case of sessile species such as shellfish, or of a particular age-range, as in the case of fish which have pelagic eggs, the total abundance of which can be measured by fine-meshed nets hauled vertically through the water column. The second is by marking or tagging, in which a known number of marked individuals are mixed into the population and the ratio of unmarked to marked individuals is subsequently measured from samples. The third is applicable to commercially exploited populations where the total annual catch is known; the mortality rate caused by exploitation is measured, based on the age-composition of the population, thus establishing the fraction which the catch is of the population. The last two methods are most generally used. Details of the methods are given by R. J. H. Beverton and S. J. Holt (1957).

The largest measured populations are of pelagic fish, particularly of the herring family and related species (Clupeoidei); one of these is the Norwegian herring, *Clupea harengus*, which contains on the order of 1,000,000,000 mature individuals and ranges over hundreds of miles of the northeast Atlantic.

Fluctuations in the size of marine fish populations may be broadly classified as long-term, with a periodicity of the order of 30–100 years, and short-term, with a periodicity of less than 10 years. The first are known from historical records of certain commercial fisheries and are probably caused by long-term climatic changes influencing water conditions, especially the current systems. The second are due primarily to variation in the size of the year-classes entering the population each year; year-class variation may be up to thirtyfold and is caused mainly by environmental conditions affecting the survival of eggs and larvae. *See* CLIMATIC CHANGE. [R.J.H.B.]

Biological species and water masses. Biogeographical regions in the ocean are related to the distribution of water masses. Their physical and ecological individuality is derived from partly closed patterns of circulation and from amounts of incident solar radiation characteristic of latitudinal belts. Each region may be described in terms of its temperature-salinity property and of the biological species which are adapted to all or part of the relatively homogeneous physical-chemical environment.

Cosmopolitan species. The discrete distributions of many species are circumscribed by the regions of oceanic convergence bounding principal water masses. Other distributions are limited to current systems. Cosmopolitan species are distributed across several of the temperature-salinity water masses or oceans; their wider specific tolerances reflect adaptations to broadly defined water types. No pelagic distribution is fully understood in terms of the ecology of the species.

A habitat is integrated and maintained by a current system: oceanic gyral, eddy, or current, with associated countercurrents. This precludes species extinction that could occur if a stock were swept downstream into an alien environment. The positions of distribution boundaries may vary locally with seasonal or short-term changes in temperature, available food, transparency of the water, or direction and intensity of currents.

Phytoplankton species are distributed according to temperature tolerances in thermal water masses, but micronutrients (for example, vitamin B_{12}) are essential for growth in certain species. The cells of phytoplankton reproduce asexually and sometimes persist in unfavorable regions as resistant resting spores. New populations may develop in prompt response to local change in temperature or in nutrient content of the water. Such species are less useful in tracing source of water than longer-lived, sexually reproducing zooplankton species.

Indicator organisms. The indicator organism concept recognizes a distinction between typical and atypical distributions of a species. The origin of atypical water is indicated by the presumed affinity of the transported organisms with their established centers of distribution.

Zooplankton groups best understood with respect to their oceanic geography are crustaceans such as copepods and euphausiids, chaetognaths (arrow worms), polychaetous annelids, pteropod mollusks, pelagic tunicates, foramaniferans, and radiolarians. Of these the euphausiids are the strongest diurnal vertical migrants (200–700 m). The vertical dimension of euphausiid habitat agrees with the thickness of temperature-salinity water masses, and many species distributions correspond with the positions of the masses. In the Pacific different species, some of which are endemic to their specific waters, occupy the subarctic mass (such as *Thysanoessa longipes*), the transition zone, a mixed mass lying between subarctic and central water in midocean and between subarctic and equatorial water in the California Current (for example, *Nematoscelis difficilis*), the barren North Pacific central (such as *Euphausia hemigibba*) and South Pacific central masses (for example, *Euphausia gibba*), the Pacific equatorial mass (such as *Eu-*

phausia diomediae), a southern transition zone analogous to that of the Northern Hemisphere (represented by *Nematoscelis megalops*), and a circumglobal subantarctic belt south of the subantarctic convergence (for example, *Euphausia lucens*).

Epipelagic fishes and other strongly swimming vertebrates are believed to be distributed according to temperature tolerances of the species and availability of food. However, distributions of certain bathypelagic fishes (such as *Chauliodus*) have been related to water mass. [E.B.]

Bibliography: R. J. H. Beverton and S. J. Holt, On the dynamics of exploited fish populations, *Fish. Invest.* (London), 19, 1957; G. E. Fogg, *The Metabolism of Algae*, 1953; H. W. Harvey, *The Chemistry and Fertility of Sea Waters*, 2d ed., 1957; Natl. Acad. Sci.—Natl. Research Council, *The Effects of Atomic Radiation on Oceanography and Fisheries*, Publ. 551, 1957; H. B. Moore, *Marine Ecology*, 1958.

Seal (zoology)

Any of a large number of aquatic carnivores of the suborder Pinnipedia, all adapted for aquatic life through the modification of the legs into flippers, the digits being fully webbed. The body is slender and the tail is short. There are three families, two of which occur off the coast of North America. The family Otariidae includes the sea lions and the fur seals. The latter are valuable fur bearers. The northern fur seal, *Callorhinus ursinus*, is now well established as a thriving and valuable fur resource, breeding in large numbers on the Pribiloff Islands, Alaska. At one time, this species became almost extinct due to unregulated hunting.

The harbor seal, *Phoca vitulina*; length to 60 in. (*From E. L. Palmer, Fieldbook of Natural History, McGraw-Hill, 1949*)

The hair seals, family Phocidae, lack the wooly underfur of the Otariidae and have other anatomical distinctions. There are many species of this cosmopolitan family. Several are important as food and skin animals to the Eskimos, but otherwise they are of little value. *See* CARNIVORA; SEA LION. [J.D.B.]

Seal, pressure

A seal is used to make pressureproof the interface (contacting surfaces) between two parts that have frequent or continual relative rotational or translational motion and are known as dynamic seals, as compared with static seals (*see* GASKET). While the pressure in seals is lower than in gaskets, the motion hinders their effectiveness so that there are more types of seals than gaskets, each type attempting to serve its environment. The materials are leather, rubber, cotton, flax, and for piston rings, cast iron. The forms of nonmetallic seals are rectangular, V-ring, and O-ring. Cartridge seals are available for rolling-contact bearings. Special seals include carbon ring and labyrinth seals for turbines (*see* STEAM TURBINE) and mechanical seals for pumps. [P.H.B.]

Seamount and guyot

An isolated submarine mountain rising 3000 ft or more above the ocean floor. The oceans cover at least 10,000 such mountains which occur as volcanic peaks on ridges, or rises, or as individual peaks.

Flat-topped seamounts are called guyots, or tablemounts (see illustration). They are present on

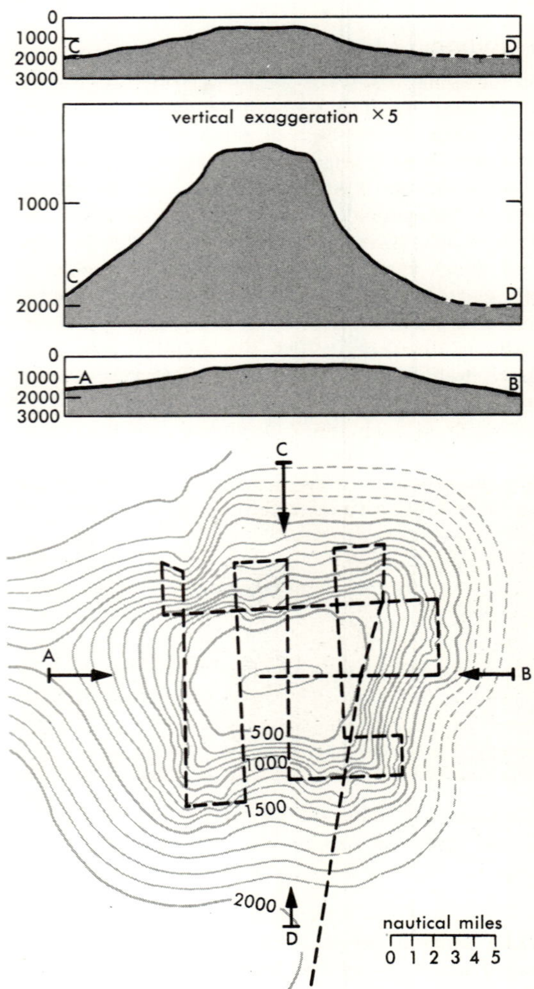

Pratt Seamount: 142°30′W 56°20′N. Plan and profiles of an isolated flat-topped seamount (guyot) in the Gulf of Alaska. Contour interval 100 fathoms.

all ocean floors but are most common in the Pacific. Bottom samples dredged from several guyots include reef corals and rounded volcanic cobbles. Both the coral and volcanic erosion debris indicate that the flat tops were once at sea level though they are now 1000–7000 ft below the ocean surface. Thus guyots are ancient islands which were truncated to sea level by erosion. *See* OCEANIC ISLANDS.

Guyots may have subsided to their present depths or they may have been relatively immobile while sea level rose. It is not certain which process dominated in general or in any particular place. A possible exception is known in the Gulf of Alaska where a line of guyots intersects the Aleutian Trench. There all guyots are about ½ mile deep except one situated on the axis of the trench which is almost 1½ miles deep. It appears that the additional mile of depth may have been produced by the subsidence of the trench under the deepest guyot. *See* OCEANIC ISLANDS; SEA LEVEL FLUCTUATIONS; SUBMARINE TOPOGRAPHY; *see also* ATOLL; CORAL REEF.

[E.L.H.; H.W.M.]

Seaplane

An airplane that takes off from and alights on water (*see* AIRPLANE). In a seaplane, the necessarily large size of the pontoons leads to the use of the underside of the fuselage as a hull, stepped for planing on the water; vehicles so designed are termed flying boats. The fluidity of the water absorbs shock; thus, even if strut-mount floats are used, shock absorbers are unnecessary. Because of spray, elevator and engine nacelle with propeller may be mounted higher than for a land-based aircraft, as illustrated. With the engine thrust well above the center of gravity, changes in thrust produce appreciable changes in trim. While at rest on the water, a seaplane is supported by buoyancy.

Four-passenger, all-metal amphibian has tricycle for use on land and stepped hull with sponsons under wings for use on water. (*Colonial Aircraft Corp.*)

While planing preparatory to take-off, it is supported partly by hydrodynamic forces on its hull and partly by aerodynamic forces on its wings. Once airborne, a seaplane is supported by wing lift. The vehicle must be stable throughout this change of support. *See* BUOYANCY; FLIGHT CHARACTERISTICS.

[F.H.R.]

Search coil

A device used for measuring the flux density in a small region of a magnetic field. The apparatus consists of a small coil connected by flexible leads to a ballistic galvanometer. If the coil is placed with its plane perpendicular to the magnetic field, the flux threading the coil is $\Phi = BA$, where A is the area of the coil and B the magnetic induction (*see* INDUCTION, MAGNETIC; MAGNETIC FLUX). When the coil is quickly turned through a quarter turn or is withdrawn from the field to a place where B is zero, the flux through the coil is changed from BA to zero. During this change, there is an electromotive force (emf) whose instantaneous value is e, and a current in the closed circuit whose instantaneous value is i, given by

$$i = \frac{e}{R}$$

where R is the resistance of the entire circuit of coil and galvanometer. If N is the number of turns in the coil

$$e = -N\frac{d\Phi}{dt}$$

and $$i = -\frac{N}{R}\frac{d\Phi}{dt}$$

Hence $$\int_0^t i\,dt = q = -\frac{N}{R}\int_\Phi^0 d\Phi = \frac{N\Phi}{R}$$

or $$\Phi = \frac{R}{N}q$$

and $$B = \frac{\Phi}{A} = \frac{Rq}{NA}$$

The deflection of the ballistic galvanometer is proportional to the charge q; hence, if the instrument is properly calibrated, the value of B can be determined (*see* GALVANOMETER). The coil is frequently permanently connected to the galvanometer, and the calibration made with the coil in place.

If the direction of the field is not accurately known, several readings can be made with the plane of the coil in different orientations. The maximum deflection will be that for which the plane of the coil was perpendicular to the field. From this observation, both the magnitude and the direction of the field are obtained. *See* FLUXMETER. [K.V.M.]

Bibliography: F. K. Harris, *Electrical Measurements*, 1952; W. C. Michels, *Electrical Measurements and Their Applications*, 1957.

Sebaceous gland

A gland which produces and liberates sebum, a mixture composed of fat, cellular debris, and keratin. When the gland arises in association with a hair follicle it forms a thickened outpushing from the side of the developing follicle near the epidermis. Central cells in these sebaceous glands form oil droplets within the cytoplasm. These cells disintegrate to liberate the sebaceous substance and are therefore of the holocrine type. The Meibomian or tarsal glands within the tarsus or supporting plate at the edge of the eyelids are sebaceous and complex tubuloacinous structures. The numerous separate glands open along the entire edge of the upper and lower lids. Retained secretions of the tarsal glands produce a cyst termed a chalozion or Meibomian cyst. *See* EPITHELIUM; GLAND. [O.E.N.]

Second (time unit)

The fundamental unit of time which, together with the centimeter and the gram, constitutes the centimeter-gram-second or cgs system of units. The official definition of the second is the fraction $1/_{31,556,925.9747}$ of the tropical year for 1900 January 0 at 12 hr, ephemeris time. The date 1900 January 0 is used by astronomers synonymously with 1899 December 31. The second was formerly defined as the 86,400th part of the mean solar day, but was redefined in 1956 by international authority, owing to variations in the duration of the day.

The second is nearly equal to the 86,400th part of the average mean solar day during the eighteenth and nineteenth centuries. The sidereal second is the 86,400th part of the sidereal day. The mean solar second is the 86,400th part of the mean solar day. *See* TIME. [G.M.C.]

Secondary emission

The emission of electrons from the surface of a solid into vacuum caused by bombardment with charged particles, in particular with electrons. The mechanism of secondary emission under ion bombardment is quite different from that under electron bombardment; the discussion here will be limited to the latter case because it is in this sense that the term secondary emission is generally used.

The bombarding electrons and the emitted electrons are referred to respectively as primaries and secondaries. Secondary emission has important practical applications because the secondary yield, that is, the number of secondaries emitted per incident primary, may exceed unity. Thus, secondary emitters are used in electron multipliers and in other electronic devices such as television pick-up tubes, storage tubes for electronic computers, and so on.

Secondary yield. The most thoroughly investigated property of secondary emission is the yield as a function of the energy of the primaries. The yield may be measured by means of the circuit shown schematically in Fig. 1. A beam of primary electrons strikes a target with an energy determined by the potential difference between the target and the cathode. The primary beam passes through a hole in the "collector" which has been made positive with respect to the target. The secondaries emitted by the target then flow to the collector, and the yield is obtained as the ratio of the secondary current i_s to the primary current i_p.

An example of the yield δ as a function of the primary energy E_0 is shown in Fig. 2 for a single crystal of magnesium oxide (MgO). The shape of the yield curve is essentially the same for all materials, that is, δ increases with increasing E_0 to a maximum value δ_m occurring at a primary energy E_{0m}, and then decreases as the primary energy increases beyond E_{0m}. The maximum yield of approximately 24 for MgO single crystals is the highest yield found so far for any material, at least in the absence of field effects.

For metals and semiconductors the maximum yield is of the order of unity and occurs at primary energies of several hundred electron volts, as may

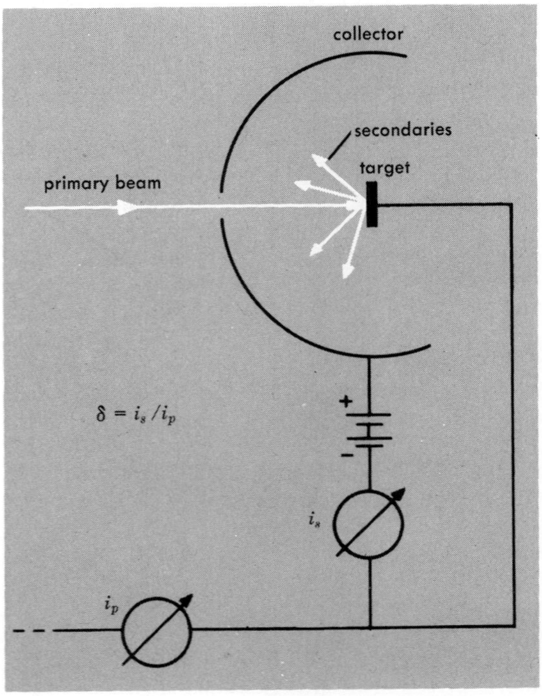

Fig. 1. Schematic circuit for measuring the secondary yield; i_p and i_s represent the primary and secondary currents.

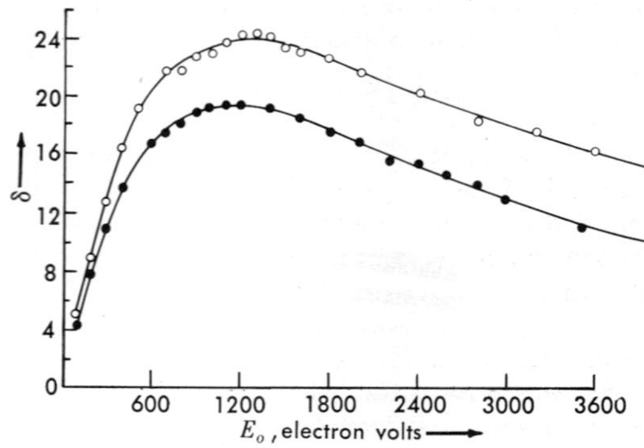

Fig. 2. The secondary yield δ as a function of primary energy E_0 for a single crystal of MgO. The lower curve represents the yield before any heating, and the upper curve that after heating the crystal to 800°C for several hours and cooling to room temperature. (*After N. R. Whetten and A. B. Laponsky, J. Appl. Phys., 28:515, 1957*)

Table 1. Values of maximum secondary yield δ_m and corresponding primary energy E_{0m} for some metals and semiconductors

Metal	δ_m	E_{0m}, ev	Semiconductor	δ_m	E_{0m}, ev
Li	0.5	85	B	1.2	150
Al	0.95	300	Si	1.1	250
Ni	1.35	550	Ge	1.15	400
Cu	1.3	600	Sb	1.3	500
Pt	1.8	700	Bi	1.15	550

be seen from Table 1. Reported values of δ_m for insulators lie in the neighborhood of 5, but it is not impossible that atomically clean surfaces of some of these materials may exhibit yields approaching the value of 24 found for MgO; in order to obtain reliable yield measurements characteristic of clean surfaces, extremely good vacuum techniques are required.

Mechanism of the process. The shape of the curve of Fig. 2 can be understood by considering the secondary emission process as a two-step mechanism. One step involves the production of internal secondaries, that is, the transfer of energy from the primary beam to the electrons in the solid. In this process, some of the electrons in the solid are raised to energy levels which lie above the energy of an electron to rest in vacuum. The diffusion toward and the escape from the surface of such relatively high-energy internal secondaries constitute the second step in the mechanism.

Both the production and the escape mechanism are extremely complex. An elementary interpretation may be given as follows. Suppose the average number of secondary electrons produced per primary by the primary beam in a slab between x and $x + dx$ below the surface of the solid is equal to $n(x)\,dx$. As the secondaries diffuse toward the surface they gradually lose energy, and one may assume for simplicity that the probability of escape of a secondary produced at a depth x is proportional to $\exp(-x/x_0)$; x_0 is a measure for the average depth from which a secondary can escape. The yield may then be written as

$$\delta = B \int_0^R n(x) e^{-x/x_0}\,dx$$

where B is a constant and R represents the penetration depth of the primaries. It is reasonable to assume that $n(x)$ is proportional to the energy loss per unit depth per primary. According to energy dissipation measurements by J. R. Young in 1956, the latter quantity is essentially constant over the primary range R. Hence, if one writes $n(x) = E_0/R\varepsilon$, where E_0 is the primary energy and ε the average energy required to produce an internal secondary, one may write

$$\delta = (BE_0/R\varepsilon) \int_0^R e^{-x/x_0}\,dx$$
$$= (BE_0 x_0/R\varepsilon)(1 - e^{-R/x_0})$$

Since the primary range R increases with primary energy approximately as $E_0^{1.35}$, this formula indeed predicts a maximum in the curve of δ versus E_0. A more detailed analysis shows that this formula describes satisfactorily the shape of the observed yield curves up to primary energies of about 4000 electron volts. From such an analysis, one may calculate x_0 and ε/B from experimental yield curves; illustrative values are given in Table 2.

Table 2. Values of x_0 and ε/B

	Platinum	Germanium	Magnesium oxide
x_0, angstroms	16	27	180
ε/B, ev	154	137	19.7

The secondary escape depth x_0 is determined by the processes by which secondaries can lose energy as they diffuse toward the surface. In metals, secondaries lose their energy mainly to the free conduction electrons; since this is an efficient process, the escape depth in metals is small and so is the yield. In insulators, the secondaries lose energy mainly to crystal lattice vibrations; this results in a relatively large escape depth and a high yield. Since the amplitude of the lattice vibrations increases with temperature, the escape depth and hence the yield of an insulator decreases slightly with increasing temperature. For metals, the yield is independent of temperature. The secondary emission properties of semiconductors are intermediate between those of metals and insulators.

The secondary yield for a given primary energy increases as the angle θ between the primary beam and the normal to the surface increases; the secondaries are then produced closer to the surface and consequently have a larger escape probability. At the same time, the energy for which the yield reaches its maximum value increases with increasing θ.

Secondary energies. A typical energy distribution of secondary electrons emitted by a silver target bombarded with primaries of 155 electron volts energy is given in Fig. 3. Note that most of the

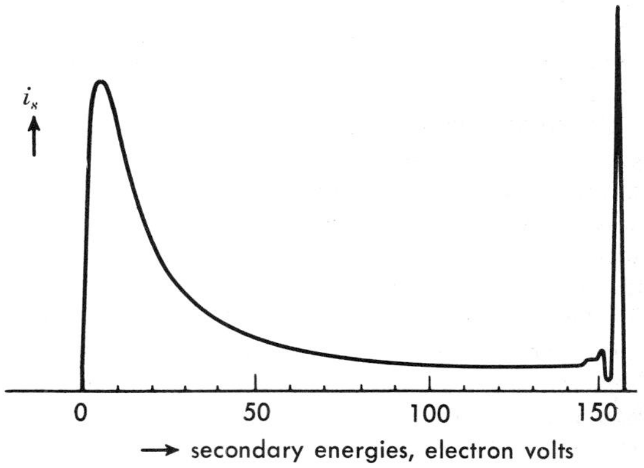

Fig. 3. The secondary current i_s as a function of the energy of the secondary electrons emitted by silver upon bombardment with primaries of 155 electron volts energy. (*After E. Rudberg, Inelastic scattering of electrons from solids, Phys. Rev., 50:138, 1936*)

secondaries have relatively low energies. A small fraction of the emitted electrons have the same energy as the incident primaries and are called reflected primaries.

Field-enhanced emission. When an insulating film deposited on a metal base is bombarded with primary electrons, the surface of the target will become positively charged if the yield is larger than unity. Thus, strong fields may be established across the insulating film, leading to field-enhanced secondary emission. In 1936 L. Malter discovered that in such cases self-sustained secondary emission may result for long periods after the primary beam is cut off. *See* FIELD-ENHANCED EMISSION. [A.J.DE.]

Bibliography: F. Seitz and D. Turnbull (eds.), *Solid State Physics*, vol. 6, 1958; A. Van der Ziel, *Solid State Physical Electronics*, 1957.

Second-order transition

A change of state through which the free energy of a substance and its first derivatives are continuous functions of temperature and pressure, but at which the second derivatives are discontinuous.

For all physical and chemical processes carried out reversibly, the free energy changes continuously. At an ordinary phase transition, such as the boiling of a liquid, the entropy S, the enthalpy H, and the volume V show sharp discontinuities when plotted as functions of the temperature T or pressure P. Because all of these functions are first derivatives of the free energy

$$S = -(\partial F/\partial T)_P \qquad H = [\partial(F/T)/\partial(1/T)]_P$$
$$V = (\partial F/\partial P)_T$$

such phase changes are usually called first-order transitions.

However, for many systems there are points at which the entropy, enthalpy, and volume are continuous, but at which temperature or pressure derivatives, such as the heat capacity $C_p = (\partial H/\partial T)_P$, the coefficient of thermal expansion $\alpha = (\partial \ln V/\partial T)_P$, and the isothermal compressibility $\kappa = (\partial \ln V/\partial P)_T$, show discontinuities. Because these correspond to second derivatives of the free energy, this phenomenon is called a second-order transition.

The illustration shows the typical thermodynamic behavior at first- and second-order transitions. The dotted vertical line for the heat capacity at the first-order transition is a Dirac delta function, that is, a line of zero width and infinite height representing a finite and nonzero area, the heat of the transition which occurs at a single temperature. The dashed lines represent metastable phases continued beyond the transition temperature (for example, superheated liquid above the boiling point and supercooled vapor below). Both the low-temperature (α) and high-temperature (β) phases show such extensions beyond a first-order transition, whereas only the β phase shows such an extension at a second-order transition.

Qualitatively all theories of second-order transitions have the following features in common: a system is capable of existing in two forms, one (α) having a lower enthalpy H and a lower entropy S than the other (β). At sufficiently low temperature, the enthalpy difference will be the dominant factor and the system will be all α, whereas at sufficiently high temperatures, it will be largely or entirely β. If the conditions were such that the α and β forms could not coexist, there would be a first-order transition at a temperature

$$T = (H_\beta - H_\alpha)/(S_\beta - S_\alpha)$$

On the other hand, if the change from α to β can take place gradually (that is, if a mixed phase including both forms can exist) and if the energy required to convert an element of the system (a molecule or group of molecules) from α to β decreases as the amount of β increases, a second-order transition will occur. This change-over from α to β in a sense catalyzes itself, so one refers to phenomena of this kind as cooperative. The temperature at which the last trace of α disappears is the λ-point or Curie point; it is of course meaningless to extrapolate the $\alpha + \beta$ curve beyond this point. The heat capacity is very large at the λ-point; it is difficult to be sure whether in some systems it may not actually become infinite; in any case the area under the curve (ΔH) is finite.

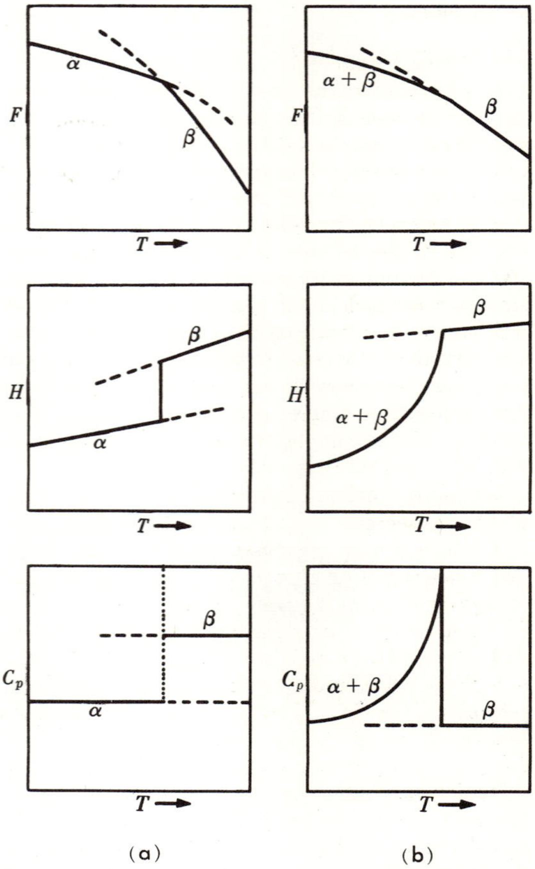

Thermodynamic behavior at (*a*) first- and (*b*) second-order transitions.

Ferromagnetism. In certain metals and alloys (iron and nickel) at low temperatures, the atomic magnets are arranged into ordered groups or domains which can orient in a magnetic field. As the temperature increases, the order within the domains decreases until at the Curie temperature, all long-range order is gone and only paramagnetic behavior remains. *See* FERROMAGNETISM.

Order-disorder in crystals. In certain solid solutions (such as β-brass, Cu-Zn), the different atoms are distributed regularly in an alternating arrangement. As the temperature increases, the two kinds of atoms exchange positions until all long-range order is lost at the Curie point, above which the arrangement is essentially random. A similar phenomenon occurs in the solid ammonium halides; each NH_4^+ tetrahedron can have two different orientations. At low temperatures, all have the same orientation; above the Curie temperature, they are distributed randomly between the two.

Liquid helium. Below 2.19°K, helium shows peculiar superfluid properties. At the lowest attainable temperature, all the molecules are in a superfluid state; as the temperature increases, more and more molecules are excited to nonsuperfluid levels until at the λ-point (2.19°K), the superfluid properties have disappeared and the helium is an ordinary liquid. *See* SUPERFLUIDITY; *see also* EQUILIBRIUM, PHASE. [R.L.S.]

Secretory structures, plant

Cells or organizations of cells which produce a variety of secretions. The process of secretion is a separation of a substance from the protoplast of a cell. The secreted substance may remain deposited within the secretory cell itself or may be excreted, that is, released from the cell. Substances may be excreted to the surface of the plant or into intercellular cavities or canals. Some of the many substances contained in the secretions are not further utilized by the plant (resins, rubber, tannins, various crystals), others take part in the functions of the plant (enzymes, hormones). Secretory structures range from single cells scattered among other kinds of cells to complex structures involving many cells and often called glands.

Glandular hairs. Epidermal hairs of many plants are secretory or glandular. Such hairs commonly have a head composed of one or more secretory cells borne on a stalk (Fig. 1*a* and *b*). The shaggy hairs of winter buds of many trees produce sticky secretions that permeate and cover the buds with a protective film. The hair of a stinging nettle is bulbous below and extends into a long fine process above (Fig. 1*c*). If one touches the hair, its tip breaks off, the sharp edge penetrates the skin, and the poisonous secretion is released.

Nectaries. Glands secreting a sugary liquid—the nectar—in flowers pollinated by insects are called nectaries (Fig. 2). Nectaries may occur on the floral stalk or on any floral organ: sepal, petal, stamen, or ovary. The nectary may be flat, depressed, or padlike. Its surface layer, one or more

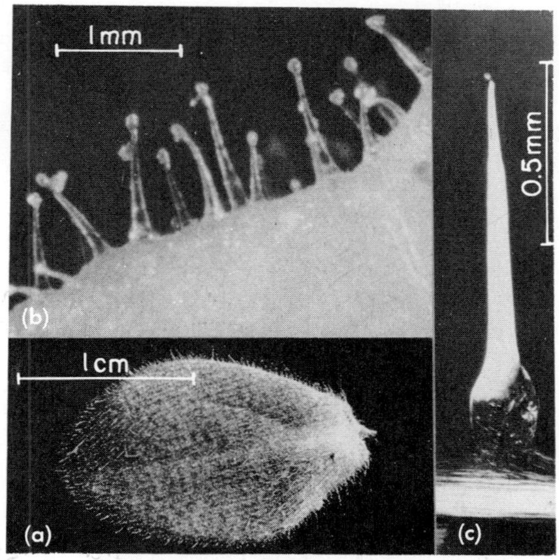

Fig. 1. (*a*) Young leaf of *Petunia* covered with glandular hairs. (*b*) Higher magnification of these hairs. Each has a head borne on a stalk. Some extraneous matter clings to the viscid hairs. (*c*) Stinging hair of nettle photographed with polarized light. The brilliantly illuminated parts of the hair are hard and stiff. The globule at the bent tip is attached by a thin wall part which breaks easily and leaves the beveled end exposed. The multicellular bulb below has soft walls.

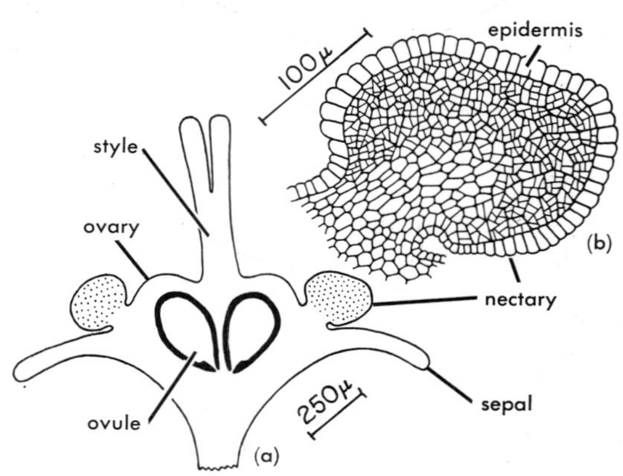

Fig. 2. (*a*) Longitudinal section of *Ceanothus* flower. The nectary is ringlike and surrounds the ovary. (*b*) Enlarged view of a section of the nectary. The epidermis and the small cells beneath it are secretory.

cells deep, consists of secretory cells which usually have dense cytoplasm. The vascular tissue, especially the sugar-conducting phloem, occurs close to the secretory tissue. The nectar may contain as much as 60% sugar.

Hydathodes. These structures discharge water—a phenomenon called guttation—through openings in margins or tips of leaves. The water flows through the xylem to its endings in the leaf, then through the intercellular spaces of the hydathode tissue toward the openings in the epidermis. Strictly speaking, such hydathodes are not glands

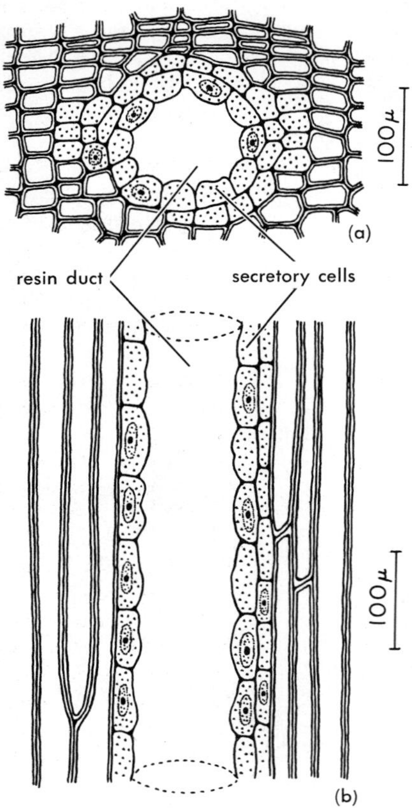

Fig. 3. Resin duct from pine wood in (a) transverse and (b) longitudinal sections.

because they are passive with regard to the flow of water. The pressure forcing the water to be discharged originates in the root. Some hydathodes, however, are like nectaries in having a layer of actively secreting cells.

Digestive glands. Some carnivorous plants have glands producing secretions capable of digesting insects and small animals. These glands occur on leaf parts modified as insect-trapping structures. In the sundews (*Drosera*) the traps bear stalked glands, called tentacles. Each gland consists of four layers of cells, the innermost being in contact with the conducting tissue of the stalk. When an insect lights on the leaf, the tentacles bend down and cover the victim with a mucilaginous secretion the enzymes of which digest the insect. In the Venus' flytrap (*Dionaea*) the insect is trapped by the sudden folding of the leaf. The upper surface of the leaf bears digestive glands and hairs acting as triggers for the folding mechanism. The traps in the pitcher plants are pitcher-shaped leaves that bear nectaries on the surface and inside the trap near its top; the digestive glands are deeper inside. Both consist of a layer of secretory columnar cells, a layer of rounded cells, and a layer of suberized cells. If an insect, attracted by the nectar, falls into the pitcher, it is prevented from escaping by rigid hairs pointing downward and is digested within the trap. There is no evidence, however, that the digested insects are essential to the growth of such plants.

Resin ducts. Resin ducts are canals lined with secretory cells that release resins into the canal (Fig. 3). The canals are intercellular spaces that originate by separation of cells. Resin ducts are common in gymnosperms and occur in various tissues of roots, stems, leaves, and reproductive structures. They may arise during normal development or as a result of injury.

Gum ducts. These ducts are similar to resin ducts and may contain resins, oils, and gums. Usually the term gum duct is used with reference to the dicotyledons, although gum ducts also may occur in the gymnosperms.

Oil ducts. Oil ducts are intercellular canals whose secretory cells produce oils or similar substances. Such ducts may be seen, for example, in various parts of the plant of the carrot family (Umbelliferae). In contrast to the oil ducts, the oil cavities of the kind found in the fruit rind and other plant parts of citrus result from a breakdown of oil-containing cells.

Laticifers. Cells or systems of cells containing latex, a milky or clear, colored or colorless, liquid.

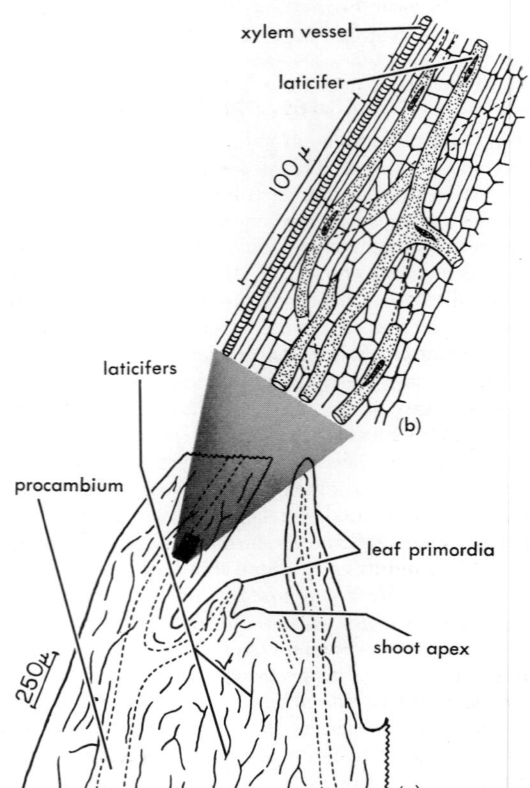

Fig. 4. Laticifers of *Nerium oleander*. (a) Longitudinal section of shoot tip. The wavy solid lines represent parts of laticiferous cells, which are arranged very irregularly. (b) An enlarged view of the area in the black rectangle in (a) showing the parts of laticifers with cytoplasm and nuclei.

Latex occurs under pressure and exudes from the plant when the latter is cut. The single-cell laticifer is often much branched (Fig. 4); a few laticifer cells originating in the embryo may branch and invade all newly produced plant parts, forming the entire laticifer system of the mature plant. Laticifers derived from more than one cell result from breakdown of walls between adjacent cells; they are sometimes called laticiferous tubes. Both kinds of laticifers have multinucleate protoplasts. The latex is the product of this protoplast, but its relation to the cytoplasm seems to be more complex than that between the vacuole and cytoplasm in ordinary cells. The latex of some plants has a high content of rubber and is used as the source of natural commercial rubber. Examples of rubber-yielding plants are the Brazilian (*Hevea*) and Indian (*Ficus*) rubber trees. *See* CYTOLOGY; INSECTIVOROUS PLANTS; PLANT METABOLISM; PLANT TISSUE SYSTEMS. [K.E.]

Bibliography: *See* PLANT ANATOMY.

Sedation

A state of decreased activity produced by various drugs. Sedation is the quiet, relaxed state of rest not necessarily accompanied by sleep. The induction of sedation depends largely on the physical environment, the condition and susceptibility of the individual, and the proper selection of a sedative which will accomplish its purpose. Any drug which will depress the activity of the central nervous system, particularly higher centers, may theoretically be used as a sedative. In practice, however, there are serious drawbacks to the use of many of these depressants, due to such factors as low margin of safety, toxicity, undesirable side effects, and addiction or habituation. *See* CENTRAL NERVOUS SYSTEM.

Barbiturates, bromides, and chloral hydrate. Of the older drugs which promote relaxation, the barbiturates, bromides, and chloral hydrate are the most commonly used. The barbiturates constitute a valuable group of medicinals, mainly because of their relatively wide margin of safety and the extent of their action. This varies over a wide range, from mild sedation to profound anesthesia, depending upon the compound, the dose, and the route of administration. Intermediate effects are principally related to induction of sleep, that is, they produce a hypnotic effect. In addition, some barbiturates have a selectively depressant effect on the motor portions of the nervous system so that they may be used as anticonvulsants in disease and toxic states. *See* ANESTHETIC.

The sedative action of a barbiturate is thought to be the result of the depression of diencephalic centers so that nerve impulses to the cortex, or higher levels of consciousness, are reduced. There is also a direct effect on the cerebral cortex, because both wakefulness and convulsions may be reduced. Increased tolerance and habituation may occur. *See* BRAIN.

Bromides depress the activities of the motor and sensory cortex by action of the bromide ion, and produce a state of dullness, unconcern, and apathy. Although still much used by laymen, bromides are being replaced as sedatives because 1–2 days of treatment are necessary to produce results and chronic poisoning, or bromism, frequently follows prolonged therapy.

Chloral hydrate depresses the cerebrum and, to a lesser extent, the spinal reflexes. Because it is inexpensive, chloral hydrate is used in many institutions for sedation and hypnosis, particularly in the treatment of drug addicts and alcoholics. It is contraindicated in patients with liver and heart disease and may be irritating to the stomach. *See* HYPNOSIS.

Other sedatives sometimes used are paraldehyde, carbromal, and chlorobutanol. They have actions similar to choral hydrate.

Tranquilizer. An entirely new field of therapy has been opened by the advent of new synthetic sedatives and, more important, by the development of a new kind of drug, the tranquilizer, or ataraxic.

The disadvantages of habituation, toxicity, and frequency of overdoses in the emotionally labile patient gave impetus to the search for new preparations. With increasing tensions, states of nervousness, and neurosis, the older preparations have been augmented by a multitude of new compounds variously described as tranquilizers, ataraxics (producing peace of mind), antidepressants, and antitension agents. *See* PSYCHIC ENERGIZER; TRANQUILIZER.

Part of this new flood stemmed from the effects produced by new antihypertension drugs, such as rauwolfia compounds, and from antiemetics such as chlorpromazine. These produced sedation and quietude when used for other purposes, especially in disturbed patients. Undesirable side effects have appeared, however, and their use cannot be indiscriminate.

Frenquel next appeared, together with other members of the group of meprobamates, such as Equanil and Miltown, which produce muscle relaxation. The number of such drugs in the United States runs to several hundred. Many are older, established preparations, some are newer forms and combinations of barbiturates, and the remainder are newer drugs of the tranquilizer variety.

There is a difference in effect between the tranquilizer and the antidepressant in that the latter acts as a controlled stimulant in states of mental depression of all kinds. Tension and nervousness may be present in some forms of depression so that sedation is also helpful.

There is also an increasing use of combined forms in one preparation. Sedatives, tranquilizers, and antidepressants are used together or with other drugs, especially those with antihypertensive, antiemetic, and antispasmodic effects. [E.G.ST.]

Sedimentary rocks

One of the three major groups of rocks that make up the crust of the earth, the other two being igneous and metamorphic. Most sedimentary rocks are layered, and, as is implied by the name, have originated by the sedimentation, or settling, of particles. Thus layering, or stratification, is the most important single characteristic of sediments and sedimentary rocks, even though there are some igneous and metamorphic rocks that show some kind of stratification or pseudostratification. *See* IGNEOUS ROCKS; METAMORPHIC ROCKS.

The distinction between sedimentary and other rocks is understood best by considering their origin. Sediments are formed at or very near the surface of the earth as a result of processes operating at the surface, at normal earth surface temperatures and pressures. Most igneous and metamorphic rocks, on the other hand, are formed as a result of conditions deep in the crust of the earth, where temperatures and pressures may be very high. Some overlapping between the three rock families exists; for example, it is difficult to classify rocks that originate as volcanic ash falls (igneous) but are then transported and become interlayered with normal sediments. It may also be difficult to distinguish between a hard, compacted sedimentary rock and a weakly metamorphosed rock.

Though sedimentary rocks account for only 5% of the earth's outer crust (a shell 10 mi thick), they make up 75% of the exposed rocks at the surface. From this relationship alone it becomes apparent that, in general aspect, sediments are distributed as a rather thin layer at the surface. The thickness of this thin layer may vary greatly from place to place; the thickness of the total sedimentary volume may be only a few tens of feet at the edges of some old igneous mountain masses such as the Ozark or Adirondack Mountains, but may be well over 30,000 ft in some places where the crust is subsiding rapidly, such as the delta of the Mississippi River (*see* DELTA; GEOSYNCLINE). Though sediments are quantitatively relatively unimportant as crustal constituents, they have been the chief means of elucidating the history of the earth and, with their contained fossils, the development and evolution of life forms. Sedimentary rocks are also important as the source of many of our major mineral resources, notably coal, oil and gas, iron ores, and limestone.

Origin. Sedimentary rocks originate primarily as the result of the fragmentation and destruction of preexisting rocks. As rocks are weathered by the action of water, wind, frost, and organic decay, large masses become mechanically broken into finer sizes and some of the constituents dissolve in rain or soil water. The solid fragments, ions in solution, and colloids in suspension, are transported, primarily by running water and secondarily by wind and ground water, from the site of weathering, the source area, to this site of deposition. Transportation of detritus may be temporarily interrupted by sedimentation in streams or lakes, resulting in river bars, alluvial fans, or lake deltas. Eventually, however, most of the material reaches the site of the lowest gravitational potential energy on the earth's surface, the bottom of the sea. *See* SEDIMENTATION (GEOLOGY). After final deposition the soft, water-saturated muds, silts, and sands become buried under successive layers of later sediment, the water is squeezed out, the sediments become compacted, and chemical changes result in cementing the original unconsolidated material to a rock. *See* DIAGENESIS.

Sedimentary petrologists commonly divide the sedimentary rocks into two large groups, the detrital and the chemical. This division is based on the differing origins of the two groups. The detrital (sometimes called clastic) rocks are formed by the sedimentation of mineral or rock fragments that were derived from the mechanical disintegration of preexisting rocks in the source area and have been transported, as solids, to the site of deposition. The chemical rocks originate as chemical precipitates at the site of deposition; they may be inorganic precipitates formed from supersaturated solutions or they may be formed by the biochemical action of organisms, as are the calcium carbonate shells of mollusks. A great many sediments are mixtures of detrital and chemical components; many chemically precipitated limestones contain some fine-sized grains of quartz and clay minerals, most of which probably originated as wind-blown or animal-carried material. Predominantly detrital rocks, such as sandstones and shales, commonly contain some amount of chemical precipitate, calcium carbonate and silica being the two most abundant; the chemical components may have been introduced at the time of deposition or during postdepositional changes (diagenesis). *See* PETROLOGY; ROCK.

The three most abundant kinds of sedimentary rocks, shale, sandstone, and limestone, together account for over 95% of all sediments. Of these, limestone, the chemical rock, composes only about 20% of the total volume of sedimentary rocks in the crust. Estimates of the relative proportions of the two detrital rocks, shale and sandstone, vary, but it appears that shales are between two and three times as abundant as sandstones.

Textural characteristics. Because the majority of sediments are dominantly mechanical mixtures of detrital mineral and rock fragments, floccules of colloidal materials such as clay, and chemically precipitated particles, it is important to determine the geometrical properties of the individual particles and their relationships to each other. Textural analysis has led to a fuller understanding of the genetic factors involved in the formation of sediment by settling of particles through a fluid (water) or gaseous (air) medium, for the textures are directly related to the hydrodynamics of the medium. A number of different textural properties have been defined.

Size. Perhaps the most important textural property is the size of individual particles and the size distribution of all the particles in the sediment. Size of particles is the basis for the division of the detrital rocks into lutite, or shale (fine); arenite, or sandstone (medium); and rudite, or conglomerate (coarse). Because it is manifestly impossible to

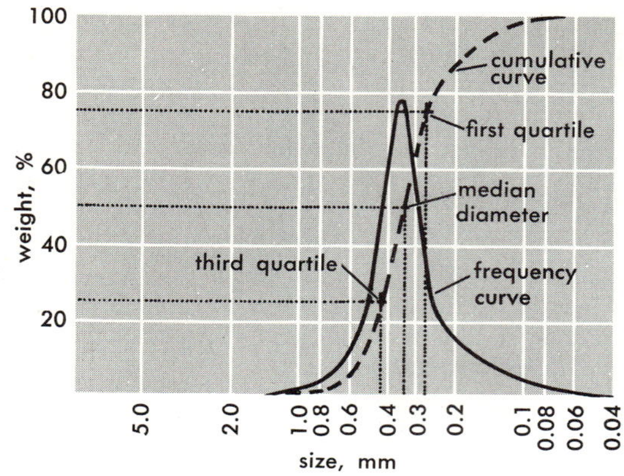

Fig. 1. Size distribution curves.

measure the sizes of all of the particles in a sediment, and one seeks to characterize a distribution of sizes, statistical analysis has been used extensively in this work. By statistical analysis the sedimentary petrographer can characterize four properties of size distributions based on counts of frequency of grains in several size grades. The first property is the calculation of some kind of average size, either arithmetical or geometrical mean, or a median (50 percentile) size. The most common measure used by geologists is median size. A second property of the size distribution is the spread or dispersion of size values about the mean or median size. Standard deviation is one measure of this spread. One of the more common measures is quartile deviation (evaluating spread between the 75 and 25 percentiles) or sorting index (Fig. 1). A third property of size distributions is the skewness or asymmetry of the distribution about the median value. This measure indicates whether a sediment has much more material finer than the median size as compared with the fraction coarser than the median size, or vice versa. A fourth property of the size distribution is kurtosis, which measures the number of grains that cluster in size around the median as compared with the number of grains that are much finer or coarser than the median. The median size of a sediment has been used to estimate the competence of a current to transport sediment: the coarser the size, the stronger the current. Sorting has been used to distinguish between beach sands, river sands, bar sands, and others, but there is some uncertainty in the validity of this interpretation. Skewness and kurtosis have been little used in the interpretation of origin of a sediment.

Shape and roundness. In addition to the size distribution, textural analysis includes the study of the shape and roundness of the particles. Shape is defined as the degree of sphericity, or approach to a sphere. Roundness is defined as the degree of sharpness of corners or edges of a particle. A particle may have many sharp, small projections and have a low roundness value and yet be very close to a sphere in shape. On the other hand, a particle may be long and rodlike, very far in shape from a sphere (low sphericity), and yet be very smooth and rounded (high roundness). Shape and roundness are related to mechanical abrasion during transportation of the detritus prior to deposition; the greater the abrasion, the higher the roundness, and, in general, the higher the sphericity (Fig. 2).

Surface configuration. Another element of texture is the surface configuration of the grains. Some sand grains show frosting and pitting, which has been interpreted as the result of being windblown and having gone through a great many collisions with other grains. Collisions under water are softened by the fluid medium and do not result in this texture.

Packing and fabric analysis. In the late 1940s and 1950s two new kinds of textural analysis were introduced, packing analysis and grain-shape fabric analysis. Packing analysis relates to the way in which the particles are arranged in the rock to give more or less dense aggregates. The density of packing is related to the weight of overlying sediments and perhaps also to lateral compressive forces operating during mountain-building episodes. Grain-shape fabric analysis is the study of the degree of preferred orientation of the long axis of elongate particles and the direction of the orientation. A rock with a high degree of preferred orientation would show all of the long particles lined up in the same direction. This direction of preferred orientation is related to the average direction of current flow of the medium from which the sediment settled. *See* PETROFABRIC ANALYSIS.

Chemical composition. Chemical composition of sedimentary (as well as other) rocks is expressed in terms of the oxides of the elements. Determinations are made normally by wet chemical analysis and the assumption is made that the elements are present as oxides, the oxygen not being determined directly. Although the chemical composition of sediments varies widely with lithology and grain

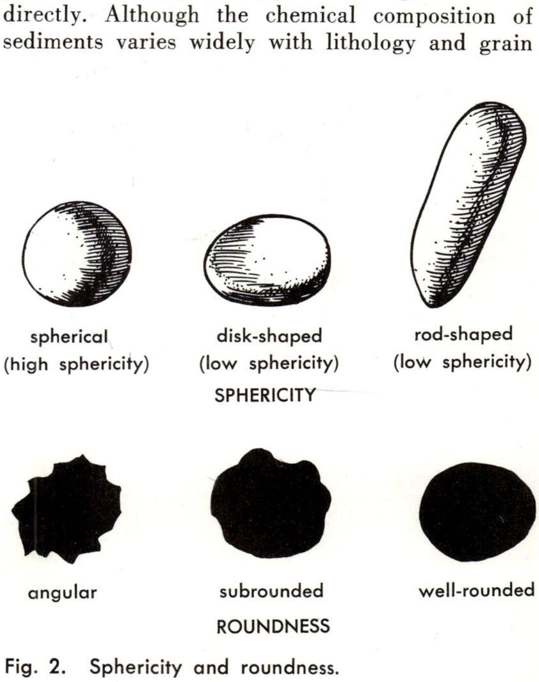

Fig. 2. Sphericity and roundness.

size, the average sedimentary rock contains about 58% silicon dioxide (SiO_2), 13% aluminum oxide (Al_2O_3), 6% calcium oxide (CaO), 5% ferrous oxide (FeO) and ferric oxide (Fe_2O_3), 5% carbon dioxide (CO_2), and smaller amounts of magnesium oxide (MgO), potassium oxide (K_2O), sodium oxide (Na_2O), and many trace elements. This average sediment may be thought of as a combination of sandstone (mainly SiO_2), shale (mainly Al_2O_3), and limestone (mainly CaO and CO_2). The average sedimentary rock differs from the average igneous rock in having much more CO_2, much lower Na_2O, and much more ferric than ferrous iron. The CO_2 is added primarily from the atmosphere, in the process of weathering. Sodium is lost, during weathering, as soluble salts, which accumulate in the oceans, and much ferrous iron is changed to ferric under the oxidizing conditions prevalent over most of the earth's surface. *See* WEATHERING PROCESSES.

Mineralogical composition. The composition of sediments is best expressed in terms of the relative abundances of minerals present. This is so because it is not always possible to calculate mineralogy from gross chemical composition without knowing what minerals are present; however, the chemical composition can easily be calculated once the mineral assemblage is known. Also, because sediments are often a mechanical mixture of minerals of many different chemical origins, the interpretation of mineralogy proceeds in a much more direct fashion than that of gross chemical composition. Although a great number of mineral species have been found in sedimentary rocks, usually in very small amounts, only some 20 minerals account for over 99% of the bulk of the sedimentary rocks. These, in general, are the minerals that are fairly stable chemically in earth surface environments. The minerals of sediments, just as the rocks themselves, are divided into two groups, the detrital and the chemical. The detrital minerals are mechanically transported and deposited, and, normally, are considered to have originated by the mechanical disintegration of the parent rock in a weathering source area of sediments. The chemical minerals are considered to have formed by precipitation, either inorganic or biogenic by the action of organisms, at the site of deposition.

Many minerals may be either or both detrital and chemical in origin. Quartz, for example, is dominantly detrital and yet it is frequently found as a chemically precipitated mineral that formed in rocks after they were deposited (*see* AUTHIGENIC MINERALS). Although calcite (calcium carbonate) is normally considered to be a chemical precipitate, it is recognized in many limestones as an essentially detrital mineral, having been transported some distance from its original place of formation and deposited mechanically in the same way as a quartz sand grain. The most abundant detrital minerals in sediments are quartz and clay minerals. Less abundant, but still quantitatively important in many sediments are feldspar, rock fragments, and coarse grained micas. A host of other detrital minerals are found in many sedimentary rocks, the sum of them rarely making up more than 1% of the rock. Most of these are grouped under the name "heavy minerals," because they have specific gravities greater than 2.85, the specific gravity of bromoform, the liquid commonly used to separate heavy from light minerals (quartz, feldspar) by a sink or float method. Some of the most common detrital heavy minerals are zircon, tourmaline, garnet, hornblende, epidote, rutile, staurolite, and magnetite. *See* MINERALOGY; PETROGRAPHY.

The major chemically precipitated minerals are the carbonates—calcite, aragonite, dolomite, and siderite. Less important are chert, gypsum and anhydrite, other saline residues, such as common rock salt (halite), and a number of phosphates, such as collophane. A number of heavy minerals may be chemical in part; zircon and tourmaline often show chemically precipitated secondary additions to the original detrital grain and anatase seems always to be formed as a chemical precipitate subsequent to deposition.

Significance of detrital minerals. The detrital minerals are significant as guides to the composition of the parent rocks of the sediment and as indexes of the degree of weathering of those parent rocks. If a source area terrain is subjected to rapid mechanical erosion and little chemical action, most of the major minerals and rock fragments will be transported as such to the depositional area. In such cases the sediment may show a high percentage of minerals that are unstable at the earth's surface. Such mineral assemblages in the sediment may differ, depending on the kinds of rocks exposed to erosion in the source area. Also low-grade metamorphic may be distinguished from high-grade metamorphic or igneous terrains. Heavy minerals are particularly useful for this purpose. If mechanical erosion is at a minimum and chemical weathering is intense in the source area, most unstable minerals will tend to dissolve or alter, leaving behind a residue rich in quartz, the common constituent of most rocks that is stable in sedimentary environments. Thus a sediment that contains only quartz as a detrital mineral may be interpreted as the product of intense chemical weathering in the source area. The ratio of quartz to feldspar in true surface sediments is a rough index of source area weathering, as quartz is stable and feldspar unstable. The higher the quartz feldspar ratio, the more intense the chemical weathering of the parent rocks. One cannot always interpret detrital mineralogy strictly in terms of the above ideas, however, for there may be changes in mineralogy after deposition. It appears that certain of the heavy minerals, in particular, olivine, augite, and other ferromagnesians, tend to dissolve if the rock has been buried a long time.

Significance of chemical minerals. The chemical precipitates are of interpretive significance as reflectors of the chemical environment of deposition. The chemical controls that determine if and what kinds of minerals are precipitated are the composition of the water solutions at the depositional site, that is, sea water, brackish water, or fresh water; the oxidation-reduction potential (Eh) and the acidity-alkalinity (pH) of the solutions; and to a lesser extent pressure and temperature. In addition to the inorganic chemical controls, there are the biological factors controlling mineral precipitates. Many invertebrates and plants, in particular mollusks and algae, are the prime agents for the pre-

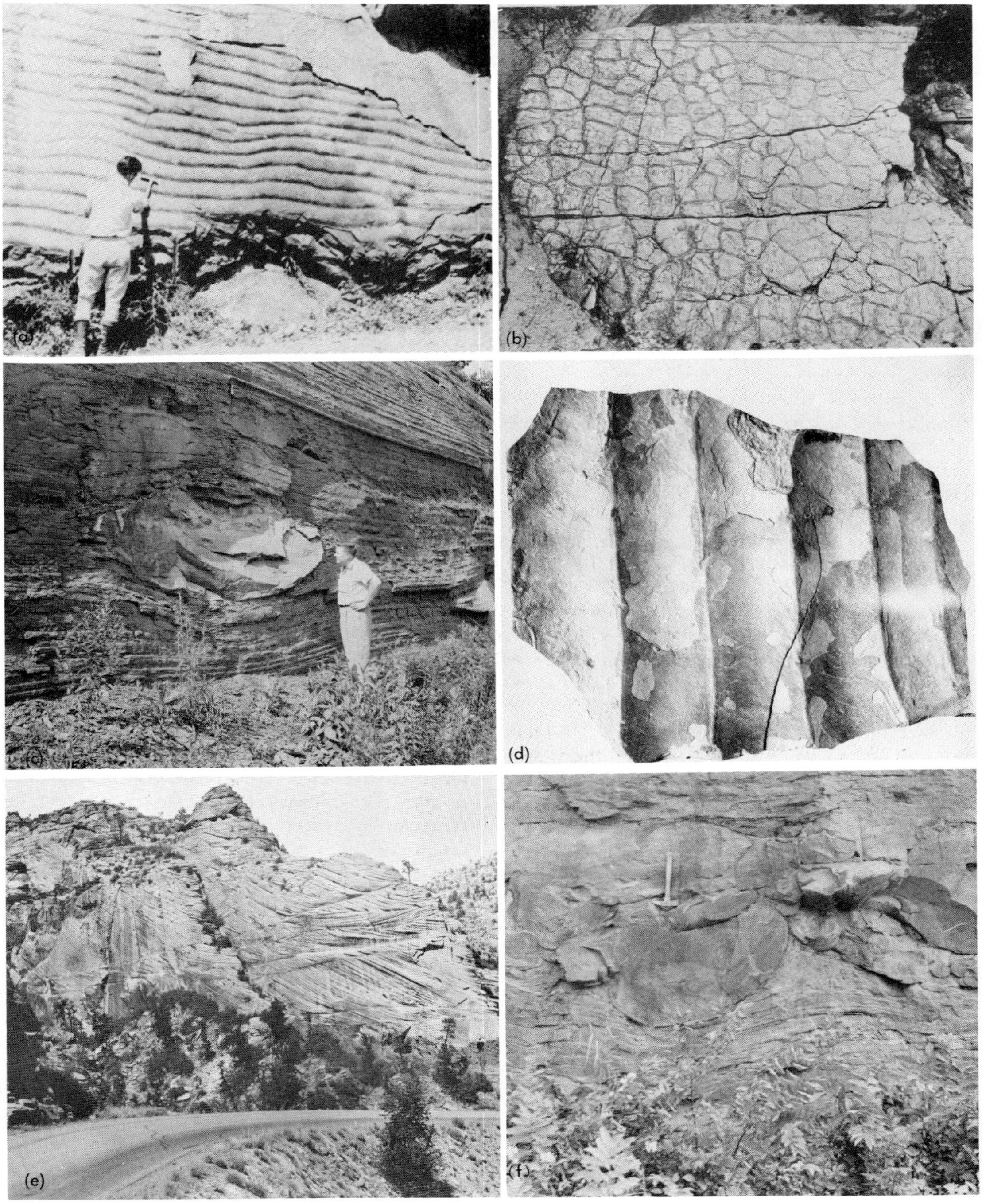

Fig. 3. Structures in sedimentary rocks. (*a*) Current ripple marks in copper ridge dolomite, south of Bluefield, Va.; (*b*) mud cracks in Mississippian limestone, northeast of Bluefield, Va.; (*c*) concretions in Pennsylvanian shale, near Montgomery, W.Va. (*Virginia Division of Mineral Resources*); (d) oscillation ripples in a Cretaceous sandstone; (e) dune bedding in the Navajo sandstone near entrance to Zion Canyon; (*f*) flow rolls in the Chemung sandstone at Chemung Nose, south of Elmira, N.Y. (*Carl O. Dunbar*).

cipitation of carbonates. Indeed, some authorities believe that originally all sedimentary carbonates were biogenic, and that lack of fossil structures is chiefly due to postdepositional recrystallization or solution and reprecipitation. Carbonate minerals can be interpreted mainly as the result of biological activity and the pH of the solutions. Iron and sulfide minerals may reflect pH, oxidation-reduction potentials, and biological activity. Chemically precipitated minerals that form postdepositionally in a rock are often called authigenic; some of the most common abundant authigenic minerals are quartz, calcite, and dolomite. *See* AUTHIGENIC MINERALS.

Clay minerals. The clay minerals occupy a position between the detrital and chemical. Basically the clays are detrital, and their fundamental crystallographic structures are normally preserved through mild weathering, transport, deposition, and diagenesis. But the clay minerals are very susceptible to exchange of alkali and alkaline earth cations with their environment, and the clay that is finally produced in a sediment may be of a composition and over-all structure quite different from that originally supplied from the parent rocks in the source area. *See* CLAY; CLAY MINERALS.

Sedimentary structures. These are the larger textural features of sediments such as bedding, ripplemarks, and concretions that were formed during or shortly after deposition, as distinct from the still larger elements of structure, folds and faults, that were produced much later than deposition (*see* STRUCTURAL GEOLOGY). Sedimentary structures include the mechanical, made by the currents that transport sediment; the chemical, produced by inhomogeneous precipitation; and the organic, produced by organisms living in the environment.

Mechanical structures. These structures include bedding of various kinds, such as cross-bedding, inclined bedding, graded bedding, and ripplemarks, and slump structures caused by small landslides more or less contemporaneous with sedimentation (Fig. 3). Fossil mudcracks, caused by temporary desiccation of a mud bottom, raindrop impressions, and frost crystal casts are included in this category. Results of work done in the late 1940s and 1950s on different kinds of sedimentary structures indicate that such structures are related to the environment of deposition, and that many of the structures, cross-bedding in particular, are a clue to the direction of sediment transport. Many different kinds of linear structures on bedding plane surfaces have been described, including rill marks and swash marks, made by the retreat of waves on a beach; groove casts and load casts, made by movement of a pebble or cobble on a muddy bottom; and more problematical structures of unknown origin. Most of these have been ascribed to current origin and the lineation parallels the current.

Chemical structures. These structures form as segregations of originally dispersed chemical substances. They include oolites and pisolites, concretions, geodes, nodules, and septaria, as well as stylolites, a solution feature. They take many forms, from highly irregular, perhaps even branched forms, to regular spheres or ellipsoids, and range in size from dumbbell-like objects 3 ft in diameter to tiny spherulites only a few millimeters in diameter. Some of the structures apparently form at the same time as the sediment, others may form very soon after sedimentation, or before compaction, and others form after compaction and consolidation of the sediment, perhaps quite late in its history.

Nodules are concretionary structures, with a great diversity of irregular shapes, composed of material different from that of the rock in which they occur. Most frequently nodules are flattened or elongated in a direction parallel to the bedding. Sometimes they tend to coalesce to form almost continuous layers. The most common nodules are of chert (silica); others are of iron oxides, phosphates, iron carbonates, and clay ironstones. They occur most typically in limestones but may also be found in shales and sandstones.

Septaria are rather large nodules, normally greater than 3–4 in., that display a system of polygonal cracks at the center and which tend to die out toward the edges of the nodule (Fig. 4). In almost all of these nodules the cracks are filled with a crystalline mineral, normally calcite. The septaria was originally a gel concretion that hardened or dehydrated on the outside first; shrinkage caused by dehydration of the gel was responsible for the cracking. Later mineral solutions filled the cracks. *See* CONCRETION; GEODE; OOLITE AND PISOLITE; STYLOLITES.

Organic structures. Preserved in sedimentary rocks are textural elements that are either the remains of organisms that lived during the time the sediment was being laid down or that have resulted from the activities of those organisms. Of greatest importance are the former, the fossils, which, in the main, are the preserved hard parts of the plants or animals. The evidences of organism activity are less frequent but consist of worm borings and tubes, faecal pellets of many different kinds, and larger structures of excretory origin, the coprolites. Fossils may consist of calcareous or phosphatic shells,

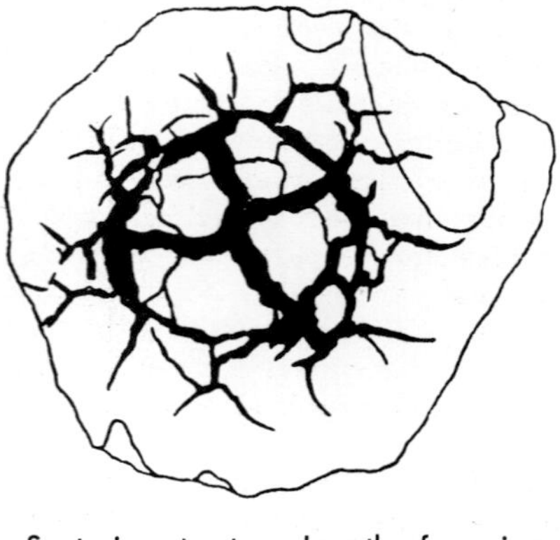

Fig. 4. Septarian structure. Length of specimen about 5 in. (*From F. J. Pettijohn, Sedimentary Rocks, Harper, 2d ed., 1957*)

of siliceous shells, of chitinous materials, or of carbonaceous films or impressions. Typically the most abundant fossils are found in limestones and dolomites. They may, in fact, constitute the bulk of the rock as in the variety of limestone called coquina. All traces of original shell materials may be gone but the shape and even fine details of the fossil may be preserved in a mold of the fossil in the surrounding rock. Algal structures, produced by the action of various kinds of lime-secreting algae are often found in limestones. Some of these structures, laminated in a variety of ways, are called stromatolites. *See* ALGAE FOSSILS; FOSSIL; STROMATOLITE.

Classification and nomenclature. Sedimentary rocks have been classified on two bases, the purely descriptive, and the genetic. Thus, in one group might be found all rocks that are colored red or all rocks that contain a certain proportion of any mineral, or, on the other hand, all rocks that were formed as river floodplain deposits. The disadvantage of the purely descriptive approach is that rocks of widely divergent origin might be lumped together. The disadvantage of the genetic approach is that the origin of the rock must be known before it can be classified; this is often one of the most difficult problems of sedimentary geology. The more fruitful approach has been to combine the two into a classification based on objective descriptive characteristics that have some genetic significance. Using this as a basis, geologists subdivide the sedimentary rocks into two broad groups, the clastic or detrital, and the nonclastic or chemically precipitated rocks. The clastic rocks are characterized by individual grains, often of heterogeneous composition, derived from the erosion of preexisting rocks; the grains may be more or less rounded by abrasion during transport from the erosional area to the site of deposition. The chemical precipitates are normally fairly homogeneous in composition and are characterized by an interlocking crystalline texture, where the crystal sizes may vary greatly. The chemical precipitates form at the site of deposition and are not derived directly from preexisting rocks.

Clastic rocks. Rocks of detrital origin are further subdivided on the basis of particle size into three classes, the rudites (conglomerates), the arenites (sandstones), and the lutites (shales). Siltstones are intermediate in grain size between arenites and lutites and may be considered as very fine-grained arenites or grouped with the lutites. Rudites are composed of particles larger than 2 mm in diameter; arenites, 2–1/16 mm; siltstones, 1/16–1/256 mm; and lutites finer than 1/256 mm (Fig. 5). In practice, since all of the particles of any rock are rarely in the same size group, the dominant particle size is selected, usually the median size.

Chemical rocks. The nonclastics are subdivided by chemical composition into the carbonates, the siliceous rocks, the evaporites, the phosphorites, ferruginous and manganiferous rocks, carbonaceous rocks, and miscellaneous rare chemically precipi-

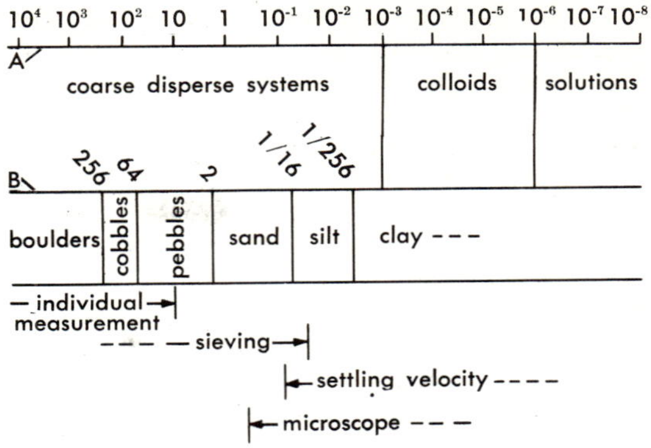

Fig. 5. Range of particle size in clastic sediments, and methods of mechanical analysis applicable to various size ranges. Scale A, logarithmic diameter in millimeters; scale B, diameter in millimeters. (*From W. C. Krumbein and L. L. Sloss, Stratigraphy and Sedimentation, Freeman, 1951*)

tated rocks. The carbonate rocks, the limestones and dolomites, are much the most important in terms of total quantity in the crust and areal extent. Some of the chemically precipitated rocks are formed directly by the agency of organisms, the most prominent being the biogenic limestones. Others are formed by inorganic precipitation, such as the iron formations. *See* ARENACEOUS ROCKS; ARGILLACEOUS ROCKS; CHERT; CONGLOMERATE; LIMESTONE; SANDSTONE; SHALE. [R.S.]

Bibliography: C. O. Dunbar and J. Rodgers, *Principles of Stratigraphy*, 1957; W. C. Krumbein and L. L. Sloss, *Stratigraphy and Sedimentation*, 1951; F. J. Pettijohn, *Sedimentary Rocks*, 2d ed., 1957; H. Williams, F. J. Turner, and C. M. Gilbert, *Petrography*, 1954.

Sedimentation (geology)

The processes that operate at or near the earth's surface to deposit rock-forming material, or sediment. Sedimentation includes the weathering processes that act mechanically and chemically to break up preexisting rocks, the processes of transportation by which the material is carried from its source to the depositional site, the processes of deposition in the sedimentary environment, and the postdepositional processes or diagenesis by which the sediment is compacted and hardened to rock.

Source of material. The raw materials of sedimentation are the products of weathering of igneous, metamorphic, and sedimentary rocks. Weathering may be primarily chemical, mechanical, or both. In chemical weathering minerals are dissolved or altered by solution and the more soluble salts are carried away by running water, leaving behind a residue of insolubles. In mechanical weathering the rock is physically disintegrated by the action of water, wind, freezing and thawing, and temperature extremes. As the whole spectrum of rock types may be involved in weathering, the detritus supplied from the source may be extremely heterogeneous in mineralogical and chemical composition. *See* WEATHERING PROCESSES.

Transportation and deposition. Material supplied by weathering in the source area is carried by water, wind, and mass movements, such as landslides, to the site of deposition. Transportation and deposition may be intermittent, and may alternate with each other until the detritus reaches its final resting place. A sand grain may be carried for a short distance and dropped temporarily many times in the course of traveling down a river to the sea. In the course of transportation the composition and texture of the sediment may be slightly or drastically altered by the transporting medium. More soluble minerals may dissolve, softer minerals may wear out by abrasion, and material may become sorted by size.

Transportation by wind. Turbulent motion of air close to the ground is responsible for lifting small particles and transporting them. Dust is carried in suspension but sand may be carried both in suspension and along the surface. Movement of sand grains may take place by saltation, a process of moving in discrete jumps, or by surface creep, in which the grains are rolled or pushed forward by the force of the wind and impact of landing grains. In saltation the grains are thrown into the air by collisions of rolling grains. Once thrown into the air the grain follows a parabolic course and comes back to the ground a short distance away. As it hits it may bounce back into the air or it may knock another grain upward (Fig. 1). Wind transportation operates over all land areas but is most effective in arid regions. Although dust may be carried to great heights by the wind, sand grains usually remain only a few feet above the surface. Wind normally transports only fine-grained sand, and the sand is well sorted. When the velocity of the wind decreases to the point where it can no longer carry particles, material is dropped, and a windblown sediment is formed. The most common windblown deposits are sand dunes, which have a variety of shapes and sizes. They are common in desert areas and along coastlines and some river valleys, where there is a ready source of sand. Loess is a windblown dust and fine silt deposit. *See* DUNE, LOESS.

Transportation and deposition by ice. Sedimentary particles may be trapped in glaciers and icebergs and transported long distances before the ice melts and the particles are deposited. The material is not sorted in size, and deposits from glaciers are commonly unstratified. Deposits from melting icebergs may be mixed with marine sediment; the resulting glaciomarine sediment may be stratified and slightly sorted by size but contains large boulders and cobbles, called erratics. The most common deposit of glaciers is till, a very heterogeneous mixture of finely ground-up rock flour, clay, sand, and pebbles and boulders. Glaciers pick up avalanche material and rock material by engulfing particles pried loose by frost action and by plucking or quarrying large blocks from the bottom over which

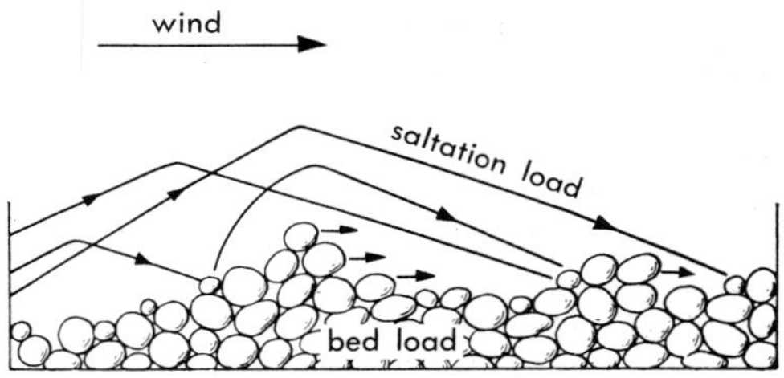

Fig. 1. Detail of saltation and bed load for wind. The range of grain sizes shown is from 0.2–2 mm. With increase of wind velocity the smaller particles of the rolling load become part of the saltation load. (*From R. M. Garrels, A Textbook of Geology, Harper, 1951*)

the glacier rides. Glacial erosion and transportation do not in general round sharp corners and edges of particles, and thus till particles tend to be sharp and angular. *See* GLACIER; TILL.

Transportation and deposition by water. Moving waters account for most of the transportation of rock materials on the earth's surface. Streams and rivers carry tremendous tonnages of materials daily from the erosion areas to the sea. The currents in the sea shift material from place to place on the sea floor. Clastic sediments commonly retain characteristics produced by the transporting currents. Thus the sorting by size in response to the current velocity and density of the water and the rounding of particles by abrasion are both properties of clastic sediments produced during transportation.

Movement of water. Water may move either by laminar or turbulent flow. When the water moves in straight lines parallel to the confining channel surfaces, without mixing of adjacent layers of water, the flow is characterized as laminar. When the water moves in irregular lines in eddies and swirls, and different layers mix, the flow is turbulent. Water in streams is dominantly turbulent, and it is this kind of flow that is responsible for erosion, abrasion, and transportation of all but the finest particles. *See* LAMINAR FLOW; TURBULENT FLOW.

Methods of transportation. Transportation by running water is accomplished in three ways. Ions and compounds are carried in solution and end up in lakes or the sea, where they may accumulate or take part in reactions that precipitate solids. Insoluble particles move either by suspension in the water or by being moved along the bottom. These are called, respectively, suspended load and traction load. The suspended load includes colloidal sizes, which need very low velocities to keep them in suspension, and larger particles, which may be suspended only temporarily by the action of high-velocity turbulent currents. The traction or bottom load moves by rolling and sliding of grains. Saltation is an important process in stream transportation, just as in wind movement (Fig. 2).

Causes of deposition. As soon as water is unable to continue transport of material, deposition starts.

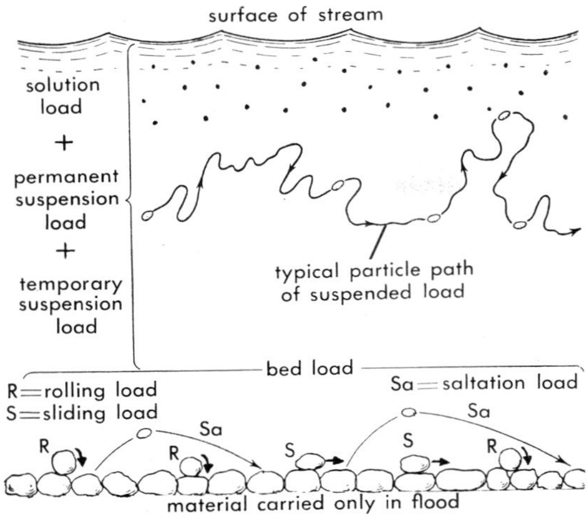

Fig. 2. Schematic diagram showing types of load carried by a stream. (*From R. M. Garrels, A Textbook of Geology, Harper, 1951*)

Also, ions carried in solution may participate in chemical reactions that produce precipitates that settle to the bottom. *See* FLUVIAL EROSION LANDFORMS; STREAM TRANSPORT AND DEPOSITION.

Flocculation and deflocculation. Colloidally suspended materials, dominantly clay minerals, are of small size and therefore need only low velocities and slight turbulence to keep them in suspension. Only when water movement practically ceases can these materials settle by the action of gravity alone. But the colloids are susceptible to flocculation, a process of coagulation or aggregation of the individual particles into larger sized clumps, or floccules. The floccules, being larger and heavier, will settle to the bottom by gravity even though some current continues. Flocculation may occur when a fresh-water suspension mixes with a salt solution, as when a river enters the sea. Deflocculation, or peptization, is the process of dispersing and breaking up the floccules into smaller colloidal-sized particles. *See* COLLOID; FLOCCULATION.

Settling velocities. When a current is no longer able to keep a particle in suspension, the particle settles to the bottom with a velocity proportional to its size, shape, density, and the viscosity of the suspended medium (in this case, water). Stokes' law states this dependence quantitatively for spherical particles at constant temperature as $v = kr^2$, where v is velocity, r is the radius of the sphere, and k is a constant. The constant $k = \frac{2}{9}(\rho_1 - \rho_2)g/\eta$, where ρ_1 and ρ_2 are the densities of the particle and fluid respectively, g is the acceleration due to gravity, and η is the viscosity of the fluid. This law holds for particles smaller than 0.08 mm in diameter, and appreciable deviations from the law are found with particles over 0.2 mm in diameter. For the larger particles, viscosity forces become unimportant and a different law, the impact law, becomes applicable. The impact law can be stated as $v = k_2\sqrt{r}$, where the constant depends on the same factors as in Stokes' law. The operation of the laws of settling velocities accounts for the size sorting that is the result of transportation and deposition by running water. Sorting by shape is also important because spherical grains will offer less frictional resistance to settling and will fall to the bottom faster than irregular ones. The fluid through which the grains settle will have its effect on settling velocities. As the viscosity and density of the fluid medium increases, settling velocities decrease and the sorting action of the current decreases. Turbidity or density currents, in which suspended fine material significantly increases the density and viscosity, are poor in ability to sort material by size; they deposit a mixture of a wide range of sizes. Glaciers may be considered as a high-viscosity transporting medium that deposits extremely poorly sorted material. *See* TURBIDITY CURRENT.

Transportation and deposition in the sea. Transportation of suspended and dissolved material in the sea is accomplished by currents caused by wind, tides, and density differences. Terrigenous sediments, derived directly from matter eroded from land surfaces and transported to the sea by rivers, are first deposited either in deltas or may be moved along the shore by littoral currents, currents that operate near the shoreline at shallow depths. Other slower moving currents may redistribute material on the continental shelves, the shallow aprons surrounding the continents. Turbidity currents distribute some terrigenous sediment to abyssal depths. Pelagic sediments are derived from extremely fine windblown mineral particles and organisms that dwell in the sunlit zone of the seas down to about 600 ft; these sediments are spread in thin layers over all of the ocean bottoms. A most important factor in marine sedimentation is the role of organisms that secrete calcium carbonate, silica, or phosphatic shells (tests). These biogenic precipitates may settle to the bottom or may be caught up in currents and transported, eventually to be deposited in some other spot. *See* DELTA; MARINE SEDIMENTS; *see also* ESTUARINE OCEANOGRAPHY.

Environments of deposition. The sedimentary environment is usually defined in terms of a complex of physical and chemical conditions associated with a particular geomorphologic unit. For example, a lacustrine environment includes all of the physical and chemical forces at work in a lake, be it the beach or the deep water center of the lake. Some environments, such as the swamps, may be fairly homogeneous; others, such as the littoral (nearshore) marine, may be very variable and heterogeneous. Other concepts of environment are used, such as tectonic, chemical, or hydrodynamic; here each controlling factor is segregated and the total environment is a combination of all of the individual components. Whereas source materials primarily influence detrital mineralogy of a sediment, the environments of deposition primarily control the textures of clastic sediments and the composition of chemical sediments.

An extensive classification of environments by W. H. Twenhofel is shown in the accompanying list.

- Continental environments
  - Terrestrial
    - Desert
    - Glacial
  - Aqueous
    - Fluvial
      - Piedmont
      - Valley flat
    - Paludal
      - Lake swamps
      - River swamps
      - Flat-land swamps
      - Paralic swamps
    - Lacustrine
      - Fresh
      - Salt
    - Spelean-cave
- Mixed continental and marine
  - Littoral
  - Delta
  - Marginal lagoon
  - Estuary
- Marine environments
  - Neritic
  - Bathyal
  - Abyssal

Sediments of physical deposition. The clastic, or detrital, sediments are those that have been deposited by mechanical action, by currents of one kind or another. It is most convenient to subdivide them on the basis of particle size, for the size and sorting are dependent on a genetic factor, the current type and strength (Fig. 3). The current regime is reflected in the sediment not only by size and sorting but by its bedding characteristics and sedimentary structures, such as cross-bedding, ripple-marks, and current lineation.

Coarse-grained clastics. Large particles, such as pebbles, cobbles, and boulders, can be carried only by high velocity or high density currents. The coarse clastics—gravels, and their indurated equivalents, conglomerates—are those whose particle size is greater than 2 mm in diameter. Rivers transport such coarse materials in times of flood; on land they may also be transported by landslides and mudflows. Submarine currents are in general unable to carry such materials, but they may be found in some sediments deposited by turbidity currents, and to a minor extent, by being rafted by ice or caught in the roots of marine plants, they may be carried far out to sea. Glaciomarine sediments commonly carry coarse detritus. *See* CONGLOMERATE; GRAVEL.

Medium-grained clastics. Sands and their consolidated equivalents, sandstones, are the medium-grained clastics, and range in size from 1⁄16 to 2 mm in diameter. Most streams and rivers, as well as near-shore littoral ocean currents, have velocities competent to carry this size particle. Windblown sands are common but tend to be fine-grained and very fine-grained sand, ¼–⅛ mm and ⅛–1⁄16 mm in diameter respectively. Sands accumulate as river bars, dunes, alluvial fans, beach deposits, barrier islands, and offshore marine bars, and are a major component of deltas. *See* SAND; SANDSTONE; SHORE PROCESSES.

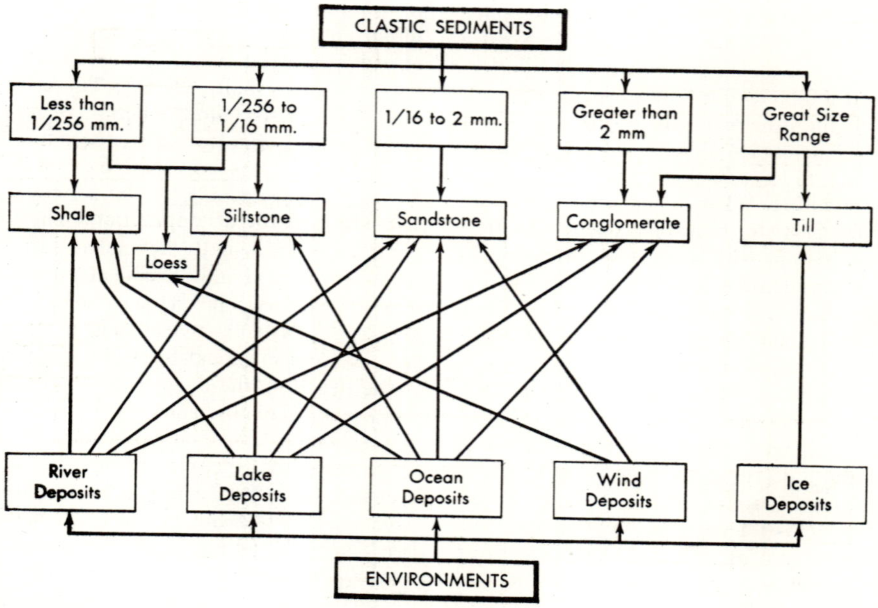

Fig. 3. Interrelation of grain size and environment for some major types of clastic sedimentary rocks. (*From R. M. Garrels, A Textbook of Geology, Harper, 1951*)

Fine-grained clastics. The fine-grained clastics are the silts and muds and their indurated equivalents, shales, siltstones, and mudstones. The particle size of silt is 1/16–1/256 mm in diameter. Clay size particles are finer than 1/256 mm. The fine-grained clastics are by far the most abundant sediment type and are found either alone or mixed with other sediment types in almost every environment of deposition. Because of their fine size, down to colloidal dimensions, they may be carried in suspension by very slowly moving currents and so may be distributed far and wide over the ocean bottoms. The clay minerals, hydrous aluminosilicates, all tend to have colloidal-sized particles and make up, on the average, about one-third of the fine clastic sediments. *See* ARGILLACEOUS ROCKS; CLAY MINERALS; SHALE.

Sediments of chemical deposition. The chemical sediments are formed by reactions between the dissolved components in the water of the environment to form precipitates. The precipitation may be inorganic or it may take place by the action of organisms, such as mollusks, which secrete calcium carbonate shells (Fig. 4). By far the most important chemical sediments quantitatively are the carbonates, dominantly calcium carbonate.

Carbonate sediments. A great many invertebrates are able to extract calcium carbonate from sea water for their shells. After death the shells accumulate to form sediment. In some areas very fine-grained calcium carbonate muds accumulate on the bottom; the mud, when lithified, becomes a dense, finely crystalline limestone. There is some dispute over the origin of the mud, some workers claiming it to be an inorganic precipitate and others finding evidence for biogenic origin in the ground up fragments of small shells and the presence of needles of aragonite, one of the crystalline forms of calcium carbonate, that are secreted by blue-green algae. A variety of calcium carbonate deposits is associated with coral or algal reefs. Globigerina and pteropod oozes are deep-sea carbonate deposits formed from the shells of the animals for whom the deposits are named. These pelagic organisms live in surface layers of the sea and, at death, sink to the bottom, where they accumulate. Chalk is a lithified deposit of this type. Carbonate sediments may contain an admixture of clay, in which case they are called marl. Dolomite is a carbonate sediment composed largely of $CaMg(CO_3)_2$, the mineral dolomite. *See* CHALK; DOLOMITE; LIMESTONE; MARL.

Siliceous sediments. The sediments composed largely or completely of chemically precipitated silica are formed mainly in environments in which the supply of clastic material is small or nonexistent. Siliceous sediments being formed today are practically limited to accumulations of tests, or shells, of silica-secreting pelagic organisms, such as the diatoms and radiolaria. Their lithified equivalents are radiolarites, radiolarian earths, diatomites, and diatomaceous earths. Inorganic precipitation of silica around hot springs forms the siliceous sinters. Chert is a siliceous sediment that occurs chiefly as nodules in carbonate rocks, probably of postdepositional origin by replacement of calcium carbonate.

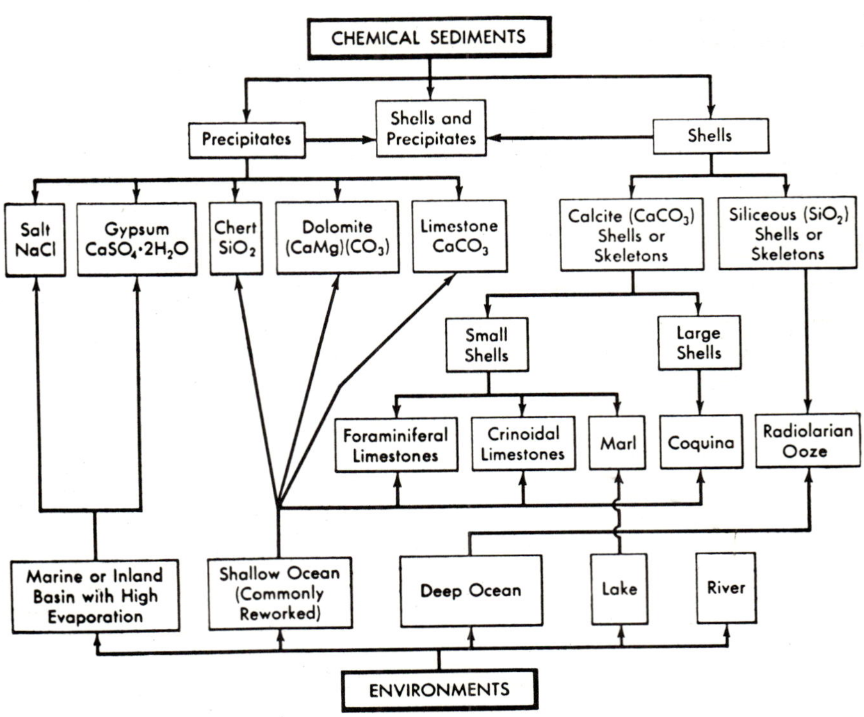

Fig. 4. Interrelation of composition and environment for some major types of chemical and organochemical sedimentary rocks. (*From R. M. Garrels, A Textbook of Geology, Harper, 1951*)

Some chert is bedded and may represent primary precipitation from sea water, either inorganically or by a biochemical mechanism, such as the formation of diatom oozes. Sediments that contain large amounts of volcanic ash commonly are siliceous, the silica being contributed from the alteration of volcanic glass. *See* CHERT; DIATOMACEOUS EARTH; RADIOLARIAN EARTH; SILICEOUS SINTER; VOLCANIC GLASS.

Ferruginous sediments. The iron-rich sediments are often formed, as the siliceous sediments are, in the absence of large amounts of clastic material. Many ferruginous sediments are formed in reducing environments where ferrous compounds predominate. Pure iron mineral sediments are rare in the geologic record, most being limited to the Precambrian. Mixtures of ferruginous minerals and clastic components give ferruginous shales and sandstones. Ferruginous limestone may be formed by the replacement of calcium carbonate by siderite, the iron carbonate. Replacement of limestone by hematite formed the Clinton iron ores, which contain oolites and fossils of hematite. Bog iron ores are formed in fresh water lakes; they are yellow and brown concretionary or irregular masses containing sand and clay.

Phosphatic sediments. Shales and limestones may contain a mixture of different calcium phosphate minerals, some of them precipitated from sea water and others formed in the sediment after deposition by replacement. Phosphate nodules are common in some limestones. Bedded phosphates occur interbedded with other marine sediments. Guano, a deposit of bird excreta, is found on some desert islands. Other phosphate accumulations are residues from the solution of a phosphatic limestone. *See* APATITE.

Manganese sediments. Manganese carbonate concretions and thin layers occur in some marine sediments, many times in association with ferruginous sediments. Manganese oxide is precipitated in the sea as finely divided pigment in some deep-sea sediments, coatings, and nodules and concretions.

Sedimentary sulfur. Native sulfur is found as small particles, lenses, and irregular bodies in some sediments, commonly the argillaceous ones. It is precipitated by the partial reduction of sulfate or partial oxidation of sulfides in solution. Sulfides are precipitated in reducing environments, such as at the bottom of the Black Sea. The sulfate-reducing bacteria play an important role in this precipitation. *See* BLACK SEA.

Carbonaceous sediments. The carbonaceous sediments are formed by the accumulation of vegetable and animal organic debris, either pure or mixed with some clastic material. The most important deposits are peats and coals. The preservation of the organic material depends on rapid burial or lack of oxygen in the environment. The original remains are partially decomposed and altered in part by bacterial action. If mixed with clastic material, the deposit becomes a carbonaceous shale or sandstone.

Decomposition. Vegetation is transformed into peat and then coal by a combination of bacterial and chemical decay and alteration. The end result of this process, called incoalation, is the reduction of proportion of oxygen and hydrogen and the enrichment of carbon. *See* DOPPLERITE.

Environments of carbonaceous deposition. Peat deposits are formed in fresh-water swamps, bogs, or moors, where vegetation is lush and growth is rapid, and where there is little bacterial activity decomposing the organic matter. Other deposits rich in organic matter are formed at the bottoms of lakes, lagoons, and some inland seas, where the combination of organic productivity and reducing environments is such as to prevent oxidation and decomposition of the organic matter. Bottom deposits of the Black Sea, whose bottom waters are reducing, are notably rich in organic material.

Classification of carbonaceous sediments. The carbonaceous sediments may be divided into the humic (coaly) and sapropelic (bituminous), with all gradations between. The humic deposits are low in hydrogen and high in oxygen, whereas the sapropelic are high in hydrogen and low in oxygen. The best known representative of the humic class is coal; true cannels exemplify the sapropelic class.

Peat deposits. Peat deposits are composed of two different types of altered plant materials, anthraxylous and attrital. The anthraxylous is composed largely of anthraxylon, the visible ligneous parts of plants, such as wood, leaves, stems, and bark, impregnated with colloidal humic decomposition products. The attrital is composed mainly of attritus, the finely divided decay-resistant plant materials, particularly the waxy, resinous, and fatty materials. Peats of one type are gradational with the other and both may grade into sapropelic deposits, such as cannel. *See* BLACK SHALE; COAL; SAPROPEL.

Diagenetic processes and lithification. After deposition the unconsolidated water-saturated sediment undergoes compaction, squeezing out of water, alteration of minerals and precipitation of new ones, and solution; the end result of the combination of these processes is a lithified sediment, a sedimentary rock. Collectively the processes are known as diagenesis. Extensive changes and modifications of mineralogy and texture of the original sediment may take place. *See* AUTHIGENIC MINERALS; DIAGENESIS.

Provenance and dispersal. The provenance of sediments includes the evaluation of the composition and location of the source rocks whose erosion produces the raw material of sedimentation. Dispersal is the process by which the sedimentary materials are carried by various transporting agents from source area to depositional area and distributed in the environments of sedimentation. The kind of source rock is of primary importance in determining the nature of the clastic material, but major modifications result from the effects of climate and topography. Only if mechanical erosion dominates will the source composition be unaltered. If, owing to climate or topography, chemical erosion is significant, the source materials may be drastically altered before they become transported to the sedimentation area. Changes in texture (size distribution, roundness, sphericity) and mineral composition may result from dispersal conditions.

Nature of the source area. If chemical weathering of source rocks is at a minimum, then reconstruction of source rock types from the sediment is relatively simple, subject mainly to modifications by dispersal. The major problems arise in reconstruction of source areas in which there has been chemical weathering. One way of assessing the amount of chemical weathering is by chemical composition. The residues of weathering tend to be enriched in silica, alumina, and iron and impoverished in alkalies and alkaline earths. Another way of assessment is by consideration of mineral stabilities under weathering conditions. The most stable minerals, quartz, muscovite, and the clay minerals, dominate the residue, whereas the most unstable ferromagnesian minerals such as olivine, pyroxenes, and amphiboles are lost. The degree of weathering may be determined by a mineral stability series, in which the most unstable species are lost first and the most stable are preserved in the residue. Olivine is one of the most unstable, followed by augite and hornblende. Kyanite, staurolite, and garnet are intermediate in stability. The most stable, besides quartz, are zircon, tourmaline, and rutile. Caution is required in using mineral stabilities for source rock evaluation, for there is evidence that over long periods of time some of the unstable minerals may disappear by solution during diagenesis, a process called intrastratal solution.

Mineralogical maturity. The maturity of a sediment is the degree to which the sedimentary material has had its more unstable minerals removed before deposition. A sediment with many unstable minerals is an immature one; one with only stable minerals is mature. Thus maturity is a measure of weathering in the source area, which is in turn a function of climate and topography.

Determining transport distance and direction. The size distribution of the particles in a clastic sediment is strongly affected by transportation agents. The maximum size and median size tend to decrease with travel downstream; thus this property can be used in a general way to deduce distance of transport. Size reduction may be the result not only of abrasion and splitting of grains but also of selective sorting, so that only finer material is carried on while coarser material is dropped. Rate of size reduction is affected by mineral composition, for the softer minerals will wear down quicker than the hard ones. Roundness and sphericity increase in downstream direction. Since the late 1930s roundness has been successfully used as an indication of abrasion during transportation (Fig. 5*a*). Relation of compositional change to transport distance can also be used (Fig. 5*b*). Feldspar, with a hardness of 6 and good cleavage, apparently wears out somewhat faster than quartz, with a hardness of 7 and no cleavage. Thus changes in quartz-feldspar ratio may indicate changes in transportation. The effect, however, seems to be small and has not as yet been fully evaluated.

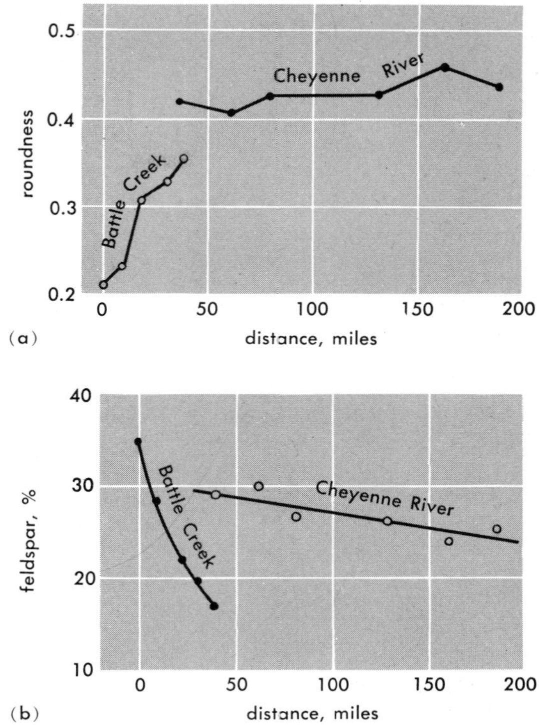

Fig. 5. Relation of textural and compositional change to distance of transport in South Dakota streams. (*a*) Increase in roundness of quartz sand in the 1- to 1.414-mm class; perfect roundness = 1.0. (*b*) Percentage of feldspar in the 1- to 1.414-mm class. (*From W. J. Plumley, Black Hills terrace gravels: A study in sediment transport, J. Geol., 56(6):526–577, 1948*)

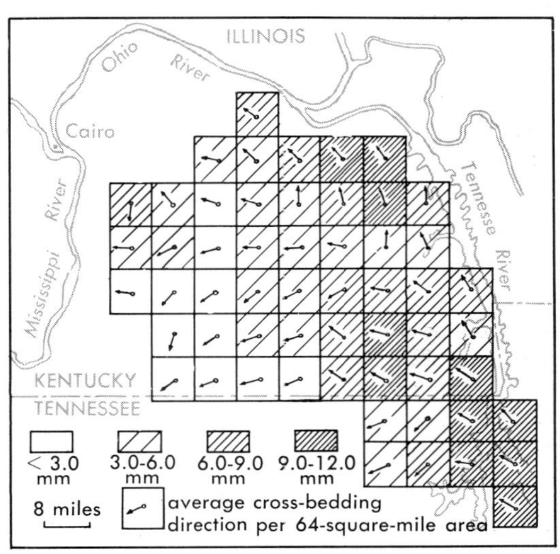

Fig. 6. Average of cross-bedding direction and median size in Lafayette gravel. (*After P. E. Potter, Petrology and origin of the Lafayette gravel: 1, Mineralogy and petrology, J. Geol., 63(1):1–38, 1955*)

Dilution by other sources. Material coming from one source area may become diluted by material from other sources as it travels downstream to the sea. Areal mapping of mineralogical and textural composition is necessary to sort out the contributions of different source areas.

Sedimentary structures and source area. A development of the 1950s has been the use of cross-bedding and other directional sedimentary structures to determine the directions of source areas (Fig. 6). Such structures are mapped to get the direction of transport at each place and thus to ascertain regional trends. The direction of the source is the upstream direction of transport. Studies of marine sedimentary structures have been used to determine directions of marine transport.

Sedimentary petrologic provinces. A sedimentary petrologic province is an area in which the sediments have a more or less uniform composition and come from the same source (Fig. 7). If two or more provinces overlap in time, the rock succession represents provincial alteration, or provincial succession. The latter may represent continuing changes in source area contributions.

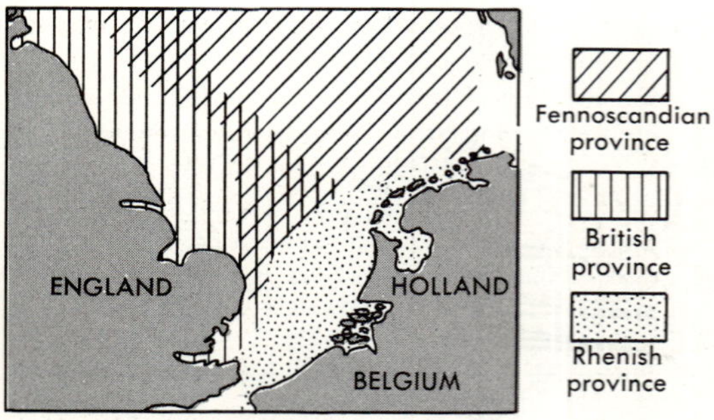

Fig. 7. Sedimentary petrologic provinces of southern North Sea. (*After J. A. Baak, in P. D. Trask, ed., Recent Marine Sediments, reprint, 1955*)

Sedimentary facies. The areal variation of lithology in a single bed or stratigraphic unit leads to the concept of facies. Facies involve particular lithologic characteristics of a bed at one place that grade into different characteristics in the same bed at other places. Thus a bed may vary in grain size from a shale facies to a sand facies, even though all of the material may come from the same source. Facies maps show the areal distribution of clastics versus chemical sediments, sand/shale ratios, or other lithologic variations in sediments of the same age. *See* FACIES (GEOLOGY).

Diastrophic movement and sedimentation. Diastrophism, earth deformation, has direct and indirect effects on sedimentation. The tectonic formation of sedimentary basins provides a place for the deposition of material. The depth of water influences sedimentation. The depth is related to downwarping of the basin and sea level, which are influenced by tectonics. The relief and climate of the source area are fundamentally controlled by earth movements. High mountains are the scene of vigorous mechanical erosion. Long continued earth stability leads to lowlands with chemical erosion predominating. Besides these general considerations, specific sedimentary associations are connected with mountain building episodes.

Flysch and molasse. The terms flysch and molasse originated as stratigraphic names in the northern Alps, where the orogenic sediments (those involved in overthrusting and recumbent folds) were the flysch, and the postorogenic sediments (later than folding and thrusting) were the molasse. Since their original use, an extension of their lithologies and orogenic and postorogenic significance has resulted in the use of flysch as synonymous with graywacke, and molasse with subgraywacke. Typically the flysch, which is interbedded shale and graywacke, increases in coarseness upward (a reflection of beginning orogeny). The molasse, mainly shale and subgraywacke sandstone, decreases in size upward (a reflection of dying-out orogeny). The molasse is a product of paralic sedimentation. *See* OROGENY; TECTONIC PATTERNS.

Paralic sedimentation. This type is defined as sedimentation taking place in areas within or peripheral to continents, where a great amount of terrigenous alluvial deposits is laid down. The deposits may be partially marine or brackish water but are dominantly terrestrial.

Geosynclinal cycle. Many geologists have recognized that geosynclines not only may have distinctive suites of rock types but that the total prism of sediments that was deposited in the geosyncline represents an evolution of rock types more or less characteristic of all geosynclines and that such a succession of rock types is linked to the tectonic development of the geosynclines. F. J. Pettijohn characterizes the major cycle as beginning with orthoquartzitic and carbonate sediments deposited on the flooded craton (the stable core of the continent) and the cratonic border of the geosyncline. Following this comes mild uplift of part of the geosyncline and deposition of cherts, black shale, and phosphorite facies. This is succeeded by strong upwarp in adjacent areas and the deposition of flysch, increasing in coarseness upward and including submarine extrusives and tuffs. The next stage is that of postorogenic molasse sedimentation and paralic sedimentation, dominated by nonmarine deposits. Finally the geosyncline is deformed and uplifted into mountain belts. There are many variations in geosynclinal development and therefore in the geosynclinal cycle; thus, no two are alike. The general concept of the cycle, however, seems to have some validity. *See* GEOSYNCLINE; *see also* EVAPORITE (SALINE); PALEOGEOGRAPHY; PALEOGEOLOGY; SEDIMENTARY ROCKS. [R.S.]

Bibliography: C. O. Dunbar and J. Rodgers, *Principles of Stratigraphy*, 1957; W. H. Twenhofel, *Principles of Sedimentation*, 2d ed., 1950; F. J. Pettijohn, *Sedimentary Rocks*, 2d ed., 1957; W. C. Krumbein and L. L. Sloss, *Stratigraphy and Sedimentation*, 1951.

Sedimentation (industrial)

A process based on the settling of solid particles through a liquid. Sedimentation is used in several ways in industrial processes. Thus, it may be used to obtain a concentrated slurry from a dilute suspension of a solid in a liquid. This is called thickening. It may also be used to remove solid particles from a liquid, to obtain a clear supernatant liquid. This is called clarification. The driving force for the process is the difference in density between the solid and liquid. Ordinarily, sedimentation is accomplished by the force of gravity, and the liquid is water or an aqueous solution. For a given density difference, the rate of settling decreases with particle size. When the particles are fine, or when the density difference is small, gravity settling may be too slow to be practicable; then centrifugal force can be used in place of gravity (*see* CENTRIFUGATION). When centrifugal force is not adequate, the more positive method of filtration may be used. All these methods of treating solids and liquids fall into the generic group of mechanical separations.

Operations based on sedimentation are important in such industrial processes as ore dressing, cement manufacture, water purification, and industrial-waste treatment.

Settling of spheres through fluids. The basic laws of sedimentation are those describing the resistance offered by a fluid to the motion of a solid particle through it. When a constant force is applied to a particle initially at rest in a fluid, the particle immediately accelerates and moves through the fluid. The motion of the particle generates a frictional resistance which in turn reduces the acceleration. When the force moving the particle and the resisting force become equal, the acceleration, by Newton's law of motion, drops to zero, and the particle continues in motion at constant velocity as long as the applied force is active. This steady velocity is called the terminal velocity. Most sedimentation processes are conducted at the terminal velocity.

The law of resistance of a solid particle moving through a fluid is complicated. The resistance depends on all these variables: the diameter and shape of the particle, the viscosity and density of the fluid, the velocity and acceleration of the particle, and the nearness of the particle to other particles and to the wall of the equipment. The basic situation is simplified when the particle is a smooth

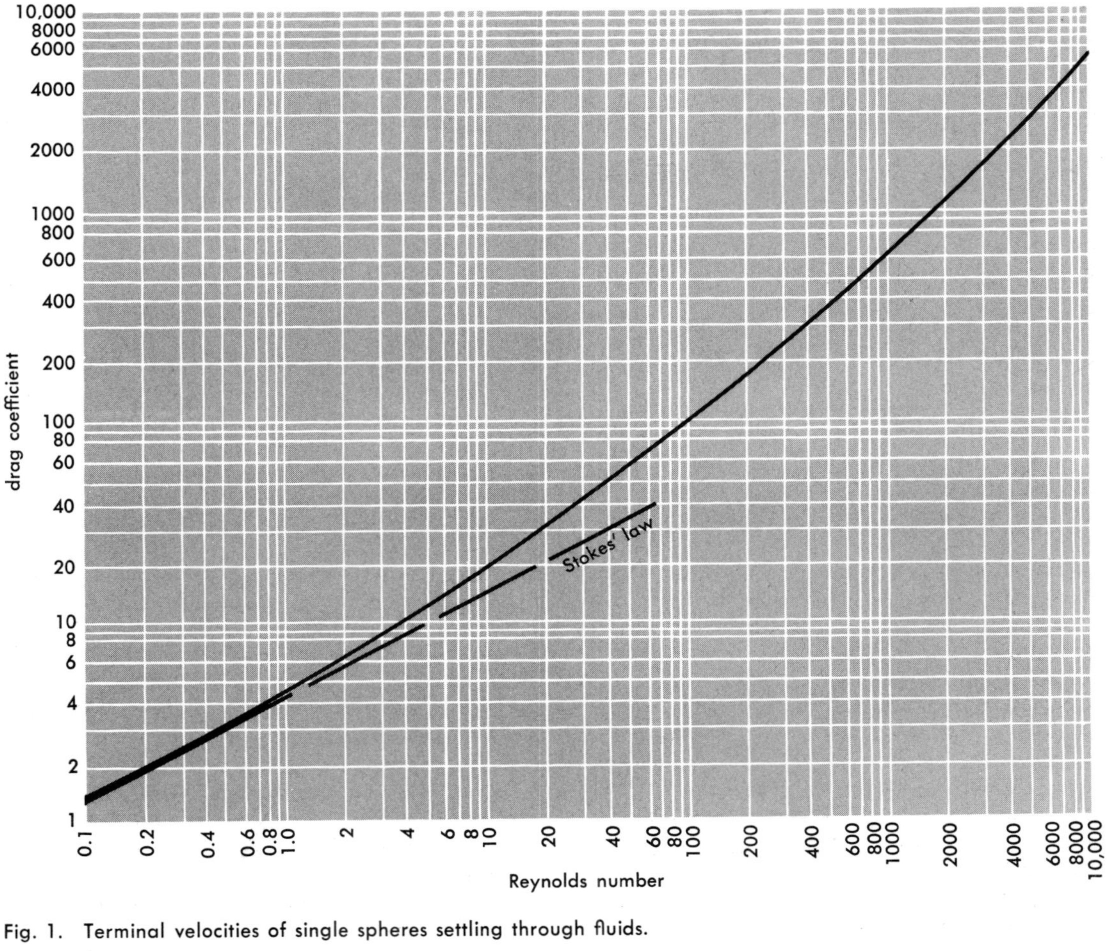

Fig. 1. Terminal velocities of single spheres settling through fluids.

sphere, when each particle is sufficiently far away from all other particles and from the wall of the equipment that its motion is unaffected by their presence, and when the particle is moving at constant velocity, that is, without acceleration. Figure 1 shows how, for this case, the terminal velocity u_t is influenced by the acceleration of gravity g, the diameter and density of the particle, D_p and ρ_p, respectively, and the density and viscosity of the fluid, ρ and μ, respectively. Figure 1 is a log-log plot of two dimensionless groups. The abscissa, $D_p u_t \rho/\mu$, is a Reynolds number, and the ordinate

$$(D_p\rho/\mu)\sqrt{gD_p(\rho_p/\rho - 1)}$$

is related to a factor called the drag coefficient. The units used in these expressions may be any self-consistent set, such as the foot-pound-second units suggested in Fig. 1.

To use Fig. 1 to predict a terminal velocity, the magnitude of the ordinate is calculated from known values of g, D_p, μ, ρ, and ρ_p. The corresponding value of the abscissa is read from the solid curve of Fig. 1, and the value of u_t, the terminal velocity, calculated from the value of the abscissa.

Stokes' law. For Reynolds numbers below about 0.1, the law of settling takes the following simple form, called Stokes' law.

$$u_t = \frac{gD_p^2(\rho_p - \rho)}{18\mu}$$

This equation is plotted as the straight dotted line in Fig. 1.

Sedimentation rates in practice. At large Reynolds numbers, which are found with large particles and small viscosities, the terminal velocity is less than that predicted from Stokes' law, and the curved line in Fig. 1 must be used. Also, in practice, the particles are usually very close together, and some are also close to the wall of the equipment. Under these hindered settling conditions, the rate of settling is considerably less than that predicted by the curve of Fig. 1. Actual rates may sometimes be estimated by applying correction factors to the correlation of Fig. 1, but for accuracy, it is best to measure the rates experimentally on the actual sludge to be settled.

Batch sedimentation of flocculated particles. Particles too small to be settled at practicable rates are often amenable to flocculation. This is accomplished by adding agents such as sodium silicate, alum, lime, and alumina. The particles agglomerate into coarse flocs, which act like single large particles, settle at a practicable speed, and leave a clear supernatant liquid behind.

A batch of flocculated pulp passes through several stages during sedimentation. Figure 2 shows the process diagrammatically. Figure 2*a* represents the original homogeneous flocculated pulp ready to settle. In Figs. 2*a* to *e*, layer *B* is a uniform suspension of the same solid concentration as that of the original pulp, and layer *A* is clear supernatant liquid. Layer *D* consists of flocs resting lightly on one another, with liquid filling the voids between the flocs. Layer *C* is a transition layer, the solid concentration of which varies continuously from that in layer *B* to that in layer *D*. As settling continues, layers *C* and *D* decrease and finally disappear. The end of this stage is shown in Fig. 2*d*. Then, a new effect, called compression, begins. During compression, the weight of the deposit breaks down the structure of the flocs, and some of the liquid in the flocs of layer *D* is expelled as small geysers. The thickness of layer *D* decreases to an equilibrium height called the ultimate height, as shown in Fig. 2*e*, and the process stops. *See* CLARIFICATION; REYNOLDS NUMBER; SEPARATION (MECHANICAL); THICKENING. [W.L.M.]

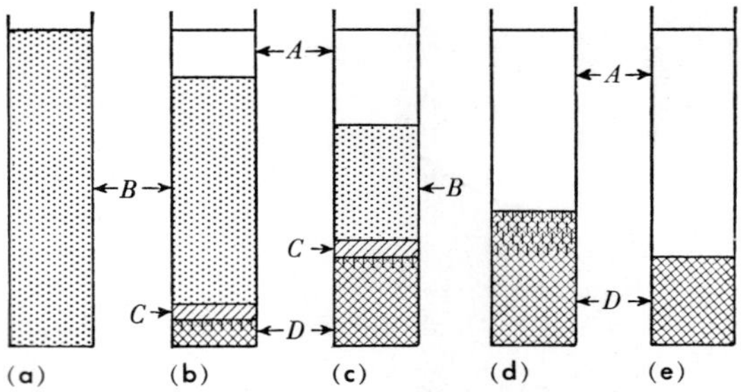

Fig. 2. (a-e) Batch sedimentation of flocculated particles. (*From W. L. McCabe and J. C. Smith, Unit Operations of Chemical Engineering, McGraw-Hill, 1956*)

Bibliography: H. S. Coe and G. H. Clevenger, *Methods for determining the capacities of slime-settling tanks, Trans. AIMME*, 55:356, 1916; W. L. McCabe and J. C. Smith, *Unit Operations of Chemical Engineering*, 1956; J. H. Perry (ed.), *Chemical Engineers' Handbook*, 3d ed., 1950.

Seed (botany)

A fertilized ovule containing an embryo which forms a new plant upon germination. Seed-bearing characterizes the higher plants—the gymnosperms (conifers and allies) and the angiosperms (flowering plants). Gymnosperm (naked) seeds arise on the surface of a structure, as on a seed scale of a pine cone. Angiosperm (covered) seeds develop within a fruit, as the peas in a pod. *See* FLOWER (BOTANY); FRUIT (BOTANY).

Seed structure. One or two tissue envelopes, integuments, form the seed coat which encloses the seed except for a tiny pore, the micropyle (Fig. 1). The micropyle is near the funiculus (seed stalk) in angiosperm seeds. The hilum is the scar left when the seed is detached from the funiculus. Some seeds have a raphe, a ridge near the hilum opposite the micropyle, and a bulbous strophiole. Others such as nutmeg possess arils, outgrowths of the funiculus, or a fleshy caruncle developed from the seed coat near the hilum, as in the castor bean. The fleshy, edible aril of the Philippine kamanchile completely encloses the seed to form a fruitlike structure.

The embryo consists of an axis and attached cotyledons (seed leaves). The part of the axis above

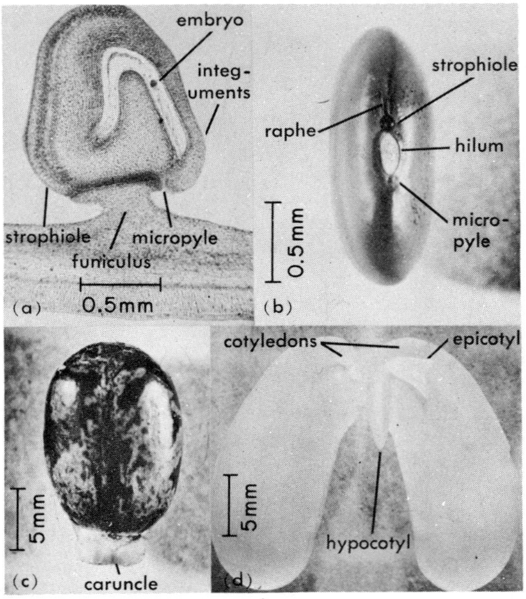

Fig. 1. (*a*) Median longitudinal section of pea ovule shortly after fertilization, showing attachment to pod tissues. (*b*) Mature kidney bean. (*c*) Mature castor bean. (*d*) Opened embryo of mature kidney bean.

the cotyledons is the epicotyl (plumule); that below, the hypocotyl, the lower end of which bears a more or less developed primordium of the root (radicle). The epicotyl, essentially a terminal bud, possesses an apical meristem (growing point) and, sometimes, leaf primordia. The seedling stem develops from the epicotyl. An apical meristem of the radicle produces the primary root of the seedling, and transition between root and stem occurs in the hypocotyl. *See* MERISTEM, APICAL; ROOT (BOTANY); STEM (BOTANY).

Two to many cotyledons occur in different gymnosperms. The angiosperms are divided into two major groups according to number of cotyledons: the monocotyledons including orchids, lilies, grasses, and sedges; and the dicotyledons such as beans, roses, and sunflowers. Mature gymnosperm seeds contain an endosperm (albumen or nutritive tissue) surrounding the embryo. In some mature dicotyledon seeds the endosperm persists, the cotyledons are flat and leaflike, and the epicotyl is simply an apical meristem (Fig. 2). In other seeds, such as the bean, the growing embryo absorbs the endosperm, and food reserve for germination is stored in fleshy cotyledons. The endosperm persists in common monocotyledons, for example, corn and wheat, and the cotyledon, known as the scutellum, functions as an absorbing organ during germination (Fig. 3). Grain embryos also possess a coleoptile and a coleorhiza sheathing the epicotyl and radicle, respectively, The apical meristems of lateral seed roots also may be differentiated in the embryonic axis near the scutellum of some grains.

Monocotyledon and dicotyledon seeds also differ in seed coat structure. In grains, or caryopsis fruits, the mature fruit wall and seed coat may be fused and the outermost endosperm cells form an aleurone layer, rich in proteins. Flour brans consist of fragments of aleurone and fruit-seed coats. Most of the thiamine and riboflavin of grains occur in these tissues and near the scutellum.

Cellular structure of the seed coat varies in the dicotyledons. An outer, water-impervious layer of wax or cuticle is usually present. The outermost cells in bean and pea seed coats, known as macrosclereids (Malpighian cells), have fluted, cellulosic wall thickenings; underlying them is a layer of osteosclereids, short bone-shaped cells. Two layers of macrosclereids occur at the hilum in peas and beans. Underlying them is a group of tracheary cells with reticulate (netted) wall thickenings, and a surrounding spongy tissue of branched cells and large air spaces. This structure appears to be well suited for water absorption during germination.

Many so-called seeds consist of hardened parts of the fruit enclosing the true seed which has a thin, papery seed coat. Among these are the achenes, as in sunflower, dandelion, and strawberry, and the pits of stone fruits such as cherry, peach, and raspberry. Many common nuts are also of this structure.

Seed dispersal. Mechanisms for seed dispersal include parts of both fruit and seed. Some dry fruits have membranous air sacs which aid in seed

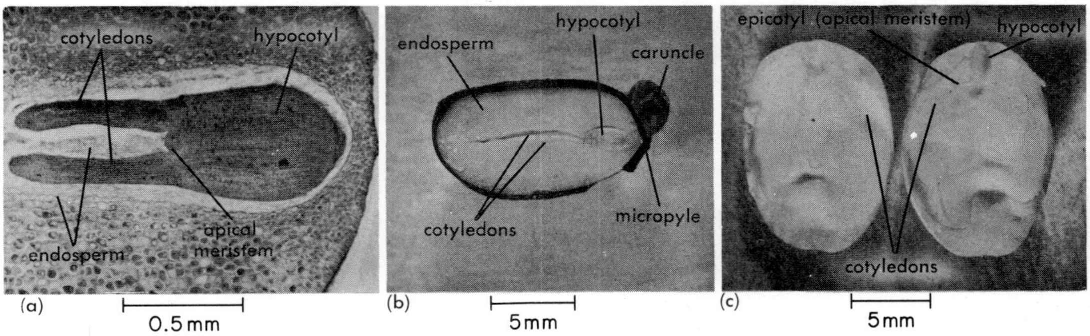

Fig. 2. (*a*) Median longitudinal section of embryo of *Garrya elliptica* (silk tassel bush) embedded in endosperm removed from mature seed. (*b*) Castor bean cut longitudinally. (*c*) Castor bean removed from seed coat and split longitudinally between cotyledons.

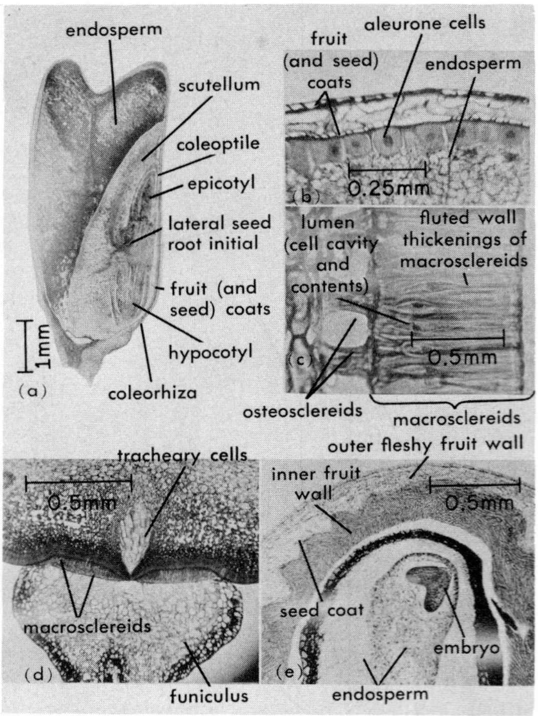

Fig. 3. (*a*) Longitudinal section of corn kernel (*from M. J. Wolf et al., Cereal Chem., 29:321–382, 1952*). (*b*) Section of outer cell layers of wheat grain. (*c*) Section of outer cells of mature pea seed coat. (*d*) Section of hilum area of nearly mature pea. (*e*) Section through portion of young raspberry drupelet (segment of berry).

dispersal by water. In cockleburrs and beggarticks the seeds are enclosed in spiny, barbed fruits which are readily carried in the fur of animals. Seed dispersal by wind is aided by hair tufts of dandelion achenes, wings of elm and maple fruits and of conifer seeds, and by seed coat hairs of willow, cottonwood, and milkweed seeds. Cotton fibers consist of greatly elongated epidermal hairs of the seed coat as do also the kapok fibers of the silk-cotton tree, *Ceiba pentandra* (*see* COTTON; KAPOK TREE).

Economic importance. Propagation of plants by seed and technological use of seed and seed products are among man's most important activities. Specializations of seed structure and composition provide rich sources for industrial exploitation other than direct use as food. Common products include starches and glutens from grains, hemicelluloses from guar and locust beans, proteins and oils from soybeans and cotton seed. Drugs, enzymes, vitamins, spices and condiments are obtained from embryos, endosperm, and entire seeds, often including the fruit coat. Most of the oils of palm, olive, and pine seeds are in the endosperm. Safflower seed oil is obtained mainly from the embryo, whereas both the seed coat and embryo of cotton seed are rich in oils. See separate articles for foods, drugs, and other products listed above; *see also* AGRICULTURE; BIOCHEMISTRY; FOOD ENGINEERING; PLANT ANATOMY; REPRODUCTION, PLANT. [R.M.R.]

Seiche

A standing wave (stationary oscillation) that occurs in enclosed or partially enclosed water bodies, such as lakes, bays, gulfs, and harbors, in which the water has a natural period of oscillation depending on the horizontal dimensions and depth of the containing basin. Seiches are commonly generated by wind, atmospheric pressure gradients, tides, or oscillations of adjacent water bodies. Relatively weak external forces can start a prolonged set of damped seiche oscillations that may reach large proportions if the periodicity of the cause approximates the natural seiche period. *See* WAVE MOTION IN LIQUIDS.

As wave length is long relative to the depth of the water body, seiche behavior conforms to long-wave theory. Thus, the velocity is given by $\sqrt{gh}$, where g is the acceleration of gravity and h, the water depth. The natural period of oscillation for uniformly deep, completely enclosed rectangular basins is given by $2l/n\sqrt{gh}$, where n is the number of nodal lines present (1 for the fundamental case or mononodal seiche and 2 for the binodal seiche). Antinodes exist at the extremities of such a basin, with the node (or nodes) in the central region.

In the accompanying profile (Fig. 1) of standing waves in the closed basins $ABB'A'$ and $ACC'A'$, vectors indicate the vertical and horizontal components of motion as the water surface oscillates from the position of the solid line to that of the broken line. (Directions are reversed in the return motion.) Points 1 and 2, below which there is maximum horizontal and no vertical motion, are called nodes. Points A, B, and C, below which there is maximum vertical and no horizontal motion, are called antinodes. The seiche in the smaller basin $ABB'A'$, having only one node, is the fundamental mode, or mononodal seiche. The seiche in the larger basin $ACC'A'$ has two nodes and is therefore a binodal seiche.

The natural period for an open-end basin is $4l/n\sqrt{gh}$, with nodal and antinodal lines at the open and closed margins, respectively. Basin irregularities and Coriolis effects may introduce errors as much as 10% of formula calculations; however, refined procedures exist which make it possible to correct seiche determinations and compensate for these errors.

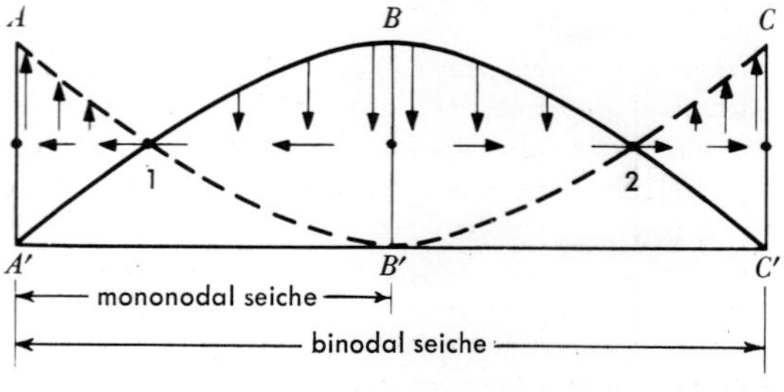

Fig. 1. Seiches in rectangular basins. Note: Figure is exaggerated in that oscillation of water surface is shown to extend to bottom of basin.

Coriolis effect. The Coriolis (deflective) effect of the earth's rotation imposes a secondary effect on the water level changes produced by a seiche. In plan views Fig. 2*a* and *b*, arrows show surface currents at ¼ and ¾ cycles respectively, after high water at the channel head. Coriolis deflection, indicated by dotted arrows, causes an increase of water level in the right half of the current in both A and B, with a corresponding decrease to the left. The dotted line is one of mean level with respect to the deflection. Water profiles along the nodal lines are shown to the left of A and B. Point O, called the amphidromic point, is thus the only position of no water change. Line ON connects points having simultaneous high water. Other such "cotidal lines" are shown in Fig. 2*c* for different phases of one seiche oscillation. The superposition of the transverse (deflective) motion on the primary longitudinal motion causes this line to rotate counterclockwise through one complete oscillation, in the Northern Hemisphere. *See* TIDE.

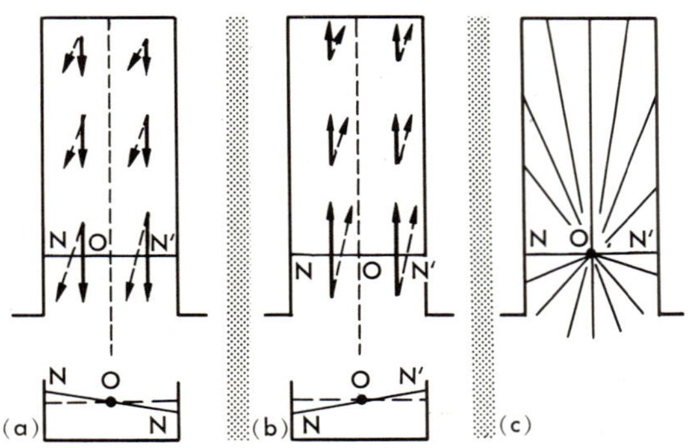

Fig. 2. Schematic representation of transverse oscillations in a bay in the Northern Hemisphere leading to the development of an amphidromic point. (*a*) Ebb current. (*b*) Flood current. (*c*) Cotidal lines.

Internal seiches. Internal seiches may develop when vertical variations of temperature or salinity produce a significant variation in density between two water layers. The zone or surface between these layers, will, if the density contrast is sufficient, behave as a free surface, similar to that between water and air. Seiches can be supported by this internal surface where the vertical water motion will be a maximum. [W.L.D.]

Seismograph

A device for detecting, amplifying, and recording motion of the ground. Ground motion varies greatly, depending on the manner of excitation. Seismic waves from explosions and earthquakes occur in the frequency range from about 100 to 1/3000 cps. Three components of ground motion are involved in the propagation of seismic waves, and the ground motion can vary by a factor of about 10^8 from the smallest to the largest recorded earthquakes. For these reasons, many different types of seismographs have been designed. Two major categories exist: pendulum seismographs and strain seismographs.

Pendulum seismographs. This type measures the relative motion between the ground and a loosely coupled inertial mass. In some instruments, optical magnification is used, whereas other devices exploit the advantages of electromagnetic transducers, photocells, galvanometers, and electronic amplifiers to achieve higher magnification.

Most widely used in seismograph stations is the electromagnetic pendulum seismograph with galvanometric registration. A coil is attached to the pendulum, and a sensitive galvanometer is connected to the coil. Motion of the coil in a magnetic field induces an electromotive force which activates the galvanometer movement. A mirror on the galvanometer deflects a light beam so as to produce a record on photographic paper. An alternate form uses a variable reluctance transducer in which movement of the pendulum varies the reluctance of a magnetic circuit. The resulting magnetic flux variations induce an electromotive force in a coil surrounding an armature in the magnetic circuit.

The equations for a pendulum-galvanometer seismograph system are (after S. K. Chakrabarty):

$$\frac{d^2x}{dt^2} + 2\epsilon_o \frac{dx}{dt} + \omega_o^2 x - \sigma_o \frac{d\theta}{dt} = -\frac{d^2Z}{dt^2}$$

$$\frac{d^2\theta}{dt^2} + 2\epsilon_g \frac{d\theta}{dt} + \omega_g^2\theta - \sigma_g \frac{dx}{dt} = 0$$

Here, x is the pendulum displacement, Z is the ground displacement, and θ is the galvanometer deflection; ϵ and ω are, respectively, damping constant and natural (circular) frequency. The subscript o refers to pendulum and g to galvanometer, σ_o is the coupling factor of galvanometer to pendulum and σ_g that of pendulum to galvanometer, and t is time. For any given ground displacement Z, θ (hence the trace motion on the seismogram) can be computed. Also, Z can be found from θ and its derivatives and integrals. By selecting appropriate values for the constants ϵ, ω, and σ, a seismograph system can be designed for specified sensitivity, frequency response, and phase distortion. For example, the frequency response R of a seismograph system is found by solving for θ in the preceding equation after setting

$$Z = C \sin \omega_e t$$

where C is the maximum amplitude and ω_e is the circular frequency of an assumed steady sinusoidal ground displacement. The frequency response is

$$R = \{[(\omega_o^2 - \omega_e^2)(\omega_g^2 - \omega_e^2) - 4\omega_e^2\epsilon_o\epsilon_g]^2 + [2\omega_e\epsilon_o(\omega_g^2 - \omega_e^2) + 2\omega_e\epsilon_g(\omega_o^2 - \omega_e^2)]^2\}^{-1/2}$$

Need for diverse types. In order to achieve adequate dynamic range and the ability to record over a wide frequency range, several different types of seismograph systems are used simultaneously. Strong motion instruments of low magnification record ground accelerations for severe shaking which would disable more sensitive apparatus. The background noise (microseisms) in the earth varies with frequency. Thus, for frequencies near ⅙ cps,

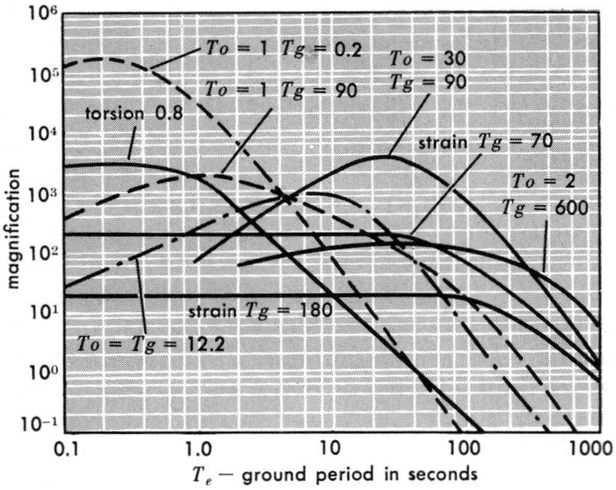

Fig. 1. Response curves for some seismograph systems in current use: *To*, pendulum period; *Tg*, galvanometer period.

Fig. 2. The Benioff vertical seismometer.

the maximum usable magnification of a seismograph located almost anywhere in the world is about 2000. For frequencies near 5 cps, magnifications of 1×10^6 are usable at certain quiet locations. For this reason, broad band instruments with uniform response over the entire seismic frequency range are not used. Rather, seismographs covering limited parts of the spectrum are designed. This not only improves the signal-to-noise ratio, but also makes possible the discrimination between seismic waves of different frequencies which arrive at the same time but are controlled by different mechanisms of propagation.

In Fig. 1 are shown the response curves for several seismograph systems in current use. The Benioff vertical seismometer (pendulum period, 1 sec) is shown in Fig. 2. The Press-Ewing vertical seismometer (pendulum period, 30 sec), is shown in Fig. 3.

Strain seismographs. The principle of the Benioff linear strain seismograph is shown in Fig. 4. Two piers, P_1 and P_2, are tied to bedrock and separated by distances of about 100 ft. L is a rigid fused-quartz tube attached to P_1 and suspended so as to have a single degree of freedom in the longitudinal direction. Strains in the ground produce proportional variations in the distance between P_1 and P_2 which may be detected by a sensitive transducer in the gap between the free end of L and P_2. Electromagnetic and variable discriminator transducers are used. The strain seismograph detects secular strains related to tectonic processes and tidal yielding of the solid earth. Strains associated with propagating seismic waves are also recorded. The response of a strain seismograph to longitudinal motions is

$$Y = -V \frac{L}{c} \cos^2 \alpha \frac{d\xi}{dt}$$

where V represents the response of the transducer; L, the length of the rod; c, the apparent surface velocity of the seismic waves; α is the angle between the direction of the rod and the propagation direction, and ξ is the ground particle displacement. Under favorable conditions of stable sites, strains as small as 10^{-9}–10^{-10} are recorded. A typical installation is shown in Fig. 5.

Seismic prospecting apparatus. In seismic exploration for petroleum, explosions provide the source of seismic waves. The frequency range is 2–20 cps for refraction shooting and 20–300 cps for reflection shooting, depending on local conditions. The principal field problem is to obtain maximum signal-to-noise ratio and to survey large areas rapidly. This requires portable or truck-mounted apparatus and the use of arrays of detectors, filtering techniques, and special recording methods.

Fig. 3. The Press-Ewing vertical seismometer.

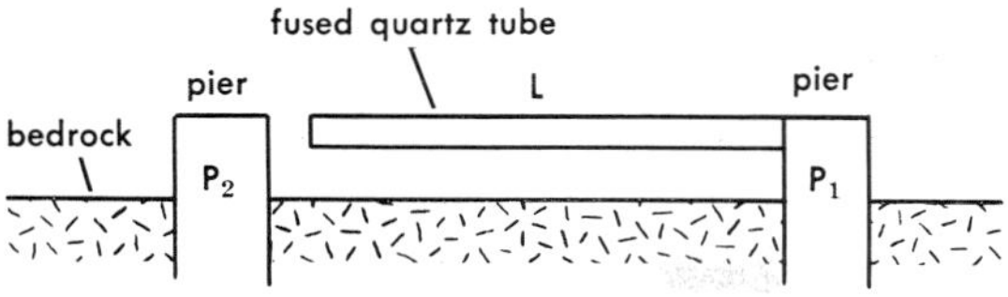

Fig. 4. Principle of the Benioff linear strain seismograph.

Fig. 5. Benioff linear strain seismograph.

The detectors (geophones), often weighing less than 1 lb, contain a mass supported by a spring and constrained to move in the vertical direction. An electromagnetic transducer transforms the mechanical motion into voltages. Special amplifiers are designed for small size, large gain, and freedom from noise; these incorporate band pass filters with adjustable low and high frequency cutoff. Automatic gain control provides usable signals from the arrival of the initial large amplitude waves until signal strength drops below the level of background noise.

Magnetic tape and drum recording are used to exploit the advantages of dynamic range and playback. Photogalvanometric recording systems are also used. *See* GEOPHYSICAL EXPLORATION; PROSPECTING; SEISMOLOGY; *see also* EARTHQUAKE.

[F.P.]

Bibliography: H. Benioff, Earthquake seismographs and associated instruments, in H. E. Landsberg (ed.), *Advances in Geophysics*, vol. 2, 1955.

Seismology

The science of earthquakes and the earth's interior as revealed by seismic waves. It is a relatively young science in which the major developments began about 1880 with the invention of a machine for measuring earthquakes (*see* SEISMOGRAPH). An earthquake or explosion represents a source of seismic waves which propagate in the earth's interior and emerge to be recorded by seismographs distributed over the earth's surface. From wave analysis it is possible to infer the sub-surface structure and sometimes the mechanism of the source.

Earthquakes. Elastic strain accumulates in the outermost several hundred kilometers of the earth because of thermal processes in the interior. Earthquakes are sudden failures of rock under strain.

Faults and fault zones. Ruptures or faults occur most frequently in belts of rapid strain accumulation in association with new mountain systems, deep sea trenches, and volcanic zones. Most occur at shallow depths (<30 km), and none occur at depths greater than 600–700 km.

In the circum-Pacific belt, for example, geodetic measurements indicate a gradual deformation, as if the Pacific Ocean basin is rotating in a counterclockwise direction with respect to the continent. Where observable, faults associated with earthquakes in this belt show movements consistent with this pattern of deformation.

The earthquake belts of the world are depicted in Fig. 1. The circum-Pacific belt is the major earthquake feature of the world. Other belts of lesser activity are the Alpide belt which runs from Burma to the Alps, the mid-Atlantic Ridge, the East African rift zone, and the Pamir-Baikal zone. Stable blocks where seismic activity is the least, are the deep sea basins and the continental shields.

Tectonic-seismic-gravity associations. Associated with the circum-Pacific belt is a remarkable arcuate tectonic-seismic structure where a deep sea trench, a seismic belt, a gravity anomaly, and a volcanic chain occur in a definite pattern. The highest seismicity is associated with this type of structure (Fig. 2). Earthquakes in Japan, the Aleutians, and the Pacific Coast of South America are examples. Block movements along faults are another tectonic-seismic feature. Earthquakes in California fall into this category in which displacements between large blocks occur along faults which are sometimes visible at the surface. The San Andreas fault of California is a classic example.

Elastic rebound theory. California block faulting gave rise to the elastic rebound theory, according to which an earthquake occurs at a point where the gradually accumulating strain exceeds the strength of the rock, producing a slip. However, this theory can account for only very shallow earthquakes because the frictional locking of fault blocks under hydrostatic pressure requires stress differences higher than the strength of the rock permits. A condition of unstable plastic deformation (creep) has been suggested as an alternative mechanism.

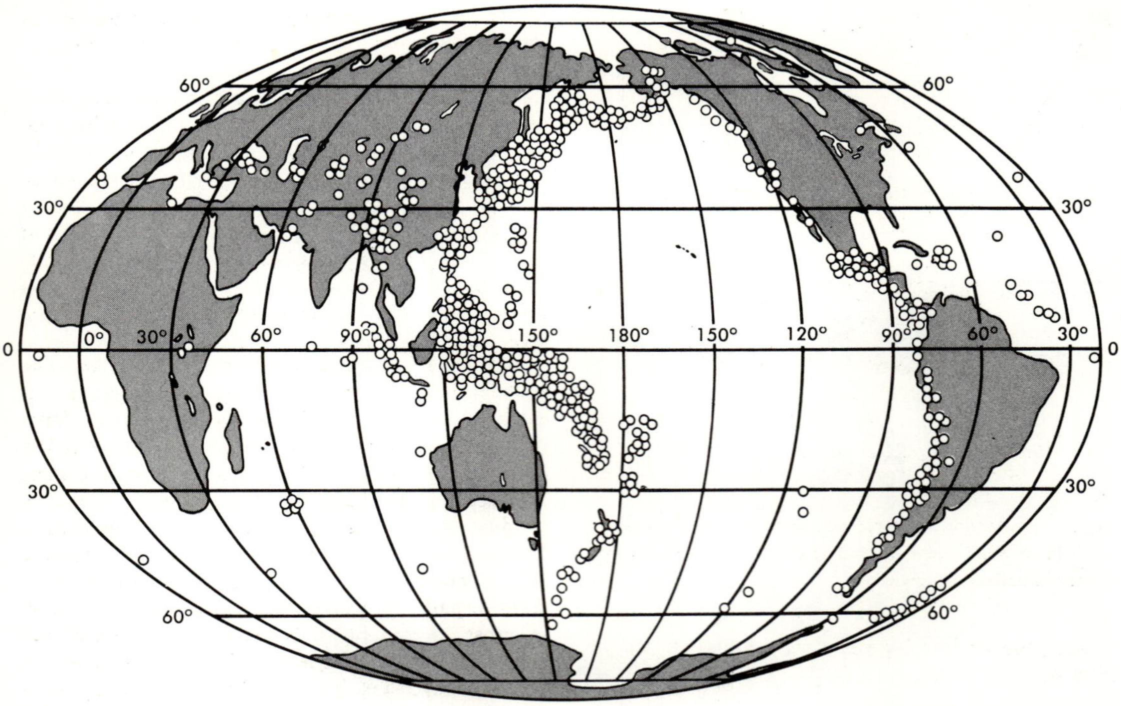

Fig. 1. Earthquake belts of the world as shown by epicenters of shallow focus earthquakes 1918–1952. (*After B. Gutenberg and C. F. Richter*)

Types of earthquakes. Most earthquakes of significant size are associated with tectonic activity and are known as tectonic earthquakes. Magma movements associated with active volcanoes may induce minor earthquakes (volcanic earthquakes). Deep focus earthquakes occur almost entirely in association with the arcuate structures of the circum-Pacific belt. Their maximum depth of about 700 km places a lower depth limit for strain accumulation under long-term stresses.

Size of earthquakes. This aspect may be characterized by intensity or magnitude. Intensity is a measure of the degree of shaking at a place as indicated by effects on people, objects, and structures. In the modified Mercalli intensity scale, 12 grades of intensity are used. Isoseismals are equal-value lines, similar to contours on a map, representing boundaries between zones of successive intensity as compiled from field investigation. Magnitude is more quantitative. It is an instrumental scale in which the logarithm (to the base 10) of the amplitude of ground velocity is added to a factor which corrects for distance and local conditions at the seismograph station. The smallest recorded earthquakes have magnitudes between 0 and 1, and the largest magnitude found was 8.6. The energy released in the seismic waves of the largest earthquakes is about 10^{25} ergs. Although the annual number of earthquakes is reckoned in the millions, the few shocks of largest magnitude each year account for most of the energy release.

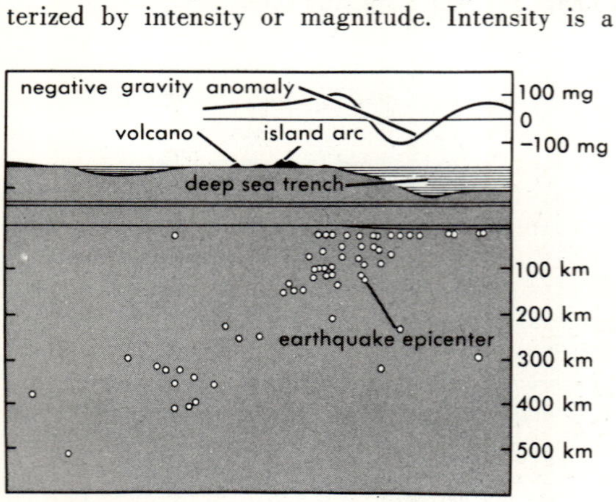

Fig. 2. Arcuate tectonic-seismic structure showing pattern of deep sea trench, seismic belt, gravity anomaly and volcanism. (*From B. Gutenberg and C. F. Richter, Seismicity of the Earth and Associated Phenomena, 2d ed., Princeton University Press, 1954*)

Seismic waves. The equations of motion for an elastic solid may be written in the form

$$\rho \frac{\partial^2 s}{\partial t^2} = (k + \tfrac{1}{3}\mu) \operatorname{grad} \theta + \mu \nabla^2 \mathbf{s}$$

where $\mathbf{s}$ is the particle displacement vector, $\theta = \operatorname{div} \mathbf{s}$ is the dilatation, ρ is the density, and μ and k are the elastic constants of rigidity and bulk modulus, respectively (*see* VECTOR ANALYSIS). Taking the divergence of the preceding equation yields a wave equation for θ.

$$\rho \frac{\partial^2 \theta}{\partial t^2} = (k + \tfrac{4}{3}\mu) \nabla^2 \theta$$

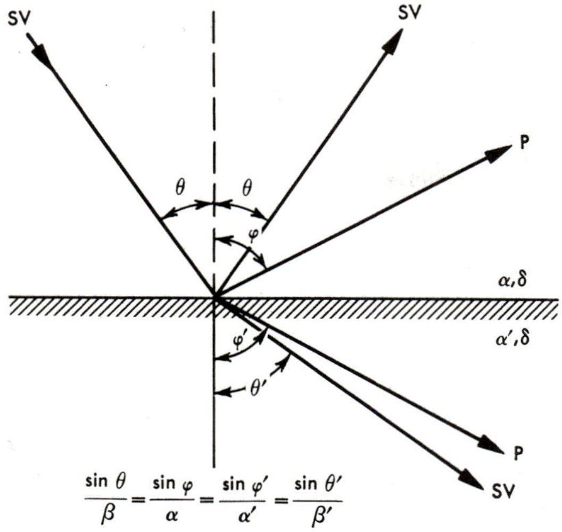

Fig. 3. Reflection and refraction of elastic waves according to Snell's law; α and β are compressional and shear velocities respectively.

Taking the curl of 1 gives a wave equation for the rotation

$$\rho \frac{\partial^2 \psi}{\partial t^2} = \mu \nabla^2 \psi$$

where ψ = curl s. These equations establish the existence of the dilatational (P) waves and rotational or shear (S) waves for an elastic solid with velocities $\sqrt{(k + \frac{4}{3}\mu)/\rho}$ and $\sqrt{\mu/\rho}$, respectively.

Most consolidated rocks found near the surface of the earth have compressional velocities in the range 2–8 km/sec. Sedimentary rocks generally fall in the velocity range 2–5 km/sec, and igneous rocks have values 5–8 km/sec. The velocity increases with pressure and compaction, and decreases with temperature. Crystalline sedimentary and igneous rocks fall in the higher velocity range, and for these rocks, the compressional velocity is about $\sqrt{3}$ times greater than shear velocity.

Shear waves are conveniently divided into SV and SH waves which are polarized respectively in vertical and horizontal planes. In general, P and SV waves incident upon a horizontal interface separating two media having different elastic constants produce two reflected and two refracted waves (Fig. 3). Snell's law is satisfied, and the phenomenon of critical reflection is possible. This may be verified, and reflection and transmission coefficients may be found by seeking solutions of the wave equations which satisfy the conditions of continuity of stress and displacement at the interface. SH waves are reflected and refracted without changing type.

If, in the solution for reflected and refracted waves, one inserts the condition that amplitudes decrease (usually exponentially) with distance from the interface or from a free surface, the existence of elastic interface (Stoneley) waves or surface (Rayleigh) waves may be established. The velocity is slightly less than the lowest velocity for shear waves near the interface or surface. *See* REFRACTION WAVES.

Another type of surface wave is composed of P and S waves traveling in a layer, confined by total reflections at its boundaries. These waves are dispersed, the frequency associated with a given phase velocity being determined by the conditions of constructive interference between successive orders of reflection. Love waves are an example in which horizontally polarized shear waves are confined between the earth's surface and the bottom of the crust.

Seismic wave transmissions in the earth. The earth may be considered as a spherically symmetrical body in which elastic velocity varies as a function of depth only. Discontinuities, such as at the bottom of the crust and the mantle-core interface, produce velocity jumps, and, in general, there is a gradual velocity variation between discontinuities. Under these circumstances, many different seismic phases occur. These represent waves which have definite paths and may have undergone conversion in crossing interfaces. Some of these phases are shown in Fig. 4. Whenever reflection or refraction occurs, conversion from P to S or S to P is possible, as is shown in the once-reflected phases PP and PS, and in the twice-reflected PPS or SPS. Similar considerations hold for core reflections P_cP and P_cS. Transmission through the fluid core is possible only for compressional waves (K), hence the phases PKP, PKS, and SKS occur. A phase traversing the inner core is $PKIKP$, where I represents the compressional wave segment through the presumably solid inner core.

An important feature is the shadow cast by the earth's core (Fig. 4). It is a consequence of the reduction in elastic velocity in the core corresponding to vanishing rigidity. The shadow zone extends from 105° to 143°. The absence of phases which have traversed the core as shear waves is the primary evidence for a fluid core.

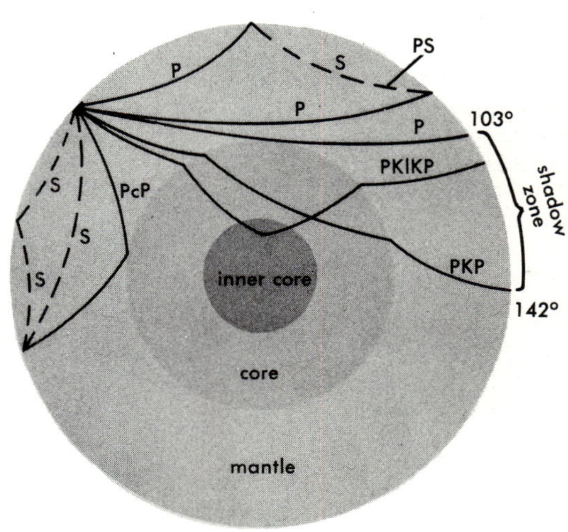

Fig. 4. Paths of seismic waves in the earth.

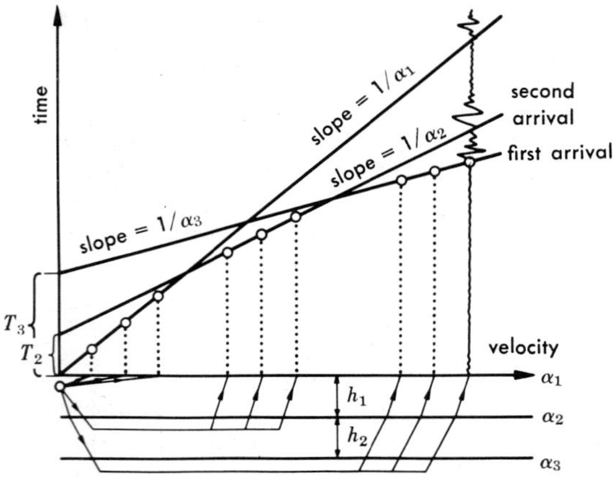

Fig. 5. Paths and travel time curves of seismic refractions used in crustal studies.

An important seismic wave is *Pn*, which follows the path shown in Fig. 5. This phase is the main source of information about crustal thickness.

For deep focus earthquakes, an additional family of phases becomes possible, such as *pP* and *sS*. These are *P* and *S* waves which have been reflected from the surface in the vicinity of the epicenter.

There are numerous channels within the earth which transmit surface waves and guided waves. In general, the waves are dispersed. Love waves and Rayleigh waves are associated with the wave guide formed by the continental and oceanic crust. *G* waves are long Love waves controlled by the velocity variation in the mantle, as is the case for mantle Rayleigh waves. Higher mode Love and Rayleigh waves exist, and these are particularly sensitive to velocity variations within the crust. *Lg* is a short-period wave guided by the gradual velocity increase with depth near the upper part of the continental crust. The *T* phase is a short period, earthquake-induced sound wave, transmitted efficiently across the deep ocean. Finally, normal modes of vibration of the earth as a whole are possible.

The seismic body waves (*P*, *S*, and composite types like *PS*) have predominant periods in the range 1–15 sec, with *P* and *S*, respectively, at the short-period and long-period end of this range. Crustal surface waves may occur with periods in the range 5–50 sec and mantle surface waves have been observed with periods of 50–500 sec. The greatest mode of a free spheroidal vibration of the whole earth is about 53 min.

Travel time curves. Curves and tables giving the travel times of the main seismic phases as a function of distance are available (Fig. 6). These have been prepared by a process of successive approximations involving a least square adjustment of epicenter, focal depth, and origin time. Input data are the arrival times of the phases as reported by the several hundred seismograph stations of the world. It is remarkable that nuclear explosions, where the origin time and location are known, have verified the *P* travel times to within 2 sec; but even this small discrepancy is explainable by the special conditions of a thin earth's crust at the detonation point. Travel time curves are used routinely by seismologists to locate earthquake epicenter, focal depth, and origin time, and to infer the elastic properties of the earth's interior.

Dispersion studies of surface waves yield data on phase and group velocity as a function of period. Experimental results for crustal and mantle Rayleigh waves are shown in Fig. 7. The corresponding wavelengths range from 50–2000 km, so that the short periods are primarily influenced by velocity variations in the crust and subadjacent mantle, whereas velocities in the deeper mantle affect the longer waves. Mantle Rayleigh waves which have circled the earth up to 7 times have been observed from large earthquakes.

Determining velocity depth variations. One of the fundamental problems of seismology is to determine the variation of elastic constants with depth in the earth. The travel times of seismic phases *P* and *S*, and the dispersion (phase and group velocity dependence on period) of surface waves have been used for this purpose.

It follows from Snell's law that for a given seismic ray in the earth

$$\frac{r \sin i}{V} = p = \frac{r_m}{V_m} = \text{constant}$$

where r is the radius to the point where the velocity is V, and i is the acute angle between the radius and the direction of the ray. At the deepest point reached by the ray, $i = 90°$; r_m and V_m are the corresponding radius and velocity. The parameter p is obtainable as a function of distance from a travel time curve (t plotted against angular distance Δ) from

$$p = dt/d\Delta$$

It can be shown that under certain restrictions on its behavior, the velocity depth function may be obtained from the following relationship

$$\pi \log \frac{r_0}{r_1} = \int_0^{\Delta_1} \cosh^{-1}\left[p\,\frac{v_1}{r_1}\right] d\Delta$$

Here r_0 is the radius of the earth. r_1 and V_1 are, respectively, the radius and velocity corresponding to the maximum depth of penetration for the ray which emerges at angular distance Δ_1.

In studying the earth's crust, one may consider flat layers and use relationships (see Fig. 5) like

$$h_2 = \frac{1}{2}\left(T_3 - 2h_1\frac{\sqrt{\alpha_3^2 - \alpha_1^2}}{\alpha_3\alpha_1}\right)\frac{\alpha_3\alpha_2}{\sqrt{\alpha_3^2 - \alpha_2^2}}$$

$$h_1 = \frac{1}{2}T_2\frac{\alpha_2\alpha_1}{\sqrt{\alpha_2^2 - \alpha_1^2}}$$

This enables one to obtain thicknesses h_1 and h_2 of the layers in the crust. The velocities are obtained from the slopes of the corresponding travel time segments. In practice, explosion-generated seismic waves are used to obtain travel time curves pertinent to studies of the earth's crust. Earthquakes

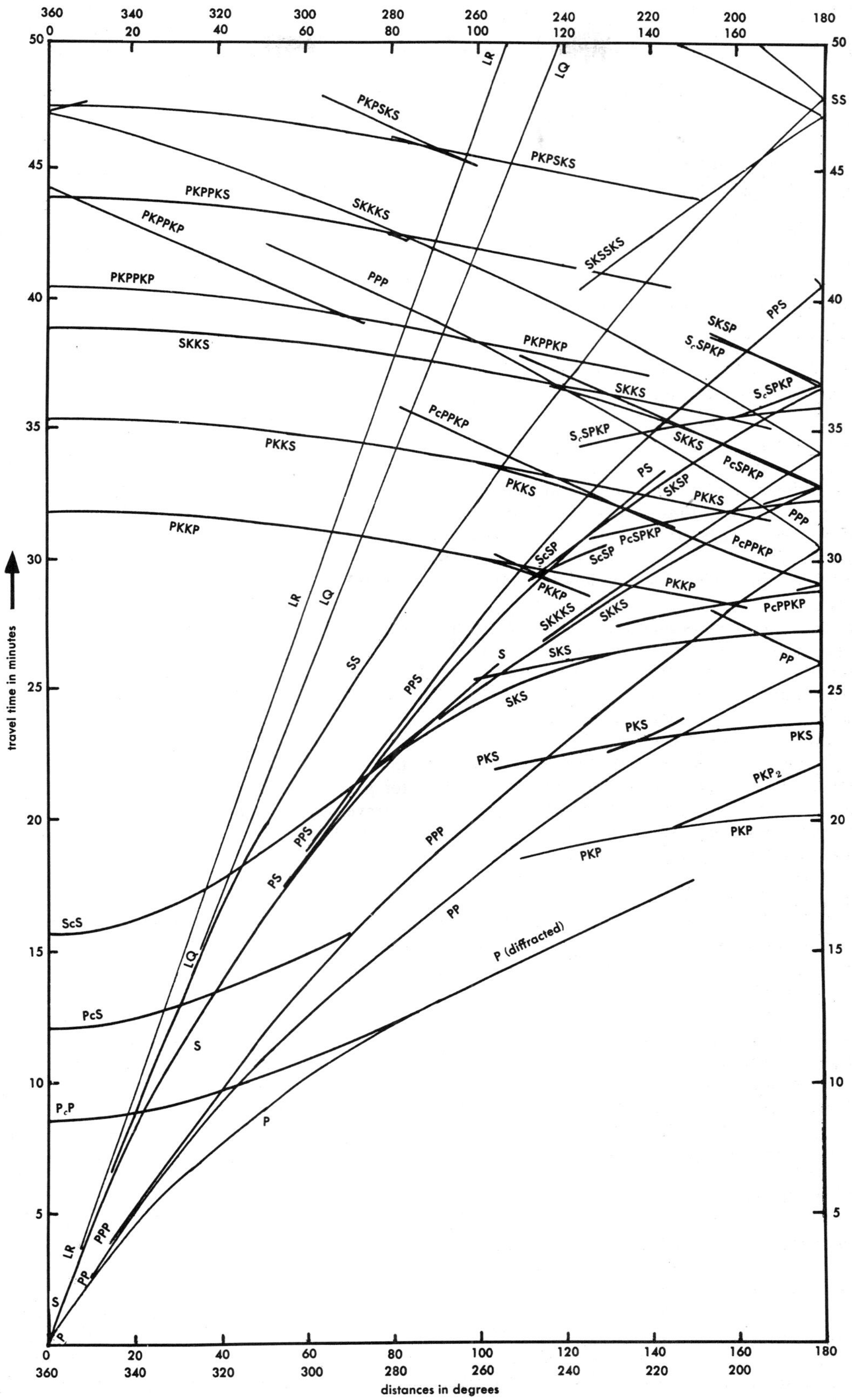

Fig. 6. Seismic travel time curves. (*After H. Jeffreys and K. E. Bullen*)

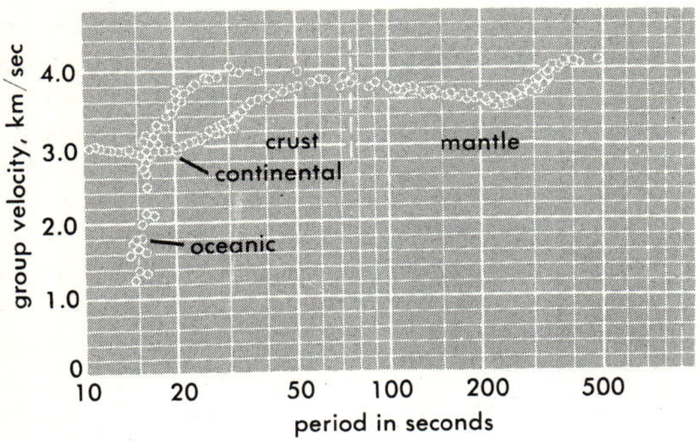

Fig. 7. Crustal and mantle Rayleigh wave group velocity curves. (*After W. M. Ewing and F. Press*)

and, more recently, nuclear explosions provide the seismic waves needed for study of the deeper layers in the earth.

Surface waves also yield velocity-depth information. The procedure is to compare experimental dispersion curves like those in Fig. 7 with theoretical curves computed on the basis of assumed velocity-depth functions. With the use of digital computers, theoretical dispersion curves may be computed for a variety of assumptions. In certain problems, for example the delineation of the Gutenberg low-velocity zone at a depth of about 150 km, surface-wave methods may be more powerful than methods using P and S data.

Seismograms. Arrival times, amplitudes, pulse shapes, dispersion, and direction of approach of seismic waves are the basic data derived from seismograms. In general, three matched seismographs are used to obtain records of up-down, north-south, and east-west motions of the ground. The character of a given earthquake recording depends on the response characteristics of the instruments; on the distance, focal depth, and magnitude of the earthquake; and in the case of surface waves, on the nature of the propagation path. Seismic phases are identified by their sequence of arrival times, character, and orbital motion. For example, Love waves would show as a transverse horizontal ground motion, whereas Rayleigh waves are polarized in the vertical plane of propagation. P waves are first arrivals; they are usually the shortest period waves on the seismogram and exhibit a longitudinal orbital motion. In general, the larger the magnitude of an earthquake, the more pronounced are the long-period components of the spectrum. Also, selective absorption of shorter periods tends to increase the predominant periods with propagation distance. Typical seismograms with phases identified are shown in Fig. 8 for several distances.

Internal constitution of the earth. The methods described earlier lead to curves and tables of elastic

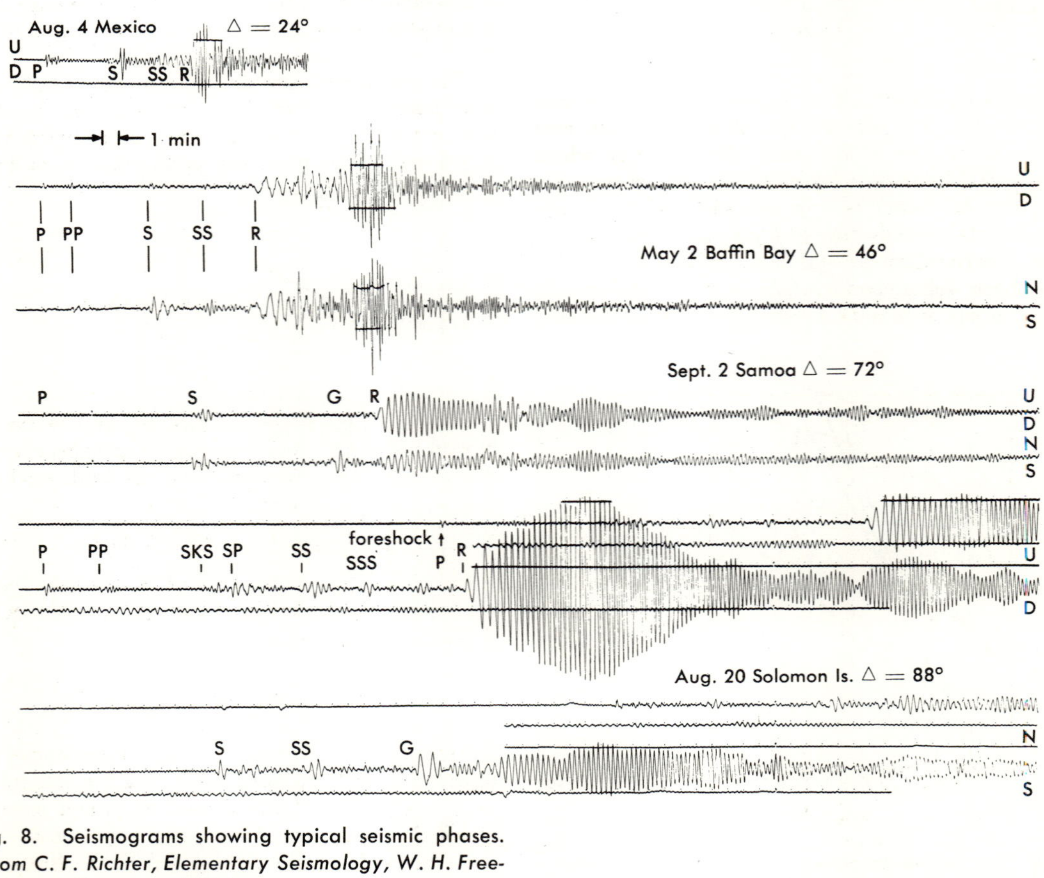

Fig. 8. Seismograms showing typical seismic phases. (*From C. F. Richter, Elementary Seismology, W. H. Freeman and Company, 1958*)

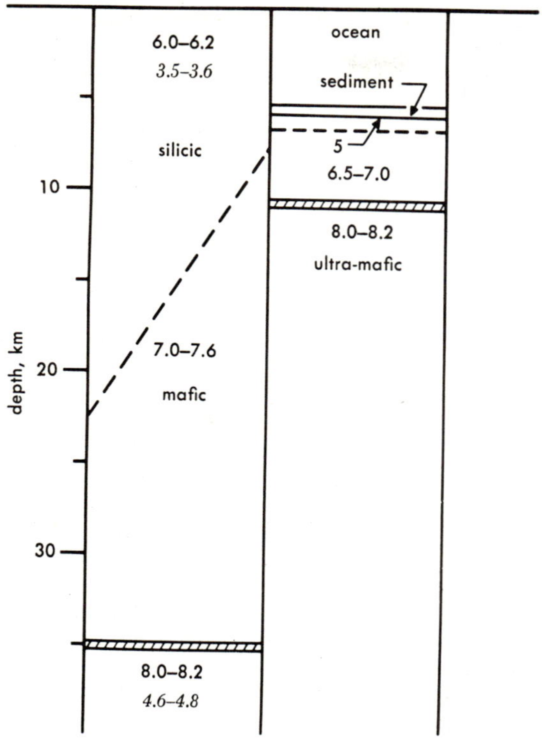

Fig. 9. Seismic velocities and crustal structure for continents and oceans. Italic numbers represent shear velocity.

velocities as functions of depth for the continental crust (Fig. 9), and for the mantle and the core (Fig. 10). The Mohorovičić discontinuity which separates the mantle from the crust is everywhere present. Its depth is approximately 35 km under continental shields and 10 km under ocean basins. The intermediate crustal layer is less definite. It has been definitely identified in many regions. In other areas, it may be present but masked because

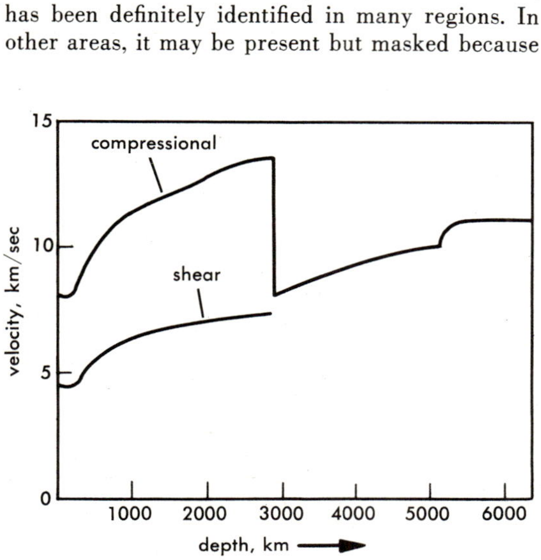

Fig. 10. Seismic velocities in the mantle and core of the earth. (*After B. Gutenberg*)

the corresponding refracted waves needed to establish its existence do not occur as first arriving waves. For this reason, its depth has been indicated as uncertain in Fig. 9. In the California-Nevada region, it is found about 20 km beneath the surface. The elastic velocity in the oceanic crust corresponds to that found in the intermediate layer of the continental crust. Under regions of high topography like the Rocky Mountains, the crust may thicken to as much as 50–55 km. If densities appropriate to the observed elastic velocities are assigned, it is found that the weight per unit area of a column of rocks extending from the surface to a standard depth in the mantle (say 60 km) is approximately the same everywhere. This is a consequence of a fundamental law of isostasy which governs crustal structure. It states that on a regional basis, the crust is in approximate hydrostatic equilibrium.

In Fig. 11 is presented another view of isostatic adjustment. In a profile across the United States from Pacific to Atlantic oceans is presented the topography, the depth to the Mohorovičić discontinuity as revealed by variations in phase velocity of earthquake-excited Rayleigh waves, and finally the variation in gravity. Gravity deficiency under the mountains reflects a deficiency of mass below, as would occur when more dense mantle rocks are replaced by less dense rocks of a thickened crust. This relationship between seismic and gravity data establishes isostatic adjustment on a regional basis.

The elastic velocity variation in the mantle and core may be used to infer something of the density variation with depth. Assuming hydrostatic stress for the interior, the pressure p at radius r satisfies the relation

$$\frac{dp}{dr} = -g\rho$$

where g is the gravitational acceleration:

$$g = \frac{\gamma m}{r^2}$$

γ being the gravitational constant, and m the included mass. Further assuming a homogeneous medium and adiabatic conditions, the bulk modulus k is

$$k = \rho(dp/d\rho) = \rho(\alpha^2 - \tfrac{4}{3}\beta^2)$$

where α and β are the P and S velocities given earlier. These three equations lead to the Adams-Williamson expression

$$\frac{d\rho}{dr} = \frac{\gamma m \rho}{r^2(\alpha^2 - \beta^2)}$$

Density-depth variation is obtained by numerical integration with the restriction that the integrated density gives the observed total mass and moment of inertia of the earth. The variation of density and pressure with depth derived by K. E. Bullen is shown in Fig. 12.

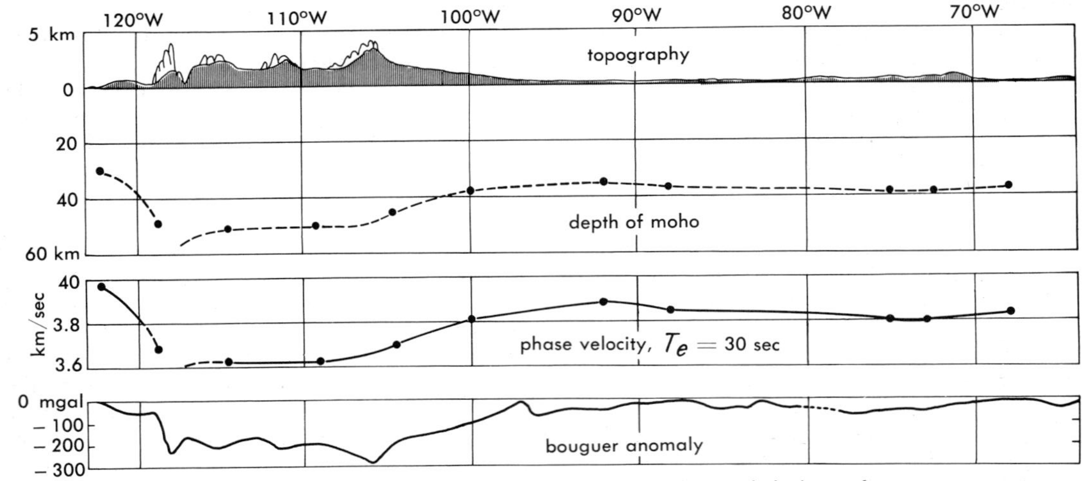

Fig. 11. Cross section of observed topography, Rayleigh wave phase velocity, Bouger gravity anomaly, and inferred crustal thickness from west to east across the United States. (*After W. M. Ewing and F. Press*)

Inferences about the physical constitution of the interior are usually made concordant with the observed seismic velocities, the relative abundance of elements, the distribution of rock types at the surface, and the composition of stony and iron meteorites. Under these restrictions, a reasonable hypothesis suggests that the crust is composed of granitic and basaltic rocks. The mantle is peridotite, and the core is iron or iron-nickel alloy. The outer core may also contain some modified ultrabasic rock. An iron core so modified is also consistent with the density distribution. It is uncertain whether the Mohorovičić discontinuity represents a composition change or a phase change. The density variation in the mantle does not permit homogeneity, that is, chemical and phase changes occur. *See* EARTH INTERIOR.

Microseisms. These are background oscillations of the ground, unrelated to earthquakes. Although always present to some extent, they often build up in intensity in microseism storms which may last from several hours to several days. Ground amplitudes during a storm may rise to 10 μ, rarely to 100 μ. Microseisms limit the sensitivity of seismographs. In the period range 2–10 sec, microseisms have their origin in energy transfer from atmospheric pressure and wind disturbances to the crust by a process which is particularly efficient for storms over the oceans. The precise mechanism is not known. One theory asserts that standing sea waves induce pressure fluctuations on the sea floor by a second-order effect. Another suggests that impulsive modification of sea waves by wind gusts initiates compressional waves in the ocean. An older theory emphasizes the role of swell impinging on steep coasts.

Microseisms in the range 2–4 sec are usually incoherent and do not propagate very far. On the other hand, 6–10 sec period microseisms can cross a continent as coherent Rayleigh waves. There have been suggestions that microseisms may be used to track severe weather disturbances. A working system has never been successfully established, although major efforts have been made.

Microseisms in the period range less than 2 sec are usually local phenomena associated with traffic, factory vibrations, and wind effects on trees and structures. Microseisms with periods longer than 10 sec have been observed and are being investigated.

Seismic prospecting. This principal method of geophysical prospecting is based on the reflection and refraction of compressional waves. The field procedure is to generate compressional waves with a near-surface explosion and to infer the depths of elastic interfaces by analyzing the travel times of reflected and refracted waves recorded by arrays of surface seismographs. Refraction shooting is the technique used for mapping a layer with a higher elastic velocity than the beds above it (see Fig. 5). The shot-detector distance must be several times greater than the depth of the layer in order that the waves occur among the earliest arriving events.

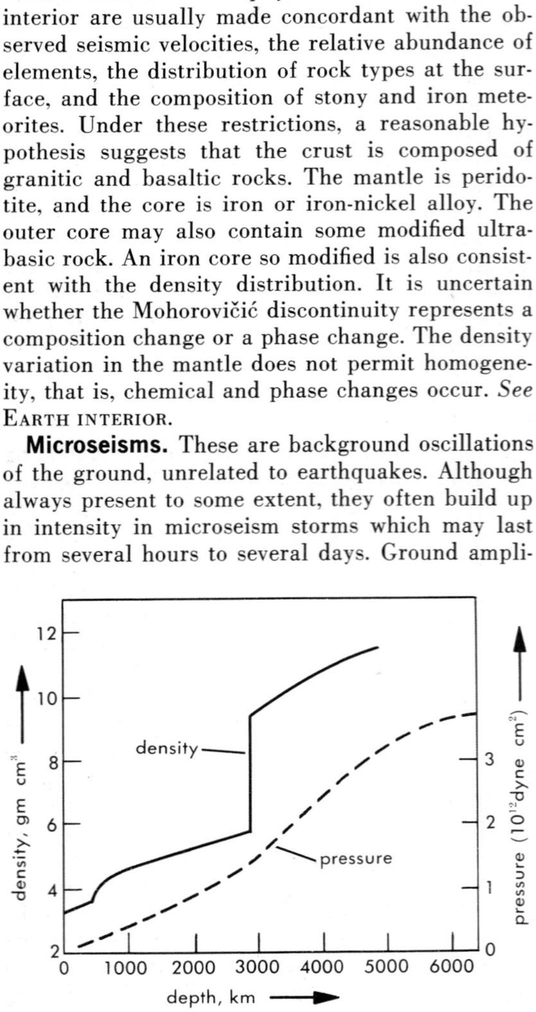

Fig. 12. Density and pressure variation in the earth. (*After K. E. Bullen*)

Reflection shooting involves near-vertical reflections, hence small shot-detector distances. In order to detect a reflection event in a background of surface waves and scattered waves, array techniques involving mixing of signals, filtering, automatic gain control, and patterns of explosions are used. *See* TERRESTRIAL GRAVITATION. [F.P.]

Bibliography: K. E. Bullen, *Introduction to the Theory of Seismology*, 2d ed., 1953; W. M. Ewing, W. S. Jardetzky, and F. Press, *Elastic Waves in Layered Media*, 1957; C. F. Richter, *Elementary Seismology*, 1958.

Seisonacea

An order of the class Rotifera which comprises a group of little-known marine animals. They form a single family with about seven species and are found only in Europe. The Seisonacea are epizoic or possibly ectoparasitic on crustacea, especially on members of the genus *Nebalia*. They have a very elongated jointed body with a small head; a long, slender neck region; a thicker, fusiform trunk; and an elongated foot, terminating in a perforated disk.

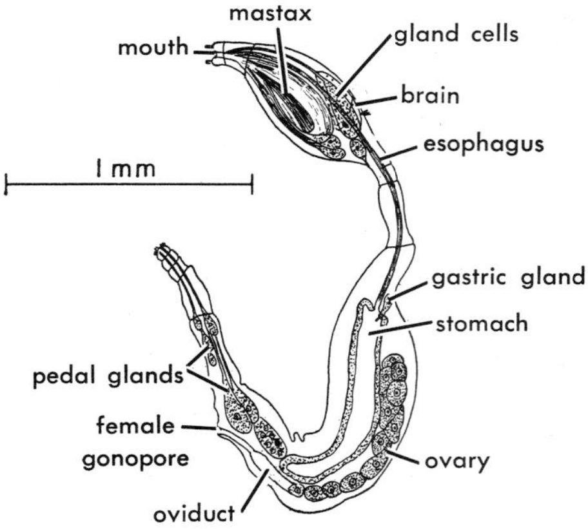

Seison, a rotifer. (After Plate, 1887)

The Seisonacea differ from other rotifers in a number of characters. They have a very reduced corona, a quite aberrant type of mastax (fulcrate), paired ovaries but no vitellaria in females, and paired testes in the males. Furthermore, the sexes are of similar size and development; only one type of egg is produced, and it requires fertilization. The Seisonacea are also larger than other rotifers, attaining sizes up to 3 mm in length. *See* ROTIFERA. [E.H.A.]

Selachii fossils

All fossil sharks, except for the very primitive cladoselachians of the Devonian and the aberrant pleuracanths of the late Paleozoic, are customarily included, with existing forms, in the order Selachii (subclass Elasmobranchii). They are differentiated from the cladoselachians by the presence of claspers and constricted rather than broad bases for the paired fins. The most ancient suborder, the Hybodontoidea, flourished in the late Paleozoic and survived into the Mesozoic, giving rise in the Jurassic to the four surviving suborders. *See* CLADOSELACHII; ELASMOBRANCHII FOSSILS; HYBODONTOIDEA; PLEURACANTHODII.

Of the surviving suborders, the Heterodontoidea and Notidanoidea are rare and unimportant except as showing primitive characters. Close relatives of *Heterodontus* (*Cestracion*), the living Port Jackson shark, appeared in the Late Jurassic; *Heterodontus*, like most modern sharks, has lost the primitive amphistylic jaw suspension of the hybodonts, but retains such features of that group as the heterodont dentition, with sharp teeth anteriorly and flattened crushing plates at the rear. Of the notidanoids, which retain the primitive jaw suspension but are not markedly primitive in other features, *Hexanchus* (*Notidanus*), the cow shark, likewise appeared in the Jurassic, but *Chlamydoselache*, only known representative of the other family of the suborder, is not known before the Miocene.

Among fossil post-Triassic sharks, as among living forms, most are included in the suborders Galeoidea and Squaloidea; both groups appear in the Jurassic and were abundant and diversified by the Late Cretaceous. All known fossils appear to be referable to families still existing, and all generally accepted family groups are represented in the fossil record with the exception of the whale sharks (Rhincodontidae) and false cat sharks (Pseudotriakidae), the minute teeth of which have so far escaped discovery. Five families, the Isuridae, Orectolobidae, Scylliorhinidae, Carcharhinidae and Squatinidae, had already developed before the close of the Jurassic. By the Late Cretaceous, 11 of the 17 families currently recognized were present, represented by species generically indistinguishable from existing types. [A.S.R.]

Selaginellales

The plant order of small club mosses, containing only one living genus *Selaginella*. Of the approximately 700 species, most live in the tropics, although a few species give the genus a world-wide distribution in moist, shady habitats. One species, the "resurrection plant," *Selaginella lepidophylla*, is tolerant of xeric (dry) conditions, reacting to the absence of water by contracting into a ball-shaped mass which unrolls when moisture becomes available. A few species are cultivated for their ornamental value. The order, Selaginellales, has many characteristics of other members of the phylum Lycopsida. It differs mainly from the order Lycopodiales in having two sizes of spores (heterosporous) and a special, small membranous outgrowth called a ligule which is borne on the upper or adaxial base of the leaf. The sex organs, antheridia (male) and archegonia (female), always develop on separate gametophytes, and the sporangia always occur on clusters of sporophylls (spore-bearing leaves) collectively called strobili. Arising at the base of each leafless branch is a rootlike organ called a rhizophore which bears adventitious roots at its tip (Fig. 1). *See* ROOT (BOTANY).

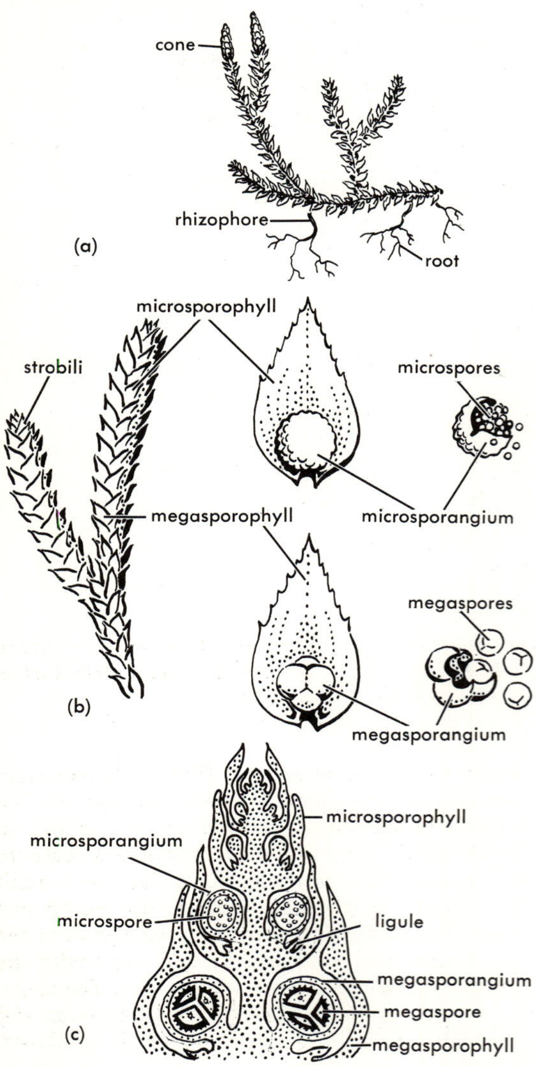

Fig. 1. (a) *Selaginella,* or small club moss, with creeping, dichotomous stems bearing numerous, tiny leaves and leafless branches, or rhizophores, which produce roots at their tips. Cones are located at the ends of the leafy branches (*from H. J. Fuller and O. Tippo, College Botany, Holt, 1949*). (b) Enlarged view of strobilus of *Selaginella* (*from W. W. Robbins, T. E. Weier, C. R. Stocking, Botany: An Introduction to Plant Science, Wiley, 1950*). (c) Diagram of a longitudinal section through a *Selaginella* cone, showing upper microsporophylls with microsporangia containing many small microspores, and lower megasporophylls with megasporangia containing 4 large megaspores (*after Lyon from H. J. Fuller and O. Tippo, College Botany, rev. ed., Holt, 1954*).

Structure. Most species of the genus are multibranched. The branches may be either dichotomous (forked) or monopodial, and some produce frondlike growths (*see* FILICALES). Many species are prostrate and creeping, whereas others are climbing and epiphytic. The stems are densely covered with closely appressed leaves which are arranged either spirally or in four rows of two alternating pairs. In the latter, each pair consists of a large and a small leaf with those on the lower, or ventral, surface being much larger than those on the upper, or dorsal, surface. This provides a more complete exposure of the leaves to sunlight. The ligule, which is characteristic of this order, is associated with both the vegetative leaves and the modified spore-bearing leaves, or sporophylls, of the strobili (cones). In some species the ligule soon withers away. Since the ligule is present in some of the treelike fossil forms, it may indicate a close relationship of the modern living species with a much larger extinct group of the Carboniferous (*see* GEOLOGY; PALEONTOLOGY).

Depending on the species, the origin and development of the stem may be either from a single apical cell or a group of cells. In maturation the development is usually exarch, or toward the center of the stem. It has been observed that in some species certain cells differentiate into true xylem vessels rather than tracheids, a condition usually found in the higher seed plants (*see* XYLEM).

Alternation of generations. In the sporophytic generation, the sporophylls closely resemble the vegetative leaves and, in some species, are so loosely arranged as to make the strobilus inconspicuous despite the fact that the presence of a

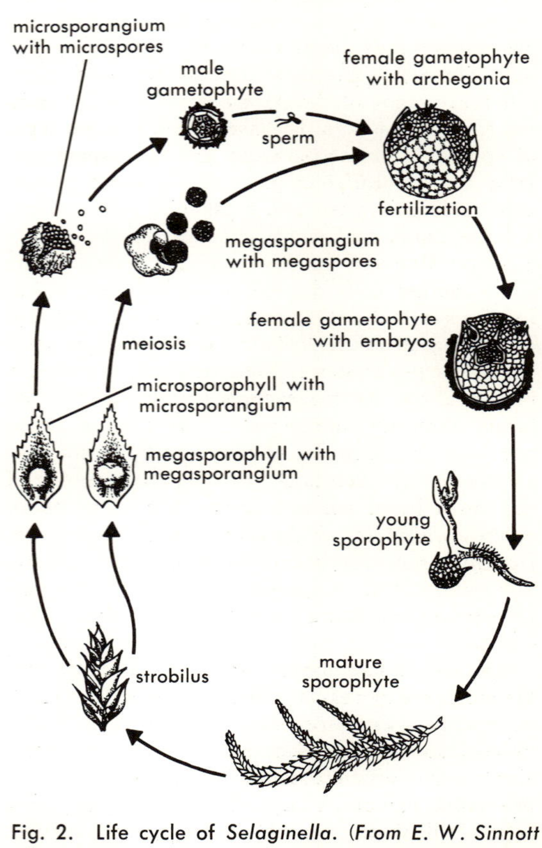

Fig. 2. Life cycle of *Selaginella.* (*From E. W. Sinnott and K. S. Wilson, Botany: Principles and Problems, 5th ed., McGraw-Hill, 1955*)

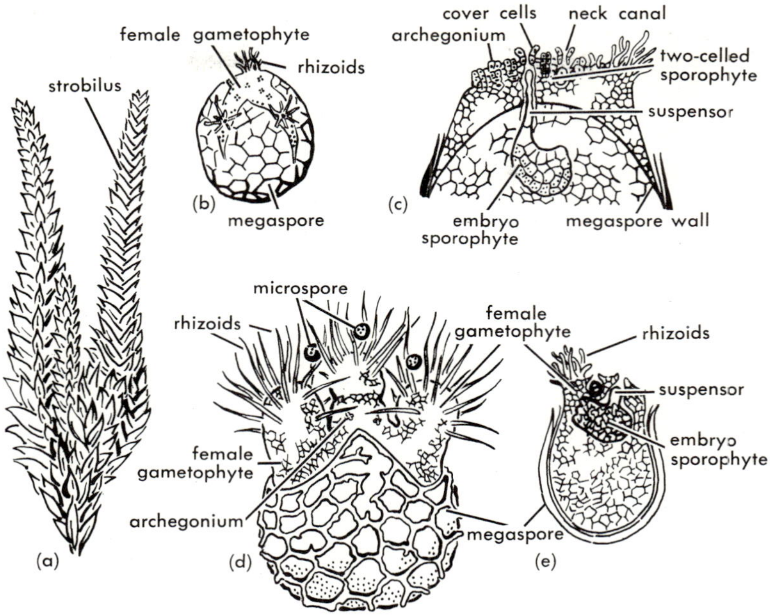

Fig. 3. Female gametophyte and embryo of *Selaginella*. (*a*) Sporophyte with strobilus. (*b–e*) Stages in development of the embryo sporophyte in the female gametophyte. (*After Bruchmann from W. W. Robbins, T. E. Weier, and C. R. Stocking, Botany: An Introduction to Plant Science, Wiley, 1950*)

strobilus is a common character of the group (Fig. 2). Usually each sporangium arises near the adaxial base of a sporophyll, but in some species a sporangium may develop from the stem in the leaf axil. Since the growth of the strobilus is apical, different stages of development are usually present from the apex to the base. There are two types of sporangia, microsporangia and megasporangia. In a microsporangium most of the spore mother cells undergo meiosis and cytokinesis and produce a large number of small microspores (*see* CYTOKINESIS; MEIOSIS). In the megasporangium, all the spore mother cells degenerate except one which, by meiosis and cytokinesis, develops a tetrad of large megaspores that distend the sporangial wall. The two types of spores (heterospores) are similar in shape, but the megaspores are about 20 times larger than the microspores. All have a thick, warty, cutinized wall. At maturity both kinds of sporangia split across the top and constriction of the unslit basal portions causes the spores to be ejected with force. This action may bring spores from different plants into proximity and, eventually, with the development of gametophytes (gamete-producing plants), this may cause cross fertilization.

Because of heterospory, remarkable changes in the morphology and physiology of the gametophyte generation take place. The microspore, at first a uninucleate cell, develops into a male, or microgametophyte which, at maturity, consists of a small prothallial cell and a large antheridium, the jacket cells of which enclose 128 or 256 spermatogenous cells, each capable of developing into a biflagellate motile sperm. Most of this development takes place within the microsporangium. These cells lack chlorophyll and are parasitic on the sporophyte. The megaspore may begin division while it is still in the tetrad. Starting with the uninucleate megaspore cell, a series of free nuclear divisions occurs until the megaspore envelope is filled with nuclei and cytoplasm. Beginning near the ridge region of the megaspore, cell walls appear and eventually the whole structure becomes cellular tissue. The megaspore wall ruptures in the region of the ridge and it is here, as the megagametophyte is produced, that the archegonia (female sex organs) appear (Fig. 3). Although it is reported that some megagametophytes produce chlorophyll and develop functional (absorbing) rhizoids, it is probable that in most cases the major part of nutrition is obtained by the megagametophytes from the parent plant. In some instances, both fertilization and the development of the zygote (fertilized egg) may occur on the parent sporophytic plant. Usually, however, the two types of spores are discharged from the sporangia, the gametophytes develop on the ground, and fertilization takes place exterior to the parent plants. The zygote divides, one of the cells producing an embryo, the other a suspensor (a chain of cells which serve to put the embryo in a favorable position in relation to its food supply). At maturity the embryo displays a feature characteristic of higher plants in that two seed leaves, or cotyledons, are borne on the axis above the hypocotyl. It is of interest that in this phase of the life cycle, the young sporophytes are parasitic upon the megagametophyte, while later the young megagametophytes may be parasitic upon parent sporophytes. The heterosporous condition of *Sel-*

aginella has long interested botanists as it implicates the evolution of seed plants. *See* LYCOPODIALES; TRACHEOPHYTA. [P.A.V.]

Bibliography: *See* LYCOPSIDA.

Selection rules (physics)

Rules summarizing the changes that take place in the quantum mechanical state describing a physical system when a transition between states occurs. From a given initial state of a system, a new state may be reached under the influence of an interaction. As in classical mechanics, certain quantities are either unchanged or are changed in generally specifiable ways during the course of the interaction. Classically, these are "constants of the motion," which are analogous to the quantum numbers characterizing the state (*see* QUANTUM NUMBERS). For a given interaction, the selection rules give the changes in the quantum numbers which must take place in order for the probability of a given transition to be nonzero. It may happen, however, that the probability vanishes for other reasons, even if the selection rule is obeyed. Selection rules are an important result of the so-called conservation laws; for a discussion of these, *see* NUCLEAR REACTION; SYMMETRY LAWS (PHYSICS).

The electromagnetic interaction with a physical system (atom, molecule, or nucleus) in which photons are emitted and the system changes state is a case of general interest for which selection rules have been worked out. For nuclei and subnuclear particles, transitions other than electromagnetic are also important.

Examples. Consider a simple atom, such as hydrogen, undergoing electromagnetic transitions. The state of the system, nucleus plus electron, is described in terms of the quantum numbers n, l, m, where n is the principal quantum number related to the energy of the system, l is the quantum number for the angular momentum of the electron, and m is the magnetic quantum number, that is, the component of the orbital angular momentum along a selected direction. The probability that a change of state will occur, accompanied by the emission or absorption of a photon, is obtained quantum mechanically by computing the integral of the product of the wave functions for the initial and final states times the function describing the electromagnetic interaction. As in classical electromagnetism, the interaction is the product of the electron momentum and the vector potential of the electromagnetic field. A multipole expansion of this potential is possible. For a definition of the term "wave function" and related information, *see* QUANTUM THEORY, NONRELATIVISTIC; *see also* QUANTUM MECHANICS.

The relative importance of successive terms in the expansion of the vector potential depends on the system in question and the wavelength involved. For atoms and visible light, only the first term is important, unless the probability of a transition involving it vanishes. This is the dipole term of the multipole expansion, and transitions produced by this part of the interaction are called dipole transitions. The dipole transition probability vanishes unless l and m change in a specified way:

$$\Delta l = (l_{\text{final}} - l_{\text{initial}}) = \pm 1$$
$$\Delta m = (m_{\text{final}} - m_{\text{initial}}) = 0, \pm 1$$

This is the dipole selection rule in the one-electron case. Since it can be shown that the photon carries off one unit of angular momentum in the dipole case, the rule is a result of the conservation of angular momentum of the combined system, atom plus radiation. No selection rule exists for n; conservation of energy merely demands that energy lost or gained by the atom equal the energy of the emitted or absorbed radiation, which can be of any amount.

Another characteristic of the state is its parity: if the wave function of the state changes sign when the coordinates are reflected through the origin, the parity is odd; if not, the parity is even. For dipole radiation, the Laporte selection rule holds: dipole transitions occur only between states of opposite parity. The selection rule on l automatically insures that the Laporte rule is obeyed in the simple case of hydrogen. *See* PARITY (QUANTUM MECHANICS).

Atomic selection rules. In the general case, when several electrons are involved, the total angular momentum quantum number for the atom J and its projection M in a suitable direction are involved. For dipole radiation, one has

$$\Delta J = 0, \pm 1 \qquad (0 \rightarrow 0 \text{ excluded})$$
$$\Delta M = 0, \pm 1 \qquad (0 \rightarrow 0 \text{ excluded if } \Delta J = 0)$$

The notation $0 \rightarrow 0$ means that no jump can occur from a state with $J = 0$ to another state with $J = 0$.

The manner in which the total angular momentum arises from the individual orbital and spin angular momenta of the electrons in the atom is not readily predictable for a given atom. However, two extreme cases are easily conceived, and many atoms appear, at least approximately, to fit one or the other. One of these is the Russell-Saunders (or LS) coupling case, where $\mathbf{J} = \mathbf{L} + \mathbf{S}$, and $\mathbf{L} = \Sigma \mathbf{l}_i$, $\mathbf{S} = \Sigma \mathbf{s}_i$. The orbital and spin angular momenta of the ith electron are designated by $\mathbf{l}_i$ and $\mathbf{s}_i$, respectively, and the sum is taken over all electrons. For this coupling case

$$\Delta S = 0$$
$$\Delta L = 0, \pm 1$$

are rules holding in addition to those on J and M, and if a single electron is making the transition, $\Delta l_i = \pm 1$ for this electron. The other extreme occurs when $\mathbf{J} = \Sigma \mathbf{j}_i$, $\mathbf{j}_i = \mathbf{l}_i + \mathbf{s}_i$ (jj coupling). In this case, L and S are not defined, and only the rules on J and M apply. In both cases, the Laporte rule holds.

Transitions which obey these rules are called allowed transitions, while any transition which does not is said to be forbidden. Actually, forbidden transitions are rarely observed in atomic spectra, since the electric dipole probabilities are very much greater than those associated with other multipoles in the expansion of the vector potential. However, if the dipole probability vanishes, even though selection rules are obeyed, transitions caused by the magnetic dipole and electric quadrupole terms are next most probable, and these "forbidden" transitions can be observed. The selection rules are as follows:

For electric quadrupole transition, $\Delta J = 0, \pm 1, \pm 2$; $0 \rightarrow 0$, $\frac{1}{2} \rightarrow \frac{1}{2}$, $0 \rightarrow 1$ excluded, no parity change.

For magnetic dipole transitions, $\Delta J = 0, \pm 1$; $0 \rightarrow 0$ excluded, no parity change.

The magnetic dipole rule is the same, therefore, as for the electric dipole, except that there is no parity change. *See* ATOMIC STRUCTURE AND SPECTRA.

Molecular selection rules. A molecule, consisting of at least two nuclei and some attendant electrons, has energy levels involving energy contributions arising from three types of motion: electronic, vibrational, and rotational. The latter two are peculiar to molecules, and quantum mechanical treatments of the rigid rotator and harmonic oscillator (which are models whose features approximate the behavior of the actual molecule) show that the transition probability integrals vanish unless

$$\Delta J = \pm 1 \qquad \Delta v = \pm 1$$

for transitions in which the electronic state of the molecule does not change. Here $Jh/2\pi$ is the angular momentum of the rigid rotator, while v is the vibrational quantum number of an oscillator with energy levels at $(v + \frac{1}{2})h\nu$, where h is Planck's constant, and ν is the classical oscillation frequency. The effect of anharmonicity of the vibration is to allow $\Delta v = \pm 2, \pm 3, \ldots$, but with rapidly decreasing intensity in the corresponding spectrum lines. These selection rules suffice for the infrared spectra of diatomic molecules, that is, in energy regions where electronic transitions do not occur, unless the orbiting electrons contribute to the total angular momentum of rotation of the molecule. The component of the electronic angular momentum along the internuclear axis is denoted as $\Lambda\hbar$ ($\hbar = h/2\pi$), and when $\Lambda \neq 0$, $\Delta J = 0, \pm 1$. In the case of changes involving electronic transitions, the changes in v are governed by a set of rules given by the Franck-Condon principle.

The bands characteristic of molecular spectra result from modifications of the electronic energy levels produced by vibrational and rotational motion, and the selection rules for the vibrational structure of the band are as in the infrared spectra. The rotational fine structure depends on the total angular momentum of the system suitably composed from nuclear rotation, electronic rotation, and electronic spin.

The quantum numbers for these angular momenta are, respectively, N, Λ, and Σ, where Σ, like Λ in the case of electronic orbital angular momentum, is actually the projection on the internuclear axis of the electronic spin rather than the spin S itself. These angular momenta may be combined in several ways to form the total angular momentum J, but only two of the cases defined by F. Hund will be mentioned here.

Hund's case (*a*): Λ combines with Σ to make a total electronic angular momentum projection Ω on the internuclear axis. The nuclear rotational angular momentum N combines with Ω to make J. The total angular momentum then takes on values

$$J = \Omega, \Omega + 1, \Omega + 2, \ldots$$

Hund's case (*b*): Λ combines with N to form K, while the spin S combines with K to form J. Thus, K has values $K = \Lambda, \Lambda + 1, \ldots$ and $J = K + S, K + S - 1, \ldots |K - S|$. The molecular selection rules concern these quantum numbers.

In addition to angular momentum selection rules, there are symmetry selection rules. The symmetries of a molecular state are many. A state has negative (−) or positive (+) symmetry depending on whether its wave function does or does not change sign on reflection of the electrons through a plane perpendicular to the internuclear axis, that is, the line joining the nuclei of the molecule. If the nuclei have the same charge, the state is even (*gerade* or *g*) or odd (*ungerade* or *u*) according as the wave function does not or does change sign on reflection of all axes. If the molecule is homonuclear, that is, has identical nuclei, its states are antisymmetric (*a*) or symmetric (*s*) according as the wave function does or does not change sign under interchange of nuclei.

The selection rules for the rotational-vibrational electronic spectra may now be written down for dipole transitions:

1. General:

$$\Delta J = 0, \pm 1 \qquad (0 \rightarrow 0 \text{ forbidden})$$
$$+ \leftrightarrow -, \text{ but } + \rightarrow + \text{ or } - \rightarrow - \text{ forbidden}$$

Where they apply:

$$g \leftrightarrow u, \text{ but } g \rightarrow g \text{ or } u \rightarrow u \text{ forbidden}$$
$$s \rightarrow s,\ a \rightarrow a, \text{ but } s \rightarrow a \text{ forbidden}$$

2. For either case (*a*) or (*b*):

$$\Delta\Lambda = 0, \pm 1$$
$$\Delta S = 0$$

3. For case (*a*) only:

$$\Delta\Sigma = 0$$
$$\Delta\Omega = 0, \pm 1$$

4. For case (*b*) only:

$$\Delta K = 0, \pm 1 \quad \text{except } \Delta K = 0 \text{ forbidden}$$

if both states have $\Lambda = 0$. *See* MOLECULAR STRUCTURE AND SPECTRA.

Nuclear selection rules. For electromagnetic transitions, nuclear selection rules are the same as for atoms, although only the rules on the total angular momentum, called I, are of practical interest. There is a difference, however, in the relative importance of the forbidden transitions. In the nuclear case, the forbidden transitions play a much more important role. It is worthwhile to indicate the selection rules for higher multipoles than those indicated for atoms. If the order of the pole is

2^l; that is, $l = 1$ for dipole, 2 for quadrupole, and so on, then the selection rules can be written, $\Delta I = \pm l,\ \pm(l-1),\ \ldots,\ 0;\ 0 \rightarrow 0$ excluded. The parity rules can be generalized for the higher multipoles as follows. If a change in parity between initial and final states is represented by +1 or "yes," and no change by −1 or "no," then the parity change must be

$$(-1)^l \quad \text{for electric } 2^l \text{ pole radiation}$$
$$(-1)^{l+1} \quad \text{for magnetic } 2^l \text{ pole radiation}$$

When nuclear particles (protons, neutrons, α-particles) are emitted, the selection rules are again consequences of conservation of angular momentum, and must be worked out taking into account the orbital angular momentum and spin of the emitted particle. *See* NUCLEAR STRUCTURE.

A new quantum number can be defined for nuclei: it is the isotopic spin, where the nuclear charge Q is related to the z-component of isotopic spin T_3 by

$$Q = e\left(\frac{A}{2} + T_3\right)$$

Here A is the nuclear mass number, and e is numerically equal to the electronic charge (*see* ISOTOPIC SPIN). The assignment of the value of the total isotopic spin T for a given nucleus is a matter of interpretation of experimental results. Obviously $T \geqq T_3$. In nuclear reactions, charge conservation is expressible in the selection rule

$$\Delta T_3 = 0$$

There is a large amount of evidence that the rule

$$\Delta T = 0$$

is also obeyed, that is, that nuclear forces are charge independent.

Beta-decay selection rules. The physical principles involved in β-decay selection rules are similar to those operating in the case of electromagnetic transitions. Thus angular momentum is conserved, and while parity is not conserved in the total state (of nucleon, electron, neutrino), there is no effect of this on the states of the nuclei themselves. The interaction is different, of course, and the vector potential of the electromagnetic case must be replaced by an interaction characteristic of the electron-neutrino field. For an extended discussion of β-decay, *see* RADIOACTIVITY; *see also* NEUTRINO.

The electron and antineutrino (or positron and neutrino) each have intrinsic spin ½ in units of $\hbar$, and so can yield a combined spin angular momentum of 0 or 1 (singlet or triplet emission). The most probable, or allowed, emissions occur with the light particles having zero orbital angular momentum, so that there must be no parity change between the states of parent and daughter nuclei. (The lack of parity conservation affects the behavior of electron and neutrino with respect to each other, but does not change the nuclear selection rule.) These statements can be expressed as selection rules, where the singlet and triplet cases are described, respectively, by the Fermi and Gamow-Teller selection rules, named after the physicists (E. Fermi, G. Gamow, E. Teller) who suggested the types of interaction leading to them.

Fermi rule:

$$\Delta I = 0 \quad \text{no parity change}$$

Gamow-Teller rule:

$$\Delta I = 0, \pm 1 \quad \text{no parity change } (0 \rightarrow 0 \text{ excluded})$$

In the case of β-decay, the interaction is not certain, as in the case of electromagnetic transitions; and for a given emitter, whether either or both of these rules is obeyed is a matter for experimental investigation.

If the electron and antineutrino are emitted with orbital angular momentum l, the transition is said to be forbidden of the lth order. This corresponds to the electromagnetic case for quadrupole, octupole, and higher-order radiation. In the case of the Fermi interaction, the selection rule is

$$\Delta I = \pm l,\ \pm(l-1),\ \ldots \quad \text{parity change } (-1)^l$$

whereas for the Gamow-Teller interaction

$$\Delta I = \pm(l+1),\ \pm l,\ \pm(l-1),\ \ldots \quad \text{parity change } (-1)^l \quad (0 \rightarrow 0 \text{ excluded})$$

Since the nuclear charge changes in β-decay, the z-component of isotopic spin also changes. The Fermi and Gamow-Teller interactions have different selection rules for the total isotopic spin. Thus

$$\Delta T = 0 \quad \text{(Fermi)}$$
$$\Delta T = 0, \pm 1 \quad \text{(Gamow-Teller)}$$

The rules stated are for nonrelativistic treatment of the interactions. Since the neutrino is always relativistic (like the photon or electromagnetic quantum, it is supposed to have zero rest mass), this can only be an approximation. Of the five possible relativistic interactions, only two are now thought to operate in nature for β-decay, because of the parity violation experiments and their interpretation. These lead, in the nonrelativistic limit, to the Fermi and Gamow-Teller selection rules. When the relativistic effects are considered, selection rules in the same order l are found to be just those already noted except for parity change: in the first-order relativistic correction, the change is opposite to that in the nonrelativistic limit. The similarity to the electromagnetic case when electric and magnetic multipoles of the same order are compared will be noted. Also, the relativistic transitions are less probable than the nonrelativistic, just as magnetic multipole transitions are less probable than electric multipole transitions of the same order. *See* MULTIPOLE RADIATION.

Hyperon selection rules. For particles assumed to be "elementary" which exceed the nucleon in mass, it is profitable to define a quantum number called "strangeness" which is conserved in hyperon decays. For a consistent discussion of these particles along with nuclei and light particles, it is also useful to introduce a nucleon (or baryon) number and a lepton number. A discussion of these matters is given elsewhere. *See* HYPERON; MESON; SYMMETRY LAWS (PHYSICS). [M.H.H.]

Bibliography: J. M. Blatt and V. F. Weisskopf, *Theoretical Nuclear Physics*, 1956; E. U. Condon and G. H. Shortley, *The Theory of Atomic Spectra*, 1935; G. Herzberg, *Atomic Spectra and Atomic Structure*, 1937; G. Herzberg, *Molecular Spectra and Molecular Structure*, vol. 1, 1950; R. G. Sachs, *Nuclear Theory*, 1955.

Selectivity

The ability of a radio receiver to separate a desired signal frequency from other signal frequencies, some of which may differ only slightly from the desired value. Selectivity is achieved by using tuned circuits that are sharply peaked and by increasing the number of tuned circuits. With a sharply-peaked circuit, the output voltage falls off rapidly for frequencies increasingly lower or higher than that to which the circuit is tuned. *See* RADIO RECEIVER. [J.MR.]

Selenium

Chemical element number 34, selenium, Se, has a chemical atomic weight of 78.96. The approximate per cent abundances of the stable isotopes in natural selenium are Se^{74}, 0.87%; Se^{76}, 9.02%; Se^{77}, 7.58%; Se^{78}, 23.52%; Se^{80}, 49.82%; and Se^{82}, 9.19%. The element was first isolated by Jons J. Berzelius in 1817.

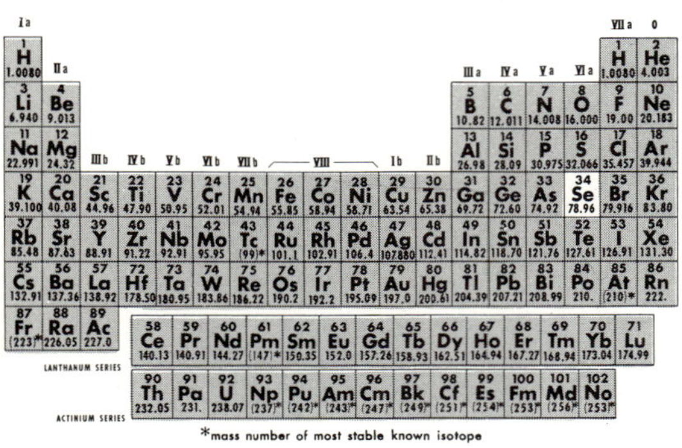

*mass number of most stable known isotope

Natural occurrence. Selenium makes up approximately 10^{-8}% of the earth's igneous rock. It occurs as the free element in Mexico, California, and Japan and as selenides in Europe, Mexico, and Argentina. The important minerals are berzelianite, Cu_2Se; tiemannite, HgSe; and naumannite, Ag_2Se. Others are crookesite, $(Cu,Tl,Ag)_2Se$; eucairite, $(Ag,Cu)_2Se$; and zorgite, a double selenide of copper and lead containing some iron and silver. Selenium sometimes occurs in conjunction with sulfur deposits and in the soil of the dry plains of the midwestern United States, where it is sometimes absorbed by plants, thereby making the herbage poisonous to grazing animals. Selenium also occurs with many sulfide ores, and often can be recovered from flue dusts obtained during the roasting of these ores. In addition, the lead chamber sludge from sulfuric acid manufacture is often rich in selenium, as is the anode mud obtained during the electrolytic refining of copper.

Preparation of the element. The extraction of selenium is usually carried out by the digestion of selenium-containing materials with hot sulfuric acid to form selenium dioxide, SeO_2, which is then purified by sublimation or by crystallization as selenious acid, H_2SeO_3. The latter can be reduced by sulfur dioxide in aqueous solution to free selenium:

$$H_2SeO_3 + 2SO_2 + H_2O \rightarrow Se + 2H_2SO_4$$

Properties of the element. There are three important allotropic modifications of selenium. Monoclinic selenium (red selenium), of which there are two crystalline forms, α and β, can be obtained by crystallizing selenium from a carbon disulfide solution of the element; monoclinic selenium melts at 144°C and has a specific gravity of 4.42; this form is metastable and transforms to gray selenium.

Metallic selenium (gray selenium) exists in two forms, A and B. B is the stable form, having a melting point of 220.2°C and a specific gravity of 4.82. Both forms are slightly soluble in carbon disulfide and are poor conductors of electricity; the electrical conductivity of the gray form is increased up to a thousandfold on exposure to light, the most effective wavelength being 7000 A.

Amorphous selenium exists in three important forms: vitreous selenium, red amorphous selenium, and colloidal selenium. Vitreous selenium can be obtained by quenching molten selenium with cold water; it is a dark material (sp gr 4.28) which softens at 50°C, changes to the metallic form between 60 and 80°C, and is fairly soluble in carbon disulfide. Red amorphous selenium is formed by precipitation with sulfur dioxide of selenious acid solutions, and is soluble in carbon disulfide and selenium oxychloride. Colloidal selenium is a red sol which can be formed by mixing dilute solutions of selenious and sulfurous acids.

The boiling point of selenium is 684.8°C. The molecular weight in carbon disulfide, liquid phosphorus, or liquid sulfur solutions corresponds to Se_8; at 900°C, the molecular weight of the vapor corresponds to Se_2, and at 2000°C, it corresponds to Se (the configuration of the Se_8 molecule is a puckered ring, similar to that of sulfur).

Uses. The uses of elemental selenium are primarily in photoelectric cells, which take advantage of the fact that the electrical conductivity of the element is approximately proportional to the square root of the intensity of light falling upon it. Selenium is also used to give glasses a reddish color, in the preparation of photographic toning baths, in the vulcanization of rubber, to clarify glasses made green by the presence of iron compounds, in the production of certain steels, in the manufacture of selenium rectifiers, and in the production of certain petroleum-cracking catalysts. Selenium compounds are extremely poisonous. *See* PHOTOVOLTAIC CELL; SEMICONDUCTOR RECTIFIER.

Selenium burns in air with a blue flame to produce selenium dioxide. The element also reacts directly with many metals and nonmetals, including hydrogen and the halogens. Selenium is not attacked by nonoxidizing acids, but dissolves in concentrated sulfuric acid, caustic alkalies, and nitric acid.

Principal compounds. Hydrogen selenide (H_2Se) is the only important compound containing only hydrogen and selenium. It is a colorless inflammable gas with an offensive odor and is more toxic than hydrogen sulfide. It has a melting point of −66°C and a boiling point of −41.5°C. Its specific gravity at its boiling point is 2.004 (liquid) and the molecule has an angular configuration with the Se—H distance being 1.6 A. It forms a hydrate, $H_2Se{\cdot}xH_2O$, which melts above 30°C. The compound can be formed by heating selenium in hydrogen, by heating iron filings with selenium to form ferrous selenide which gives hydrogen selenide on treatment with acid, or by the hydrolysis of aluminum selenide:

$$Al_2Se_3 + 6H_2O \rightarrow 2Al(OH)_3 + 3H_2Se$$

The compound is less stable toward heat than the sulfide and gives a weakly acidic solution in water which can precipitate the selenides of many metal ions. Two series of selenides, acid selenides, M^IHSe, and normal selenides, $M_2{}^ISe$, are known. Normal alkali and alkaline-earth selenides in aqueous solution dissolve selenium to form polyselenides, $M_2{}^ISe_x$, in a manner similar to the formation of polysulfides. Most of the heavy metal selenides are only very slightly soluble in water.

Halogen compounds. Selenium monochloride, Se_2Cl_2, is a reddish-brown liquid, melting at −85°C, boiling at 130°C with decomposition, and having a specific gravity of 2.9. It is formed from the elements and decomposes in water as follows:

$$2Se_2Cl_2 + 3H_2O \rightarrow H_2SeO_3 + 3Se + 4HCl$$

Selenium tetrachloride, $SeCl_4$, is a yellowish-white solid of cubic structure, which sublimes at 180°C and melts at 305°C; it is formed by the action of excess chlorine on selenium or by the action of PCl_5 on SeO_2, and it decomposes in water to give selenious acid and hydrogen chloride; it can form salts of H_2SeCl_6. Selenium monobromide, Se_2Br_2, is a dark red liquid melting at −46°C, boiling at 227°C with decomposition, and having a specific gravity of 3.6; it is formed from the elements. Selenium tetrabromide, $SeBr_4$, is an orange-red solid, melting with decomposition at 74°C, which is also formed from the elements and which can form the acid H_2SeBr_6 and its salts. Selenium hexafluoride, SeF_6, is a colorless gas melting at −39°C and boiling at −34.5°C. Selenium tetrafluoride, SeF_4, is a colorless liquid melting at −13.2°C and boiling at 93°C. Selenium oxychloride, $SeOCl_2$, is a yellow liquid melting at 8.5°C, boiling at 176.4°C with decomposition, and having a specific gravity of 2.42 at 22°C; it is formed by the partial hydrolysis of the tetrachloride:

$$SeCl_4 + H_2O \rightarrow SeOCl_2 + 2HCl$$

Selenium oxybromide, $SeOBr_2$, is a reddish-yellow solid melting at 41.6°C and boiling at 217°C with decomposition. Selenium oxyfluoride, $SeOF_2$, is a colorless liquid melting at 4.6°C, boiling at 124°C, and having a specific gravity of 2.67. The oxybromide is formed by treatment of the oxychloride with sodium bromide, and the oxyfluoride is formed from the oxychloride and silver fluoride. The oxyfluoride readily decomposes glass to give SiF_4.

Oxides. The important oxides of selenium are the dioxide, SeO_2, and the trioxide, SeO_3. The dioxide melts at 345°C, sublimes at 315°C, and has a specific gravity of 3.95. The crystal contains chains of alternating selenium and oxygen atoms in a tetrahedral-like orientation, with the Se—O distance being 1.73 A. It is formed from the elements and reacts with water to give selenious acid. The trioxide is a white, deliquescent solid formed by passing oxygen and selenium vapor through a glow discharge at reduced pressures. It melts at 118°C and decomposes to SeO_2 above 180°C. It has two crystalline modifications corresponding to the cubic and asbestoslike forms of sulfur trioxide. It reacts vigorously with water to form selenic acid.

Oxy acids. The important oxy acids of selenium are selenious acid, H_2SiO_3, and selenic acid, H_2SeO_4. Selenious acid can be formed as water-soluble, hexagonal prisms by evaporation of an aqueous solution of SeO_2. It forms both acid selenites ($KHSeO_3$) and normal selenites (K_2SeO_3) upon treatment with bases, as well as superacids, such as $KH_3(SeO_3)_2$, and heteropolyacids with vanadium, uranium, and molybdenum oxides. Selenic acid is produced by the reaction of the trioxide with water, by the oxidation of selenious acid with permanganate or electrically, and by certain special reactions such as the oxidation of selenium in water by chlorine:

$$Se + 4H_2O + 3Cl_2 \rightarrow H_2SeO_4 + 6HCl$$

The pure acid melts at 58°C and forms hydrates with one and four water molecules. The concentrated acid is a strong oxidant which chars organic matter and dissolves copper and gold to give $CuSeO_4$ and $Au_2(SeO_4)_3$, respectively. The dilute acid dissolves zinc to give hydrogen, but not iron, which becomes coated with a thin protective layer of selenium. The acid resembles sulfuric acid in many of its chemical properties, and correspondingly, selenates resemble sulfates. For example, calcium selenate forms a hemihydrate, $(CaSeO_4)_2{\cdot}H_2O$, which is similar to plaster of paris. Barium selenate and lead selenate are only slightly soluble in water, and double salts are formed by certain heavy metal selenates with alkali metal selenates.

Nitride. Selenium nitride, Se_4N_4, having a structure similar to that of S_4N_4 can be formed by passing dry ammonia into a benzene solution of selenium oxychloride or by the action of liquid ammonia on selenium tetrabromide in CS_2.

Mixed compounds. Mixed oxides of sulfur and selenium can be prepared by dissolving the free element of one in the acid or oxide of the other (for example, blue $S{\cdot}SeO_3$, green $Se{\cdot}SO_3$). Carbon selenides, such as carbon oxyselenide, O=C=Se, can be prepared by heating selenium in dry carbon monoxide. It is monomeric, melts at −122.2°C, boils at −20°C, is colorless, has a foul odor, and is stable in the cold. Carbon sulfoselenide, S=C=Se, can be prepared from carbon disulfide and hot ferrous selenide. It is a yellow liquid which boils at 84°C, melts at −85°C, is a strong lacrimator, and is decomposed by light. Carbon diselenide, Se=C=Se, is readily obtained by the action of methylene dichloride on hot selenium. It is a yellow liquid which boils at 124°C, melts at −45.5°C, does not burn in air, and is slowly decomposed by light.

Organic compounds. The important organic selenium compounds are summarized in the table.

Organic selenium compounds

Type	Example	Properties
Dialkyl selenides, R_2Se	$(CH_3)_2Se$	Bp 58°C, more reactive than ethers
Monoalkyl selenides, RSeH (alkyl selenomercaptans)	CH_3SeH	Bp 12°C
Monoaryl selenides, RSeH (aryl selenomercaptans)	C_6H_5SeH	Bp 183.6°C
Diaryl selenides, R_2Se	$(C_6H_5)_2Se$	Prep. from diaryl sulfones and Se
Cyclic selenoethers, $(CH_2)_nSe$	$(CH_2)_4Se$	5-Membered ring; bp 135°C
Selenophene compounds	$(CH)_4Se$	5-Membered diene ring; bp 108°C
Selenonium compounds, R_3SeX	$(CH_3)_3SeCl$	Saltlike compounds; R = alkyl or aryl
Polyselenides, R_2Se_2, R_2Se_3	$(CH_3)_2Se_2$	Bp 156°C
	$(C_2H_5)_2Se_3$	Bp 100°C at 26 mm
	$(C_6H_5)_2Se_2$	Mp 63.5°C
	$(C_2H_5)_2Se_8$	Bp 100°C at 26 mm
Organic selenium halides	$(CH_3)SeCl$	
	C_6H_5SeCl	
	$(C_6H_5)CH_3SeBr_2$	Mp 115°C (dec.)
	$C_6H_5SeBr_3$	Mp 105°C
Selenoxides, R_2SeO	$(C_6H_5)SeO$	Mp 113–114°C
Organic selenium hydroxides	$(CH_3)_3SeOH$	Strong base
	CH_3SeOH	Selenenic acid
	$(CH_3)_2Se(OH)_2$	Dec. to selenoxide
	$CH_3Se(OH)_3$	Dec. to RSeOOH (seleninic acid)
Selenones, R_2SeO_2	$(C_6H_5)_2SeO_2$	Mp 155°C
Selenious esters, R_2SO_3	$(CH_3)_2SeO_3$	Bp 60°C at 15 mm
	$(C_2H_5)HSeO_3$	
	$C_2H_5SeO_2Cl$	Bp 175°C
Seleninic acids, $RSeO_2H$	$C_6H_5SeO_2H$	Mp 170°C
Selenic esters, R_2SeO_4	$(CH_3)_2SeO_4$	Bp 100°C at 15 mm
Selenonic acids, $RSeO_3H$	$C_6H_5SeO_3H$	Mp 142°C
Seleno ketones, R_2CSe or dimer	$[(CH_3)_2CSe]_2$	Bp 220–230°C

See SULFUR; TELLURIUM. [S.K.]

Bibliography: J. W. Mellor, *A Comprehensive Treatise on Inorganic and Theoretical Chemistry*, vol. 10, 1930; H. Remy, *Treatise on Inorganic Chemistry*, vol. 1, 1956; N. V. Sidgwick, *The Chemical Elements and Their Compounds*, vol. 2, 1950.

Semaeostomeae

An order of the class Scyphozoa including most of the common medusae, such as *Aurelia aurita*. The umbrella of these medusae is more flat than high and is usually domelike. The margin of the umbrella is divided into many lappets. Sensory organs are situated between the lappets. The tentacles arise between the lappets in *Pelagia*, on the exumbrella in *Aurelia*, and on the subumbrella in *Cyanea*. They are generally well developed and very long except in a few forms such as *Aurelia*. The oral arms are well developed and either curtain-

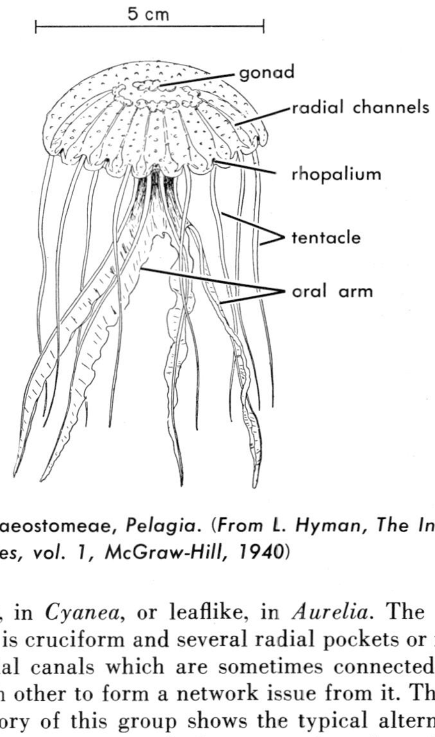

Semaeostomeae, *Pelagia*. (From L. Hyman, *The Invertebrates, vol. 1*, McGraw-Hill, 1940)

like, in *Cyanea*, or leaflike, in *Aurelia*. The stomach is cruciform and several radial pockets or many radial canals which are sometimes connected with each other to form a network issue from it. The life history of this group shows the typical alternation of generations, with forms passing through the stages of planula, scyphopolyp, strobila, and ephyra (*see* METAGENESIS). The Semaeostomeae are distributed mostly in temperate zones and are generally coastal forms with a few exceptions such as *Pelagia* and *Sanderia*. Some of them, such as *Dactylometra* and *Sanderia*, are known to have violent poison on their tentacles. This poison is not only harmful to the skin but also to the nerves, which it paralyzes temporarily. *Cyanea* is used as bait in fishing. Several fossils of this group were found in the strata of the Jurassic period. *See* SCYPHOZOA. [T.U.]

Semiconductor

A solid crystalline material whose electrical conductivity is intermediate between that of a metal and an insulator. Semiconductors exhibit conduction properties that may be temperature-dependent, permitting their use as thermistors (temperature dependent resistors) or voltage-dependent, as in varistors. If the electrical connections to the semiconductor are nonohmic, then over-all nonlinear voltage-current characteristics are obtained and the device may become, for example, a diode rectifier. Semiconductor devices called transistors exhibit amplification properties and are rapidly replacing vacuum tubes in selected applications. *See* DIODE, SEMICONDUCTOR; SEMICONDUCTOR RECTIFIER; THERMISTOR; TRANSISTOR; VARISTOR; *see* also CRYOSAR; PHOTOELECTRIC DEVICES.

CONDUCTION IN SEMICONDUCTORS

The electrical conductivity of semiconductors ranges from about 10^3 to 10^{-9} ohm^{-1} cm^{-1}, as compared with a maximum conductivity of 10^7 for good conductors and a minimum conductivity of 10^{-17} ohm^{-1} cm^{-1} for good insulators. *See* ELECTRICAL CONDUCTIVITY OF METALS; INSULATOR, ELECTRIC.

The electric current is usually due only to the motion of electrons, although under some conditions, such as very high temperatures, the motion of ions may be important (*see* IONIC CRYSTALS). The basic distinction between conduction in metals and in semiconductors is made by considering the energy bands occupied by the conduction electrons.

A crystalline solid consists of a large number of atoms brought together into a regular array called a crystal lattice. The electrons of an atom can each have certain energies, so-called energy levels, as predicted by quantum theory. Because the atoms of the crystal are in close proximity, the electron orbits around different atoms overlap to some extent and the electrons interact with each other; consequently the sharp, well-separated energy levels of the individual electrons actually spread out into energy bands. Each energy band is a quasi-continuous group of closely spaced energy levels. *See* BAND THEORY OF SOLIDS.

At absolute zero temperature, the electrons occupy the lowest possible energy levels, with the restriction that at most two electrons may be in the same energy level. In semiconductors and insulators, there are just enough electrons to fill completely a number of energy bands, leaving the rest of the energy bands empty. The highest filled energy band is called the valence band. The next higher band, which is empty at absolute zero temperature, is called the conduction band. The conduction band is separated from the valence band by an energy gap which is an important characteristic of the semiconductor. In metals, the highest energy band that is occupied by the electrons is only partially filled. This condition exists either because the number of electrons is not just right to fill an integral number of energy bands or because the highest occupied energy band overlaps the next higher band without an intervening energy gap. The electrons in a partially filled band may acquire a small amount of energy from an applied electric field by going to the higher levels in the same band. The electrons are accelerated in a direction opposite to the field and thereby constitute an electric current. In semiconductors and insulators, the electrons are found only in completely filled bands, at low temperatures. In order to increase the energy of the electrons, it is necessary to raise electrons from the valence band to the conduction band across the energy gap. The electric fields normally encountered are not large enough to accomplish this with appreciable probability. At sufficiently high temperatures, depending on the magnitude of the energy gap, a significant number of valence electrons gain enough energy thermally to be raised to the conduction band. These electrons in an unfilled band can easily participate in conduction. Furthermore, there is now a corresponding number of vacancies in the electron population of the valence band. These vacancies, or holes as they are called, have the effect of carriers of positive charge, by means of which the valence band makes a contribution to the conduction of the crystal. *See* HOLES IN SOLIDS.

The type of charge carrier, electron or hole, that is in largest concentration in a material is sometimes called the majority carrier and the type in smallest concentration the minority carrier. The majority carriers are primarily responsible for the conduction properties of the material. Although the minority carriers play a minor role in electrical conductivity, they can be of great importance in rectification and transistor actions in a semiconductor.

Electron distribution. The probability f for an energy level E to be occupied by an electron is given by the Fermi-Dirac distribution function (*see* FERMI-DIRAC STATISTICS):

$$f = \left[1 + \exp\left(\frac{E - W}{kT}\right)\right]^{-1}$$

where k is the Boltzmann constant and T is the absolute temperature. The parameter W is the Fermi energy level; an energy level at W has a probability of ½ to be occupied by an electron. The Fermi level is determined by the distribution of energy levels and the total number of electrons.

In a semiconductor, the number of conduction electrons is normally small compared with the number of energy levels in the conduction band, and the probability for any energy level to be occupied is small. Under such a condition, the concentration of conduction electrons is given by

$$N_n = \frac{2}{h^3}(2\pi m_n kT)^{3/2} \exp\left[\frac{(W - E_c)}{kT}\right]$$

where h is Planck's constant, E_c is the lowest energy of the conduction band, and m_n is called the effective mass of conduction electrons. The effective mass is used in place of the actual mass to correct the coefficient in the equation and to bring the results in line with experimental observations. This correction is necessary because the theory leading to these equations is based upon electrons moving in a field free space, which is not the exact picture. The electrostatic Coulomb potential throughout the crystal is varying in a periodic manner, the variation being due to the electric fields about the atomic centers. The concentration of holes in the valence band is given by

$$N_p = \frac{2}{h^3}(2\pi m_p kT)^{3/2} \exp\left[\frac{(E_v - W)}{kT}\right]$$

where m_p is the effective mass of a hole and E_v is the highest energy of the valence band.

Mobility of carriers. The velocity acquired by charge carriers per unit strength of applied electric field is called the mobility of the carriers. The velocity in question is the so-called drift velocity in the direction of the force exerted on the carriers by the applied field. It is added to the random thermal velocity. In semiconductors the carrier mobility normally ranges from 10^2 to 10^5 $cm^2/(sec)(volt)$. A material's conductivity is the product of the charge, the mobility, and the carrier concentration.

Electrons in a perfectly periodic potential field can be accelerated freely. Impurities, physical defects in the structure, and thermal vibrations of the atoms disturb the periodicity of the potential field in the crystal, thereby scattering the moving carriers. It is the resistance produced by this scattering that limits the carriers to only a drift velocity under the steady force of an applied field.

Intrinsic semiconductors. A semiconductor in which the concentration of charge carriers is characteristic of the material itself rather than of the content of impurities and structural defects of the crystal is called an intrinsic semiconductor. Electrons in the conduction band and holes in the valence band are created by thermal excitation of electrons from the valence to the conduction band. Thus an intrinsic semiconductor has equal concentrations of electrons and holes. The intrinsic carrier concentration, N_i, is determined by

$$N_i = \frac{2}{h^3}(2\pi kT)^{3/2}(m_n m_p)^{3/4} \exp\left(-\frac{E_G}{2kT}\right)$$

where E_G is the energy gap. The carrier concentration, and hence the conductivity, is very sensitive to temperature and depends strongly on the energy gap. The energy gap ranges from a fraction of 1 ev to several electron volts. A material must have a large energy gap to be an insulator.

Impurity semiconductors. Typical semiconductor crystals such as germanium and silicon are formed by an ordered bonding of the individual atoms to form the crystal structure. The bonding is attributed to the valence electrons which pair up with valence electrons of adjacent atoms to form so-called shared pair or covalent bonds. These materials are all of the quadrivalent type, that is, each atom contains four valence electrons, all of which are used in forming the crystal bonds. *See* CRYSTAL STRUCTURE.

Atoms having a valence of 3+ or 5+ can be added to a pure or intrinsic semiconductor material with the result that the 3+ atoms will give rise to an unsatisfied bond with one of the valence electrons of the semiconductor atoms, and 5+ atoms will result in an extra or free electron that is not required in the bond structure. Electrically, the 3+ impurities add holes and the 5+ impurities add electrons. They are called acceptor and donor impurities, respectively. Typical valence 3+ impurities used are boron, aluminum, indium, and gallium. Valence 5+ impurities used are arsenic, antimony, and phosphorus.

Semiconductor material "doped," or "poisoned," by valence 3+ acceptor impurities is termed *p*-type, whereas material doped by valence 5+ donor material is termed *n*-type. The names are derived from the fact that the holes introduced are considered to carry positive charges and the electrons negative charges. The number of electrons in the energy bands of the crystal is increased by the presence of donor impurities and decreased by the presence of acceptor impurities. Let N be the concentration of electrons in the conduction band and let P be the hole concentration in the valence band. For a given semiconductor, the relation $NP = N_i^2$ holds, independent of the presence of impurities. The effect of donor impurities tends to make N larger than P, since the extra electrons given by the donors will be found in the conduction band even in the absence of any holes in the valence band. Acceptor impurities have the opposite effect, making P larger than N. *See* ACCEPTOR ATOM; DONOR ATOM.

At sufficiently high temperatures, the intrinsic carrier concentration becomes so large that the effect of a fixed amount of impurity atoms in the crystal is comparatively small and the semiconductor becomes intrinsic. When the carrier concentration is predominantly determined by the impurity content, the conduction of the material is said to be extrinsic. There may be a range of temperature within which the impurity atoms in the material are practically all ionized, that is, they supply a maximum number of carriers. Within this temperature range, the so-called exhaustion range, the carrier concentration remains nearly constant. At sufficiently low temperatures, the electrons or holes that are supplied by the impurities become bound to the impurity atoms. The concentration of conduction carriers will then decrease rapidly with decreasing temperature, according to either $\exp(-E_i/kT)$ or $\exp(-E_i/2kT)$ where E_i is the ionization energy of the dominant impurity.

Physical defects in the crystal structure may have similar effects as donor or acceptor impurities. They can also give rise to extrinsic conductivity.

Hall effect. Whether a given sample of semiconductor material is *n*- or *p*-type can be determined by observing the Hall effect. If an electric current is caused to flow through a sample of semiconductor material and a magnetic field is applied in a direction perpendicular to the current, the charge carriers are crowded to one side of the sample, giving rise to an electric field perpendicular to both the current and the magnetic field. This development of a transverse electric field is known as the Hall effect. The field is directed in one or the opposite direction depending on the sign of the charge of the carrier. *See* HALL EFFECT.

The magnitude of the Hall effect gives an estimate of the carrier concentration. The ratio of the transverse electric field strength to the product of the current and the magnetic field strength is called the Hall coefficient, and its magnitude is inversely proportional to the carrier concentration. The coefficient of proportionality involves a factor which depends on the energy distribution of the carriers and the way in which the carriers are scattered in their motion. However, the value of this factor normally does not differ from unity by more than a factor of two. The situation is more complicated when more than one type of carrier is important for the conduction. The Hall coefficient then depends on the concentrations of the various types of carriers and their relative mobilities.

The product of the Hall coefficient and the conductivity is proportional to the mobility of the carriers when one type of carrier is dominant. The proportionality involves the same factor which is contained in the relationship between the Hall coefficient and the carrier concentration. The value obtained by taking this factor to be unity is referred to as the Hall mobility.

MATERIALS AND THEIR PREPARATION

Elemental semiconductors. The group of chemical elements which are semiconductors includes germanium, silicon, gray (crystalline) tin, selenium, tellurium, and boron. Germanium, silicon, and gray tin belong to group IV of the periodic table and have crystal structures similar to that of diamond. Germanium and silicon are two of the best known semiconductors. They are used extensively in devices such as rectifiers and transistors. Gray tin is a form of tin which is stable below 13°C. White tin, which is stable at higher temperatures, is metallic. Gray tin has a small energy gap and a rather large intrinsic conductivity, about 5×10^3 ohm^{-1} cm^{-1} at room temperature. The *n*-type and *p*-type gray tins can be obtained by adding aluminum and antimony, respectively.

Selenium and tellurium both have a similar structure, consisting of spiral chains located at the corners and centers of hexagons. The structure gives rise to anisotropy of the properties of single crystals; for example, the electrical resistivity of tellurium along the direction of the chains is about one-half the resistivity perpendicular to this direction. Selenium has been widely used for making rectifiers and photocells.

Semiconducting compounds. A large number of compounds are known to be semiconductors. Copper(I) oxide (Cu_2O) and mercury(II) indium telluride ($HgIn_2Te_4$) are examples of binary and ternary compounds. The series zinc sulfide (ZnS), zinc selenide (ZnSe), zinc telluride (ZnTe), and the series zinc selenide (ZnSe), cadmium selenide (CdSe), and mercury(II) selenide (HgSe) are examples of binary compounds consisting of a given element in combinations with various elements of another column in the periodic table. The series magnesium antimonide (Mg_2Sb_2), magnesium telluride (MgTe), and magnesium iodide (MgI_2) is an example of compounds formed by a given element with elements of various other columns in the periodic table (*see* PERIODIC TABLE).

A group of semiconducting compounds of the simple type AB consists of elements from columns symmetrically placed with respect to column IV of the periodic table. Indium antimonide (InSb), cadmium telluride (CdTe), and silver iodide (AgI) are examples of III-V, II-IV, and I-VI compounds, respectively. The various III-V compounds are being studied extensively, and many practical applications have been found for these materials. Some of these compounds have the highest carrier mobilities known for semiconductors. The compounds have zincblende crystal structure which is geometrically similar to the diamond structure possessed by the elemental semiconductors, germanium and silicon, of column IV, except that the four nearest neighbors of each atom are atoms of the other kind. The II-VI compounds, zinc sulfide (ZnS) and cadmium sulfide (CdS), are used in photoconductive devices. Zinc sulfide is also used as a luminescent material. *See* LUMINESCENCE; PHOTOCONDUCTIVITY.

Binary compounds of the group lead sulfide (PbS), lead selenide (PbSe), and lead telluride (PbTe) are sensitive in photoconductivity and are used as detectors of infrared radiation. The compounds, bismuth telluride (Bi_2Te_3) and bismuth selenide (Bi_2Se_3), consisting of heavy atoms, are found to be good materials for thermocouples used for refrigeration or for conversion of heat to electrical energy. *See* THERMOELECTRICITY.

The metal oxides usually have large energy gaps. Thus, pure oxides are usually insulators of high resistivity. However, it may be possible to introduce into some of the oxides impurities of low ionization energies and thus obtain relatively good extrinsic conduction. Copper(I) oxide (Cu_2O) was one of the first semiconductors used for rectifiers and photocells; extrinsic *p*-type conduction is obtained by producing an excess of oxygen over the stoichiometric composition, that is, the 2-to-1 ratio of copper atoms to oxygen atoms. A number of oxide semiconductors can be obtained by replacing some of the normal metal atoms with metal atoms of one more or less valency. The method is called controlled valence. An example of such a semiconductor is nickel oxide containing lithium.

Preparation of materials. The properties of semiconductors are extremely sensitive to the presence of impurities. It is therefore desirable to start with the purest available materials and to introduce a controlled amount of the desired impurity. The zone refining method is often used for further purification of obtainable materials. The floating zone technique can be used, if feasible, to prevent any contamination of molten material by contact with crucible. See ZONE REFINING.

For basic studies as well as for many practical applications, it is desirable to use single crystals. Various methods are used for growing crystals of different materials. For many semiconductors, including germanium, silicon, and the III-V compounds. the Czochralski method is commonly used. The method of condensation from the vapor phase is used to grow crystals of a number of semiconductors, for instance, selenium and zinc sulfide. For materials of high melting points, such as various metal oxides, the flame fusion or Vernonil method may be used. See CRYSTAL GROWTH.

The introduction of impurities, or doping, can be accomplished by simply adding the desired quantity to the melt from which the crystal is grown. Normally, the impurity has a small segregation coefficient, which is the ratio of equilibrium concentrations in the solid and the liquid phases of the material. In order to obtain a desired impurity content in the crystal, the amount added to the melt must give an appropriately larger concentration in the liquid. When the amount to be added is very small, a preliminary ingot is often made with a larger content of the doping agent; a small slice of the ingot is then used to dope the next melt accurately. Impurities which have large diffusion constants in the material can be introduced directly by holding the solid material at an elevated temperature while this material is in contact with the doping agent in the solid or the vapor phase.

RECTIFICATION IN SEMICONDUCTORS

In semiconductors, narrow layers can be produced which have abnormally high resistances. The resistance of such a layer is nonohmic; it may depend on the direction of current, thus giving rise to rectification. Rectification can also be obtained by putting a thin layer of semiconductor or insulator material between two conductors of different material.

Barrier layer. A narrow region in a semiconductor which has an abnormally high resistance is called a barrier layer. A barrier may exist at the contact of the semiconductor with another material, at a crystal boundary in the semiconductor, or at a free surface of the semiconductor. In the bulk of a semiconductor, even in a single crystal, barriers may be found as the result of a nonuniform distribution of impurities. The thickness of a barrier layer is small, usually 10^{-3}–10^{-5} cm.

A barrier is usually associated with the existence of a space charge. In an intrinsic semiconductor, a region is electrically neutral if the concentration n of conduction electrons is equal to the concentration p of holes. Any deviation in the balance gives a space charge equal to $e(p-n)$, where e is the charge on an electron. In an extrinsic semiconductor, ionized donor atoms give a positive space charge and ionized acceptor atoms give a negative space charge. Let N_D and N_A be the concentrations of ionized donors and acceptors, respectively. The space charge is equal to $e(p-n+N_D-N_A)$.

A space charge is associated with a variation of potential. A drop in potential, $-\Delta V$, increases the potential energy of an electron by $e\Delta V$, consequently every electronic energy level in the semiconductor is shifted by this amount. With a variation of potential, the electron concentration varies proportionately to exp (eV/kT) and the hole concentration varies as exp $(-eV/kT)$. A space charge is obtained if the carriers, mainly the majority carriers, fail to balance the charge of the ionized impurities.

A conduction electron in a region where the potential is higher by ΔV must have an excess energy of $e\Delta V$ in order for it to have the minimum energy on reaching the low potential region. Electrons with less energy cannot pass over to the low potential region. Thus a potential variation presents a barrier to the flow of electrons from high to low potential regions. It also presents a barrier to the flow of holes from low to high potential regions.

Surface barrier. A thin layer of space charge and a resulting variation of potential may be produced at the surface of a semiconductor by the presence of surface states. Electrons in the surface states are bound to the vicinity of the surface, and the energy levels of surface states may lie within the energy gap. Surface states may arise from the absorption of foreign atoms. Even a clean surface may introduce states which do not exist in the bulk material, simply by virtue of being the boundary of the crystal.

The surface is electrically neutral when the surface states are filled with electrons up to a certain energy level ϵ in the energy gap E_G, which is the energy difference between the bottom of the conduction band E_C and the top of the valence band E_V. If the Fermi level W in the bulk semiconductor lies higher in the energy gap, more surface states would be filled, giving the surface a negative charge. As a result the potential drops near the surface and the energy bands are raised for n-type material (Fig. 1). With the rise of the conduction band, the electron concentration is reduced and a positive space charge due to ionized donors is obtained. The amount of positive space charge is

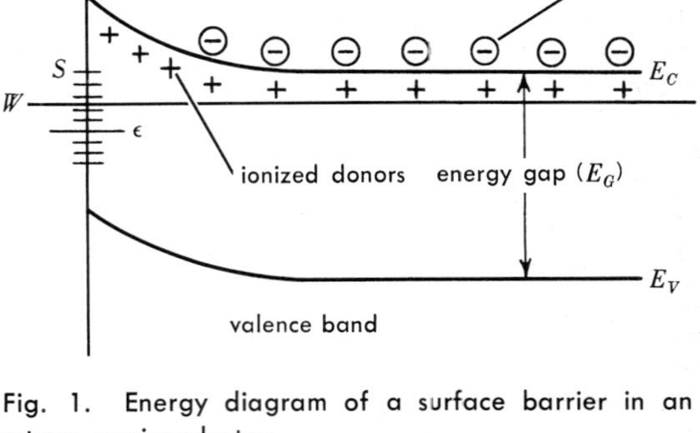

Fig. 1. Energy diagram of a surface barrier in an n-type semiconductor.

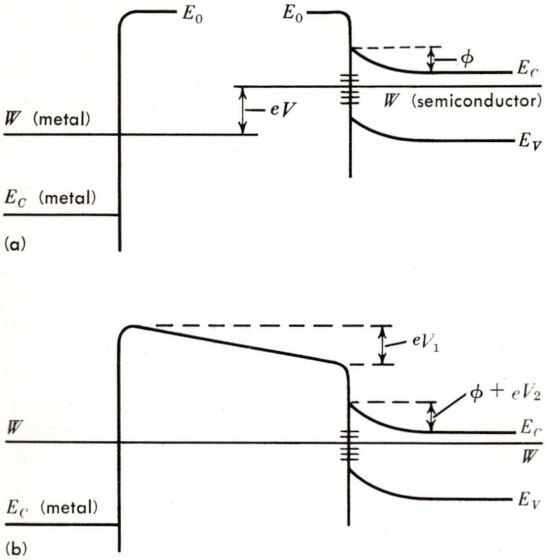

Fig. 2. Energy diagrams for a metal (*left*) and an n-type semiconductor (*right*). E_0 is the potential energy of an electron outside the material, E_C is the energy at the bottom of the conduction band, and E_V is the energy at the top of the valence band. (*a*) Semiconductor and metal isolated. (*b*) Semiconductor and metal in electrical contact, $eV_1 + eV_2 = eV$.

equal to the negative surface charge given by the electrons in the surface states between ϵ and the Fermi level.

Contact barrier. The difference between the potential energy, E_0, of an electron outside a material and the Fermi level in the material is called the work function of the material. Figure 2 shows the energy diagram for a metal and a semiconductor, the work functions of which differ by eV. Upon connecting the two bodies electrically, charge is transferred between them so that the potential of the semiconductor is raised relative to that of the metal, that is, the electron energy levels in the semiconductor are lowered. Equilibrium is established when the Fermi level is the same in the two bodies. In this case, the metal is charged negatively and the semiconductor is charged positively. The negative charge on the metal is concentrated close to the surface, as is expected in good conductors. The positive charge on the semiconductor is divided between the increase of space charge in an extension of the barrier and the depopulation of some of the surface states. The charging of the semiconductor is brought about by a change of eV_2 in the barrier height ϕ. The sum of eV_2 and the potential energy variation eV_1 in the space between the two bodies is equal to the original difference eV between the work functions.

With decreasing separation between the two bodies, the division of eV will be in favor of eV_2. However, if there is a very large density of states, a small eV_2 gives a large surface charge on the semiconductor due to the depopulation of surface states. It is possible that eV_2 is limited to a small value even at the smallest separation, of the order of an interatomic distance in solids. In such cases, the barrier height remains nearly equal to the value ϕ of the free surface, irrespective of the body in contact. This situation has been found in germanium and silicon rectifiers. Before the explanation was given by J. Bardeen, who postulated the existence of surface states, it had been assumed that the height of a contact barrier was equal to the difference of the work functions. *See* WORK FUNCTION (ELECTRONIC).

Single-carrier theory. The phenomenon of rectification at a crystal barrier can be described according to the role played by the carriers. Where the conduction property of the rectifying barrier is determined primarily by the majority carriers, the single-carrier theory is employed. Such cases are likely to be found in semiconductors with large energy gaps, for instance, oxide semiconductors. Figure 3 shows the energy diagrams of metal-semiconductor contact rectifiers under conditions of equilibrium. The potential variation in the semiconductor is such as to reduce the majority carrier concentration near the contact. If the energy bands were to fall in the case of an n-type semiconductor or to rise in the case of a p-type semiconductor, the majority carrier concentration would be enhanced near the contact, and the contact would not present a large and rectifying resistance. It is clear that in the cases shown in Fig. 3, the minority carrier concentration increases

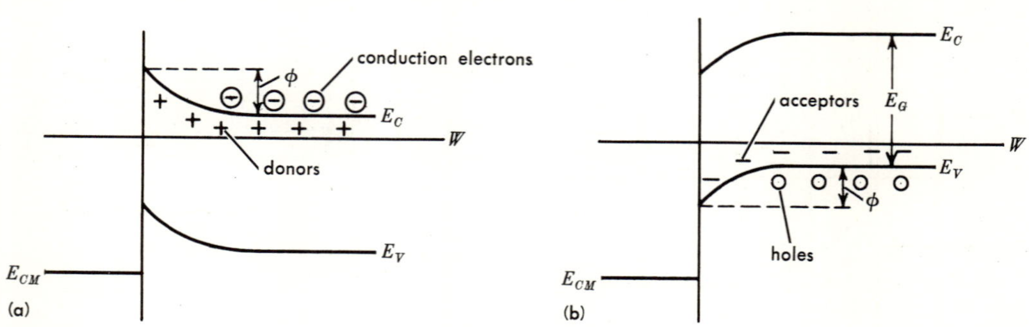

Fig. 3. Energy diagrams of a metal-semiconductor rectifying contact. (*a*) n-Type semiconductor. (*b*) p-Type semiconductor.

near the contact. However, if the energy gap is large, the minority carrier concentration is normally very small, and the role of minority carriers may be still negligible even if the concentration is increased.

Under equilibrium conditions, the number of carriers passing from one body to the other is balanced by the number of carriers crossing the contact in the opposite direction, and there is no net current. The carriers crossing the contact in either direction must have sufficient energies to pass over the peak of the barrier. The situations under applied voltages are shown in Fig. 4 for the case of an n-type semiconductor. When the semiconductor is made positive, its energy bands are depressed and the height of the potential barrier is increased, as shown in Fig. 4*a*. Fewer electrons in the semiconductor will be able to cross over into the metal, whereas the flow of electrons across the contact from the metal side remains unchanged. Consequently, there is a net flow of electrons from the metal to the semiconductor. The flow of electrons from the metal side is the maximum net flow obtainable. With increasing voltage, the current saturates and the resistance becomes very high. Figure 4*b* shows the situation when the semiconductor is negative under the applied voltage. The energy bands in the semiconductor are raised. The flow of electrons from the semiconductor to the metal is increased, since electrons of lower energy are able to go over the peak of the barrier. The result is a net flow of electrons from the semiconductor to the metal. There is no limit to the flow in this case. In fact, the electron current increases faster than the applied voltage because there are increasingly more electrons at lower energies. The resistance decreases, therefore, with increasing voltage. The direction of current for which the resistance is low is called the forward direction, while the opposite direction is called the reverse or blocking direction. A general expression for the current can be written in the form

$$j = enC\left(\exp\frac{-\phi}{kT}\right)\left[\exp\left(\frac{eV}{kT}\right) - 1\right]$$

where j is the current density, n is the carrier concentration in the bulk of the semiconductor, ϕ is the barrier height, and V is the applied voltage taken as positive in the forward direction. The factor C depends on the theory appropriate for the particular case.

Diffusion theory. When there is a variation of carrier concentration, a motion of the carriers is produced by diffusion in addition to the drift determined by the mobility and the electric field. The transport of carriers by diffusion is proportional to the carrier concentration gradient and the diffusion constant. The diffusion constant is related to the mobility, and both are determined by the scattering suffered by moving carriers. The average distance traveled by a carrier in its random thermal motion between collisions is called the mean free path. If barrier thickness is large compared to mean free path of carriers, motion of carriers in the barrier can be treated as drift and diffusion. This viewpoint is the basis of the diffusion theory of rectification. According to this theory, the factor C in the preceding equation depends on the mobility and the electric field in the barrier.

Diode theory. When the barrier thickness is comparable to or smaller than the mean free path of the carriers, then the carriers cross the barrier without being scattered, much as in a vacuum tube diode. According to this theory, the factor C in the rectifier equation is $v/4$, where v is the average thermal velocity of the carriers.

Two-carrier theory. Often the conduction through a rectifying barrier depends on both electron and hole carriers. An important case is the p-n junction between p- and n-samples of semiconductor material. Also in metal-semiconductor rectifiers, the barrier presents an obstacle for the flow of majority carriers but not to the flow of minority carriers, and the latter may become equally or more important.

Rectification at p-n junctions. A p-n junction is the boundary between a p-type region and an n-type region of a semiconductor. When the impurity content varies, there is a variation of electron and hole concentrations. A variation of carrier concentrations is related to a shift of the energy

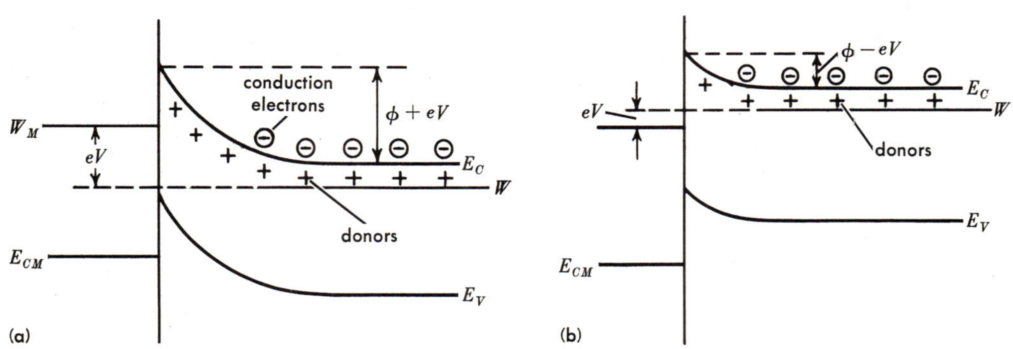

Fig. 4. Energy diagrams of a rectifying contact between a metal and an n-type semiconductor under an applied voltage V. (*a*) Positive semiconductor. There is a net flow of electrons from metal to semiconductor. (*b*) Negative semiconductor. There is a net flow of electrons from semiconductor to metal.

bands relative to the constant Fermi level. This is brought about by a variation of the electrostatic potential which requires the existence of a space charge. If the impurity content changes greatly within a short distance, a large space charge is obtained within a narrow region. Such is the situation existing in a rectifying *p-n* junction.

When a voltage is applied to make the *n*-region negative relative to the *p*-region, electrons flow from the *n*-region, where they are abundant, into the *p*-region. At the same time, holes flow from the *p*-region, where holes are abundant, into the *n*-region. The resistance is therefore relatively low. The direction of current in this case is forward. Clearly, the resistance will be high for current in the reverse direction.

With a current in the forward direction, electrons in the *n*-region and holes in the *p*-region flow toward the junction and there must be continuous hole-electron recombination in the neighborhood of the junction. The minority carrier concentration in each region is increased near the junction due to the influx of the carriers from the other region. This phenomenon is known as carrier-injection. When there is a current in the reverse direction, there must be a continuous generation of holes and electrons in the neighborhood of the junction, from which electrons flow out into the *n*-region and holes flow out into the *p*-region. Thus current through a *p-n* junction is controlled by the hole-electron recombination or generation in the vicinity of the junction.

The transistor consists of two closely spaced *p-n* junctions in a semiconductor with an order *p-n-p* or *n-p-n*.

Contact rectification. If the height of a rectifying contact barrier is high, only a very small fraction of majority carriers can pass over the barrier. The fraction may be so small as to be comparable with the concentration of the minority carriers, provided the energy gap is not too large. The current due to the minority carriers becomes appreciable if the barrier height above the Fermi level approaches the energy difference between the Fermi level and the top of the valence band (see Fig. 3).

The concentration of minority carriers is higher at the contact than in the interior of the semiconductor. With a sufficiently high barrier, it is possible to obtain at the contact a minority carrier concentration higher than that of the majority carriers. The small region where this condition occurs is called the inversion layer.

As in the case of a *p-n* junction, a forward current produces injection of minority carriers. With the presence of an inversion layer, the injection can be so strong as to increase appreciably the conductivity in the vicinity of the contact. Ordinarily, contact rectifiers consist of a semiconductor in contact with a metal whisker. For large forward currents, the barrier resistance is small, and the resistance of the rectifier is determined by the spreading resistance of the semiconductor for a contact of small area. By increasing the conductivity in the vicinity of the contact where the spreading resistance is concentrated, carrier injection may reduce considerably the forward resistance of the rectifier. [H.Y.F.]

Bibliography: F. S. Brackett (ed.), *The Present State of Physics*, AAAS Publ. 35, 1954; W. C. Dunlap, *An Introduction to Semiconductors*, 1957; H. K. Henisch, *Rectifying Semi-conductor Contacts*, 1957; F. Seitz and D. Turnbull (eds.), *Solid State Physics*, vol. 1, 1955; W. Shockley, *Electrons and Holes in Semiconductors*, 1950; H. C. Terrey and C. A. Whitmer, *Crystal Rectifiers*, 1948.

Semiconductor rectifier

Formerly called metallic rectifier, this rectifier uses a semiconductor material to obtain its asymmetric conducting properties—low resistance to current flow in one direction, high resistance in the opposite direction. A semiconductor is an electronic conductor with resistivity in the range between metals and insulators. The rectifying action takes place at the junction between a semiconductor having a deficiency of electrons and another semiconductor or a metal which has a supply of free electrons. The low-resistance direction is toward the electrode with free electrons. The elementary device consisting of the electrodes with the junction and any contact terminals is called a cell.

Semiconductor rectifiers are used for converting alternating current to direct current in applications such as battery charging, electrolytic production of metals and gases, and driving dc motors. The cells are rated on the basis of the forward current and the reverse, or inverse, voltage (rms or peak) to which they are subjected in rectifier circuits. The operating characteristics are usually shown on a volt-ampere curve, which gives the forward voltage drop as a function of the current and the reverse current as a function of the inverse voltage. The ratings are determined largely by temperature limits. The forward and reverse resistances are decreased as the temperature is increased; the effect on the reverse resistance is greater. The circuit symbol is shown in Fig. 1. Cells can be connected in parallel to obtain higher output currents, and in series for higher output voltages. The most frequently used rectifier circuits are the single-phase and 3-phase bridge (also called 6-phase double-way). *See* RECTIFIER.

Fig. 1. Circuit symbol of semiconductor rectifying element.

There are two general classes of semiconductor rectifiers: (1) polycrystalline, in which the semiconductor consists of many crystals; copper oxide and selenium rectifiers are in this class; (2) monocrystalline, in which the semiconductor is a slice of a single crystal; germanium and silicon rectifiers are in this class. The monocrystalline rectifiers have a higher current density and a higher ratio of forward to reverse current.

Polycrystalline rectifiers. The rectifying junction, also called the barrier layer, is between the semiconductor and a metallic conductor. Figure 2 shows enlarged sections of copper oxide and selenium cells. Their approximate volt-ampere characteristics are shown in Fig. 3.

In a copper oxide cell, the rectifying junction is between copper and cuprous oxide (semiconductor), which is produced on the copper by oxidation at a high temperature. Cells have been made in

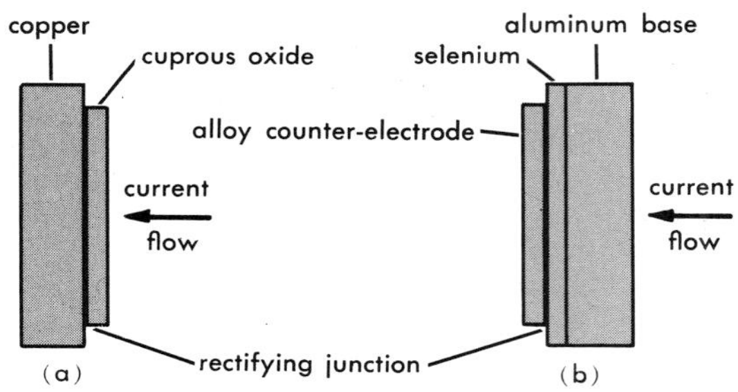

Fig. 2. Cross sections of (*a*) copper oxide and (*b*) selenium rectifying cells.

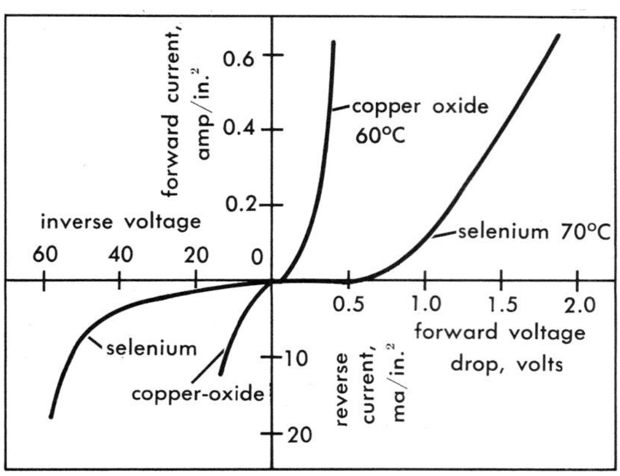

Fig. 3. Typical volt-ampere characteristics of copper oxide and selenium rectifying cells. The voltages and current densities are instantaneous values.

ratings from a few milliamperes to about 6 amp average dc, convection cooled, with current densities of about 0.1–0.2 amp/in.^{2} The ratings are more than doubled by forced-air cooling. The highest peak-inverse voltage rating is about 11 volts for low-voltage cells and 20 volts for high-voltage cells. The operating temperature is usually kept below 60°C to prevent excessive aging.

A selenium cell is made by depositing selenium on an aluminum plate and following with a heat treatment. A eutectic alloy is then deposited over the selenium, and the cell is electroformed by applying a reverse voltage, which produces the rectifying junction between the selenium (semiconductor) and the alloy coating. Cells are made for cur-

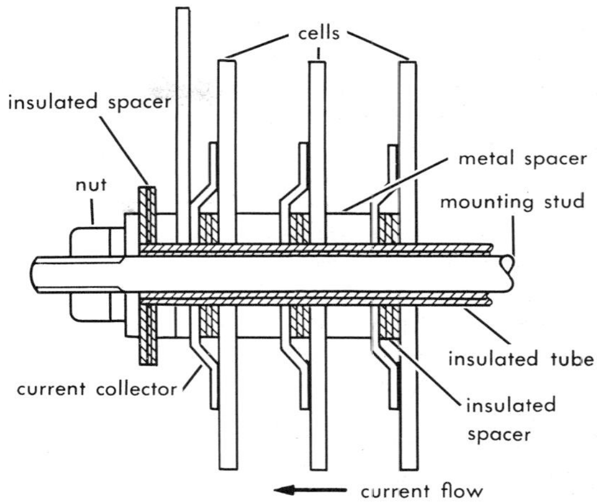

Fig. 4. Selenium rectifier stack.

rent ratings from a fraction of an ampere to 15 amp average dc, convection cooled, with densities of about 0.16–0.3 amp/in.^{2} The ratings can be increased 2 to 3 times by forced-air cooling. Peak-inverse voltage ratings are available from about 20–75 volts. The maximum operating temperature range is 85–100°C.

Copper oxide and selenium cells are assembled in stacks and given a protective coating against atmospheric conditions. A typical assembly of a selenium stack is shown in Fig. 4. Both types of cells are subject to aging, which gradually increases the forward resistance and lowers the output voltage by about 15% for copper oxide rectifiers and 5–10% for selenium rectifiers. The approximate range of over-all efficiencies of 6-phase rectifier units at rated load is, after aging, 65–75% for copper oxide and 75–85% for selenium rectifiers.

Monocrystalline rectifiers. The rectifying junction is within a slice of a single crystal of the semiconductor. For a discussion of conductivity types, doping of semiconductors, and definitions of terms, *see* SEMICONDUCTOR.

If a semiconductor crystal is doped to make one side *n*-type and the other side *p*-type, the *p-n* junction has rectifying properties, as illustrated in Fig. 5. If a cell is connected in a circuit with polarities, as shown in Fig. 5*a*, the potential gradient

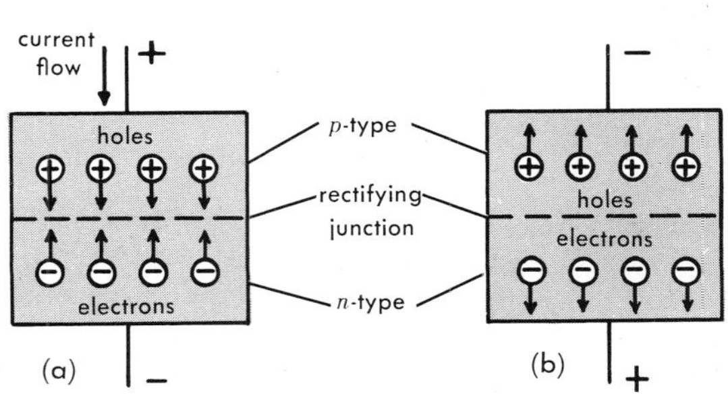

Fig. 5. Simplified representation of operation of monocrystalline rectifier. (*a*) Conducting. (*b*) Blocking.

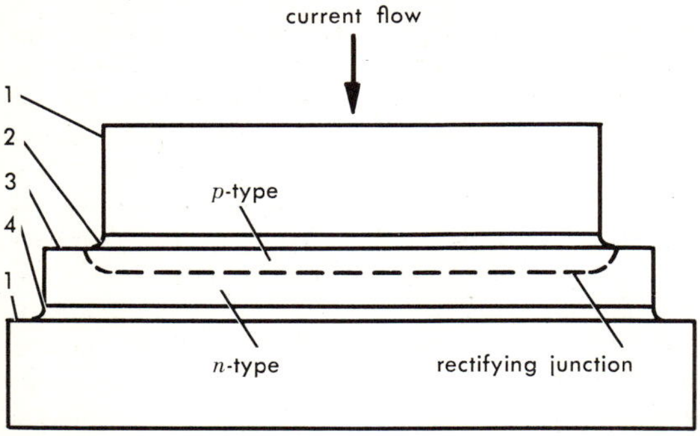

1 molybdenum, 0.030″ thick

2 soldering and p-doping material, indium for germanium, aluminum for silicon

3 germanium or silicon wafer 0.015″ thick

4 soldering material, tin for germanium, silver-lead-antimony alloy for silicon

Fig. 6. Typical construction of germanium and silicon rectifying cells.

will drive the charge carriers across the junction, producing a current. If the polarities are reversed, as shown in Fig. 5*b*, the charge carriers are drawn away from the junction and the current stops, except for a small leakage current. This is a simplified explanation of the operating principle. *See* JUNCTION DIODE.

The crystal is grown from a melt of highly purified germanium or silicon and is sliced into wafers which are used for making rectifying cells. Figure 6 shows an enlarged cross section of a typical germanium or silicon cell. The wafer is *n*-type, obtained by doping the melt from which the crystal is grown with a 5-valence element, such as arsenic. The wafer is soldered to the molybdenum terminals by heating the assembly. The soldering alloy on one side has a 3-valence doping element which converts the *n*-type semiconductor on that side to a *p*-type, thus producing a *p-n* junction. Other techniques and materials have also been used.

The cell is sealed hermetically in a case under a dry gas, for protection against moisture. One side is soldered to the bottom of the case for good thermal and electrical conductivity; the other side is soldered to a conductor, which is insulated from the case. A typical encapsulation of a silicon cell is shown in Fig. 7. For dissipating the heat produced by the losses, the case is attached to an air- or water-cooled heat sink; in some designs it is an integral part of the case.

The rectifier ratings are determined by the operating temperature limits of the junctions, about 65–85°C for germanium and 175–200°C for silicon cells. No aging effects have been observed at these temperatures. Because of the small size of the cells, germanium and silicon rectifiers have only a limited overload capacity for short periods. Excessive forward currents or inverse voltages can destroy the junction and will usually result in a short circuit through the cell. Adequate protection against overloads and fault currents is a vital part of a rectifier unit. As protection against voltage surges, it is the practice in many applications to limit the operating peak inverse voltage (PIV) to one-half or less of the rated PIV.

Germanium and silicon rectifiers are available in current ratings from a fraction of an ampere to several hundred amperes average dc per cell, and PIV up to about 300 volts for germanium and 1000 volts for silicon cells. The highest PIVs are for the smaller cells. The approximate average current densities corresponding to the maximum ratings are 300–600 amp/in.^{2} for germanium and 700–1300 amp/in.^{2} for silicon. A cell of a given current rating has a range of PIV ratings so as to limit the magnitude of reverse current. These ratings are determined by testing and grading during manufacture on the basis of the inverse characteristic. Figure 8 shows approximate volt-ampere characteristics for cells of medium and high current ratings. There can be large variations in the reverse characteristics.

When rectifier cells are connected in parallel, their forward characteristics have to be matched, or some other means provided for balancing the currents. When they are connected in series, it is the usual practice to use shunting resistors or other means for dividing the inverse voltage. The dc output voltage can be adjusted by transformer taps, an induction regulator, or an adjustable reactor in the ac circuit. The over-all efficiency of 6-phase ger-

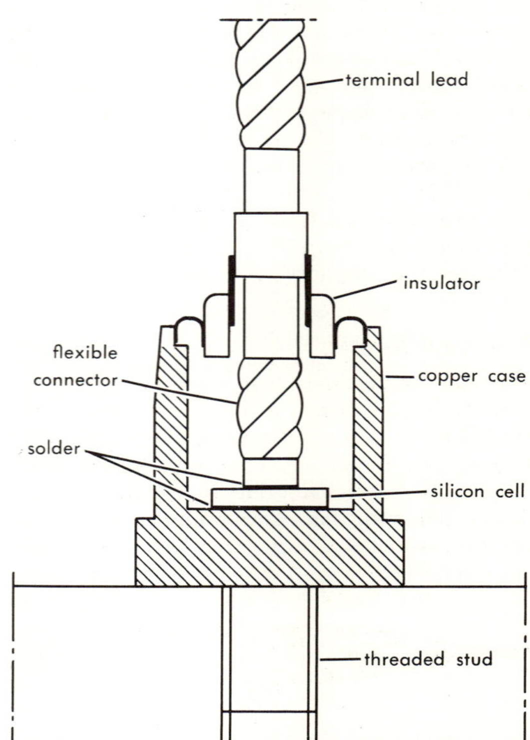

Fig. 7. Assembly of silicon rectifier.

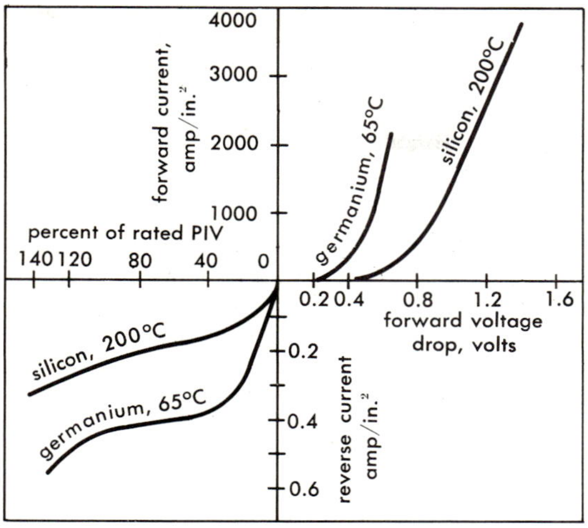

Fig. 8. Volt-ampere characteristics of germanium and silicon rectifying cells. The voltages and current densities are instantaneous values.

manium or silicon rectifier units at rated load is usually in the range of 90–97%.

The commercial application of germanium rectifiers was started in about 1950; of silicon, in about 1954. Before the introduction of medium- and larger-size silicon cells, there were many applications of germanium rectifiers, including a number of high-current installations for electrolytic service, one of which has a capacity of 140,000 amps, 250 volts dc. Because of its higher operating temperature and higher available voltage ratings, the silicon rectifier has become the preferred type, particularly for dc voltages above 50 volts. Its application has grown rapidly and has been extended into fields which were formerly monopolized by mercury-arc rectifiers. *See* CONTROLLED RECTIFIER. [H.W.]

Bibliography: AIEE, *Rectifiers in Industry*, Publ. T-93, 1957; W. C. Dunlap, Jr., *An Introduction to Semiconductors*, 1957; H. K. Henisch, *Rectifying Semi-Conductor Contacts*, 1957; S. P. Jackson, *Selection and Application of Metallic Rectifiers*, 1957.

Semionotiformes

Actinopterygian fishes first occur in the Upper Permian but are best developed in the early Mesozoic. Modern gars belong to this order of otherwise archaic forms which are also known as the Ginglymodi. The scales are thick, more or less rhomboidal, and have an enamel-like surface. The structure is diverse but modern forms are elongate

Spotted gar, *Lepisosteus productus.* (After G. B. Goode, *Great International Fisheries Exhibition, London, 1883, U.S. Natl. Museum Bull. 27, 1884*)

and have produced bony jaws with sharp teeth and an abbreviated heterocercal caudal fin. The swim bladder is highly vascularized. The single Recent family Lepisosteidae contains one genus, *Lepisosteus*, with seven species (see illustration). They are restricted to lowland fresh and brackish waters of North and Central America. Gars are despised by fishermen because of their predation on more valuable fishes. They are themselves caught with difficulty and are almost worthless as food. The roe is highly toxic when eaten by warm-blooded animals. *See* ACTINOPTERYGII; SWIM BLADDER. [R.M.B.]

Senescence

The study of the biological changes related to aging, with special emphasis on plant, animal, and clinical observations which may apply to man.

Aging process. From the time the ovum is fertilized to form a new individual until the individual dies, the processes of aging are at work. Body cells, in most cases, like the beginning embryo, are first relatively unspecialized, or undifferentiated. They are literally bursting with life and their primary initial function seems to be that of growth, then reproduction. Following this period of activity, the average cell then enters a variable period of time during which it will function as one of a number of similar units so that a particular function or group of metabolic requirements may be carried out. Inevitably, at some point certain changes occur; these may be almost imperceptible at first but they mark the onset of a period of decline which will end in the reduction of activity and, finally, in cessation of biologic function, death. *See* DEATH.

In a sense, all recognizable stages or changes during the lifetime of a cell or tissue indicate senescence, since aging is a function of time, and time ultimately limits all living things. More properly, however, senescence includes the gradual decrease in functioning of many general and specific characteristics of both cellular and intracellular components of the body over the organism's lifetime.

Aging does not occur at a uniform rate, nor does it occur at the same rate in different tissues, or for that matter, even in different individuals. Therefore the concepts of senescence, although they mark the influence of time, cannot be directly correlated with its actual passage.

Physiological changes. Another aspect which must be considered in aging is that the changes which indicate deterioration or degeneration are not actually causes of senescence, but are effects of more subtle disturbances of metabolism or regulation which, for the most part, are recognized only when they become accentuated. The long, intricate chain of events which ultimately produces a recognizable alteration is usually so intertwined with other contemporary changes that it is hard to disentangle a single causative thread. For the most part the present course is to compare the normal aging tissue, if possible, with abnormal tissue from either individuals who show severe evidence of senescence or from individuals whose tissues show changes much earlier or later than the average. This immediately brings up the question of what is abnormal, or pathologic. This may be easy to define in extreme cases but, as the range narrows toward normalcy, there will be less and less agreement and more divergent views on interpretation, unless large groups of tissues, organs, or individuals are considered so that certain patterns appear even in the borderland regions.

Certain generalizations regarding senescence may be made if it is kept in mind that these apply to large groups and not necessarily to a specific individual.

In older persons there is usually a specific atrophy of certain tissues, notably those of the sex organs, the skin, the skeletal framework, and the connective elements of the body. Even in those organs where there is an adequate function, age produces a decrease in the size of the organ cells, or parenchyma. There may also be a proportionate increase in the amount and density of the noncellular components. Thus, in a senescent liver, for example, the cells are smaller than those in the liver of a young adult. The intercellular connective tissue appears more prominent, but usually this is due only to a relative shift in volumes, although an actual increase in connective tissue may occur under the influence of further aging. *See* ATROPHY.

It is believed that at least part of this cellular atrophy is due to the accumulation in the immediate environment of the cell of materials which in a more vigorous individual would be carried off or metabolized. Biochemical investigations are shedding some light on this phenomenon, both from the standpoint of the type and quantity of material accumulated and also on the measurable decreases in cellular activity that occur first.

Secondary alterations. The great integrating and regulating systems of the body show the effects of aging and, in turn, produce secondary alterations on other organs and tissues. For example, gradual accumulation of fatty deposits in blood vessels commonly results in a hardened, thickened, abnormal wall with decreased elasticity. Blood passing to an organ through such a vessel may not furnish nourishment or remove wastes as effectively as formerly. This may be due to a decrease in actual blood flow because of partial obstruction by the arteriosclerotic plaque, or it may be due to the possibly injurious effects of an intermittently higher blood pressure, since the wall cannot stretch to dissipate some of this force.

Changes in the endocrine system and in the nervous system, also integrative and regulatory in nature, are routinely seen in elderly persons. The complex interrelationships of the hormones of the body may be altered so that both specific target tissues and general body cells are adversely affected, albeit at a slow rate.

The nervous system is somewhat unique in that its functional units, the neurons, do not multiply or regenerate after embryonic life. In addition, they are also highly susceptible to damage caused by a decrease in, or lack of, oxygen and by other factors. Nervous system atrophy is expected in most elderly persons, but this does not necessarily relate directly to severe or pronounced mental effects. It does mean, though, that such atrophy, when combined with changes in blood vessels, glands, and other organs, may hasten the appearance of the composite picture we call senility.

Tissue changes. Perhaps the most marked changes in senescence are seen in the various intercellular tissues of the body, the bony matrix, blood-forming centers, connective tissue and joints, and similar components in which there is a very low content of living cells. Much of this material accumulates with age as the result of cellular secretion, metabolic activity, and physicochemical processes which are only beginning to be understood. Such material is known to decrease in lability and rate of turnover as old age advances. Since most of the parenchymal cells of the body which still function are held in place by some form of this intercellular substance, it is easy to see that these cells might also be adversely affected by decreased metabolism of the supporting structures.

In certain tissues, senescence is marked by typical changes in pigmentation, such as the brown atrophy of the heart, liver, or kidney, and, of course, the accumulation of lipid materials in blood vessel walls and other places. *See* PIGMENTATION.

Changes in the colloidal systems. Many senescent alterations are believed to be due to the gradual change in certain physical or chemical properties of body colloidal systems. This is seen most clearly in the relation of hydrophilic colloidal particles to the surrounding aqueous phase of such a system, but analogous situations probably occur in the aging of elastic fibers in which collagen replaces the formerly resilient material in tissues like blood vessels, the skin, lungs, and musculoskeletal structures.

Scope of senescence. In consideration of the scope of senescence, it must be realized that there is but a hairline of difference between normal or average aging changes and the development of those diseases and disturbances to which the elderly are most subject. Such diseases, however, lie more properly in the realm of geriatric medicine. The changes mentioned in this article apply, with allowances for differences in life span and other species characteristics, to most animal tissues with any degree of specialization. *See* GERONTOLOGY.

Although cumulative injury and repair of body elements is excluded in a theoretical consideration of senescence, the practical aspects cannot be ignored in an evaluation of aging. It is also most difficult to quantify such repetitive injury in terms of a single individual.

Metabolic and physiologic changes which occur with age are of great import. The obvious effects of even such mundane factors as decreased physical activity, improper elimination, and faulty diet do not create the best nutritional situation to facilitate the fight against deleterious influences.

With an increase in age, particularly beyond the middle years, there is thought to be a gradually progressive dessication, or drying, of the body components. There is also a gradual decrease in tissue resilience, elasticity, and tensile strength. Muscular effectiveness drops off relatively early, and is followed in later years by variable reductions in the integrity of the nervous system. The latter produces, quite often, changes in vision, hearing, coordination, reception of sensory stimuli, and mental processes, particularly those of memory and concentration.

Nonspecific changes. Other important, nonspecific changes in body cells include a gradual decrease in the rate of normal cell division, so that normal or excessive repair proceeds more slowly. There is also a corresponding reduction in the quantity or quality of cell secretions, including the important portions of the digestive system, the pancreas, liver, and the vital endocrine glands.

Most body tissues decrease in activity with advancing age; this may be demonstrated repeatedly by comparisons of basal metabolic tests of groups of different age medians.

It has been aptly stated that all of the many factors which constitute the aging process tend to deplete body reserves so that the aging individual gradually becomes less able to withstand the accidents or the unusual stresses of living and finally is unable to provide adequate maintainence of the functions and structures necessary for life.

[E.G.ST.]

Senile dementia

A psychosis characterized by intellectual deterioration, impairment of judgement, and gross emotional instability. *See* PSYCHOSIS.

The psychosis is largely due to metabolic changes in the senile brain. Atrophy and reduction in the number of cells, especially in the frontal lobes, are characteristic. Psychological factors, particularly threats to the security of the aged patient, are important. E. Gruenberg showed that general physical illness is often a precipitating or aggravating factor.

The symptoms are impaired abstract thinking and a return to what K. Goldstein called concrete thinking. Patients are not able to handle ordinary problems of living, are oversensitive, irritable, anxious, and often quite suspicious of being mistreated. Frequently, they suffer from paranoid delusions and various hallucinations. The behavior of some patients becomes quite aggressive, with a regression to infantile sexual patterns. The differential diagnosis against cerebral arteriosclerosis is not always possible. *See* DELUSION; HALLUCINATION; PARANOID STATE.

Treatment is symptomatic. Patients need much psychological support, and the prevention of severe threats, such as illness without care; the older individual needs to have a feeling of security and stability. The use of tranquilizing drugs is in order. Prolonged hospitalization may be indicated, but should be recommended only after careful evaluation of the patient. *See* TRANQUILIZER. [F.C.R.]

Bibliography: J. R. Ewalt, E. A. Strecker, and F. G. Ebaugh, *Practical Clinical Psychiatry*, 8th ed., 1957; A. P. Noyes and L. C. Kolb, *Modern Clinical Psychiatry*, 5th ed., 1958.

Sensation

A term commonly used to refer to the subjective experience resulting from stimulation of a sense organ, for instance, a sensation of warm, sour, or green. As a general scientific category, the study of sensation is the study of the operation of the senses.

Sense receptors. Sense receptors are the means by which information presented as one form of energy, for example, light, is converted to information in the form used by the nervous system, that is, impulses traveling along nerve fibers. The receptors may conveniently be classified on the basis of their location and stimulus source, in a modification of the system introduced by C. S. Sherrington.

Pain does not fit neatly into the following classification, its receptors apparently being distributed throughout most of the body.

Exteroceptors. These are receptors at the surface of the body which transmit information about the external environment. They may be further classified as teleceptors (the eye, ear, and nose) which are concerned with the more distant environment, and proximoceptors (involved in taste and in cutaneous sensations) which report primarily on the contiguous external environment. *See* HEARING; SMELL; VISION.

Proprioceptors. Proprioceptors are found in subcutaneous tissues, muscles, tendons, joints, and in the labyrinth of the ear. They signal the spatial position and movements of the body and its members, as well as muscular tension.

Interoceptors. Interoceptors are located in the visceral organs and generally yield diffuse, poorly localized sensations. They are responsible for the feelings of distention and temperature which can arise from some of these organs. The interoceptors are also involved in the characteristic sensations accompanying hunger or nausea.

Specific irritability. Specific irritability of receptors is essential for the ability to discriminate among different kinds of stimulation. Each sense is specialized for the detection of one form of energy, called the adequate stimulus for that sense. While the eye can be stimulated by sufficiently intense mechanical energy, from a blow, for instance, it responds to a very much smaller intensity of its adequate stimulus, which is light. The touch receptors of the finger-tips, unstimulated by the glare of the noonday sun, respond to the minute skin deformations resulting from the gentle strokes of a feather.

An additional principle beyond that of specific irritability of receptors is needed in order to account for the ability to discriminate among different kinds of stimulation. The need for the further principle arises because the nature of the adequate stimulus has no effect on the characteristics of the resulting neural impulses which transmit the stimulus information through the nervous system. Consequently, there is nothing distinctive about the neural impulses to indicate what sort of stimulation initiated them.

Johannes Muller's doctrine of specific nerve energies, also known as Muller's law, explains why the messages from the different senses are not confused with each other, even though they consist of similar nerve impulses. There is discrimination among the adequate stimuli for the several senses because the corresponding neural impulses travel over different nerves and arrive at different destinations within the nervous system.

One demonstration of this principle is the fact that stimulation of a sensory nerve by any means results in a sensation appropriate to that nerve. Stimulation of the optic nerve, even if mechanical, electrical, or thermal, always yields visual sensations.

Properties common to the senses. Each sense has mechanisms and characteristics peculiar to itself, but all display the phenomena of absolute threshold, differential threshold, and adaptation.

Absolute threshold. Not until sufficient stimulation impinges on a receptor can the presence of a stimulus be detected. The quantity of stimulation required is known as the absolute threshold. The magnitude of the absolute threshold is affected by many factors. Some of the more general ones are mentioned below under strength of response.

Differential threshold. Not until a sufficient change occurs in some aspect of a stimulus can the change be detected. The magnitude of the change required is called the differential threshold.

A generalization known as Weber's law, which is found empirically to be only an approximate rule of thumb, states that the stimulus increment which can barely be detected (the just noticeable difference, or j.n.d.) is a constant fraction of the initial magnitude of the stimulus. The Weber fraction is $1/588$ for the frequency of a moderately loud auditory tone 3 octaves above middle C, $7.5/1$ for the intensity of a weak tone 3 octaves below middle C. The values of the Weber fractions for other stimuli fall between these values.

The size of the Weber fraction is influenced in many ways. It may be different for different senses, such as smell and temperature, although there is much overlap in this regard. For one sense, it varies with such qualitative factors as hue or pitch. With all other factors held constant, the Weber fraction may remain fairly constant throughout the middle range of stimulus intensities, but its values at the extremes of intensity may differ a hundredfold. Rate of stimulus change, stimulus duration or size, and a host of other factors may have a bearing in any particular case. If the limitations of Weber's law are kept in mind, however, it is still useful to remember that the "resolving power" of a sense is roughly proportional to the magnitude of the stimulus. *See* TEMPERATURE SENSES.

Special procedures known as psychophysical methods have been developed for measuring absolute and differential thresholds, and for relating magnitude of sensation to magnitude of stimulation. *See* PSYCHOPHYSICAL METHODS.

Adaptation. Under steady stimulation, there is a decrease in sensitivity of the corresponding sense, as indicated by a shift in the absolute threshold and in the magnitude of sensation. After the stimulation ceases, sensitivity increases. An obvious example of visual adaptation occurs when one goes from bright to dim surroundings or vice versa. Depending on the sense involved and the amount and duration of the change in stimulation, the shift in sensitivity may be by a factor of as much as 100,000, and may require as little as a fraction of a second, or more than an hour.

Strength of response. The magnitude of sensory response to stimulation may be affected by the size or duration of the stimulus. Stimuli which are very small or very short often, although not invariably, yield less response.

An important influence on the strength of response is the level of the absolute threshold. A stimulus below absolute threshold occasions no response; a stimulus slightly above threshold arouses a weak response; and a stimulus well above threshold yields a strong response. While there are many influences on thresholds in each sense department, some are of considerable generality.

The nature of the stimulus has a profound effect on the absolute threshold, since receptors are not equally sensitive to the entire range of their adequate stimuli. The absolute threshold of the human eye for greenish-yellow light is $1/1000$ of that for red light, and the threshold concentration of one odorous substance may be millions of times greater than that of another.

Receptor condition may affect absolute threshold, not only through adaptation, but also because of impaired functioning. For instance, vitamin A deficiency can lead to so-called night blindness, and normal aging processes may affect hearing thresholds. *See* VITAMIN A.

Condition of the central nervous system also plays a role in sensitivity. The nervous system is continuously active, and evidence is accumulating that, for at least some sensory systems, the level of activity in the central nervous system has rather direct effects on sensitivity. *See* RETICULAR FORMATION (BRAIN).

The locus of stimulation, for senses in which this can vary, will affect the absolute threshold. Some areas of the tongue are more sensitive to bitter stimuli, others to salty. Looking directly at a visual stimulus is the best procedure if one's purpose is to resolve fine detail, but a poor procedure if one wishes to detect a faint pinpoint of light in the dark, because of differences among receptors in different parts of the retina of the eye.

Species differences in sensitivity have been examined by means of behavioral tests and by recording of electrical activity evoked in the nervous system by sensory stimuli. The common cat can respond to auditory tones of frequencies three times higher than the upper limit for man, and there is electrophysiological evidence that pure water has a distinctive taste to the dog. Lest it be thought, however, that man invariably makes a poor showing when his sensitivities are compared to those of other species, let it be said that under suitable conditions man's eye and ear can detect stimuli which are nearly as weak as are detectable by physical means.

Spatial localization. With fairly good accuracy, man can localize visual objects, sounds, and cutaneous contacts, and can discriminate the spatial orientation of his body and its members. With rather poor accuracy, he can localize many of the stimuli originating within his own body.

With the exception of hearing, in which sense localization depends on differences in the acoustic stimuli reaching the two ears, there appears to be a common principle involved in giving spatially separated receptors their different local signs. Stimulation at different points on the receptive surface results in peaks of electrical activity at different loci in the brain. In no sense is there anything like a private wire from each sensory cell to a corresponding point in the brain. In fact, there are so many opportunities for a signal to go astray on its way from the receptor to the brain that it is surprising that spatial discrimination is as good as it is. Nevertheless there is clear evidence that, by a combination of anatomical and functional arrangements, spatial differences at the receptor level are translated into topologically similar spatial differences in brain activity.

Compensation for sensory deficiencies. Although compensation for deficiencies in such senses as pain, proprioception, or taste is important to the individuals concerned, much more effort has been devoted to compensating for deficiencies in sight or hearing, the major channels through which sensory information about the environment is transmitted.

There are two types of solution to these problems. In one, devices are used which alter the input of stimulation in a way which compensates directly for the sensory defect. Common examples are the use of eyeglasses to correct for refractive errors of the eye, and the use of hearing aids to amplify sound, either as compensation for lowered auditory sensitivity or as a means of bypassing the affected part of the ear. *See* EYE GLASSES; HEARING AID.

Sometimes such direct compensation is not possible, and there is recourse to a different sense as a communication channel. The Braille system of printing for use by the blind illustrates the substitution of tactual for visual information, and lip-reading is a means of substituting visual information for the speech sounds no longer available to the deaf. There might be more such substitutions available if the possibilities were more fully explored. *See* ITCH; PAIN, CUTANEOUS; PAIN, DEEP; PARESTHESIA; SENSE, CHEMICAL; SOMESTHESIS; STEREOGNOSIS; TASTE; TICKLE; TOUCH. [J.F.H.]

Sense, chemical

The senses of smell (olfaction) and taste and the so-called common chemical sense comprise the three chemical senses. The sense cells of olfaction and taste are specialized receptor neural elements. The receptors of common chemical sense appear to be undifferentiated, free nerve endings. These are distributed throughout the moist, mucous membranes of land dwelling animals and over the skin of aquatic vertebrates. In order of thresholds, olfaction is the most sensitive, taste is intermediate, and common chemical sensitivity is the least sensitive.

The chemical senses, with their relatively simple morphology (structure and form) are often classified as lower senses when compared to the higher senses of vision and hearing. The chemical senses mediate the selection and acceptance of foodstuffs, the avoidance of irritants and, in lower organisms especially, the detection of enemies and prey and the selection of mates. Their role in controlling general behavior appears to be more important in lower organisms than in man.

The common chemical senses are considered to be the sensitivity to mildly irritating chemicals like dilute solutions of alkali, acids, and salts. Such chemical sensitivity is distinct from tactile sensitivity and, in the mouth and nose, is distinct from taste and smell. Some authorities claim that chemical sensitivity can be demonstrated as an entity distinct from pain and touch. However, other authorities are of the opinion that this has not been clearly shown. *See* CHEMORECEPTION; SENSATION; SMELL; TASTE. [C.P.]

Bibliography: S. S. Stevens (ed.), *Handbook of Experimental Psychology*, 1951.

Sense organ

A structure which is a receptor for external or internal stimulation. A sense organ is often referred to as a receptor organ. External stimuli affect the sensory structures which comprise the general cutaneous surface of the body, the exteroceptive area, and the tissues of the body wall or the proprioceptive area. These somatic area receptors are known under the general term of exteroreceptors. Internal stimuli which originate in various visceral organs such as the intestinal tract or heart affect the visceral sense organs or interoceptors. A receptor structure is not necessarily an organ; in many unicellular animals it is a specialized structure within the organism. Receptors are named on the basis of the stimulus which affects them.

Photoreceptors. Those structures which are sensitive to light and in some instances are also capable of perceiving form, that is, of forming images, are called photoreceptors. Light-sensitive structures include the stigma of phytomonads, photoreceptor cells of some annelids, pigment cup ocelli and retinal cells in certain asteroids, the eyespot in many turbellarians, and the ocelli of arthropods. The compound eye of arthropods, mollusks, and chordates is capable of image formation and is also photosensitive. *See* PHOTORECEPTION.

Phonoreceptors. Structures which are capable of detecting vibratory motion or sound waves in the environment are phonoreceptors. The most common phonoreceptor is the ear, which in the vertebrates has other functions in addition to sound perception. Among the fishes, it has been demonstrated that the air bladder and lateral line organs, which in certain species have receptors (neuromasts) in their canals, serve as sound receptors. Sound perception in snakes, which lack a middle ear and are sensitive to air-borne sounds, is by bone conduction which makes them sensitive to ground vibrations. Sensory hairs or hair sensilla and tympanal organs occur in insects. Other organs which may function as sound receptors are the statocysts of certain crustaceans, although they are primarily statoreceptors. *See* PHONORECEPTION.

Statoreceptors. Structures concerned primarily with equilibration, such as the statocysts found throughout the various phyla of invertebrates and the inner ear or membranous labyrinth filled with fluid, are statoreceptors. Halteres are unique structures found in the Diptera which aid in orientation of these insects. *See* EQUILIBRIUM, BIOLOGICAL.

Olfactoreceptors. The sense of smell is dependent upon the presence of olfactory neurons in the olfactory epithelium of the nasal passages among the vertebrates. Olfactory hairs extend from each olfactoreceptor cell, giving it a brushlike appearance. Jacobson's organ in amphibians may play a minor role in olfaction. Little is known about the sensation of smell among the invertebrates. The auricle of planarians has olfactory receptors and the rhinophores of certain land mollusks are regarded as smell receptors. Odor sensitivity has been observed among many of the invertebrates, but no specific structure has been implicated. The most widely studied invertebrate group is the insects in which three types of structures have been described: the sensilla placodea, sensilla basiconica, and the sensilla coeloconica. *See* SMELL.

Gustatoreceptors. One of the best-known senses is that of taste, which is mediated by the taste buds. In most vertebrates these taste buds occur in the oral cavity, on the tongue, pharynx, and lining of the mouth, but among certain species of fish, the body surface is supplied with taste buds as are the barbels of the catfish. Again, among the invertebrates, the most studied group has been the insects and most of the experimental work has been based on behavioral studies. However, because sensilla are known to occur on the antennae, mouthparts, tarsus, and tibia of many species, and on the ovipositor of a few forms, it has been postulated that these structures are involved. Contact stimulation appears to activate the receptors. *See* TASTE.

Cutaneous receptors. The surface skin of vertebrates contains numerous varied receptors associated with sensations of touch, pain, heat, and cold. Tangoreceptors are associated with the phenomena of touch and pressure. Because sensory endings of both tangoreceptors and algesireceptors (pain receptors) terminate in the skin, it is often difficult to distinguish between the two. Among the vertebrates, cutaneous receptors are Grandry's corpuscle, Herbst corpuscle, the bulb of Krause, Merkel's corpuscle, Meissner's corpuscle, the Pacinian corpuscle, and the end organ of Ruffini. The end bulb of Krause is a thermoreceptor and is stimulated by cold (frigidoreceptor), whereas the end organs of Ruffini are believed to be influenced by heat (caloreceptor). Sensory and labial pits of certain reptiles are temperature sensitive. Many of the invertebrates are temperature sensitive, especially during developmental life cycles. Thermal perception among insects has been attributed to the antennae and mouthparts as well as the cerci and tarsi. The ampullae of Lorenzini found in elasmobranchs may have a thermosensory function according to recent studies. *See* PAIN, CUTANEOUS; SENSATION; SENSE, CHEMICAL; TOUCH. [C.B.C.]

Sensible temperature

An estimate of the degree of human comfort which is felt under various combinations of atmospheric temperature, humidity, air movement, and radiation to and from surroundings. It is expressed by the temperature at which the same degree of comfort would be felt in air at standard humidity, air movement, and radiation. The human body produces heat constantly, and to avoid chilling or overheating, it must lose heat at approximately the rate of production, which depends on muscular activity. Sensible temperature measures the cooling power or thermal acceptance of the atmosphere. *See* TEMPERATURE.

Heat is lost by conduction to cooler air, by evaporation of perspiration into unsaturated air, and by radiative exchange with the surroundings. Air motion (wind) affects the rate of conductive and evaporative cooling of skin but not of lungs. Radiative losses occur only from skin, and depend primarily on sunshine intensity and on the temperature of sky and surroundings. Clothing affects all three avenues of heat loss from skin. Heat can also be lost by conduction to water, to the ground, or to another surface in substantial contact with the body. Hence any sensible temperature or cooling-power formula applies to a man at a given rate of activity, with specified clothing or no clothing.

Many cooling-power formulas have been proposed, as well as such names as effective, equivalent, operative, or comfort temperature; or comfort, sultriness, or wind-chill index. Most have been derived from special instruments such as Hill's katathermometer, Dorno's (Davos) frigorimeter, K. Buettner's frigorigraph, P. A. Siple's cylinder, and C. P. Yaglou's globe. Others are based on experiments with human subjects; most used are Yaglou's effective temperature (given graphically in the annual Guide of the American Society of Heating and Air-Conditioning Engineers) and the operative temperature of C. E. A. Winslow, L. P. Herrington, and A. P. Gagge. [A.C.]

Bibliography: C. E. A. Winslow and L. P. Herrington, *Temperature and Human Life*, 1949.

Sensitivity

The ability of the output of a device or system to respond to an input stimulus. Mathematically, sensitivity is expressed as the ratio of the response or change induced in the output to a stimulus or change in the input. If the sensitivity varies with the level of the input signal, then sensitivity is usually expressed in terms of the derivative of the output with respect to the input at a specified input level.

The reciprocal sensitivity is called the scale factor or figure of merit, and it represents the conversion factor by which the output indicator or scale reading must be multiplied to obtain the magnitude of the input. Occasionally the scale factor is called the sensitivity through loose usage.

Sensitivity is closely related to noise. Quite often the limiting factor in increasing the sensitivity of devices or systems is the inherent noise level. For instance, the noise level for an electric galvanometer or a gravitational weight balance arises from the random or Brownian movement of the air molecules surrounding the apparatus. In electrical circuits, the noise level arises from the random movement of electrons in resistors. *See* BROWNIAN MOVEMENT; NOISE, ELECTRICAL.

Examples of sensitivity would be visual contrast sensitivity, which is the ability of the eye to distinguish between the luminances of adjacent areas; galvanometer sensitivity, expressed as the ratio of the scale deflection in millimeters per microampere input; and radio receiver sensitivity, which is actually expressed in terms of reciprocal sensitivity in the form of antenna voltage in microvolts necessary to cause a specified output.

The sensitivity of a radio receiver is a measure of its ability to reproduce weak broadcast signals with satisfactory output volume. The sensitivity of a television camera tube determines its ability to deliver a usable picture signal under poor lighting conditions. [J.MR.]

Sensory learning

A term used to describe learning situations in which a person or an animal is trained to respond to changes in or differences between some aspects of a physical stimulus presented to one of the sense organs. These studies are not always easily distinguished from other learning experiments, such as those of conditioning and problem solving. *See* REFLEX, CONDITIONED; PROBLEM SOLVING (PSYCHOLOGY). The basis for the sensory learning class lies in the convenience of relating a number of studies which use more or less similar experimental procedures, and which provide information about the organism's ability to discriminate sensory cues, to learn, and to remember.

Sensory learning is extensively used in animal experiments. An animal is almost always required to raise a leg or to press a lever, or to go from one place to another in order to approach a reward, such as food or water, or to avoid an electric shock. Since the occurrence of the reward or the punishment depends upon the presence of a specific sensory stimulus, the animal's task is to respond selectively to a certain stimulus, such as the presence of a high tone, or to choose one stimulus among others, for example, to choose a red rather than a green color. The animal is said to have learned the sensory discrimination, or acquired the discriminative habit, when it makes the appropriate movements to the specified sensory stimulus. Results of such studies provide information concerning both the animal's sensory capacity and learning ability.

Sensory learning studies. These may be divided into two groups. The first group includes those studies on the normal animal's sensory capacities. Using the method just described, the lower limit of an animal's capacity to perceive a sensory stimulus is determined by gradually decreasing the strength of the stimulus until the animal fails to respond. The animal's differential sensitivity, that is, its ability to perceive the difference between two stimuli, is determined by gradually decreasing the difference between two stimuli until the animal ceases to respond differentially to them.

Most experiments on sensory learning belong to the second group. They are studies dealing with the localization of brain centers which correlate first with the memory of a learned sensory discrimination, and second with the capacity to learn a sensory discrimination. The procedure common to the former is first to train an animal to sensory stimuli. After the animal learns a problem, such as a color discrimination, a relevant neural structure, like the cortical visual area, is surgically destroyed. If the animal performs successfully on the problem after this operation the damaged cortical region is judged not critical for maintaining this habit. If the animal fails on the problem, this may mean either that the operation impairs the sensory capacity, that is, the animal cannot see the stimuli; or that the operation causes amnesia, that is, the animal can see the stimuli but cannot remember the habit. To determine which alternative applies, the animal is retrained on the same discrimination. If it relearns the problem, the inference is made that the cortical damage does not affect the animal's sensory capacity, but does affect retention of the learned response. On the other hand, if the animal fails to relearn, it may be inferred that sensory capacity is sometimes impaired. However, many factors, such as method of training, degree of motivation, and so on, greatly influence learning, and therefore a failure in relearning after brain operation is not always correlated with a loss of sensory capacity. The animal may still see the stimuli, have the capacity to learn, and yet may not relearn because of behavioral deficits resulting from changes in motivation, attention, and the like.

A somewhat different procedure is to destroy the neural structure first and then to train an animal on a sensory discrimination. This method is used to localize in the brain the initial learning capacity. Most of the studies on human subjects with brain injury are of this type. A retarded rate of learning by an animal or man with brain injury indicates a decreased learning capacity. The destroyed brain region is then correlated with a specific sensory learning. The same difficulties with respect to differentiating sensory and learning capacities discussed above also apply to this procedure.

Sensory learning in normal animals. These studies have been undertaken to determine the limits and differential sensitivity of animal senses. Results from studies on vision show that chimpanzees and monkeys have visual acuity, detail vision, and color vision approximate to that of man (*see* VISION). Lower mammals like cats, dogs, and rats have poorly developed color vision or none at all; but chickens, pigeons, and some fish can discriminate colors. Rats also have poor visual acuity and weak detail vision. Experiments on audition show that chimpanzees and monkeys can hear slightly higher tones than man, while cats, dogs, and rats can hear considerably higher tones. All these animals have about the same sensitivity to sound intensity as man (*see* HEARING). The method of sensory learning has also been used to study the skin senses, muscle senses, and smell of many mammals. Almost all of these studies show that an animal can learn to choose one stimulus against another stimulus but they do not indicate the limits and the range sensitivity. Monkeys, cats, and rats can learn to discriminate differences in weights, surface roughnesses, shapes of objects, and temperature. Chimpanzees have not been trained on temperature discrimination but have been studied on the other tasks (*see* SOMESTHESIS). Monkeys and rats can learn to discriminate between two odors, and dogs can localize meat by odors so faint that the human observer cannot detect the scent (*see* SENSE, CHEMICAL).

Studies on taste have taken advantage of an animal's natural preference for sweet, weak salt, and weak bitter solutions, and his natural aversion to strong sweet, salt, bitter, and sour solutions. The lower limits of taste sensitivity have been determined by letting the animals select the preferred solution. These studies, like most studies of human sensation, are not included under the heading of sensory learning since no training is required, at least in regard to the discrimination itself.

Neural mechanisms of sensory learning. These are studied by testing the learning or memory of a sensory habit in man and animals with brain lesions. The lesions are usually on those areas of the cerebral cortex or within the subcortex, that are known on the basis of other evidence to be related to the sense used in the learning problem. Although there are many subcortical centers in the brain for each of the senses, these centers have not been as extensively studied as have the cortical sensory areas. The cortical areas on the surface layer of the brain have been demonstrated either anatomically or physiologically to receive connections from the peripheral sense organs. Figure 1 shows the approximate location of the anatomically determined visual, auditory, somatic, and olfactory cortical areas in man. Somatic area II is the only second somatic sensory area identified in man. The darkly shaded areas in Fig. 1 are anatomically determined regions. The surrounding lightly shaded areas, when stimulated electrically, arouse various sensations in conscious human patients. Figure 2 shows the physiologically determined sensory areas in rat, rabbit, cat, and monkey. These are identified by recording changes of electric potentials on the cortical surface when the peripheral sense organs are stimulated. With few exceptions each sense is represented twice on the cortex. In the somatic

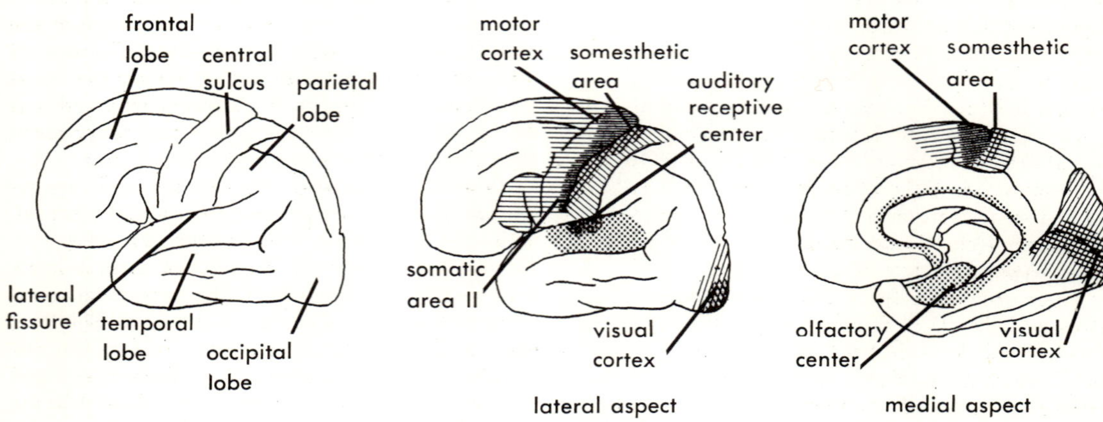

Fig. 1. Diagram of the cortical areas of the human brain. (*Modified from S. W. Ranson and S. L. Clark, Anatomy of the Nervous System, 9th ed., Saunders, 1953*)

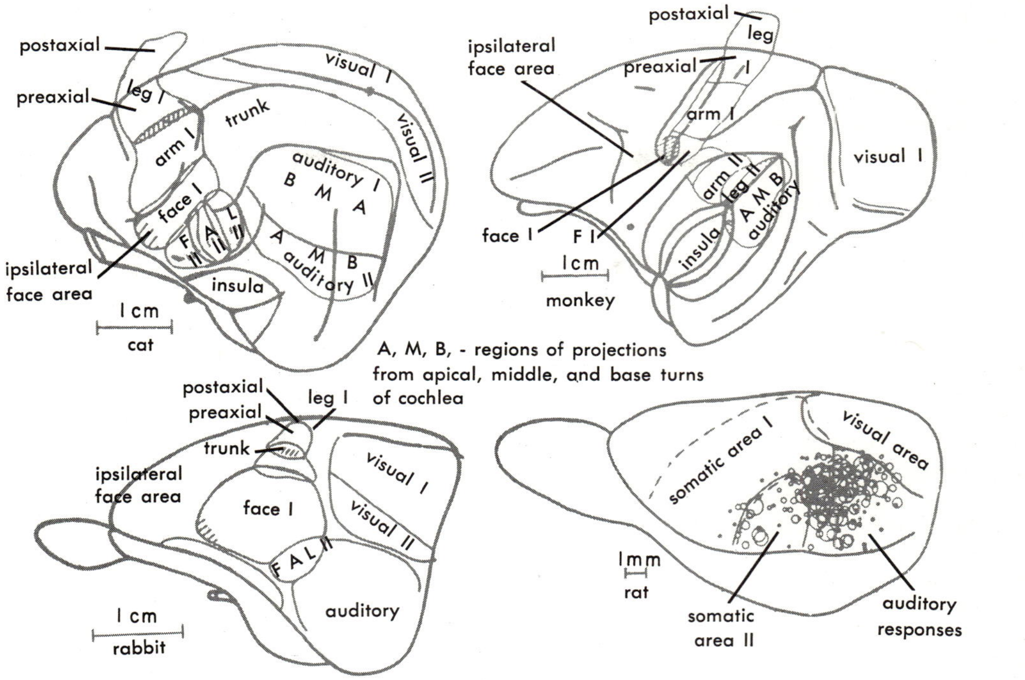

Fig. 2. Lateral view of left hemisphere of brain showing physiological maps of sensory areas of cortex. *(From C. N. Woolsey in Biology of Mental Health and Disease, Hoeber, 1952)*

areas of the cat, monkey, and rabbit (Fig. 2), the parts of the body are separately outlined in both somatic areas I and II. The leg area of somatic area I of cat and monkey extend into the medial surface.

Visual learning. Complete bilateral destruction of the cortical visual areas abolishes practically all visual sensation in man. Partial damage of the visual areas causes a severe sensory loss in a limited area of the visual field. The brain lesions in man usually involve both cortical and subcortical structures. It is not known whether a man with a lesion confined to visual area only can be trained to perceive light intensities. Complete removal of the visual area in monkeys, cats, dogs, and rats causes amnesia of all learned visual discriminations, but the animals can relearn to discriminate between two lights of different intensity. This relearning is probably based on total energy of light rather than on light brightness, that is, energy per unit area. These results indicate that only the memory, not the capacity for intensity learning is localized in the visual area of these animals. Cats can also relearn to discriminate different speeds of movements. Rats with a small portion of the visual area left intact remember form discriminations, such as the difference between pictures of a triangle and a circle.

There are only a few studies on the effect of destroying subcortical centers on visual learning. In rats and monkeys destruction of the superior colliculi, which are midbrain centers controlling visual reflexes, does not affect either the memory of, or the ability to learn, a visual intensity and form discrimination. Lesions in thalamic nuclei (part of the thalamus in the forebrain), other than the anterior nuclear group, cause a slower rate of learning the intensity and form discriminations in rats.

In man unilateral lesions in the posterior parietal cortex, a region outside of the visual area, are responsible for the clinical syndrome, visual agnosia (*see* AGNOSIA). The patient can see an object but not say what it is or what its use might be. A man with a damaged posterior parieto-temporal cortex also learns complex visual perceptual problems more slowly than normal subjects. In monkeys, bilateral ablation of temporal cortex causes an amnesia of learned color and form discriminations, but relearning is possible. Cortical lesions outside the visual area increase the rat's tendency to generalize from one visual stimulus to another. The rats respond more frequently to other, somewhat different, form discriminations as if they are the originally learned one. The common features of these results are that with lesions outside the visual area there is no change of visual capacity and that the neural mechanisms affected pertain to the memory or the meaning of complex visual habits.

Normal chimpanzees, cats, pigeons, and rats immediately recognize with one eye a visual discrimination that has been learned with the other eye. This interocular transfer does not occur in animals that are reared from birth in darkness or in an environment with no other visual stimuli than diffuse light. Such animals can, however, relearn the discrimination with the originally untrained eye and they do so at a faster rate than they showed in the origin of learning. The right and left halves of the brain normally are interconnected by the optic chiasma and the corpus callosum. Cats in which these fiber tracts have been severed do not show interocular transfer. They can relearn the visual discrimination through the untrained eye at a normal rate. This shows that when the major fiber connections between the two cerebral hemispheres are cut, the memory of a visual discrimination learned through one eye is confined to the same side of the brain. In addition, infant chimpanzees reared with only one eye exposed to normal daylight and the other eye to diffuse light also fail to show immediate interocular transfer. It seems that the experience of seeing through both eyes is essential for the fiber connections between the two hemispheres to function normally.

Auditory learning. After bilateral ablation of the auditory cortical areas, cats, dogs, and rats cannot remember discriminations of changes in sound intensity or sound frequency, but can relearn such discriminations at about the normal rate. Cats with these lesions also lose the habit of localization of sound in space. The animal relearns this problem only when the two sounds are separated by a wider angle than that which can be discriminated by normal cats. If the cortical lesion does not destroy all of the auditory cortex, the animals usually remember these problems to some extent. When the lesion includes all the physiologically identified auditory areas, plus some of the surrounding cortex, somatic area II, or the temporal and insular area, a cat may fail to remember but can still relearn a frequency discrimination. Clearly, the memory of these auditory problems is localized in the auditory cortex, but not the capacity to learn them.

The auditory areas are necessary, however, for both the memory of and the capacity to learn a tonal pattern discrimination. A cat with auditory cortices removed cannot be retrained to differentiate two patterns of three tones such as high-low-high from low-high-low.

Of the many subcortical centers in the auditory system, only the inferior colliculi of the mesencephalon have been studied. After destruction either of these structures alone or in addition to an auditory cortical lesion, cats show amnesia for learned intensity discrimination. They are able to relearn such a problem only when the difference in intensities is much larger than the ones they learned before the operation.

Somesthetic learning. In man, a complete destruction of the somatic area I will cause temporary loss of all types of sensation. After some time, feelings of dull pain, pressure, and temperature gradually return in that order. The sense of movement and sense of position in space are permanently impaired or may be totally lost. There is also permanent loss of the ability to discriminate two touch stimuli on the skin as well as the power to differentiate on the basis of touch the superficial quality and form of objects. If the cortical lesion involves a part of the somatic area in one side of the brain, for example, the arm area, only a part of the body opposite to the side of lesion (such as the arm) loses the sensations. Patients with incomplete lesions of this type usually recover the sensations faster and more completely than do people with total destructions. Damage of somatic area II in man does not result in any detectable sensory disturbance. A sensory loss similar to that following removal (ablation) of somatic area I may also occur in human subjects who have cortical damage in the posterior parietal lobe. The clinical syndrome, astereognosis, is caused by a lesion either in the somatic area or in the posterior parietal region (*see* ASTEREOGNOSIS). Patients with these types of brain injury can feel an object such as an apple and can describe it as smooth, cool, and round, but cannot recognize it. They usually have impaired tactile localization and tactile discrimination of two points.

After bilateral ablation of the cortical somatic area I, chimpanzees and monkeys do not remember, but can relearn, weight, roughness, and simple somatic object discriminations. The threshold of roughness and weight discriminations is sometimes slightly increased, for example the animals can only relearn coarse differences in weight and roughness. Rats retain form and temperature discriminations after such a lesion. Both cats and rats lose roughness discrimination and relearn it at an increased threshold. Destruction of the physiologically defined somatic area II has little effect on the memory of somatic discriminations in monkeys and rats, but impairs the ability to relearn a roughness discrimination in cats. When the lesion includes both somatic areas I and II, the effect is like the removal of area I alone in monkeys, and more severe than area I lesion in cats and rats.

Chimpanzees lose the ability to relearn somesthetic discriminations after a cortical lesion including the somatic area I plus the rest of the parietal lobe. Monkeys with such a lesion are able to relearn with difficulty the easy discriminations, such as large differences in weights and simple objects. Cortical lesions outside the somatic areas, either in the posterior parietal or in the parietotemporal region also impair the memory of somatic discrimination in monkeys, an effect similar to that of destroying somatic area I alone. These experimental results indicate that the memory of, and the capacity to learn, somatic problems do not reside exclusively in the somatic areas but also in other temporoparietal areas as well.

Similar to the evidence on interocular transfer, the fiber connectives between the two sides of the brain are necessary for transfer of somesthetic learning from one paw to the other in cats. After section of corpus callosum, cats fail to transfer roughness, softness, and form discriminations from the trained to the untrained paw. They can relearn these problems with the untrained paw at a normal rate.

Olfactory learning. Neither the central neural connections nor cortical areas of smell have been well established. The few studies on olfactory learning indicate that, following lesions in a large number of neural structures, such as septum, amygdaloid, hippocampus, cerebral cortex, and anterior thalamic nuclei, rats retain olfactory discriminations. Only after cutting the olfactory tract is the ability to discriminate odors lost and then it is not a memory loss, but a permanent loss of all olfactory sensation. After bilateral ablation of anterior temporal cortex, however, monkeys forget a discrimination of orange and vanilla odors. They can relearn it only after prolonged training.

Studies on cortical mechanisms of taste have not involved the method of sensory learning as defined above. It will be merely stated in passing that bilateral ablation of physiologically defined cortical taste area in rats causes a decreased discriminatory ability; that is, the animals accept stronger sweet, salt, bitter, and sour tasting solutions than they would normally. In monkeys a similar decreased discriminatory ability occurs after destruction of the anterior insular and frontotemporal region. This area has not been clearly demonstrated as the cortical taste area in monkeys. *See* BRAIN; LEARNING THEORIES. [K.L.C.]

Bibliography: C. T. Morgan and E. Stellar, *Physiological Psychology*, 2d ed., 1950.

Separation (chemical and physical)

A method used in chemistry to purify substances or to isolate them from other substances, for either preparative or analytical purposes. In industrial applications, the ultimate goal is the isolation of a product of given purity, whereas in analysis, the primary goal is the determination of the amount or concentration of that substance in a sample. In principle it is always more convenient to carry out quantitative determinations directly on portions of the original sample. In cases where the analytical methods available are not sufficiently selective to permit this direct approach, it is necessary to employ preliminary separations to reduce the concentration of, or to remove completely, those substances which interfere in the final estimation.

Although special considerations arise in a comparison of separation methods for engineering or laboratory analytical purposes because of differences in the scale of operations, the various separation processes are based on the same principles. There are three factors of importance to be considered in all separations: (1) the completeness of recovery of the substance being isolated, (2) the extent of separation from associated substances, and (3) the efficiency of the separation. The recovery factor or yield R_A of a separation of substance A is defined as

$$R_A = \frac{Q_A}{(Q_A)_0}$$

where Q_A and $(Q_A)_0$ are the amounts of A after and before the separation.

The degree of separation $S_{B/A}$ of two substances A and B is given by the separation factor R_B for B with respect to A, and is defined as

$$S_{B/A} = \frac{(Q_A)_0 Q_B}{(Q_B)_0 Q_A} \cong \frac{Q_B}{(Q_B)_0} \cong R_B$$

Although complete separations are usually preferred, they are not always necessary in analytical applications. The degree of purity will depend upon the choice of the method of final estimation. Sometimes merely a reduction in the quantity of foreign substance present is enough to simplify the subsequent analytical task.

The third factor, efficiency, is a measure of the amount of work required to obtain a given amount of product with a prescribed purity. This consideration is of much greater consequence in industrial separations in which both the scale and the cost of the operation are important.

There are many types of separations based on a variety of properties of materials. Among the most commonly used properties are those involving solubility, volatility, adsorption, and electrical and magnetic effects, although others have been used to advantage. The most efficient separation will obviously be obtained under conditions for which the differences in properties between two substances undergoing separation are at a maximum.

The common aspect of all separation methods is the need for two phases. The desired substance will partition or distribute between the two phases in a definite manner, and the separation is completed by physically separating the two phases. The ratio of the concentrations of a substance in the two phases is called its partition or distribution coefficient.

In analytical work the original phase is usually a liquid, that is, a solution of the sample, and the separation is brought about through the addition or formation of a solid, a liquid, or a gaseous second phase. Although the actual separation of the phases may be physical in nature, chemical reactions are usually required to convert or modify the substance to a form which permits the formation of the new phase or the partition of the substance to the second phase. In some separation methods this step may also be accomplished by physical means.

If two substances have very similar distribution coefficients, many successive steps may be required for a separation. The resulting process is called a fractionation.

Based on the nature of the second phase, the more commonly used methods of separation are classified as follows:

1. Methods involving a solid second phase include precipitation, electrodeposition, chromatography (adsorption), ion exchange, and crystallization. These methods involve a solid second phase either through the formation of a slightly soluble product, deposition as a metal on the surface of an electrode, or by physical or chemical adsorption on a suitable solid material.

2. The outstanding method involving a liquid second phase is solvent extraction, in which the original solution is placed in contact with another liquid phase immiscible with the first. Separations are achieved as a result of differences in the distribution of solutes between the two phases. Solid materials may also be separated by extraction with organic solvents.

3. Methods involving a gaseous second phase include gas evolution, distillation, sublimation, and gas chromatography. Mixtures of volatile substances can often be separated by fractional distillation. *See* EXTRACTION; MASS-TRANSFER OPERATION. [G.H.MO.]

Bibliography: I. M. Kolthoff and E. B. Sandell, *Textbook of Quantitative Inorganic Analysis*, 3d ed., 1952.

Separation (mechanical)

A group of industrial operations by means of which particles of solid or drops of liquid are removed from a gas or liquid, or are separated into individual fractions, or both. Mechanical separations are differentiated from a second large class of separations based on mass transfer, in which homogeneous mixtures and solutions are divided into fractions by vaporization, condensation, precipitation, and diffusion. *See* MASS-TRANSFER OPERATION.

Particle sizes in mechanical separations may range from large chunks of crushed rock and ore to fine dusts and fogs of particle sizes as small as 0.1 μ (1 μ is 10^{-4} cm). The objective may be to eliminate droplets of fog from a stream of process gas, to remove completely solids suspended in a liquid, to separate a mixture of particles by size only, or to sort into pure fractions a mixture of solid particles differing in chemical composition. Examples of these applications are the precipitation of acid mist in a contact sulfuric acid plant; the removal of dust from air, as in an ordinary vacuum cleaner; the removal of solids from a liquid suspension in a filter; the separation by size of a mixture of sand and gravel in a screening plant; and the sorting of the valuable ore from the worthless gangue in ore dressing.

The techniques of mechanical separations are based on physical differences among the particles such as size, shape, density, wettability, and electrical and magnetic properties. Such techniques are applicable to separation of liquids from gases, solids from liquids, and solids from solids. The general methods are to use a sieve, septum, or membrane, such as a screen or filter cloth, which retains one component and allows the other to pass; to use the differential velocities of particles or drops of different sizes or densities through liquids or gases; to use centrifugal force in place of the force of gravity in utilizing density differences or to develop pressure for filtration through a septum; and to use differences in the electrical or magnetic characteristics of the substances.

Separation of solids by size alone is called classification. When the solids are segregated in accordance with their chemical composition, the operation is often called sorting. Both classification and sorting may be achieved in the same equipment. Such units are sorting classifiers. *See* CLASSIFICATION, MECHANICAL.

Because of the differences in the properties of the materials, the wide range in the sizes of the particles, and the variety of the tasks performed by mechanical separators, many separation devices have been invented. Some of the more important methods are shown in the table. The techniques are grouped according to the phases involved.

Types of mechanical separator

Materials separated	Separators
Liquid-gas	Settling chambers, cyclone separators, electrostatic separators, impingement separators
Solid-gas	Bag filters, settling chambers, cyclone separators, electrostatic separators, impingement separators
Liquid-liquid	Gravity decanters, liquid cyclones, centrifugal decanters
Solid-liquid	Filters, clarifiers, thickeners, liquid cyclones, filtering centrifuges
Solid-solid	
Without sorting	Screens, hydraulic classifiers, centrifugal classifiers
With sorting	Hydraulic separators, air separators, jigs and cones, flotation, electrostatic separators, magnetic separators, sink-and-float

Methods using septums. In liquid-solid filters and bag filters, solids are removed from liquids or gases by forcing the fluid by hydrostatic pressure through a filter medium or septum which retains the solids as a cake on the septum and allows the fluid to pass through. The cake is then removed, continuously or periodically, from the septum. The same method is used in filtering centrifuges. In screening, perforated membranes or sieves are used as septums that pass smaller particles and retain larger ones. *See* CENTRIFUGATION; FILTRATION; SCREENING.

Methods based on fluid mechanics. All other methods listed in the table are based on the movement of particles or drops through fluids. Even in the operations listed under solid solid, although the objective is to sort or separate one size or kind of solid from another, the techniques require the presence of a fluid, which is added deliberately to suspend the solid for further manipulation. After the solids have been separated by size or kind, they are removed from the fluid, which is discarded or reused.

Gravity, unaided by other forces, is relied upon to treat relatively coarse particles and drops in settling chambers, gravity decanters, most clarifiers and thickeners, hydraulic classifiers, sink-and-float hydraulic separators, and jigs and cones. Gravity is also the active force in flotation. Here, air bubbles are selectively absorbed by one of the finely divided solid components suspended in a liquid. The effective density of the aerated particles is so reduced that they float to the top of the liquid, and removal of the froth containing these particles completes the separation. *See* CLARIFICATION; FLOTATION; SEDIMENTATION (INDUSTRIAL).

Sink-and-float is an example of hydraulic methods. It is used where the two solids have a considerable difference in density. The two materials, screened to about the same particle size, are suspended in a rising stream of liquid having a density larger than that of one of the solids and smaller than that of the other. The heavier solid sinks through the liquid and settles on the bottom of the equipment. The lighter solid rises through the liquid and floats out of the top of the separator with the liquid. The liquid is then removed from the lighter solid in an external settler and returned for reuse.

Centrifugal force is used in place of gravity when the force of gravity is too weak to give rapid separations. It is used in all centrifuges and also in air separators and cyclone separators. In the latter, a swirling rotary motion is imparted to the liquid or gas carrying the solid or drops. The particulates are thrown out by centrifugal force to the wall of the cyclone, where they are collected and removed. *See* DUST AND MIST COLLECTION.

Impingement separators act on the principle that when a fluid, ordinarily a gas, is given a sharp change in direction, any particulates therein, because of their inertia, do not conform to the new direction of flow, but continue to move nearly in their original direction. Then they strike the solid surfaces of the equipment, which may consist of a bed of solid shapes, a series of baffles, a bed of liquid-covered fibers, or other collecting surfaces. The particles striking these surfaces coalesce and are removed from the gas stream. Many entrainment separators use this principle. The oil-covered glass-fiber units used in household air filters are of this type.

Electrostatic and electromagnetic forces are used to drive very small particles of solid or liquid through gases or liquids at rates impossible to achieve by other driving forces. Electrostatic precipitators are used to remove dust and fog completely from streams of gas. Electromagnetic and electrostatic separators are used to separate substances differing in electric properties and which, because they have nearly the same density, could not be separated by gravity or centrifugal force. *See* ELECTROSTATIC PRECIPITATOR; MAGNETIC SEPARATION METHODS; SEPARATION (CHEMICAL AND PHYSICAL); THICKENING; UNIT OPERATIONS.

[W.L.M.]

Bibliography: W. L. McCabe and J. C. Smith, *Unit Operations of Chemical Engineering*, 1956; J. H. Perry (ed.), *Chemical Engineers' Handbook*, 3d ed., 1950; A. F. Taggart (ed.), *Handbook of Mineral Dressing*, 1945.

Sepioidea

An order of the molluscan subclass Coleoidea. The members of this order are abundant as shore- and bottom-dwelling forms in marine environments. *Sepia*, *Spirula*, and *Sepiola* are common examples of this order. *Voltzia palmeri*, reported from the Upper Jurassic, is the oldest fossil sepioid. In *Sepia*, one pair of tentacles is elongate and capable of retraction into pits at their base. The eye is highly developed in this group, an internal shell is present which in some species (*Spirula*) is partly external, the fins are separated posteriorly, and chromatophores are present in the dermis. The shell of *Sepia* is the cuttlebone of commerce. *See* COLEOIDEA; DECAPODA (MOLLUSCA). [C.B.C.]

Sepiolite

A complex hydrated magnesium silicate ($Mg_3Si_4O_{11}\cdot 4H_2O$) with a type of fibrous crystallization remotely related to that of the amphiboles. In some instances the fibrosity is pronounced, but the typical, more familiar occurrences are massive interlacings of disoriented fibers in aggregates so porous that they float, leading both to the name sepiolite (alluding to cuttlefish bone) and the German name Meerschaum or "sea foam." Sepiolite crystallizes in the monoclinic system. It has a hardness of 2.0–2.5 and a specific gravity of 1–2. The mineral is an alteration product of massive serpentine, and is common, but best known from Asia Minor and Greece.

The soft stone is easily carved, takes a high polish with wax, and hardens when warmed—properties which have led to widespread use for pipes and cigarette holders. [W.F.BR.]

Septibranchia

A small order in the class Pelecypoda, the bivalve mollusks. The anterior and posterior adductor muscles are about equal in size, and the foot is long and slender. The mantle remains partially open and there are two siphons. The gills have disappeared as respiratory organs and have become a pumping mechanism by being transformed into a muscular septum. Oxygen is obtained through the walls of the suprabranchial chamber, the water being conveyed through pores in the muscular septum.

All species in this order are carnivorous and all are marine. They are found in waters varying from a few fathoms to great depths. *See* PELECYPODA.

[W.J.C.]

Septic tank

A single-story settling tank in which settled sludge is in immediate contact with sewage flowing through the tank while solids are being decomposed by anaerobic bacterial action. Such tanks have limited use in municipal treatment, but are the primary resource for the treatment of sewage from individual residences. There are probably well over 4,000,000 septic tanks in use on home disposal systems in the United States. Septic tanks are also used by isolated schools and institutions and for sanitary sewage treatment at small industrial plants.

Home disposal units. Septic tanks have a capacity of approximately 1 day's flow. Since sludge is collected in the same unit additional capacity is provided for sludge. One formula for sludge storage that has been used is $Q = 17 + 7.5y$, where Q is the volume of sludge and scum in gallons per capita per year, and y is the number of years of service without cleaning. About one-half of a 500-gal tank is occupied by sludge in 5 years in an ordinary household installation. The majority of the states require a minimum capacity of 500 gal in a single tank. Some states require a second compartment of 300 gal capacity. Single- and double-compartment tanks are shown in Figs. 1 and 2. Such units are buried in the ground and are forgotten until the system gives trouble due to clogging or overflow. Commercial scavenger companies are available in most areas. A tank truck equipped with pumps is brought to the premises and the tank content is pumped out and taken to a sewer manhole or a treatment plant for disposal. In rural areas the sludge may be buried in an isolated place.

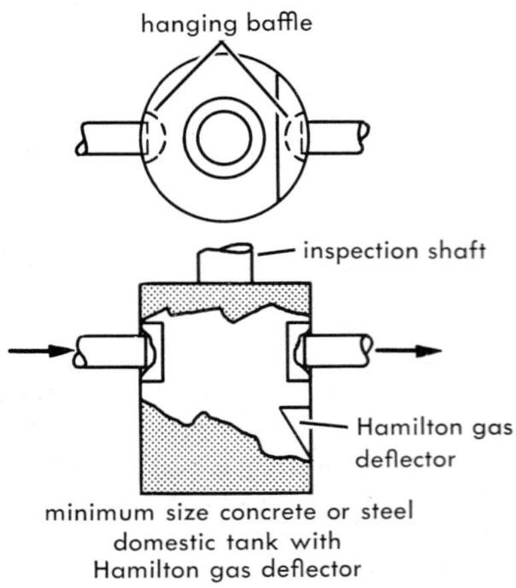

Fig. 1. Circular household septic tank. (*From H. E. Babbitt and E. R. Baumann, Sewerage and Sewage Treatment, 8th ed., Wiley, 1958*)

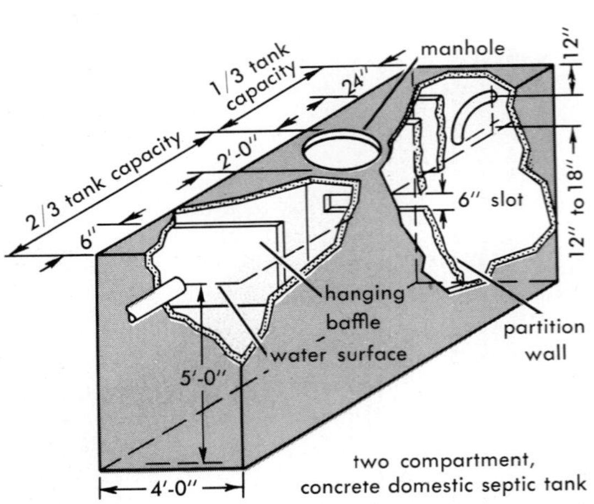

Fig. 2. Two-compartment rectangular household septic tank. (*From H. E. Babbitt and E. R. Baumann, Sewerage and Sewage Treatment, 8th ed., Wiley, 1958*)

Municipal and institutional units. These are designed to hold 12–24 hours' flow, with additional sludge capacity provided. Provision is made for sludge withdrawal about once a year. Desirable features of design are (1) watertight and corrosion-resistant material (concrete, and well-protected metal have been used); (2) the tank must be vented; (3) manhole openings in the roof of the tank to permit inspection; (4) baffles at the inlet and the outlet to a depth below the probable scum line, usually 18–24 in. below the water surface; (5) sludge draw-off lines—although seldom used, they should be designed so that they can be rodded or unplugged by some positive mechanism; (6) hoppers or sloped bottoms so that digested sludge can be withdrawn as required; (7) provision for safe handling of septic tank effluent by disposal underground or by chlorination before discharge to a stream, or both.

Tank efficiency. Septic tank effluent is dangerous and odorous. It will contain pathogenic bacteria and sewage solids. Particles of sludge and scum are trapped in the flow and will cause nuisance at the point of discharge unless properly handled. Efficiency in removal of solids is less than that for plain sedimentation. While 60% suspended solids removal is used theoretically, it is seldom obtained in practice. Improvement is noted when tanks are built with two compartments. Shallow tanks give somewhat better results than very deep tanks. *See* SEWAGE TREATMENT. [W.T.I.]

Bibliography: U.S. Robert A. Taft Sanitary Engineering Center, *Studies on Household Sewage Disposal Systems*, U.S. Public Health Serv., pts. 1, 2, and 3, 1949, 1950, and 1954.

Septicemia

A condition in which infection is disseminated throughout the body in the circulating blood by microorganisms or their poisonous products. The condition has also been called blood poisoning. Bacteria may be recovered by culture of the blood in bacteremia. Virus may also invade the blood stream as in viremia. Frequently there is a recognizable primary focus of infection. For example, in diphtheria, *Corynebacterium diphtheriae* is found in the nose and throat from where the diphtheria toxin is circulated in the blood throughout the body. The principal manifestations are fever, chills, generalized aching, and severe prostration. Prior to the availability of the sulfonamide drugs and antibiotics the fatality rate in septicemia was high; with present-day methods of treatment these infections can often be brought under control. *See* BACTERIOLOGY, MEDICAL; DIPHTHERIA. [P.B.BE.]

Sequoia

The giant sequoia or big tree (*Sequoia gigantia*). occupies a limited area in California and is said to be the oldest and most massive of all living things. The leaves are evergreen, scalelike, and overlapping on the branches. In height sequoia is a close second to the redwood (300–330 ft) but the

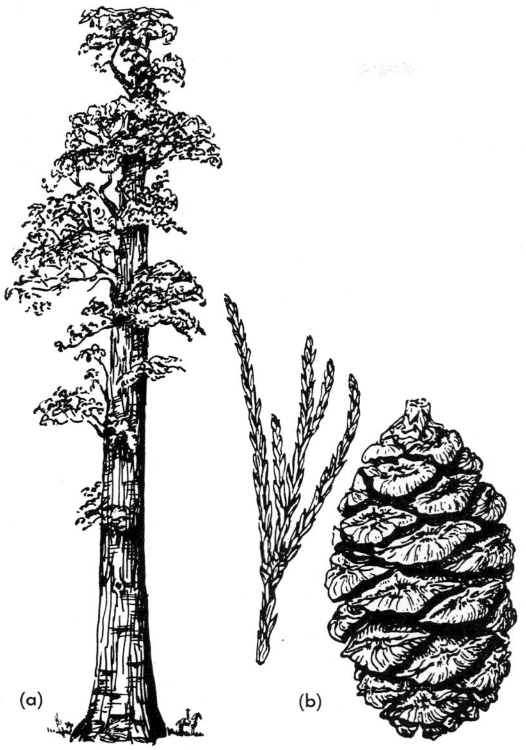

(*a*) *Sequoia gigantia*. (*b*) Overlapping scalelike leaves on branch and a cone. (*USDA*)

trunk is more massive (*see* REDWOOD). Sequoia trees may be 27–30 ft in diameter 10 ft from the ground. The stump of one tree showed 3400 annual rings. *See* STEM (BOTANY). The red-brown bark is 1–2 ft thick and spongy (*see* BARK). Vertical grooves in the trunk give it a fluted appearance. The heartwood is dull purplish-brown, lighter, and more brittle than that of the redwood (*see* XYLEM). Because the loss in felling the trees is so great, the logs so difficult to handle, and the wood so brittle, sequoia is almost no longer marketed. The wood and bark contain much tannin, which is probably the cause of the great resistance to insect and fungus attack. The most magnificent trees are within the General Grant and Sequoia National Parks, where there are many individual trees 25 ft in diameter, containing 500,000 board ft and up, and probably 3000 years old. The largest specimen weighs about 6000 tons. Giant sequoia is sometimes grown in the eastern United States and responds well to cultivation in the British Isles and in Central Europe. *See* FOREST AND FORESTRY; TREE.

[A.H.G.]

Series

The indicated sum of a succession of numbers or terms. Series are used to obtain approximate values of infinite repeating decimals, to solve transcendental equations, to obtain values of logarithms or trigonometric functions, to evaluate integrals, and to solve boundary value problems.

For a finite series, with only a limited number of terms, the sum is found by addition. For an infinite series, with an unlimited number of terms, a sum or value can be assigned only by some limiting process. When the simplest such process yields a value, the infinite series is convergent. There are many tests for convergence which enable one to learn whether a sum can be found without actually finding it.

If each term of an infinite series involves a variable x and the series converges for each value of x in a certain range, the sum will be a function of x. Often the sum is a given function of x, $f(x)$, for which the series having terms of some given form is desired. Thus the Taylor's series expansion

$$f(x) = \sum_{n=0}^{\infty} f^{(n)}(a) \frac{(x-a)^n}{n!}$$

can be found for a large class of functions, the analytic functions, and represents such functions for sufficiently small values of $|x-a|$. For a much less restricted type of function on the interval $-\pi < x < \pi$, a Fourier series expansion of the form

$$\tfrac{1}{2}A_0 + \sum_{n=1}^{\infty} (A_n \cos nx + B_n \sin nx)$$

can be found.

Finite series. Here the problem of interest is to determine the sum of the first n terms,

$$S_n = u_0 + u_1 + u_2 + \cdots + u_{n-1}$$

when u_n is a given function of n. Examples are the arithmetic series, with $u_n = a + nd$ and $S_n = (n/2)[2a + (n-1)d]$, and the geometric series, with $u_n = ar^n$ and $S_n = a(1-r^n)/(1-r)$. *See* PROGRESSION (MATHEMATICS).

If v_n is any function such that $v_{n+1} - v_n = u_n$, then $S_n = v_n - v_0$. For methods of solving this difference equation $\Delta v_n = u_n$, *see* INTERPOLATION. For example, if u_n is any polynomial of the nth degree, then v_n will be a polynomial of the $(n+1)$st degree. In particular,

$$1 + 2 + 3 + \cdots + n = \frac{n(n+1)}{2}$$

$$1^2 + 2^2 + 3^2 + \cdots + n^2 = \frac{n(n+1)(2n+1)}{6}$$

$$1^3 + 2^3 + 3^3 + \cdots + n^3 = \frac{n^2(n+1)^2}{4}$$

As an example with u_n a rational function of n, from

$$\frac{-1}{n+2} - \frac{-1}{n+1} = \frac{1}{(n+2)(n+1)}$$

it may be concluded that if

$$u_n = \frac{1}{(n+1)(n+2)} \quad \text{then} \quad S_n = 1 - \frac{1}{n+1} = \frac{n}{n+1}$$

Convergence and divergence. An infinite series is the indicated sum of an unlimited number of terms

$$u_0 + u_1 + u_2 + \cdots + u_n + \cdots$$

or more briefly $\sum\limits_{n=0}^{\infty} u_n$ or simply Σu_n

read "sigma of u_n." The sum S_n of the first n terms is known as the nth partial sum. Thus S_n is the finite sum

$$\sum_{k=0}^{n-1} u_k$$

If, as n increases indefinitely or becomes infinite, the partial sum S_n approaches a limit S, then the infinite series Σu_n is convergent. S denotes the sum or value of the series. For example, if $|r| < 1$,

$$S = \Sigma ar^n = \frac{a}{1-r} \quad \text{since} \quad S_n = a\frac{1-r^n}{1-r}$$

and

$$S = \Sigma \frac{1}{(n+1)(n+2)} = 1 \quad \text{since} \quad S_n = 1 - \frac{1}{n+1}$$

If, as n becomes infinite, the partial sum S_n does not approach a finite limit, then the infinite series Σu_n is divergent. For example, $\Sigma 1$ diverges, since here $S_n = n$ becomes infinite with n. Also $\Sigma(-1)^n$ diverges, since here $S_n = \frac{1}{2}[1 - (-1)^{n+1}]$ which is alternately one and zero.

It follows from the definition of S_n that $S_{n+1} - S_n = u_n$. For a convergent series, one may take limits in this equality and so deduce that

$$\lim_{n\to\infty} u_n = S - S = 0$$

Thus, if as n becomes infinite, either u_n approaches a nonzero limit or u_n fails to approach a limit, the series must diverge. This checks the earlier conclusion about $\Sigma 1$, with $\lim u_n = 1$, and also that about $\Sigma(-1)^n$, where $u_n = (-1)^n$ does not approach a limit.

A convergent series remains convergent if a finite number of terms is added, removed, or changed either at the beginning or distributed throughout the series. This changes all S_n after a certain point by the same finite constant. Thus the limit S is changed by this same constant, but there is no change in the fact of approach to a limit.

Positive series. These are series each of whose terms is a positive number or zero. For such series, the partial sum S_n increases as n increases. If for some fixed number A, no sum S_n ever exceeds A, the sums are bounded and admit A as an upper bound. In this case, S_n must approach a limit, and the series is convergent. If every fixed number is exceeded by some S_n, the sums are unbounded. In this case, S_n must become positively infinite and the series is divergent. The tests for convergence of positive series are tests for boundedness, and this is shown by a comparison of S_n with the partial sums of another series or with an integral.

The integral test. Let the function $f(x)$ be positive and always decrease as x increases for x greater than m, some fixed positive integer. Then the series $\Sigma f(n)$, with $u_n = f(n)$, converges if the integral

$$\int_m^{\infty} f(x)\,dx$$

converges. The series diverges if the integral diverges. The principal application of this test is that with $f(x) = 1/x^p$, where p is any positive constant. The result is that the series

$$\sum_{n=1}^{\infty} \frac{1}{n^p}$$

converges if p is greater than 1 and diverges if p is less than or equal to 1. Such divergent series as

$$\sum_{n=1}^{\infty} \frac{1}{n} \quad \text{or} \quad \sum_{n=1}^{\infty} \frac{1}{\sqrt{n}}$$

illustrate that series may diverge and still have $\lim u_n = 0$. The slow divergence of

$$\sum_{n=1}^{\infty} \frac{1}{n}$$

may be seen from the fact that over 10^{433} terms must be taken to make the partial sums exceed 1000.

Comparison tests. Let k be any positive constant, Σu_n a positive series to be tested, and Σc_n a positive series known to be convergent. Then if $u_n \leqq kc_n$ for all n greater than some fixed integer m, the series Σu_n converges.

If Σd_n is a positive series known to be divergent, and $u_n \geqq kd_n$ for all n greater than some fixed integer m, then the series Σu_n is divergent.

As corresponding tests involving limits, if

$$\lim_{n\to\infty} \frac{u_n}{c_n} = L$$

where L is a finite limit, then Σu_n converges. But if

$$\lim_{n\to\infty} \frac{u_n}{d_n} = L \quad (L > 0, L = +\infty)$$

then the series Σu_n is divergent.

The ratio test. For positive series, the simple ratio test is based on a consideration of

$$\lim_{n\to\infty} \frac{u_{n+1}}{u_n} = t$$

If t is less than unity, the series converges. If t is greater than unity, or if the ratio becomes positively infinite, the series diverges.

If $t = 1$, no conclusion can be drawn directly. But in many such cases

$$\frac{u_{n+1}}{u_n} = 1 - \frac{b}{n} + \frac{c}{n^2} + \cdots$$

may be written. In this case the series converges if $b > 1$, and diverges if $b \leqq 1$.

Cauchy's test, which is related to the ratio test but depends on a single term, is as follows. If for a positive series

$$\sqrt[n]{u_n} \leqq r < 1 \quad \text{for all } n \text{ greater than } m$$

the series converges. If $\sqrt[n]{u_n} \geqq 1$ for an infinite number of values of n, the series diverges.

Alternating series. These are series whose terms are alternately positive and negative. For such a series, if each term is numerically less than the preceding term, and

$$\lim_{n\to\infty} u_n = 0$$

the series converges. An example is

$$\Sigma(-1)^n \frac{1}{n+1}$$

For such a series, the difference between S_n, the nth partial sum, and S, the sum of the series, is numerically less than the first unused term, u_n.

Absolute convergence. For any series Σu_n which may have both positive and negative terms, the series of absolute values, $\Sigma|u_n|$, is a positive series whose convergence may be proved by one of the tests for positive series. If $\Sigma|u_n|$ converges, then Σu_n necessarily converges and is said to converge absolutely. The sum of an absolutely convergent series is independent of the order of the terms.

Conditional convergence. A series which converges but which does not converge absolutely is said to be conditionally convergent. For such a series, a change in the order of the terms may change the sum or cause divergence. In fact, by a suitable rearrangement, any sum may be obtained. The series $1 - \frac{1}{2} + \frac{1}{3} - \cdots$ is conditionally convergent with sum $\log_e 2$. The rearrangement $1 + \frac{1}{3} - \frac{1}{2} + \frac{1}{5} + \frac{1}{7} - \frac{1}{4} + \cdots$ obtained by taking blocks of two positive terms and then one negative term is conditionally convergent with sum $\frac{3}{2} \log_e 2$.

Operations on series. Two convergent series may always be added termwise to give a convergent series. If $\Sigma u_n = S$ and $\Sigma v_n = T$, then $\Sigma(u_n + v_n) = S + T$. If both series are absolutely convergent, then the double series $\Sigma\Sigma u_m v_n$, where m and n each run from 1 to infinity, converges to the product ST absolutely, and so does any rearrangement. In particular, this is true for the Cauchy product, with $u_0 v_n + u_1 v_{n-1} + \cdots + u_{n-1} v_1 + u_n v_0$ as its $(n+1)$st term.

If Σu_n and Σv_n each converge, and if the Cauchy product series converges, its sum is ST. This will necessarily be the case if at least one of the series converges absolutely.

In any convergent series, parentheses may be inserted to form a new convergent series, with the same sum. But the removal of parentheses may convert a convergent series to a divergent one, for example $\Sigma(1-1) = 0$ becomes $\Sigma(-1)^n$, which diverges.

Power series. These are series with $u_n = a_n x^n$. For such a series, it may happen that

$$\lim_{n\to\infty} \left|\frac{a_{n+1}}{a_n}\right| = A$$

If $A = 0$, the series converges for all values of x. If $A \neq 0$, the series converges for all x of the interval $-1/A < x < 1/A$. It will diverge for all x with $|x| > 1/A$. For any power series, the interval of convergence is related in this way to a number A given by the superior limit of

$$\sqrt[n]{|a_n|}$$

Similar remarks apply to the series with $u_n = a_n(x-c)^n$. Here the interval of convergence is $|x-c| < 1/A$, where A may be

$$\lim \left|\frac{a_{n+1}}{a_n}\right| \quad \text{or} \quad \lim \sqrt[n]{|a_n|}$$

if these limits exist. In any case, A is equal to the superior limit,

$$\overline{\lim} \sqrt[n]{|a_n|}$$

One of the most important power series is the binomial series:

$$1 + mx + \frac{m(m-1)}{1\cdot 2}x^2 + \cdots + \frac{m(m-1)(m-2)\cdots(m-n+1)}{n!}x^n + \cdots$$

When m is a positive integer, this is a finite sum of $m+1$ terms which equals $(1+x)^m$ by the binomial theorem. When m is not a positive integer, the interval of convergence is $-1 < x < 1$, and for x in this interval, the sum of the series is $(1+x)^m$. *See* BINOMIAL THEOREM.

The sum function of a power series is continuous inside the interval of convergence. If the series converges at either end of the interval, the function is continuous at this end. Inside the interval of convergence, a power series may be integrated termwise. At any point inside the interval, it may be differentiated termwise. For example, from the series

$$\frac{1}{1-x} = 1 + x + x^2 + \cdots + x^n + \cdots \quad \text{for } -1 < x < 1$$

differentiation gives

$$\frac{1}{(1-x)^2} = 1 + 2x + \cdots + nx^{n-1} + \cdots$$

Integration gives

$$\log_e(1-x) = -x - \frac{x^2}{2} - \frac{x^3}{3} - \cdots - \frac{x^{n+1}}{n+1} - \cdots$$

Since this converges when $x = 1$, $\log_e 2 = 1 - \frac{1}{2} + \frac{1}{3} - \frac{1}{4} + \cdots$.

Taylor series. Let the power series $\Sigma a_n(x-c)^n$ have the sum function $f(x)$. Then $a_0 = f(c)$, $a_n = f^{(n)}(c)/n!$, and the series is the Taylor series of $f(x)$ at $x = c$. Thus, every power series whose interval of convergence has positive length can be put in the form

$$f(x) = f(c) + f'(c)\frac{x-c}{1!} + \cdots + f^{(n)}(c)\frac{(x-c)^n}{n!} + \cdots$$

where $f(x)$ is the sum function.

The Maclaurin series is the special case of Taylor series with $c = 0$

$$f(x) = f(0) + f'(0)\frac{x}{1!} + \cdots + f^{(n)}(0)\frac{x^n}{n!} + \cdots$$

Remainder term. For any function which is finite together with all of its derivatives for $x = c$, the difference between the function and the first $(n+1)$ terms of its Taylor series is the remainder R_n. For a suitable value x_1 in the interval $c < x_1 < x$,

$$R_n = f^{(n+1)}(x_1)\frac{(x-c)^{n+1}}{(n+1)!}$$

which is Lagrange's form for R_n. This gives $x_1 = c + \theta h$, where $h = x - c$, and θ is a suitable value between 0 and 1. It is true that, if $h \to 0$, $\theta \to 1/(n+2)$. Cauchy's form for R_n is

$$R = f^{(n+1)}(c + \theta h)(1 - \theta)^n \frac{h^{n+1}}{n!}$$

This may be used to prove that the binomial series converges to $(1+x)^m$, by showing that $R_n \to 0$ as $n \to \infty$. Lagrange's form for R_n may be used to show that

$$e^x = 1 + x + \frac{x^2}{2!} + \cdots + \frac{x^n}{n!} + \cdots$$

$$\sin x = x - \frac{x^3}{3!} + \frac{x^5}{5!} - \cdots$$

$$\cos x = 1 - \frac{x^2}{2!} + \frac{x^4}{4!} - \cdots$$

are each Maclaurin series valid for all values of x.

Operations on power series. Two power series $\Sigma a_n(x-c)^n$ and $\Sigma A_n(x-c)^n$ may be added or multiplied by the Cauchy product rule. The resultant series will converge in an interval at least as large as the smaller of the intervals of convergence of the two given series. The first series may be divided by the second, provided that the divisor series is not zero at $x = c$, to give a series for the quotient with some nonzero interval of convergence. A Taylor series with constant term c may be substituted in another series about $x = c$. A power series may be inverted, that is if $y = \Sigma a_n(x-c)^n$, with $a_1 \neq 0$, there is an expansion $x = \Sigma b_n(y - a_0)^n$ where $b_0 = c$, and the other b_n may be found by substitution.

Complex series. The series Σz_n, in which the general term is the complex number $x_n + iy_n$, with $i^2 = -1$, is said to converge to $A + iB$ if $\Sigma x_n = A$ and $\Sigma y_n = B$. Thus the tests for real series may be made on Σx_n and Σy_n, or the positive series $\Sigma|z_n|$ may be tested. If this converges, the given complex series converges absolutely, and necessarily converges.

Complex power series. The series with $u_n = a_n z^n$, where the $a_n = p_n + iq_n$ are now complex numbers, and the complex variable $z = x + iy$, are of great importance in the theory of analytic functions of a complex variable. More generally, power series occur in $(z - c)$, with $u_n = a_n(z-c)^n$, where c is any complex number. There is always a radius of convergence $R = 1/A$, where A may be

$$\lim \left|\frac{a_{n+1}}{a_n}\right| \quad \text{or} \quad \lim \sqrt[n]{|a_n|}$$

if these limits exist. In any case, A is equal to the superior limit

$$\overline{\lim}\ \sqrt[n]{|a_n|}$$

If A is finite and not zero, so is R, and there is a circle of convergence $|z - c| < R$ within which the series converges. If $A = 0$, the series converges for all values of z. If $A = \infty$, $R = 0$, and the series converges for no value except $z = c$. For example, $\Sigma n!z^n$ converges for $z = 0$, but for no other values of z; $\Sigma(z-3)^n/2^n$, with $A = \frac{1}{2}$, $R = 2$, converges [to $2/(5-z)$] for $|z - 3| < 2$; $\Sigma z^n/n!$, with $A = 0$, $R = \infty$, converges (to e^z) for all values of z.

Uniform convergence. Let each term of a series be a function of z, $u_n = g_n(z)$. Let S_n be the sum of the first n terms, and S the sum to which the series converges for a particular value of z. Then $R_n = S - S_n$ is the remainder after n terms, and for the particular value of z, lim R_n must equal zero. If, for a given range of z, it is possible to make $R_n(z)$ arbitrarily small for sufficiently large n without specifying which z in the range is under consideration, the series converges uniformly. The Weierstrass comparison test may be used to test for uniformity. If ΣU_n is a convergent series of positive constants, or a uniformly convergent series of positive terms, and $|u_n| \leqq U_n$ in the range considered, then Σu_n converges uniformly in the range. For a uniformly convergent series, if each $u_n(z)$ is continuous, the sum function $S(z)$ is continuous. A uniformly convergent series may be integrated termwise. If the differentiated series converges uniformly, the series may be differentiated termwise.

Analytic functions. A function $f(z)$ is analytic in a two-dimensional region if, at each point of the region, the derivative $f'(z)$ exists. A uniformly convergent series of analytic functions has a sum function which is analytic and, in particular, may be integrated or differentiated termwise. These results all apply to any power series whose radius of convergence R is not zero, since every power series $\Sigma a_n(z-c)^n$ converges uniformly in any circle $|z - c| < R_1$, where $R_1 < R$. Thus the sum function $f(z) = \Sigma a_n(z-c)^n$ is an analytic function of z for any z such that $|z - c| < R$. For such z, the function $f(z)$ possesses derivatives of every order, each expandable in a series obtained by termwise differentiation. Each such series has the same radius of convergence R. The power series is necessarily the Taylor series of the function $f(z)$, so that $a_n = f^{(n)}(c)/n!$. If a function is single-valued and analytic in a two-dimensional region of the complex plane, and c is any point inside this region, then there is a Taylor expansion $f(z) = \Sigma f^{(n)}(c)(z - c)^n/n!$. Its radius of convergence R is at least as great as the largest circle with center at c, all of whose interior points are interior points of the given region of analyticity.

Fourier series. Let $f(x)$ be a periodic function of period T, so that $f(x+T) = f(x)$. Then the Fourier series for $f(x)$ is

$$A + \sum_{n=1}^{\infty} (A_n \cos n\omega x + B_n \sin n\omega x)$$

where $\omega = 2\pi/T$ and

$$A = \frac{1}{T}\int_a^{a+T} f(x)\,dx \qquad A_n = \frac{2}{T}\int_a^{a+T} f(x) \cos n\omega x\,dx$$

$$B_n = \frac{2}{T}\int_a^{a+T} f(x) \sin n\omega x\,dx$$

That is, A is the average of $f(x)$, A_n is twice the average of $f(x) \cos n\omega x$, and B_n is twice the average of $f(x) \sin n\omega x$ over any interval of length T. Because of the periodicity, it is immaterial which interval, or value of a, is used. If, in each interval of period T, the graph of $f(x)$, which need not be continuous, is made up of arcs which collectively have finite length, then the Fourier series necessarily converges to $f(x)$ at each point where $f(x)$ is continuous. At each point of discontinuity, the series converges to $\frac{1}{2}[f(x-) + f(x+)]$, the average of the right- and left-hand limits. If $f(x)$ is continuous at all points, and has a uniformly bounded derivative, the series will converge uniformly for all values of x.

A Fourier series may always be integrated termwise, but it may not be permissible to differentiate such a series termwise unless $f'(x)$ satisfies some condition for development in a Fourier series.

Fourier sine series. On the interval $0 < x < L$, the Fourier sine series

$$f(x) = \sum_{n=1}^{\infty} B_n \sin n\omega x$$

where

$$\omega = \pi/L \quad \text{and} \quad B_n = \frac{2}{L}\int_0^L f(x) \sin n\omega x\,dx$$

may be considered as the Fourier series for the function $F(x)$ which is odd, $F(x) = -F(-x)$, of period $2L$, $F(x+2L) = F(x)$, and which equals $f(x)$ on the interval $0 < x < L$. Thus if $f(x)$ was an odd function for $-L < x < L$, the sine series is valid inside this interval.

Fourier cosine series. On the interval $0 \leqq x \leqq L$, the Fourier cosine series

$$f(x) = A + \sum_{n=1}^{\infty} A_n \cos n\omega x$$

where $\quad \omega = \pi/L \qquad A = \dfrac{1}{L}\displaystyle\int_0^L f(x)\,dx$

and $\quad A_n = \dfrac{2}{L}\displaystyle\int_0^L f(x) \cos n\omega x\,dx$

may be considered as the Fourier series for the function $F(x)$ which is even, $F(x) = F(-x)$, of period $2L$, $F(x+2L) = F(x)$, and which equals $f(x)$ on the interval $0 < x < L$. Thus if $f(x)$ was an even function for $-L \leqq x \leqq L$, the cosine series is valid inside this interval. The sum of the sine or cosine series is always $\frac{1}{2}[F(x-) + F(x+)]$ for each value of x, so that the sum is $F(x)$ at each point of continuity.

Cesaro summability. Let S_n be the sum of the first n terms of a series Σu_n, and form the sequence

$$C_1 = S_1/1, \quad C_2 = \tfrac{1}{2}(S_1 + S_2), \quad \ldots, \quad C_n = \frac{1}{n}\sum_{k=1}^{n} S_k$$

If

$$\lim_{n\to\infty} C_n = L$$

exists, the series is said to be summable in the sense of Cesaro, or $C(1)$ to L. A convergent series with sum S is necessarily summable $C(1)$ to S, but a divergent oscillating series may be summable $C(1)$. For example, if $u_n = (-1)^n$, the series is $1 - 1 + 1 - \cdots$ which diverges. But $S_k = 1$ and 0 alternately, and $L = \frac{1}{2}$. Thus the series is summable $C(1)$ to $\frac{1}{2}$, a value also made plausible from consideration of the limit as $x \to 1$ of the identity

$$\frac{1}{1+x} = 1 - x + x^2 - x^3 + \cdots$$

There are other more elaborate methods of summability by which sums can be assigned to certain divergent series.

The Fejér theorem. There are continuous functions whose Fourier series do not converge. But, for any continuous periodic function, the theorem of Fejér asserts that the Fourier series is always summable $C(1)$ to the function for every value of x.

Convergence in the mean. Let $f(x)$ be of period T, and

$$T_n = a + \sum_{k=1}^{n-1} (a_k \cos k\omega x + b_k \sin k\omega x)$$

be any trigonometric sum of order $n-1$. Then T_n may be regarded as an approximation to $f(x)$, and the degree of the approximation may be measured by the average of the square of the error,

$$E_n = \frac{1}{T}\int_a^{a+T} [f(x) - T_n(x)]^2\,dx$$

Then

$$E_n = \frac{1}{T}\int_a^{a+T} [f(x)]^2\,dx - A^2 - \tfrac{1}{2}\sum_{k=1}^{n-1}(A_k^2 + B_k^2)$$
$$+ (A-a)^2 + \tfrac{1}{2}\sum_{k=1}^{n-1}(A_k - a_k)^2 + (B_k - b_k)^2$$

This shows that, for all trigonometric sums of given order, S_n, the one formed with Fourier coefficients, $a = A$, $a_k = A_k$, $b_k = B_k$, makes the average error least. Moreover, for any function such that

$$\int_a^{a+T} [f(x)]^2\,dx$$

is finite, as n becomes infinite the limit of E_n, the average squared error for S_n, is zero. Thus

$$\lim_{n\to\infty} \int_a^{a+T} [f(x) - S_n(x)]^2\, dx = 0$$

This is the condition for the trigonometric sums S_n to converge in the mean to $f(x)$.

Integration. If the sequence $S_n(x)$ converges in the mean to $f(x)$, and $g(x)$ is any fixed function, it is true that

$$\lim \int S_n(x) g(x)\, dx = \int f(x) g(x)\, dx$$

for the same limits on the integrals in both members. That is, the series with partial sums $S_n(x)$ may be integrated termwise, and the same is true of the series with partial sums $g(x)S_n(x)$. This fact may be used to derive the formulas for the Fourier coefficients in terms of integrals which were given above.

Again, if $T_n(x)$ converges in the mean to $g(x)$, it is true that

$$\lim \int S_n(x) T_n(x)\, dx = \int f(x) g(x)\, dx$$

with the same limits on the integrals in both members. From this Parseval's identity may be deduced,

$$\int_a^{a+T} f(x)g(x)\, dx = AA' + \tfrac{1}{2} \sum_{k=1}^{\infty} (A_kA_k' + B_kB_k')$$

where A', A_k', B_k' are Fourier coefficients for $g(x)$.

Examples of Fourier series. For $-\pi < x < \pi$, there is a series

$$\sin ax = \frac{2 \sin a\pi}{\pi}\left(\frac{\sin x}{1^2 - a^2} - \frac{2 \sin 2x}{2^2 - a^2} + \frac{3 \sin 3x}{3^2 - a^2} - \cdots\right)$$

The analogous expansion for $\cos ax$, with $x = 0$ and $a = z/\pi$ leads to

$$\csc z = \frac{1}{z} - \frac{2z}{z^2 - \pi^2} + \frac{2z}{z^2 - 2^2\pi^2} - \frac{2z}{z^2 - 3^2\pi^2} + \cdots$$

The expansion for $\cos ax$ is valid for $x = \pi$. With this and $a = z/\pi$, it gives

$$\cot z = \frac{1}{z} + \frac{2z}{z^2 - \pi^2} + \frac{2z}{z^2 - 2^2\pi^2} + \frac{2z}{z^2 - 3^2\pi^2} + \cdots$$

The last two expansions hold for any complex z for which no denominator is zero.

Infinite product for the sine. A series for $\log_e [(\sin z)/z]$ may be found from the expansion of $\cot z - 1/z$ by integration. This leads to

$$\sin z = z\left(1 - \frac{z^2}{\pi^2}\right)\left(1 - \frac{z^2}{2^2\pi^2}\right)\left(1 - \frac{z^2}{3^2\pi^2}\right) \cdots$$

the infinite product for the sine. Putting $z = \pi/2$ leads to

$$\frac{\pi}{2} = \frac{2 \cdot 2 \cdot 4 \cdot 4 \cdot 6 \cdot 6 \cdots}{1 \cdot 3 \cdot 3 \cdot 5 \cdot 5 \cdot 7 \cdots}$$

which is Wallis's product. Equating the z^3 term on the right with the term $-z^3/3!$ in $\sin z$ gives

$$1 + \frac{1}{2^2} + \frac{1}{3^2} + \frac{1}{4^2} + \cdots = \frac{\pi^2}{6}$$

This is a special case of

$$\sum_{n=1}^{\infty} \frac{1}{n^{2k}} = \frac{2^{2k-1}\pi^{2k}}{(-1)^{k-1}(2k)!} B_{2k}$$

where the B_{2k} are rational fractions, the Bernouilli numbers. $B_2 = 1/6$, $B_4 = -1/30$, $B_6 = 1/42$. For n odd, B_n is zero unless $n = 1$, and $B_1 = -1/2$. These B_n are the coefficients in

$$\frac{x}{e^x - 1} = \sum_{n=0}^{\infty} B_n \frac{x^n}{n!}$$

They occur in such expansions as

$$\tan x = \sum_{k=1}^{\infty} \frac{(2^{2k} - 1)2^{2k}(-1)^{k-1}}{(2k)!} B_{2k}x^{2k-1}$$

$$\cot x = \frac{1}{x} + \sum_{k=1}^{\infty} \frac{(-1)^k 2^{2k}}{(2k)!} B_{2k}x^{2k-1}$$

$$\csc x = \frac{1}{x} + \sum_{k=1}^{\infty} \frac{(2 - 2^{2k})(-1)^k}{(2k)!} B_{2k}x^{2k-1}$$

Stirling's formula for m!. The expansion

$$\log_e (m!) = \log_e \sqrt{2\pi} + \left(m + \frac{1}{2}\right)(\log_e m) - m + \sum_{r=1}^{n} \frac{B_{2r}m^{-2r+1}}{2r(2r - 1)} + R_n$$

leads to a divergent series if R_n is omitted and $n \to \infty$. But the series is asymptotic in the sense that R_n is always numerically less than the last term in the sum which involves B_{2n}. Thus for large m, a few terms of the sum give a good approximation. This leads to Stirling's approximation to $m!$, $m! \sim \sqrt{2\pi m}\, m^m e^{-m}$. For $m \to \infty$, the absolute error becomes infinite, but the percentage error is of the order of $e^{1/12m} \sim 1/12m$ which is small even for moderate m.

Applications of series. Series sometimes appear in disguised form in arithmetic. Thus the approximation of a rational number by an infinite repeating decimal is really a geometric series, $1/3 = 0.\dot{3}$ being a series with $u_n = 3/10^{n+1}$ with sum $1/3$. It is possible to use $1 = 0.\dot{9}$ to find the fraction represented. For example $0.\dot{2}8571\dot{4}$ equals $285714/999999 = 2/7$.

Roots are often conveniently found by the binomial series. For instance, for small x, $(1 + x)^{1/n} \sim 1 + (x/n)$, $\log_e (1 + x) \sim x$.

Solution of equations. Sometimes algebraic or transcendental equations are best solved by reverting series. By translations of x and y, $y = y_0 + Y$, $x = x_0 + X$, it is possible to make $Y = 0$ correspond to $X = 0$ near the points of interest. Also, a change of scale, $X = kX'$, makes the first coefficient one. Then with new notation, $y = x + bx^2 + cx^3 + dx^4 + ex^5 + \cdots$, with x and y small. The reverted series is $x = y - by^2 + (2b^2 - c)y^3 - (5b^3 - 5bc + d)y^4 + (14b^4 - 21b^2c + 6bd + 3c^2 - e)y^5 + \cdots$. For example, with $d = 0$, $e = 0$, $y = -q$, this gives a series solution of the cubic equation $cx^3 + bx^2 + x + q = 0$ for small values of q.

Tables. The values of logarithms for tables may be computed by judicious use of the series

$$\log_e \frac{1+x}{1-x} = 2\left(x + \frac{x^3}{3} + \frac{x^5}{5} + \cdots\right)$$

The expansions

$$\sin x = x - \frac{x^3}{3!} + \frac{x^5}{5!} - \cdots$$

$$\cos x = 1 - \frac{x^2}{2!} + \frac{x^4}{4!} - \cdots$$

$$\sin^{-1} x = x + \frac{1}{2}\frac{x^3}{3} + \frac{1 \cdot 3}{2 \cdot 4}\frac{x^5}{5} + \frac{1 \cdot 3 \cdot 5}{2 \cdot 4 \cdot 6}\frac{x^7}{7} + \cdots$$

or

$$\tan^{-1} x = x - \frac{x^3}{3} + \frac{x^5}{5} - \frac{x^7}{7} + \cdots$$

are useful in computing tables of trigonometric functions. The last series makes it possible to compute π easily by using relations like

$$\frac{\pi}{4} = 2 \tan^{-1} \tfrac{1}{3} + \tan^{-1} \tfrac{1}{7}$$

or

$$\frac{\pi}{4} = 4 \tan^{-1} \tfrac{1}{5} - \tan^{-1} \tfrac{1}{239}$$

Integrals may often be found from series. For example

$$\int_0^x e^{-x^2}\,dx = x - \frac{x^3}{3} + \frac{x^5}{5 \cdot 2!} - \frac{x^7}{7 \cdot 3!} + \cdots$$

makes it possible to evaluate the probability integral. For large values of x it is easier to use the divergent, but asymptotic expression

$$\int_0^x e^{-x^2}\,dx =$$

$$\frac{\sqrt{\pi}}{2} - e^{-x^2}\left(\frac{1}{2x} - \frac{1}{2^2x^3} + \frac{3}{2^3x^5} - \frac{3 \cdot 5}{2^4x^7} + \cdots\right)$$

Ordinary differential equations. An ordinary point of a linear differential equation is a value of x at which all the coefficients are analytic, with the first coefficient not zero. For x near such a value, the complete solution may be expressed in terms of Taylor series. For example, $x = 0$ is an ordinary point of

$$(1 - x^2)\frac{d^2y}{dx^2} - 2x\frac{dy}{dx} + n(n+1)y = 0$$

which is Legendre's equation. For n zero or an integer, one solution is

$$P_n(x) = \sum_{k=0}^{\infty} (-1)^k \frac{(2n-2k)!}{2^n k!(n-k)!(n-2k)!}$$

the Legendre polynomial of degree n.

Fourier-Legendre series. If, for $-1 < x < 1$, the function $f(x)$ satisfies the conditions for expansion in a Fourier series, there is an expansion

$$f(x) = \sum_{n=0}^{\infty} a_n P_n(x)$$

where

$$a_n = \frac{2n+1}{2}\int_{-1}^{1} f(x)P_n(x)\,dx$$

The sum to $(n+1)$ terms gives the polynomial of the nth degree best approximating $f(x)$ in the sense of least square error.

Regular singular point. Let a differential equation have the form

$$A_2\frac{d^2y}{dx^2} + A_1\frac{dy}{dx} + A_0y = 0$$

where $A_2(x)$ is analytic and not zero at x_0, $(x - x_0)A_1(x)$ and $(x - x_0)^2A_0(x)$ are each analytic at x_0. Then if x_0 is not an ordinary point, it is a regular singular point. Near such a point, at least one solution may be found in the form $(x - x_0)^s$ times a Taylor series, where s is some real or complex value. For example, for n zero or any positive integer, the Bessel function of order n

$$J_n(x) = \sum_{k=0}^{\infty} \frac{(-1)^k x^{n+2k}}{2^{n+2k}k!(n+k)!}$$

is a solution of Bessel's differential equation

$$\frac{d^2y}{dx^2} + \frac{1}{x}\frac{dy}{dx} + \left(1 - \frac{n^2}{x^2}\right)y = 0$$

Partial differential equations. Certain boundary-value problems in partial differential equations may be solved by the use of series. Thus, in two dimensions, Laplace's equation in polar coordinates is

$$\frac{\partial^2 U}{\partial r^2} + \frac{1}{r^2}\frac{\partial^2 U}{\partial \theta^2} + \frac{1}{r}\frac{\partial U}{\partial r} = 0$$

Let the values on the boundary of a circular region $r = a$ be given in the form $f(\theta)$, and let the Fourier expansion of this function of period 2π be

$$f(\theta) = A + \sum_{n=1}^{\infty} (A_n \cos n\theta + B_n \sin n\theta)$$

Then the solution of Laplace's equation with $U(a,\theta) = f(\theta)$ is

$$U(r,\theta) = A + \sum_{n=1}^{\infty} \left(\frac{r}{a}\right)^n (A_n \cos n\theta + B_n \sin n\theta)$$

Again, the solution of Laplace's equation

$$\frac{\partial^2 U}{\partial x^2} + \frac{\partial^2 U}{\partial y^2} = 0$$

with $U(0,y) = 0$, $U(L,y) = 0$, $U(x,+\infty) = 0$, $U(x,0) = f(x)$, with Fourier sine expansion

$$f(x) = \sum_{n=1}^{\infty} B_n \sin \frac{n\pi x}{L}$$

is given by

$$U(x,y) = \sum_{n=1}^{\infty} B_n e^{-n\pi y/L} \sin \frac{n\pi x}{L}$$

See FOURIER SERIES AND INTEGRALS. [P.FR.]

Bibliography: P. Franklin, *Differential and Integral Calculus*, 1953; W. Kaplan, *Advanced Calcu-*

lus, 1952; I. S. Sokolnikoff and R. M. Redheffer, *Mathematics of Physics and Modern Engineering*, 1958.

Series circuit

An electric circuit in which the principal circuit elements have their terminals joined in sequence so that a common current flows through all of the elements. The circuit may consist of any number of passive and active elements, such as resistors, inductors, capacitors, electron tubes, and transistors.

The algebraic sum of the voltage drops across each of the circuit elements of the series circuit must equal the algebraic sum of the applied voltages. This rule is known as Kirchhoff's second law and is of fundamental importance in electric circuit theory. *See* KIRCHHOFF'S LAWS OF ELECTRIC CIRCUITS.

When time-varying voltages and currents are involved, it is necessary to employ differential or integral equations to express the summation of voltages about a series circuit. If they are varying sinusoidally with time, functions of a complex variable are used in place of the calculus. *See* ALTERNATING-CURRENT CIRCUIT THEORY; CIRCUIT, ELECTRIC; DIRECT-CURRENT CIRCUIT THEORY.

[R.L.R.]

Serine

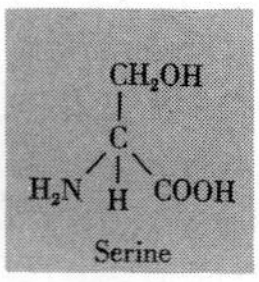

Serine

Physical constants of the L isomer at 25°C:

pK_1 (COOH): 2.21; pK_2 (NH_3^+): 9.15

Isoelectric point: 5.68

Optical rotation: $[\alpha]_D(H_2O)$: −7.5; $[\alpha]_D(5\,N\,HCl)$: +15.1

Solubility (g/100 ml H_2O): 5.02 (DL)

An amino acid. The amino acids are characterized physically by the following: (1) the pK_1, or the dissociation constant of the various titratable groups; (2) the isoelectric point, or pH at which a dipolar ion does not migrate in an electric field; (3) the optical rotation, or the rotation imparted to a beam of plane-polarized light (frequently the D line of the sodium spectrum) passing through 1 decimeter of a solution of 100 grams in 100 ml; (4) solubility. *See* EQUILIBRIUM, IONIC; ISOELECTRIC POINT; OPTICAL ACTIVITY; SPECTROPHOTOMETRIC ANALYSIS.

Serine reacts with periodate to yield glyoxylate, ammonia, and formaldehyde. Serine is a biosynthetic precursor of several important metabolites: glycine, cysteine, choline (and hence betaine), and the side-chain of tryptophan. D-Serine is a constituent of the antibiotic, polymyxin. Serine originates, biosynthetically, from 3-phosphoglyceric acid (*see* AMINO ACIDS). Since glycine and serine are interconverted rapidly, the pathway glyoxylate → glycine → serine can furnish some serine also. *See* CYSTEINE; GLYCINE; POLYMYXIN; TRYPTOPHAN.

During metabolic degradation, serine deaminase (dehydrase) nonoxidatively attacks serine to yield pyruvate and ammonia. Another major pathway starts with transfer of the hydroxymethyl group to tetrahydrofolic acid, yielding glycine. Finally, a transaminase for serine exists, forming β-hydroxypyruvate. This compound may be convertible to glyceric acid, or the actual pathway may be the reversal of the biosynthetic one, involving phosphorylated intermediates.

[E.A.AD.]

Serology

The division of biological science concerned with certain properties of serum, specifically, the antigen-antibody reactions. Although the term serology is properly used in connection with any of these reactions, it is often used in a more limited sense to denote merely the laboratory diagnostic tests, especially those for syphilis. While serological reactions take place in many natural processes of immunity and are also utilized in the laboratory for evaluating the immune state, they are not limited to the products of pathogenic microorganisms and can be utilized for a variety of other scientific purposes, such as comparative taxonomy and study of the structure of proteins and polysaccharides. Serology, although useful in the study of immunology, is not synonymous with the latter, which includes many factors of resistance other than those associated with the serum reactions. *See* ANTIGEN-ANTIBODY REACTION; IMMUNITY; POLYSACCHARIDE; PROTEIN; SYPHILIS.

[H.P.T.]

Serotonin

Serotonin, 5-hydroxytryptamine, is a derivative of tryptophan, an indole-containing amino acid. It is widely distributed among all animal species examined, particularly in the blood and gastric mucosa of mammals. It is found in small amounts in the brain of mammals, birds, and reptiles. The occurrence of similar compounds in plants and bacteria testifies to their general biological importance, while the occurrence in the urine of dog and man of 5-hydroxyindoleacetic acid identifies another pathway in the metabolism of tryptophan. Serotonin plays an important role in brain and nerve function and possesses many significant pharmacological properties. *See* BRAIN; NERVOUS SYSTEM; TRYPTOPHAN.

Metabolism. The formation of serotonin in the animal involves first the hydroxylation of tryptophan by the enzyme tryptophan hydroxylase, which has been isolated from a bacterium, *Chromobacterium violaceum*. The enzyme involved in the second step in the synthesis, 5-hydroxytryptophan decarboxylase, is found in relatively large amounts in the liver, kidney, and stomach. It is highly specific in its action, which is restricted to only one of the stereoisomers, 5-hydroxy-L-tryptophan. The vitamin pyridoxine is required for this reaction. The low levels of serotonin in pyridoxine-deficient chicks may well be associated with the neurological symptoms characteristic of pyridoxine deficiency. The removal of serotonin from the tissues (by its catabolism) involves its oxidative deamination to 5-hydroxyindoleacetic acid, which is excreted in the urine. The enzymes that catalyze these terminal reactions are known, and some of their properties have been described. *See* ENZYME.

Function in brain. The brain normally contains about 0.55 μg of serotonin per gram, mainly in a bound form which protects it from the highly active enzyme, monoamine oxidase. On treatment with certain drugs, brain cells partially lose their capacity to retain serotonin, which is then liberated and rapidly catabolized. Intraventricular administration of serotonin produces a lethargic state and other largely depressant conditions in the nervous system, suggesting that the compound plays an important role in brain function, possibly as a neurohumoral agent.

Pharmacological action. The interaction of serotonin with certain alkaloids is of considerable clinical importance, because of the use of such drugs as tranquilizing agents in the treatment of nervous and mental disorders. Of these, reserpine and some of its derivatives are of particular importance because, when administered to experimental animals, they release serotonin from its bound form in various tissues (brain, intestine, and blood platelets); thus serotonin mediates the prolonged tranquilizing effects of these drugs. *See* TRANQUILIZER.

[H.H.MI.]

Serpentine

The name applied to the hydrous magnesium silicate mineral assemblage of the rock serpentinite. Serpentinization alters host rocks to compositions approximating $3MgO{\cdot}2SiO_2{\cdot}2H_2O$, and the names serpentine, chrysotile, antigorite, and others are applied with regard to degree of crystallinity achieved locally in given serpentinized groundmasses. *See* SERPENTINITE.

The state of the solid matter in massive serpentine has certain aspects in common with synthetic polymers. It is permeated with crystalline nucleations insufficiently articulated to meet the criteria of true crystals. Associated with the massive assemblage, substantially the same compositions, possibly under the influence of some mechanical stress, become organized into varying degrees of improved alignments and have received the various names of the serpentine minerals group.

The most important serpentine mineral is the fibrous variety, chrysotile (Fig. 1), which accounts for as much as 95% of the asbestos of commerce. Although chrysotile is not a crystal in the sense of having plane-bounding faces, it nevertheless has a crystal structure consisting of regularly arrayed sheets of component atoms wrapped about a less well-organized or possibly even unoccupied core. The high tensile strength of these tubular fibers permits the shredding, spinning, and fabrication of useful asbestos products (Fig. 2). *See* ASBESTOS.

Other named varieties, such as antigorite, bastite, marmolite, and picrolite, consist of comparable regularly-arrayed sheets, more or less corrugated and stacked colinearly to present a more platy, or at least lathlike, aspect. Some of these platy varieties

Fig. 1. Veins of chrysotile asbestos in serpentine. (Canadian Johns-Manville Co., Ltd.)

Fig. 2. Silky fibers of asbestos-bearing rock. (Canadian Johns-Manville Co., Ltd.)

may involve a pseudomorphism after host crystals which imposed some control onto the product to which they altered. The platy varieties are not currently known to have any value in themselves.

Serpentine is easily cut and polished for ornamental stone. It may be blackish-green through leek green to nearly white, or brownish or yellowish, and exhibits attractive textures because of the disseminated filamentous crystalline nucleations. Serpentinized carbonate rocks sometimes show an attractive clouded green color and are then called verde antique, or serpentine marble. [W.F.BR.]

Serpentinite

An abundantly occurring, fine-grained massive rock, generally considered to have been derived by pneumatolytic or hydrothermal processes from preexisting basic and ultrabasic igneous intrusions. The inferred processes are called serpentinization and the product serpentine. The serpentinization of metamorphic rocks, and of limestones and dolomites, is common. The composition of serpentine approximates $3MgO{\cdot}2SiO_2{\cdot}2H_2O$. The groundmass itself and the varietal associated minerals of the same composition are called the serpentine minerals. *See* METAMORPHIC ROCKS; PNEUMATOLYSIS.

Serpentine differs from the more conventional rocks—those composed of separable grains of various minerals—for serpentine has no idiomorphic crystal grains of its own. When grains are apparent, they are relicts from the preexisting crystalline rock which was altered to serpentine while retaining the old grain boundaries (called pseudomorphs after the older grain). Pseudomorphs are most frequent after chrysolite (olivine), common after amphibole and pyroxenes, and have been noted after other minerals.

Massive serpentinites often form layers, dikes, or necks (composed more or less of pure serpentine) in various positions in crystalline schists. Many serpentinites develop by the low-grade metamorphism, or hydrothermal alteration, of olivine-rich rocks, especially peridotites and dunites in mountain ranges. Experiments have shown that at low pressure, magnesian olivine is stable only above 400°C when it is in contact with water vapor; below that temperature it changes into serpentine and brucite. Iron-rich olivine is stable in the presence of water at much lower temperatures, and does not readily change into serpentine.

In many rocks where serpentine has formed pseudomorphs after olivine, traces of the original mineral may still be present as mesh structure. Other rocks are composed wholly of fibrous chrysotile serpentines and still others are made up of flaky antigorite which developed under stress conditions. Sometimes the flakes and blades of antigorite are arranged at random in sheaflike bundles. See SERPENTINE. [T.F.W.B.]

Serum

The thin, yellowish fluid residue of the blood that remains after the blood cells and clotting elements have been removed. (Plasma is the fluid portion of the blood without the blood cells but with the fibrinogen and the other clotting factors, such as prothrombin and calcium.) The serum contains numerous soluble organic and inorganic substances such as glucose, albumin, certain globulins, salts, and buffer systems. Many of the globulins present are either antibodies or contain antibodies. These have been formed by various body tissues in response to antigenic stimulation by bacteria and other foreign materials, notably proteins. See ANTIBODY; ANTIGEN; BLOOD; FIBRINOGEN; GLOBULIN; GLUCOSE.

The complex and technical field of serology is based upon the occurrence and alteration of the substances present in serum. Common serologic tests include blood typing and diagnostic tests for certain diseases, such as syphilis. See SEROLOGY; SYPHILIS.

Serum also refers to any watery fluid produced by secretion of, or damage to, body cells. It includes the watery secretions of the mouth, lacrimal glands, nose, and portions of the digestive tract. It also includes the clear material found in blisters and seeping from the skin after an abrasion. Finally, a serous fluid is normally secreted by serous membranes such as the peritoneum and pericardium; one apparent function is to lubricate opposing surfaces to reduce friction. [E.G.ST.]

Servomechanism

A special type of closed-loop control system capable of producing a controlled output motion at a high energy or power level in response to a low energy level input signal. The block diagram of Fig. 1 shows the basic elements of a single-loop servomechanism and the manner in which they interact with each other. In this simple example, each connection between elements is such that only a unidirectional cause and effect action takes place in the directions shown by the arrow.

The servomechanism provides the amplification and controllable power needed to move large masses rapidly in the presence of large load disturbances. In addition, a servomechanism employs negative feedback, which makes possible a comparison of the reference input signal R with the measured value of the controlled output motion C'. A low energy level error signal E is thereby produced. The servoamplifier and servomotor respond to the error signal to cause the output motion C to approach the desired value.

The reference input signal R frequently exists in the form of an electric voltage or current. It may also exist in the form of any one of many

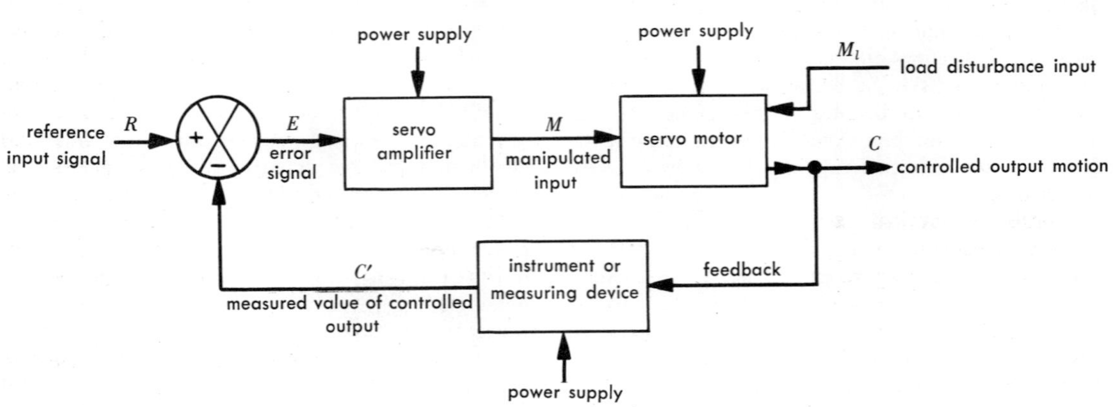

Fig. 1. Block diagram showing the basic elements of a servomechanism.

different physical phenomena, such as a hydraulic or pneumatic pressure or rate of flow, mechanical motion or force, heat flow rate or temperature, or intensity of nuclear radiation. The output of a servomechanism, however, is always a motion of a shaft, lever, linkage, or similar mechanical element. Thus a servomechanism is a special type of closed-loop control system and most of the important basic concepts involved in the operation of closed-loop control systems are directly applicable to servomechanisms. For discussion of closed-loop control systems, *see* CONTROL SYSTEMS.

SERVOMECHANISM SYSTEMS

When the measuring device generates a signal C', representing a measured value of the position of the output member of the servomotor, a position servomechanism results, and the output C is a position which changes in response to changes in the reference input signal R.

When the measuring device generates a signal C', representing a measured value of the velocity of the output member of the servomotor, a velocity servomechanism results, and the output C is a time rate of change of position which changes in response to changes in the reference input R.

When the measuring device generates a signal C', representing a measured value of the acceleration of the output member of the servomotor, an acceleration servomechanism results, and the output C is a time rate of change of velocity which changes in response to changes in the reference input R.

To attain the best possible performance from the servomechanism, the measuring device must have a high degree of accuracy in the static sense (steady-state) and it must have negligible dynamic lag in comparison with the dynamic response characteristics of the rest of the system. The quality of this type of control system can never exceed that of the instrument which measures the output motion.

Problems of greatest concern in the design and operation of servomechanisms are usually those related to (1) system speed of response, (2) closed-loop stability, (3) sensitivity to load disturbances including noise in the input signal, (4) power-supply requirements, (5) peak signal levels which may occur throughout the system, and (6) operation under conditions in which nonlinear effects are important, such as limiting or saturation of components, on-off or relay type of amplification, coulomb or static friction, and nonlinear valve characteristics.

Electromechanical servomechanism. In an electromechanical system, the reference input signal is an electric voltage or current, and the servoamplifier, the servomotor, and the measuring device are electric, electronic, electromechanical, or mechanical devices.

The input signal may be an alternating current or voltage (ac) signal or a direct current or voltage (dc) signal. The measured value of the output, sometimes called the feedback signal, must be of the same type as the input signal. The difference between the reference input signal and the feedback signal is usually obtained by means of an electrical summing network. The sign or phase of the feedback signal must be negative with respect to the input signal in order to obtain the desired difference with a summing network. With ac signals, the frequency of the carrier must be well above the highest frequency anticipated in the motion of the output shaft of the servomechanism, and the feedback signal carrier must be carefully phased with respect to the input signal. Harmonic distortion in either signal must be kept to a minimum.

The servoamplifier is often electronic but may be a magnetic amplifier, a relay type of amplifier, or any combination of these types. In addition to providing needed power amplification, the servoamplifier may incorporate dynamic characteristics, sometimes referred to as compensation, required for accurate, stable operation of the complete servomechanism. Because the servoamplifier plays roughly the same role in a servomechanism that a controller plays in many types of automatic control systems, the servoamplifier is sometimes referred to as the controller. The electric output power supplied by the servoamplifier to the servomotor may be ac or dc in accordance with the type of servomotor employed.

The servomotor may be a conventional type of ac or dc motor, but more often it is of special design to meet demanding requirements of speed of response, peak load, small size, minimum weight, and high accuracy.

The instrument, or measuring device, is usually a transducer operating at a low energy level and capable of producing an electric output signal that is an accurate measure of output shaft position, velocity, or acceleration. *See* TRANSDUCER.

Because of the ready availability of electric power, simplicity, ease of manufacture and maintenance, and low cost of electromechanical servomechanisms, they are widely used in applications involving the control of low to medium mechanical power level output motions.

Hydraulic servomechanism. In a hydraulic system the servomotor includes a hydraulic motor, converting high-pressure hydraulic power to the mechanical power required in controlling the output motion of the servomechanism. When the input signal to the servomechanism is electrical and the servoamplifier is electric or electronic, the system is an electrohydraulic servomechanism. The hydraulic servomotor involves, in addition to the hydraulic motor, a means of modulating the flow of hydraulic power to the motor in response to the low energy level mechanical or electrical output signal from the servoamplifier.

For low to medium power level hydraulic drives, valve-controlled hydraulic servomotors are most frequently employed. The control valve is often referred to as a servovalve and, in an electrohydraulic servomechanism, an electromechanical transducer sometimes called a force motor, or torque motor, is employed to actuate the servovalve. High-pressure hydraulic power from a suitable source, usually at constant pressure, must be supplied to a valve-controlled servomotor.

For higher power level hydraulic drives, pump displacement controlled hydraulic transmissions are often used, in which the flow of power to the hydraulic motor is modulated by varying the displacement of a variable displacement hydraulic pump. High energy level mechanical power from a suitable source, usually at constant speed, must be supplied to drive the shaft of the pump. The displacement control member of the pump, usually a shaft or lever, is often actuated by a low power level electric or hydraulic actuator.

A complete hydraulic servomechanism is often designed as a single unit and referred to as a hydraulic amplifier or hydraulic actuator when it is used as a component of another control system.

Hydraulic servomechanisms are frequently used in applications involving the control of medium to high mechanical power level output motions because of their relatively small size, light weight, and fast response characteristics. They are sometimes required in systems where the use of electricity is forbidden by fire or explosion hazards.

Pneumatic servomechanism. In a pneumatic system the servomotor includes a pneumatic motor, which converts pneumatic power to the mechanical power required in controlling the output motion of the servomechanism. When the input signal to the servomechanism is electrical and the servoamplifier is electric or electronic, the system is an electropneumatic servomechanism. The pneumatic servomotor includes, in addition to the pneumatic motor, a control valve to modulate the flow of pneumatic power to the motor in response to the low energy level mechanical or electrical output signal from the servoamplifier. Pneumatic power from a suitable source, usually at constant pressure, must be supplied to the servomotor.

A complete pneumatic servomechanism is often designed as a single unit, and it is referred to as a pneumatic amplifier or pneumatic actuator when it is used as a component of another control system.

Pneumatic servomechanisms are frequently used in applications involving the control of medium mechanical power level output motions because of their relatively small size, light weight, and fast response characteristics. They are sometimes required in systems where the use of electricity is forbidden, because of fire or explosion hazards, and they can be designed to operate over a wide range of working temperatures.

SERVOMECHANISM COMPONENTS

Various devices or elements are required to form the complete closed loop of a servomechanism. In many instances, the components employed in servomechanisms are commercially available devices which have been developed for general industrial and laboratory uses. In some instances, however, the components required are highly specialized devices, developed to meet specific and often exacting requirements. Great care must be taken to understand thoroughly the performance characteristics of servomechanism components as individual devices and as interacting components in a given system. The use of components that are incompatible with each other results in inferior performance of the complete servomechanism.

Measuring devices. These components are the instruments, transducers, or pickups employed to measure the output position, velocity, or acceleration of servomechanisms. It is usually desirable to determine the value of the output quantity without appreciably affecting the output, that is, without imposing a noticeable load on the output shaft of the servomotor. The accuracy of the measuring device must be at least as good as the accuracy required of the complete servomechanism. Some measuring devices can be used as an error detector. The term error detector usually means measurement of input, measurement of output, and determination of the difference between measured input and measured output. Although measuring devices are expected to be quite sensitive to the output quantity to be measured, they should be relatively insensitive to all other effects which might cause an erroneous or noisy error signal in the servomechanism.

Electrical measuring devices. The output of these devices is an electric signal representing the position, velocity, or acceleration of the output shaft of the servomechanism. In these devices one or more electrical quantities are influenced by the output condition to be measured. These devices, used as elements in an electric circuit, develop a usable electric signal, which is a measure of the output condition. The various kinds of electrical measuring devices may be classified according to the predominant electrical quantity sensitive to the output condition to be measured.

Variable resistance devices. The variation of the servomechanism output can be used to cause a variation in the resistance of a variable resistance element. A useful electric signal can, in some instances, be derived directly from the variable resistance element. For example, if it is supplied with a constant current from a suitable electric energy source, the voltage across the variable resistance element is directly proportional to its resistance; or if a constant voltage is supplied to the variable resistance element, the current flowing through it is inversely proportional to its resistance.

More frequently, however, the variable resistance element is combined with other circuit elements into networks, making it easier to generate a drift- and noise-free electric signal as a measure of the servomechanism output. A bridge-type network is often used, in which one or more variable resistance elements are sensitive to the servomechanism output and one or more variable resistance elements are sensitive to the reference input (*see* BRIDGE

CIRCUIT). These combine to form an error detector whose output is an electric signal that is a measure of the difference between the reference input and the controlled output of the servomechanism. The most commonly employed types of variable resistance elements are precision potentiometers and strain gage pickups. These devices are usually employed for position measurement. *See* POTENTIOMETER (VARIABLE RESISTOR); STRAIN GAGE.

Variable inductance devices. Sometimes referred to as differential transformers, these employ the variation of servomechanism output to cause variation of the mutual inductance between a primary winding and one or more secondary windings, thereby varying the voltages induced in the secondary windings (*see* TRANSFORMER).

Linear differential transformers employ a movable slug of magnetic material with fixed primary and secondary coils. The slug moves along a straight line inside the coils as in Fig. 2. The secondary coils are connected in series opposing. The output voltage from the secondaries is, therefore, theoretically zero when the slug is centered, and it varies linearly with slug position when it is moved from its center position.

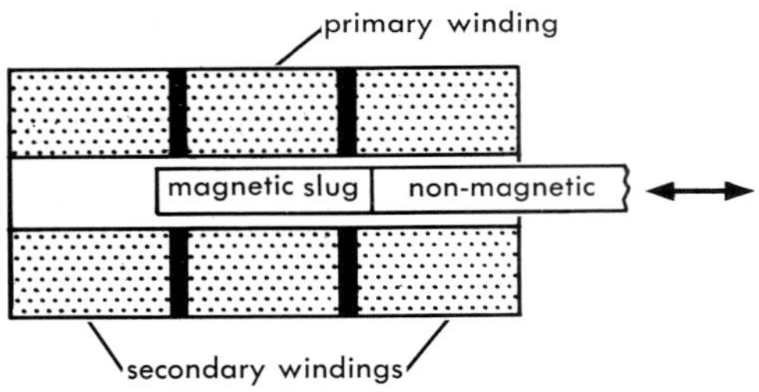

Fig. 2. Schematic of a linear differential transformer. (*From J. E. Gibson and F. B. Tuteur, Control System Components, McGraw-Hill, 1958*)

E pickoffs employ a movable arm of magnetic material and an E-shaped magnetic core containing a primary and two secondary windings as in Fig. 3. The operation of this device is analogous to that of the linear differential transformer. The motion of the arm is limited by the air gaps between the arm and the E-shaped core; therefore this device is usually employed to measure only small linear or angular motions.

To measure large angular motions, a considerably more sophisticated type of variable transformer, called a synchro, is used. A synchro has a single primary winding on a rotor, which can rotate through a full 360°, and a 3-winding stator or secondary. The signals induced in the stator windings vary with the position of the rotor and can therefore be used to measure rotation of the rotor. *See* SYNCHRO.

Variable capacitance devices. If one or more variable capacitors are coupled to the output of the servomechanism, the output motion results in change in capacitance. When incorporated into a bridge-type network, operating with an alternating-

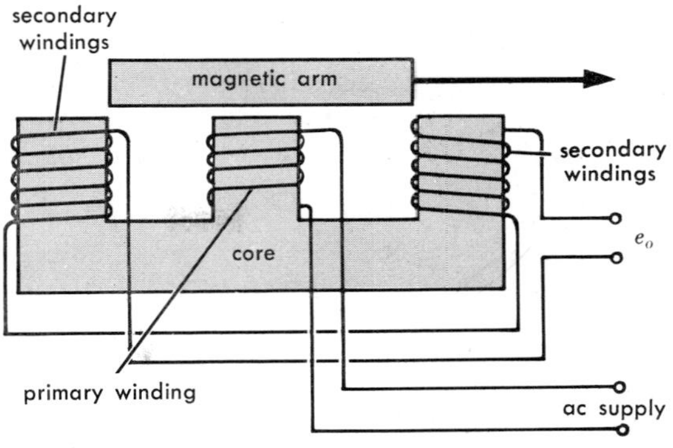

Fig. 3. E pickoff.

current supply, the variation of capacitance of one or more arms of the bridge results in a varying alternating-current signal, which is a measure of the output position of the servomechanism.

Electromagnetic devices. These devices involve relative motion between magnetic fields and electric conductors. The voltage induced in a conductor moving with its axis normal to the lines of magnetic flux is proportional to the strength of the field and to the velocity of the relative motion. Perhaps the simplest form of such a device consists of a permanent-magnet slug moving inside a coil of wire. The output voltage across the terminals of the coil is directly related to the velocity of the slug. This device, therefore, measures linear velocity. A conventional alternating-current generator, which employs the same basic principle but involves rotary motion, is sometimes used for measuring angular velocity. The major drawback of this type of tachometer is that both frequency and voltage of the ac output are proportional to shaft speed. This type of output signal is especially difficult to use at low speeds.

Another simple type of electromagnetic measuring device is the rotating-magnet speedometer, in which a permanent-magnet armature on a rotating shaft rotates inside a concentric cup mounted on nearly frictionless bearings. The cup is made of a material that is a good conductor of electricity. A torque is produced by the eddy currents induced in the cup by the motion of the field of the rotating permanent magnet. This torque is proportional to the speed of the rotor shaft, and may be used as a measure of rotor speed.

A somewhat more sophisticated type of electromagnetic measuring device is the ac drag-cup tachometer, which is basically a two-phase motor having a drag-cup rotor and arranged so that one of the fields is excited with a constant ac voltage. An ac signal voltage is generated in the other field and is proportional to the speed of rotation of the drag-cup armature. For further discussion, *see* TACHOMETER.

Mechanical measuring devices. These devices usually yield a force or motion type of signal representing the position, velocity, or acceleration of the output of the servomechanism. Such devices involve one or more of a number of mechanical principles, which are influenced by the condition to be measured. The various types of mechanical measuring devices can be classified according to the predominant mechanism which is used to measure the output condition.

Linkages and gear trains. These are often used to amplify or attenuate the motion to be measured and to compare it with a reference input motion. For examples of such mechanisms, *see* GEAR TRAIN; LINKAGE, MECHANICAL.

Springs. Displacement may be measured in terms of the force required to produce that displacement of a spring. Thus the force in the spring is a measure of the relative positions of its ends. A single spring may be used as an error detector, the motion of one end corresponding to the reference input, and the motion of the other corresponding to the output shaft of the servomechanism.

Viscous drag dashpots. Velocity may be measured in terms of the force required to produce that velocity in a dashpot. Thus the force at the terminals in such a device is a measure of the relative velocities of its ends, and a single dashpot may be used as an error detector of velocity just as a spring measures errors of position.

Governors and gyroscopes. These devices measure acceleration in terms of the force required to produce that acceleration of a mass (for linear motion) or inertia (for angular motion). A fly-ball governor is an example of the measurement of angular velocity by means of a moving mass. The mass is mounted to the rotating shaft so that it rotates about the center line of the shaft and experiences a radial acceleration which varies as the second power of the shaft speed. Thus the radial force on the fly weight is a measure of the square of the shaft speed. *See* GOVERNOR.

Perhaps the most complicated type of mechanical measuring device based on inertial effects is the gyroscope, which consists of an inertia rotating at high speed and mounted in gimbals so that it can undergo angular motions about any axis. Motion of the gyroscope about any axis that does not coincide with the axis of rotation results in a precessive torque about an axis that is mutually perpendicular to the spin axis and the axis of the imposed motion. This torque is proportional to the product of the imposed angular velocity and the sine of the angle between the spin axis and the axis of the imposed motion. *See* GYROSCOPE.

Servoamplifiers or controllers. This portion of the servomechanism closed-loop system reacts to the low energy level error signal and delivers a higher energy level signal to the servomotor or actuator. The input of the servoamplifier usually must be very sensitive to the error signal without loading the summing network or error detector. It must operate on readily available sources of power to provide the power gain required to drive the servomotor. Relatively fast dynamic response is usually expected in the servoamplifier (as compared to the dynamic responses of many of the other elements in the servomechanism loop), and special dynamic characteristics, sometimes referred to as compensation, are also frequently included to provide required closed-loop performance. Internal feedback within the amplifier is often employed to achieve the necessary compensation or to minimize the sensitivity of the amplifier to such undesirable disturbances as supply-voltage or temperature variations and drift effects caused by aging components.

Electric and electronic amplifiers. The required power gain is usually achieved through the use of variable-resistance elements, such as vacuum tubes, gas tubes, transistors, and relays, which require relatively small amounts of energy to vary their resistance. However, these devices are relatively wasteful of power, because they are made up of dissipative devices. Therefore, when large amounts of power are required to drive the servomotor, more efficient devices, such as field-controlled generators, are employed to provide the desired power gain by varying the rate of generation of electric energy. When the amplifier is direct-coupled, so that it can transmit steady direct-current signals, it is referred to as a dc amplifier. In some instances, it is desirable to employ high-gain ac amplifiers because of their lower sensitivity to drift and aging effects. Modulators and demodulators are then employed to convert dc signals to ac signals and vice versa, if necessary. When internal feedback is employed to provide compensation, it is applied only to those portions of the amplifier system which involve dc signals. Drift of dc amplifiers is often minimized through the use of chopper stabilization. *See* DIRECT-COUPLED AMPLIFIER; POWER AMPLIFIER; VIBRATOR.

Electric and electronic servoamplifiers are often provided with means to adjust their gain and dynamic response characteristics so that standard amplifiers may be employed for a wide variety of systems. If the output impedance of the servoamplifier is not sufficiently low relative to the input impedance of the servomotor it drives, the servomotor may interact with the servoamplifier and change its characteristics appreciably.

Hydraulic and pneumatic amplifiers. These usually achieve the required power gain through the use of variable orifices in the form of special control valves, often referred to as servovalves. However, these devices are dissipative. When relatively large amounts of power are required to drive the servomotor, more efficient devices, such as variable displacement pumps, are employed to provide the required power gains by varying the rate of generation of hydraulic power. The system shown in Fig. 4 illustrates how amplification is attained through the use of one variable orifice and one fixed orifice. This valve, referred to as a flapper valve, is frequently used in industrial pneumatic controllers and in the first stages of hydraulic servovalves.

A four-way valve, employing four variable orifices, is shown in Fig. 5 coupled to a hydraulic ram to form a hydraulic amplifier. The position x_1 of the valve spool is the input and the position x_2 of the ram is the output. Low energy level inputs are achieved through careful design of the valve to minimize fluid flow forces.

The integration which exists between the input x_1 and the output x_2 of the system shown in Fig. 5 can be eliminated by the use of mechanical feedback (Fig. 6) and by considering the input to be

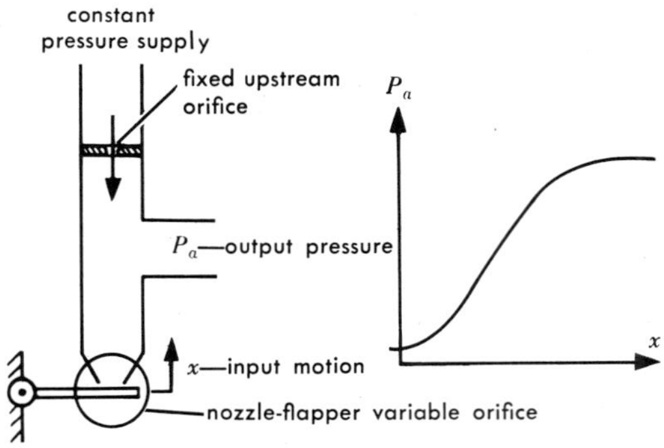

Fig. 4. Simple nozzle-flapper amplifier.

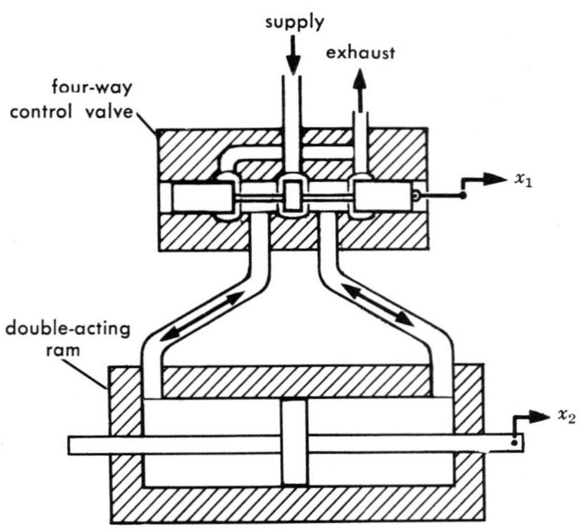

Fig. 5. Hydraulic amplifier with four-way valve.

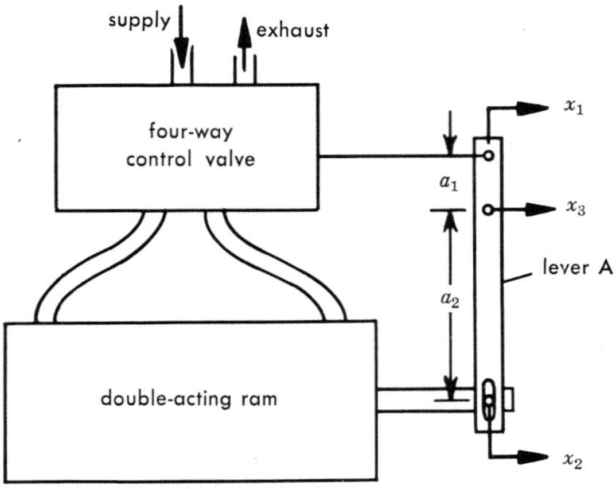

Fig. 6. Mechanical feedback on a hydraulic amplifier with four-way valve.

the motion x_3 of the intermediate point on the lever A. The steady-state gain is determined by the location of the intermediate point having motion x_3 and is equal to $(a_1 + a_2)/a_1$.

The dynamic characteristics of hydraulic and pneumatic amplifiers are often strongly influenced by the type of loads imposed on their outputs, especially when large masses are involved, making it necessary to include the effects of fluid compressibility. As with electronic amplifiers, several stages of amplification may be incorporated in a single amplifier. The last stage of valving is often considered to be part of the servomotor, resulting in a valve-controlled hydraulic or pneumatic servomotor.

Relay amplifiers. These special types of amplifiers, based on variable resistance or variable impedance devices, have only two operating impedance levels—nearly infinite (open circuit) and nearly zero (closed circuit). The degree of sophistication of relay amplifiers varies from simple on-off control involving a single relay to pulse-width modulation systems involving several relays. Many types of relays involve mechanically operated switches and have limited life for reliable operation. Magnetic amplifiers employ relays consisting of saturable reactors, sometimes referred to as transductors, having magnetic cores made of material with a sharp saturation characteristic and ac and dc windings such that small changes in the dc magnetization current result in large changes (from nearly infinite to nearly zero) of the impedance of the ac windings. No moving parts are involved in magnetic amplifiers, and they are well suited for driving 2-phase ac servomotors, because they operate with a dc input signal and an ac output. *See* MAGNETIC AMPLIFIER; SATURABLE REACTOR.

Servoamplifier compensation. Certain dynamic control characteristics are incorporated into the servoamplifier to attain required closed-loop performance. An uncompensated amplifier is sometimes called a proportional controller, because it produces an output signal that is proportional to its input signal with negligible dynamic lag between output and input. To maintain zero steady-state error of the complete servomechanism, integral compensation is introduced, and the amplifier output includes a component that is proportional to the time integral of the input. Derivative compensation is sometimes needed for system dynamic stability; the servoamplifier output then includes a component that is proportional to the derivative of the input. Thus a servoamplifier with proportional plus integral plus derivative action is described by the following differential equation

$$M = k_c E + k_i \int E\,dt + k_d \frac{dE}{dt}$$

where E represents the input, M the output, and k_c, k_i, and k_d the constants of the proportional, integral, and derivative terms, respectively. Similarly, many other types of compensation may be incorporated to meet closed-loop system requirements. Compensation is sometimes accomplished by employing a passive network in series with an uncompensated amplifier. Another way to accomplish compensation is through the use of negative feedback around a high-gain amplifier. The choice of amplifier compensation depends upon the characteristics of the components in the rest of the system and over-all system requirements and specifications. In many instances, the final choice is made on the basis of trial and error and experience. Many servoamplifiers and controllers are designed so that many different kinds of compensation can be achieved with a minimum of modification.

Servomotors (actuators). These are the "muscles" of servomechanism systems. Often referred to as the final control element of a system, a servomotor normally operates at a higher power level than any of the components preceding it. It is often required to move large masses or inertias and overcome large friction forces and external load forces. Over-all system dynamic response is often limited by the characteristics of the servomotor, and the servomotor and its power supply often account for most of the volume and weight of the complete servomechanism.

Electric servomotors. These are employed to directly produce limited linear motions and angular motions from small fractions of a radian to continuous rotation. Solenoids (moving slug), voice coils (moving coil), and torque motors (moving rotor), all operating with electromagnetically induced forces, are employed to provide limited linear or limited angular motions at relatively high force levels, such as those required for stroking servovalves and actuating brakes and clutches. For continuous rotary motion, a wide variety of ac and dc motors may be employed (*see* ALTERNATING-CURRENT MOTOR; DIRECT-CURRENT MOTOR). Field-controlled dc servomotors often have a center-tapped field winding to operate from the push-pull output of an electronic servoamplifier, as shown in Fig. 7*a*. These can be driven in either direction. When large controlled currents are available, as from a field-controlled generator, armature-controlled dc motors with constant field excitation are employed, as in the Ward-Leonard system shown in Fig. 7*b*. When high starting torques are required, universal motors having the field windings in series with the armature may be employed as either ac or dc servomotors. A typical circuit is shown in Fig. 7*c*. When low power level ac servomotors are required, two-phase induction motors are employed. One field winding is kept at constant voltage (Fig. 7*d*).

Hydraulic and pneumatic servomotors. These are employed to produce limited linear motions and

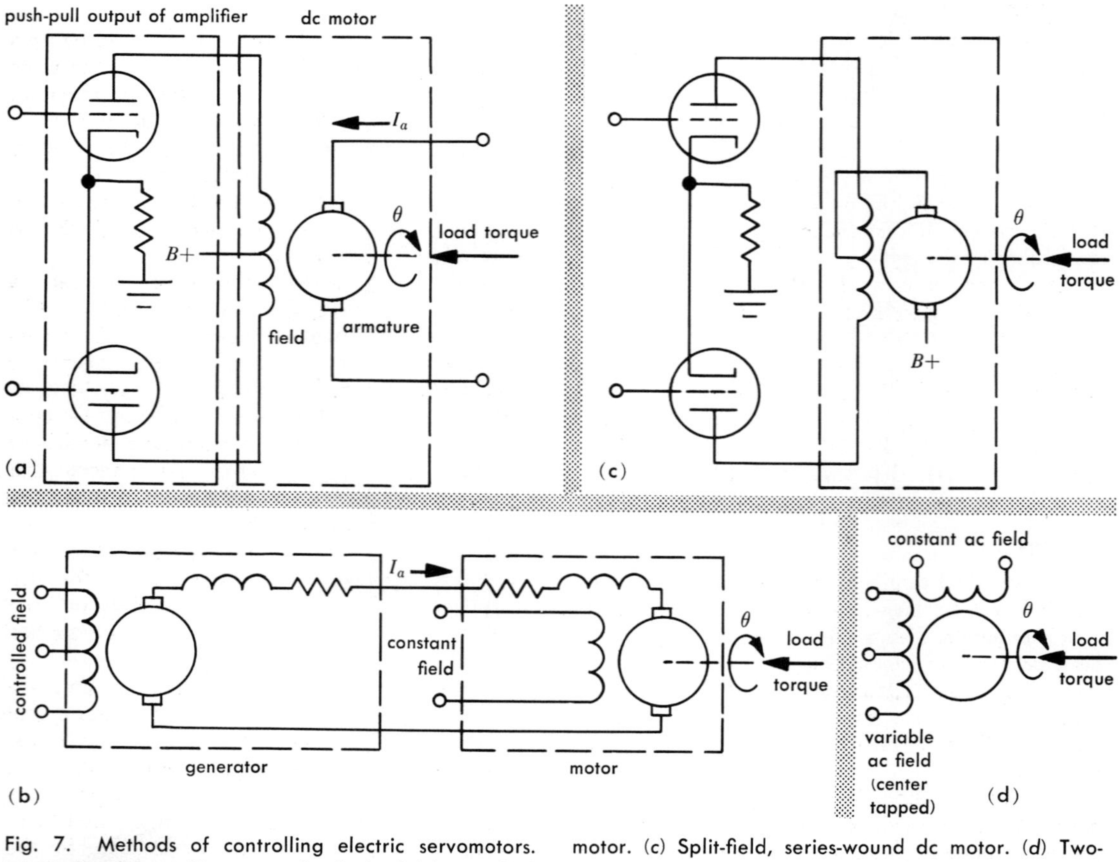

Fig. 7. Methods of controlling electric servomotors. (*a*) Push-pull amplifier control of dc field-controlled servomotor. (*b*) Ward-Leonard system for control of dc motor. (*c*) Split-field, series-wound dc motor. (*d*) Two-phase ac servomotor.

angular motions of from small fractions of a radian to continuous rotation. Figure 8 illustrates the simplest kind of hydraulic and pneumatic motor. A closely fitted piston in a cylinder is driven by fluid flow under pressure, which exerts a net force to move the ram (piston) and drive the load. Although the servovalve was discussed in the section on hydraulic and pneumatic amplifiers as a component of hydraulic amplifiers, it is customary to include the last stage of valving, which supplies controlled fluid power to the output motor, with the motor and designate the combination as a valve-controlled servomotor. When rotary output motion is desired, rotary units, such as multipiston, gear, and vane motors, as well as turbines, may be employed as output motors in servomechanisms. *See* HYDRAULIC ACTUATOR.

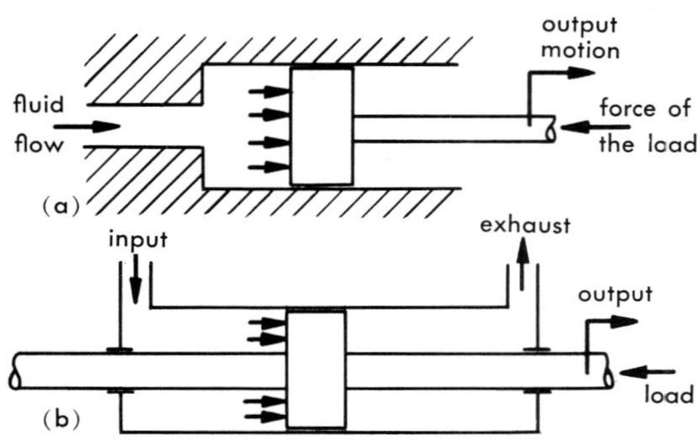

Fig. 8. Ram-type fluid motors. (*a*) Single-acting. (*b*) Double-acting.

When large power levels are involved and it is important to minimize power dissipation in a hydraulic system, a variable displacement pump is used to control the flow of hydraulic fluid to the motor. A hydraulic amplifier is usually employed to position the displacement control lever of the pump, but sometimes an electric servomotor is employed instead. [J.L.SH.]

Bibliography: J. F. Blackburn, G. Reethof, and J. L. Shearer, *Fluid Power Control*, 1960; J. E. Gibson and F. B. Tuteur, *Control System Components*, 1958; J. G. Truxal (ed.), *Control Engineers' Handbook*, 1958.

Set theory

A mathematical term referring to the study of collections or sets. Consider a collection of objects (such as points, dishes, equations, chemicals, numbers, or curves). This set may be denoted by some symbol, such as X. It is useful to know properties that the set X has, irrespective of what the elements of X are. The cardinality of X is such a property.

Cardinality of sets. Two sets A and B are said to have the same cardinal, written $C(A) = C(B)$, provided there is a one-to-one correspondence between the elements of A and the elements of B. For finite sets this notion coincides with the phrase "A has the same number of elements as B." However, for infinite sets the above definition yields some interesting consequences. For example, let A denote the set of integers and B the set of odd integers. The function $f(n) = 2n - 1$ shows that $C(A) = C(B)$. Hence, an infinite set may have the same cardinal as a part or subset of itself.

Subset. A is called a subset of B if each element of A is an element of B, and it is expressed as $A \subset B$. The collection of odd integers is a subset of the reals. Also, each set is a subset of itself.

The continuum hypothesis. An infinite set is called uncountable if it cannot be put in a 1-1 correspondence with the positive integers. Here is one of the unsolved problems of set theory. It is of particular interest since so many mathematicians have tried unsuccessfully to solve it. If X is an uncountable subset of the reals R, is $C(X)$ equal to $C(R)$? The conjecture that the answer is in the affirmative is called the continuum hypothesis.

Comparing cardinality of sets. One says that $C(A) \leqq C(B)$ if there is a 1-1 correspondence between the elements of A and a subset of the elements of B. One useful theorem that can be proved states that any two sets A, B are comparable; that is, either $C(A) \leqq C(B)$ or $C(B) \leqq C(A)$ (possibly both). Another theorem states that if $C(A) \leqq C(B)$ and $C(B) \leqq C(A)$, then $C(A) = C(B)$. Each of these results may be proved by using well orderings of A and B.

Ordering. An ordering is one way of setting up a 1-1 correspondence between two sets of the same cardinality. A relation $<$ is an order relation for a set X if it satisfies the following conditions:

1. If x_1, x_2 are two elements of X, either $x_1 < x_2$ or $x_2 < x_1$ (any two elements are related).
2. $x_1 \nless x_1$ (no element is less than itself).
3. If $x_1 < x_2$ and $x_2 < x_3$, then $x_1 < x_3$ (the order relation is transitive).

The following are examples of ordered sets: a horizontal line where $<$ means "is to the left of"; the reals where $<$ means "is less than"; the collection of words in the dictionary where ordering is alphabetical.

Well ordering. An ordering of a set is called a well ordering if it satisfies the additional condition:

4. Each subset Y of X has a first element; that is, there is an element y_0 of Y such that if y' is another element of Y, $y_0 < y'$.

Used above was the natural convention that each set contains at least one element and not the artificial convention that there is an "empty" set. Some authors use an "empty set" but its introduction is not necessary.

The natural ordering of the positive integers is a well ordering, but neither the natural ordering of the integers nor of the reals is a well ordering. A well ordering for the integers is 0, 1, 2, . . . , −1, −2, ... Since a well ordering of the reals cannot be written down, one might guess that there is none. This guess is shown to be false by the theorem that states that any set X has a well ordering. In proving this, one considers the collection Z of all subsets of X, selects a point $x_\alpha = f(z_\alpha)$ from each element z_α of Z, and well orders X so that if S_α is the set of all elements that precedes x_α, then $x_\alpha = f(Z - S_\alpha)$.

Some of the theorems proved by well ordering are so strange that their truths do not seem intuitively obvious. Well ordering is also used to construct pathological examples which serve as counterexamples to various conjectures. These counterexamples are useful since they show that the conjectures are false and it is useless to try to prove them.

Formation of sets. One approved method of forming a set is to consider a property P possessed by certain elements of a given set X. The set of elements of X having property P may be considered as a set Y. The expression $p \in X$ is used to denote the fact that p is an element of X. Then $Y = \{p/p \in X$ and p has property $P\}$. Another approved method is to consider the set Z of all subsets of a given set X. It may be shown in this case that $C(X) < C(Z)$.

Paradoxically, it is not permissible to regard the collection of all sets as a set. If such a collection X were called a set, and Z were used to denote the set of all subsets of X, one would arrive at the absurdity that $C(X) < C(Z)$.

Operations with sets. In set theory, one is interested not only in the properties of sets but also in operations involving sets: addition, subtraction, multiplication, and mapping.

Sum or union. The sum of A and B ($A + B$ or $A \cup B$) is the set of all elements in either A or B, that is, $A + B = \{p/p \in A \text{ or } p \in B\}$.

Intersection, product, or common part. The intersection of A and B ($A \cdot B$, $A \cap B$, or AB) is the set of all elements in both A and B, that is, $A \cdot B = \{p/p \in A \text{ and } p \in B\}$. If there is no element which is in both A and B, one says that A does not intersect B and writes $A \cdot B = 0$.

Difference. The expression $A - B$ is used to denote the collection of elements of A that do not belong to B, that is, $A - B = \{p/p \in A \text{ and } p \notin B\}$. If $A \subset B$, it is expressed as $A - B = 0$.

An example. If a person were to squirt some black ink on a plane, the set A of points in the dark spot would be an example of a point set. Suppose a set B is determined by squirting some red ink on the plane. Then $A + B$ designates the set of points covered by ink, $A \cdot B$ designates the set covered by both kinds of ink, and $A - B$ designates those covered by black but not red ink (Fig. 1).

Boolean algebra. By using the previous notation, it follows that the sets of Fig. 1 satisfy some of the familiar laws of algebra as

$$A + B = B + A$$
$$A \cdot (B + C) = A \cdot B + A \cdot C$$

However, other identities are not so familiar:

$$X - (A + B) = (X - A) \cdot (X - B)$$
$$X - A \cdot B = (X - A) + (X - B)$$

See BOOLEAN ALGEBRA.

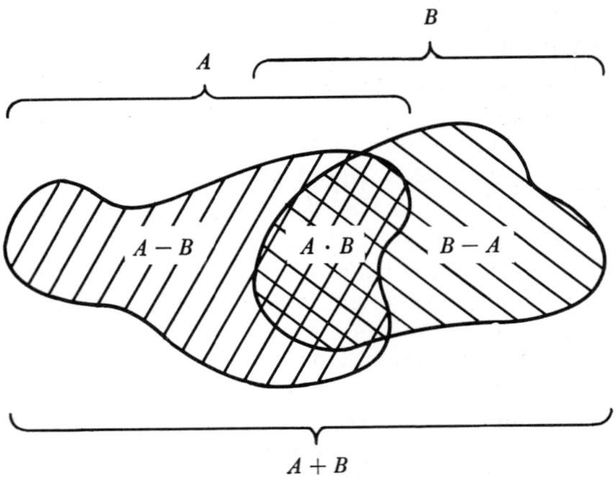

Fig. 1. Sets of point on a plane.

Transformations. A transformation of a set X into a set Y is a function that assigns a point of Y to each point of X. Such a transformation is also called a mapping. The transformation shown in Fig. 2 is the vertical projection of a set X onto a segment Y. The point assigned to X under a transformation f is called the image of x and denoted by $f(x)$. Also, the set of all points x mapped into a particular point y of Y is called the inverse of y and denoted by $f^{-1}(y)$. The inverse of y shown in Fig. 2 is the sum of three segments. The equation $f(x) = x^2$ represents a transformation that takes each real number into its square, for example, 2 to 4, −5 to 25, and so on. Rotations, congruences, and similarities are examples of transformations from geometry. However, in general, a transformation may change both the size and shape of an object.

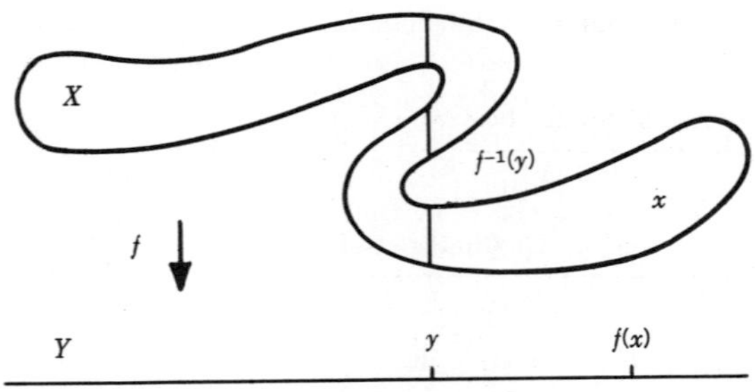

Fig. 2. Vertical projection of a set X onto a segment Y.

Topology. Topology is one of the branches of mathematics that makes extensive use of set theory. Here, not only does one have sets of points for consideration but also collections of interiors of spheres or neighborhoods. These neighborhoods enable one to study limit points and the continuity of transformations. See CONFORMAL MAPPING; RING THEORY; TOPOLOGY. [R.H.BI.]

Bibliography: E. Kamke, *Theory of Sets*, trans. from 2d German ed., 1950.

Sewage

A combination of (1) liquid wastes conducted away from residences, institutions, and business buildings; and (2) the liquid wastes from industrial establishments with (3) such surface, ground, and storm water as may find its way or be admitted into the sewers. Category (1) is known as sanitary or domestic sewage; category (2) is usually referred to as industrial waste; category (3) is known as storm sewage.

Relation to water consumption. Sewage is the waste water reaching the sewer after use; hence it is related in quantity and in flow fluctuation to water use. The quantity of sewage is generally less than the water consumption since some portion of water used for fire fighting, lawn irrigation, street washing, industrial processing, and leakage does not reach the sewer. These losses are compensated for partly by the addition of water from private wells, ground-water infiltration, and illegal connections from roof drains. Water consumption increases with size of community served and many other community characteristics. Characteristics of each city must be studied and analyzed for specific information. As a general average estimate, communities with population under 1000 use about 60 gal per capita per day (gcd), while communities of 100,000 use about 140 gcd. In a 1957 study of large cities of the United States, the median consumption was 154 gcd, and the median population was 658,000 (Fig. 1). An accepted unit flow for domestic sewage as shown in the table is 100 gcd.

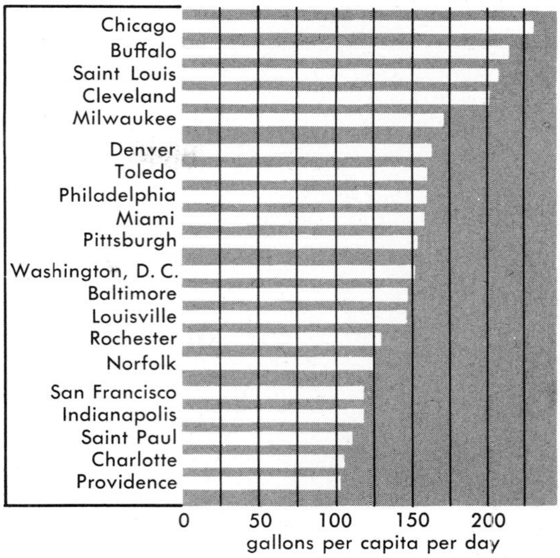

Fig. 1. Estimated water use in 20 major cities. (*Research Division, New York University, College of Engineering, New York*)

Infiltration of ground water should be held to a minimum. It may be expected to be equal to or less than 30,000 gal per day per mile of sewer including house connections. Much depends on the quality of sewer construction. Water may enter through poorly made joints, and, in quantity, through poorly constructed, leaky manholes and illegal and abandoned sewers. Sewers in wet ground with a high water table will have more infiltration. Sewers under pressure may have infiltration or leakage to the surrounding ground. The danger of ground water pollution from leaky sewers should be avoided. *See* SEWAGE COLLECTION SYSTEMS.

Fluctuations in sewage flow are related to water use characteristics but tend to dampen out since there is a time lag from the time of use to appearance in the sewer mains and trunks (Fig. 2). Hourly, daily, and seasonal fluctuations affect design of sewers, pumping stations, and treatment plants.

Rates of sewage flow from various sources[a]

Character of district	Gal per capita per day	Gal per acre per day	Source of sewage	Gal per capita per day
Domestic			Trailer courts	50[b]
Average	100		Motels	53[b]
High-cost dwellings	150	7,500	State prisons	
Medium-cost dwellings	100	8,000	Maximum	280[c]
Low-cost dwellings	80	16,000	Average	176
			Minimum	104
Commercial			Mental hospitals	
Hotels, stores, and office buildings		60,000	Maximum	216[c]
			Average	123
Markets, warehouses, wholesale districts		15,000	Minimum	38
			Grade school	4.4[d]
			High school	3.9[d]
Industrial				
Light industry		14,000		

[a] From H. E. Babbitt and E. R. Baumann, *Sewerage and Sewage Treatment*, 8th ed., Wiley, 1958.
[b] From report of State Sanitary Engineers, *Public Works*, p. 108, March, 1957.
[c] From J. C. Frederick, *Public Works*, p. 112, April, 1957.
[d] Average of 4.4 gal per day per pupil between 7:30 A.M. and 5:30 P.M. The average for the high school is spread over more hours per day. From C. H. Coberly, *Public Works*, p. 143, May, 1957.

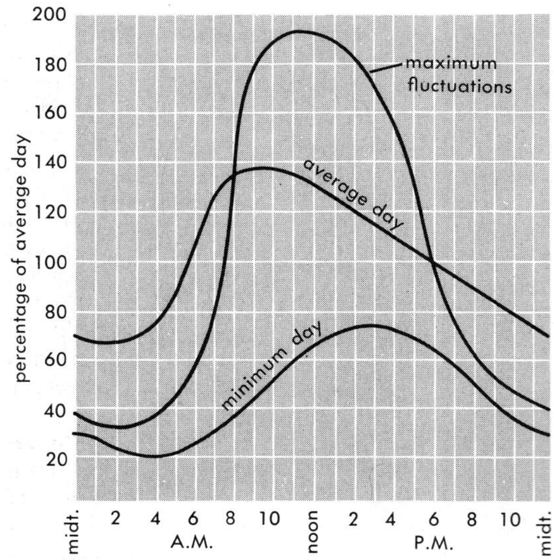

Fig. 2. Typical hourly variations in sewage flow. (*From H. E. Babbitt and E. R. Baumann, Sewerage and Sewage Treatment, 8th ed., Wiley, 1958*)

Daily and seasonal variations depend largely on community characteristics. Weekend flows may be lower than weekday. Industrial operations of seasonal nature influence the seasonal average. The seasonal average and annual average are about equal in May and June. The seasonal average is about 124% in late summer and may drop to about 87% at the end of winter. Peak flows may reach 200% of average at the treatment plant and may be over 300% of average in the laterals. Laterals are designed for 400 gcd and mains and trunks for 250 gcd.

Design periods. These are dependent on the proposed sewer construction. Lateral sewers may be designed for ultimate flow of the area to be sewered. Mains may be designed for periods of 10–40 years. Trunk sewers may be planned for long periods with provision made in design for parallel or separate routings of trunks of smaller size to be constructed as the need arises. Economics, available funds, and engineering judgment affect selection of the design periods. Appurtenances may have a different life, since replacement of mechanical equipment will be necessary. A span of 20–25 years is often selected and a time table of additions during that period is then scheduled in the over-all improvement plan.

Storm sewage. Storm sewage is liquid flowing into sewers during or following a period of rainfall and resulting therefrom. An estimate of the quantity of storm sewage is necessary in sewage design.

Estimating quantity of storm sewage requires a knowledge of intensity and duration of storms, distance water travels before reaching sewer, permeability and slope characteristics of the surface over which water flows to sewer inlet, and shape and amount of area to be drained to inlet. These general considerations are included in the equation $Q = CAIR$ expressing the runoff from a watershed having no retention or storage of water. Q is expressed in cubic feet per sec (cfs), A is area, I is the relative imperviousness of the surface expressed as a decimal, R is the rate of rainfall in inches per hour. C is a coefficient permitting the expression of the factors in convenient units; in the above units it may be taken as 1 so that the equation becomes $Q = AIR$.

Time of concentration is a combination of the theoretical time required for a drop to run from the most distant point to the inlet and time from sewer inlet to point of concentration. The inlet time may range from 3 min for a steep slope on an impervious area to 20 min on a city block. The time of flow is assumed to be the velocity in the full flowing sewer divided by the length of sewer from inlet to point of concentration. Flood crest and storage time while the sewer is filling are usually neglected, the effect being that of assuming a larger rate of flow and therefore providing a safety factor in design.

Values of I, the runoff coefficient, range from 0.01 in wooded areas to 0.95 on roofed surfaces. A common value used in residential areas with considerable land in lawn, garden, and shrubbery is 0.30–0.40. In built-up areas, values of 0.70–0.90 may be used.

Rainfall intensity values are selected on the basis of frequency and duration of storms. In some sewer design it is necessary to select a value for the expected occurrence of maximum runoff. This is done by using one of the several formulas which will allow a prediction of R for 5, 10, or 15 years. The element of calculated risk is combined with engineering judgment in deciding which R to choose. For lesser structures in residential areas a 5-year frequency may be used with reasonable safety. Where failure would endanger property, the 10-, 15-, or 25-year frequency of occurrence provides a more conservative design basis; 50-year frequency may be selected where flooding could cause lasting damage and disrupt facilities. In such instances cost-benefit studies may be made to guide the selection of a suitable frequency. *See* HYDROLOGY.

Pumping sewage. Not all sewage will flow by gravity without unnecessary expense in circuitous routing or deep excavation, therefore, pumping stations may be advantageous. Pumping stations may be required in the basements of large buildings. Pumping stations are provided with two or more pumps of sufficient capacity so that with one unit out of service the remaining unit or units will pump the maximum flow. Motive power is required from at least two sources, usually electric motors and auxiliary fuel-fed motor drive. Care must be taken to have motive power above flood level and protected from the elements. Screening is usually required ahead of pumping stations, unless the pumps themselves are self-cleaning. Many states require that pumping units be installed in a dry well and that sewage be confined to a separate wet well. Buildings above ground should fit the surroundings. Small pumping stations are often one unit and made fully automatic so that minimum attendance is required. Safety measures must be considered. Centrifugal pumps are used almost exclusively in larger stations. Air ejector units may be installed in smaller stations.

Examination of sewage. Sewage is actually water with a small amount of impurity in it. Examination of sewage is required to know the effects of these impurities. Various tests are used to aid in determining the characteristics, composition, and condition of sewage. These include physical examination, solids determination, tests for determining the oxygen requirement of organic matter, chemical and bacteriological tests, and examination under the microscope.

Physical tests for turbidity, odor, color, and temperature are made. Normal fresh sewage is gray and somewhat opaque, has little odor, and has a temperature slightly higher than the water supply. Decomposition of organic matter darkens the sewage, and odors are characteristic of stale or septic sewage.

Tests for residue or solid matter provide an indication of the types of solids, the strength of the sewage, and the physical state of the solids. Total solids determinations measure both suspended and dissolved solids. A sample of the sewage is filtered. The suspended solids can be determined by drying the material recovered on the filter. The dissolved solids can be determined by evaporation of the filtered portion. Heating the solids residue until organic matter gasifies separates volatile solids from fixed solids or inorganic ash. Loss on ignition represents the volatile or organic fraction and is a good measure of sewage strength.

Measurement of the part of the suspended solids heavy enough to settle is made in an Imhoff cone. The settleable-solids test is useful in determining sludge-producing characteristics of sewage.

Tests for organic matter are made principally to determine the oxygen requirement of sewage. These tests include the biochemical oxygen demand test (BOD), the chemical oxygen demand test, the oxygen consumed test, and the relative stability test. Organisms in sewage require oxygen for growth and the BOD measures the amount of dissolved oxygen required for decomposition of organic solids for a measured time at a constant temperature. The standard measurement is made for 5 days at 20°C and is a good measure of sewage strength. Since the BOD measurement includes both biological and chemical oxygen requirement, another test, the chemical oxygen demand, is sometimes used to measure the chemical oxygen requirement. Sewage is heated in the presence of an oxidizing agent such as potassium dichromate. The oxygen requirement is that of chemical digestion since all organisms have been killed. This test has somewhat limited use. The oxygen consumed test uses potassium permanganate as the oxidizing agent. The result offers some index of the readily oxidizable carbonaceous material. The relative stability test indicates when the oxygen present in plant effluent or polluted water is exhausted. The data express as a percentage the approximate amount of oxygen available in water in relation to the amount required for complete stability. The test is a color test using methylene blue. Reducing agents, precipitation of color, concentration of dye, amount of dissolved oxygen in the sample, and other factors affect the reliability of this test, and it is considered generally as a rough or screening test of the condition of plant effluent. Tests for nitrogen include those for free ammonia, albuminoid ammonia, organic nitrogen, nitrites, and nitrates. The latter are indications of oxidation change and stabilization and are used in checking condition of plant effluent.

Bacteriological tests are made primarily to determine the presence of organisms of the coliform group. The organisms exist in the intestines of warm-blooded animals and are used as an index of the presence of fecal material. The coliform test is made on chlorinated effluents to determine the efficiency of chlorination. Occasionally other bacteriological determinations are made in special studies to determine the presence of organisms of the *Salmonella* group or dysentery group in polluted water and sewage.

Microscopic tests are not normally made on raw sewage. They are used as part of plant operator control in treatment processes. Examinations for the presence of algae, protozoans, bacteria, fungi, rotifers, and worms are made when necessary. *See* BACTERIOLOGY; BIOLOGY; PROTOZOA. [W.T.I.]

Bibliography: APHA-AWWA-FSIWA Joint Committee, *Standard Methods for the Examination of Water, Sewage, and Industrial Wastes*, 10th ed., 1955; W. T. Ingram, *Water Fluoridation Practices in Major Cities of the United States*, pt. 1, Research Division, New York University College of Engineering, 1958; F. A. Kristal and F. A. Annett, *Pumps*, 2d ed., 1953.

Sewage collection systems

Systems of pipes and conduits, together with control devices, pumping stations, and appurtenances used for the collection and transfer of waste waters. Waste waters may include sanitary sewage (that is, the liquid wastes of residential premises), industrial wastes, storm waters, and ground-water infiltration. *See* SEWAGE.

Sewer pipe. Sewer pipe is manufactured from a number of different materials, such as vitrified clay, concrete, asbestos cement, corrugated iron, cast iron, and steel. Plastics, bituminous wood fiber, and wood stave pipe also are used. All sewer pipes may be surcharged or filled at some time and must be capable of withstanding some hydraulic pressure. Pressure lines connected to pump discharge, or lines carried under roads or streams, that is, inverted siphons, are designed for the particular condition, and materials which withstand pressure are selected.

Sewers are laid underground and must be able to withstand external pressures such as those caused

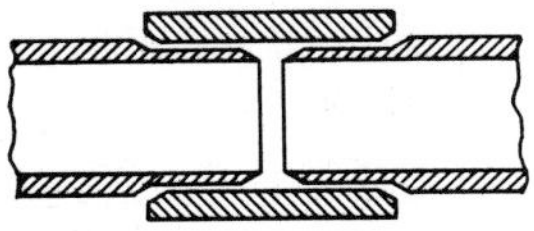

Fig. 1. Diagram of joint for asbestos-cement pipe. (*From H. E. Babbitt and E. R. Baumann, Sewerage and Sewage Treatment, 8th ed., Wiley, 1958*)

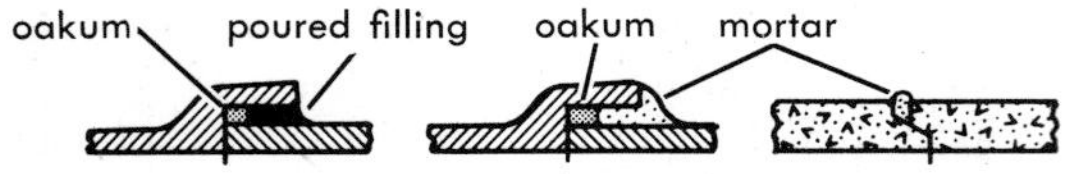

Fig. 2. Types of joints for bell-and-spigot pipes. (*From H. E. Babbitt and E. R. Baumann, Sewerage and Sewage Treatment, 8th ed., Wiley, 1958*)

by soil, water, and the extra weight of traffic. Large-diameter reinforced concrete sewers or conduits may be constructed in place.

Pipes are made in various lengths. Pipe joints are of many types. An asbestos-cement pipe joint is shown in Fig. 1. The bell-and-spigot and ring types are common (Fig. 2). Jointing materials include cement, mortar. asphalt, plastics, sulfur, rubber, rings, and plastic gasket.

Clay sewer pipe is manufactured in strengths and with dimensions as provided in American Society for Testing Materials (ASTM) Specifications C13, C261, C200, and C268. Safe supporting strengths are specified in Tables 1 and 2.

Concrete sewer pipe is manufactured as provided in ASTM Specifications C14, C76, and C362. Asbestos-cement pipe is required to meet Federal Specifications SS-P-331A for gravity flow and ASTM Specification C296 where pressure pipe is required. *See* STRENGTH OF MATERIALS.

Precast concrete pipe is made by spinning or tamping semidry concrete against a mold. The centrifugal process was introduced in the United States in the 1920s. Reinforced concrete pipe is also made by the centrifugal method. Reinforcing is placed as a single, double, or elliptical cage (Fig. 3). It may also be a steel plate cylinder.

Special conditions of soil, construction, geology, pressure, and capacity may require sewers to be built in place. Concrete construction of circular, elliptical, egg shape, and horseshoe shape requires special forming of wood or metal (Fig. 4). Machine tamping and vibration equipment are used to produce a dense, impervious concrete shell. Concrete in pipe is designed to withstand 3000–4000 psi. Required thickness of shell and amount of reinforcing are designated in specifications.

Both internal and external corrosion of sewer pipe can occur. Metals are attacked unless corrosion-resistant alloys are added. Organic materials in sewage are attacked by bacteria and other microorganisms and form acids which attack concrete and metals. Protective coatings of asphalt compounds and, within the last 5 years, plastics and epoxy resins have been applied to interior pipe surface. Exterior corrosion may occur due to

Table 1. Crushing strength requirements for standard strength clay sewer pipe

Size, in.	Average strength, min, lb per linear ft	
	Three-edge-bearing method	Sand-bearing method
4	1000	1500
6	1100	1650
8	1300	1950
10	1400	2100
12	1500	2250
15	1750	2625
18	2000	3000
21	2200	3300
24	2400	3600
27	2750	4125
30	3200	4800
33	3500	5250
36	3900	5850

Table 2. Crushing strength requirements for extra strength clay sewer pipe

Nominal size, in.	Average strength, min, lb per linear ft	
	Three-edge-bearing method	Sand-bearing method
6	2000	3000
8	2000	3000
10	2000	3000
12	2250	3375
15	2750	4125
18	3300	4950
21	3850	5775
24	4400	6600
27	4700	7050
30	5000	7500
33	5500	8250
36	6000	9000

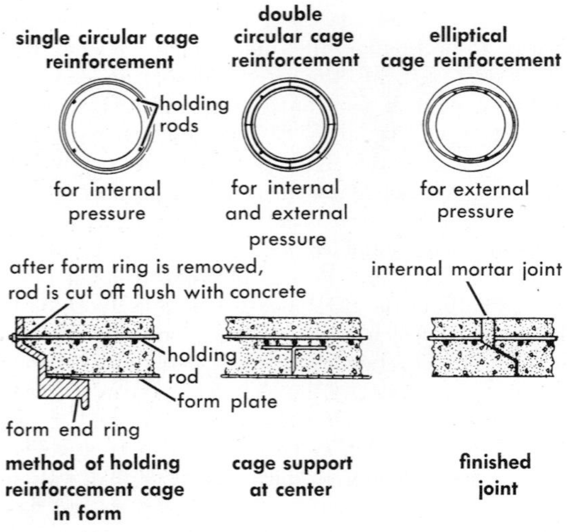

Fig. 3. Methods of reinforcing concrete pipe. (*From H. E. Babbitt and E. R. Baumann, Sewerage and Sewage Treatment, 8th ed., Wiley, 1958*)

electrolysis, bimetallic corrosion, and electrochemical and bacterial attack. *See* Corrosion.

Sewage flow. This must be known or estimated to complete design. Storm water from roof drainage or ground and street surfaces is excluded.

Velocity. Velocity of flow must be maintained at a rate sufficient to carry contained sewage solids.

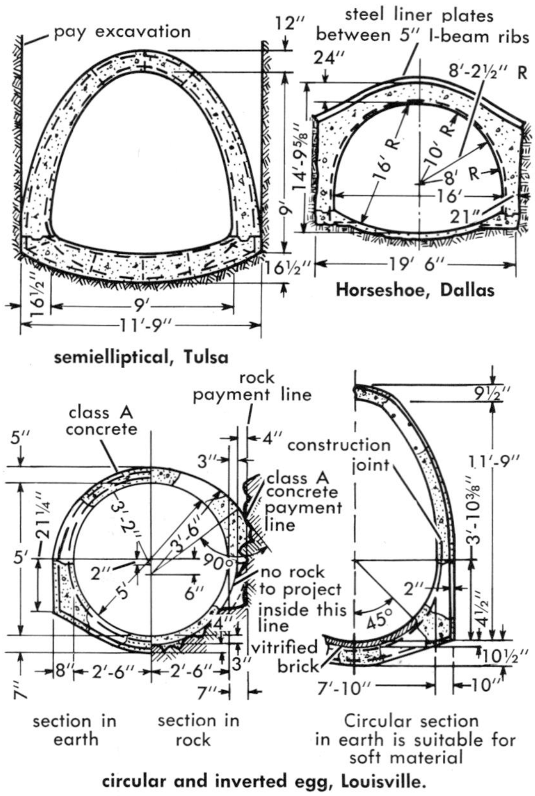

Fig. 4. Sewer shapes. (*From E. W. Steel, Water Supply and Sewerage, 3d ed., McGraw-Hill, 1953*)

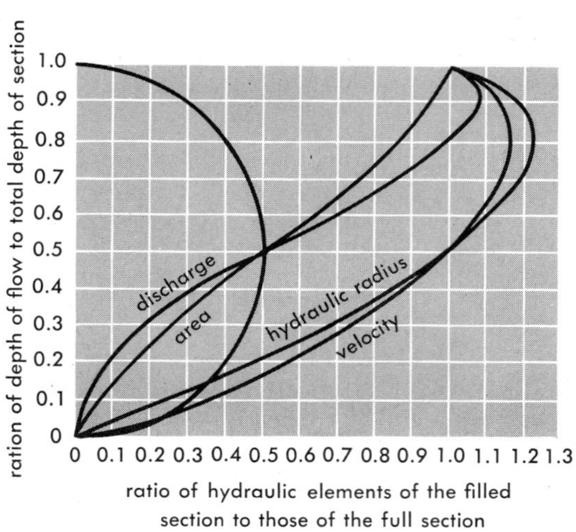

Fig. 5. Hydraulic elements of a circular pipe. (*From E. W. Steel, Water Supply and Sewerage, 3d ed., McGraw-Hill, 1953*)

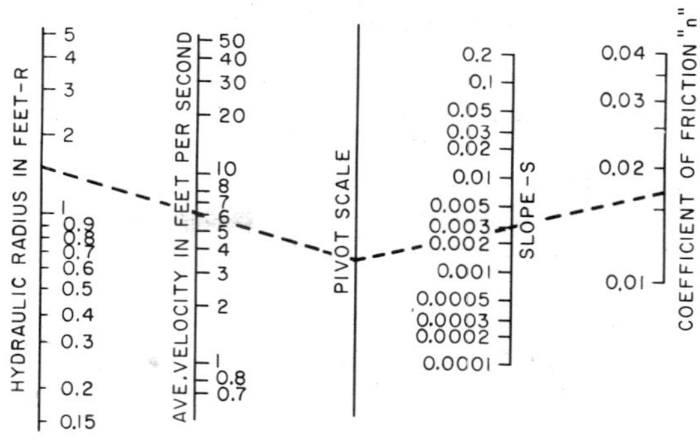

Fig. 6. Nomogram based upon Manning formula. (*From E. W. Steel, Water Supply and Sewerage, McGraw-Hill, 1953*)

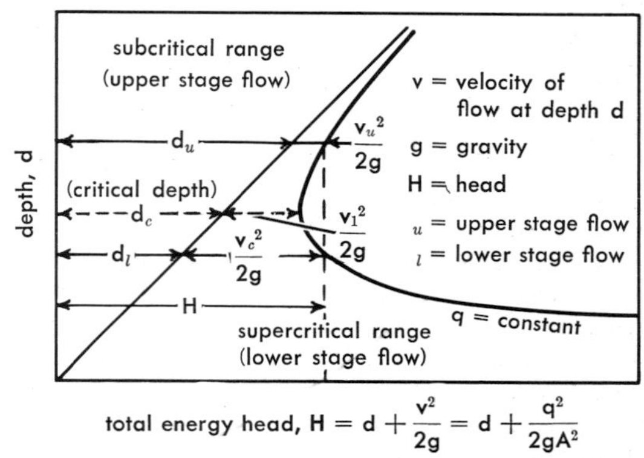

Fig. 7. Depth of flow versus total energy head. (*American Society of Civil Engineers, Manual of Design and Construction of Sanitary and Storm Sewers No. 37, 1959*)

For sanitary sewage the minimum velocity is 2 ft/sec and for storm water the minimum velocity is 2.5 ft/sec. In a circular pipe the velocity is the same whether flowing half full or full; at one-fourth depth it is about two-thirds that when flowing full. The maximum flow occurs at about 0.9 depth and the maximum velocity occurs at about 0.8 depth (Fig. 5). A pipe carrying design flow when flowing full provides about 8% safety factor in handling peak flows. Tables, charts, and monographs have been constructed to aid in the solution of pipe flow problems (Fig. 6).

Transitions. Changes in direction, grade, elevation, and pipe size and the union of two or more sewers into a common trunk are carried out at manholes and junction points. Inlet and exit losses introduced in such structures must be included in computation of the hydraulic grade line. At the end of a sewer pipe or at a marked change in slope, the flow of liquid is no longer uniform. Velocity may be decreasing as in the case of a reduced slope, or increasing with increase in slope or free fall at end of line. These points of change are called transition points. Design of sewers must provide for transition changes in hydraulic gradient. *See* Fluid-flow principles; Fluid mechanics.

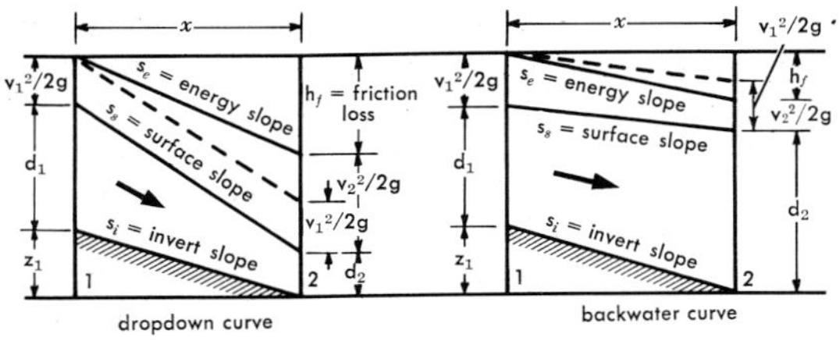

Fig. 8. Dropdown and backwater curves. (*From H. E. Babbitt and E. R. Baumann, Sewerage and Sewage Treatment, 8th ed., Wiley, 1958*)

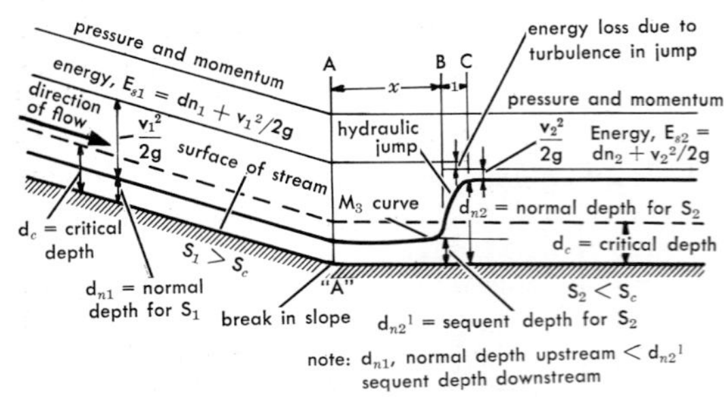

Fig. 9. Free hydraulic jump. (*From H. E. Babbitt and E. R. Baumann, Sewerage and Sewage Treatment, 8th ed., Wiley, 1958*)

Critical depth. Energy is minimal for a given cross section and discharge. Under critical flow condition (Fig. 7) there is only one depth, velocity, and hydraulic gradient that will satisfy the energy equation of flow in open channels, expressed in terms of depth. This relationship may be solved graphically or by nomograph when channels are irregular in shape or when the hydraulic radius and area are not conveniently expressed in terms of depth.

At conditions of flow other than critical, alternate stages at which the same flow may take place are possible. These stages are the normal depth or upper-stage, and lower depth or lower-stage flow. Variations in velocity occur at dropdown curves, backwater curves, and hydraulic jumps. The dropdown curve occurs near the free outlet end of a sewer where the velocity of flow is increasing. The backwater curve is caused by an obstruction in the sewer such as a dam or by discharge into a body of water whose surface is above the normal level of flow in the sewer. A backwater curve will also result from flattened grade. The velocity of flow decreases in the backwater curve. Dropdown and backwater curves are illustrated in Fig. 8.

The hydraulic jump occurs when the depth of flow changes abruptly from the lower stage to the upper stage (Fig. 9). For most sewer shapes the length of transition for the several conditions mentioned is calculated by trial and error, since the formulas are complex.

Sewer appurtenances. Sewer appurtenances are manholes, inlets, regulators, inverted siphons, and outfalls. A manhole is an opening from the ground or street surface permitting a man to enter to make examinations and repairs (Fig. 10). Manholes are spaced along the sewer at 300- to 500-ft intervals or placed at any other point where access is necessary.

Inlets. An inlet provides an opening for storm water and is usually placed at the curb line of the street. The structure below the inlet is called a catch basin (Fig. 11). A short length of sewer connects the inlet or catch basin to a manhole on the sewer.

The capacity of inlets is determined by complicated analytical methods. The length of opening is a function of the amount of storm water, gutter shape, and depth. *See* HIGHWAY ENGINEERING; HYDRAULICS.

Regulators. A regulator is a device designed to divert sewage flow from one sewer to another chan-

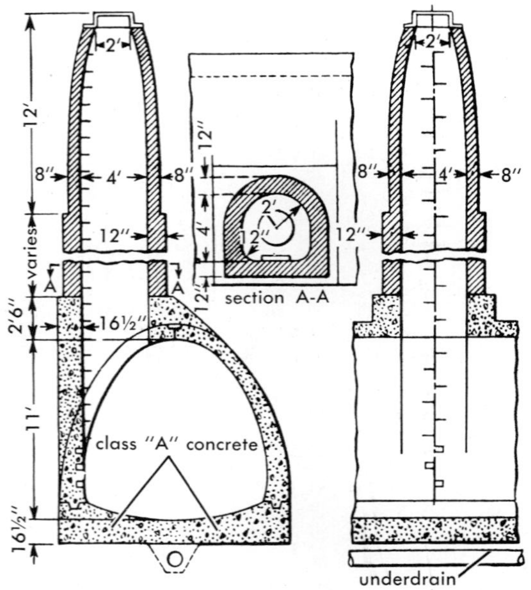

Fig. 10. Deep manhole with access to large sewer. (*From E. W. Steel, Water Supply and Sewerage, 3d ed., McGraw-Hill, 1953*)

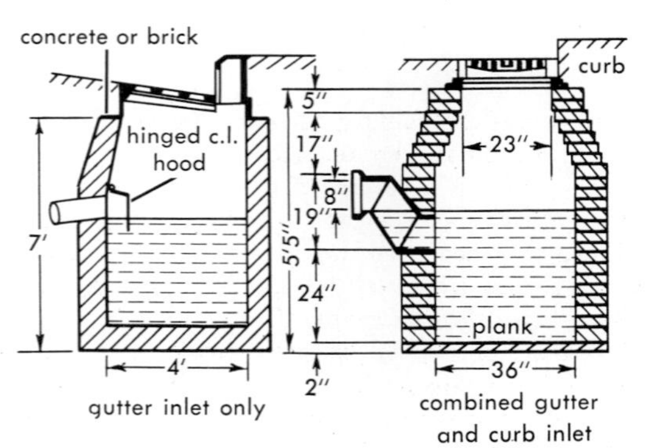

Fig. 11. Types of catch basins. (*From E. W. Steel, Water Supply and Sewerage, 3d ed., McGraw-Hill, 1953*)

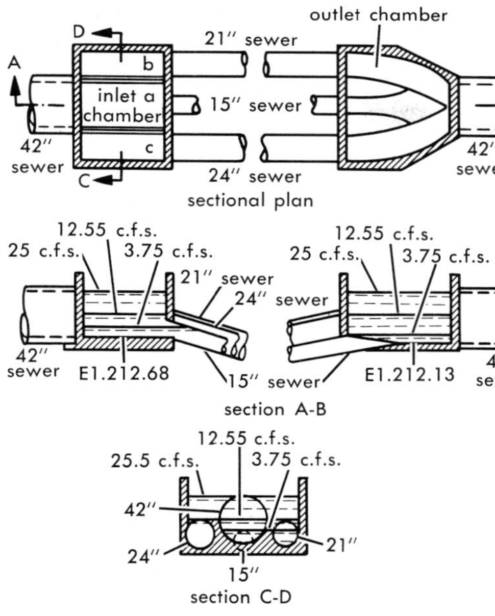

Fig. 12. An inverted siphon. (*From E. W. Steel, Water Supply and Sewerage, 3d ed., McGraw-Hill, 1953*)

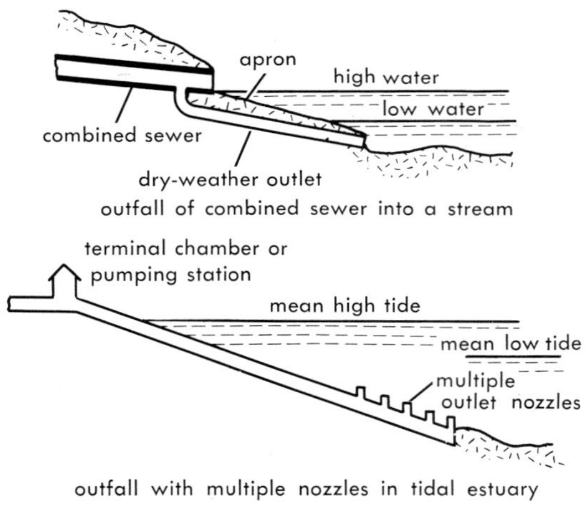

Fig. 13. Sewage outfalls. (*From G. M. Fair and J. C. Geyer, Elements of Water Supply and Waste-Water Disposal, Wiley, 1958*)

nel. It is most frequently used in combined sewer systems to regulate the amount of storm water permitted to flow to a sewage treatment plant. Types of regulators include side flow weirs, leaping weirs, and float-operated gates and valves. Practice varies, but regulators are usually designed to divert twice or three times the dry weather flow.

Inverted siphons. An inverted siphon is a length of sewer set below hydraulic grade line so that it flows under pressure between an inlet chamber and an outlet chamber (Fig. 12). It is usually constructed with two or more pipes of smaller diameter to regulate velocity at minimum and maximum flow and to avoid clogging and reduce cleaning.

Outfall. The outfall is a structure designed to admit treated or untreated sewage to a receiving body of water (Fig. 13). It may be submerged or partially submerged. Storm waters may flow entirely above water level. The simplest form is a headwall supporting a pipe end equipped in some instances with a tide gate to prevent backflow under high water conditions. Treatment plant outfalls frequently are carried into deep water and the piping ends in a series of outlets, rosette-shaped or in fingers, to provide dispersion of the effluent flow. Openings may be turned upward to prevent clogging by shifting bottom deposits.

Sewer system design. A comprehensive study of the community or area to be sewered is made for the purpose of estimating the flow that must be handled by the system at some future period of time, such as 10 or 20 years, or the period of ultimate development. Decisions must be made concerning the type of system to be constructed, separate or combined.

Investigations. Major factors affecting the quantity and flow patterns of sanitary sewage are (1) population and population increase; (2) population density and density change; (3) water use, water demand, water consumption; (4) industrial requirements present and future; (5) commercial requirements present and future; (6) expansion of service geographically; (7) ground-water geology of area; (8) topography of area.

A preliminary layout of sewers and tentative selection of sizes, grades, and location follows. Physical characteristics of the areas to be sewered are determined and attention is given to elevation and plan location of roads, streets, water courses, buildings, basements, underground utilities, and geology. The preliminary report includes a plan of the proposed system together with an estimate of its cost. After the preliminary design is accepted final design begins. Field work is required to establish location and elevation of all existing structures that may affect the design. Borings are made, if necessary, to determine soil and foundation characteristics along the route of sewers and system structures. Final plans and profiles, specifications, and cost estimates are prepared. The project is then ready for the letting of bids and construction.

Plans for construction of sewers usually require review and approval by a supervising state agency such as the health department. The engineer must become familiar with specific regulations and legal requirements applicable to the approval of plans for sewers in the state in which work is to be done.

Sanitary sewer system design. The completed design will include a general map of the whole area, showing location of all sewers and structures and the drainage areas; detailed plans and profiles of sewers, showing ground levels, size of pipe, and slope and location of appurtenances; detailed plans of all appurtenances and structures; a complete narrative report with necessary charts, graphs, and tables to make clear the exact nature

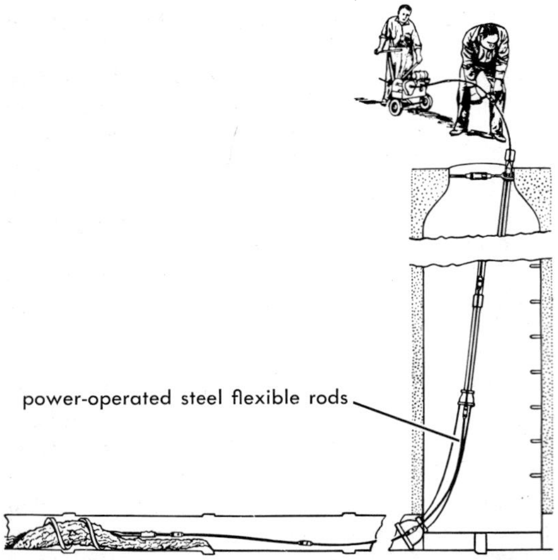

Fig. 14. Typical sewer-cleaning operation. (*From E. W. Steel, Water Supply and Sewerage, 3d ed., McGraw-Hill, 1953*)

of the project; complete specifications; and a confidential estimate of costs made available to the authority or owner.

Extensive plans require tabulation of data beginning at the upper end of the system and proceeding downstream from manhole to manhole bringing into the computation each addition to flow from connecting sewers. Details of such tabulations appear in standard textbooks.

Storm sewer system design. As in sanitary sewer design an estimate must be made of the amount of water to be carried by the system. Factors which must be considered include topography, surface permeability, rainfall intensity and duration, time of runoff, and time of concentration in sewers. The method of analysis is discussed in textbooks on hydrology, hydraulics, and sewerage. The procedure for design and report is essentially that described for sanitary sewers.

Combined systems. Combined system design considers factors for both sanitary and storm sewer flow. Provision must be made for handling dry weather or sanitary flow at proper velocity in sewers that may carry large quantities of water following rainfall. Design is complicated by the need for diversion of waters not flowing to the sewage treatment plant. Structures for diversion are located at or near appropriate water courses, and the effects of discharging polluted water, a combination of sanitary and storm waters, must be fully investigated.

Sewerage system construction. Many features of sewer construction are no different from other types of construction (*see* CONSTRUCTION ENGINEERING). Excavation of trenches and laying of pipe in trenches and tunnels require some variation in construction methods. To construct a sewer it is necessary to remove pavement and ground overburden; use sheeting and bracing to support vertical side walls or the sides and crown of tunnels; dewater the trench; protect all adjacent pipes and structures both in and on the ground; backfill, tamp, and settle soil over finished pipe; and replace pavement.

Maintenance. Roots and accumulations of debris must be removed from sewers periodically (Fig. 14). Special equipment is used to clean out sections of pipe between manholes. An instrument designed to cut roots and scrape sand is installed at a manhole and is attached to rods or cable placed in the sewer. This is pushed or pulled forward. Accumulations are raised to the surface by bucket or similar equipment. Safety precautions must be taken by inspectors and workmen to avoid the hazards of sewer gases and insufficient oxygen. [W.T.I.]

Bibliography: ASCE-FSIWA Joint Committee, *Design and Construction of Sanitary and Storm Sewers;* H. E. Babbitt and E. R. Baumann, *Sewerage and Sewage Treatment*, 8th ed., 1958; C. V. Davis (ed.), *Handbook of Applied Hydraulics*, 2d ed., 1952; G. M. Fair and J. C. Geyer, *Elements of Water Supply and Waste-Water Disposal*, 1958; M. Romanoff, *Underground Corrosion*, Nat. Bur. Standards Circ. 579, 1957; H. Rouse (ed.), *Engineering Hydraulics*, 1950; Storm Drainage Research Committee, *The Design of Storm-Water Inlets*, 1956.

Sewage disposal

All waste waters are eventually discharged into surface- or ground-water courses, which constitute the natural drainage of an area. Most waste waters contain offensive and potentially dangerous substances, which can cause pollution and contamination of the receiving water bodies. Contamination is defined as the impairment of water quality to a degree that creates a hazard to public health. Pollution refers to the adverse effects on water quality that interfere with its proper and beneficial use.

In the past, the dilution afforded by the receiving water body was usually great enough to render these waste substances innocuous. Since the turn of the century, however, the dilution of many rivers has been inadequate to absorb the greater waste discharges caused by the increase in population and expansion of industry.

The principal sources of pollution are domestic sewage and industrial wastes. The former includes the used water from dwellings, commercial establishments, and street washings. The industrial wastes constitute the acids, chemicals, oils, and animal and vegetable matter carried by the cleaning or used process waters from factories and plants. For a discussion of sources of wastes, *see* SEWAGE.

Regulation of water pollution. This is primarily a responsibility of the state, in cooperation with the federal and local governments. The health departments of many states are given statutory power and responsibility for the control of water pollution, and they have established specific water quality standards. There are two basic types of standards—stream standards, dealing with the quality of the receiving water, and effluent standards, referring to strength of wastes discharged. Both types are based on the capacity of the receiving waters to absorb waste substances and on the beneficial uses made of the water.

The self-purification capacity is determined by the available dilution, the biophysical environment of the stream, and the strength and characteristics of the wastes. Beneficial uses include drinking, bathing, recreation, fish culture, irrigation, industrial uses, and disposal of wastes without creation of pollution.

Adjustment of these conflicting interests and equitable distribution of water resources is complex from the technical, economic, and political viewpoints. These considerations have led to the establishment of interstate commissions, which provide a means of coordinated control of the larger rivers.

Water-quality criteria deal with the physical, chemical, and biological parameters of pollution. The most common standards are concerned with physical appearance, odor production, dissolved-oxygen concentration, pathogenic contamination, and potentially toxic or harmful chemicals. For a discussion of these characteristics, *see* SEWAGE. The allowable quantity and concentration of these characteristics and substances vary with the water usage.

Absence of odor and unsightliness and the presence of some dissolved oxygen are common minimum standards. Preliminary or primary treatment of waste waters is usually required for the maintenance of these standards. Highest quality waters require clarity, oxygen saturation, low bacteriological counts, and absence of harmful substances. In these cases, intermediate or complete treatment may be required. *See* SEWAGE TREATMENT; WATER ANALYSIS; WATER PURIFICATION.

Stream pollution. Biological, or bacteriological, pollution is indicated by the presence of the coliform group of organisms. While nonpathogenic itself, this group is a measure of the potential presence of contaminating organisms. Because of temperature, food supply, and predators, the environment provided by natural bodies of water is not favorable to the growth of pathogenic and coliform organisms. Physical factors, such as flocculation and sedimentation, also help remove these bacteria. Any combination of these factors provides the basis for the biological self-purification capacity of natural water bodies.

When subjected to a disinfectant such as chlorine, bacterial die-away is usually defined by Chick's law, which states that the number of organisms destroyed per unit of time is proportional to the number of organisms remaining. This law cannot be directly applied in natural streams because of the variety of factors affecting the removal and death rates in this environment. The die-away is rapid in shallow, turbulent streams of low dilution, and slow in deep, sluggish streams with a high dilution factor. In both cases, higher temperatures increase the rate of removal.

The concentration of many physical characteristics and chemical substances may be calculated directly if the relative volumes of the waste stream and river flow are known. Chlorides and mineral solids fall into this category. Some substances in waste discharges are chemically or biologically unstable, and their rates of decrease can be predicted or measured directly. Sulfites, nitrites, some phenolic compounds, and organic matter are examples of this type of waste.

These simple relationships, however, do not apply to the concentration of dissolved oxygen. This factor depends not only on the relative dilutions, but also upon the rate of oxidation of the organic material and the rate of reaeration of the stream.

Nonpolluted natural waters are usually saturated with dissolved oxygen. They may even be supersaturated due to the oxygen released by green water plants under the influence of sunlight. When an organic waste is discharged into a stream, the dissolved oxygen is utilized by the bacteria in their metabolic processes to oxidize the organic matter. The oxygen is replaced by reaeration through the water surface exposed to the atmosphere. This replenishment permits the bacteria to continue the oxidative process in an aerobic environment. In this state, reasonably clean appearance, freedom from odors, and normal animal and plant life are maintained.

An increase in the concentration of organic matter stimulates the growth of bacteria and increases the rates of oxidation and oxygen utilization. If the concentration of the organic pollutant is so great that the bacteria use oxygen more rapidly than it can be replaced, only anaerobic bacteria can survive and the stabilization of organic matter is accomplished in the absence of oxygen. Under these conditions, the water becomes unsightly and malodorous, and the normal flora and fauna are destroyed. Furthermore, anaerobic decomposition proceeds at a slower rate than aerobic. For maintenance of satisfactory conditions, minimal dissolved oxygen concentrations in receiving streams are of primary importance. *See* BACTERIAL METABOLISM; WATER MICROBIOLOGY.

Figure 1 shows the effect of municipal sewage and industrial wastes on the oxygen content of a stream. Cooling water, used in some industrial processes, is characterized by high temperatures, which reduce the capacity of water to hold oxygen in solution. Thermal pollution, however, is significant only when large quantities are concentrated in relatively small flows. Municipal sewage requires oxygen for its stabilization by bacteria. Oxygen is utilized more rapidly than it is replaced by reaeration, resulting in the death of the normal aquatic life. Further downstream, as the oxygen demands are satisfied, reaeration replenishes the oxygen supply.

Any organic industrial waste produces a similar pattern in the concentration of dissolved oxygen.

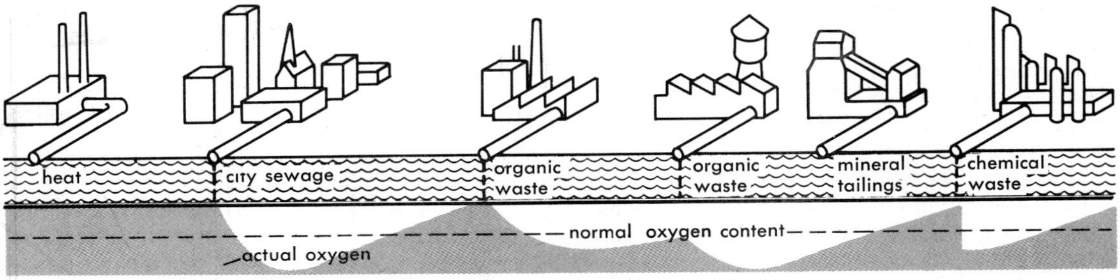

Fig. 1. Variation of oxygen content of polluted stream.

Certain chemical wastes have high oxygen demands which may be exerted quickly, producing a sudden drop in the dissolved oxygen content. Other chemical wastes may be toxic and destroy the biological activity in the stream. Strong acids and alkalies make the water corrosive, and dyes, oils, and floating solids render the stream unsightly. Suspended solids, such as mineral tailings, may settle to the bed of the stream, smother purifying microorganisms, and destroy breeding places. Although these latter factors may not deplete the oxygen, the pollutional effects may still be serious.

Deoxygenation. Polluted waters are deprived of oxygen by the exertion of the biochemical oxygen demand (BOD), which is defined as the quantity of oxygen required by the bacteria to oxidize the organic matter. The rate of this reaction is assumed to be proportional to the concentration of the remaining organic matter, measured in terms of oxygen. This reaction may be expressed as follows

$$\frac{dL}{dt} = -K_1L$$

which integrates to

$$L_t = L_0e^{-K_1t}$$

or

$$y = L_0(1 - e^{-K_1t})$$

in which L_t is BOD remaining at any time, t; L_0 is ultimate BOD; y is BOD exerted at end of time, t; K_1 is coefficient defining the reaction velocity; and t is time. The coefficient is a function of temperature

$$K_T = K_{20} \cdot 1.047^{T-20}$$

in which T is temperature in degrees centigrade; K_T is value of the coefficient at temperature T; and K_{20} is value of the coefficient at 20°C.

The BOD of a waste is determined by a standard laboratory procedure and is reported in terms of the 5-day value at 20°C. From a set of BOD values determined for any time sequence, the reaction velocity constant K_1 may be calculated. Knowledge of this coefficient permits determination of the ultimate BOD from the 5-day value in accordance with the above equations. For municipal sewages and many industrial wastes the value of K_1 at 20°C is between 0.15 and 0.75 per day. A common value for sewage is 0.4 per day.

The coefficient determined from laboratory BOD data may be significantly different from that calculated for stream BOD data. The determination of the stream rate may be made from a reexpression of the above equations as follows

$$K_r = \frac{1}{t} \log \frac{L_A}{L_B}$$

L_A is the BOD measured at an upstream station and L_B the BOD at a station downstream from A, and t is the time of flow between the two stations. Values of K_r range from 0.10 to 3.0 per day. The difference between the laboratory rate K_1 and the stream rate K_r is due to the turbulence of the stream flow, biological growths on the stream bed, insufficient nutrients, and inadequate bacteria in the river water. These factors influence the rate of oxidation in the stream as well as the removal of organic matter. Such processes as flocculation, sedimentation, and scour of the organic material in the river affect the removal rate but do not necessarily influence the rate of oxidation and the associated dissolved oxygen concentration. Field surveys are usually required to determine the pollution assimilation capacity of a stream.

When a significant portion of the waste is in the suspended state, settling of the solids in a slow-moving stream is probable. The organic fraction of the sludge deposits decomposes anaerobically, except for the thin surface layer which is subjected to aerobic decomposition due to the dissolved oxygen in the overlying waters. In warm weather, when the anaerobic decomposition proceeds at a more rapid rate, gaseous end products, usually carbon dioxide and methane, rise through the supernatant waters. The evolution of the gas bubbles may raise sludge particles to the water surface. Although this phenomenon may occur while the water contains some dissolved oxygen, the more intense action during the summer usually results in depletion of dissolved oxygen.

Reoxygenation. Water may absorb oxygen from the atmosphere when the oxygen in solution falls below saturation. Dissolved oxygen for receiving waters is also derived from two other sources: that in the receiving water and the waste flow at the point of discharge, and that given off by green plants. The latter source is restricted to daylight hours and the warmer seasons of the year and, therefore, is not usually used in any engineering analysis of stream capacity.

Unpolluted water maintains in solution the maximum quantity of dissolved oxygen. The saturation value is a function of temperature and the concentration of dissolved substances, such as chlorides. When oxygen is removed from solution, the deficiency is made up by the atmospheric oxygen, which is absorbed at the water surface and passes into solution. The rate at which oxygen is absorbed, or the rate of reaeration, is proportional to the degree of undersaturation and may be expressed as follows:

$$\frac{dD}{dt} = -K_2D$$

in which D is dissolved oxygen deficit, t is time, and K_2 is reaeration coefficient.

The reaeration coefficient depends upon the ratio of the volume to the surface area and the intensity of fluid turbulence. An approximate value of the coefficient may be obtained from the following formula:

$$K_2 = \frac{D_L U^{1/2}}{H^{3/2}}$$

in which D_L is coefficient of molecular diffusion of oxygen in water, U is average velocity of the river flow, and H is average depth of the river section.

The effect of temperature on this coefficient is identical with its effect on the deoxygenation coefficient. A common range of K_2 is from 0.20 to 5.0 per day. Many waste constituents, such as surface active substances, interfere with the molecular diffusion of oxygen and reduce the value of the reaeration rate from that of pure water. Winds, waves, rapids, and tidal mixing are factors which create circulation and surface renewal and enhance reaeration.

Oxygen balance. The oxygen balance in a stream is determined by the concentration of organic matter and its rate of oxidation, and by the dissolved oxygen concentration and the rate of reaeration. The simultaneous action of deoxygenation and reaeration produces a pattern in the dissolved oxygen concentration known as the dissolved oxygen sag. The differential equation describing the combined action of deoxygenation and reaeration is as follows:

$$\frac{dD}{dt} = K_1L - K_2D$$

This equation states that the rate of change in the dissolved oxygen deficit D is the result of two independent rates. The first is that of oxygen utilization in the oxidation of organic matter. This reaction increases the dissolved oxygen deficit at a rate that is proportional to the concentration or organic matter L. The second rate is that of reaeration, which replenishes the oxygen utilized by the first reaction and decreases the deficit. Integration of this equation yields

$$D_t = \frac{K_1L_0}{K_2 - K_r}(e^{-K_rt} - e^{-K_2t}) + D_0e^{-K_2t}$$

L_0 and D_0 are the initial biochemical oxygen demand and the initial dissolved oxygen deficit, respectively, and D_t is the deficit at time t. The proportionality constants K_1 and K_2 represent the coefficients of deaeration and reaeration, respectively, and K_r the coefficient of BOD removal in the stream.

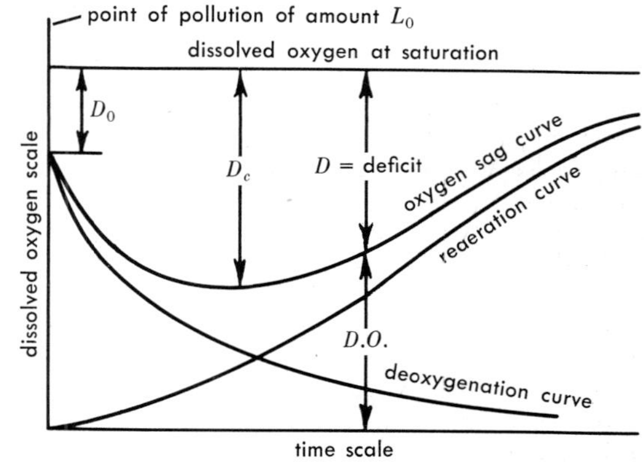

Fig. 2. Typical dissolved oxygen sag curve and its components.

Figure 2 shows a typical dissolved oxygen sag curve resulting from a pollution of amount L_0 at $t = 0$. The sag curve is shown to result from the deoxygenation curve and the reaeration curve. A point of particular significance on the sag curve is that of minimum dissolved oxygen concentration, or maximum deficit. At this location, the rate of change of the deficit is zero, which results in the numerical equality of the opposing rates of deoxygenation and reoxygenation. The balance at this critical point may be written

$$K_2D_c = K_1L = K_1L_0e^{-K_rt_c}$$

where the BOD at the critical point has been replaced by its equivalent at zero time (the location of the waste discharge). The value of the time t_c may be calculated from the equation

$$t_c = \frac{1}{K_2 - K_r} \log \frac{K_2}{K_r}\left[1 - \frac{D_0(K_2 - K_r)}{K_1L_0}\right]$$

Allowable pollutional load. The pollutional load L_0 that a stream may absorb is a function of the dissolved oxygen deficit D_c, the coefficients K_1, K_r, and K_2, and the initial deficit D_0. The dissolved oxygen deficit is usually established by water pollution standards of the health agency, and the initial deficit is determined by upstream pollution. The engineering problem is usually associated with the assignment of representative values of the coefficients K_1, K_r, and K_2 for a given flow and temperature condition.

Seasonal temperatures influence the saturation of oxygen and the rates of deaeration and reaeration. Variation in stream flow with the seasons affects the dilution factor. The most critical conditions occur during the summer when the stream runoff is low and the temperatures are high.

Pollution in lakes and estuaries. In lakes self-purification is slower than in streams because of the low rates of dispersion of the waste waters. The turbulence characteristic of flowing rivers is not present, and mixing depends primarily on winds, waves, and currents. Waste-water outfalls are designed to take advantage of the dispersion induced by these factors and to prevent the development of concentrated sewage fields.

In estuaries, the dispersion of waste waters is complicated by the tides, which carry various portions of the pollutant back and forth over many cycles, and by the difference of density in fresh water, waste water, and salt water. The equation defining the oxygen balance must be modified to allow for the greater time that an average particle of pollution is detained within the estuary; the flushing mechanism of such bodies is therefore of primary concern. *See* ESTUARINE OCEANOGRAPHY.

Each estuary presents problems of density currents, configuration, and exchange that distinguish it from others. Field measurements of salinities, currents, and cross sections, in addition to the measurement of physical, chemical, and biological characteristics, are necessary to evaluate the pollution capacity of these watercourses. Dilution and dispersion in ocean waters is complicated by many of the same factors as in estuaries. The death rates of the coliform bacteria are greater in sea water. The outfalls must be designed and located to promote effective dispersion and to prevent the accumulation of sewage fields.

Oxidation ponds and land disposal. The forces of natural purification are utilized in shallow ponds, called oxidation ponds. Successful operation of these basins usually requires relatively high temperatures and sunshine. Carbon dioxide is released by means of the bacterial decomposition of the organic matter. Algae growth develops, consuming the carbon dioxide, ammonia, and other waste products and releasing oxygen under proper climatic conditions. Oxidation ponds are efficient and relatively economical.

Instead of relying on the algae as a primary source of oxygen, mechanical aeration of the pond contents may be employed. Lagoons aerated in this manner are not as susceptible to climatic conditions as the oxidation ponds.

Land disposal of sewage is occasionally practiced by surface or flood irrigation. The former is the discharge of sewage upon the ground, from which it evaporates and through which it percolates. However, a significant portion remains which must be collected in surface drainage channels. Although this method is not particularly efficient for domestic sewage, a modification of it, spray irrigation, has been successfully employed in the treatment of a few industrial wastes. In flood irrigation, all the sewage is permitted to seep through the ground and is usually collected in underdrains. This method takes advantage of the mechanical filtration and biological purification afforded by the soil. Unless the sewage is treated before irrigation, odors and clogging usually occur and possible contamination of ground or surface water can result. *See* SANITARY ENGINEERING. [D.J.O.]

Bibliography: G. M. Fair and J. C. Geyer, *Water Supply and Waste-Water Disposal*, 1954; H. W. Streeter and E. B. Phelps, *A Study of the Pollution and Natural Purification of the Ohio River*, Public Health Bull. 146, 1925; U.S. Public Health Service, *Oxygen Relationships in Stream*, Tech. Rept. W58–2, 1958.

Sewage solids

A semiliquid mass, called sludge, removed from the liquid flow of sewage.

Solids treatment. This depends on the source of the solid and its characteristics. Solids are removed as screenings, grit, primary sludge, secondary sludge, and scum.

Screenings. Screenings are putrescible and offensive. They are either ground and returned to the sewage, ground and transferred to the digestor, incinerated, or buried. The quantity of screenings is variable and is dependent on sewage characteristics (*see* SEWAGE). Coarse screenings will vary from 0.3 to 5 ft^3/1,000,000 gal. Fine screenings will range from 5 to 35 ft^3/1,000,000 gal. Grit collection has a wide variation. Normally the volume will be between 1 and 10 ft^3/1,000,000 gal.

Sludge. Sludge will vary in amount and characteristics with the characteristics of sewage and plant operations. The following refers to a reasonable normal for each source of sludge. Primary sludge is composed of gray, viscous, identifiable solids, putrescible, odorous, 2500 gal 95% moisture content per 1,000,000 gal. Trickling filter sludge is black, dark brown, granular or flocculent, partially decomposed, not highly odorous when fresh, 500 gal 92.5% moisture content sludge per 1,000,000 gal. Activated sludge is dark to golden brown, partially decomposed, granular flocculent, earthy odor when fresh, 13,500 gal 98% moisture content per 1,000,000 gal.

Digestion. Digestion is the anaerobic decomposition of organic matter resulting in partial gasification, liquefaction, and mineralization. Sludges (except chemical sludges) from treatment processes can be digested provided there are no substances such as cyanides and chromium, toxic to organisms present in the sludge. Sludges are transferred to separate digestion tanks except where Imhoff-type tanks or septic tanks are in use.

The digestion process is a progressive decomposition of organic matter, which comprises about 70% of the total sludge weight. Carbohydrates are attacked first and organic acids are formed. This stage is known as acid fermentation. Organisms living in acid environment continue digestion during a second stage, known as acid digestion, when organic acids and nitrogenous materials are at-

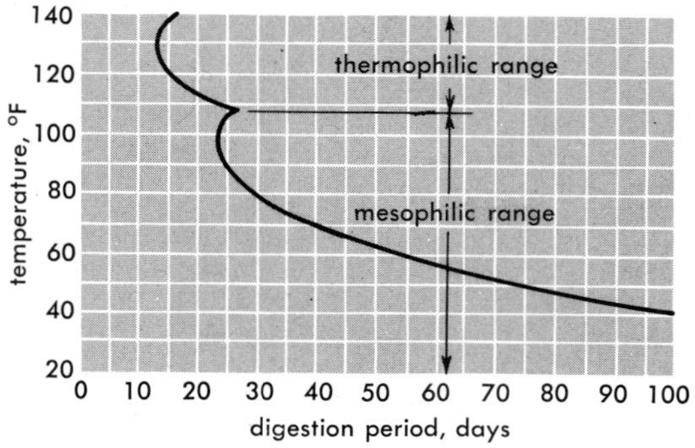

Fig. 1. **Effect of temperature on time of digestion of seeded sludge.** (*From E. W. Steel, Water Supply and Sewerage, 3d ed., McGraw-Hill, 1953*)

tacked. During the third stage—a period of digestion, stabilization, and gasification—proteins and amino acids are subject to bacterial action. Volatile acids are reduced and pH rises. The final stage is referred to as alkaline digestion. The principal gas produced during this stage is methane. *See* AMINO ACIDS; CARBOHYDRATE; FERMENTATION; METHANE; PROTEIN.

In a single tank all three stages proceed simultaneously. Fresh solids mixed with well-digested materials providing balance and holding the pH above 7.0 offer a fairly ideal condition. Liquefied materials, excess liquor (or supernatant liquor), and digested solids are removed, making room for fresh material. After the balance has been obtained it is possible to continue the operation if fresh solids are held to less than 4% of the tank solids measured on a dry weight basis.

Sludge-digestion tanks are circular or rectangular, heated or unheated units. Most states have established schedules of capacity requirements for different types of sludge on heated and unheated bases. Imhoff tanks are unheated and require sludge capacity of 3–4 ft^3 per capita. Primary sludges require 2–3 ft^3 heated and 4–6 ft^3 per capita unheated. Filter and primary sludge mixed runs from 3–5 ft^3 heated and 6–10 ft^3 per capita unheated. Activated sludge requires 4–6 ft^3 heated and 8–12 ft^3 per capita unheated. Heated tanks provide controlled temperature for thermophilic (110–140°F) or mesophilic digestion. Temperatures around 100°F are optimum (Fig. 1).

Provision is made for manipulation of the sludge, and the system may include preheater and heater equipment, recirculation pumps with sludge suction at several levels, supernatant liquor drawoff at several levels, gas dome or collector, stirring mechanism, sludge rakes, and drawoff. Covers may be fixed or floating (Fig. 2).

Multistage digestion occurs when two digestors or more are placed in series, the sludge drawoff of the first being connected to the second and continuing. In this system flexibility in manipulating and mixing sludges and in controlling supernatant liquor is possible.

Supernatant liquor. This is the liquid fraction in a digestor. It is offensive in odor, high in solids and in BOD (biochemical oxygen demand). It is discharged to the incoming sewage and treated in the primary sedimentation unit. It is withdrawn in small quantities from a level having fewer solids. Activated sludge with high moisture content is sometimes settled in a preliminary operation before discharge to digestor. The volume may be reduced as much as 50% and the decant liquor is less objectionable than the supernatant.

Sludge gas. Sludge-gas production under good operating conditions is about 12 ft^3/lb of volatiles destroyed. The gas is 60–70% methane and 20–30% carbon dioxide with minor amounts of impurities such as hydrogen sulfide. Gas has a fuel value

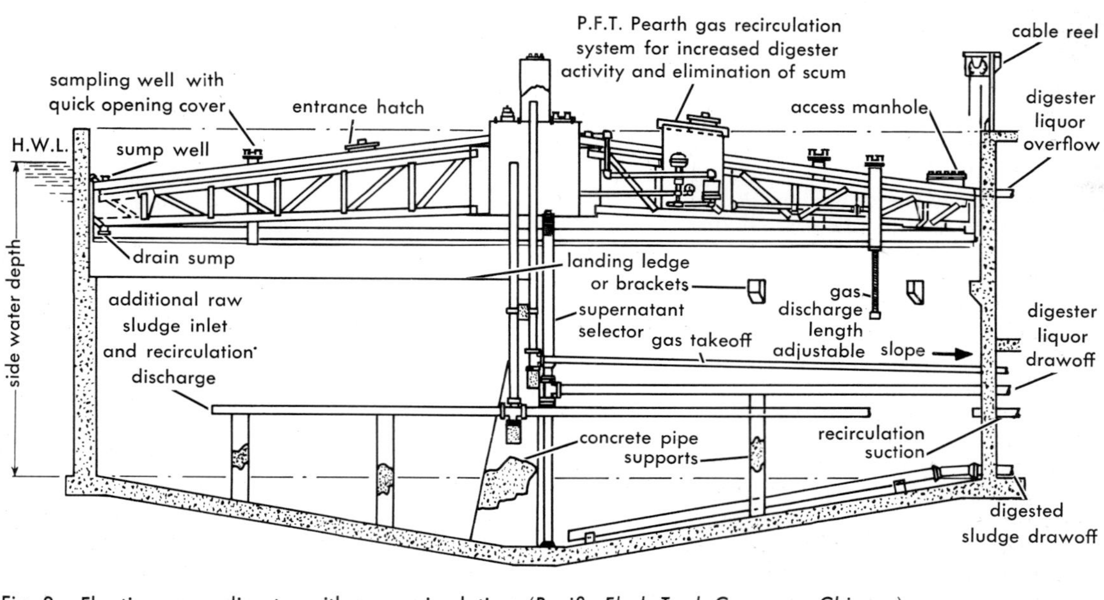

Fig. 2. Floating cover digestor with gas recirculation. (*Pacific Flush Tank Company, Chicago*)

of 600–700 Btu/ft^3 and is used at the plant to operate auxiliary engines and provide heat for sludge-heating systems. Excess gas is burned.

Sludge-drying beds. These are provided at smaller plants to handle sludge removed from digestors without further treatment. Drying beds should have an area of 2–3 ft^2 per capita. Covered beds require about three-quarters of that area. Beds consist of up to 12 in. of coarse sand over 12-in. gravel packed around underdrains. Sludge is drawn to a depth of 9–12 in. and allowed to drain and dry. A well-digested granular sludge drains easily and reduces to a depth of 3–4 in. when dry (60–70% moisture content). Sludge is removed from the bed and eventually may be used as humus material. It has little or no odor.

Sludge processing. Sludge processing may be required if the sludge is to be disposed of by other methods. Elutriation, or washing of sludge with plant effluent, removes undesirable amino-ammonia nitrogen and reduces or eliminates the need for conditioning chemicals. Lime or ferric chloride may be used to prepare sludge for vacuum filtration. Filter cake containing 70–80% moisture is more easily handled. In some plants built in the 1950s raw sludge is conditioned and processed on various filters without digestion. Such sludge is offensive and is handled in the same manner as screenings.

Filter rates expressed as lb/(ft^2)(hour) may be taken generally at 3.5 and will range from 2.5 for fresh activated sludge to 8 for primary digested sludge. Filters are revolving drums covered with wire, plastic, or cotton cloth, or flexible metal spring-like coils. Drums revolve at 1.5–9 min per revolution, passing through a basin of sludge. Vacuum within the drum picks up sludge against the media and separates water from the solids. The filtrate is returned to sewage flow or to elutriators.

Solids disposal. Sewage solids must be disposed of without nuisance or hazard to health. Burial, incineration, and drying for use as fertilizer are means of final disposal. Lagooning may be used and in sea-coast cities sludge may be taken out to sea.

Burial. Screenings are handled by hand in small plants. At one time special screening pits were prepared and the material was placed and covered until it had composted. The recovery does not justify the work and generally screenings are burned or placed in a sanitary landfill. The fill may be either on the plant premises or a part of municipal refuse disposal.

Incineration. Incineration of sludge has developed as a means of disposal in larger plants. Incineration introduces problems of air pollution (*see* AIR POLLUTION CONTROL). Incineration of sludge requires auxiliary heat because the moisture content is high. Gas or oil or digestor gas may be used as fuel. Incinerators used to burn sludge are generally multiple hearth. Sludge is fed to the top hearth and as it dries it is dropped down to the next hearth by agitator arms. Water is driven off and volatile gases are released by the heat. The gases are ignited by the furnace temperature. To avoid excessive odors, the temperature should be maintained at 1200–1400°F. Ash residue is inert and may be used for fill or cover on sanitary landfill. Since the volume of treatment-plant sludge is small, plant incinerators are not operated daily. Auxiliary fuel is required to preheat the combustion chamber. If there is no digestor gas fuel available, the costs of sludge incineration can be excessive. Sludge cake can be mixed with refuse and burned in municipal incinerators, if the two facilities are adjoining.

Drying. Drying of sludge is substituted for incineration in some plants. The dried sludge can be used for fertilizer by enriching it with chemicals, particularly potash, which is lacking in most digested sludges. In this process water is driven off without burning the material. Sludges from drying beds may be stockpiled. After a year or so these sludges are earthlike and may be used as a soil conditioner or as a soil builder when preparing new land areas over sanitary landfill and on sand. Sludges from beds and filter cake may be put into sanitary landfill. The fill is compact but burnable.

Flash driers operate by mixing a portion of dried sludge with the incoming wet sludge cake. A high-velocity, high-temperature gas stream evaporates the water. The dried material is then passed through a cyclone separator and is carried to storage, which may be at a fertilizer plant. Municipalities having refuse incinerators at the sewage treatment site provide a ready source of heat that can be used for sludge drying.

Spray driers have some usefulness in handling liquid sludges. The sludge suspension is ejected under high pressure into a heated chamber. The sudden pressure release atomizes the suspension and water is quickly driven off as the material falls to the bottom. The material is removed by a separator. Heat requirements are high and the method has the limitations of fuel cost if no source of waste heat is available.

Land disposal. Disposal on land has limited application. There are a few locations where sludge may be taken to fields and plowed under. Occasionally liquid sludge may be applied to gardens and lawns around a treatment plant. The practice of land disposal has certain public-health dangers and must be closely supervised. [W.T.I.]

Bibliography: ASCE-FSIWA Joint Committee, *Sewage Treatment Plant Design*, 1959; H. E. Babbitt and E. R. Baumann, *Sewerage and Sewage Treatment*, 8th ed., 1958; G. M. Fair and J. C. Geyer, *Elements of Water Supply and Waste-Water Disposal*, 1958; E. W. Steel, *Water Supply and Sewerage*, 3d ed., 1953.

Sewage treatment

Any process to which sewage is subjected in order to remove or alter its objectionable constituents and thus render it less offensive or dangerous. These processes may be classified as preliminary, primary, secondary, or complete, depending on the degree of treatment accomplished. Preliminary treatment may be the conditioning of industrial waste prior to discharge to remove or to neutralize substances injurious to sewers and treatment processes, or it may be unit operations which prepare the water for major treatment. Primary treatment is the first and sometimes the only treatment of sewage. It is the removal of floating solids and coarse and fine suspended solids. Secondary treatment utilizes biological methods of treatment, that is, oxidation processes following primary treatment by sedimentation. Complete treatment removes a high percentage of suspended, colloidal, and organic matter.

Septic tanks and Imhoff tanks are considered secondary treatment methods because sedimentation is combined with biological digestion of the sludge. *See* IMHOFF TANK; SEPTIC TANK.

Coarse solids removal. This is accomplished by means of racks, screens, grit chambers, and skimming tanks. Racks are fixed screens comprised of parallel bars placed in the waterway to catch debris. The bars are usually spaced 1 in. or more apart. Screens are devices with openings usually of uniform size 1 in. or less placed in the line of flow. Screens may be fixed or movable and vary in construction as bar screens, band screens, or cage screens. Such screens are hand cleaned or mechanically cleaned (Fig. 1). Grit chambers remove inorganic solids but may also trap heavier particles of organic nature such as seeds (Fig. 2). Grit chambers are designed so that the flow in the chamber is at 1 ft/sec or more. At less than that velocity, organic material also settles. Removal of grit is done either by hand or mechanically. Devices are added to mechanically cleaned units which wash most of the organic material out of the grit. Skimming chambers are devices for removing floating solids and grease. Air has been used to coagulate greases which then float and are skimmed off mechanically or by hand.

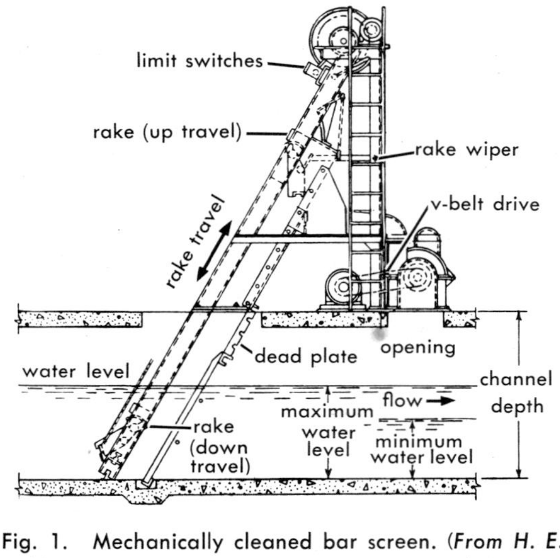

Fig. 1. Mechanically cleaned bar screen. (*From H. E. Babbitt and E. R. Baumann, Sewerage and Sewage Treatment, 8th ed., Wiley, 1958*)

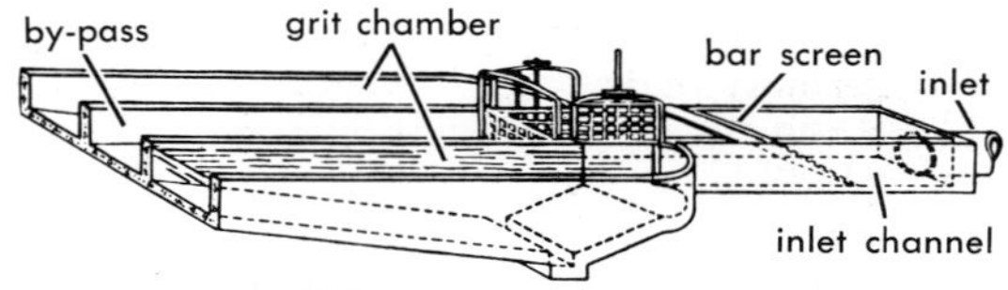

Fig. 2. Diagram of a grit chamber. (*From Bull. 58 Eng. Extension Dept. Iowa State College, 1953*)

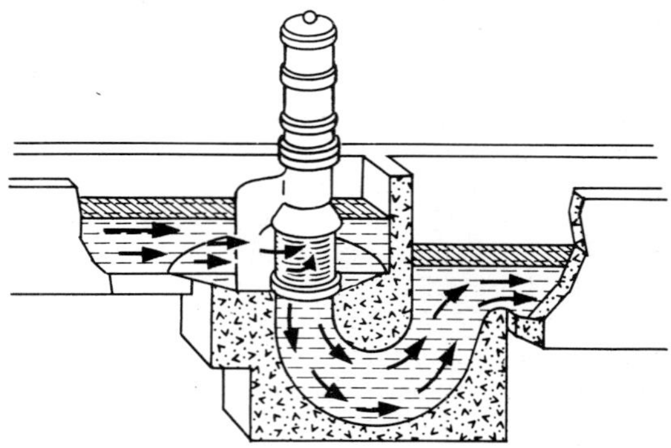

Fig. 3. A comminutor in place. (*Chicago Pump Company*)

Fine solids removal. This is accomplished by screens with very small openings $\frac{1}{16}$ or $\frac{1}{32}$ in. wide, by sedimentation, or by both.

Fine screens are set in the line of flow and are operated mechanically. Band screens, drum screens, plate screens, and vibratory screens are in use and the finer particles of floating solids are removed as well as coarse solids passing a rack. In some treatment plants screenings are passed through a grinder and returned to the flow so that they will settle out in the sedimentation tank. Another device, the comminutor, barminutor, or griductor, has high-speed rotating edges working in the flow of sewage (Fig. 3). These blades cut, chop, and shred the solids and they pass on to the sedimentation unit.

Sedimentation. Sedimentation has one objective, the removal of settleable solids. Some floating materials are also removed by skimming devices, called clarifiers, built into sedimentation units. The basins are either circular or rectangular. In the circular unit sewage flows in at the center and out over weirs along the circumference (Fig. 4). In the rectangular tanks sewage flows into one end and out the other (Fig. 5).

The efficiency of a settling basin is dependent on a number of factors other than particle size, specific gravity, and settling velocity. Concentration of suspended matter, temperature, retention period, depth and shape of basin, baffling, total length of flow, wind, and biological effects all have an effect on solids removal. Density currents and short-circuiting may negate theoretical detention

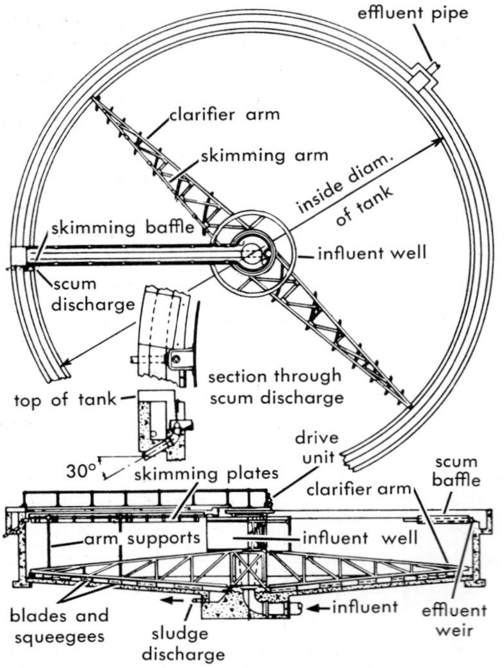

Fig. 4. Typical circular clarifier. (*From H. E. Babbitt and E. R. Baumann, Sewerage and Sewage Treatment, 8th ed., Wiley, 1958*)

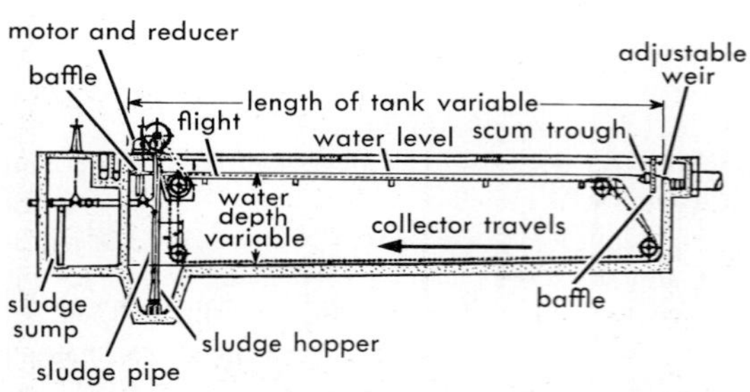

Fig. 5. Longitudinal section of typical rectangular clarifier. (*From E. W. Steel, Water Supply and Sewerage, 3d ed., McGraw-Hill, 1953*)

computations. Improper baffling may have the effect of reducing the effective surface area and creating dead or nonflow areas within the tank. In general a settling tank of good design with surface settling rates of 1000 gal/(ft^2)(day) and a 2-hour detention period will remove 50–60% of the suspended solids and at the same time remove 30–35% of the biochemical oxygen demand (Fig. 6).

The settling velocity of a particle is a function of specific gravity of the particle, specific gravity of water, viscosity of liquid, and particle diameter. Settling rates of particles larger than 0.1 mm are determined empirically. Sizes less than 0.1 mm settle in accordance with Stokes' law. Theoretically if the forward motion of the water is less than the vertical settling rate of the particle, a particle at the surface will settle some distance below the surface in a given time interval. After that time interval the surface layer of water could be removed and it would contain no solids. The term surface settling rate is introduced as a practical measure of the rate of flow through the basin, if the rate of flow is equal to the surface area times the settling velocity of the smallest particle to be removed. Hence the selection of an overflow or surface settling rate expressed as gallons per day per square foot of surface area establishes a relationship between flow and area.

Flocculent suspensions have little or no settling velocity. These may occur in raw sewage but occur more frequently in secondary settling of effluents from activated sludge units. Such suspensions may be removed by passing the inflowing water upward through a blanket of the material (Fig. 7). Theoretically there is a mechanical sweeping action in which smaller particles are attached to larger particles which then have sufficient weight to settle. Another type of treatment for such material is provided by an inner chamber equipped with baffles which rotate and stir the liquor and aid the formation of larger and heavier floc (Fig. 8). The same purpose is also achieved by agitation with air. Some of the settled sludge is raised by airlift and mixed in with the material, thus forming a mixture with improved settling characteristics.

Sedimentation basin design. Practical considerations and engineering judgment must be applied in designing sedimentation basins. Depth is usually held at 10 ft sidewall depth or less. The surface area requirement is usually 600 gal/(ft^2)(day) for primary treatment alone and 800–1000 for all other tanks. The detention period is normally 2 hours. These three parameters of design must be adjusted since each is dependent on the other for a given design flow (average daily flow at a plant). When mechanical sludge removal equipment is used, the tank dimensions are usually sized to a conventional equipment specification. Rectangular tanks are built in units with common walls between units and

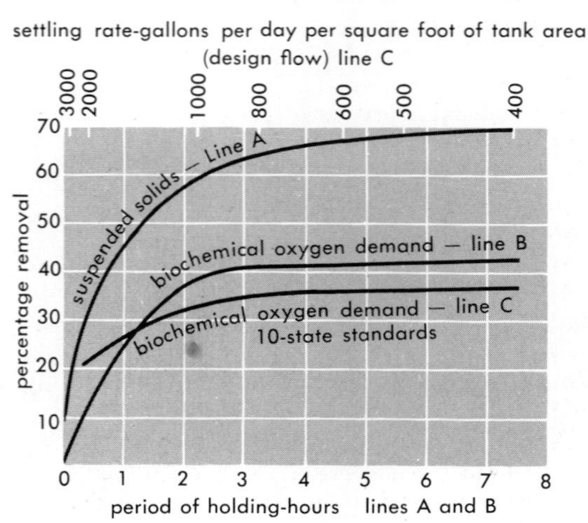

Fig. 6. Probable performance of sedimentation basins. (*From H. E. Babbitt and E. R. Baumann, Sewerage and Sewage Treatment, 8th ed., Wiley, 1958*)

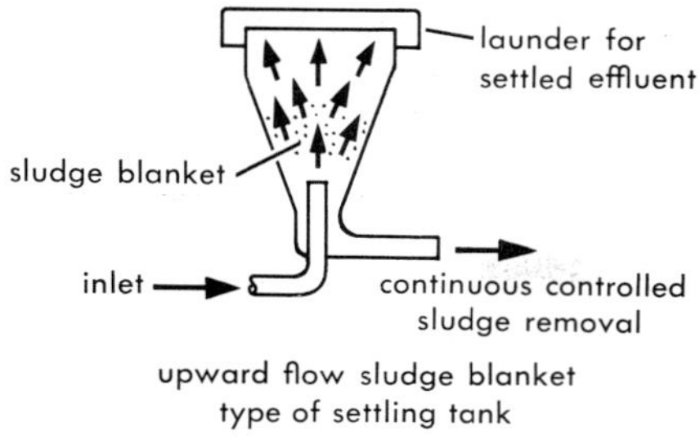

Fig. 7. Diagram of a vertical-flow sedimentation tank. (*From H. E. Babbitt and E. R. Baumann, Sewerage and Sewage Treatment, 8th ed., Wiley, 1958*)

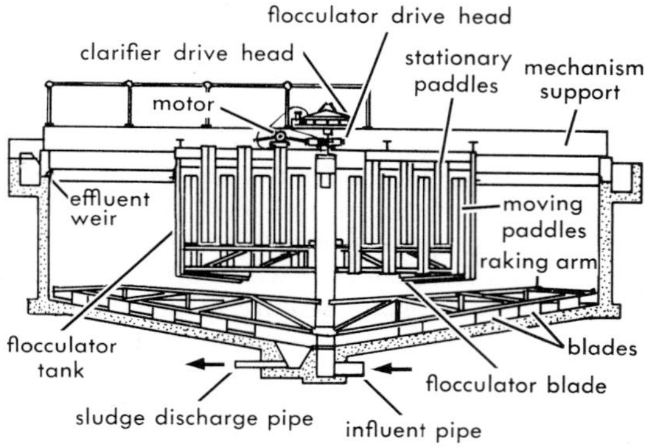

Fig. 8. The Dorr clariflocculator. (*From E. W. Steel, Water Supply and Sewerage, 3d ed., McGraw-Hill, 1953*)

unit width up to 25 ft. The length-width ratio, frequently determined by economical design dimension, should not be greater than 5:1. The minimum length should be 10 ft. Final sizing may be fitted to convenient equipment dimensions.

Sludge removal on a regular schedule is mandatory in separate sedimentation tanks. If sludge is not removed, gasification occurs and large blocks of sludge begin to appear on the surface. These must then be removed by scum removal mechanism or broken up so that they will settle. In circular tanks radial blades move the sludge to a center sludge hopper. In rectangular tanks the hopper is located at the inlet end and blades on a traveling chain move sludge in reverse of sewage flow. The heavier solids settle at the inlet and have a short travel path. These same blades may rise to the surface and move scum with the sewage flow to the outlet end where it is held by a baffle and removed by some form of scum removal device. Sludge removal mechanisms are often operated intermittently by time-clock relay mechanisms.

Appurtenances in the form of skimmers, scrapers, and other mechanical devices are many. Manufacturers have variants to offer and competition is keen. Manufacturers' literature should be studied carefully and specifications should be carefully written to procure equipment meeting the requirements of engineering design.

Detention periods are theoretical. The actual flowthrough time is influenced by the inlet and outlet construction. On circular tanks inlets are submerged. Water rises inside a baffle extending downward to still the currents. Rectangular tank inlets may be submerged, or more commonly, sewage is brought to a trough which has a weir extending the width of the tank. The flow then moves forward with less short-circuiting. The outlet device on circular tanks is nearly always a circumferential weir adjusted to level after installation. The weir may be sharp-edged and level or provided with a sawtoothlike series of v-notches. On rectangular tanks, in order to provide enough weir length, a device known as a launder is used. A launder is a series of fingerlike shallow conduits set to water level and receiving flow from both sides of the conduit. Each of the fingers is connected to a common exit trough. The normal weir loading should not exceed 10,000 gal/linear ft of weir per day in small plants, or 15,000 in units handling more than 1,000,000 gal per day (1.0 mgd).

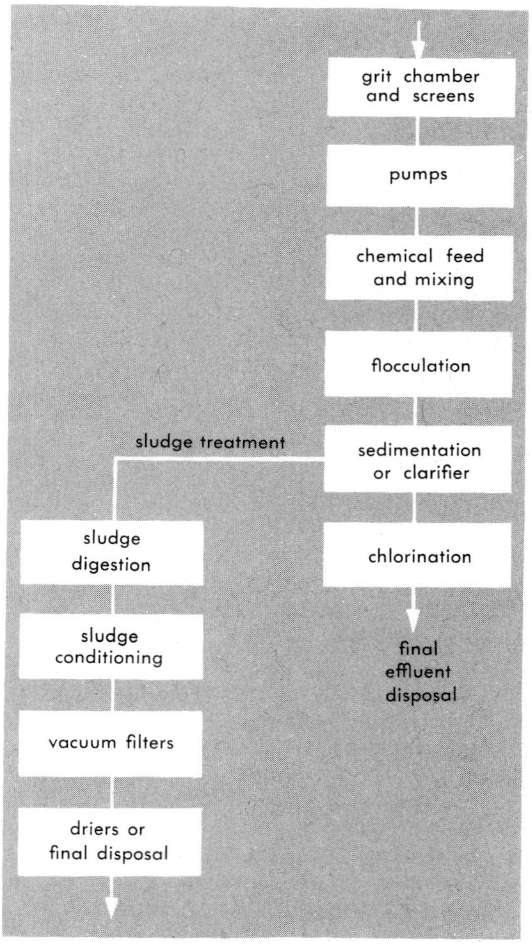

Fig. 9. Flow-through diagram of a chemical treatment plant. (*From H. E. Babbitt and E. R. Baumann, Sewerage and Sewage Treatment, 8th ed., Wiley, 1958*)

Chemical precipitation. Many attempts have been made to utilize chemical coagulants in the flocculation of sewage. The process, if used, is similar to that used in water treatment (*see* WATER TREATMENT). The cost of chemicals and the somewhat intermediate treatment obtained with chemicals have kept this process out of general use. Its principal use today is in the preparation of sludge for filtration. Various steps in chemical precipitation are shown in Fig. 9. Alum, ferric sulfate, ferric chloride, and lime are used to form an insoluble precipitate which adsorbs colloidal and suspended solids. The entire floc settles and is removed as sludge. Sixty-four patents for proprietary chemical treatment processes were granted in the United States from 1873 to 1935. The Guggenheim process employs ferric chloride and aeration. The Scott-Darcy process employs ferric chloride made by treating scrap iron with chlorine solution.

Oxidation processes. These are secondary treatment processes, although a few activated sludge plants have been built without primary sedimentation. Oxidation process methods are (1) filtration by intermittent sand filters, contact filters, and trickling filters; (2) aeration by the activated sludge process or by contact aerators; (3) oxidation ponds. There are three basic oxidation methods, all depending on biological growth. Each provides a method of bringing organic matter in suspension or solution in sewage into immediate contact with a population of microorganisms living under aerobic conditions. The processes are called filtration, activated sludge, and contact aeration.

Filtration. Intermittent sand filters are sand beds provided with underdrains. Sewage is dosed intermittently by siphon or by pump, at rates from 20,000 gal per acre per day (gad) to a maximum of 125,000 gad when operated as a secondary treatment process. Rates may go to 500,000 gal per acre per day (0.5 mgad) when operated as a tertiary process. Beds are usually 2½–3 ft deep and are constructed with 6–12 in. of gravel at the bottom. The sand is sized to a uniformity coefficient of 5.0 or less (3.5 preferred) with effective size of 0.2–0.5 mm. The uniformity coefficient is the ratio between the sieve size that will pass 60% to the effective size. The effective size is the sieve size in millimeters that permits 10% of the sand by weight to pass. A mat of solids is formed in the surface layer of sand and must be removed periodically. The dry surface mat can be scraped clean, but periodically the top 6 in. or so of mat must be removed and replaced. Plants with sand filters operate at better than 95% removal of biochemical oxygen demand (BOD).

Trickling filters are beds of media, usually rock, over which settled sewage is sprayed. Microorganisms form a slime layer on the media surface and the water passes down over the surface in a thin film. Nutrients from the sewage are adsorbed in the slime layer and absorbed as food by organisms. Filters are ventilated through the underdrainage system or by other means, and thus oxygen, sewage, and organisms are brought together. Plants with trickling filtration have been operated at 90–95% efficiency of BOD removal.

Filter media include various materials such as stone, crushed rock, ceramic shapes, slag, and plastics. Stone and crushed rock which do not fragment, flour, or soften on exposure to sewage are preferred media. Rock sizes range from 1 to 6 in.; however, current practice employs sizes between 2- and 4-in. nominal diameter. Plastic corrugated sheets have been employed on very deep filters. Pretreatment of sewage is normally required. When the waste contains a concentration of dissolved solids, as with milk waste, without any great concentration of settleable solids, the waste may be applied directly to the filter. Some advantage is gained by preaeration so that the waste applied to the filter has some dissolved oxygen.

Filters are classified as standard or low-rate filters, high-rate filters, and controlled filters. The filter introduced in the United States early in the twentieth century was a bed of stone 6–8 ft deep with a distribution system of fixed nozzles. This type of filter is called a standard or low-rate filter. The allowable organic loading is about one-third that of a high-rate filter having 3- to 6-ft depth introduced during the decade 1930–1940 and developed with many variations of recirculation and application of sewage since that time. In 1956 a new approach was taken with the introduction of controlled filtration on sectionalized units comprising a deep filter. The loading rate with no recirculation on such filters is 10–12 times that of low-rate filters.

Low-rate filters are dosed at a rate of 1–4 mgad by siphon through nozzles so spaced that water reaches every part of the filter surface during a

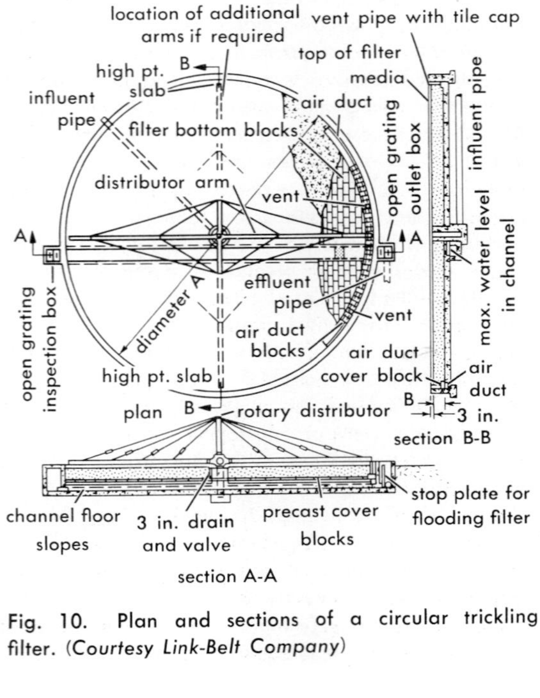

Fig. 10. Plan and sections of a circular trickling filter. (*Courtesy Link-Belt Company*)

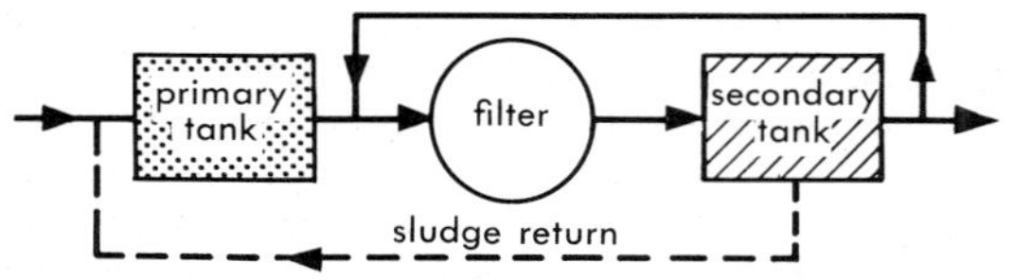

Fig. 11. Flow diagram of a single-stage high-rate trickling filter plant. (*From Am. Soc. Civil Engrs., Sewage Treatment Plant Design, Manual of Practice 36, 1959*)

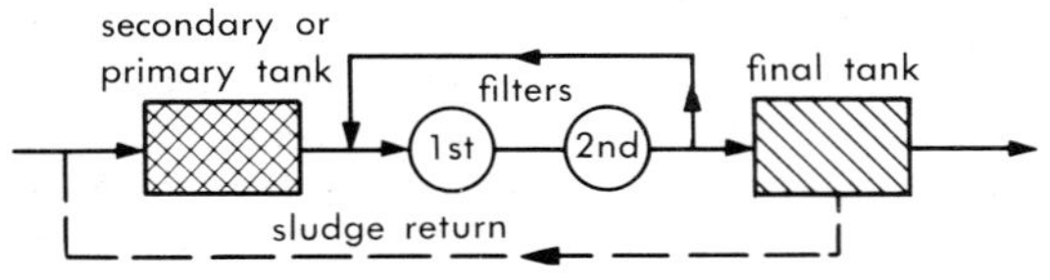

Fig. 12. Flow diagram of a two-stage high-rate trickling filter plant. (*From Am. Soc. Civil Engrs., Sewage Treatment Plant Design, Manual of Practice 36, 1959*)

dosing cycle. The application of water by this method is intermittent. The rotary distributor (Fig. 10) may also be operated by siphon. This type of distributor has two or four radial arms supported on a center pedestal. Hydraulic force of water passing through the nozzles fixed to the arm causes the arm to rotate. The distributor may be operated in continuous rotation by feeding from a weir box. In either case the filter is sprayed as the arm passes over a given section and the dosing is intermittent with a short time interval between doses. With the fixed nozzle method the interval may be 5 min, but with the rotary distributor the dosing interval may be no more than 15 sec.

High-rate filters depend on recirculation. The hydraulic loading rate is about 20 mgad with a range of 9–44 mgd. Rotary distributors are used. Pumps pick up settled effluent and return it. Filters are often set up as primary and secondary filters with recirculation of water to each. Several alternative flow arrangements are demonstrated in Figs. 11 and 12. Recirculation ratios range from 1:1 to about 5:1. Final sedimentation is required for both low- and high-rate filters as filter slime and organic debris are washed free.

Aeration. Aeration is accomplished in tanks in which compressed air is diffused in liquid by various devices: filter plates, filter tubes, ejectors, and jets; or in which air is mixed with liquid by mechanical agitation. The high degree of treatment possible with conventional activated sludge, 95–98% BOD removal, has made it a popular method of treatment. Sewage organisms seeded in sludge which has passed through treatment are returned to incoming sewage and mixed thoroughly with the liquor. In this way the biota, oxygen supply, and sewage are brought together. Contact aeration utilizes air diffusion to keep a biota suspension thoroughly mixed; however, the biota are also maintained in active growth on plates of impervious material such as cement-asbestos suspended in the mixed liquor of the aeration tank. Slime growth forms on the plates and liquid passing by the plates furnishes the plate biota with a source of nutrients.

Activated sludge process, the conventional process, requires an aeration period of 4–8 hours. Much of the oxidation takes place in the first 3 hours of detention. Aeration tanks are usually long, narrow, rectangular tanks with porous plates or diffusers along the length to keep the liquor well agitated throughout (Fig. 13). Widths are 15–30 ft and depths about 15 ft. Length-width ratio is about 5:1.

Air requirements are 0.2–1.5 ft^3 air/gal of sewage treated. It is necessary to maintain dissolved oxygen (DO) levels at 2 ppm or higher.

Mechanical aeration is done in square or rectangular aeration tanks, depending on the mechanism. In the Simplex method liquor is drawn by impeller up a draft tube and expelled over the tank surface (Fig. 14). In the Link-Belt unit, brushes introduce a spiral motion with considerable agitation. The period of aeration may be up to 8 hours with this method (Fig. 15).

Modifications of the aeration process are many and include modified aeration, step aeration, tapered aeration, stage aeration, biosorption, bioactivation, dual aeration, and others.

Recirculation of sludge is one of the essentials of the process. About 25–35% of the sludge settled

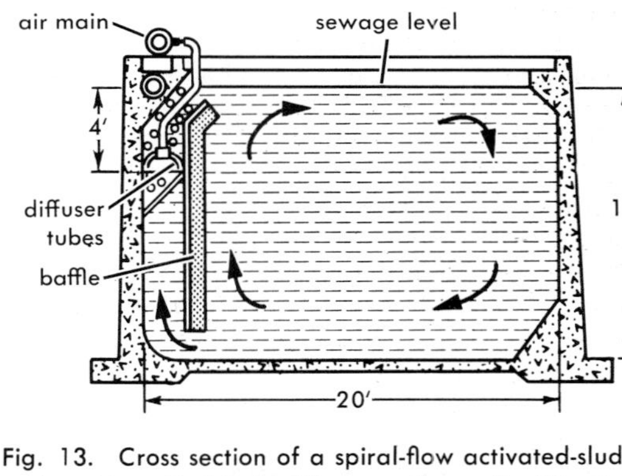

Fig. 13. Cross section of a spiral-flow activated-sludge tank with cylindrical diffusers. (*From E. W. Steel, Water Supply and Sewerage, 3d ed., McGraw-Hill, 1953*)

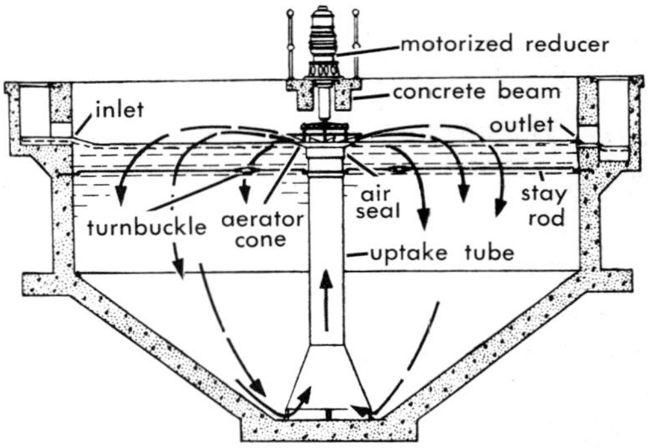

Fig. 14. Simplex aerator. (*From E. W. Steel, Water Supply and Sewerage, 3d ed., McGraw-Hill, 1953*)

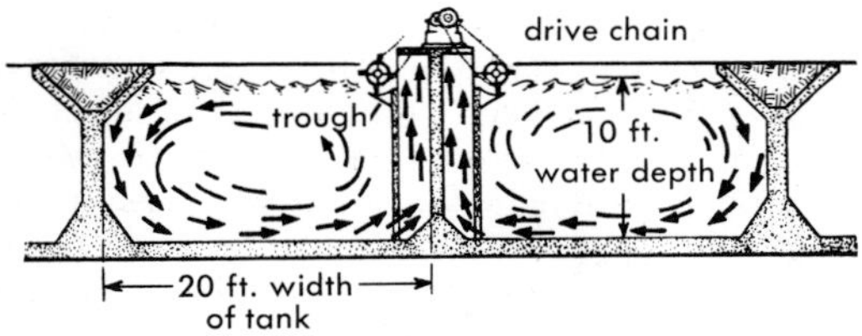

Fig. 15. Cross section of a Link-Belt mechanical aerator. (*From E. W. Steel, Water Supply and Sewerage, 3d ed., McGraw-Hill, 1953*)

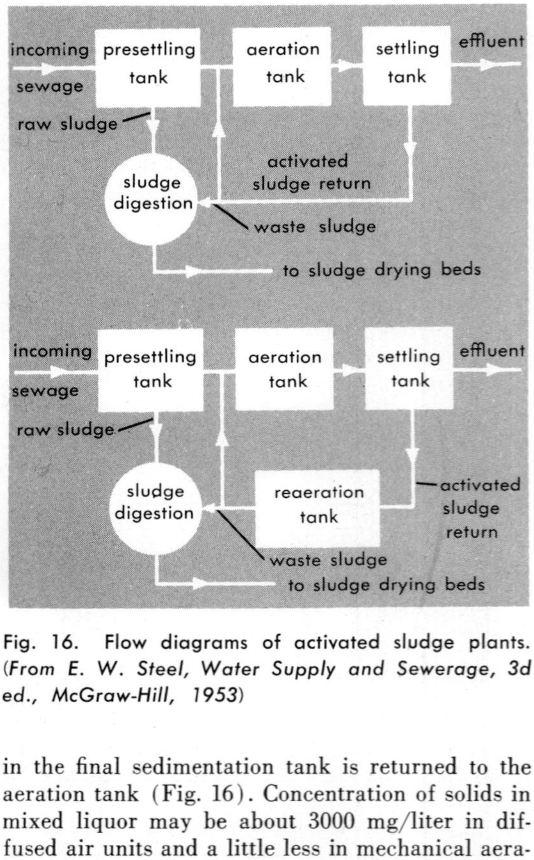

Fig. 16. Flow diagrams of activated sludge plants. (*From E. W. Steel, Water Supply and Sewerage, 3d ed., McGraw-Hill, 1953*)

in the final sedimentation tank is returned to the aeration tank (Fig. 16). Concentration of solids in mixed liquor may be about 3000 mg/liter in diffused air units and a little less in mechanical aeration units. The ratio of sludge volume settled to suspended solids is an important operating index and is known as the Mohlmann index.

$$\text{Mohlmann index} = \frac{\text{volume of sludge settled in 30 min, \%}}{\text{suspended solids, \%}}$$

A good settling sludge has an index below 100. Sludge age, another important factor, is the average time that a particle of suspended solids remains under aeration and is the ratio of the dry weight of sludge in the tank in pounds to the suspended solids load in pounds per day.

Contact aerators provide an aeration period of 5 or more hours. Aeration is usually preceded by preaeration of the raw sewage before primary settling. The preaeration lasts 1 hour. Loadings are based on two factors: pounds per day per 1000 ft^2 of contact surface (6.0 or less), and pounds per day per 1000 ft^2 per hour of aeration (1.2 or less). Air supply of 1.5 ft^3/gal of flow is required. The process has an over-all plant efficiency of about 90% BOD removal.

Chlorination. Chlorination of treated sewage has one major purpose: to reduce the coliform group of organisms. Sufficient chlorine to satisfy demand and provide a residual of 2.0 mg/liter should be added. The following magnitude of dosage is possible: primary effluent, 20 mg/liter; trickling filter plant effluent, 15 mg/liter; activated sludge plant effluent, 8 mg/liter; sand filter effluent, 6 mg/liter. The contact period should be at least 15 min at peak hourly flow.

Oxidation ponds. These are ponds 2–4 ft in depth designed to allow the growth of algae under suitable conditions in sewage media. Oxygen is absorbed from the air, but the conversion of CO_2 to O_2 by *Chlorella pyrenoidosa* and other algae provides an additional source of oxygen of great value. The development of scientific design data for oxidation ponds is in progress. Ponds should be preceded by primary treatment. A loading of 17 lb BOD/(day)(acre) is an early figure. More recently 50 lb BOD/acre has been employed. BOD removal efficiency may range from 40 to 70%.

[W.T.I.]

Bibliography: ASCE-FSIWA Joint Committee, *Sewage Treatment Plant Design*, Am. Soc. Civil Engrs. Manual of Practice 36, 1959; H. E. Babbitt and E. R. Baumann, *Sewerage and Sewage Treatment*, 8th ed., 1958; G. M. Fair and J. C. Geyer, *Water Supply and Waste-Water Disposal*, 1954; L. Metcalf and H. P. Eddy, *American Sewerage Practice*, vol. 3, 3d ed., 1935; E. W. Steel, *Water Supply and Sewerage*, 3d ed., 1953; J. Wilford and T. P. Conlon, Contact aeration sewage treatment plants in New Jersey, *Sewage and Industrial Wastes*, 29(8):845, 1957.

Sewing machine

A mechanism that stitches cloth, leather, pages of books, and other material that is to be joined by means of a double-pointed needle or an eye-pointed needle. In ordinary two-threaded machines, a lock stitch is formed (Fig. 1). An eye-pointed reciprocating needle carries an upper thread through the layers of fabric, forming a loop beneath the material. A shuttle carrying a bobbin of under thread passes through the loop. Alternatively, a rotary hook takes the loop of upper thread and passes it around the bobbin of under thread. The needle withdraws, and a thread take-up lever pulls the stitch tight.

The machine carries out these necessary motions and also feeds the material past the needle inter-

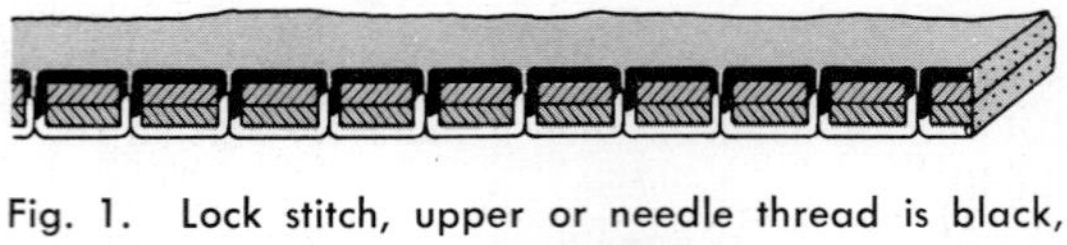

Fig. 1. Lock stitch, upper or needle thread is black, under or bobbin thread is white.

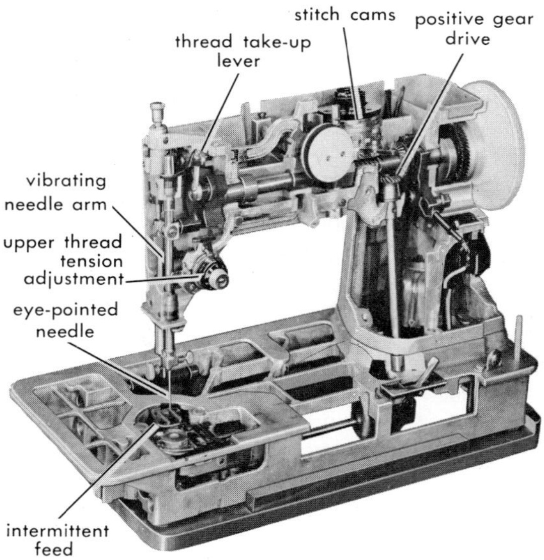

Fig. 2. Modern domestic sewing machine has positive gear drive throughout for slip-free operation. (*Singer Mfg. Co.*)

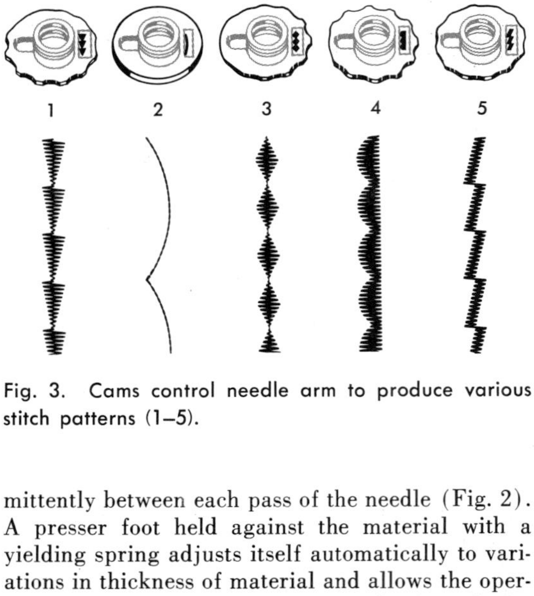

Fig. 3. Cams control needle arm to produce various stitch patterns (1–5).

mittently between each pass of the needle (Fig. 2). A presser foot held against the material with a yielding spring adjusts itself automatically to variations in thickness of material and allows the operator to turn the material as it feeds through the machine. A cluster of cams, any one of which can be selected to guide the needle arm, makes possible a variety of stitch patterns including buttonhole stitches and other standard and decorative stitches (Fig. 3). [F.H.R.]

Sex determination

The genetic constitution of the organism is the basic determinant of sex in nearly all forms of life.

Definition of sexuality. The essential feature of sexuality is the capacity for forming haploid gametes (each containing only one chromosome of a kind) which later unite in the process of syngamy or fertilization, resulting in a zygote with the diploid chromosome number. Haploid gametes may appear similar morphologically or they may differ in size and shape. Sexual reproduction in a simple form takes place in the pond scum, *Spirogyra* (Fig. 1). Cells in adjacent filaments of this green alga first become connected by conjugation tubes. The contents of a single cell, including a nucleus with each chromosome represented once, comprise a gamete here. Since both conjugating cells are of the same size and shape, they are called isogametes. Syngamy results when the cellular material of one conjugating cell passes through the conjugation tube to fuse with the cellular material of the other isogamete. Union of the two haploid nuclei of the isogametes produces a diploid zygospore which, upon undergoing the two meiotic divisions, germinates to form new filaments composed of haploid cells. Sexual differentiation occurs in higher organisms with larger female gametes or eggs that unite with smaller male gametes or spermatozoa (Fig. 2).

Sexual bipotentiality. All plants and animals, whether simple or near the top of the evolutionary scale, exhibit a fundamental sexual bipotentiality. That is, both male and female sex determiners are present in each sex. This is seen in monoecious plants, where a single individual produces both male and female gametes (Fig. 3). Most familiar

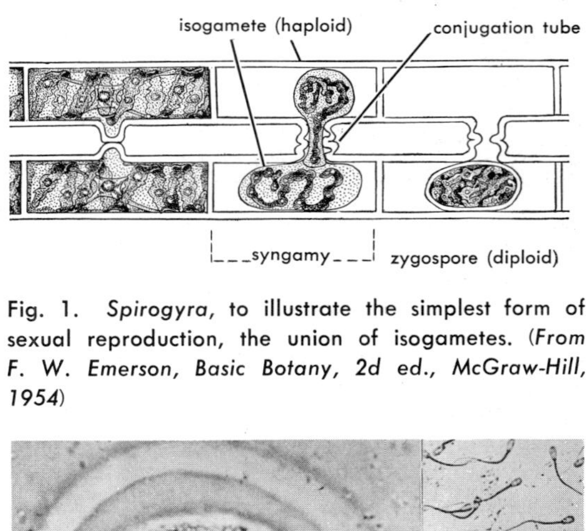

Fig. 1. *Spirogyra*, to illustrate the simplest form of sexual reproduction, the union of isogametes. (*From F. W. Emerson, Basic Botany, 2d ed., McGraw-Hill, 1954*)

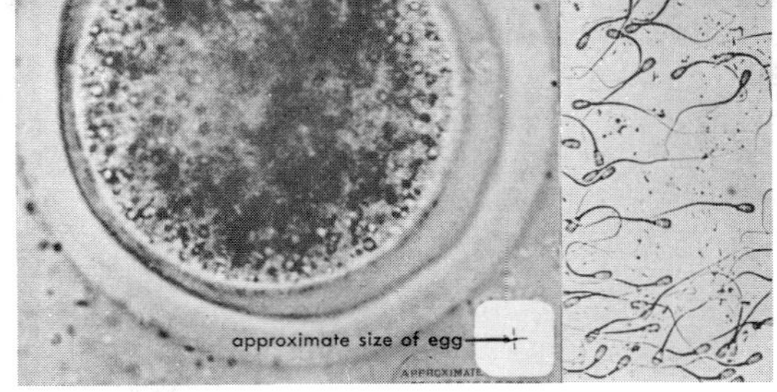

Fig. 2. The bovine egg and sperm. (*By C. C. Hartman and W. H. Lewis from R. C. Cook, A one-celled cow, J. Heredity, 23(5):193–199, 1932*)

Fig. 3. (*a*) A dioecious plant, *Rumex hastatulus*, with stamens (male organs) and pistil (female organs) on separate plants (*photograph from B. W. Smith*); (*b*) a monoecious plant, *Ornithogalum tempskyanum*, with both stamens and pistil in the same flower structure (*from L. F. Randolph and J. Mitra, J. Heredity, 48*(*5*): *213–216, 1957*).

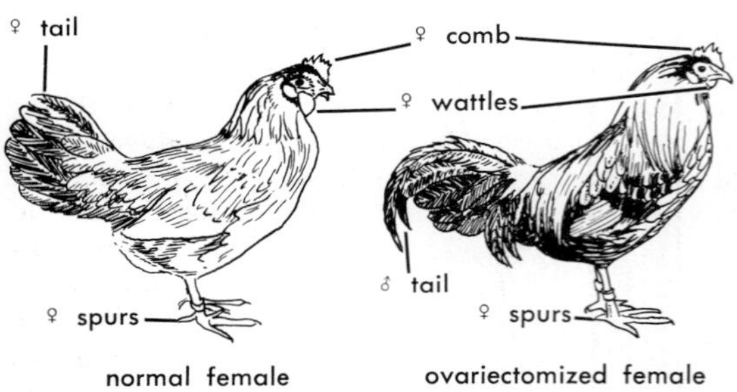

Fig. 4. Normal and ovariectomized hens. (*After Finlay, Brit. J. Exptl. Biol., 2:439, 1925*)

seed plants, such as the lily, garden pea, and corn, are monoecious in that they produce both male organs, called stamens and anthers, and female organs, known as carpels, enclosing ovaries. These occur on one individual plant, and often within the same flower structure. The date palm, willow, poplar, and papaya, on the other hand, are dioecious, bearing male flowers and female flowers on separate plants. Although the sexes are distinct individuals in most animals, there are several large groups of hermaphroditic animals, in which a single individual produces both egg and sperm. These groups include most flatworms, the earthworm, leeches, the garden snail, the oyster, and a few species of fish. However, because of protandry, or the temporal spacing of the production of male versus female gametes, many hermaphroditic animals and monoecious plants are not self-fertilizing.

In animals and plants where the sexes are separate individuals, this sexual bipotentiality shows itself in abnormal cases of partial sex reversal. A Brown Leghorn hen, with the functional left ovary removed, develops malelike plumage and characteristic male tail feathers (Fig. 4). Both male and female sex hormones are present in each sex in vertebrates, and it is a matter of the quantity of the various kinds of sex hormones present which directs the course of sexual differentiation. The original capacity for producing the several sex hormones depends on the genetic constitution of the sex in healthy individuals. *See* ANDROGEN; ESTROGEN.

MECHANICS OF SEX DETERMINATION

Sex determination has a genetic basis, just as do other morphological and physiological characters of an organism. A special pair of chromosomes often differing in size and shape characterizes the two sexes in some plants and in many animals. Sex determination may depend upon a pair of alleles located in the sex chromosomes in some organisms or upon a balance between the male and female tendency genes located not only in the sex chromosomes but also in one or more other chromosome pairs (autosomes)—the genic balance concept. These mechanisms are discussed in the following paragraphs. *See* GENE; GENE ACTION.

Homogametic and heterogametic sex. In some species the special paired sex chromosomes are of unequal size in one sex (heterogametic sex) and the same size in the other sex (homogametic sex).

Heterogametic male, homogametic female. The members of the pair of sex chromosomes are of unequal sizes in men, whereas they are the same size in women (Fig. 5). The two sex chromosomes in women are called X chromosomes. A man possesses an X chromosome and a smaller, Y chromosome. These unequal chromosomes are homologous (have similar genes) for only a small part of their length. However, this homology makes them partners at the two divisions of meiosis preceding gamete formation. Since half the sperm of a man

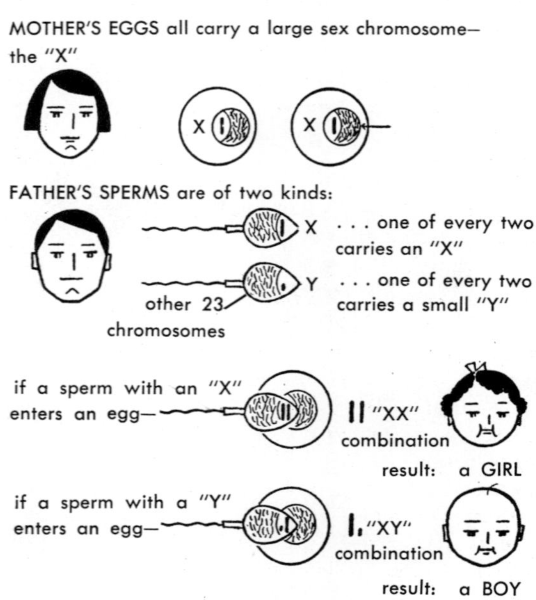

Fig. 5. Sex determination in man. (*From A. Scheinfeld, Human Heredity Handbook, Lippincott, 1956*)

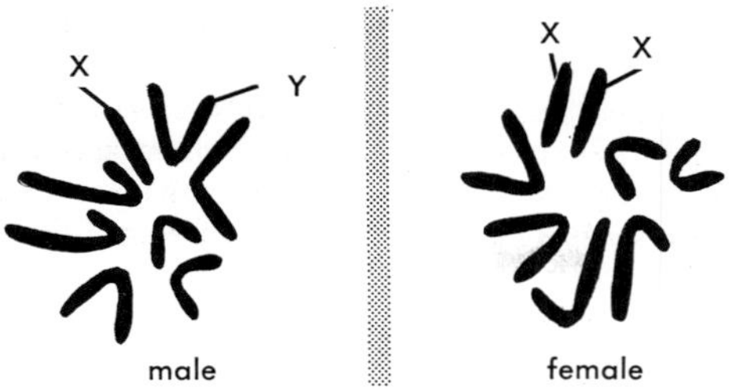

Fig. 6. Sex chromosomes in male and female *Drosophila stonei;* metaphase stage from larval brain cell.

Fig. 7. Sex chromosomes of male *Melandrium album;* the anaphase stage. (*After Belar, in E. W. Sinnott, L. C. Dunn, and T. Dobzhansky, Principles of Genetics, 4th ed., McGraw-Hill, 1950*)

contains an X and the other half a Y chromosome, the male in this species is the heterogametic sex. All of the ova of a woman contain an X chromosome, so the female human belongs to the homogametic sex. Union of an X-bearing sperm with an X-bearing egg results in a female offspring. Union of a Y-bearing sperm with an X-bearing egg produces a male offspring. Sex determination in man thus occurs at the time of fusion of gametes, that is, at syngamy. *See* CHROMOSOME; SYNGAMY.

The XY male and XX female type of sex determination has also been found in (1) such animals as the cat, sheep, hamster, rat, field mouse, mole, ferret, and several marsupials; (2) such insects as Neuroptera (ant lions, lace wing flies), some Heteroptera (true bugs), some Dermaptera (earwigs), some Coleoptera (beetles), and nearly all *Drosophila* (vinegar flies) (Fig. 6); (3) over 50 species of dioecious plants including *Elodea gigantea, Rumex acetosa, Melandrium album* (a pink) (Fig. 7), *Humulus lupulus* (a hop), *Gingko biloba* (the maidenhair tree), *Populus balsamifera* (a poplar), and *Salix andersoniana* (a willow).

A variation of the general male heterogamety, female homogamety type of sex determination occurs in some of the Orthoptera (certain grasshoppers and certain mantids) and some of the Heteroptera. The males of some of these species have one X chromosome but lack a Y chromosome (XO), the female possessing two X chromosomes (XX). The diploid chromosome number in these forms is one less in the male than in the female.

Heterogametic female, homogametic male. The members of the pair of sex chromosomes are of unequal size in the female whereas they are of equal size in the male. Female heterogamety characterizes the members of two diverse animal groups: the Lepidoptera (butterflies and moths), and the birds. Here the female produces half X-bearing eggs, half Y-bearing eggs. To this group also belong the blood fluke, *Schistosoma douthitti,* and among plants the strawberry, *Fragaria elatiar.*

Sex chromosomes morphologically similar. Although a search has been made, no special sex chromosomes are found in any of the hermaphroditic groups of animals nor have they been verified in the many plants that are habitually monoecious. However, not all separate-sexed species of plants and animals show X and Y chromosomes differing from one another in size or shape. Among vertebrate animals, only birds and mammals possess sex chromosomes that are morphologically distinct from the other chromosomes (autosomes). O. Winge found in the guppy, *Lebistes reticulatus,* that maleness is dependent upon perhaps a single gene in the Y chromosome; femaleness being its allele in the X chromosome. Morphologically the X and Y are similar and homologous for a great part of their length since crossing over, or exchange, occurs between them. The result is the occasional transfer of sex-linked mutants controlling the brilliant body pigmentation from the Y chromosome, where they are inherited from father to son, to the X chromosome. A stock of *Lebistes* was found in which a pair of genes not on the X or Y chromosomes came to assume control of sex determination. In the new stock, the female was the heterogametic sex and the male the homogametic sex. Unlike sex alleles result in a female; like sex alleles, in a male. M. Gordon found that female heterogamety is the rule for domesticated aquarium stocks of the platyfish, *Xiphophorus* (*Platypoecilus*) *maculatus,* but certain wild stocks of this species from Mexico show male heterogamety.

The mosquito, *Culex molestus,* also shows a simple type of sex determination; the male is heterozygous for a single dominant gene, the female is homozygous for its recessive allele, and no differences in the chromosomes of the two sexes are apparent. Conversely, the axolotl, *Ambystoma mexicanum,* has an XY female and XX male, though the X and Y cannot be distinguished cytologically. In spinach, *Spinacea oleracea,* the male plant produces X- and Y-bearing microspores in equal numbers. The female plant produces X-bearing megaspores only. Although the sex chromosomes do not differ in size and shape, they do divide ahead of the other chromosomes (autosomes), a common characteristic of sex chromosomes at meiosis.

Genic balance concept. A simple XX-XY distinction between the sexes does not account for all the genetic facts known about sex determination.

In *Lymantria dispar,* the gypsy moth, with female heterogamety, the X chromosome bears the male determiner or determiners either as a single allele or a group of closely linked genes. The genetic basis for femaleness is carried in the Y chro-

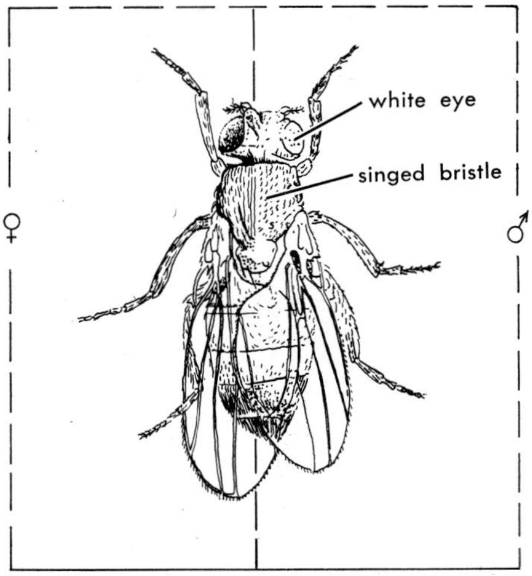

Fig. 8. A gynandromorph in *Drosophila melanogaster* caused by loss of one of two X chromosomes (on the right side) in an early embryo; the recessive sex-linked mutant's white eye and singed bristles are carried by the single X chromosome present in all cells on the right side. (*From J. T. Patterson, Journal of Experimental Zoology*)

mosome. There are a number of European and Japanese races of the gypsy moth differing in the strength or potency of the male and female determiners. When a "strong" Japanese male, X_SX_S, is crossed with a "weak" European female, X_WY_W, all the female offspring, X_SY_W, are intersexual, showing mixtures of male and female primary and secondary sexual characters. The reciprocal cross of strong female with weak male gives normal-sexed offspring which, if inbred, produce half their male offspring intersexual in the next generation. R. B. Goldschmidt's finding of these intersexual forms in *Lymantria* led to his formulation of the genic balance theory of sex determination: (1) each sex possesses both male and female potentialities; and (2) sex is determined by a quantitative relation between male- and female-determining genes in an individual. Within the European weak races of the gypsy moth, the action of the male-determining gene in one X chromosome is weaker than the action of the female-determining gene in the weak Y so that X_WY_W results in a female. The action of two doses of male-determining genes in X_WX_W outweighs the female-determining material in the cytoplasm initiated by the action of the Y_W female gene in the egg before meiosis, with the result that X_WX_W is a male. Similarly, in the strong races, X_SY_S gives a female and X_SX_S gives a male. However, in the female hybrid between a strong male and a weak female, X_SY_W, the male-determining potency of the strong X outweighs the female-determining potency of the weak Y, and the result is intersexuality. The sex genes in the X and Y chromosomes of different races of *Lymantria* behave as series of multiple alleles of different potencies.

The genic balance theory has also been shown to apply to sex determination in *Drosophila melanogaster*. Here the main female-determining genes are in the X chromosome and the male-determining genes in the autosomes. The Y chromosome of the male contains male fertility genes but no sex determiners. The decision of sexuality in diploid *Drosophila* depends on the dosage of the X chromosome. In early 2X 2A embryos, whenever one X chromosome is lost, the tissue derived from cells with 1X 2A undergoes male differentiation; that derived from the original 2X 2A cells becomes female. Thus a sex mosaic, or gynandromorph, is formed (Fig. 8).

The proof that male-determining genes are in the autosomes came from the discovery by C. B. Bridges of intersexes in the progeny of triploid female *Drosophila* and diploid males. These sterile intersexes, possessing 2X chromosomes and three of each autosome (2X 3A), show varying mixtures of male and female parts. An extra set of autosomes shifts sexual development strongly in the male direction. The ratios of the numbers of X and autosome sets in the various sexual forms in *Drosophila melanogaster* are given in the table.

Ratio of X to autosomes in Drosophila melanogaster

Sexual form	Chromosome composition		Ratio, X to A
Female tissue	1X 1A	Haploid tissue, female; haploid individual	1.0
Diploid female	2X 2A		1.0
Triploid female	3X 3A		1.0
Tetraploid female	4X 4A	(Occasionally found)	1.0
Superfemale	3X 2A	(Sterile)	1.5
Diploid male	1X 2A		0.5
Tetraploid male	2X 4A	(Rarely identified)	0.5
Supermale	1X 3A	(Sterile)	0.33
Intersex	2X 3A		0.67
Intersex	3X 4A	(Rarely identified)	0.75

To locate further female-determining sex determining genes in the X chromosome of *Drosophila*, a stock was used containing an X chromosome broken previously by means of x-rays somewhat to the left of its middle. The point of breakage was determined accurately by studying the broken X chromosome cytologically in the giant cells of the larval salivary gland. By special crosses, either the left-hand fragment or the right-hand fragment of this X chromosome was added to the 2X 3A intersex chromosome set. Individuals with either fragment plus 2X 3A developed as weakly fertile females, though individuals with the right fragment only plus 2X 3A sometimes retained vestiges of male sex combs. Since a shorter left- than right-hand fragment plus 2X 3A produced a weakly functional female, female determiners must be more concentrated to the left of the middle of the chromosome, although a number of regions throughout the X chromosome must contain female-determining genes.

This method has been used in an attempt to locate the male-determining genes in the large autosomes (II and III chromosomes). The tiny chromosome IV has no male-determining genes in it, since it may be present once or three times in the diploid sexes without affecting sex. Short sections of broken II or III chromosomes have been added to the 2X 3A intersex set. Although the experiments covered the entire length of these large autosomes in short sections, no pronounced shift in the male direction was observed in the hyperintersexes. Further experiments adding long fragments of the II or III chromosomes either to the 2X 3A intersex set or to the 3X 3A triploid female set failed to produce a shift in the male direction. The results show that there must be a minimum of two major male-determining genes in the autosomes of *Drosophila*. The male tendency genes must be located on both II and III chromosomes.

In the silkworm, *Bombyx mori*, a very strong female determiner has been located in a small section of the Y chromosome.

Sex balance in the dioecious plant, *Rumex acetosa*, is similar to the *Drosophila* type. A ratio of 1X to 2 autosome sets (2A) makes a male plant; 2X 2A makes a female plant. By studying a series of plants carrying a definite autosome in triplicate but diploid for the other autosomes, it has been determined that three of the autosomes had a net male-promoting effect, whereas two of them had a net female-promoting effect.

A strong male-promoting gene necessary for anther formation, male fertility genes, and a female suppressor gene in the Y chromosome of male plants have been found in the dioecious plant *Melandrium album*. By crossing triploid and tetraploid *Melandrium* strains, the intersexual XXX Y and XXXX Y plants were obtained. The male tendency of one Y was unable to outweigh fully the female tendency of four or even three X chromosomes.

Females diploid, males haploid. In the insect order Hymenoptera (sawflies, bees, ants, and wasps), females are biparental, developing from fertilized eggs. Males have a mother only, arising parthenogenetically from unfertilized eggs. Thus the female begins development with a diploid chromosome set; the male, with a haploid set. P. W. Whiting, in his study of the tiny parasitic wasp, *Habrobracon juglandis*, found that although no sex chromosomes are identifiable as such, femaleness is dependent upon heterozygosity at a sex locus. At least nine sex alleles have been found which behave as a series of multiple alleles. The combination of any two different alleles results in a diploid female; a single allele occurs in a haploid male. In sex mosaics or gynandromorphs of this species, female tissue arises from diploid cells heterozygous for a sex locus; male tissue arises from haploid cells. Homozygosity of two sex alleles can occur in inbred crosses, and gives rise to weak, usually inviable, diploid males. Sex determination in the honeybee, *Apis mellifera*, has been found to be similar to that in *Habrobracon*, being dependent

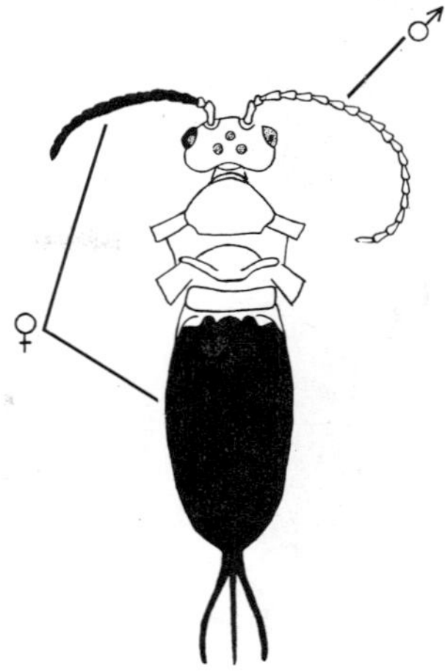

Fig. 9. Haplodiploid gynandromorph in *Habrobracon juglandis*. (From P. W. Whiting, *J. Heredity*, *34*(12): 355–366, 1943)

upon a series of 11 multiple alleles. Diploid males homozygous for a sex allele always die, but male diploid tissue is found in certain mosaic bees.

SEX IN MICROORGANISMS

Sexual reproduction has long been known in thallophytic plants. Some species of bread molds, such as *Mucor*, are homothallic—the hyphae, or filaments, of a single strain undergoing conjugation with one another. Other species are heterothallic, the filaments of one strain requiring conjugation with filaments which are of a physiologically different strain.

Complex systems of mating types have been discovered in *Paramecium* and other protozoa, as well as in some green algae. In *P. aurelia* there are 17 known mating types. Animals of the same mating type cannot mate with one another, but must mate specifically with animals of a certain other mating type. Thus type I mates with type II, type VII with type VIII, and so on. Type XIV has not been found; type XIII mates, rather poorly, with type II. These mating types fall into two varieties. The determination of the mating type in variety A is primarily by means of a genetic locus in the macronucleus; in variety B the determination is cytoplasmic. When no cytoplasmic exchange occurs between conjugating animals of variety B, each exconjugant remains of the same mating type as before; but when extensive cytoplasmic exchange takes place during mating, the exconjugant individuals will be exclusively of one parental type or the other. In variety A, however, the segregation and recombination of the mating-type locus produces Mendelian frequencies of like and unlike exconjugants. Thus, the cross Mt-I × Mt-II produces both exconjugants of type I, one of type I and one of type II, or both of type II in predictable frequencies. In this and similar mating-type phenomena, a multiplicity of mating types is not to be regarded as necessarily a multiplicity of sexes, since a mat-

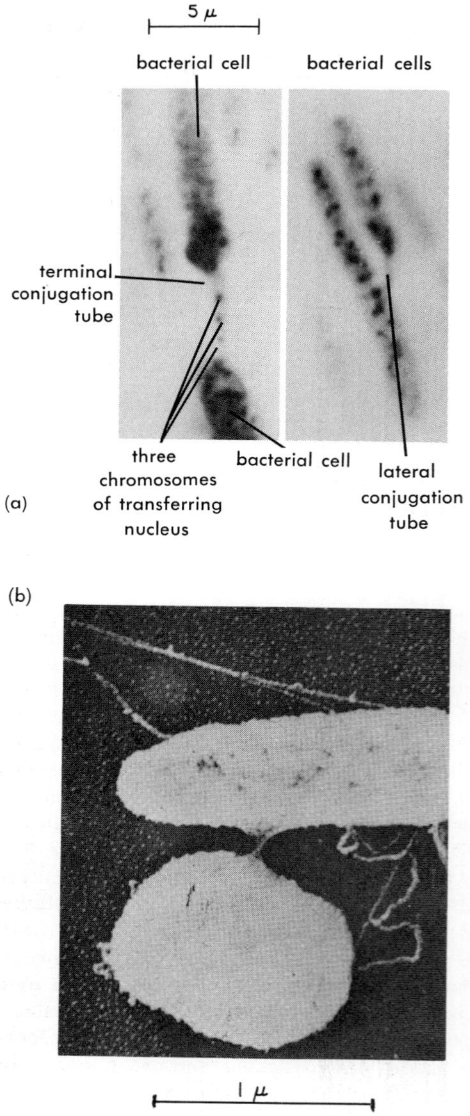

Fig. 10. Conjugation in bacteria. (*a*) *Bacillus megatherium* (*from E. D. Delamater, Cold Spring Harbor Symposia Quant. Biol., vol. 16, 1951*); (*b*) *Escherichia coli* showing the 500-A bridge between the long narrow K-12 Hfr cell of one mating type and CF^- cell of the opposite mating type (*from T. F. Anderson, E. L. Wollman, and F. Jacob, Annales de l'Institut Pasteur, 93:950, 1957*).

ing type involves sexuality plus a sexual attraction mechanism.

Nuclei of certain bacteria, yeasts, and blue-green algae have been studied cytologically and photographed. The chromosomal behavior in these forms is in many ways similar to that found in higher animals and plants. Sexual reproduction has been shown by cytological means to occur in the large bacterium, *Bacillus megatherium*, and by electron micrographs in *Escherichia coli*, the common colon bacillus (Fig. 10). In *E. coli* sexuality can be demonstrated also through the occurrence of genetic recombination into a single bacterial strain of marker traits introduced separately into a mixture of two bacterial strains, according to the work of J. Lederberg. Here not only + and − strains but also + and + strains will undergo recombination. Two − strains will yield no recombinants; that is, they do not undergo sexual reproduction. *See* ENDOCRINE GLAND; GENETICS. [S.B.P.]

Bibliography: E. Allen (ed.), *Sex and Internal Secretions*, 2d ed., 1939; H. Beale, *The Genetics of Paramecium Aurelia*, 1954; F. A. E. Crew, *Sex Determination*, 2d ed., 1946; M. J. D. White, *Animal Cytology and Evolution*, 2d ed., 1954; E. L. Wollman and J. Jacob, Sexuality in Bacteria, *Sci. Am.*, July, 1956.

Sex-influenced inheritance

That part of the inheritance pattern on which sex differences operate to promote character differences. These characters are expressed differently in the two sexes, even when their genotypes are identical. Some confusion exists between this term and sex-limited inheritance. Historically, differences in horns in sheep first illustrated this mode of inheritance. Such characteristics as baldness in man, mahogany coat color in cattle, and eosin and some other eye colors in *Drosophila* all show influence of sex. The effects frequently are attributed to hormones secreted by the sex cells, but often hormones cannot be demonstrated. The differences found within the individual body cells of the sexes may be the responsible factors. The subject has attracted attention because of its relation to important economic and esthetic characters. *See* GENE ACTION.

Breeds of sheep are differentiated by presence of horns in both sexes, by presence of horns in males but not in females, and by absence of horns in either sex. Where present, the horns in the females are smaller than those in the males. This holds true even for sheep of the Hebrides, where rams may have four horns and females two horns. Early castrations of horned males result in hornless males. B. L. Warwick and P. B. Dunkle's studies suggest that the major inheritance for hornlessness is found at one locus in the genome. Three alleles appear to exist at this locus. For breeds where both sexes are horned, their inheritance is represented as H^1H^1. Purebreds having horned males and hornless females are considered as *hh* and pure hornless breeds as *HH*. Presence of horns or hornlessness comes about as a consequence of the cooperation of these genes with sex in the embryological development of the sheep.

Sex-limited inheritance differs from sex-influenced in that the inheritance finds expression only in one sex. A set of three allelic genes (+, *Hr*, and *trans*) in the third chromosome of *Drosophila* illustrates this case. When the wild-type alleles ++ are present, the resulting progeny are normal males or females. *Hr* acts on diploid females causing them to become converted into hermaphrodites with parallel development of both male and female reproductive systems. The males are completely unaffected by this inheritance. When the other allele, *trans/trans*, is homozygous in any diploid female, it causes the whole female reproductive system to convert into a fully developed male system. Milking goats offer a like example where inheritance that will affect only one sex converts the females into hermaphrodites.

Another gene (*Ne*) located at a different place in the third chromosome of *Drosophila*, causes diploid females to have only male progeny without regard to their male mates. This gene, as with those described above, does not affect the males in any way. Under normal conditions, *Drosophila*, like man, produces families with approximately equal numbers of males and females.

These genes, sex-limited in their actions, are interesting in that man, among other species, displays hermaphroditic and other types of deviation from normal males and females which may be due to similar genes. Sex-limited inheritance may be more common than formerly thought. Sterility of several types and in several species may be caused by genes acting in this sex-limited manner. Historically, the term sex-limited inheritance has often included much that was really sex influence. *See* GENETICS. [J.W.G.]

Bibliography: E. Allen (ed.), *Sex and Internal Secretion*, 1939; J. W. Gowen and R. H. Nelson, Predetermination of sex, *Science*, 96:558–559, 1942; J. W. Gowen and S. T. C. Fung, Determination of sex through genes in a major sex locus in *Drosophila melanogaster*, *Heredity*, 11:397–402, 1957; A. L. Rae, The genetics of the sheep, *Advances in Genet.*, 8:190–253, 1956; B. L. Warwick and P. B. Dunkle, Inheritance of horns in sheep, *J. Heredity*, 30:325–329, 1939.

Sex-linked inheritance

The transmission to successive generations of differences that are due to genes located in the sex chromosomes. Because of their ready identification and straightforward expression, sex-linked genes have frequently provided the basis for crucial observations and experiments in genetics. The best known sex-linked traits in man concern color vision and hemophilia. *See* COLOR VISION; HUMAN GENETICS.

Sex-linked genes are distinguished as such not by the traits they control but by characteristic patterns of transmission with respect to the sex difference, paralleling in all respects the distribution of the sex-determining X or Y chromosomes. Except for the sex-determining factors themselves, sex-linked genes are no more likely than nonsex-linked genes to be concerned with differences in the primary or secondary sexual traits that differentiate male from female. The following tabulation is an example of sex-linked inheritance in *Drosophila* where X is the X chromosome, Y the Y chromosome, and *W*, *w* are sex-linked genes concerned with eye pigment.

Parents:	X^wX^w White-eyed female	×	X^WY Red-eyed male
	Eggs: All X^w		Sperm: $\frac{1}{2}X^W$, $\frac{1}{2}Y$
Offspring:	$\frac{1}{2}X^wX^W$, Red-eyed females		$\frac{1}{2}X^wY$ White-eyed males

Parents:	X^WX^w Red-eyed female	×	X^wY White-eyed male
	Eggs: $\frac{1}{2}X^W$, $\frac{1}{2}X^w$		Sperm: $\frac{1}{2}X^w$, $\frac{1}{2}Y$

Offspring:

$\frac{1}{4}X^WX^w$	$\frac{1}{4}X^wX^w$	$\frac{1}{4}X^WY$	$\frac{1}{4}X^wY$
Red-eyed females	White-eyed females	Red-eyed males	White-eyed males

Sex linkage is termed complete in the case of genes located in the differential segment of the X chromosome which is not represented by a homologous counterpart in the Y. Sex linkage is termed incomplete or partial if genes are located in homologous X- and Y-chromosome segments that pair and exchange parts during meiosis. Contrary to most textbooks, cases of completely Y-linked genes are rare or doubtful and evidence for partially sex-linked genes in man is inconclusive.

Complete linkage. The pattern of transmission of completely X-linked genes is similar in different groups of organisms except that its phase with respect to sex may be reversed depending upon which of the two sexes possesses two X chromosomes—the male (birds, butterflies, moths, and some fishes), or the female (most other animals, including vertebrates, and dioecious plants). Recessive X-linked traits are expressed more frequently in the heterogametic sex, which, being XY or X—, possesses only one X chromosome, than they are in the homogametic sex with two X chromosomes, because in the homogametic sex they are masked by dominant counterparts, in other words, alleles. This is true both in individual family pedigrees and in populations of diploid organisms, where the expected fraction of recessive individuals within the heterogametic sex is the square root of the fraction within the homogametic sex. Completely X-linked genes at different chromosome loci recombine with each other by crossing over in the homogametic sex. *See* MENDELISM; RECOMBINATION, GENETIC.

Partial linkage. Where partially sex-linked genes occur, two homologous sets are present in both sexes, and the traits are seen among offspring in the Mendelian ratios characteristic of autosomal nonsex-linked genes, except that partially sex-linked differences show nonindependent segregation with respect to the sex difference, and crossing over in the heterogametic sex is necessary in order for recombination with sex to occur.

Sex-linked genes in the X chromosome of *Drosophila* furnished the basis for the first correct explanation of genetic linkage, the first linkage maps, the first detailed and critical proof that inherited differences are located in chromosomes, and the first proof that mutations are produced by x-rays. The sex chromosomes are the only chromosomes in man identified as carriers of specific genes, and until recently the only proved cases of genetic linkage in man involved sex-linked genes.

The term sex-linked was introduced by T. H. Morgan to distinguish differences due to genes located in the sex chromosomes from sex-limited differences whose expression is limited to one or the other sex, for example, differences in milk yield in mammals, but whose chromosomal basis may be other than sex-linked. *See* GENETICS; HUMAN GENETICS; SEX DETERMINATION; SEX-INFLUENCED INHERITANCE. [D.D.P.]

Bibliography: T. H. Morgan, Sex-limited and sex-linked inheritance, *Am. Naturalist,* 48:577, 1914; N. E. Morton, Further scoring types in sequential linkage tests, with a critical review of autosomal and partial sex linkage in man, *Am. J. Human Genet.,* 9:55–75, 1957; E. W. Sinnott, L. C. Dunn, and T. Dobzhansky, *Principles of Genetics,* 5th ed., 1958; C. Stern, The problem of complete Y-linkage in man, *Am. J. Human Genet.,* 9:147–169, 1957.

Sextant

A navigation instrument used for measuring angles, primarily altitudes of celestial bodies. The original sextant had arcs of 60°, or one-sixth of a circle, from which the instrument derived its name. Because of the double reflecting principle used, such an instrument could measure angles as large as 120°.

In modern practice, the name sextant is commonly applied to all instruments of this type regardless of the length of the arc, which is seldom 60°. However, the terms octant, quintant, and quadrant are applied occasionally to instruments having arcs of 45°, 72°, and 90°, respectively.

Development. John Hadley, an Englishman, and Thomas Godfrey, an American, are credited with inventing the marine sextant independently. The sextant has remained virtually unchanged since each of these men designed a similar instrument in 1730. Various less accurate, more cumbersome angle-measuring devices preceded the sextant, including the common quadrant, astrolabe, cross-staff, backstaff or sea quadrant, and nocturnal. Tycho Brahe (1546–1601) invented several instruments with arcs of 60°, which he called sextants. In 1700, Sir Isaac Newton designed an instrument incorporating the double reflecting principle, but his invention was not made public until after sextants had been constructed by Hadley and Godfrey.

Modern instruments may be grouped in two classes, marine sextants and air sextants.

Marine sextant. This is an instrument designed for use by mariners. It utilizes the visible sea horizon as the horizontal reference. It is equipped with a removable telescope to magnify the image of the horizon, and with shade glasses or filters to reduce glare. The marine sextant has been widely used as a symbol of navigation.

Fig. 1. A marine sextant.

Fig. 2. A periscopic sextant for use in aircraft.

Air sextant. An instrument designed for use in aircraft is called an air sextant. Such sextants vary widely in design, but all modern instruments have built-in artificial horizons. A bubble is most commonly used for this purpose, but pendulums and gyroscopes have also been used. Most modern air sextants are periscopic, to permit observation of celestial bodies without need of an astrodome in the aircraft. Figure 2 illustrates a modern periscopic sextant.

Sextant altitude corrections. Altitudes measured with the sextant are subject to certain corrections. These may be classified as those related to: (1) inaccuracies in reading, (2) inaccuracies in horizontal reference, (3) bending of ray of light from the body, (4) adjustment to equivalent reading of center of body, and (5) adjustment to equivalent reading at center of earth. [A.B.M.]

Bibliography: N. Bowditch, *American Practical Navigator*, U.S. Navy Hydrographic Office, H. O. 9, 1958; J. C. Hill, II, *Dutton's Navigation and Piloting*, 1958.

Sexual dimorphism

A fundamental phenomenon of life is bisexuality, in which females produce relatively large, nonmotile eggs and males produce small, motile spermatozoa which fertilize eggs. In some organisms, such as the green alga *Chlamydomonas*, or echinoderms, such as the sea urchin or the sea star, the functional separation of the sexes is not marked by obvious morphological differences. Commonly, however, the sexes can be diagnosed by differences of form or pattern. Such diagnostic differences between the sexes comprise the phenomenon of sexual dimorphism.

Examples among animals. Sexual dimorphism ranges from differences detectable only with care, to differences so extreme that careful study has been required to assign the sexes to the same species. A very minor dimorphism occurs in snakes, in which the main external difference between the sexes is in the tapering of the tail posterior to the cloaca. This is abrupt in females, gradual in males. Moderate dimorphism is well known in man, where secondary sex characters include differences in bodily proportions and contours, in distribution of subcutaneous fat, in voice, and in other dimorphic characters. In deer, the males are marked by antlers. Plumage differences are common in birds. Thus the male red-winged blackbird presents a striking pattern, whereas the female is a dull, gray bird ordinarily recognized only by experts.

More extreme sexual dimorphism is common among the Rotifera. These animals are, on the average, the smallest metazoans, yet males are usually much smaller than females and may lack various structures and organs. In *Epiphanes*, males are about half the length of females, and they lack a digestive system. In more extreme cases, males may be only one-twelfth the length of females, and only the reproductive structures are well developed. *See* ROTIFERA.

A striking example is afforded by the echiuroid worm, *Bonellia viridis* (Fig. 1). The female is complex, with a bulbous body about 1 in. long and a long, bifid proboscis. Yet the male is less than 1 mm long, has rudimentary organ systems, and resembles a large, ciliated protozoan. Sex determination has been thoroughly studied in *Bonellia*. A larva which develops in isolation becomes a female, but one which settles on the proboscis of an established female becomes a male. *See* ECHIUROIDEA.

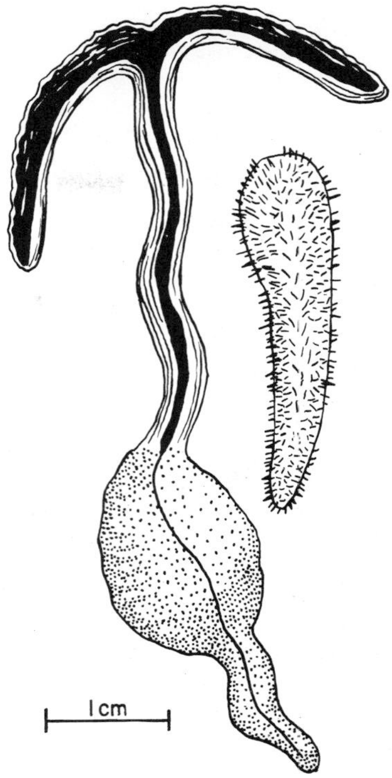

Fig. 1. *Bonellia*. On the left is a female, on the right a male, drawn at a much greater magnification. The ciliatelike male is less than 1 mm in length. (*From E. O. Dodson, Genetics, Saunders, 1956*)

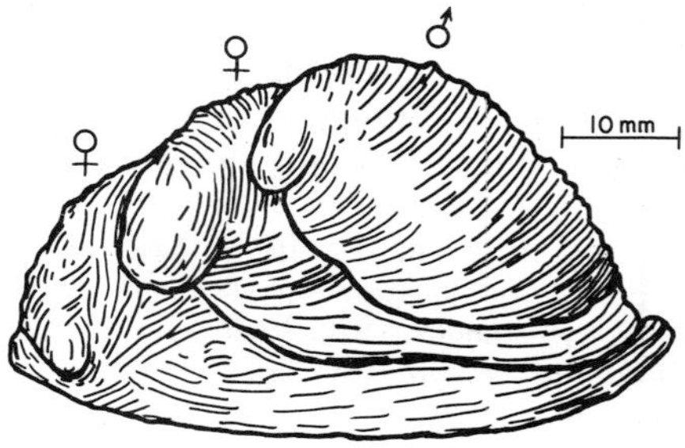

Fig. 2. *Crepidula*. Females are large, males smaller, because of protandry. (*Courtesy of W. R. Coe and the Quart. Rev. of Biol., vol. 19, 1946*)

Dimorphism and protandry. A type of dimorphism which is common among mollusks is associated with protandry. A protandrous animal develops first as a male, then later becomes a female. A simple dimorphism of size results, small individuals being male and large ones female. *Crepidula*, the slipper shell, is a good example. The shipworm, *Teredo*, transforms so early that females comprise about 90% of a typical colony. The males, however, are very prolific sperm producers. In oysters, protandrous individuals live side by side with those of permanent sex. Protandry is common in the west-

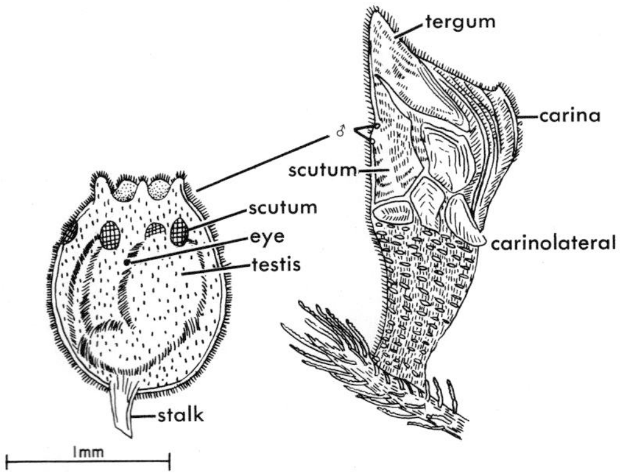

Fig. 3. *Scalpellum vulgare*, a stalked barnacle. The large individual is a female bearing two complemental males, one of which is shown in detail at the left. (*After Darwin and Smith, from L. A. Borradaile and F. A. Potts, The Invertebrata, 2d ed., Cambridge Univ. Press, 1953*)

ern oyster, *Ostrea lurida*, but less common in the eastern oyster, *Crassostrea virginica*. The occurrence of protandrous and permanently sexed individuals in one species provides an opportunity to study the genetic basis of protandry.

Protandry occurs even among the Chordata. It is common among tunicates, and occasionally occurs in toads. Near each testis is a small potential ovary. If the toad is castrated surgically or by parasitic infection, the potential ovaries enlarge and become functional.

Sexual dimorphism among arthropods. Sexual dimorphism of all degrees occurs in arthropods. Crabs show very moderate dimorphism, males having a narrow abdomen, females a broad one. Also, male arthropods are generally smaller than females. Some copepods, such as *Cyclops* and *Calanus*, are moderately dimorphic. Males are somewhat reduced in size, and their antennae are modified for clasping the females. However, *Chondracanthus*, a parasite upon gills of marine fish, has large females so strongly modified for parasitism that few typical copepod characters remain in adults. Yet the male, minute and permanently attached near the genital pore of the female, shows fairly typical copepod structure.

An interesting type of dimorphism occurs among stalked barnacles. These are hermaphroditic, but they may bear complemental males. These are always smaller than the host and may be reduced in structure, lacking digestive and other systems. Relatively complete complementals live at the edge of the mantle cavity of the host, whereas in species with very degenerate complementals a highly protected position inside the mantle cavity of the host is taken. In the most extreme cases, the testes of the host are not developed, so that the sexes are separate and highly dimorphic.

An outstanding insect example is the markedly dimorphic royal pair in a termite colony. After their nuptial flight and the founding of their colony, they lose their wings. The king remains fairly typical otherwise, but the queen is grossly modified, her abdomen becoming an enormous reproductive sac. She is completely dependent upon workers which feed her. There are two kinds of supplemental males and females. The first is winged, and serves chiefly for establishment of new colonies. The second is wingless, and becomes actively reproductive only if the royal pair dies. In addition to these, the colony swarms with sterile workers and soldiers.

Examples among plants. In the great majority of flowering plants, a single flower includes both stamens which produce pollen, and pistils in which the ovules develop. Just as in hermaphroditic animals, sexual dimorphism is out of the question here. However, staminate and pistillate flowers are separate in some species, and thus sexual dimorphism can occur. In some, both kinds of flowers occur on the same plant, which is then monoecious but dimorphic. Perhaps the best example is Indian corn, the tassels being staminate and the ears pistillate inflorescences. Other examples include the oak, hickory, beech, and birch trees. In other cases, the two kinds of flowers occur on different plants, and hence they are dioecious as well as dimorphic. Examples include the date palm, hemp, box elder, willow, and poplar. Only males exist in the Lombardy poplar, so that vegetative propagation is obligate.

Significance of sexual dimorphism. The significance of this widespread phenomenon is not easily determined, and probably no single explanation embraces all cases. Sexual dimorphism was explained by Darwin on the basis of sexual selection. This aspect of Darwinian theory has not been widely accepted, and it seems unlikely that sexual selection has produced most cases of sexual dimorphism. However, it is probable that dimorphism has selective value based upon recognition of the sexes and stimulation of sexual responsiveness.

In some cases, sexual dimorphism seems to be a step toward elimination of the male and establishment of obligate parthenogenesis. The three orders of Rotifera exemplify this. In the Seisonidea, the sexes are about equally developed, although males are somewhat reduced in size. In the Monogononta, reduction of males ranges from moderate in *Epiphanes* to extreme in *Asplanchna*. Finally, in the Bdelloidea, males are unknown and reproduction is exclusively parthenogenetic. *See* BDELLOIDEA; MONOGONONTA; SEISONACEA.

In other instances, sexual dimorphism increases the probability of successful fertilization. Thus, in *Bonellia*, in which males exist only as parasites within the females, the availability of sperm is assured to the host female. Again, sexual dimorphism may play a role in the promotion of cross fertilization. This is well exemplified by the stalked barnacles. In general, these are hermaphroditic and colonial, and cross fertilization is the rule. Some species, however, are solitary, and it is in these that

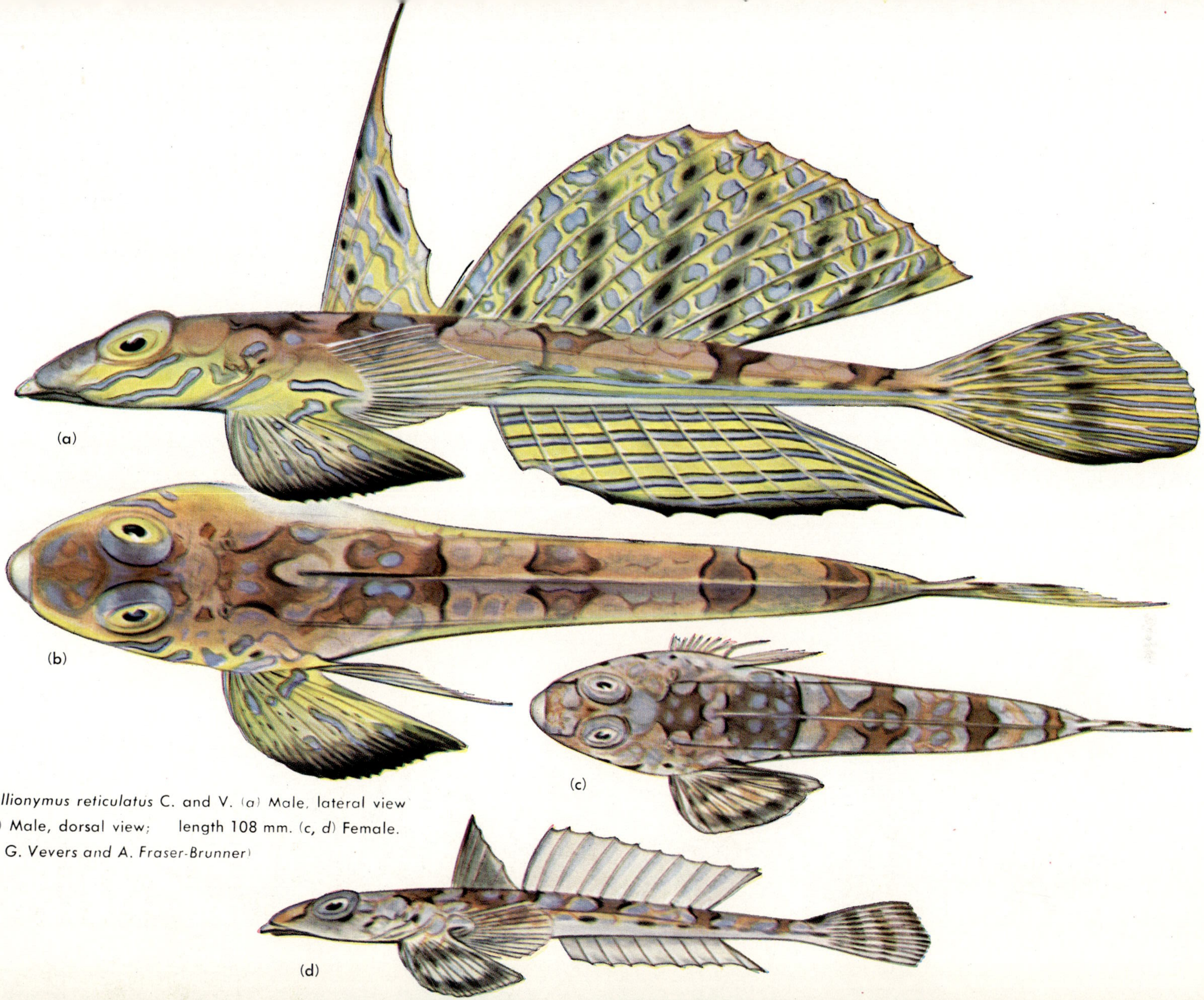

Callionymus reticulatus C. and V. (*a*) Male, lateral view (*b*) Male, dorsal view; length 108 mm. (*c, d*) Female. (*H. G. Vevers and A. Fraser-Brunner*)

Fig. 4. Corn, a monoecious, dimorphic plant. The ears are the pistillate, the tassels the pollinate, inflorescences. The background is a 4 × 8 ft board. (*DeKalb Agricultural Association*)

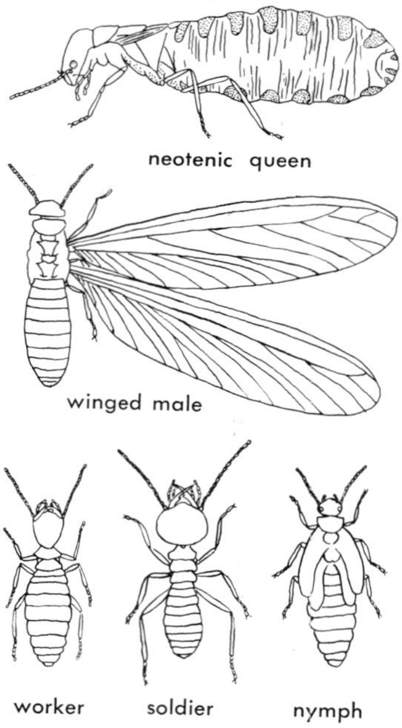

Fig. 5. Polymorphism in a termite, *Hamitermes silvestri*. (*After Tillyard, from L. A. Borradaile and F. A. Potts, The Invertebrata, 2d ed., Cambridge Univ. Press, 1953*)

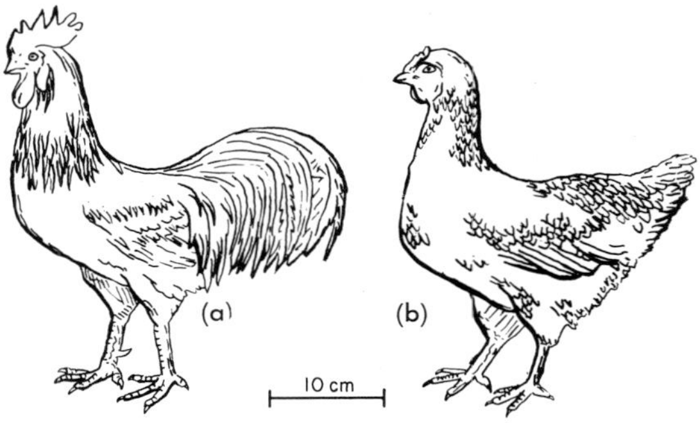

Fig. 6. (*a*) Cock-feathering. (*b*) Hen-feathering. (*From E. O. Dodson, Genetics, Saunders, 1956*)

complemental males are best developed. The aptness of this mechanism for promoting cross fertilization in a solitary, hermaphroditic species is obvious. Whether self-sterility genes exist in these animals is unknown.

Another type of sexual dimorphism, polymorphism, is exemplified by termites. In this group, it appears to be one aspect of a complex polymorphism based upon the caste society of these insects. *See* SOCIAL INSECTS.

Physiology. The physiology of establishment of sexual dimorphism has been well studied in respect to feathering patterns in chickens. A cock-feathered chicken has long, pointed, fringed feathers in the neck and tail regions, whereas a hen-feathered chicken has a more uniform coat of shorter, rounded, fringeless feathers. In most breeds, this sex difference is quite regular, but in Campines, Hamburgs, and some other breeds, males may be either cock-feathered or hen-feathered, and Sebright bantams are always hen-feathered regardless of sex. Experimental crosses show that hen-feathered cocks carry a dominant gene, *Hf*. Chickens of genotypes *HfHf* or *Hfhf* are hen-feathered irrespective of sex, whereas chickens of genotype *hfhf* show typical dimorphism. Chickens of any genotype, if castrated, develop poulard feathering, which is similar to cock-feathering. If castrated chickens are artificially supplied with sex hormones, the result depends both upon the genotype and the hormone. Homozygous *hfhf* chickens will, following the next molt, develop hen feathering in the presence of female sex hormone or cock feathering in the presence of the male sex hormone. However, chickens of genotypes *HfHf* and *Hfhf* develop hen feathering in the presence of either sex hormone. These experiments show that cock feathering or, more accurately, poulard feathering is a neutral state from which dimorphic deviations may be induced by an appropriate combination of genotype and sex hormones. *See* SEX DETERMINATION. [E.O.D.]

Seymouriamorpha

A group of Late Carboniferous and Permian tetrapods which almost exactly bridge the gap in skeletal structure between labyrinthodont amphibians and cotylosaurian "stem" reptiles. Hence, in default of knowledge of their embryology, they cannot be positively assigned to either of the two classes. In vertebral construction, their expanded

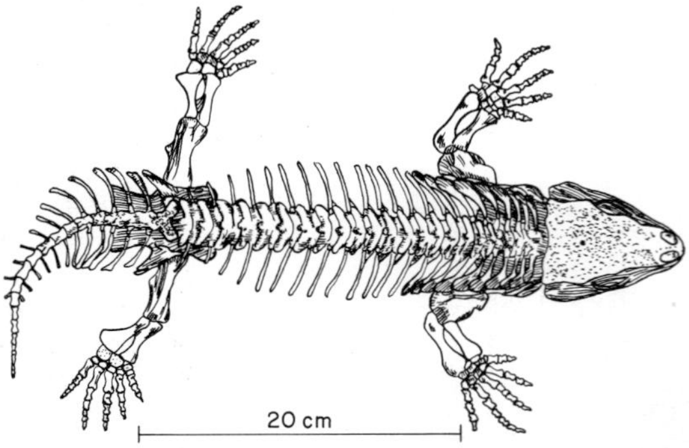

Dorsal view of *Seymouria*; about 20 in. long. (*After Williston, as used in A. S. Romer, Vertebrate Paleontology, 2d ed., Univ. of Chicago Press, 1945*)

neural arches match those of the cotylosaurs, but the intercentra are larger than in typical reptiles. In the skull such features as the presence of an intertemporal bone and a deep otic notch placed high in the cheek are primitive amphibian characters, but features nevertheless to be expected in an ancestral reptile. Representative of the group is *Seymouria* of the Texas Early Permian. *See* COTYLOSAURIA; LABYRINTHODONTIA. [A.S.R.]

Shad

Any of a variety of fresh-water and marine fishes of the order Isospondyli, and related to the herrings.

Perhaps best known is the American shad, *Alosa sapidissima*, a laterally compressed, silvery fish, which attains a length of 2½ ft and sometimes

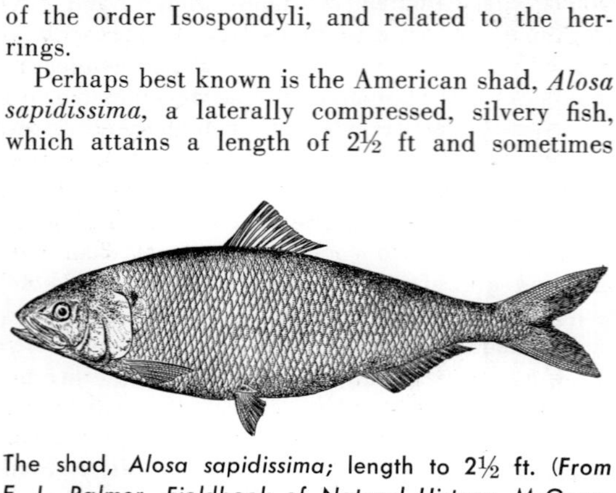

The shad, *Alosa sapidissima*; length to 2½ ft. (*From E. L. Palmer, Fieldbook of Natural History, McGraw-Hill, 1949*)

weighs over 13 lb. This deep sea fish is native to the Atlantic Coast, but it has been successfully introduced along the Pacific Coast. To spawn it migrates into fresh water where it is caught in large quantities, being highly favored for both its flesh and its roe. Dams, pollution, and overfishing have reduced or eliminated it entirely in many streams where it was once abundant.

The gizzard shad, *Dorosoma cepedianum*, is a very abundant fresh-water shad, especially in the southern and midwestern parts of the United States, where it is important as a forage fish. There are several related species in American waters. *See* CLUPEIFORMES; HERRING. [J.D.B.]

Shadow

A region of darkness caused by the presence of an opaque object interposed between such a region and a source of light. A shadow can be totally dark only in that part called the umbra, in which all parts of the source (or sources) are screened off. With a single source of appreciable breadth, there is a region of the shadow called the penumbra which is illuminated by only part of the source. For example, the shadow of the moon shown in the diagram has an umbra which barely touches the earth during a total eclipse of the sun. The eclipse is only partial for an observer situated in the penumbra. *See* ECLIPSE, ASTRONOMICAL.

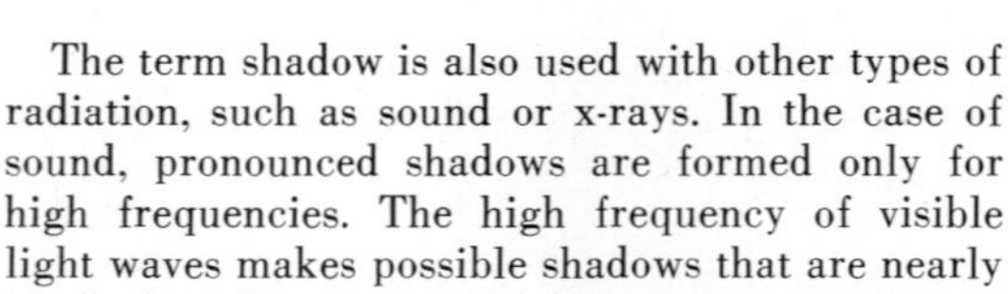

Shadow of the moon at the time of an eclipse. (Relative sizes are not to scale.)

The term shadow is also used with other types of radiation, such as sound or x-rays. In the case of sound, pronounced shadows are formed only for high frequencies. The high frequency of visible light waves makes possible shadows that are nearly black (*see* DIFFRACTION). X-ray photographs are shadowgrams in which the bones and tissues appear by virtue of their different opacities, or degrees of absorption (*see* RADIOGRAPHY). The bending of light rays by reflection or refraction may also produce shadow patterns. An important practical example is the schlieren method of observing flow patterns in wind tunnels. *See* SCHLIEREN PHOTOGRAPHY. [F.A.J.]

Shadowgraph of fluid flow

A simple method of making visible the disturbances that occur in fluid flow at high velocity. The three principal methods of optical fluid flow measurements, schlieren, interferometer, and shadowgraph, depend on the fact that light passing through a flow field of varying density is retarded differently

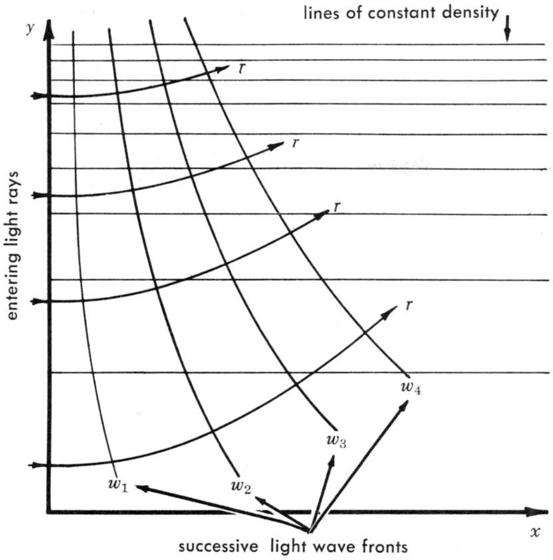

Fig. 1. Density gradient in fluid flow refracts light rays.

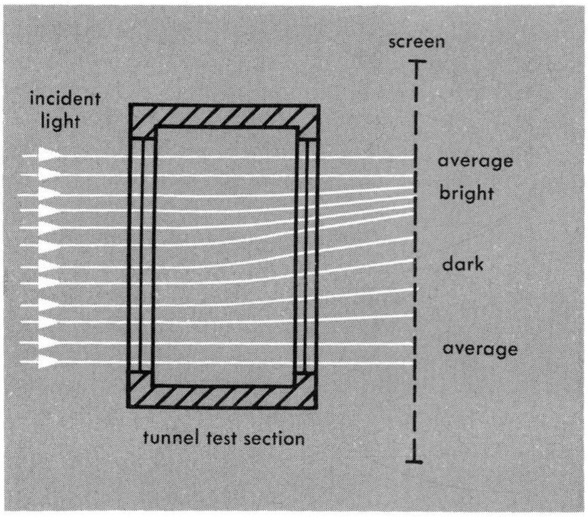

Fig. 2. Shadowgraph produced by light passing through fluid in test section indicates flow pattern.

through the field, resulting in a turning of the wave fronts, that is, a refraction of the rays, and in a relative phase shift along different rays. The first of these, the refraction of the rays, is the basis for shadowgraph flow visualization.

Figure 1 shows light crossing a flow field of varying density in the y direction, as represented by the lines of constant density. The light can be considered to be acting as a wave entering at the left. The lines marked w are the wave fronts at successive times, and the lines orthogonal to them and marked r are the rays of light. Because large density gradients have a greater effect on the velocity of the light, the wave front turns as shown. The ray of light is turned through the same angle. It can be shown that for plane flow, where conditions are the same along any x direction, the deflection of the ray is proportional to the density gradient. For three dimensional flow, as in a wind tunnel, the final deflection depends upon all of the density gradients encountered.

The application of the shadowgraph to a wind tunnel is shown in Fig. 2. The rays enter normal to the side wall through the window at the left. (The window is optical glass of high quality to reduce refraction and to assure that observed effects are caused by the flow in the test section.) As they pass through the test section, they encounter a change in density of the fluid in the tunnel, and are deflected. The light rays then fall on a screen where they are observed or photographed. Where the rays have crowded together, the screen is brightly illuminated, and where the rays diverge, the screen is dark. Where the spacing is unchanged, the illumination is normal, even though there has been a change of density along the path of the ray.

The light need not be parallel when it enters the test section, so the slit source and lens systems of the other methods are not required. This simplicity makes the shadowgraph system considerably less expensive than other methods; it is often used where the finer resolution of the other systems is not required or desirable. *See* INTERFEROMETRY; SCHLIEREN PHOTOGRAPHY; SHOCK-WAVE DISPLAY. [R.F.C.]

Bibliography: J. V. Charyk and M. Summerfield (eds.), *High Speed Aerodynamics and Jet Propulsion*, vol. 9, 1954.

Shaft balancing

When a body, unconstrained by bearings, rotates in space, its center of gravity lies in the axis of rotation. However, when a body (rotor or shaft) is rotating in bearings, the axis of rotation is determined geometrically by the bearings, and the center of gravity does not usually lie in that axis but describes a small circle about the axis of rotation. The resulting centrifugal force causes vibration and vibratory forces on the bearings that may spread throughout the foundation. The centrifugal force F is $m\omega^2 r$ where m is the mass of the body, r its radius, and ω is its rotational speed. Because F is proportional to the square of rotational speed, the effect is unimportant for slow-speed machines and of overwhelming importance for high-speed rotors. For a 300 rpm rotor the eccentricity of the center of gravity which causes a centrifugal force equal to the weight (and hence makes the machine jump from its bearings) is about 0.4 in., for a 3000 rpm machine that distance is 0.004 in., and for a 30,000 rpm machine it is 0.00004 in.

Balancing machine. After a rotor has been manufactured it is placed in a balancing machine. This device consists of bearings with small lateral stiffness that present little resistance to vibration, and the machine is equipped with sensitive vibration-measuring instruments. The rotor is spun and the resulting bearing vibration is an indication of the unbalance or eccentricity of the center of gravity of the rotor. This is compensated for by placing small correction weights or drilling small indentations at appropriate locations on the rotor.

Balancing machines have been developed to great accuracy and also have been made automatic. Small electric rotors for domestic appliances are balanced in such an automatic machine; the entire process of measuring and locating the eccentricity and of placing the correction weights takes a small fraction of a minute. Automobile front wheels are balanced on similar machines. Rotors for gyroscopes for navigational purposes running at speeds of 20,000 rpm and higher are balanced on machines that reduce the eccentricity of the center of gravity to a fraction of a millionth of an inch.

Unbalance in turbines. A vexing problem in steam and gas turbines is that of heat distortion of the rotor, in which the temperature rise causes a bending of the rotor with consequent displacement of its center of gravity. Such rotors cannot be balanced, and the original design must be such as to preclude heat distortion.

Another type of unbalance is hydro- or aerodynamic unbalance which occurs in propellers, pumps, and turbines as the result of slight differences in the angle of pitch of individual blades. This type, being independent of rotational speed, is of relatively large importance in low-speed machinery and can be corrected by changing the pitch angles of the blades. [J.P.D.H.]

Shafting

The machine element that supports a roller and wheel so that they can perform their basic functions of rotation. Shafting, made from round metal bars of various lengths and machined to dimension the surface, is used in a great variety of shapes and applications. Because shafts carry loads and transmit power, they are subject to the stresses and strains of operating machine parts. Standardized procedures have been evolved for determining the material characteristics and size requirements for safe and economical construction and operation.

Types. Most shafting is rigid and carries bending loads without appreciable deflection. Some shafting is highly flexible; it is used to transmit motion around corners.

Solid shafting. The normal form of shafting is a solid bar. Solid shafting is obtainable commercially in round bar stock up to 6 in. in diameter; it is produced by hot-rolling and cold-drawing or by machine finishing with diameters in increments of ¼ in. or less. For larger sizes, special rolling procedures are required, and for extremely large shafts, billets are forged to the proper shape. Particularly in solid shafting, the shaft is stepped to allow greater strength in the middle portion with minimum diameter on the ends at the bearings. The steps allow shoulders for positioning the various parts pressed onto the shaft during the rotor assembly.

Hollow shafting. To minimize weights, solid shafting is bored out or drilled, or hollow pipes and tubing are used. Hollow shafts also allow internal support or permit other shafting to operate through the interior. The main shaft between the air compressor and the gas turbine in a jet aircraft engine is hollow to permit an internal speed reduction shaft with the minimum requirement of space and weight. A hollow shaft, to have the same strength in bending and torsion, has a larger diameter than a solid shaft, but its weight is less. The center of large shafts made from ingots are often bored out to remove imperfections and also to allow visual inspection for forging cracks.

Functions. Shafts used in special ways are given specific names, although fundamentally all applications involve transmission of torque.

Axle. The primary shafting connection between a wheel and a housing is an axle. It may simply be the extension of a round member from each side of the rear of a wagon, and on the end of each the hub of a wagon wheel rotates. Similarly, railroad car axles are large, round bars of steel spanning between the car wheels, supporting the car frame with bearings on the axle outside the wheels. Axles normally carry only transverse loads, as in the examples above, but occasionally, as in rear automobile housings, the axles also transmit torsion loads.

Spindle. A short shaft is a spindle. It may be small, or slender, or tapered. A spindle is capable of rotation or of having a body rotate upon it. It is similar to an arbor or a mandrel, and usage defines the small drive shaft of a lathe as a live spindle. The term originated from the round tapering stick on a spinning wheel on which the thread is twisted.

Head. A short stub shaft mounted as part of a motor or engine or extending directly therefrom is a head shaft. An example is the power take-off shaft on a tractor.

Counter shaft. A secondary shaft that is driven by a main shaft and from which power is supplied to a machine part is called a counter shaft. Often the counter shaft is driven by gears, and thus rotates counter to the direction of the main shaft. Counter shafts are used in gear transmissions to obtain speed and torque changes in transmitting power from one shaft to another.

Jack shaft. A countershaft, especially when used as an auxiliary shaft between two other shafts, is termed a jack shaft.

Line shafting. One or more pieces of shafting joined by couplings is used to transmit power from, for example, an engine to a remotely located machine. A single engine can drive many lines of shafting, which, in turn, connect in multiple fashion to process equipment machines. Belts operate on pulleys to transmit the torque from one line to another and from the shafting to the machines. Clutches and couplings control the transfer of power from the shafting.

The delivery of power to the machines in a shop has generally been converted from line shafting to individual electric motor drives for each machine. Thus, in a modern processing plant line shafting is obsolete. *See* BELT DRIVE; PULLEY. [J.J.R.]

Shale

A fine-grained laminated or fissile sedimentary rock made up of silt- and clay-sized particles. The average shale consists of about one-third quartz, one-third clay minerals, and one-third miscellaneous minerals, including carbonates, iron oxides, feldspars, and organic matter. The mineral composition of shales is poorly known because ordinary microscopy is unsatisfactory with such fine-grained materials, and x-ray diffraction methods do not always give analyses of all of the material.

Chemical composition. The chemical composition of shales, because of the difficulty of mineral identification, has been studied more than mineral composition. The average shale, as calculated by F. W. Clarke, consists of about 58% silicon dioxide, SiO_2, 15% aluminum oxide, Al_2O_3, 6% iron oxides, FeO and Fe_2O_3, 2% magnesium oxide, MgO, 3% calcium oxide, CaO, 3% potassium oxide, K_2O, 1% sodium oxide, Na_2O, 5% water, H_2O, and lesser amounts of other metal oxides and anions. Chemical composition varies with grain size, the coarser fractions having more silica and the finer having more alumina, iron, potash, and water. Shales normally contain a large amount of quartz silt, up to 60% in some analyses. Some shales have abnormally high silica content that cannot be ascribed to silt content. Most of the excess silica is in the form of very finely crystalline quartz, chalcedony, or opal, probably the result of large numbers of diatoms or volcanic ash in the sedimentary environment. Iron-rich shales may contain much pyrite or siderite, or iron silicates, all of which imply at least mildly reducing conditions in the original environment. Calcareous shales may contain several carbonate minerals and may be fossiliferous, grading laterally into limestones in some cases. High lime content may also be associated with the presence of gypsum. Potash is almost always more abundant than soda, possibly as a result of fixation in illitic clay minerals. Shales rich in organic matter signify lack of oxygen in the sedimentary environment. Organic matter may be carbonaceous, in which oxygen is chemically bound to carbon and hydrogen, or petroliferous, in which oxygen is essentially absent and the compounds are hydrocarbons of various kinds. The color of shales is not in general due to distinctive chemical composition but rather to pigmentation by small quantities of colored material. Red color in redbeds is associated with pigmentation by ferric iron, black color with carbonaceous material, green or dark gray with ferrous iron.

Fissility. The tendency for shales to split along bedding planes is called fissility; it is a product of the subparallel orientation of the individual grains of micaceous minerals along bedding planes. Part of this orientation may be introduced during sedimentation by slow settling of particles, but most of the effect is due to post-depositional compaction and reorientation of the platy minerals under pressure.

Lamination in shales, which ranges from 0.05 to 1.0 mm, is due to grain size variations, mineralogical variations, or color variations resulting from minor amounts of mineral or organic matter pigment. The laminations may be due to seasonal climatic factors that affect the source of supply and to sedimentation currents, similar to factors that influence the formation of varves in varved clay. Lack of lamination is chiefly the result of reworking of soft sediment by bottom-dwelling scavenging organisms. Such reworking can also result in mottled mixtures of clay and silt. *See* CLAY; CLAY MINERALS; VARVE.

Classification. No general agreement on a classification scheme for shales has been reached. On the basis of the composition of the silt fraction, W. C. Krumbein and L. L. Sloss have divided shales into groups roughly paralleling sandstone types: quartzose shale (orthoquartzite), feldspathic shale (arkose), chloritic shale (graywacke), and micaceous shale (subgraywacke). Their interpretation of the origin of these groups roughly is the same as for corresponding sandstone groups, stipulating that the shales were deposited in quieter waters than the sandstones. Shales may also be subdivided on the basis of origin, as was done by F. J. Pettijohn, into residual (from reworked soils), transported, and hybrid. The hybrid shales are mixtures of clastic and nonclastic materials (carbonate, organic matter, iron oxides). The hybrid shales represent slow rates of clastic sedimentation that gave the opportunity for chemical and biochemical processes to dominate the sedimentary environment. In this group belong the black shales, which may contain up to 25% organic matter and may even grade laterally into purely organic beds such as coal. Black shales are commonly phosphatic; any fossils present are thin carbonized impressions or pyritized. The faunas represented may be dwarfed or depauperate. Any calcareous shells are thin, normally of the *Lingula* type. Siliceous, alumina-rich, and iron-rich shales represent special conditions under which the appropriate elements are incorporated into the clay minerals or precipitated as oxides. *See* ARGILLACEOUS ROCKS; ARGILLITE; BENTONITE; BLACK SHALE; LOESS; REDBED; SEDIMENTARY ROCKS. [R.S.]

Shaped charge

A high explosive device that exploits the so-called Munroe effect. Charles E. Munroe, a leader in the development of explosives in the United States, made this novel discovery. He found that when letters are inscribed on the base of an explosive charge, an exact image is engraved in a steel plate upon which the charge is detonated.

This shaped or hollow-charge effect results from the interaction of the shock wave (detonation wave) in the exploding charge with the surface of

Fig. 1. Block of explosive (right) with words "Munroe Effect" molded on the surface, and a steel plate (left) which was placed in contact when the charge was exploded. (*Trojan Powder Company*)

Fig. 2. A heavy steel disk engraved by a charge similar to that used in Fig. 1. (*Trojan Powder Company*)

the concavity. Because the detonation front encloses the cavity, it produces in it a convergent shock wave. Extremely high pressures and temperatures result. If the cavity is lined with a thin sheet of metal, the liner is driven inward and extrudes as a very thin, extremely high-velocity, liquid jet.

The shaped charge has been used in many armor-piercing weapons, notably the bazooka of World War II fame. Since the war, special shaped charges have been developed for piercing the steel casings in oil wells. The so-called Jet Tapper also exploits the same principle to tap a blast furnace.

Experimental and theoretical studies of the shaped-charge effect have advanced the understanding of high-speed impacts such as those produced by meteorites on the earth. Shaped-charge techniques for producing high-speed pellets with velocities of about 7000 m/sec have also made possible the investigation of atmospheric erosion (ablation) on meteorites and missiles. Besides cavities, various other means are used to shape the detonation wave in order to produce such phenomena. *See* EXPLOSION AND EXPLOSIVE; EXPLOSIVE FORMING.

[W.E.G.]

Shaper

A machine tool for cutting flat or flat contoured surfaces by reciprocating a single-point tool across the workpiece. A shaper is usually used for small pieces requiring a short cutter stroke rather than for large pieces requiring longer strokes provided by a planer.

Shapers are classified as either horizontal or vertical depending on the plane of motion of the reciprocating ram. The latter type is often referred to as a slotter. The maximum stroke provided by the ram designates the size of the machine.

On horizontal shapers, a movable tool head mounted on the end of the ram permits angular cuts as illustrated. A hinged clapper box provides tool relief on the return stroke, while a movable table holds the workpiece.

Horizontal ram type shaper. (*Rockford Machine Tool Co.*)

Vertical machines hold the tool firmly on the ram while the workpiece is fed into the tool by moving the table. *See* MACHINING OPERATIONS.

[A.T.]

Shark

Any of about 200 species of fishes with cartilaginous skeletons, heterocercal tails, very hard placoid scales, and gills in separate clefts in pairs of five, except in one species with six and one with seven. The body is typically spindle-shaped. Sharks are further characterized by the presence of spiracles, a spiral valve, and a ventral mouth, usually equipped with several rows of teeth. These fishes are looked upon with disfavor by most commercial fishermen because of the damage they do to nets and because of their predatory feeding habits.

Kinds of sharks. Sharks are almost all marine. One small species, however, lives in Lake Nicara-

The mackerel shark, *Lamnia nasus;* length to 12 ft. (*From E. L. Palmer, Fieldbook of Natural History,* McGraw-Hill, 1949)

gua, and another small form is found in mountain lakes in the Philippine Islands. Occasionally a marine shark will wander upstream some distance into fresh water. Sharks are the largest of all fishes and include the largest living animals except the whales. *Caracharodon carcharias,* the man eater or great white shark, grows to a length of 30 ft. This is the shark usually responsible for attacks on humans in Australian waters. The two largest sharks are plankton feeders, straining food from the sea by means of their gill rakers. These are *Cetorhinus maximus,* the basking shark, which grows to a length of 40 ft, and the whale shark, *Rhineodon typus,* which reaches a length of 50 ft. Most other sharks are predaceous and feed on shrimp, squid, lobsters, crabs, and fishes. Some of the larger species prey upon seals and sea lions.

Economic importance. Except in Australia where it is a favored fish, shark meat is not utilized to any great extent as human food. Shark fins, however, are prized by some Oriental peoples, and sharks are used for animal food and fertilizer. Sharks have large livers which are rich in oil, and there is a substantial fishery for shark liver oil. This oil is rich in vitamin A and is processed for use as a human and animal vitamin supplement.

Shark skin is utilized commercially as an abrasive, under the name shagreen. It is also tanned into a serviceable leather.

The various species of small sharks called dogfishes are commonly used for dissection in vertebrate anatomy courses as an example of the primitive vertebrate structure.

Reproduction. Sharks produce large eggs containing much stored yolk. Primitive species lay eggs, sometimes in elaborate, leathery cases. Most sharks, however, retain the eggs in the uterus until they complete development, giving birth to living young. In all instances shark eggs are fertilized internally, the males having a pair of copulatory organs called claspers as modifications on their pelvic fins. *See* CHONDRICHTHYES. [J.D.B.]

Shear

A straining action wherein tangentially applied forces produce a sliding or skewing type of deformation. A shearing force acts parallel to a plane as distinguished from tensile or compressive forces, which act normal to a plane. Examples of force systems producing shearing action are (1) forces transmitted from one plate to another by a rivet that tend to shear the rivet, (2) forces in a beam that tend to displace adjacent segments by transverse shear, and (3) forces acting on the cross section of a bar that tend to twist it by torsional shear (Fig. 1). Shear forces are usually accompanied by normal forces produced by tension, thrust, or bending.

Shearing stress is the intensity of distributed tangential force expressed as force per unit area. Shear stresses on mutually perpendicular planes at a point in a stressed body are equal. When no normal stresses exist, the state of stress is pure shear, which induces both normal and shear stresses on oblique planes. Elements oriented 45° to planes of pure shear are subjected to biaxial tension and compression equal to the shear stress (Fig. 2*a*). Similarly, pure shear is induced by equal and op-

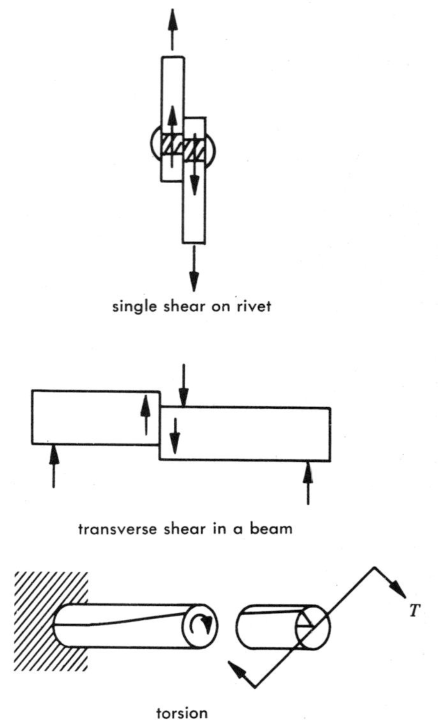

Fig. 1. Shearing actions.

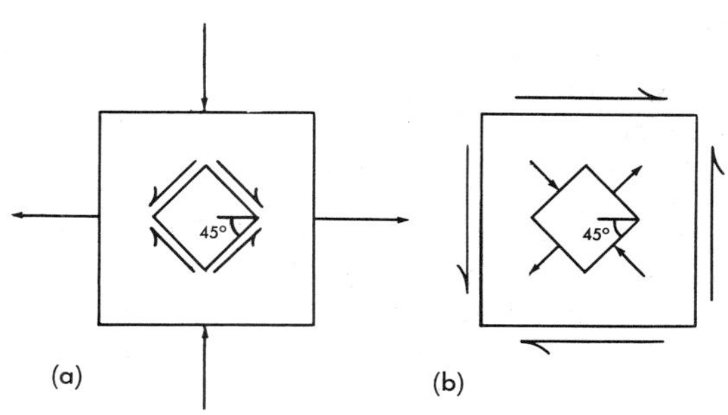

Fig. 2. State of pure shear. (*a*) Shear produces biaxial tension and compression. (*b*) Biaxial stresses induce shear.

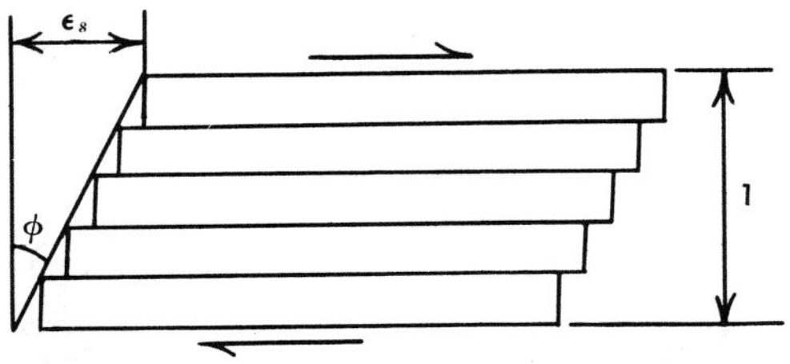

Fig. 3. Shear strain.

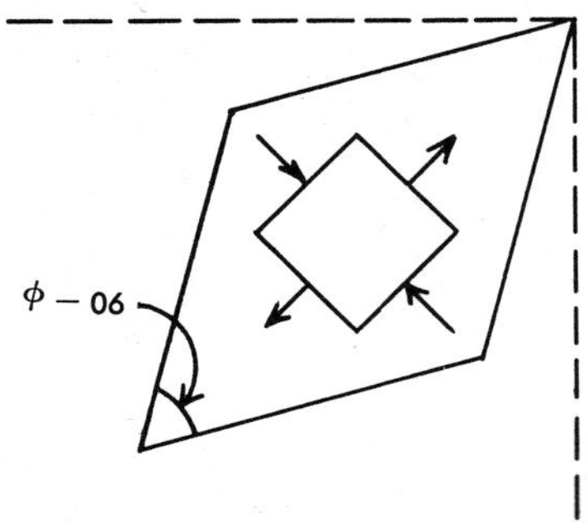

Fig. 4. Shear distortion.

posite biaxial stresses (Fig. 2*b*). Pure shear is the maximum shear stress at the point. Under combined stress, the shear stress is found by principal stress analysis.

Shearing strain is the displacement ϵ_s of two parallel planes, unit distance apart, which accompanies shear stresses acting on these planes. Shearing distortion is visualized as sliding without separation or contraction of all planes parallel to the shear forces, like cards in a pack (Fig. 3). For planes unit distances apart

$$\epsilon_s = \tan \phi \approx \phi \text{ radians}$$

where ϕ is the small angle of distortion.

Modulus of rigidity, designated E_s, G, or N, is the shearing modulus of elasticity, which according to Hooke's law is the constant of proportionality between shearing stress S_s and shearing strain S_s/ϕ during elastic behavior.

After shear distortion, the angles of a rectangular element are altered by the shear strain ϕ (Fig. 4). Changes in length of diagonals must be consistent with the biaxial strains, which leads to the relationship

$$E_s = \frac{E}{2(1 + \mu)}$$

where E is Young's modulus, and μ is Poisson's ratio. The value of μ can be found when E_s and E are experimentally determined. *See* STRESS AND STRAIN. [W.J.KR.]

Shear center

The point in the plane through a section of a structural member at which a shear force can be applied without producing a rate of twist of that section. The shear center of a section through a member such as an I beam that has two planes of symmetry coincides with the geometric center or centroid of the section. When such a member is loaded transversely it bends without twisting. However, if the structural member is unsymmetrical or has but one plane of symmetry and is loaded elsewhere than through that plane, as in an open section made of thin material for resisting bending in aircraft construction, the load force and the reaction force of the structural member constitute a couple that, in general, causes the member to twist. The load can be applied through a unique point in the plane of the section such that the moments in the plane of the section are balanced and the beam is not twisted. This unique point is called the shear center.

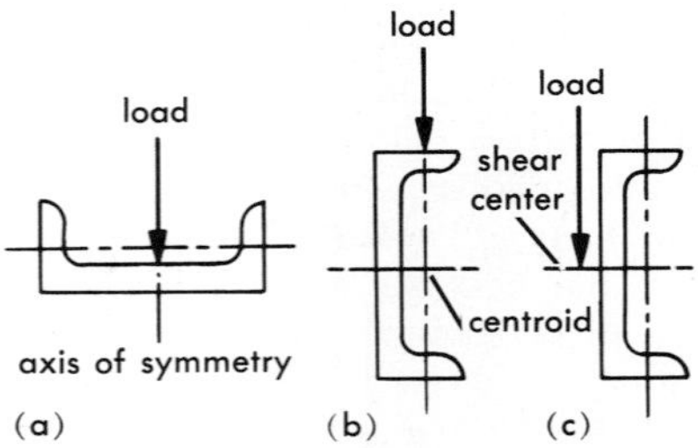

Three conditions of channel section under load. (*a*) Channel resists bending when load is along axis of symmetry. (*b*) Channel bends even with load through centroid if load is not along axis of symmetry. (*c*) Channel resists bending when load is through shear center.

Location of the shear center for a section depends only on the section dimensions. The member is only subject to twisting when subjected to shear; shear center is of no significance for a beam in pure bending. *See* BEAM; LOADING, TRANSVERSE; SHEAR. [W.J.KR.]

Bibliography: D. J. Peery, *Aircraft Structures*, 1950.

Sheep

The sheep-production branch of animal husbandry involves breeding, feeding, and management for commercial purposes (see Table 1 and Fig. 1).

Domestication of sheep began about 6000 B.C. in Asia Minor. Domestic sheep (Table 2) descended from the Asiatic moufflon, *Ovis orientalis*, the European moufflon, *O. musimon*, and the Asiatic urial, *O. vignei* (Fig. 2). Although sheep breeds are found throughout the world, they predominate in the marginal, sparsely settled, temperate areas of the Western Hemisphere.

Successful breeding of sheep requires knowledge of their physiology of reproduction which varies with breed and type, systems and standards of selection for genetic improvement, and systematic flock management. *See* BREEDING (ANIMAL).

Physiology of reproduction. This aspect of sheep husbandry pertains to the phenomena associated with the reproductive cycle. Ewes are usually bred

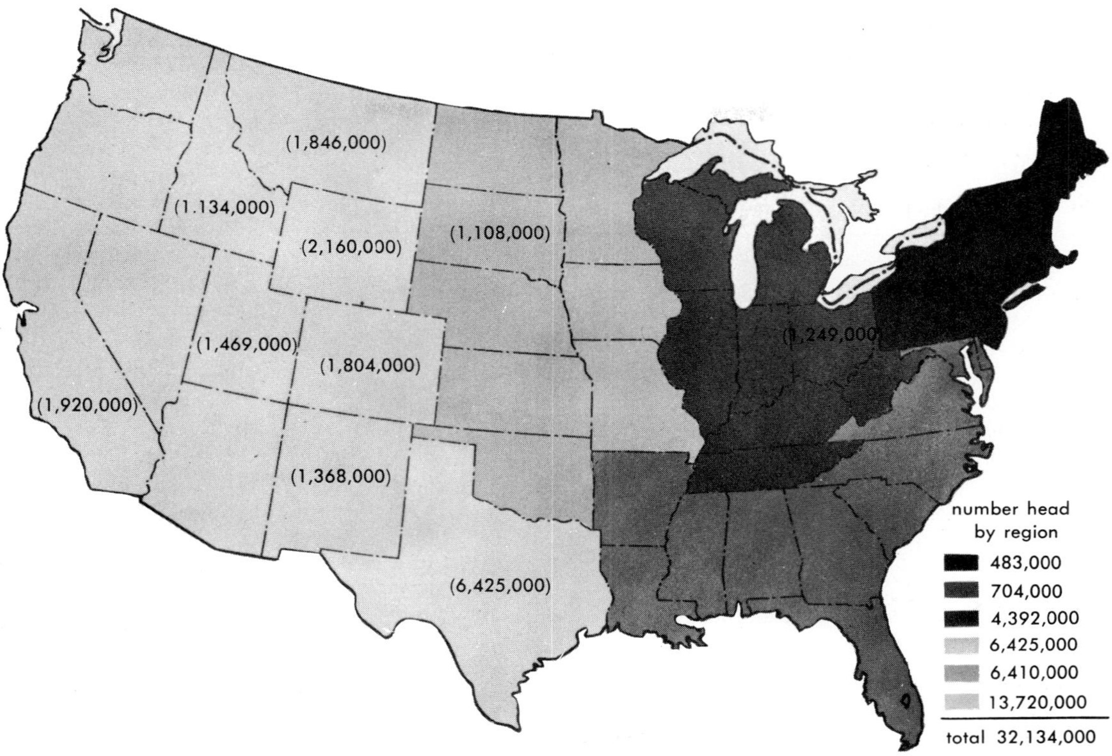

Fig. 1. Average number of sheep and lambs (1947–1956) on farms in the United States by regions. States having more than 1,000,000 head indicated in parentheses. (*From Livestock and Meat Statistics, USDA Statist. Bull. 230, July, 1958*)

to lamb at 2 years of age. Ewe lambs benefiting from optimum conditions can be bred as lambs, but a low lamb crop will usually result. Such ewes are temporarily stunted, but they are comparatively more productive on a lifetime basis. Wool production is not affected.

Gestation periods vary in number of days with breeds and types. Breed averages are: fine wools, 148–152 days; medium wools, 144–148 days; long wools, 146–149 days. Individuals within breeds also vary.

Estrus in ewes is difficult to detect visually. It lasts from 1 to 3 days. Ovulation occurs toward the end of the heat period. The estrous cycle averages 17 days.

All breeds, with the exception of the Rambouillet, Merino, and Dorset, have a definite breeding season. Medium-wool breeds, long-wool breeds, and crossbred types start exhibiting estrus in August. Their normal breeding season lasts through December. Attempts to induce estrus by hormones for extraseasonal breeding have been unsuccessful. Because Rambouillet, Merino, and Dorset ewes come into heat throughout the year, nonseasonal lambs can be produced.

Growthy ram lambs are fertile at 6–7 months of age. Sperm motility and concentration are about the same in the semen of lambs and mature rams. Rams breed throughout the year. Range operators use three yearling or older rams per 100 ewes. Under the same conditions, five ram lambs may be used.

In pasture- or hand-mating, a ram lamb can serve up to 30 ewes, or an adult can serve 50–60 ewes without endangering fertility. Where single rams are used for matings, assurance of a lamb crop is gained by replacing the first ram with a second after the first or second estrous cycle.

Genetic improvement by selection. Selection of lambs on the basis of production is replacing empirical methods. Records of economic factors are utilized for selection and culling. These factors are sometimes combined into a selection index. In the more complex fine-wool and dual-purpose breeds, economic factors include fleece production as measured by staple length and fleece weight; wean-

Table 1. Number and value of sheep and lambs on farms in the United States on January 1, 1958

Period	No. of head, × 1000	Farm value, × $1000
1947–1956 (average)	32,134	$562,525
1957	30,840	461,361
1958	31,328	601,929

SOURCE: *Livestock and Meat Statistics*, USDA Statist. Bull. 230, July, 1958.

Table 2. Prominent breeds of sheep in the United States classified according to grade of wool

Fine wool	Medium wool	Crossbred wool	Long wool	Carpet wool	Fur
Merino Rambouillet	Cheviot Dorset Hampshire Oxford Shropshire Southdown Suffolk	Columbia Corriedale Panama Romeldale Targhee	Cotswold Lincoln Romney	Karakul Navajo	Karakul lamb

ing weight, adjusted for type of birth, inbreeding, and age of dam; and such related factors as face covering, smoothness in body and neck, type, and condition.

Ram progeny performance-testing provides an additional source of production-tested sires. In Texas, where this program has been conducted since 1950, the average gain for Rambouillet ram lambs has increased from 0.37 to 0.60 lb per day; grease fleece weight from 13.8 to 19.5 lb per shearing; and staple length from 3.34 to 4.20 in. per fiber.

Flock management. Breeding performance of ewes is affected by flock management. Body weight, obesity, nutrition, and age of dam are important factors. Ewes that are heaviest as yearlings maintain that standing throughout their lifetime. When compared with ewes of lighter body weight, heavier ewes have a higher conception rate, have more live lambs born per ewes bred, and wean more pounds of lamb. Obese ewes do not follow this trend.

Ewes are kept more profitably at maintenance weight, except just prior to breeding and parturition. About 2–6 weeks prior to breeding, ewes may be conditioned, or flushed, by giving them a good feed supplement, or by placing them in a lush, unused pasture. This stimulates a higher ovulation rate.

Age is a factor in breeding efficiency of ewes. Older ewes are more prolific and breed more easily than yearlings. Yearling ewes tend to produce single lambs, whereas older ewes have an increasing tendency to twin.

Sheep nutrition. Under range conditions, sheep depend primarily on range plants for their sustenance. Sheepmen minimize feeding operations but generally provide supplements to overcome range deficiencies (*see* NUTRITION).

In the intermountain area, sheep generally graze in the spring and fall in home pastures in the foothills. During the winter months, when snow covers the ground, feeding becomes necessary. Where alfalfa hay is abundant, it may be the sole feed. Approximately 6 weeks prior to lambing, it is a good practice to feed a pelleted protein supplement in addition to hay. This might consist of barley, oats, dehydrated molasses, beet pulp, soy-bean meal, and

Fig. 2. Examples of prominent breeds of sheep. (*a*) Rambouillet ram (*Sheep and Goat Raiser*). (*b*) Dorset ram (*Continental Dorset Club*). (*c*) Hampshire ram. (*d*) Southdown ram (*Southdown Assn.*). (*e*) Suffolk ewe and ram (*photograph by A. L. Henley*). (*f*) Columbia ram (*photograph by A. Sponagel*). (*g*) Lincoln ram (*Cook and Gormley*). (*h*) Navajo ram (*Southwestern Range and Sheep Breeding Lab., Fort Wingate, N.M.*).

Fig. 3. Ram progeny groups on performance test at Texas Agricultural Experiment Station.

molasses. In parts of this area, the ranges are deficient in phosphorus. The inclusion of monosodium phosphate in the supplement in such deficient areas improves productivity in ewes in terms of weight gains, wool yield, and lamb crop.

In the Southwest, ranges are subject to periodic droughts and feeding is entirely different from that of the intermountain region. Here a very common year-around practice is to feed free choice a 25% salt and 75% cottonseed meal mixture. Salt acts as a regulator of feed intake. An abundance of water and a source of roughage are required for such a system. Water should be at a sufficient distance from the feeders to require the ewes to travel back and forth, hence restricting intake and encouraging more efficient range utilization. This supplement carried thousands of sheep through one of the most disastrous droughts in the history of the Southwest. The average intake of pregnant lactating ewes is from 0.073 lb to 0.347 lb of cottonseed meal and 0.024 lb to 0.116 lb of salt.

A urea-molasses liquid supplement self-fed to pregnant and lactating ewes compares favorably with cottonseed cake as a supplement to dry range forage. This consists of a mixture of 88% molasses, 10% urea, and 2% phosphoric acid, fish oil, and trace minerals. The protein equivalent of this mixture is 30%.

Other feed additives include such antibiotics as chlortetracycline and oxytetracycline. These are fed in lamb fattening rations at the rate of 10–15 mg/lb of feed. In wether lambs a hormone is often administered, this being done by implanting it between the skin layers of the ear. Diethylstilbestrol pellets weighing 3–6 mg are recommended. No ill effects on carcass grades or secondary sex characteristics are noted with implants of such weight. These treatments produce faster feed lot gains and improve feed efficiency. When excessive estrogens are fed or administered, deleterious effects are observed, such as poor carcass quality, prolapse of the rectum, and abnormal development of secondary sex characteristics (*see* ESTROGEN).

Economic results are obtained by feeding therapeutic amounts of certain antibiotics to feed lot lambs (*see* ANTIBIOTIC). This economy is gained through the reduction of digestive disorders, such as diarrhea and enterotoxemia (*see* ENTERIC BACILLI). Further, the antibiotic serves as a feed restrictor so that lambs can be self-fed on complete rations. Rate of gain is improved as well as increase in feed efficiency.

Tranquilizers have not been studied extensively as feed additives (*see* TRANQUILIZER). Increased daily gains of 12–17% by feed lot lambs have been reported from the use of hydroxyzine. Texas workers fed this tranquilizer in a total mixed ration at levels of 1.5 g up to 25 g/ton of mixed feed. Response was essentially the same for all levels.

Pelleting complete diets or supplemental rations is advantageous if the costs are not exorbitant. These advantages include increased daily gain, increased intake of high-roughage feed, prevention of separation of feed ingredients, improved efficiency, reduced storage space, reduced loss from storage and wind, and savings of time and labor in feeding operations. However, operators report that up to 5% of feed lot lambs initially refuse pelleted feeds. The operator is required to give special attention to these lambs.

Parasite control. Parasites, external and internal, find sheep highly satisfactory hosts. Correct and thorough use of insecticides and anthelmintics prevents serious economic losses in meat and wool (*see* INSECTICIDE).

External parasites. These live in the surface layers of the skin and in the wool fiber follicles. Dips, sprays, and smears are effective measures for prevention and cure. Such parasites include lice, *Bovicula ovis;* sheep bots, *Oestrus ovis;* screw worms, *Callitroga hominivorax;* fleece worms, *Phormia regina;* sheep keds, *Melophagus ovinus;* and scab mites, *Psoroptes ovis* (*see* ANOPLURA; MYIASIS). There are others of lesser importance. Control measures for these are shown in Table 3.

Scab mites, keds, and lice spend their whole lives on sheep. Elimination must be thorough and must reach all of the animals.

Bots, screw worms, and fleece maggots are parasites only during a part of their development. Here prophylactic measures for reducing fly strike may be fairly effective. Maggots of screw worms cannot enter the skin unless it is broken. Therefore, fresh cuts in animals should be prevented whenever possible, or prompt treatment should be given.

Internal parasites. These live within the body and organs of animals. Those of greatest significance are the eastern, or large, stomach worm, *Haemonchus contortus;* the brown stomach worm, *Ostertagia circumcincta;* the thread-necked worm, *Nematodirus* spp.; the stomach hair worm, *Trichostrongylus axei;* the nodular worm, *Oesophagostomum columbianum;* the hookworm, *Bunostomum trigocephalum;* the broad tapeworm, *Moniezia* spp.; and the fringed tapeworm, *Thysanosoma actinoides* (*see* CESTODA; NEMATODA).

Table 3. Control measures for external parasites

Parasite	Treatment	Remarks
Sheep and goat lice	As a dip, use one of the following: (1) one lb derris or cube powder per 100 gal water; (2) 0.25% DDT, TDE, methoxychlor, toxaphene, or chlordane*; (3) 0.03% gamma BHC, or lindane; (4) 5.0% rotenone per 100 gal water	To use (2) as a spray, double the indicated concentration; to use (3) as a spray, double the indicated concentration; *do not use on animals under one month of age, or on animals in poor condition*
Sheep bots, nose fly	As a smear, use pine tar repellent; irrigate nasal passages with 1 fl oz of a 3% solution of saponated cresol under 35–45 lb pressure	To prevent infestation, smear nostrils at weekly intervals during fly season
Screw worms	As a smear, use EQ335 remedy diluted with 9 parts of water in infested area	Active toxic agent in EQ335 is lindane, which should not be used on animals under 30 days of age
Fleece worms, wool maggots	As a smear, use EQ335 as for screw worms; as a spray, use 0.5% toxaphene, or 0.5% chlordane*	
Sheep keds, ticks	Same as for sheep and goat lice; as a dust, use 1.5% dieldrin	
Scab mites and mange	As a dip, (1) use a 0.06% suspension of gamma BHC or lindane, prepared with a wettable powder; or (2) use either 1.5% polysulfides of lime-sulfur, or 0.05–0.07% nicotine	Do not use (1) on animals under 30 days of age, or on thin animals; temperature of dip (2) should be 95–105°F

* Chlordane accumulates in body fat; no more than four applications should be made each season.

Fortunately, the drug, phenothiazine, is effective for all these parasites except tapeworms. Sheep are highly tolerant of phenothiazine. Such other effective agents as copper sulfate and nicotine sulfate (Cu-Nic), carbon tetrachloride, tetrachlorethylene, and hexachlorethane have a narrow margin of safety.

Therapeutic dosages of phenothiazine consist of 2.5 oz for sheep weighing more than 60 lb; only 1.5 oz is required for those under that weight. Drenches are more commonly applied than boluses or capsules.

Prophylactic quantities of phenothiazine can be ingested where a mixture of 9 parts salt and 1 part phenothiazine is kept free choice before range sheep or goats. This is not a substitute for drenching where the need is evident.

Parasitologists generally concede that the broad tapeworms are not detrimental to sheep or goats (*see* PARASITOLOGY). Lead arsenate is the preferred treatment in doses of 0.5 g for animals weighing less than 60 lb and 1.0 g for those over that weight. Lead arsenate can be purchased in combination with phenothiazine. No effective treatment has been found for the fringed tapeworm (*see* TAPEWORM).

Sheep that are provided with adequate feed are more resistant to worm infection than those in poorly nourished condition. Overstocking of ranges and pastures should be avoided.

Judging. Fat lambs and breeding sheep comprise the two general judging classifications. Fat lambs are judged as slaughter animals; breeding sheep are judged as perpetuating stock. In each classification, judging is based on over-all phenotypic (physical) merit, in which differences among the sheep under consideration are rationalized and systematically arrayed according to accepted standards of excellence (*see* GENETICS). Sex and breed characteristics are described in literature published by the various breed associations, each association setting for its specific breed the accepted standards of excellence for conformation and fleece.

Fat lambs. These are judged on the basis of conformation, finish, and quality. In conformation, they should be rectangular, moderately low-set, and blocky. The well-balanced body should be deep and uniform in width; ribs well-sprung, close together, and long; loin thick and broad; back strong and straight; hips and shoulders smooth; rump long, level, wide, and square to the dock; twist deep and plump; leg large, meaty, and thick; neck short and thick; and shoulder smooth on top.

Finish gives a thick and plump condition. Covering should be uniform and firm over loin, back, ribs, and shoulders. A full dock, plump neck, and shoulder vein indicate fatness. The lamb must be handled in determining true finish and conformation, since elaborate dressing of the fleece for show purposes can be visually deceiving.

Quality concerns refinement of head and bone structure, and freedom from body folds. Refinement does not sacrifice size and growthiness. Quality is a major factor in dressing percentage; hence, ample finish, light pelt, freedom from paunchiness, and a trim, straight underline are desired.

Breeding sheep. These are generally shown in full fleece, and therefore require careful handling to determine body width, depth, and smoothness. Breed characteristics become important factors, and breeding sheep may be classified according to wool type: medium wool, medium wool dual-purpose, and fine wool. Rams should be rugged with strong bones, and ewes more refined than rams, showing femininity about the neck and head.

Fig. 4. Fat lambs are judged as slaughter animals.

Body conformation among most of the medium-wool breeds is similar, with emphasis placed on width and depth, compactness, low-setness, size of leg, and strength of back. Uniformity of width, depth of body, and balance are important in both ewes and rams. Depth of body is measured from back to belly. Degree of low-setness and compactness should not be carried to excess. Size and ruggedness are desirable.

Constitution implies hardiness or ability of sheep to thrive under adverse conditions. It is generally agreed that hardiness is indicated by strength and straightness of legs, fullness of heart girth, width of chest, roominess of middle, full arch of rib, and strength of head.

Quality in medium-wool breeds is denoted by refinement of head and bone, pinkness of skin, and desirable fleece. Fleeces of the medium-wool breeds vary in importance according to association regulations. In general, however, heavy emphasis is placed upon conformation. Factors such as malocclusion and wool-blindness are serious defects. Cryptorchidism is intolerable.

For the medium-wool dual-purpose types, about 65% of the emphasis is placed on conformation and 35% on wool. The fleece is of significant importance, and should be staple in length, free from black fiber, dense, and tight. Grade of wool varies with the breed. Breechiness, excessive variation in fiber diameter, and medullation are discriminated against.

In judging fine-wool breeds, emphasis is still placed upon conformation, but the fleece is much more important than with medium-wool breeds. About 50–60% of the income from fine-wool sheep is derived from lambs, with the remaining 40–50% from wool.

Length of staple is the most important single factor of the various physical properties of the wool. For a 12-month growth, at least 2.5 in. is expected. Uniformity of length is important. Density of fleece is an indicator of light-shrinking wool and heavy fleece weight. An open fleece is undesirable, and is indicated by loose, ropy locks and large, exposed areas in the flanks and armpits.

Purity is extremely important, meaning that the fleece should have no dark or black spots or single, pigmented fibers. The fleece itself should be bright and should display a distinct and uniform crimp. Belly wool should be dense and not carry high into the side wool.

The fine-wool sheep breeder compromises between mutton and wool production. He tends to demand open-faced sheep and does not tolerate body and neck folds, malocclusion, or weak pasterns. *See* FIBER, NATURAL; GOAT; MOHAIR; WOOL. [T.D.W.]

Bibliography: *See* AGRICULTURAL SCIENCE (ANIMAL).

Sheepshead

An American fresh-water fish, *Aplodinotus grunniens*, of the family Sciaenidae. The sheepshead, or fresh-water drum, is also known as the white perch, especially in the middle Mississippi Valley; in Louisiana, it is called the gaspergou. It has perhaps the widest natural range of any American fresh-water fish, being found from Guatemala northward into the Hudson Bay drainage. It attains considerable size, formerly 100 lb or more, but most sheepshead now caught weigh under 10 lb. It is a solid, silvery fish, with a short, square tail and a long, low dorsal fin.

The sheepshead is a good food fish in most waters, although in some localities, at least seasonally, it is not considered desirable. It is found in great abundance in certain waters, notably Lake Erie and Lake Winnebago in Wisconsin. Like others of the croaker family, it produces a distinct purring croak, clearly audible under water. Its otoliths are frequently carried as lucky stones. *See* PERCH; PERCIFORMES. [J.D.B.]

Sheet metal forming

The shaping of metal that is initially in thin sheets by applying local pressure. Sheet metal parts (including strip, plate, and tubing) are fabricated by a variety of processes, the nomenclature of which is not standardized. Many of the processes are so interrelated that classification into major groups is difficult; consequently, the major classes considered herein (bending, deep drawing, shearing, spinning, and squeezing) are purely arbitrary. With few exceptions (such as forming brittle metals or thick plate) the operations are performed at room temperature.

Bending. The curving of sheet, strip, or sections to produce a permanent deformation is considered a bending operation. Strictly speaking, bending refers to a curving process whereby the outer fibers are lengthened while the inner fibers are compressed or shortened. If the curving leaves the outer fibers essentially unstressed while the inner fibers are compressed, the process is shrinking; and if both the outer and inner fibers are stretched, but not to the same degree, the process is stretching. Any of the three methods can be used to produce a bend.

The minimum bend radius to which a sheet can be bent is limited by cracking in the bend or at the edges of the bend. For a specific material, no definite value of minimum bend radius may be specified since the width of the bend, the edge conditions, and the method of accomplishing the bend (which influences friction forces) have a significant effect. The type of metal and the degree of prior strain hardening also affect the minimum bend radius because the unit strain in the outer fibers increases as the bend radius decreases.

If a part is bent in one direction only, the radius of curvature will increase upon release of the bending forces because of elastic recovery of the material. Accurate determination of this springback must be made by experiment; values determined by mathematical analysis are usually lower than those encountered in practice, particularly for sharp bends. Correction for springback is usually made by overbending so that, on release of the forming force, the metal returns to the desired shape.

The equipment used for bending depends upon the nature of the bend and the shape of the section being bent. Some parts are bent along straight lines and others along curves; all the bends may be in the same direction or they may be in different directions. The quantity of parts to be produced is an important factor, in some instances, in the selection of the optimum equipment. Thus, different types of bending equipment are required.

For forming long sections in large quantities, roll forming is the most economical method. A roll forming machine consists of numerous small power driven rolling mills mounted on a frame. Each successive roll bends the strip a little farther than the preceding roll; the number of stations, or passes, and the design of the roll profiles depend upon the shape of the final cross section desired. In drawbench forming the rolls are not power driven, but the strip is pulled through the rolls.

Parts containing straight bends are formed on mechanical presses using special purpose dies (die bending) if quantities are large, or on press brakes using general purpose dies if small quantities of large, extended parts are to be produced. Press formed parts are often called stampings.

The aircraft industry uses the Guerin process extensively to form parts with straight and curved flanges. In this process, the upper platen of a hydraulic press is equipped with a contained, thick rubber pad, which acts primarily as the female die. Various metallic, or even nonmetallic, materials are used for the male die, or formblock. Usually several different formblocks are placed on a table acting as the lower platen of the press; thus, several different parts are formed simultaneously with low tooling costs.

A modification of this process, called the Wheelon process, allows higher pressures than are possible by the rubber pad forming method. In the Wheelon process, a rubber bag (or fluid cell) replaces the rubber pad in the upper platen of the press. By inflating the bag with hydraulic fluid, a rubber throw pad in contact with the blank transmits the pressure and forces the blank to bend over the formblock. In both processes, there is no control over the movement of the blank during the early stages of the forming, and wrinkling may occur in certain shrinking operations.

Roll bending, which is confined to bends of large radius, is accomplished by three parallel rolls, one of which has a movable axis. The movable roll may be replaced with a cam, a roller, or a forming block.

Wiper-type bending is a method in which the stock is forced to assume the curvature of a formblock by pressing it progressively against the block with a roll, slide block, or a wiping block. This method permits control of the cross section of the piece being formed, as is necessary in bending tubing without its flattening at the bend. It also allows forming to smaller bend radii because controlled compressive stresses in the outer fibers permit greater plastic deformation (material can be extended a greater amount if it is simultaneously compressed in a lateral direction).

Stretch forming or wrap forming equipment superimposes longitudinal tension on the bending moment. This is achieved by mechanical or hydraulic means and causes tensile stresses of different magnitude in both the outer and inner portions of the bend. Because of the absence of compressive stresses in the inner fibers, springback and wrinkling are minimized, but the metal does become thinner. Stretch forming may be combined with wiping to obtain some of the advantages of each.

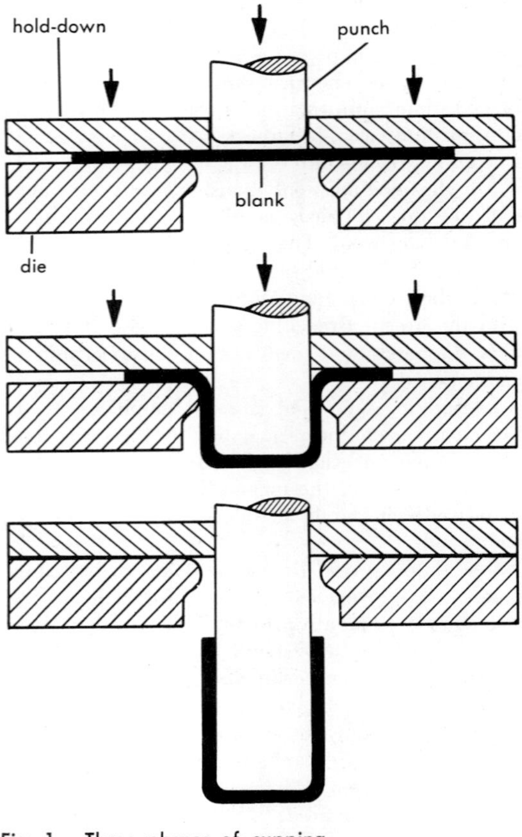

Fig. 1. Three phases of cupping.

Deep drawing. The press forming of cup- or box-shaped parts from sheet metal by controlled plastic flow is termed deep drawing or shell drawing. The first operation is cupping; it consists of forcing a properly shaped blank through a die as shown in Fig. 1. The part need not be forced all the way through the die, in which case a flanged cup can be produced. The hold-down (blank holder or pressure plate) is required on relatively thin sheets to prevent wrinkling (or buckling), which tends to occur because of circumferential stresses that are present during the drawing operation. The hold-down pressure may be exerted by springs or hydraulic or air pressure. Excessive pressure is not desirable as the increased frictional forces would limit the severity of the draw (ratio of depth of cup to diameter of cup, or percentage reduction from diameter of blank to diameter of cup). Lubrication, die finish, die radius, punch radius, and the properties of the metal being drawn are also factors which affect the maximum severity of draw (which may be as high as 75%).

If deep parts, stepped parts, cups and boxes with recessed bottoms, conical or hemispherical shapes are required, one or more redrawing operations are needed (Fig. 2). Depending upon the properties of the metal and the severity of draw in each operation, interstage annealing may be required to restore ductility to the metal. If a tapered wall thickness or a thin wall and thick bottom are desired, an operation known as ironing is used (Fig. 3). In ironing, the inside diameter remains constant and the clearance between punch and die is less than the desired thickness of the cup. Consequently, as the cup is forced through the die, the wall is thinned or ironed, resulting in an extension of the cup. By tapering the punch, a tapered wall is obtained, such as is required in cartridge cases.

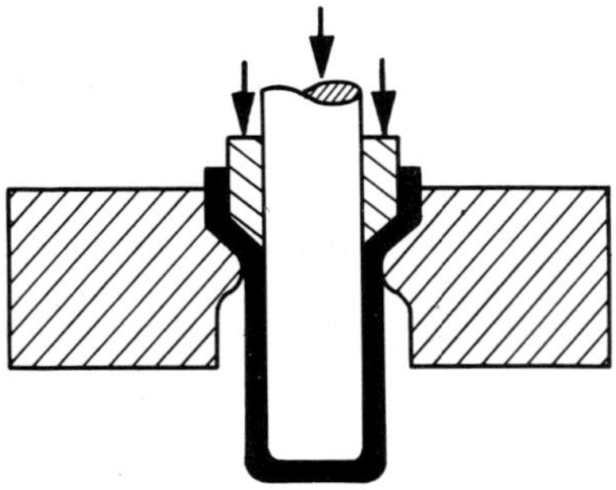

direct redraw

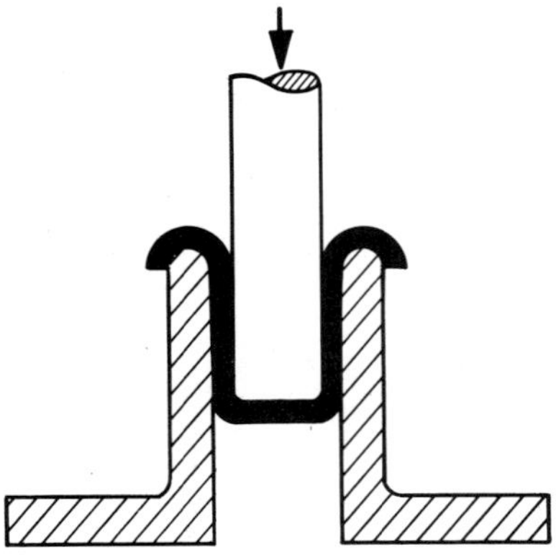

reverse redraw

Fig. 2. Types of redrawing operations.

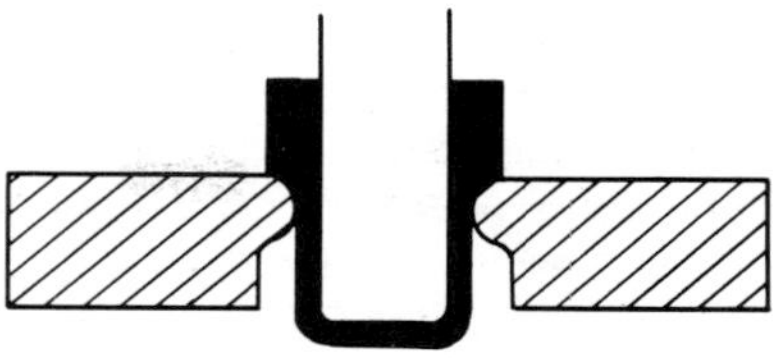

Fig. 3. Ironing produces a wall of varying thickness.

Bulging out the parallel walls of a deep-drawn cup into a more complex shape may be accomplished by pressing a rubber punch into the cup or by forcing a fluid under pressure into the cup, thereby causing the cup to assume the shape of a surrounding die.

The Guerin process can also be used to form shallow cups. A modification of this process (known as the Marform or Hidraw process) uses a rubber pad for the female die, a solid punch, and a solid blank holder (Fig. 4*a*). The blank holder and punch are independently actuated by hydraulic pressure. Higher pressures are obtained than those in the Guerin process, permitting deeper draws without wrinkling. Another modification (termed the Hydroform and Corform processes) uses an oil-backed rubber diaphragm, a solid punch and a solid blank holder (Fig. 4*b*). The pressure of the oil can be varied and acts as the top blank holder as well as a variable pressure female die. The blank holder is stationary, being mounted on the press base. In another flexible die method (the Hydrodynamic process) the female die is a machined metal die, and water pressure forces the blank into the die cavity (Fig. 4*c*). High pressure water passes through a hole in a spring-actuated solid pressure pad and acts as a fluid punch of continually changing shape on the underside of the blank. Explosive forming is also a flexible punch process. Sheet metal is forced into an open female die by a shaped charge or by transmitting the explosive force through water. In all these processes the stresses are more uniform than those in conventional drawing; this uniformity permits deeper draws and more complex shapes.

Deep drawn parts may be formed on drop hammers (*see* FORGING) or by stretch forming. In the latter process, the die and hold-down of conventional press forming equipment are replaced by two plane surfaced, straight edge jaws. The stretch die is then forced against the clamped blank, stretching the metal and causing it to assume the contour of the stretch die.

Another cupping technique uses the frictional forces developed between the blank and the die and

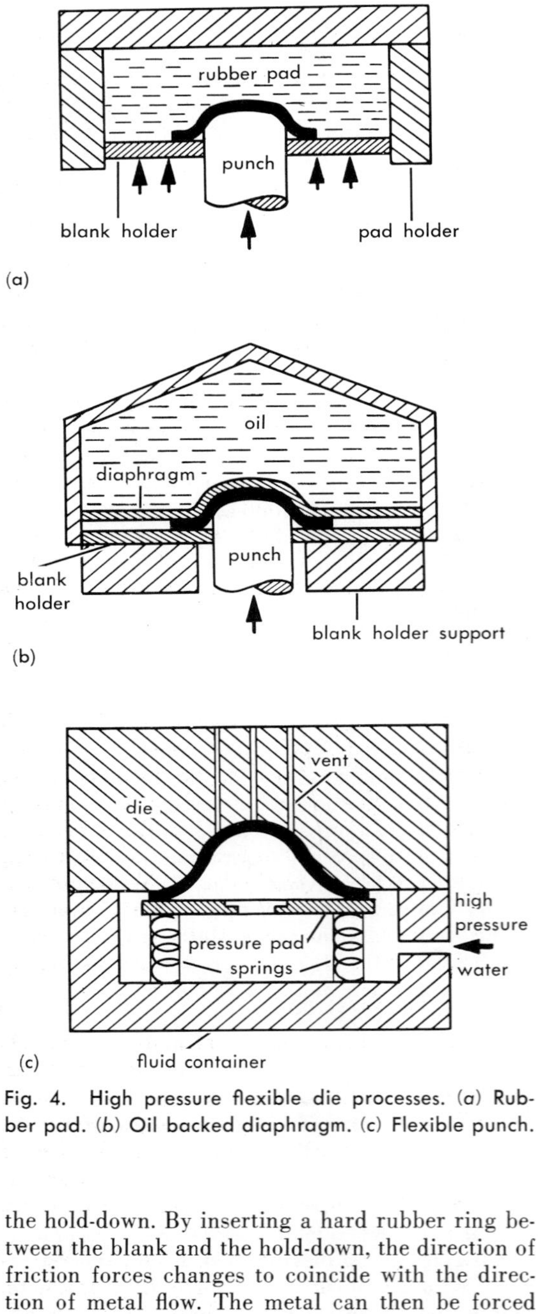

Fig. 4. High pressure flexible die processes. (*a*) Rubber pad. (*b*) Oil backed diaphragm. (*c*) Flexible punch.

the hold-down. By inserting a hard rubber ring between the blank and the hold-down, the direction of friction forces changes to coincide with the direction of metal flow. The metal can then be forced into the die without the aid of a punch, although a supplementary punch improves the quality of the part and facilitates the action of the friction element. In operation, pressure is applied to the die and transmitted through the blank to the rubber ring, compressing the latter and forcing it to move in the direction of least resistance toward the die cavity. The metal also moves toward the die cavity if the frictional forces between the rubber and the blank are large enough. Lubrication is important in this process because frictional forces between the blank and die, and between the rubber ring and blank-holder, must be minimized.

Spinning. Parts of more or less rotational symmetry can be formed by the spinning process. The basic process consists of rapidly rotating a form block, against which a circular metal blank is rigidly held. The blank is progressively pressed around the block by a bar or roller type tool (Fig. 5). The process can be carried out on an ordinary (but rigid) lathe or on special purpose spinning or rolling machines. Spun parts are commercially competitive with die formed parts, particularly for small quantities, large sizes, and complex contours. Tubing and tubular parts can be spun to expand or reduce the diameter, generally at the end of the part.

In ordinary spinning any change in thickness of the metal as a result of the spinning operation is unintentional; however, depending upon the shape of the part and pressures exerted by the spinning tool, thinning or thickening of the metal may occur. A modification of spinning combines features of spinning with those of rolling and in some cases, extrusion. It is a roll forming operation for making conical, tubular, or curvilinear surfaces of revolution. In cold power spinning (also called Hydrospin or Floturning), a thick blank (disk, tube, or other shape) is cold-rolled under high pressure against a rotating mandrel, and the metal is, in effect, extruded in an axial direction. High strength, close tolerances, and excellent surface finish are obtained by this process.

Squeezing. A variety of press forming operations involve squeezing or compressing the metal to produce small over-all deformation but, generally, rather high local deformation. These operations are termed stamping; however, stamping in a more general sense includes all press forming of sheet metal parts.

Common operations of this type are sizing, coining, swaging, and embossing. The first three processes are basically cold die forging operations in which the metal is confined and made to flow plas-

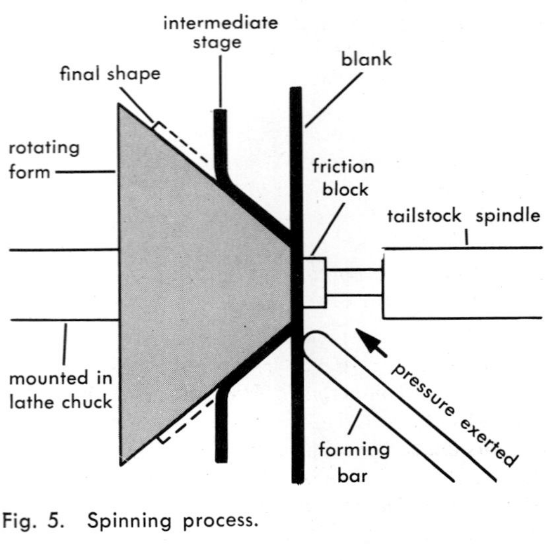

Fig. 5. Spinning process.

tically into the punch or die impressions, or both. Sizing is an operation in which the part (previously formed to almost finished shape) is pressed between male and female dies to improve its dimensional accuracy. In coining operations a pattern is pressed into one or both surfaces of the metal (as it is in a coin), to change the thickness of the piece at many points. Swaging is used extensively to displace metal around a hole, thereby increasing the local thickness of thin sheet for subsequent threading operations. Embossing is a form of shallow drawing or bending in which a pattern or recess is pressed into one surface of the metal and the reverse side assumes a negative configuration. The metal thickness remains essentially the same at all points.

Shearing. Shearing is an operation which involves a complete separation of the metal along one or several planes. The shearing tool operates much like a pair of scissors. Although shearing processes differ in detail, the mechanism of shearing is basically the same: two square sliding edges deform the metal severely along a narrow line; fracture begins at each surface and progresses through the cross section, the fracture surfaces meeting if clearance between shearing edges is correct. To obtain a clean cut the shearing tool should have small clearance and sharp edges, and the metal should be brittle. To reduce the total shearing force, the tools are designed so that deformation and fracture occur only over a short length of the metal at any one time. This is accomplished by cutting an angle on the face of the punch (called the angle of shear) or on the die.

If the shape which is sheared from the sheet is retained, and the remainder discarded, the operation is blanking. As it is usually desirable that the blank remain flat, the angle of shear is on the die, and the end of the punch is flat. On the other hand, if the sheared shape is the scrap and the hole is the objective of the operation, the process is piercing, punching, or perforating (the latter term generally referring to a large number of holes). In these processes, the shear angle is on the surface of the punch and the die surface is flat. In a parting operation neither piece is scrapped and no shear angle can be used.

If the metal is not entirely separated, the operation is one of slitting. Notching is the cutting of indentations in the edges of the sheet. Trimming is usually the final shearing operation which cuts an article to finished dimensions. The trimming of heavy blanks to obtain smooth, square edges is shaving. *See* METAL FORMING. [R.L.F.]

Bibliography: E. V. Crane, *Plastic Working of Metals and Non-metallic Materials in Presses*, 3d ed., 1944; V. N. Krivobok, G. Sachs, et al., *Forming of Austenitic Chromium-Nickel Stainless Steels*, 2d ed., 1954; G. Sachs, *Principles and Methods of Sheet-metal Fabrication*, 1951; G. Sachs and K. R. Van Horn, *Practical Metallurgy*, 1940.

Shellac

The name applied to the lac resin (secreted by the lac insect) when used in flake (or shell) form. Shellac varnish is a solution of shellac in denatured alcohol.

Shellac varnish is widely used in wood finishing where a fast-drying, light-colored, hard finish is desired. Shellac varnish is not water-resistant and is not suitable for exterior coatings, but is widely used for finishing floors and for sealing materials (for example, asphalt, coal tar, oil stains, and resinous knots) which are soluble in mineral spirits but not in alcohol.

Natural shellac has a brown color and is often referred to as orange shellac. Bleaching with chlorine produces bleached or white shellac. Bleached shellac is slightly less durable and hard than orange shellac but is lighter in color.

Shellac consists of a number of acids and esters and about 3% shellac wax, which is removed to make dewaxed shellac. Shellac varnishes are usually sold on the basis of the number of pounds of shellac dissolved in a gallon of alcohol, as 3-, 4-, or 5-lb cuts. Shellac varnish should be freshly prepared, because it deteriorates on aging.

In addition to floor and furniture finishing, which consume the largest quantities of shellac varnishes, there are other uses such as the modification of lacquers and inks. Water-thinned products may be made from alkaline solutions of shellac. Shellac is an important ingredient in a number of antifouling paints, where it helps to regulate the decomposition of the film to make fresh toxic material available at the coating surface. *See* HEMIPTERA; SURFACE COATING. [F.SD.]

Shielding, electrical

A means of avoiding pickup of undesired signals or noise, suppressing radiation of undesired signals, and confining wanted signals to desired paths or regions. These shielding objectives cannot be realized without some degree of modification of the electric and magnetic fields involved, although the effects of these modifications may not seriously interfere with wanted objectives.

A change in an electric field gives rise to magnetic effects, similarly, a change in a magnetic field gives rise to electric effects. As a result, electromagnetic shielding is more commonly encountered than either electrostatic shielding or magnetostatic shielding.

A shield is a sheet, tube, screen, grid, or other object, usually of conducting material. Sometimes a shield is magnetic, or both magnetic and conducting, or even laminated. Fields may be largely confined to or suppressed from a specified region by shields.

The ratio of the unwanted signal in a communication circuit when the source of shielding is present to the same signal when the shielding is absent is called the shield factor.

Electrostatic shielding. Virtually complete shielding from external electrostatic fields is obtained by totally enclosing the space to be shielded with a highly conducting surface, usually grounded. Conversely, no change in electrostatic conditions within the shield can appreciably influence the state of conductors, circuits, or fields external to the shield. In effect, a highly conducting shield substantially terminates electrostatic fields from either within or without.

Magnetostatic shielding. Substantially complete shielding from an external and virtually constant magnetic field is obtained by shunting the flux around the space to be shielded through a low-reluctance path. Conversely, a constant magnetic field enclosed by the shield is largely constrained from reaching the region outside the shield. Ideally, the shield should present a complete magnetic path to the flux; that is, the space to be shielded should be surrounded by a ferromagnetic material, preferably of high permeability. The shield factor may be reduced by increasing the thickness of the shield, by increasing its permeability, or by using several nested shields. Relatively thick, single-layer shields are not economical.

Electromagnetic shielding. The magnetostatic shielding such as that described above is likewise effective when the magnetic field is changing with time. A coil, for example, might be surrounded by a shield of this type. Such a shield would increase the inductance of the coil, illustrating that effective shielding necessarily modifies the fields. This kind of magnetic shielding is necessary for shielding constant or slowly varying fields.

A different yet highly effective method of realizing electromagnetic shielding is by means of an electrostatic shield. For frequencies above the upper audio range, shield factors of the order of 10^{-4} are readily obtainable, as, for example, between electric wave filters mounted in close proximity. Here the shield might be a closed box made of copper, about 1/16 in. thick. The main precautions are good soldered joints and absence of small cracks. A magnetic field impinging upon the copper sets up eddy currents, which oppose the penetration of the flux. The better the conductivity of the eddy-current paths, the more perfect the shielding action. In contrast to a magnetostatic shield, this type of shield reduces the inductance of a shielded coil.

In situations where space occupied by the shield is at a high premium, a laminated shield consisting of layers alternating between low-reluctance and high-conductance materials can be used.

Substantial electrostatic shielding may be obtained without appreciable magnetic shielding by making the electrostatic shield nonmagnetic and shaped so that eddy-current paths are interrupted.

Shielded wires and cables. The cable sheath, which encloses many communication lines within a telephone cable, constitutes one of the most striking examples of electromagnetic shielding. The sheath is grounded at one or more points along its length. Even with high-resistance grounds, the cable conductors are well shielded from external fields. At higher carrier frequencies, the shielding is so effective that significant conductor noise does not appreciably exceed resistance noise.

At moderately high frequencies, a coaxial line is an excellent example of a shielded wire, but at sufficiently low frequencies, the shielding effect is negligible. For this and other reasons, shielded twisted wires are preferred for transmitting television signals at video frequencies. At low frequencies, where a nonmagnetic shield is ineffective, transpositions act to attenuate noise, interference, and unwanted radiation. *See* NOISE, ELECTRICAL. [H.S.BL.]

Shigella

The dysentery bacilli, a genus of the family Enterobacteriaceae (*see* ENTEROBACTERIACEAE). Members of the *Shigellae* are the causative agents of bacillary dysentery (*see* BACILLARY DYSENTERY). The *Shigellae* are not motile and flagella are absent. The microorganisms do not produce carbon dioxide from sugars. The methyl red test is positive, and the Voges-Proskauer test is negative (*see* IMViC TEST).

Approximately 30 varieties of *Shigella* can be distinguished by cultural and serological methods. This is important for epidemiology and medicine. Serological differentiation depends on somatic, or O, antigens (*see* SALMONELLA). The classification scheme of the Enterobacteriaceae Subcommittee of the International Association of Microbiologists distinguishes the following four subgroups: A, *Sh. dysenteriae;* B, *Sh. flexneri;* C, *Sh. boydii;* D, *Sh. sonnei.*

Only ten varieties or types are found with frequency. These are: Type A1 of subgroup A, commonly known as Shiga's bacillus; type A2, *Shiga ambigua* or Schmitz' bacillus; seven types of subgroup B known as Flexner's bacillus; a single type of subgroup D known as Sonne's bacillus. In the United States and most of western Europe, only serotypes of group B and *Sh. sonnei* are of high incidence.

Diarrheal infection is also caused by *Bacillus* (or *Shigella*) *alkalescens* and *B.* (or *Sh.*) *dispar.* These two bacteria are taxonomically intermediate between *Shigella* and *Escherichia* (*see* ESCHERICHIA). *See also* BACTERIOLOGY, MEDICAL. [A.J.W.]

Bibliography: P. R. Edwards and W. H. Ewing, *Identification of Enterobacteriaceae*, 1955; A. J. Weil and I. Saphra, *Salmonellae and Shigellae*, 1953.

Shiner

Any of several small fishes, usually of the genus *Notropis*, family Cyprinidae. Shiners range from the Rocky Mountains eastward, and from Canada southward into Central America. There are about 50 species of *Notropis*, most of which are under 4 in. long, although a few species may reach a length of 6, or occasionally, 8 in. They usually are

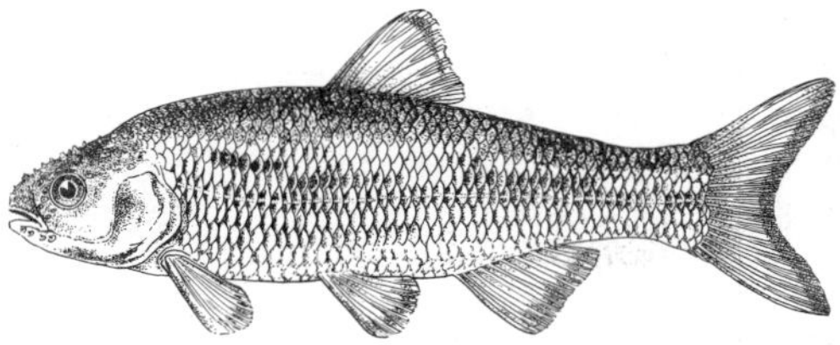

The common shiner, *Notropis cornutus;* length to 8 in. (From E. L. Palmer, *Fieldbook of Natural History,* McGraw-Hill, 1949)

the most common group present, and are the basic forage fish in most waters within their range. Although their diet and habits vary greatly, most of them eat plankton.

They are called shiners because of their silvery sides, which are quite evident as they twist and turn while feeding, usually in schools.

Some other small Cyprinidae are called shiners, the best known being the relatively large golden shiner, *Notemigonus chrysoleucas.* This is a common minnow, 8–9 in. in length. It is successfully cultured as a bait fish. *See* CYPRINIFORMES. [J.D.B.]

Ship, merchant

A power-driven vessel employed in commercial transport on the oceans and large bodies of inland water such as the Great Lakes. The relatively small craft used for inland waterway transportation are not commonly referred to as ships. For a description of these, *see* INLAND WATERWAYS TRANSPORTATION.

The principal classes of merchant-ship transportation include general cargo, bulk dry or liquid cargo, and passenger. There may also be mail, express, or other special classes, such as refrigerated cargo. Generally, cargo is transported for much less cost per ton-mile by water than by rail, truck, or air.

General cargo includes miscellaneous goods packed in boxes, bales, crates, cases, bags, cartons, barrels, containers, or drums. It may also include lumber, motor vehicles, pipe, steel, and machinery. The vessels engaged primarily in the transportation of general cargo are called freighters.

Examples of bulk dry cargo are ore, coal, sugar, cement, and grain. These items are poured or otherwise loaded into the ship's cargo compartments without being boxed, bagged, or hand stowed. This class of cargo may also be transported in freighters. However, vessels designed specifically for the ore or coal trade are referred to as ore carriers and colliers.

The principal bulk liquid cargoes are petroleum and its by-products. Others include wine, fruit juices, and molten asphalt. These liquids are transported in large tank spaces bounded by the vessel's main structural transverse and longitudinal bulkheads, outside plating, and deck. Such vessels are called tankers.

A passenger ship, as defined by international safety of life at sea rules, is one that carries more than 12 passengers on international voyages. Some passenger ships, such as those navigating solely on the Great Lakes, are excepted from these rules. Most seagoing passenger vessels transport considerable freight. Others, particularly fast liners having large passenger capacity, may transport only express, mail, baggage, and passengers' motor cars.

DESIGN CONSIDERATIONS

With very few exceptions, merchant ships conform to classification society rules. Classification societies are civilian organizations in the principal maritime nations that issue rules for the construction, equipping and maintenance of merchant ships. Compliance with their rules assures owners and insurers of the vessel's strength and seaworthiness.

Dimensions. The breadth of normal seagoing vessels is usually from 10 ft to 20 ft more than 10% of the length in feet between perpendiculars (LBP); for example, a 500-ft LBP ship might have $(0.1 \times 500 \text{ ft}) + 20 \text{ ft} = 70$ ft breadth. The length between perpendiculars, a classification society dimension, is defined as approximately the distance from the bow to the rudder, but not less than 96% of the load waterline length (the immersed hull length when floating at the maximum permissible draft). The depth of seagoing merchant vessels from the keel to the strength deck (usually the uppermost continuous deck) is rarely less than one-fourteenth of the LBP and for freighters and passenger vessels is commonly one-tenth to one-thirteenth. For Great Lakes vessels, which operate under different conditions, the depth to LBP ratio may safely be about one-eighteenth. The breadth is also less than that of ocean vessels of the same length. For the same cargo capacity and speed, a relatively long, slender vessel such as is found on the Great Lakes is more costly to construct, but requires slightly less propulsive power and in bad weather need not slow down as much as a shorter vessel to avoid damage. For additional information, *see* SHIP DESIGN.

The maximum permissible draft (submerged depth of vessel) is limited by international load line regulations. The draft to which general freight and bulk cargo vessels may be loaded depends principally upon the ship's depth below the freeboard deck, her length, and other factors including the length of superstructure (deck erections) such as a poop, bridge, forecastle, or shelter deck. In normal full scantling (maximum strength) cargo vessels the freeboard deck is the uppermost deck that extends continuously all fore and aft. The freeboard deck of shelter deck freighters is the deck below the uppermost continuous deck. The maximum draft of typical full scantling 500-ft general cargo freighters and bulk carriers is approximately three-quarters of the depth from keel to the freeboard deck. For 250-ft vessels this ratio increases to nearly seven-eighths. The maximum draft of shelter deckers is greater in relation to the depth to the freeboard deck but is less in relation to the depth to the uppermost continuous deck. The draft of tankers is slightly greater than that of corresponding full-scantling general cargo freighters. Safety regulations usually restrict the draft of passenger vessels to less than cargo vessel drafts.

If the draft at the bow is greater than at the stern a vessel is said to trim by the bow (or head). If greater at the stern she trims by the stern. When fully loaded with homogeneous cargo, fuel, and water (no water ballast), well-designed ships do not trim by the bow, but 2- or 3-ft trim by the stern is usually acceptable except in restricted depths of water.

Bulkheads. Ships are divided into watertight compartments by transverse bulkheads (walls) to reduce the extent of sea-water flooding in case of damage and to stiffen the hull structure. A normal 250-ft general-cargo freighter has not less than four such bulkheads and if 500 ft long, eight bulkheads are typical. Passenger vessels usually have more bulkheads, as required by the international safety regulations. With the exception of bulk-liquid-cargo vessels, an inner bottom is fitted between the peak (endmost) bulkheads. An inner bottom is horizontal plating, usually from 3 to 5 ft above the keel, extending from side to side of the ship. It provides tank space below for fuel oil, fresh water, and sea-water ballast, and also minimizes chances of serious sea-water flooding in case of minor bottom damage, which may happen, for example, in grounding.

Stability. Both the breadth and depth affect transverse stability against capsizing, so, in ship design, the one must be adjusted in relation to the other. An unusually deep ship must also be unusually broad. Very little stability is risky and too much results in objectionable violent rolling in heavy weather.

As fuel and fresh water are consumed en route, the reduction in weight aboard at low levels in the hull reduces stability, possibly risking capsizing if the ship is damaged by collision. In the case of passenger vessels particularly, sea-water ballast may be pumped aboard to replace this reduced weight, preferably into low-level compartments not ordinarily used for fuel oil.

Speed factors. Speed in knots is the velocity in nautical miles (6076.1 ft) per hour. The underwater shape of slow seagoing vessels (those in which the speed in knots is equal to 50–60% of the square root of the waterline length in feet) has about 30% of underwater parallel body (no change in shape) and rather bluff forward and after bodies. High-speed vessels (those in which the speed in knots exceeds the square root of the waterline length in feet) have no parallel underwater body and both ends are quite fine, the forebody waterlines being straight or even hollowed near the bow.

The increased construction cost of weight saving aluminum alloy in a high-speed vessel's structure, furniture, and equipment is often more than justified by the resultant reduction in horsepower and fuel consumption. The situation pyramids as a reduction in propulsion machinery weight and reduced fuel weight aboard further reduces the displacement.

The outstanding modern high-speed liner is the 990-ft over-all length, quadruple-screw, 2000-passenger American ship SS *United States* (Fig. 1). During several days of her trans-Atlantic maiden voyage she averaged over 36 knots, an unprecedented merchant-ship speed.

Propulsion. Cargo vessels rarely exceed about 25,000 shaft horsepower (shp) and one propeller is usual. Higher-powered vessels may have two or four propellers. Three are possible but unusual. Steamships of less than about 3000 shp normally have reciprocating steam engines but for higher powers, steam turbines are used. Diesel engines up to 15,000 brake horsepower per engine are also in wide use, particularly in foreign-built ships. Nuclear power, gas turbines, free piston engines, and other power units are also used to propel merchant ships. Sails are almost extinct.

High-revolution propellers are inefficient. Therefore, if the power plant operates at high revolutions

Fig. 1. SS *United States*. (*Mariners Museum, Newport News, Virginia*)

per minute (rpm), it either drives the propeller shaft through reduction gears, or powers a direct-drive high-rpm electric generator. This, in turn, drives the propeller shaft through a low rpm reversible rotation electric motor.

The propulsion machinery of most passenger vessels and general cargo freighters is located between amidships (midlength of the LBP) and about three-fourths of the LBP from the bow. A four- or six-hold general cargo freighter normally has two or three holds forward of and two or three holds abaft (sternward from) the machinery space. A five-hold freighter usually has three holds forward of and two holds abaft the machinery. The shaft tunnel and shape of holds near the stern, however, make general cargo stowage difficult. There is a trend toward locating the machinery and the deck houses above it further aft. *See* MARINE MACHINERY; PROPELLER, MARINE; SHIP PROPULSION.

Size-speed relationship. Other factors being equal, a large, fully loaded cargo vessel (general or bulk cargo) will transport cargo at less cost per ton-mile than a small, fully loaded vessel. However, a fully loaded moderate capacity cargo vessel will transport cargo at less over-all cost per ton-mile than the same quantity of cargo in a partially loaded much larger vessel of the same type.

Deadweight is the weight of cargo, fuel, fresh water, stores, passengers, crew, and baggage aboard a ship. General-cargo transoceanic freighters rarely have over 15,000 long tons deadweight capacity, and freighters of 8000 to 11,000 tons capacity predominate. In ship design and operation the long ton, 2240 lb, is used. Larger vessels would have less frequent sailings for a given annual cargo and often there is not enough cargo to load fully even an 8000-ton deadweight ship of this type. Tankers and bulk cargo vessels over 100,000 long tons deadweight capacity are in ocean service, but tankers in the 20,000–40,000-ton class are most common.

High speed is an important advantage in war time, and in peace time it attracts the patronage of both passengers and cargo shippers. If two ships of the same cargo capacity are both fully loaded, the cost of shipment per cargo ton-mile is greater for a high speed vessel. Nevertheless, the faster ship may earn larger dividends if her speed attracts full loads of high-quality cargo and a slower ship does not. Most modern cargo vessels over 400 ft in length have speeds of about 14–16 knots. Relatively few make over 18 knots. American vessels are a little faster, on the average, than those of other nations.

Cargo handling. The cargo handling time for general cargo freight is much longer than for bulk cargo, and this involves costly idle ship time (spending instead of earning) while in port. For a vessel of the same basic dimensions and speed, a general-cargo freighter can transport far more weight or volume of payload cargo than a containerized cargo vessel. Bulk cargo transportation is the most economical per ton-mile even though the vessel is in ballast without cargo on the return passage. For relatively long distances general-cargo (not containerized) freighters provide more economical transportation than containerized cargo vessels. This is not the case for short distances. As an extreme example, on a 10-mile passage a ferry could transport loaded truck vans (equivalent of containerized cargo) much more cheaply and more quickly than if even more truckloads of cargo were unloaded onto the ferry and then reloaded onto trucks at the other end of the passage. For a 10,000-mile passage, however, it is cheaper to transfer the cargo to a ship and transport only the payload cargo, even though considerable time and money is spent loading and unloading.

Size restrictions. Canal locks and harbor water depths restrict the dimensions of the larger classes of vessels. The Panama Canal recommends that vessels not exceed 900-ft length over-all, 104-ft breadth, and 35-ft draft. The Suez Canal limits the draft to 35 ft, but the length and breadth are unrestricted. The maximum suggested for the St. Lawrence Seaway is 715-ft length over-all, 72-ft breadth. The minimum channel depth is 27 ft, so about 25–26 ft is the maximum draft. The minimum clear-

Typical large merchant ship characteristics*

Characteristic	Passenger and freight	General cargo	Ocean tanker	Ocean ore carrier	Great Lakes self-unloader
Over-all length, ft	730	490	716	664½	666¼
Length between perpendiculars, ft	658	459	685	631½	640
Beam, ft	93	66	93	87	72
Depth, ft	64	41	48⅔	45¾	36
Summer keel draft, ft	32¾	30 11/12	36⅚	34 1/12	25 5/12
Lightweight	21,080	5,822	12,460	10,505	7,850
Deadweight capacity	14,310	11,788	37,745	31,560	20,040
Displacement, full load	35,390	17,610	50,205	42,065	27,890
Shaft horsepower, normal	34,250	10,000	21,300	12,500	7,000
Shaft horsepower, maximum	37,675	11,000	23,430	13,750	7,700
Service speed, knots	23	17	18	15.5	14.8
Passengers	1,175	12	4	4	6
Officers and crew	640	50	51	46	41

* Weights are in long tons (2240 lb). Lightweight is the weight of the empty ship with no deadweight aboard. The full-load displacement equals the weight of the water displaced by the lightweight plus the deadweight.

ance under bridges along the Seaway is 120 ft. All of these dimensions except the bridge clearance can be slightly exceeded in an emergency.

Although large bulk-cargo vessels have economic advantages, if the draft exceeds about 38 ft (few harbors have over 40-ft depth) operations may be restricted. Tankers of great draft, such as those of 50,000 to 100,000 tons deadweight capacity, may have to unload partially into a smaller tanker or into an offshore pipeline before entering many harbors. Representative data on merchant ships are given in the table on the previous page.

TYPES OF VESSELS

Passenger vessels. These vary from small inter-island or cross-channel ships to liners over 1000 ft in length. The speed range varies from about 12 to over 30 knots and the passenger capacity from 13 to over 2000. Three separate classes of passengers are provided for in most large liners, for example, first, cabin, and tourist classes. Some vessels are "one class."

The larger passenger vessels provide staterooms with 1–4 berths, each room with bath and toilet in first-class quarters. A few rooms are connecting and suites may include a living room, dressing room, and even a private outdoor veranda. On most large and many small passenger vessels, air conditioning is customary in all staterooms, public spaces, and in officer and crew quarters. Large liner passenger spaces and facilities may include a dining room and galley, lounge and observation spaces, cocktail or bar room, open and closed promenade deck, movie theater, library, smoking room, writing room, ballroom, children's play room, shopping center, restaurant, gymnasium, swimming pool, game deck area, beauty shop, barber shop, and doctor's office. Some of the public spaces such as the movie theater or swimming pool may be made available to two classes of passengers at different hours. A few staterooms are arranged to permit assignment to either of two classes, depending upon the number of passengers booked for each class. One galley may serve separate dining rooms for two classes. A typical freight and passenger vessel of about 100-passenger capacity usually has only a dining room and galley, lounge, cocktail room, card and game room, library, and possibly a swimming pool.

The number of officers and crew for a 500-ft, 13-passenger vessel is 55–60, and for 100-passenger ships, 100–125. For typical transoceanic vessels of over 1000 passengers the number of officers and crew is slightly over half the number of passengers.

International safety of life at sea regulations include rules for the fireproof construction of ships. Carelessness may result in a stateroom fire. To confine the fire the joiner bulkheads are usually fireproof and the furniture at least fire resistant. The 1948 regulations permit automatic fire alarms and sprinkler systems instead of fireproof bulkheads. Fireproof or fire resistant paint is now used in quarters of the highest class of passenger vessels. Draperies, carpets, and bedding may be treated to make them fire-resistant.

General-cargo freighters. An inner bottom is required for general-cargo freighters of over about 300-ft length and is customary in smaller vessels. Transverse framing is usual for the main hull structure, although some have longitudinal bottom framing. (Framing is discussed later in this article.) The inner bottom forms the lowest level of the cargo spaces (holds) in which general cargo is stowed. All except very small freighters have one or two decks below the strength deck and the cargo on these intermediate decks is said to be stowed in the " 'tween decks."

The cargo is loaded and discharged through large rectangular deck openings (hatches) over each

Fig. 2. The general-cargo freighter *Old Colony Mariner*. She has a top speed of approximately 22.5 knots.

cargo space. Mechanically operated hatch covers are now used to close the openings. The hatch covers in the 'tween decks are strong enough to support cargo which may be stowed on them. The topside hatch covers are weathertight. Hatch sizes vary, but the width is usually 35–50% of the ship's breadth and the length 50–60% of the hold length. The main transverse bulkheads form the ends of the cargo spaces and in sizable transoceanic freighters are usually from 40 to 70 ft apart. Some have one or more tanks abreast the shaft tunnel in the after holds for bulk edible liquid cargo.

Shelter deck (complete superstructure above the freeboard deck) and full-scantling minimum freeboard freighters predominate, the former being popular for voluminous (low density) cargo. A forecastle, midship deckhouse, and often a poop are usual. Long bridges for cargo stowage are occasionally fitted. Unless the propulsion machinery is aft (a growing trend), the midship deckhouse is above the machinery space. A typical modern transoceanic general cargo freighter is shown in Figure 2.

In some ports general cargo is transferred between the ship and shore by cranes located on the pier, but usually the ship's booms are used. The booms are raised or lowered by adjustable wire rigging (topping lifts) led from the mast or kingpost (the equivalent of a mast) to the boom ends. Two adjustable ropes (vangs) from each boom end to the deck are used to swing the booms or hold them in a fixed position over the pier or the hatch (Fig. 3). A wire rope (cargo fall) leads over sheaves from a steam or electric winch to the outer end of each boom and terminates in a cargo hook. Cargo

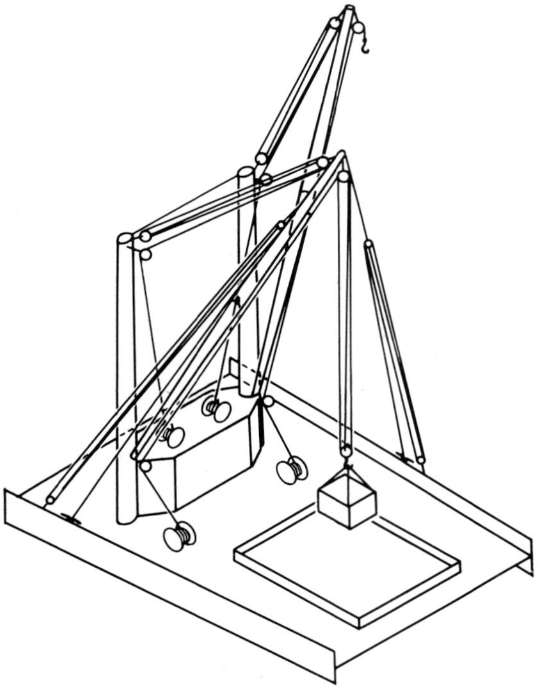

Fig. 3. Swinging-boom cargo handling.

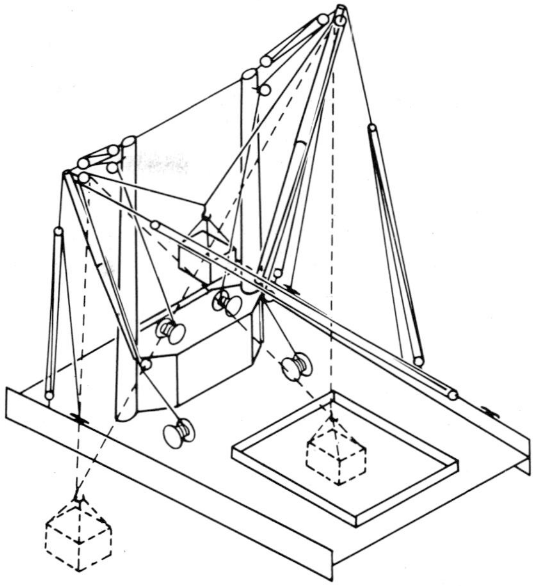

Fig. 4. Burtoning.

can be hoisted using one boom and then slowly shifted to or from the ship by swinging the boom. This is customary only for occasional very heavy units of cargo (over 5 or 10 long tons). Usually two booms are used with one boom end over the hatch and the other over the pier, as shown in Fig. 4. The vangs secure the booms in these positions. The cargo falls from both booms are joined to one cargo hook. When loading, cargo is hoisted by the pier boom until well above the hatch level. Then the hatch boom cargo fall is rapidly wound in while the pier boom cargo fall is skillfully unwound at such a rate that the cargo is suspended by both boom falls and moves horizontally well above the deck until over the hatch. Thereafter both boom falls are unwound, and the cargo is lowered into the cargo space. This process, called burtoning, is reversed when discharging cargo from the ship to the pier. Burtoning is usual for cargo loads up to 5 long tons and a few vessels can burton 10-ton loads. The cargo is often piled together in a large net which is emptied and returned for the next load. Many large vessels have one pair of booms at each end of each hatch. This expedites cargo handling but increases the stevedoring cost. Most freighters have one heavy lift boom of 30–50 tons capacity for occasional units of heavy cargo. A few freighters have cranes mounted on the ship instead of masts, booms, and winches.

Uniform classes of cargo such as cartons of canned goods may be stacked on inexpensive small flat platforms called pallets and hoisted aboard. The loaded pallets are efficiently handled aboard ship (some by fork lift) and remain under the cargo until after discharge.

Containerized-cargo vessels. Containers consist of weatherproof boxes (usually metal), strengthened to withstand stacking and motion at sea. They hold general cargo, preferably loaded by the shipper but occasionally by stevedores on the pier. Except for customs inspection, the containers are delivered unopened to the consignee. They vary from small sizes such as 6-ft cubes, some of which are collapsible, to boxes about 8 ft square and 35 ft long resembling large truck vans. The principal advantage of containers is rapid cargo handling. If the container is loaded aboard and discharged from the ship by being rolled on highway wheels or casters, the vessel is called a roll-on, roll-off ship corresponding to a ferry boat carrying truck vans. Loading ramps are necessary on the pier.

Container ships, as distinguished from roll-on, roll-off ships, lift the wheelless containers on and off the vessel with cranes. For a given payload cargo capacity, both roll-on, roll-off and container ships are more costly to build than conventional freighters, but both the cargo handling cost and the idle ship time in port are much reduced.

Large container ships usually load or discharge a full cargo in 8 or less hours, compared to several days for the same amount of cargo in conventional freighters. Roll-on, roll-off ships may have even shorter turnaround time in port. Containerized cargo vessels make more voyages per year than conventional freighters. Container transportation is most efficient with vessels designed specifically for the job.

The class of cargo that can be containerized is limited by the container size. The quantity of pay load cargo that a fully loaded container ship can load is considerably less than that of a similar size conventional freighter. For roll-on, roll-off ships it may be as little as 25% of the pay load cargo capacity of a general-cargo freighter of the same dimensions and speed. A large number of containers (with or without wheels) are necessary ashore as well as enroute at sea, and to maintain the required number of empty containers available in each port for reloading, the ships must be fully loaded with containers on every passage, whether they are full of cargo or empty.

Because fully loaded roll-on, roll-off ships cannot carry enough cargo to immerse them as deeply as the maximum draft of general-cargo freighters of the same dimensions, it is practical to have sideports (doors in the vessel's sides) above the waterline through which roll-on, roll-off containers can be rolled aboard and discharged using pier ramps. Ships of this class also usually have a transom stern (cut off square like a rowboat) fitted with doors through which containers are rolled on wheels. Roll-on, roll-off ships may have several cargo decks, the containers being shifted from the loading deck to other decks by either elevators or sloping ramps.

Efficient container ships are designed to load and discharge the containers mechanically and to guide them mechanically as they are lowered into or unloaded from their stowed position. To permit rapid movement of containers to or from the stowed-aboard ship location, container ships may have unusually large hatches. By use of gantry cranes and guides much like elevator tracks, the containers are stowed directly below and only within the hatch area. In some cases additional containers may be carried above the topside deck.

Bulk-liquid-cargo vessels. These ships, commonly called tankers, have the propulsion machinery aft, with a small (little used) general dry cargo space near the bow. The bulk liquid cargo spaces are between these areas. The total length of the liquid cargo space is, typically, 60% of the LBP (see Fig. 5). The length of the cargo tanks and the forward hold is adjusted so that the center of gravity of the weight of the loaded ship is in the same fore and aft location as the center of volume (equal upward force) of the displaced sea water. As a result the draft is the same at the bow and stern. This basic principle applies to other classes of ships as well.

An inner bottom is fitted beneath the machinery space and either an inner bottom or deep tank under the forward dry cargo space. Tankers have one or two, and large vessels have three, longitudinal bulkheads in the cargo tank spaces. The individual tank lengths (fore and aft) rarely exceed about 40 ft. To minimize damage from the liquid swashing as the ship rolls and pitches, and to provide for cargo expansion if the temperature rises, the tanks are each filled to about 98% of capacity. There is only one deck and no inner bottom beneath the cargo tanks. This main deck, the bottom and side plating, and the bulkheads are framed longitudinally. Longitudinal framing consists of

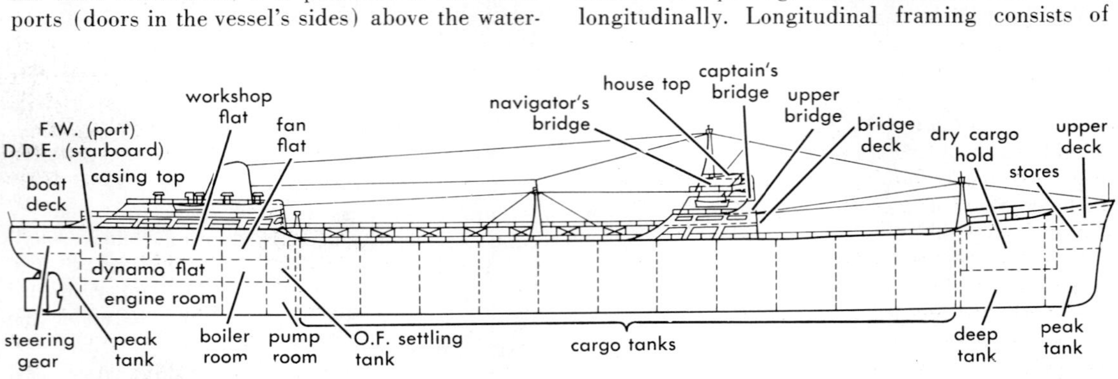

Fig. 5. Cross-sectional view of the supertanker *Spyros Niarchos*, which has 47,122 tons deadweight capacity. The cargo space is divided into 33 cargo tanks totaling 2,289,370 ft.^{3} (From A. C. Hardy, *Merchant Ships: World Built*, John De Graff, 1957)

Fig. 6. The tanker *Esso Gettysburg*. She is of the single-deck type with forecastle, bridge, and poop-deck houses. The deadweight is 37,689 tons and cargo capacity is 318,000 barrels. One of the fastest oil tankers to fly the American flag, she is capable of maintaining a speed of 18 knots. (*From A. C. Hardy Merchant Ships: World Built, John De Graff, 1958*)

shapes such as angle or tee bars 2–3 ft apart running fore and aft. Widely spaced (such as 10 ft) transverse structural members between transverse bulkheads support longitudinals. In transverse framing shapes run athwartship or vertically instead of fore and aft; supporting members run fore and aft. Either longitudinal or transverse framing may be used for the side shell. Some vessels transport special classes of liquid cargo in tanks that are not a part of the hull.

Conventional tankers (Fig. 6) have a poop, forecastle, and either a bridge or deck house slightly forward of amidships. The pilot house, radio room, and quarters for the navigating officers are in the midship deck house. The remainder of the quarters are in the poop and in deck houses above it. Tankers are rarely in port more than one day at the end of each passage, so as an inducement for continuity of crew service the quarters of tankers are generally much more livable than those of freighters.

The deadweight capacity of sizable tankers is approximately three times the light (empty) weight of the ship. Bulk-liquid or bulk-dry-cargo vessels are normally fully loaded on one passage and on the return passage carry sea water ballast to provide suitable draft. A typical large tanker can transfer over 2000 tons of liquid cargo to or from shore in an hour. Shore pumps load and ship pumps discharge the liquid.

Bulk-dry-cargo vessels. The density of bulk dry cargo varies widely. A ship loaded to her maximum draft with iron ore utilizes a much smaller volume of cargo space than if transporting an equal weight of coal or sugar. Although there are exceptions, most vessels of this type are designed for a specific trade, such as iron ore or coal. If the cargo space for a maximum permissible load of heavy ore extended across the full hull width, the ore would occupy only a small portion of the depth from the inner bottom to the topside deck. The cargo weight thus concentrated low in the vessel would result in excessive stability and if the ore is poured into the ship, it would not self stow uniformly throughout the hold. In a severe storm, bulk cargo in a partially filled space may shift to one side, causing the vessel to list heavily or even capsize.

For dense cargo then, the bottom of bulk-dry-cargo spaces is much higher above the keel than is a general-cargo vessel's inner bottom, and the cargo is stowed between two longitudinal bulkheads outboard of which are spaces sometimes used as water ballast tanks. The height of the cargo space bottom is such as to result in acceptable stability. The cargo space width between the longitudinal bulkheads is also controlled by the density of the cargo. The cargo space corners under the deck often have sloping plating (inboard to the hatch edges) at the angle of repose of the cargo. This eliminates empty hold space after the cargo is poured to the level of the hatches. The corners at the bottom of the holds may also be sloped. This expedites discharging as the cargo slides either to a relatively small area all of which can be reached through the hatch by shore-based grab buckets, or into hoppers over a self-unloading conveyor belt.

Conventional bulk-dry-cargo vessels have a poop and a forecastle. The propulsion machinery is aft. For ocean service there is either a short bridge or midship deckhouse for the pilot house, radio room and quarters for the navigating officers, or these spaces may be located at the forward end of the poop. Great Lakes vessels have no midship deckhouse, the pilot house and deck officers' quarters being located on the forecastle. As in the case of tankers, the port turnaround time is short, so highly livable crew quarters are customary.

Bulk dry cargo is loaded by pouring it into the cargo spaces through spouts, or by belt conveyors. It is discharged from the ship by shore-based grab buckets or by belt conveyors installed in the ship. Up to 3000–5000 tons of bulk dry cargo can be loaded or discharged per hour. Cement can be moved rapidly between the ship and pier by blowing it through tubes with compressed air.

Other types of vessels. In addition to the larger classes of merchant ships and inland waterway craft there are fishing vessels, ice breakers, dredges, light vessels, ferries, tugs, salvage ships, fireboats, patrol boats, pilot boats, and other craft. *See* SHIPBUILDING. [S.A.V.]

Bibliography: D. Arnott (ed.), *The Design and Construction of Steel Merchant Ships*, 1955; H. C. Hanson, Fishing vessels of the Pacific Coast of America, *Soc. Naval Architects Marine Engrs. Trans.*, 62:30–42, 1954; J. J. Henry, Modern ore carriers, *Soc. Naval Architects Marine Engrs. Trans.*, 63:57–111, 1955; H. F. Johnson, Development of ice breaking vessels for the U.S. Coast Guard, *Soc. Naval Architects Marine Engrs. Trans.*, 54:112–151, 1946; G. M. Phannemiller, Modern design and construction methods as applied to 95-ft patrol boats, *Soc. Naval Architects Marine Engrs. Trans.*, 62:643–687, 1954; C. D. Roach, Tugboat design, *Soc. Naval Architects Marine Engrs. Trans.*, 62:593–642, 1954; H. F. Robinson, J. F. Roeske, and A. S. Thaeler, Modern tankers, *Soc. Naval Architects Marine Engrs. Trans.*, 56:422–471, 1956.

Ship, naval

The present-day American naval fleet bears little resemblance to the one with which the United States finished World War II. Battleships, once the most important surface component, have been inactivated. Smaller and faster cruisers now serve as fleet flagships. Aircraft carriers, handling as many as 70 heavy jet fighters, have become the core of surface naval operations. Guided-missile launchers are replacing conventional gun batteries for armament. Above all these startling changes looms an even greater one—the conversion to nuclear power for ship propulsion. Nuclear submarines came first, and by 1962 the U.S. Navy will have nuclear-powered aircraft carriers, cruisers, and destroyers in service.

This article discusses the major surface units of the modern navy. For information on other important naval vessels, *see* LANDING SHIPS AND CRAFT; SUBMARINE; *see also* ANTISUBMARINE WARFARE; ARMAMENT, NAVAL.

Aircraft carriers. The largest warships of modern navies are aircraft carriers. They are designed to support and operate aircraft in all weather conditions and to engage in naval operations against enemy ships, aircraft, and shore installations. An aircraft carrier is in fact a floating and mobile air station. Her flight deck provides runways, and her island is a control tower; a large hangar area below the flight deck contains maintenance and repair facilities, and tanks built into the hull carry fuel for aircraft. In the U.S. Navy, provision for all these features for about 70 heavy jet aircraft and a 4000-man crew results in 1000-ft ships with displacements of 65,000–85,000 tons.

U.S. Navy carriers are generally being provided with guided-missile batteries for defense against attacking aircraft. Principal reliance for self-defense, however, is placed on her own aircraft and the more extensive guided-missile batteries of her escorts.

Aircraft launching and recovery. The outstanding feature of modern attack aircraft carriers is the angled flight deck. A 700-ft landing strip makes an angle of 7–10° with the axial portion of the deck, whose length approximates 1000 ft. Thus two separate aircraft strips are provided. Launching of aircraft is accomplished by steam catapults located at the forward ends of the axial and angled strips (Fig. 1). Powered by steam from the ship's boilers, these massive "slingshots" can impart an end speed of about 140 knots to a 70,000-lb aircraft, and develop upwards of 60,000,000 ft-lb of energy in a single stroke. Automatic positioners for final spotting of aircraft on catapults are being incorporated in modern carriers to expedite launching operations. By launching (taking off) only from the axial strip and recovering (landing) on the angled one, safety of operations is greatly enhanced. Aircraft not successfully arrested on landing can either fly off or be allowed to crash into the sea over the forward end of the angled deck without harm to other aircraft awaiting catapulting on the forward portion of the axial deck.

Aircraft to be recovered approach the carrier from the stern, lower a tail hook from the underside of the fuselage, and touch down on the angled strip. Engagement of the tail hook with one of the several horizontally stretched cables, which are lifted mechanically just before the plane lands, gives complete stoppage in about 300 ft. This "run-

Fig. 1. Fighter aircraft, F9F-8, being launched from starboard catapult of USS *Forrestal*, CVA 59. (*U.S. Navy Official Photo*)

out" is decelerated smoothly by hydraulic plungers, located below the flight deck, to which the cables are attached. A mirror landing system furnishes the incoming pilot with an optical indication that he is in a correct glide path for his approach. For foul weather or night operations, an electronic carrier control approach system is provided to assist in the landing of aircraft.

Large hydraulically operated deck edge elevators, having capacities of over 80,000 lb, are provided to transport aircraft between the flight and hanger decks. Extensive shop and repair facilities are available in the hangar space. Outlets for fueling, electrical service, and compressed air for engine starting are situated at convenient locations on both the flight and hangar decks.

Bomb magazines below decks, connected with the flight deck by special elevators, supply aircraft ammunition. Ready rooms for briefing pilots, air operations centers, combat intelligence facilities, and electronic equipment for the control of aircraft operations and the receipt of aerological information are among the principal special features of aircraft carriers.

Advantages of nuclear power. Nuclear propulsion is incorporated in the USS *Enterprise*, CVA(N)65, whose displacement of 85,000 tons makes her the largest warship ever constructed. She is scheduled for completion in 1962. Eight pressurized water reactors will provide dry saturated steam to geared turbines, which will develop over 250,000 shaft horsepower and propel the ship at a trial speed in excess of 30 knots (Fig. 2). A dramatic increase in endurance will be achieved with this nuclear power plant, enabling the *Enterprise* to cruise many times around the world without refueling. Other substantial advantages which accrue with the use of nuclear power for aircraft carriers are (1) freedom in location of island without regard to uptake positions, (2) elimination of smoke nuisance from aircraft operations, and (3) greater resistance to battle damage by elimination of uptakes.

Fig. 2. USS *Enterprise*, CVA(N)65. Artist's conception of the first nuclear-powered aircraft carrier as she will appear when completed in 1962. The *Enterprise* will be 1100 ft long and will cruise at over 30 knots. Note angled flight deck. (*U.S. Navy Official Photo*)

Cruisers. These are medium-sized warships designed to engage in operations against aircraft, guided missiles, and surface or subsurface opposition singly or in support of other forces. With the passing of the battleship as queen of the battle force, the cruiser has become the heavy surface combatant. From the role of scout and screen ship for the battle line, the cruiser has evolved into a fast, lightly armored, and heavily armed defender of the aircraft carrier, now the core of modern task forces. Cruisers also serve as flagships for major fleet commands. U.S. Navy planning envisages cruisers in the 14,000- to 17,000-ton displacement range, equipped with guided-missile armament, and propelled by nuclear machinery.

Guided-missile ships. The initial effort in this plan by the U.S. Navy, as well as other navies, is the conversion of existing cruisers to guided-missile ships. Under a program which started in 1956 to provide surface-to-air guided missile capability, several U.S. Navy light cruisers of the 14,000-ton *Cleveland* class have been converted. This conversion consists primarily of removing the after 6-in. turrets, the after 5-in. mounts, and all 3-in. mounts and installing aft either a short-range Terrier or a long-range Talos missile launcher. The resultant armament is, therefore, a combination of conventional guns forward and a guided-missile launcher aft (Fig. 3). Some of these ships are, in addition, being provided with extensive facilities to make them suitable for flagship service with the major fleets. Space and weight for these additions are being made available by additional removal of conventional guns.

The Italian Navy is engaged in a major modernization of the 11,000-ton *Garibaldi* to provide a guided-missile system aft in combination with new 5.3-in. and 3-in. guns forward and in the waist (the middle part of the ship).

The French Navy has considered providing a guided-missile system aft in such cruisers as *Colbert*, whose construction was completed in 1958. A mixed armament of guns and guided missiles was contemplated.

A program for converting U.S. Navy 17,000-ton heavy cruisers of the *Baltimore* and *Oregon City* classes to guided missile ships started in 1958. The first Terrier guided-missile cruisers of the U.S. Navy, *Boston* and *Canberra*, had been converted from these classes in 1955. In the new conversion designs, all conventional guns will be removed. Armament will ultimately consist of Talos guided missile systems for long-range attack against aircraft, Tartar systems for close-in air defense, large missiles for long-range attack against land targets, and antisubmarine weapons consisting of shipboard weapon launchers and helicopters. With this varied armament, these ships will be the most powerfully armed of any warships in modern navies.

Fig. 3. *Little Rock,* CLG4. Artist's conception of U.S. Navy conversion of ex-CL92 with Talos launcher aft and fleet flagship facilities forward. Terrier ships will have similar appearance. (*U.S. Navy Official Photo*)

The Long Beach. The first nuclear-propelled surface ship to be constructed for the U.S. Navy is the cruiser *Long Beach* (Fig. 4). Her armament is similar to that of the *Baltimore* and *Oregon City* conversions. but carried in a lighter hull of 14,000 tons full-load displacement. Thus she is a true prototype for the force of cruisers toward which present planning is projected. In her machinery spaces, two pressurized-water reactors furnish saturated steam to conventional geared turbines. Performance includes a trial speed of over 25 knots, and a greatly extended endurance. In the *Long Beach,* powerful guided-missile armament coupled with nuclear propulsion typify the ultimate in current U.S. Navy planning for cruisers.

Destroyer types. These are small, high-speed, unarmored warships ranging in size from 1700 to 7500 tons. Among the navies of the world, they are variously and with little uniformity designated destroyers, destroyer escorts, and frigates. A frigate in one navy may be considered a destroyer escort in another, and a destroyer in a third. Within these broad categories there are many specialized ships in which a single function is optimized over the general purpose functions of most destroyer types. For example, some U.S. Navy destroyers are specially fitted as radar pickets, and some British ones as aircraft fighter directors. Because of their small size and high utility, destroyer types constitute the largest number of warships in most navies.

Destroyers. These vessels are designed to screen other surface units against attack by enemy air, subsurface, or surface units; to operate offensively against enemy aircraft, submarines, and light forces; and to carry out light bombardment and amphibious support tasks. In 1958, the U.S. Navy was embarked on a destroyer construction program based on the *Forrest Sherman,* DD 931, whose commissioning took place in 1955 (Fig. 5). With a full-load displacement of 3500 tons and a speed in excess of 30 knots, this ship represents a sub-

Fig. 4. *Long Beach,* CGN 9. Artist's conception of U.S. Navy prototype for a guided-missile, nuclear-powered cruiser. Two Terrier launchers forward, one Talos launcher aft, and provision for long-range surface-to-surface missiles amidships comprise the main features of her armament. Note absence of smokestacks. (*U.S. Navy Official Photo*)

Fig. 5. USS *Forrest Sherman,* DD 931. This destroyer was commissioned in 1955 and has since become the prototype vessel for a construction program. (*U.S. Navy Official Photo*)

stantial improvement over previous designs. Her seakeeping qualities and maneuverability are excellent, and her armament is heavy and varied. Antiaircraft guns of the most advanced types, torpedoes for attack against surface ships, hedgehogs (short-range antisubmarine weapons) and depth charges combined with modern sonar (underwater detection gear) for use against submarines, all make this ship a powerful offensive and defensive warship.

In 1958 the Italian Navy was also embarked on a construction program for fleet destroyers comparable to *Forrest Sherman. Impetuoso,* completed in 1957, is the class leader, with a displacement of 3600 tons and speed of 34 knots. *Impavido,* laid down in 1957, is the newest addition to the class. Equipment plans call for a surface-to-air guided missile system, some conventional antiaircraft guns, antisubmarine torpedoes, and an Italian-developed three-barrelled antisubmarine howitzer.

In 1957, the U.S. Navy initiated a program for the construction of guided-missile destroyers, designated the DDG 2 class (Fig. 6), on the concept of producing a ship similar to the DD 931 class, but with improved armament. The improvement in antiaircraft armament consisted primarily of substituting a twin Tartar launcher for the after 5-in., 54-caliber mount. Advanced sonar and antisubmarine weapons were also provided. As a result of these extensive changes, the ship was enlarged over DD 931, and has a full-load displacement of 4500 tons, an increase of some 650 tons. It is interesting to note that the first destroyers constructed for the U.S. Navy during the Spanish-American War displaced only 420 tons.

Destroyer escorts. Vessels intended primarily to screen convoys from enemy submarines are called destroyer escorts. Their mission includes the tasks of detecting and destroying enemy submarines threatening the convoy, and the capability of limited self-defense against enemy air and surface attack. In time of war, such ships are needed in large quantities to ensure adequate transport of strategic materials across the oceans. Destroyer escorts, therefore, are designed under the criteria of minimum size, complexity, and cost to facilitate rapid construction.

In 1958, the *Courtney* class, DE 1021, represented the latest in destroyer escort designs in the U.S. Navy (Fig. 7). In her 1950-ton hull, a vessel of this class carries antiaircraft guns and depth charges. Her machinery can propel her at over 22 knots, which provides an adequate speed margin for screening slow merchant ships.

In 1958 the Royal Canadian Navy commissioned the leader of the new *Restigouche* class of antisubmarine escorts. Fully designed in Canada with some British assistance, she has the mission of detecting and destroying fast modern submarines. Her displacement of 2600 tons, speed in excess of 25 knots, and powerful armament of antiaircraft guns, depth charges, mortars, and acoustic homing torpedoes make her an extremely effective escort whose capabilities approach that of a destroyer. Flush main deck, low and spacious bridge, the extensive use of aluminum, and depth-charge mortars (Limbo) located in an after well below the main deck are interesting features. The addition of antisubmarine helicopter facilities is a possibility.

Frigates. These are destroyer types which are not amenable to precise definition. In European navies frigates are generally small, short-range, antisubmarine ships of about 2000–3000 tons displacement, while in the U.S. Navy the most modern frigates are in fact destroyer flotilla leaders whose 7000 tons approach light cruiser size.

The new British *Whitby* class consists of antisubmarine frigates of over 2800 tons displacement

Fig. 6. DDG 2. Artist's conception of U.S. Navy guided-missile destroyer. The 435-ft vessel makes over 30 knots, and uses the Tartar guided-missile system. (*U.S. Navy Official Photo*)

Fig. 7. USS *Evans,* DE 1023. U.S. Navy destroyer escort of DE 1021 class. The 310-ft ship has a speed of over 22 knots. (*U.S. Navy Official Photo*)

Fig. 8. HMS *Torquay*, British antisubmarine frigate of the *Whitby* class. (*British Joint Services Mission, Washington, D.C.*)

and speed in excess of 30 knots (Fig. 8). They are very similar to Canada's destroyer escort, *Restigouche*, in size, armament, and general features. Special attention was given in these designs to seakeeping qualities and maintaining speed in a seaway. High freeboard forward, a long forecastle, and an unusual hull form were adopted to obtain good rough weather performance. Having been primarily designed for locating and destroying modern submarines, *Whitby* is equipped with the latest underwater detection devices and antisubmarine weapons, including Limbo launchers and torpedoes.

Revolutionary advances in armament and propulsion are being effected in U.S. Navy frigates. The DLG 16 class, designed in 1957, is a guided-missile frigate of almost 7000 tons displacement, long range, and speed of over 30 knots. Her mission is to screen fast task forces against enemy air, submarine, and surface threats, and to operate offensively against naval forces and shipping. Installation of two Terrier launching systems make her the first destroyer to carry missiles both forward and aft. Advanced underwater detection devices, antisubmarine weapons, torpedo tubes, and helicopters give this ship an impressive antisubmarine capability. Close-in protection is provided by two twin rapid-fire gun mounts.

Fig. 9. DLG(N). Artist's conception of U.S. Navy nuclear-powered guided-missile frigate. Note absence of smokestacks. (*U.S. Navy Official Photo*)

The DLG(N) incorporates nuclear propulsion for the first time in a ship of frigate size, and is the third surface ship in the U.S. Navy to have nuclear propulsion, *Long Beach*, CG(N)9, and *Enterprise* being the forerunners. Two pressurized-water-type reactors furnishing steam to geared turbines permit great endurance without refueling. This design incorporates essentially the same armament as DLG 16, but in a somewhat larger hull of about 7600 tons (Fig. 9).

Mine sweepers. These are small ships equipped to remove, or render harmless, naval mines. U.S. Navy types are representative of those of many countries because large numbers of these designs were constructed for NATO navies under the mutual defense assistance programs.

Ocean mine sweepers (MSO). These are ships of approximately 720 tons, equipped for magnetic, acoustic, and moored mine sweeping (Fig. 10). They are intended for heavy-duty sweeping in company with fleet units. These ships are unique in their non-magnetic features, which minimize the chances of the inadvertent detonation of magnetic mines. Wooden hulls fitted with machinery and equipment made of non-magnetic materials reduce the magnetic attraction of the ships and permit them to move safely through magnetic mine fields (*see* DEGAUSSING).

Fig. 10. MSO 495. U.S. Navy ocean mine sweeper. (*U.S. Navy Official Photo*)

Coastal mine sweepers (MSC). These are ships of about 370 tons, designed for magnetic, acoustic, and moored mine sweeping in coastal waters. They are roughly equivalent to the YMS type widely used during World War II. Like the ocean mine sweepers, these ships have wooden hulls and are of nonmagnetic construction.

Mine sweeping boats (MSB). These are 88,000-lb craft for sweeping inshore waters; they are equipped to handle magnetic, acoustic, and moored mines. These boats are wooden hulled, are of nonmagnetic construction, and feature gas turbine drive for mine sweeping generators.

Mine sweeping launches (MSL). These are the smallest of the family of mine sweepers. They are 36 ft long, displace about 21,000 lb, and are used in clearing all types of mines in assault sweeping operations.

Motor torpedo boats. These are fast vessels designed for submarine chasing in narrow or sheltered waters, sneak torpedo attacks on large surface ships, and the destruction by gunfire of minor seagoing traffic. Emphasizing speed and maneuverability, such vessels, designated PT in the U.S. Navy, carry more armament per ton of displacement than any combatant ship.

Four experimental PT boats, completed in 1951, represent the most advanced type in the U.S. Navy. If additional construction were undertaken, these would probably be the basis for a prototype. The four boats (PT 809, 810, 811, and 812) have aluminum hulls, and initially had identical main propulsion units consisting of four Packard gasoline engines and identical gasoline-engined auxiliary generators (Fig. 11). The hulls, however, differ considerably in their lines and arrangements. In continuation of the experimental function of these craft, several have been re-engined, one to diesel drive and another to a combination of diesel engine and gas-turbine drive.

Fig. 11. PT 811. U.S. Navy motor torpedo boat, completed in 1951. The vessel has a speed of 42 knots and can house as many as four torpedoes. (*U.S. Navy Official Photo*)

Other navies have shown more continued interest in motor torpedo boats. For example, in Great Britain, greater emphasis has been placed on this type for the detection and chasing of submarines in narrow waters, and an active development and construction program has been maintained. The *Brave* class of fast patrol boats is the latest design, embodying gas-turbine drive for three propellers, with additional thrust obtained from the gas-turbine exhaust. The immediate predecessor, the *Dark* class, featured aluminum construction and diesel drive. This design represented a departure from earlier torpedo boats, such as were used during World War II and which had wooden hulls and gasoline engines.

Naval auxiliaries. These are service and logistic ships that support naval operations. Included in this broad class are a large number of specialized ships, such as tenders, reefers, oilers, ammunition ships, cargo ships, and icebreakers, all of which are basically merchant types adapted to naval service. Some of the most important U.S. Navy types are discussed here; auxiliaries of other navies are comparable.

Tenders. These are floating and mobile repair supply stations. They have shops capable of accomplishing ship repairs, which do not require heavy

Fig. 12. USS *Mauna Kea,* AE 22. U.S. Navy ammunition ship. She is 502 ft long, and has a full-load displacement of 17,400 tons. (*U.S. Navy Official Photo*)

shipyard facilities such as drydocks or large-capacity cranes. Medical and dental services, and laundry and dry cleaning facilities, are available on them as well as limited resupply of provisions, spare parts, ammunition, and fuel. Although the basic design of the various types of tenders and repair ships is essentially the same, minor variations have been introduced to suit particular functions. For example, submarine tenders have torpedo shops, periscope repair shops, submarine battery repair and charging facilities, and battery water stowage. Destroyer tenders, on the other hand, include in their special capabilities torpedo stowage and repair facilities, but none of the other facilities just enumerated.

Ammunition ships. These closely resemble commercial cargo carriers, radar systems and armament being the only external indications of their special capabilities (Fig. 12). Diversification of stowage is one of their principal features. Each cargo hold is fitted with adjustable metal stowage systems which permit the efficient stowage of any type or size of ammunition. Large-capacity elevators deliver ammunition to the main deck quickly and without the use of booms or rigging, and fork lift trucks handle it fore and aft. A major feature is the capability for transferring ammunition to another ship at sea while underway. This is done by means of winches which automatically adjust for the motions of both ships.

Oilers. These are similar to large commercial tankers, but have several specialized military fea-

Fig. 13. USS *Neosho,* AO 143. U.S. Navy oiler. She is shown in center conducting simultaneous refueling at sea of an aircraft carrier and a destroyer. (*U.S. Navy Official Photo*)

tures. Probably the most important is the capability of rapidly discharging fuel to other ships while at sea and underway (Fig. 13). For this task they are provided with large pumps and special rigging for supporting hoses that connect with receiving ships. Another military requirement which complicates naval oilers is the diversification of cargo, because the same tanker must supply fuel oil, diesel oil, lubricating oil, gasoline, and fuel for jet aircraft. Speed requirements for fleet support are also somewhat greater than for commercial practice, and armament for limited self-defense is a military feature found in commercial tankers only in wartime. *See* GUIDED MISSILE; MARINE MACHINERY; REACTOR, SHIP PROPULSION; SHIP DESIGN; SHIP PROPULSION. [A.M.MO.]

Bibliography: R. Blackman (ed.), *Jane's Fighting Ships, 1959–1960*, 61st ed., 1959.

Ship design

The design of a ship involves a selection of the features of form, size, proportions, and other arrangements which are open to choice, in combination with those features which are imposed by circumstances beyond the control of the design naval architect. Each new ship should do some things better than any other ship. This superiority must be developed in the evolution of the design, in the use of the most suitable materials, in the application of the best workmanship, and in the application of the basic fundamentals of naval architecture and marine engineering.

As ships have increased in size and complexity, plans for building them have become more detailed and numerous. The intensive research since the period just prior to World War II has brought about many technical advances in the design of ships. These changes have been brought about principally by the development of welding techniques, developments in main propulsion plants, advances in electronics, and changes in materials and methods of construction. *See* SHIPBUILDING.

All ships have many requirements which are common to all types, whether they are naval, merchant, or special-purpose ships. The first of such requirements is that the ship must be capable of floating when carrying the load for which it was designed. A ship floats because as it sinks into the water it displaces a certain amount of water, and the pressure of the water produces an upward force which is called the buoyancy (*see* BUOYANCY). The buoyancy force is equal to the weight of the water displaced by the ship and is called the displacement. Displacement is equal to the underwater volume of the ship multiplied by the density of the water in which it is floating. When floating in still water, the weight of the ship, including everything it carries, is equal to the buoyancy or displacement. The weight of the ship itself is called the lightweight. This weight includes the weight of the hull structure, fittings, equipment, propulsion machinery, piping and ventilation, cargo or other handling equipment, and other items which are required for the efficient operation of the ship. The load which the ship carries in addition to its own weight is called the deadweight. This includes cargo, passengers, crew and effects, stores, fresh water, feed water for the boilers in the case of steam propelling machinery, and other weights which may be part of the ship's operational load. The sum of all these weights plus the lightweight of the ship gives the total displacement; that is,

$$\text{Displacement} = \text{lightweight} + \text{deadweight}$$

One of the first things which a designer must do is to determine the weight and size of the ship and decide upon a suitable hull form to provide the necessary buoyancy to support the weight.

Owner's requirements. Ships are designed, built, and operated to fulfill the requirements and limitations specified by the operator and owner. These owner's requirements denote the essential considerations which are to form the basis for the design. They may be generally stated as (1) a specified minimum deadweight carrying capacity, (2) a specified measurement tonnage limit, (3) a selected speed at sea, or a maximum speed on trial, and (4) maximum draft combined with other draft limitations.

In addition to these general requirements, there may be a specified distance of travel without refueling and maximum fuel consumption per shaft horsepower hour limitation, as well as other items which will influence the basic design. Apart from these requirements, the ship owner expects the designer to provide a thoroughly efficient ship. Such expectations include (1) minimum displacement on a specified deadweight carrying capacity, (2) maximum cargo capacity on a minimum gross tonnage, (3) appropriate strength of construction, (4) the most efficient type of propelling machinery with due consideration to weight, initial cost, and cost of operation, (5) stability and general seaworthiness, and (6) the best loading and unloading facilities as well as ample accommodations for stowage.

Design procedure. From the specified requirements, an approach is made to the selection of the dimensions, weight, and displacement of the new design. This is a detailed operation, but some rather direct approximations can be made to start the design process. This is usually done by analyzing data available from an existing ship which is closely similar. For example, the design displacement can be approximated from the similar ship's known deadweight of, say, 11,790 tons and the known design displacement of 17,600 tons. From these figures, a deadweight-displacement ratio of 0.67 is obtained. Thus, if the deadweight for the new design is, for example, 10,000 tons, then the approximate design displacement will be 10,000/0.67 or 15,000 tons. This provides a starting point for the first set of length, beam, and draft dimensions, after due consideration to other requirements such as speed, stability, and strength.

Beam is defined as the extreme breadth of a ship at its widest part, while draft is the depth of the lowest part of the ship below the waterline.

Length and speed. These factors are related to the hull form, the propulsion machinery, and the propeller design. The hull form is the direct concern of the naval architect, while the propulsion machinery and propeller design are of indirect concern. The naval architect has considerable influence on the final decisions regarding the efficiency, weight, and size of the propulsion machinery and the size and efficiency of the propeller, as both greatly influence the design of the hull form. *See* MARINE MACHINERY; PROPELLER, MARINE.

Speed has an important influence on the length selected for the ship. The speed of the ship is related to the length in terms of the ratio $V/\sqrt{L}$, where V is the speed in knots and L is the effective waterline length of the ship. As the speed-length ratio increases, the resistance of the ship increases. Therefore, in order to obtain an efficient hull form from a resistance standpoint, a suitable length must be selected for minimum resistance. Length in relation to the cross-sectional area of the midship or maximum section, that is, the fineness of the underwater form (the so-called prismatic coefficient), is also very important insofar as resistance is concerned. Fast ships require fine (slender) forms or relatively low fullness coefficients as compared with relatively slow ships which may be designed with fuller hull forms.

Beam and stability. A ship must be stable under all normal conditions of loading and performance at sea. This means that when the ship is inclined from the vertical by some external force, it must return to the vertical when the external force is removed. Stability may be considered in the transverse or in the longitudinal direction. In surface ships, longitudinal stability is of much less concern than transverse stability. Submarines, however, are concerned with longitudinal stability in the submerged condition.

The transverse stability of a surface ship must be considered in two ways, first at small angles of inclination, called initial stability, and second at large angles of inclination. Initial stability depends upon two factors, (1) the height of the center of gravity of the ship above the base line and (2) the underwater form of the ship. The center of gravity is the point at which the total weight of the ship may be considered to be concentrated (*see* CENTER OF GRAVITY). The hull form factor governing stability depends on the beam B, draft d, and the proportions of the underwater and waterline shape. For a given location of the center of gravity, the initial stability of the ship is proportional to B^2/d. Beam, therefore, is a primary factor in transverse stability.

At large angles of heel (transverse inclination), freeboard is also an important factor. Freeboard is the amount the ship projects above the waterline of the ship to certain specified decks (in this case, to the weatherdeck to which the watertight sides extend). Freeboard affects both the size of the maximum righting arm (Fig. 3) and the range of stability, that is, the angle of inclination at which the ship would capsize if it were inclined beyond that angle.

Depth and strength. A ship at sea is subjected to many forces because of the action of the waves, the motion of the ship, and the cargo and other weights which are distributed throughout the length of the ship. These forces produce stresses in the structure, and the structure must be of suitable strength to withstand the action. The determination of the minimum amount of material required for adequate strength is essential to attaining the minimum weight of the hull. The types of structural stress experienced by a ship riding waves at sea are caused by the unequal distribution of the weight and buoyancy throughout the length of the ship. The structure as a whole bends in a longitudinal plane, with the maximum bending stresses being found in the bottom and top of the hull girder. Therefore, depth is important because as it is increased, less material is required in the deck and bottom shell. However, there are limits which control the maximum depth in terms of practical arrangement and efficiency of design.

Hull form. The desired hydrostatic and hydrodynamic characteristics which the new ship form should have are formulated by considering resistance and propulsion, stability, volumetric requirements, steering and maneuverability, performance in rough weather, and the freeboard necessary for seaworthiness and safety. After these factors have been decided, the work on the drawing of the hull form is started.

The geometry of the ship is usually delineated on a drawing which is commonly referred to as the lines drawing of the ship, or simply the lines. Every set of lines consists of three plans or views. These are the elevation, or profile of the ship; a view looking down upon the vessel, known as the half-breadth plan; and a transverse view, known as the body plan (Figs. 1 and 2). Ordinarily, only one side of a ship's form is delineated, since ships are normally symmetrical about their longitudinal center plans. By means of these projections, taken in conjunction with each other, it is possible to determine fully the relative positions in three-dimensional space of all points and lines considered to be on the vessel with the same accuracy as though they were being examined and measured on the ship itself. It is from lines such as these that all the calculations that determine the geometric properties of the ship are made.

In selecting the dimensions of a ship and the fullness of the form to provide the necessary displacement, it is quite possible to have many different forms. The form with least resistance to propulsion is the one that is most desired, since it will require the least powerful machinery and expenditure of fuel.

The total resistance of a ship is usually divided into two principal parts, frictional and residual.

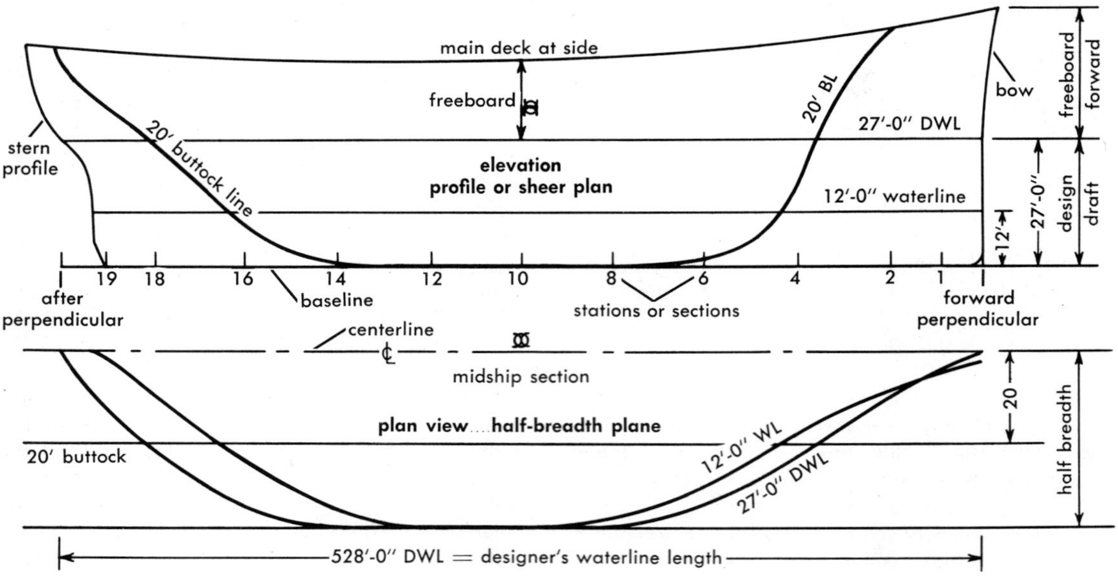

Fig. 1. Profile view (*above*) and half-breadth plan (*below*) of a ship. These views, along with the transverse view (Fig. 2) make up the lines of a vessel.

The residual resistance is made up of wave-making resistance and separation and eddying resistance. The latter two can be reduced to very small amounts by good hull design and attention to detail fairing of endings and shapes. The frictional resistance depends on the wetted surface area of the hull, and little can be done to reduce this resistance beyond the reduction of the wetted area. Wave-making resistance, however, is related to the features of the hull form, and much can be done in the way of selecting the proper hull length and fineness to minimize wave-making resistance. The beam of the ship in relation to its draft has some effect on resistance, and generally the resistance increases with the beam-to-draft ratio. Little can be done in the way of reducing the beam, however, as it is generally fixed by the stability requirements. The shape of the waterlines, particularly the surface waterline, is important in the reduction of wave-making resistance. High-speed ships require waterlines with small angles of slope to the centerline, while slower ships may have larger waterline angles without undue waste of power. Other features of form which have a bearing on wave-making resistance are the relative fullness of the ends of the hull form and the parallel middle body portion, the use of a bulbous bow at properly related speeds, and fineness of the hull form. For a discussion of resistance, *see* SHIP PROPULSION.

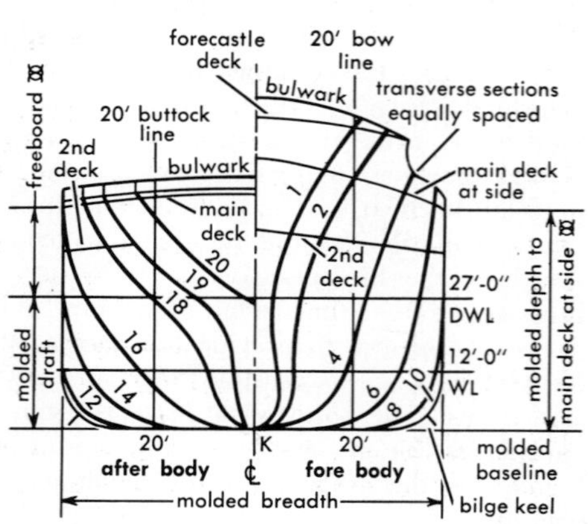

Fig. 2. Transverse view, also called the body plan, of a ship.

After the lines have been completed, they are usually sent to a model basin (test facility) for the construction of a scale model and for testing in a basin. These tests are generally carried out in still water. The power determined from these experiments represents the power required to drive the ship under ideal sea conditions, and allowances must be made for many departures from these ideals, such as probable rough water performance and increased resistance as a result of fouling. Rough water performance is a matter which has not lent itself to precise testing in model basins, although much work is being done in this area. With the new test techniques and facilities now available at many model basins, such as the wave-making machine at the David Taylor Model Basin in Carderock, Maryland, and the research work which is being carried on, designers should soon have a better insight into the technical elements of hull form which affect performance in rough water.

Stability of ships. A ship may be inclined in any direction. Any inclination may be considered as being composed of an inclination in the transverse plane and an inclination in the longitudinal

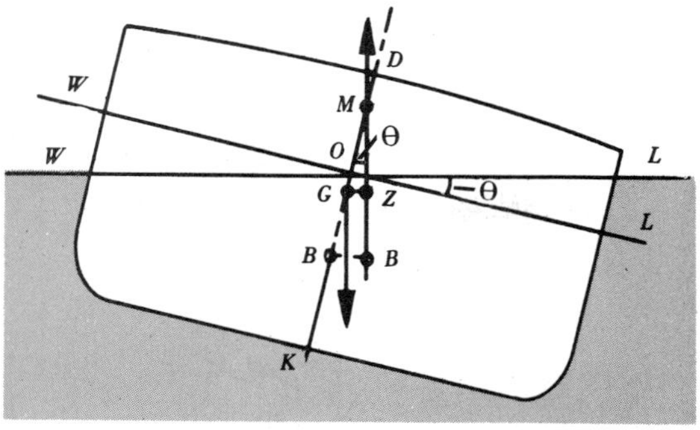

Fig. 3. Diagram of an inclined ship, showing new center of buoyancy B'.

plane of the ship. The transverse inclination is called heel or list, and the longitudinal inclination is called trim. In ship calculations each is treated separately.

If, under the action of wind, waves, or shifting weights, a ship is inclined from a waterline WL through a small angle θ to a new waterline $W'L'$, as shown in Fig. 3, the displacement weight is not changed. Therefore, the buoyancy force remains constant and acts vertically (normal to the water surface) through a new center of buoyancy B'.

The angles WOW' and LOL' are known as the emerged and immersed wedges, respectively. The point M at which the vertical through the new center of buoyancy B' intersects the ship's centerline KOD is known as the metacenter. The height of the metacenter above the center of gravity is termed the metacentric height and is designated GM. The height of the metacenter above the center of buoyancy B is known as the metacentric radius and is usually designated as BM. The arm GZ drawn through G, the center of gravity, at right angles to the new vertical through B', is known as the righting lever or righting arm GZ. The point K can be any convenient reference point, but is usually taken at the top of the keel amidships.

The two forces of weight and buoyancy, being equal in amount and acting through the centers G and B', respectively, form a couple equal to Δ (displacement) $\times\ GZ$ foot-tons, known as the righting moment. If M is above G, this couple is positive and tends to restore the ship to its original position. If M coincides with G, there is no righting moment and the ship is in neutral equilibrium with no tendency to move in either direction. If M is below G, the couple is negative and tends to increase the inclination until the ship either capsizes or reaches a new position of equilibrium. The ship is said to have negative GM when M is below G and positive GM when M is above G.

Referring to Fig. 3, it is seen that for small inclinations

$$\begin{aligned} \Delta\,(GZ) &= \Delta\,(GM) \sin\theta \\ &= \text{righting moment, ft-tons} \end{aligned}$$

where Δ is the displacement, GZ is the righting arm, GM is the initial metacentric height, and θ is the small angle of heel (extending up to 10°).

Again referring to Fig. 3, it is seen that

$$GM = KB + BM - KG$$

where KB is the vertical height of the center of buoyancy above the molded base line (point K), KG is the vertical height of the center of gravity above the molded base line, and BM is the metacentric radius.

In order to find the initial metacentric height GM, it is necessary to determine the value of BM. The value of KB is obtained by integrating the volume of the buoyancy to the several waterlines of a body plan together with the respective vertical moments of the buoyancy above the base line K. These and other conventional calculations concerning the displacement and other hydrostatic properties of the underwater portion of the ship determine a set of curves.

It is sufficient, for purposes of brevity, to accept the condition that BM is a firm geometric property of the displaced volume of the ship and the waterline at which it is floating. The classic expression for this property is

$$BM = \frac{I}{V}$$

where I is the transverse moment of inertia of the waterline and V is the displaced volume to the waterline in question.

From this relationship, the value of BM is obtained directly by calculation from the hull form of the ship. Since KB can be obtained by direct calculation also, the precise location of M for any waterline for any given ship hull form can be determined. Thus, the value of KM is readily determined by calculation. The value of KG, which specifies the location of the center of gravity, is a much more difficult and illusive problem to solve.

Unless one has access to reliable data from other ships, it is difficult to make the preliminary weight estimate and to locate the position of the center of gravity of a new ship before detailed plans have been drawn. After the value of KG has been estimated, the beam, depth, and draft of the ship and its general design features are adjusted, as necessary, in order to find the location of M and G where the desired stability is attained.

Metacentric height. One of the fundamental features of any design is the metacentric height. It should have a value such that it will meet the following requirements:

1. It must be large enough in passenger ships to prevent capsizing or an excessive list in case of flooding.

2. It must be large enough to prevent listing to unpleasant or dangerous angles if, for example, all passengers crowd to one side of the ship.

3. It must be large enough to minimize the possibility of a serious list under pressure from strong beam winds.

4. It must be small enough to prevent violent rolling in waves.

The naval architect is confronted with the problem of adjusting the principal dimensions and characteristics of a design so as to provide enough GM to comply with requirements 1, 2, and 3, yet not too much, so that requirement 4 is also satisfied.

Damage stability considerations may occasionally require excessive metacentric heights. In view of this, several formulas have been devised to establish the maximum GM that need not be exceeded in the interest of safety. One such formula for passenger vessels of over 50-ft beam states that GM in feet need not exceed the value given by

$$GM = 0.06B - \frac{350}{(B/10)^4}$$

in which B is the beam of the ship in feet.

Longitudinal stability and trim. Longitudinal stability is a measure of a ship's ability to return to its position of equilibrium after being inclined forward or aft by external forces. This statement is analogous to the definition of transverse stability. Trim is a measure of a ship's fore and aft inclination when it is floating in equilibrium, free of any external inclining forces. It is important to understand the distinction between problems involving stability and those involving trim, since both relate to the longitudinal inclination. Stability deals with the behavior of a ship when it is temporarily inclined from its position of equilibrium by external forces. Trim problems are concerned with more or less permanent positions of equilibrium assumed by the ship as the result of its particular internal weight distribution.

It should be pointed out that the angles generally dealt with in transverse inclination are much larger than those involved in longitudinal inclination. The former are sometimes as high as 30–40° from one side to the other, but in the case of longitudinal inclination, angles greater than 10–12° seldom occur in connection with stability, and the angles are seldom over 2–3° in trim problems. This means that approximations can often be made in the longitudinal case which would not be admissible in the transverse case.

Stability of submarines. The stability and trim of a submarine is a unique problem involving three main phases: (1) operation on the surface, (2) period of submergence or surfacing, and (3) underwater operation. While operating on the surface, the stability and trim, and the calculations concerning them, are the same as those for surface ships.

The condition during the period of submergence can best be understood by keeping in mind the fact that while the actual weight of the submarine remains unchanged, an amount of buoyancy exactly equal to the reserve buoyancy, while the submarine is operating on the surface, must be lost (Fig. 4). This loss of buoyancy is brought about by flooding the main ballast tanks with sea water. Buoyancy below the surface condition waterline is lost, but is replaced simultaneously by buoyancy of exactly equal amount that was reserve buoyancy above the waterline. There is, therefore, a rise in the center of buoyancy on submerging. The center of gravity of the submarine remains fixed and the weight remains unchanged.

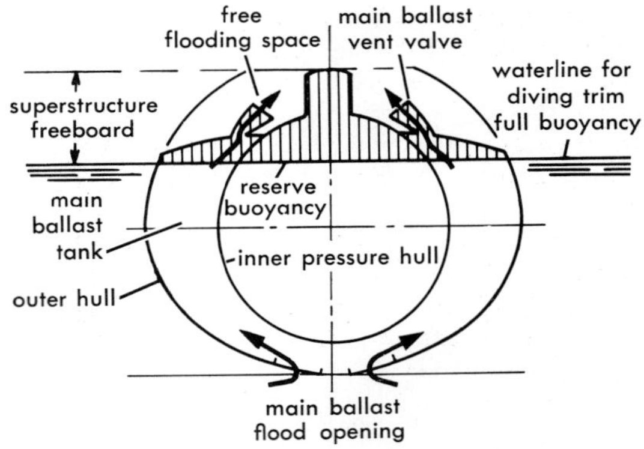

Fig. 4. Sectional view of double-hulled submarine.

As the submarine dives, the waterplane (the effective area of the surface waterline at which the vessel floats on the surface of the water) disappears and its moment of inertia becomes zero, so that $BM = I/V = 0$. The metacenter M coincides with the center of buoyancy B. The submerged submarine is held upright by the fact that the center of gravity G always lies below the combined center of buoyancy and metacenter. This gives the submerged submarine what is known as pendulum stability, with G always below B, as illustrated in Fig. 5. Any moment of weight forward or aft of the submerged center of buoyancy causes the submarine to trim in that direction until the center of gravity is directly under the center of buoyancy. A small offset may result in a large trim angle unless the necessary correction is made to the position of G. When the submarine is inclined submerged by some external moment, either in heel or trim, there is always a pendulum-type restoring moment acting to level it off at an equilibrium attitude. When the ship is under way, this moment is superposed on the hydrodynamic moments. A submarine in diving trim, that is, in a condition so that it can submerge with neutral buoyancy and zero trim, must be floating at a par-

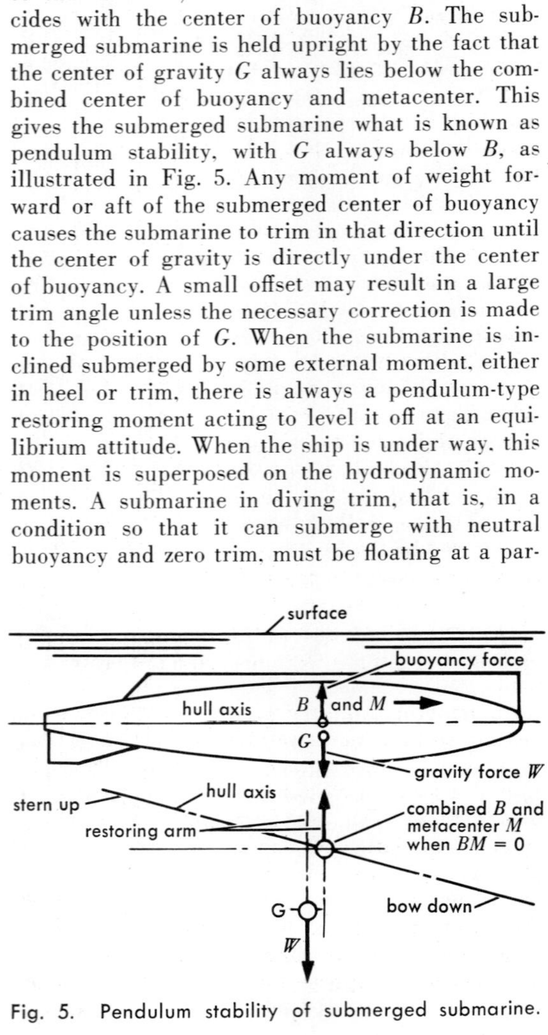

Fig. 5. Pendulum stability of submerged submarine.

ticular waterline such that the volume of the main ballast tanks equals the volume of reserve buoyancy and that the center of volume of the main ballast tanks lies directly under the center of volume of the reserve buoyancy. *See* SUBMARINE.

Ship motion. A ship is considered to have six freedoms of motion in a seaway: rolling, pitching, heaving, yawing, swaying, and surging. Rolling is transverse angular rotation about a horizontal fore and aft axis; pitching is vertical angular rotation about a transverse horizontal axis; heaving is a vertical linear movement of translation; yawing is a horizontal angular rotation about a vertical axis, that is, a transverse horizontal swing of the bow and stern; swaying is a transverse linear movement of translation, that is, a sideways movement; surging is a fore and aft movement of translation.

Of these motions, rolling is of most concern. The next two, pitching and heaving, are of somewhat less concern, but are still important considerations, especially when attempting to drive a ship in a rough sea.

Large amplitudes of pitching and heaving occur when there is synchronism with the period of encounter of the waves with the ship. The worst motions and the greatest loss in speed are experienced when the lengths of the waves are between about three-quarters of the length of the ship and one and one-quarter times the length of the ship. For the comfort of the passengers and crew, these motions should be reduced as much as possible. However, the only practical action which is feasible is to alter speed or course, or both; aside from this, there is little that can be done.

The rolling motion of a ship also is greatest when the rolling period is in synchronism with that of the waves. This can also be minimized by changing course or speed, or both.

The rolling period of a ship is a function of the beam and the initial stability *GM* of the ship. One of the cheapest and most effective ways of increasing the resistance to rolling is to fit bilge keels. The bilge keel, shown in Fig. 2, is a flat surface fitted and located along the bilge of the ship and normal to the shell. As the ship rolls, the flow is interrupted by the flat surface of the bilge keel and the water is forced to flow around the outside edges. This force opposes the rolling motion of the ship. The bilge keel has been shown to be more effective in reducing rolling when the ship has ahead motion.

An important roll-reducing development which has gained favor is the so-called activated fin. This consists of a fin projecting from each side of the ship at a location close to amidships and the turn of the bilge. The fins have the appearance of balanced rudders, and may or may not be retractable. The action of the fin as a hydrofoil is controlled by an automatic gyrocontrol device. When the ship is rolling and has headway, the angle of attack on the fin is controlled so as to provide moments which oppose the rolling motion of the ship. Full-scale trials have shown good results in reducing the amplitudes of roll to one-quarter or one-half of those experienced without the fins.

Structural design. After having established the principal dimensions, form, and general arrangement of the ship, the designer undertakes the problem of providing a structure capable of withstanding the forces which may be imposed upon it. The hull of a steel merchant ship (Fig. 6) is a complex structure, unique in the field of engineering structures in that it is primarily a plate structure, depending for its major over-all strength on the plating of the shell, decks, and in most cases, also on the inner bottom and longitudinal bulkheads. The framing members, each of which has its own function to perform, are designed primarily to maintain the plate membranes to the planned contours and their positions relative to each other when subjected to the external forces of water pressure and breaking seas, as well as to the internal forces caused by the services for which the ship is designed. Unlike most other large engineering structures, the forces supporting the ship's hull as well as the loads which may be imposed upon it vary considerably, and in many cases, cannot be determined accurately. As a result, those responsible for the structural design of ships must be guided by established standards.

Basic considerations. The problem of the development of a satisfactory structure generally involves the following considerations:

1. It is necessary to establish the sizes of, and to combine effectively, the various component parts so that the structure, with a proper margin of safety, can resist the major over-all stresses re-

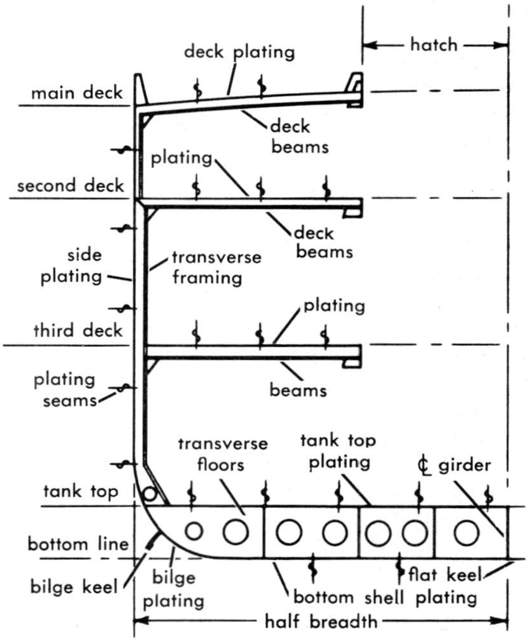

Fig. 6. Schematic midship section showing structural arrangement of plating and transverse framing in a steel merchant ship.

sulting from longitudinal and transverse bending.

2. Each component part must be so designed that it will withstand the local loads imposed upon it from water pressure, breaking seas, the weight of cargo or passengers, and other superimposed loads such as deckhouses, heavy machinery, masts, and so on, including such additional margins as sometimes may be required to meet unusually severe conditions encountered in operation. *See* STRESS AND STRAIN; STRUCTURAL DEFLECTIONS.

Rules of classification societies. The various classification societies have continued to modify and improve their rules to keep pace with the records of service experience, an increasing amount of research, and the constantly growing understanding of the scientific principles involved. In the modern rules of the societies, the designer has available to him formulas and tables of scantlings, dimensions of framing shapes, and thicknesses. These are directly applicable to practically all the ordinary types of sea-going merchant vessels being built today, and contain a flexibility of application to vessels of special types.

The design of structural features of a merchant ship is greatly influenced by the rules of classification societies; in fact, the principal scantlings of most merchant ships are taken directly from such rules. Scantlings are defined as the dimensions and material thicknesses of frames, shell plating, deck plating, and other structures, together with the suitability of the means for protecting openings and making them sufficiently watertight or weathertight. The classification society rules contain a great deal of useful information relating to the design and construction of the various component parts of a ship's structure. Scantlings can be determined directly from the tables given in these publications. In many cases, a good conception of the usual "good-practice" construction can also be gleaned from the sketches and descriptive matter available from the classification societies. *See* SHIP, MERCHANT.

Load lines and freeboard. The limitations on the drafts of passenger and cargo ships have been imposed by the International Conventions on Safety of Life at Sea (1929, 1948) and the Load Line Convention (1930), and have been established by law. Load lines, or drafts, must be marked on each side at midlength of the summer load line of the ship, with a particular form of mark also established by law. The marks show the deepest draft to which the vessel may be lawfully submerged in salt or fresh water, and in various zones and seasons of the year.

Freeboard. The fundamental part of these regulations involves freeboard, which is the distance measured vertically downward from the upper edge of the deck line to the upper edge of the summer load line. The deck line is defined as the line of intersection of the top of the freeboard deck amidships with the outer surface of the side plating.

In the determination of freeboard, there are three principal considerations:

1. The geometry of the vessel; that is, the shape and dimensions of the watertight hull and weathertight superstructures.

2. The structural strength of the vessel and provisions for making the hull strong and tight.

3. The standard of subdivision in the case of a passenger vessel; that is, the degree to which the vessel is subdivided into compartments by means of watertight bulkheads and decks, and the number of compartments that in damaged condition may be flooded before reserve buoyancy and stability disappear. Reserve buoyancy is the volume or potential buoyancy of the watertight structure of the vessel above the load line.

The minimum freeboard permissible from the first two considerations is computed from a booklet covering the load-line regulations of the United States. This minimum freeboard can be achieved in practice only if the standards of structural strength and protection of the openings are fully complied with. If the strength of the ship is less than standard, the freeboard must be increased accordingly. In the case of a classed vessel, the strength standards of the American Bureau of Shipping and other classification societies are accepted.

The third principal consideration, subdivision, applies only in case the vessel is designed to carry more than 12 passengers. When the subdivision draft is less than the draft corresponding to the minimum freeboard obtained from the load-line regulations, the subdivision draft becomes the maximum permissible draft. The subdivision draft is the draft to which the standard of watertight subdivision for passenger ships is keyed, that is, the spacing of watertight transverse bulkheads is a function of a specific limiting draft, designated the subdivision draft.

Tonnage. As vessels grew in size, they varied greatly in dimensions. Some were long and narrow, and some short and wide. Some were shallow, and some relatively deep. It developed that the length, or even length and breadth, did not adequately express the relative size of vessels, but that size was a matter of carrying capacity. Tonnage measurements go back to the early Egyptian, Phoenician, and Chinese times. There was confusion as to whether to express carrying capacity in cubic measure or weight, and this confusion persists to some extent even to the present day. It has only been since about 1860 that the word "tonnage" has been generally accepted as a measure of cubic capacity or volume of the vessel in units of 100 ft^3. Tonnage is a term applied to the internal volume of a ship in tons of 100 ft^3.

Gross tonnage is the entire internal volume, except for certain exempted spaces. Net tonnage is the tonnage remaining after the nonearning spaces, such as machinery spaces and crew spaces, have been deducted from the gross tonnage. Net tonnage was originally intended as a measure of a ship's earning ability, and is the basis upon which tolls for canal transit, port charge, and so on are calculated.

All of the principal maritime governments have their rules describing how tonnage is to be measured. The Suez and the Panama Canal authorities also each have their own rules.

Safety of life at sea. The safety and welfare of passengers and crew, the provision of adequate lifesaving appliances, and the like, in all ships are under government jurisdiction. The safety regulations represent the accumulated experience and the lessons learned from many serious casualties involving loss of life which have occurred to ships at sea, including such major accidents as the loss of the *Titanic* in 1912 and the burning of the *Morro Castle* in 1934. Since shipping is largely an international business, it is necessary that safety regulations be international in character. The Safety of Life at Sea Conventions of 1914, 1929, and 1948 and the Load Line Convention of 1930 represented major steps toward improvement in safety of life at sea.

Standards of subdivision. If the underwater shell of a ship is torn open by collision or other cause, that portion of the ship which has access to the sea will fill with water to the level of the sea outside. This will affect the sinkage and trim and the transverse stability of the ship.

The ship settles bodily into the water by an amount which will depend on the quantity of water which enters the ship. At the same time, unless the flooded compartments are near amidships, the ship will trim by the bow or by the stern as the case may be. If the flooding is not symmetrical about the fore and after centerline, the ship will take a list to port or starboard. The effect of sinkage, trim, and list may be to reduce the freeboard to the deck to which the watertight bulkheads are carried. If this deck is brought below the level of the sea, water may enter the undamaged compartments and cause progressive flooding until the ship is lost.

The flooding of one or more compartments will generally change the transverse stability of the ship. The net effect may be either a gain or loss of stability, depending on the proportions of the ship and the length of the flooded spaces. If there is a loss of stability, the ship may become unstable and take a severe list or even capsize.

The prevention of foundering, loss due to bodily sinkage, or capsizing lies in fitting transverse bulkheads so spaced that the leakage water is confined to an amount which will not sink or trim the ship sufficiently to immerse the tops of the bulkheads, and so that the lost waterplane is confined to an extent which will not cause sufficient loss of stability to result in an excessive angle of list. The amount of the leakage water also depends on the permeability of the space.

Permeability of volumes. Permeability (μ) refers to the extent to which leakage water can permeate flooded spaces. It is expressed as a percentage of the total volume. For instance, an empty hold can take nearly its entire volume of sea water, the only water-excluding space being the structure of the ship itself. The permeability in this compartment would be about 98%. At the other extreme, a ballast tank entirely filled with water would not admit any sea water if opened to the sea; its permeability would be zero. Spaces used for various purposes may have any degree of permeability between these two extremes.

It is customary to use an average of permeability for each of the three major divisions of a ship: (1) the spaces forward of the machinery space, (2) the machinery spaces, and (3) the spaces aft of the machinery space.

Spacing of bulkheads. From the point of view of safety when flooded, very long compartments may be a menace; if the bulkheads were extremely far apart, the ship might be unable to survive any damage beyond that with which the pumps could cope. On the other hand, however, having closely spaced bulkheads may interfere with cargo handling and stowage, passenger accommodations, and machinery arrangements; if the bulkheads are extremely close together, the ship might be rendered useless from a commercial point of view. Somewhere between lies the proper spacing of the subdivision bulkhead.

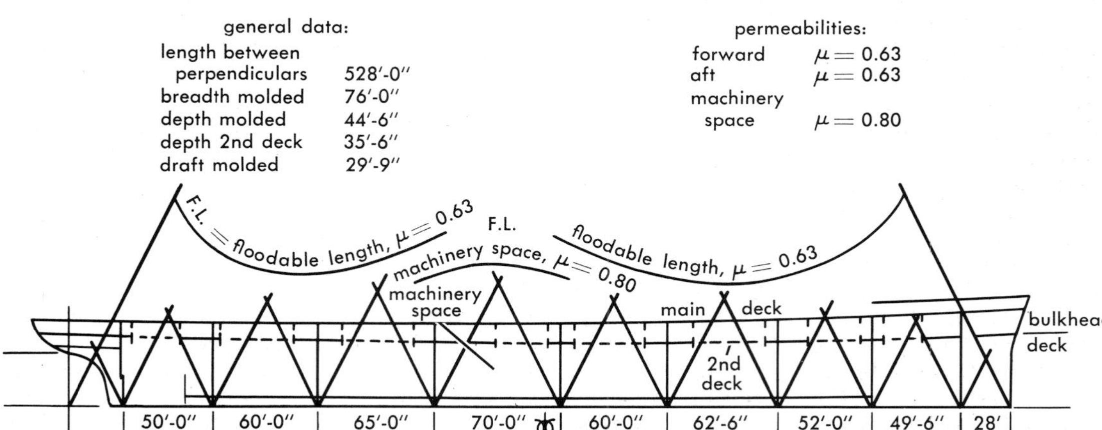

Fig. 7. Subdivision diagram and floodable length curve.

At the Safety of Life at Sea Conventions of 1929 and 1948, a system of regulations and formulations was set up whereby the spacing of subdivision bulkheads of sea-going ships carrying more than 12 passengers is fixed by a standard of comparison to the floodable length. The floodable length is shown in Fig. 7 as a curve drawn upon the profile of a ship in which, at each point, the ordinate equals the maximum portion of the length of the ship, having its center at the point in question, which can be flooded without the ship being submerged beyond the margin line.

Legislation now in force for safety of life at sea deals with many other matters in addition to subdivision. The broad principle, with regard to locations of watertight bulkheads upon which the Convention regulations are based, is that the minimum standard for passenger ships shall be that a ship shall be capable of floating with any one compartment flooded. This standard is increased as the length of the ship is increased. A closer spacing of bulkheads than that which would give a one compartment standard is adopted in the longer ships. It is also reasonable to increase the standard of safety in ships carrying a larger number of passengers than in those carrying a few passengers. Thus, bulkhead spacing is reduced as the ship becomes more and more a passenger ship.

When dealing with the problem of flooding, it is important to ensure that the ship has ample stability. One of the provisions of the Convention is that the stability of a new ship shall be tested by an inclining test after completion.

There are also regulations concerning precautions against fire. In passenger ships, the risk of fire spreading throughout the ship is reduced by dividing the ship into vertical zones by means of fire-resisting bulkheads spaced not more than 131 ft apart.

Regulations of the U.S. Coast Guard govern the safety features of ships carrying the American flag. These regulations are more stringent in many instances than those of the International Conventions. The Coast Guard regulations are largely based upon Senate Report 184, which followed the loss of the *Morro Castle* in 1934.

Warship design. Merchant ships are designed principally for civilian use, while warships are designed for military use. Both, of course, must be designed to meet their specified requirements in the most efficient manner possible—one for commercial efficiency or profit, the other for military or combat efficiency.

The attainment of such efficiency is a matter of technical knowledge and organization. In the United States, the Bureau of Ships of the Department of the Navy is the technical bureau responsible for the design, construction, and maintenance of naval ships. The Office of the Chief of Naval Operations prepares the characteristics and determines the missions for new naval vessels.

As in the case of a merchant ship, the most efficient warship is one in which the desired capabilities are obtained in the lightest ship of smallest practicable dimensions. Materials of high strength, efficiently arranged, are used in order to reduce the hull and machinery weight as much as possible. Thus the necessary military offensive and defensive equipment is allowed the remaining permissible weight to the greatest possible degree. Ship design is, then, a matter of balancing the need against the cost of attaining it to decide whether the resulting ship can meet its competitors on an equal or better status. *See* HYDRODYNAMICS; HYDROFOIL CRAFT; SHIP, NAVAL. [J.C.N.]

Bibliography: D. Arnott (ed.), *Design and Construction of Steel Merchant Ships*, 1955; L. Attwood, H. S. Pengally, and A. J. Sims, *Theoretical Naval Architecture*, 1953; K. C. Barnaby, *Basic Naval Architecture*, 2d ed., 1954; J. P. Comstock, *Introduction to Naval Architecture*, 1942; W. Muckle, *Modern Naval Architecture*, 1956; H. E. Rossell and L. B. Chapman (eds.), *Principles of Naval Architecture*, 2 vols., 1939; Soc. Naval Architects Marine Engrs. *Historical Transactions, 1893–1943*, 1945; D. W. Taylor, *The Speed and Power of Ships*, rev. ed., 1933; W. P. A. Van Lammeren, L. Troost and J. G. Koning, *Resistance, Propulsion and Steering of Ships*, 1948.

Ship propulsion

The earliest propulsion of boats and vessels by paddles or oars was succeeded or augmented by the use of sails. Power propulsion, using paddle wheels, began in the early 1800s when the steam engine was adapted to marine use. Screw propellers came into use about 1850 and offered great advantages over paddle wheels. The diesel engine developed rapidly after 1892 and now supplies a large portion of the horsepower for ship propulsion. Nuclear power has been successfully used for both submarine and surface ship propulsion and appears to offer great possibilities for future development.

This article discusses resistance of ships, model testing, powering of ships, ship vibrations, steering and maneuverability of ships, and ship trials. For related information *see* HYDROFOIL CRAFT; MARINE ENGINE; MARINE MACHINERY; PROPELLER, MARINE; REACTOR, SHIPBUILDING; SHIP DESIGN; SUBMARINE.

Resistance of ships. In order to move a ship through the water, resistance must be overcome. The total resistance of ships to propulsion is made up of skin frictional resistance, wave-making resistance, eddy resistance, and air resistance. These various types of resistance vary with speed, as well as with form and condition of the hull.

Skin frictional resistance is the drag of the water on the surface of the ship's hull, and it is generally the largest factor in the total resistance. Wave-making resistance occurs as a result of the energy required to set up the system of waves around the hull as the ship moves through the water. This type of resistance is usually the second largest, but for high-speed ships may be the largest factor in the total resistance. Eddy resistance refers to the resistance created by hull appendages, such as struts, as they move through the water. It is distinguished from turbulence eddies near the ship's hull which are associated with skin resistance. For a well-designed ship, eddy resistance is a minor factor. Air resistance is the term applied to the total resistance offered by the air as the ship moves. It varies with conditions but is not a major factor.

For a general understanding of the resistance of ships to propulsion, consider first conventional ships, meaning those which proceed through the water as opposed to hydrofoil craft, which skim the surface, or submarines, which travel beneath the surface. The area of that part of the hull of a ship which is immersed and in contact with the water is called the wetted surface; it is a significant factor in resistance of all ships at all speeds. Length is also an important factor in determining both resistance and the maximum speed at which a conventional ship can be driven economically. For example, the resistance to propulsion per ton of ship is very nearly the same for the longer SS *United States* at 35 knots as for the shorter Coast Guard cutter *Ingham* at 20 knots. Under such conditions these two ships are traveling at corresponding speeds, that is, speeds proportional to the square roots of their lengths. The ratio $V/\sqrt{L}$, where V is speed in knots and L is the length of the hull at the waterline in feet, is called the speed-length ratio and it is the same for both ships at corresponding speeds.

The relationships which exist between speed, resistance to propulsion, length, and weight of conventional ships of various sizes may be shown on a single graph (Fig. 1). It may be noted in Fig. 1

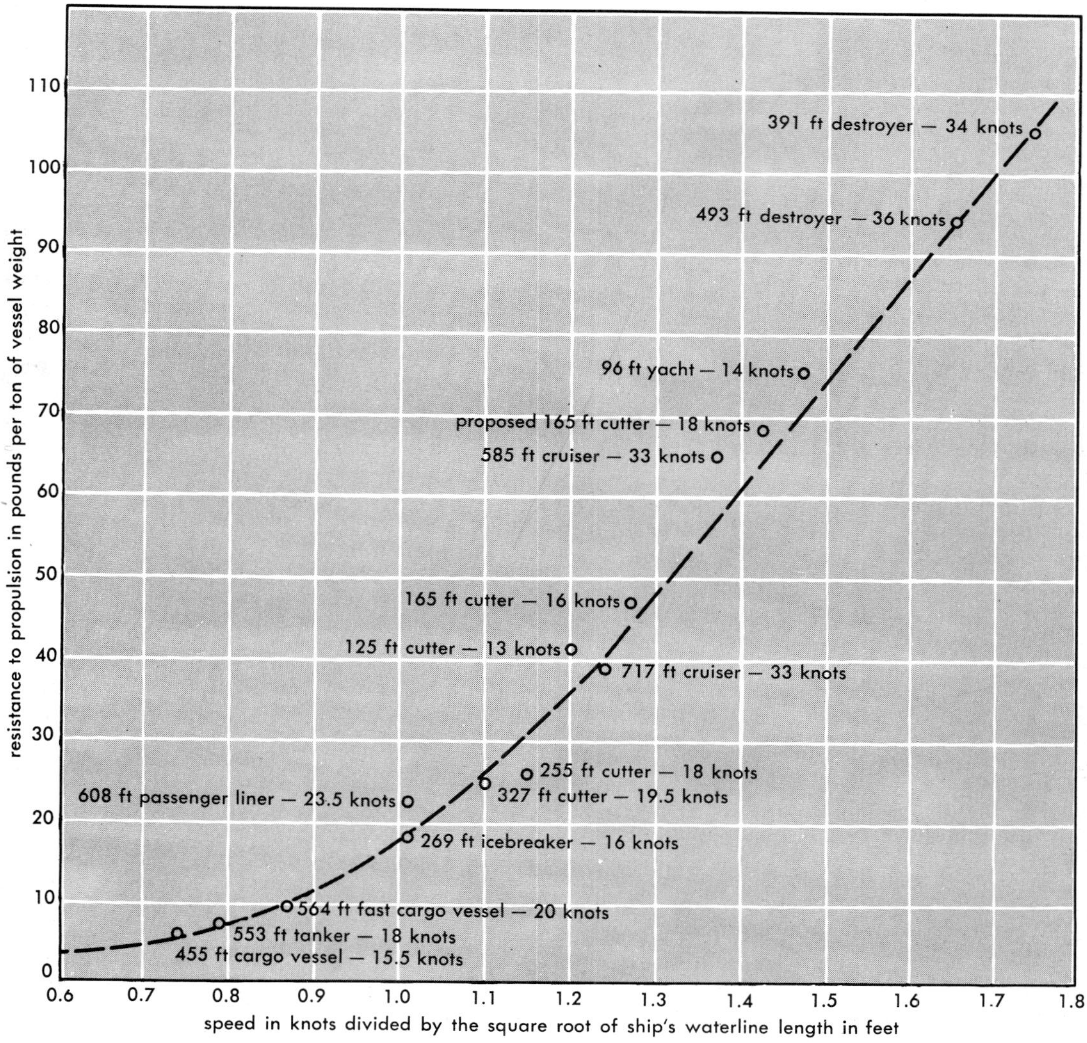

Fig. 1. Graphic representation of the relationship between speed, resistance to propulsion, length, and weight of a ship. (*U.S. Coast Guard*)

that all ships plotted fall near a single smooth curve. It will also be noted that large cargo vessels with great carrying capacity, ruggedness, durability, and economy of operation all fall at the lower left of the curve. At the other end of the scale are lightly constructed vessels with shorter life expectancy, little carrying capacity for their size, and high cost of operation. In general, as the curve ascends, the weight, size, and cost of the propelling plant become an increasingly greater part of the total weight, size, and cost of the whole ship. For a given type ship, cost considerations make it impractical to go above some point on the curve. For cargo vessels that point is well down on the curve. A passenger ship must move faster to be successful, so it can afford to move up the curve above the cargo vessel. Near the top of the curve are naval vessels whose speed is one of their most important characteristics.

Reynolds number. In 1883 Osborn Reynolds introduced a dimensionless ratio bearing his name that finds important application as a criterion in fluid-flow work. The Reynolds number is the ratio of the inertia force to the viscous frictional force in a fluid system. It is commonly written $VL\rho/\mu$, where V is a typical velocity of flow, L is a typical length, ρ is the density of the fluid, and μ is its coefficient of absolute viscosity. Since skin frictional resistance particularly involves both of the forces whose ratio is expressed by the Reynolds number, this criterion has special application to skin resistance (*see* REYNOLDS NUMBER).

Froude number. The waves created by the motion of a surface ship or model introduce gravity-restoring forces. It is thus apparent that the acceleration of gravity is one of the governing variables where wave-making resistance is involved. The Froude number, introduced by William Froude, is a dimensionless ratio of inertia force to gravity force, which finds important application as a criterion in such cases. The ratio is commonly written $V/\sqrt{gL}$, where g is the acceleration of gravity and V and L are as indicated for the Reynolds number. When testing a surface ship model in a towing tank, the value of the Froude number for the model must be kept the same as for the full-size ship. *See* FROUDE NUMBER.

Origin of ship waves. The pressure which exists at a submerged point ahead of a ship which is moving steadily ahead in still water is directly proportional to the depth of the water at the point. As the ship approaches the point, the water is disturbed and pressure variations occur. If the point is well submerged, the pressures are everywhere exactly balanced by opposing pressures from adjoining layers and there will be no tendency for vertical movement of water at the point. At the surface, however, the upper layer is not acted upon by an adjoining water layer and the unbalanced pressures from below cause upward accelerations of the water. At higher speeds of the ship, greater pressures are created and greater upward accelerations occur. As the ship continues ahead at a steady speed, pressures and velocities of the water continue to vary in accordance with Bernoulli's principle; that is, as velocities increase, pressures decrease (*see* BERNOULLI'S THEOREM). With decreasing pressures, as the bow cuts through the water, downward accelerations are produced and the water, due to its own inertia, will overtravel, thus producing a phase lag. The net result of these actions is to start a bow wave train. The crest of the bow wave train becomes higher and further abaft the stem as the speed of the ship increases.

Figure 2 shows the general characteristics of a bow wave train as sketched by Froude. Note that succeeding waves along the length of the ship are of decreasing height but of approximately the same length as the initial bow wave. Wave length is directly related to wave height and varies with the speed of the ship.

Froude reasoned that just as the bow wave train starts with a crest due to the increased pressures in the vicinity of the bow, so should a stern wave train start with a trough due to the decreased pressures in that area as the water rushes in to fill the space behind the moving ship. Since the length of the bow wave varies with the speed of the ship, it appears that succeeding waves along the length of the ship may arrive at the stern in such phase as to accentuate the trough of the stern wave train, or they may arrive in some opposing phase. This is illustrated in Fig. 3 where, in the upper sketch, the profile of a ship and bow wave are shown as well as the initial undisturbed stern trough. The ship in this sketch has an imaginary parallel middle body of sufficient length to permit the disappearance of the bow wave train before it reaches the stern trough. In the lower sketch the actual ship and wave profiles are shown and, for this case, the resulting accentuated stern trough is indicated. At other speeds the phase relationship would vary and the stern trough would be reduced rather than accentuated.

The ship waves considered here should not be confused with ocean waves which are created by the effects of wind. Such ocean waves also significantly affect the resistance to propulsion of ships.

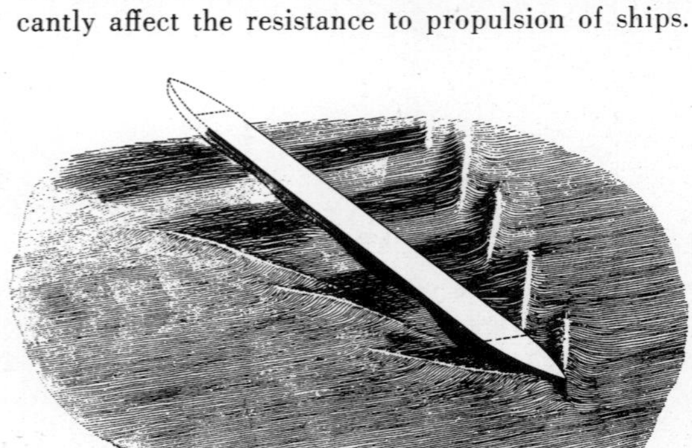

Fig. 2. Froude's sketch of characteristic bow wave train. (*From H. E. Rossell and L. B. Chapman, Principles of Naval Architecture, vol. 2, Soc. Naval Architects Marine Engrs., 1939*)

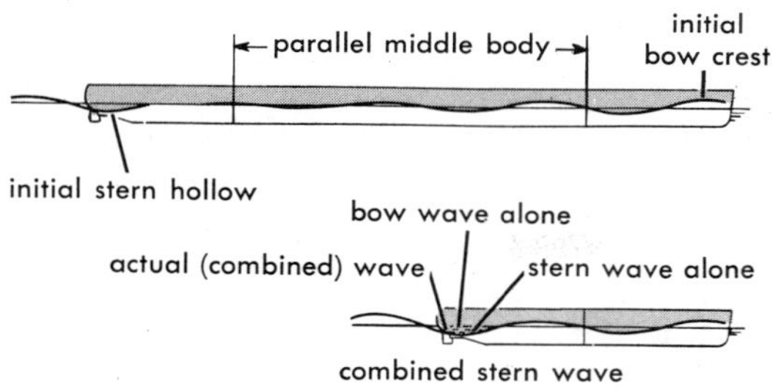

Fig. 3. Natural stern wave train and wave interference. (*From H. E. Rossell and L. B. Chapman, Principles of Naval Architecture, vol. 2, Soc. Naval Architects Marine Engrs., 1939*)

See OCEAN WAVES; *see also* WAKE (SEAGOING VESSELS).

Resistance of surface vessels. The total resistance to propulsion of surface ships is a function of both Reynolds number, which governs skin friction resistance, and Froude number, which governs wave-making resistance. Hence, correlation between the total resistance of a ship and that of a geometrically similar model can be effected only if both the Reynolds number and the Froude number are made equal. In general, this is impossible to achieve since the controlling variables would demand that the model be moved faster for the same Reynolds numbers and slower for the same Froude numbers. In practice the model is towed at speeds corresponding to those of the full-size ship and the total resistance determined.

In 1932 K. E. Schoenherr plotted the test data from many earlier authorities on a single graph and drew a smooth average curve through them. Choice of his formulation and friction coefficients or those of other authorities gives results which vary somewhat, all of which indicates the complexity of the problem and its experimental rather than scientifically exact nature. Continuing efforts toward better utilization of the work of authorities in model work in different countries are reflected by the agreements reached at the International Towing Tank Conference in Madrid in 1957.

Resistance of submarines. When submarines operate on the surface, the principles which apply to surface vessels are used for the determination of resistance. When sufficiently submerged, so that wave-making resistance becomes negligible, inertia and viscous frictional forces are involved and the Reynolds number applies. The remaining resistance of deeply submerged submarines does not change with speed since wave-making resistance is not involved and the flow pattern remains the same at all speeds.

Model testing and towing tanks. The reliable prediction of resistance to propulsion of a full-size ship from model tests requires thorough appreciation of the many factors involved as well as refined techniques. Uncertainties of varying magnitude, particularly in self-propelled models, may arise even with very careful work. Larger models, 20 ft or more in length, rather than smaller types, 4–6 ft in length, may be used to advantage in reducing the uncertainties.

Towing tanks for testing ship models and for other purposes have been in use at various locations throughout the world since about 1872. Many new facilities have been added since World War II as a result of a world-wide increase in research activity. For example, 25 new towing tanks in addition to an even larger number of various types of basins, water tunnels (used primarily for testing propellers), and flow channels are under construction or have been completed in various countries since World War II.

One of the world's most extensive testing facilities is at the David Taylor Model Basin at Carderock, Md., near Washington, D.C. (Fig. 4). These facilities are primarily used to serve the ship design needs of the U.S. Navy, but are accessible to commercial interests upon payment of cost. Other such facilities, serving the needs of both naval and

Fig. 4. David Taylor Model Basin—aerial view of grounds and buildings. The long building houses the towing basins described in the text. (*U.S. Navy*)

commercial shipping, are found in several locations in the United States as well as in many other countries of the world. The low-speed towing tank, or deep-water basin, at Carderock is 2775 ft long, 51 ft wide and 22 ft deep. The high-speed basin is 2968 ft long, 21 ft wide and 10 ft deep for part of its length and 16 ft deep for the remainder. Ship models are fastened to, and controlled from, metal towing carriages. In addition, the installation at Carderock contains a shallow-water basin, a turning basin, a test pond, a circulating water channel, variable-pressure water tunnel, wind tunnels, shops, offices, laboratories, and miscellaneous buildings. *See* TOWING TANK; WATER TUNNEL.

Figure 5 shows a model of the rotating arm and maneuvering basins at the David Taylor Model Basin, completed in 1959. Such facilities are used to study the course-keeping or steering and turning qualities of ships by model tests. Seakeeping qualities are also studied. Means are provided for producing not only regular head and following waves, but also irregular and short-crested waves at various angles of encounter. Facilities of this type are also found in other leading maritime nations of the world.

Powering of ships. Many methods have been used to estimate the power required to drive a given ship at a given speed. Early methods were empirical in nature. However, the early assumptions that resistance is wholly frictional and that power varies as the cube of the speed were fair approximations, since most early ships were of slow speed and ordinary proportions. In England a method which is referred to as the admiralty coefficient method continues to be used. This method also assumes that resistance is all frictional but does not use wetted surface directly. In the United States the admiralty coefficient method is often used for preliminary estimating. For all ships, and particularly those of higher speeds and fine (slender) proportions, the more accurate and generally used method is to determine the resistance of a model, or similar ship, and then deduce the resistance of the ship in question by application of Froude's law of comparison. In the case of similar ships, if trial results are known for one ship, then the power requirements for another similar one may be deduced by utilizing the trial results in the same manner as model results are used.

Marine engineering equipment. In propulsion of a ship, the total resistance is most commonly overcome by using a power plant to turn a suitable screw propeller, though in the case of sailing vessels the resistance is overcome by the action of the wind on sails. Marine power plants, like power plants on shore, are designed to meet the particular problem at hand. Many factors are considered in selecting the type of equipment to be used, including the size and speed of the ship, the purpose for which it will be used, the weight and initial cost of the machinery, the space occupied, reliability, fuel consumption, maintenance cost, and the characteristics of the propeller to be used. For an extended discussion of various types of marine power plants and equipment, *see* MARINE ENGINE; MARINE MACHINERY.

Ship vibrations. Like other structures, a ship's hull will vibrate if subjected to an exciting force. When the frequency of the exciting force is equal, or nearly equal, to one of the natural frequencies of the ship, a condition of resonance occurs and the resulting vibration may be very noticeable and undesirable. The hull may vibrate as a whole in either a vertical, horizontal, or torsional fashion,

Fig. 5. Architect's model showing arrangement of the rotating arm basin (right) and the maneuvering and seakeeping basin (left) of the David Taylor Model Basin. (*David Taylor Model Basin*)

or in a combination of these. In addition, the various parts of the ship all have natural frequencies which make local vibrations possible and probable. These include drumhead-type vibration of bulkheads or vibration of the light upperworks of the ship.

Exciting forces for the production of ship vibrations commonly originate with the propellers or with the operating machinery. The frequency of vibration does not necessarily equal the speed of rotation of the propeller or some item of operating machinery; it may be some multiple or combination of multiples of such speeds. Vibration-generating equipment which is capable of operating at various frequencies and amplitudes is useful in the location and correction of hull vibrations. For more details on causes and cures of vibrations, *see* MECHANICAL VIBRATION; VIBRATION; VIBRATION DAMPING.

Steering and maneuverability. For turning and course-keeping, ships are commonly equipped with a rudder located at the stern, its midposition being in the vertical centerline plane. When the ship is moving straight ahead and the rudder is moved and held at some angle from its midposition, the force exerted by the rudder causes the ship to turn about a vertical axis, to change trim about a transverse axis, and to heel over about a longitudinal axis. The turning is the most obvious effect, though heeling is easily observed at high speeds. With steady engine speed the center of gravity (CG) of the ship traverses a path which becomes circular as soon as the turning becomes uniform, as indicated in Fig. 6. The diameter $2R$ of the turning circle, the advance, and the transfer, as shown in Fig. 6, all vary with rudder angle and speed of the ship.

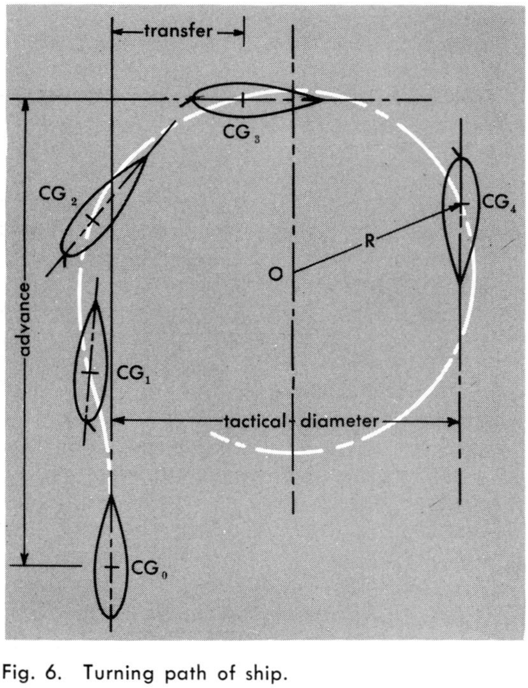

Fig. 6. Turning path of ship.

For the same rudder angle the diameter of the turning circle is less at lower than at higher speeds. At full speed and maximum rudder angle, the diameter of the turning circle varies with the type of ship and rudder but is commonly four to six times the vessel's length.

Ships are said to be directionally stable or unstable according to the size of the force (due to rudder angle) required to maintain a straight course—the smaller the rudder angle required, the more stable the ship. Referring to Fig. 6, the initial effect of applying right rudder is to push the ship bodily left, away from the intended direction of turning. As the rudder force continues to act, the turning moment forces the water pressure distribution around the hull to be altered. This alteration produces a resultant or net pressure force on the hull which acts either to assist or to resist the turning of the ship. It assists the turning if its effective point of application is forward of the center of gravity of the ship; if aft of the center of gravity, it resists. The latter case is the more directionally stable since in resisting the turn the ship tends to maintain a straight course and is better able to do so with smaller rudder angle than if directionally unstable.

Maneuverability is much affected by the number of screws a ship carries, twin screws having great advantage over a single screw. In addition, maneuverability is more important in some ships than in others, the demands of the particular service being a determining factor. For example, merchant ships operate on long runs which require little maneuvering, while warships should be able to maneuver readily.

Factors affecting steering. Steering of a ship is assisted by high length-breadth ratio of ship, low length-draft ratio, trim (greater draft) by the stern, a rudder of large area located in the propeller slip stream, a smooth sea, and deep water. In shallow water, the under-hull pattern of flow is disturbed, making steering erratic. Ocean waves often affect steering adversely and, to meet incipient fall-offs from the course, early and quick rudder action is necessary.

With rudder applied to a moving ship, the net resultant pressure on the two sides of the rudder acts at the center of pressure. Because of fluid friction the resultant pressure is at a slight angle from the normal to the plane of the rudder and may be broken up into two components, one of which is parallel and the other normal to the direction of motion of the ship. The parallel component is called the drag while the normal or effective steering component is called the lift of the rudder. The lift of a rudder is influenced by area, area orientation, and rudder angle; rudder outline or shape has less influence. For rectangular rudders the ratio of the depth (span) to the width (chord), the latter being the dimension in the direction of water flow, is called the aspect ratio. For other rudder shapes the aspect ratio is the ratio of the square of the span to the lateral area. Tests demonstrate that lift

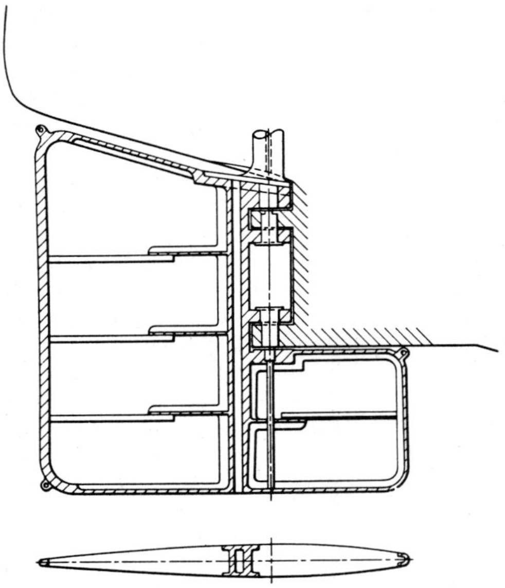

Fig. 7. Semibalanced rudder for twin-screw ship. (*From H. E. Rossell and L. B. Chapman, Principles of Naval Architecture, vol. 2, Soc. Naval Architects Marine Engrs., 1939*)

of rudders of the same size varies considerably with aspect ratio. The greater the aspect ratio the smaller the critical angle, or angle at which the flow pattern on the downstream side of the rudder changes from streamlined smooth flow to irregular turbulent flow, with resulting fall-off in lift. Rudders are normally designed to be most effective at rudder angles of 30–35°, and their range of motion is usually limited to these values.

Rudder types. A rudder is said to be balanced when the location of the center of pressure coincides with the turning axis of the rudder. Since the position of the center of pressure shifts with rudder angle, thorough balance for all angles is not attained. Common practice is to design for balance at 15° rudder angle, thus reducing power requirements for the steering engine. In ships with a cruiser-type stern, a semibalanced rudder (Fig. 7) is sometimes used because it fits the design well. A rudder which is balanced for the ahead direction will be greatly unbalanced for motion astern, and maximum rudder torque may occur when going astern.

Figure 8 shows a spade rudder as installed on a single-screw passenger ship. The rudder stock of this type rudder, having no support at the bottom, must withstand bending as well as torsional stress, necessitating a heavy design. Twin rudders of this type are sometimes installed on twin-screw ships, one in each propeller slip stream; such an installation gives good maneuverability at all speeds. They are also used in multiple, both ahead and astern of propellers, on river towboats where great steering power is required. There is no fixed rule for determining area of rudder since requirements for maneuverability vary widely with different types of vessels. The selection is made by comparison with similar vessels of desirable and proven maneuvering qualities.

Automatic steering devices. Because the gyrocompass is more sensitive in detecting initial deviations from the course than is the human eye, automatic steering devices offer an improvement over hand steering. The improvement may be greater than sometimes realized, since reduction of rudder drag reduces fuel used, reduction of rudder angle required reduces the work of the steering engine, and reduction of deviations from course reduces distance traveled. *See* GYROCOMPASS.

Backing power of ships. By international law, oceangoing ships have for years been required to have sufficient power for astern operation to secure proper control of the ship in all normal circumstances. In addition, maximum safe speed under conditions of reduced visibility is an important operational consideration and is closely related to stopping ability. The problem varies with type of power plant. For example, reciprocating steam engines offer excellent stopping and maneuvering ability since they can be quickly reversed and can apply full-power torque at any shaft speed from full ahead to full astern. Direct- or geared-drive diesel reversing engines have nearly equal torque characteristics for ahead or astern operation. With the ship underway, however, special arrangements for changing the direction of engine rotation must be made because of the drag or opposing torque of the propeller. Diesels or turbines with ac or dc electric drives show variable characteristics according to the particular power plant, and their stopping and maneuvering characteristics range from good to excellent. Geared turbine drive power

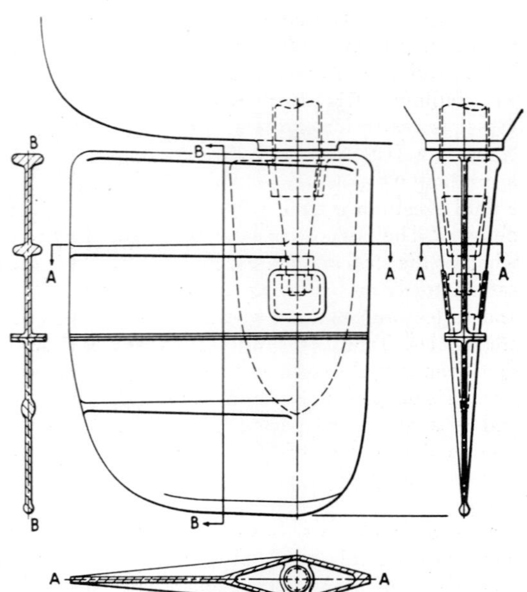

Fig. 8. Spade rudder. (*From H. E. Rossell and L. B. Chapman, Principles of Naval Architecture, vol. 2, Soc. Naval Architects Marine Engrs., 1939*)

plants are the most common type of steam plant for ships and they are usually designed to produce 80% of normal ahead torque at an astern propeller speed of 50% of normal ahead speed.

Actual tests of the geared turbine drive plant on the 34,100-ton tanker *Esso Suez* demonstrated that if the engines were put full astern while going ahead at 17 knots the ship traveled nearly 1 nautical mile before stopping. This and other tests show that quick stopping ability from higher-ahead speeds for vessels of low resistance per square foot of midship area is not possible even with large backing power. It appears therefore that head reach alone is not the best criterion for determination of backing power since it may depend more on the type of vessel than on astern power. Maneuverability around docks and ability to stop in pilot waters from medium ahead speeds are better criteria. It should also be noted that application of backing power to a propeller shaft does not mean that a corresponding effective astern thrust is developed immediately. The change from ahead to astern thrust is a gradual process and the propeller characteristics have a bearing on the changing relationship between torque and thrust.

Ship trials. The number and type of trials given new ships varies with the particular contract and type of ship. As the various items of auxiliary machinery and equipment are installed in the ship they are given individual installation tests. When the main machinery is installed and properly connected with its auxiliaries and other equipment necessary to its proper functioning, it is usual practice to hold a dock trial. The machinery is warmed up and run very slowly until it is assured that no troubles exist to prevent continuing the trial. Speed is increased to the highest safe speed while the ship is tied up to the dock, including astern operation, and the trial usually lasts 4–6 hours. Builders usually give new vessels an underway builder's trial to assure themselves that the vessel is ready for the trials required by the contract. New U.S. Navy ships are required to undergo a preliminary acceptance trial prior to delivery and a final acceptance trial within a specified time after delivery. The purpose of such trials is to determine whether the ship and her machinery and equipment satisfactorily meet the requirements of the contract and to test and observe operating capabilities and efficiencies. Trials of similar purpose are held on new commercial vessels built for the U.S. Maritime Commission and these trials are followed, at the end of a guarantee period, by a thorough inspection to determine defects or deficiencies. Private owners and others may prescribe trials of varying scope.

It is desirable to reduce uncertainties to a minimum during trials by choosing minimum wind and sea conditions along with a clean hull and propeller. The program for trials will depend on total time available but it should be complete in all details and prepared and approved in advance. The personnel taking active part should be familiar with and practiced in their duties. In conducting acceptance trials, the progressive speed or standardization trials are usually held first, followed by maneuvering trials, steering-gear tests, anchor windlass tests, miscellaneous tests, and finally, the fuel-consumption trials.

Standardization trials. These are conducted over the best available nautical mile course at specified displacement. The measured mile course approximately 5 miles ESE of Rockland, Maine, shown in Fig. 9, is the best available course on the Atlantic Coast of the United States. The courses and distances indicated in Fig. 9 give ample time for steadying of conditions before and during traversal of the measured mile. Runs at slow speed are made first with at least two runs, one north and one south, for each speed. Propeller rpm is kept as nearly constant as possible for each speed and the

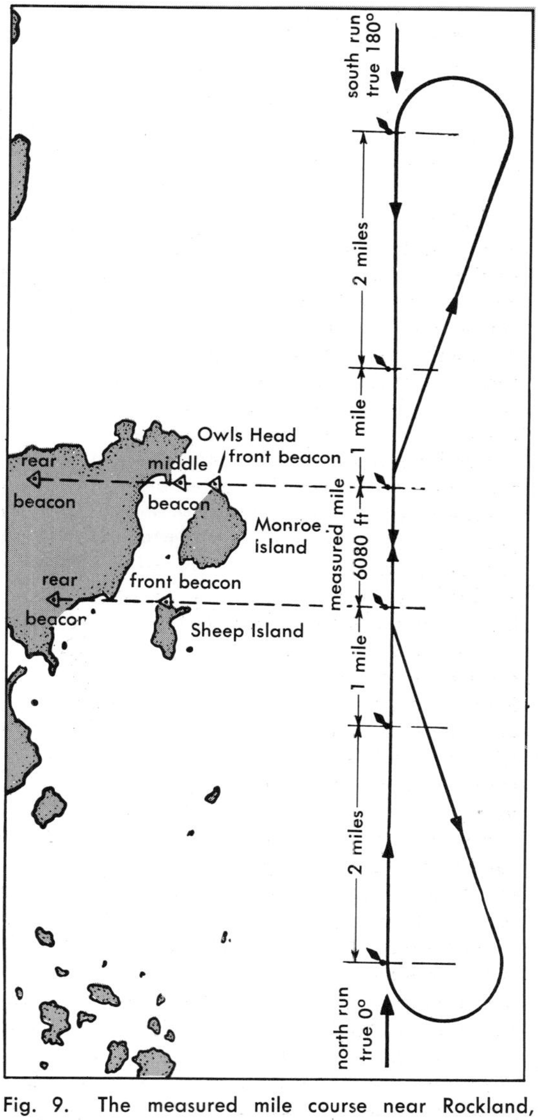

Fig. 9. The measured mile course near Rockland, Maine. *(From J. M. Labberton and L. S. Marks, Marine Engineers' Handbook, McGraw-Hill, 1945)*

number of speeds checked should be at least six (more for large speed ranges). In addition to recording the specific readings and data for each run, certain computations should be made and plots maintained in order that deviations from a fair curve, indicating inaccuracies in measurements, may be taken into account.

Maneuvering and steering trials. Maneuvering trials include tests to determine turning circle diameter with full rudder, ability to go from full speed ahead to full astern, ability to develop full astern power without undue heating (turbine drives), ability to go from full speed astern to full ahead, time required to go from full speed ahead to dead stop, and time required to change course a certain number of degrees with a specified rudder angle. Steering-gear tests check the time required to change the rudder from full right to full left (or vice versa) when going full speed ahead and when going full speed astern. Operation of the emergency or hand steering arrangements is also checked.

Fuel consumption trials. These are conducted in deep water in the open sea to determine the economy of operation of the machinery. For commercial vessels the guarantee usually applies at normal operating shaft horsepower, while for naval vessels guarantees apply at about 5-knot intervals from approximately 10 knots to full power. Trial runs are made at all guarantee points and at full power. Instruments and equipment used during the trials, such as meters, gages, and tanks, are tested or calibrated before the trials. Special fuel-oil meters are usually installed and these are calibrated both before and after the trials. Samples of the fuel used during the trial are analyzed for heat value. The time required for fuel consumption trials is from 4 to 8 hours.

Analysis of trial results. In analyzing trial results disturbing influences such as wind and current effects must be considered. Careful analysis reveals that the effects of wind and current are not eliminated by averaging the results of several runs in opposite directions over the measured mile. Since the effect of current is to increase or decrease speed over the ground without affecting resistance, it follows that current has no effect on thrust, torque, or rpm. Wind, however, affects resistance; therefore differences in readings of thrust, torque, and rpm for runs in opposite directions reflect the effects of wind alone. With the effects of wind and current separated, it is possible to arrive at the true relation of shaft horsepower to speed. Disturbing effects such as variations from designed steam pressure, temperature, and condenser vacuum also enter fuel-consumption trials. Correction factors are applied to take care of such effects. *See* HYDRODYNAMICS. [K.K.C.]

Bibliography: J. M. Labberton and L. S. Marks, *Marine Engineers' Handbook*, 1945; H. E. Rossell and L. B. Chapman, *Principles of Naval Architecture*, vol. 2, 1939.

Ship salvage

The act of saving or recovering a ship, its cargo, or its equipment from the sea. It is also called marine salvage. Rescue of passengers and crew is usually termed lifesaving. Several situations may present themselves to the salvage crew:

1. The ship may be helpless or damaged but still afloat. In this case, it may be towed to port by rescue or salvage tugs. If the ship is a derelict, members of the salvage crew must board it to attach the towline, and possibly to steer it under tow.

2. The ship may be aground and resting at a higher level than when floating freely. Because of unrepaired damage, water may enter the ship when it is floated.

3. The ship may be aground and resting at a lower level than when floating freely, but with the main hull not completely submerged.

4. The ship may be sunk, with the main hull reasonably intact but lying completely below the surface. The ship may be bottom up, lying on its side, or right side up.

Operating procedures. A ship-salvage project involves a number of tasks, several of which may be carried on simultaneously. The first step is locating the wreck when it has foundered unnoticed (perhaps at night or in a storm), when it is an abandoned derelict, or when it has sunk out of sight. Next, further motion and damage to the wreck must be prevented by laying out anchors or moorings and possibly by temporary, controlled internal flooding. Then a salvage plan is formulated. The method of lowering, lifting, moving, or unwatering the ship is chosen after preliminary calculations have been made of the buoyancy or lifting forces needed. The method chosen must enable the ship to be held under control during all parts of the operation. The plan includes an estimate of the salvage time and cost, as well as the setting up of a base of operations.

On the scene, the wreck must be sealed watertight (airtight if compressed air is to be used in the unwatering process). Doors, hatches, and covers built into the ship must be secured, and necessary plugs, patches, and other temporary closures provided. If the ship has to be lowered or moved horizontally, obstructions between it and deep water must be removed.

Floating the wreck. Salvaging ships that are on rocks, beaches, or shoals by floating them must be done in several stages. Rocks projecting into the ship bottom must be carefully blasted out so that the hull can be moved horizontally. The wreck must be lifted bodily with jacks and improvised launching ways inserted under it. The wreck is then eased into deep water. Sometimes a cradle must be built under the lifted ship and the ship rolled or skidded until it can be launched or pulled into protected deep water. A deep channel, into which the ship can be pulled or slid, may need to be dredged or excavated alongside (Fig. 1). A stranded ship is pulled off a beach with heavy winches or capstans

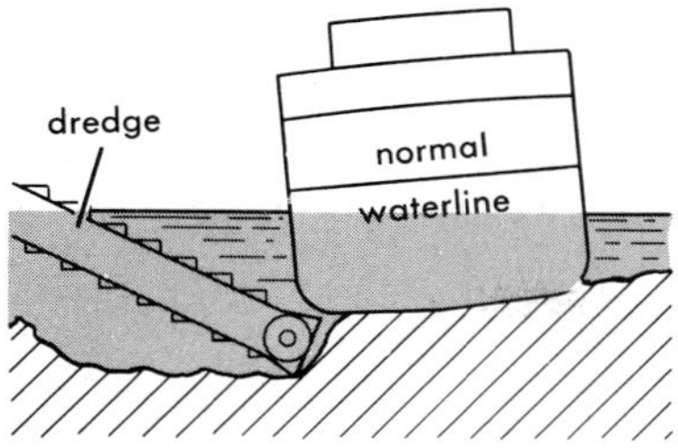

Fig. 1. Dredging a channel alongside a grounded ship.

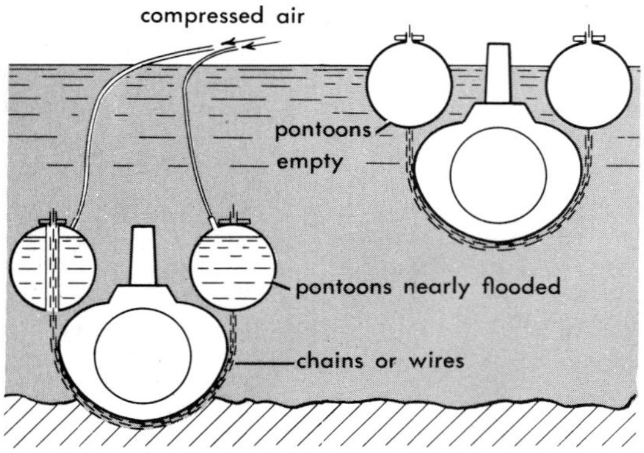

Fig. 2. Lifting a sunken submarine by pontoons.

pulling steadily on wires or chains strung between the ship and large anchors placed well offshore in deep water. Motive power may be on the beach.

Sunken ships may be raised by several methods. One, used when the ship's bottom is intact, is to pump out the water to lift the ship with its own buoyancy. If the main hull is not too far under water, its sides are extended upward with watertight walls, or watertight trunks are built around the hatches or other deck openings. All permanent closures are secured; temporary closures are added as required; wooden, steel, or concrete patches are fitted over holes in the hull; the ship is lightened as much as possible; and pumping follows.

If the upper boundaries of the hull are strong and intact and the lower ones open or damaged, as in a ship lying bottom up, the water may be blown out by compressed air. The upper compartments are sealed off, and air is then blown through lines and valves until enough water is forced out for the ship to float.

Another method is to lift the wreck bodily with large cylindrical tanks called pontoons. As is shown in Fig. 2, they are used in pairs, the connecting wires or chains passing under the ship's hull. When water is blown out of them, they rise and the ship is lifted.

A similar method is to lift the wreck by securing other craft to it, then pumping water out of these other craft. The rise of the tide also can be employed, and sometimes both other craft and the tide are used. A wreck in deep water may be lifted progressively by wire or cable slings passing under the ship's hull and connecting pairs of barges or lighters. The wreck is moved progressively into shallower water with each lift.

A wreck is righted from an upside-down position or from a horizontal position on its side by blowing water out of its low side or by pulling on tackles secured to brackets built out from the ship's side and extended to powerful winches and anchorages (Fig. 3).

The most sensitive and dramatic part of the entire salvage operation is the movement of the wreck. The salvage crew must maintain a maximum of control during the entire moving phase, for an error can greatly complicate or even doom the entire salvage operation.

After a wreck is floated it may be lightened and run up on a nearby beach for temporary repairs at low tide. Afterward it can be floated off when some of the buoyancy lost by damage has been regained. This makes it more seaworthy for towing or propelling to a repair yard.

Other salvage techniques. Salvaging a shallow wreck which has more or less disintegrated, such as the old USS *Maine* in Havana harbor, is accomplished by building a dam of earth and piling or of steel sheet piling around the wreck and pumping out the water inside. This makes possible a careful examination of the wreckage without the inevitable disturbance and additional damage from salvage operations. Certain historic ships sunk in lakes have been salvaged by pumping out the entire lake.

At times it is possible to salvage only part of a wreck, such as that carrying the propelling machinery. This can be later joined to new parts to make a ship which still has many years of useful

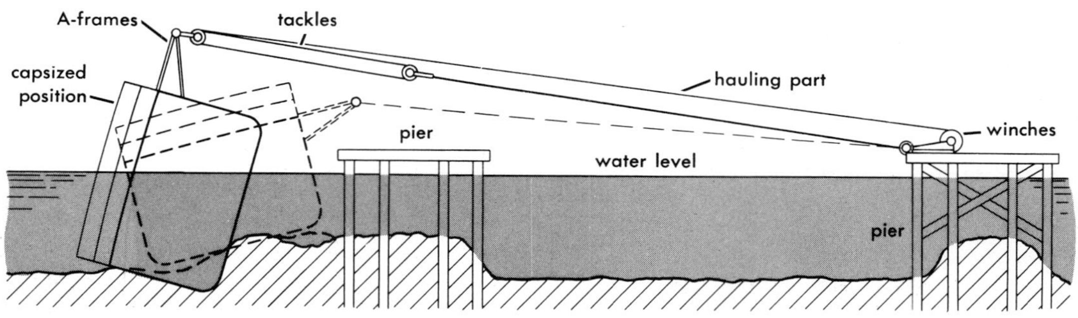

Fig. 3. Righting a capsized ship by tackles and winches.

life. In more extreme cases, only the equipment and machinery are removed, while the hull is left to the elements. [H.E.S.]

Bibliography: W. A. Sullivan, Marine salvage, *Trans. Soc. Naval Architects Marine Eng.*, 56:104–148, 1948.

Shipbuilding

The building of ships is one of mankind's oldest occupations and one of the most important of the world's heavy industries.

In the 12 years ending with 1957, 6228 ships (of over 1000 gross tons) carrying a total of over 64,000,000 tons deadweight were built, and in 1957, 794 ships carrying over 10,000,000 tons deadweight were delivered. The 10 major shipbuilding countries, in order of the total deadweight tonnage built in each during the same 12 years, are the United Kingdom, Japan, West Germany, Sweden, the United States, Netherlands, France, Italy, Denmark, and Norway.

A ship must be designed and built to endure the worst that a storm at sea can do to her, and at the same time the ship and all her appurtenances, from the main propelling machinery to the galley range, must continue to operate efficiently.

Most ships are custom-built to the requirements of their particular trade. Large numbers of identical ships, such as the Hog-Island freighters of World War I and the Liberty freighters of World War II, are built only to meet a war emergency.

The building of a ship may be divided into six phases: design, work prior to keel-laying, work on the ways, launching, outfitting, and sea trials.

The design of ships and sea trials are covered elsewhere (*see* SHIP DESIGN; SHIP PROPULSION). This article covers the remaining four phases, and is, in general, applicable to the building of both merchant and naval ships.

Shipyard layout and facilities. A shipyard site must be on deep water accessible to the ocean (or, in the United States, to the Great Lakes), and must have railroad connections for the delivery of materials. Figure 1 shows a large American shipyard.

Fig. 1. A large American shipyard. (*Newport News Shipbuilding and Dry Dock Co.*)

The principal facilities in a shipyard are offices for administration and design, storage areas and warehouses for the storage of steel and other materials, ship sheds and working areas for lay-off, fabrication and subassembly, the building ways on the water front on which the ships are built, shops for outfitting work, and piers at which the ships can lie during outfitting after launching.

An extensive materials-handling system is also needed. This includes cranes in all storage and working areas, heavy-lift cranes at the building ways and on the outfitting piers, a yard railway system, and a fleet of trucks.

Shop facilities include a mold loft, a bending slab with furnaces, a joiner shop, a machine shop, a pipe shop, a sheet-metal shop, an electrical shop, and perhaps a foundry.

Yard facilities include an extensive compressed-air system, and may also have a propane-gas system as well as the usual distribution of water and electricity. Some shipyards have a pickling plant to remove mill scale from steel plates; others remove mill scale by sand-blasting. Many shipyards have one or more drydocks used both for new construction and for ship repairs.

Ship steel. The great majority of merchant vessels are made of mild steel (MS), made to specifications laid down by classification societies. For a discussion of classification societies, *see* SHIP, MERCHANT.

For naval vessels special steels are used extensively both for strength and for protection. The two most commonly used are high-tensile steel (HTS), about one-third stronger than mild steel, and special treatment steel (STS), about twice as strong as mild steel.

Aluminum is being increasingly used in shipbuilding, principally for deckhouses and for such items of equipment as lifeboats, hatch covers, and so on. Certain magnesium-bearing aluminum alloys show excellent resistance to corrosion at sea, and can be welded with joints nearly as strong as the material itself.

Nomenclature. Before the construction of the steel hull can be described, some description of this structure is necessary. The construction varies considerably in different types of ships. For the purposes of this article the structure of a conventional cargo ship is used.

The keel consists of a flat keel and a vertical keel (see Fig. 2). The flat keel is simply the center strake (row of plates) of the bottom shell plating. The vertical keel is a vertical plate on the centerline of the bottom of the ship, running the entire length of the ship. The bottom shell and the side shell form the outer watertight envelope of the ship. Floors are vertical plates extending across the bottom of the ship, square to the centerline. Longitudinals are similar vertical plates running more or less parallel to the centerline. The inner bottom is precisely that—an inner bottom a few feet above the bottom shell, serving as protection in case of minor bottom damage, and as a floor to the holds

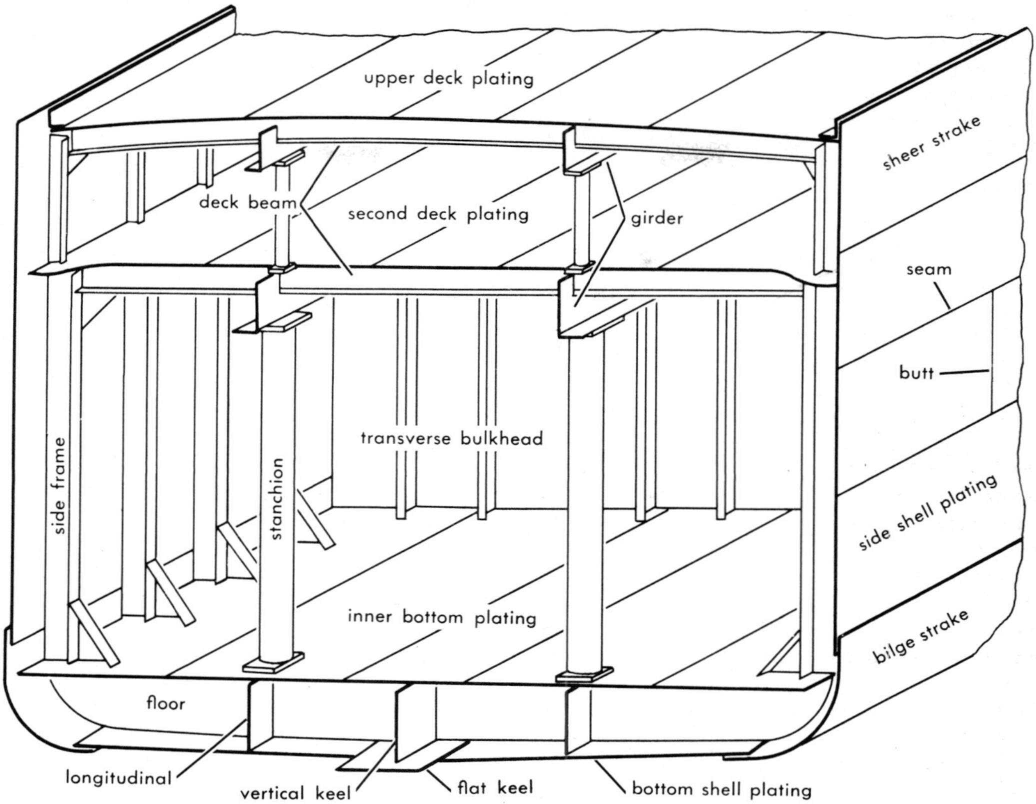

Fig. 2. Section through the structure of a simple cargo vessel.

and machinery space. Decks are horizontal surfaces corresponding to the floors of a building. Bulkheads are vertical divisions, either transverse or longitudinal (fore-and-aft), corresponding to partitions in a building and dividing the ship into compartments. Side frames are the stiffeners holding the side shell against the pressure of the sea. Deck beams and bulkhead stiffeners are self-explanatory. The deck beams are supported by girders, and the girders by stanchions, or pillars. Seams are the joints between the long edges (usually fore-and-aft) of plates, and butts are the joints between the short edges of plates. Seams connect strakes, and butts connect the plates within a strake.

Besides the trades common to other types of construction, shipbuilding involves two trades peculiar to itself. The shipfitters build the steel structure, including laying off and fabricating the individual members, subassembly, and erection on the shipway. The shipfitters are supported by the other trades of shipwrights, erectors, welders, and so forth.

The shipwrights' responsibility is to see that the structure is straight and true and of the designed dimensions. Their work starts with the laying down of the keel blocks, and continues throughout the steel work.

The many other trades involved in shipbuilding, such as riveting, welding, joiner work, machine-shop work, sheet-metal work, pipefitting, and electrical work, are essentially similar to the same trades in other industries.

Electric arc welding has largely supplanted riveting in shipbuilding for both steel and aluminum structure.

Work prior to keel-laying. In order that work may proceed efficiently after keel-laying, the job must be well advanced prior to this date. Structural design should be nearly finished, most of the steel should be ordered, and over half of it received in the shipyard.

Prior to keel-laying, the work consists of preparing working drawings (this continues right up to completion), ordering steel and other material, laying down the lines in the mold loft, making molds or templates, fabricating the steel, and assembling it into subassemblies as large as can be handled by the cranes.

Working drawings are made in the design division of the shipyard from the contract plans and the specifications. For a large passenger or naval ship several thousand separate drawings may be made.

Mold loft. This is a room with a smooth wood floor large enough to lay down the lines of the ship full-scale (Fig. 3). From these full-scale lines and the working drawings of the structure, molds or templates, made of thin wood or of heavy paper,

Fig. 3. A mold loft floor. (*Newport News Shipbuilding and Dry Dock Co.*)

Fig. 4. Photo lay-off. (*Newport News Shipbuilding and Dry Dock Co.*)

are developed for each plate, frame, or other structural piece. These molds carry all information necessary to finish the metal member ready for assembly. The templates are carried to a lay-off area where they are laid on the piece of steel to be used and the information transferred to, or laid off on, the steel.

Photo lay-off. This is a process in which, instead of using templates, special drawings are made on which is shown the same information which would have been given on a template. These drawings are photographed on negatives. Then in a darkened projection chamber the pattern is projected directly onto the steel and the steel laid off using the projected image (Fig. 4).

As soon as steel and templates are available, fabrication (the making of the individual members) is begun. The members are in turn assembled into subassemblies, ready for erection, as near the building ways as practicable. By the time the keel is laid, half of the structure may be in subassemblies ready for erection.

Preparation of the shipway. The area to be occupied by the ship during building must be clear, and provision must be made to carry the weight of the ship. This often requires a foundation of piles, capped at grade level either by cap timbers or by a solid concrete slab, which in turn supports the keel blocks, shores, cribbing, and other support for the ship until its weight is transferred to the launching ways.

The launching arrangements must be planned before the keel is laid. If the ship is to be end-launched, the keel blocks are carefully set to the designed height and slope. If the ship is to be side-launched, or built in a graving dock, the keel track is level.

Spauls, of either steel or wood, cut to the shape of the bottom of the ship, are set up and leveled at right angles to the centerline of the ship. The spauls form a cradle on which the bottom shell is laid.

On the ways. The work on the ways consists of erecting the structure, previously subassembled as far as practicable, riveting or welding it together, testing the watertight parts such as the shell plating, watertight bulkheads, and tank boundaries, and making a start at installing the propelling machinery. The outfitting trades also are started while the ship is on the ways, but most of this work is done after launching.

Laying the keel. The first section of the flat keel is placed on the keel blocks. This is usually a fairly large subassembly which includes the first section of the keel (Fig. 5).

When enough of the bottom shell has been laid on the keel blocks and spauls, subassemblies consisting of sections of floors, longitudinals, and inner bottom are laid on the bottom shell, aligned, and welded. This is followed by transverse bulkheads, side shell and side framing, longitudinal bulkheads, and decks.

Foundations for propelling machinery are completed early in the construction, so that the large

Fig. 5. Laying the keel of an oil tanker. (*Newport News Shipbuilding and Dry Dock Co.*)

and heavy items of machinery can be installed before the ship is closed in overhead. If this cannot be done, structure may be left loose to permit placing the machinery later.

Testing. All watertight or oiltight structures are tested for tightness. The boundaries of tanks which are to carry oil or water are tested by filling with water to the test head, usually some distance above the top of the tank. An equivalent air test is used as an alternative to this method, especially in naval work. Tight structure other than tanks is usually checked by a hose test.

As welding progresses the shipwrights keep a close check on the shrinkage caused by welding, and adjustments are continually made to keep the structure straight and true and to ensure that the ship maintains the intended over-all dimensions.

When the welding of the structure from the machinery space to the stern is practically complete, the shaft line (centerline of the line shaft which extends from the propelling machinery to the propeller) is run as a straight line between points located in the machinery space and at the intended location of the propeller. The aftermost section of the line shaft, called the tail shaft, and the propeller, are usually fitted before launching.

The degree of completion at launching depends on many factors: how badly the shipway is needed for the next ship, the relative ease of handling materials on the shipway as compared with the outfitting dock, and so on. In general, the ship is launched as soon as the hull is sufficiently complete

Fig. 7. Side launching of the 22,000-ton ore carrier *Ernest T. Weir.* (*American Shipbuilding Co.*)

Fig. 8. Floating-out launching of the *SS United States.* (*Newport News Shipbuilding and Dry Dock Co.*)

Fig. 6. End launching of the 60,000-ton tanker *Sansinena.* (*Newport News Shipbuilding and Dry Dock Co.*)

to stand the very considerable strain of launching.

Launching. Ships are launched in three ways: end launching (Fig. 6), side launching (Fig. 7), and floating out (Fig. 8).

In end-launching shipways, the ship is built more or less at right angles to the shore line, with launching ways parallel to the centerline of the ship. The ways extend far enough out into the water to prevent the ship from tipping down over the way-ends when launched. The launching ways usually have a slope of about ⅝ in. per foot. There are usually two, but may be three or four, launching ways.

A variation of the end-launching shipway is called a semisubmerged shipway, in which the outboard end of the launching ways is protected by side walls and a gate exactly like a graving-dock gate. (For information on graving docks and other types of drydocks, *see* DRYDOCKING.) The entire length of ways is thus in the dry during construction, and the ship can be located near the outboard end of the ways. When the ship is ready for launching, the lower end of the shipway is flooded to the level of the sea and the gate is removed.

In side-launching shipways, the ship is built parallel to the shore line, and the many launching ways are square to the centerline of the ship. Side-launching ways are given a slope of about 1½ in. per foot, much steeper than for end launchings, because in a short distance the ship must attain a speed which will carry it clear of the way-ends. A side-launching of a large ship is a spectacular sight (Fig. 7).

Very large ships are often built in graving docks, usually built for this purpose rather than for regular drydocking. The ship is launched simply by flooding the dock, removing the gate, and towing the ship to the outfitting pier (Fig. 8).

End launching. In preparing for an end launching, the first step is to haul the ground ways under the ship and set them to the required height and slope. Unless a semisubmerged shipway is used, the outboard or underwater part of the ground ways must be set by divers. A layer of launching grease, consisting of a thick base coat and a thin slip coat, separates the ground ways from the sliding ways, which are placed next. The grease is kept from being squeezed out before the launching by grease irons, about ½ in. thick, laid between the ground ways and the sliding ways and not removed until just before the launching. The area of the sliding ways is arranged to give a pressure of about 2 tons/ft^2 on the grease. Next comes a layer of wedges, which are not set up hard until just before the launching. A flat layer of timber is fitted above the wedges, and the space between this timber and the ship is fitted with packing.

As the ship enters the water there comes a time when the buoyancy of the stern will cause the stern to lift and the entire ship to pivot about the forward end of the ways. This throws a large pivoting pressure onto the forward end of the ways, which is taken by a special construction called a fore poppet.

The launching procedure consists of withdrawing the grease irons a few hours before the launch, setting up hard on the wedges, knocking out all shores and cribbing, and removing the keel blocks. At this point the ship is resting on the grease and is held only by triggers or some other restraining device. Then, with a signal to the sponsor who is waiting to christen the ship the moment it starts, the triggers or other devices are released and the ship slides down the ways.

If the water into which the ship is launched is limited in extent, the ship's speed must be checked promptly after she leaves the ways. The usual method of doing this is to arrange heavy piles of chain connected to the vessel by wire rope so that they will begin to be dragged by the ship as soon as she is water-borne.

If there is unlimited water the ship is simply let run until it is taken in tow by tugs.

Side launching. In a side launching the initial procedure is much the same as in an end launching. The principal difference in procedure is in the method of restraint and release. There is a trigger (dog-shore, or spur-shore) at each of the many launching-ways, and these must be released simultaneously. One of the simplest and most successful ways of doing this is to arrange a series of guillotines, which are released electrically by pressing a single button, and each of which cuts a manila line and thus releases its trigger.

Because of the great resistance to motion sideways through the water, there is no snubbing problem in a side launching; on the contrary, the problem is to make sure that the vessel clears the way-ends. The vessel also takes a large angle of list away from the shipway as it leaves the ways, and the stability must be such as to prevent capsizing.

Floating out. If a ship is launched by simple flotation, most of the launching problems disappear. All that is necessary is to determine that the vessel has sufficient stability at all phases of the operation, and that when the bow lifts (it usually lifts first) the resulting concentration of pressure at the stern is adequately provided for.

Outfitting. When the ship moves to the outfitting piers, the outfitting period starts in earnest. Many trades must work simultaneously: joiner work, deck covering, pipefitting, machinery installation, heating, ventilation and air-conditioning, piping, wiring, painting, and the installation of all kinds of equipment. The most careful planning of sequence of operations is required to ensure, for instance, that piping systems are completed before installing the joiner ceiling that covers them up.

As a result of the fact that the ship must be complete and self-contained in all respects, the amount of work in any one trade on a large passenger ship is enormous. The electrical system, for example, includes a generating plant in the engine room, power distribution of all electrically driven equipment, lighting all over the ship, an interior communication system, often a dial telephone system, and ship-to-ship and ship-to-shore communication.

During the outfitting period the installation of the propelling machinery is completed and thoroughly tested at the pier by a dock trial, in which the ship is held fast by special mooring lines and the machinery is run up to full torque (full twisting force in the shaft). Full power cannot be obtained on a dock trial because the designed torque will not turn the propellers at the designed revolutions per minute with the ship tied to the pier. The anchor windlass and the steering gear are installed and tested, and the cargo handling gear is tested to the satisfaction not only of the builder but also of regulatory authorities.

When the joiner work is finished, the deck covering is laid, and the painters are through, the ship is ready for her sea trials and delivery to her owners. [J.P.C.]

Bibliography: D. Arnott (ed.), *The Design and Construction of Steel Merchant Ships*, 1955; H. F. Garyantes, *Handbook for Shipwrights*, 1944; J. M. McNeill, Launch of the Queen Mary, *Trans. Inst. Naval Architects*, vol. 77, 1935; H. E. Rossell and L. B. Chapman (eds.), *Principles of Naval Architecture*, vol. 1, 1955.

Shipworm

Any of 12 or more species of the family Teredidae, class Pelecypoda, phylum Mollusca.

Best known is the common shipworm, *Teredo navalis*, which is virtually worldwide in distribution. There are two other genera, *Bankia* and *Xylophaga. Bankia gouldi*, the estuarine shipworm, is common along the Atlantic coast in waters of reduced salinity, such as Chesapeake Bay, but gives way to *Teredo* in fully saline waters. There appears to be little difference in the structure or life histories of the different species. Most published information deals with *Teredo navalis*. Shipworms are highly specialized mollusks having slender bodies with a small shell anteriorly located.

Prior to the use of steel hulls for ships and the development of antifouling paints and creosote treatment of wood, the destruction done to ships and wharves by the shipworm was colossal. Even now, with the life history understood and adequate treatment available, damage to waterfront timbers is said to exceed $50,000,000 each year in the United States alone.

The fundamental structure and life history of the shipworm is similar to that of the oyster and marine clam. Eggs are shed freely into the water, where a veliger larva develops quickly into a minute bivalve larva, almost identical with the young oyster. This larva appears to have a chemical affinity for wood, and will leave other objects if it happens to settle on them. The larva enters wood at right angles to the grain, but turns and bores with the grain except when encountering a knot or the burrow of another shipworm.

Boring is accomplished by turning the body to the right and left in 180° arcs, cutting being done by the two small shells. Borings are swallowed and pass through the intestine. There is some indication that the cellulose in the borings is partially utilized as food, but this has been disputed. The small opening made by the larva upon entering the piece of wood is not enlarged, so that the growing shipworm is trapped for life within its burrow. Shipworms in temperate waters attain a length of a little less than 1 ft. Small excurrent and incurrent siphons protrude from the original opening. The siphons are supported and protected by a pair of shell-like flaps, called pallets, which can be closed. The burrow is lined with a limy shell as the animal grows.

Food consists primarily of microorganisms drawn in by the incurrent siphons. Sexes are separate. Young males later turn into females. *See* PELECYPODA. [J.D.B.]

Shock absorber

Effectively a rubber spring which, together with the mass of the mechanism it supports, has a natural frequency that is sufficiently different from the forcing frequency of the system to bring about a minimum amplitude of vibration and have suffi-

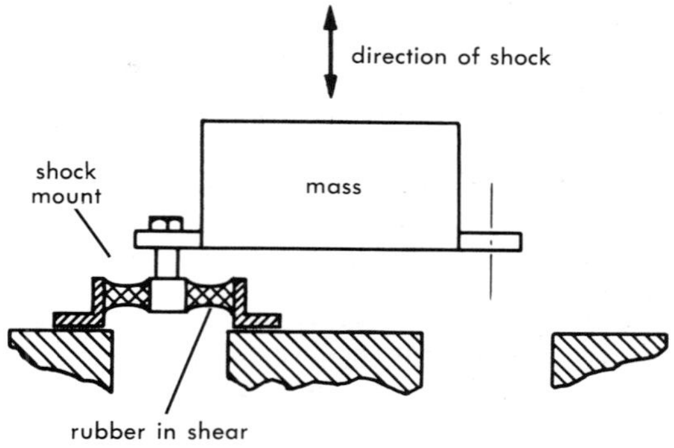

Shock absorbing mount.

cient static load-carrying capacity to support the load, as illustrated. The primary function of a shock absorber is to prevent impact or instantaneous loads from being transmitted from the base or frame of a machine to a delicate mechanism mounted on such a base. The rubber is usually stressed in shear for optimum performance and life. *See* SPRING (MECHANICAL). [L.S.L.]

Shock absorber cord

A spring support that is stretched along its axis. The cord is usually made of rubber or synthetic elastomer, the essential characteristic being a high ratio of allowable stress to modulus of elasticity. An application was in airplane shock absorbers where a multistrand cord is stretched in tension. Nylon and similar synthetic fibers have appreciable elasticity so that they stretch elastically under sudden shock loading, whereas a rope of inelastic fiber might break. Some elasticity is desirable in safety ropes as used by rock climbers so that, in an emergency, the rope acts as a shock absorber. [F.H.R.]

Shock isolation

The application of isolators to alleviate the effects of shock on a mechanical device or system. Although the term shock has no universally accepted definition in engineering, it generally denotes suddenness, either in the application of a force or in the inception of a motion. *See* SHOCK WAVE.

Shock isolation is accomplished by storing energy in a resilient medium (isolator, cushion, and so on) and releasing it at a slower rate. The effectiveness of an isolator depends upon the duration of the shock impact. An isolator may be effective in one case where there is a high *G* loading with a short duration, 0.001 millisecond (msec) or less, but may magnify the shock where there is a lower G loading but a longer duration (0.001–0.015 msec). The quantity G is equal to the so-called limit acceleration a divided by the acceleration of gravity g and is discussed later. Most shock isolators, also known as shock mounts or shock absorbers, that are available commercially are effective for the 0.001-msec or less interval.

Rubber is the most common material used in commercial shock isolators. Rubber isolators are generally used where the shock forces are created through small displacements. For larger displacement shock forces, such as those experienced by shipping containers in rough handling conditions, thick cushions of felt, rubberized hair, sponge rubber, cork, or foam plastics are used. Shock isolation systems which use the various cushion materials are generally custom designed to the particular application and cannot be considered from the standpoint of standardized isolators, but rather from that of the basic principles involved.

Absorption of shock. The shock load must be divided between the case, the shock cushion, and the equipment. The case, since it must withstand effects of rough handling such as sliding and dropping, is by necessity rigid. The more rigid the case the closer to a 1:1 ratio will be the transfer of the shock from outside to inside. The absorption of the shock is primarily between the cushion and the equipment. To dissipate the energy of a 1-ft drop with a cushion having a linear spring rate would require a thickness of 2 ft of cushion. Since cushions of such thickness are not feasible, the equipment itself must withstand part of the shock. The cushion that is needed to dissipate the energy from various heights of drop with the equipment sharing part of the load may be determined by consideration of the principles involved.

Limit acceleration. When a body moving with velocity v has to be stopped to complete rest, a deceleration (negative acceleration) must be applied. In order to make the stopping process smooth, a maximum value for the acceleration a is prescribed as a limit. Usually the full amount of the limit acceleration cannot be attained for the entire duration of the stopping process. However, an ideal process can be imagined with constant limit acceleration a. Deviation from this ideal case will be considered later.

Such uniformly decelerated motion is exactly the reverse process of uniformly accelerated motion. The same formulas apply

$$S = \frac{v^2}{2a}$$

where S is the total distance traveled, and v is the initial and final velocity. When the velocity to be stopped is produced by a free fall, the necessary height of drop H must be

$$H = \frac{v^2}{2g}$$

where g is the gravitational acceleration. From these formulas

$$v^2 = 2aS = 2gH$$

Since the velocity at the end of the free fall is the same as at the beginning of the stopping process,

$$2aS = 2gH$$

$$S = H\frac{g}{a}$$

It is customary to give the limit acceleration in the form of a G value, defined by $G = a/g$. With this notation, the stopping distance can be written as

$$S = \frac{H}{G} \tag{1}$$

Force-distance diagrams. It is useful to analyze the ideal case again from another aspect. The energy E stored in the falling body after fall from height H is $E = WH$, where W is the weight of the moving body. The same amount of energy must be consumed during the stopping period. Therefore,

$$E = GWS$$

The active force is GW in this stopping phase. Equating the last two equations for the energy yields Eq. (1). As represented in Fig. 1, the energy can be visualized as the area under the curve in a force-distance diagram. The curve is a horizontal line in the case of constant deceleration.

Whatever device is used for checking the velocity, it is hardly possible to obtain a constant deceleration of exactly the limit value. Therefore, allowances have to be made for practical considerations.

If the opposing force is provided by an ordinary spring, no constant force is produced. The force F is built up gradually with the compression of the spring, as shown in Fig. 2. The area under the force-displacement curve is in this case a triangle, representing the energy

$$E = \tfrac{1}{2}GWS_1$$

The kinetic energy from the fall is the same as before, $E = WH$. By equating the last two expressions for the energy, the stopping distance for the spring of Fig. 2 is

$$S_1 = \frac{2H}{G} \tag{2}$$

This means a distance twice as long as that for the ideal case of Fig. 1 is needed. This less favorable condition is caused by the fact that the permitted maximum force is used only at the end of the process. The diagram of Fig. 2 is not "filled up" completely; it is only half-filled.

Preloaded springs have a better-filled diagram, as shown in Fig. 3. An inherent disadvantage of the better-filled-up spring diagrams is the steep slope at the beginning, which means that the spring system must be designed for a predetermined

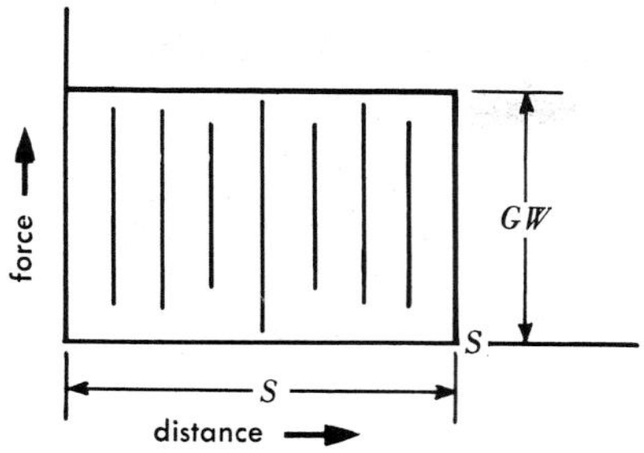

Fig. 1. Force-distance diagram for ideal case.

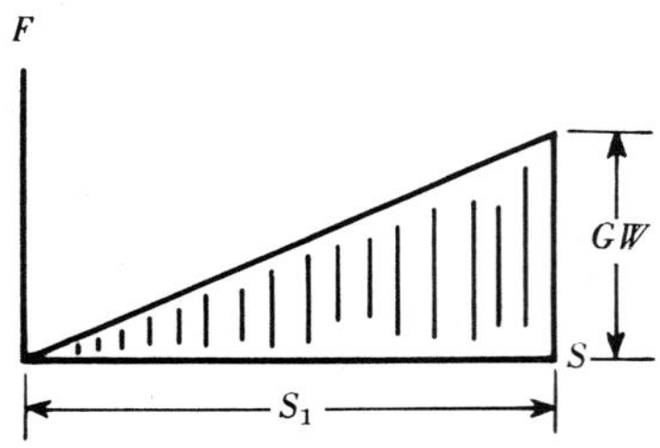

Fig. 2. Force-distance diagram for ordinary spring.

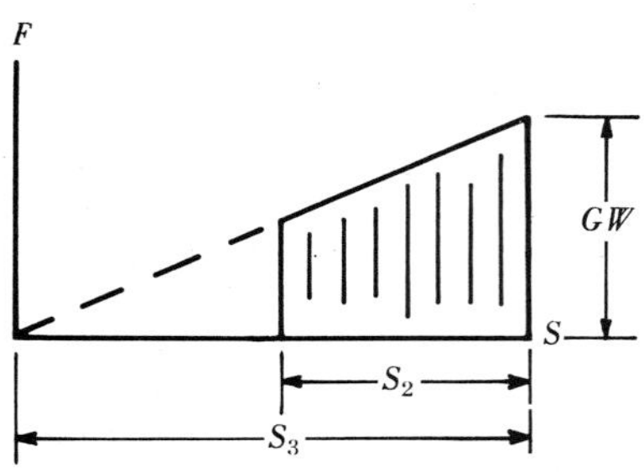

Fig. 3. Force-distance diagram for preloaded spring.

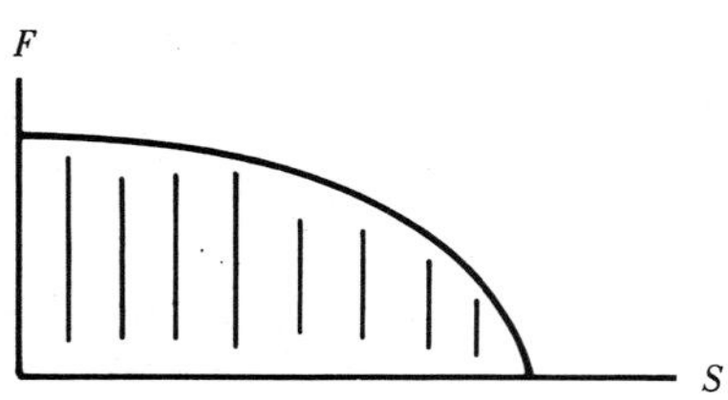

Fig. 4. Force-distance diagram for pure viscous damping.

shock. This design does not help in isolating small shocks. It has not been proven which shocks do the most damage—the large shock that happens occasionally or the small, repeated shock that occurs almost continuously. From the isolation standpoint, the spring characteristic of an isolator should start up with a moderate slope and continue with gradually sharper increase of force. This is why a system representing a completely filled diagram cannot be applied, and therefore a certain deviation from the ideal condition represented by Eq. (1) is unavoidable.

Damping forces are helpful for attaining longer deceleration force at the beginning of the stopping process. Pure viscous damping alone would result in an elliptical diagram like the one shown in Fig. 4.

Since the high velocity at the beginning produces large opposing forces, spring force and damping effects can be combined to make a well-filled diagram, as shown in Fig. 5.

Practical shock absorbers usually fall between the two cases of completely filled diagrams and half-filled diagrams. The stopping distance provided therefore lies between

$$S = \frac{H}{G} \quad \text{Eq. (1)} \quad \text{and} \quad S_1 = \frac{2H}{G} \quad \text{Eq. (2)}$$

The value given by Eq. 2 is considered as conservative and therefore is recommended, since a certain tolerance is necessary.

Using Eq. 2 and assuming that the equipment can withstand $30G$, the operating deflection of the cushion under a 24-in. drop must be

$$S = \frac{2 \times 24}{30} = 1.6 \text{ in.}$$

Since S is the operating distance, the total thickness of cushion will depend on the compression ratio of the material.

The preceding equation will provide the operating thickness needed to bring a given mass to rest with a linear spring cushion, leaving the equipment to share $30G$ of the impact load. It must be remembered that cushions, springs, and so forth store energy and, depending upon their inherent friction, will return this energy. Thus, in an ideal spring system, no energy is absorbed. However, by adding

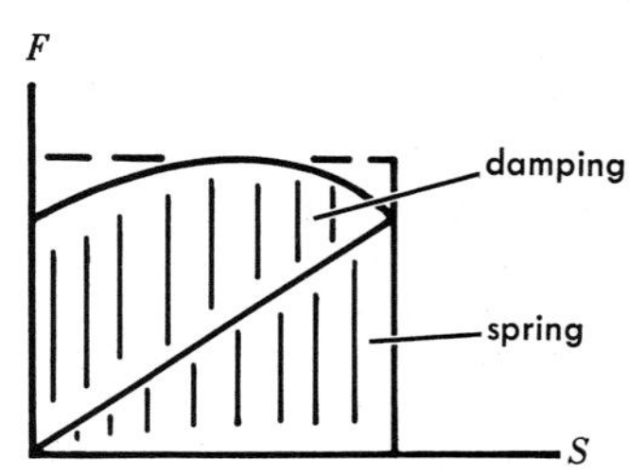

Fig. 5. Force-distance diagram for combined effects of spring force and damping.

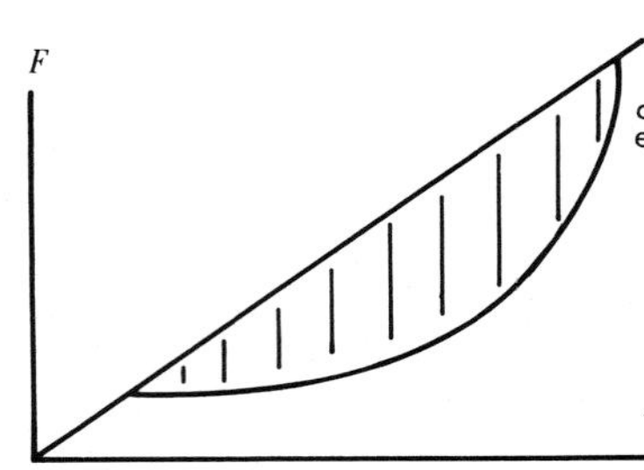

Fig. 6. Absorbed energy for ideal spring system with damping.

damping or by using a material with inherent damping, there will be provided a resilient medium that possesses characteristics as shown in Fig. 6. The shaded area under the curve is the absorbed energy.

To accomplish such a system requires engineering information on the cushioning materials under dynamic conditions. Once their behavior is known, the materials that have the desired force-displacement characteristics can be selected, and the damping forces may be added to provide a shock-absorbing system approaching the effectiveness of the one shown in Fig. 6. *See* DAMPING; RUBBER; SPRING (MECHANICAL). [K.W.J.]

Bibliography: C. E. Crede, *Vibration and Shock Isolation*, 1951; C. E. Crede and C. M. Harris (eds.), *Handbook of Shock and Vibration Control*, 1960.

Shock syndrome

A state of collapse marked by failure of the peripheral circulation, with reduction of effective blood volume and blood pressure. *See* CARDIOVASCULAR SYSTEM.

There are several varieties of shock, each seen in a fairly specific emergency; each has its own characteristics as well as the general pattern of the shock reaction. The term is also used, sometimes inaccurately, to denote other superficially similar states, such as emotional, insulin, and electrical shock.

Pallor, apprehension, cold skin, sweating, low blood pressure, and rapid, shallow pulse and breathing are typical findings. Any form of severe stress may induce this state; the most common causes are hemorrhage, physical injury with severe pain, and burns.

Previously used terms such as primary and secondary shock appear to be outmoded. The former referred to a syndrome which could best be exemplified by mere fainting, from whatever cause. The latter, secondary shock, corresponds more closely to the current use of the expression shock syndrome. This implies a progressive deterioration which may be fatal if not corrected and which will cause actual morphologic changes in certain organs and tissues that are especially vulnerable to decreased blood volume or pressure.

Blood loss, plasma loss (as in burns), and water and electrolyte loss (for example, from excessive vomiting) are common causes of the decrease of blood volume. Peripheral inadequacy of blood volume, with decreased pressures, commonly results from heart failure, severe toxic states, obstruction of major vessels, and from poisoning by certain substances such as drugs that cause dilation of blood vessels. [E.G.ST.]

Shock wave

A fully developed compression wave of large amplitude. Shock waves arise from sharp and violent disturbances generated from a lightning stroke, bomb blast, or other form of intense explosion, and from steady supersonic flow over bodies. The abrupt nature of a shock wave can best be visualized from a schlieren photograph or shadowgraph of supersonic flow over objects (*see* SCHLIEREN PHOTOGRAPHY; SHADOWGRAPH OF FLUID FLOW). Such photographs show well-defined surfaces in the flow field across which the density changes rapidly, in contrast to waves within the range of linear dynamic behavior of the fluid (*see* WAVE MOTION IN FLUIDS). Measurements of fluid density, pressure, and temperature across the surfaces show that these quantities always increase along the direction of flow, and that the rates of change are usually so rapid as to be beyond the spatial resolution of most instruments. These surfaces of abrupt change in fluid properties are called shock waves or shock fronts.

Shock waves in supersonic flow may be classified as normal or oblique according to whether the orientation of the surface of abrupt change is perpendicular or at any angle to the direction of flow. A schlieren photograph of a supersonic flow over a blunt object is shown in Fig. 1. Although this photograph was obtained from a supersonic flow over a stationary model in a shock tube, the general shape of the shock wave around the object is quite typical of those observed in a supersonic wind tunnel, or of similar objects (or projectiles) flying at supersonic speeds in a stationary atmosphere. The shock wave in this case assumes an approximately parabolic shape and is clearly detached from the blunt object. The central part of the wave, just in

Fig. 1. Schlieren photograph of a supersonic flow over a blunt object. (*Avco Research Laboratory*)

front of the object, may be considered an approximate model of the normal shock; while the outer part of the wave is an oblique shock wave of gradually changing obliqueness and strength. For a discussion of experimental techniques for supersonic flow, *see* BALLISTIC RANGE; WIND TUNNEL.

Normal shock wave. The changes in thermodynamic variables and flow velocity across the shock wave are governed by the laws of conservation of mass, momentum, energy, and also by the equation of state of the fluid. Thus, for the case of normal shock, the sketch in Fig. 2 illustrates a steady flow across a stationary wavefront. The mass flow and momentum equations are the same as for an acoustic wave. However, in a shock wave, changes in pressure and density across the wavefront can no longer be considered small. As a consequence, the velocity of propagation of the shock wave relative to the undisturbed fluid is

$$u_1^2 = \frac{\rho_2 (p_2 - p_1)}{\rho_1 (\rho_2 - \rho_1)} \tag{1}$$

In addition, conservation of thermal and kinetic energy across the shock front requires that,

$$h_1 + \tfrac{1}{2} u_1^2 = h_2 + \tfrac{1}{2} u_2^2 \tag{2}$$

where h is the specific enthalpy (or total heat per unit mass) of the fluid. By eliminating u_2 and u_1 with the aid of Eq. (1) and the conservation of mass, the energy equation becomes

$$h_2 - h_1 = \frac{1}{2}\left(\frac{1}{\rho_1} + \frac{1}{\rho_2}\right)(p_2 - p_1) \tag{3}$$

If the thermodynamic properties of the fluid are known, specific enthalpy h can be expressed as a function of pressure and density, or any other pair of thermodynamic variables. Equations (1) and (3), together with the appropriate equation of state of the fluid, are known as the Rankine-Hugoniot equations for normal shock waves. From this set of equations, all thermodynamic variables behind the shock front (denoted by subscript 2) can be expressed as functions of the propagation velocity of the shock wave and the known initial state of the fluid (denoted by subscript 1). For example, if the fluid is a perfect gas of constant specific heats, enthalpy h can be written as

$$h = C_pT = \frac{\gamma}{\gamma - 1}\frac{p}{\rho} = \frac{a^2}{\gamma - 1} \tag{4}$$

where γ is the ratio of specific heats, and a is the adiabatic speed of sound given by $(\gamma RT)^{1/2}$. For this case, the pressure and density ratios across the shock front are given by

$$\frac{p_2}{p_1} = \frac{2\gamma M_1^2 - (\gamma - 1)}{\gamma + 1} \tag{5}$$

$$\frac{\rho_2}{\rho_1} = \frac{u_1}{u_2} = \frac{(\gamma + 1)M_1^2}{2 + (\gamma - 1)M_1^2} \tag{6}$$

The temperature ratio, deduced from the perfect gas law, is

$$\frac{T_2}{T_1} = \frac{[2\gamma M_1^2 - (\gamma - 1)][2 + (\gamma - 1)M_1^2]}{(\gamma + 1)^2 M_1^2} \tag{7}$$

These expressions show that the ratios of all thermodynamic variables and flow velocities across the shock depend on only one parameter for a given gas, which is the Mach number M_1 of the flow relative to the shock front, where $M_1 = u_1/a_1$, or what amounts to the same thing, the velocity of the shock wave divided by the speed of sound for the gas into which the shock propagates. Because of this, the magnitude of M_1 is often used as a measure of the strength of the shock. For comparison with the amplitude of acoustic waves, sound waves correspond to values of $M_1 \cong 1.001$ or less.

The results of Eqs. (5), (6), and (7) have been derived for gases of constant specific heats. From molecular and atomic physics, it is well known that when a gas is heated to high temperatures, vibrational excitation, dissociation, and ionization take place, with accompanying changes in heat capacities of the gas. Therefore, for strong shock waves, the appropriate expression for the specific enthalpy h, and the equation of state, which takes into account these phenomena, must be used in place of Eq. (4) to obtain the shock wave solution from Eqs. (1) and (3). The ratios p_2/p_1, ρ_2/ρ_1 and T_2/T_1 for normal shock waves in air at standard atmospheric density are plotted in Figs. 2, 3, and 4. The approximate solutions, as given by Eqs. (6) and (7), hold only for the weaker shock waves ($M_1 < 6$), even though the pressure ratio is relatively insensitive to the changes in heat capacities of the gas.

Because the Rankine-Hugoniot equations did not impose any limit on the value of M_1, there remains the question of whether a shock wave can propagate into an undisturbed gas at a speed some-

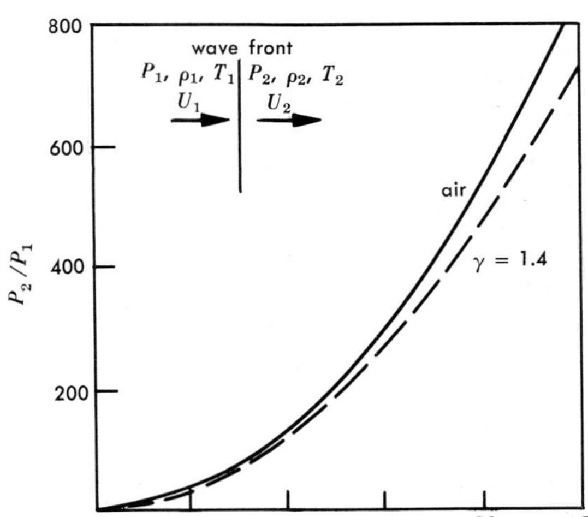

Fig. 2. Pressure ratio across normal shock wave in air at standard atmospheric density.

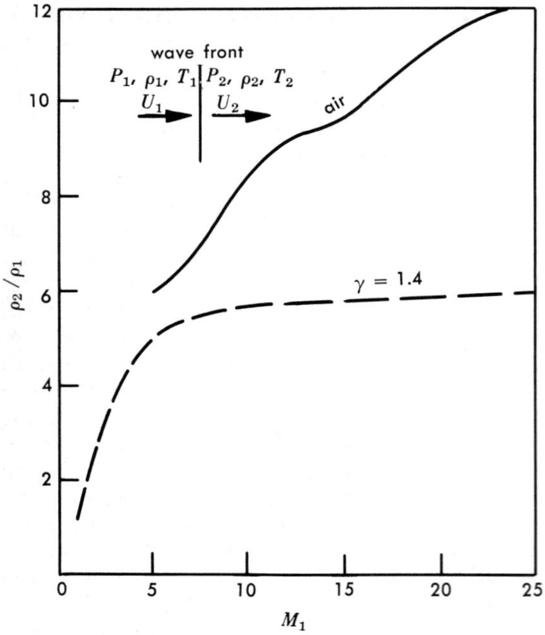

Fig. 3. Density ratio across normal shock wave in air at standard atmospheric density.

what lower than the speed of sound of this gas, as would be the case for $M_1 < 1$. Although this question cannot be answered by the First Law of Thermodynamics, an examination of the change in specific entropy across the shock front provides an answer. Thus

$$\Delta S_{12} = S_2 - S_1 = \int_1^2 \frac{dE + p\,dv}{T} \tag{8}$$

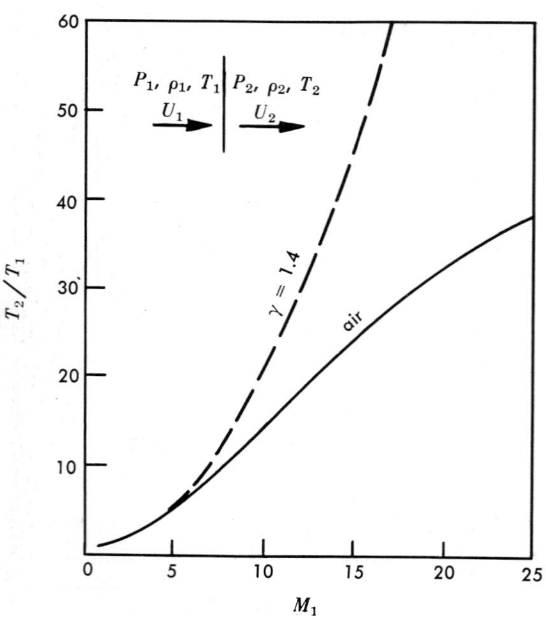

Fig. 4. Temperature ratio across normal shock wave in air at standard atmospheric density.

which, for gases of constant specific heats, becomes

$$\frac{\Delta S_{12}}{R} = \frac{1}{\gamma - 1} \ln \frac{[2\gamma M_1^2 - (\gamma - 1)]\,[2 + (\gamma - 1)M_1^2]^\gamma}{(\gamma + 1)^{\gamma+1} M_1^{2\gamma}} \tag{9}$$

showing that entropy change ΔS_{12} assumes a negative value when M_1 is noticeably less than unity. This violates the Second Law of Thermodynamics which states that the entropy accompanying any naturally occurring processes always tends to increase. Therefore, one may conclude that shock waves always travel at supersonic speeds relative to the fluids into which they propagate.

Oblique shock wave. The changes in flow variables across an oblique shock wave are also governed by the laws of conservation of mass, momentum, and energy in a coordinate system which is stationary with respect to the shock front. In this case, the problem is slightly complicated by the fact that the flow velocity will experience a sudden change of direction as well as magnitude in crossing the shock front. Thus, if β_1 and β_2 denote the acute angles between the initial and final flow velocity vectors and the shock surface (Fig. 5), then in crossing the oblique shock, the flow will be deflected by a finite amount $\theta = \beta_1 - \beta_2$.

The oblique shock solution can be obtained directly from the complete set of conservation equations. However, the solution already obtained for normal shock waves provides the following simplifying information.

The rate of mass flow per unit area across the shock wave is determined by the normal component of the flow velocity (Fig. 5). Thus, for conservation of mass across the shock,

$$\rho_1 u_1 \sin \beta_1 = \rho_2 u_2 \sin \beta_2 \tag{10}$$

On the other hand, conservation of the parallel component of momentum across the shock front requires that

$$\rho_1 u_1^2 \sin \beta_1 \cos \beta_1 = \rho_2 u_2^2 \sin \beta_2 \cos \beta_2 \tag{11}$$

Equations (10) and (11) show that

$$u_1 \cos \beta_1 = u_2 \cos \beta_2 \tag{12}$$

which is equivalent to the statement that the tangential component of the flow velocity must remain unchanged in crossing the oblique shock wave. Therefore, the resultant flow across the oblique shock shown in Fig. 5 will be identical to what an observer would see if he moved at a uniform velocity $u_1 \cos \beta_1$ along the surface of a normal shock wave propagating at a velocity $u_1 \sin \beta_1$. Such a translation of the frame of reference in the direction parallel to the shock front should not change the strength of the shock wave; thus, the changes in thermodynamic variables across the shock should depend only on the velocity component normal to the shock wave. The substitution of M_1 in Eqs. (5), (6), (7), and (9) with $u_1 \sin \beta_1$ gives the corresponding expressions for oblique

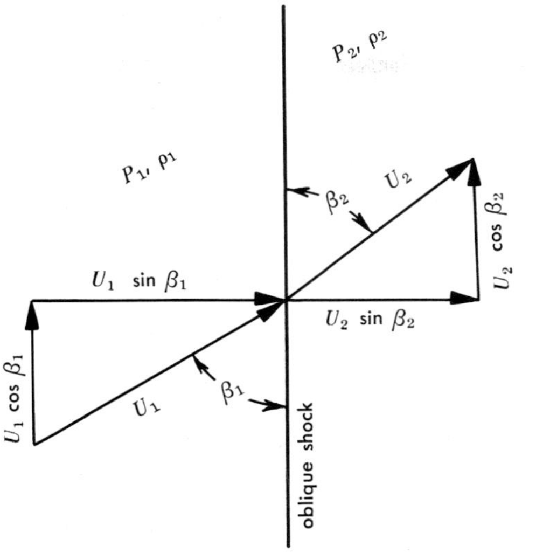

Fig. 5. Flow across an oblique shock wave.

shock waves in gases of constant specific heats. Again thermodynamic considerations show that the normal component of the flow velocity into the oblique shock wave must be at least sonic. Therefore, in a supersonic stream of Mach number $M_1 > 1$, the value of β_1 must lie within the range

$$\sin^{-1}\frac{1}{M_1} \leq \beta_1 \leq \frac{\pi}{2} \quad (13)$$

The lower limit corresponds to the Mach angle of acoustic waves, while the upper limit corresponds to that of the normal shock wave. At both limits, the flow deflection angle $\theta = \beta_1 - \beta_2$ will be zero. For any intermediate value of β_1, the value of β_2 can be obtained from Eqs. (10) and (12), thus

$$\beta_2 = \tan^{-1}\left(\frac{\rho_1}{\rho_2}\tan\beta_1\right) \quad (14)$$

The flow deflection angle θ for oblique shock waves in air at normal density is plotted in Fig. 6 as a function of M_1 and β_1. For a given Mach number M_1, the flow deflection angle first increases with the wave angle β_1, reaches a maximum value $\theta_{\max}$, and then decreases toward zero again as the wave angle approaches that of the normal shock. Conversely, for any given flow deflection angle $\theta < \theta_{\max}$ such as would be produced by sudden introduction of a wedge or an inclined plane surface into the initially uniform supersonic stream, there exist two possible values of β_1. The higher value of β_1 corresponds to the stronger shock. If the wedge angle or the inclination of the plane surface so introduced exceeds the value of $\theta_{\max}$ for the given M_1, the shock wave will either become detached from the obstructing object or will form a more complicated pattern. The question of exactly what shock-wave pattern to expect from a given situation is complicated by interaction of the resultant shock wave and flow pattern on the over-all boundary condition as well as on the local state of the flow.

Bomb blast. When energy is suddenly released into a fluid in a concentrated form, such as by a chemical or a nuclear explosion, the local temperature and pressure may rise instantly to such high values that the fluid tends to expand at supersonic speed. When this occurs, a blast wave forms and propagates the excess energy from the point of explosion to distant parts of the fluid. If the point of explosion is far away from any fluid boundary, the blast wave assumes the form of an expanding spherical shock wave followed by a radially expanding fluid originating from the point of detonation. The changes in thermodynamic variables across the spherical shock are the same as those for a normal shock propagating at the same instantaneous velocity. However, because of the continuous expansion and the finite amount of energy available from the explosion, both the strength of the shock and the specific energy of the expanding fluid must decay with time. The decay of a blast wave goes through three principal stages. A strong shock period begins immediately after the formation of the blast wave, during which the shock strength decays rapidly with distance from the point of detonation. During this stage, the shock velocity decays with the inverse 3/2 power of the distance, and the over-pressure behind the shock decays with the inverse cube of the distance. The second stage is a transition period, during which the strong spherical shock gradually changes into an acoustic wave. During the last stage or residual acoustic decay period, the acoustic wave carries the sound of explosion great dis-

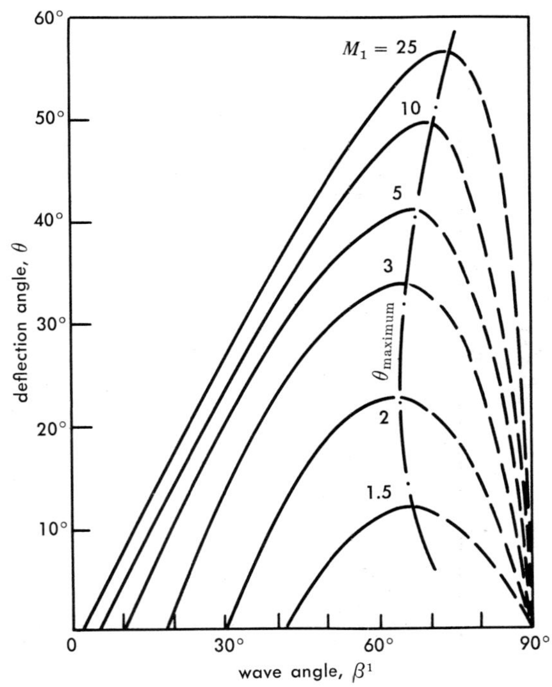

Fig. 6. Oblique shock solution for air at standard atmospheric density.

tances from the point of detonation. As characteristic of sound propagation in three dimensions, the over-pressure carried by the spherical acoustic wave decays inversely with distance, and the velocity of propagation remains constant. [S.C.L.]

Bibliography: H. W. Liepmann and A. Roshko, *Elements of Gasdynamics*, 1957; G. Taylor, The formation of a blast wave by a very intense explosion, *Proc. Roy. Soc. (London)*, *A*201:175–86, 1950.

Shock-wave display

The making visible of shock waves for purposes of photography. Shock waves accompany sparks, explosions, or high-speed flows past objects in fluids. In general these waves are not directly visible, and the display of their presence requires the use of special techniques. Changes in pressure, density, and velocity of the fluid, as well as changes in its chemical composition and in the internal energy of its molecules can be used for this purpose. The method best suited to display a shock wave depends upon the strength of the shock wave, ambient conditions in the fluid, and the constituents of the fluid. *See* SHOCK WAVE.

Refraction methods. Shock wave display techniques are used most extensively in aerodynamic test facilities (*see* BALLISTIC RANGE; WIND TUNNEL). Of the techniques, the most widely used are the schlieren and shadowgraph techniques; optical interferometry is also used (*see* INTERFEROMETRY; SCHILIEREN PHOTOGRAPHY; SHADOWGRAPH OF FLUID FLOW). All three methods are based on the variation of the index of refraction with density in the medium traversed by light. In the schlieren technique, the field displayed on the final image corresponds to the first derivative of the density normal to the schlieren knife orientation. In the shadowgraph technique, the change of illumination depends on the second derivative of the density. These two methods have the advantage of simplicity of experimental set-up, and are usually preferred because they readily display even weak disturbances such as nearly sonic shocks.

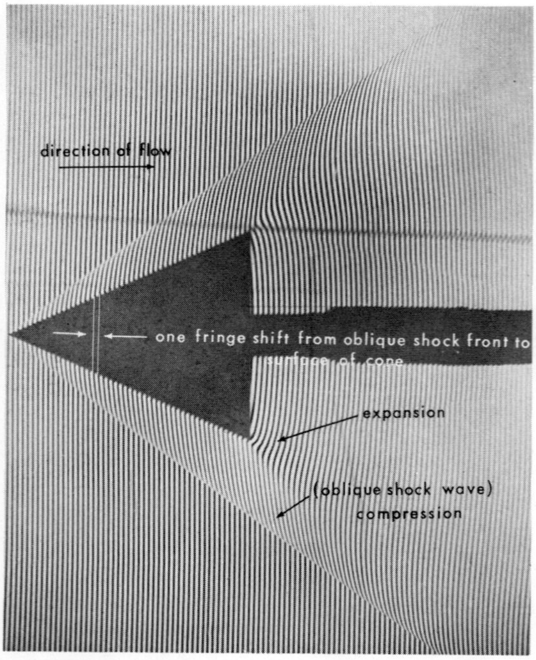

Fig. 1. Interferogram of shock wave formed around 45° cone at Mach number approximately 4. (*U.S. Naval Ordnance Laboratory*)

The interferometric technique permits a direct measurement of the density change across the shock wave. The change is presented as displacement of interference fringes (Fig. 1). The method is best suited for obtaining quantitative information on shock waves and flow fields. The limit of sensitivity of this technique is determined by the fringe displacement that can be measured accurately, usually considered to be 0.1 of their spacing. For air, a 10-cm path, and light of 5000 A wavelength, this limit corresponds to a density change of 2.2×10^{-6} g/cm^3. The type of instrument most commonly used is the Mach-Zehnder interferometer. Its design provides for a convenient adjustment of the virtual location of the fringes to the object under study, and the large separation of the two interfering light beams makes it adaptable to even large test facilities.

Absorption methods. For the visualization of shock wave phenomena in gases of low density, the above methods become too insensitive. Under such conditions, techniques utilizing the absorption of radiation or corpuscular rays provide better means for detection of shock waves. Best suited for the radiation absorption techniques are the spectral regions of strong continuous absorption, which for most gases are in the ultraviolet, or the region of soft x-rays (*see* X-RAY OPTICS). For investigations in air, the oxygen absorption continuum between 1750 and 1300 A and soft x-rays up to 13 A can be used. The x-ray technique can easily be adapted to studies in high-density fluids by utilizing x-rays of appropriately higher energy.

Corpuscular-ray absorption techniques use monoenergetic particles such as electrons, protons, or α-particles. Their attenuation results from true absorption, scattering, and slowing down. By suitable selection of the initial energy of the particles, the technique can be adjusted to flow studies over a certain range of conditions.

Most important is the method utilizing electrons with energies of 10–30 kev. The lower limit of usefulness of the absorption techniques for shock wave display is usually correlated with a 10% change in absorption. For air and a 10-cm absorbing path, a shock wave of strength 6:1 can be recognized by a 10% change in absorption if the density ahead of the shock wave is about 2.5×10^{-6} g/cm^3 for 13 A x-rays, 2.5×10^{-7} g/cm^3 for 10-kev electrons, and 10^{-7} g/cm^3 for the absorption at 1470 A of the oxygen molecules.

Equipment needed for absorption techniques is basically simple, consisting of a radiation or particle source with associated components, and a receiver. The range of wavelengths and energies involved requires that windows, if needed, and optical components be made of low absorbing material, that the radiation and particle path outside the test chamber be enclosed, and that the housing be evacuated or filled with a nonabsorbing atmosphere. The receiver may be a sensitized screen, photographic film of appropriate characteristics, photoelectric receiver, ionization chamber, or Geiger counter. *See* MICRORADIOGRAPHY.

Glow methods. Two other methods for visualization of especially low density flows are the glow discharge and afterglow techniques. In these techniques the gas, prior to entering the test chamber, is subjected to a strong electrical discharge which excites it to luminescence. The light emission accompanying the discharge is used in the first method; the second method utilizes the glow persisting after the exciting discharge has been cut off. Shock waves become visible because density changes affect the luminous intensity as well as the spectral distribution of the glow (Fig. 2). The useful range of the glow methods covers a density range from 2.5×10^{-4} g/cm^3 to 2.5×10^{-7} g/cm^3. The gas is usually excited in an insulated section which may be a separate discharge tube ahead of the wind tunnel, or the wind-tunnel nozzle proper. Depending upon the mode of excitation, which may be on electrodeless discharge, or a high-frequency condensed discharge between suitably placed electrodes, and also depending upon the operating pressure level and the type of gas to be excited, the power requirements range from a fraction of 1 kilowatt to several kilowatts. The flow can be observed directly or it can be photographed.

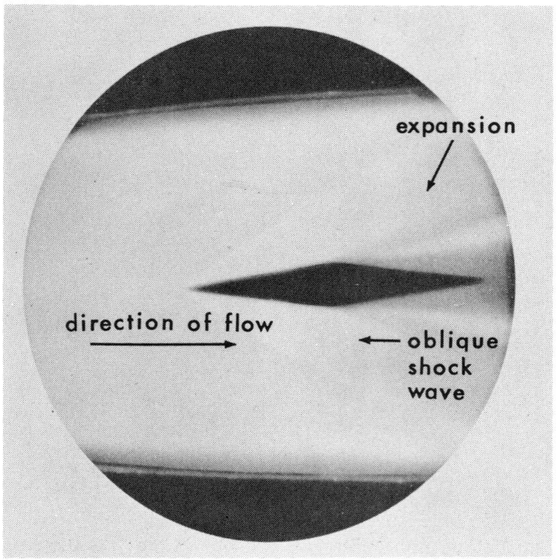

Fig. 2. Nitrogen afterglow around two-dimensional airfoil; Mach number approximately 2.6. (*Princeton University Press*)

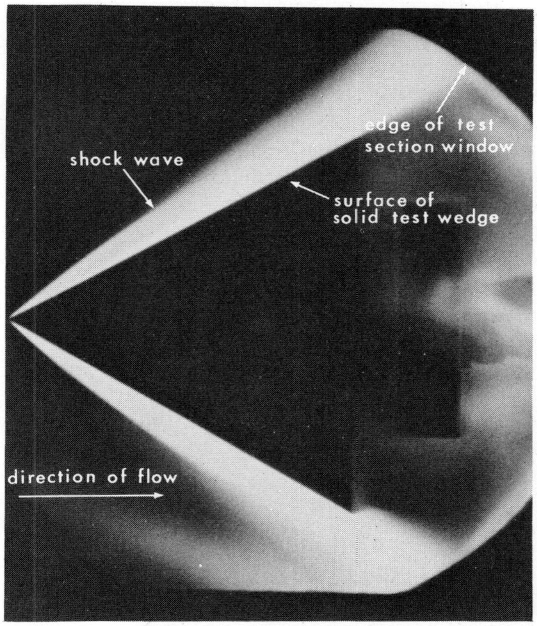

Fig. 3. Luminous flow of xenon over 60° *wedge;* Mach number approximately 12. (*U.S. Naval Ordnance Laboratory*)

At flow conditions involving sufficiently high stagnation temperatures, the gas molecules become electronically excited, and luminosity will be observed behind the shock front (Fig. 3). For air, this occurs at velocities approaching those of intercontinental ballistic missiles reentering the atmosphere. The luminous phenomenon is pronounced in the case of a nuclear bomb where a radiation front expands from the fireball. This front is characterized by a radiative process which advances faster than the pressure shock wave. This condition exists until the temperature behind the radiation front has fallen to about 300,000°K; thereafter, the velocity of the pressure shock wave exceeds that of the radiation front. The gas behind the pressure shock wave continues to remain luminous until, by further expansion, its temperature has decreased to only a few thousand degrees Kelvin; the luminous front then ceases to exist. [E.M.WR.]

Bibliography: J. V. Charyk and M. Summerfield (eds.), *High Speed Aerodynamics and Jet Propulsion*, vol. 9, 1954; H. C. Wolfe (ed.), *Temperature: Its Measurement and Control in Science and Industry*, vol. 2, 1955.

Shore processes

The geologic processes which shape, or alter, the shore features of coastlines, particularly the mechanics of waves, currents, and tides as these relate to the sediments and organisms of the continental shelf and nearshore waters. The action of waves, winds, and currents is effective in keeping sediment in motion in shallow water and along the shores of oceans, bays, and lakes. This action may erode the coast and transport unconsolidated material, usually sand, along the coast. The formation of sandy beaches along the shores is the most common manifestation of shore processes.

Erosional and depositional nearshore processes play an important role in determining the configuration of coastlines. Erosion is usually dominant off headlands and along coastal sections backed by alluvium and other unconsolidated material; whereas, deposition is most common along indentations between headlands. The over-all effect of such processes is usually a straightening and smoothing of the coastline. However, this is not always the case; differential wave erosion may cause a rapid erosion of material between headlands and thus cause irregularities in the coastline. *See* COASTAL LANDFORMS.

Whether deposition or erosion will be predominant in any particular place depends upon a number of interrelated factors: the amount of available beach sand and the location of its source; the configuration of the coastline and of the adjoining ocean floor; and the effects of wave, current, wind, and tidal action. The establishment and persistence of natural sand beaches are often the result of a delicate balance among a number of these factors, and any changes, natural or man-made, tend to upset this equilibrium.

Waves. Waves and the currents which they generate are the most important factor in the transportation and deposition of nearshore sediments. Waves are effective in moving material along the bottom and in placing it in suspension for weaker currents to transport. In the absence of beaches, the direct force of breaking waves erodes cliffs and sea walls.

Wave action along most coasts is seasonal in nature in response to the changing wind systems over the waters where the waves are generated. The height and period of the waves depend on the speed and duration of the winds generating them and the fetch, or length, over which the wind blows. Consequently, the nature and intensity of wave attack against coastlines varies considerably with the size of the water body, as well as with latitude and exposure. Waves generated by winter storms in the Southern Hemisphere of the Pacific Ocean may travel more than 5000 miles before breaking on the shores of California, where they are common summer waves for the Northern Hemisphere.

The profiles of ocean waves in deep water are long and low, approaching a sinusoidal form. As the waves enter shallow water the wave velocity and length decrease, the wave steepens, and the wave height increases, until the wave train consists of peaked crests separated by flat troughs. Near the breaker zone the process of steepening is accelerated, so that the breaking waves may attain a height several times greater than the deep-water wave. This transformation is particularly pronounced for long-period waves from a distant storm. The profiles of local storm waves and the waves generated over small water bodies such as lakes show considerable steepness even in deep water, so that the shallow-water steepening is not as pronounced as in the case of ocean swell.

The shallow-water transformation of waves commences at the depth where the waves "feel bottom." This depth is equal to one-half the deep-water wave length, where the wave length is the horizontal distance from wave crest to crest. The deep-water wave length is given by the relationship $L = gT^2/2\pi$, where g is the acceleration of gravity and T is the wave period in seconds. Upon entering shallow water, waves are also subjected to refraction, a process in which the wave crests tend to parallel the depth contours. For straight coasts with parallel contours, this decreases the angle between the approaching wave and the coast, and causes a spreading of the energy along the crests. The wave height is decreased by this process, but the effect is uniform along the coast (Fig. 1). A submarine canyon or depression causes waves to be refracted, or bent, in such a manner that waves over the canyon will diverge and decrease in height and the line of wave crests will be convex toward the shore. Waves will converge on either side of the canyon over a ridge, causing the wave height to increase and the line of wave crests to be concave toward the shore. The amount of wave refraction and consequent change in wave height and direction at any point along the coast is a function of wave period, direction of approach, and the configuration of the bottom topography.

Waves from distant storms may have periods as great as 20 seconds or more when they reach ocean coasts. Since refraction commences when waves

Fig. 1. Longshore currents are generated when waves approach the beach at an angle. In this photograph at Oceanside, California, the longshore current is flowing toward the observer. (*Photograph from Department of Engineering, University of California, Berkeley*)

reach a depth equal to one-half the wave length, these long waves will be refracted by topographic features on the ocean floor in depths as great as 300 m. Thus, the formation of beaches and the effect of waves on a coastline may be influenced by the topography of the bottom many miles from the coast. When submarine ridges cause a concentration of wave energy at certain points along a coast, severe erosion and damage to coastal structures often results. This occurs periodically along the California coast when the wave period, deep-water direction of approach, and height are such as to focus energy on coastal structures. *See* OCEAN WAVES; SEA STATE; WAVE MOTION IN LIQUIDS.

Currents in the surf zone. When waves break so that there is an angle between the crest of the breaking wave and the beach, the momentum of the breaking wave has a component along the beach in the direction of wave propagation. This results in the generation of longshore currents that flow parallel to the beach inside of the breaker zone (Fig. 2). After flowing parallel to the beach as longshore currents, the water is returned seaward along relatively narrow zones by rip currents. The net onshore transport of water by wave action in the breaker zone, the lateral transport inside of the breaker zone by longshore currents, the seaward return of the flow through the surf zone by rip currents, and the longshore movement in the expanding head of the rip current all constitute the nearshore circulation system. The pattern that results from this circulation commonly takes the form of an eddy or cell with a vertical axis. The positions of the rip currents are dependent on the submarine topography and configuration of the coast, and the height and period of the waves. Periodicity or fluctuation of current velocity and direction is a characteristic of flow in the nearshore system. This variability is primarily due to the grouping of high waves followed by low waves, a phenomenon which gives rise to a pulsation of water level in the surf zone called surf beat.

Water in the surf zone has a slow net offshore flow near the bottom between rip currents. This flow between rip currents does not appear to extend through the breaker zone; rather, water outside the surf zone moves toward the shore. There is no evidence of strong undertow in the surf zone, other than the instantaneous motion occurring as the backwash flows under a wave breaking on the face of a steep beach. The principal danger to swimmers is from rip currents, which may carry them seaward rather unexpectedly. Studies by D. L. Inman and W. H. Quinn show that longshore currents commonly flow with velocities between 15 and 75 cm/sec, although flows of 125 cm/sec have been measured.

In many cases where waves break along a straight beach with parallel bottom contours, it is possible to estimate the velocity of the longshore current from a consideration of wave and beach characteristics. However, where the beaches are not straight, or where offshore topography causes differential wave refraction, the longshore currents are dependent on the gradient of breaker height along the beach, as well as on the angle between the breaking wave and the beach. As has been shown by F. P. Shepard and D. L. Inman, for such cases, the positions of the circulation cells are largely dependent upon the location of zones of wave convergence and divergence, and quantitative prediction of longshore velocity cannot be made.

Points, breakwaters, and piers all influence the circulation pattern and alter the direction of the currents flowing along the shore. In general, these obstructions determine the position of one side of the circulation cell. In places where a relatively straight beach is terminated on the down-current side by points or other obstructions, a pronounced rip current extends seaward. During periods of large waves having strong diagonal approach, these rip currents can be traced seaward for one or more miles.

Beach types. A knowledge of types of beaches and their configuration is essential to the understanding of beach and nearshore processes. Beaches consist of transient clastic material (unconsolidated fragments) which reposes near the

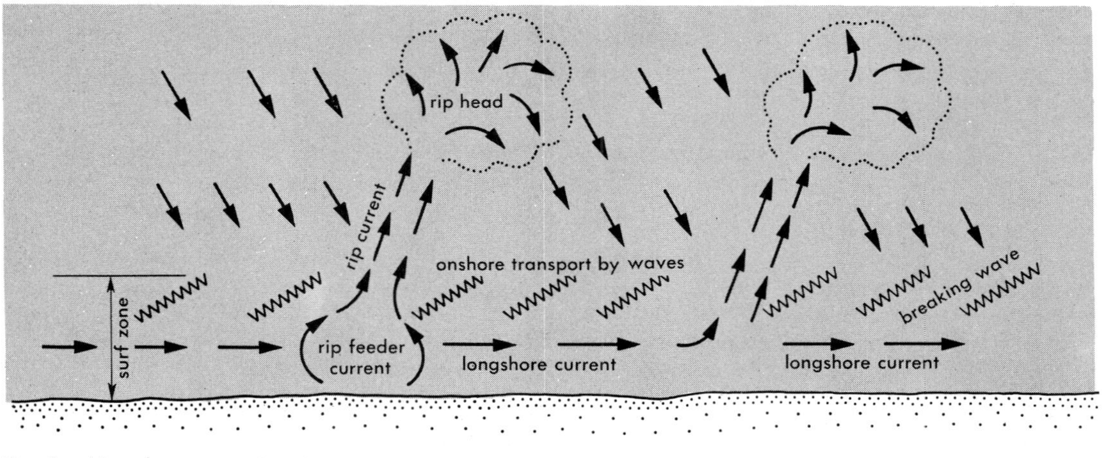

Fig. 2. Nearshore current system.

interface between the land and the sea and which is subject to wave action. The material is in dynamic repose rather than a stable deposit, and thus the width and thickness of beaches is subject to rapid fluctuations, depending upon the amount and rigor of erosion and transportation of beach material. Beaches are essentially long rivers of sand that are moved by the littoral processes, and are derived from the material eroded from the coast and brought to the sea by streams.

The geometry of beaches is also dependent on coastal history, and there is a close relationship between beach characteristics and type of coast. Long straight beaches are typical of low sandy coasts; shorter crescent-shaped beaches and small pocket beaches are more common along mountainous coastlines. The coast may be cliffed as shown in Fig. 3, or it may contain a ridge of wind-blown sand dunes and be backed by marshes and water. Along many low sandy coasts, such as the East and Gulf coasts of the United States, the beach is separated from the mainland by water or by a natural coastal canal. Such beaches are called barrier beaches. A beach that extends from land and terminates in open water is referred to as a spit, while a beach that connects an island or rock to the mainland or another island is a tombolo.

While differing in detail, beaches the world over have certain characteristic features which allow application of a general terminology to their profile (Fig. 3). The beach or shore extends landward from the lowest water line to the effective limit of attack by storm waves. The region seaward is termed the offshore; that landward, the coast. The beach includes a backshore and foreshore. The backshore is the highest portion and is only acted upon by waves during storms. The foreshore extends from the crest of the berm to the low-water mark, and is the active portion of the beach traversed by the uprush and backwash of the waves. The foreshore consists of a steep seaward dipping face related to the size of the beach material and the rigor of the uprush; and a more gentle seaward terrace, sometimes referred to as the low-tide terrace or step, over which the waves break and surge. In some localities, the foreshore face and terrace merge into one continuous curve; in others, there is a pronounced discontinuity at the toe of the beach face. The former condition is characteristic of fine sand beaches and of coasts where the wave height is equal or greater than the tidal range; the latter is typical of coastlines where the tidal range is large compared with the wave height, as along the Patagonian coast of South America and portions of the Gulf of California.

The offshore zone frequently contains one or more bars and troughs that parallel the beach; these are referred to as longshore bars and longshore troughs. Longshore bars commonly form on the bottom at the plunge point of the wave, and their position is thus influenced by the breaker height and the nature of the tidal fluctuation.

Beach cycles. Waves are effective in causing sand to be transported laterally along the beach by longshore currents and in causing movements of sand from the beach foreshore to deeper water and back again to the foreshore. These two types of transport, although interrelated, are more conveniently discussed in separate sections. The offshore and onshore transport of sand is closely related to the beach profile and to the cycles in beach width, and will be discussed here and under the mechanics of beach formation.

Along most coasts there is a seasonal migration of sand between the beaches and deeper water, in response to the changes in the character and direction of approach of the waves. In general, the beaches build seaward during the small waves of summer and are cut back by high winter storm waves. There are also shorter cycles of cut and fill associated with spring and neap tides, and with nonseasonal waves and storms. According to D. L. Inman and G. A. Rusnak, bottom surveys indicate that most offshore-onshore interchange of sand occurs in depths less than 10 m, but that some effects may extend to depths of 30 m.

Figure 3 shows the profile of a typical summer beach which has been built seaward by low waves. During stormy seasons the beach foreshore is eroded, frequently forming a beach scarp. Subsequent low waves build the beach foreshore seaward again. The beach face is a depositional feature and its highest point, the berm crest, represents the maximum height of the uprush of water on the

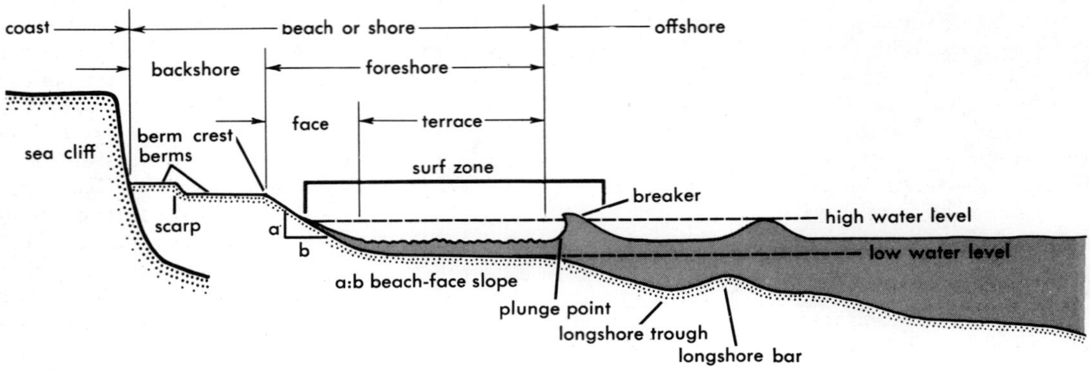

Fig. 3. Beach profile.

Fig. 4. The effect of headlands on the accretion of beach sand is shown in this photograph of Point Mugu, California. The point forms a natural obstruction which interrupts the longshore transport of sand, causing accretion and a wide beach to form (foreground). The regularly spaced scallops are beach cusps. (*Photograph from Department of Engineering, University of California, Berkeley*)

face of the beach. The height of wave uprush usually varies between one and three times the height of the breaking wave. Since the height of the berm depends on wave height, the higher berm, if it is present, is sometimes referred to as the winter or storm berm, and the lower berm as the summer berm.

The entire beach may be cut back to the country rock during severe storms. Under such conditions, the waves erode the coast and form the well-known sea cliffs and wave-cut terraces, which are frequently preserved in the geologic record and serve as markers to the past relations between the levels of the sea and the land.

A series of regularly spaced scallops, called beach cusps, sometimes forms on the beach foreshore (Fig. 4). Beach cusps consist of short transverse valleys formed in the beach face and separated by ridges with cuspate points. The spacing between cuspate points ranges from several to several hundred meters, while the depths of the valleys range from a few centimeters on fine sand beaches to several meters on pebble and cobble beaches. The formation of cusps is related to the volume and vigor of wave uprush on the beach face, and they occur with greatest frequency during neap tides when the fluctuations in water level are minimal.

Mechanics of beach formation. Wherever there are waves and an adequate supply of sand or coarser material, beaches form. Even man-made fills and structures are effectively eroded and reformed by the waves. The initial and most characteristic event in the formation of a new beach from a heterogenous sediment is the sorting out of the material, coarse material remaining on the beach and fine material being carried away. Concurrent with the sorting action, the material is rearranged, some being piled high above the water level by the uprush of the waves to form the beach berm, some carried back down the face to form the foreshore terrace. In a relatively short time, the beach assumes a profile which is in equilibrium with the forces generating it.

The mechanics of accretion and erosion of the beach face appear to be connected with the differential between the speed and amount of water in the uprush and backwash over the beach face. Percolation of the uprush into a permeable beach reduces the amount of flow in the backwash, and is thus conducive to deposition of the sand transported by the uprush. If, in addition, the beach is dry, this action is accentuated. Coarse sands are more permeable, and consequently more conducive to deposition and the formation of steep beach faces. Large waves elevate the water table in the beach. When the beach is saturated, the backwash has a higher velocity, a condition which is conducive to erosion. From the foregoing it follows that the slope of the beach face, sometimes referred to as the foreshore slope, increases both with increasing sediment size and with decreasing wave height and intensity. Some typical average values for the face slopes of ocean beaches are given below.

Size of beach sediment, mm		*Slope of beach face, degrees*
Very fine sand	1⁄16–1⁄8	1
Fine sand	1⁄8–1⁄4	3
Medium sand	1⁄4–1⁄2	5
Coarse sand	1⁄2–1	7
Very coarse sand	1–2	9
Granules	2–4	11
Pebbles	4–64	17
Cobbles	64–256	24

As mentioned in the discussion of currents in the surf zone, there is an onshore transport of water associated with wave motion near the breaker zone. This shoreward transport of water causes fluctuations in sea level and a rise in the sea surface in the surf zone. The "piling up" of water in the surf zone is compensated by a seaward return flow, part of which is in the form of rip currents. Theory and observations suggest that the onshore transport of water is a maximum at the water surface, and that in addition to rip currents, this transport is compensated by a net seaward return flow along the bottom in the surf zone. The magnitude of net offshore drift along the bottom usually does not exceed 10 cm/sec. The convergence of this net offshore drift along the bottom inside the surf zone, with the net onshore drift outside the surf zone, may account for the tendency of longshore bars to form beneath breaking waves.

The instantaneous particle motion associated with shallow-water progressive waves is oscillatory in nature, the motion under the crest being in the direction of wave propagation, while that under the wave trough is in the opposite direction. As waves approach the breaker zone, a differential develops between the magnitude of the crest and trough orbital velocities; the velocity under the crest exceeds that of the trough and becomes of shorter duration. In general, the differential between crest and trough orbital velocities increases as the wave height and frequency decrease. However, the picture of wave motion in shallow water is considerably complicated by currents and by longer-period waves or surges, such as surf beat.

The relative increase in the onshore velocity under the wave crest, as compared with the offshore velocity of the wave trough, probably accounts for the shoreward migration of sand during periods of low waves, particularly when the waves have long periods. On the other hand, a greater amount of material is maintained in suspension above the bottom when the waves are high and have short periods, and for this condition the differential between onshore and offshore orbital velocities is much less. Thus, the net offshore drift along the bottom may result in a net offshore transport of material when the waves are high and have short periods.

High short-period waves also augment beach erosion by increasing the velocity of the currents in the nearshore circulation system. Sand is transported along the beach by longshore currents and through the breaker zone by rip currents to deeper water, where it is deposited. It is probable that a significant net seaward transport of sediment by diffusion occurs between the surf zone, where the concentration of suspended sediment is high, and the offshore waters, where the concentration is relatively low.

Oscillatory ripples, which form on the sandy bottom, are an important factor in the mechanism of transportation and sorting of beach sands. Because turbulence and lift force are most intense at the crest of the ripple, only the coarsest material is deposited there. The fine material is placed in suspension and removed from the area, while the coarse material moves with the ripple toward the beach. As a consequence of the sorting action by ripples and by the uprush and backwash on the beach face, beach sands are the best sorted sedimentary deposits.

Fig. 5. Laminations in the beach face at La Jolla, California. Black laminae are heavy minerals and the light are quartz. Scale is 15 cm long. (*From D. L. Inman, Beach Erosion Board, U.S. Eng. Corps Tech. Memo. 39, 1953*)

The beach face is frequently characterized by laminations (closely spaced layers) that show slight differences in the size, shape, or gravity of the sand grains. The laminations parallel the beach face and represent differential sorting within the lamina as it is deposited by the uprush and backwash (Fig. 5).

Longshore movement of sand. The movement of sand along the shore occurs in the form of bed load (material rolled and dragged along the bottom) and suspended load (material stirred up and carried with the current). Bed load transportation results in part from the longshore component of the wave uprush on the beach face and in part from the fluid drag of the longshore current generated by the wave action in the surf zone. Suspended load transportation occurs primarily in the surf zone where the turbulence and vertical mixing of water are most effective in placing sand in suspension, and where the longshore currents which transport the sediment laden waters have the highest velocity. Concentrations of suspended sand in the surf zone, as high as 30 g/liter, have been measured along the coast of the Black Sea, suggesting that suspension is one of the most important processes in littoral drift. On the other hand, it is probable that suspension becomes less important for very coarse sand and cobble beaches.

The volume of littoral transport along oceanic coasts is usually estimated from the observed rates of erosion or accretion, most often in the vicinity of coastal engineering structures such as groins or jetties. In general, beaches build seaward upcurrent from obstructions and are eroded on the current lee, where the supply of sand is diminished. Such observations indicate that the transport rate varies from almost nothing to several million cubic meters per year, with average values commonly falling between 150,000 and 600,000 m^3/year. Along the shores of smaller bodies of water, such as the Great Lakes of the United States and the Mediterranean Sea, the littoral transport rate can be expected to vary from about 7000 to 150,000 m^3/year. In general these are conservative estimates, since the volume of material moved commonly exceeds that indicated either by deposition or erosion.

The large quantity of sand moved along the shore and the pattern of accretion and erosion that occurs when the flow is interrupted pose serious problems for the coastal engineer. The problem is particularly acute when jetties are constructed to stabilize and maintain deep navigation channels through sandy beaches. A common remedial procedure is to dredge sand periodically from the accretion on the up-current side of the obstruction and deposit it on the eroding beaches in the current lee. Recent trends include the installation of permanent sand by-passing plants which continually remove the accreting sand and transport it by hydraulic tube to the beaches in the lee of the obstruction.

Source of beach sediments. The principal sources of beach and nearshore sediments are the rivers, which bring large quantities of sand directly to the ocean; the sea cliffs of unconsolidated material, which are eroded by waves; and material of biogenous origin (shell and coral fragments and skeletons of small marine animals). Occasionally, sediment may be supplied by erosion of unconsolidated deposits in shallow water. Beach sediments on the coasts of Holland are derived in part from the shallow waters of the North Sea. Windblown sand may be a source of beach sediment, although winds are usually more effective in removing sand from beaches than in supplying it. In tropical latitudes many beaches are composed entirely of grains of calcium carbonate of biogenous origin. Generally the material consists of fragments of shells, corals, and calcareous algae growing on or near fringing reefs. The material is carried to the beach by wave action over the reef. Some beaches are composed mainly of the tests of foraminifera that live on sandy bottom offshore from the reefs.

Streams and rivers are by far the most important source of sand for beaches in temperate latitudes. Cliff erosion probably does not account for more than about 5% of the material on most beaches. Wave erosion of rocky coasts is usually slow, even where the rocks are relatively soft shales. On the other hand, retreats greater than 1 m a year are not uncommon in the unconsolidated sea cliffs. C. Lyell in 1873 showed that the Ganges and Brahmaputra Rivers carry a volume of sediment into the Bay of Bengal each year that is 780 times greater than the material eroded by wave action from the 36-mile stretch of cliffs in the vicinity of Holderness, England. The Holderness cliffs, which are on the exposed North Sea coast, are noted for their rapid rate of erosion. According to J. A. Steers, they are 40 ft high and recede at the rate of about 7 ft per year.

Surprisingly, the contribution of sand by streams in arid countries is quite high (Fig. 6). This is because arid weathering produces sand-size material and results in a minimum cover of vegetation, so that occasional flash floods may transport large volumes of sand. The maximum sediment yield occurs from drainage basins where the mean annual precipitation is about 30 cm/year as has been shown by W. B. Langbien and S. A. Schumm.

Following initial deposition at the mouths of streams entering the ocean, much of the sand-size fraction of terrestrial sediments is carried along the coast by longshore currents. The sand carried by these littoral currents may be deposited in continental embayments, or it may be diverted to deeper water by submarine canyons which traverse the continental shelf and effectively tap the supply of sand. Recent observations by H. W. Menard suggest that most of the deep sediments on the abyssal plains along a 250-mile section of the California coastline are derived from two submarine canyons, Delgada Submarine Canyon in northern California and Monterey Submarine Canyon in central California.

Fig. 6. Sand deltas, such as Rio de la Concepcion on the arid coast of the Gulf of California, are important sources of sand for beaches. (*Photograph from Scripps Institution of Oceanography, University of California, La Jolla*)

Biological effects. The rigor of wave action and the continually shifting substrate make the sand beach a unique biological environment. Because few large plants can survive, the beach is occupied largely by animals and microscopic plants. Much of the food supply for the animals consists of particulate matter that is brought to the beach by the nearshore circulation system and trapped in the sand. The beach acts as a giant sand filter that strains out particulate matter from the water that percolates through the beach face.

Since the beach-forming processes and the trapping of material by currents and sand are much the same everywhere, the animals found on sand beaches throughout the world are similar in aspects and habits, although, according to E. Dahl, different species are present in different localities. In addition, since the slopes and other physical properties of beaches are closely related to elevation, the sea animals also exhibit a marked horizontal zonation. Organisms on the active portion of the beach face tend to be of two general types, insofar as the procurement of food is concerned: those that burrow into the sand, using it for refuge while they filter particulate matter from the water through siphons or other appendages that protrude above the sandy bottom; and those that remove organic material from the surface of the sand grains by ingesting them or by "licking." There are usually few species, but those which are present may be very abundant; for example, *Thoracophelia nucronata*, a beach worm which ingests sand grains, was estimated to have a population of 100,000/m^3 in the fine sand beach at La Jolla, California, and the bean clam *Donax gouldii* has peak populations of over 25,000/m^2 on the same beach.

A black layer is frequently found at depths of 5–75 cm below the surface of the beach foreshore. Chemically this is a reducing layer which has pH values greater than 8.0. J. R. Bruce has shown that the discoloration is caused by presence of ferrous sulphide which oxidizes to a reddish-yellow on exposure to air. The formation of the layer is apparently related to the activity of bacteria on the organic material in the beach. This reducing layer is conducive to the deposition of calcium carbonate and may play an important role in cementing the beach sand and forming beach rock and nodules.

In tropical seas the entire shore may be composed of the cemented and interlocking skeletons of reef-building corals and calcareous algae. When this occurs, the nearshore current system is controlled by the configuration of the reefs that the organisms form. Where there are fringing reefs, breaking waves carry water over the edge of the reef, generating currents that flow along the shore inside of the reef, and then flow back to sea through deep channels between reefs. Under such conditions beaches are usually restricted to a berm and foreshore face bordering the shoreward edge of the reef. *See* CORAL REEF. [D.L.I.]

Bibliography: D. W. Johnson, *Shore Processes and Shoreline Development*, 1919; P. H. Kuenen, *Marine Geology*, 1950; National Research Council, *Treatise on Marine Ecology and Paleoecology*, Geol. Soc. Am. Mem. 67, vol. 1, 1957; F. P. Shepard, *Submarine Geology*, 1948; U.S. Beach Erosion Board, Corps of Engineers, *Shore Protection Planning and Design*, Tech. Rept. 4, 1957.

Short circuit

A term commonly used to describe an electrical connection of negligible impedance connected across a pair of terminals.

Common use implies an undesirable condition arising from electrical insulation failure due to improper operation, mechanical damage, or damage by natural causes such as lightning and rain. Protection against such short circuits is a major field in electrical power engineering (*see* ELECTRIC PROTECTIVE DEVICES). The location of short circuits, or faults as they are often called, is made by electric bridge measurements. *See* CIRCUIT TESTING, ELECTRICAL.

Although short circuits are undesirable on power transmission lines they are often used to advantage on high-frequency lines. For instance, a stub transmission line, one-quarter wavelength long and short-circuited at one end, acts as an insulator at the opposite end and therefore is used as a support for a high-frequency transmission line. In a similar manner shorting bars are used to tune transmission lines. *See* CIRCUIT, ELECTRIC; TRANSMISSION LINES. [R.L.R.]

Shoulder

The area of union between upper limb and trunk in tetrapods and man. The bony framework consists of an anterior clavicle (collarbone) and a posterior scapula (shoulder blade). They join laterally to form the shoulder joint, that is, the articulation with the humerus, or upper arm bone. A wide range of motion of the arm is obtained by

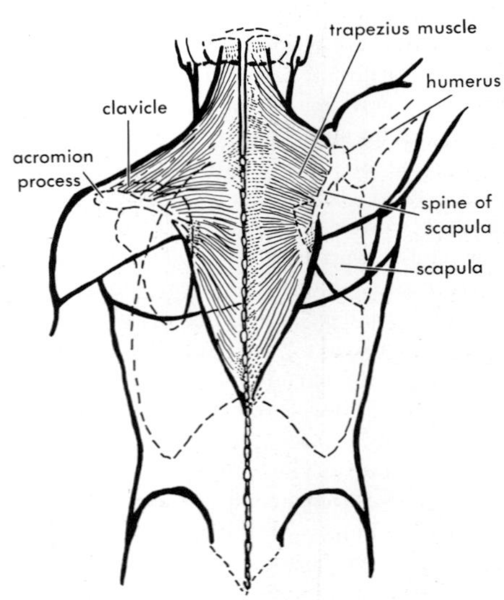

Human shoulder, back view. (*From J. M. Dunlop, Anatomical Diagrams, Macmillan, 1935*)

a ball-and-socket joint, the movement being effected by anterior, posterior, superior, and inferior muscle groups. Stability and strength of the joint, however, are poor. The axilla, or armpit, carries the major blood vessels and nerves to and from the arm and shoulder region, including the important brachial plexus. *See* BONE; SKELETAL SYSTEM. [E.G.ST.]

Shovel, power

A power shovel is a digging machine, usually self-propelled on crawler, rubber tire, or sometimes, rail mountings. It is equipped with a

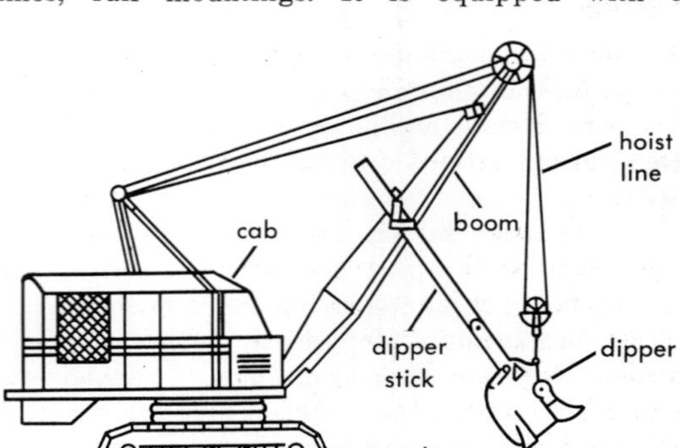

Principal parts of a crawler shovel. (*Cummins Engine Co., Ltd.*)

shovel boom, dipper attached to the front end of a dipper stick, dipper trip mechanism, padlock (dipper sheave block), crowding mechanism and cables all carried on a fully revolving superstructure. So-called commercial sizes have capacity ratings from ¼–2½ yd^3; special sizes up to 40 yd^3 have been produced. *See* BULK-HANDLING MACHINES; CONSTRUCTION EQUIPMENT. [D.O.H.]

Bibliography: R. L. Peurifoy, *Construction Planning, Equipment and Methods*, 1956; U.S. Department of Commerce, Power Cranes and Shovels, Commercial Standard CS90, 1958.

Shrew

Any of several small to very small mammals of the order Insectivora, a group which also includes the moles. Shrews occur on all continents except Australia. In the United States 24 species are recognized. These animals are generally considered to be the most primitive of living placental mammals. They have small, sharp pointed teeth, soft velvety fur, long snouts, five toes, and small eyes.

The short-tailed shrew, *Blarina brevicauda*; length 5 in. (*From E. L. Palmer, Fieldbook of Natural History*, McGraw-Hill, 1949)

Shrews feed either on the ground or in animal burrows, eating various small animals, especially insects, but sometimes killing rodents larger than themselves. The strength of the shrew is remarkable. In relation to its size, the shrew is reputed to be by far the strongest of all mammals. Shrews are shy animals and are seldom seen, even where common. *See* INSECTIVORA. [J.D.B.]

Shrike

Any member of the perching bird family Laniidae, having 67 known species distributed throughout the Northern Hemisphere, Africa, and New Guinea. Two closely related species occur in the United States.

The northern shrike, *Lanius excubitor*, and the loggerhead shrike, *L. ludovicianus*, are similar in appearance, the latter lacking the faint, wavy bars across the breast characteristic of the northern shrike. They are bluish-gray birds, white below, with black masks, black wings, and a long black tail bordered with white. The black mask and short, hooked beak distinguish them from the mockingbird, which they resemble. Shrikes are often called butcher birds from their habit of impaling

The migrant shrike, *Lanius ludovicianus*; length 9⅕ in. (*From E. L. Palmer, Fieldbook of Natural History*, McGraw-Hill, 1949)

insects, small birds, and other food animals on thorns and fence barbs. *See* MOCKINGBIRD; PASSERIFORMES. [J.D.B.]

Shrimp

Any of several marine animals belonging to the suborder Natantia, order Decapoda, class Crustacea, phylum Arthropoda.

Economic importance. Shrimp are among the most valuable marine invertebrates. They support an important fishery in the Gulf of Mexico and substantial fisheries off the Atlantic and Pacific coasts. In 1954, the United States and Alaska shrimp fishery accounted for 268,000,000 lb, worth

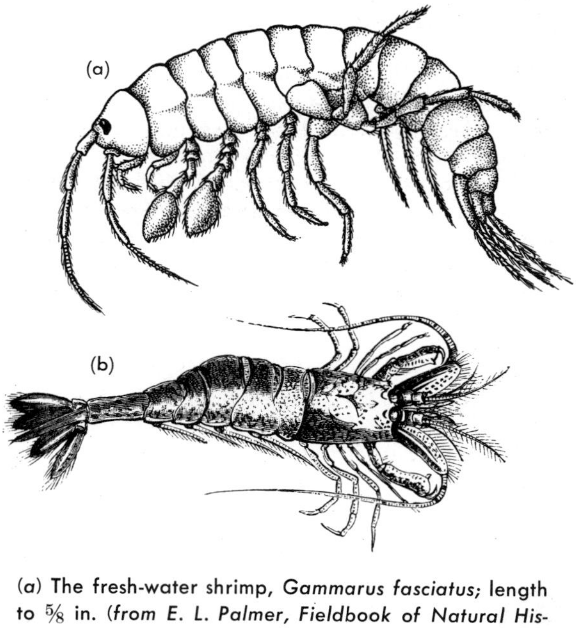

(*a*) The fresh-water shrimp, *Gammarus fasciatus*; length to ⅝ in. (*from E. L. Palmer, Fieldbook of Natural History*, McGraw-Hill, 1949). (*b*) *Crago septemspinousus*; length to 2½ in. (*from P. M. Duncan, ed., Cassell's Natural History*, Cassell).

$61,000,000. The Gulf of Mexico annually produces about 90% of the commercial catch (*see* FISHERIES CONSERVATION). Approximately 99% of the commercial harvest is taken by trawls. In addition to the direct commercial value of shrimp, they are significant as food for fishes. The fresh-water prawns, which are closely related to the shrimp, are also significant in the aquatic food chain although they are of little commercial value.

Characteristics. In the United States at least ten species of shrimp are commercially important; 4 in the Gulf of Mexico, 1 in the Atlantic, and 5 in the Pacific. Other species enter the commercial catch in lesser quantities. The common or white shrimp, *Penaeus setiferus*, is the most important and best-known species.

Shrimp are in the same order with lobsters and crayfish and differ from them primarily in having an elongated abdomen with the sixth pair of abdominal appendages (the uropods) and telson formed into an efficient swimming organ (*see* CRAYFISH; LOBSTER). The thorax is usually compressed, and the legs are long. Shrimp are quite similar to crayfish in their general anatomy, excepting the characteristics listed above. Closely related are the prawns, which occur in both fresh and salt water; in fact, the names shrimp and prawn are frequently used interchangeably. There are about 185 genera in the suborder, and several hundred species. They are world-wide in their distribution.

Reproduction and development. Like other Arthropoda, the young of shrimp pass through a series of stages, or instars, in their development. In the genera *Penaeus* and *Lucifer* the eggs hatch into a very primitive stage, called the nauplius, and gradually change to the adult form. In other genera the eggs hatch into a more advanced larval stage.

In *Penaeus setiferus* the male attaches a spermatophore (packet of sperm) to the female prior to emission of the eggs, which are fertilized as they are laid. Each female will produce about 500,000 eggs at each spawning, and probably spawns more than once each season. In contrast to most decapods (including several species of shrimp) the female does not carry the eggs but deposits them directly into the water. There are at least 10 larval stages. Most spawning takes place at sea, with the young moving into estuarine waters soon after, or during, the last larval stages, when they are about 50 millimeters long. After a period in estuarine waters, during which time the smallest shrimp are the closest to shore, they gradually move back into the open sea. Spawning appears to be continuous from March through September. Other than onshore-offshore movements related to development, the movements of the schools of shrimp appear to be wanderings in search of food instead of true migrations. Most shrimp are omnivorous, capturing such food as they can with their pinchers. The genus *Crago*, the most important Pacific Coast group, is made up of species which are primarily scavengers.

Size and distribution. Shrimp and prawns vary in size from fresh-water forms less than 1 in. long to tropical marine species 2 ft in length. The fresh-water prawns of the family Palaemonidae are widespread in the lakes and streams of the eastern United States. There are also marine species in the family. One form in the lower Mississippi Valley is about 4 in. long, large enough to be trapped by the river dwellers for food. *See* DECAPODA (CRUSTACEA). [J.D.B.]

Shrink fit

A shrink or heavy force fit has considerable negative allowance so that the diameter of a hole is less than the diameter of a shaft that is to pass through the hole. Shrink fits are used for permanent assembly of steel external members, as on locomotive wheels. The difference between a shrink fit and a force fit is in method of assembly. Locomotive tires, for instance, would be difficult to assemble by force, whereas a shaft and hub assembly would be convenient for force fit by a hydraulic press. In shrink fits, the outer member is heated, or the inner part is cooled, or both, as required. The parts are then assembled and returned to the same temperature. *See* ALLOWANCE; FORCE FIT. [P.H.B.]

Shunting

The act of connecting one device to the terminals of another so that the current is divided between the two devices in proportion to their respective admittances. Shunting is widely used in ammeters, galvanometers, and other current-measuring instruments to bypass part of the current around the instrument so as to change the measuring range. Resistors are frequently shunted across tuned circuits to broaden the tuning characteristics.

Shunting is equivalent to connecting in parallel. Shunting one resistor with another gives a lower resistance for the combination, whereas shunting one capacitor with another gives a total capacitance equal to the sum of the individual values. *See* ALTERNATING-CURRENT CIRCUIT THEORY; DIRECT-CURRENT CIRCUIT THEORY. [J.MR.]

Sideband

The frequency band located either above or below the carrier frequency within which fall the frequency components of the wave produced by the process of modulation (*see* CARRIER; MODULATION). Apart from the carrier, all components of an amplitude-modulated sinusoidal carrier, when taken together, form a pair of sidebands extending on either side of the carrier frequency in mirror symmetry and containing all the frequency components of the modulating wave. The sidebands above and below the carrier frequency are called upper sideband and lower sideband, respectively. *See* AMPLITUDE MODULATION. [H.S.BL.]

Siderite

The mineral form of ferrous carbonate, often containing appreciable amounts of magnesium and manganese substituting for iron.

Siderite may occur in sedimentary deposits or in hydrothermal veins. It may be formed by the action of iron-bearing solutions on limestones. The equilibrium replacement of calcium for iron has been determined and found to increase as a function of temperature. It is found in England, Greenland, Spain, and North Africa. In the United States it occurs in Connecticut, Pennsylvania, and the various mining districts of the Middle and Far West.

Siderite has hexagonal (rhombohedral) symmetry and the same structure as calcite. Individual crystals are often rhombohedral in shape, sometimes with curved faces. Massive varieties also occur. Siderite is often brownish and sometimes gray or greenish. The specific gravity is 3.9 and the hardness is 4. The stability of siderite is dependent on the partial pressure of oxygen. *See* CARBONATE MINERALS; IRON. [R.I.H.]

Siderocapsaceae

A family of aquatic bacteria of the order Pseudomonadales, found in iron-bearing waters. It is a heterogeneous group of gram-negative organisms, possessing in common the ability to deposit iron or manganese compounds around the cells. Cell morphology may be apparent only after treatment with dilute acid. A poorly studied group, only a few members of the family have been cultivated or obtained in pure culture. The species are organized into ten genera, seven containing encapsulated cells and three containing nonencapsulated ones. *See* PSEUDOMONADALES.

The spherical cells of *Siderocapsa* occur in masses which are surrounded by a common capsule, whereas the cells of *Siderosphaera*, though similar, always occur in pairs. The ellipsoidal cells of *Sideronema* occur in chains surrounded by a heavy gelatinous capsule, as do the rod-shaped cells of *Ferribacterium*. The coccoid to rod-shaped cells of *Sideromonas* produce well-defined capsules which may fuse to form zoogloea-like masses.

In *Naumanniella*, a marginal thickening, the torus, completely surrounds the cell, whereas in *Ochrobium* the torus remains open, being horseshoelike.

The spherical cells of *Siderococcus* measure less than a micron in diameter and lack a gelatinous capsule. The rod-shaped cells of the genera *Siderobacter* and *Ferrobacillus* also lack capsules. *Siderobacter* occurs in neutral or alkline waters, whereas *Ferrobacillus* occurs in acid mine wastes. *Ferrobacillus ferroxidans* is a strict autotroph, and may be identical with *Thiobacillus ferroxidans*. [R.S.W.]

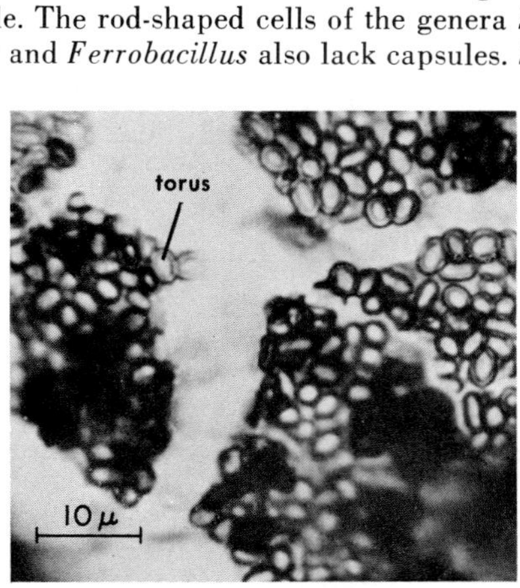

Photomicrograph of *Naumanniella* species showing the torus.

Sight

The special sense perceived and transmitted by the visual apparatus. This includes the eye, the optic tracts, and the intracerebral pathways which carry sensory impulses to both unconscious and conscious levels of perception.

Light entering the eye through its aperture, or pupil, is focused by the crystalline lens and fluids so that the rays fall on or near the retina. Light- and color-sensitive receptors, the rods and cones, are stimulated in a biochemical reaction involving rhodopsin, or visual purple, and other substances. In some manner these, in turn, cause nerve impulses to be relayed to the brain in such a way that intensity, patterns, and different wavelengths (color) of light are perceived so that the objects from which light originates or is reflected are seen by the observer. *See* EYE; VISION. [E.G.ST.]

Signal generator

A piece of electronic test equipment that delivers a sinusoidal output of accurately calibrated frequency. The frequency may be anywhere from audio to microwave, depending upon the intended use of the instrument. The frequency and the amplitude are adjustable over a wide range. The oscillator must have excellent frequency stability, and its amplitude must remain constant over the tuning range.

The Wien-bridge oscillator is commonly used for frequencies up to about 200 kc. For a radio-frequency signal generator up to about 200 Mc, a resonant circuit oscillator is used (such as a tuned-plate tuned-grid, Hartley, or Colpitts). Beyond this range vhf and microwave oscillators are used.

Many signal generators contain circuitry that allows the output to be either amplitude or frequency modulated. The most common forms of amplitude modulation are sinusoidal, square-wave, or pulse. The frequency is either kept constant, is sinusoidal-modulated, or is swept linearly across a band of frequencies. For example, for testing broadcast receivers, it is important to sweep the generator frequency over a range of ± 10 kc at a low rate, say 60 times a second. *See* OSCILLATOR. [J.MI.]

Bibliography: F. E. Terman and J. M. Pettit, *Electronic Measurements*, 2d ed., 1952.

Signal tracer

A device for the systematic test analysis and the location of the point of improper operation of a radio receiver or audio amplifier. The servicing of a radio receiver can be a time-consuming task, particularly if the cause of the failure is an uncommon occurrence. The signal tracer simplifies the task and combines into one instrument the several instruments required for receiver servicing.

The general function of the device is to inject into the antenna a signal which can be traced at various points through the circuit by using a vacuum-tube voltmeter. The absence of a signal at points in the circuit permits tracing the trouble to the malfunctioning circuit component.

The usual signal tracer includes a tunable rf signal generator and modulator to provide the signal and an rf amplifier and audio amplifier for tracing the signal. Thus, the functioning of the receiver can be traced from the antenna terminals through the rf section of the receiver, the converter stage, the i-f amplifier, and the audio amplifier. The local oscillator of the receiver can also be checked. The gain of each stage is checked by measuring the output voltage and comparing this with the input voltage. A low value of gain could indicate a weak tube or an improperly functioning component.

The usual signal tracer includes several additional features. One of these is a means for locating noisy potentiometers, such as the volume control. If a high dc voltage is placed across the component and the audio amplifier and a loudspeaker are connected across the component, the noise, being accentuated by the high voltage, will be heard through the loudspeaker. The loudspeaker is a built-in feature which is also used to test the output circuits of radios that were brought in for servicing without their own loudspeakers. Another feature provides a means for measuring the power consumption of the receiver. This measurement can often provide a clue as to the underlying reason for component failure. *See* RADIO RECEIVER. [H.F.K.]

Signal-to-noise ratio

The ratio, at some location, of some measure of the desired signal to the same measure of the total noise, abbreviated S/N. This is a primary consideration in the design of any communication system, and is a measure of the efficiency of the system. Distortion of the signal by noise causes errors. The objective of engineering is the maintenance of error-free communications over the smallest possible S/N value.

In radio communications, it is common to rate a receiver in noise factor (NF). The value of the factor is stated in decibel units of the noise contribution of the receiver circuitry over that of a theoretically perfect receiving device. Good engineering design can maintain a figure of approximately 2.5 db NF in the vhf band. *See* NOISE, ELECTRICAL. [W.LS.]

Significant figures

The numbers considered here are real numbers in decimal form. All numbers are considered as positive although the theory applies also to negative numbers. Every digit in a number is significant unless its sole purpose is to fix the position of the decimal point. Thus the significant figures in 31.50 are 3, 1, 5, and 0; in 0.0315 only 3, 1, and 5 are significant; and in 3005 both zeros, 3, and 5 are significant.

A number having n significant figures has n-figure accuracy. Thus 31.50 and 3005 have four-figure accuracy but 0.00315 has three-figure accuracy. Evidently the number 3150 may have four-figure or three-figure accuracy. The so-called powers-of-ten notation is used to remove ambiguity. In the powers-of-ten notation a number is written in the form $a \times 10^k$ where k is an integer and a equals 1 or lies between 1 and 10. In this notation all figures in a are significant. Thus 3150 would be written 3.150×10^3 to indicate four-figure accuracy and written 3.15×10^3 to indicate three-figure accuracy.

The process of writing a number having n-figure accuracy and differing as little as possible from a number having more than n significant figures is called rounding off the latter number to n figures; when there is a choice, an even digit is used in preference to an odd one. Thus 2.614, 2.185, 2.155, 3.2451 rounded off to three figures are 2.61, 2.18, 2.16, and 3.25.

An approximate number must be used generally to express the magnitude of a measured quantity and the number of significant figures in it is the index of its accuracy. The *absolute error* in an approximate number is the numerical difference between the number and a number considered exact. The *relative error* is the absolute error divided by the exact value. The relative error is independent of the unit of measure and the position of the decimal point. It is the true index of accuracy. The error of a number having n-figure accuracy is not greater than ½ in the nth place. Hence, if the first significant figure is $k \neq 0$, the relative error is not greater than $1/(2k \cdot 10^{n-1})$.

The following facts should be observed in computation. Subtractions should be avoided. Note the loss of accuracy in $86.345 - 86.300 = 0.045$. Absolute errors are the basis of judging the accuracy of a sum. Thus to add 8.6, 32.431, and 0.5751, find the sum of 8.6, 32.43, 0.58 to get 41.61, write 41.6 and realize that this may be in error by 0.1 since $0.05 + .005 + .005 = 0.06$. A product or quotient is rounded off to the number of significant figures in the least accurate factor. Thus, the product of 0.75×3.1416 is 2.4. If E_{rel} is the relative error in u then the relative error in a power of u, u^P, is PE_{rel} and in a root of u, $\sqrt[n]{u}$, it is E_{rel}/n. [L.M.K.]

Bibliography: J. B. Scarborough, *Numerical Mathematical Analysis*, 4th ed., 1958.

Silencer

A device employed to reduce the sound radiated from a noise source such as an automobile exhaust, air compressor intake, or gun muzzle. It is always associated with noise sources in which the transmission of sound accompanies a flow of gas or liquid.

The term silencer denotes complete elimination of noise, a situation which rarely is realized. A more generally employed term is muffler. For a more detailed discussion of the design of such noise attenuating devices, *see* MUFFLER. [W.J.G.]

Silicate

A salt or ester of a silicic acid. The alkali-metal silicates are water soluble, and sodium silicate is a syrupy liquid known as water glass. It is used as an adhesive for corrugated boxes and as a waterproofing and fireproofing agent.

Silicon dioxide is not soluble in water. When solutions of soluble silicates are acidified, a gelatinous precipitate comes down which can best be described as a hydrous oxide and represented by the formula $(SiO_2)_x(H_2O)_y$. This hydrous oxide is weakly acidic and will dissolve in alkaline solutions. Aqueous solutions of monosilicic acid, H_2SiO_3, can be prepared by special techniques.

The basic unit of silicates is the SiO_4^{4-} group which exists in orthosilicates such as Zn_2SiO_4. Other discrete silicate anions contain several of these SiO_4^{4-} groups in groups or rings linked by Si—O—Si bonds. Some of these anions are $Si_2O_7^{6-}$, $Si_3O_9^{6-}$, $Si_4O_{12}^{8-}$, and $Si_6O_{18}^{12-}$. Extended chains and sheets of variable sizes are also known. Mica falls in this latter category. In some of these naturally occurring polymers, aluminum atoms replace some of the silicons in the network. *See* CEMENT; GLASS AND GLASS PRODUCTS; SILICATE MINERALS; SILICON. [E.E.WR.]

Silicate minerals

All silicates are built of a fundamental structural unit, the so-called SiO_4 tetrahedron. The crystal structure may be based on isolated SiO_4 groups or, since each of the four oxygen ions can bond to either one or two silicon (Si) ions, on SiO_4 groups shared in such a way as to form complex isolated groups or indefinitely extending chains, sheets, or three-dimensional networks. Mixed structures in which more than one type of shared tetrahedra are present also are known. *See* SILICON.

Classification. Silicates are classified according to the nature of the sharing mechanism, as revealed by x-ray diffraction study, and an abbreviated form of such a classification is given below. The sharing mechanism gives rise to a characteristic ratio of Si to O, but it is possible for oxygen ions that are not bonded to Si to be present in the structure, and sometimes some or all of any aluminum present must be counted as equivalent to Si. The constitution and classification of the silicates were controversial before the advent of x-ray structure analysis methods in 1912. The silicates were then usually considered as salts of silicic acids, many of them hypothetical, and a chemical classification as orthosilicates, metasilicates, and the like was applied.

Silicate structures and their characteristic Si/O ratios

Type	Nature of Si-O linkage	Si/O ratio	Examples
Nesosilicates	Isolated SiO_4 groups	1:4	Olivine, garnet
Sorosilicates	Isolated compound groups, Si_2O_7, Si_6O_{18}, and so on	2:7, 6:18, and so on	Thortveitite, beryl
Inosilicates	1-Dimensional extended chains and bands	1:3, 4:11, and so on	Amphiboles, pyroxenes
Phyllosilicates	2-Dimensional extended sheets	2:5	Mica, clays, talc, chlorite
Tectosilicates	3-Dimensional networks	1:2	Feldspars, feldspathoids, zeolites

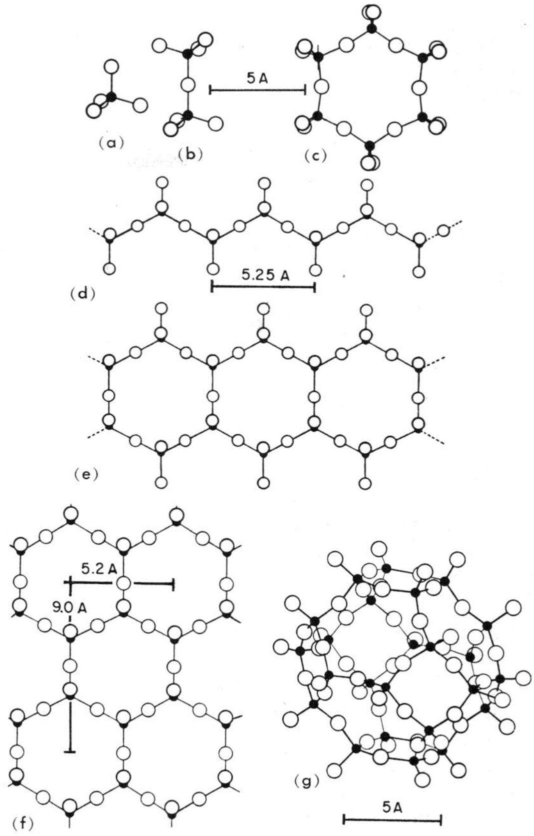

Forms of silicon-oxygen linkage. (*a*) Nesosilicate $(SiO_4)^{4-}$. (*b*) Sorosilicate $(Si_2O_7)^{6-}$. (*c*) Sorosilicate, or cyclosilicate $(Si_6O_{18})^{12-}$. (*d*) Inosilicate $(SiO_3)^{2-}$, showing chain structure of pyroxenes. (*e*) Inosilicate $(Si_4O_{11})^{6-}$, showing band structure of amphiboles. (*f*) Phyllosilicate $(Si_2O_5)^{2-}$, showing extended sheets. (*g*) Tectosilicate, showing three-dimensional structure of lazurite, a feldspathoid. *(From W. L. Bragg, Atomic Structure of Minerals, Cornell, 1937)*

Structure. A dominant feature of the crystal chemistry of the silicates that in large part determines the chemical complexity of these species is the dual role played by aluminum in the crystal structure. The radius ratio of this ion with oxygen is near the critical value between 4-coordination and 6-coordination, and the aluminum (Al) ion can occur in one or the other or both roles simultaneously. When in 4-coordination, the trivalent Al ion substitutes for the quadrivalent Si ion, introducing a valence deficiency of one unit and requiring a concomitant substitution of another cation elsewhere in the crystal structure to provide valence compensation. This mechanism usually involves the coupled substitution of a divalent for a monovalent cation, as of calcium (Ca) for sodium (Na), less frequently of a trivalent for a divalent ion. Other less common mechanisms involve coupled omissions or substitutions among the anionic units of structure. In some silicates, such as those with a silicon-oxygen framework based on a polymorph of silica, the serial substitution of Al for Si is compensated by the entrance of a cation, such as Na, into vacant interstices of the crystal structure.

The detailed crystallographic and physical properties of the various silicates are broadly related to the type of silicate framework that they possess. Thus, the phyllosilicates as a group typically have a platy crystal habit, with a cleavage parallel to the plane of layering of the structure, and are optically negative with rather high birefringence. The nesosilicates, based on an extended 1-dimensional rather than 2-dimensional linkage of the SiO_4 tetrahedra, generally form crystals of prismatic habit; if cleavage is present, it will be parallel to the direction of elongation. The tectosilicates commonly are equant in habit, without marked preference for cleavage direction, and tend to have a relatively low birefringence.

Important minerals. Silicate minerals comprise the bulk of the outer crust of the earth and form in a wide range of geologic environments. Many silicates are of economic importance. Among the clays, feldspars, and refractory minerals, andalusite and wollastonite are used in the ceramic industries, mica as an electrical insulating agent, asbestos and exfoliated vermicularite as thermal insulating agents, and garnet as an abrasive. Talc is a constituent of facial powder. Other silicates are important as ore minerals, beryllium being obtained from beryl, zirconium and hafnium from zircon, and thorium from thorite. Some silicates such as jadeite and nephrite are prized as ornamental materials, and peridot, garnet, tourmaline, and aquamarine are well-known gem stones. *See* CLAY, COMMERCIAL; CLAY MINERALS; GLASS AND GLASS PRODUCTS; SILICATE PHASE EQUILIBRIA.

For discussions of certain silicate mineral groups, *see* AMPHIBOLE; ANDALUSITE; CHLORITE; CHLORITOID; EPIDOTE; FELDSPAR; FELDSPATHOID; GARNET; HUMITE; MICA; OLIVINE; PYROXENE; SCAPOLITE; SERPENTINE; ZEOLITE. [C.FR.]

Silicate phase equilibria

Silicate phase equilibria studies define the conditions of temperature, composition, and pressure at which silicates can stably coexist. Silicate phase equilibria relations are used by geologists, ceramists, and cement manufacturers to explain the variation of composition of silica-bearing minerals, as well as their number and order of appearance in rocks, slags, glasses, and cements. They are also useful to interpret the chemistry of refractories, boiler scale deposits, and welding fluxes.

Silica itself makes up nearly 60% by weight of the earth's crust. The next most abundant oxides in decreasing order are Al_2O_3, CaO, Na_2O, FeO, MgO, K_2O, and Fe_2O_3; all of these occur principally combined with silica as silicates. Free silica and the hundreds of silicate minerals comprise nearly 97% of the earth's crust. The study of silicate phase equilibria was initiated by geologists seeking to apply the phase rule of J. Willard Gibbs to these abundant natural substances. Most of the work since 1907 has been done at the Geophysical Laboratory, Carnegie Institution of Washington. *See* EQUILIBRIUM, PHASE; SILICATE MINERALS.

Silicate phase equilibria are determined in systems generally specified by naming the components involved. Components are the smallest number of independently variable chemical constituents necessary to express the composition of each phase present. A binary system has two components, which may be simple oxides such as Na_2O and SiO_2, or complex compounds such as Na_2SiO_3 and $Na_2Si_2O_5$. In either case, the preferred order of writing the components or oxides follows the rule that the oxides are grouped according to increasing valence of the cation, and then in alphabetical order, as K_2O-Na_2O-CaO-Al_2O_3-SiO_2 for a quinary system and $K_2O \cdot Al_2O_3 \cdot 6SiO_2$ or $KAlSi_3O_8$ for a complex compound.

Equilibrium relations between molten silicate liquids and crystalline silicates, called liquidus equilibria, are most often determined. Equilibria between crystalline silicates in the absence of any liquid silicate melt, called subsolidus equilibria, are determined only infrequently.

Historically, research has been from binary or ternary systems toward more complicated systems. More than 500 systems have been studied to date. The most complicated studies involve limited portions of a system with six components. Specific portions of multicomponent systems are determined because of their importance. For example, in 1913, the first complete silicate phase equilibrium study, the binary system $NaAlSi_3O_8$ (albite)-$CaAl_2Si_2O_8$ (anorthite) representing the very abundant rock-forming plagioclase feldspars, was determined by N. L. Bowen, one of the most famous pioneering experimental petrologists. However, this binary system was but a subsystem from the very complicated quaternary system Na_2O-CaO-Al_2O_3-SiO_2 which has not yet been completely determined.

Silicate crystal structures in general require at least 30% by weight of silica; hence the less siliceous portions of systems are often not determined. The limited number of possible silicate crystal structures determines the possible ratios of oxides to silica in silicates. Certain elements substitute for each other in silicate crystal structures, and this reduces the possible varieties of silicates so that those making up the overwhelming bulk of rocks fall into less than a dozen major groups, called rock-forming silicates. Typical of such groups are the feldspars and feldspathoids in which alkali ions substitute for one another and aluminum and ferric iron ions substitute in part for silicon ions, and the olivine and pyroxene groups in which divalent iron, magnesium, manganese, and calcium ions substitute for each other. Most experiments on silicate phase equilibria have been and continue to be on the relations within and between these important groups of minerals. *See* FELDSPAR; FELDSPATHOID; OLIVINE; PYROXENE.

Dynamic and static methods. Both methods are used to determine equilibrium in silicate systems. Dynamic methods like those used for metals and alloys, such as measuring the heat effects of phase changes on heating or cooling, are not particularly satisfactory with silicates. These methods require large samples of silicates that are difficult to prepare, and require also that equilibrium be reached quickly. Silicates in general are slow to react, and supercooling or superheating of hundreds of degrees before reaction occurs is common. Many silicates react sluggishly at temperatures of 1000°C or higher. Crystals in liquid may persist metastably for weeks, nucleation of new phases may require months, and certain crystallographic processes, such as the ordering of aluminum and silicon atoms into different structural sites, may take a lifetime. Subsolidus reactions are extremely sluggish and may require geologic time before equilibrium is attained or even approached. It is this fact that permits the geologist and petrologist to determine the past thermal history of a rock; the higher-temperature products are preserved for study, even through very long cooling intervals, by the slow rate of attaining equilibrium.

Most silicate phase equilibria are determined by the static method of holding a sample under controlled conditions until equilibrium is attained, then quenching the sample for examination. Charges of 10–100 mg of pure components or glass, gel, or crystals of known composition are wrapped in platinum foil and held in air in furnaces in which the temperature is carefully regulated. Controlled partial pressures of oxygen or inert gases are required to study systems with components, such as FeO, MnO, Fe_2O_3, and TiO_2, that are oxidized or reduced by contact with air, or that alloy with platinum. The samples are quenched rapidly to room temperature by dropping them into water or mercury. Most silicate liquids with a ratio of oxygen to aluminum plus silicon of 2:1 or 3:1 are viscous and quench to glass. Crystals grown from liquid during the quench are usually fine-grained and fibrous, and are easily recognized. Certain displacive polymorphic transitions in silicate crystals cannot be quenched and are studied by special techniques, but in general the crystals stable at high temperature can be quenched and studied at room temperature. The composition, amount, and number of glassy and crystalline phases are determined by optical microscopy, x-ray diffraction and thermal analysis.

Establishment of equilibrium. Equilibrium is established when the products obtained by heating a sample to a given temperature are identical with the products of cooling a sample to that temperature; no requirement is made regarding the texture, shape, or grain size of the products. Another criterion used in recognizing equilibrium is that no change of the sample can be observed after holding the charge at a given temperature for very long periods of time. Particularly useful for this purpose are thin, sharp, crystal fragments that sinter and become rounded on melting, or splinters of glass that devitrify to crystals before larger fragments are affected. When the same products are obtained regardless of starting materials, equilibrium is thought to be attained. At equilibrium, the products are usually uniformly distributed in the sample, and there is consistency between the equilibrium results of the system studied and those portions common to other equilibrium systems. Finally, the results of equilibrium are consistent with the trends shown by natural occurrences of the same minerals.

Diagrammatic representation. The use of diagrams to present silicate phase equilibria is customary, as such diagrams express quantitatively the amount and composition of each phase present at any bulk composition in the system at any temperature. Even when molten, silicates have very low vapor pressures, and the vapor pressure and composition of the vapor are practically unaffected by large variations of temperature. Therefore the vapor phase is not considered and the phase diagrams are drawn isobarically at a total pressure of 1 atm. However, diagrams involving only a single component are usually drawn with temperature plotted against total pressure.

Binary systems. Binary systems are plotted with composition (usually in percentage by weight) against temperature at constant pressure. An example of such a binary system with complete solid solution without a maximum or minimum is the system $NaAlSi_3O_8$-$CaAl_2Si_2O_8$ (Fig. 1). The region labeled liquid plus feldspar is a two-phase field where feldspar, whose composition lies along the lower curve, coexists with liquid, whose composition lies on the upper curve. A line connecting the liquid and feldspar that coexist at a given temperature is called a tie line. The curve separating the liquid field from the liquid-plus-feldspar field is called the liquidus. The curve separating the liquid-plus-feldspar field from the feldspar field is called the solidus.

A composition such as 50% by weight $NaAlSi_3O_8$ and 50% by weight $CaAl_2Si_2O_8$, or more simply $Ab_{50}An_{50}$, is all liquid at temperatures above 1450°C. The first crystals to appear in it on cooling would be of composition $Ab_{18}An_{82}$ on reaching the liquidus at 1450°C. If equilibrium is maintained during further cooling, these crystals will react continuously with the liquid and change in composition toward $Ab_{50}An_{50}$ along the solidus and at the same time increase in amount. Simul-

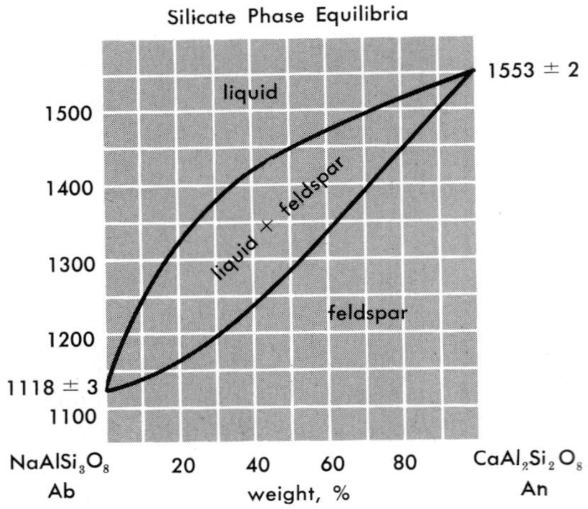

Fig. 1. System $NaAlSi_3O_8$-$CaAl_2Si_2O_8$. (*After N. L. Bowen, Am. J. Sci., ser. 4, 35:577–599, 1913; melting point of* $NaAlSi_3O_8$ *from J. W. Greig and T. F. W. Barth, Am. J. Sci., ser. 5, 35A:94, 1938*)

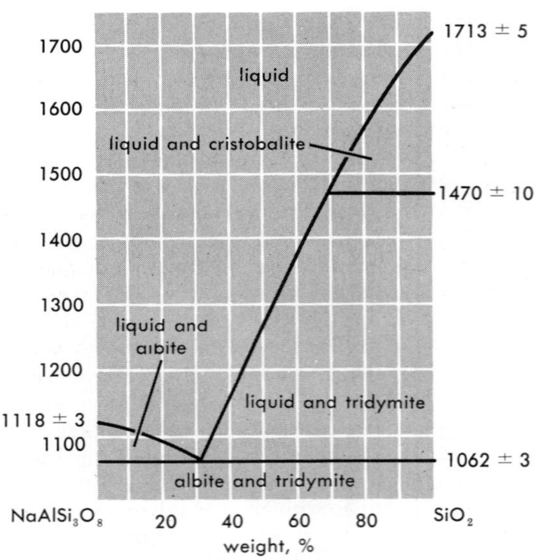

Fig. 2. System $NaAlSi_3O_8$-SiO_2. (*After J. F. Schairer and N. L. Bowen, Am. J. Sci., 254:161, 1956*)

taneously the composition of the liquid will change along the liquidus toward $Ab_{86}An_{14}$. The ratio of liquid to crystals can be found at any temperature by the lever rule, which states that the amounts of the two phases are inversely proportional to the distances of their compositions from the bulk composition. At 1285°C, the last liquid of composition $Ab_{86}An_{14}$ will react with crystals and the entire mass will consist of feldspar of $Ab_{50}An_{50}$ composition.

Fractional crystallization occurs when crystals are removed from contact with the liquid for any reason such as by crystal settling or by protection by armoring of the core by a rim of different composition. The magnitude of the change produced in the remaining liquid can be determined from the diagram if the composition and amount of crystals isolated are known. Thus, the equilibrium diagram can be used to determine the consequences of disequilibrium.

The diagram for the binary system $NaAlSi_3O_8$-SiO_2 shows no solid solution in the crystalline phases (Fig. 2). In most systems of this type, solid solution in the crystalline phases does exist, but is too limited to be observed by ordinary techniques or to be shown on the scale of the diagram.

In the two-phase region of this system, a crystal of composition of the appropriate components exists with liquid until, upon heating, it dissolves in the liquid at the temperature appropriate for the particular bulk composition chosen. When cooled to the temperature of the eutectic, the first crystal of the composition of the second component appears, and the temperature remains constant until all the liquid has crystallized. The bulk composition is then represented as an aggregate of crystals of both components. Only at the eutectic composition and at the pure components themselves do crystals melt directly to a liquid without an intervening stage of crystals and liquid. Contrary to the behavior of systems of metals, mixtures at the composition of the eutectic in silicate systems rarely form intimately intergrown aggregates with so-called eutectoid texture. Discrete crystals of both phases appear instead.

Many other types of binary systems are known, and one additional type is described below.

Ternary systems. Ternary systems are plotted in triangular coordinates with the three components at the apexes of an equilateral triangle, and temperatures on some surface such as the liquidus are indicated by contours projected onto the diagram, as in the system $KAlSiO_4$-$NaAlSiO_4$-SiO_2 (Fig. 3). The fields of primary crystallization of the various solid phases are indicated, as are the compositions and melting temperatures of the end-members of solid solutions, intermediate compounds, and pure components of the system. Binary systems appear as lines on ternary diagrams; the system $NaAlSi_3O_8$-SiO_2 discussed above appears as a segment of the boundary $NaAlSiO_4$-SiO_2.

Isothermal sections. The use of isothermal sections is another graphic method used to represent ternary systems. Phase relations and compositions are shown in detail at a specified temperature. An isothermal section at 600°C for the system $KAlSiO_4$-$NaAlSiO_4$-SiO_2 is a typical example (Fig. 4).

More complex systems are represented by tetrahedrons or by various projections.

Pseudobinary system. A pseudobinary system is shown in the temperature-composition diagram for $KAlSi_3O_8$-$NaAlSi_3O_8$ (Fig. 5). These components make up the abundant alkali feldspars. The system is pseudobinary; the equilibria represented by the heavy lines are truly binary, but the equilibria represented by the light lines are ternary and represent a section through regions. the compositions of the phases of which do not lie in the plane of the diagram. The ternary nature of

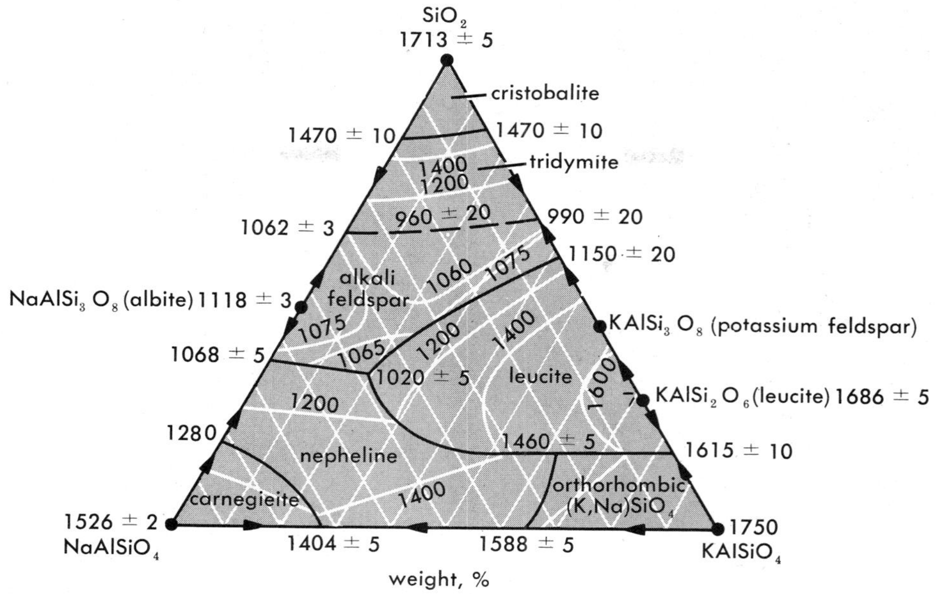

Fig. 3. Liquidus of the system $NaAlSiO_4$-$KAlSiO_4$-SiO_2. (*After J. F. Schairer, J. Geol., 58:514, 1950*)

this system results from the incongruent melting of solid solutions at and near $KAlSi_3O_8$ composition to leucite, $KAlSi_2O_6$, and a liquid whose composition is indicated by the intersection of the liquidus surface and the prolongation of the line drawn from leucite composition through the bulk composition of the feldspar under consideration. These relations are apparent in the illustration of the system $KAlSiO_4$-$NaAlSiO_4$-SiO_2 (Fig. 3).

The truly binary portion of the system $KAlSi_3O_8$-$NaAlSi_3O_8$ is an example of another major group of binary systems with a minimum in the liquidus and solidus, or with a maximum. The one crystal in the solid solution series at the composition of the minimum melts directly to a liquid without an intervening region of liquid and crystals.

The approximate limits of a two-phase subsolidus region are shown, within which a homogeneous alkali feldspar of intermediate composition will separate into two alkali feldspar solid solutions. The curve defining this region is the locus of points representing pairs of compositions that coexist, and is called a solvus.

A subsolidus, nonquenchable polymorphic transition from monoclinic to triclinic symmetry is also shown.

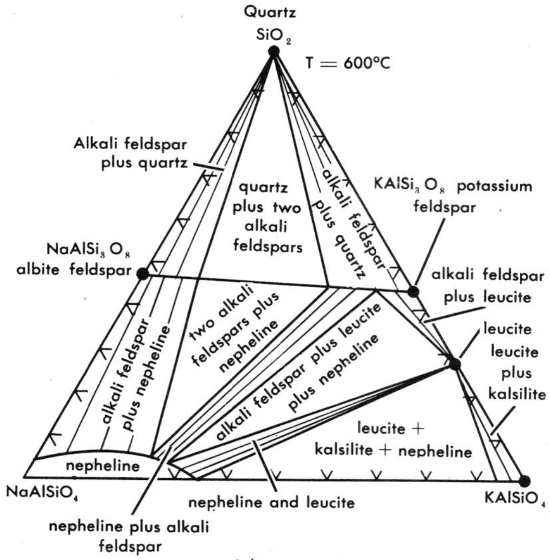

Fig. 4. Isothermal section at 600°C in the system $NaAlSiO_4$-$KAlSiO_4$-SiO_2, in part schematic. (*After O. F. Tuttle and J. V. Smith, Am. J. Sci., 256:587, 1958*)

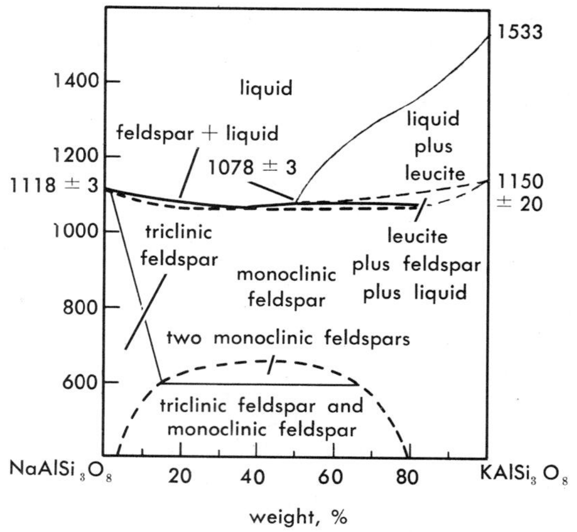

Fig. 5. System $NaAlSi_3O_8$-$KAlSi_3O_8$. Heavy solid or dashed curves refer to binary equilibrium. Light solid or dashed curves refer to ternary equilibrium. (*After J. F. Schairer, J. Geol., 58:515, 1950; N. L. Bowen and O. F. Tuttle, J. Geol., 58:497, 1950; W. S. MacKenzie, Am. J. Sci., Bowen vol., 331, 1952*)

Applications. Silicate phase diagrams permit the user to trace the paths of crystallization of a given composition both when equilibrium is maintained and when disequilibrium occurs. These paths place limits on possible relations and thereby restrict interpretation of natural occurrences.

There are several important generalizations which result from the determination of hundreds of phase diagrams for silicates. The most important of these is the generalization that solid solutions are very prevalent among silicates, and wide variations of composition are possible within a given mineral group. An important consequence is that the composition of the crystal or crystals growing from silicate melt changes as the temperature in the system varies. The liquids also change composition during this process, and if liquids are separated from crystals, fractionation and differentiation of silicate magma can proceed very extensively. Furthermore, careful study of the compositions of coexisting mineral phases can, after comparison with experimental results, indicate details of the temperatures and pressures at which the minerals were formed.

The extent of solid solution in a mineral group varies with the temperature and pressure; the limits are indicated by the appearance of another crystalline phase. High temperature favors greater solid solution in general, while high pressure decreases the amount of solid solution. The extent of solid solution may be used in certain cases where the pressure of formation is known or can be approximated to determine the temperature of formation. The method requires that two minerals with a common component coexist at equilibrium. If they are cooled so that they do not react, then their compositions by comparison with the appropriate solvus will indicate the temperature at which the assemblage formed. Solvi are known in the pyroxene, feldspar, nepheline, and other abundant mineral groups, so this method has wide geologic application. *See* GEOLOGIC THERMOMETRY.

Since 1950 a few experimental studies of the distribution of trace elements between coexisting silicates have been made. The results suggest that the minor elements may also be useful to evaluate the temperature and pressure of formation of mineral assemblages.

Another generalization is that there are frequently sharp changes in the number and relative amounts of the phases present as a given composition is cooled from an all-liquid to an all-crystalline state. Other sharp changes occur even in the crystalline state. A great many points have been established, most of which are of peritectic (reaction) points rather than eutectic points. A decrease in temperature may cause an assemblage of minerals and liquid to transform completely into other minerals plus liquid, to several new minerals, or to a single mineral of intermediate composition. For example, siliceous liquid and leucite, $KAlSi_2O_6$, react with each other on cooling to form potassium feldspar, $KAlSi_3O_8$, at 1150°C.

Phase equilibria studies of aluminosilicates have shown that when incongruent melting occurs, the liquid formed is richer in silica than the newly formed crystal in most cases. This generalization is of considerable significance in petrogenesis, as it indicates a reason for the progressive enrichment in silica of late magmatic differentiates. *See* MAGMA.

Phase equilibria studies of silicate systems have also shown that the hypothesis of the immiscibility of granitic and basaltic magmas is not useful geologically. Liquid immiscibility has been shown to occur experimentally, but only in compositions that are not found in nature.

The available phase diagrams show that the silicate liquids that crystallize last and at the lower temperatures, either by equilibrium or disequilibrium crystallization, tend to be enriched in ferrous iron, silica, and alkali aluminosilicates with a molecular ratio of alkalis to alumina of unity. Further crystallization tends to remove the ferrous iron, and the last liquid remaining very closely approaches the system $KAlSiO_4$-$NaAlSiO_4$-SiO_2 shown above. This is known as petrogeny's residua system. Two major types of rocks are represented by this system. The most abundant are the granites, represented by the subsystem $KAlSi_3O_8$-$NaAlSi_3O_8$-SiO_2. Analyses of hundreds of granites plot in or close to the low-temperature "valley" on the liquidus of this system running from the $KAlSi_3O_8$-$NaAlSi_3O_8$ binary toward SiO_2. This constitutes a major reason for believing most granites originate from silicate magmas.

The second major type of rocks represented are the silica-deficient feldspathoidal syenites of the subsystem $KAlSiO_4$-$NaAlSiO_4$-$KAlSi_3O_8$-$NaAlSi_3O_8$. More experimental detail is required to understand all of the geologic consequences of this system, but there is excellent agreement between synthetic and natural assemblages of phases. A major problem remains in finding the reason why certain magmas differentiate toward the silica-deficient subsystem. The solution to this problem undoubtedly involves many additional components. It is apparent, however, that the join $KAlSi_3O_8$-$NaAlSi_3O_8$, where binary, is a barrier or hump across which liquids do not move as a consequence of equilibrium crystallization.

The concept of incompatible mineral assemblages is another major consequence of studies of silicate phase equilibria. Nepheline, $NaAlSiO_4$, is incompatible at equilibrium with quartz, SiO_2, as they react to form albite feldspar, $NaAlSi_3O_8$. Similarly leucite, $KAlSi_2O_6$, and quartz are incompatible and react to form potassium feldspar, $KAlSi_3O_8$. Tens of incompatible assemblages are now known to petrologists. The recognition of any of these assemblages in the field indicates unusual conditions in which equilibrium was not attained.

Since 1948, silicate phase equilibria in systems with oxidizable components have been intensively studied to determine the effect of variations of oxygen pressure upon the phase relations. The results show that when the oxygen partial pressure is changed, the temperature at which minerals containing an oxidizable component can exist is changed. The magnitude of the changes is quite remarkable; a small fraction of an atmosphere of oxygen may change the stability limits of an iron-bearing silicate by as much as 300°C. This principle, of course, is fundamental to the operation of blast furnaces. In nature, changes of partial pressure of oxygen are expected as gases are expelled from crystallizing magma, and these changes are now believed to affect profoundly the course of differentiation of the magma. The trend in silicate research is toward more detailed study of this effect, particularly as it affects the silica content of the residual liquid.

Much geological field evidence indicates that the temperatures required to obtain silicate liquids in experimental work are hundreds of degrees higher than those determined by various methods to have existed in magmatic assemblages or observed in volcanic eruptions. Silicate phase equilibria studies of increased complexity indicate that as the number of components is increased, the temperatures required for liquid to form are lowered. The magnitude of this effect is inadequate to explain the geological discrepancy. Geologists have suggested that volatile components, notably H_2O, cause the great lowering of liquidus temperatures. This has indeed been shown to be the case, but experimentally the volatile components must be contained by high external pressure. Pressures of H_2O of 15,000 psi lower the temperatures of the liquidus as much as 700°C; simultaneously, several per cent of H_2O dissolve in the molten silicate. Reactions are greatly accelerated, and important groups of hydrous minerals become stable. So spectacular have been the results that most current research in silicate phase equilibria for geological applications is conducted under high pressures of volatile components. For additional detail, *see* HIGH-PRESSURE PHENOMENA.

Thermodynamic theory is sufficiently developed to permit silicate phase equilibria to be calculated when values of the necessary physical-chemical parameters are available and ideal solution occurs. Accurate measurements of all these parameters are difficult, and the results of most phase equilibria studies are not precise enough to allow calculation of thermodynamic parameters. Some simple subsolidus and liquidus relations have been rigorously calculated. In general, however, the quality of calculations of complex phenomena is poorer than rough experimental work, the usual difficulties of establishing equilibrium notwithstanding.

See CEMENT; CERAMIC TECHNOLOGY; GLASS AND GLASS PRODUCTS; LITHOSPHERE, GEOCHEMISTRY OF; PETROLOGY; PORCELAIN; REFRACTORY; SILICATE. [D.B.ST.]

Bibliography: Wilhelm Eitel, *The Physical Chemistry of the Silicates*, 1954; E. M. Levin, H. F. McMurdie, and F. P. Hall, *Phase Diagrams for Ceramists*, 1956; L. H. Ahrens et al. (eds.), *Physics and Chemistry of the Earth*, vol. 3, 1959; J. F. Schairer, Melting relations of the common rock-forming oxides, *J. Am. Ceram. Soc.*, 40:215–235, 1957.

Siliceous sinter

A porous silica deposit formed around hot springs. It is white to light gray, and sometimes friable. Geyserite is a variety of siliceous sinter formed around geysers. The siliceous sinters are deposited as the hot subterranean waters cool after issuing at the surface and become supersaturated with silica that was picked up at depth. The sinters are frequently deposited on algae that live in the pools around the hot springs. *See* GEYSER; THERMAL SPRING. [R.S.]

Silicoflagellata

A class of unicellular flagellate microorganisms of the phylum Chrysophyta, which are a part of the marine plankton. Their exoskeletons, siliceous and coarsely perforate, resemble those of the Radiolaria, with which they have been grouped. They are usually subpyramidal or hemispherical in shape,

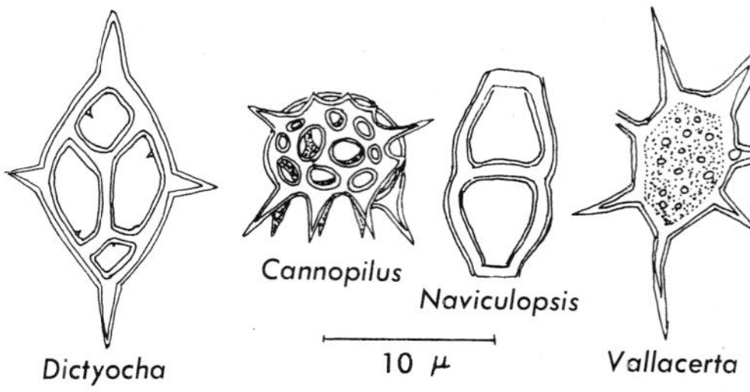

Fossil and modern Silicoflagellata: *Dictyocha*, Cretaceous to Recent; *Cannopilus*, Miocene; *Naviculopsis*, Eocene to Miocene; *Vallacerta*, Upper Cretaceous.

delicately filigreed, and range in size from 10 to 150 microns. Two families and eleven genera of silicoflagellates have been described from siliceous sedimentary rocks ranging in age from Upper Cretaceous to Recent, in association with abundant diatoms and siliceous sponge spicules. One genus, *Dictyocha*, lives in the ocean today. *See* CHRYSOPHYTA; MICROPALEONTOLOGY. [D.J.J.]

Silicon

Element number 14, silicon, Si, is the most abundant electropositive element found in the earth's crust. It is a bluish-gray metalloid which has a decided metallic luster and is exceedingly brittle. It crystallizes in the diamond lattice, has a specific gravity of 2.42, melts at 1420°C, and boils at 3500°C. It has an atomic weight of 28.06. The element is usually tetravalent in its compounds, although sometimes divalent, and is decidedly electropositive in its chemical behavior. It forms a great variety of inorganic and organic compounds.

Uses. Elementary silicon is used as an alloying constituent to strengthen aluminum, magnesium, copper, and other metals. It also has a deoxidizing effect on steel, and in much larger proportion it also confers chemical inertness on ferrous alloys.

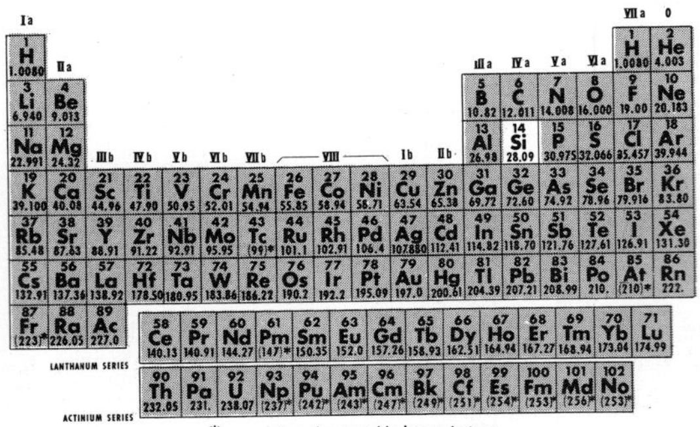

*mass number of most stable known isotope

Silicon of 99% purity is used as the starting material for silicone resins, oils, and elastomers, and high-purity silicon is used in semiconduction devices such as rectifiers, transistors, and solar batteries.

Silicon dioxide, besides its use in tremendous quantities in the ceramic industries, is used as the raw material for making elementary silicon and for silicon carbide. Sizable crystals of it are used for piezoelectric crystals, those thin wafers of quartz which control the frequency of radio oscillators by vibrating at a very exact frequency. Fused quartz sand becomes silica glass, used in chemical laboratories and plants as well as an electrical insulator. A colloidal dispersion of silica in water is used as a coating agent and as an ingredient in certain polishes.

There is no widespread use of the hydrides of silicon. Small amounts are made and decomposed in the preparation of very pure transistor-grade silicon. There are some organochlorosilanes (such as CH_3SiHCl_2) which contain silicon-hydrogen bonds, and these are valued because they may be converted into silicone polymers which retain the reactive silane groups. These special silicones are used to impart water-repellent films to textiles and leather.

The organochlorosilanes are used for the manufacture of silicone polymers, which may be resinous (cross-linked structures of high molecular weight), fluid (linear structures with blocking-groups at the ends), or elastomeric (linear structures of very high molecular weight, connected at intervals by the action of curing agents). Such silicone materials are used for electrical insulation, as mold-release agents, for water-repellent coatings, as hydraulic fluids, in polishes and lubricants, in cosmetics, and in a host of other applications. The silicate esters are used as sources of pure silica, in paint formulations, and as heat-transfer fluids.

Properties of the element. Naturally occurring silicon contains 89.6% of the isotope of mass number 28, 6.2% of silicon-29, and 4.2% of silicon-30. In addition to these stable, natural isotopes, artificially radioactive isotopes of masses 27 and 31 are known. The expected stability of an even-even isotope provides only a partial explanation for the prevalence of silicon-28 in the universe, and there must be some further explanation which necessarily becomes an important part of any theory of the genesis of the elements. *See* ELEMENTS (COSMIC ABUNDANCE); ELEMENTS AND NUCLIDES (ORIGIN).

Elementary silicon has the physical properties of a metalloid, resembling germanium below it in group IV of the periodic table, and, to a lesser extent, arsenic and boron in the diagonal relationship. It shows an appreciable electrical conductivity, but this is clearly semiconductance in the sense that it increases with rise of temperature. In very pure form silicon is an intrinsic semiconductor, although the extent of its semiconduction is greatly increased by the introduction of minute amounts of impurities. Elements of the third group, such as boron, introduce atoms with a deficiency of electrons into the crystal structure and produce the so-called *p*-type silicon, which conducts an electric current by migration of electron vacancies or "holes." Similarly, the semiconductance of pure silicon is greatly increased by the introduction of group V elements such as phosphorus or arsenic, but in this case the increased current is carried by migration of extra electrons and the solid solution is called *n*-type silicon. *See* SEMICONDUCTOR.

Silicon resembles the metals in its chemical behavior, and is commonly assigned an electronegativity of 1.8 in the Pauling scale. Close investigation of its relative electronegativity in the sp^3 hybrid state shows it to be about as electropositive as tin, and decidedly more positive than germanium or lead. In keeping with this rather metallic character, silicon forms tetra-positive ions and a variety of covalent compounds in which it is the positive partner of a dipole; it appears as a negative ion in only a few silicides, and, of course, as a positive constituent of oxy acid or complex anions. Its chemical behavior is characterized further by a very high heat of oxidation, so that natural silicon is found in the completely oxidized state as the dioxide or as the silicate minerals. To indicate the extent of its affinity for oxygen and the halogen elements, Table 1 gives the heats of formation of some representative compounds of silicon in comparison with the corresponding compounds of carbon and germanium. Despite this high heat of oxidation, however, silicon apparently remains unoxidized at room temperature for periods of many years; it has been shown to have a high activation energy for oxidation, and hence the rate of reaction becomes appreciable only at temperatures in excess of 1200°C. Crystals of the element are insoluble in single dilute or concentrated acids, but a mixture of concentrated nitric and hydrofluoric acids will dissolve the element slowly by converting it to the dioxide and dissolving the dioxide as tetrafluoride. The crystalline element is slowly soluble in concentrated solutions of sodium and potassium hydroxide, in which it liberates hydrogen and forms solutions of the corresponding alkali silicates. Finely divided silicon produced by reduction of its compounds below the melting point of the element is correspondingly more active toward the same reagents. Several series of hydrides are formed, a va-

Table 1. Heats of formation of silicon, carbon, and germanium compounds*

Carbon compound	kcal/mole	Silicon compound	kcal/mole	Germanium compound	kcal/mole
$CO_2(g)$	94.4	$SiO_2(s)$	201.3	$GeO_2(s)$	128.3
		SiC	1.43	(forms no carbide)	
$CH_4(g)$	17.9	$SiH_4(g)$	11.9	$GeH_4(g)$	
CI_4	−50. (approx)	$SiI_4(s)$	27.7	GeI_4	
$CBr_4(g)$	12.0	$SiBr_4(l)$	91.5	$GeBr_4$	
$CCl_4(l)$	33.3	$SiCl_4(l)$	149.1	$GeCl_4(l)$	130.
$CF_4(g)$	162.5	$SiF_4(g)$	361.3	GeF_4	

* (*g*) gas; (*l*) liquid; (*s*) solid.

riety of halides (some of which contain silicon-to-silicon bonds), and also many series of oxygen-containing compounds which may be either ionic or covalent in their properties.

Natural occurrence. Silicon occurs in many forms of the dioxide and as almost numberless variations of the natural silicates. For a discussion of the structures and compositions of the representative classes, *see* SILICATE MINERALS. For a discussion of many other silicates, synthetic to a varying degree, *see* CERAMIC TECHNOLOGY. In abundance, silicon exceeds by far every other element except oxygen. It constitutes 27.72% of the solid crust of the earth, whereas oxygen constitutes 46.6%, and the next element after silicon, aluminum, accounts for 8.13%. In fact, all the solid material of the earth's crust, insofar as it has been available for investigation, is known to be silicious except for the carbonate and phosphate rocks, which of course are not igneous or original. This tremendous abundance of silicon makes it of particular interest as a chemical raw material, and the versatility of its chemical behavior has encouraged more and more uses to be developed through intensive research.

Preparation of the element. In the laboratory, free silicon can be obtained by reducing potassium fluosilicate with metallic potassium at elevated temperatures, followed by washing out the resulting potassium fluoride with water. A somewhat more convenient method consists of the reduction of finely ground or precipitated silicon dioxide with magnesium powder at red heat, followed by dissolution of the resulting magnesium oxide in dilute acid and by washing the resulting brownish powder free of soluble material. Neither of these reactions gives a very pure product, since contamination by the original starting materials or by excess reducing agent usually results. Commercial reduction of the dioxide is accomplished electrothermally with carbon at a temperature considerably above the melting point of the silicon

$$SiO_2 + 2C \rightarrow Si + 2CO$$

Since quartz, sand, and other natural forms of silica always contain some oxides of iron, aluminum, titanium, and other materials, the silicon so obtained usually is of about 98% purity. A typical analysis of such a commercial silicon is given in Table 2. The impurities may be reduced by grinding the silicon to 50–100 mesh size and leaching exhaustively in a dilute mixture of hydrofluoric and sulfuric acids. Very pure silicon for use as a semiconductor may be prepared by purifying the tetrachloride exhaustively by distillation and then reducing it with exceedingly pure distilled zinc, followed by distillation of the zinc chloride and fusion of the spongy or powdered silicon. Further purification then is accomplished by zone melting and solidification or by growing single crystals from a large bath of melted material. *See* ZONE REFINING.

Table 2. Composition of a commercial silicon

Si	98.53%	Ca	0.12%	Other metals	0.08
Fe	0.56	Mn	0.04%	Oxygen	0.34
Al	0.31	Ti	0.02		

Principal compounds. Silicon forms useful and important compounds with hydrogen, carbon, the halogen elements, nitrogen, oxygen, and sulfur. In addition, useful organosilicon derivatives have been prepared.

Hydrides. The hydrides of silicon are named silanes, the compound SiH_4 being called monosilane, Si_2H_6 disilane, Si_3H_8 trisilane, and so on. Compounds in which oxygen atoms alternate with silicon atoms in the principal part of the structure are called siloxanes, and those with nitrogen between silicon atoms are called silazanes. All other covalent compounds of silicon are considered for the purpose of nomenclature to be derived from these silanes and modified silanes and are named according to substituent groups and their placement along the principal silicon-containing chain or ring. Thus $(C_2H_5)_3SiOH$ is triethylsilanol, $[(CH_3)_2SiO]_4$ is octamethylcyclotetrasiloxane, $C_6H_5SiCl_3$ is phenyltrichlorosilane, CH_3SiHCl_2 is methyldichlorosilane, and so on.

The hydrides of silicon were first investigated thoroughly by Alfred Stock, who prepared them by the reaction of magnesium silicide (from the reaction of silica with excess magnesium at minimum temperature) with dilute aqueous hydrochloric or phosphoric acid. He obtained the saturated series of silanes given in Table 3, all of which could now be derived more readily from the corresponding chlorides by reduction with lithium aluminum hydride in ether solution

$$2Si_2Cl_6 + 3LiAlH_4 \rightarrow 2Si_2H_6 + 3LiCl + 3AlCl_3$$

Table 3. Hydrides of silicon and some derivatives

Name	Formula	Melting point, °C	Boiling point, °C
Monosilane	SiH_4	−185.	−112.
Disilane	Si_2H_6	−133.	−14.5
Trisilane	Si_3H_8	−117.	+52.9
Tetrasilane	Si_4H_{10}	−90. (approx)	109. (approx)
Chlorosilane	SiH_3Cl	−118.	−30.4
Dichlorosilane	SiH_2Cl_2	−122.	+8.3
Trichlorosilane	$SiHCl_3$	−127.	31.8
Bromosilane	SiH_3Br	−94.	1.9
Dibromosilane	SiH_2Br_2	−70.1	66. (approx)
Tribromosilane	$SiHBr_3$	−73.5	111.8
Iodosilane	SiH_3I	−57.	45.4
Diiodosilane	SiH_2I_2	−1.0	149.5
Triiodosilane	$SiHI_3$	8.	220.
Disiloxane	$H_3SiOSiH_3$	−144.	−15.2

The products must be handled in a well-designed vacuum system, because all the silanes are readily oxidized by air and form spontaneously flammable or explosive mixtures with air. All are also attacked by water in the presence of even minute traces of hydroxyl ion to evolve hydrogen and form silicic acid or hydrated silica. The reaction with water is further accelerated by larger amounts of inorganic or organic base and becomes a dependable method for the quantitative determination of hydrogen bonded directly to silicon, since each silicon-hydrogen bond evolves one molecule of H_2.

The hydrolysis of silanes is believed to take place through the coordination attachment of a hydroxyl ion to silicon, followed by the loss of a hydride ion and its consequent combination with a proton of water to form molecular hydrogen. A similar process takes place in the hydrolysis of silicon halides, ejecting a halide ion and producing the corresponding hydrohalogen acid along with silicic acid or silica. It is further believed that molecules of water may coordinate in the same way as hydroxyl ions in the initial phase of the reaction and that the demonstrated ability of silicon to expand its valency group to accommodate six covalent neighbors allows this process to go on rapidly. The corresponding mechanism is, of course, not available to carbon, which lacks *d*-orbitals for the expansion of its covalency shell beyond four atoms or groups.

The silicon hydrides decompose thermally at elevated temperatures to liberate hydrogen and deposit spongy, brownish silicon. The most stable hydride, monosilane, decomposes rapidly at 500°C and slowly but appreciably at 250°C, especially in the presence of solid materials such as silica gel or activated alumina. The gas-phase reactions of the silanes are correspondingly limited because of this competing decomposition but all undergo such reactions as addition to the double bond of olefins at slightly elevated temperatures or in the presence of catalysts which generate free radicals. Thus monosilane adds readily to ethylene at room temperature under irradiation with ultraviolet light to form ethylsilane

$$SiH_4 + H_2C{=}CH_2 \rightarrow C_2H_5SiH_3$$

A corresponding reaction occurs in a general way whenever any compound containing a silicon-hydrogen bond is mixed with an alkene or an alkyne under appropriate conditions of concentration, temperature, and catalyst, so that organosilanes and their derivatives become available by this route.

The silanes are oxidized rapidly or explosively by the halogens, but react in controllable fashion with hydrogen halides in the presence of the corresponding aluminum halide to yield halogen-substituted silanes. By control of the concentrations and conditions, progressive substitution with halogen is possible.

Of the substituted hydrides, trichlorosilane (silicochloroform, $SiHCl_3$) is perhaps the best known because it can be prepared readily by the action of hydrogen chloride on elementary, crystalline, or reduced silicon at a temperature of 300–450°C. Since it boils at 31.8°C, it is easily separated from the tetrachloride and other chlorosilanes by distillation. It shows the typical reactions of the silicon-hydrogen bond, as well as those of a silicon chloride.

Silicon carbide. The reduction of silica with excess carbon under appropriate conditions gives silicon carbide, SiC, which crystallizes in a number of forms but is best known in the cubic-diamond form with spacing a_0 of 4.35 A (compared with 3.56 A for diamond). In the pure form, silicon carbide is colorless, but the commercial product known as Carborundum is black and has a bluish or greenish iridescence. The carbide is not easily oxidized by air except above 1000°C, and retains its physical strength up to this temperature. For these reasons it is a favorite structural refractory material for the ceramic arts. It also is extremely hard, with a Mohs hardness in excess of 9, and so has found wide application as an abrasive. It melts in the neighborhood of 2700°C and is quite unreactive except toward oxidizing agents at temperatures up to 1000°C. Chlorine will convert the hot carbide to a mixture of carbon tetrachloride and silicon tetrachloride. Although the bond energy of the carbon-silicon bond was long believed to be in the neighborhood of 58 kcal/mole, it is now believed to be more nearly 75 kcal/mole. *See* ABRASIVE.

Silicon halides. Silicon tetrachloride, $SiCl_4$, is perhaps the best-known covalent compound of silicon. It is readily available commercially. It can be prepared by chlorinating elementary silicon, or by the action of chlorine on a mixture of silica with finely divided carbon, or by the chlorination of silicon carbide. It is a volatile liquid which fumes in moist air and hydrolyzes rapidly to silica and hydrochloric acid; the mechanism of this reaction, which so sharply distinguishes the tetrachlorides of silicon and carbon, has been given above. Silicon tetrachloride reacts readily with alcohols and glycols to form the corresponding ethers, which may also be considered to be esters of silicic acid.

$$SiCl_4 + 4C_2H_5OH \rightarrow 4HCl + Si(OC_2H_5)_4$$

Table 4. Halides of silicon

Name	Formula	Melting point, °C	Boiling point, °C
Tetrafluorosilane (silicon tetrafluoride)	SiF_4	−95.7	−65 at 1810 mm
Tetrachlorosilane (silicon tetrachloride)	$SiCl_4$	−70.	+57.6
Tetrabromosilane (silicon tetrabromide)	$SiBr_4$	+5.	153.
Tetraiodosilane (silicon tetraiodide)	SiI_4	121.	290.
Hexachlorodisilane	Si_2Cl_6		147.
Octachlorotrisilane	Si_3Cl_8		216.
Decachlorotetrasilane	Si_4Cl_{10}		150 at 15 mm Hg
Hexabromodisilane	Si_2Br_6	95.	265.
Hexafluorodisiloxane	Si_2OF_6	−47.8	−23.3
Hexachlorodisiloxane	Si_2OCl_6	−28.1	+137.
Hexabromodisiloxane	Si_2OBr_6	+27.9	118 at 15 mm Hg

It also reacts with Grignard reagents and with the alkyls of zinc, lithium, and mercury to produce organic derivatives of silicon, and has been used as a starting material for the commercial preparation of such compounds.

Some of the principal halides of silicon, together with their physical properties, are listed in Table 4. Although the iodides decompose readily with the liberation of iodine to limit their utility in some reactions, all show the characteristic properties of the silicon-halogen bond such as ready hydrolysis and reaction with alcohols and Grignard reagents. Silicon tetrafluoride complexes readily with hydrogen fluoride to produce fluorosilicic acid, H_2SiF_6, which forms a well-known series of fluorosilicate salts. The chlorides, bromides, and iodides do not form complex acids of this type, but hexacovalent silicon is known in some chelate compounds containing bidentate groups. Mixed halides containing more than one kind of halogen are known, and pseudohalides containing cyanate, isocyanate, or thiocyanate groups can readily be prepared by exchange reactions or by reaction of the silicon chlorides with silver cyanate, isocyanate, or thiocyanate.

Silicon nitride. The action of nitrogen on elementary silicon at 1300°C or above produces a refractory silicon nitride of the composition Si_3N_4. The same substance results from the thermal decomposition of ammonia addition compounds of monosilane or silicon tetrachloride, probably by way of the intermediate silicon imide, $Si(NH)_2$, which, like the dioxide, is polymeric. The nitride is inactive chemically and has some use as a refractory.

Silicon oxides. Silicon dioxide is perhaps best known as one of its crystalline modifications called quartz, colorless crystals of which are also known as rhinestones and Glens Falls diamonds. Purple or lavender-colored quartz is called amethyst, the pink variety is rose quartz, and the yellow type, citrine. At 530°C the ordinary or α-quartz changes over reversibly to β-quartz, which has a lower density, and at 870°C β-quartz changes to a different crystal modification, β-tridymite. At a still higher temperature, 1470°C, β-tridymite becomes a third modification called β-cristobalite. Both β-tridymite and β-cristobalite have lower temperature α modifications which have lower optical symmetry, and all the forms can be maintained at room temperature if chilled rapidly from the stable equilibria. Each appears to have its own melting point, but the usual melting point for silica is that of β-cristobalite, about 1710°C. Rapid chilling of the liquid produces silica glass, often incorrectly called quartz glass, a vitreous modification of SiO_2 which has a very low coefficient of expansion and is, of course, isotropic. This silica glass finds many uses because of its resistance to thermal shock and its low electrical conductivity; it is useful in the temperature range below 1000°C, but above this temperature it begins to devitrify and at 1250°C will crystallize quite rapidly.

Because rock crystal has been collected and admired for thousands of years, large and perfectly formed natural crystals of quartz now are very rare. With the growth of radio broadcasting and the electronics industry, piezoelectric crystals cut from perfect specimens of quartz have been used in increasing quantities, to the point of scarcity of natural crystals. As the supply diminished, considerable effort was devoted to the problem of growing crystals of quartz by artificial means. Some success has been achieved by growing the crystals hydrothermally from a solution of silica glass in water containing an alkali or a fluoride, the whole being maintained at a temperature of 200–300°C or higher in a high-pressure bomb. The vitreous silica has a higher solubility in an aqueous medium at the operating temperature than does quartz and so will dissolve in the solution and crystallize out upon a tiny seed crystal of quartz. The solubility of quartz in water and in solutions of hydroxides or fluorides does not decrease sharply as the critical point of water is reached; in fact, the solubility in pure water above the critical point is ample for growing crystals.

A number of natural noncrystalline varieties of silicon dioxide also are known, such as the hydrated silica known as opal and the dense unhydrated variety known as flint. Onyx and agate represent still other semiprecious forms. *See* GEM; QUARTZ.

Silicon monoxide, SiO, is a brownish powder which can be obtained by heating a mixture of finely divided silica and elementary silicon in the absence of air above 1400°C. It is much more volatile than the dioxide, and is used to form a durable coating on lenses and mirrors and in electron microscopy for shadowing. There is considerable doubt that the monoxide persists as a stable substance at temperatures below 1300°C; it seems to disproportionate into crystallites of silicon and silica if it cools slowly. The monoxide oxidizes readily to dioxide, which of course does not interfere with its use as a lens coating. At a temperature of 300–400°C it also is reactive to halogens and behaves in general as a readily oxidized lower-valent compound of silicon would be expected to do. The only other known compounds of divalent silicon are some nonvolatile liquid lower chlorides, which are also covalently unsaturated and will react (for example) with methyl chloride to form methylchlorosilanes.

Sulfides. Although no monosulfide of silicon has been isolated, silicon disulfide is known in the form of long, flexible acicular crystals which appear to be infinite chains of SiS_4 tetrahedrons. The crystals are mechanically strong in the direction of the long axis but hydrolyze readily to produce silica and hydrogen sulfide.

Esters of silicic acid. As indicated above, the reaction of ethyl alcohol with silicon tetrachloride produces ethyl silicate, a colorless, volatile liquid of pleasant odor which boils at 168°C. Because this can be distilled to a high degree of purity and is made from materials which contain no alkalies or other ionic impurities, it is a favored source of pure silica to be used for the preparation of silicate phosphors and similar materials. It hydrolyzes very slowly in pure water but more rapidly in the presence of a small amount of dilute acid as catalyst, to form hydrated silica and ethanol. The hydrolysis can also be controlled with limited water to give condensed ethyl polysilicates, which have been used as paint vehicles for the protection of masonry.

The reaction of methanol with silicon tetrachloride produces methyl silicate, a liquid of camphor-like odor which boils at 121°C. This might be expected to have the same uses as ethyl silicate, but when made in the manner indicated, it is an extremely dangerous substance to handle. It causes perforating ulceration of the cornea and eventual blindness. The same methyl silicate can be made by the direct action of methanol on elementary silicon at elevated temperatures and in the presence of a copper catalyst, and this product (made in the absence of halogen-containing substances) has been found to have no adverse physiological effect upon experimental animals.

Higher alkyl silicates, as well as some aromatic silicates, are items of commerce and find some use as heat-transfer media because of their long liquid range and their considerable thermal stability. All will hydrolyze in the presence of strong acids and bases, however.

Organic compounds. Important organosilicon derivatives include tetraalkyls and tetraaryls, halides, hydrides, and the organosiloxanes.

Tetraalkyls and tetraaryls. Organic compounds of silicon which contain direct silicon-carbon bonds have been known since about 1860, and some of them show remarkable thermal stability and resistance to oxidation. The tetraalkyl- and tetraarylsilanes can be made by the action of Grignard reagents, zinc alkyls, lithium alkyls, or other alkylating agents on silicon tetrachloride, followed by hydrolysis of the reaction mixture and distillation of the desired compound. A list of some representative compounds, together with their physical properties, is given in Table 5. Of the compounds listed, tetraphenylsilane has the most striking stability; it can be distilled in air at 428°C, and remains quite unchanged. Tetramethylsilane does not decompose until a temperature of about 600°C is reached, but it does oxidize at 350°C or higher.

Organosilicon halides and hydrides. The organic compounds of silicon which contain both directly attached organic groups and silicon-chlorine or silicon-hydrogen bonds are of greater interest than the tetraalkyl silanes because they lend themselves so well to further reactions and hence may serve as chemical intermediates for a wide variety of organosilicon products. The alkyl and aryl chlorosilanes represent one class of such compounds, and these find application as intermediates for the preparation of silicone polymers. In general, organosilicon halides may be made by the action of the classical organometallic reagents, such as lithium alkyls and Grignard reagents, on the halides of silicon according to the general equations

$$SiX_4 + yRMgX \rightarrow R_ySiX_{4-y} + yMgX_2$$
$$Si_2X_6 + yRMgX \rightarrow R_ySi_2X_{6-y} + yMgX_2$$

This substitution method has extreme flexibility and so is widely applicable to laboratory and commercial preparations. For quantity production of a limited number of organochlorosilanes, however, it has given way to the direct synthesis from alkyl or aryl chlorides and elementary silicon

$$2RCl + Si \xrightarrow[\text{Heat}]{\text{Catalyst}} R_2SiCl_2 \text{ and other products}$$

For example, methyl chloride reacts readily with elementary commercial silicon in the presence of copper powder as a catalyst at a temperature in the vicinity of 300°C to produce principally dimethyl-

Table 5. Organosilicon compounds of the type, R_4Si

Compound	Formula	Melting point, °C	Boiling point, °C
Tetramethylsilane	$(CH_3)_4Si$		26.5
Trimethylethylsilane	$(CH_3)_3SiC_2H_5$		62.
Dimethyldiethylsilane	$(CH_3)_2Si(C_2H_5)_2$		95.8
Methyltriethylsilane	$CH_3Si(C_2H_5)_2$		127.
Tetraethylsilane	$(C_2H_5)_4Si$		153.
Triethylvinylsilane	$(C_2H_5)_3SiC_2H_3$		146.
Diethyldiphenylsilane	$(C_2H_5)_2Si(C_6H_5)_2$		297.
Tetrapropylsilane	$(n\text{-}C_3H_7)_4Si$		212.
Tetrabutylsilane	$(n\text{-}C_4H_9)_4Si$		157 at 22 mm Hg
Tetraphenylsilane	$(C_6H_5)_4Si$	233.	428.
Tetrabenzylsilane	$(C_6H_5CH_2)_4Si$	127.	

Table 6. Some organochlorosilanes, R_xSiCl_{4-x}

Name	Formula	Melting point, °C	Boiling point, °C
Methyltrichlorosilane	CH_3SiCl_3	−77.8	65.7
Dimethyldichlorosilane	$(CH_3)_2SiCl_2$	−76.1	70.0
Trimethylchlorosilane	$(CH_3)_3SiCl$	−57.7	57.3
Methylphenyldichlorosilane	$(CH_3)C_6H_5SiCl_2$		82.5
Ethyltrichlorosilane	$C_2H_5SiCl_3$	−105.6	97.9
Diethyldichlorosilane	$(C_2H_5)_2SiCl_2$	−96.5	129.
Triethylchlorosilane	$(C_2H_5)_3SiCl$		143.5
Ethylphenyldichlorosilane	$(C_2H_5)C_6H_5SiCl_2$		230.
Vinyltrichlorosilane	$C_2H_3SiCl_3$		92.
Divinyldichlorosilane	$(C_2H_3)_2SiCl_2$		119.
Propyltrichlorosilane	n-$C_3H_7SiCl_3$		122.7
Dipropyldichlorosilane	$(n$-$C_3H_7)_2SiCl_2$		175.
Butyltrichlorosilane	n-$C_4H_9SiCl_3$		148.9
Decyltrichlorosilane	n-$C_{10}H_{21}SiCl_3$		183 at 84 mm Hg
Phenyltrichlorosilane	$C_6H_5SiCl_3$		201.5
Diphenyldichlorosilane	$(C_6H_5)_2SiCl_2$		305.2
Triphenylchlorosilane	$(C_6H_5)_3SiCl$	88.	378.
Benzyltrichlorosilane	$C_6H_5CH_2SiCl_3$		216.
Dibenzyldichlorosilane	$(C_6H_5CH_2)_2SiCl_2$	51.	243 at 100 mm Hg
Naphthyltrichlorosilane	α-$C_{10}H_7SiCl_3$		168 at 22 mm Hg

dichlorosilane, with some methyltrichlorosilane, trimethylchlorosilane, methyldichlorosilane, and a number of other less important products. From this reaction mixture the pure methylchlorosilanes are distilled and used as intermediates for silicone oils, resins, and rubber, since they hydrolyze readily to form the corresponding organosiloxane polymer (silicone) and hydrochloric acid. Because methyl chloride may be made from methanol and hydrochloric acid, the raw materials are seen to be silicon and methanol, which in turn are made from silica, coal, and water as ultimate raw materials. Similarly, phenylchlorosilanes may be made by the action of chlorobenzene on elementary silicon, with silver or copper as catalyst and at temperatures of 350–500°C. The distilled products are cohydrolyzed with methylchlorosilanes to make methyl phenyl silicone polymers which show greater resistance to oxidation than the methyl silicones themselves.

A great variety of organochlorosilanes are known, and many also are available commercially. Table 6 lists a number of the more important ones. By appropriate procedures of synthesis, functional groups may be attached to the organic portions of these organochlorosilanes, and a growing technology of so-called organofunctional silicones is developing. It appears that organosilicon groups may be introduced into a large variety of organic polymers, dyes, drugs, and other products, with results yet to be determined.

The reaction of organochlorosilanes and related substances with alcohols produces the corresponding esters or ethers of the type $R_xSi(OR')_y$. These hydrolyze also to organosiloxanes, but in a controllable way that makes them desirable for some applications. A similar reaction of organochlorosilanes (and notably the methylchlorosilanes) with hydroxyl groups of other organic structures takes place readily and attaches organosilicon entities to the structures. Thus the vapor-phase reaction of methylchlorosilane with cellulose results in chemical attachment of a highly water-repellent film to the cellulose fiber, and the coating cannot be removed by organic solvents or by washing with aqueous detergents.

Organosiloxanes. The organohalosilanes hydrolyze readily to form organosiloxanes, all of which are polymeric. These cyclic and linear polymers are known as silicones because they were first regarded by F. S. Kipping to be analogs of the or-

Table 7. Some organosiloxanes

Name	Formula	Melting point, °C	Boiling point, °C
Hexamethylcyclotrisiloxane	$[(CH_3)_2SiO]_3$	64.	134.
Octamethylcyclotetrasiloxane	$[(CH_3)_2SiO]_4$	17.5	175.
Decamethylcyclopentasiloxane	$[(CH_3)_2SiO]_5$	−38.	210.
Dodecamethylcyclohexasiloxane	$[(CH_3)_2SiO]_6$	−3.	245.
Hexamethyldisiloxane	$(CH_3)_3SiOSi(CH_3)_3$		100.5
Octamethyltrisiloxane	$(CH_3)_8Si_3O_2$	−80.	153.
Decamethyltetrasiloxane	$(CH_3)_{10}Si_4O_3$	−70.	194.
Hexaethylcyclotrisiloxane	$[(C_2H_5)_2SiO]_3$	14.	117. at 10 mm Hg
Octaethylcyclotetrasiloxane	$[(C_2H_5)_2SiO]_4$	−50.	159. at 10 mm Hg
Hexaphenylcyclotrisiloxane	$[(C_6H_5)_2SiO]_3$	190.	295. at 1 mm Hg
Octaphenylcyclotetrasiloxane	$[(C_6H_5)_2SiO]_4$	201.	335. at 1 mm Hg

ganic ketones, but they are decidedly polymeric in composition. The individuality of silicon gives them properties which are very different from those of the ketones.

Table 7 lists a number of pure organosiloxanes and some of their physical properties. In general, the siloxanes are chemically inactive, being unchanged by dilute or even moderately concentrated acids and by most ionic reagents. The siloxane chain is attacked by strong alkalies, however, with the formation of alkali metal salts of the corresponding organosilicates. Hydrofluoric acid also will attack the siloxane chain to produce monomeric organofluorosilanes and water.

Organic groups of the organosiloxanes show a wide variation of chemical inertness or reactivity. The methyl groups in a methyl polysiloxane remain attached at temperatures up to 500°C, where the siloxane chain depolymerizes into volatile, cyclic structures, and phenyl groups are equally firmly attached to silicon. In the presence of oxygen, however, methyl groups begin to be oxidized to formaldehyde at 300°C or higher, whereas phenyl groups are capable of withstanding temperatures of 400–450°C. Higher alkyl groups oxidize at a rate which increases with the chain length, so that the higher alkyl silicones have little applicability at elevated temperatures. In general, negative groups such as halogen, OH, COOH, and NH_2 as substituents in the organic part of the molecule decrease the stability of the carbon-silicon bond through their inductive effect upon its polar properties.

Among the interesting physical properties of the methyl and methyl phenyl polysiloxanes is an unusually small dependence of viscosity, dielectric constant, and compressibility upon temperature. The very low temperature coefficient of viscosity of methyl silicone oils, therefore, makes them useful as hydraulic fluids, lubricants, and dielectric fluids, so that they come into service both at very low and very high temperatures in comparison with their organic counterparts. *See* CARBON; GERMANIUM; SILICONE RESINS. [E.G.R.]

Bibliography: R. K. Iler, *The Colloid Chemistry of Silica and Silicates*, 1955; R. R. McGregor, *Silicones and Their Uses*, 1954; H. W. Post, *Silicones and Other Organic Compounds of Silicon*, 1948; E. G. Rochow, *An Introduction to the Chemistry of the Silicones*, 2d ed., 1951; N. V. Sidgwick, *The Chemical Elements and Their Compounds*, vol. 1, 1950; R. B. Sosman, *The Properties of Silica*, ACS Monograph 37, 1927; A. E. Stock, *Hydrides of Boron and Silicon*, 1933.

Silicon diode

A small silicon rectifier of either the point-contact, bonded-contact, or junction type. It is distinguished from a silicon rectifier by size only, the latter term being used for units of relatively large power-handling capacity. Point-contact types have found application in microwave detectors and mixers. The bonded and microjunction types have shown a marginal life and stability. Silicon diodes contrasted with germanium diodes are capable of operating at higher temperatures and therefore at higher power levels. Operation at 200°C and voltages up to 1000 volts are possible. *See* JUNCTION DIODE; POINT-CONTACT DIODE. [L.P.HU.]

Silicone resins

Polymers composed of alternating atoms of silicon and oxygen with organic substituents attached to the silicon atoms.

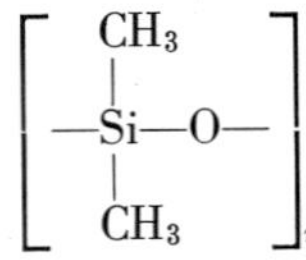

Silicones, also called organopolysiloxanes, may be liquids, greases, resins, or rubbers. The distinguishing characteristics of silicone polymers are their resistance to water and oxidation, and their stability at high temperatures.

Preparation. Silicones are obtained by the condensation of hydroxy organosilicon compounds formed by the hydrolysis of organosilicon halides. The required halide can be prepared by a direct reaction between silicon and an alkyl halide, or the reaction between a silicon halide and a Grignard reagent.

$$\left.\begin{array}{l} CH_3Cl + Si \xrightarrow[300°C]{Cu,CuO} \\ \\ SiCl_4 + CH_3MgBr \rightarrow \end{array}\right\} \begin{array}{l} CH_3SiCl_3,\ (CH_3)_2SiCl_2 \\ (CH_3)_3SiCl,\ (CH_3)_4Si \end{array}$$

After separation of the reaction products by distillation, organosilicon halides can be polymerized by carefully controlled hydrolysis.

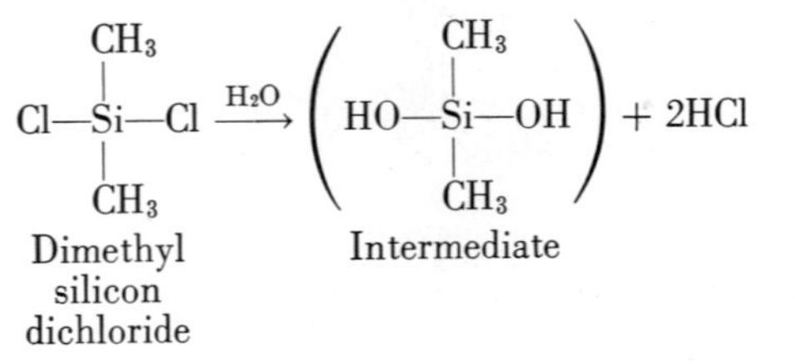

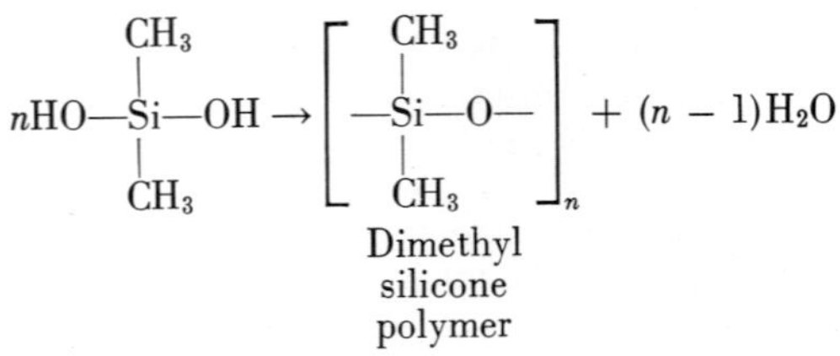

The difunctional products, such as the dimethyl silicon dichloride, are clearly the most important, because they can yield polymers of very high molecular weight. By means of the process known as equilibration, a mixture of the mono- and trimethyl silicon halides can be converted to the dimethyl derivative.

$$CH_3SiCl_3 + (CH_3)_3SiCl \xrightarrow{AlCl_3} (CH_3)_2SiCl_2$$

By this process the dimethyl compound can be obtained in high yield from the products of the initial reaction.

The molecular weight and structure may be varied by including in the original polymerization mixture small amounts of monofunctional $(CH_3)_3SiCl$ or the trifunctional $(CH_3)SiCl_3$ along with the difunctional $(CH_3)_2SiCl_2$. The monofunctional compound is a chain-stopper and the proportion used directly determines the upper limit of the molecular weight. The trifunctional product gives cross-linking and the proportion used directly determines the amount of cross-linking that can be obtained in the system.

Uses. Variations of the organo portion of the resins can be made. The dimethyl silicones are the most commonly used. Copolymers of the dimethyl and diphenyl derivatives are also employed. In recent years other variations have received attention. Vinyl silicon trichloride is employed as a treating agent for glass fiber in order to cause unsaturated polyesters and epoxy resins to adhere more firmly. Amino derivatives have great affinity for certain metals.

The liquids, generally dimethyl silicones of relatively low molecular weight, have low surface tension, great wetting power for metals, and very small change in viscosity with temperature. They are used as hydrolytic fluids, antifoaming agents, and treating agents for leather and textiles, and in cosmetic preparations.

The greases are particularly desired for applications requiring effective lubrication at very high and at very low temperatures.

Silicone resins are frequently selected for coating applications in which thermal stability in the range 300–500°C is required. The dielectric properties of the polymers make them suitable for many electrical applications, particularly in electrical insulation exposed to high temperatures.

Silicone rubbers are compositions containing high-molecular-weight dimethyl silicone linear polymer, finely divided silicon dioxide as the filler, and a peroxidic curing agent. It has been suggested that cross-linking takes place through reaction of the peroxide with two methyl groups on adjacent chains, either to form a dimethylene bridge or to replace the two methyl groups by a single oxygen bridge. The silicone rubbers have the remarkable ability to remain flexible at very low temperatures and to remain stable at high temperatures. *See* PLASTICS FABRICATION; POLYMER PROPERTIES; RUBBER; SILICON. [J.A.M.; L.M.H.]

Silk

The lustrous fiber produced by the larvae of silkworms; also the thread or cloth made from such fiber. The United States is the greatest importer and consumer of silk. Silk will probably always be prized by the consumer even though certain man-made fabrics now have some qualities that were formerly possessed only by silk. The silk industry grosses about $500,000,000 a year.

Japan, the first country to use scientific methods in cultivating the silkworm, has always ranked highest in the production of fine silk, although satisfactory types are made in other silk-producing countries. The domestication and cultivation of the silkworm, which began in Japan about 3000 B.C., requires extreme care and close supervision, and the reeling of the filament from the cocoons can be done successfully only by skilled operators whose training is the result of many generations of experience.

Life cycle of the silkworm. Since the discovery many years ago that the filament composing the cocoon of the silkworm can be unwound and constructed into a beautiful and durable fabric, silkworms have been bred for the sole purpose of producing raw silk. The production of cocoons for their filament is called sericulture. The cocoon of the silkworm is the second stage of development of the life cycle of *Bombyx mori*, a species which spins a thread of high quality. In sericulture, all four stages of the life cycle of this moth are important, because some of the better cocoons must be set aside to permit full development, thus supplying eggs for another hatching. By scientific breeding, silkworms may be hatched three times a year; under natural conditions, breeding occurs only once a year. The life cycle includes (1) the egg, which develops into the larva or caterpillar, the silkworm; (2) the silkworm, which spins its cocoon for protection, and to permit development into the pupa or chrysalis; (3) the chrysalis, which emerges from the cocoon as the moth; and (4) the moth, of which the female lays eggs, thus continuing the life cycle.

Within three days after emerging from the cocoons, the moths mate and the female lays 350–400 eggs. The adults seldom fly, do not eat, and live only a few days. Each healthy egg hatches into a grub, or larva, about ⅛ in. long. The larva requires careful nurturing for approximately 20–32 days. Dur-

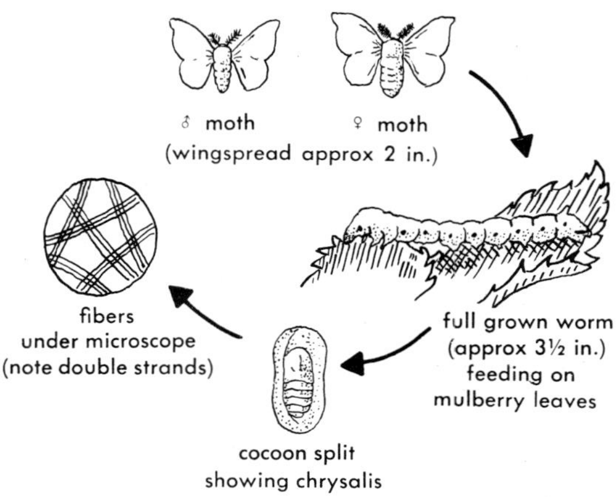

Life cycle of the silkworm. (*From H. H. Manchester, The Story of the Silkworm, rev. ed., Cheney Brothers, 1924*)

ing this period, the tiny worm has a voracious appetite, requiring 5 daily feedings of chopped mulberry leaves. After four changes of skin, or moltings, the worm reaches full growth in the form of a smooth, grayish-white caterpillar about 3½ in. long. After about 6 weeks, its interest in food ceases. It shrinks somewhat in size and acquires a pinkish hue, becoming nearly transparent. A constant restless rearing movement of the head indicates that the worm is ready to spin its cocoon. Clusters of twigs or straw are provided for this purpose.

The silk worm begins to secrete a proteinlike substance through its spinneret, a small opening under its jaws. With a bending motion a filament is spun around the worm in the form of the figure eight. The silkworm is hidden from view within 24 hours; in 3 days, the cocoon is completed. It is about the size and shape of a peanut shell. The filament is in the form of a double strand or fibroin, which is held together by a gummy substance called sericin, or silk gum. Chemically, the silk fibroin and sericin are composed of approximately 95% protein and 5% wax, fats, salts, and ash. The liquid substance hardens immediately on exposure to the air. If left undisturbed, the chrysalis inside the cocoon develops into a moth within 2 weeks. To emerge, the moth must break through the top of the cocoon by excreting an alkaline liquid that dissolves the filament. As this cutting through damages the cocoon so that the filament cannot be unwound in one long thread, the growers terminate the life cycle at this point by a process known as stoving or stifling. The cocoons are heated to suffocate the chrysalis, but the delicate silk filament is not harmed. *See* LEPIDOPTERA.

Filature operations. The cocoons are delivered to a factory, called a filature, where the silk is unwound from the cocoons and the strands are collected into skeins. Some of the cocoons are produced scientifically in such factories. They are sorted according to color, size, shape, and texture, as all these affect the final quality of the silk.

After the cocoons have been sorted, they are put through a series of hot and cold immersions, because the sericin must be softened to permit the unwinding of the filament in one continuous thread. Raw silk consists of about 80% fibroin and 20% sericin. In this step, only about 1% of the sericin is removed because this silk gum is a needed protection during the further handling of the delicate filament.

The process of unwinding the filament from the cocoon is called reeling. The care and skill used in the reeling operation prevent defects in the raw silk. As the filament of a single cocoon is too fine for commercial use, 3–10 strands are usually reeled at a time to produce the desired diameter of raw silk thread. The cocoons float in water, bobbing up and down as the filaments are drawn upward through porcelain eyelets and rapidly wound on wheels or drums while the operator watches to detect flaws. As the reeling of the filament from each cocoon nears completion, the operator attaches a new filament to the moving thread. Skilled operators have an uncanny ability to blend the filaments, always retaining the same diameter of the rapidly moving silk strand. The sericin acts as an adhesive, aiding in holding the several filaments together while they are combined to form the single thread.

The usable length of the reeled filament is from 1000–2000 ft. The remaining part of the filament is valuable raw material for the manufacture of spun silk.

The term reeled silk is applied to the raw silk strand that is formed by combining several filaments from separate cocoons. It is reeled into skeins, which are packed in small bundles called books, weighing 5–10 lb. These are put into bales, ranging in weight from 135–145 lb. In this form, the raw silk is shipped to all parts of the world.

From the filature, the books of reeled silk go to the throwster where it is transformed into silk yarn, also called silk thread, by a process known as throwing. Persons engaged in this work are called throwsters. Silk throwing is analogous to the spinning process that changes cotton, linen, or wool fibers into yarn. Unlike those fibers, however, the manufacture of silk yarn does not include carding, combing, and drawing out, the usual processes for producing a continuous yarn. The raw silk skeins are sorted according to size, color, and length or quantity, then soaked in warm water with soap or oil. This softening of the sericin aids in handling the thread. After being mechanically dried, the skeins are placed on light reels, from which the silk is wound on bobbins.

During this winding operation, single strands may be given any desired amount of twist. If two or more yarns are to be doubled, they are twisted again in the same or in a reverse direction, depending on the kind of thread to be made. To equalize the diameter, the thread is run through rollers. It is then inspected and packaged ready for shipment to manufacturers for construction into fabric.

Wild silk. Wild or tussah silk may be distinguished from cultivated silk by its coarse thick form, which appears flattened. Cultivated silk is a narrow fiber with no markings. Wild silk is a broader fiber with fine, wavy longitudinal lines running along its surface, giving it a dark hue under a microscope. *See* FIBER, NATURAL. [M.D.P.]

Sillimanite

A nesosilicate mineral, composition Al_2SiO_5, crystallizing in the orthorhombic system. It commonly occurs in slender crystals or parallel groups and is frequently fibrous, hence the synonym, fibrolite. There is one direction of perfect cleavage; the luster is vitreous and the color brown, pale green, or white. The hardness is 6–7 on Mohs scale and the specific gravity is 3.23. *See* SILICATE MINERALS.

Sillimanite, kyanite, and andalusite are all polymorphic forms of Al_2SiO_5. They are metamorphic minerals, found in highly aluminous gneisses and schists. Each mineral is stable under different conditions but the transitions from one to another are so sluggish that they may coexist in the same rock. Sillimanite is less common than the others and is found in the highest-grade metamorphic rocks associated with quartz, corundum, garnet, and muscovite. It is rarely found as a contact metamorphic mineral. *See* ANDALUSITE; KYANITE. [C.S.HU.]

Silurian

A geologic period of time during which a system of rocks, the Silurian system, was deposited. The strata are identified in terms of the principal forms of life extant at that time as determined by means of the fossils.

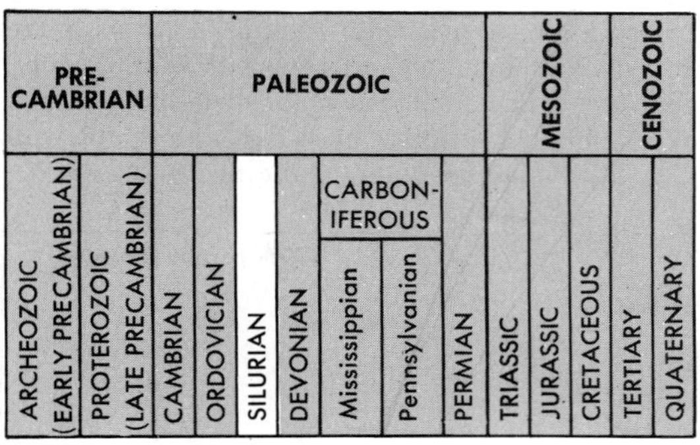

This system was named by R. I. Murchison for exposures in Wales, the name being derived from an ancient Celtic tribe (Silures) who occupied this region during the Roman conquest. Murchison described the stratigraphic character and fossil content of these rocks in his monograph, *The Silurian System*, in 1839. In this and later publications the Silurian system was defined to include some older strata which were later removed to the Ordovician system by C. Lapworth in 1879. *See* ORDOVICIAN.

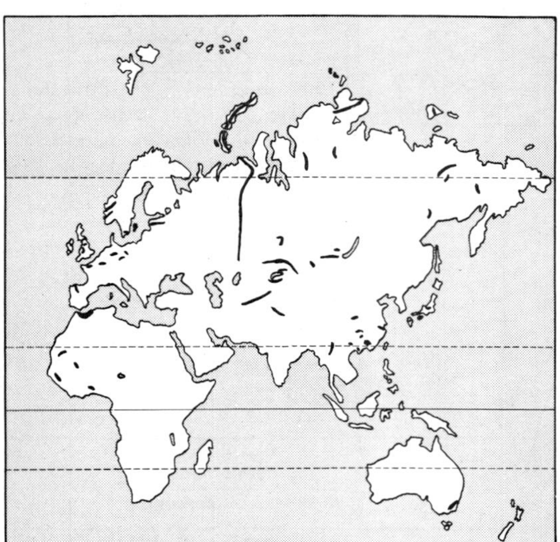

Fig. 1. Generalized distribution of Silurian outcrops in Eastern Hemisphere.

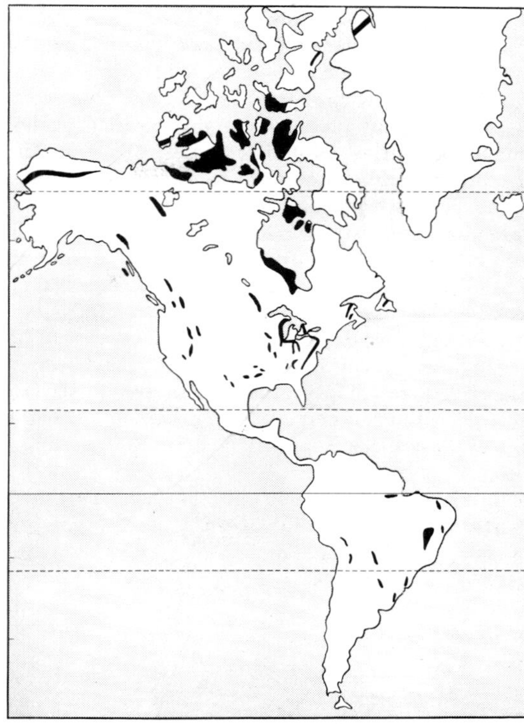

Fig. 2. Generalized distribution of Silurian outcrops in Western Hemisphere.

Silurian series. Silurian rocks are present in many parts of the world (Figs. 1 and 2), but three of the best known sections are in Great Britain, eastern United States, and central Bohemia. In the type region of the British Isles the Silurian is divided into four series, in ascending order: Llandoverian, Wenlockian, Ludlovian, and Downtonian. The Silurian rocks of the eastern United States are divided into three series, the Lower or Medinan (sometimes Alexandrian), Middle Silurian or Niagaran series, and the Upper Silurian or Cayugan series. In Bohemia the series components are designated e-1, e-2, and f-1 rather than by geographic names, a method that dates back to the classic work of Joachim Barrande (*Système Silurien du centre de la Bohême*, 1865–1907).

In addition to the areas mentioned above, the island of Gotland in the Baltic Sea has a well-known Silurian section and several of the fossils from this island were described by the famous Swedish naturalist, Karl von Linné (Linnaeus). Other regions that have been studied in some detail include Poland; southern Norway; the Urals, U.S.S.R; China; and Australia.

United States	Great Britain	Bohemia
Cayugan	Downtonian	f-1
	Ludlovian	e-2
Niagaran	Wenlockian	
Medinan	Llandoverian	e-1

Silurian facies. Silurian deposits are represented by both terrestrial and marine strata, but the latter are predominant. The marine strata are composed largely of carbonate facies, either limestone or dolomite, but there are sandstones and shales, including dark graptolite shales. The marine strata commonly contain a large invertebrate fossil fauna. This is especially true of the limestones and dolomites. Terrestrial Silurian deposits are much less common. Strata of this type are present in southeastern New York and eastern Pennsylvania, mostly as conglomeratic sandstones bearing a sparse eurypterid fauna. Westward these strata grade into a typical marine facies as shown in Fig. 3. Terrestrial Silurian strata have been reported on the southern tip of Africa, although the age of these rocks is in question. *See* EURYPTERIDA; GRAPTOLITHINA.

In late Silurian times the seaways of eastern North America began to dry up, leading to the formation of salt beds. Thick beds of Cayugan salt underlie parts of New York, Pennsylvania, and Michigan, and in places are mined on a large scale. Lower and Middle Silurian strata bear extensive iron deposits, mostly in the form of hematite. Rich iron deposits of this type are mined in the area around Birmingham, Ala. Some oil and gas are obtained from beds of Silurian age in Ohio, western New York, and other areas. *See* EVAPORITE (SALINE).

Climate. The climate during the Middle Silurian must have been mild. The invertebrate faunas have a cosmopolitan aspect. Similar and even identical species have a wide geographic distribution from equatorial regions into latitudes north of the Arctic Circle. Moreover, extensive Niagaran carbonate deposits containing numerous organic reefs are known from strata at least as far north as 50°N lat. In the latter part of the Silurian period more severe conditions appear to have prevailed. The extensive salt deposits of the Upper Silurian indicate increasing aridity and probably decreasing temperature, most likely associated with the Caledonian Mountain building which affected

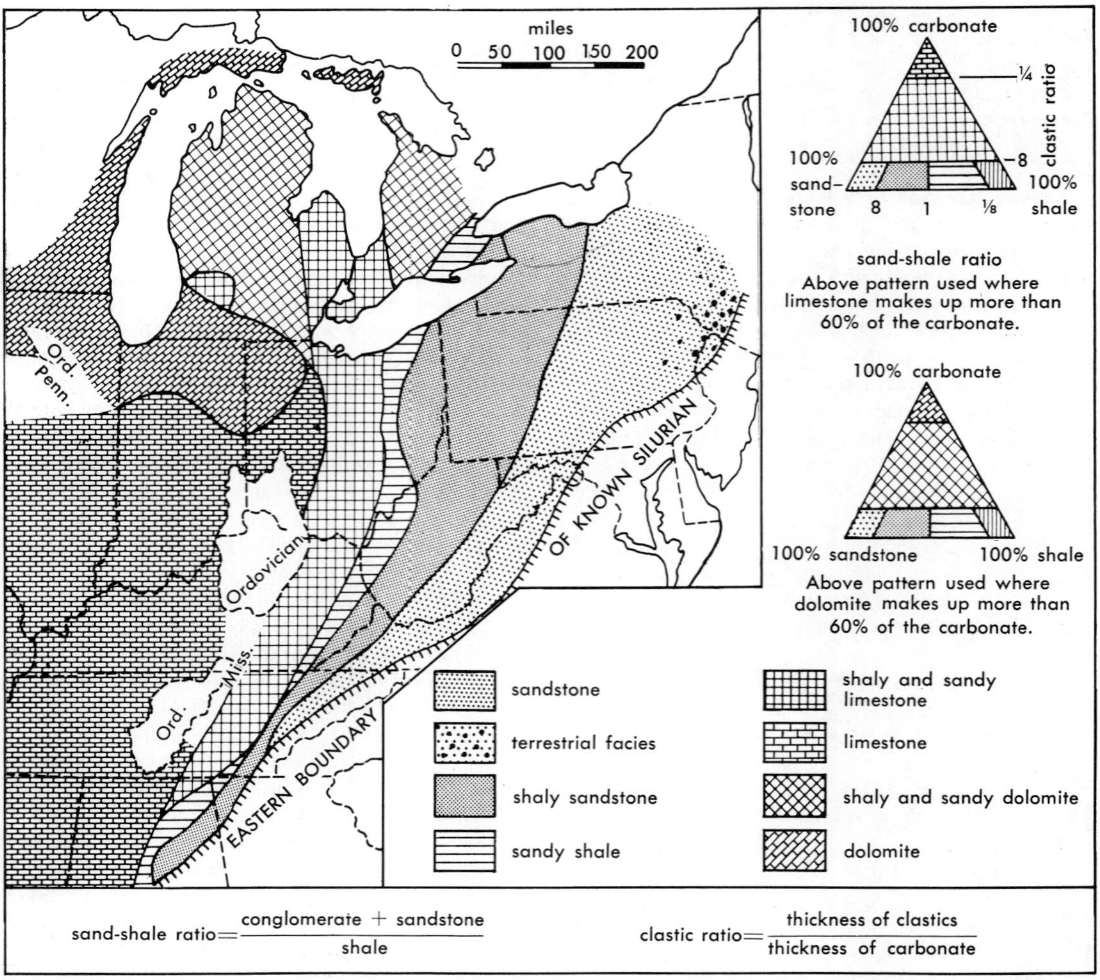

Fig. 3. Lithofacies map of Lower Silurian (Median) series in the eastern United States. Pure shale facies are not present. Triangular diagrams show how lithofacies map was constructed. (*After T. W. Amsden, Bull. Am. Assoc. Petrol. Geologists, 39(1):60–74, 1955*)

parts of arctic North America, Scandinavia, France, Germany, northern Africa, and Siberia.

Silurian life. Silurian rocks contain a prolific invertebrate fossil assemblage. Among the best represented groups were the brachiopods belonging to the class Articulata (Pygocaulia). Pentameroid brachiopods such as *Conchidium* and *Pentamerus* and the dalmanelloids such as *Rhipidomelloides* were especially abundant. Another common group, the spire-bearing brachiopods, became numerous for the first time. The coral faunas also were prolific and included both solitary and colonial tetracorals. The tabulate corals were even more numerous. Such genera as *Favosites* and *Halysites* are among the more common of Silurian fossils. Locally the corals were associated with the stromatoporoids, Bryozoa, and other sedentary reef-building organisms. The straight-shelled nautiloid cephalopods were abundant, but other mollusk groups were well developed only in local areas. Crinoids (echinoderms) became common for the first time. Complete articulated crowns are seldom found, but the stems and isolated plates are abundant. The latter are so concentrated in some beds that they become the dominant rock constituent (encrinite). Trilobites and graptolites are present although fewer species are found than in Ordovician rocks. The eurypterids are an extinct group of Arachnoidea that flourished during the Silurian and Devonian periods; a few of the Silurian species attained a length of almost 9 ft, making them the largest of all known Arthropoda. Scorpions with a body form similar to living species have been found in Middle Silurian strata. If these were air-breathers they represent the oldest known terrestrial animals; however, the evidence concerning their mode of respiration is inconclusive.

Primitive fish are the only known Silurian vertebrates. One group, the Agnatha (ostracoderms), had no jaws and were early relatives of the modern cyclostome fish. A second group, the Placodermi, had a primitive jaw apparatus and made their appearance late in the Silurian, although they did not become common until the Devonian period. Most of these early fish had the head and front part of the body protected by an armor of bony plates. *See* OSTRACODERM; PLACODERMI.

Calcareous algal structures are present in many Silurian beds. These primitive aquatic plants are present in ancient Precambrian strata, and are locally common in many Paleozoic rocks. Of greater significance is the presence of vascular plants belonging to the Pteridophyta (Tracheophyta). These plants, which are representatives of the Psilopsida and Lycopsida, undoubtedly lived on land and are the oldest known terrestrial plants.

[T.W.A.]

Bibliography: C. O. Dunbar, *Historical Geology*, 1949; R. C. Moore, *Introduction to Historical Geology*, 2d ed., 1958; E. Neaverson, *Stratigraphical Palaeontology*, 2d ed., 1955; C. Schuchert, *Stratigraphy of the Eastern and Central United States*, Historical Geology of North America, vol. 2, 1943; H. Termier and G. Termier, *Histoire Geologique de la Biosphere*, 1952.

Silver

Chemical element number 47, silver, Ag, is a gray-white, lustrous metal with atomic weight 107.88. Chemically, it is one of the heavy metals and one of the noble metals; commercially it is a precious metal. Copper, silver, and gold make up group Ib

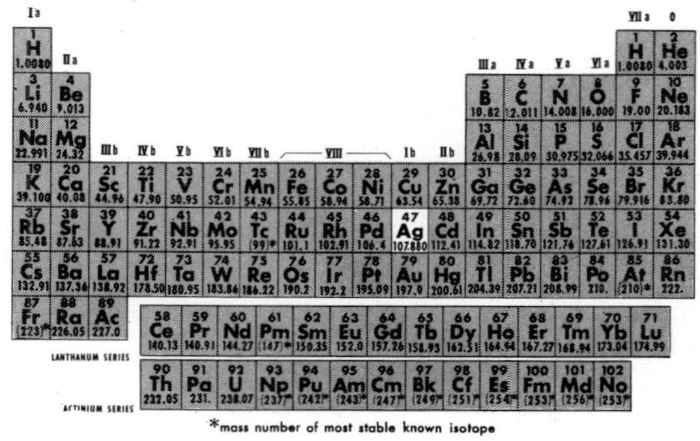

*mass number of most stable known isotope

of the periodic table of elements. Silver has been known as a metal since very ancient times; it was mentioned in the books of the Egyptian king Menes, about 3600 B.C., who set its value at two-fifths that of gold. Nineteen isotopes of silver have been reported, with atomic masses ranging from 102 to 115. Ordinary silver is made up of the two isotopes of masses 107 and 109.

Uses of the metal. In most of its uses, silver is alloyed with one or more other metals (*see* SILVER ALLOYS). The chief use of silver is in coins. Widespread use of copper, silver, and gold for this purpose has caused them to be known as the coinage metals. Silver also has well-known uses in jewelry and silverware. It is used in some fuses and medical instruments, in silver solder, and in corrosion-resistant storage batteries. Alloys in which silver is a minor ingredient include dental amalgam and metals for engine pistons and bearings. Silver has some germicidal properties, and it has been used in a process for sterilization of water.

Occurrence. Silver is a rather rare element, ranking sixty-third in order of abundance. It constitutes about $1 \times 10^{-6}\%$ of the earth's crust. Sometimes it occurs in nature as the free element (native silver) or alloyed with other metals. Norway has the world's most important deposit of native silver; one piece weighing over 1500 lb has been found there. For the most part, however, silver is found in ores containing silver compounds. The principal silver ores are argentite, Ag_2S, cerargyrite or horn silver, AgCl, and several minerals in which silver sulfide is combined with sulfides of other metals: stephanite, $5Ag_2S \cdot Sb_2S_5$; polybasite, $9(Cu_2S, Ag_2S) \cdot (Sb_2S_3, As_2S_3)$; proustite, $3Ag_2S \cdot As_2S_3$; and pyrargyrite, $3Ag_2S \cdot Sb_2S_3$. About three-fourths of the silver produced is a by-product of the extraction of other metals. In addition to this new metal, substantial quantities of silver have been recovered from foreign coins, which were replaced with others containing less or no silver. The recovery of silver from industrial scrap, including photographic residues, is also important. For descriptions of the commercial extraction processes, *see* SILVER METALLURGY.

Silver metal. Pure silver is a white, moderately soft metal, somewhat harder than gold. When polished, it has a brilliant luster and reflects 95% of the light falling on it. Silver is the best conductor of heat and electricity among the metals, and it is second to gold in malleability and ductility. Its density is 10.5 times that of water, so that 1 ft^3 of silver weighs 655 lb. Silver melts at 961°C and boils at about 2200°C. Gold and silver may be mixed to form true solutions (alloys) in any proportions. The quality of silver, its fineness, is expressed as parts of pure silver per 1000 parts of total metal. Commercial silver is usually 999 fine.

Chemical properties. Although silver is the most active chemically of the noble metals, it is not very active in comparison with most other elements. It does not oxidize at all readily (as iron does when it rusts), but it reacts with sulfur or hydrogen sulfide to form the familiar silver tarnish. Electroplating silver with rhodium will prevent this discoloration. The tarnish may be removed from silver articles by abrasion with a silver cream or polish, which also removes the very thin surface layer of silver that has combined with sulfur. Tarnish may be removed chemically by heating the article in a dilute solution of sodium chloride (table salt) and sodium hydrogen carbonate (baking soda) or placing the tarnished article in contact with a more active metal such as aluminum, which reacts with the sulfur and restores the silver to the metallic state. Silver itself does not react with dilute nonoxidizing acids (hydrochloric or sulfuric acids) or strong bases (sodium hydroxide). However, oxidizing acids (nitric or concentrated sulfuric acids) dissolve it by reaction to form the unipositive silver ion, Ag^+. This ion, which is present in solutions of all simple, soluble compounds of silver, is rather easily reduced to the free metal, as in the deposition of silver mirrors by organic reducing agents. Electroplating of silver involves reduction of complex silver ions. The Ag^+ ion is colorless, but a number of silver compounds are colored because of the influence of their other constituents. Oxygen dissolves in silver to a surprising extent, about 20 parts of oxygen to 1 of silver by volume at the melting point of silver. Even after cooling, the silver retains 0.75 part of oxygen by volume.

Silver compounds. Silver is almost always monovalent in its compounds, but an oxide, fluoride, and sulfide of divalent silver are known. Some coordination compounds of silver, also called silver complexes, contain divalent and trivalent silver. Some of the more important compounds are listed in the table.

Compounds of silver

Name and formula	Uses	Properties, remarks
Silver nitrate (lunar caustic), $AgNO_3$	Medicinal; preparation of silver compounds, silver mirrors, inks	Colorless, very soluble compound; stains skin; poisonous internally; easily reduced to metallic silver
Diammine silver hydroxide, $[Ag(NH_3)_2]OH$		Soluble coordination compound, formed by adding ammonium hydroxide to silver salt solutions; on standing, forms highly explosive "fulminating silver"
Silver cyanide, AgCN	Electroplating	Used with excess sodium or potassium cyanide in electroplating to form complex ions $[Ag(CN)_2]^-$ and $[Ag(CN)_3]^{--}$, which are reduced to metallic silver
Silver chloride, AgCl	Photography; ionization detector for cosmic rays	White, insoluble compound; dissolves in ammonium hydroxide to give $[Ag(NH_3)_2]^+$ complex ions
Silver bromide, AgBr	Photography	Light yellow, insoluble compound, more resistant to dissolving than is AgCl
Silver iodide, AgI	Cloud seeding; photography	Yellow, insoluble compound, more resistant to dissolving than is AgBr; unit crystals almost identical with those of ice in cloud seeding
Silver sulfide, Ag_2S		Least soluble of all silver salts; black; main component of silver tarnish

Although silver does not oxidize when heated, it can be oxidized chemically or electrolytically to form silver oxide or peroxide, a strong oxidizing agent. Because of this activity, silver finds considerable use as an oxidation catalyst in the production of certain organic materials. A silver oxide or peroxide anode in conjunction with a zinc cathode in an alkaline electrolyte constitutes an electric battery which will give a large output per unit weight or volume and therefore finds application in special military devices where weight and space are at a premium. This type of battery can be recharged a few times and is therefore a storage battery, but it can withstand only a limited number of cycles of charge and discharge and also has a rather limited shelf life. These features, as well as the cost, restrict its more general use. For a discussion of the important photographic uses of silver compounds, *see* PHOTOGRAPHIC MATERIALS.

Monovalent silver forms a large number of stable coordination compounds. These are often two-coordinate, having two ionic or molecular groups attached to a central Ag^+ ion, as in $[Ag(NH_3)_2]^+$ or $[Ag(CN)_2]^-$. Three-coordinate complexes, such as $[AgCl_3]^{2-}$, are also known, and four-coordinate complexes like $[AgCl_4]^{3-}$ and $[Ag(CN)_4]^{3-}$ probably occur in solution. Silver cyanide, AgCN, is a long-chain coordination compound, made up of alternate silver and cyanide ions. Divalent silver can be stabilized against decomposition by complexing the Ag^{2+} ion with the organic compounds *o*-phenanthroline, pyridine, and α, α'-dipyridyl. The trivalent Ag^{3+} ion can be stabilized through complexing with ethylenedibiguanide. All the coinage metals, that is, copper, silver, and gold, complex more readily with substances that can provide nitrogen, sulfur, or halogen atoms for attachment to the metal than they do with oxygen-providing substances. Complexes of silver with hydroxide ion, for example, are not very stable compared with the hydroxide complexes of zinc, which is a good coordinator with oxygen. Accordingly, silver oxide dissolves only slightly in strong solutions of sodium hydroxide, whereas zinc hydroxide dissolves through coordination with hydroxide, displaying the property known as amphoterism.

Most silver compounds are poisonous and silver nitrate is classed as a hazardous chemical.

Analytical methods. Solutions containing silver ion may be readily identified by precipitation of silver chloride upon addition of hydrochloric acid or a soluble chloride salt. The precipitate may be distinguished from those of lead and monovalent mercury by its ability to dissolve when excess ammonium hydroxide is added, and to reprecipitate when nitric acid is added. Quantitatively, either silver chloride or silver bromide may be conveniently precipitated, dried, and weighed. Silver ion may also be reduced by electrolysis and weighed as metallic silver. Standard potassium thiocyanate solution may be used to analyze for silver volumetrically. *See* COPPER; ELECTROPLATING OF METALS; GOLD. [W.E.C.]

Bibliography: J. C. Bailar, Jr. (ed.), *The Chemistry of the Coordination Compounds*, 1956; N. A. Lange and G. M. Forker (eds.), *Handbook of Chemistry*, 9th ed., 1956; N. V. Sidgwick, *The Chemical Elements and Their Compounds*, vol. 1, 1950; M. C. Sneed, J. L. Maynard, and R. C. Brasted (eds.), *Comprehensive Inorganic Chemistry*, vol. 2, 1954.

Silver alloys

Combinations of silver with one or more other metals. Pure silver is very soft and ductile but can be hardened by alloying. Copper is the favorite hardener and normally is employed in the production of sterling silver, which must contain at least 92.5% silver, and also in the production of coin silver, which, in the United States contains 90% silver, the balance being copper. Pure silver melts at 960°C, but this is lowered by the addition of copper to a minimum of 779°C at 28% copper. This silver-copper eutectic and modifications containing zinc, tin, and cadmium are widely used for brazing purposes, where strong joints, having relatively good corrosion resistance, are required. The high electrical and thermal conductivities of pure silver, slightly exceeding those of copper, along with its resistance to oxidation and moderate cost, have led to the large use of silver for electrical contacts. Because of the tendency of silver to form sulfide films, it is desirable to employ voltages in excess of about 12 volts and reasonable pressures in using such contacts. Frequently, the silver for contacts is alloyed with 10% copper, with a small amount of cadmium, or, better still, with cadmium oxide, which improves the behavior of the contact material under many conditions.

Noble-metal alloys. Silver may be alloyed with gold or palladium in any ratio, producing soft and ductile alloys; certain of these intermediate alloys are useful for electrical contacts, where resistance to sulfide formation must be achieved. This requires about 40–50% palladium or slightly more gold.

In recent years, silver has sold for a little less than $1 per troy ounce or about 3 cents per gram. It has a density of 10.5 so that pure silver costs about $5 per cubic inch. Sterling and coin silver cost slightly less. In an early effort to reduce the cost of silver articles, the technique of bonding silver to copper and rolling the composite material into sheet was developed in England, particularly in the vicinity of Sheffield; the product was therefore known as Sheffield plate. However, with the development of the method of electrodepositing silver from a double cyanide solution by G. R. and H. Elkington in 1844, the production of Sheffield plate almost disappeared and was replaced by electroplated silver. This was the first commercial use of electroplating, and basically the same solutions are still employed. The Elkingtons found that the most satisfactory base metal upon which silver could be plated was the ancient Chinese alloy known as paktong, which was found to contain nickel, copper, and zinc. It is still made under the general name of German silver or nickel silver and still constitutes the most satisfactory base for silver electroplate.

Technical alloys. Silver has proved to be a useful component for high-duty bearings in aircraft engines, where it may be overlaid with a thin layer of lead and finally with a minute coating of indium.

The fact that silver is readily oxidized and reduced is a limitation to its use for certain electrical purposes, particularly where silver conductors are in contact with vulcanized fiber or similar materials and are subjected to a reasonable voltage gradient. Under these circumstances, the silver tends to migrate and develop conducting paths within the insulation, which are difficult to detect.

Specially developed alloys of silver with tin, plus small percentages of copper and zinc in the form of moderately fine powder, can be mixed with mercury to yield a mass which is plastic for a time and then hardens, developing relatively high strength. This material was developed specifically for dental use and is generally known as amalgam, although the term amalgam actually includes all the alloys of mercury with other metals. Dental amalgam is the most widely used tooth-filling material and is one of the most useful of the dental materials. It possesses some limitations in strength and also discolors in the mouth, but silver has some germicidal effect which probably is helpful in this instance in preventing further decay, which is particularly important in children's teeth.

Silver is a component in many of the colored gold alloys used not only for jewelry but also for dentistry; when added to the gold-copper alloys or the palladium-copper alloys, a large improvement results in the strengths obtainable by heat treatment. The addition of platinum or palladium to some of these alloys further augments their strength or hardenability and also renders them more suitable for rubbing electrical contacts. The softness of silver, its insolubility in iron, and its freedom from oxidation at high temperatures make silver powder useful in preventing the threads in high temperature bolting from sticking at temperatures up to about 650°C. In the absence of silver, such bolts may bond together after a short time so that they cannot be unscrewed. Where silver is to be used at high temperatures, the pure metal should be employed; it has been so used for electrical windings in motors operating as high as 500°C. At moderate temperatures, silver coatings are frequently useful in electrical devices to ensure good electrical contact in bolted joints and to provide a low-resistivity coating in equipment which will be operated at high frequency where the skin conduction is of primary importance. Thin silver electroplate also may be applied to copper and copper parts to facilitate soldering, particularly after the parts have been stored for some time and have become oxidized and difficult to solder in the absence of the silver coating.

Because silver does not oxidize on heating, it is used in applying electrically conducted coatings to ceramics; a paste containing finely divided silver plus certain additives is applied to the ceramic. Upon heating to redness, the silver coating becomes firmly bonded to the ceramic. With care, the product can be soldered; the method is widely used in making connections to ceramic capacitors and similar devices. Similar coatings are used for decorating glassware. These may be coated with additional silver by electrodeposition and finally with rhodium to prevent tarnishing.

Because of its resistance to acetic acid and many other organic materials, as well as to alkalies, silver-lined equipment finds considerable use in the chemical industry—in autoclaves, piping, and similar pieces of equipment, some of it very large. In high-temperature applications, the high solubility of oxygen in silver and the rapid diffusion of oxygen through it must be recognized. Silver, alloyed with a base metal, will suffer internal oxidation when heated in air, and a silver coating applied to a base metal, such as iron, affords no useful protection at high temperatures, because the oxygen will go through the silver, oxidizing the iron at the interface. Furthermore, silver will be damaged if heated in an oxidizing atmosphere and then in a reducing atmosphere, particularly if the latter contains hydrogen. *See* SILVER; SILVER METALLURGY.

[E.M.WI.]

Silver chloride electrode

An electrode made of silver, covered or intimately mixed with silver chloride. One method of preparation consists of coating a silver wire or a silver-plated noble metal with silver chloride by electrolysis as an anode in a chloride solution. A second method consists of three steps: (1) pasting silver oxide or oxalate on a platinum helix, (2) reducing the oxide or oxalate to silver by heating to 500°C, and (3) chlorodizing part of the silver to silver chloride. A third method consists of pasting an intimate mixture of silver oxide and silver chlorate on a platinum helix and heating to 500°C. Although these systems are frequently referred to as silver chloride electrodes, they are, in reality, not electrodes until immersed in a chloride solution. The standard potential of the silver chloride electrode is −0.2224 volt relative to the normal hydrogen electrode at 25°C. The potential of the electrode is a logarithmic function of chloride ion; that is, it is reversible to chloride ion. This electrode finds wide use in the study of chloride systems, as a replacement for calomel in half-cells, and as the inner electrode in some glass electrodes. *See* CALOMEL ELECTRODE; ELECTRODE POTENTIAL; GLASS ELECTRODE; HYDROGEN ELECTRODE. [W.J.H.]

Silver metallurgy

The extraction from ores, refining, and preparation of silver. Silver is widely distributed in nature and occurs in both native and combined forms. The most common minerals are argentite, Ag_2S; cerargyrite, $AgCl$; and stephanite, Ag_5SbS_4. Some ores are treated for silver alone (or for combined silver-gold values); however, the greater part of silver is produced as a by-product of the smelting of base metal ores, notably those containing copper, lead, or zinc sulfides. The world's annual production of silver is about 250,000,000 fine ounces. Among the principal silver-producing countries (with 1957 production in millions of ounces) are Mexico, 46; U.S.S.R., estimated 25; Canada, 24.2; Peru, 23.0; and Australia, 15. The U.S. Treasury price of silver has been fixed since 1946 at $0.905 per ounce.

Cyanidation. As applied to silver ores, this process differs only in detail from the cyanidation of gold ores (*see* GOLD METALLURGY). Dilute cyanide solutions, containing dissolved air or other weak oxidizing agents, dissolve not only metallic silver but also silver sulfide and chloride. The silver cyanide complex $Ag(CN)_2^-$ is formed in each case. The precipitation procedures, involving reduction with zinc dust and subsequent refining of the precipitate, also resemble those for gold.

Refining of electrolytic slimes. Nearly all base metals which are refined electrolytically (for example, copper and nickel) contain some silver which is usually insoluble in the electrolyte and collects on the anode or as an anode slime. Recovery of the silver and other precious metals from this slime is commonly effected by (1) roasting to oxidize base metals, (2) water-leaching the calcine to remove soluble metals, (3) smelting the residue with a suitable flux to slag off remaining base metals, and (4) electrolytic refining (using either the Balbach-Thum or Moebius process) to separate the gold and silver. *See* GOLD.

Parkes process. This is employed almost universally to recover by-product silver from lead ores, nearly all of which contain some silver that collects in the lead bullion during smelting. In the Parkes process, zinc metal is stirred into the molten lead and combines with the silver to form an alloy which collects on the surface and is removed by skimming.

Amalgamation. This process, when applied to silver ores, is similar in principle to that used with gold ores. Although now largely superseded by cyanidation, it is of considerable historical interest and was used from the sixteenth to the nineteenth centuries to produce vast amounts of silver in Mexico and South America. *See* SILVER; SILVER ALLOYS. [J.H.]

Bibliography: W. H. Dennis, *Metallurgy of Non-Ferrous Metals*, 1954.

Silverfish

A small, primitive, wingless insect, *Lepisma saccharina*, of the order Thysanura. It is found throughout the United States and is a well-known household pest. It is about ½ in. long, covered with scales, and silvery gray in color. It is readily recognized by the three long appendages at the end of the abdomen. Silverfish eat food of high starch content, and may damage clothing, wall paper, books, and book bindings.

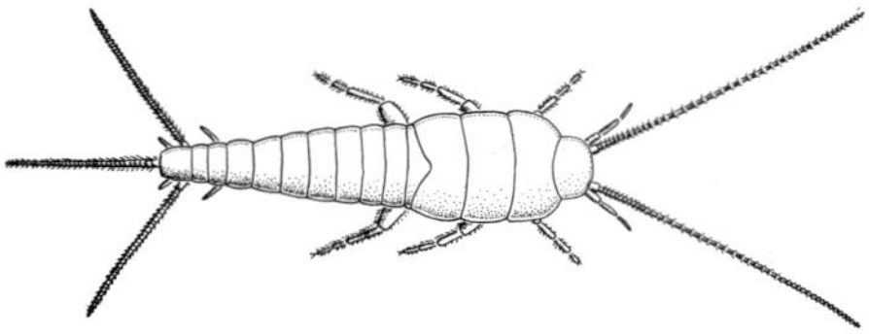

The silverfish, *Lepisma saccharina*; length to ⅓ in. (From E. L. Palmer, *Fieldbook of Natural History*, McGraw-Hill, 1949)

There is no metamorphosis, the young being similar to the adult. A few eggs at a time are laid in cracks or folds of food material. Development takes about 2 years in temperate regions and the life span is somewhat longer than that of most insects.

A closely related species, the firebrat, *Thermobia domestica*, can be distinguished by the dark and light mottlings on its back. This species prefers warm, even hot, situations, and frequents such places as bakeries and furnace rooms. *See* THYSANURA. [J.D.B.]

Silviculture

Silviculture and forest management are closely related branches of forestry that deal with the treatment and planned use of forests. This article discusses both silviculture and forest management.

SILVICULTURE

Silviculture is the theory and practice of controlling the establishment, composition, and growth of stands of trees for any of the goods and benefits that they may be called upon to produce. In practicing silviculture, the forester draws upon knowledge of all natural factors that affect trees growing upon a particular site and guides the development of the vegetation, which is either essentially natural or only slightly domesticated, along the lines that will best meet the economic demands of society in general and ownership in particular (*see* FOREST ECOLOGY).

The techniques proceed on the assumption that the natural vegetation of any site normally tends to extend itself to occupy all available growing space, making the fullest possible use of the available growth factors, such as light and moisture. The forester attempts, usually by cutting in the act of harvesting useful wood products, to create vacancies in the forest vegetation that will provide environments that are favorable either to the establishment of new, desirable trees (reproduction cuttings), or to the enhanced growth of those that remain (intermediate cuttings).

Intermediate cuttings. These are made during the life of a particular crop of trees to correct defects in its composition, to provide income, and to increase the amount or value of the timber produced. Under intensive management, these may include various kinds of release cuttings designed to free desirable trees not past the sapling stage from the competition of taller or faster-growing trees that either are of undesirable species or are defective specimens of otherwise acceptable species. If the trees to be removed in such cuttings are not merchantable, they may be killed either by girdling or by various tree-killing chemicals. *See* PLANT TRANSLOCATION (ORGANIC SOLUTES).

Most of these techniques of killing unwanted trees involve the poisoning or removal of a band of bark and cambium to halt the downward movement of carbohydrates from the leaves to the roots (*see* PLANT ANATOMY). The roots eventually starve, thus causing the death of the whole tree. Water-soluble poisons can also be introduced into the wood (xylem) and are then carried upward to kill the crown (*see* XYLEM). Some chemicals can be sprayed onto the foliage and branch surfaces, from either the ground or the air. Under the proper conditions, this treatment kills undesirable deciduous species without significant harm to conifers beneath them.

Improvement cuttings are made for the same general purposes as release cuttings but in stands beyond the sapling stage. The varied assortment of material eliminated is often merchantable; such

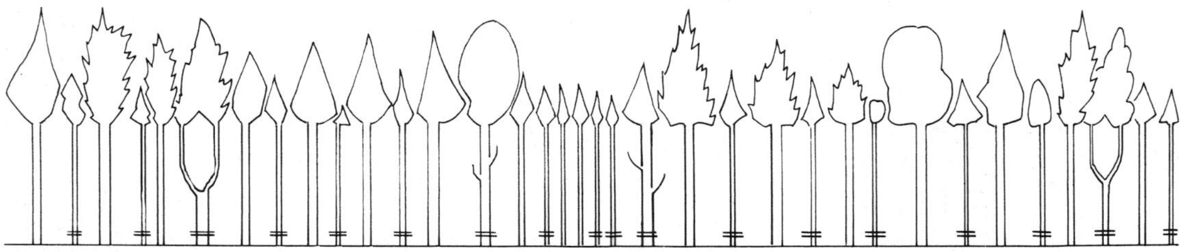

Fig. 1. A previously untreated, even-aged stand marked (by horizontal lines near the bases of the stems) for a typical thinning. In general the least vigorous trees would be removed but there are several instances in which thrifty trees of poor quality are cut to free trees of good form but lower vigor.

cuttings frequently constitute the first step in bringing previously untreated stands under management. Salvage cuttings are somewhat comparable operations conducted to save useful material from trees that have been killed or injured by the unexpected attacks of various damaging agencies.

Thinnings are made in immature stands to stimulate the growth of the trees that remain and to increase the total production (Fig. 1). Basis for their conduct is found in the fact that new stands start with many hundreds or thousands of trees per acre and ultimately decline to no more than several hundred as a result of severe competition. The crowns and roots of individual trees expand and the individuals that grow fastest in height overtop and ultimately kill the laggards. Artificial thinning represents a judicious acceleration and guidance of this process. So long as the trees of a stand are sufficiently numerous to be capable of expanding to occupy the available growing space completely, the gross volume of raw wood produced by a stand of given composition and age on a particular site remains nearly independent of the number of trees per acre. Because a given volume of wood has greater value and utility if produced on a small number of large trees rather than a large number of small trees, it is generally desirable to reduce the stocking periodically to the smallest number of good trees that will fully occupy the site. This process can enhance the growth of the remaining trees in diameter, but not in height. Thinning also increases the actual yield of a stand through the harvest of surplus merchantable trees that might otherwise be killed by competition and lost to decay. The practice often imitates the natural process by eliminating the smaller trees in favor of the larger. However, if the smaller trees happen to be more desirable than the larger trees and are of sufficient vigor, it is possible to alter the pattern of removals to rescue them. Thinning enables the forester to guide an existing stand along lines of optimum development and to secure current income from an otherwise immature crop (Fig. 2).

Pruning is the removal of branches from the stems of living trees, ordinarily for the purpose of increasing the volume of clear (knot-free) wood. The practice is best confined to high-quality trees destined to be favored by thinning as part of the final crop.

Reproduction cuttings. These are made when the prospective growth of an old stand is deemed less valuable than that of a new stand. In American practice, new stands are usually sought from natural seeding. Artificial planting is used mainly in situations in which there is little or no possibility of establishing a desirable stand by natural

Fig. 2. (*a*) Thinning in 60-year-old Douglas-fir, western Oregon (*Crown Zellerbach, Amer. Forest Products Ind.*). (*b*) Thinning of loblolly pine stand for pulpwood and saw logs (*South Carolina Forestry Commission, Amer. Forest Products Ind.*). (*c*) Eastern white pine plantation, 25 years old, after pruning and first thinning, Connecticut (*Yale Univ. School of Forestry*).

Fig. 3. Application of the clearcutting method of natural reproduction in old-growth Douglas-fir in the Pacific Northwest. (*Amer. Forest Products Ind.*)

seeding; however, it can also be used where natural regeneration is regarded as too slow or if selected superior strains are to be introduced.

Successful natural regeneration demands basic knowledge of the ecological requirements for germination and survival of the desired species (*see* DENDROLOGY). An abundant source of seed is essential. The condition of the seed bed and the arrangement of the remaining trees must also be adjusted to provide an adequate number of spots where the environment is more favorable to establishment of the desired species than to any others. The native flora of any forest region includes species adapted to colonize almost any kind of vacancy that might be created in the forest by natural disturbances, great or small. The forester attempts to simulate the kind of disturbance that will lead to establishment of the particular species desired.

Reproduction cuttings are usually classified according to the degree of exposure caused by removal of tree cover, the number and arrangement of trees left for seed on the cutover area, and the age distribution of the new stands created. In the clearcutting method all the trees on the area where regeneration is sought are cut, exposure is complete, and adjacent uncut stands are relied upon as a source of seed (Fig. 3). The seed-tree method differs only in that a limited number of trees are temporarily left on the area to provide seed, but the degree of exposure is scarcely diminished. These two methods are successful only for wind-disseminated species and where environmental conditions are highly favorable to the establishment of seedlings. They are, for example, used in the management of the southern pines and the coastal form of Douglas-fir. In the shelterwood method enough trees are left on the cutting area to reduce the degree of exposure significantly and also to provide a substantial source of seed (Figs. 4*a*, 5). In this method the growth of a major portion of the preexisting crop continues, and the old trees

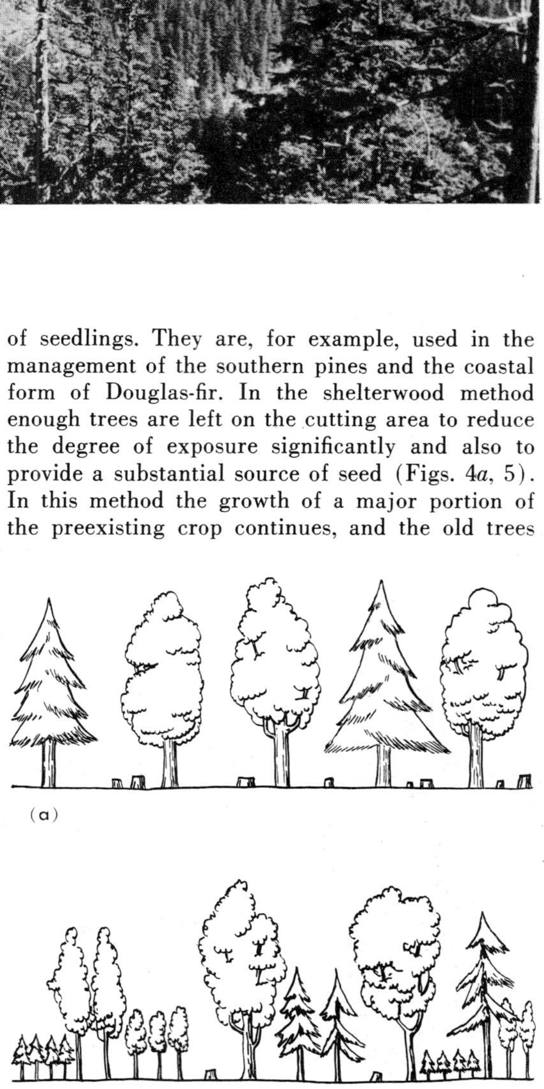

Fig. 4. (*a*) An even-aged stand after a uniform shelterwood cutting. (*b*) An uneven-aged stand after a single-tree selection cutting.

Fig. 5. Uniform shelterwood cutting in 80-year-old stand of oak and other hardwoods, Connecticut. (*Yale Univ. School of Forestry*)

are not entirely removed until the new stand is well established. The three methods just described lead to the creation of essentially even-aged stands.

The various modifications of the selection method, however, lead to the creation of uneven-aged stands characterized by the presence of three or more distinct age classes (Figs. 4*b*, 6). Reproduction cuttings under the selection method occur in the same stand at more or less equal intervals of time and are not concentrated near the end of the life of the crop as in the even-aged stands. Both the shelterwood and selection methods are subject to a very large amount of variation and modification; they are both applied in one form or another to the management of almost all species and forest types, except the few species that need protection from extreme exposure to become established.

The so-called coppice methods involve establishment of new stands from sprouts arising from the stumps or roots of cut trees. They remain important in America only in the reproduction of stands of aspen, but are commonly used in many parts of the world where forests of sprouting hardwoods are managed primarily for the production of quick crops of small fuel wood.

Clearcutting is not necessarily crude silviculture; it may even represent the final harvesting operation following a series of frequent intermediate cuttings. Conversely, light periodic partial cuttings do not invariably represent highly efficient practice but may involve "high-grading" in which the few good trees of an otherwise untended stand are cut to leave it dominated by undesirable trees.

In the application of partial cutting, use is often made of the concept of financial maturity as a guide in determining which trees to cut and which to reserve for additional growth. According to this concept the market value of each standing tree is regarded as a capital investment, and an attempt is made to predict the increase in value of the tree between the cutting immediately in prospect and the next cutting. If this increase in value does not represent an acceptable rate of compound interest on the investment represented by the present value of the tree, it is marked for immediate cutting. Under this concept, fast-growing trees of good quality are reserved while poor, unthrifty trees are likely to be cut. *See* FOREST SEEDING AND PLANTING.

Accessory practices. The application of silviculture involves a number of accessory practices other than cutting. Fire, like the axe, is an agency that can be used wisely to build the forest or unwisely in its witless destruction. In localities of high fire risk, it is often mandatory to dispose of the slash (logging debris) after cutting by burning it either broadcast as it lies or after it has been piled. This not only reduces the potential fuel, but it may also help the establishment of seedlings by baring the mineral soil or reducing the physical barrier represented by the slash. Slash disposal is most often

Fig. 6. (*a*) Selection cutting in ponderosa pine stand, eastern Oregon (*U.S. Forest Service*). (*b*) Natural reproduction of redwood after selection cutting, northern California (*Amer. Forest Products Ind.*).

necessary where the cutting has been very heavy or where the climate is so cold or dry that decay is slow. Deliberate prescribed burning of the litter beneath existing stands of fire-resistant species is sometimes carried out even in the absence of cutting to reduce the fuel for wild fires, to kill undesirable understory species (including those representing stages in plant succession later than those desired), to enhance the production of forage for wild and domestic animals, and to improve seedbed conditions. Prudence demands that such burning be prescribed only by experts under carefully chosen weather conditions and with ample provision for controlling the fires (*see* FOREST FIRE CONTROL). The practice has been used mainly in stands of fire-resistant 2- and 3-needled pines, primarily in the southeastern United States, where it closely imitates the effect of natural fires. Heavy disk-plows, modified bulldozers, and similar equipment also find limited use in the elimination of undesirable vegetation and preparation of the soil for establishment of new crops by planting or natural seeding.

Silvicultural practices may be conducted not only for the growth of wood products, but also to develop forest stands of form and composition consistent with wise management of water, wildlife, and other forest resources (Fig. 7). For example, the yield of water from a forested watershed can

Fig. 7. Stand of yellow poplar, oak, and other hardwoods with an understory of hemlock after several partial cuttings on a municipal watershed in Connecticut. (*Yale Univ. School of Forestry*)

Fig. 8. Two age classes of loblolly pine developed from natural seeding. Stand in background is about 80 years old and that in foreground is about 10 years old. (*Amer. Forest Products Ind.*)

be enhanced by frequent cuttings that interrupt the crown canopy and thus reduce the amount of precipitation that is intercepted and lost through evaporation. Periodic cuttings of the proper sort can also be used to provide a continuous supply of low food plants for wildlife (*see* CONSERVATION OF RESOURCES).

Silvicultural systems. Integrated schedules of treatment for stands are called silvicultural systems. They cover both intermediate and reproduction cuttings, but are classified and named in terms of the general method of reproduction cutting contemplated. Such programs are evolved for particular situations and kinds of stands with due regard for all the significant biological and economic considerations involved. These considerations include the desired uses of the land, kind of wood products sought, prospective costs and returns of the enterprise represented by management of the stand, funds available for long-term investment in stand treatments, harvesting techniques and equipment employed, reduction of losses from damaging agencies, and the natural requirements that must be met in reproducing the stand and fostering its growth.

FOREST MANAGEMENT

This phase of forestry deals with the application of business methods and technical forestry principles to the operation of particular forest properties. In the broadest sense, it involves the integrated application of all pertinent knowledge drawn from the natural and social sciences to the conduct of enterprises involving ownership of forest land. Technological aspects of the subject are described in the following paragraphs.

Rotations and cutting cycles. The rotation is the period required to establish and grow a timber crop to a specified condition of maturity. The basic objective is to use the rotation that will give the maximum average annual profit per acre for the whole rotation. Important among the factors that fix this rotation length are (1) the inevitable decline in physiological efficiency and growth associated with increased size and age of trees; (2) increasing incidence of loss caused by damaging agencies, especially heart-rotting fungi; (3) degree of attainment of the tree sizes required for efficient manufacture of the products desired; (4) the concept of financial maturity; and (5) attainment of seed-bearing age. In North American forests, economic rotations in managed forests tend to be 50–120 years long, involving attainment of final tree diameters of 18–30 in., although management exclusively for small products such as pulpwood may involve shorter rotations and smaller final diameters.

The cutting cycle is the planned interval between harvesting operations in a particular stand or group of stands. Under the least intensive kind of management, it is equal to the rotation. In uneven-aged stands consisting of a wide variety of intermingled age classes, it is usually a relatively rigid fraction of the rotation ranging from about 1/10–1/3, depending on the intensity of management. In even-aged stands the interval between cuttings tends to fluctuate according to a pattern related to the need for tending the crop at different stages of development, although the desirability of systematic programming imposes a certain degree of regularity.

Regulation of cuts. One of the peculiar features of the managed forest is the fact that living trees constitute both the productive machinery (that is, growing stock or forest capital) and the product. Careful judgment must be exercised in determining which and how many trees should be reserved for growth and which represent a surplus over the essential growing stock and may thus be regarded as product available for harvest. The ideal objective is to create a growing stock containing the correct distribution of ages, species, tree diameters, and classes of tree quality to provide a steady and optimum yield of product trees in perpetuity. Ideally the capacity of the manufacturing facilities dependent on the particular forest for raw material is in balance with this sustained yield. Under such circumstances, the economic security of the enterprise tends to depend upon the accuracy with which the volume of growing stock and the volume of cut stock are kept in balance.

In its simplest terms, a forest capable of providing an optimum sustained yield would have an equal area occupied by even-aged stands of the desired composition representing each year of age from 1 year to that corresponding to the chosen rotation age. For example, a forest of 80,000 acres arranged on this ideal basis with a 80-year rotation would have 1000 acres each of stands 1 year old, 2 years old, 3 years old, and so on up to 1000 acres of 80-year-old stands ready for immediate cutting (Fig. 8). The approach embodied here is regulation by area, and it can be applied only to forests consisting of stands that are essentially even-aged. Yield tables are often available to provide a basis for estimating the volumes of timber present in each age class (*see* FOREST MENSURATION). The main component of the annual cut is the volume present in the one final, mature age class, although substantial additional volumes may be harvested from thinnings and other intermediate cuttings in the immature age classes.

An alternative technique is regulation of cut by volume. Basically this also involves the ideal forest with all necessary age classes represented on equal areas, but no attention is paid either to the ages of trees or to the area occupied by different age classes. Instead, the appropriate distribution of numbers of trees, with respect to their diameter (Fig. 9), or volume or basal area, with respect to diameter or age class, is derived from either yield tables or various hypotheses about the straight-line relationship that the logarithm of the average number of trees per acre bears to tree diameter in forests with the correct distribution of age classes. Advantage is taken of the fact that, in forests with the correct distribution of age, volume, and tree di-

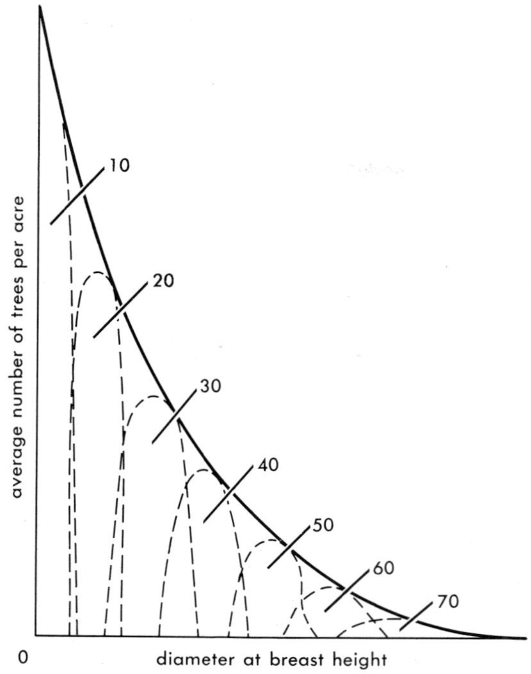

Fig. 9. Graph showing (by the solid line) the distribution of numbers of trees with respect to diameter in a stand or forest in which the representation of age classes has been adjusted to guarantee a perpetual and uniform sustained yield. The dashed lines under the curve illustrate how this diameter distribution is derived from the summation of nearly bell-shaped curves defining the diameter distributions of 7 even-aged stands varying from 10 to 70 years of age.

ameter, the cut that can be made in a particular period without jeopardizing future sustained yield is equal to the growth occurring in the same period. This growth can be estimated both from extrapolation of direct measurements of growth in the period of equal length just past and from use of yield tables.

The application of either general method of regulating the cut is complicated by differences in the growing capacity of various sites, composition, annual market requirements, silvicultural methods, and harvesting techniques, but most of all by the fact that the ideal distributions can at best be only approximated even after decades of carefully planned cutting. [D.M.S.]

Bibliography: *See* FOREST AND FORESTRY.

Similitude

The use in scientific studies and engineering designs of the corresponding behavior between large and small objects of similar nature. Two structures behave similarly if they are geometrically, kinematically, and dynamically similar. For geometric similarity the ratios of critical dimensions, such as ratios of diameters to lengths, must be equal. For kinematic similarity, corresponding velocities and velocity gradients must be in the same ratios at corresponding locations. For dynamic similarity ratios of forces acting within the two structures, such as viscosity and inertia, must be equal.

As an example, cavitation occurs in a flowing liquid when the total pressure falls below the vapor pressure. To test a hydraulic turbine for cavitation may be expensive and cumbersome, so a small model is built and tested. In scaling the model, the geometrical dimensions of the prototype are reduced, and a fluid is used with a correspondingly scaled vapor pressure, or the operating pressure is scaled to preserve the relations between the characteristics that affect the behavior of the turbine and the model. *See* DIMENSIONAL ANALYSIS; DYNAMIC SIMILARITY; FLUID MECHANICS; MODEL THEORY.

Model tests in wind and water tunnels, towing tanks, dynamometers, antenna test ranges, and plasma reacting with magnetic fields are predicated on similitude relations between the model being tested and the full-sized object being studied. The use of small models greatly increases the speed with which design changes can be explored. Where explosive reactions may occur or where the structure is tested to failure, for example, the use of small models reduces hazards. Considerable power is saved and other economies are achieved in the use of small models. [F.H.R.]

Simple machine

Any of several elementary machines, one or more of which will be found in practically every machine. The group of simple machines usually includes only the lever, wheel and axle, pulley (or block and tackle), inclined plane, wedge, and screw. However, the gear drive and hydraulic press may also be considered as simple machines. The principles of operation and typical applications of simple machines depend on several closely related concepts. *See* EFFICIENCY; FRICTION; MECHANICAL ADVANTAGE; POWER; WORK.

Two conditions for static equilibrium are used in analyzing the action of a simple machine. The first condition is that the sum of forces in any direction through their common point of action is zero. The second condition is that the summation of torques about a common axis of rotation is zero. Corresponding to these two conditions are two ways of measuring work. In machines with translation, work is the product of force and distance. In machines with rotation, work is the product of torque and angle of rotation. *See* BLOCK AND TACKLE; GEAR DRIVE; HYDRAULIC PRESS; INCLINED PLANE; LEVER; SCREW; WEDGE; WHEEL AND AXLE.

Work is the product of a force and the distance through which it moves. For example, the work done in raising a 10-lb object 15 ft is 150 ft-lb. In this example the work done on the weight goes into increasing the potential energy of the object. Work and energy, both potential and kinetic, have the same units, and in general the purpose of a machine is to convert energy into work.

For rotating machines, it is more convenient to consider torque and angular displacements than force and distance. Work is then expressed as the product of the torque and the angle (in radians) through which the object rotates while acted on by the torque. Torque, in turn, is the force exerted at a given radius from an axis of rotation. Thus, a 10-lb force at the end of a 15-ft crank exerts a torque of 150 lb-ft.

Power is the rate of doing work. For example, one horsepower is arbitrarily defined as 550 ft-lb per sec, which equals 33,000 ft-lb per min.

[R.M.PH.]

Simulator

Any device in which a physical or a conceptual process or a mechanical, electronic, biological, or social system or combination thereof is represented in such a way that the phenomenon can be imitated and thus studied experimentally.

In a sense all analog computers, particularly those in which there is a one-to-one correspondence between the problem parameters and variables and the computer parameters and variables, are simulators. An example of simulation in this respect is the study of dynamical systems by a network analyzer, in which there exists a correspondence between mass, damper, and spring on the one hand and inductors, resistors, and capacitors on the other.

More suggestive examples are the submarine, flight, and wind-tunnel drive simulators. In the first two, the actual physical equipment and instrumentation is modeled for training use. Trainee personnel experience rolling, pitching, and yawing motions, observe instruments, and manipulate controls which, through analog equipment, solve various force and moment equations and, through servomechanism means, alter the behavior of the model.

Digital computers have been used to simulate complex industrial processes. An example is the simulation of a factory in the production of an item made up of many parts. Given the number of each part required, the machines required in making the part, the time required on each machine, and the productive capacity of the machines, the computer can be programmed so as to schedule a sequence of machine operations which will tend to minimize over-all production time. Various guiding rules, which are believed to establish schedules of an optimal nature or which must be stated to preclude unworkable sequences of operation, are included in the program.

Another type of digital-computer simulation is management gaming. Various controllable aspects of a business enterprise, such as advertising budget, research budget, production volume, and product price, may be determined within limits by the human "player." These decisions are fed into the computer, which then simulates the operation of the business and prints out reports indicating, for example, sales volume, current inventory, statement of profit and loss, and production capacity for another period.

Digital computers may simulate, in a way reminiscent of direct analog simulation, various processes that are thought by some to be inherently digital or symbolic in character. Examples are learning, concept formation, and heuristic reasoning processes. However, there is no wide agreement on the validity of the several analogies. *See* ANALOG COMPUTER; DIGITAL COMPUTER. [R.J.N.]

Bibliography: C. C. Gotlieb and J. N. P. Hume, *High-Speed Data Processing*, 1958; E. M. Grabbe (ed.), *Automation in Business and Industry*, 1957; G. R. Stibitz and J. A. Larrivee, *Mathematics and Computors*, 1957.

Sine wave

A wave having a form which, if plotted, would be the same as that of a trigonometric sine or cosine function. It generally results from the solution of a problem having a one-dimensional space coordinate, such as the transverse vibrations of a string, longitudinal vibrations of a bar, or the propagation of plane waves of electromagnetic radiation or sound.

The sine wave may be thought of as the projection on a plane of the path of a point moving around a circle at uniform speed. For example, in part a of the illustration, assume that the moving point travels around the circle at constant speed v. The projection onto a plane would trace back and

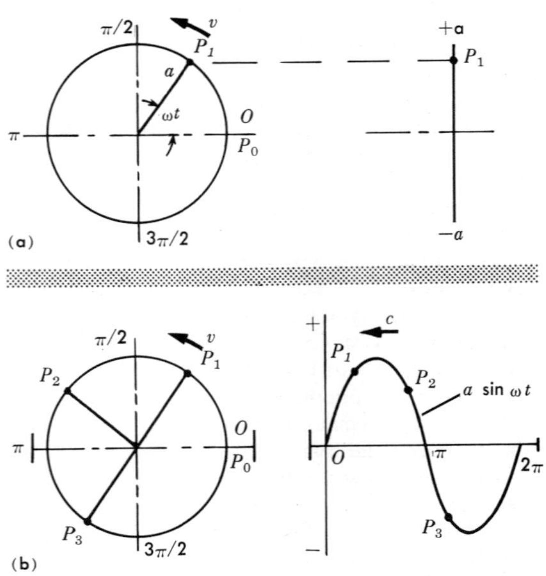

Simple harmonic motion and sine wave. (*a*) Point P moves around circle at constant tangential speed v; its projection moves up and down between limits $+a$ and $-a$. (*b*) Same conditions as in (*a*), except that plane on which projection is plotted is traveling to left at speed c, thereby generating a sine wave. Angular speed is ω rad/sec; total angle traced out from zero position equals ωt rad.

forth on the line indicated as the point went around the circle. If the plane is now moved to the left at constant speed c, as in part b of the illustration, the resulting trace has the form of the graph of the sine function.

The sine wave trace of part b describes a body moving in simple harmonic motion. This motion changes in magnitude and time so that it repeats itself exactly, as long as uniform speed is maintained. It is characteristic of one-dimensional vibrations and one-dimensional waves having no dissipation. *See* HARMONIC MOTION.

The sine wave is the basic function employed in harmonic analysis. It can be shown that any complex motion in a one-dimensional system can be described as the superposition of sine waves having certain amplitude and phase relationships. The technique for determining these relationships is known as Fourier analysis. *See* FOURIER SERIES AND INTEGRALS; *see also* WAVE EQUATION; WAVEFORM; WAVE MOTION. [W.J.G.]

Single crystal

In crystalline solids, the atoms or molecules are stacked in a regular manner, forming a three-dimensional pattern which may be obtained by a three-dimensional repetition of a certain pattern unit called a unit cell (*see* CRYSTAL STRUCTURE; CRYSTALLOGRAPHY). When the periodicity of the pattern extends throughout a certain piece of material, one speaks of a single crystal. A single crystal is formed by the growth of a crystal nucleus without secondary nucleation or impingement on other crystals.

Growth techniques. Among the most common methods of growing single crystals are those of P. Bridgman and J. Czochralski. In the Bridgman method, the material is melted in a vertical cylindrical vessel which tapers conically to a point at the bottom. The vessel then is lowered slowly into a cold zone. Crystallization begins in the tip and continues usually by growth from the first formed nucleus. In the Czochralski method, a small single crystal (seed) is introduced into the surface of the melt and then drawn slowly upward into a cold zone. Single crystals of ultrahigh purity have been grown by zone melting. Single crystals are also often grown by bathing a seed with a supersaturated solution, the supersaturation being kept lower than is necessary for sensible nucleation.

When grown from a melt, single crystals usually take the form of their container. Crystals grown from solution (gas, liquid, or solid) often have a well-defined form which reflects the symmetry of the unit cell. For example, rock salt or ammonium chloride crystals often grow from solutions in the form of cubes with faces parallel to the (100) planes of the crystal, or octahedra with faces parallel to the (111) planes. The growth form of crystals is usually dictated by kinetic factors and does not correspond necessarily to the equilibrium form. *See* CRYSTAL GROWTH; CRYSTALLIZATION; ZONE REFINING.

Physical properties. Ideally, single crystals are free from internal boundaries. They give rise to a characteristic x-ray diffraction pattern. For example, the Laue pattern of a single crystal consists of a single characteristic set of sharp intensity maxima. *See* X-RAY DIFFRACTION.

Many types of single crystal exhibit anisotropy, that is, a variation of some of their physical properties with the direction along which they are measured. For example, the electrical resistivity of a randomly oriented aggregate of graphite crystallites is the same in all directions. The resistivity of a graphite single crystal is different, however, when measured along different crystal axes. This anisotropy exists both for structure-sensitive properties, which are strongly affected by crystal imperfections (such as cleavage and crystal growth rate), and structure-insensitive properties, which are not so affected (such as elastic coefficients). The anisotropy of a structure-insensitive property is described by a characteristic set of coefficients which can be combined to give the macroscopic property along any particular direction in the crystal. The number of necessary coefficients can often be reduced substantially by consideration of the crystal symmetry; whether anisotropy, with respect to a given property, exists depends on crystal symmetry.

The structure-sensitive properties of crystals (for example, strength and diffusion coefficients) appear to be governed by defects, often on an atomic scale, in the crystal. *See* CRYSTAL DEFECTS. [D.T.]

Bibliography: H. E. Buckley, *Crystal Growth*, 1951; A. Holden, Preparation of metal single crystals, *Trans. Am. Soc. Metals*, 42:319–346, 1950; F. Seitz and D. Turnbull (eds.), *Solid State Physics*, vol. 4, 1957, and vol. 6, 1958.

Sink flow

A point in three-dimensional flow, into which fluid is presumed to flow uniformly from all directions. The strength of a sink is defined as the volume per unit time flowing into the point. A sink may also be defined as a negative source. *See* SOURCE FLOW.

In two-dimensional flow, in which all flow occurs in parallel planes that have identical flow patterns, a sink is a straight line into which fluid flows uniformly from all directions at right angles to the line. It appears as a point on the customary two-dimensional flow diagram.

By analogy, flow of ground water into a well point closely approximates three-dimensional sink flow. The concept of the sink is useful in building up complex flow patterns when used in conjunction with sources, doublets, and uniform flow. [V.L.S.]

Sintering

The welding together and growth of contact area between two or more initially distinct particles at temperatures below the melting point of the substance. Sintering may take place at room temperature in some materials, but the technologically important cases are those occurring at elevated temperatures. Since the rate of sintering is greater with smaller particles than with large, the process is most important with powders, as in powder metallurgy and in the firing of ceramic oxides. Some writers extend the term to include situations in which some liquid is present; the term then can be applied to all ceramic firing operations.

Sintering is observed macroscopically as an increase in strength, electrical conductivity, and density. Often the first two can be observed before there is an increase in density and an accompanying shrinkage.

Although sintering will occur in loose powders, it is enhanced greatly by compacting the powder. This may be done before or during the heat treatment; the latter is called hot pressing (*see* CERAMIC TECHNOLOGY). Generally, after precompaction, the greater the density before heating, the greater will be the fired or sintered density for a given heat-treatment.

It has been observed that sintering begins in many materials at a temperature about half the melting temperature (on an absolute temperature scale), and this temperature has been called the Tammann temperature. However, this rule is followed only very roughly. The onset of sintering is affected by the particle size of the powder, small amounts of impurities, and other factors.

Originally the term sintering was used to refer to processes in which it was thought no liquid whatsoever was present. However, since it is not absolutely certain that there is no liquid in some industrially important cases of sintering and since, in any case, there are important ceramic processes in which liquid is clearly present, it has become customary in recent years to consider that sintering sometimes occurs in the presence of a liquid. This is often referred to as wet sintering, the process with no liquid being called dry sintering.

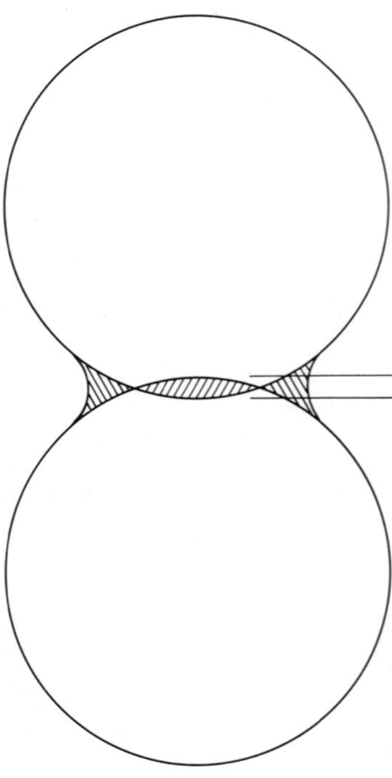

Sintering of two spheres, initially in tangential contact, showing the resulting shrinkage. The material in the lens where the particles overlap is removed to form a neck between the particles.

It is generally conceded that the driving force in sintering is the surface energy, which decreases because the total surface area decreases as sintering proceeds. When two spheres originally in tangential contact sinter, they move closer together and the material near the point of contact (shaded area in the illustration) must be moved to the neck between the spheres. This process of material transport is not yet completely understood. Movement of liquids could take place by viscous flow (as when two drops of water coalesce to form one) and it appears that this does occur in sintering of glass. However, in crystalline materials such a process seems less plausible and it is thought that here defects in the crystal structure (that is, missing atoms) move toward the point of contact and material moves in the opposite direction. It seems reasonable that the rate of sintering will depend on the distance material must diffuse and this agrees with the fact that fine powders sinter much more readily than do coarse powders.

Another aspect of the effect of particle size is the phenomenon of grain growth. Besides reducing their surface energy by growing together, the grains of a powder can reduce their surface energy by increasing in size. This second process competes with sintering, and when exaggerated grain growth occurs, it seems to do so very rapidly and stop all further shrinkage. The smaller the initial particle size of the powder, the larger the grains will be after the sudden growth.

Sintering in the presence of a liquid also involves movement of material, partly by viscous flow of the liquid and partly by ionic migration through it. It seems that the solubility of the solid in the liquid is greater at the points of contact, because the pressure is higher there. Once in solution, the ions diffuse away from the points of contact between particles and reprecipitate elsewhere, filling the voids. *See* CERMET; POWDER METALLURGY.

[M.C.M.]

Sinus

A space in an organ, tissue, or bone. Usually, reference is to the paranasal sinuses of the face. In man, four such sinuses, lined with ciliated mucous membrane, communicate with each nasal passage through small apertures. The ethmoid and sphenoid sinuses are located centrally between and behind the eyes. The frontal sinuses lie above the nasal bridge, and the maxillary sinuses are contained in the upper jaw beneath the eye sockets. In addition,

Unsintered, as-pressed "green" briquet of bronze bearing metal. White areas are metallic tin powder; black areas are graphite; reddish areas are copper. (*Greenback Industries*)

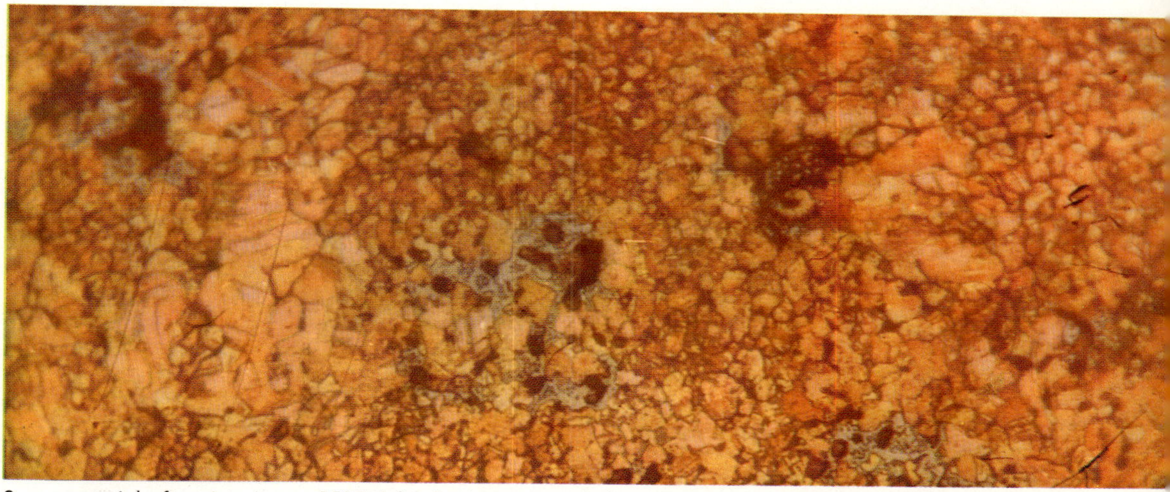

Same material after sintering at 1525°F for 5 minutes in endothermic atmosphere. White areas are undiffused tin; reddish areas are high copper; larger black areas are voids; smaller black areas are graphite. (*Greenback Industries*)

Same material after sintering at 1550°F for 15 minutes in a dissociated ammonia atmosphere. Entire microstructure shows large yellow-brownish equiaxed grains, with normal black voids throughout. (*Greenback Industries*)

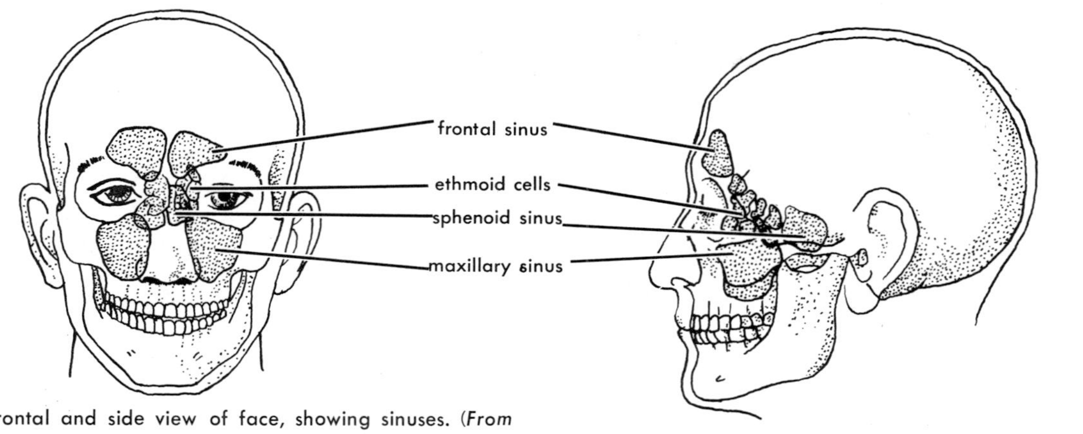

Frontal and side view of face, showing sinuses. (*From N. L. Hoerr and A. Osol, eds., Gould Medical Dictionary, 2d ed., Blakiston–McGraw-Hill, 1956*)

the mastoid portion of the temporal bones contains air cells lined with similar epithelium.

These accessory air sinuses vary in size and shape with the individual and with age. They are present in the infant only as minute cavities, if at all, and usually develop to adult size as the skull bones enlarge after eruption of the permanent teeth.

Fluid or inflammatory products that accumulate in the sinuses usually alter voice resonance and produce headache or pain. They are frequently the result of acute or chronic infections, or sinusitis. The membranes are well supplied with blood, lymphatics, and nerves, particularly those nerves sensitive to pain and pressure. Various theories concerning the function of the sinuses are prevalent, but none is satisfactory. In man, erect posture decreases natural drainage of the paranasal sinuses. [E.G.ST.]

Sinusitis

Inflammation of the mucous membranes lining the paranasal sinus. Abundant mucus is produced, which cannot drain into the nose because the narrow ducts are obstructed by the swollen mucous membrane. Pressure rises in the sinus, causing pain in the form of localized headaches. The localization varies according to the sinus involved. Extension of the infection to the neighboring structures becomes dangerous because of the close relationship to the brain. In chronic sinusitis polyps can develop which protrude into the nose. [E.WE.]

Siphonaptera

An order of insects commonly known as the fleas. These animals are of importance because they are bloodsucking pests of man and animals and transmit serious diseases from animals to man.

Fleas in the adult stage are recognized with ease. They are small, about ⅛ in. in length, dark brown in color, laterally flattened, and with three pairs of legs modified for jumping. The body is more or less oval in shape and armed with spines and setae, adapting the flea for living among the hairs of animals. The head is provided with mouthparts modified for sucking blood.

Life history. Fleas have a complete metamorphosis, that is the life history involves four distinct phases, the egg, larva, pupa, and adult. Each adult female flea lays a number of eggs, 400 or more in some species, over a long period of time. These eggs may be laid while the flea is on or in the nest or sleeping place of its host. Large numbers may occur on mats or cushions where cats and dogs rest or sleep. The egg, which is white and ovoid, usually hatches in a few days into a slender, very active, segmented larva which is yellowish-white in color and characterized by well defined, transverse rows of setae. Blood, host feces, and adult flea feces, as well as vegetable matter, make up the larval diet. The blood of the host is essential, but the larvae are not usually parasitic and the blood is obtained largely from adult flea feces. The length of time spent in the larval stage is highly variable and depends on conditions of nutrition, temperature, and humidity, as well as the species of flea involved. After maturity, the larva spins a silken cocoon which is covered with particles of dirt and nest debris. Molting takes place in the cocoon to form the pupal stage, a nonfeeding, resting period which may last only a few days to nearly a year, depending on circumstances. The adult flea emerges from the pupa. The life cycle of a dog flea may be completed in as little as two weeks under favorable conditions. The adult fleas, both male and female, feed exclusively on the blood of the host. They do not live on the host all the time, but visit it to feed, and many are found in the litter and debris of the nest.

Medical importance. During the two decades following the outbreak of World War II, there was great activity in the study of fleas and the world flea fauna became very well known, with specialists recognizing about 1400 species and subspecies. The great majority of these are obscure ectoparasites of mammals, and to a lesser extent birds, and do

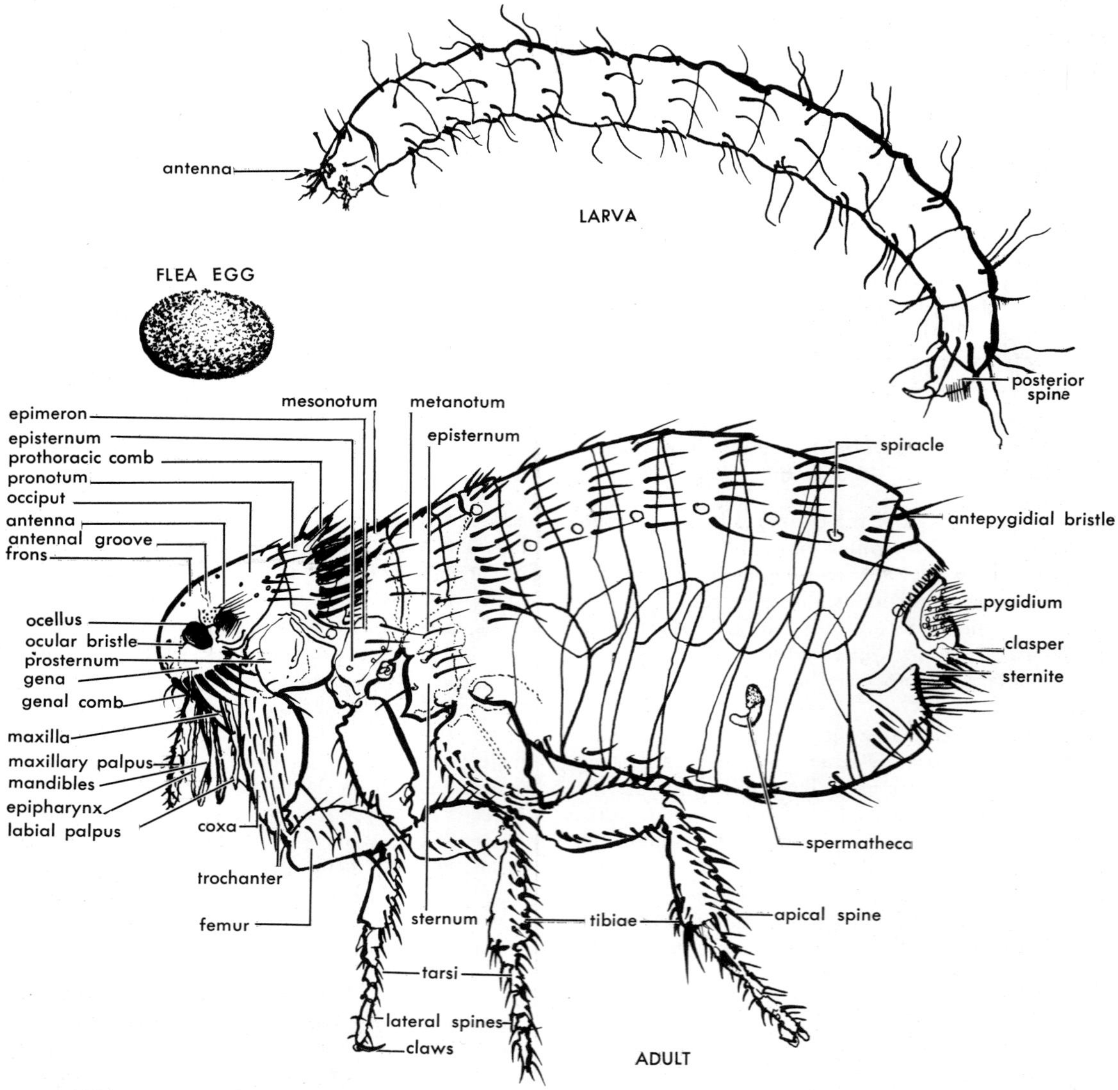

Fig. 1. Siphonaptera, various stages of the life cycle of the dog flea, *Ctenocephalides canis* (Curtis). (*From E. O. Essig, College Entomology, Macmillan, 1942*)

not affect man. Host specificity is exhibited in that, ordinarily, a certain taxonomic group of fleas will parasitize a specific taxonomic group of hosts. Nevertheless, some species may leave their favored hosts and attack man, and these species are the pests and disease carriers. Bubonic plague, which was responsible for millions of deaths in India during the first two decades of this century, is a disease of rats. It is transmitted between rats and from rats to man by fleas, particularly the oriental rat flea, *Xenopsylla cheopis*. This is a cosmopolitan species occurring on rats throughout the world. Sylvatic plague is the term used to denote plague occurring among the wild small animals of western United States, such as ground squirrels, chipmunks, and prairie dogs. Although many species of fleas are capable of transmitting sylvatic plague, human cases are rare, perhaps because man does not often come in contact with the animal and insect reservoirs. Murine or endemic typhus fever, another disease transmitted from domestic rats to man by rat fleas, was of great public health significance in southeastern United States during the period 1944–1950, when thousands of cases occurred. Tularemia, or rabbit fever, a disease of small wild animals, particularly of western and midwestern United States, may be transmitted to man by fleas. However, this disease, which is not common, is more often contracted while skinning infected animals. Flea allergy is the term applied by physicians to the severe reactions which sometimes result from flea bites. This disease is regarded as a major dermatological problem in California. Often the urticarial papular eruption resulting from flea

Important species of Siphonaptera

Scientific name	Common name	Importance
Ctenocephalides canis	Dog flea	Common household pest; intermediate host of dog tapeworm; readily attacks man
Ctenocephalides felis	Cat flea	Often abundant in households
Ctenopsyllus segnis	Mouse flea	Cosmopolitan, common on mice and rats; transmits plague among rodents
Ceratophyllus gallinae	European chicken flea	Infests fowl
Ceratophyllus niger	Western chicken flea	Fowl
Diamanus montanus	Ground squirrel flea	Transmits sylvatic plague and tularemia among rodents
Echidnophaga gallinacea	Sticktight flea, chicken flea	Serious pest of domestic fowl
Nosopsyllus fasciatus	European rat flea	May transmit bubonic plague from rodent to man
Pulex irritans	Human flea	Attacks man; not as common as the cat and dog flea; may transmit plague and endemic typhus
Tunga penetrans	Chigoe, jigger, sand flea, burrowing flea	Female burrows into skin of domestic animals and man
Xenopsylla cheopis	Indian rat flea, tropical rat flea	Transmits bubonic plague and endemic typhus

bites in allergic persons is mistakenly diagnosed as hives and attributed to some food. The fleas involved are those species that most often occur as household pests, the dog flea (*Ctenocephalides canis*), the cat flea (*Ctenocephalides felis*), and the human flea (*Pulex irritans*). In tropical America, a flea known as the "chigoe," "nigua," or sand flea (*Tunga penetrans*) can penetrate into the skin between the toes of man and cause festering sores which may lead to tetanus and gangrene. The sticktight, or chicken flea (*Echidnophaga gallinacea*), is similar to the nigua but attacks the heads of poultry, cats, and dogs, and sometimes annoys man. Fleas are also intermediate hosts of cat and dog tapeworms, and very rarely children may become infected by accidentally swallowing dog or cat fleas.

Control of fleas. Bubonic plague and murine typhus fever are controlled by measures directed against rats, such as poisoning, trapping, and ratproofing buildings, and by the application of a 10% DDT dust to rat runs and harborages to kill rat fleas. Dog and cat fleas are resistant to DDT but often may be controlled by applying a 4% malathion dust. To kill immature stages, cellars and outbuildings should be sprayed with creosote oil, and living quarters should be treated with flaked naphthalene at the rate of 5 lb per room. Carbolated petroleum jelly and derris powder are helpful in ridding poultry, dogs, and cats of sticktight fleas. In man, the nigua should be removed with a sterilized needle and the wound treated with iodine. Shoes should be worn as a prophylactic measure wherever this flea is common. The immature stages of the nigua breed in the ground or sand frequented by domestic animals, particularly pigs. It may be destroyed by spraying lightly with creosote oil. *See* INSECTA. [I.FO.]

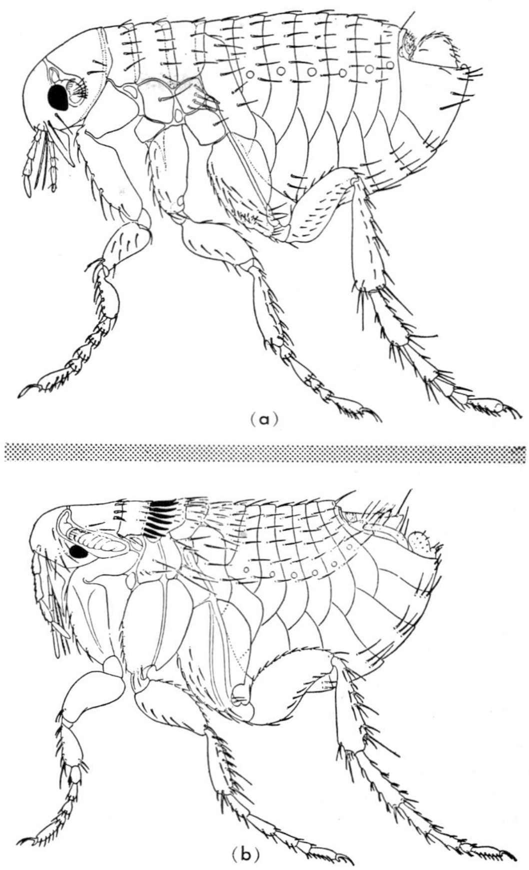

Fig. 2. (*a*) Human flea, *Pulex irritans* L. (*b*) European rat flea, *Nosopsyllus fasciatus* (Bosc). (*From E. O. Essig, College Entomology, Macmillan, 1942*)

Bibliography: I. Fox, *Fleas of Eastern United States*, 1940; H. E. Ewing and I. Fox, *The Fleas of North America*, USDA Misc. Publ. 500, 1943; C. A. Hubbard, *Fleas of Western North America*, 1947; G. P. Holland, *The Siphonaptera of Canada*, Canada Dept. Agr. Tech. Bull. 70, 1949.

Siphonophora

An order of the class Hydrozoa of the phylum Coelenterata, characterized by an extremely complex organization of components of several different types, some having the basic structure of a jellyfish, others of a polyp. The components may be connected by a stemlike region or may be more closely united into a compact organism. The medusalike components include swimming bells, a gas-filled float, bracts which are tough gelatinous bodies presumed to be protective, and gonophores which produce the eggs and sperms. Polyplike components include gastrozooids with mouth and one long hollow tentacle; dactylozooids without mouth and often with one tentacle; and gonozooids which bear the gonophores.

Most siphonophores possess a float and are animals of the open seas. Best known is the Portuguese man-of-war, *Physalia*, with a float as much as 40 cm long, and tentacles which extend downward for many meters. These animals may be swept shoreward and make swimming not only unpleasant but dangerous. *Velella*, the by-the-wind sailor or purple sail, has a flap which extends above water. Some siphonophores propel themselves by jets of water produced by the swimming bells. *See* COELENTERATA; HYDROZOA. [S.CR.]

Sipunculoidea

A phylum of marine worms which dwell in burrows, secreted tubes, or adopted shells. The body is greatly elongated in the dorsoventral axis. As a result, both mouth and anus occur close together at one end of the body, and the alimentary tract is a long, twisted loop which extends ventrally and shows little differentiation along its course. The jawless mouth, surrounded by tentacles, is situated in an eversible proboscis, or introvert, which is projected by hydrostatic pressure generated by the strongly muscular body wall. The proboscis can be withdrawn by special retractor muscles which are anchored in the trunk wall and extend to its tip. Species that inhabit snail shells have developed special holding devices at their ventral extremity.

The body cavity is an extensive schizocoele which shows no evidence of segmentation but is filled with fluid and traversed by fibers. There is also a lacunar system of vessels forming a ring at the base of the tentacles. The vessels extend into the tentacles and also into one or two large, saclike polian vesicles lying along the esophagus. The fluid of both systems contains abundant pink corpuscles with a respiratory pigment, hemerythrin; it also contains peculiar ciliated bodies, termed urns, which are budded from the walls of the polian vesicles and thought to have a stirring and scavenging function. Amebocytes also occur.

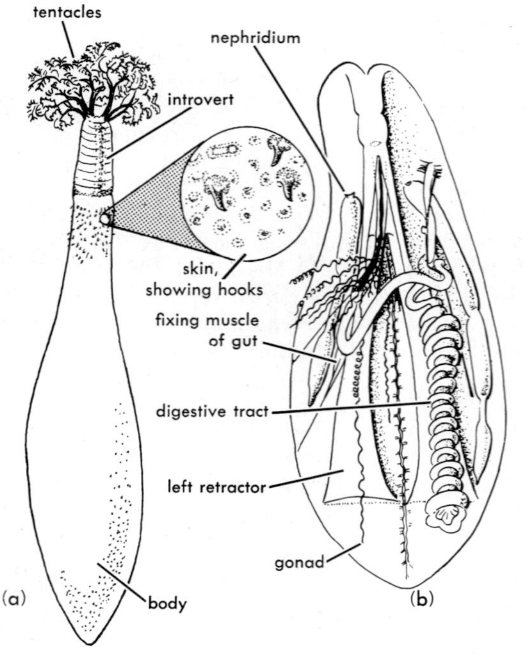

Two examples of Sipunculoidea. (*a*) *Dendrostoma petraeum*. (*b*) *Dendrostomum pyroides* (after Chamberlin).

A pair of giant metanephridia whose ventral pores open to the exterior serve in excretion. Assisted by cilia along the tube walls, the metanephridia expel waste-filled coelomic cells called excretophores. They also serve, perhaps even more importantly, as genital canals.

The nervous system includes a dorsal brain that is continuous, by right and left connectives, with a nonsegmented ventral cord which extends to the posterior tip of the worm. Frontal and nuchal organs with apparently sensory functions and often pigmented eyes also are present near the brain.

In this group of animals, the sexes are separate. The gonads, which arise from the coelomic lining near the origins of the ventral proboscis retractor muscles, release reproductive cells which mature chiefly while free in the coelomic fluid.

In early development cleavage is spiral. The blastopore becomes the mouth and an anus forms anew and is located dorsally. A free-swimming trochophore larva is formed bearing a preoral ring of small cilia, the prototroch, and a postoral ring of longer locomotory cilia, the metatroch.

There are about a dozen genera, including *Phascolosoma* and *Dendrostoma*, with probably more than 200 species. Sipunculids are widely distributed. All are bottom-dwellers, feeding by means of tentacles and cilia on the organic matter in the ingested mud and sand. *See* CLEAVAGE, EMBRYONIC; SCHIZOCOELA; *see also* CELL LINEAGE. [F.A.B.]

Siren (acoustics)

An apparatus for generating sound by the mechanical interruption of the flow of fluid (usually air) by a perforated rotating disk or cylinder. The disk may be so shaped that the fluid under pressure causes the disk to rotate, or conversely, the rotation of the disk may cause the flow of the fluid. The resulting frequency is the product of the speed of rotation of the disk by the number of perforations. Sounds of predetermined frequency ratio can be had from simultaneous use of two or more rows of holes in the same disk; this feature was exploited in early experiments on musical tones. Before the advent of the jet engine, the siren was the most powerful of man-made steady sound sources, and as such it found application to fog signaling and as an emergency warning signal. [R.W.Y.]

Sirenia

An order of mammals, popularly called sea cows, that are adapted to aquatic life. There are two genera of living sirenians, the manatees (*Triche-*

chus) of the Atlantic basin, and the dugong (*Dugong*) of the Pacific and Indian Oceans. Fossil remains of several other genera are known from various parts of the world. Sirenians were already specialized for aquatic life in the Eocene, and the modern forms are almost as modified as the whales. The body is torpedo-shaped, the front limbs are flipperlike, the hind limbs are completely absent, and the tail is in the form of a broad horizontal fin. The ribs and limb bones lack a marrow cavity and are remarkable for their extreme density. Sirenians are strictly herbivorous. Their closest living relatives are believed to be the elephants. *See* MAMMALIA. [D.D.D.]

Sirenia fossils

Fossil representatives of this order, like the living forms, were large, ponderous animals completely adapted to aquatic life. From *Prorastomus* (Eocene of Jamaica) and *Protosiren* and *Eotheroides* (Eocene of Egypt) to the living *Trichechus* (manatee) and *Dugong* (sea cow), sirenians are characterized by reduction of hind limbs, correlated with a change to fish-shaped bodies, and by development of paddlelike front limbs. They have very heavy and massive bones, especially the ribs, and low-crowned, generally bilophodont cheek teeth. Later forms tend to have reduced numbers of simplified teeth. The recently exterminated *Hydrodamalis* of the Bering Islands had no functional teeth. Throughout most of their history sirenians seem to have preferred shallow, brackish water habitats such as estuaries and lagoons, although some fossil forms have been found in deposits representing former open-sea reefs and shelf areas.

The two principal groups of sirenians that are distinguishable in the later Cenozoic are the Pacific and Indian Ocean Dugongidae, with strongly deflected rostrum and greatly reduced dentition, and the Atlantic Trichechidae, with little-deflected rostrum and bilophid cheek teeth. These families are not distinguished with certainty in the early Tertiary fossils. Dugongids show a wider distribution than manatees; for the fossil dugongids *Halitherium, Halianassa, Felsinotherium,* and others are known from Tertiary marine rocks in Africa, Europe, and the Americas. Dugongids may have had much of their evolutionary development in the Western Hemisphere but are now restricted to the Eastern Hemisphere. Fossil forms of unquestioned trichechid affinity like *Ribodon, Potamosiren,* and *Trichechus* are known only from the upper Miocene and Pleistocene rocks of the Atlantic border of the Americas.

Sirenians represent an early Eocene, or earlier, differentiation from a land-mammal group that was also ancestral to Proboscidea, Desmostylia, and perhaps Perissodactyla and certain enigmatic orders of large extinct mammals in Africa and South America. *See* DESMOSTYLIA; PERISSODACTYLA FOSSILS; PROBOSCIDEA FOSSILS. [D.E.S.]

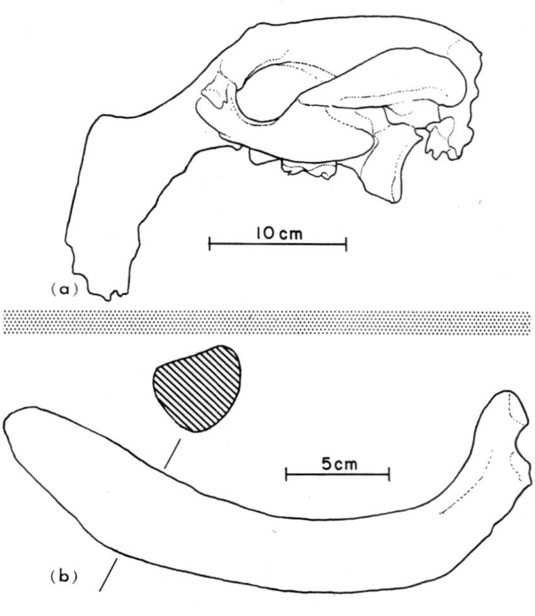

Caribosiren, Oligocene dugong from Puerto Rico. (*a*) Side view of the cranium. (*b*) A rib.

Sirius

Alpha Canis Majoris, the brightest star visible from Earth. From the Northern Hemisphere, Sirius appears in the south in midwinter. The apparent magnitude, -1.5, makes it a rival of the planets in brightness. The distance, 2.7 parsecs, makes it the sixth nearest star to the Sun. Absolute magnitude $+1.4$ and spectral type A1 are representative of early type A stars, with an effective temperature near 10,000°K, strong hydrogen lines, and large Balmer continuous absorption. Special interest attaches to the system of Sirius because of the close companion, of the ninth magnitude (more than 10^4 times fainter), the white dwarf α CMaB. The period of the binary system is 50 years, the mass of α CMaA is 2.28 M⊙, that of α CMaB, 0.98 M⊙. Stars of solar mass usually have spectral type G, and yellow color, but α CMaB is bluish and is reported to show hydrogen lines; that is, it is hotter, and much smaller than our Sun. Until an accurate determination is made, the temperature and radius of α CMaB are uncertain. The theoretical radius is only 5900 km, as contrasted to that of the primary, which is about 10^6 km. *See* STAR; WHITE DWARF STAR. [J.L.GR.]

Sirocco

A southerly or southeasterly wind current from the Sahara or from the deserts of Saudia Arabia which occurs in advance of cyclones moving eastward through the Mediterranean Sea. The sirocco is most pronounced in the spring, when the deserts are hot and the Mediterranean cyclones are vigorous. It is observed along the southern and eastern coasts of the Mediterranean Sea from Morocco to Syria as a hot, dry wind capable of carrying sand and dust great distances from the desert source. The sirocco is cooled and moistened in crossing the Mediterranean and produces an oppressive, muggy atmosphere when it extends to the south coast of Europe. Rain which falls in this air is often discolored by the dust or sand which is precipitated along with the water drops. Under sunny conditions, the sirocco can produce temperatures in excess of 100°F in southern Europe. Various other names are used to denote the sirocco in specific localities; notably, the term khamsin in Egypt. *See* AIR MASS; WIND.

[F.S.]

Sisal

This plant, *Agave sisalana*, is a species of the Amaryllis family (Amaryllidaceae) and a native of Mexico and Central America. The best supply is produced in Yucatan, where it is known as henequen. It is also grown in Florida, Central America, the West Indies, East Indies, Hawaii, and Africa. It is very drought-resistant and will thrive in habitats too arid for other species in this plant family. Little cultivation is required. The leaves contain coarse, stiff, yellow fibers which are removed by hand or by means of a coarse rasp. After cleaning and drying, the fibers are baled for shipping. This fiber is used in making sisal binder twine, for rope, for bristles of inexpensive brushes, and as a substitute for horsehair in upholstering. Sisal may be mixed with hemp, but it is not used for ship cordage because it disintegrates in salt water. A large amount of sisal fiber is imported into the United States from Mexico and the East Indies. [P.D.S.]

Diseases of sisal. Sisal is subject to two bole (stem) rots. *Aspergillus niger* has been associated with one which is a soft wet rot. The other, a dry rot, has been attributed to *Fusarium solani*. The

Fig. 1. Bole rot of sisal caused by *Fusarium solani*. (*a*) General view of diseased plants. Split boles showing (*b*) advanced necrosis and (*c*) early infection (see point of pencil).

Fig. 2. Leaf spot of sisal caused by *Colletotrichum agaves*.

fungus enters through the exposed base of harvested leaves and spreads inwardly (Fig. 1).

Leaf spots are caused by the fungi *Microdiplodia agaves*, *Marssonina agaves*, and *Colletotrichum agaves* (Fig. 2).

Nutritional diseases include leaf banding, chlorotic leaf spot, and tip withering and mottling.

The cause of the destructive "red rot" disease is suspected to be a virus. *See* PLANT VIRUS.

The fiber of the closely related plants *A. fourcroydes* and *A. cantala* is sometimes called sisal fiber; these species are affected by many of the above diseases. *See* LILIALES; PLANT DISEASE.

[T.E.SU.]

Size reduction

That operation in which a body of matter is divided into smaller bodies. The term is generally applicable to solid matter, but it may be applied to liquids and gases under some conditions. There is a companion term, size enlargement, which involves the creation of larger pieces by bonding together small pieces or by adding new material to small pieces. For information on other processes *see* CRYSTALLIZATION; EXTRUSION; SINTERING.

Applications. Size reduction methods are used on quarry rock employed in roadbuilding; on ore to facilitate separation of the valuable ore particles from gangue; on coal so that it will be easy to handle and burn; on minerals such as cement raw mix or phosphates so as to create uniform mixtures which can react chemically; on pigments and fillers to provide smooth paints of good hiding power; on grain so as to improve product texture; on chemicals to facilitate handling, reactivity, or mixing. In the liquid and gas phases, size reduction covers emulsification, aerosols, and gas dispersion.

In function, size reduction creates a particulate group in which the maximum particle size, the surface area, or the size distribution may be important. To achieve the desired end in any product, equipment needs to be chosen, its manner of operation set, and the cost of energy and of maintenance considered. At times the product may be injured

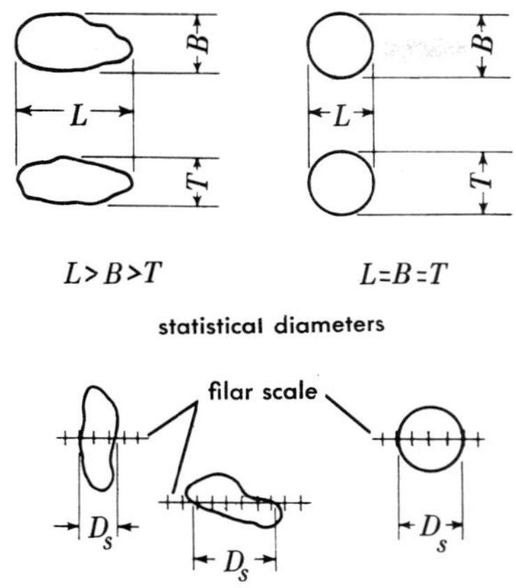

Fig. 1. Dimensions of particles. (*From R. E. Kirk and D. F. Othmer, eds., Encyclopedia of Chemical Technology, Interscience, 1956*)

by contamination developing during the operation.

Particle size measurement. The individual particle is a unit generally held together by molecular forces in a compact state. It may be heterogeneous and may contain different crystals closely joined, it may be an essentially continuous phase within which are pores or particles, it may be a crystal, or it may be a glass. A particle should be distinguished from an agglomerate, in which a number of individual particles are more loosely bonded together. These may range from lumps of clay or dried filter cake particles, to lightly sintered metal powders. There is a border area in which it is difficult to make a distinction between particles and agglomerates. This difference and the significant shapes and dimensions of the individual particle are shown in Fig. 1. Because of the nature of the individual particle, it is generally difficult to assign it a true diameter and consequently almost impossible to calculate its surface area. There are, however, properties which can be measured to define a diameter or to yield surface area. For example, a particle that is not too long will upend and pass through a square screen aperture with its intermediate and small dimensions controlling separation; or a particle will settle in a fluid at a definite rate based on its weight and its surface resistance, from which the diameter of an equivalent sphere can be calculated. Likewise, surface area may be measured directly, for example, by adsorption; or it can be calculated from a diameter. It is important that the criterion for measurement be stated when particle size values are given, for there are substantial differences between these measurements. Being relative, they still make it possible to do precise work and to draw valuable conclusions within these limitations.

The particles in size reduction occur as a granular or powdery mass containing a large number of individual particles, agglomerates, or both, usually in a wide range of sizes. Accurate measurement depends on taking a representative sample and reducing it to a measurable amount without impairing its representative nature and then, using statistics, evaluating it to secure reproducible results. The results desired may be a maximum surface, a maximum diameter, or a distribution of sizes by numbers of particles or weights of material occurring in each size range measured. The chief methods are listed in the table.

Methods of particle size measurement

Test method	Approximate useful range	Property measured
Sieves	A few inches to 50 microns, μ	Distribution of particles by weights in sieve ranges; sieve aperture controls passage of the small cross section of particle
Microscope		Distribution of particles by measurement of each particle; any one of several measured diameters—usually misses the shortest
Visible light	100–0.3 μ	
Ultraviolet	100–0.05 μ	
Electron	100–0.005 μ	
Elutriation	100–5 μ	Distribution of particles by separation according to settling rate in fluid; recovered fractions may also be measured in size
Sedimentation		
Regular	50–1 μ	Distribution based on settling in fluid medium
Centrifugal	50–0.1 μ	No recovered fraction
Ultracentrifuge	Molecular	Diameter based on Stokes' law for equivalent spheres or on a composite of forces in ultracentrifuge
Turbidimeter	50–0.3 μ	Surface area function directly; with sedimentation, yields distribution based on surface-settling rate
Sorption	50 μ to finest sizes	Surface area by direct measurement
Permeability	A few hundred μ to fractional μ	Surface area or a calculated surface mean diameter from measurement of a voids/surface-resistance relation
X-ray	1–2 μ to several hundredths micron	An average crystal diameter by x-ray diffraction
Electron counter	50 μ to fractional μ	Distribution of particles by frequency for a function of cross section

Particle size data obtained from some of these methods of test may be represented in units of surface area, such as square meters per gram; or data may be simply converted to an equivalent average diameter, such as a surface mean diameter of x microns. More significant information is available in the distribution of size. This may be expressed in numbers, such as that diameter at which 10% by weight of the total sample is larger and 90% smaller. The data can be plotted graphically in terms of cumulative per cent oversize against the corresponding sizes. Such typical curves are shown

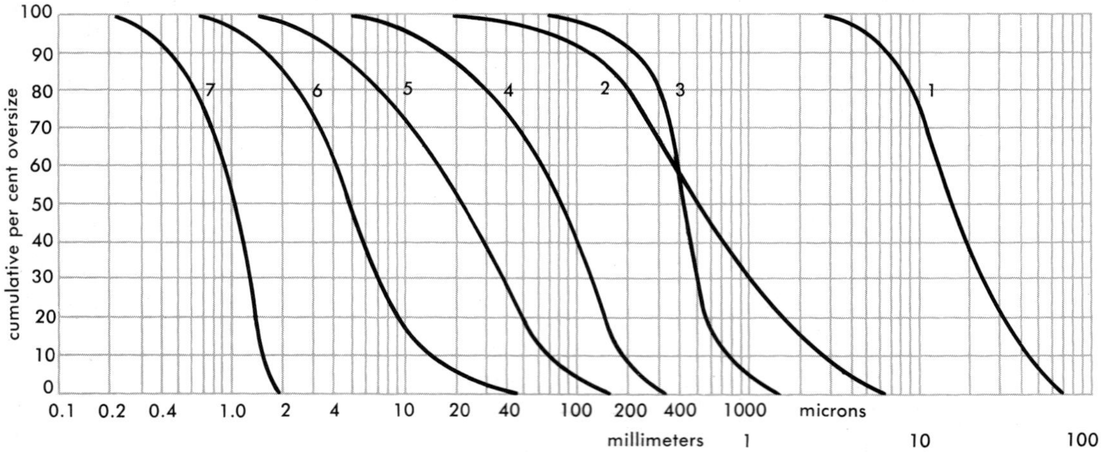

Fig. 2. Particle size distribution curves for typical materials. Curves: 1, coarse aggregate; 2, fine aggregate; 3, filter sand; 4, powdered coal; 5, portland cement; 6, mineral fillers; 7, pigments. (*From R. E. Kirk and D. F. Othmer, eds., Encyclopedia of Chemical Technology, Interscience, 1956*)

in Fig. 2. A plot of the slopes of such an S curve yields a modal curve wherein the per cent in each unit of diameter is plotted against that diameter. Both types of curve are basic and independent of the diameters of measurements, provided that an adequate set of such points is employed. On the other hand, there are in the literature some deceptive curves based on arbitrary and variable intervals. When cumulative per cent undersize is plotted against the reciprocal diameter, the area beneath the curve represents surface; when per cent per unit of diameter is plotted against logarithm of diameter, a similar surface value results. These are respectively the Gates and Roller diagrams.

One object of grinding is to produce material in the desired particle sizes. If a maximum size is important, some control such as a sieve or a classifier is employed to return oversize to the mill and to accept only undersize in the product. If freedom from fines is important, only a limited control is possible, since particles normally fracture into a random group of sizes. This has been the subject of statistical study by several workers, but no single theoretical treatment has been generally accepted. Properties and preparation of the feed, the kind of mill action, the prevention of overgrinding by classification, and recirculation all play a part. Usually it is unprofitable to remove the fines and reject them, though this is sometimes necessary and justified.

Energy requirements. Energy requirements are important because of their bearing on cost of grinding. It is usually difficult to separate the useful work from the necessary effort incidental to running a grinding operation. One theory relates necessary energy to the production of new surface and another to the work done in deforming a particle to the point of fracture. A third relation recognizes both factors and is intermediate between them. Energy requirements increase greatly in the production of fine products and often limit the warranted size reduction.

Processes and devices. Well-known metal- and wood-working devices give some control in size reduction. Examples are cutters, shears, slitters, files, grinders, saws, shaving machines, turning devices, milling equipment, and chippers. Decorticating machines for removing hulls from grain are in this category of controlled functional operation. So, too, are shot towers and flakers. Flaking may be done on a drying belt or on chill rolls. Extrusion and sintering are operations used for controlled size enlargement.

Other methods permit less control of product size or of range of sizes. One technique consists of shattering by thermal shock. For example, a calcined ore, while still hot, may be quenched in water so that the shrinkage stresses cause rupture. This action is greater on larger pieces than it is on fine particles. Thermal shock does, however, have a place in fine pulverization if the unit particles of a mass are small and the bonds cementing the particles are weak. The effect of the shock is still greater if the bonding layers are transformed or weakened by the heat so as to be subject to the attack of water. Another technique consists of superheating wet material in an autoclave and suddenly releasing pressure. Grain is puffed and wood is explosively disintegrated into its fibrous structure in this treatment. Equipment for explosive shattering is illustrated in Fig. 3.

Sprays are employed for breaking a fluid stream, with or without contained solids, into droplets of generally controlled size. Such droplets may be dried to solids (*see* DRYING); they may be combustible, and when mixed with air may form a smoothly burning mixture; or they may be molten, and when cooled or quenched yield fine particles, such as some commercial metal powders, free-flowing salts, and the like.

A specific class of devices is known by the term colloid mill. These devices employ sharp shear gradients in a fluid medium to rupture solid particles which are generally friable, to break down

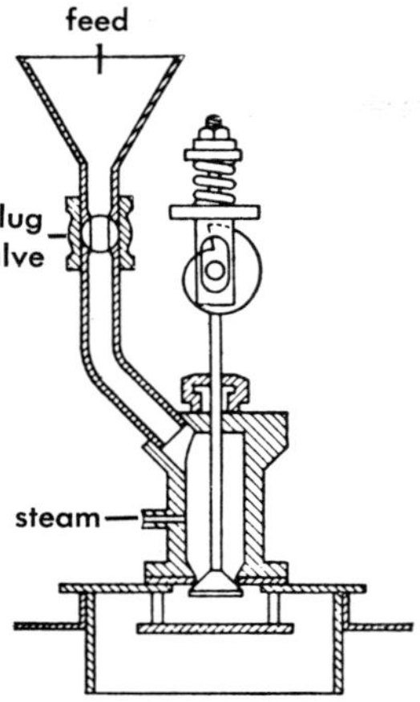

Fig. 3. Sectional diagram of explosive-shattering machine. (*U.S. Bur. Mines, 402, 1938; from J. H. Perry, ed., Chemical Engineers' Handbook, 3d ed., McGraw-Hill, 1950*)

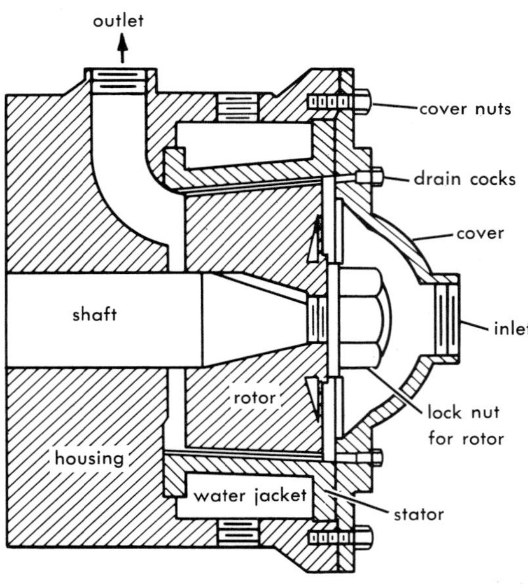

Fig. 4. Charlotte colloid mill. (*From J. H. Perry, ed., Chemical Engineers' Handbook, 3d ed., McGraw-Hill, 1950*)

agglomerates, and to create emulsions or to make the suspended liquid droplets finer. Because of the speeds, the contact surfaces are subject to wear from the more abrasive particles. When they are used to break emulsions into finer particulates, they are called homogenizers. Colloid mills (Fig. 4) generally contain a rotor in a stator ring. The shear is developed in liquid pumped across these closely spaced faces. Static devices may also be used; a partly closed valve through which feed is pumped is one simple type. *See* CLASSIFICATION, MECHANICAL; CRUSHING AND PULVERIZING; GRINDING; METAL FORMING; UNIT OPERATIONS. [L.T.W.]

Bibliography: J. M. Dalla Valle, *Micromeritics, The Technology of Fine Particles*, 2d ed., 1948; G. Herdan, *Small Particle Statistics*, 1953; R. E. Kirk and D. F. Othmer (eds.), *Encyclopedia of Chemical Technology*, vol. 12, 1954; J. H. Perry (ed.), *Chemical Engineers' Handbook*, 3d ed., 1950; L. T. Work, *Size Reduction, Ind. and Eng. Chem., Chem. Eng. Revs.*, 51(3), pt. II:395–396, 1959.

Skarn

A rock term generally reserved for rocks composed entirely, or almost so, of lime-bearing silicates, and derived from nearly pure limestones and dolomites into which large amounts of silicon, Si; aluminum, Al; iron, Fe; and magnesium, Mg, have been introduced. *See* METAMORPHIC ROCKS; PNEUMATOLYSIS.

Skarn is an old term in mining, and was originally used by Swedish miners to designate dark silicate minerals occurring in seams or masses adjacent to an ore-bearing vein. Now it is applied to a coarse-grained rock or association of minerals formed by reaction between hot silica-rich solutions or acid gases and a limestone or dolomite. The reaction rock is made up of coarse masses of minerals of various kinds, depending upon the temperature and the composition of the reacting gases or solutions. The most important skarn minerals are andradite-garnet, hedenbergite-diopside, iron-rich hornblende, and actinolite-tremolite. Various ore minerals, oxides, and sulfides, as well as fluorite, often occur in connection with skarn. The common occurrence of fluorite, CaF_2, supports the idea that silica and metal halogenides have reacted in the limestone. Thus the formation of andradite may be explained as follows:

$$2FeF_3 + 3SiO_2 + 6CaCO_3 \rightarrow \underset{\text{Andradite}}{Ca_3Fe_2Si_3O_{12}} + \underset{\text{Fluorite}}{3CaF_2} + 6CO_2$$

Not only are fluorides active, but also chlorides may be introduced, in which case $CaCl_2$ is formed and reacts with formation of scapolite. Large masses and dikes of scapolite are common in many skarn rocks.

Oxidic ore is very common in skarns and forms in many places useful deposits of skarn ore. Ore formation will take place especially if silica is not introduced:

$$2FeF_3 + 3CaCO_3 \rightarrow Fe_2O_3 + 3CaF_2 + 3CO_2$$

Thus were formed, for example, the famous hematite deposits in Elba containing large masses of andradite plus the calc-iron silicate lievrite at the contact between granite and limestone. Among other classical contact deposits with skarn formation are those at Campiglia Maritima, Tuscany; Banat, Rumania; Franklin Furnace, New Jersey; and Clifton-Morenci, Arizona. *See* ORE AND MINERAL DEPOSITS. [T.F.W.B.]

Skeletal system

Those vertebrate structures composed of bone, cartilage, or a combination of both which provide a framework for the animal body and serve as attachment for muscles. Relative movements of bony parts are permitted by the joints (articulations) between them. Although the skeleton of vertebrates is characteristically internal (endoskeleton), many species also possess external or exoskeletal structures of bony or horny material to provide support or protection, as the shell of the turtle. Such exoskeletal structures and their derivatives comprise the dermal skeleton. Invertebrates which possess skeletal structures have an exoskeleton composed of chitin.

EMBRYOLOGY

Bone is one of the connective tissues which as a major histological subdivision is characterized by the presence of cells specialized to produce an intercellular substance. The histological classification of connective tissues is based mainly upon the character of the intercellular substance, although the cells of each type of tissue are also distinctive. Osseous tissue is distinguished by its hard, impermeable matrix through which the cells maintain an intricate canalicular system. *See* BONE.

Skeletal histogenesis. All connective tissues, including the osseous variety, are derived from the mesodermal germ layer. In early embryonic development just prior to the appearance of skeletal anlage, the mesoderm is distributed both compactly as well-defined paraxial columns (somites) and lateral plates, and diffusely as a syncytium, the mesenchyme. The vertebrae originate from the somites, but most of the skeletal elements of the head, girdles, and limbs arise locally from the mesenchyme. The first step in the histogenesis of a skeletal element involves an intensive proliferation of mesenchymal cells at sites of prospective bone centers. At this preliminary stage, though consisting simply of compactly arranged mesenchymal cells, each model in external form and relative position approximates its definitive derivative (*see* CONNECTIVE TISSUE). The second step in the histogenesis of skeletal elements involves the differentiation of mesenchyme into cartilage or bone. In either case, mesenchymal cells on the surface of each skeletal element differentiate into a sheet of fibroblasts. The latter elaborate collagenous and elastic fibers that are woven into a snug-fitting sleeve called the fibrous perichondrium if it covers cartilage or periosteum if it covers bone. In instances where skeletal elements are arranged end to end as in the limbs or vertebral column, the fibroelastic investment is continuous from one element to the next in the series; thus, at sites of articulation the sleeving constitutes a joint capsule.

Joints. The fate of the mesenchyme within a joint may be correlated with the degree of movement of the developed joint. The mesenchyme retains its original delicate syncytial character in the joints which are presumptively freely movable (the diarthroses). By the time such a joint is subjected to movement, its mesenchyme has differentiated into a synovial membrane that secretes a small amount of fluid and forms a lining on all aspects of the joint interior except the articulating surface, which remains cartilaginous throughout life. At the sites of the presumptively immovable joints (the synarthroses), the mesenchyme differentiates into a dense fibrous connective tissue (syndesmosis), a fibrous or hyaline cartilage (synchondrosis), or into bone (synostosis). An amphiarthrosis is a partially movable joint which has a minute joint cavity supported by a dense covering of fibrous connective tissue or fibrocartilage.

Mesenchyme differentiation. The mesenchymal model may differentiate into cartilage and remain cartilaginous throughout life as in the case of the entire skeleton of elasmobranchs, or it may differentiate directly into bone by a process called intramembranous or membrane bone formation. The so-called dermal bones that form, for example, the roof of the cranium in the bony fishes and in all higher vertebrates are derived exclusively by intramembranous ossification. Incidentally, the membrane bones may be regarded as part of an external skeleton, which also includes various scales of epidermal origin, and serves to protect the animal from the exterior as a sort of biological suit of mail. As a third alternative in the histogenesis of a skeletal element, the model may differentiate into cartilage which subsequently is gradually replaced by bone, a process called endochondral or intracartilaginous ossification. *See* SCALE (ZOOLOGY).

Formation of osseous elements. The majority of the osseous elements of the vertebrate endoskeleton are formed by a combination of the intramembranous, or periosteal, and endochondral processes. Therefore, to describe bone formation in somewhat more detail, the development of certain long bones of a mammalian limb will be considered. All major ossification centers are established by the same sequence of events. Following their differentiation from the mesenchyme, chondroblasts first proliferate intensively for a brief period and thereafter show progressive maturation changes among which are enlargement and the elaboration of glycogen and alkaline phosphatase. Next, the matrix about the mature chondrocytes undergoes calcification, a change followed by disintegration of the sealed-off cells. The empty chambers (lacunae) are invaded by a capillary bud which carries with it bone-forming (osteogenic) and blood-forming (hematopoietic) cells. The osteogenic cells, called osteoblasts, apply themselves to the calcified cartilaginous walls of the lacunae and proceed to elaborate bone matrix, ossein. Even before a colony of osteoblasts is established at the center of the cartilage model, an osseous "waistband" appears that represents the product of activity of the perichondrial osteoblasts which subsequently become the periosteal osteoblasts. The waistband of periosteal bone thickens and widens toward the bone ends, and simultaneously the endochondral ossification center spreads toward the ends. A reserve of chondroblasts near one or both ends of the growing bone by continuous proliferation maintains a cartilaginous front against the invading files of osteoblasts. This cartilaginous disk, called the epiphyseal plate, is present throughout the animal's growth period; in fact, its presence is indicative of the bone's capacity to grow in length. It is important to understand that the normal epiphyseal plate represents a kinetic balance between the mitotic rate of the reserve chondroblasts and death rate of the enlarged chondrocytes in the zone of calcification (Fig. 1). It is only by invading the spaces created by disintegration of the chondrocytes that osteoblasts can add osseous matrix to the length of the bone. Uncalcified cartilage has the consistency of firm jelly and yields to growth pressures. Thus, though surrounded by matrix, the chondroblasts can increase both in number and in size. Such a process, by which a structure increases in size from within, is known as interstitial growth. Osseous matrix normally is mineralized soon after deposition and as such is as rigid and impermeable as concrete. Thus, new bone formation can take place only by appositional growth, or growth on a surface such

Fig. 1. Metaphysis of a long bone, shown semischematically.

as that provided by the empty cartilage lacunae or on the trabecular surface as indicated in Fig. 1. Once imbedded in the mineralized ossein, the cells can neither proliferate nor enlarge. The term osteocyte applies to the imbedded bone cells, whereas osteoblast refers to the free, actively osteogenic cell on the trabecular, endosteal, or periosteal surfaces. Although imbedded, the osteocyte, by a fine canalicular system that ramifies through the matrix, maintains a life line for metabolic exchanges with the regional, usually Haversian, blood vessels. In contrast to bone matrix, uncalcified cartilage matrix without a canalicular system is permeable enough to satisfy the metabolic needs of the chondrocytes.

Ossification centers. Although the sequence of cellular activities is the same at all centers of endochondral ossification, wide variation is encountered in the growth rate and duration of activity at each center. Such variation is reflected in the size of each center's contribution to the formation of the skeleton. The shaft or diaphysis of a long bone such as the femur is formed from a single center called the diaphyseal or primary or main center; the ends of the femur, the head and condyles, are formed from separate centers called the secondary or epiphyseal centers, proximal and distal, respectively. The diaphyseal center of a given long bone is established much earlier, grows more rapidly, and ceases activity later than the small epiphyseal centers. Though adjoining the epiphysis, the epiphyseal plate "faces" the diaphyseal cavity and contributes to the length of the diaphysis only. A bone's period of growth in length ends with closure of its epiphyseal plate or plates. The time of epiphyseal plate closure, as well as the time of appearance of diaphyseal and epiphyseal centers, is characteristic for each bone. This information has been compiled in detail for the skeletal system of man and the common laboratory animals. It constitutes a chronology of events in the ossification of the skeletal system and a basis upon which skeletal age determinations may be made, information which is often useful as a diagnostic procedure in certain thyroidal disorders of infants and children. Skeletal aging has also been used to evaluate the skeletal maturation-promoting effect of certain endocrine factors such as thyroxin and the steroid hormones.

Endocrine and dietary factors. A number of dietary and endocrine factors have a pronounced effect on the skeletal system, especially during periods of rapid development. The growth hormone of the pituitary gland is the most effective skeletal growth-promoting agent. Thyroxin is the most effective skeletal maturation-promoting agent. The sex hormones, especially testosterone, are moderately effective in promoting both growth and maturation of the skeletal tissues. Parathormone promotes bone resorption. Vitamin A has an effect similar to that of parathormone, vitamin C is necessary for the formation of ossein, and vitamin D is necessary for the proper mineralization of ossein.

Furthermore, the dietary level of proteins, calcium, and phosphorus must be above a certain minimal level for proper bone formation. *See* PARATHYROID GLAND; THYROID GLAND.

MORPHOGENESIS OF THE SKELETON

Axial skeleton. The notochord represents the most primitive form of axial support and may be regarded as the most constant and characteristic feature of the vertebrates. In *Amphioxus* and the cyclostomes the notochord forms a permanent structure, essentially a rather stiff but elastic rod lying ventral to, and coextensive with, the neural tube. In all higher vertebrates the notochord is supplanted by the vertebral column.

Vertebral column. Though enormous diversity is encountered in the structure of vertebrae of different species, all vertebrae are homologous. The basic plan in the formation of the vertebra is clearly demonstrable in elasmobranchs. From its ventromedial aspect each somite gives forth a mesoblastic growth that establishes a platelike structure, the sclerotome, over the lateral aspect of both the notochord and neural tube. In the course of development the posterior half of each sclerotome takes on a more dense arrangement and so is differentiated from the anterior half. Each half becomes further subdivided into dorsal and ventral arcualia. In elasmobranchs the pair of arcualia that appears in the denser posterior half of the sclerotome, known as the basidorsal and basiventral basalia, has a larger share in the formation of the vertebra than the anterior pair, the interdorsal and interventral interbasalis (Fig. 2*a*). Thus, the basidorsal forms the neural arch and the basiventral, the transverse process. The centra are formed as mesoblasts from the basalia bilaterally invade the notochordal sheath to establish intersegmental rings. These thicken at the expense of the notochord and, together with the neural and hemal arch, undergo chondrification. At segmental levels the notochord persists and, in fact, indents the ends of the centra which, because of their biconcave shape, are said to be of the amphicoelous type.

In the bony fishes and tetrapods, considerable variation is encountered in the particular contribution each of the four pairs of arcualia makes to the final product. Perhaps the most consistent contribution is that of the basidorsals to neural arch formation. Some of these variations may be discerned in Fig. 2*b–g*. The degree to which the interbasals participate in vertebral formation is often reflected in the shape of the definitive centrum. The amphicoelous type, characteristic of elasmobranchs, is also found in the amphibians of strictly aquatic habitat. In Anura intervertebral rings, presumably derived from the interbasalia, completely crowd out the notochord and subsequently fuse with the basalia derivative in front or behind to form respectively a procoelous or opisthocoelous type of centrum. In most reptiles the centra are of the procoelous type; in birds they are heterocoelous (saddle-shaped); and in mammals the acelous, or flat-surfaced, type occurs. Since contour of the ends of the centra determines the shape of the intervertebral joints, the terms used to describe the centra are also used to describe the intervertebral joints.

Tail. The formation of the caudal vertebrae follows the plan outlined for vertebrae in general. Some modifications do occur; for instance, in elas-

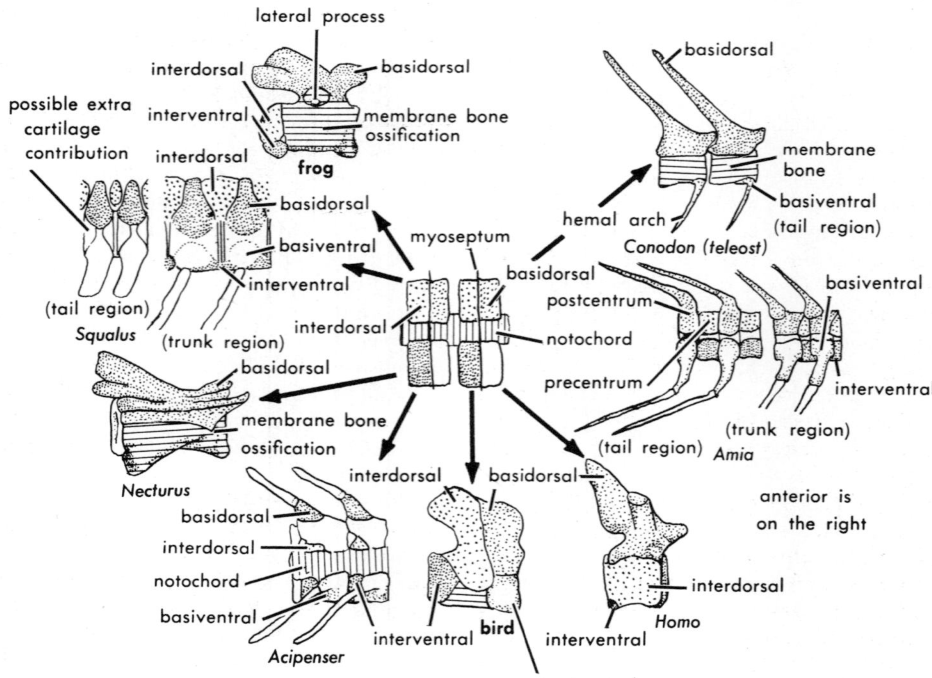

Fig. 2. Proportion of the vertebra formed by dermal bone and each of the arcualia. (*From O. E. Nelsen, Comparative Embryology of the Vertebrates, Blakiston, 1953*)

mobranchs and the bowfin, *Amia*, two vertebrae develop per muscle segment, a condition known as diplospondyly. That is, two centra and a duplicate set of dorsal and ventral acualia correspond to one pair of myomeres and spinal nerves. The original single set of basalia lengthens, then divides transversely. Thereafter a centrum forms in relation to each new set of basalia. Diplospondyly facilitates lashing the tail from side to side, the movement by which fish propel themselves. Rib processes are reduced or absent on caudal vertebrae, but the ventral or hemal arches derived from the basiventrals are well developed and serve to protect the caudal vessels.

Sternum. A true sternum, which develops only in the higher vertebrates, is of paired origin and is closely associated with the ventral end of the ribs, if not actually derived from them. It belongs to the axial rather than to the appendicular skeleton. The sternum of modern Amphibia has become secondarily isolated, a change associated with the reduction of the ribs. In mammalian development the primordia of the sternum first appears independent of the ribs as a pair of mesenchymal sternal bands. The bands secondarily establish continuity with the rib blastema and move toward the ventromedian line, where they develop centers of chondrification and fuse with one another. The portion of the sternum attached to the ribs is called the mesosternum, rostral to which is the pre- or episternum and caudad is the meta- or xiphisternum. Each cartilaginous segment of the mesosternum, known as a sternabra, develops a right and left center of ossification.

The manubrium, or uppermost piece of the sternum, is usually ossified from three centers, the body of the sternum from the several sternebral centers, and the xiphisternum from a single center.

Ribs. Although absent in the *Amphioxus* and the cyclostomes, ribs occur in all other vertebrates. Two varieties called dorsal and ventral, or pleural, ribs are distinguished. Both types develop as rodlike extensions from the basiventral arcualia into the transverse septa (myosepta, myocommata): the dorsal set, along lines of intersection with the horizontal septa; the ventral set, in the parietal wall of the coelom (Fig. 3). Ventral ribs are found in bony fishes whereas dorsal ribs are found in elasmobranchs and tetrapods. No modern amphibian has ribs typical of the higher vertebrates. The typical rib of the higher vertebrates is bifurcated; the capitulum, or ventral fork, is articulated to the centrum below; and the tuberculum, or dorsal fork, to the extremity of the transverse process, or diapophysis, of the neural arch above. The tetrapod rib chondrifies and later ossifies separately, though during the mesenchymal model or blastema stage the rib capitulum and vertebral centrum are in continuity. The gastralia or abdominal ribs of the alligators and crocodiles are accessory bony elements of dermal origin and bear only a topographic relationship to the true ribs.

Skull. The skull of vertebrates is a composite structure. The two rather strikingly different sub-

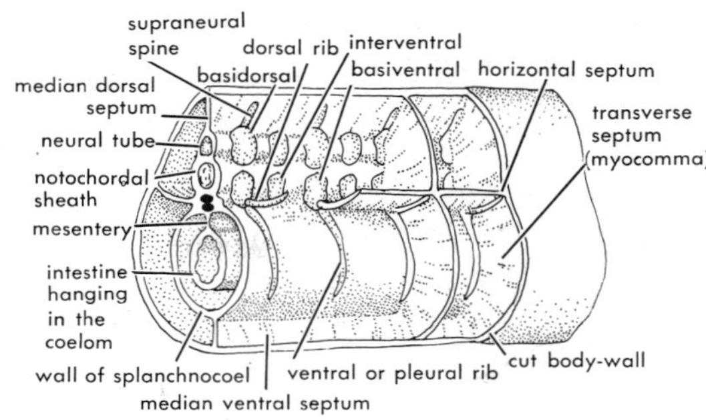

Fig. 3. Principal septa in relation to axial structures and ribs of the vertebrate trunk. (*From E. S. Goodrich, Studies on the Structure and Development of Vertebrates, vol. 1, Dover, 1958*)

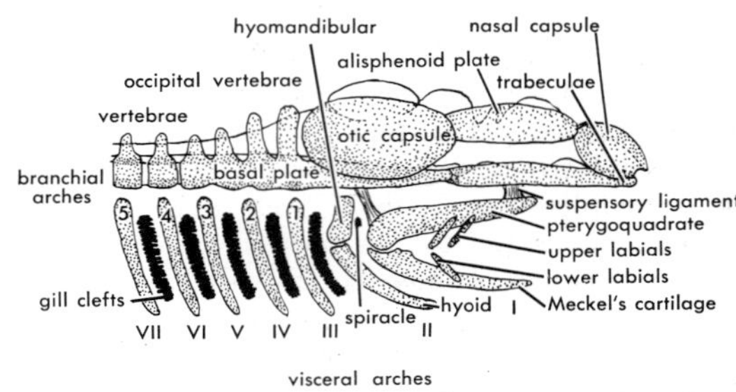

Fig. 4. Diagram of the early elasmobranch chondrocranium in side view. (*From J. S. Kingsley, Comparative Anatomy of Vertebrates, Blakiston, 1912*)

divisions are the neurocranium, which houses the brain and organs of special sense, and the splanchnocranium, which supports the gut near its oral end. Although the skull is modeled in mesenchyme as the initial step in its differentiation, it is only with chondrification of the model that the architectural plan becomes evident.

Chondrocranium. In all vertebrates from elasmobranchs through the mammals, the chondroneurocranium is constructed according to the same basic plan. The median or axial portion of the skull base is established by a series of bilateral bars. The more prominent of these are the trabeculae cranii that extend rostrally from the level of the pituitary gland and the parachordals that flank the notochord directly behind the pituitary gland. Developing caudal to and in series with these elongated cartilages is a chain of small stout bars, the occipital sclerotomes, homologous to those that form the vertebrae. The occipital sclerotomes together with the parachordals fuse to form the basal plate. Lateral arctiform extensions of the occipital sclerotomes form the posterior wall of the cranium. Elsewhere, the side walls and peripheral portion of the skull base are derived from the capsules that house the organs of special sense. Anteriorly to posteriorly these bilateral cartilaginous vesicles are the nasal capsules which fuse with the ends of the trabeculae cranii to form the ethmoidal plate, the optic capsules, and the auditory capsules (Fig. 5).

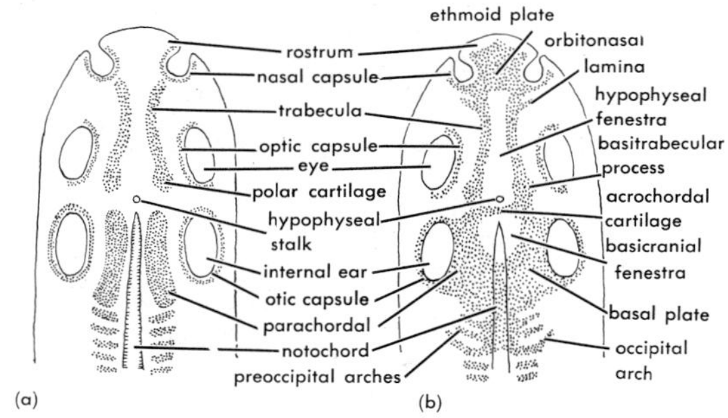

Fig. 5. Diagrams of the development of the chondrocranium. (*a*) Early stage. (*b*) Later stage. (*From L. H. Hyman, Comparative Vertebrate Anatomy, 2d ed., Univ. of Chicago Press, 1942*)

Roofing over of the side walls to form a complete cartilaginous brain cap or calvarium occurs only in the elasmobranchs and in ganoid fishes such as *Amia*. In all higher vertebrates only the occipital sclerotomes extend themselves sufficiently to establish a complete cartilaginous arch over the brain and this is limited to the posterior pole of the cranium. The large defect in the calvarial portion of the chondrocranium is closed by membrane bones. Since in higher vertebrates the cartilage is almost entirely replaced by bone, the distinction between bone of dermal origin and that formed by the endochondral process must be kept in mind when studying the comparative embryology of the skull. No attempt will be made here to describe the very numerous ossification centers that occur in each of the dermal and cartilaginous elements.

Splanchnocranium. As a series of cartilaginous arches in the wall of the foregut, the splanchnocranium develops independently from the chondroneurocranium. These visceral arches strengthen the wall of the pharynx without interfering with its expansion-contraction movements. The first visceral arch, the mandibular arch, supports the jaws; its dorsal component is called the palato- or pterygoquadrate, and its ventral component, Meckel's cartilage. The first and second (hyoid) arches usually become attached to the chondrocranium above, while the more posterior visceral arches, the true branchial arches of fish, remain free from the skull and become much modified and reduced in air-breathing vertebrates. Each branchial arch, visceral arches 3 to 7, typically consists of a short chain of rodlike elements which, beginning with the most dorsal, are named pharyngeal, epibranchial, ceratobranchial, and hypobranchial. The last element is joined to its contralateral fellow in the midventral line by an unpaired basibranchial or copula. Branchial rays for gill support radiate outward from their attachments on the epibranchial and ceratobranchial elements (Fig. 6). *See* RESPIRATORY SYSTEM.

With respect to modification of the primitive cartilaginous splanchnocranium, as represented in the elasmobranchs and outlined above, consideration should be given to the fate of the original elements, the addition of dermal or membrane bones, and the changes associated with jaw suspension.

In elasmobranchs the dorsal and ventral components of the first visceral arch meet those of the opposite side in the midline below the level of the ethmoidal plate and nasal capsules to form the teeth-bearing upper and lower jaws. Both jaws are secured behind by ligaments to the second visceral arch which in turn is suspended by a ligament from the chondrocranium (hyostylic mode of suspension). An additional suspensory ligament is usually present more rostrally.

In bony fishes and all higher vertebrates the palatoquadrate bars fail to meet in front to form a complete upper jaw arch. The teeth for biting then derive their support from new elements ossified in membrane. Meckel's cartilage also becomes reduced in importance and is superseded by dermal bones.

Aside from the dermal bones that supplant the palatoquadrate, there are two important derivatives of the primitive upper jaw in tetrapods, named the epipterygoid, equivalent to the alisphenoid of mammals, and the quadrate. Both of the latter are ossified from endochondral centers. The quadrate usually becomes a fixed and integral part of the skull and, except in mammals, bears the

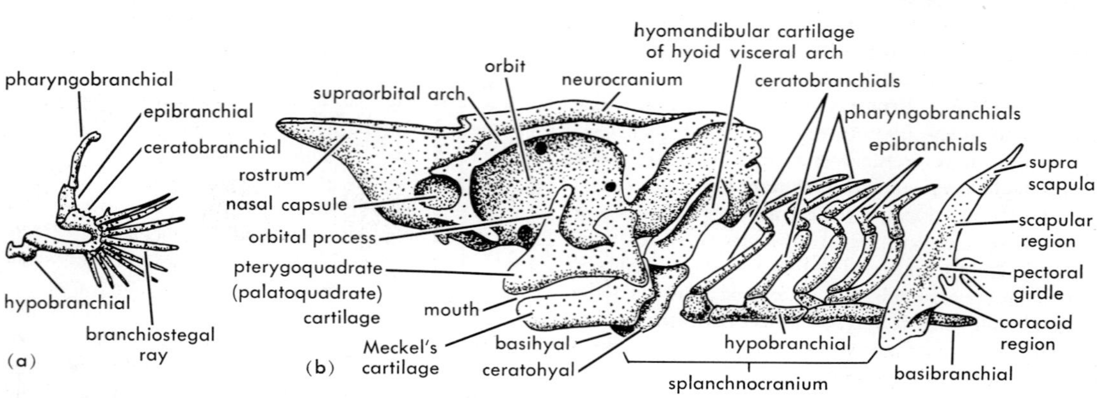

Fig. 6. (*a*) Chondrocranium and (*b*) gill supports in an adult elasmobranch. (*From O. E. Nelsen, Comparative Embryology of the Vertebrates, Blakiston, 1953*)

articulating surface for the lower jaws, the latter constituting an autostylic type of jaw suspension. The principal derivative of Meckel's cartilage is the articular bone, appropriately named since it articulates with the quadrate and so brings the lower jaw into the autostylic suspension. The hyoid apparatus and laryngeal cartilages are derived from the ventral components of the hyoid arch and the successive branchial arches. The dorsal element of the hyoid arch becomes the sound-transmitting ossicle, or columella auris.

The splanchnocranium of larval anurans undergoes a radical reconstruction during metamorphosis, as a result of which the relationships are more typical of tetrapods in general than before. The first visceral arch undergoes the most conspicuous modification through which the small ventral mouth of the tadpole is converted into the wide slitlike terminal mouth of the frog.

In mammals, derivatives of the primitive cartilaginous splanchnocranium have been reduced to a few elements crowded back behind the jaws. The most distinctive of these derivatives are the minute middle ear ossicles that transmit sound from the eardrum to the organ of hearing. The three ossicles are named malleus, incus, and stapes and are derived from Meckel's cartilage, the palatoquadrate, and the dorsal part of the hyoid arch, respectively. *See* EAR.

Appendicular skeleton. The appendicular skeleton consists of fins or limbs and their supporting girdles, used primarily for locomotion. Appendages are found in almost every class of vertebrate and occur as unpaired median fins, paired fins, and paired limbs.

Median fin development. In elasmobranchs the formation of the median fin becomes evident with the appearance of a narrow fold of the integument consisting of epidermis and a layer of mesenchyme along the dorsomedian line of the trunk. A similar fold develops along the ventromedian line from just behind the cloaca to the tail where the two longitudinal folds become continuous with one another. While these folds continue to lengthen along their whole extent, at certain locations the mesenchymal core becomes particularly thickened. Eventually the fin folds regress everywhere except at these restricted sites of thickening, which then represent the anlage of the definitive dorsal, caudal, and postanal fins. Bilaterally the muscle buds from the several myotomes spanned by the developing fin extend into its platelike mesenchymal core. These buds become detached and form the radial muscles, each of the latter in the adult having arisen from one bud and, therefore, one body segment. A skeletal radial develops between right and left members of each pair of radial muscles. The skeletal elements first appear as bandlike mesenchymal condensation which subsequently chondrify. In bony fishes median fins develop in the same fashion as in the elasmobranchs except that they become ossified in the former.

Paired appendages. It seems to be generally agreed that the paired appendages have evolved from bilateral longitudinal fin folds. The paired lateral (pectoral and pelvic) fins of fish develop in the same manner as that of the median fins described above. The girdles supporting the lateral fins arise by extension inward of the base of the jointed radiales, the base having been formed by the fusion of the proximal row of radiales into a single bar (Fig. 7).

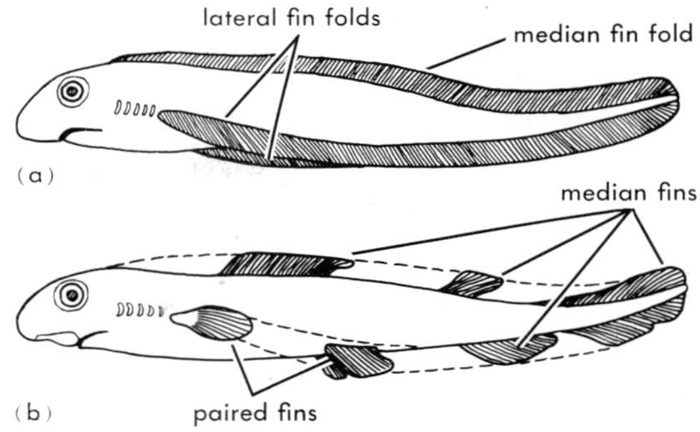

Fig. 7. Diagrams to illustrate the fin-fold theory. (*a*) Early stage. (*b*) Later stage. (*After Wilder from L. H. Hyman, Comparative Vertebrate Anatomy, 2d ed., Univ. of Chicago Press, 1942*)

Attempts to derive the tetrapod limb (cheiropterygium) from the lateral fin (icthyopterygium) have not been altogether successful. Perhaps the most revealing comparison is that between the limb of an ancient tetrapod, *Eryops*, and the pectoral and pelvic fins of a primitive aquatic fish, the extinct crossopterygian, *Sauripterus*. Both have a single proximal segment, the tetrapod's humerus or femur; two intermediate elements, the tetrapod's radius and ulna or tibia and fibula; and about a dozen marginal radials, which is decidedly less than the number of skeletal elements in the tetrapod's pentadactyle foot (Fig. 8).

Pectoral girdle. The pectoral or shoulder girdle is the internal archlike supporting structure for

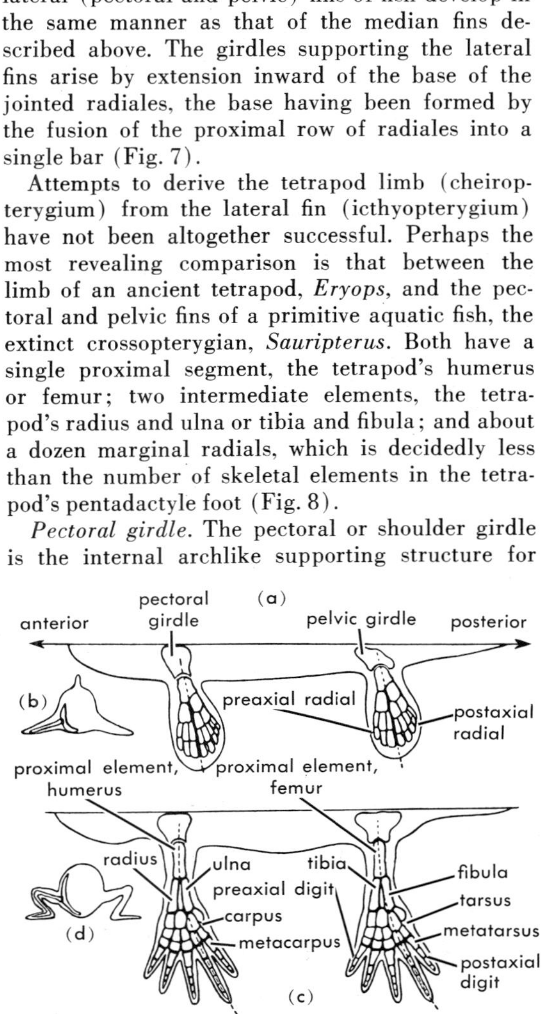

Fig. 8. Comparison of the primitive appendages of fish and tetrapod. (*a*) Primitive fishlike form. (*b*) Same in transverse section. (*c*) Primitive tetrapod. (*d*) Same in transverse section. (*From E. S. Goodrich, Studies on the Structure and Development of Vertebrates, vol. 1, Dover, 1958*)

the anterior pair of appendages. In elasmobranchs the pectoral girdle is a simple U-shaped cartilaginous bar formed by the right and left basal radials which had become displaced ventromedially until their ends met and fused at the midline. The point at which the fin is attached, situated near the bend of the U bilaterally, is called the glenoid fossa. The horizontal portion of the girdle that connects between glenoid fossae is the coracoid region; the portion that extends above the glenoid fossa on each side with a free tip is the scapular region. The latter may be subdivided into scapula proper nearest the fossa and suprascapula more distally. In the bony fishes the latter elements which are preformed in cartilage become overlaid by dermal bones. Thus, on each side the clavicle appears over the coracoid, the cleithra over the scapula. Beyond the scapula additional dermal elements appear that unite the pectoral girdle to the skull.

In tetrapods two divergent lines in the evolution of the shoulder girdle have been traced from ancient reptilian stocks, depending upon whether the coracoid region is ossified from a single center, often called the precoracoid center, or from two centers, the anterior and posterior coracoids. The former has led to reptiles and birds; the latter, to mammals. However, except in the monotremes in which the two coracoid centers appear, only the posterior coracoid has persisted in mammals and even at that has been reduced to a small projection, the coracoid process of the scapula. The scapula, on the other hand, is always well developed, particularly in placental mammals. As in bony fishes, the pectoral girdle of primitive tetrapods is strengthened by the presence of dermal bones, including paired clavicles ventrally, paired cleithra dorsally, and an unpaired ventromedian interclavicle. The cleithrum disappears except in frogs and certain reptiles. The clavicle persists in all but the more highly specialized forms such as the tailed amphibians, snakes, alligators, ungulates, whales, and a few carnivores. In birds the clavicle is fused with its contralateral mate and as such is called the furcula or, more popularly, the wishbone. The interclavicle, absent in most tetrapods, persists in certain reptiles and uniquely in the monotremes.

Pelvic girdle. The pelvic girdle serves the same function with respect to the posterior appendages as the pectoral for the anterior. Furthermore, the basic elements of the two girdles and their plans of development are very much alike. In elasmobranchs the unsegmented cartilaginous pelvic arch bears a socket, the acetabulum, on each side where the pelvic fins are attached. The horizontal portion of the arch that connects between acetabular fossae is called the ischiopubic bar; the free dorsal projections of the arch are the iliac processes. In bony fishes the pelvic arch is similar in form to that of elasmobranchs but is ossified. The pelvic girdle in all fishes remains free from the vertebral column, wheras in tetrapods it gains additional strength by attachment to the vertebral column. Characteristically, the ischiopubic bar develops two centers; the pubis is ossified from the anterior one and the ischium from the posterior. The ilium is ossified from a center in the iliac process. The three centers (pubic, ischial, and iliac) converge upon the acetabulum and, until they actually fuse, the boundary between them is clearly delineated in the floor of the fossa as a triradiate cartilage. Dermal bones have no share in the formation of the pelvis. Attachment of the pelvis to the vertebral column is provided through the ilium, the medial edge of which meets the sacral ribs on either side.

[D.G.W.]

ANATOMY

The vertebrate skeleton is primarily an endoskeleton of cartilage and bone. There are two types of bone: cartilage (endochondral, or replacing bone) and dermal (membrane, or investing bone). The first type is preformed in cartilage and ossifies later, while the second type ossifies directly from membrane without any cartilaginous predecessor. Otherwise they are similar. Dermal bone occurs only in the skull and shoulder region. An exoskeleton of hair, claws, nails, hoofs, horns, antlers, feathers, scales, and bony ossicles in the skin may also be present.

The skeleton consists of the axial skeleton, comprising the skull, vertebral column, and associated structures, and the appendicular skeleton or appendages. The skull is subdivided into the neurocranium and the branchiocranium, also known as the splanchnocranium or viscerocranium.

Axial skeleton. Skeletal organization is quite uniform except for variations in the number, size, and arrangement of the bones. The neurocranium consists of the braincase and three pairs of capsules associated with the organs of smell, sight, equilibrium, and hearing (Fig. 10). The olfactory capsules develop around the olfactory lobes, the optic capsules around the eyeballs, and the otic or auditory capsules around the ear. The olfactory and otic capsules unite with the rest of the neurocranium, but the optic capsules remain free to permit eye movements. The neurocranium has openings (foramina) for the passage of blood vessels and nerves.

The jaws, hyoid, and gill (branchial) arches or their derivatives form the branchiocranium (Fig. 10). The jaws are closely associated with the neurocranium, while the other branchial arches are indirectly related to it. The jaws constitute the first pair of branchial arches, the hyoid the second pair, and the true gill arches the remaining pairs.

Fig. 9. Skeletal systems. (*a*) Man (*from N. L. Hoerr and A. Osol, eds., Gould Medical Dictionary, 2d ed., Blakiston–McGraw-Hill, 1956*). (*b*) Frog (*from T. I. Storer and R. L. Usinger, General Zoology, 3d ed., McGraw-Hill, 1957*). (*c*) Bird (*from W. H. Atwood, Comparative Anatomy, 2d ed., Mosby, 1955*). (*d*) Fish (*after Dean from A. S. Romer, The Vertebrate Body, 2d ed., Saunders, 1955*).

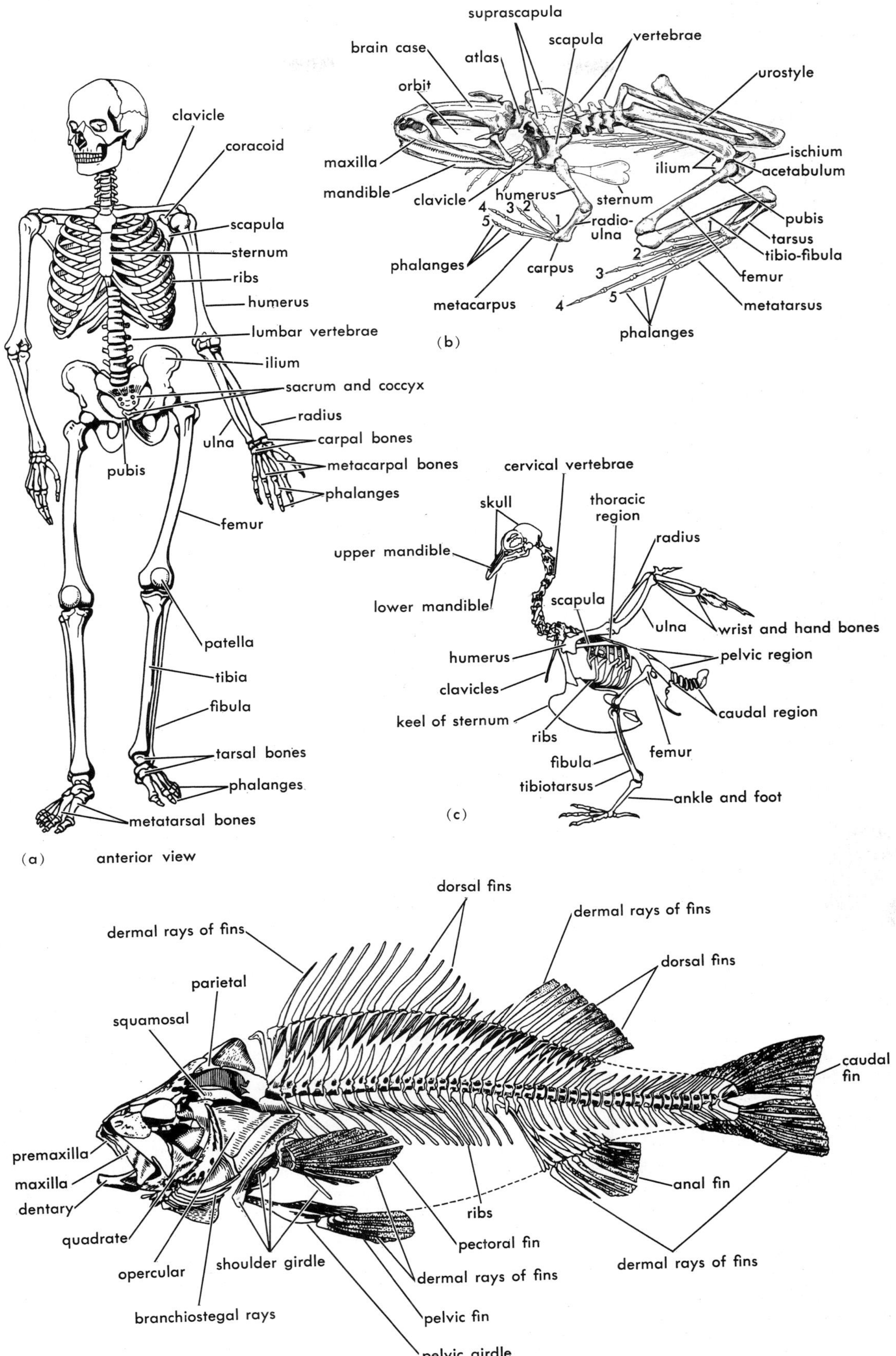
clavicle
coracoid
scapula
sternum
ribs
humerus
lumbar vertebrae
ilium
sacrum and coccyx
radius
ulna
carpal bones
metacarpal bones
pubis
phalanges
femur
patella
tibia
fibula
tarsal bones
phalanges
metatarsal bones
(a)
anterior view
suprascapula
brain case
atlas
scapula
vertebrae
urostyle
orbit
maxilla
mandible
clavicle
humerus
sternum
ilium
ischium
acetabulum
pubis
tarsus
tibio-fibula
femur
metatarsus
radio-ulna
carpus
phalanges
metacarpus
phalanges
1
2
3
4
5
(b)
cervical vertebrae
skull
thoracic region
upper mandible
radius
lower mandible
scapula
ulna
wrist and hand bones
humerus
pelvic region
clavicles
caudal region
keel of sternum
ribs
femur
fibula
tibiotarsus
ankle and foot
(c)
dorsal fins
dermal rays of fins
dermal rays of fins
dorsal fins
parietal
squamosal
caudal fin
premaxilla
maxilla
dentary
quadrate
opercular
shoulder girdle
branchiostegal rays
ribs
pectoral fin
dermal rays of fins
pelvic fin
pelvic girdle
anal fin
dermal rays of fins
(d)

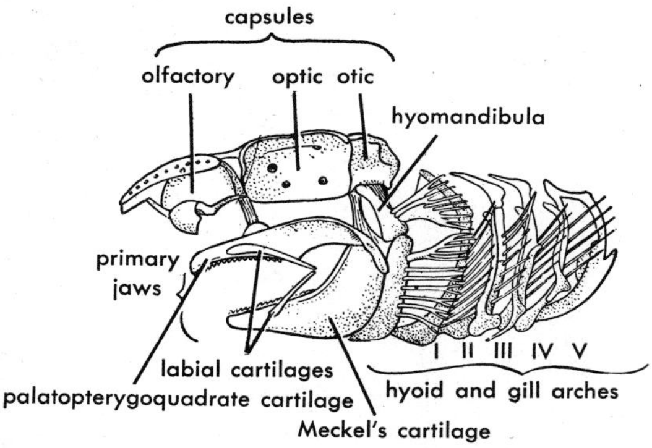

Fig. 10. Skull of a shark. (*From W. K. Gregory, Our Face from Fish to Man, Putnam, 1929*)

The relation of the jaws to the neurocranium varies in different vertebrates. In the early fishes, Acanthodii, the jaws were probably attached to the neurocranium by ligaments (autodiastylic suspension) and the remaining branchial arches bore gills. Most fishes have hyostylic suspension; that is, the jaws are attached to the neurocranium by the hyomandibular. In some fossil fishes and a few living ones the upper jaw, in addition to the hyomandibular support, is directly attached to the neurocranium (amphistylic suspension). Lungfishes (Dipnoi), chimaeras (Holocephali), and tetrapods (amphibians, reptiles, birds, and mammals) have the upper jaw fused with the neurocranium without hyoid support (autostylic suspension).

The branchial arches are divided into movably articulated parts. The first or mandibular arch consists of paired palatoquadrate (upper jaw) and Meckel's (lower jaw) cartilages. Usually the cartilage bones of the primary (inner) jaws are ensheathed in dermal bones forming secondary (outer) jaws.

The hyoid or second arch consists of paired hyomandibular or epihyal, ceratohyal, and hypohyal elements, plus a single basihyal connecting the opposite halves of the arch ventrally. This is the condition in certain fossil fishes such as the Pleuracanthodii and Acanthodii, in fossil and living bony fishes, and in lungfish.

The gill arches, usually branchial arches 3–7, are composed of paired pharyngobranchial, epibranchial, ceratobranchial, and hypobranchial elements. A median unpaired basibranchial, the copula, connects the halves of each arch ventrally. Attached to the epibranchials and the ceratobranchials are the gill or branchiostegal rays supporting the gills. The functional gill arches of most fishes have this essential arrangement although the number of arches may be reduced and more than one arch may attach to a single basibranchial element. In the tetrapods the branchial elements behind the jaws disappear or are changed into ear bones, the hyoid apparatus, or laryngeal cartilages.

The Chondrichthyes, that is, the sharks, rays, and chimaeras, have a cartilaginous skull, but adult Osteichthyes, or bony fishes, and tetrapods have an ossified skull. Some bony fishes have more than 150 separate skull bones, but during evolution the number was reduced to approximately 27 bones in the human skull.

Crossopterygians. In *Eusthenopteron* (Fig. 11), a generalized fossil crossopterygian fish (the group believed ancestral to tetrapods), the neurocranium has 5 unpaired bones—basioccipital, interparietal, parasphenoid, sphenethmoid, and vomer—and 28 paired bones—frontal, lateral extrascapular, medial extrascapular, nasals 1–3, opisthotic, parietal, postnasal, postparietal, postrostral 1–2, prootic, rostral, septomaxillary, supratemporal, and tabular. The total is 33 bones. In addition, there are 8 circumorbital bones—jugal, lacrimal, postorbital, and prefrontal, all paired; and 6 laterotemporal bones—quadratojugal, preopercular, and squamosal, all paired. The sphenethmoid, parasphenoid, basioccipital, prootic, opisthotic, and vomer are probably cartilage bones; the others are dermal bones.

The branchiocranium of *Eusthenopteron* contains 98 bones. The primary (inner) upper jaws consist of 16 paired bones—ectopterygoid, epipterygoid, palatine, pterygoid, quadrate, and 3 suprapterygoids. The secondary (outer) upper jaws consist of 4 paired bones—maxillary and premaxillary. There are only 2 bones (paired articulars) in the primary lower jaw but 16 paired bones (angular, coronoid 1 and 2, dentary, postsplenial, prearticular, splenial, and surangular) in the secondary lower jaw. The quadrate and articular are the only cartilage bones in the mandibular arch. The crossopterygian hyoid and gill arches are similar to those of typical fishes. The gills are covered by an operculum of paired opercular and subopercular bones and the "throat" by 5 paired marginal gular bones; 4 pairs of pharyngeal bones may have been present. *Latimeria,* the only living crossopterygian fish, is very similar to *Eusthenopteron.*

Primitive tetrapods. A generalized primitive tetrapod condition is represented by *Eryops,* a Permian labyrinthodont amphibian. The neurocranium (Fig. 12*a*) may be divided into the following regions: roof, circumorbital, laterotemporal, ear, braincase floor, snout, and occipital. All the bones except the basioccipital, basisphenoid, parasphenoid, and sphenethmoid are paired. There is one tripartite occipital condyle, composed of the basioccipital and the two exoccipital bones.

The primary upper jaw consists of the ectopterygoid, epipterygoid, palatine, pterygoid, and quadrate bones; the secondary upper jaw is composed of the premaxillary and maxillary bones. The bones of the primary lower jaw are the mentomeckelian and articular, while the secondary lower jaw is composed of the angular, coronoid, dentary, intercoronoid, postsplenial, prearticular, precoronoid, splenial, and surangular. All jaw bones are paired. The teeth are greatly infolded (hence the name labyrinthodont) like those of the

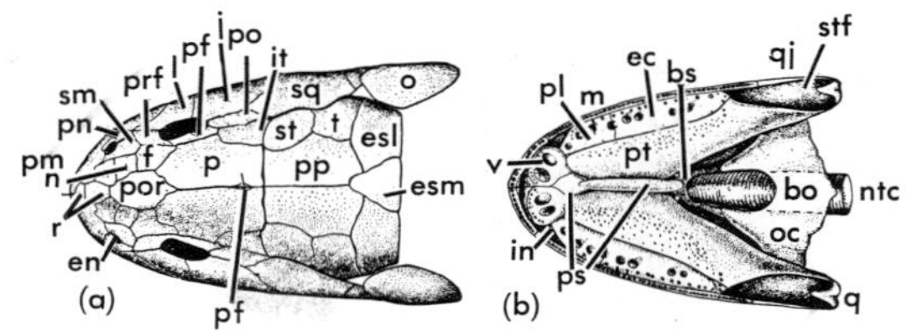

Fig. 11. Skull of *Eusthenopteron,* a Devonian crossopterygian fish. (*a*) Dorsal aspect. (*b*) Ventral aspect. (*From A. S. Romer, The Vertebrate Body, 2d ed., Saunders, 1955*)

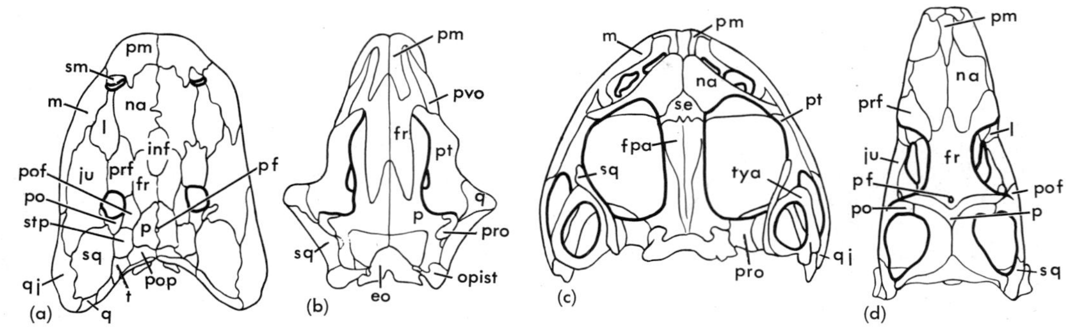

Fig. 12. Skulls of fossil and living amphibians and reptiles. (*a*) *Eryops.* (*b*) The mudpuppy, *Necturus.* (*c*) The bullfrog, *Rana.* (*d*) The lizard, *Iguana.* (*Modified from F. G. Evans, The morphological status of the modern Amphibia among the Tetrapoda, J. Morphol., 74:43–100, 1944*)

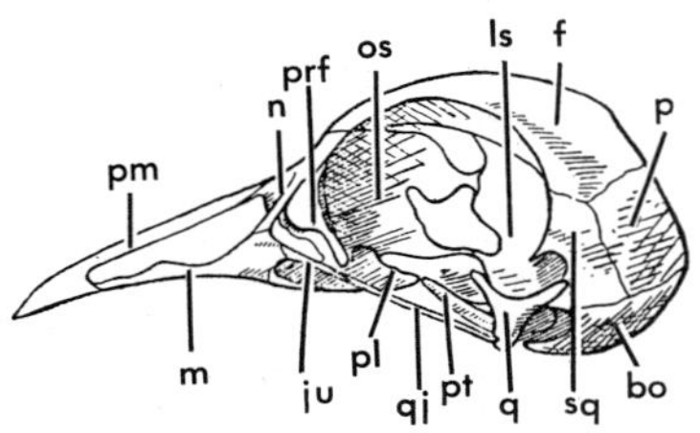

Fig. 13. Skull of *Columba,* the pigeon. (*From G. R. deBeer, Vertebrate Zoology, Sidgwick and Jackson, 1928*)

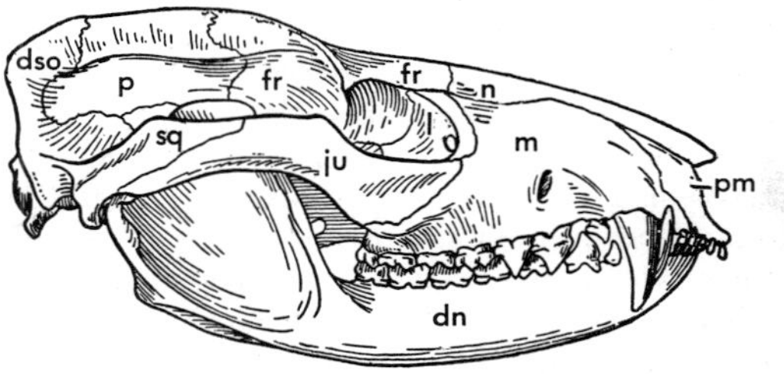

Fig. 14. Skull of *Didelphis,* the modern opossum. (*From W. K. Gregory, Our Face from Fish to Man, Putnam, 1929*)

bo, basioccipital
bs, basisphenoid
dn, dentary
dso, dermosupraoccipital
ec, ectopterygoid
en, external naris
eo, exoccipital
esl, lateral extrascapular
esm, medial extrascapular
f, frontal
fpa, frontoparietal
in, internal naris
inf, internasofrontal
it, intertemporal
ju, jugal
la, lacrimal
ls, latersphenoid
m, maxillary
n, nasal
ntc, notochord
o, opercular
oc, otic capsule
opist, opisthotic
os, orbitosphenoid
p, parietal
pf, parietal foramen
pl, palatine
pm, premaxillary
pn, postnasal
po, postorbital
pof, postfrontal
por, postrostral
pp, post parietal
prf, prefrontal
pro, prootic
ps, parasphenoid
pt, pterygoid
pvo, prevomer
q, guadrate
qj, quadratojugal
r, rostral
s, squamosal
se, sphenethmoid
sm, septomaxillary
st, supratemporal
stf, subtemporal fossa
t, tabular
tya, tympanic annulus
v, vomer

contemporaneous crossopterygian fishes. Except for the absence of gills, the skull is similar to that of crossopterygians.

Amphibia. In the living amphibians (Fig. 12*b,c*), all specialized, the number of skull bones is reduced from 40 in *Eryops* to 18 in some of the legless Apoda (*Ichthyophis*) and the bullfrog *Rana* and to 15 in the mudpuppy *Necturus.* Typical living amphibians have lost the internasofrontal, postparietal, jugal, lacrimal, postorbital, supratemporal, tabular, otic, basisphenoid, basioccipital, and supraoccipital bones from the neurocranium. The bones usually absent in the jaws are the ectopterygoid, epipterygoid, intercoronoid, postsplenial, precoronoid, splenial, and surangular. Two exoccipital condyles are present for articulation with the vertebral column. Much of the reduction in number of skull bones is due to fusion. The teeth are simple cusps.

The hyoid apparatus is largely cartilaginous. In *Ichthyophis* it consists of a single basibranchial and four paired ceratobranchials. In the bullfrog *Rana* it is a flat cartilaginous plate with paired cartilaginous ceratohyals and ossified thyroid processes. The mudpuppy *Necturus* has a pair of hypohyal and ceratohyal elements plus two basibranchial, two pairs of hypobranchial, and three pairs of ceratobranchial elements. All are cartilaginous.

Reptiles. The skull of some Permian cotylosaurs, the most primitive reptiles, is very similar to that of labyrinthodont amphibians, and one of them, *Seymouria*, has been classified as both a reptile and an amphibian. The skull of the living lizard, *Iguana* (Fig. 12*d*), has all the neurocranial bones found in *Eryops* except the internasofrontal, postparietal, quadratojugal, supratemporal, supraoccipital, and sphenethmoid. The prootic, opisthotic, and exoccipital bones have fused. The only *Eryops* bones lost from the jaws are the mentomeckelian, intercoronoid, precoronoid, and splenial. Except for the advanced mammal-like reptiles, there is a single tripartite occipital condyle in reptiles. The columella auris of lizards consists of an ossified proximal part, the stapes or columella, and a distal cartilaginous part, the extracolumella. In many respects living reptiles are more primitive than contemporaneous Amphibia.

The hyoid apparatus of *Iguana* consists of the hyoid and first 2 branchial arches. There are 1 basihyal, 2 hypohyals, 2 ceratohyals, 2 ceratobranchials, and 2 epibranchials. Only the first pair of ceratobranchials is ossified.

Teeth are absent in turtles, but conical single-cusped teeth are present in *Sphenodon*, Crocodilia, and lizards. Snakes have curved teeth which are grooved or hollow in poisonous species. The mammal-like reptiles have different kinds of teeth and some dinosaurs such as *Anatosaurus* and *Edmontosaurus* had many leaflike teeth, approximately 2000, closely pressed together to form grinding structures. Snakes have a special joint in the braincase to permit wide opening of the mouth, and the opposite sides of the lower jaw are joined by an elastic ligament. Thus, a snake can swallow prey with a greater diameter than its own body.

Birds. The adult bird skull (Fig. 13) is fused into four complexes: (1) a thin, rigid, continuous neurocranium, (2) a vertically hinged upper beak supported on a basically reptilian upper jaw and palate, (3) a mandible of fused reptilian elements, and (4) a jointed hyobranchial chain. In the pigeon *Columba livia*, the neurocranium consists of paired basitemporal, ectoethmoid or turbinal, exoccipital, frontal, jugal, lacrimal, laterosphenoid, nasal, orbitosphenoid, parietal, prefrontal, quadratojugal, and squamosal bones, plus single basioccipital, basisphenoid, mesethmoid, parasphenoid, and supraoccipital bones. The basisphenoid and mesethmoid bones form the interorbital septum. There is a single occipital condyle. The upper beak consists of the paired maxillary, palatine, premaxillary, most of the beak, pterygoid, and quadrate bones. The lower beak is double-jointed as in lizards and snakes and consists of the paired articular, angular, surangular, and dentary bones. No living birds have teeth, but they were present in some early fossil ones such as *Archaeopteryx* and *Ichthyornis*.

The pigeon hyoid apparatus (hyoid and third gill arches) consists of three single median elements: the entoglossal cartilage or fused ceratohyals anteriorly, the basihyals in the middle, and the basibranchial or third arch posteriorly. A pair of short anterior horns (the cornu) and long posterior horns (the third ceratobranchials and epibranchials) are also present.

Mammals. Adult mammal skulls typically lack paired prefrontal, postfrontal, postorbital, quadratojugal, and septomaxillary bones. The reduction in total bone number is due to fusion, and in the primitive opossum, *Didelphis*, there are 42 separate bones (Fig. 14). Paired frontal, jugal, lacrimal, nasal, parietal, periotic, and squamosal bones plus unpaired basioccipital, basisphenoid, dermosupraoccipital, ethmoid, presphenoid, and vomer form the neurocranium. The paired premaxillary, maxillary, pterygoid, and palatine bones form the upper jaw, and the dentary, the lower jaw. A diagnostic mammalian feature is the presence of a tympanic and three paired auditory ossicles, the malleus, incus, and stapes. Auditory bullae are usually present and there is one occipital condyle. In the nose there are three paired turbinate bones. The hyoid apparatus consists of paired ceratohyals and thyrohyals plus a single basihyal.

In the adult human skull (Fig. 15) the pre- and basisphenoids fuse to form the sphenoid; the basi-, ex-, and supraoccipitals to form the occipital; the periotic, squamosal, and tympanic to form the temporal; the premaxillary and maxillary to form the maxilla. The upper and middle turbinates or conchae fuse with the ethmoid. The basi-, cerato-, and thyrohyals form the hyoid bone.

Mammals typically have two sets of teeth, the milk or baby teeth and the permanent incisors, canines, premolars, and molars. Teeth are absent in some mammals, such as the anteaters, and sec-

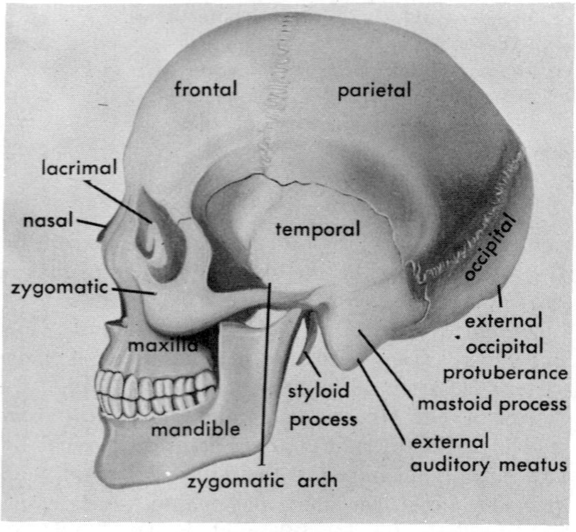

Fig. 15. Adult human skull. (*From F. G. Evans, Atlas of Human Anatomy, Littlefield, Adams, 1957*)

ondarily increased in number but reduced to simple cusps in others, as the toothed whales. *See* DENTITION.

Vertebrae. The term vertebrate refers to the vertebral column composed of vertebrae. A typical vertebra consists of a body or centrum and a neural arch surrounding the spinal cord lying dorsal to the centrum. The neural arch has spinal and transverse processes for the attachment of muscles and ligaments and articular processes, the zygopophyses, for articulation with adjacent vertebrae. In certain fossil reptiles the bodies of the cervical vertebrae also have articular processes, exopophyses. The bodies of the caudal vertebrae often have ventral hemal arches surrounding the blood vessels to the tail. In the tails of reptiles and mammals the hemal arches, chevrons, are attached to small elements, intercentra, wedged in ventrally between successive centra. Fish have little regional differentiation in the vertebral column, but tetrapods typically have cervical, thoracic, lumbar, sacral, and caudal vertebrae. Legless forms lack sacral vertebrae. The first two cervical vertebrae, the atlas and axis, of amphibian fossils and amniotes are specialized for support and movements of the head. Tetrapods have fibrocartilaginous intervertebral disks between the bodies of most of the movable vertebrae.

There are seven cervical vertebrae in all mammals, including man and giraffe. Exceptions are the manatee. *Trichechus*, and two-toed sloth, *Cholaepus*, which have six, and the three-toed sloth, *Bradypus*, which has nine neck vertebrae. *Elasmosaurus*, a fossil marine reptile, had 76 neck vertebrae. The Xenartha (sloths, armadilloes, and anteaters) have an extra pair of zygapophyses on the posterior dorsal and lumbar vertebrae. The number of separate vertebrae varies from over 200 in some Apoda and pythons to only 6 in some frogs.

Ribs and sternum. Two types of ribs, dorsal and ventral, occur in vertebrates. Dorsal ribs lie in the septum dividing the trunk musculature into dorsal and ventral masses; ventral or pleural ribs lie in the septa dividing the trunk musculature into segments. Many fish have both types of ribs, but only dorsal ribs occur in tetrapods. The earliest tetrapods had ribs on all of the presacral vertebrae, but in living forms, especially birds and mammals, movable ribs are usually restricted to the thoracic region. In tetrapods with functional hind legs, one or more sacral ribs articulate with the pelvis. Cervical ribs, when present, are usually short, whereas thoracic ribs are typically long and curved. In turtles the trunk vertebrae and ribs are immovably fused to the carapace or upper shell. In certain mammals (living armadilloes and fossil glyptodonts) there is considerable fusion of the trunk vertebrae.

Most tetrapods have a sternum which in amniotes is connected with the ventral ends of the ribs via costal cartilages. The sternum in living amphibians is mostly cartilaginous and incorporated with the pectoral girdle. It does not connect with the ribs, which are short and straight, but some of

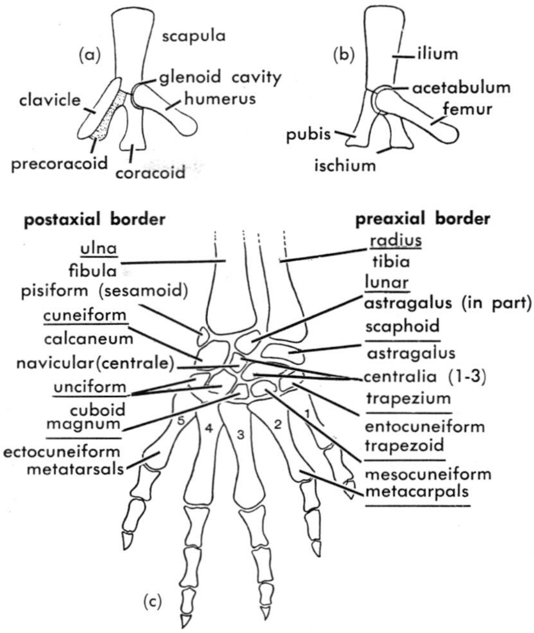

Fig. 16. Diagram of (*a*) the pectoral and (*b*) the pelvic girdles and (c) generalized hand and foot of a tetrapod. Bones of forelimb are underlined. (*From T. J. Parker and W. A. Haswell, A Text-book of Zoology, vol. 2, 6th ed., Macmillan, 1951*)

the fossil amphibians, Labyrinthodontia, have long curved ribs which may have articulated with a sternum via costal cartilages. The sternum is absent in the legless living amphibians, turtles, legless lizards, and snakes. The bird sternum is an ossified quadrangular bone with a prominent median keel in flying forms and penguins. The keel is absent in ostriches and related forms. The mammalian sternum is an ossified median structure of several parts which may remain separate or fuse.

Appendicular skeleton. This consists of the skeleton of the pectoral (shoulder) and the pelvic (hip) girdles and of the paired appendages (Fig. 16). The girdles provide attachment and support for the appendages, which may be fins, flippers, legs, wings, or arms. The pectoral girdle differs from the pelvic one in consisting of both dermal and cartilage bones and, except in skates and pterodactyls, in having no cartilaginous or bony connection with the vertebral column. The pectoral girdle is located close to the head, in bony fishes and some early tetrapods actually being attached to the skull, while the pelvic girdle is associated with the outlets of the digestive and reproductory systems. However, in some bony fishes the pelvic fins are in the throat region anterior to the pectoral fins.

Pectoral and pelvic girdles. In bony fishes the pectoral girdle consists of a scapulocoracoid bone (sometimes the two elements are separate) which is almost covered externally by dermal bones named, in a dorsoventral order, post-temporal, supracleithrum, cleithrum, postcleithrum, and clavicle. The clavicles of the opposite halves of the girdle unite in a symphysis. The post-temporal bone attaches the girdle to the back of the skull. In tetrapods the connection with the skull is soon lost and the dermal elements are reduced progressively in size and number while the cartilage bones increase.

The *Eryops* pectoral girdle consists of an ossified scapulocoracoid, cleithrum, clavicle, and interclavicle. In salamanders only the scapula is ossified. Frogs have an ossified scapula, coracoid, clavicle, and omosternum. Living Amphibia lack the cleithrum and interclavicle. The Apoda have no girdles.

The *Eryops* pelvic girdle consists of ossified ilium, pubis, and ischium meeting in the acetabulum. Living Amphibia have a cartilaginous pubis. The frog has a saberlike, forwardly directed ilium; in others it is vertical. The ilium articulates with a single sacral vertebra in living Amphibia, but some fossil forms have two or three sacral vertebrae.

The lizard *Iguana* has an ossified scapula, coracoid, clavicle, and interclavicle attached to a cartilaginous sternum and ribs. The pelvic girdle consists of an ossified ilium, ischium, and pubis. Both girdles are absent in snakes and legless lizards. Turtles are the only vertebrates with the pectoral girdle inside the ribs.

Birds have a swordlike scapula, an enlarged coracoid, and clavicles joined by an interclavicular element, the furcula, to form the wishbone. The bird pelvis consists of expanded ilium, articulating with several sacral vertebrae, and a slender pubis paralleling the expanded ischium. The pubic symphysis is typically lacking, but ostriches and rheas have both a pubic and an ischiatic symphysis.

Mammals typically have a clavicle, a scapula, and no coracoid except in monotremes (egg-laying mammals). Horses have only the scapula. The three pelvic bones are usually fused into a single pelvic bone. Monotremes, marsupials, insectivores, many rodents, hoofed mammals, and carnivores have both a pubic and an ischiatic symphysis.

Appendages. The skeleton of the paired appendages (fins) of fish consists of cartilaginous or bony fin rays. Tetrapod limbs are all modifications of a single basic pattern (Fig. 16). Most variation occurs in the number and shape of hand and foot bones. The forearm and leg bones fuse in frogs, and the ulna and fibula are small splints in hoofed mammals.

Eryops has 12 carpal (wrist) bones, but in living frogs (Salientia) and salamanders (Caudata) the number varies from 3 to 9. *Iguana* has 9 carpals. Some primitive fossil amphibians such as *Trematops* have 13 tarsal bones, but living Caudata have 3–12 and Salientia from 5 to 7 tarsals. *Iguana* has 3 tarsals. Birds have 4 carpals, the 2 distal ones fused with the metacarpals. The tarsals of birds are fused with the leg and metatarsal bones with the main joint between them. Mammals have 8–11 carpals and usually 7 tarsals.

Five is probably the original tetrapod number of metacarpals, metatarsals, and digits. The number of digits in living forms is 4 in Amphibia, 5 in reptiles, 3 in birds, 2 in even-toed mammals (Artiodactyla) 4 on the front feet and 3 on the hind feet in tapirs and rhinoceros (Perissodactyla), 1 on each foot in horses (Perissodactyla), and 5 in most other forms. Some fossil marine reptiles (Ichthyosaurs) had 8 digits in their flippers.

The number of bones (the phalanges) in each digit is written as a formula beginning with the first digit. *Eryops* has a phalangeal formula of 2-2-3-2-2 in the hand or manus, but the formula is 2-2-3-2 in typical Caudata and 2-2-3-3 in *Rana*. *Iguana* has the normal reptilian formula of 2-3-4-5-3. The formula in birds and mammals is 1-2-1-0-0 and 2-3-3-3-3, respectively.

The phalangeal formula in the foot is 2-3-4-4-3 in *Trematops*, 2(1)-2-3-3-2 in most Caudata, and 2-2-3-4-3 in *Rana*. *Iguana* has the typical reptilian formula of 2-3-4-5-4. Turtles frequently have a reduction in phalangeal number, but the ichthyosaurs and plesiosaurs have the phalangeal number increased. Birds have a formula of 2-3-4-5-0 and mammals the same one as the hand. [F.G.EV.]

PHYSIOLOGY OF BONE

Bone is a highly specialized form of connective tissue, composed of branching cells in a calcified intercellular substance, which forms the skeleton or framework of the body of most vertebrate animals. It has long been recognized that the gross structure of bone is well adapted to its mechanical functions of providing the skeletal support of the body and of protecting the vital organs of the cranial and thoracic cavities. It has more recently been appreciated that bone tissue has important physiologic functions, relating to the life and health of the whole organism, and that its microscopic and submicroscopic structure is equally adapted to these functions.

Cellular components. The cellular components of bone are associated with specific functions: osteoblasts, located on bone surfaces, with the formation of bone; osteocytes, in cavities or lacunae within bone, with the maintenance of the integrity of bone as a living tissue; and osteoclasts, on the surfaces of bone, with its destruction or resorption. These cells, having common ancestors, are closely interrelated. During active growth, frequent transformation occurs from one form to another, depending upon the functional activity the cells are called upon to perform.

Membranes. The connective tissue surrounding bone is the periosteum. In the young, rapidly growing animal this includes an inner layer of osteoblasts, actively engaged in the formation of bone. In the adult these cells lose the characteristics of osteoblasts; this layer is activated, however, following injury and again forms new bone. The bone marrow cavities are lined with a thin layer of connective tissue cells, the endosteum, also having the capability of forming bone and of participating in the repair of bone following a fracture.

Interstitial substance. In addition to being calcified, bone has a fibrillar structure similar to that of ordinary connective tissue; the fibers are mainly those of collagen. The ground substance, filling the spaces between fibers and mineral, is semifluid and is characterized by its content of complex sugars. Since it communicates freely with the intercellular fluids of the body, and hence with the blood, it plays an important part in the physiologic activity of bone. The total organic matter in bone makes up 35% or more of its dry, fat-free weight.

Bone marrow. In the higher animals, the bone marrow is the principal, or perhaps only, source of new red blood cells. It also participates in the formation of bone; its reticular cells, under certain circumstances, undergo transformation into the cells of bone. The endosteum is a condensed peripheral layer of the stroma of the bone marrow. *See* HEMATOPOIESIS.

Minerals. The mineral of bone, often called the bone salt, accounts for as much as 65% of its dry, fat-free weight, and is responsible for its hardness. The fact that the bone salt is difficultly soluble is of far-reaching physiologic significance. This property not only determines the deposition of the bone mineral, but it also preserves the structure and rigidity of the bones.

The mineral of bone, while not constant in composition, is now well recognized by means of chemical analysis, x-ray diffraction, and other methods as a member of the apatite series. It corresponds closely in composition and crystal structure to hydroxyapatite, $3Ca_3(PO_4)_2 \cdot Ca(OH)_2$. The unit of the mineral of bone, at the ultrastructural level, is a microcrystal of hydroxyapatite of colloidal dimensions, only a few hundred angstrom units (A) in length and with a thickness of 20–50 A.

Growth and remodeling. Most of the bones of the body are laid down first in the embryo as cartilage models, which are later replaced by bone, originating in ossification centers. A portion of the cartilage model persists as the epiphyseal cartilage near the ends of the long bones. This serves a double purpose; it forms a union between the ends of the bone or epiphyses and the shafts or diaphyses, giving structural stability while permitting growth, and it provides a growth apparatus within which growth in length persists until adult life is reached. As long as growth continues it is under regulation by the growth hormone of the pituitary gland; when growth stops, the epiphyseal cartilage disappears. In the rat, the epiphyseal cartilages of the long bones persist, and growth continues throughout the life of the animal.

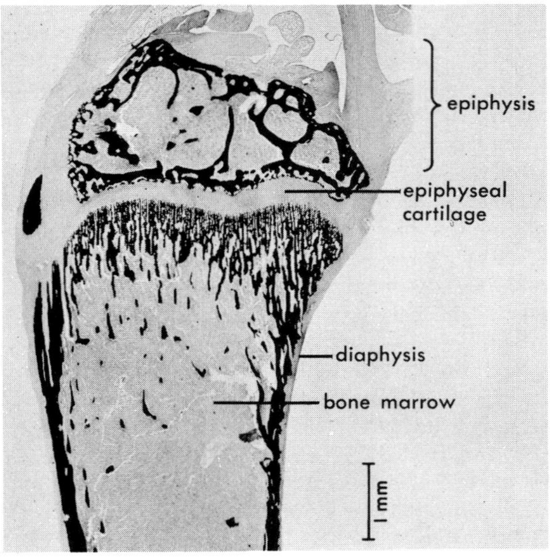

Fig. 17. Sagittal section of head of tibia of normal rat, age 7 weeks. Cut, without decalcification, and treated with silver nitrate to illustrate distribution of calcified tissues, which stain black.

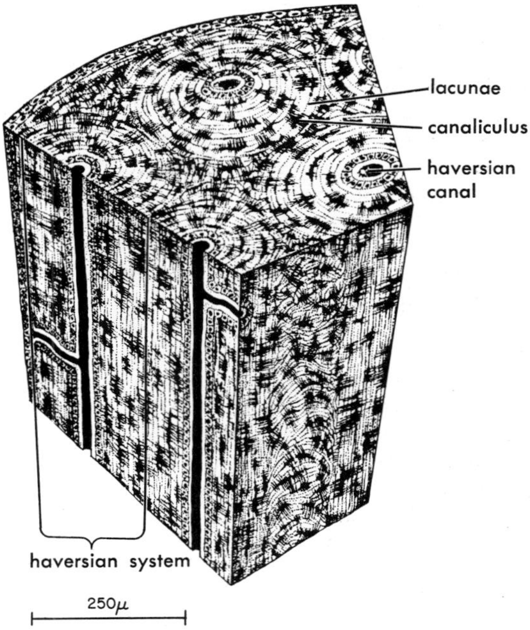

Fig. 18. Three-dimensional diagram of the microscopic structure of the Haversian systems, or osteons, of bone. Illustrates central canals, lacunae, canalicules, and concentric layers of osteons.

Growth in diameter of bones occurs by another process. The long bones are gradually resorbed from within the marrow cavity, while new bone is added on the external, periosteal surface. While growth continues, the bones are subject to extensive remodeling, permitting increase in size with only minor changes in external configuration.

Haversian system. The unit of structure of bone at the microscopic level is the Haversian system, or osteon, which is peculiar to bone (Fig. 18).

The osteon, when fully formed, is an irregularly cylindrical and branching structure, with thick walls and a narrow lumen, the Haversian canal. The canal carries blood capillaries and venules. The walls have a lamellar structure, in concentric layers. Between the lamellae are large numbers of lacunae, housing the osteocytes and interconnected with one another and with the lumen of the canal by means of branched canalicules. Blood reaches the bone through the capillaries. These, together with the lacunae and their interconnecting canalicules, make possible the deposition of mineral in the organic matrix and interchange of ions and metabolites with the fluids of the body.

Internal or Haversian remodeling, which continues throughout life, is closely related to the regulation of the exchange of mineral between the blood and the bones. It consists in the formation of tunnels, or resorption cavities, to be subsequently filled in by new Haversian systems, by deposition of concentric layers of bone on the walls of the tunnel, continuing until only the usual small canal remains. During this remodeling the new bone is reactive and readily accessible to the circulation over periods of months or more.

Calcification of the skeleton. For the bones to become hard the mineral portion must be deposited in a soft organic matrix; failure of such deposition results, in the infant, in rickets. This deposition of a complex calcium phosphate, mainly hydroxyapatite, is calcification.

Calcification of the skeleton begins in the embryo, as cartilage models are transformed into bone; it continues throughout life, as new bone replaces old. It depends upon two processes: initiation of crystal formation, known as crystal seeding, and crystal growth. Crystal growth is well described from the contributions of crystallography. Crystal seeding, in a calcifiable tissue, is not well understood; it is not known why bone and some parts of cartilage calcify, while other tissues, resembling the organic matrix of bone in chemical composition, remain free of deposits of mineral.

For the skeleton to calcify normally, a continuous supply of calcium and phosphate must be carried in the blood to the sites of calcification, at concentrations adequate to ensure their combination in the form of the bone mineral. These conditions define the humoral factor in calcification. The humoral factor is of course common to all parts of the body, but not all parts of the body calcify; there is also a local factor that determines where calcification will occur. The nature of this local factor is not well understood; it has been attributed to the collagen of the bone matrix, or to the ground substance, or to a chemical or physical association of the two. The osteoblasts of growing bone are rich in an enzyme, alkaline phosphatase, believed to play a part either in calcification or in the formation of a calcifiable matrix. No way has yet been found to assess the property of calcifiability in a tissue, except the observation of calcification itself.

Bone as a reservoir of minerals. Of the numerous elements present in the body the skeleton serves to store large amounts of calcium, magnesium, sodium, and phosphorus, releasing these to meet other needs of the organism. In addition, a number of other elements are found in small quantities in the mineral of bone; some as contaminants, owing to their accidental presence in the body; others in storage, to meet the needs of metabolism.

Ion transfer between blood and bone. In order for the skeleton to serve as a source of mineral elements for the remainder of the body, there must exist mechanisms for the transfer of these elements between blood and bone, in both directions. Except in the case of calcium, in which there is a special mechanism to regulate its concentration in the fluids of the body, transfers between blood and bone are for the most part effected by the movement of positively or negatively charged ions in accord with physicochemical forces tending to maintain a steady state. Thus, if the blood is depleted of sodium, an element essential to life, the ions of this element are transferred from the bone to blood by diffusion, until equilibrium or a steady state is reached; if an excess of sodium is absorbed into the blood, a portion of it is excreted, while another part is stored in the bones.

Regulation of blood calcium. The major portion of the calcium in the blood plasma is in one of two forms: free calcium ions, Ca^{++}, and calcium held in combination with proteins. These two fractions are in equilibrium with each other. Only the free calcium ions are physiologically active and are subject to transfer between bone and blood. There is no appreciable amount of calcium in the red blood cells. The blood plasma of a man weighing 70 kilograms contains a total of approximately 275 milligrams of calcium, and an amount equal to one-third of this or more moves back and forth between blood and bone every minute. For this to be accomplished without large fluctuations in the concentration in the blood, a dual mechanism is required. One part of the mechanism, dependent only upon ion transfer, does not differ materially from that which restores sodium to the blood as needed, and depends upon a passive physicochemical process. The other mechanism, which controls the fine adjustment of the calcium level in the blood plasma, requires biologic activity mediated by the parathyroid glands, four tiny bodies embedded in the thyroid gland. The parathyroid glands monitor the calcium ion concentration in the blood plasma, and keep this up to a normal level, approximately 10 milligrams per 100 milliliters of plasma, by releasing additional amounts of calcium from the bones as needed. This part of the dual mechanism operates as a feedback; a reduction in the concentration of calcium ions in the blood plasma provides the stimulus for increased activity of the parathyroid glands. Such increased activity releases additional parathyroid hormone into the blood, and this acts directly on bone to transfer more calcium to the blood.

Vitamin D and bone. Vitamin D, best known for its protection of infants against rickets, plays an important part in the mineralization of the skeleton and in the regulation of the concentration of calcium ions in the blood plasma. It promotes calcification chiefly by aiding in the absorption of calcium in the gastrointestinal tract; it acts together with the parathyroid hormone in mobilizing calcium from the bones, and thus in maintaining the level of this element in the blood. *See* VITAMIN.

Hormones and bone. Reference has been made above to the regulation of growth of the bones by the growth hormone of the anterior lobe of the pituitary gland, and of the regulation of the concentration of calcium in the blood by the hormone of the parathyroid glands. Other hormones, notably those of the thyroid gland and of the cortex of the adrenal glands, as well as the male and female sex hormones, influence the growth and metabolism of the skeleton, although for the most part their effects are not specifically directed to the bones as target organs. See separate articles on the various endocrine glands.

Of particular interest is the effect of the female sex hormones, estrogens, in birds. Before and during the egg-laying cycle the bird produces a secondary system of spongy bone, filling much of the bone marrow cavities, especially of the long bones; this serves as a reservoir of calcium for the requirements of the egg shell, and is accompanied by other phenomena incident to ovulation. A similar effect is produced in either male or female birds by administration of the female sex hormones; the male sex hormones have no such effect. Alone among the mammals, the mouse forms bone in the marrow cavities under the influence of administered estrogens.

Repair of bone. Mammals have inherited from lower vertebrates an extraordinary capacity to repair injury and replace missing parts of the skeleton. Instead of being patched together by scar tissue, as is usual in most tissues, bone repair is ordinarily complete, new bone formation being an automatic reaction to injury of bone tissue.

The healing of a fracture begins with a blood clot in and around the fracture gap, followed by formation of a fibrocartilaginous callus. New bone formation then begins from the cells of the periosteum and of the endosteum. This replaces the fibrocartilaginous callus, and forms the bony callus. When the fracture gap is closed, by apposition of new bone from the separated fragments, bony union occurs. Following this there is reorganization of the callus with reshaping of the new bone.

In modern surgery much use is made of bone grafts. Transplants of bone from the same individual often survive and produce new bone, uniting with the bone in the site of implantation. Implants of frozen-dried, devitalized, or preserved bone from other species do not survive, but they serve to fill gaps and to lead to new bone formation by the host. Bone banks, in which bone from various sources is stored and preserved in various ways, are now in common use.

Radioactive isotopes and bone. Neither radiation itself nor the metabolism of radioactive isotopes has any place in the normal physiology of bone. Beginning shortly before World War II, and greatly accelerated by the researches incident to atomic energy, these subjects have assumed increasing importance, with the skeleton playing a major role in relation to them. Radiation, external and internal, is of importance because of its physiologic effects; the use of radioactive elements in tracer amounts is frequent in biologic investigations.

Internal radiation. Excessive doses of radiation applied to the surfaces of the body may have damaging effects upon the bones; for the most part, these are overshadowed by effects upon the soft parts of the body. When certain radioactive elements are taken into the body, they may accumulate in the bones and may injure the bone tissue and the bone marrow. Accidental poisoning, occurring in radium-dial workers, and resulting in malignant tumors of bone, was attributed to radium as early as 1929. Of current interest is the fission product, strontium-90, liberated in the atmosphere in the testing of atomic bombs; this radioactive isotope also may cause malignant tumors in bone.

Isotopes as tracers. The use of radioactive phosphorus, P^{32}, in the study of the metabolism of bone was reported in 1935; radioactive calcium, Ca^{45}, was not available in quantities adequate to meet the needs in biology until 1948. Another isotope of calcium, Ca^{47}, is also in use, and strontium, in the form of Sr^{85}, has been used as a substitute for calcium. The use of these radioactive isotopes has permitted a kinetic analysis of the movement and accumulation of the bone-seeking elements in the body. Methods are available for calculating the rates of accretion, resorption, and exchange reactions in the skeleton. *See* BONE (BIOPHYSICS); SPECIALIZED TISSUE. [F.C.M.]

Bibliography: G. H. Bourne (ed.), *The Biochemistry and Physiology of Bone*, 1956; G. R. deBeer, *Vertebrate Zoology*, 1928; F. G. Evans, The morphological status of the modern Amphibia among the Tetrapoda, *J. Morphol.*, 74:43–100, 1944; F. G. Evans, *Atlas of Anatomy, Simplified*, 1958; W. H. Flower, *An Introduction to the Osteology of the Mammalia*, 1870; W. K. Gregory, *Our Face from Fish to Man*, 1929; W. K. Gregory, *Evolution Emerging*, 2 vols., 1951; F. C. McLean and M. R. Urist, *Bone: An Introduction to the Physiology of Skeletal Tissue*, 1955; W. F. Neuman and M. W. Neuman, *The Chemical Dynamics of Bone Mineral*, 1958; S. H. Reynolds, *The Vertebrate Skeleton*, 1897; A. S. Romer, *Osteology of the Reptiles*, 1956; A. S. Romer, *The Vertebrate Body*, 2d ed., 1955; A. S. Romer, *Vertebrate Paleontology*, 2d ed., 1945.

Skeletal system disorders

The response of the bones to disease and injury is similar to that of other tissues of the body but is modified to some extent by the unique structure of bone.

Characteristics of bone. Bone is a living tissue which participates actively in the metabolic processes of the body. There is constant interchange between the calcium and phosphorus in the blood and that in the bone salt. The concentration of calcium and phosphorus in the blood is in turn controlled chiefly by the hormone of the parathyroid gland. Bone is composed of four main elements. Connective tissue fibers, similar to those in tendons and ligaments, give to it tensile strength and resiliency. These are impregnated with bone salt, a complex mineral compound which is similar to, perhaps identical with, the naturally occurring mineral hydroxyapatite, $Ca_{10}OH(PO_4)_6$, and which imparts compressive strength and hardness. Specialized bone cells (osteoblasts, osteoclasts, osteocytes) control the relationship between the connective tissue fibers and the bone salt, and direct the local metabolic processes of bone. Blood vessels permeate the substance of even the densest portions of bone, providing nutriment and assisting in repair if damage occurs. *See* PARATHYROID GLAND.

Disease of the skeletal system may be classified under five main headings: congenital and hereditary defects, fractures, infection (osteomyelitis), metabolic disease, and tumors. *See* BONE; BONE BIOPHYSICS.

Congenital and hereditary defects. In dyschondroplasia and achondroplasia, bone growth is both retarded and irregular. *See* ACHONDROPLASIA; DYSCHONDROPLASIA. In hereditary multiple exostoses, irregular outgrowths of cartilage and bone appear at the growing ends of long bones. These outgrowths usually increase in size until skeletal maturity is reached. Osteogenesis imperfecta is characterized by a deficiency in the number of bone forming cells (osteoblasts). The bones are fragile and susceptible to fracture to a degree depending upon the severity of the disease. In severe cases the infant's skeleton is so delicate that it is crushed during passage through the birth canal. In osteopetrosis the bones are excessively heavy and dense, differing markedly from the thin delicate bones of osteogenesis imperfecta. The defect is also due to a deficiency of osteoblasts, and although increased quantities of calcium are present the bones are excessively brittle and, as in osteogenesis imperfecta, are susceptible to multiple fractures.

Rarer congenital defects include abnormalities of shape or structure involving specific bones such as abnormally shaped vertebrae (platyspondylia), abnormally long bones of the hands and feet (arachnodactyly), deformity and shortening of the tibia (osteochondrosis deformans tibia), localized regions of abnormal density in bones (osteopoikilosis), localized deficiency of bone formation in the skull and in the clavicle (cleidocranial dysostosis), deformities of growth of the bones of the lower arm (Madelung's deformity), and enlarged pointed skull associated with defective separation of fingers and toes (acrocephalosyndactylism).

In rare individuals skeletal growth proceeds at a greatly retarded rate but is normal in all other respects, resulting in a tiny but perfectly proportioned individual known as an ateliotic dwarf.

Fracture. Bone fracture may be either traumatic, if resulting simply from application of excessive force, or pathological, if occurring through a region of bone rendered weak as a result of preexisting disease such as a tumor, cyst, or osteogenesis imperfecta. The fracture may be classified as to severity, being termed simple if it is a clean break without penetration of the overlying skin, compound if the overlying skin is penetrated, and comminuted if the bone is shattered into several fragments.

When fracture occurs the broken ends become displaced, misaligned, or both. Other tissue, such as muscle or tendon, may become interposed between them and result in numerous blood vessels being broken and in the formation of a blood clot of varying size. This clot was previously thought to provide a necessary framework upon which healing could take place. This is no longer believed to be true. Not only is the clot unnecessary; it can, if large enough, seriously impede healing. The fracture may heal rapidly and normally (normal union), its rate of healing may be retarded (delayed union), or it may never heal properly (nonunion).

Healing. Normal healing requires alignment of the fragments and close apposition of the broken ends with removal of any intervening tissue, proper splinting to prevent motion of the broken ends during the healing process, adequate local blood supply at the fracture site, and good general bodily health. If these requirements are not properly satisfied, bony unification of the broken ends is retarded or, in extreme cases, prevented.

Two types of non-union are recognized, fibrous union and nearthrosis (new joint formation). In fibrous union, the broken ends are united by a solid but flexible connective tissue scar. In nearthrosis, a cystic space resembling a joint cavity develops between the poorly joined ends.

Osteomyelitis. Infection of a bone and its marrow cavity is osteomyelitis and may be either acute or chronic. Although use of antibiotics in recent years has led to decrease in its incidence and severity, it is still a disease of major importance. Osteomyelitis may result from local injury to a bone or, more commonly, infection may be carried to bone via the blood stream. Although any bone in the body may be involved, the long bones of the arms and legs are most commonly affected.

The organism responsible for most cases of acute osteomyelitis is the staphylococcus. Tuberculosis is the most common cause of chronic osteomyelitis. Syphilis, fungus disease, and leprosy occur but are less common.

The outcome of osteomyelitis, once established, depends upon the virulence of the causative organism and the ability of the body to resist infection. Thus, in mild cases, healing may occur without residual damage. In more severe cases a portion of bone or the entire bone may die. Such a dead portion of bone, termed a sequestrum, is gradually absorbed over a period of months or years, or it may be so large as to require removal by operation. In order to maintain its strength and form during this process, the bone forms a supporting shell around the sequestrum known as an involucrum.

Complications of osteomyelitis include extension into an adjacent joint cavity to produce arthritis, spread of the infection to other organs via the blood stream, and, in growing children, retardation of the growth of the involved bone.

Metabolic disease. Metabolic disease of bone is characterized by diminished calcium content and increased fragility of the bones. In adults this is known as osteomalacia. In infants and children there is, in addition, retardation of skeletal growth and the condition is known as rickets. A special form of osteomalacia known as generalized osteitis fibrosa (von Recklinghausen's disease, hyperparathyroidism) is caused by increased activity of the parathyroid glands (usually from a parathyroid tumor).

Most metabolic disease of bone may be traced to disparity between the body's requirements for calcium and phosphorus and the amounts of these substances that enter the blood stream daily. This disparity may be caused by diminished absorption or increased bodily requirements.

Diminished absorption may result from insufficient calcium and phosporus in the diet or failure to absorb that eaten because of vitamin D deficiency or disease of the digestive system such as celiac disease, sprue, or steatorrhea. Increase in bodily requirements is seen in chronic kidney disease and in overactivity of the parathyroid glands; in both cases the kidneys excrete abnormally large quantities of calcium and phosphorus.

Tumors. Bone tumors occur most frequently in the adolescent and young adult, differing from tumors of most other organs which increase in frequency with advancing age. They may be benign or malignant (cancer). The latter may be primary in bone or metastatic (spread from elsewhere in the body).

Of the many types of benign bone tumors that have been described, the three most common are the osteochondroma composed of bone and cartilage; the chondroma composed of cartilage; and the giant cell tumor composed of fibrous tissue and large cells with multiple nuclei.

More than 25 different types of primary bone cancer have been described. The 3 most common are myeloma, characterized by overgrowth of plasma cells; sarcoma, arising from cartilage, osteoblasts, or fibrous tissue; and Ewing's tumor, composed of cells of unknown origin.

Metastatic bone cancer, although more common than combined benign and malignant primary tumors, is of lesser importance than the latter because it is frequently only an incidental terminal event in cancer already widespread throughout the body. Exceptions to this are some cases of cancer of the thyroid, breast, prostate, and lung, which may first spread to the bones before involving other organs of the body. *See* ONCOLOGY; SKELETAL SYSTEM; SPECIALIZED TISSUE. [W.R.AD.]

Skin

The external covering, or integument, of the body. In vertebrates, there are two layers, the outer epidermis composed of flattened epithelial cells, and the inner corium or dermis, formed by a fibrous yet elastic connective tissue in which there is a rich supply of blood vessels, nerves, and accessory structures.

The outer portion of the epidermis consists of many layers of dying or dead cells; the inner portion is usually a single layer of germinating cells, the stratum germinativum, which is in contact with the underlying corium.

In many vertebrates distinct types of epidermal variations are found—the feathers of birds, the scales of fish and reptiles, and the hair of mammals. In addition, the epidermis also may specialize to form different types of glands, including the sweat glands and oil or sebaceous glands of man and other animals.

The integument has many important functions: protection; pliability and extensibility, which allow movement; regulation of body fluids; waste excretion; regulation of body temperature; development of sense organs in the skin; and the formation of specialized accessory structures. *See* INTEGUMENT. [E.G.ST.]

Skin disorders

These include localized and generalized skin disorders as well as those of primary occurrence in the skin and those secondary to involvement of other tissue. Only the most common skin diseases will be mentioned. *See* SKIN.

Many authorities, particularly histopathologists, classify skin disorders on the basis of the layer or layers of the skin which are primarily or principally affected. Others categorize the same lesions on an etiologic basis. Still others attempt a clinical systemization. Probably the most useful systems are combinations of all three points of view.

There are many rather common skin disorders that so far have offered little evidence of their causes. Common examples include psoriasis, parapsoriasis, and lichen planus. The skin is the major recipient of damage from trauma of all kinds. Sunburn, frostbite, lacerations, hematomas, thermal burns, and chemical irritations and burns are commonplace.

Skin conditions occurring in regions which have a large number of oil, or sebaceous, glands usually produce an increase in secretion so that an oily skin is characteristic. Acne is the most common example and probably results from endocrinologic imbalance and other factors. Rosacea is a chronic condition seen in the middle-aged in which the middle third of the face is commonly involved. Redness, papules, and oiliness are present. The cause is unknown, but emotional upsets are often involved. Seborrheic dermatitis is an inflammatory condition of the skin, particularly the scalp, in which greasy scales are formed. A mild form is the ordinary dandruff that occurs in many people. *See* ACNE.

Well-known diseases such as tuberculosis, syphilis, and the less common sarcoidosis and lupus erythematosus have characteristic patterns of skin alteration that accompany pathologic changes in other organs. *See* SYPHILIS; TUBERCULOSIS.

Infectious etiology. The skin is the site of direct and indirect involvement in many infections. The most common diseases produced by the pyogenic, or pus-forming, bacteria include erysipelas, caused by *Streptococcus hemolyticus*; impetigo, caused by either streptococci, staphylococci, or both; folliculitis, which results from staphylococcal invasion of the superficial skin; and furunculosis, or boils, caused by several bacteria or combinations. *See* STAPHYLOCOCCUS; STREPTOCOCCUS.

Viral infections account for many of the infectious exanthemas such as measles, chicken pox, and smallpox, and also for other disorders. They are responsible for the formation of warts and may be implicated in certain other skin tumors. Viruses are also the agents of herpes simplex (fever blisters) and herpes zoster (shingles). *See* ANIMAL VIRUS; CHICKENPOX AND SHINGLES; MEASLES; SMALLPOX.

Fungus infections include the superficial types, such as tinea or ringworm, moniliasis or athlete's foot, and similar disorders. The infrequent but potentially dangerous systemic fungus infections such as actinomycosis, blastomycosis, and histoplasmosis often are marked by skin lesions. *See* ACTINOMYCOSIS; BLASTOMYCOSIS; HISTOPLASMOSIS; CANDIDIASIS.

Parasites commonly cause skin conditions. Head and body lice, crab lice, and scabies or "the itch" are seen frequently in large metropolitan clinics.

Noninfectious etiology. The term eczema is generally used to include any noninfectious, inflammatory lesion in which there are papules, blisters, and oozing of serum from the affected parts. Contact dermatitis, neurodermatitis, infantile and senile eczemas, and similar conditions are usually included, but the category is used in many ways by various authorities. *See* ECZEMA.

The erythemas are disorders in which the prominent feature is involvement of the vasculature of the affected skin. Such involvement often produces a characteristic swelling, or edema, in these disorders and includes urticaria, or hives, which results from any of the allergic agents. Other forms of erythema, which simply means redness, are erythema multiforme and erythema nodosum.

A special group of skin manifestations is that of the drug eruptions. Many individuals appear to be especially sensitive to certain medications and these conditions are quite common. Aspirin, the barbiturates, the antibiotics, especially penicillin forms, certain laxatives containing phenolphthalein, and many others produce lesions which are fortunately quite often of a typical pattern. *See* ANTIBIOTIC; BARBITURATES.

Skin lesions are common in many psychic disorders. These vary from simple pruritus (itching) to widespread, compound lesions of great severity. In some neuroses there may be marked tendencies toward chronic excoriation, or self-induced trauma of the skin; this is distinct from dermatitis factitia in which a surface wound is made in an attempt to gain attention, or for some other ulterior motive. The subtle relationships between mind and body are far from understood and in the future many diseases of now unknown cause may be established as psychosomatic disorders. *See* NEUROSIS.

As the largest organ of the body, and because of its functions and close relationships to other organ systems, the skin may reflect a great many changes induced by diseases of those systems. Endocrine disturbances, vitamin deficiencies, cardiovascular diseases, diabetes, nephritis, various malignancies, and certain blood dyscrasias may produce such varied symptoms as sweating, scaling, jaundice, hemorrhages, and others. *See* ENDOCRINE SYSTEM; VITAMIN.

Primary tumors of the skin are common and most often are of the benign varieties. Skin cancer, however, has one of the highest rates of incidence of any malignancy. Its superficial location and relative ease of treatment also produce one of the highest rates of cure, if diagnosis is made early.

There are many disorders of pigmentation, ranging from the local appearance of freckles or birthmarks to pigmentary changes that reflect hereditary factors or metabolic dysfunction. [E.G.ST.]

Skin diving

Skin diving or free diving is that technique of diving which employs swim fins, face mask, and commonly, but not always, Self-Contained-Underwater-Breathing-Apparatus (SCUBA), such as the Aqualung (Fig. 1). Originally, the term referred to the act of diving with only a swim suit, fins, and mask; the diver held his breath for the duration of his stay underwater (usually 1–2 min). However, with the almost universal acceptance of SCUBA by the skindiver, and since this equipment also requires the diver to wear fins and face mask, the term is commonly used when referring to diving with various types of SCUBA and accessory equipment such as exposure suits, camera, and scientific equipment. The term applies to most types of SCUBA diving, regardless of purpose, so long as the diver is in no way connected to the surface by lines or air hoses.

Fig. 1. A diver-scientist, equipped with an Aqualung and exposure suit, examining organisms in a rock bottom environment. Direct observations of this type using SCUBA provide valuable information to depths of 50 m. (*Photo by R. F. Dill, U.S. Navy Electronics Laboratory*)

Free diving or SCUBA diving is used to some extent in most of the marine scientific (Fig. 2) and engineering investigations currently being undertaken in all environments, including the Arctic. The underwater observer is limited to about 50 m of water depth for physiological reasons when obtaining observations which require him to think clearly and to solve problems while submerged. Below this depth, when using compressed air as a breathing gas, he is limited, not by his equipment, but by the complex temporary changes which take place in his body chemistry while breathing gas (air) under high pressure. Nitrogen enters the tissue of the body at higher rates and reaches higher concentrations when breathed under pressure, the solubility going up with increased partial pressure. Under these higher concentrations in the body tissues, nitrogen has a narcotic effect which dulls the diver's reactions to simple problems, a definite danger if an unforeseen occurrence takes place, and reduces his ability to concentrate on his scientific observations. In addition to the narcotic effect, the diver must, upon returning to the surface, allow his body to desaturate from its excess nitrogen content (decompress). Too rapid a return to the surface would allow the excess nitrogen, which is in a supersaturated state in the cells and blood, to pass beyond the "bubble point." Small bubbles of gas form in the body tissue, causing the "bends" or caisson disease, one of the worst maladies of diving. Tables containing permissible rates of ascent and time allowable at different depths are available from the U.S. Navy Experimental Diving Unit, Washington, D.C. Breathing a mixture of helium instead of nitrogen with oxygen can increase the diver's ability to think underwater, but owing to the expense and difficulties of handling the specialized diving gear and breathing mixtures needed for this type of deep diving, it is not at the present time practical for SCUBA use.

Fig. 2. Diving scientist about to drop into a submarine canyon to make observations of the rock formations and organisms they contain. The picture is taken at the rim of Scripps Submarine Canyon in southern California. In this area the wall just below the diver drops as a sheer cliff to a depth of 180 ft before it reaches the bottom of the canyon. (*Photo by R. F. Dill, U.S. Navy Electronics Laboratory*)

By using SCUBA, trained divers are no longer restricted to the sea surface but are able to investigate underwater problems first hand. Free diving or SCUBA diving is used as a tool in solving offshore problems, in much the same way as is oceanographic instrumentation for underwater investigation. *See* SCUBA; UNDERWATER PHOTOGRAPHY. [R.F.DI.]

Skin effect

The crowding of high-frequency electric current into a thin surface layer of a conductor.

For a steady unidirectional flow of electricity, the current is uniformly distributed over the cross section of a uniform conductor, that is, the current density (current per unit area) is the same at all points in a cross section. For an alternating current, there is no longer this simple uniformity, but rather the current density is greater near the outer surface than at the center. The magnitude of the nonuniformity increases as the frequency rises. For low frequencies, the effect is very small, but at frequencies for which the wavelength within the conducting material is comparable with the dimensions of the conductor, or smaller, the entire current may be considered to be within a relatively thin surface layer.

The skin effect is largely due to self-induced electromotive forces (emf's) which are different for different paths within the conductor. These emf's increase with frequency, since they depend upon the rate of change of flux. The emf's are smallest for paths that link the smallest flux. The internal linkages decrease as the frequency increases, and for infinite frequency there would be no internal linkages. For this condition, the skin effect may be called complete.

The skin effect results in a resistance for a conductor that is greater for an alternating current than for a direct current, since the effective cross section of the conductor is decreased. Also, the resistance varies with frequency, increasing as the frequency rises. Again, this behavior follows from the decrease in the effective area of the conductor. [K.V.M.]

Bibliography: S. S. Attwood, *Electric and Magnetic Fields*, 3d ed., 1949; M.I.T. electrical eng.

staff, *Electric Circuits*, 1940; L. Page and N. I. Adams, *Principles of Electricity*, 3d ed., 1958; E. R. Peck, *Electricity and Magnetism*, 1953.

Skin friction

A type of friction force which exists at the surface, or skin, of a solid body immersed in and in motion with velocity u_1 relative to a much larger volume of fluid, as illustrated. The magnitude of skin friction per unit surface area, the shear stress τ_w, was equated by Isaac Newton to the rate of distortion of an adjacent fluid element, $(\partial u/\partial y)_w$, times a transport property of the fluid called the viscosity coefficient μ (*see* VISCOSITY OF GASES). Such flow distortion is significant only in a thin boundary layer, which may be laminar or turbulent and outside of which, with y greater than $\delta(x)$, the motion is essentially inviscid.

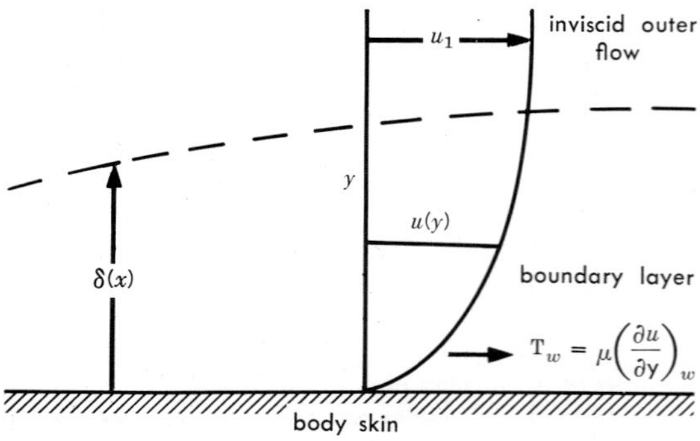

Boundary-layer velocity profile.

The skin friction force contributes directly to the drag of the body; it also contributes indirectly because, by its action, the inviscid outer flow may be modified with effect upon the pressure distribution. Conveniently, although far from completely, it has been possible to relate the skin friction quantitatively to conditions of the outer flow, such as its velocity, viscosity, density, and Mach number, and to body size and shape.

Because skin friction does work upon the fluid, its action in gases at high speeds tends to produce high temperatures in the boundary layer, which lead to the problem of cooling the skin. *See* AEROTHERMODYNAMICS; BOUNDARY-LAYER FLOW; HYPERSONIC FLIGHT. [B.M.L.]

Bibliography: L. Prandtl, *Essentials of Fluid Dynamics*, 1952; H. Schlicting, *Boundary Layer Theory*, 1959.

Skin test

A procedure for evaluating the immunity status, involving the introduction of a reagent into or under the skin. The table shows representative intracutaneous diagnostic skin tests.

Certain toxic antigens applied in minute doses will give a visible, but readily tolerated, lesion if less than a threshold amount of antibody is

Representative intracutaneous diagnostic skin tests

Name of test and use	Material and administration	Readings and interpretation
Schick test (for determining susceptibility to diphtheria); positive test presumptive of lack of immunity	0.1 ml diluted diphtheria toxin intradermally; 0.1 ml heated toxin in opposite arm as control	Score as positive or negative after 48 hours; a positive reaction shows edema and usually scaling for 7 days; the control permits evaluation of sensitivity to bacterial protein
Dick test (for determining susceptibility to scarlet fever); positive test presumptive of lack of immunity to the erythrogenic toxin	0.1 ml diluted streptococcus erythrogenic toxin intradermally	Read between 18–24 hours; positive test requires an erythema over 10 mm in diameter
Frei test (for lymphogranuloma venereum infection)*	0.1 ml chick-embryo culture of antigen (*Miyagawanella lymphogranulomatosis*) intradermally; 0.1 ml normal yolk sac material on opposite arm as control	Read at 48 and 96 hours; positive reaction requires papule greater than 6 mm, with lesser reaction for control; reaction may be negative during first 1–6 weeks of infection. Hypersensitivity may persist for life
Tuberculin test (for tuberculosis infection)*	Either purified protein derivative (PPD) or old tuberculin (OT); 0.1 ml of suitable dilution intradermally	Read in 36–48 hours, noting diameter of redness and swelling; hypersensitivity may be persistent
Ducrey test (for venereal disease due to *Haemophiluo ducreyi*)*	0.1 ml of killed culture intradermally	Read in 47 hours; positive test requires an area of induration in excess of 7 mm
Brucellergen test (for *Brucella* infection)*	0.1 ml of an extract of *Brucella*	Read after 48–72 hours; positive test requires erythema with an induration of at least 1 cm; should be evaluated in connection with other clinical data
Trichinella test (for trichinosis)*	0.1 ml of worm extract intradermally, with saline control	A positive reaction usually appears as a wheal with pseudopods within 20 min; some individuals may exhibit a delayed reaction, evident only after 24 hours

* In all these instances, within the serologic specificity of the test, a positive reaction is presumptive evidence of past or present infection with the specific microorganism.

present in the skin. Examples are the Schick test for diphtheria antitoxin and the Dick test for scarlet fever antitoxin. If sufficient antibody is present, the toxin will be largely or completely neutralized, and the lesion will be minimal or absent. False positive reactions occur in allergies to the reagent toxin, but these can be controlled with inactivated toxins. Other tests involve substances that are not reactive with normal individuals but give a visible skin reaction in the presence of antibodies acquired as a result of hypersensitivity or allergy to an infecting organism. A positive reaction is thus presumptive for previous contact with a specific infectious agent, for example, the tuberculin and Mantoux tests for tuberculosis, and the Mallein test for glanders. Because of persisting, often life-long, hypersensitivity a positive skin test may be elicited perhaps years after active infection has ceased. In a few instances, antibody may be injected as a reagent and the neutralization of toxins present in the skin observed, as in the Schultz-Charlton blanching test in scarlet fever. *See* BRUCELLOSIS; DIPHTHERIA; GLANDERS; LYMPHOGRANULOMA VENEREUM; SCARLET FEVER; SOFT CHANCRE; TRICHINOSIS; TUBERCULOSIS. [H.P.T.]

Bibliography: M. J. Pelczar and R. D. Reid, *Microbiology*, 1958.

Skink

Any of about 600 species of smooth-scaled lizards of the family Scincidae, with 30 recognized genera, 3 of which are found in the United States. They are most common in the Eastern Hemisphere, especially in the East Indies and Australia.

The blue-tailed skink, *Plestiodon fasciatus;* length to 10 in. (*From E. L. Palmer, Fieldbook of Natural History, McGraw-Hill, 1949*)

Most skinks produce living young, but a few lay eggs. Within the family there is a strong tendency toward reduction of the limbs, and there are many legless forms. Others have weak or vestigial limbs. Skinks also show a strong tendency to break off their tails voluntarily when attacked by an enemy, and eventually to regrow a stubby tail to replace the lost one.

Best known is the genus *Eumeces*, represented in the United States by 15 species, most of which are brightly-colored terrestrial forms, but some live in trees. The hard, slick, shiny appearance, caused by the flat, smooth, and overlapping scales, readily distinguishes the skinks from all other lizards. *See* LIZARD; REPTILIA. [J.D.B.]

Skull

The bones of the head which form the cranium and the face. In man the eight cranial bones which form a hollow, protective brain case include the occipital, sphenoid, ethmoid, and frontal, as well as the paired lateral, temporal, and parietal bones. Fourteen facial bones include the vomer, mandible (lower jaw), and the paired nasal, lacrimal,

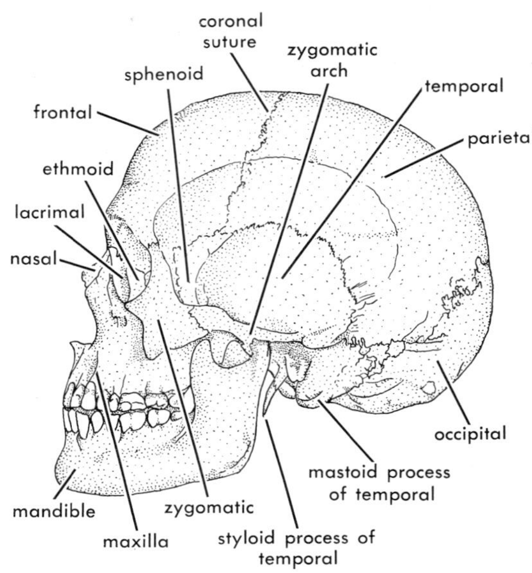

Lateral view of human skull. (*From W. T. Foster, Anatomy, Foster Art Service*)

zygomatic (cheek), palatine, maxillae (upper jaw), and the inferior turbinates of the nasal passage. Skull articulations are generally fixed, serrated sutures. The skull base contains openings, or foramina, for blood vessels and nerves. Certain bone segments are hollow and form the air sinuses such as frontal, maxillary, and mastoid.

The development of the skull affords one of the major clues to classification of the vertebrates because quite typical features are seen in each major subdivision of the phylum. *See* ANIMAL EVOLUTION. [E.G.ST.]

Skunk

Any of several American carnivorous mammals of the family Mustelidae, characterized by scent glands which produce a liquid of strong odor, sprayed at enemies as a defensive measure. All the skunks are gaudily marked in black and white, a pattern often described as an example of warning coloration (*see* PROTECTIVE COLORATION). Except for coyotes and great horned owls, which prey upon them regularly, skunks are carefully avoided by most other animals. Their diet is primarily small animals, including insects, but they eat considerable plant material. They are beneficial because of their destruction of rats, mice, and insects. The fur is durable and beautiful.

The common skunk, *Mephitis mephitis;* length 30 in.

There are three genera of skunks: *Mephitis*, the common skunks; *Spilogale*, the spotted skunks; and *Conepactus*, the rooter skunks. The first two are common over much of the United States; the latter is southern in distribution. *See* CARNIVORA; SCENT GLAND. [J.D.B.]

Skutterudite

A mineral with composition (Co,Ni)AS_3 (cobalt and nickel arsenides), an ore of cobalt and nickel. It crystallizes in the isometric system, and crystals may show the cube, octahedron, and dodecahedron but are rare. More commonly the mineral is massive with metallic luster and tin white color. The hardness is 5½–6 and the specific gravity 6.6. The name smaltite, with composition given as $CoAs_2$, is common in mineralogical literature, but it has been shown that this mineral exists only as a mixture having skutterudite as a principal ingredient. Skutterudite is found in veins formed at moderate temperatures and is associated with cobaltite, niccolite, arsenopyrite, native silver, and bismuth. Famous localities are at Freiberg, Annaberg, and Schneeberg in Germany, and at Cobalt, Ontario. *See* COBALT; NICKEL. [C.S.HU.]

Slate

A group name for various very fine-grained rocks derived from mudstones, siltstones, and other clayey sediments as a result of low-degree regional metamorphism. Highly characteristic of slates is the perfect fissility or slaty cleavage which is a regular and perfect planar schistosity, the slates themselves thus grading into phyllites. *See* CLEAVAGE, ROCK; METAMORPHIC ROCKS; PHYLLITE; SCHISTOSITY, ROCK.

Development of slaty cleavage. The manner in which slaty cleavage develops is best illustrated by ordinary argillaceous sediments and by some fine volcanic tuffs. Other types of sediments, calcareous or quartzitic, may occasionally show a similar cleavage but of a less perfect type. Original semispherical bodies (for example round fossils) which have been sheared during the deformation of the rock afford by their distortion some measure of the deformation. In a usual roofing slate the ratio of the semiaxis $a:b:c$ is frequently about 1.5:1:0.4. There has been, therefore, a great compression of the rock in the direction perpendicular to the cleavage planes and a certain elongation along the cleavage planes in the direction of the dip (Figs. 1 and 2). The mineral constituents are finegrained and cannot be discerned by the naked eye. With the microscope it can be seen that a large part of the rock is made up of thin flaky plates of muscovite or chlorite set in subparallel position. These minerals have recrystallized, that is, they have grown during the low-grade metamorphism suffered by the slates. It is of general interest in the study of the fabric of rocks that crystals grow with the axis of best thermal conductivity perpendicular to the isotherms, and, more importantly, with the axis of largest compressibility along the pressure gradient. Therefore, the newly formed crystalloblasts of muscovite, chlorite, stilpnomelane (all being phyllosilicates exhibiting the characteristic sheetlike nature in their crystal structure), all orient themselves, as they grow, in flaky crystals sitting on the planes of slaty cleavage, thus highly accentuating the cleavage. One may distinguish between, on the one hand, paracrystalline and precrystalline deformation commonly identified by undeformed aligned flakes of mica and chlorite which crystallized during or later than the slip movements; and on the other hand, postcrystalline

Fig. 1. Cleavage in banded slate. Folded rock structure near Walland, Blount County, Tennessee. (*USGS*)

Fig. 2. Detail of cleavage in folded structure beds showing small horizontal bed of quartzite in westwardly overturned plications and quartz vein in slaty cleavage dipping east. Scale shown on photograph is 4 in. Rutland County, Vermont. (*USGS*)

movements partially aligning favorably situated flakes of mica and chlorite which become bent and twisted in the process. *See* SILICATE MINERALS.

Mineral constituents. The chief minerals of slates are muscovite (as sericite), chlorite, and quartz. Common accessories are tourmaline, rutile, epidote, sphene, hematite, ilmenite. Stilpnomelane has been shown to be a major constituent in many slatelike rocks in southern New Zealand. Stilpnomelane has a position between chlorite and the clay minerals, and may have a wider general distribution in low-metamorphic rocks than has usually been recognized.

Uses. Slates are widely used for roofing purposes. They are easily prepared and fixed, are weatherproof and durable, and in many areas are cheaper and better than any other thatching materials.

The active slate-producing districts in the United States are the Monson (Maine) district; the New York–Vermont district (including Washington and Rutland Counties; the Lehigh district, Peach Bottom district, and Berks County, all in Pennsylvania; Harford County, Md.; and Buckingham and Albemarle Counties, Va. There are also important quarries in England (Devon, the Lake district), North Wales, Scotland (Ballachulish), Ireland (Kilkenny), France (the Ardennes), Bohemia, Germany (near Coblenz).

The making of slates is still performed by hand using chisel and mallet. Big slabs are split into separate slates, the thickness of which varies with the size required and the quality of the rock. An average roofing tile of the best kind of rock is about 5 mm thick. The slates are afterward trimmed to size either by hand or by means of machine-driven rotating knives. *See* QUARRYING; STONE AND STONE PRODUCTS. [T.F.W.B.]

Sleep

Although sleep involves changes in almost every system of the body, it is most conspicuously characterized by (1) a relaxed position of the musculature, (2) relative immobility, (3) elevation of sensory and motor thresholds, and (4) termination through spontaneous awakening or premature arousal by any sufficiently compelling stimuli. The last item of definition is particularly important since it allows normal sleep to be clearly differentiated from comatose states, anesthetized states, and hypnotic trances. These latter sleeplike states, although they share a great number of the physiological characteristics of sleep, neither allow the affected organism to awaken spontaneously nor permit simple arousal by a wide range of nonspecific stimuli. *See* HYPNOSIS.

Insomnia. Insomnia or, more correctly, hyposomnia represents one of the commonest disruptions of the normal sleep-wakefulness cycle. This disruption is characterized by the lessening of either the depth, the duration, or both depth and duration of the individual's typical sleep period. A shorter duration of sleep is what is most typically referred to in the lay use of the term insomnia. Although insomnia can be categorized in many different ways, one of the commonest breakdowns is according to the time during the usual sleep period when the wakefulness occurs. Thus, predormitional insomnia refers to wakefulness when it persists during the beginning of an attempted sleep period, while intermittent and terminal insomnia refer to periods of wakefulness occurring during and at the end of the usual sleep period respectively. The potential causes of insomnia are too extensive to permit individual listing. Any form of pathology that prohibits complete relaxation can lead to insomnia. There are, in addition, some medical conditions not involving apparent subjective discomfort that also predispose to insomnia, such as, hypertension, cardiac dysfunction, uremia, and poorly controlled diabetes. These conditions presumably cause insomnia by producing a level of central nervous system activity that is not compatible with sleep. Any process that contributes input to the central nervous system (CNS) favors wakefulness. This is clearly the mechanism by which various forms of environmental stimulation contribute to insomnia. The commonest cause of insomnia, "an overactive mind," represents intrinsic CNS activity which in itself is incompatible with the development of sleep. *See* METABOLIC DISORDERS; NERVOUS SYSTEM.

Hypersomnia. The opposite of insomnia is hypersomnia which is excessive depth, frequency, or duration of sleep. This condition is a much rarer disruption of the normal, sleep-wakefulness rhythm than is insomnia. Hypersomnia must raise the suspicion of endocrine or neural pathology until proven otherwise. However, cases having a purely psychological etiology have been reported. The best known cause of hypersomnia is epidemic or lethargic encephalitis or, as it is popularly called, sleeping sickness. This infectious disease is presumably caused by a virus invasion of the nervous system, and autopsies reveal diffuse and variable small lesions throughout the CNS. The basal ganglia and paraventricular gray matter in the midbrain are particularly prone to involvement. Not all cases of this form of encephalitis show somnolence and many symptoms other than hypersomnia occur. Epidemics of this disease were reported in various parts of the world until 1926, when they apparently stopped. *See* BRAIN; SLEEPING SICKNESS, AFRICAN.

Duration of sleep. The duration of sleep, even under normal circumstances, varies considerably from individual to individual and from time to time in the same individual. However, averages from large samples reveal systematic curves of total time spent in sleep by individuals of different ages. The average for the newborn is about 18 hours; by 5 years of age only 12 hours of the 24-hour period are spent in sleeping; and by the late teens most individuals have arrived at the typical adult pattern of sleeping about 9 hours. That cultural factors may help determine the duration of sleep is suggested by the fact that a study of Japanese children reveals that they average about 1 hour less sleep than American children of comparable age. Data on total sleep time obscure the important additional fact that the number of periods of sleep also changes as a function of age. In the adult and older child, all the needed sleep is customarily obtained in one nighttime period. Such individuals are said to be in a monophasic sleep cycle. Younger children and infants, so-called polyphasic sleepers, cumulate their needed quota of sleep in two or more sleep periods, depending on their age. Many lower animals, even when fully matured, persist in the polyphasic sleep pattern. However, there are species exceptions to this rule. For example, canaries and some species of snakes have a monophasic sleep pattern.

Depth of sleep. The depth of sleep is apparently not a simple curvilinear function across the sleep period as might be imagined from superficial observation of a sleeping individual. Measurement of variables that are thought to change with depth of sleep show that the depth of sleep fluctuates in an irregular manner even when the individual is apparently sound asleep. Depending on which physiological criteria are used estimates vary as to just when during the sleep period the greatest depth of sleep occurs. It seems that no single physiological index is an infallible measure of depth of sleep. Contrary to public myth, there is no convincing evidence to indicate the first few hours of sleep to be necessarily the most restful. For example, heart-rate deceleration and rise in electrical skin resistance suggest that the maximal depth of sleep occurs during the third quarter of an 8-hour sleep period, while auditory thresholds vary in a manner to suggest that maximum depth of sleep is present at scattered times over the course of the entire first half or more of a night's sleep. Comparably, motility of the sleeper is reported to be less during the first half than the second half of a night's sleep. On the other hand, lowered blood pressure and elevated carbon dioxide in alveolar air do suggest that the deepest sleep is during the earliest hours of the sleep period. Perhaps depth of sleep would be more accurately measured by some method of summing the various indices.

Physiological changes. The physiological changes accompanying sleep are more numerous than those just mentioned as indices of depth of sleep. There are, for example, widespread changes in the body musculature. The eyes are commonly diverged laterally. This is thought to represent the position of the eyeball when all extraocular muscles are equally relaxed. At the same time, the pupils are constricted, which is the opposite of what would be expected of the iris muscle if it were simply reacting to being in the dark.

The general, skeletal musculature is clearly more relaxed in sleep than during the waking state, but it is far from being flaccid or completely lacking in tone. Sleepers can maintain various postures during sleep; the tightly clenched fist of the sleeping infant is a dramatic example of the fact that some considerable degree of muscular tonus is not incompatible with sleep. The well-controlled sphincters of the adult sleeper are further evidence in this direction. Clearly, the musculature during sleep is far removed from the totally noninnervated state seen in the flaccid paralytic. Coincident with the relative decrease of normal tonicity or hypotonia that is present, reflex thresholds are high. Of particular interest in this respect is the report that in some individuals, during very deep sleep, the Babinski reflex, or extension of the great toe upon stroking the sole of the foot, becomes positive, presumably indicative of a low level of cerebral cortical activity. *See* REFLEX, UNCONDITIONED.

Sensory thresholds are also elevated during sleep. In contradistinction to comatose states, however, a sufficiently intense stimulus can be perceived by the sleeper and premature arousal results.

Core body temperature, typically at its peak in the late afternoon, falls gradually during sleep, reaching its minimum during the early morning hours. Most typically, this diurnal temperature curve starts its rising phase somewhat prior to awakening.

The electroencephalogram also varies systematically with sleep. Although there is variability in details from record to record, there is, among other reported changes, a typical shift from the predominant alpha rhythm (waves of approximately 10/sec) seen in the resting individual over to the larger and slower (0.5–5/sec) delta waves that are characteristic of sleep. This shift reverses itself on awakening. *See* ELECTROENCEPHALOGRAPHY.

Although it is generally agreed that respiratory changes are a common concomitant of sleep, the pattern of change varies considerably between individuals and even within individuals. The most constant finding is increased regularity of breathing. Whatever the pattern of respiration may be in a given case, it is common to find increased partial pressure of carbon dioxide both in alveolar air and in the serum along with pH shifts of the blood. These observations have led to the suggestion that during sleep the respiratory center of the medulla, along with the rest of the nervous system, is less efficient and thus performs its homeostatic duties less accurately.

This by no means exhausts the list of physiological changes that have been reported to occur during sleep. Practically every system in the body has been found to change in some manner.

Evolutionary theory. The most comprehensive and widely accepted theory of sleep and wakefulness is N. Kleitman's evolutionary theory. Its point of departure is the manner in which the polyphasic sleep cycle gives way to the monophasic pattern as a function of both phylogenetic and ontogenetic development. According to this theory, the wakefulness associated with polyphasic sleep which is characteristic of lower animals and younger members of higher animal groups is to be viewed as a "wakefulness of necessity." Such short-term wakefulness is said to be maintained by a subcortical mechanism which is in turn activated by simple afferent impulses. Particular emphasis is given to the importance of sensory impulses arising in the viscera and the muscles, tendons, and joints which contribute to the activity of the subcortical mechanisms. Contrastingly, the longer-term wakefulness associated with monophasic sleep cycle which is characteristic of the adult forms of the higher animals is regarded as a "wakefulness of choice." This more advanced form of wakefulness is considered to be learned and to represent the organism's adaptation to the 24-hour light-dark cycle. As such, it is thought to involve a cortical mechanism which, as a result of experience, comes to exert some control over the more primitive subcortical mechanism. Experiments involving prolonged wakefulness, as well as the observation of individuals' sleep habits when day and night cues are absent, indicate that the monophasic rhythm, once established, can cycle itself autonomously for a considerable period of time without benefit of the cues which helped establish it.

Neural mechanism. Knowledge of the neural mechanism responsible for sleep and wakefulness is in substantial agreement with the evolutionary theory. The most important system in the CNS in this regard is the ascending reticular system. This system is subcortical and for present purposes is best regarded as a mechanism for maintaining wakefulness. It acts to maintain the higher levels of the nervous system, particularly the cerebral cortex, in a state of activity compatible with the performance of its waking functions. The reticular system itself is capable of being activated by at least two routes, by any sensory input to the CNS or by direct pathways from the cortex itself. It is thought that in the absence of such support, the reticular system is unable to maintain a suitable level of nonspecific activity in the higher nervous system. In such a case drowsiness and eventually sleep would thus result. Certain localized sleep and wakefulness centers appear to be located either in the thalamus, or the hypothalamus, or in both, but their relation to the reticular mechanism is not clear. Removal of the so-called wakefulness center in the region of the mamillary bodies, in the hypothalamus of the brain, results in somnolence and this may mean that the center is simply an area having particularly strong arousal effects on the reticular system. Alternatively the somnolence might be the indirect result of the fall in body temperature that accompanies such a lesion. There is some evidence to suggest that the hypothalamic sleep center functions by having an inhibitory effect on either the waking center or on the reticular system itself. *See* BODY RHYTHM. [R.A.M.]

Bibliography: N. Kleitman, *Sleep and Wakefulness*, 1939.

Sleeping sickness, African

An endemic and occasionally epidemic infectious human disease which is progressive and usually fatal if untreated. It is caused by either of two protozoan species, *Trypanosoma gambiense* or *T. rhodesiense*, and transmitted by the bite of the tsetse fly (*Glossina*). The disease is also known as African trypanosomiasis, *maladie du sommeil*, and *Schlafkrankheit*. *See* DIPTERA; TRYPANOSOMATIDAE.

Symptoms. Clinically it is marked by an initial skin lesion, often unperceived; fever, particularly at the outset; generalized enlargement of the lymph nodes; skin rashes; cardiovascular disturbances, edemas, and a variety of neurological manifestations of the central nervous system which become more marked as the disease progresses.

Epidemiology. Trypanosomiasis exists only on the African continent and a few adjacent islands, and even there is restricted within the tropics, roughly from 15° North to 20° South latitude. Similar diseases exist enzootically, that is, in animals. The trypanosomiases, human and animal, have their own history, but of more interest is their continuing influence on the history of tropical Africa. It is estimated that in about half of an area of some 4,500,000 square miles draft animals are virtually absent, and until the advent of mechanical power both agriculture and transport had to depend on human power. Both the Arab and the Portuguese conquests of tropical Africa were stopped chiefly, it is believed, through decimation of horses by "tsetse disease." The protein malnutrition of tropical Africa can be traced in part to the partial

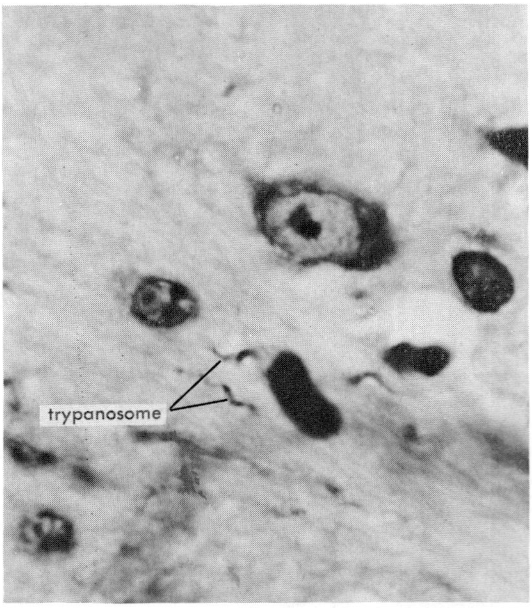

Trypanosomes in brain substance. *T. rhodesiense*, 15–30 microns long by 1.5–3 microns wide and with a serpentine outline, is visible in the tissue, where it is readily distinguished from the irregularly rounded brain-cell nuclei.

vegetarianism enforced on a society in which it is impossible to rear the common meat-producing animals. The problem is by no means solved; in 1956 it was stated by the British Medical Journal that "the tsetse fly infests 280,000 out of a total of 640,000 square miles of East Africa; in the last 25 years measures to control it have produced a net gain of only 3% in land cleared permanently of the fly." Human trypanosomiasis has been much reduced, but the measures employed are far from optimal since they are often temporary and of a type requiring short-term renewal. In part they depend on closing large tracts of land to human settlement, a procedure which, never unobjectionable, remains feasible only so long as the continent is underpopulated and underdeveloped.

Treatment. Treatment with organic arsenical compounds, such as tryparsamide or melarsen BAL, may be effective at all stages of the disease; the nonmetallic compounds such as antrypol (Bayer 205) and pentamidine now in wide use are virtually useless against pronounced brain and spinal cord involvement. The last two compounds are used prophylactically to prevent infection. Such chemoprophylaxis has proved a satisfactory, if temporary, control measure in certain areas. Eradication of tsetse also will control human trypanosomiasis; the problem is complex, and the required measures vary according to the habitat of the species of *Glossina* in cause. *See* PARASITOLOGY, MEDICAL.

[D.W.]

Bibliography: D. Weinman, *The Trypanosomiases of Man*, vol. 5, 1950.

Slide rule

A mechanical analog computing aid which is used extensively for multiplication and division and to a lesser degree for looking up functions. In its most common form a slide rule consists of a body formed from two parallel members rigidly fastened together, a slide which can be moved left or right between the body members, and a transparent indicator which carries a hairline and can be moved left or right over the face of the body and the slide. Scales are provided on the body and the slide as shown in the illustration.

The C and D scales, used for multiplication and division, are graduated from 1 on the left to 10 on the right, with intermediate numbers distributed logarithmically. Multiplication on a slide rule is based upon the fact that the product of two numbers can be obtained by adding the logarithms of the two numbers and then taking the antilogarithm (*see* LOGARITHM). To perform multiplication, the index on the slider is aligned with the graduation on the body that represents the multiplicand, and the indicator hairline is placed over the position on the slider that represents the multiplier. The product appears under the hairline on the body. The illustration shows the positions of the body, slider, and indicator for performing the multiplication of 1.5 by 2.

Division, which is accomplished by subtraction of the logarithm of the divisor from that of the dividend, is performed in reverse sequence to multiplication. Thus, in the illustration the indicator would be set over the dividend (3) as read on the D scale, and then the slider would be moved until the divisor (2) as read on the C scale also appeared under the hairline. The quotient (1.5) then appears on the D scale opposite the index on the slider.

Numbers for which graduations do not exist must be estimated by eye, and a rough mental calculation must be performed to determine the position of the decimal point in the result obtained.

Many slide rules also carry scales from which sines, cosines, tangents, natural logarithms, logarithms to the base 10, squares, and cubes can be read. The S scale on the body of the rule illustrated is calibrated from approximately 6 to 90°, with the angles placed opposite the corresponding sines as read on the D scale. Thus the sine of 17.5 degrees is 0.3. Logarithms to the base 10 can be read using the L and D scales.

The most common rule has a 10-in. scale, but larger and smaller straight rules and circular rules

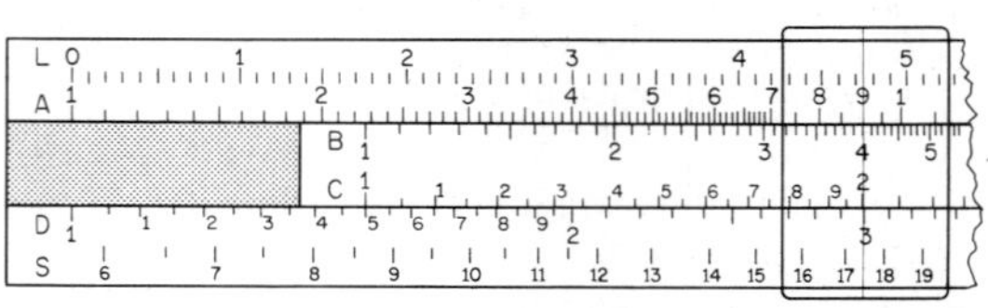

Left half of slide rule showing multiplication of 1.5 by 2.

are used. Circular rules offer a convenience in multiplication. For example, multiplication of 2 by 6 using the left-hand C index is not possible with a straight rule. Instead, one must reverse ends and use the right-hand C index. With a circular rule this problem does not arise, because the scale is continuous.

In addition to the usual type of slide rule, many special rules have been devised to mechanize particular computations. Examples include slide rules for carrying out Ohm's law calculations and for finding the reactance of a given inductance or capacitance at a prescribed frequency. These may be thought of as mechanized nomographs. *See* NOMOGRAPH. [W.W.S.]

Slider crank mechanism

A four-bar linkage, most widely used to convert reciprocating to rotary motion (as in an engine), or to convert rotary to reciprocating motion (as in pumps), but with numerous other applications. The principal parts of the mechanism are named in Fig. 1. Positions at which reversal of the slider takes place are called dead centers. When the crank and connecting rod are extended in a straight line and the slider is at its maximum distance from the axis of the crank shaft, the position is top dead center (TDC); when the slider is at its minimum distance from the axis of the crank shaft, the position is bottom dead center (BDC).

For a given crank throw, or radius, the action of the slider approaches simple harmonic motion as the length of the connecting rod is increased. Maximum accelerations occur at reversal of the slider. For constant angular velocity of the crank, the slider acceleration at top dead center is somewhat greater, and at bottom dead center somewhat less, than accelerations that would occur if motion were simple harmonic.

While the idea of combining a crank with a connecting rod is quite old (fifteenth century), the crank was not successfully applied to a steam engine until 1780; and the completion of the linkage to include a slider had to wait for a satisfactory means of making metal guides for the slider (about 1820–1830) and for satisfactory lubrication.

Many attempts were made during the next fifty years to produce rotary motion directly and thus eliminate the need for the slider crank mechanism in prime movers. Hundreds of different rotary engine designs (many of which employed slider crank linkage in a form not recognized by the inventor) were proposed, but for a prime mover that depends for its operation upon such a fluid as steam or air within a chamber, no arrangement has been found superior to the conventional one described here.

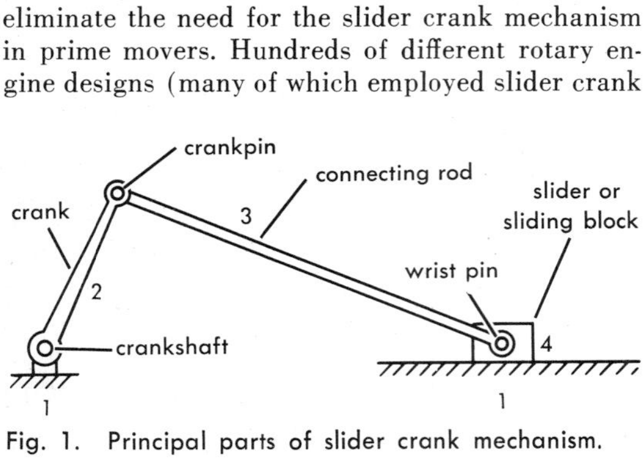

Fig. 1. Principal parts of slider crank mechanism.

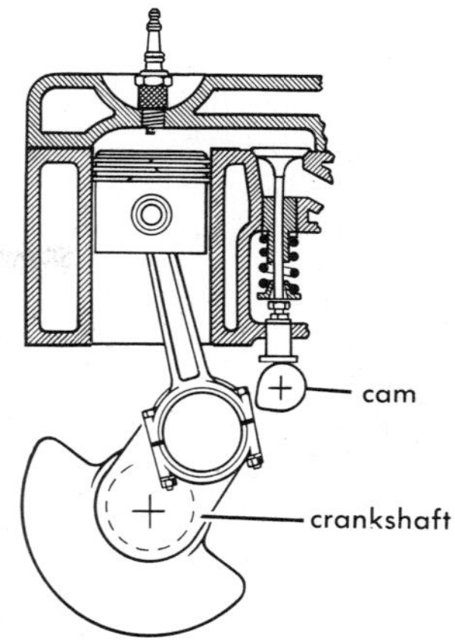

Fig. 2. Slider crank mechanism in internal combustion engine (*The Texas Co.*)

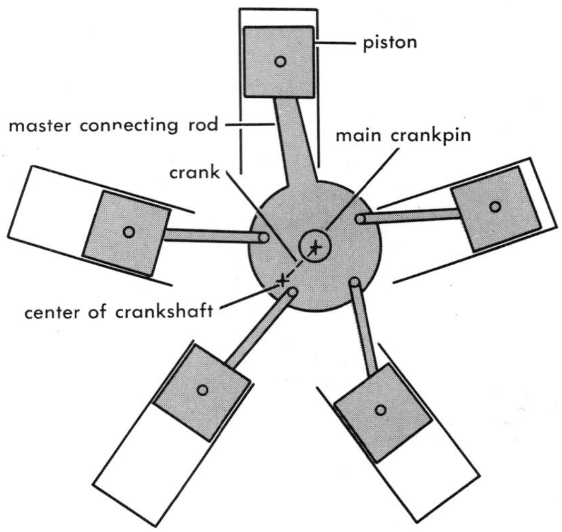

Fig. 3. Master connecting rod, piston, and crankshaft in radial engine constitute slider crank mechanism.

Internal combustion engine. The conventional internal combustion engine employs a trunk piston arrangement, in which the piston becomes the slider of the slider crank mechanism (Fig. 2). While satisfactory for engines of moderate lifespan, the reversal of side thrust on the piston twice during each revolution, complicates the problem of keeping the piston tight enough in the cylinder to contain the working medium in the combustion space. Because of angularity of the connecting rod, most wear occurs at the lower end of the cylinder. In some large low speed engines, a crosshead, differing in arrangement but similar in principle to the crosshead of a steam engine, is used to reduce cylinder wear.

Radial engines for aircraft employ a single master connecting rod to reduce the length of the crankshaft (Fig. 3). The master rod, which is connected to the wrist pin in a trunk piston, is part of a conventional slider crank mechanism. The other pistons are joined by their connecting rods to pins on the master connecting rod. There is only one slider-crank mechanism in this engine, the crank pin of the master connecting rod being the only one that follows a circular path.

Steam engine. The slider of the slider crank mechanism in a conventional steam engine is the crosshead, to which is attached the piston rod. The crosshead thus guides one end of the piston rod in a straight line, while the piston guides the other end, producing no side thrust on the cylinder. In a horizontal engine the weight of the piston causes wear on the lower half of the cylinder, but with satisfactory lubrication, wear is slight.

A reciprocating engine will not start when it is on dead center. In multicylinder engines, the cranks are arranged so that all cylinders cannot be at dead center simultaneously. In the locomotive engine, for example, this is accomplished by setting the cranks of the two cylinders 90° apart. Single cylinder engines are usually moved by hand, or jacked, off dead center for starting.

Reciprocating pumps and compressors. The crank shaft is driven usually through belting by an electric motor, or it may be driven directly or through belting by an internal combustion engine. Portable reciprocating compressors use trunk pistons; stationary compressors generally employ crossheads and guides.

Toggle mechanism. A force, acting through a limited distance, may be greatly multiplied by a toggle mechanism. Punch presses and stone crushers frequently use a toggle, sometimes called a knuckle joint. A modified slider crank linkage may be used in the mechanism, although a rocker may be substituted for the slider. In Fig. 4, force F acts

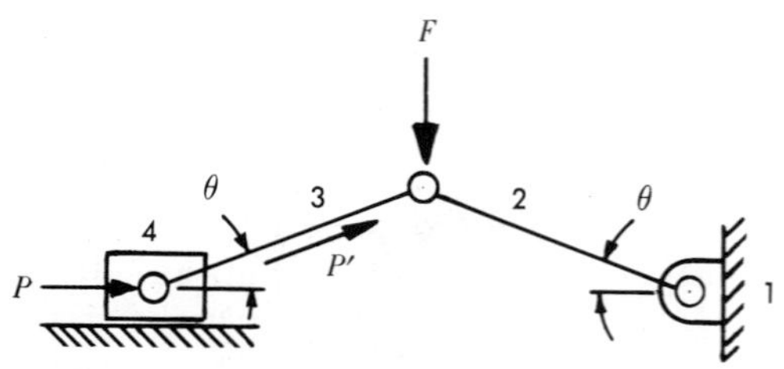

Fig. 4. Toggle mechanism.

on the crank pin joint, producing a reaction P at the slider, as well as a reaction on the crank shaft.

By constructing force polygons, the reaction along link 3, $P' = F/(2 \sin \theta)$, and $P = P' \cos \theta$. Solving for P in terms of F gives $P = (F \cos \theta)/(2 \sin \theta)$. For a finite value of F, when θ approaches 0 degrees, P approaches infinity. P is limited by the yielding of the links and pins of the mechanism. [E.S.F.]

Sliding pair

A system of two adjacent links in a mechanism in which one link is constrained to move in a particular path with respect to the other link. The lower pair, or closed pair, is completely constrained by the design of the links of the pair. A turning pair is always a lower pair, and the connection between links is equivalent to a pin joint, in which a pin is encircled by a properly fitted bushing. The two links thus turn about each other. A closed sliding pair has the sliding block of one link constrained

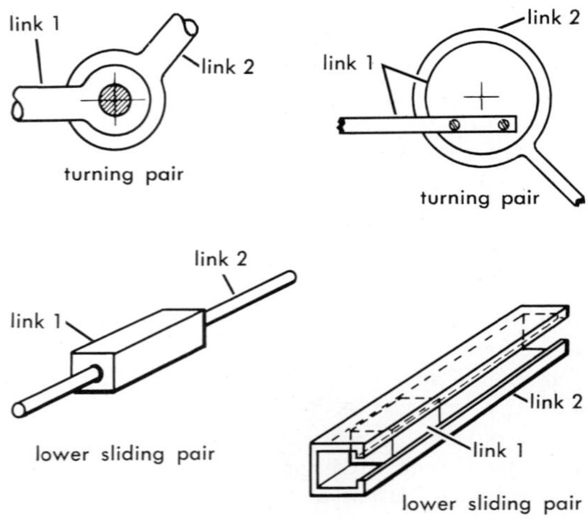

Fig. 1. Lower pairs of mechanical links.

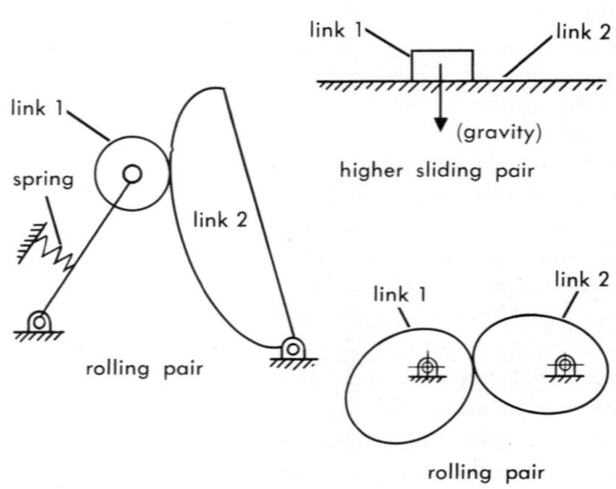

Fig. 2. Higher pairs of mechanical links.

by a rod or guide on the other link (Fig. 1). A higher pair, or open pair, requires an auxiliary force to maintain contact between links. The force may be that of gravity or a spring, or, in the case of gearing or conjugate rolling surfaces, result from fixed centers of rotation (Fig. 2). [E.S.F.]

Slip

The difference between the operating speed of an induction motor and its synchronous speed (the speed of the rotating field). Slip s is usually expressed as a decimal fraction of synchronous speed n_s

$$s = \frac{n_s - n}{n_s}$$

where n is the rotor, or operating, speed. *See* INDUCTION MOTOR; SYNCHRONOUS SPEED. [A.F.P.]

Slip rings

Electromechanical components which, in combination with brushes, provide a continuous electrical connection between rotating and stationary conductors. Typical applications of slip rings are in electric rotating machinery, synchros, and gyroscopes. Slip rings are also employed in large assemblies where a number of circuits must be established between a rotating device, such as a radar antenna, and stationary equipment.

Electric rotating machines. Slip rings are used in wound-rotor induction motors, synchronous motors and alternators, and rotary converters to connect the rotor to stationary external circuits. These slip rings are usually constructed of steel with the cylindrical outer surface concentric with the axis of rotation. Insulated mountings insulate the rings from the shaft and from each other. Conducting brushes are arranged about the circumference of the slip rings and held in contact with the surface of the rings by spring tension. A typical assembly is shown in Fig. 1. Other arrangements, such as concentric slip rings mounted on the face of an insulating disk, may be employed in special cases. Alternating-current windings normally require one slip ring per phase except that a single-phase winding requires two slip rings. Two slip rings are required for rotating dc field windings. *See* ELECTRIC ROTATING MACHINERY. [A.R.E.]

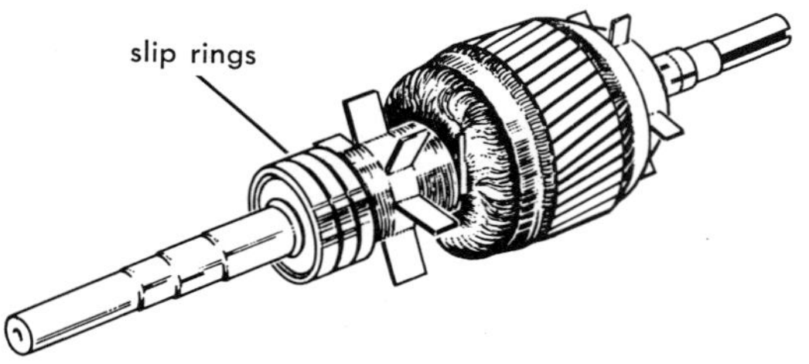

Fig. 1. Rotor of electric rotating machine showing slip rings.

Slip-ring assemblies. These integral mechanical structures contain a plurality of slip rings which, in combination with self-contained brushes, provide continuous electrical connection between electric and electronic equipment mounted on stationary and rotating platforms.

Slip-ring assemblies are designed for a wide range of electric circuits. The same assembly may have circuits for power up to several hundred kilowatts; high voltage to 50 kilovolts, power pulses for radar transmitters, and data signals, including radio frequencies to 100 megacycles (Mc), strain-gage signals and thermocouple signals. Ring surface speeds range from a few feet per minute to over 15,000 ft per minute.

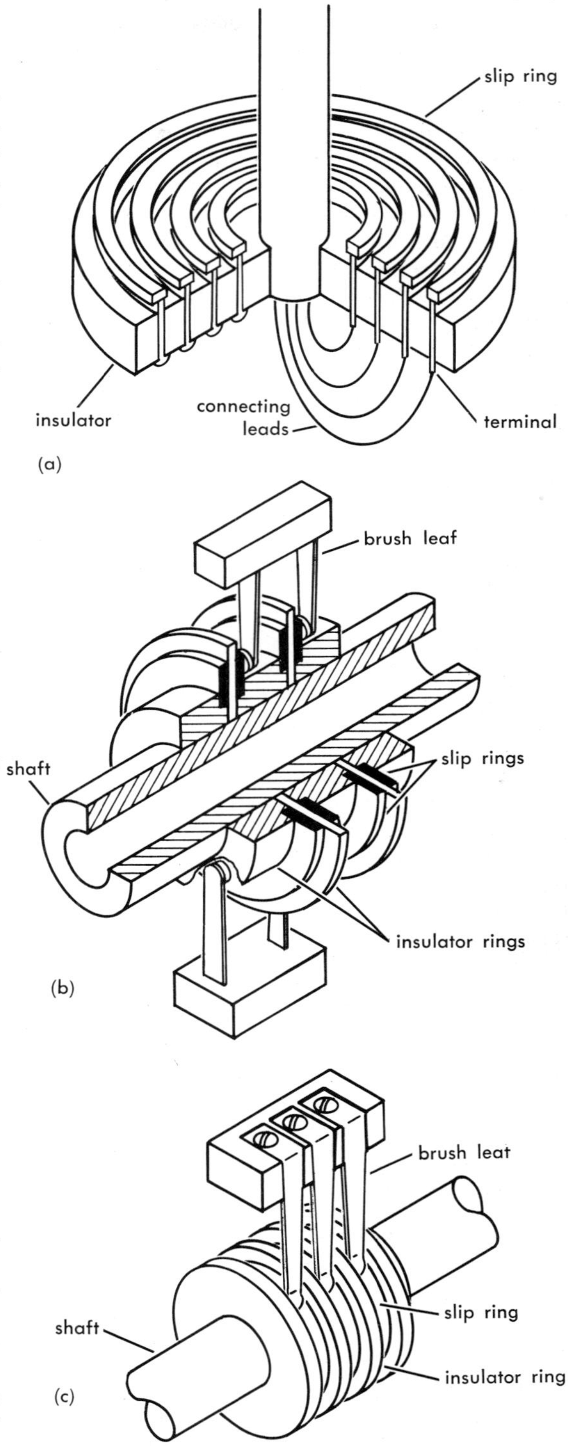

Fig. 2. Slip-ring assembly configurations. (*a*) Concentric-ring. (*b*) Back-to-back. (*c*) Drum.

Slip rings for slip-ring assemblies are made usually of coin silver, stamped from hard rolled sheet or cut from drawn tubing, or made of strip silver overlay on copper, formed into a ring and silver brazed. Fine silver rings may be electroformed onto the insulating material. Surface finish is machined or mill rolled from 4 to 16 microinches. Brushes for slip-ring assemblies are graphite combined with copper or silver in proportions suitable for the application, and may be welded or brazed to a spring-temper leaf. Leaf brush pressure is from 1½ to 3 oz.

Large assemblies, such as those used with radar search antennas, are fabricated from individual insulators and conducting materials arranged in either the concentric-ring configuration (Fig. 2*a*) or the back-to-back ring configuration (Fig. 2*b*). Small assemblies, such as those used to transmit signals through gimbals or high-speed turbines, may be fabricated in the back-to-back ring configuration or the drum configuration (Fig. 2*c*).

Other manufacturing methods employ casting the individual rings into filled epoxy resins, electroforming fine silver into grooves machined on filled epoxy resin tubes, or molding the individual rings with electrical grades of thermosetting resins. The casting, electroforming, and molding methods have the advantage of low tolerance build-up, which is important for the synchro sizes.

Background noise for strain-gage signals should not be greater than a few microvolts; intercircuit interference (crosstalk) should not be greater than 70 decibels (db) down at 30 Mc; insertion loss at 30 Mc no greater than 0.5 db, and brush contact resistance approximately 0.005 ohms.

Synchro slip rings are made from fine silver or gold alloy. Brushes are also made from precious metal alloys, usually in the form of hard-drawn, spring-temper wires. Surface speeds are usually low. [W.F.MA.]

Slope

The trigonometric tangent of the angle α that a line makes with the x axis (*see* ANALYTIC GEOMETRY); the slope of a plane curve C at a point P

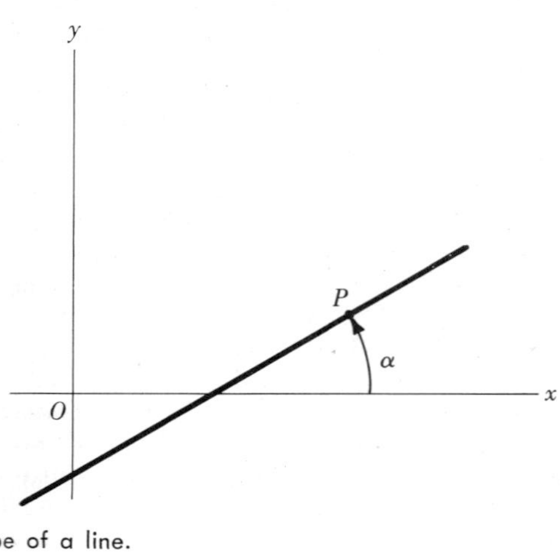

Slope of a line.

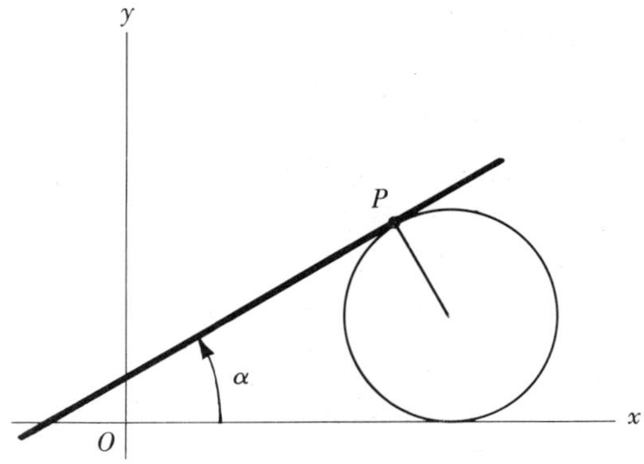

Slope of a curve.

of C is the slope of the line that is tangent to C at P. If $y = f(x)$ is an equation in rectangular coordinates of curve C, the slope of C at $P(x_0,y_0)$ is the value of the derivative $dy/dx = f'(x)$ at P, denoted by $f'(x_0)$, and hence an equation of the nonvertical tangent to C at P is

$$y - y_0 = f'(x_0)(x - x_0)$$

See CALCULUS, DIFFERENTIAL AND INTEGRAL. [L.M.BL.]

Sloth

Any of several species of arboreal mammals of the family Bradypodidae, order Edentata, found in the forests of tropical America. There are two genera: the two-toed sloths of the genus *Choloepus*, having two toes on the forefeet and three on the hind feet; and the three-toed sloths, genus *Bradypus*, with three toes on each foot. Virtually all living mammals have seven cervical vertebrae, but *Bradypus* has nine and *Choloepus* has only six.

The three-toed sloth, *Bradypus griseus;* length to 20 in. (*From P. Martin Duncan, ed., Cassell's Natural History, Cassell & Company, Ltd., London*)

Sloths are remarkably modified to hang upside down on the underside of limbs, moving with fantastic slowness. They are primarily nocturnal and feed almost entirely upon the leaves and young twigs of the cecropia tree. *See* EDENTATA. [J.D.B.]

Slug (unit)

A unit of mass in the British absolute system of units. By definition, a force of 1 lb acting on a body of mass 1 slug produces an acceleration of 1 ft/sec^2. The weight of 1 slug is 32.174 lb at sea level and 45° latitude, and there are 14.594 kg in a slug. *See* MASS; UNITS, SYSTEMS OF.

Slug (zoology)

Any of several mollusks of the order Pulmonata, class Gastropoda, with the shell reduced or lacking. Slugs are similar in all details to the better-known land snails except for the lack of a shell.

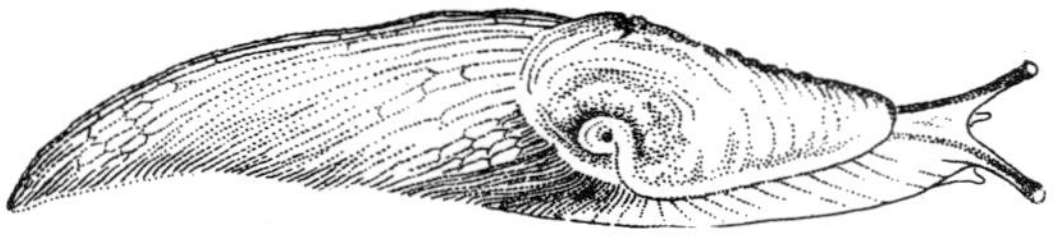

The field-gray slug, *Deroceras agreste;* length to 1½ in. (*From E. L. Palmer, Fieldbook of Natural History, McGraw-Hill, 1949*)

The most common forms in the United States are members of the genus *Limax,* all native to Europe, but now widespread in this country. The largest, and probably the most common, is *Limax maximus,* a gray species marked with alternate rows of black spots and stripes. It is about 4 in. long. These animals are elongate, fusiform (spindle-shaped), and have two pairs of antennalike tentacles; the posterior pair is prominent. They travel fairly rapidly on a slime track. Slugs are sometimes destructive to gardens and fields of vegetables, but most of the time are only a nuisance because of their slime tracks.

There are five other species in the United States. *Limax flavus,* a large brown species, is one of the more common. Others are blackish or gray, and usually only about 1 in. long.

Certain marine forms of gastropods lack a shell, and are sometimes called sea slugs, but more often are known as nudibranchs. *See* MOLLUSCA; MUSSEL; NUDIBRANCH; SNAIL. [J.D.B.]

Smallpox

An acute, infectious, viral disease characterized by severe systemic involvement and a single crop of skin lesions which proceeds through macular, papular, vesicular, and pustular stages. Smallpox is also known as variola major. A mild form, variola minor, also occurs.

Infectious agent. Variola virus is about 200 × 300 millimicrons in size. It withstands drying for months at room temperature, and at lower temperatures for years. In the dry state, it resists heating at 100°C for 5–10 min. However, when moist, it is destroyed in 10 min at 60°C. At room temperature, it is destroyed by exposure to 1% phenol, 50% alcohol, or 0.01% potassium permanganate for 1 hour. In vivo, variola grows only in man and monkeys. It produces characteristic lesions on the chorioallantoic membrane of embryonated egg. *See* CULTURE, EMBRYONATED EGG.

Antigenically, variola major virus is indistinguishable from that of variola minor, and shows only slight differences from vaccinia virus (used for vaccination) and from cowpox virus. *See* ANTIGEN.

Pathogenesis. The virus enters through mucous membranes of the upper respiratory tract and propagates in lymphoid or other tissues. After a 12-day incubation period, the disease begins with 1–5 days of fever and malaise. During this period of time the virus multiplies, circulates in the blood stream, and localizes in the epidermis and produces skin lesions. These are papular (small, circumscribed, solid elevations) for 1–4 days, vesicular (large, blisterlike) for 1–4 days, and pustular for 2–6 days. They form crusts which fall off 2–4 weeks after the first lesions and leave pink scars which fade slowly. Permanent scarring occurs because there is involvement of all skin layers, including necrosis of the corium. The nature and extent of the rash parallels the severity of the disease. Vaccinated contacts may develop variola sine eruptione, with all the prodromal (early) symptoms but with no rash or further progress of illness. In severe cases, the rash is hemorrhagic. With discrete rash the case mortality rate is about 5%; with confluent rash, it is over 40%.

Variola minor or alastrim has less severe prodromal symptoms, and the lesions are more discrete, smaller, and more superficial. Variola major in vaccinated persons may produce a similar syndrome. However, upon transmission of variola major virus to contacts, severe smallpox may result; with variola minor, the illness is always mild.

Laboratory diagnosis. Laboratory diagnosis demands careful choice of tests, depending upon the stage of illness. Among these are microscopic examination of material from skin lesions to detect cytoplasmic inclusions (Guarnieri bodies) which consist of masses of elementary bodies or infectious virus particles; isolation of virus from lesion material inoculated into chick embryo chorioallantois; and complement-fixation tests for virus or for serum antibody. Laboratory tests do not distinguish variola major from variola minor. *See* ANTIBODY; COMPLEMENT-FIXATION TEST.

Epidemiology. Smallpox has had world-wide distribution, but with vaccination it has been virtually eliminated from some countries. It is transmitted by contact with oral or nasal secretions of infected persons or with material from skin lesions. Because the virus survives drying, it also has been transmitted through infected bedclothes or on cotton imported from distant countries.

All human beings are susceptible. Children are born with maternal antibodies, but these are soon lost. Control measures in addition to vaccination include rigorous quarantine regulation of travelers and of those in contact with known or suspected cases, and strict isolation procedures with patients. *See* ANIMAL VIRUS.

Immunization. Immunization by introducing living vaccinia (cowpox) virus under the skin dates from K. Jenner's work, published in 1798. The original source of vaccinia virus is uncertain; the strains now widely used for vaccination have been passed for many years by dermal inoculation in calves, sheep, or rabbits. Primary vaccination is recommended in infants 4–6 months old, with revaccination at regular intervals or in the face of possible exposure. In countries where smallpox is endemic, revaccination is recommended at intervals of 1 year or more often. In the United States, revaccination every 7 years is recommended.

Vaccination responses must be interpreted with care. An immediate or immune reaction, typical in immune persons, may also occur in a fully susceptible person from vaccine inactivated by heat while moist. This may occur in normal tropical temperatures, for example. In doubtful instances, revaccination with a new lot of vaccine is recommended.

Complications of vaccination may include generalized vaccinia; bacterial contamination; in rare instances, a postvaccinal encephalitis in which the central nervous system is attacked; and progressive vaccinia, which occurs in children incapable of making antibody (*see* AGAMMA-GLOBULINEMIA). Generalized vaccinia may develop in children with eczematous skin conditions. Except under great risk of smallpox, vaccinia virus should not be introduced into such a child's family, by vaccination of the child or his siblings, because the mortality case rate in generalized vaccinia is 30–40%. *See* ECZEMA.

Precautions against bacterial contamination of the vaccination site are important; and no dressing should be applied, because tetanus infections may occur in covered vaccination lesions.

Primary vaccination in infancy has the lowest incidence of encephalitic complications and provides early protection against smallpox. [J.L.M.]

Smell

One of the chemical senses. It is known technically as olfaction.

Anatomy. In mammals the olfactory receptors are located in the upper parts of the nasal cavity. In man they are located in a small patch in each nasal cavity, consisting of about 2.5 cm^2 of mucous membrane, colored with a yellow pigment. The olfactory cells are long and ovoid in shape, terminating at the distal end in several delicate hairs projecting into and possibly through the mucous covering the nasal epithelium (Fig. 1). The proximal end consists of fine unmyelinated nerve fibers which pierce the bony cribriform plate to enter directly the olfactory bulb of the brain. *See* BRAIN.

Nerve endings of the trigeminal nerve fibers, utilized in common chemical sensitivity, are widely distributed throughout the nasal and olfactory area. Many odorants, for example, ammonia, stimulate both the olfactory and free nerve endings.

Odorous molecules may be carried to the olfactory region by slight eddy currents during quiet

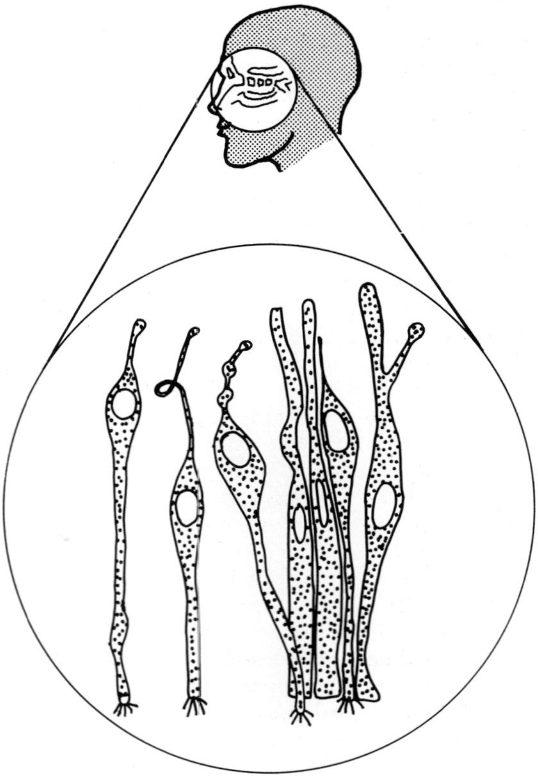

Fig. 1. Olfactory cells, showing sensory hairs, and sustentacular cells from the olfactory epithelium of man. (*S. S. Stevens (ed.), Handbook of Experimental Psychology, Wiley, 1951*)

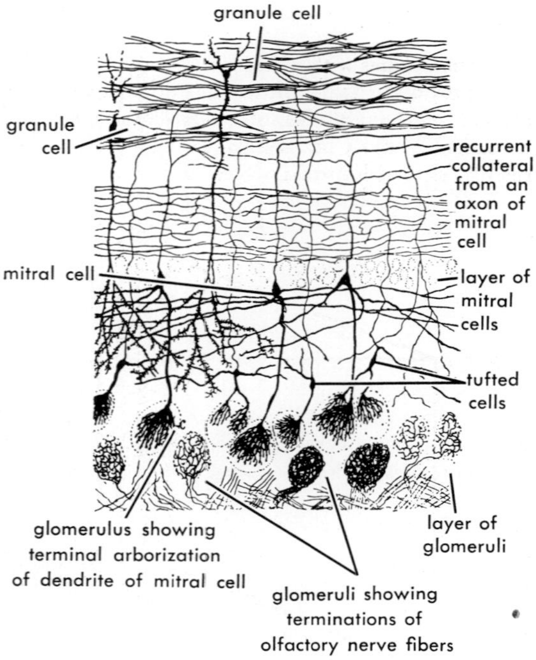

Fig. 2. Histological section of the olfactory bulb of a kitten, Golgi Method. (*S. S. Stevens (ed.), Handbook of Experimental Psychology, Wiley, 1951*)

respiration, but vigorous sniffing will produce a surge or turbulence which brings the odor to the olfactory area. A common cold, which results in congestion of the mucous membrane, effectively inhibits or eliminates olfaction.

In vertebrates olfactory nerve fibers entering the olfactory bulb end in a series of intricate basketlike terminations, or clusters, called glomeruli, where the first synapse occurs (Fig. 2). The primary sense cells synapse with either the large mitral cells or the tufted cells. Most of the mitral cell axons form the lateral olfactory tract running to the higher olfactory centers on the ventral surface of the brain. Most of the tufted cell axons enter the anterior part of the anterior commissure and terminate in the olfactory bulb of the opposite side.

Although neurologists designate a rather widespread area of the brain as rhinencephalon, or smell brain, there is increasing evidence that much of this structure is concerned more with the emotional responses than with purely olfactory sensitivity. Details of the higher neural centers for olfaction have yet to be elaborated.

Odor qualities and stimuli. The odor vocabulary is rich with words attempting to describe the great variety of olfactory qualities. Odor terminology, however, suffers in that most are object names and most attempts at odor classification are purely psychological. One of the best-known classifications is that of H. Henning, in which there were six main odor qualities; fruity, flowery, resinous, spicy, foul, and burnt. Henning attempted to systematize the relations among these odor sensations by a diagram, the smell prism (see Fig. 3), but he did not equate these with six primary receptor cell types, or areas.

Recent studies with electrophysiological methods have provided further information on the sensory mechanisms underlying different odor qualities. Electrical activity in the olfactory bulb can be readily detected by fine wire electrodes inserted

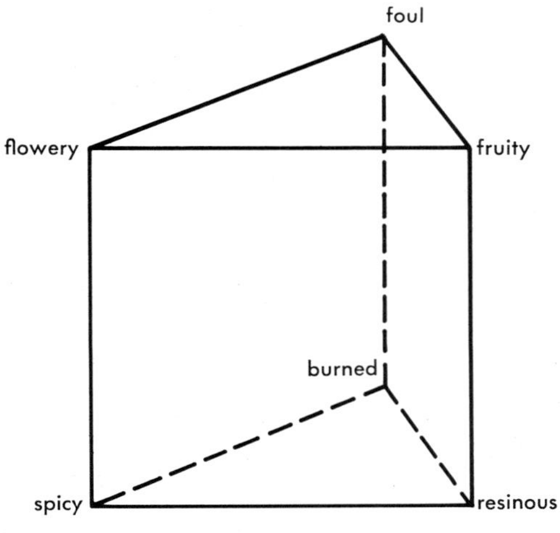

Fig. 3. The smell prism, according to H. Henning. (S. S. Stevens (ed.), *Handbook of Experimental Psychology*, Wiley, 1951)

into the bulb itself. Two forms of activity have been observed: one, a wavelike response, may be recorded from the surface of the bulb, and the other is an impulse activity which reflects the olfactory bulb's mitral cell discharges following stimulation. In the rabbit those portions of the bulb toward the anterior, or oral, region are more sensitive to water soluble substances whereas the fat soluble substances appeared to stimulate the more posterior, or aboral, parts of the olfactory bulb. In addition, different mitral cells could be characterized according to the groups of chemicals which stimulated them. However, no clear primary types emerged from such studies. It appears unlikely that the olfactory system is characterized by a few primary receptor cell types.

Sensitivity. It is generally accepted that an odorous substance must be slightly volatile. Apparently, fat and water solubility are also necessary but are not in themselves sufficient, since many inodorous substances have both these properties. The unique chemical or physical property for odor is yet to be defined.

Most odorous compounds are organic. Both the arrangement and structure of the molecule as well as the presence of certain groups within the molecule appear to influence odor. In spite of the relative inaccessibility of the olfactory end organs, odor materials can be detected at extremely low concentrations. It has been estimated that olfaction is 10,000 times more sensitive than taste. Threshold concentrations for well-known odorants, such as ethylmercaptan, have been cited in the range of 4×10^{-8} mg/liter of air. Tables of threshold values prepared by different investigators show major discrepancies when compared.

Differential sensitivity and factors influencing sensitivity. Differential sensitivity for olfaction is relatively low, ranging from a fraction ($\Delta I/I$) equal to 1 for weak stimuli and equal to ⅕ for the more intense odors. The fraction refers to the necessary increment in intensity or strength of stimulus in order that the stimulus be perceived as just noticeably greater.

Stimulus variables that affect olfactory sensitivity are difficult to investigate because the stimulation pathway is so indirect. Temperature and humidity influence the strength of odors largely because these factors influence volatility or the transport of the odorous particles from the source to the observer. The most common deviation in sensitivity is an acquired anosmia, absence of sense of smell, following nasal infection. This may be differential, that is to say, only the odor for certain substances may be reduced. Paranosmia, or change in odor quality, has also been described.

Changes in sensitivity correlated with different phases of human menstrual cycle have been reported, particularly for certain odorants related to steroids or sex hormone types of substances. Changes in olfactory acuity in relation to hunger and appetite have also been recorded by some workers. Finally ageusia, a condition somewhat like taste blindness, is reported for certain floral odors. Further work on genetically determined anosmias or partial anosmias is needed (*see* HUMAN GENETICS).

Adaptation and odor interactions. The adaptation to odors is a striking feature of olfactory sensitivity. Nearly everyone is familiar with the fact that odors rapidly disappear upon continued exposure; the stronger the odor, the greater the adaptation. Some odors will produce cross adaptation. For example, camphor elevates the threshold for eucalyptol and eugenol. This is not due to fatigue, because the threshold for benzaldehyde has been little affected. Benzaldehyde has little effect on the threshold of camphor, eucalyptol, or eugenol.

The ancient art of perfumery and the modern science of odor control attest to the realities of odor mixing and blending. Masking one odor by another occurs when the intensity of one is substantially greater than that of the other. Odor neutralization by a chemical interaction between two odors is often the basis for odor control. Nonchemical. purely physiological or psychological neutralization, often called compensation, has been reported by some workers but has not been generally accepted. When two different odors of equal intensity are presented simultaneously, the components may be readily identified. The more they resemble each other, the greater will be the tendency for blending. Although a successful perfume is a total fused impression, the trained chemist is usually able to discriminate the component notes.

Theories. There is an extensive literature on theories of olfaction, ranging from the purely chemical to the purely physical. The factors considered are the chemical or physical nature of an adequate stimulus and the nature of the receptor; for example, the receptor may be composed of only a few basic kinds of endings or many specialized different endings. Though many theories are provocative, there is, as yet, insufficient ground for accepting one theory over the other.

Behavioral effects. The distinctive and unique character of different foods and flavors is due largely to the sense of smell. Flavor is a composite of the common chemical sense, texture, pressure, temperature, and taste, as well as olfaction, but the aroma of roast beef or the delicate bouquet of a wine result largely from olfactory stimulation. The practical importance of flavor has been recognized in recent years in the food industry where it is now common practice to maintain flavor panels for the psychophysical assessment of quality and quantity of food flavor. There is no chemical or objective substitute for the human nose and palate and certainly no substitute for the human observer in assessing the total effect of food or drink. Panel techniques have been developed to assess detectability of different odorants or taste stimuli, for matching or describing the different qualities of flavor present, and for rating the affective or hedonic effect, that is, its pleasantness or unpleasantness (*see* MOTIVATION).

The importance of olfaction for the control of sexual behavior appears more obvious in lower animals than in man. Numerous experiments have demonstrated that the odor of female moths will attract large swarms of males of the same species. and it has recently been demonstrated that the honeybee's ability to discriminate friend from foe depends upon odor cues. Many wild animals utilize odor glands whose secretions assist in the detection and attraction of mates of the opposite sex. Perhaps rudiments of such mechanisms underlie some of man's reactions and behavior. It is of interest that the basic component in many perfumes is musk or the musklike odors which chemically are related to certain of the sex hormones. Some recent observations appear to show that sensitivity to certain musk perfumes is substantially affected by variations in the menstrual cycle. *See* SENSE, CHEMICAL; TASTE. [C.P.]

Smelt

Any of several fishes of the family Osmeridae. The smelts are found along both the Atlantic and Pacific coasts of the United States. They are small, slender, slightly compressed, green-backed, silver-sided fish, with adipose fins and deeply forked caudal fins. Best known is the eastern, or Atlantic,

The smelt, *Osmerus mordax;* length to 14 in. (*From E. L. Palmer, Fieldbook of Natural History, McGraw-Hill, 1949*)

smelt, *Osmerus mordax*, common from the Gulf of St. Lawrence southward to Maryland. It lives in the shallow sea and moves into fresh water to spawn. It is of considerable commercial importance, and is an excellent food fish in spite of its small size. Most adults are about 8 in. long and weigh only 4–5 oz. This species has been introduced into the Great Lakes where it has become abundant and has assumed major importance, both as a food fish and as a forage fish for larger species. *See* CLUPEIFORMES. [J.D.B.]

Smithsonite

The natural form of zinc carbonate. Smithsonite may contain small amounts of iron, calcium, cobalt, copper, manganese, cadmium, and magnesium. It occurs chiefly as a secondary mineral formed by the oxidation of primary zinc ores, such as sphalerite.

Smithsonite has hexagonal (rhombohedral) symmetry and has the calcite structure-type. It is rarely found in well-formed crystals and is more often in compact or porous masses. Its color is often dirty white, but a variety of tints, including greens, blues and browns, may be found. depending on the impurities present. For pure smithsonite the hardness is about 4 or 4½ and the specific gravity is 4.43.

Very pure smithsonite can be synthesized by heating zinc oxide to moderate temperatures at elevated pressures of carbon dioxide. The thermal stability of smithsonite at various pressures of carbon dioxide has been found by experiment to range from 350 to 450°C between 15,000 and 50,000 psi.

Smithsonite has been referred to as calamine, but the term is confusing as it has also been used as a synonym for hemimorphite and hydrozincite. Smithsonite occurs at many localities in Europe and Africa and throughout the United States. Large deposits occur at Monarch and Leadville, Colorado. *See* CARBONATE MINERALS; ZINC. [R.I.H.]

Smog

Air pollution consisting of smoke and fog. This portmanteau word was coined and first used publicly in England in 1905. The characteristics of Los Angeles and London smogs, the two principal types, are listed in the accompanying table.

Smog types*

Characteristic	Los Angeles	London
Temperature	75–90°F	30–40°F
Relative humidity	<70%	85% (plus fog)
Type of inversion	Subsidence	Radiation
Wind speed	<5 mph	Calm
Visibility	<½–1 mile	<100 yards
Maximum occurrence	Midday; Aug.–Sept.	Early morning; Dec.–Jan.
Major fuels used	Petroleum	Coal and petroleum
Principal components	Organic matter, nitrogen oxides, O_3, CO	Sulfur compounds, aerosols, CO
Reactions	Photochemical and thermal	Thermal
Principal effects on humans	Temporary eye irritation	Bronchial irritation; coughing; increased respiratory mortality

* After L. H. Rogers, Report on photochemical smog, *J. Chem. Educ.*, June, 1958.

See ATMOSPHERIC POLLUTION. [E.W.HE.]

Smoke

Fumes and smoke are dispersions of finely divided solids or liquids in a gaseous medium. The particle-size range is 0.01–5.0 microns (μ). (1 μ = 0.001 mm = 0.00004 in.) Typical dispersions are smokes from incomplete combustion of organic matter such as tobacco, wood, and coal; soot or carbon black; oil-vapor mists; chemical fumes such as sulfur trioxide (SO_3) and phosphorus pentoxide (P_2O_5) mists, ammonium chloride (NH_4Cl), and metal oxides; and the products of hydrolysis of metal chlorides such as titanium tetrachloride ($TiCl_4$), stannous chloride ($SnCl_2$), and aluminum chloride ($AlCl_3$) by moist air. Oil-vapor and P_2O_5 mists (formed by burning phosphorus in moist air) have been extensively used in military operations to produce screening smokes.

Air pollution in many industrial centers has stimulated development of methods for sampling and identification of the components of smokes and fogs (when both are present the term smog has been used). The data obtained are being used as the basis for regulatory and corrective legislation to eliminate or greatly reduce the amount of smoke and fumes which are hazardous to the health of humans, cause extensive damage to vegetation, and corrode many structural materials. Thus in several cities such as Pittsburgh, Pennsylvania, and St. Louis, Missouri, the combustion of coal for space heating and for manufacture of steam and power is regulated so that only negligible amounts of smoke and fly ash are discharged into the atmosphere. The smog problem in Los Angeles, California, has been particularly difficult to solve. The recurrent eye-irritating smog is due to the formation of peroxides from certain hydrocarbons and nitric oxide in the presence of sunlight and to the temperature inversion (a blanket of stagnant or slow-moving air which is warmer than the layer of air close to the ground). The atmospheric inversion condition combined with the effect of the surrounding foothills of the Sierra Nevada sharply reduces the flow of air in the area.

The contaminants discharged daily into the atmosphere of the industrial area in and around Los Angeles is estimated to be more than 2000 tons. In many instances, such as the discharge of sulfur compounds from petroleum refineries, the recovery of the discharged material is industrially profitable. Many other contaminants can be eliminated at their sources by suitable collection devices such as centrifugal collectors (cyclones) for dusts, electrostatic precipitators and cloth filters for finer particles, and wet collectors for mists and fogs. Efficient designs of furnaces for both domestic and industrial use are available to greatly reduce smoke and fly ash. In industrial plants where recovery or collection is technically difficult or very costly, atmospheric dilution of the smoke by increase in height of the smoke stack and by proper location of the plant with respect to ground contours, elevations, and meteorological conditions is usually quite effective. *See* AIR POLLUTION CONTROL; DUST AND MIST COLLECTION; SCREENING SMOKE.

[H.H.ST.]

Bibliography: L. C. McCabe (ed.), *Air Pollution*, Proc. 1950 U.S. Tech. Conf. on Air Pollution, 1952.

Smut (microbiology)

A term applied both to fungi of the order Ustilaginales of the subclass Heterobasidiomycetidae and to the diseases which they cause. Plants of several families are subject to smuts, but smuts of cereals are of greatest economic importance. In the United States, stinking smut, or bunt, destroys about 1.3% of the wheat crop annually. Loose and covered smuts of oats, caused by *Ustilago avenae* and *U. kolleri*, destroy 40–50 million bushels of oats each year. Barley smuts, caused by *U. nuda*, *U. nigra*, and *U. hordei*, account for annual losses of 2–5 million bushels, and corn smut may ruin 2% of the total crop. *See* FUNGI; HETEROBASIDIOMYCETIDAE; PLANT DISEASE.

The life history of corn smut, *U. maydis*, is representative of smuts belonging to the family Ustilaginaceae. Infection is initiated by basidiospores, or buds derived from them. Two compatible cells fuse, and their nuclei pair. A mycelium, each cell of which contains two nuclei—one of each compatibility—develops from the fusion cell and invades the host. Eventually, gall-like swellings are produced, and most of the cells of the mycelium develop into teleutospores. The two nuclei in each teleutospore fuse. Each gall now consists of a powdery mass of teleutospores enclosed by a thin layer of host tissue, rupture of which exposes the spores to be distributed by wind or water. Immediately, or after overwintering, a teleutospore germinates; a tubelike epibasidium develops, and the nucleus undergoes reduction division. Cross walls divide the epibasidium into four cells, each with a single haploid nucleus. Successive crops of basidiospores are budded from each cell.

Tilletia caries or *T. foetida*, causal agents of bunt of wheat, are representative of the Tilletiaceae. The binucleate mycelium initiates infection of wheat seedlings. Few symptoms of disease appear until flowering, when the ovaries are filled with teleutospores, forming "smut balls." At germination, the teleutospore forms an epibasidium which bears elongate basidiospores at its tip. These basidiospores fuse in pairs, often while still attached to the epibasidium, and nuclear pairing takes place. Binucleate spores are discharged from this fusion cell. Because the haploid phase of these smuts is so short relative to the binucleate phase, some authorities consider them to be the most advanced fungi.

The control measure most effective against all types of smuts is the development and planting of resistant varieties of grain. Bunt of wheat is also controlled by treating seed with organic mercurial fungicides to kill teleutospores adhering to the seed coat. *See* BASIDIOMYCETES; USTILAGINALES.

[R.M.P.]

Bibliography: *Plant Diseases*, USDA Yearbook of Agr., 1953; J. C. Walker, *Plant Pathology*, 2d ed., 1957.

Snail

A term which, in its broadest sense, applies to all the members of the class Gastropoda, in the phylum Mollusca. In the more commonly used sense, the word is applied only to the land and fresh-water gastropods with a shell, thus excluding the marine forms and the slugs (*see* GASTROPODA). There are about 33,000 living species of gastropods, and about 5000 belong to the order Pulmonata, a group which includes most of the fresh-water and land snails and slugs (*see* PULMONATA).

Economic importance. Generally speaking, snails are of little direct importance to man. However, several are considered food animals; others are agricultural pests, sometimes causing notable damage to garden and field crops. A few are hosts in one phase of the life cycle of parasites which affect man or his domestic animals. Snails are often important in the aquatic food chain, and some land forms are significant in the diet of shrews and other small mammals (*see* MAMMALIA; SHREW).

The most common of the snails used for food in Europe is *Helix pomatia*, considered a gourmet's delicacy in France and Belgium. It is frequently found in specialty shops in the United States.

The giant African land snail, *Achatina fulica*, which was brought to the islands of the South Pacific by the Japanese during World War II as a food source, has developed into an agricultural pest of considerable significance. This rapidly multiplying and highly destructive animal, the size of a man's fist, has been carried from island to island by ships. A constant vigil is maintained at ports of entry to prevent its establishment in the United States.

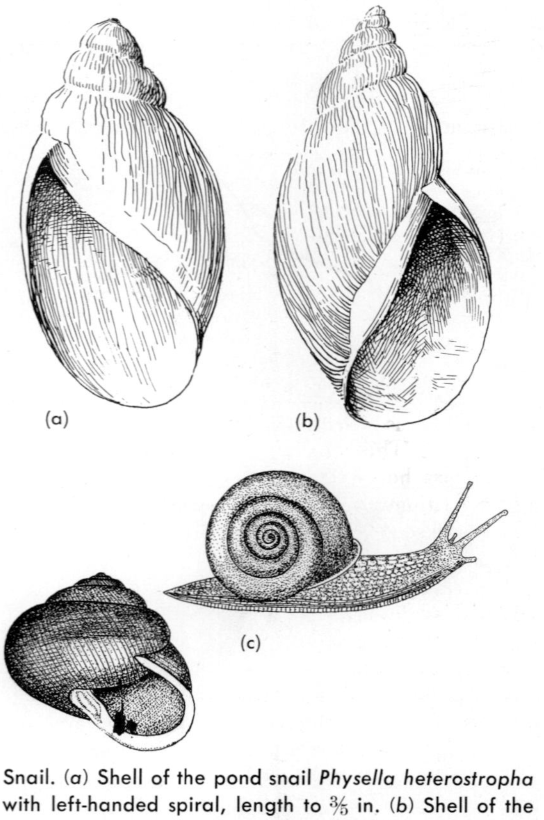

Snail. (*a*) Shell of the pond snail *Physella heterostropha* with left-handed spiral, length to ⅗ in. (*b*) Shell of the pond snail *Stagnicola palustris elodes* with right-handed spiral, length to 1⅕ in. (*c*) The white-lipped land snail, *Triodopsis albolabris;* diameter to 1¼ in. (*From E. L. Palmer, Fieldbook of Natural History, McGraw-Hill, 1949*)

Of the parasites of importance to man which are transmitted by snails, perhaps the best known is the sheep liver fluke (*see* SHEEP). This is a destructive parasite of sheep in which the larvae develop and undergo a cycle of rapid reproduction in the liver of certain pond snails. Other trematode parasites in which the snail is a secondary host include the liver fluke of man, occurring primarily in Japan and China, and the lung fluke, widespread in tropical countries and also known in the United States (*see* TREMATODA). The most important trematode parasites of man are the species of *Schistosoma* (formerly called *Bilharzia*), a genus of blood flukes which causes much suffering through wide sections of the world. These parasites are all carried by snails. The form in the Nile Valley is the best known. The disorder caused by the animals of this group is called schistosomiasis.

Swimmer's itch, prevalent in many northern and western states, as well as in Canada and Europe, is a severe itch not unlike poison ivy in its initial symptoms, caused by the penetration into the skin of man and other mammals by the cercariae (free-swimming, mature larvae) of certain snail-carried bird flukes (*see* CERCARIA). The cercariae do not continue to live in man, but they will burrow into the skin in great numbers and cause intense discomfort. The disorder is technically referred to as cercarial dermatitis and children are especially susceptible. Regular applications of copper sulfate to kill the snail host is the practice in many areas.

Characteristics. A typical fresh-water snail has a univalve shell, with a conspicuous anterior head and a long, ventral muscular foot. The foot and head are bilaterally symmetrical, but the shell is coiled and asymmetrical. There is an almost endless variety of shapes, patterns, and colors of shells. The vital organs are located in the body above the foot. The Pulmonata are so named because they have lost their gills and instead have an empty gill chamber which acts as a respiratory organ, the lung. The lung is supplied with a network of blood vessels along the inner wall of the adjacent mantle. This device functions properly only when moist; hence land snails are usually found in damp locations. The fresh-water species of Pulmonata are secondarily aquatic, being derived from land forms.

The fresh-water snails in the groups Ctenobranchia and Apsidobranchia have an internal gill and separate sexes. Included here are the viviparous snails and apple snails.

Reproduction. Most snails are hermaphroditic and reproduce by means of reciprocal fertilization, so that in mating each animal fertilizes the eggs of the other. The gelatin-covered eggs are laid in bundles or bunches in a suitable site.

Feeding habits. Food is usually vegetation, and some species are voracious feeders. Many aquarists have found that snails, kept to help in aquarium sanitation, destroy so much vegetation that keeping plants in the aquarium becomes a problem. Many tropical fish fanciers now try to maintain snail-free aquaria for this reason. There are a few carnivorous species, feeding on a variety of small animals. Many of the common aquarium species will destroy the eggs of fishes.

Land snails are active primarily at night or in damp weather. They leave a typical track or path of slime secreted by a gland on the ventral side of the head. Movement over this track is by muscular contractions of the foot. Some of the land forms have an amazing capacity to cling to life after months of inactivity because of dryness. They may appear quite dead but resume activity when introduced into a favorable habitat. *See* MOLLUSCA; SLUG (ZOOLOGY). [J.D.B.]

Snake

Any of about 2500 species of reptiles of the suborder Serpentes (Ophidia), order Squamata. Snakes are distinguished from lizards, to which they are most closely related, by several characteristics, all more or less associated with their loss of legs. Snakes have no legs, except for vestigial femurs and pelvic girdle elements found in a few primitive forms. The lower jaw is loosely attached to the skull, and the mandibles are held together by an elastic ligament, so that each of the four half-jaws may move independently. The absence of a sternum or pectoral girdle, and the highly elastic body wall of snakes, along with the free movement of the jaws, enables them to swallow prey much larger in diameter than themselves. Snakes have

Snake. (Top) Rattlesnake, *Crotalus horridus;* length to 5½ ft. (Center) Garter, *Thamnophis ordinatus;* length to 36 in. (Bottom) Black, *Coluber constrictor;* length to 6 ft. (From E. L. Palmer, *Fieldbook of Natural History,* McGraw-Hill, 1949)

also lost the urinary bladder, ear openings, and eyelids; the eyes are covered by a transparent scale. The left lung is frequently reduced in size. Most snakes have small, sharp, conical teeth, usually replaced readily when lost. In some species, front or rear fangs are developed in association with poison glands.

Quite a number of snakes are poisonous, some being extremely dangerous, whereas others may be only mildly venomous. Contrary to popular belief, snakes are neither slimy nor cold, but are covered with scales, which may be smooth or rough. They are completely dry-skinned, and have a body temperature at or near that of the environment. As cold-blooded animals, that is, animals not able to regulate their own body temperature, they become inactive and hibernate during cold weather. They cannot tolerate high temperatures, and those species that live in hot, exposed areas like the desert are generally nocturnal, as are many species in more tolerable climates.

American folklore is filled with many superstitions concerning snakes, associated with a long, unfounded fear of all species. Most snakes are beneficial, being among the most effective of all natural controls for rodents.

Feeding habits. All snakes are predatory, catching and swallowing various animals. Some of the smaller species eat only worms and insects, but the larger ones are capable of overpowering animals of considerable size, including pigs, deer, and occasionally man.

Size. In size, snakes range from 4-in. long burrowing animals superficially resembling worms to the great reticulated python, *Python reticulatus*, said to reach a length of 32 ft, and the anaconda, *Eunectes murinus*, reliably reported up to 33 ft long. Accounts of much larger individuals of both species as well as others are given, but reports have not been supported by specimens.

Reproduction. Most snakes lay eggs, which are almost always abandoned by the parents immediately. However, there are reports of one or two species which brood their eggs. A number of snakes retain the eggs in the uterus until they hatch, thus appearing to give birth to living young.

Distribution. Snakes occur throughout the world except in the colder regions. They are somewhat more abundant in tropical and semitropical regions, especially in the number and variety of species.

Most snakes are terrestrial, but several are arboreal. Some snakes are aquatic, occurring both in the ocean and in fresh water, and a few are burrowing forms. See BLACKSNAKE; BOA; COBRA; COPPERHEAD; CORAL SNAKE; COTTONMOUTH; GARTER SNAKE; GREEN SNAKE; HOG-NOSED SNAKE; PYTHON; RATTLESNAKE; WATER SNAKE; *see also* SQUAMATA. [J.D.B.]

Snap ring

A form of spring used principally as a fastener. Piston rings are a form of snap ring used as seals. The ring is elastically deformed, put in place, and allowed to snap back toward its unstressed position into a groove or recess. The snap ring may be used externally to provide a shoulder that retains a wheel or a bearing race on a shaft, or it may be used internally to provide a construction that confines a bearing race in the bore of a machine frame, as illustrated. The size of the ring and its recess determines its strength under load. Sufficient clearance and play is needed in the machine so that the ring can be inserted and seated. [L.S.L.]

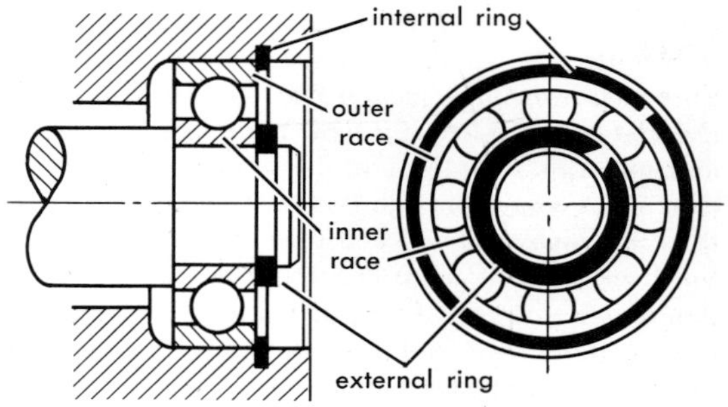

Snap rings hold ball bearing race in place. Internal ring supports any axial thrust of the axle. External ring aligns the inner race against the shoulder of the shaft.

Bibliography: P. F. Rossman, Designing snap ring fastenings, *Machine Design*, 13(5):49–51, 106–108, 1941.

Snipe

A name applied to various shore birds, but most commonly meaning the Wilson's snipe, *Capella gallinago*. This bird is easily recognized in flight by its zig-zag course and its short orange tail. Formerly considered a game bird, it has been protected since about 1940. The Wilson's snipe has an extremely long bill and is short-legged, somewhat like the woodcock. It frequents wet meadows and is Holarctic in distribution, nesting southward to California and Iowa. *See* CHARADRIIFORMES; SANDPIPER. [J.D.B.]

The Wilson's snipe, *Capella gallinago*; length to 11¾ in. (*From E. L. Palmer, Fieldbook of Natural History, McGraw-Hill, 1949*)

Snow

The most common form of frozen precipitation, usually flakes of starlike crystals, matted ice needles, or combinations, often rime-coated; also transitions between these forms, or combinations of them with ice columns, plate crystals, snow pellets, snow grains, or ice pellets. Photographs by W. A. Bentley, mostly of carefully selected six-pointed star crystals and hexagonal ice plates, show an infinite variety of beautiful symmetrical patterns. *See* PRECIPITATION (METEOROLOGY).

Geographical distribution. Snow falls at sea level generally poleward from latitudes 35°N or S, or even closer to the Equator in the interior of continents and at high elevations. On west coasts it commonly falls only poleward of latitude 45°. The elevation of snow on mountains and highlands (snow line) decreases poleward from the subtropics down to sea level in polar regions. Glaciers develop where the annual accumulation of snowfall exceeds melting, runoff, and evaporation; and along seacoasts, as in Greenland, large ice masses break off glaciers to form icebergs.

The greatest annual snowfalls in the United States attain several hundred inches in some locations. Examples are in the Sierra Nevadas, Cascade Range, and mountains of Colorado; other places with heavy amounts are northern New York near Lake Ontario and eastward, in Michigan along the southern shore of Lake Superior, and in northern New England.

Specific gravity of snow. When newly fallen, snow has a gravity ratio commonly about 1:10 but varying greatly. Wet snow is heavier, but dry, fluffy snow may have a specific gravity of 1:30 or less. The density of the snow cover increases by packing, sometimes by melting and refreezing, also because snow changes to ice granules through transfer of water by sublimation from small to larger ice particles. Deep in glaciers the entrapped air is compressed and density approaches that of pure ice.

Thermal effects of snow cover. Snow, especially when new, is a good insulator because air is entrapped by the flakes. Thus it often protects low vegetation, such as grains, from freezing. Air temperatures tend to be low over snow because it is a good reflector of sunshine.

Formation of snow flakes. U. Nakaya, who studied the forms, growth, and properties of snowflakes, found in the laboratory that star crystals could be grown only in a narrow temperature range of about −14°C to −17°C. Above −7°C only ice needles formed. Plates and columns formed predominantly between about −10°C and −22°C with relative humidity 100–110%, whereas other types grew mostly when humidity was 110–130%. B. J. Mason later reported (1959) the following experimentally determined relationships between temperature (°C) and the type of ice crystals grown on a fiber in a cold chamber: 0 to −3, thin hexagonal plates; −3 to −5, needles; −5 to −8, hollow prisms; −8 to −12, hexagonal plates; −12 to −16, dendritic crystals; −16 to −25, plates; −25 to −50, hollow prisms. *See* CLOUD PHYSICS; GLACIER; HAIL; HYDROLOGY; PRECIPITATION (CHEMISTRY); SUBLIMATION. [J.R.F.]

Bibliography: W. A. Bentley and W. J. Humphreys, *Snow Crystals*, 1931; U. Nakaya, *Snow Crystals: Natural and Artificial*, 1954.

Snow field

A term usually applied to mountain and glacial regions to refer to an area of snow-covered terrain with definable geographic margins. Where the connotation is very general and without regard to geographical limits, the term snow cover is more appropriate; but glaciology requires more precise terms with respect to snow-field areas. These terms differentiate according to the physical character and age of the snow cover. Technically, a snow field can only embrace new or old snow (material from the current accumulation year). Anything older is categorized as firn or ice. The word firn is a derivative of the German adjective fern, meaning "of last year" and hence refers to hardened snow not yet metamorphosed to ice which has been retained from the preceding year or years. Thus, by definition, a snow field comprised of firn should be called a firn field. Another term familiar to glaciologists is névé field, from the French word névé, a mass of hardened snow of glacier origin. In English, rather than a specific word for the material itself, a descriptive phrase is used such as: consolidated granular snow not yet changed to glacier ice. It is becoming most acceptable to use the French term névé when specifically referring to a geographical area of snow fields on glaciers (that is, an area covered with perennial "snow" and embracing the entire zone of annually retained accumulation). For reference to the compacted remnant of the snow pack itself, which is retained at the end of an annual melting period on a snow field or névé, it is appropriate to use the German term firn. *See* GLACIATED TERRANE; GLACIER.

[M.M.M.]

Bibliography: L. De Vries, *German-English Science Dictionary*, 1948; N. P. Larousse, *Dictionnaire Encyclopédique*, 1956; M. M. Miller, The terms "Névé" and "Firn," *J. Glaciol.*, 2 (12), 1952.

Snow gage

Instrument for measuring snow samples to yield data for assessing snow and water-equivalent resources. Such samplers were developed, beginning about 1910, by a group under the leadership of J. E. Church in Nevada. Here the need was great to know the proportions of water stored in blankets of winter snow in the mountains for later use to support life and agriculture in the oases and settlements of the desert basins.

Early gages were developed to obtain complete samples, of diameter 1½ in., to depths exceeding 20 ft. Various lengths of steel tubing were connected by screw couplings so that the combined length could be adjusted to sample any depths of snow encountered.

A later sampler is made of aluminum tubing in lengths of 30 in. graduated so that, when coupled together in order, the correct depth in inches from bottom edge of cutter can be read on the outside of the tube. Each length of tubing has staggered slots ⅛ in. wide to facilitate removal of snow core and allow measuring depth of snow in inches. Later refinements include a horizontal support and scales for weighing the tube empty or loaded. Scales with a circular dial were developed so that the pointer could be set at zero with the empty sampler tube and so graduated that when weighing the loaded sample tube the pointer would indicate actual equivalent water in inches depth. These samplers and scales are called the Mt. Rose sampler and scales.

In the 1920s G. D. Clyde of Utah adopted less-expensive sample tubes that cut a snow core with diameter of 1.485 in. Because the weight of a cylinder of water of that diameter is exactly 1 ounce avoirdupois per inch of length or 1 lb for 16 in., this adoption permits the use of commercial weighing scales for the snow samples. *See* SNOW SURVEYING. [H.P.BO.]

Snow line

A term generally used to refer to the elevation of the lower edge of a snow field. In mountainous areas, it is not truly a line but rather an irregular, commonly patchy border zone, the position of which in any one sector has been determined by the amount of snowfall and ablation. These factors may vary considerably from one part to another. In regions where valley glaciers descend to relatively low elevations, the summer snow line on intervening rock ridges and peaks is often much higher than the snow line on the glaciers, and in most instances it is more irregular and indefinite. If by the end of summer it has not disappeared completely from the bedrock surfaces, the lowest limit of retained snow is termed the orographical snow line, because it is primarily controlled by local conditions and topography. On glacier surfaces it is sometimes referred to as the glacier snow line or névé line (the outer limit of retained winter snow cover on a glacier). Year-to-year variation in the position of the orographical snow line is great. The mean position over many decades, however, is important as a factor in the development of nivation hollows and protalus ramparts in deglaciated cirque beds. The average regional level of the orographical snow line in any one year is the regional snow line. Since the regional snow line is controlled entirely by climate and not influenced by local conditions, glaciologists sometimes call it the climatological snow line. The average position of the climatological snow line over a decade or more may be termed the mean climatological snow line. On broad ice sheets and glaciers, the climatological snow line is the same as the névé line. The average position of this, over a decade or more of observation, may be termed the mean névé line. *See* GLACIATED TERRANE; GLACIER; SNOW FIELD. [M.M.M.]

Snow surveying

Application of snow gaging or sampler measurements to stream-flow forecasting. Most of these techniques have been developed since about 1911 under the direction of J. E. Church in western Nevada. Here precipitation in the form of rain is so small in quantity that practically all the useful water supply results from melting of snow on the eastern slope of the Central Sierra.

A snow course is customarily established and marked by signs, originally placed on trees, at each end of the course and at a height to be above the deepest expected snow. The locations where samples are to be taken are measured in line by tape usually at intervals of 50 or 100 ft for control in repeated measurements. The snow thus measured in successive months and also in following years affords a method of comparison of seasonal development of snow accumulation and of quantity of water equivalent in different years. The comparison of such snow survey measurements with stream flow over a period of years will show some relationship between snow water equivalent and stream flow, and after a number of years this results in development of approximate forecasting. The average of a number of samples is a far better indicator than measurements at different times at only one location. The minimum number of samples sufficient to give fairly dependable results for a snow course is probably 10 or 15.

In the Central Sierra it has been found that seasonal snow survey results prove, in general, that high altitude snow water equivalent greatly exceeds low altitude water equivalent. In analyzing

Field observers recording snow-gage measurements at a marked location along a snow-course line in the Nevada highlands. (*From Univ. Nevada Agr. Expt. Sta. Bull. 184, 1949*)

results, above 7000 ft is considered high altitude and below 7000 ft low. In adopting an adjusted water equivalent to represent a given watershed, relative areas of high level and low level should be taken into account.

A graphical diagram can also help in the analysis desirable for forecasting stream flow. In such a diagram, the vertical scale indicates adjusted value of water equivalent in inches for the whole basin. The horizontal scale represents actual runoff (at gaging station) in thousand acre feet. Each identified year of accumulated data from past experience is represented by a spot or small circle placed in correct position for adjusted snow water content and actual resulting runoff in thousand acre feet. It has been found that in the Central Sierra a straight line or regression line can be located on such a diagram based on data accumulated over a good many years. The regression line can be used as a guide for making the forecast of expected runoff.

All of the 11 states from the Rocky Mountains to the Pacific Coast are using snow surveying to aid in determining their probable summer water supply. [H.P.BO.]

Bibliography: H. P. Boardman, Snow surveys for forecasting stream flow in western Nevada, *Univ. of Nev. Agricultural Experiment Station, Bulletin No. 184*, September, 1949; R. K. Linsley, Jr., M. A. Kohler, and J. L. H. Paulhus, *Hydrology for Engineers*, 1958.

Soap and detergent

Substances which alter the surface tension of a solvent when dissolved therein. They are widely used for cleaning, washing, and textile processing. Sales of synthetics are about 3,000,000,000 lb per year, and of soap over 1,000,000,000 lb per year. The annual value of soap and detergent sales is $1,000,000,000. The annual per capita consumption in the United States is about 29 lb of soap and synthetics. The industry is one of the largest of the chemical process industries.

The marked imbalance of polarity within the molecules of soaps and detergents causes them to have unusual solubility and phase characteristics in both polar and nonpolar solvents. This behavior is responsible for their usefulness in detergency, wetting, solubilization, dyeing, and many other processes of industrial and household importance.

The basic feature of molecular structure which gives rise to these properties is the location of a highly polar function (hydrophilic) at or near the end of a long hydrocarbon chain (hydrophobic,

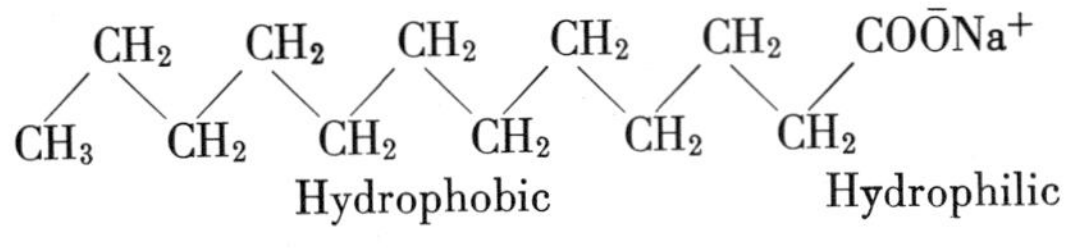

lipophilic). At the interface between two phases, there exists a concentration of detergent higher than that in the system as a whole, because of the differing attractions of the phases for the polar and nonpolar portions of the molecule. The interfacial tension is lowered. This accounts for the description of these substances as surface-active agents (surfactants). At an agitated air-solution interface, the excess concentration leads to the phenomenon of sudsing and lathering. At an oil-water interface, emulsification can result. If the interface occurs on a substrate such as cloth, glass, or metal, introduction of a surfactant causes a change in the angles of contact between the phases. *See* SURFACE-ACTIVE AGENT.

Soaps. Soaps are the alkali metal or (substituted) ammonium salts of straight-chain carboxylic acids, the molecules of which contain 10–18 carbon atoms. Metallic soaps are the alkaline-earth or other metal salts; they are extremely insoluble in water, and find application in nonaqueous systems (for example, in additives to lubricating oils, rust inhibitors, and napalm). Although soaps are a special class of detergent, the latter term is commonly restricted to synthetic materials.

Soaps are prepared from naturally occurring triglycerides (animal and vegetable fats) by hydrolysis of the ester linkages. Formerly, hydrolysis was achieved by the action of alkalies (NaOH, KOH) and water on fats in a soap kettle at high temperatures.

$$\underset{\text{Fat}}{C_3H_5(OOCR)_3} + \underset{\substack{\text{Caustic}\\\text{soda}}}{3NaOH} \rightarrow \underset{\text{Soap}}{3NaOOCR} + \underset{\text{Glycerol}}{C_3H_5(OH)_3}$$

A modern development of increasing importance is the direct hydrolysis of fats by water at high temperature. This permits isolation and rectification of the fatty acids, which are then neutralized to form soaps. This development is the basis of a continuous process for soap manufacture. Recovery of the by-product glycerol is important in the economics of soap manufacture.

Synthetic detergents. All the ingenuity of modern organic chemistry has been used to modify both the polar group and the hydrophobic group of soaps. This results in the synthetic detergents. A principal advantage of synthetics lies in their resistance to formation of insoluble soap curd. Soaps give curd by interaction with metal-ion impurities in natural water, especially those of calcium and magnesium. Alkaline-earth and heavy metal salts of anionic synthetics are generally much more soluble than those of soaps. Another advantage is resistance to acidity by many synthetic surfactants. Carboxylate soaps are hydrolyzed readily at low pH to give quite insoluble acid soap precipitates. Many additional special properties (germicidal action, fabric softening, and optimum hydrophilic-lipophilic balance) are obtained by modification of the polar and nonpolar groups of soaps.

As sources of the hydrocarbon moiety, synthetic materials (for example, polypropylene benzene from petrochemicals and products of the oxo and

Fischer-Tropsch processes) may now compete with natural materials such as tallow, grease, coconut and other nut oils, marine oils, and tall oil. Thus, a variety of chain lengths and branched structures not commonly found in nature is available.

Polar groups most commonly used to replace carboxylate are derived from sulfuric acid. Examples are alkyl sulfates, $ROSO_3Na$, prepared from fatty alcohols and sulfuric acid; alkane sulfonates, RSO_3Na, prepared from alkyl halides and sodium sulfite; and alkylaryl sulfonates, such as R—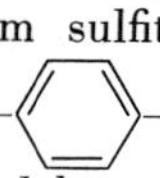—SO_3Na, prepared by sulfonation of alkyl benzenes. Fatty acid amides and esters prepared from derivatives of taurine ($H_2NCH_2CH_2SO_3H$) and isethionic acid ($HOCH_2CH_2SO_3H$) were early examples of synthetics. Alkane phosphonates represent another type. All these are examples of anionic surfactants.

Even greater modification of the polar group may be made by changing the sign of the charge on the surface-active ion. A well-known example of the class of cationic detergents, or invert soaps, is provided by the quaternary ammonium halide $C_{16}H_{33}N(CH_3)_3^+Br^-$.

In another variation, the nonionic detergents, the polar group is a nondissociated hydrophilic group, usually containing a multiplicity of oxygen functions (ether, alcohol) which engage in hydrogen bonding with water. An example is the ester made from a fatty acid and sugar. Another common type results from polymerization of several ethylene oxide units on an alcohol,

$$R{-}O{-}(CH_2CH_2O{-})_nH$$

Detergent action. The nature of the soap-water system has been investigated extensively. The phase relationships of anionic and cationic detergents are similar in a general way, and even nonionic detergents show many of the same properties.

A very dilute soap solution exists as a solution of an electrolyte that is extensively hydrolyzed. Sodium soaps are not very soluble at room temperature; the much more soluble potassium and substituted ammonium soaps are used in soft soap and liquid soap formulations. The solubility of a soap increases gradually with temperature, until suddenly, over a small temperature range, a very marked solubility increase occurs. This range is the Krafft temperature; additional amounts of soap introduced at this point do not go into true aqueous solution but exist as the colloidally dispersed phase called micelles. Micelles represent an association of the surface-active ions of which the hydrophobic groups are on the interior, and the polar groups are in contact with the water. The critical concentration at which micelles begin to form extensively may be taken as the limit of simple solubility of the soap in water. Various physical properties of the solution, which vary with soap concentration, reach more nearly constant values at this point. These include surface tension, conductivity, and

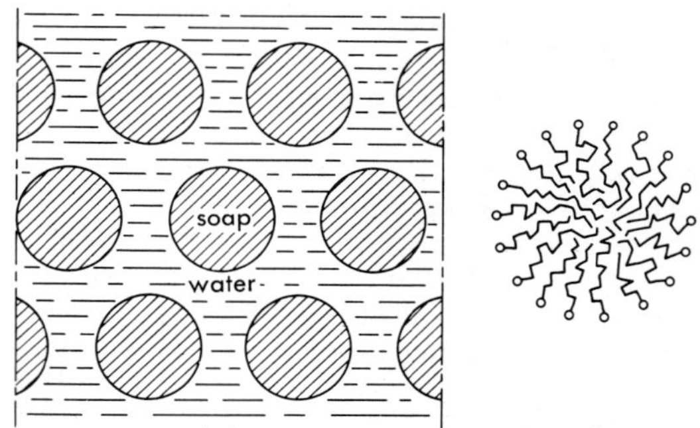

(a) (b)

Fig. 1. (*a*) Schematic structure of middle soap. (*b*) Cross section of a cylinder. (*From V. Luzzati, Nature, 180:600, 1957*)

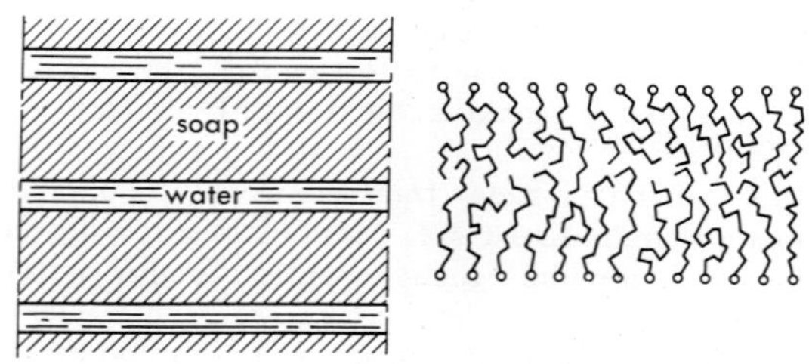

(a) (b)

Fig. 2. (*a*) Schematic structure of neat soap. (*b*) Section of a soap double layer. (*From V. Luzzati, Nature, 180:600, 1957*)

washing ability. Water-insoluble substances such as hydrocarbons may be solubilized into the interior of micelles. *See* HYDROLYSIS; MICELLE.

Such dilute soap solutions are called nigres. At sufficiently high soap concentrations, a viscous middle phase appears, presumably by coalescence of the micelles into long chains. At still higher soap concentrations, the less viscous neat phase has another type of order; great sheets of soap molecules are present. Middle and neat phases are anisotropic, or liquid crystalline phases.

The effect of salts on soap-phase behavior is very great and is of considerable importance in the chemistry of the soap kettle. Neat, the goal of the kettle boiler, is salted out (pitched) with sodium chloride when the desired stage of fat hydrolysis is reached. In subsequent treatment of the neat, as the concentration of water becomes low, the various waxy and solid crystalline phases appear. These, in the molded, roll-dried, or spray-dried forms, are the soap bars, flakes, and powders of commerce. Strictly anhydrous soap is not encountered in practice; it has none of the desired soapy properties.

The well-known household detergent washing powders are commonly obtained by spray drying a slurry of detergent, such as alkyl sulfate or alkylaryl sulfonate, mixed with sodium sulfate, alkaline builders, such as silicates, phosphates (especially $Na_5P_3O_{10}$), and a number of minor additives. *See* QUATERNARY AMMONIUM SALTS. [J.T.Y.]

Soapstone

A soft talc-rich rock. Soapstones are rocks composed of serpentine, talc, and carbonates (magnesite, dolomite, or calcite). They represent original peridotites which were altered at low temperatures by hydrothermal solutions containing silicon dioxide, SiO_2; carbon dioxide, CO_2; and other dissolved materials (products of low-grade metasomatism). Among the rock products thus formed are antigorite schists, actinolite-talc schists, and talc-carbonate rocks. To the last belongs the true soapstone, but the whole group of rock may loosely be referred to as soapstones because of their soft, soapy consistency. Such rocks were selected by prehistoric men for making primitive vessels and pots, and also for making rough carvings for ornamental purposes. *See* METASOMATISM; TALC.

[T.F.W.B.]

Social animals

Those animals whose social behavior may be defined as any behavior stimulated by or acting upon another animal of the same species. In this broad sense, almost any animal which is capable of behavior is to some degree social. Even those animals which are completely sedentary, such as adult sponges and sea squirts, have a tendency to live in colonies and are social to that extent. Social reactions are occasionally given by other species than the animal's own, for example, the relations between domestic animals and man.

Organization. Animals vary in the amount and kind of social behavior they show. This in turn determines the degree of social organization of which they are capable. The vast majority of lower animals merely form temporary aggregations based on the two most primitive forms of social behavior, sexual behavior and shelter-seeking. Among Protozoa, paramecia may come together in pairs to conjugate, or huddle together in large groups to minimize unfavorable conditions. The mere proximity of other animals of the same species is enough to create a more favorable environment under many situations. Such primitive aggregations produced by sexual behavior and shelter-seeking provide the basis for the evolution of more complex types of social behavior and organization.

The higher animals have evolved more complex types of social behavior and developed more permanent and more complex organizations. In addition to sexual behavior, their reproduction involves the care of the young, and this care-giving or epimeletic behavior is the foundation of social life in the insect societies. Ants, for example, form a permanent group of adults and young. There is a division of labor by castes, so that the sterile females, or workers, care for the young, and the fertile females, or queens, produce all the eggs. *See* SOCIAL INSECTS.

Allelomimetic and agonistic behavior. Care-giving plays an important but not predominant role in the social life of vertebrates. With their well-developed eyes, many vertebrates are capable of allelomimetic behavior, the tendency for animals to mutually mimic each other's actions. This behavior produces long-lasting cohesive groups, such as schools of fish, flocks of birds, or herds of mammals. Similar groups are formed by squids, made possible by the highly developed eyes of these mollusks. The higher animals are also capable of agonistic behavior, including fighting, escape, and defense. This results in dominance organization and social hierarchies, and also leads to the division of living space into territories. A complex vertebrate society may show all of these advanced types of social organization as well as the more primitive ones. *See* SOCIAL HIERARCHY.

General classification. There are three general types of social animals: those which are found only in temporary aggregations, including all the lower invertebrates; those which live in permanent groups bound together by care-giving behavior, such as the insect societies; and those whose permanent groups are organized through mutual imitation and agonistic behavior (often in addition to care-giving), such as many vertebrate societies. Man belongs to a fourth type in which a highly developed communication system has been added to the above social processes.

FISH

Just as mammals are dominant on land and birds in the air, fish are the dominant vertebrates in water habitats. Fish have competed successfully with both the mollusks and arthropods, formerly dominant in the ocean, and only a few of the large-sized cephalopods and crustaceans remain. Fish of larger sizes compete most successfully, and the smaller fishes have not displaced the small invertebrates.

The sharks, rays, and other elasmobranch fishes originated in salt water and never successfully moved into fresh-water habitats. On the other hand, the modern bony fishes first evolved in fresh water and have since moved back to populate the ocean.

Adaptation and behavior. The behavioral capacities of fish reflect their watery habitat. Those living in clear, well-lighted water usually have large eyes adapted for vision over short distances. Their chemical senses are highly developed, and fishes are able to taste the water in a variety of ways. Some fishes even have taste buds on the outside of the body. The nostrils also are used to sample the water, and are highly developed in sharks. The lateral lines are important sense organs not found in other vertebrates. They have the function of detecting small currents and pressure changes in the water, and a fish deprived of these lines is quiet and unresponsive to most changes in its environment. Fish have a well-developed sense of hearing but require no external organs for the concentration and transmission of sound. In some, the hearing mechanism is connected to the swim bladder. A few slow-moving, bottom-living fish such as catfish have special feelers or tactile organs. *See* SWIM BLADDER.

The motor organs of fish are primarily adapted for rapid movement through the water. Lateral movement of the tail is transformed by the flexible tail fin into forward propulsion. Other fins are primarily used for steering. Compared to other vertebrates, fish are poorly equipped with manipulative organs.

As to psychological capacities, fish are capable of rapid conditioning. Trout in hatcheries quickly learn to rise for food at any splashing noise. Fish can learn simple mazes and are able to find their way back to their home ranges quickly under natural conditions. On the other hand, much of their social behavior, particularly that connected with mating and nest building, is largely organized by instinct.

Social behavior. The social behavior of the three-spined stickleback, *Gasterosteus aculeatus*, has been thoroughly studied. Except during the breeding season, these small, fresh-water fish live in schools. When the mating season arrives, the males leave the school, change color, and select territories on a sandy and weedy bottom area. The male builds an elaborate nest mound and makes a tunnel through it. Eventually the females approach in schools, and the male goes toward them in a zigzag courtship dance. If the female is ready to spawn, she follows him to the nest and lays the eggs, which are immediately fertilized by the male. This accomplished, he drives her away and stays in the nest, fanning the eggs with his fins. The young hatch in 7–8 days and soon emerge in a swarm around the nest. If one wanders away the male picks it up in his mouth and brings it back to the school. He guards them for a few weeks, then leaves the territory and the young go on their way in the school. Such behavior is typical of many bony fishes, and even the river dogfish, *Amia calva*, one of the most primitive of the group, builds a nest and guards the young fry. *See* TERRITORIALITY.

Social life. Compared with birds and mammals, fish lack some of the basic types of social behavior. As water-living animals they have no problem of waste disposal and consequently have no special patterns of eliminative behavior. Fish never feed their young after hatching, and their care-giving behavior is limited to offering shelter and driving off predators. The connection between offspring and parents is brief at best, and the young show little care-soliciting behavior. Much of the social life of fishes is concentrated around the school, in which mutual imitation or allelomimetic behavior coordinates the group and provides mutual safety. Other social life centers around mating and nest building.

Most fish do not swim around in the water at random but live in definite localities. Fresh-water fish taken away from their home ranges will return rapidly if it is at all possible. In long-lived species the same individual may be found in the same spot

Fig. 1. Allelomimetic behavior in fishes. (*From J. P. Scott, Animal Behavior, University of Chicago Press, 1957*)

year after year. Definite territories are set up during the breeding season by many fishes and probably are important at other times as well.

Migration. Fish also show remarkable examples of migration. Salmon typically migrate in schools from their fresh-water spawning grounds to the ocean and return at maturity several years later to the same stream. They recognize the stream through their chemical senses which detect slight differences in the water. The salmon must therefore have learned to recognize this water in early life. The Atlantic eels migrate from their spawning grounds in the Sargasso Sea to the coasts of Europe and North America, returning at maturity in a trip of several thousand miles. In either case the fish show remarkable powers of orientation.

Communication. Studies with hydrophones show that many species of fish make use of auditory communication, although these noises are seldom apparent to land-living animals. The toad fishes emit warning signals when other fish approach their territories. Other species make noises which are useful in coordinating a school, either at night or in the semidarkness of the ocean bottom.

Breeding behavior. In many fish such as trout and salmon, parental care is reduced, and the importance of the school increased. In the Alpine char of Europe, the male sets up a breeding territory on the gravel bottom of a clear lake or stream. A female enters the territory and digs a nest by lying on her side and flipping up her tail to move the small stones by suction. After courtship the eggs are deposited in the nest, and the female covers them with gravel. Any eggs which are left uncovered are eaten by the parents. Usually the same female digs several nests and covers them after all the eggs have been laid. The females remain a few days in the vicinity of the nests, defending them as a territory. The male shows no further reaction to the nest but will court other females if they enter his territory. The parents have no protective reactions to the young fry, and eat them if they appear. The young fish must depend upon the school for safety.

The school. The school is perhaps the most typical part of the social life of fishes. It is usually an association between newly hatched young, but many species such as herrings and mackerel spend most of their lives in such a group. Even at spawning the school of herrings stays together and sheds sperm and eggs without parental care. The members of a typical school are highly allelomimetic, reacting to the movements of animals on either side. If one fish finds a source of food, others close by come immediately. If it reacts to a source of danger, others react in turn, so that the whole group is thrown into violent turmoil. In the resulting disturbance a predator is likely to be confused and his vision reduced. There is no evidence of leadership in such groups, nor that one fish consorts with particular individuals. At the other extreme, many species of small fish such as guppies, which normally live in muddy and weedy localities, do not form schools at all.

Social life. The social life of the bony fishes is therefore chiefly organized around places rather than individuals. The relationships set up during mating are very brief, and there are no known instances of long-lasting associations between parents and offspring. It is possible that such relationships might exist in some of the elasmobranch fishes whose young are born alive, but little is known concerning them. Social hierarchies are formed in laboratory aquariums by such species as the Siamese fighting fish and may have some importance in nature. However, most agonistic behavior is concerned with guarding territories or young individuals. Fish living in large schools do not exhibit such behavior and the group is organized simply on the basis of allelomimetic behavior and social attraction. The limitations of social organization in fish are in part a reflection of their limited manipulative abilities. These in turn limit the care which can be given to other individuals, and are in part a result of the limitations imposed on the sense organs by water living, which makes the recognition of individuals difficult. Even within these limitations, fish have developed a remarkable variety of social organizations. New and interesting facts are continually being discovered as modern scientific instruments make it possible to study fish in their natural habitats.

BIRDS

Birds are highly social animals. They move so rapidly and freely that their social behavior is not always apparent to the casual observer, but when their actions are studied more carefully it is found that birds develop a great variety and complexity of social organization. Their biological differentiation of labor is simpler than that of insects, since there are only three principal types of individuals: males, females, and young. However, they exhibit all the basic types of social behavior in well-developed forms and are capable of considerable differentiation of behavior on the basis of learning and experience.

Habitat. Birds are basically aerial animals, and various species have populated the entire surface of the earth, from polar regions to the Equator, and from deserts to midocean. Their aerial habitat gives them extraordinary freedom of movement, and many species have developed extraordinary capacities for migration. It also makes them greatly dependent upon weather conditions, and one of the characteristics of the social behavior of birds is regular seasonal change. In keeping with their habitat, birds have highly developed eyes which are the dominant sense organs. Hearing comes next and with this an extraordinary sense of balance. The senses of smell and taste are relatively unimportant.

The wing is, of course, the chief locomotor organ for all flying birds. This leaves the beak and claws to serve other functions such as manipulation and prehension, and their modification is one of the chief anatomical characteristics of the different families of birds. Even with their limited motor organs, some birds are able to perform highly complicated manipulative acts, such as nest construction by orioles.

Social behavior. In their psychological capacities birds vary continuously from such species as turkeys, which seem to have almost all their behavior organized by heredity into rigid instincts, to birds such as crows and jackdaws, which are highly adaptable and organize a great deal of their behavior on the basis of learning and experience.

Coloration. With their great mobility, birds frequently encounter other birds of different species, which creates a problem of separation between different societies. The bright colors and plumage changes found in many birds are closely related to social behavior and serve to differentiate between species, the sexes, and adults and young. Display behavior, emphasizing the plumage, is a prominent part of sexual behavior in many species. This often forms a key stimulus, or releaser, which limits the mating reaction to one species.

Communication. Display is a type of communication. In addition, birds have a wide variety of vocal signals; best-known of these is the song of the

Fig. 2. Strutting or display behavior of the male sage grouse. (*From J. P. Scott, Animal Behavior, University of Chicago Press, 1957*)

perching birds, now shown to be a territorial signal. In some species the song is in part learned; in others it is developed even by birds reared in isolation. Certain species, such as crows and parrots, have powers of vocal imitation superior to those of mammals, but even when human words are learned, there is no evidence that the birds do more than learn a complicated trick. Such parroting is found in nature in the songs of mockingbirds. Vocal and visual signals are important mechanisms in the social organization of birds but are not used as a true language.

Effect of aerial life. Social behavior is affected still more directly by aerial life. One major adaptation is rapid development of the young. The young of the song sparrow are ready to fly 17 days after hatching and are completely independent of the parents by 30 days. Even large birds such as gulls and geese reach adult size within the short breeding season of a few months. This means that in many birds there is little opportunity for a parent-offspring relationship to develop. The most prominent social relationship of most birds is consequently the male-female relationship or the pair formation.

Imprinting. Most highly social animals have a period early in life when the primary social relationships are formed on the basis of early social experience. This primary socialization or imprinting takes place very rapidly in precocious birds such as ducks and geese, but is spread over a longer period in birds which are hatched in an immature state. The parasitic cowbirds and cuckoos do not become imprinted by their foster parents but join their own kind after leaving the nest.

Localization. Another fundamental social process is that of becoming attached to particular places or home sites, or localization. Many species of birds have two homes, one during the breeding season and another at other times. Migrations between the two may cover thousands of miles, and require extraordinary capacities for orientation, the basic nature of which is the subject of much research.

Social organization. A few examples will illustrate some of the basic types of social organization and its variability in birds. Perching birds such as song sparrows and the various species of blackbirds typically spend the winter in the south and feed and fly in flocks. In the spring the males migrate northward and set up territories, each defended by a single male, which sits in the middle and sings as a warning to trespassers. Each male is dominant on its own territory. Females arrive somewhat later and usually one settles with each male. Both parents care for the eggs and young but the female usually takes a larger share in both incubation and feeding. In the autumn, the birds again unite into flocks and migrate southward. There is often a highly developed coordination in the flocks, but no leadership by a single individual.

Gallinaceous birds. In gallinaceous birds such as chickens and their wild relatives, the grouse family, the birds become attached to particular places, but these are not defended as territories. Whenever they gather into flocks they form strong dominance hierarchies and indeed these were first discovered in the domestic fowl. The hierarchy among males determines which male will be able to do most of the mating. There are no lasting sexual relationships, and mating is typically polygynous. Care of the young is entirely by females.

Doves and pigeons. The most extreme cases of pair formation are found in doves and pigeons, where the mated pair forms a lasting bond during the breeding season. In addition, the two sexes are morphologically very similar, and the care of the

Fig. 3. The gaping reaction in young desert horned larks, an example of care-soliciting (et-epimeletic) behavior in these animals. (*From J. P. Scott, Animal Behavior, University of Chicago Press, 1957*)

Fig. 4. Allelomimetic behavior in a male elk herd in Jackson Hole, Wyo. (*From J. P. Scott, Animal Behavior, University of Chicago Press, 1957*)

eggs and young is shared equally. There is a very close coordination of activities during incubation; one parent leaves and the other sits on the nest in regular rotation. Territory is established only in the immediate area around the nest, and fighting is much milder than in either of the above types of birds. There is no leadership in the flocking behavior.

Ducks and geese. As a final example, ducks and geese show strong pair formation which is often renewed season after season. Furthermore, the relation between parents and offspring is not broken off at migration, and the birds frequently migrate in family groups with a definite leader. They may meet in flocks of thousands at favorable resting places during migration, but each group sorts itself out and continues as a unit. A territory is set up only in relation to the nests themselves.

Social structure. There are numerous other kinds of bird societies, and many of them are still incompletely described and studied. The previous examples illustrate certain important features of the social life of birds, each one of which may vary from a highly complex development to none. Flocking behavior is typical of birds. In the smaller flying birds extraordinarily complex and rapid maneuvers based on allelomimetic behavior are seen; in some larger birds there appears considerable leadership. In contrast, the hawks and owls are almost entirely solitary in their movements. Territory is another important feature of bird life, ranging from the large breeding territories of the perching birds to the small nest area in gulls and other colonial sea birds. In still others there are no definite territories. Pair formation can result in male-female relationships lasting for life, or can be a casual sexual encounter. The result of combining these basic variable phenomena is a large variety of bird societies, each of which functions effectively for the particular species concerned as long as environmental conditions are reasonably constant. When the social structure is rigid, and dependent on the hereditary nature of the species, altered environmental conditions may result in extinction. When social structure is flexible, as it appears to be in mallard ducks, for example, the species may flourish in a variety of environmental conditions.

Knowledge of the social organization of birds has important practical applications in the conservation and protection of these animals for hunting and other purposes, as well as great theoretical interest for the student of animal and human societies. Just as mammals do, birds show all basic types of social adaptation, expressed in many sorts of social relationships and in complex social organizations. Unlike mammals, their behavior, and hence their social organization, is much more definitely organized by heredity or instinct. It is perhaps because of this limitation that true language, permitting a more advanced type of social organization, first appeared in a mammal rather than in the highly vocal birds.

MAMMALS

Mammals are warm-blooded vertebrates which typically live on the land surface of the globe.

Several orders, however, have semiaquatic forms, such as muskrats and beavers among rodents, and otters and seals among carnivores. One entire order, the whales, has become completely water-living, and another, the bats, has developed the power of active flight and has taken to the air.

A common characteristic of the entire group is the mammary gland, used for the feeding of the new-born young. Because this is developed only in females, there is a general tendency for a differentiation of labor in which the females care for the young and the males specialize in fighting behavior. In the subclass Marsupialia, the nipples are enclosed in a brood pouch, in which the young spend most of their early life. The early ingestive behavior of mammals, or nursing, thus has a high degree of social significance.

In addition to their great variety of habitats, mammals eat a great variety of foods, and occupy many different ecological niches. Unlike birds, there are almost as many mammals which are nocturnal as those which are chiefly active during the day.

Behavioral capacities. Depending on its habitat, a given species may use one or more of the sense organs more than the others. Some nocturnal rodents such as mice and rats use the tactile and auditory senses a great deal more than the eyes, whereas a diurnal tree-living animal such as a squirrel may chiefly use its eyes. All major sense organs are well developed in each species.

Mammals show a great variety of motor capacities. The limbs can be highly developed for digging as they are in moles and badgers, or used almost entirely for locomotion as they are in the hoofed mammals. An arboreal primate such as the chimpanzee has all four limbs developed for manipulation and prehension. Even the tail can be used for prehension by some of the monkeys. At the other extreme, the hind limbs of whales are externally invisible and the fore limbs are used only as flippers.

Mammals have relatively large and complex brains, with a large development of the cerebral cortex. There is also a considerable capacity for behavioral adaptation through learning. Unlike birds, there are relatively few instances of elaborate patterns of behavior which are completely fixed by heredity. In general, mammals are highly adaptable, both from the evolutionary viewpoint and in their capacity to adjust to daily changes in the environment.

Social behavior of the American elk. This is one of the largest living species of the deer family, standing as high as a horse. The eyes, ears, and nose are all well developed. The males grow large antlers which are used chiefly in the rutting season and are shed in the early spring. Elk were originally plains-living animals, but are now restricted by hunting to the open areas in the Rocky Mountains. They are highly social animals and live in large herds.

Social life is closely related to the seasons. In the autumn the male herds break up. Each male attempts to round up a herd of females with whom he mates as each comes into estrus. If another male

Fig. 5. Agonistic behavior in a herd of buffalo; two males sparring. (*From J. P. Scott, Animal Behavior, University of Chicago Press, 1957*)

appears, the two begin a furious pushing contest, ending when one is exhausted and driven off. *See* ESTRUS.

As winter comes on, the males and females separate into different herds and migrate to lower altitudes. Hundreds of individuals may move together, producing a migration trail which looks almost as if it were a roadway.

In the spring there is a return migration to the higher altitudes. Each band separates from the large winter herd and goes back to the locality from which it came. Males go higher than females, which stay together and bear their calves. As is true with many members of the deer family, the young calves will "freeze," or stay quiet, while the mothers leave to graze or browse. Usually at least one female stays in close proximity to the young animals. The calves at first obtain most of their nourishment by nursing from their mothers. When alarmed, they call to the mother. They grow rapidly during the summer and follow their mothers throughout the first year. The seasonal round of behavior begins again with the rutting season in the autumn.

Types of social behavior in elk. The most highly developed type of social behavior is allelomimetic, in which each animal does the same things as those nearby, all responding to each other. This results in the formation and continued existence of a herd. Also of importance in these herbivorous animals is ingestive behavior. Adults spend a great deal of time browsing and grazing. A more highly social sort of ingestive behavior is the nursing of the calves. Related to this is the care-giving or epimeletic behavior of the mothers and the care-soliciting behavior of the young. Agonistic behavior, consisting of fighting and escape behavior, is common in the males during the rutting season. As is true of most of the herd animals, fighting is individual and groups never combine against one animal. Sexual behavior occurs at the same time, but only for a brief period with each female. Investigative behavior is common, and the animals are continually raising their heads to look around. There is no special pattern of eliminative behavior and relatively little social shelter-seeking.

Social relationships. Social behavior results in the formation of social relationships. This is partly based on biological differentiation into three types, males, females, and young, each with its characteristic behavior. There are six possible combinations between them, and each kind of social behavior may result in the formation of a social relationship within a particular combination.

The sexual relationship developed between males and females is a brief one, lasting only as long as the male is able to maintain his dominant position in relation to a herd of females. It has no importance at other times of the year, although it may play some part in holding the male group together.

Both male-female and female-female relationships can involve dominance and subordination. However, as is the case in many grazing animals, whose food is scattered in such a way as to prevent competition, dominance has little importance except in the breeding season. Even here a male may be able to hold a dominant position only for a portion of the season.

Social behavior and relationships in mammals

Relationship	Type of social behavior	Typical combinations formed
Care-dependency	Care-giving Care-soliciting Ingestive (especially nursing) Shelter-seeking Eliminative (in some species)	Adult-young (especially female-young)
Sexual	Sexual Investigatory Eliminative (in some species)	Male-female
Dominance-subordination	Agonistic	Male-male (but also in all other combinations)
Leader-follower	Allelomimetic	Female-young (usually weaker in other combinations)
Mutual care	Care-giving (grooming)	Adult-adult
Mutual defense	Agonistic	Male-male

In female-young associations the care-dependency relationship is important and long-lasting. As is the case in other herd animals, females will allow only their own young to nurse, and a close association is built up throughout the nursing period. This relationship lasts into later life and forms the foundation of leadership within the herds. The young born in the same year and within the same herd also form a lasting relationship with each other. This group is particularly important for the males when they separate from the females as adults. In the female herds a leader-follower relationship develops, first between mother and offspring and eventually between older and younger females. This basically allelomimetic relationship is one of the two most important social relationships developed in an elk society.

Unlike the situation among many birds and fishes, the migration trails are apparently learned from one generation to the next and have been altered within historical times. The animals become localized, that is, they become attached to the area in which they were brought up, but there is no defense of territory. Strange herds mingle with each other without fighting, and agonistic behavior is important only in the rutting season.

Other ungulates. The social behavior of other even-toed ungulates follows a pattern similar to that of the elk. The red deer of Scotland are quite similar to them, and even more distantly related animals such as goats, sheep, and bison have the same general characteristics. There is much emphasis on ingestive behavior, because enormous quantities of vegetation must be eaten. A typical grazing animal follows a regular cycle of grazing and stops to chew its cud every few hours. Among the more social species this is always done in concert with others, so that allelomimetic behavior is also highly important. Under natural conditions the result is the enormous herds of ungulates which were once seen on the plains of North America and South Africa.

The predominant social relationship of the even-toed ungulates is the mother-young relationship. Its formation has been studied in detail in sheep. The mother sheep will allow only its own lamb to nurse and will drive all others away. The behavior of the mother toward its own lamb is fixed within the first 4 hours after birth. The result of this close and definite relationship is that leadership is developed by females, and social inheritance tends to be matrilineal.

Social behavior of wolves. These carnivorous animals are especially interesting because of their wide distribution and their close association with man in fact and legend. The first domestic animal, the dog, was derived from them and has almost identical patterns of social behavior.

Under ordinary conditions the food supply of a carnivore is not regularly available in any one area. Consequently the chief activity of wolves is hunting, a form of investigatory behavior. Much of this is done at night. Having spent the day in or near the den, the wolf pack assembles toward evening and hunts far away, usually returning by morning. The hunting range may be 20–50 miles across. In their hunting activities, they show a great deal of allelomimetic behavior, both in searching for and pulling down game, but there is little evidence of consistent leadership by one individual.

Mating and care of young. Mating behavior takes place in January or February. The entire estrous cycle of a female extends for 6 weeks or more, during which time there is long preliminary courtship as well as actual mating. A peculiarity of sexual behavior is the sexual tie, in which the male and female remain coupled for as long as ½ hour. It is not known whether the female mates with more than one male, but in related animals which do not run in packs, such as foxes and coyotes, a female typically mates with one male, and the two form a lasting association.

The cubs are born in early spring and are constantly attended by the mother during the first few days of life. As they grow older the mother must leave them to feed and hunt. There is usually a good supply of food at the den, either carried back or vomited by the adults. The cubs begin to eat freshly vomited food at about three weeks of age and are completely weaned at 6 or 7 weeks. However, they are still unable to hunt and remain dependent on the adults until approximately 6 months of age.

Thus the litter is in constant association from birth, but parents and other adults leave them for long periods each day. The closest social relationships are made with litter mates, and this is the foundation for the new pack. The young litter leaves the old den at about 1 year of age, to settle down in a new territory. It is not known whether they are joined by other litters. Most wolf packs consist of five or six adults plus any young animals running with the pack but not old enough to leave it.

Dominance. Agonistic behavior within the pack is organized into a definite dominance order. Occupying such large ranges, there is no possibility of guarding the boundaries as territories, although some packs regularly hunt over circular runways. Wolves set up scent posts which are regularly visited and at which the animals urinate and defecate. It is possible that in this way a wolf can tell whether a range is occupied and so avoid it. The den area is a definite territory; a strange wolf coming to the den is attacked and driven off.

Wolves are relatively long-lived animals, and social relationships, including the sexual one, are quite persistent. The allelomimetic relationship between litter mates is a most important one, and also important is the dominance-subordination relationship. The relationships between parents and offspring are relatively unimportant.

Cat family. By contrast, members of the cat family tend to emphasize the mother-offspring relationship. This is associated with a different method of hunting: lying in wait for the prey and pursuing it for a short distance rather than ranging widely. In addition, cats are less omnivorous and more dependent on meat than members of the dog family. The young develop rapidly and are assisted in their first hunting by the mother. With the exception of lions, cats tend to be solitary in their hunting activities and show very little allelomimetic behavior.

Seals. Alaskan seals show highly organized social behavior upon their breeding grounds. The males take up territories along the shore in the late spring and defend them against males and other mammals. As females arrive, the males guard them as a harem. A few days after arrival the females produce their pups and shortly come into estrus, mating only once. Females leave their pups in groups while they swim off to feed, but males do not feed in this period. Females do not breed until 4 or 5 years of age, and males are not able to maintain harems till about 10 years. Large groups of immature males stay outside the territories of the harem males. After the young are mature enough to leave the breeding grounds, the seal herds break up and become almost entirely aquatic until the new breeding season.

Breeding grounds relations. Two social relationships are prominent on the breeding grounds: the dominance relationship of the harem males, and the mother-offspring relationship. The sexual relationship has little social importance. In general, the social behavior of carnivores is related to their food habits. Investigatory behavior is always prominent. Possibly because of this, carnivores have a strong reputation for intelligence. Some are strongly allelomimetic, but many are almost solitary in their activities. The care of the young is greatly emphasized, because successful hunting depends upon size and experience. Agonistic behavior is also of great importance, as might be expected in a hunting animal. The parent-offspring relationship is important in early life because of the difficult manner of obtaining food. In some species, but not all, the sexual relationship is an important one, and there is a tendency toward the development of lasting bonds between males and females.

Social behavior of rodents. Among mammals, rodents are good builders and diggers, and this is reflected in their social life. Perhaps the highest development of rodent society is found in the prairie dog, a large ground squirrel of the western plains. These animals dig numerous burrows and pile the dirt in mounds which are the outward signs of a prairie dog town. A town is subdivided into territories, each inhabited by a coterie of prairie dogs. A coterie includes one or two male and two or three female adults, which produce numerous young and guard the territory of perhaps half an acre. Mating takes place in February and March, at which time there is a good deal of fighting between males and some reorganization of territories. Each female in the territory retires to a particular burrow and guards it as long as the young are below ground. Males must stay in other holes. When the young come out in April the females relax and all is again peaceful within the coterie.

The young learn the territorial boundaries by being challenged as they cross them. There is some social organization throughout the entire town, and any prairie dog that sees danger gives a warning signal which alerts all the rest. In June and July the adults leave the territory to construct new burrows on the outskirts of the colony. Thus there is a regular cultural inheritance. The young learn the territory from the parents and inherit the burrows. This system provides for colonization and also means that the experienced adult animals live on the outskirts of the colony, the part most exposed to danger.

In behavior typical of most other rodents, the principal type of social behavior is the care of the young in the nest, a relationship which tends to break up as the young leave. The male-female relationship is more important than in ungulates but is relatively unstable. The young are bound together chiefly by the territorial system. There is almost no allelomimetic behavior and consequently very little coordinated group activity. In the less social rodents such as mice, an entire society may consist of a mother and her young.

Social behavior of primates. Among primates, social behavior is characterized by an emphasis on sexual behavior. Adult males and females constantly associate with each other throughout life. There is also a great deal of emphasis on care-giving behavior, particularly by females but also by males. The young are carried by the female for at least a year and may associate with her for much longer. The result is the formation of the typical primate group of males, females, and young, of which the human nuclear family is an example.

The baboons and their relatives, the rhesus monkeys, exhibit one type of primate society. A group of two or more males, several females, and assorted young occupy an area of about 2 square miles. The adult males have a strict dominance order among themselves but combine with each other to attack predators which threaten the group. In a well-organized band, females in the period of estrus, which lasts for several days, go from male to male within the group without exciting fighting. Mothers carry their young and pay close attention to them. Males respond if the young are hurt or give a call of distress. The young baboons spend a great deal of time in play with each other and thus presumably set up the relationships which will maintain the group when they are adults.

The size and composition of primate societies varies from species to species. In the gibbons, in which the two sexes are much alike in physique and both are highly aggressive, the typical group includes only one adult male and female. The forest-living great apes, such as chimpanzees and gorillas, live in small groups of eight or nine, mostly young, with one or two each of adult males and females. In contrast, the macaques may form groups as large as 150. Thus the ground-living primates form larger groups than those which live in trees. However, there is a great variety of social organization and behavior in primates, and no one species can be taken as an exact model for primitive human behavior.

Development of social organization. The postnatal development of each species is closely related to the social organization typical of the adults. Every highly social animal has a short period early in life when it readily forms attachments to any animal with which it has prolonged contact. This critical period for primary socialization can be demonstrated by removing young from their natural parents and rearing them by hand. Experience before this period has no effect; during the period the young animal transfers all its native social behavior to man; thereafter this process becomes increasingly difficult, although captivity and hand feeding have some effect even on adults.

Process of socialization. The process of socialization begins almost immediately after birth in ungulates like the sheep, and the primary relationship is formed with the mother. In dogs and wolves the process does not begin until about 3 weeks of age, at a time when the mother is beginning to leave the pups. Consequently the strongest relationships are formed with litter mates, thus forming the foundation of a pack. Many rodents stay in the nest long after birth; primary relationships are therefore formed with nest mates. Young primates are typically surrounded by a group of their own kind, but because they are carried for long periods the first strong relationship tends to be with the mother.

Mammalian society. In summary, mammals may develop all types of social behavior to a high degree, but not necessarily in every species. The above examples include four highly developed societies from representative orders; in other species certain types of social behavior may be completely absent. Mammals have great capacities for learning and adaptation, which means that social relationships are often highly developed on the basis of learning and habit formation as well as on the basis of heredity and biological differences. The resulting societies tend to be malleable and variable within the same species and to show considerable evidence of cultural inheritance from one generation to the next. Mammalian societies have been completely described in relatively few forms, and new discoveries will probably reveal the existence of an even greater variety of social organization.

Basic human social organization and behavior obviously differs from that of all other primates, although it is related to them. At the same time, the range of variability of human societies as seen in the nuclear family does not approach that in mammals as a whole. Human societies are characterized by the presence of all fundamental types of social behavior and social relationships rather than by extreme specialization. [J.P.S.]

Bibliography: W. C. Allee, *Cooperation among Animals with Human Applications*, 1951; E. A. Armstrong, *Bird Display and Behaviour*, 1947; M. E. Brown (ed.), *The Physiology of Fishes*, vol. 2, 1957; C. R. Carpenter, A field study of the behavior and social relations of howling monkeys, *Comp. Psychol. Monographs*, 10(2):1–168, 1934; L. R. Dice, *Natural Communities*, 1952; M. M. Nice, Studies in the life history of the song sparrow, *Trans. Linnaean Soc. N.Y.*, vols. 4 and 5, 1937–1943; J. P. Scott, *Animal Behavior*, 1957; N. Tinbergen, *Social Behaviour in Animals*, 1953; A. Wolfson (ed.), *Recent Studies in Avian Biology*, 1955.

Social hierarchy

Social hierarchy was first adequately described in flocks of domestic hens. The fighting behavior of each individual hen is organized so that in any given pair, hen A always pecks hen B and never the reverse. In this way an entire flock can be described in terms of social rank, or a "peck order." This type of social organization is an important general phenomenon among higher animals.

Dominance-subordination relationships. A social hierarchy is composed of many dominance-subordination relationships. When two hens are brought together for the first time they usually fight. The next time they come together the fighting has a shorter duration. One hen forms a habit of winning and the other of losing. Eventually the relationship is reduced to a threat or peck by the dominant hen and avoidance or submission to the peck by the subordinate one. The hens thus form a social relationship based on agonistic behavior, which is defined as any behavior connected with conflict between two individuals, and including fighting, escape, and submissive behavior.

Possible arrangements of dominance hierarchies in a group of three animals, assuming that each relationship shows either a complete dominance or complete peacefulness, are shown: (1) straight-

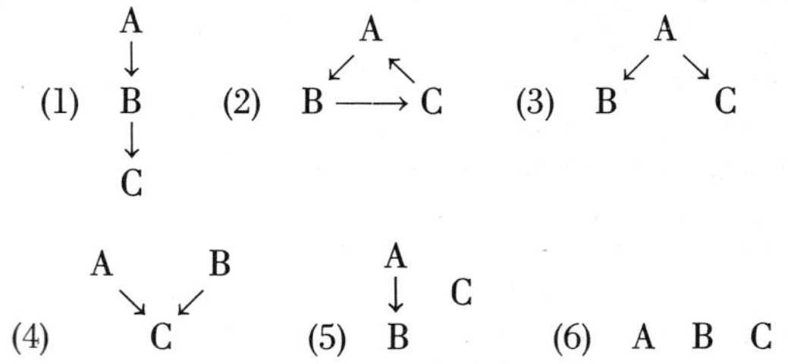

line dominance order, commonly found in flocks of hens in aggressive breeds; (2) triangular order, less common; (3) one dominant and two subordinate, common in mouse groups; (4) two dominant animals and one subordinate; (5) one dominance relationship, C is peaceful with both; (6) all animals peaceful, with no dominance order, frequently found in young animals reared together.

Since most of the lower invertebrates do not exhibit agonistic behavior, social hierarchies have so far been found only in arthropods and vertebrates. In arthropods, the best examples are found in crayfish and other Crustacea. While fighting occurs in some insects, it is doubtful whether true dominance-subordination relationships are formed. In the higher social insects, such as ants, bees, wasps, or termites, no fighting takes place within the colony. On the other hand, dominance-subordination relationships have been described in all classes of vertebrates, although they have little importance among amphibia, which are poorly equipped for fighting behavior. The social hierarchy is thus a typical attribute of vertebrate societies, reaching its highest development in birds and mammals.

Dominance organization. Dominance organization results in the control and reduction of fighting behavior within a group. Hens in an organized flock gain more weight and lay more eggs than those in a flock which is continually disturbed by the introduction of new members. At the same time, animals at the top of a dominance hierarchy frequently have the advantage over low-ranking individuals in competition for food, territory, and mates. The nature of competition depends on the genetic constitution of the species. Deer and many similar grazing mammals fight over mates, but not over food or territory. Chickens compete for food, but this is important only if the supply is limited or concentrated in one place. The net result of a social hierarchy depends on the nature of the animals involved plus the nature of their environment.

Under natural conditions dominance organization is frequently set up in a much less violent manner than that described above. Young animals accept dominance from older ones on the basis of threats. Young goats reared together may either develop peaceful relationships or develop dominance relationships on the basis of more or less playful fighting. In any case, a naturally formed group tends to be better organized. Violent and bloody fighting is likely to occur only where strange animals are brought together and kept under confined conditions.

Hereditary influences. Hereditary differences in agonistic behavior have a profound effect upon dominance relationships. In any combat, size is usually an advantage, as is aggressiveness. The small, unaggressive animal tends to come out at the bottom of the dominance order. Excessively aggressive animals, such as gamecocks or terrier dogs, are frequently unable to form dominance relationships, as the winner usually kills the loser. In contrast, unaggressive animals form flexible hierarchies or no hierarchies at all. In unaggressive breeds of dogs, any animal may be left in possession of a bone, depending on which one gets it first. In moderately aggressive breeds they tend to form rigid hierarchies like those of hens.

Range of dominance. Dominance relationships may vary a great deal in the same animal group. In a flock of goats, dominance relationships range from those which are completely peaceful through varying degrees of dominance. The most extreme dominance relationships are almost as peaceful as relationships in which there is no dominance, because the subordinate animal avoids the dominant one on all occasions. Dominance can be seen only when the two are forced together. The greatest amount of fighting occurs in a relationship of unsettled dominance, in which the animals fight every time they meet, but neither wins.

Dominance hierarchy. The dominance hierarchy can be either extremely important or completely nonexistent, depending on the species of animal involved. Well-developed dominance relationships tend to be relatively permanent, persisting over periods of years in long-lived animals. Such social organization may have important effects on the control of fighting within a species, the division of food, territory, and mates, and upon the selection of surviving individuals. At the same time it should be remembered that this is only one aspect of social organization and that important relationships which do not involve fighting, such as care-dependency and leader-follower relationships, also exist in animal societies. *See* BEHAVIOR, ONTOGENY OF; SOCIAL ANIMALS; SOCIAL INSECTS. [J.P.S.]

Bibliography: W. C. Allee, A. E. Emerson, O. Park, T. Park, and K. Schmidt, *Principles of Animal Ecology*, 1949; J. P. Scott, *Animal Behavior*, 1958.

Social insects

Social organization has evolved independently several times in unrelated groups of the class Insecta. The best examples of these social insects are the termites of the order Isoptera, and the social wasps, bees, and ants of the order Hymenoptera. Characteristically, social insects are differentiated and specialized in structure, function, and behavior into castes, which live in colonies and exhibit group integration and division of labor. The principal castes are the reproductives, which include the king and queen, and the sterile, the workers and soldiers. The reproductive caste performs the fundamental function of reproduction, distribution of the species accompanied by swarming or colonizing flight, choosing the site for the new colony, excavation of the first galleries, feeding, and care of the first young of the new colony. In termites, and some higher ants, the sterile castes consist of workers and soldiers represented by both sexes; but among social wasps, bees, and ponerine ants (Ponerinae), the workers are the only sterile caste, consisting of nonreproductive females. The function of the workers, which form a majority in the population, is that of tending the eggs, young nymphs or larvae, feeding and cleaning other castes, and constructing and repairing the nest, and gathering food. The soldiers are structurally modified to defend the community against predators.

Social insects arose from subsocial insects showing some sort of a communal mode of life and division of labor. The termites arose from ancestral cockroaches. The social wasps belong to the family Vespidae and are related to potter wasps *Eumenes* and *Odynerus*, which are essentially solitary in habit. Ants form a family Formicidae and arose from a solitary wasp, but there is no living representative of this ancestral family. The bethylid wasp (*Scleroderma*) is its closest living relative. The social bees evolved from solitary bees, which in turn evolved from sphecoid wasps.

Caste determination. In the two main orders of insects having social organization, caste determination is due to different factors. Extrinsic factors, nutrition, and genetics are involved.

In the order Hymenoptera (wasps, bees, and ants), caste differentiation has both a genetic and nutritional basis. The differences between male and female have a genetic basis. The male develops from an unfertilized egg, and therefore contains one set of chromosomes, and is haploid. The queens, workers, and soldiers are female, develop from a fertilized egg and are diploid, containing two sets of chromosomes. The differences between the worker and the queen are correlated with differences in the quality and quantity of food. It is well known that in the honeybee, the worker and drone larvae are fed with royal jelly for the first 3 days and bee bread (honey and pollen) for another 3 days. The queen larvae are fed entirely with royal jelly.

In termites, the order Isoptera, caste differentiation is due not to genetic but to extrinsic factors. The most adequate and experimentally supported theory is that of ectohormone inhibition (*see* INSECT PHYSIOLOGY), which states that the reproductives and the soldiers give off secretions or ectohormones containing inhibiting substances, passed to the nymph through mutual feeding (trophallaxis) within the colony. Thus the reproductives and soldiers exert a sphere of inhibiting influence against the development of like forms; the substance is caste specific. When, because of increase in population, some undifferentiated nymphs fall beyond this influence or the substance, being limited, is no longer distributed by trophallaxis in effective amounts, then certain additional members of the caste may differentiate. It is believed that the initial trend of development is toward reproductives. Individuals inhibited against such development tend to become soldiers, and if inhibited in this direction remain undifferentiated nymphs functioning as workers or become workers.

Reproduction and fecundity. Among the social insects, reproduction is essentially similar in the various groups. Fecundity and colony formation, however, are quite variable.

Reproduction. A single egg-laying queen is typical for most colonies of social insects. In some social wasps (*Polistes annularis* and *Nectarina*) and ants, however, the colony is built and maintained by several queens. It has been reported that one nest of the ant *Formica exectoides* contained 1407 queens. In the majority of termite nests, a single queen and king occur with or without supplementary reproductives, but several instances have been reported of two queens and a king, and two kings and a queen.

Fecundity. This is an important variable which determines the potential size of a colony. The queens of primitive social wasps, bees, ants, and termites lay few eggs per day. The fecundity of specialized social insects, however, has developed enormously. An ant queen has been recorded as laying 341 eggs per day. Army ant (*Eciton*) queens produce as many as 20,000 in a few days, and 3021 eggs per day have been counted for a honeybee queen. A queen of a certain tropical species of termite has a capacity of 6000–7000 eggs per day for 15–50 years. Such queens attain an enormous size. Old queens of *Odontotermes obesus* in India for example, reach a length of 5 in. or more.

Colony foundation. Colony foundation, in the social insects, is the result of many factors such as temperature, moisture, or overcrowding. During this phase of colony foundation, the interesting phenomenon of swarming may be observed in some groups.

In social wasps (*Polistes*, *Vespula*) and bumblebees, the impregnated queen hibernates over the winter and founds a new colony in spring. She settles in a suitable location and constructs the first shelter, lays the eggs in it, and rears the first brood. The first brood consists of workers only, which take over such duties as foraging, nest building, and care of young. In late summer, males and young queens appear in the nest. Workers, males, and old queens die as winter approaches, leaving only the young fertilized queen to survive and perpetuate the colony.

In higher ants, the queen founds a new colony by a mating flight from the old one. The young males and females emerge from exit holes for the mating flight. The queen copulates with many males which die after mating and never take part in colony foundation. The queen drops to the earth, pulls off her wings, digs a hole in the soil, imprisons herself and lays the eggs. She feeds the larvae with a secretion derived from the absorption of her wing muscles. The first brood hatched are small workers. The queen produces eggs for 12–17 years from sperms stored in the spermatheca at the time of her mating flight. As the colony grows, young winged queens and males are produced.

In some genera of ants (*Formica*, *Eciton*) "budding" of the old colony takes place when a portion of the worker population migrates with one or more fertile queens to form a new colony.

The propagation of colonies by the honeybee is by swarming. Each swarm consists of an old queen and a mass of about 35,000 workers, which fly out of the parental nest, find a new nesting site and establish a new colony. Swarming is controlled by temperature and crowding within a colony. During swarming, the bees give out an odor from their scent glands which serves as a guide in the formation of clusters. In an old colony, mating of the queen takes place during the nuptial flight in which drones from many colonies take part. The drone dies after mating.

In termites, the male and female play equal parts in colony foundation. They leave the old colony in the form of a mass exodus, the so-called swarming or colonizing flight. Swarming is controlled by a certain combination of temperature and moisture. The flight usually takes place either just before or more often just after a rainfall. The workers prepare exit holes for emergence of the winged reproductives. The reproductives become photopositive and fly short distances in all directions, for termites are weak fliers. After swarming for a brief period, they lose their wings and pairs are formed on the ground, where the females attract the males by means of their abdominal scent glands. They become photonegative once again, and together dig a hole in the soil or dead wood and block the entrance. After some time they copulate, and the first nymphs develop into workers. Copulation occurs at short intervals throughout the life of the queen.

Colony age. The age of the colony varies in different groups of social insects. Communities of social wasps (*Polistes*), bumblebees, and others living in temperate latitudes are short-lived, existing only through the summer season. Colonies of tropical species tend to live longer.

The longevity of members of the honeybee hive varies under different conditions. Normally the queen survives 3–4 seasons, although two cases have been reported where the queen lived 7–9 years. The drones survive for 3–4 months. Workers live for 6–10 weeks in summer and 6 months or longer in winter.

The colony of the ant *Formica ulkei* lives for almost 25 years. Forel has mentioned that one nest of *Formica pratensis* lived for 40 years. The longevity of all caste members of the same colony has never been determined in natural colonies. In an artificial nest of *Monomorium pharaonis* females lived 39 weeks, males not longer than 2–3 to 4–5½ weeks, and workers lived 9 weeks.

In species of termites which produce supplementary queens and kings, the colony is potentially immortal. In those species which do not produce supplementary sexuals, the colony is short-lived. F. N. Ratcliffe mentions that a colony of *Nasutitermes graveolus* at Darwin, Australia, has been in existence for over 30 years and a mound of *Tumulitermes triodiae* for over 150 years. This record is remarkable since this species does not produce any supplementary reproductives.

Population size of colonies. The colony size of social wasps varies from 25–100 individuals in the species *Polistes* to about 400 in *Vespa maculata* and 11,000 in *Vespa vulgaris.* In modern hives of honeybees a typical colony consists of about 30,000 individuals, although the population may drop to 10,000 or less during winter and rise to 70,000 during more favorable seasons. Mature colonies of bumblebee contain 1000–2000 individuals. The size of ant colonies differs for different species from 20 individuals in *Sysphincta pergandei* to 237,000 in *Formica exsectoides.* Population of fully grown termite colonies show a wide range from several hundreds in *Kalotermes* and *Reticulitermes,* to some millions in mound-building species in the tropics.

Food. Nutritional requirements of social insects vary considerably. Larvae of social wasps feed mainly on chewed insects, while adults eat a variety of food including fruit juices, syrups, and other sweets, nectar, caterpillar juices, and animal food. The bees are specialized pollen and nectar feeders. Most of the primitive ants are carnivorous, while others live on special food such as honey dew secreted by aphids and caterpillars and sweet liquid from flower nectaries and fruits. Higher ants are vegetarians and have developed harvesting habits of collecting and storing seeds and grasses. A few ants are fungus growers. Termites are specialized cellulose feeders, utilizing mainly wood and fibrous products as food materials. Some of the primitive termite species depend upon flagellate protozoa to digest the cellulose for them (*see* MASTIGOPHORA). The most specialized species have lost such dependence upon symbionts and probably produce the necessary enzymes themselves.

Among social insects, the young receive food from workers or queen. The young in turn give off secretions which are licked by the workers. This mutual exchange of food between young and adult, known as trophallaxis, is very important among social insects. It serves as a bond of solidarity and permits the colony to expand in numbers without the members losing recognition and thus becoming hostile to each other.

Cannibalism is very common in ants, termites, and wasp colonies. It probably regulates the population of a colony and conserves valuable food products.

Nest. The social insects control the environment in their nests. This control helps the population to live under stable conditions and avoid external environmental fluctuations of temperature and humidity. The nest also gives protection from predaceous animals and provides for storage of food and cultivation of fungi.

The social wasps and bees build comb nests. The social wasps make circular or hexagonal cells of the comb from a mixture of chewed old wood, tough plant fibers, and salivary secretion; this mixture is reduced to a pulp mass of paper consistency. The cells are arranged in vertical rows and open downward. The cells may be naked, as in *Polistes,* or covered with an envelope, as in *Dolichovespula.* These nests are found in dark places such as hollows of trees, between adjacent walls, under stones, caves, and branches of trees.

The honeybee comb mass is made of wax. The cells are of various sizes and arranged in regular patterns. Cells of the worker brood and those which store pollen are smaller than those which house drones and store honey.

Ants nest in all sorts of places such as hollows of trees or cavities in plant stems, seeds, nuts and galls. Some such as the carpenter ants excavate galleries in wood. The majority of ants are earth-dwellers in subterranean passageways or mounds. The earth nest may be small and simple or may be quite large and elaborate, consisting of large numbers of tunnels and galleries. Certain chambers in such underground nests serve as brood chambers, while others serve for storage of food.

The primitive termites are wood-dwelling and build nests in dry or damp wood. The higher termites are earth-dwelling and build carton nests, mounds, or subterranean tunnels. The nests have an enclosed system of tunnels which is capable of maintaining the high degree of humidity necessary for the survival of the soft-bodied termite population.

The insect society, which has all the attributes of the organism (growth, reproduction, replication, symmetry, regeneration, homeostasis, adaptation, ontogeny, and phylogeny), has sometimes been referred to as a supraorganism. The unit of natural selection here is the entire colony instead of the individual member. *See* HYMENOPTERA; ISOPTERA. [K.K.]

Bibliography: W. C. Allee et al., *Principles of Animal Ecology*, 1949; C. D. Michener and M. H. Michener, *American Social Insects*, 1951; W. M. Wheeler, *The Social Insects*, 1928.

Soda niter

A nitrate mineral having chemical composition $NaNO_3$ (sodium nitrate). Soda niter is by far the most abundant of the nitrate minerals. It crystallizes in the rhombohedral division of the hexagonal system. It sometimes occurs as simple rhombohedral crystals but is usually massive granular. The mineral has a perfect rhombohedral cleavage, conchoidal fracture, and is rather sectile. Its hardness is 1½ to 2 on Mohs scale, and its specific gravity is 2.266. It has a vitreous luster and is transparent. It is colorless to white, but when tinted by impurities, it is reddish brown, gray, or lemon yellow. *See* NITRATE MINERALS.

Soda niter is a water-soluble salt found principally as a surface efflorescence in arid regions, or in sheltered places in wetter climates. It is usually associated with niter, nitrocalcite, gypsum, epsomite, mirabilite, and halite.

The only large-scale commercial deposits of soda niter in the world occur in a belt roughly 450 miles long and 10–50 miles wide along the eastern slope of the coast ranges in the Atacama, Tarapaca, and Antofagasta deserts of northern Chile. The deposits consist of a thin bed of nitrates and associated minerals, varying from a few inches to a few feet in thickness, overlain by a shallow overburden of sand and gravel. The crude soda niter, known as caliche, is about one-fourth sodium nitrate, admixed with other salts, notably bloedite, anhydrite, gypsum, polyhalite, halite, glauberite, and darapskite, together with minor amounts of various iodates, chromates, and borates. Because of the presence of the iodate minerals lautarite and dietzeite, these deposits yield most of the world's supply of iodine.

The origin of these deposits is controversial. It is generally agreed that the nitrates were transported by ground water and deposited by evaporation. The source of the nitrate has been attributed to (1) guano; (2) nitrogen fixation by electrical storms; (3) the bacterial fixation of nitrogen from vegetable matter; and (4) a volcanic source in nearby Triassic and Cretaceous rocks. The last source seems the most probable. *See* NITROGEN CYCLE.

Chilean nitrate had a monopoly of the world's fertilizer market for many years, but now occupies a subordinate position owing to the development during World War I and subsequent years of synthetic processes for nitrogen fixation. Production of Chilean nitrate in 1957 totaled 1,456,000 tons. The 1957 sodium nitrate and potassium nitrate prices were $34.48 and $43.56, respectively, per short ton. *See* FERTILIZER.

Small deposits of soda niter similar to those in Chile are found in Bolivia, Peru, North Africa, Egypt, U.S.S.R., India, and western United States. [G.S.]

Sodalite

A mineral tectosilicate of the feldspathoid group, crystallizing in the isometric system, with chemical composition $Na_4Al_3Si_3O_{12}Cl$. Crystals, which are rare, are usually dodecahedrons. Sodalite is most commonly massive or granular. There is poor dodecahedral cleavage. The hardness is 5–6 and the specific gravity 2.2–2.4. The luster is vitreous and the color, usually blue, may also be white, gray or green. Sodalite has been cut and polished for use as an ornamental stone. *See* FELDSPATHOID; SILICATE MINERALS.

Sodalite is a relatively rare mineral found in nepheline syenites and leucite-bearing rocks. It is thus associated with nepheline, leucite, cancrinite, and feldspar. It is chiefly a primary mineral, although some has formed by the alteration of nepheline. Transparent crystals have been found in the lavas of Vesuvius, and the massive blue variety occurs in Ontario, Quebec, and British Columbia, Canada; and in Litchfield, Maine. [C.S.HU.]

Sodium

With an atomic number of 11 and an atomic weight of 22.997, sodium, Na, stands below lithium and above potassium in group Ia of the periodic table of the elements. It is a soft, low-melting, reactive metal with a specific gravity of 1.0. It is by far the most important alkali metal commercially, being produced at the rate of over 250,000,000 lb annually. The metal was named by Sir Humphry Davy, who first isolated it by electrolysis in 1807.

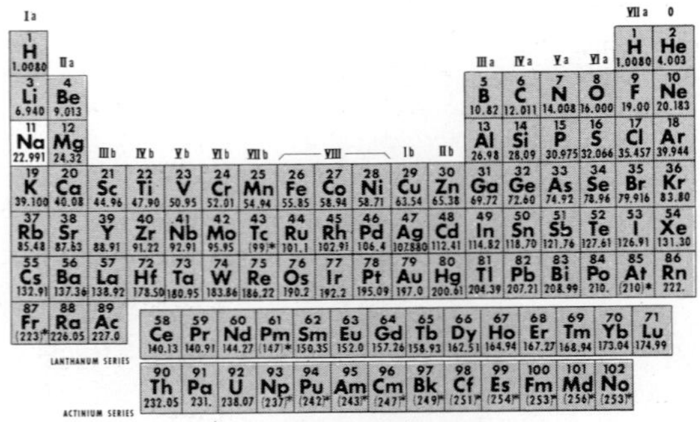

*mass number of most stable known isotope

Uses. The largest single use for sodium metal, accounting for about 60% of total production, is in the synthesis of tetraethyllead, an antiknock agent for automotive gasolines. In this process, the reaction of a sodium-lead alloy with ethyl chloride gives tetraethyllead.

$$4PbNa + 4C_2H_5Cl \rightarrow (C_2H_5)_4Pb + 3Pb + 4NaCl$$

The unreacted lead is recycled in the process.

A second major use is in the reduction of animal and vegetable oils to long-chain fatty alcohols; these alcohols are raw materials for detergent manufacture. This use has been decreasing, in favor of production of such alcohols by high-pressure catalytic hydrogenation.

Another major use is in the reduction of titanium and zirconium halides to the respective metals. Here the use of sodium is increasing at the expense of magnesium as the preferred reducing agent in such operations.

Sodium metal is also used in making sodium hydride, sodium amide, and sodium cyanide, as discussed in the section on inorganic reactions. It is also used in the synthesis of "isosebacic acid," as described under organic reactions. The use of liquid sodium metal as a heat-transfer agent in nuclear reactors is also becoming increasingly important.

Sodium chloride is used in the manufacture of sodium hydroxide, sodium carbonate, sodium sulfate, and sodium metal. In sodium sulfate manufacture, hydrogen chloride is the co-product. In metallic sodium manufacture, chlorine gas is the co-product.

Rock salt is used in curing fish, in meat packing, in curing hides, and in making freezing mixtures. Food preparation, including canning and preserving, consumes much salt. Table salt accounts for only a small percentage of sodium chloride consumption, most of it going into the industrial uses outlined above.

Sodium hydroxide is perhaps the most important industrial alkali. Its major use is in the manufacture of chemicals, about 30% going into this category. The next major use is the manufacture of cellulose film and rayon, both of which proceed through soda cellulose (the reaction product of sodium hydroxide and cellulose); this accounts for about 25% of the total caustic soda production. Soap manufacture, petroleum refining, and pulp and paper manufacture each account for a little less than 10% of total sodium hydroxide uses.

Sodium carbonate finds its major use in the glass industry, which takes about one-third of total production. Approximately another third goes into the manufacture of soap, detergents, and various cleansers. The manufacture of paper and textiles, nonferrous metals, and petroleum products accounts for much of the balance.

The major consumer of sodium sulfate (salt cake) is the kraft pulp industry. Increasing quantities of sodium sulfate are being used in the manufacture of flat glass. Other uses of salt cake are in detergents, ceramics, mineral stock feeds, and pharmaceuticals.

Occurrence. Sodium ranks sixth in abundance among all the elements in the earth's crust, which contains 2.83% of sodium in combined form. Only oxygen, silicon, aluminum, iron, and calcium are more abundant. Sodium is, after chlorine, second most abundant element in solution in sea water.

The important sodium salts found in nature include sodium chloride (rock salt), sodium carbonate (soda, trona), sodium borate (borax), sodium nitrate (Chile saltpeter), and sodium sulfate. Sodium salts are found in sea water, salt lakes, alkaline lakes, and mineral springs. Rock salt deposits occur where salt lakes and ancient seas have existed. In the United States, salt domes are the major commercial source of sodium chloride as a raw material for sodium metal manufacture. These deposits are found in Virginia, West Virginia, Michigan, along the south shore of Lake Erie, and along the Gulf Coast of Texas and Louisiana. The salt is removed from the earth both by underground mining and by pumping water down to force up salt brines.

Physical properties. The physical properties of metallic sodium are summarized in the table.

Chemical properties. For convenience, the reactions of sodium are divided into inorganic reactions and organic reactions.

Inorganic reactions. Sodium reacts rapidly with water, and even with snow and ice, to give sodium hydroxide and hydrogen. The reaction liberates sufficient heat to melt the sodium and ignite the hydrogen.

When exposed to air, freshly cut sodium metal loses its silvery appearance and becomes dull gray because of the formation of a coating of sodium oxide. Sodium probably oxidizes to the peroxide, Na_2O_2, which reacts with excess sodium present to give the monoxide, Na_2O. When sodium reacts with oxygen at elevated temperatures, sodium superoxide, NaO_2, is formed; this reacts with more sodium to form the peroxide. Sodium also forms an ozonide, NaO_3, when ozone is passed into a solution of sodium in liquid ammonia.

Sodium does not react with nitrogen, even at very high temperatures. Sodium and hydrogen react

Physical properties of sodium metal

Property	Temperature °C	Temperature °F	Metric (scientific) units	British (engineering) units
Density	0	32	0.972 g/cm³	60.8 lb/ft³
	100	212	0.928 g/cm³	58.0 lb/ft³
	800	1472	0.757 g/cm³	47.3 lb/ft³
Melting point	97.5	207.5		
Boiling point	883	1621		
Heat of fusion	97.5	207.5	27.2 cal/g	48.96 Btu/lb
Heat of vaporization	883	1621	1005 cal/g	1809 Btu/lb
Viscosity	250	482	3.81 millipoises	4.3 kinetic units
	400	752	2.69 millipoises	3.1 kinetic units
Vapor pressure	440	824	1 mm	0.019 lb/in.²
	815	1499	400 mm	7.75 lb/in.²
Thermal conductivity	21.2	70.2	0.317 cal/(sec)(cm²)(°C)	76 Btu/(hr)(ft²)(°F)
	200	392	0.93 cal/(sec)(cm²)(°C)	46.7 Btu/(hr)(ft²)(°F)
Heat capacity	20	68	0.30 cal/(g)(°C)	0.30 Btu/(lb)(°F)
	200	392	0.32 cal/(g)(°C)	0.32 Btu/(lb)(°F)
Electrical resistivity	100	212	965 microhm-cm	
Surface tension	100	212	206.4 dynes/cm	
	250	482	199.5 dynes/cm	

above about 200°C to form sodium hydride. This compound decomposes at about 400°C and cannot be melted. Sodium hydride can be formed by the direct reaction of hydrogen and molten sodium or by hydrogenating dispersions of sodium metal in hydrocarbons. Sodium reacts with carbon with difficulty, if at all, and this reaction has not been adequately studied.

At room temperature, fluorine and sodium ignite, dry chlorine and sodium react slightly, bromine and sodium do not react, and iodine and sodium do not react. However, moisture or elevated temperatures will speed the reactions enormously.

Sodium reacts with ammonia, forming sodium amide and liberating hydrogen. The reaction may be carried out between molten sodium and gaseous ammonia. Alternatively, sodium metal reacts with liquid ammonia (−30°C) in the presence of catalysts of finely divided metals. Sodium amide has been used in indigo manufacture (in the condensation of sodium phenylglycinate to sodium indoxyl, which in turn is oxidized by air to indigo). Sodium reacts with ammonia in the presence of coke to form sodium cyanide.

Carbon monoxide reacts with sodium, but the resulting carbonyl, NaCO, is stable only at liquid ammonia temperatures. At high temperatures, sodium carbide and sodium carbonate are formed from carbon monoxide and sodium.

The reactions of sodium with various metal halides to give the metal plus sodium chloride are very important. Thus, titanium tetrachloride is reduced to titanium metal. Similarly, the halides of zirconium, beryllium, and thorium can be reduced to the corresponding metals by sodium. The interaction between sodium and potassium chloride is used in the commercial production of potassium metal.

Organic reactions. Sodium does not react with paraffin hydrocarbons but does form addition compounds with naphthalene and other polycyclic aromatic compounds and with arylated alkenes. It reacts with acetylene, replacing the acetylenic hydrogens to form sodium acetylides. Sodium adds to dienes, the reaction which forms the basis of the buna synthetic rubber process used by both Germany and the U.S.S.R. in World War II. The addition of sodium to butadiene can also be controlled to give a disodium butadiene dimer, which can be carbonated to give a 10-carbon atom dibasic acid known as "isosebacic acid," an important ingredient in plasticizers for vinyl chloride polymers.

The reaction of sodium with alcohols is similar to, but less rapid than, the reaction of sodium with water. Brick sodium, molten sodium, or sodium dispersed in hydrocarbons may be used in the reaction with alcohols, and the alcoholate (alkoxide) products may be handled in solution form, in slurry form in hydrocarbons, or as dry, free-flowing powders.

Sodium reacts with organic halides in two general ways. One of these involves condensation of two organic, halogen-bearing compounds by removal of the halogen, allowing the two organic radicals to join directly. *See* WURTZ REACTION. The second type of reaction involves replacement of the halogen by sodium, giving an organosodium compound. Alternatively in this second class of reaction, a metal alloyed with sodium may replace the halogen after the halogen has been removed by sodium; the reaction of sodium-lead alloy with ethyl chloride to give tetraethyllead is an example of this type of reaction.

Sodium can effect the reduction, condensation, and alkylation of carbonyl compounds. For these purposes sodium can be used in the form of the metal, as an alloy, as the alkoxide, the amide, the hydride, or as an organosodium compound to effect various specific reactions. Most reactions of sodium with carbonyl compounds proceed through an intermediate formed by reaction of an active hydrogen atom with sodium. This intermediate then reacts with other molecules of the original compounds or with other active compounds present in the reaction mixture.

Metallurgical extraction. Raw sodium chloride either enters the plant as a brine or is fed into a dissolver tank to produce a brine. This brine is treated with sodium hydroxide and ferric chloride, and then with barium chloride to remove, among other impurities, any sulfate. The refined brine is then evaporated and the dried salt stored.

Practically all of the metallic sodium is made at the present time in the Downs cell, using a bath consisting of molten sodium chloride plus calcium chloride. The presence of 58–59% of calcium chloride depresses the melting point of pure NaCl from 800°C to 575–585°C. The lower working temperature simplifies cell construction because of less severe operating conditions.

The Downs cell is a large, refractory-lined, steel vessel. Graphite anodes project up from the bottom of the cell. These central anodes and the surrounding cylindrical steel cathode define an annular electrolysis zone. The gaseous product of the electrolysis, chlorine, rises into the top of the cell and is removed. Since the product, sodium metal, is lighter (specific gravity 0.88) than the molten bath (sp gr 2.1), it rises through the bath and is collected

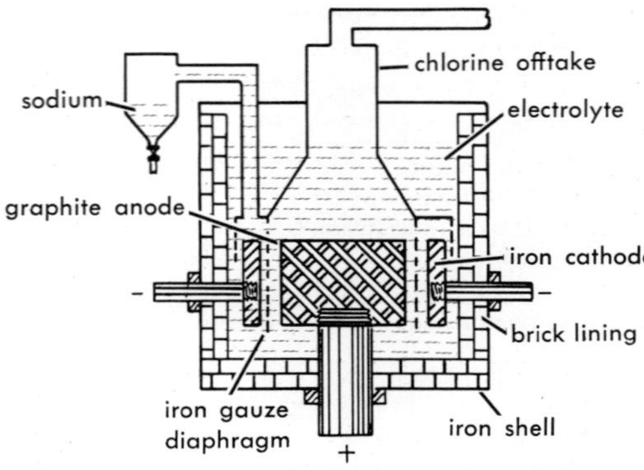

Schematic diagram of Downs cell for sodium production. (*C. L. Mantell, Industrial Electrochemistry, McGraw-Hill, 1950*)

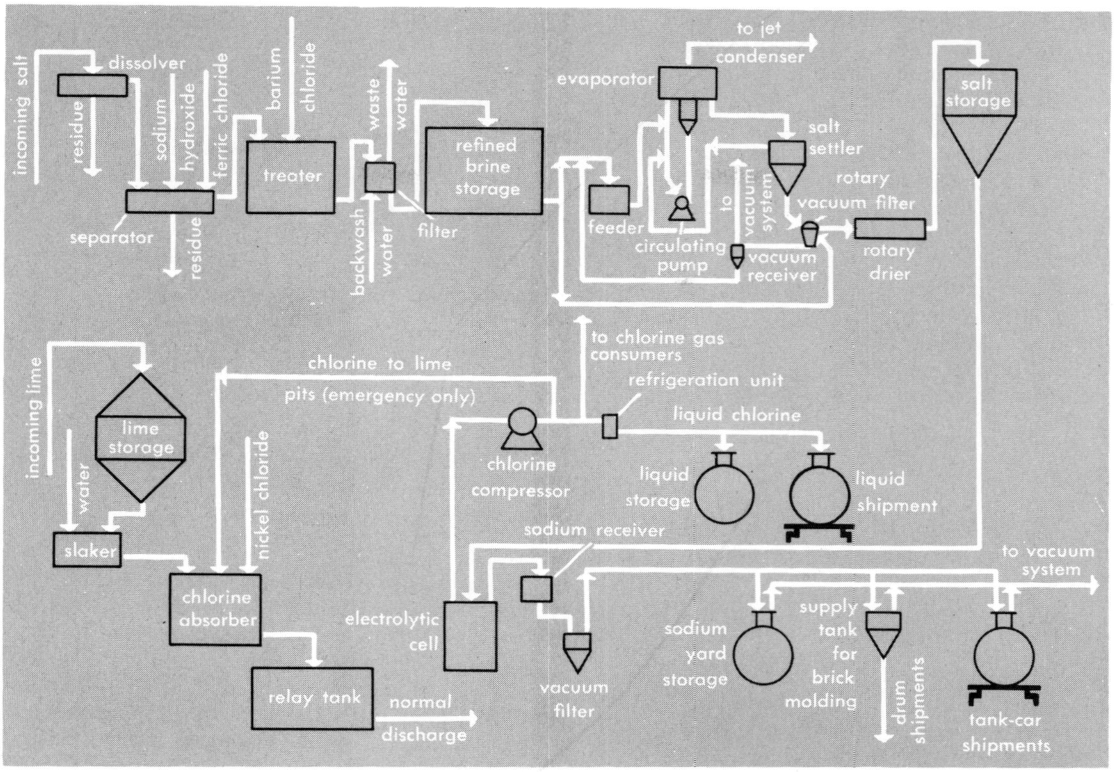

Process flow sheet of modern electrolytic sodium plant. (*US Industrial Chemicals Company*)

under the inverted collector ring and forced by buoyancy up the sodium collector pipe. Some cooling occurs in this riser pipe, and calcium metal, formed as a co-product with the sodium, precipitates out and falls back into the cells where it redissolves in the bath mixture. The sodium issuing from the receiver is 99.8% pure.

Availability. Sodium is available in two grades: a commercial grade analyzing 99.8% sodium, and a special nuclear reactor grade manufactured, loaded, and shipped under argon gas after being passed through porous metallic filters to remove oxides.

Sodium is available in containers ranging from 1 lb to 80,000 lb net weight. For laboratory use, 1-lb bricks are available in hermetically sealed, key-opening metal cans. Bricks in 1-, 2½-, 5-, 12- or 24-lb sizes are shipped in steel pails of about 25 lb net weight and in steel drums of about 300 lb net weight. Sodium is also available cast-solid in steel drums. Large consumers purchase sodium cast-solid in special railway tank cars of 80,000 lb net weight. Hot oil is pumped through channels in the walls of the special car at the destination and the molten cargo is withdrawn by vacuum.

World sodium production in 1958 was about 300,000,000 lb, and United States production accounted for about 260,000,000 lb of this total. The price of sodium in tank car lots was 16 cents per lb.

Handling. Sodium and other alkali metals are no more difficult to handle than many chemicals used every day. There are some hazards involved, however: (1) fire and explosion from hydrogen evolved if the sodium comes in contact with water; (2) caustic soda burns from the residue of a sodium-water reaction; (3) eye injuries from small pieces of sodium or caustic soda; (4) burns about the body from clothing ignited by clinging particles of burning sodium or flammable dispersed sodium; (5) irritation of the eyes and mucous membranes of the nose and throat because of breathing fumes from burning sodium; (6) flesh burns from contact with metallic sodium.

Safe handling of sodium demands that users observe the following precautions:

(1) Keep water away to avoid evolution of hydrogen which can be ignited by the heat of reaction. (2) Use only dry nitrogen (containing less than 1.0 mg/ft^3 water) to blanket sodium. (3) Use dry sodium carbonate (soda ash), dry sodium chloride, or dry graphite as fire-extinguishing agents. Do not use chlorinated hydrocarbons, carbon dioxide, foam, water, or soda-acid. (4) Know the physical and chemical properties of sodium and consider the potentialities of the reaction. Make initial trials of the reaction on a small scale. Control the reaction rate by adding sodium in small increments and by diluting the reactants with inert solvents. (5) Plan the disposal of sodium residues in advance. Dispose of scrap promptly, not allowing it to accumulate. (6) Store sodium in airtight, well-labeled, rust-free metal containers, away from water. (7) Use shal-

low metal pans on bench tops which can contain spillage in case of accident. (8) Wear clothing suited to the quantity of sodium being handled and to the temperature level at which it is being handled. This means goggles and gloves in most cases. It means full protective clothing for sodium at 1000°F under pressure. It means an impervious apron when dispersions of sodium in hydrocarbons are being handled.

Principal compounds. Sodium chloride, or common salt, NaCl, is not only the form in which sodium is found in nature, but (in purified form) is the most important sodium compound in commerce as well. World production is about 60,000,000 tons, of which the United States produces about 20,000,000 tons.

Sodium hydroxide, NaOH, is also commonly known as caustic soda. It readily absorbs water from the atmosphere and must be protected in storage and handling. It is corrosive to the skin and must be handled with extreme care to avoid caustic burns.

Most sodium hydroxide is produced by the electrolysis of sodium chloride solutions in one of several types of electrolytic cells. An older process is the soda-lime process whereby soda ash is converted to caustic soda

$$Na_2CO_3 + Ca(OH)_2 \rightarrow 2NaOH + CaCO_3$$

Sodium carbonate, Na_2CO_3, is best known under the name soda ash because sodium carbonate occurs in (and once was extracted from) plant ashes. Most sodium carbonate is produced by the Solvay or ammonia-soda process. In an initial reaction, salt is converted to sodium bicarbonate

$$NaCl + NH_3 + CO_2 + H_2O \rightarrow NaHCO_3 + NH_4Cl$$

which precipitates and is then separated. Heating of sodium bicarbonate produces sodium carbonate

$$2NaHCO_3 \rightarrow Na_2CO_3 + CO_2 + H_2O$$

The carbon dioxide is recycled in the process. Despite its apparent simplicity, the carrying out of this process requires much technical skill.

Sodium sulfate, Na_2SO_4, is also known in the anhydrous form as salt cake. The decahydrate, $Na_2SO_4 \cdot 10H_2O$, is known as glauber salt.

Most sodium sulfate is produced synthetically as a by-product or co-product in various industries. The Mannheim furnace process for hydrochloric acid manufacture is one important source of salt cake,

$$2NaCl + H_2SO_4 \rightarrow Na_2SO_4 + 2HCl$$

The neutralization of sulfuric acid in rayon and cellophane plants and in other inorganic and organic chemical processes accounts for much production.

A lesser amount of sodium sulfate is produced from natural sources, such as salt deposits in Wyoming, and lake brine in California.

Analytical methods. The determination of the sodium ion in solution is complicated by the high solubility of most sodium salts in water. Thus, quantitative determination of sodium relies heavily on gravimetric techniques using double uranyl salts, such as zinc uranyl acetate, as precipitants in most cases. The yellow color imparted to a flame by sodium ions serves as a sensitive qualitative test for the presence of sodium. *See* ALKALI METALS.

[M.SI.]

Bibliography: American Chemical Society, Handling and uses of the alkali metals, *Advances in Chem. Ser.*, 19, 1957; C. B. Jackson, *Liquid Metals Handbook, Sodium-NaK Supplement*, 3d ed., 1955; R. N. Lyon, *Liquid Metals Handbook*, 2d ed., Navexos P-733 (rev.), 1954; J. W. Mellor, *A Comprehensive Treatise on Inorganic and Theoretical Chemistry*, vol. 2, 1922; Marshall Sittig, *Sodium: Its Manufacture, Properties, and Uses*, 1956.

Sodium-vapor lamp

An electric discharge lamp of monochromatic yellow light used primarily for street lighting. The construction of a sodium-vapor lamp is shown in the illustration. The arc tube, in which the arc takes place, is made of glass. The outer bulb has three walls for better thermal insulation. The electrodes are coils of tungsten, which can be preheated before starting the arc by passing an electric current through each coil. The arc tube contains a small amount of sodium and some neon gas to facilitate starting.

In starting, the electrodes are heated by an electric current, and a neon glow forms between the starting electrode and the adjacent main electrode (*see* NEON GLOW LAMP). As the neon glow fills the arc tube, the resistance between main electrodes decreases, and an arc forms. The sodium vaporizes from the heat of the arc and electrodes, and sodium ions carry the arc during operation.

During the starting period, the lamp color is essentially red, the characteristic color of the neon glow. As the lamp warms up and the sodium vaporizes, lamp color gradually shifts to the intensely

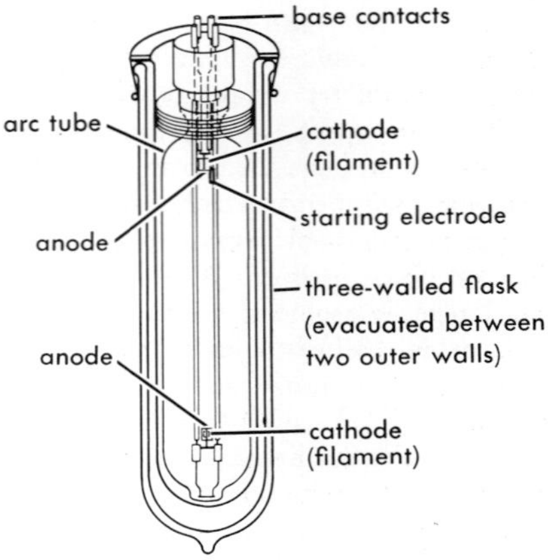

Ten-thousand-lumen sodium-vapor lamp. (*From Illuminating Engineering Society, IES Lighting Handbook, 2d ed., 1952*)

yellow color that is characteristic of sodium radiation. Sodium radiation is confined to a single band in the yellow region of the spectrum. This results in high luminous efficiency, since the human eye is quite sensitive to yellow light, but extremely poor rendition of colors. Because of their poor color-rendering properties, sodium lamps are limited to tasks where color is unimportant and efficiency is a primary consideration. In the United States, sodium-lamp lighting is restricted primarily to street lighting and some building floodlighting, and these are being superseded by other types of lamps, such as the yellow-coated mercury lamp (*see* MERCURY-VAPOR LAMP). Sodium-vapor lamps, however, are still popular in Europe.

The narrow band of radiation from sodium lamps suits them ideally to optical work requiring monochromatic light. A special sodium laboratory arc lamp is available for these applications.

[A.MA.]

Bibliography: Illuminating Engineering Society, *IES Lighting Handbook*, 2d ed., 1952; P. H. Moon, *Scientific Basis of Illuminating Engineering*, 1936.

SOFAR

The name associated with explosive signals which follow transmission paths through deep ocean layers. Sounds originating moderately deep in the ocean are refracted so that they propagate to large distances with small losses. For example, a small explosive charge set off at 3000- or 4000-ft depth is detectable many hundreds of miles away. If such an explosive signal is received at two or more points, the geographical position of the explosion can be calculated. This signaling technique has been used to locate aviators who have been forced down in the open ocean, and explains the origin of the word SOFAR, which stands for SOund Fixing And Ranging.

The refraction which makes possible this long-range propagation occurs because the velocity of sound has a minimum value at some depth (about 3600 ft in the North Atlantic). Because of refraction, a large number of rays can propagate without striking either the ocean bottom or surface. Since the sound is confined to a channel, the intensity of the sound carried by these rays decays with the inverse first power of the distance instead of as the inverse square power as it would in the absence of refraction. *See* UNDERWATER SOUND.

The signal received from an explosion through the SOFAR channel has a very special character. Because only certain rays which leave the signal source are intercepted by a receiver at a distant point, a signal from an explosion consists of a succession of pulses, each one having followed a different ray. These become closer together and more intense as time goes on, finally ceasing abruptly. The first received signal follows the ray that makes full excursions to the top or bottom, the last pulse being the one that goes along the channel axis. Although the latter pulse has gone the shortest distance, it arrives last because it has followed a route where the velocity of sound is smallest.

For a discussion of position fixing by the use of radio waves rather than sound waves, *see* HYPERBOLIC NAVIGATION SYSTEM. [R.W.MO.]

Bibliography: W. M. Ewing, W. S. Jardetzky, and F. Press, *Elastic Waves in Layered Media*, 1957.

Soft chancre

A world-wide venereal disease, also known as chancroid, caused by *Haemophilus ducreyi* (Ducrey's bacillus). This species requires X factor (a hemin factor in blood), but not V factor (phosphopyridine nucleotide). The lesion of soft chancre usually appears on the genitalia 4–10 days after contact. It is an inflamed, edematous ulcer or chancre that must be distinguished from the less inflamed, indurated "hard chancre" of syphilis. It may be followed by swelling and ulceration of the inguinal lymph nodes. Diagnosis is by culture on infusion agar containing 20% rabbit blood or by microscopic observation of the organism in material from the lesion. The tetracycline antibiotics or the sulfa drugs are effective treatment. *See* ANTIBIOTIC; HEMOPHILIC BACTERIA; SYPHILIS. [W.F.V.]

Soil

Freely divided rock-derived material containing an admixture of organic matter and capable of supporting vegetation. Soils are independent natural bodies, each with a unique morphology resulting from a particular combination of climate, living plants and animals, parent rock materials, relief, the ground waters, and period of time. Soils support plants, occupy large portions of the earth's surface, and have shape, area, breadth, width, and depth. Soil, as used here, differs in meaning from the term as used by engineers, where the meaning is unconsolidated rock material. *See* PEDOLOGY; SOIL MECHANICS.

This article is divided into five parts: origin and classification of soils, physical properties of soil, chemistry of the soil, soil management, and soil erosion.

ORIGIN AND CLASSIFICATION OF SOILS

Soil covers most of the land surface as a continuum. Each soil grades into the rock material below and into other soils at its margins, where changes occur in relief, ground water, vegetation, kinds of rock, or other factors which influence the development of soils. Soils have horizons, or layers more or less parallel to the surface and differing from those above and below in one or more properties, such as color, texture, structure, consistency, porosity, and reaction. The horizons may be thick or thin. They may be prominent, or so weak that they can be detected only in the laboratory. The succession of horizons is called the soil profile. In general, the boundary of soils with the underlying rock or rock material occurs at depths ranging from 1 to 6 ft, though the extremes lie outside of

Fig. 1. Photograph of a soil profile showing horizons. The dark crescent-shaped spots at the soil surface are the result of plowing. The dark horizon lying from 9 to 18 in. below the surface is the principal horizon of accumulation of organic matter that has been washed down from the surface. The thin wavy lines were formed in the same manner.

this range. Figure 1 is a photograph of a soil profile showing horizons.

Origin of soils. Soil formation proceeds in stages, but these stages may grade indistinctly from one into another. The first stage is the accumulation of unconsolidated rock fragments, the parent material. Parent material may be accumulated by deposition of rock fragments moved by glaciers, wind, gravity, or water, or it may accumulate more or less in place from physical and chemical weathering of hard rocks. *See* WEATHERING PROCESSES.

The second stage is the formation of horizons. This stage may follow or go on simultaneously with the accumulation of parent material. Soil horizons are a result of dominance of one or more processes over others, producing a layer which differs from the layers above and below.

Major processes. The major processes in soils which promote horizon differentiation are gains, losses, transfers, and transformations of organic matter, soluble salts, carbonates, silicate clay minerals, sesquioxides, and silica. Gains consist normally of additions of organic matter, of oxygen and water through oxidation and hydration, but in some sites slow continuous additions of new mineral materials take place at the surface or soluble materials are deposited from ground water. Losses are chiefly of materials dissolved or suspended in water percolating through the profile or running off the surface. Transfers of both mineral and organic materials are common in soils. Water moving through the soil picks up materials in solution or suspension. These materials may be deposited in another horizon if the water is withdrawn by plant roots or evaporation, or if the materials are precipitated as a result of differences in pH (degree of acidity), salt concentration, or other conditions in deeper horizons.

Other processes tend to offset those that promote horizon differentiation. Mixing of the soil occurs as the result of burrowing by rodents and earthworms, overturning of trees, churning of the soil by frost, or shrinking and swelling. On steep slopes the soil may creep or slide downhill with attendant mixing. Plants may withdraw calcium or other ions from deep horizons and return them to the surface in the leaf litter.

The kinds of horizons present and the degree of their differentiation, both in composition and structure, depend on the relative strengths of the processes. In turn, these relative strengths are determined by the way man uses the soil as well as by the natural factors of climate, plants and animals, relief and ground water, and the period of time during which the processes have been operating.

Composition. In the drier climates where precipitation is appreciably less than the potential for evaporation and transpiration, horizons of soluble salts, including calcium carbonate and gypsum, are normally found at the average depth of water penetration.

In humid climates some materials normally considered insoluble may be gradually removed from the soil or at least from the surface horizons. A part of the removal may be in suspension. The movement of silicate clay minerals would be an example. The movement of iron oxides is accelerated by the formation of chelates with the soil organic matter. Silica is removed in appreciable amounts in solution or suspension, though quartz sand is relatively unaffected. In warm humid climates free iron and aluminum oxides and silicate clays accumulate in soils, apparently because of low solubility relative to other minerals. *See* CHELATION.

In cool humid climates, solution losses are evident in such minerals as feldspars. Free sesquioxides tend to be removed from the surface horizons and to accumulate in a lower horizon, but mixing by animals and falling trees may counterbalance the downward movement.

Structure. Concurrently with the other processes, distinctive structures are formed in the different horizons. In the surface horizons, where there is a maximum of biotic activity, small animals, roots, and frost action keep mixing the soil material. Aggregates of varying sizes are formed and bound by organic matter, microorganisms, and colloidal material. The aggregates in the immediate surface tend to be loosely packed with many large pores among them. Below this horizon of high biotic activity, the structure is formed chiefly by volume changes due to wetting, drying, freezing, and thawing. Consequently, the sides of any one aggregate, or ped, conform in shape to the sides of adjacent peds.

Water moving through the soil usually follows the root channels and the ped surfaces. Accordingly, materials that are deposited in a horizon commonly coat the peds. In horizons that have received clay from an overlying horizon, the peds usually have a coating or varnish of clay making the exterior unlike the interior in appearance. Peds formed by moisture or temperature changes normally have the shapes of plates, prisms, or blocks.

Soil horizons. Pedologists have developed sets of symbols to identify the various kinds of horizons commonly found in soils. The nomenclature originated in Russia where the letters A, B, and C were applied to the main horizons of the black soils of the steppes. The letter A was used to designate the dark surface horizon of maximum organic matter accumulation. The letter C was used to designate the unaltered parent material, and the letter B was used for the intermediate horizon. The usage of the letters A, B, and C spread to western Europe, where the intermediate or B horizon was a horizon of accumulation of free sesquioxides or silicate clays or both. Thus, the idea developed that a B horizon is a horizon of accumulation. Some, however, define a B horizon by position between A and C. Subdivisions of the major horizons have been shown by either number or letter subscripts, for example, B_t or B_2. No internationally accepted set of horizon symbols has been developed. In the United States the designations shown in Fig. 2 have been widely used since about 1935, but will doubtless be modified in the future. Lower-case letters are being added to numbers in B horizons to indicate the nature of some material that has accumulated. Generally, h is being used to indicate translocated humus, t for translocated clay, and fe or ir for translocated iron oxides. Thus, B_{2t} indicates the main horizon of clay accumulation.

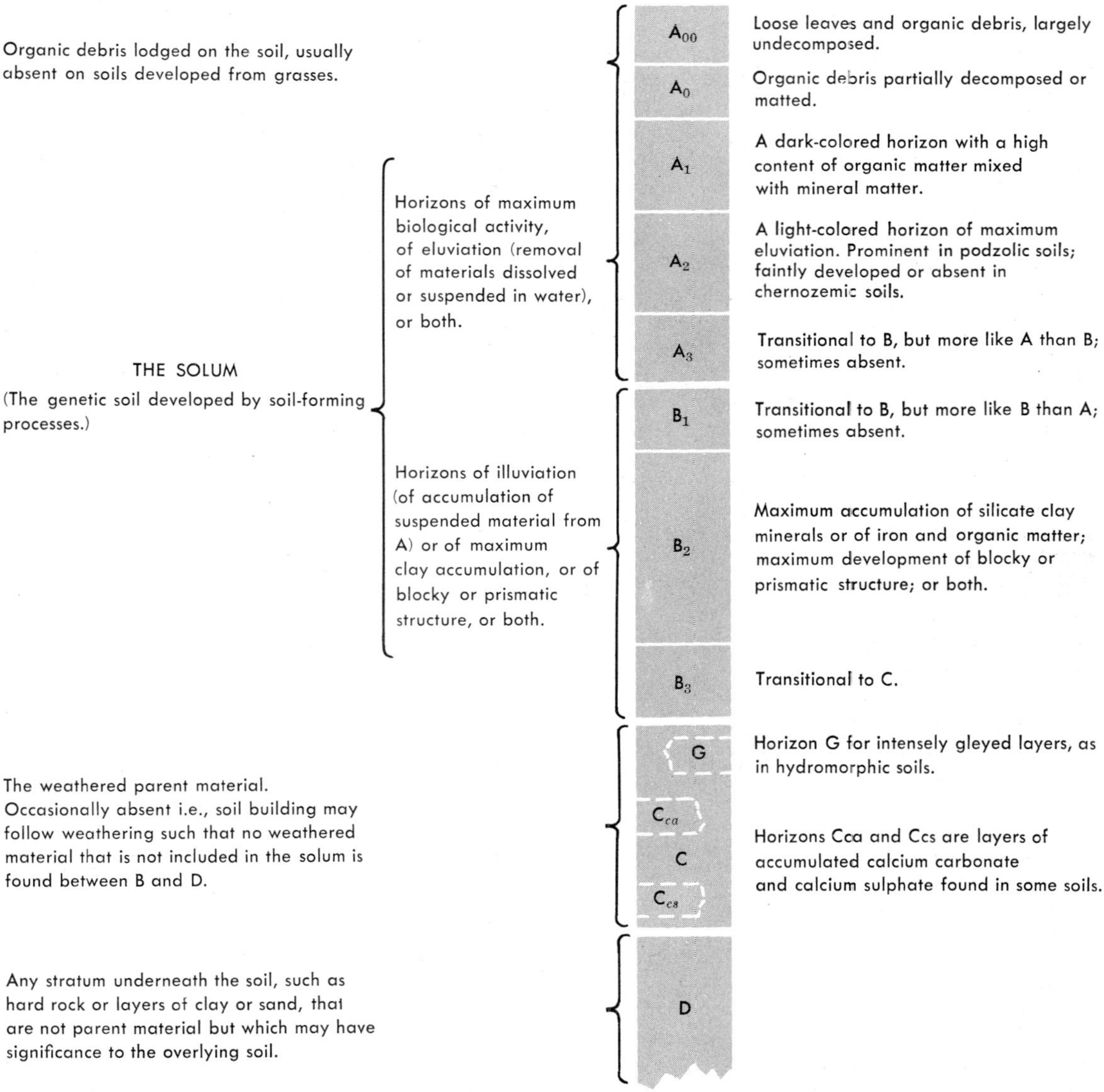

Fig. 2. A hypothetical soil profile showing all the principal horizons. Not all of these horizons are present in any one profile, but every profile has some of them. (*From USDA, Soil Survey Manual, Handbook 18, 1951*)

Soil classification. Systems of soil classification are influenced by concepts prevalent at the time a system is developed. Since ancient times, soil has been considered as the natural medium for plant growth. Under this concept, the earliest classifications were based on relative suitability for different crops, such as rice soils, wheat soils, and vineyard soils.

Early American agriculturists thought of soil chiefly as disintegrated rock, and the first comprehensive American classification was based primarily on the nature of the underlying rock.

In the latter part of the nineteenth century, Russian students noted relations between the steppe and black soils and the forest and gray soils. They developed the concept of soils as independent natural bodies formed by the influence of environmental factors operating on parent materials over time. The early Russian classifications considered soil chiefly in relation to water table, climate, native vegetation, and color. Black, brown, red, and gray colors were first recognized, so such names as gray desert, gray forest, and brown forest were considered appropriate. The vegetative part of the name related the soil to its appropriate climate and native vegetation. These were considered in the United States as great soil groups (zonal soils) and as soil types in Europe. Soil type in the classification system in the United States is comparable to a plant species in the botanic classification. *See* SOIL (GREAT SOIL GROUPS); SOIL, ZONAL DISTRIBUTION.

Many systems of classification have been attempted but none has been found markedly superior; most systems have been modifications of those used in Russia. Two bases for classification have been tried. One basis has been the presumed genesis of the soil. Climate and native vegetation were given major emphasis. The other basis has been the observable or measurable properties of the soil. To a considerable extent, of course, these are used in the genetic system to define the great soil groups. The morphologic systems, however, have not used the genesis as such, but have attempted to use properties that are acquired through soil development.

The principal problem in the morphologic systems has been the selection of the properties to be used. The color, tried in the earliest systems, produces groupings of soils of unlike genesis.

The Soil Survey staff of the U.S. Department of Agriculture and the land-grant colleges recently undertook to revise the classification scheme. Although the new system has been widely tested, only time can tell whether it will be more useful than earlier systems. As knowledge of soil genesis increases, modifications of classification systems will continue to be necessary.

The system being developed in the United States differs from earlier systems in that it may be applied to either cultivated or virgin soils. Previous systems have been based on virgin profiles, and cultivated soils were classified on the presumed characteristics or genesis of the virgin soils. The new system has many categories, based on both physical and chemical properties.

Order. In the highest category 10 orders are recognized. These are distinguished chiefly by differences in kinds and amounts of organic matter in the surface horizons, kinds of B horizons resulting from the dominance of various specific processes, evidences of churning through shrinking and swelling, base saturation (see later discussion of chemical properties), and lengths of periods of dryness during which the soil is without available moisture. The properties selected to distinguish the orders have a strong climatic bias, joining in broad climatic zones the soils that have distinct horizons.

Suborder. This is the next category, which distinguishes soils that show evidences of waterlogging during some seasons from those which do not. In orders having wide climatic ranges suborders are distinguished by characteristics which narrow the ranges; in those orders having narrow climatic ranges suborders are defined largely in terms of chemical composition. Wide differences in the content of quartz, organic matter, allophane (amorphous clays), and free sesquioxides are examples.

Great groups. The great groups constitute the next category. In this category are grouped soils having the same kinds of horizons in the same sequence. Exceptions are made for surface horizons that are apt to be mixed by plowing or lost rapidly by erosion when the soils are cultivated.

The great groups are subdivided into subgroups that show the central properties of the great group, and intergrade subgroups that show properties of more than one great group.

Family. The families are defined largely on the basis of physical and mineralogic properties of importance to plant growth. The two lowest categories, the series and type, are unchanged from earlier systems.

Series. The soil series is a group of soils having horizons similar in differentiating characteristics and arrangement in the soil profile, except for texture of the surface portion, and developed in a particular type of parent material. The series is named after the place where the soil was first recognized and defined; for example, Fayette is the name of one soil series.

Type. The soil types within a series differ primarily in the texture of the surface or plow layer. The type name is formed by adding the textural class of the surface soil to the series name. Thus, Fayette silt loam is a soil type of the Fayette series.

Many practical classifications have been developed on the basis of interpretations of the usefulness of soils for specific purposes. An example is the capability classification, which groups soils according to the number of safe alternative uses, risks of damage, and kinds of problems that are encountered under use.

Soil surveys. Soil surveys include those researches necessary (1) to determine the important characteristics of soils, (2) to classify them into defined types and other classificational units, (3) to establish and map the boundaries between kinds of soil, and (4) to correlate and predict adaptability of soils to various crops, grasses, and trees; behavior and productivity of soils under different management systems; and yields of adapted crops on soils under defined sets of management practices. Although the primary purpose of soil surveys is to aid in agricultural interpretations, many other uses have been made ranging from suburban planning, rural zoning, and highway location, to tax assessment and location of pipe lines and radio transmitters.

Soil surveys were first used in the United States in 1898. Over the years the scale of soil maps has been increased from ½–1 in. to the mile to 3–4 in. to the mile for mapping humid farming regions, and up to 16 in. to the mile for maps in irrigated areas. After the advent of aerial photography, planimetric maps were largely discontinued in favor of aerial photographic mosaics. The United States system has been used, with modifications, in many other countries. *See* AERIAL PHOTOGRAPH.

Two kinds of soil map are made. The common map is a detailed soil map, on which soil boundaries are plotted from observations made throughout the surveyed area. Reconnaissance soil maps are made by plotting soil boundaries from observations made at intervals. The maps show differences in the soils that are of significance for present or foreseeable uses.

The units shown on soil maps usually are phases of soil types. The phase is not a part of the natural classification. It may be a subdivision of any of the natural classification units, according to some feature that is of significance for use and management of the soil, but not in relation to the natural landscape. The presence of loose boulders on the surface of the soil makes little difference in the growth of a forest, but it is highly significant if the soil is to be plowed. Phases are most commonly based on slope, erosion, presence of stone or rock, or differences in the rock material below the soil itself. If a legend identifies a phase of a soil type, the soils so designated on a soil map are presumed to lie within the defined range of that phase in at least 85% of the area involved. Thus, the inclusion of soils having other type characteristics and occupying up to 15% of the area is tolerated in the mapping.

If the pattern of occurrence of two or more series is so complex that it is impossible to show them separately, a soil complex is mapped, and the

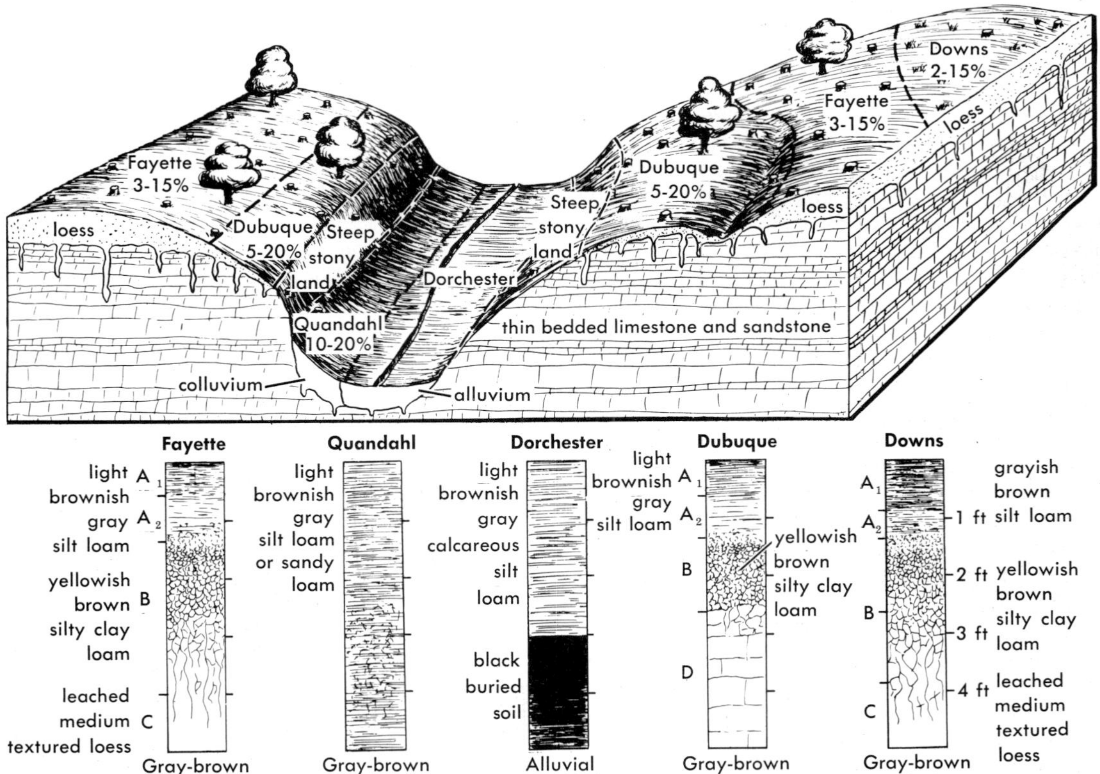

Fig. 3. Sketch showing the relation of the soil pattern to relief, parent material, and native vegetation on a farm in northeastern Iowa. Soil slope gradient expressed as percentage. (*From R. W. Simonson, F. F. Riecken, and G. D. Smith, Understanding Iowa Soils, Brown, 1952*)

legend includes the word "complex." Thus the phrase Fayette-Dubuque complex indicates that the two series occur in the area and that each represents more than 15% of the total of the area.

At times the significance of the difference between series is so slight that the expense of separating them is unwarranted. In such a case the names of the series are connected by a conjunction, and such names occur as Fayette or Downs silt loams. In this kind of mapping unit, the soils may or may not be associated geographically.

It is possible to make accurate soil maps only because the nature of the soil changes with alterations in climatic and biotic factors and in relief and ground waters, acting on parent materials over long periods of time. Boundaries between kinds of soil are made where such changes become apparent. On a given farm the kinds of soil usually form a repeating pattern related to the relief (Fig. 3).

Because concepts of soil have changed over the years, maps made 30 to 50 years ago may use the same soil type names as maps made in recent years, but with different meanings. The older maps must therefore be interpreted with caution.

[G.D.S.]

Bibliography: H. Jenny, *Factors of Soil Formation*, 1941; G. W. Robinson, *Soils*, 3d ed., 1951; USDA, *Soils and Men: The Yearbook of Agriculture*, 1938; USDA, *Soil Survey Manual*, Handbook 18, 1951.

PHYSICAL PROPERTIES OF SOIL

The physical properties of the soil are important in agriculture because of their influence on plant growth and on the management requirements of the land. They influence the plant from seeding to maturity by regulating the supply of air, water, and heat, first to the seed and then to the root system in each soil layer or horizon. The absorption of essential nutrients by plant roots is dependent upon an available supply of oxygen, water, and heat. Thus, physical properties indirectly regulate the nutrition of plants and their response to liming and fertilization. The more favorable the supply of air, water, and heat in each soil layer, the greater is the potential rooting zone for plants.

Physical properties of the soil also determine the kind, amount, and ease of tillage, the runoff and erosion potential, and the choice of crops which can or should be grown on a given soil.

Many people use the word tilth in referring to the physical condition of the soil. Tilth has been defined as the physical condition of the soil in its relation to plant growth. The physical condition of the soil is controlled by, or results from, whatever set of physical properties the soil has at any given time.

Soil physics is that branch of soil science which is concerned with the study of the physical properties of the soil. The physical properties of the soil include texture, particle density, structure, bulk density, porosity, water, air, temperature, consistency, compactibility, and color. The amount of water, air, and heat in the soil at any one time is an important aspect of any soil and the conductivity of these constituents in the soil is equally important. All these properties are interrelated.

The four major components of the soil are inorganic particles, organic matter, water, and air. The proportions of these components vary greatly from place to place in a field, from one layer or horizon to another, and in different parts of the world. The amounts of air, water, and heat in the soil change from day to day and from season to season.

Soil texture. About half the total volume of mineral soils consists of solid matter, of which 80–99% is inorganic and 1–20% is organic material. The inorganic portion consists of rock and mineral particles of many sizes and shapes. They are classified into five major size groups called separates. The two largest separates are stone and gravel. Stone particles are greater than 3 in. and gravel particles are 2 mm to 3 in. along their greatest diameter. Sand has particles between 0.05 and 2.00 mm in diameter. Silt has particles 0.002–0.05 mm in diameter. Clay, the smallest of the soil particles, has a diameter less than 0.002 mm.

After separating the coarser separates by screening, the amount of silt and clay is determined from measurements of the settling velocities of the individual particles, which have been well dispersed in water with the aid of a dispersing agent. The size of the particles is calculated from their settling speed. The stone, gravel, and sand separates in a soil can be seen with the naked eye. Clay can be examined only with an electron microscope.

Textural class. Determination of the particle size distribution in a soil is called mechanical analysis. The texture of a soil is determined by its content of sand, silt and clay. The percentages of sand, silt and clay in the 12 textural classes are shown in Fig. 4. With this triangle the textural class of a soil can be determined from its percentage of

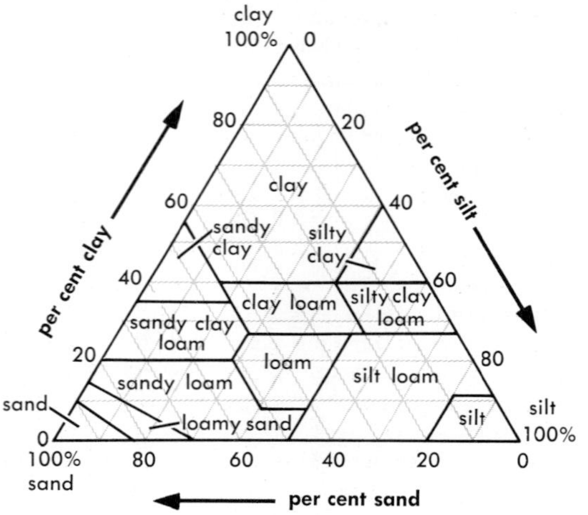

Fig. 4. Triangle showing percentage of sand, silt and clay in each textural class.

sand, silt, and clay. The textural class is combined with the series name of a soil to give the soil type, such as Sassafras sandy loam, Miami silt loam, or Houston clay loam.

Large particles. The stone, gravel, and larger sand particles usually are present as separate particles. They may be rounded, angular or platelike in shape. They are composed of rock fragments and of primary minerals such as quartz. Soils with large amounts of stone, gravel, and sand have low plant-nutrient and water-holding capacities, and permit rapid air, water, and heat movement. Sand imparts a grittiness to the feel of a soil.

Clay particles. The clay fraction controls most of the important properties of a soil. In soils of the cold and temperate regions, clay is composed chiefly of secondary crystalline alumina silicates. These consist of the kaolinite, illite, and montmorillonite groups of clay minerals. Hydrated oxides of iron and aluminum are the main components of the clay in the more highly weathered soils typical of many parts of the tropics. *See* CLAY MINERALS.

Because of their extremely small size and platelike shape, clay particles have a very large specific surface which is responsible for the great adsorptive capacity of clay soils for water, gases, ions, and organic molecules. Clays are well known for their plasticity and stickiness when wet. They also expand or swell with wetting and contract or shrink upon drying. Movement of air and water through clay soils is often very slow owing to the smallness of the pores between the clay particles.

Silt particles. Silt particles exhibit some of the properties of sand and clay. They are usually angular in shape, with quartz being the dominant mineral. Many silt particles have a coating of clay particles. Without this clay coating, silt has a floury or talcum-powder feel when dry and loose. Soils with high silt contents and moderate amounts of clay may have very poor air and moisture relations and may be very difficult to manage. They are often very easily eroded. The loam soils generally have the most desirable texture for crop growth and ease of management.

It is seldom feasible to try to change the texture of a soil. The texture of surface soils may change as a result of removal of the smaller particles by wind and water erosion or by eluviation (movement within the soil).

Organic matter. The organic matter in the soil is made up of the partially decomposed remains of plant and animal tissue, together with the bodies of living soil microorganisms and plant roots. Many good and some bad effects may accompany the decomposition of organic matter. During decomposition of organic matter by the soil microorganisms, gluelike soil aggregate bonding substances are produced. With the knowledge of the great importance of these natural soil conditioning materials, the chemical industry has produced a number of synthetic soil conditioners, Krilium being one of the best known. *See* SOIL MICROBIOLOGY; SOIL MICROORGANISMS.

Much of the organic matter of the soil has colloidal properties. It has two to three times the adsorptive capacity for water, gases, ions, and other colloids as the same amount of clay. Its superior water and nutrient holding capacity makes it an ideal substitute for clay in improving droughty, infertile sandy soils and a substitute for sand in improving tight, sticky, or hard and lumpy clay soils.

Density of particles. The inorganic soil particles may consist of many kinds of minerals with a wide range in particle density. The average particle density for most mineral soils varies between 2.60 and 2.75 g/cm^3. The average density of humus particles ranges between 1.2 and 1.4 g/cm^3. For general calculations the average particle density of soil is taken to be 2.65 g/cm^3. Based on these values, the plowed layer weighs about 2,000,000 lb per acre. The pycnometer is used to determine soil particle density. *See* DENSITY MEASUREMENT.

Soil structure. Soil structure refers to the arrangement of soil particles into aggregates of different sizes and shapes. Pure sands have a single-grain structure (Fig. 5). Because of the adhesive and cohesive properties of clay and organic matter, the inorganic and organic particles combine to form the following types of structure in the A and B horizons of most soils: platy, prismatic, columnar, blocky, nuciform, granular, and crumb (Fig. 6).

These types of structure have been developed from the bonding together of individual particles (accretion) or the breakdown of large massive mixtures of gravel, sand, silt, clay, and organic matter (disintegration). The formation, or genesis, of a given type of structure and the stability of the

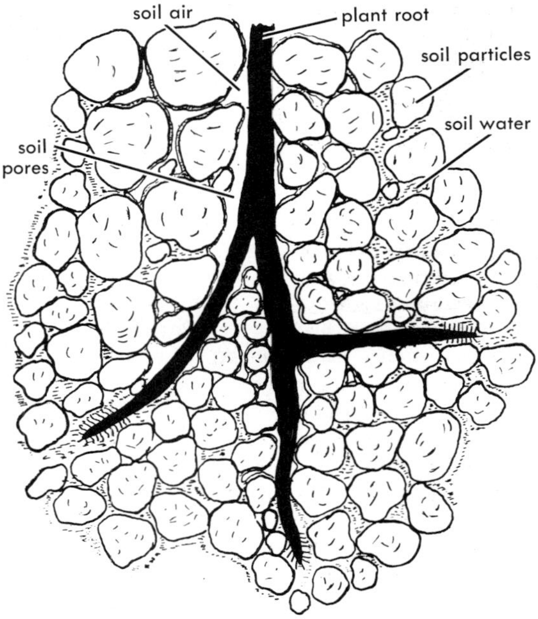

Fig. 5. Portion of surface soil with granular structure.

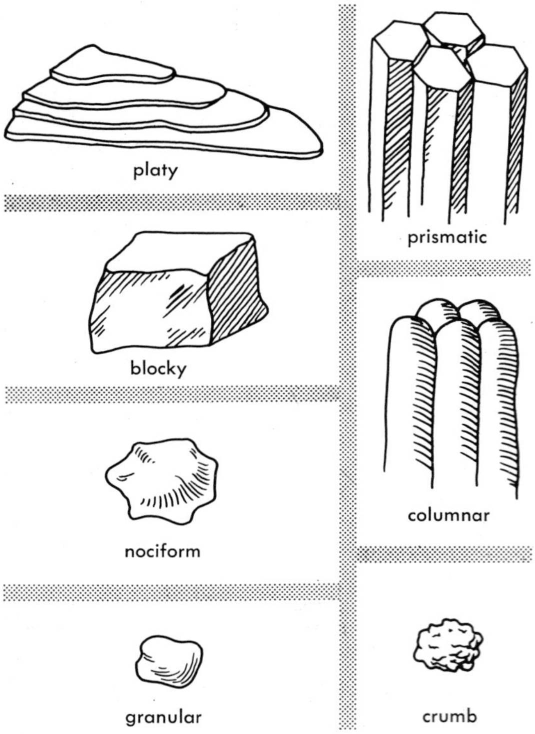

Fig. 6. Types of soil structure.

aggregate produced appear to be associated with (1) the contraction and expansion resulting from hydration and desiccation of the clay-organic matter upon wetting and drying and from freezing and thawing; (2) the physical activity of roots and animals in the soil; (3) the influence of decomposing organic matter and of the slimes and mycelia of the microorganisms that provide bonding substances with which aggregates are held together; and (4) the effect of adsorbed cations, some of which bring about flocculation and others dispersion of the colloidal matter. *See* COLLOID.

The prism, block, and sometimes the platelike types of structure are found mostly in subsoils. Granules and crumbs are found in largest numbers in surface soils. Compacted layers in the soil often have a platy structure.

The size, shape, and arrangement, particularly the amount of overlap of soil aggregates and individual sand, gravel, and stone particles are extremely important because they largely determine the size, shape, arrangement, and continuity of pores in the soil.

There are a number of ways of studying and describing the structure of a soil. The most direct is visual examination of an undisturbed section of soil. Much can be learned about the size, shape, and arrangement of the soil particles and the pore space in a soil by close inspection of each horizon with the naked eye or with a magnifying glass.

A second method is to measure how much of the soil has been aggregated into granules or crumbs with diameters above a given dimension, 0.25 mm being the most common. In well granulated soils, 70–80% of the total mass may be aggregated into granules or crumbs greater than 0.25 mm, as determined by wet-sieving a sample of soil through a 60-mesh screen. Aggregation values of 40–50% are more commonly found in soils under ordinary management. Sandy soils or clay soils having poor structure may have only 10–20% aggregation. Except for very sandy soils, such a low amount of aggregation usually forecasts a physical condition very unsatisfactory for plant growth.

Bulk density. Bulk density is the mass (weight) of a unit volume of dry soil, usually expressed in grams per cubic centimeter. It is determined by the particle density and by the arrangement of the particles.

The soil structure is the major factor in accounting for changes in the bulk density of a soil from time to time or from layer to layer in the profile. Soils with many particles closely packed together have high bulk densities and correspondingly low total pore space. Bulk density is a measure of the amount of compaction in soils. Severe compaction results from excessive traffic by farm machinery in intensively cultivated soils, trampling of cattle in heavily grazed pastures, and foot traffic on lawns and recreational areas. Bulk densities may be as high as of 1.7–2.0 g/cm^3 in such compacted layers. Bulk densities of uncompacted, porous soils are about 1.2–1.3 g/cm^3. In undisturbed forest or grassland soils it may be 0.9–1.0 g/cm^3. High amounts of organic matter will lower the bulk density.

Porosity. The voids or openings between the soil particles are spoken of collectively as the pore space. It makes up roughly one-half the volume of the soil. In very loose, fluffy soils with low bulk density it may occupy 60–65% of the total volume. In very compact soil layers it may be reduced to 35–40%. It is calculated as follows:

$$\%\ \text{Total pore space} = 100 - \frac{(\text{bulk density})}{(\text{particle density})} \times 100$$

Pore space in a soil is occupied by air and water in reciprocally varying amounts. Very dry soils have most of their pore spaces filled with air. The opposite is true for very wet soils. There is considerable variation in the size, shape, and arrangement of pores in the soil.

The effective size of a pore, from the standpoint of moisture retention, can be determined by the amount of force required to withdraw water from the pore. These suction values, expressed in centimeters of water, can be translated into equivalent pore diameters using the capillary rise equation

$$r = \frac{2T}{hdg}$$

where r is radius of pore in centimeters, T is surface tension of water, d is density of water, g is acceleration of gravity, h is suction force minus centimeters of water.

Most soils have an assortment of large and small pores. A sufficient number of large pores connected with each other is needed for rapid intake and distribution of water by and in the soil and for disposal of excess water by drainage into the substratum or into artificial drains. When empty of water they serve as air ducts. Soils with insufficient functional macroporosity lose a great deal of the rain and irrigation water as runoff. They drain slowly and often remain poorly aerated after wetting. One of the first effects of compaction is to reduce the size and number of the larger pore spaces in the soil.

The primary purpose of the small or capillary pores is to hold water. Loose, droughty, coarse sandy soils have too few small pores. Many tight clay soils would be improved if some of their small pores were converted into larger pores.

Soil water. The movement and retention of water in the soil is related to the size, shape, continuity and arrangement of the pores, the moisture content of the pores, and the total surface area of the soil particles. Movement and retention of water may be characterized by the energy relationships or forces which control these two phenomena.

Water retention. Some water is held in the soil pores by the force of adhesion, the attraction of solid surfaces for water molecules. Most of it is held by the force of cohesion, the attraction of water molecules for each other.

Water held by these two forces keeps the smaller pores full of water and also can maintain relatively thick films on the walls of large pores. Not until the pores in one layer of soil are filled with all the water they can hold does water move into the layer below.

Water is found in the soil in both the liquid and vapor state. The air in all the pores is saturated with water vapor when liquid water is present.

The liquid water may be characterized by the suction force or tension with which it is held in the soil by adhesion or cohesion. These suction or tension values may be expressed in any of four units as (1) height in centimeters of a unit water column whose weight just equals the force under consideration; (2) pF, the logarithm of the centimeter height of this column; (3) atmospheres; or (4) pounds per square inch (psi). For example, 1000 cm water tension = pF_3 = 1 atm = 14.7 psi. The moisture content of the soil is determined by drying the soil at 110°C and then dividing the weight of water lost by the weight of oven-dry soil. This value times 100 equals the percentage moisture in the soil on a dry weight basis. The percentage moisture on the volume basis for a given depth of soil is calculated as follows

$$\% \text{ Soil moisture (wet basis)} \times \text{bulk density} \times \text{depth soil (in.)} = \text{in. of water per in. soil depth}$$

Tensiometers and electrical resistance blocks are used to measure the moisture content of the soil in situ. Thus the changes in moisture content can be traced at a given point in the soil within the effective range of each instrument.

Retentive capacity for water. There are several important soil moisture equilibrium points. Water remaining at oven dryness is held at tensions above 10,000 atmospheres. The hygroscopic coefficient is a rough measurement of the water held by adhesion at a tension of about 31 atmospheres. The wilting point or wilting percentage represents that moisture content or moisture tension (15 atmospheres) at which plant roots cannot absorb water rapidly enough to offset losses by transpiration and the plant wilts, first temporarily and then permanently. Certain plants of desert and dry farming regions are able to stay alive and even grow on water held at tensions up to 25–30 atm by the soil. The field capacity of any soil layer represents the maximum amount of water it can hold against the force of gravity when free drainage is provided. It is best measured by obtaining the moisture content of the soil layer in question 24–48 hours after a thorough wetting of the soil. A definite tension value cannot be assigned to this equilibrium point. The maximum retentive capacity is the moisture content of a soil when all of its pores are filled with water.

The moisture in a soil which is available for plant use is that held between field capacity and the wilting percentage. This is called the available soil moisture. Sandy loams hold 1–1½, loams 1½–2, and clay loams 1¾–2 in. of available water per foot of soil. The retentivity of soils of different textures for moisture at different tensions is shown in Fig. 7. These are called soil moisture tension curves. Much of the water in sandy soils is held at low tensions. The opposite is true for clay soils.

Water movement. Water moves in the soil as a gas and as a liquid. Vapor transfer takes place by diffusion in response to a vapor pressure gradient. Vapor movement is through air-filled pores from a moist to a dry and from a warm to a cool layer.

Liquid movement may be expressed by the equation $V = Ki$. V is the volume of water crossing unit area perpendicular to the flow in unit time. The proportionality factor K is the hydraulic conductivity or the permeability of the soil to water. It is controlled by the size, shape, arrangement,

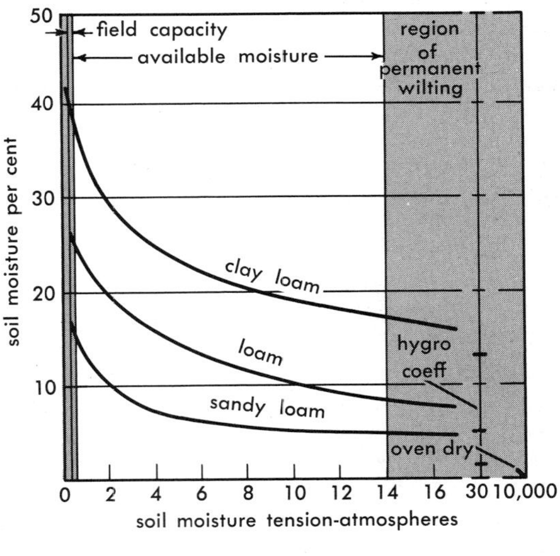

Fig. 7. Soil moisture tension curves.

and moisture content of the soil pores. The value i is the water moving force. It has two force components, the force of gravity and a suction or tension gradient force. The force or pull of gravity is of constant magnitude, and always acts in a downward direction. The suction or tension gradient force may vary both in magnitude and direction. The flow of water in unsaturated soil is equal to the gradient or difference in tension between two points in the soil. It is in the direction of the increase in tension. This is also called capillary adjustment. It accounts for the movement of water toward roots which have depleted the supply of water held at low tension in the soil at the soil root interface. Movement of water to the root system is not usually sufficient to meet the demands of the plant for water. It is believed that plants are able to satisfy their tremendous water needs only by an extension of their root system into fresh supplies of water held at low tension in hitherto untapped or recently refilled soil pores. It is important, therefore, that soil structure be such as to permit the rapid extension of the root system through the whole soil mass.

Soil air and aeration. Soil air differs from the atmosphere above the soil in that it usually contains 5–100 times as much carbon dioxide (0.15–0.65%), slightly less oxygen, and is saturated with water vapor. In deep, poorly drained soil layers or in heavily manured soils the CO_2 content of the soil area may reach 10% and the O_2 content decrease to 1%. In waterlogged soils, methane and hydrogen sulfide may be present in amounts toxic to plant life.

Aeration refers to the movement of gases in and out of each soil layer or horizon. The movement of gases within the soil as well as to and from the atmosphere is by diffusion in a direction determined by its own partial pressure. The rate of diffusion of each gas in and out of the soil depends on differences in concentration of each gas in the soil and the atmosphere, and on the ability of the soil to transmit the gases. Diffusion or aeration is proportional to the volume of air-filled pores in any soil layer.

Soil temperature. The temperature of field soils shows rather definite changes at different depths, at different times during the day and night and at different seasons of the year. These changes are determined by the amount of the radiant energy that reaches the soil surface and by the thermal properties of the soil. Only that part of the heat energy which is absorbed causes changes in soil temperature. Dark-colored soils capture a much higher proportion of the radiant energy than do light-colored soils. The insulating effects of vegetative cover and of surface mulches keep the soil cooler than a bare, fallow soil. The energy absorbed by the soil surface is disposed of by radiation to the atmosphere, heating of the air above the soil by convection, increasing the temperature of the surface soil, or by conduction to the deeper soil layers.

Consistency and compactibility. As the moisture content of a soil changes from air dryness to saturation its consistency varies from a state of hardness or brittleness, to loose, soft or friable, to tough or plastic, to sticky or viscous. The reaction of the soil to physical manipulation, such as tillage, is primarily an expression of the properties of cohesion, adhesion and plasticity. These properties are largely determined by the structure, organic matter content, kind and amount of clay, nature of adsorbed bases, and the moisture content which regulates the thickness of the water films around the soil particles. Tillage should be done only after the soil attains a soft, friable condition, when it breaks apart or can be worked into granules 1–5 mm in diameter. This is a very desirable range of particle size for good seed and root bed conditions.

Each soil has a critical moisture range, often near field capacity, at which pressure by foot or machinery traffic results in maximum compaction. Bulk density, permeability, porosity and penetrometer measurements are used to indicate the degree of compaction as found in traffic pans on the surface soil at the plow sole or in natural hardpans. Soil compaction is a very serious problem because it reduces the permeability of the soil to air and water, which in turn reduces root activity.

Soil color. Soil color is influenced by and indicates the kind of parent material, chemical composition, organic matter content, and drainage, aeration or state of oxidation of a soil. A blotched or mottled yellow, gray, and blue subsoil indicates poor drainage, aeration or oxidation. A clear red, yellow, or brown color indicates good drainage. The color of some soils is inherited from the parent material. Organic matter gives a brown to black color to that horizon where it is concentrated. Color of soils is determined by comparison with standard colors of known hue, value, and chroma in the Munsell soil color charts. [R.B.A.]

Bibliography: L. D. Baver, *Soil Physics*, 3d ed., 1956; F. E. Bear (ed.), *Soil Sci.*, 68(1):1–112, 1949; B. T. Shaw (ed.), *Soil Physical Conditions and Plant Growth*, vol. 2, American Society of Agronomy Monograph, 1952.

CHEMISTRY OF THE SOIL

Elemental composition. The elemental compositions of soils vary over wide ranges, and only a few

Table 1. Average content of less abundant constituents in surface soil

Constituent	Podzol	Gray-brown podzolic	Red-yellow podzolic	Prairie	Latosol
MnO	0.03	0.07	0.51	0.12	0.66
TiO_2	0.61	0.79	1.02	0.71	4.3
CaO	0.95	0.59	0.22	0.88	0.66
MgO	0.49	0.40	0.16	0.68	0.84
K_2O	1.75	1.65	0.88	1.97	0.42
Na_2O	0.53	1.39	0.16	1.03	0.08
P_2O_5	0.12	0.15	0.17	0.22	0.43
SO_3	0.11	0.17	0.07	0.17	0.24

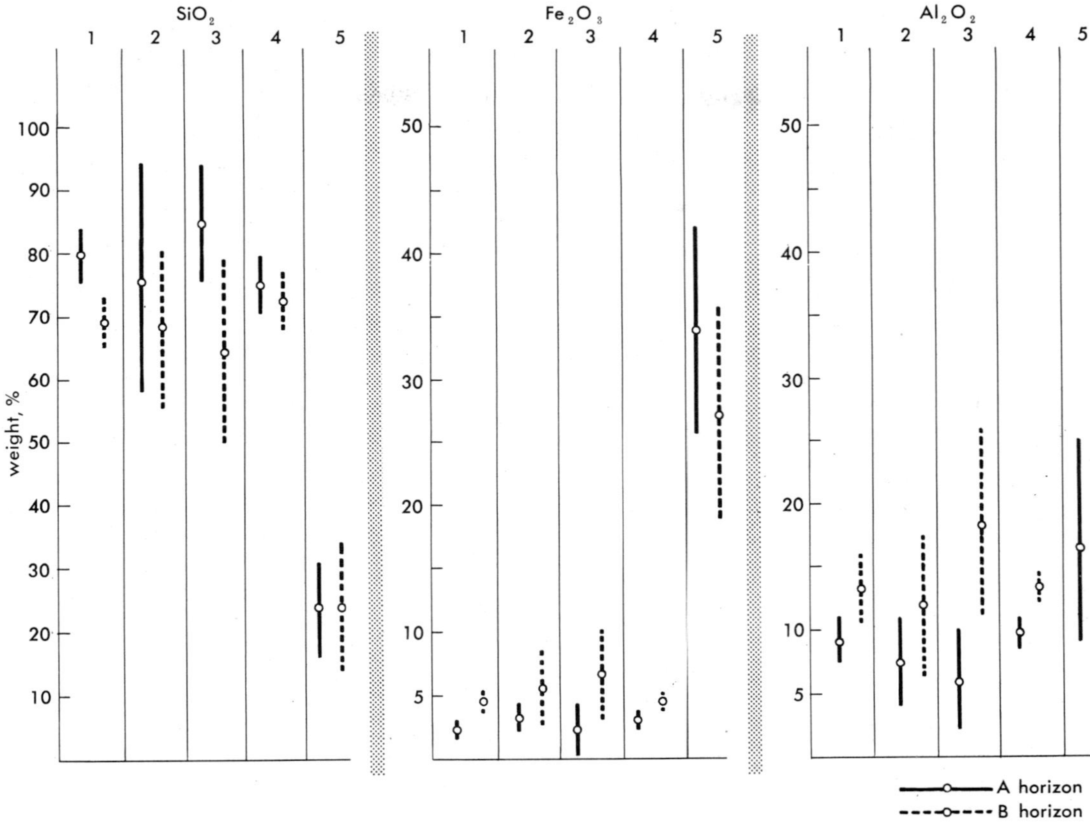

Fig. 8. Mean values and standard deviations of SiO_2, Fe_2O_3, and Al_2O contents of five great soil groups. Key: (1) podzols (means of 7 soils); (2) gray-brown podzols (23 soils); (3) red-yellow podzols (35 soils); (4) prairie (11 soils); (5) latosols (7 soils).

generalizations can be made. Soils containing more than 20% organic matter are termed organic; those with less, mineral. Organic soils such as peats and mucks may contain as much as 95% carbonaceous material. Carbon, oxygen, and hydrogen are the most abundant elements in soil organic matter. Nitrogen, phosphorus, and sulfur are important constituents. Carbon/nitrogen ratios vary between 10 and 20, the lower value being characteristic of upland soils of temperate regions. Soil organic matter contains about one-tenth as much phosphorus or sulfur as it does nitrogen.

At least 95% of the mass of most upland soils is mineral. The oxides of hydrogen, silicon, aluminum, and iron are the most abundant components of mineral soils. Except for calcareous soils and some latosols (lateritic soils), the oxides of silicon, SiO_2, aluminum, Al_2O_3, and iron, Fe_2O_3, make up at least 90% of the elemental composition of the mineral portion. Silicon, aluminum, and iron are present as primary clay and oxide minerals.

The chemical composition in a given soil varies with depth. Because of accumulations of clay and oxide minerals in deeper layers, Al_2O_3 and Fe_2O_3 content usually is higher in B than in A horizons, while that of SiO_2 is lower. Organic matter content decreases rapidly with depth.

Figure 8 shows mean values and standard deviations of the SiO_2, Al_2O_3 and Fe_2O_3 content of the A and B horizons of certain zonal soils. Average surface soil content of compounds present in smaller amounts is shown in Table 1. Cations are present as exchangeable ions or in mineral lattices. Ranges of trace element contents of soils are shown in Fig. 9.

Soil minerals. The minerals in soils are derived from parent rock and soil materials. These materials are mixtures of minerals which are broken down into separate minerals by physical and chemical weathering processes.

Primary minerals in sand and silt. Most of the common rock-forming minerals in soils occur in sand and silt fractions. As a result of weathering, as well as differences in the parent rock, their proportions vary tremendously from soil to soil. Quartz, accumulated at the expense of less resistant pyroxenes, amphiboles, micas, and feldspars, predominates in the sand and silt of soils in humid regions (Table 2). In soils of subhumid and arid regions, contents of nonquartz minerals in sand-size fractions are greater, averaging 20% and 37%, respectively.

Minerals in clay fractions. Clay and sesquioxide minerals make up the bulk of the clay-size fraction

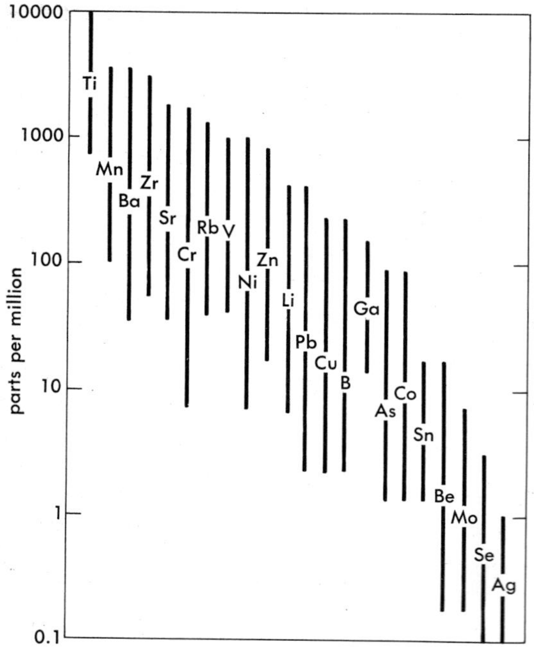

Fig. 9. Range of contents of some trace elements commonly found in mineral soils.

of soils, though others may be present. The proportions of various mineral species vary from soil to soil and within a particular profile. In the following list mineral compositions of clay fractions are given in the order of their increasing resistance to alteration. Minerals near the top of the list generally are found in clays of slightly weathered soils and those near the bottom of the list in soils which have been exposed to drastic weathering.

Mineral	*Soils*
Feldspars	Glacial soils of southern Canada
Muscovite-illite	Subhumid and arid soils; glacial soils of temperate regions
Vermiculite–interstratified clays	Subhumid and arid soils (from weathering of micas); red-yellow podzolic soils
Montmorillonite-beidellite	Volcanic ash, limestone, and basic rock soils; soils weathered from micas in temperate regions
Kaolinite-halloysite	Red-yellow podzolic soils, some latosols
Gibbsite-allophane	Red-yellow podzols, latosols
Hematite-goethite	Red-yellow podzolic soils, 5–30%; latosols, 30–75%
Anatase	Up to 35% in some latosols

Table 2. Percentage of primary minerals in topsoil

Topsoil	Quartz	Feldspars	Micas
Coarse sand	92.7 ± 6.5	3.2 ± 5.5	1.6 ± 2.5
Fine sand	86.4 ± 11.0	4.5 ± 3.6	1.4 ± 2.2
Silt	56.4 ± 15.8	8.0 ± 6.8	8.9 ± 15.0

In zonal soils of humid-cool to subhumid-temperate areas, illite is the predominant mineral. Mixtures of kaolin-vermiculite-interstratified minerals occur in humid-temperate regions. Kaolins with admixtures of smaller amounts of gibbsite and hematite predominate in warm-humid regions. Under tropical conditions, the proportions of sesquioxide minerals are high and in extreme cases only these minerals remain. Arid, basic conditions favor the formation of montmorillonoids. These also occur where soil development is retarded, as by imperfect drainage.

Cation exchange. Many small soil particles, both mineral and organic, possess net negative charges. The charges are balanced by cations which exist in more or less diffuse swarms near the surfaces of the particles. The balancing ions are called exchangeable cations and are in kinetic equilibrium with the soil solution. Their quantity, usually expressed as milliequivalents (meq) per 100 g of dry soil, is the cation exchange capacity (CEC).

Lattice substitution in clays. In clay minerals, as in micas, there is extensive proxying of ions in octahedral and tetrahedral lattice positions, with a resulting unbalance of charge. The net charges of clays are negative. They are large for micas and vermiculite (135–270 meq per 100 g), intermediate for montmorillonoids (80–135 meq per 100 g) and quite small for kaolins (2–10 meq per 100 g). In clay mineral micas and in vermiculite, much of the lattice charge is balanced by cations which occur between fixed lattice layers. Interlayer potassium ions in micas are not displaced under ordinary conditions, and are not counted with the exchangeable cations, though they are bound by the same kinds of forces. The cation exchange capacities of mica clays vary between 20 and 40 meq per 100 g.

The common interlayer ions of vermiculite are calcium and magnesium, though vermiculitelike minerals with interlayer aluminum exist in many soils. Rates of ion exchange are sluggish for such minerals, particularly when ammonium or potassium is one of the exchanging ions, and the cation exchange capacity is not well defined.

Cations balancing the isomorphous substitution charges of montmorillonoids are almost completely exchangeable, and there are excellent correlations between the exchange capacities of these minerals and their chemical compositions.

Isomorphous substitution also appears to contribute to the cation exchange capacity of kaolins, though the charge here is small.

Lattice termination in clays. Lattice terminations can result in further development of negative charge, particularly when the pH is above 6. Protons may ionize from SiOH groups around the edges of clay particles and edge aluminum ions may adsorb OH^- or may shift in coordination number from 6 to 4 as pH is raised. Maximum development of edge charge occurs at around pH 10. *See* SILICATE MINERALS.

The magnitude of the negative charge which can develop on lattice terminals depends on edge area. In meq per 100 g, it is about 20 for montmorillonite, 10 for illite, and 2–10 for kaolins, depending on particle size and crystal form.

Allophanes have cation exchange capacities (at pH 7) of 60–120 meq per 100 g. Probably this results largely from coordination shifts involving aluminum.

Organic matter. The organic matter in soils contains carboxyl, phenol, enol, and imide groups. Ionization of protons from these results in negatively charged particles, the charge increasing with the extent of ionization. The apparent ionization constant for carboxyl, COOH, groups of soil organic matter is about 5. Other acid groups ionize appreciably only when pH is above 7.

The carboxyl content of soil organic matter varies between about 150 and 275 meq per 100 g. Organic matter can bind as much as 400 meq of cations per 100 g at high pH.

Cation exchange capacities of soils. The cation exchange capacities of soils usually are measured by treatment with a salt solution, such as neutral normal ammonium acetate, to achieve saturation with one species of cation. This is an arbitrary procedure, since the quantity of cation adsorbed after such treatment is not a unique value characteristic of the soil alone, but depends as well on the concentration, the ion composition, and the pH of the saturating solution. Cations of a saturating solution displace positive ions which are bonded electrostatically to soil particles and also displace protons which ionize from weakly acidic groups on clays and organic colloids. The proportion of the latter which is replaced depends particularly upon the pH of the saturating solution, becoming larger as pH is raised.

Cation exchange capacities of soils can be generalized by referring to information on mineralogical makeup and organic matter percentages. For given mineral types CEC varies directly with amount of clay and organic matter (Fig. 10); kaolins contribute little, montmorillonoids a great deal, and micas are intermediate.

Exchangeable cations. Exchangeable cations are those balancing the negative charges on soil particles. Experimentally, they are the cations displaced upon leaching with a salt solution. These can be divided into two groups, the exchangeable metal cations (Ca, Mg, K, Na, Al are most abundant) and exchangeable hydrogen. The former are bonded largely through electrostatic forces, the latter almost entirely to weakly acidic spots.

To a first approximation, exchangeable metal cations can be regarded as neutralizing isomorphous substitution charges and those weakly acidic groups which are ionized under the prevailing conditions.

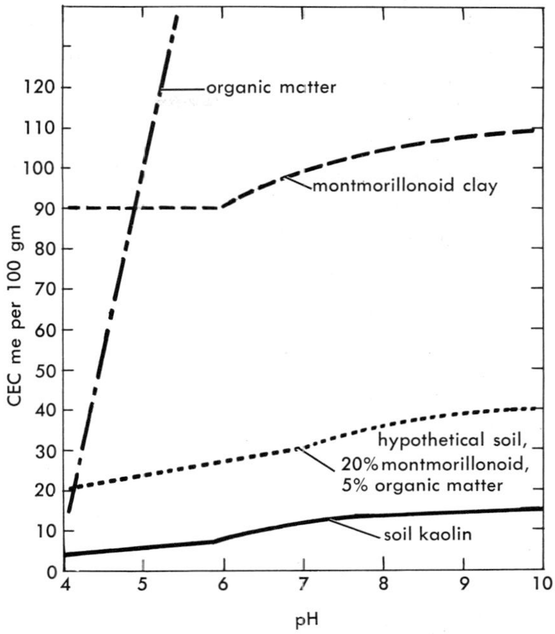

Fig. 10. Idealized cation exchange capacities of clays and organic matter.

Exchangeable hydrogen presents more difficulty. As determined experimentally in most procedures, it is the sum of hydrolyzable metal cations displaced, particularly Al, and of ionization of weakly acidic groups caused by contact with the displacing solution. This is incorrect, and exchange acidity can be resolved into its two major components by appropriate measurements.

Soils which are less weathered because of youth, low rainfall, temperate or cold climate have as exchangeable cations largely calcium and magnesium. Highly weathered soils, unless derived from basic parent material, have large proportions of their permanent charges (20–95% under native conditions) countered by exchangeable aluminum. Some soils of dry areas contain large amounts of exchangeable sodium.

Exchangeable cation populations of some representative soils are given in Table 3. The common exchangeable cations differ in their affinity for ion exchange spots on soil clays and organic colloids.

For permanent charges:

$$Al > Ca \gtreqless Mg > K > Na \gtreqless H$$

For weak-acid charges of clays:

$$H > Ca \gtreqless Mg > K > Na$$

For carboxyl groups of organic matter:

$$H \gg Ca \gtreqless Mg \gg K > Na$$

Ion exchange equilibria can be described fairly accurately by means of mass-action expressions.

Table 3. Exchangeable cations of some representative soils (meq per 100 g)

Soil	CEC_7	Ca	Mg	K	Na	H
Chernozem	55	46	6	1	2	0
Prairie	22	13	7	1.5	1	0
Alkali	21	12	2	0.3	6.7	0
Podzol	17.7	2.0	4.2	0.1	0	11.4
Red-yellow podzolic	8	2.0	1.0	0.1	0	4.9

Anion exchange. Ion exchange refers to the association of ions with solid surfaces in such a way that they can be replaced stoichiometrically by other ions of the same charge. Anion exchange in soils has a different connotation. Anion exchange does occur in many soils, but is often complicated by the occurrence of other reactions which also result in the lowering of solution concentrations of anions.

Clay and oxide minerals in soils can develop positive charges on certain exposed surfaces. This can happen through proton acceptance, hydroxyl ionization, or perhaps other events. The positive charges of soil particles increase as the pH is lowered, and in the case of some latosols, positive charge may exceed negative charge when pH is below 4. All clay minerals, however, as well as temperate region soils, possess net negative charges at every value of pH.

The development of positive charges on soil minerals may result in the exchange adsorption of anions. Halide, nitrate and sulfate ions, as well as phosphate, fluoride and carboxylic acid anions, are sorbed and are mutually replaceable. Capacity for anion sorption by this mechanism varies from near zero at neutrality for all minerals to as much as 1 meq per 100 g for kaolins, and 20 meq per 100 g for kaolin-iron and aluminum oxide combinations in latosols. Montmorillonoids and other clays with large lattice charges do not sorb small anions via this mechanism unless salt contents are very high. Halides and similar anions are negatively adsorbed by such clays under most circumstances.

Certain anions, such as phosphate and fluoride, coordinate strongly with ferric iron and aluminum and are sorbed by soils through another mechanism. Phosphate ions, for example, can react with clay and oxide minerals in soils to form basic iron and aluminum phosphates similar to strengite, variscite and palmerite. Phosphate is displaced from such minerals by hydroxyl, fluoride, or other ions which coordinate strongly with ferric iron and aluminum. Such reactions are viewed as involving solution-precipitation rather than anion exchange.

Soils containing kaolin clays and free iron and aluminum oxides can fix large amounts of phosphate through decomposition-precipitation reactions. It appears that aluminum phosphate complexes are more prevalent in soils than are ferric compounds. *See* ION EXCHANGE.

Acid soils. Soils with pH less than 7 are termed acid. Acid pH usually results from the presence of exchangeable hydrogen and aluminum ions, the former bonding to weakly acidic exchange spots, the latter to permanent lattice charges. Clays with electrostatically bonded exchangeable hydrogen are unstable, and decompose to yield silicic acid and sufficient Al or Mg to counter the exchange spots occupied initially by H ions.

The pH of acid soils varies between 3 and 7, the reaction depending on the ion saturation and the soluble electrolyte content. In the absence of excessive amounts of soluble salts or acids, a pH of less than 4 is rare.

Soil acidity can be discussed in terms of buffer curves relating soil pH to the amount of a basic metal cation such as Ca neutralizing exchange spots. Idealized examples are shown in Fig. 11.

Montmorillonoids, vermiculites, and illites are similar in their neutralization behavior. Permanent lattice charge accounts for the bulk of their exchange capacity and Al for their acidity. Buffer curves for kaolins have two sections, one indicating displacement of exchangeable Al by Ca, the other reflecting the development of weak-acid charge. Below pH 7, organic colloids behave as weak acids with dissociation constants of around 10^{-5}.

Percentage base saturation, reflecting the proportion of soil CEC countered by basic cations, has no general significance, since CEC at an arbitrary pH such as 7 consists of permanent and weak-acid components, their proportions varying from soil to soil. In the absence of organic colloids and with Ca and Mg as exchangeable ions, pH 6 corresponds closely to 100% base saturation of permanent charge.

Soil acidity often is characterized through the measurement of exchangeable hydrogen. More meaningful determinations are for exchangeable Al and for weak-acid charge at a given pH.

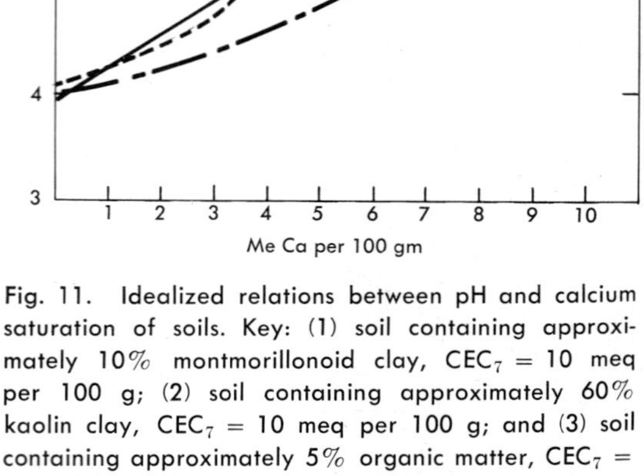

Fig. 11. Idealized relations between pH and calcium saturation of soils. Key: (1) soil containing approximately 10% montmorillonoid clay, CEC_7 = 10 meq per 100 g; (2) soil containing approximately 60% kaolin clay, CEC_7 = 10 meq per 100 g; and (3) soil containing approximately 5% organic matter, CEC_7 = 10 meq per 100 g.

Many soils of humid regions are too acid for optimum plant growth. Application of calcium and magnesium carbonates is a remedial measure.

Calcareous soils. Soils containing accumulations of calcium and magnesium carbonate are referred to as calcareous. Carbonates may occur throughout the profile or may be concentrated in certain horizons, their distributions reflecting the nature of the parent material and the weathering regime. Percentages of carbonates vary from less than 1 to 70.

Calcareous soils are saturated largely with calcium and magnesium. In contrast, with acid soils, pH and concentrations of ions in the soil solution are controlled not by ion exchange equilibria, but by the calcium carbonate-CO_2-H_2O system.

Equilibria in this system can be described by the equation

$$\text{pH} - \tfrac{1}{2}\,\text{pCa} = 4.85 - \tfrac{1}{2}\log \text{pCO}_2$$

pCO_2 is the partial pressure of carbon dioxide in the atmosphere in contact with a $CaCO_3$ water system, while pCa is the negative logarithm of the molar activity of calcium in the soil solution. The manner in which pH and solution calcium concentration vary with CO_2 partial pressure is shown in Fig. 12.

Since soil-air CO_2 content varies from 0.003%, the content in the atmosphere, to nearly 20%, calcium solubility and the reaction of calcareous soils can vary widely.

Salted soils. Soils of dry areas often contain sufficiently large amounts of soluble salts, exchangeable sodium, or both to have peculiar chemical and physical properties. Such soils are referred to as salted.

The U.S. Salinity Laboratory classification of salted soils is based on the electrical conductivity of a saturation extract rather than on salt percentage as such. Soils yielding saturation extracts with specific conductivities greater than 4 millimhos are saline. The concentration of a solution of this conductivity varies with the nature of the salt, averaging 45 meq per liter for the electrolytes commonly found in salted soils. The lower-limiting salt percentages of saline soils vary from around 0.1 for coarse-textured soils to 0.25 for heavier ones.

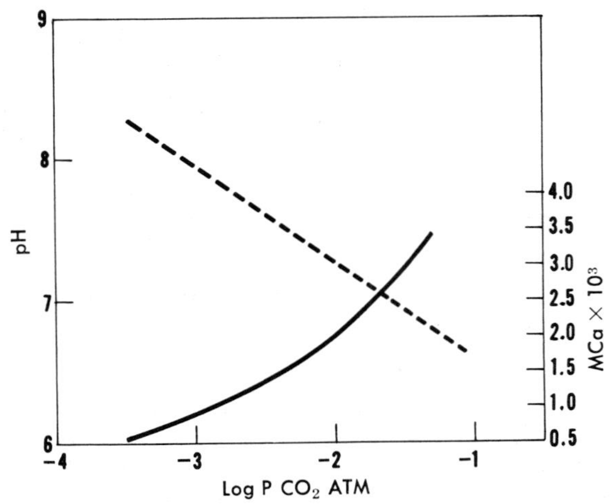

Fig. 12. Calcium solubility and pH of calcareous soil related to CO_2 partial pressure.

Table 4. Four categories of salted soils

	Exchangeable cations saturation extract	Exchangeable sodium percentage	pH	SAR
Saline	>4	<15	7.5–8.5	10–14
Saline-alkali	>4	>15	7.5–8.5	20–70
Nonsaline-alkali	<4	>15	8.5–10	14–40
Degraded alkali	<4	>15	6.5–7.5	

The second classification criterion for salted soils is their percentage sodium saturation. If this is above 15, the soil is alkali.

Salted soils can be placed into four groups (see Table 4). The salts in salted soils are largely chlorides and sulfates of calcium, magnesium, and sodium. The proportions of the cations in solution are controlled by ion exchange equilibria, as follows (ESP is exchangeable sodium percentage; SAR is sodium adsorption ratio):

$$\text{ESP} = -0.0126 + 0.01475\ \text{SAR}$$

$$\text{SAR} = \frac{1}{2}\,\frac{\text{concentration sodium}}{[\text{concentration (calcium + magnesium)}]}$$

$$\text{ESP} = \frac{\text{exchangeable sodium}}{\text{exchangeable (calcium + magnesium)}}$$

Saline-alkali soils are changed into alkali soils when salts are leached away. Similarly, normal or saline soils can be converted to alkali soils through the use of irrigation water containing excessive amounts of sodium.

Reclamation of salted soils involves the removal of excess salts by leaching and of excess sodium through ion exchange with calcium from gypsum or from soil carbonates. The latter are dissolved by the addition of acidifying materials such as sulfuric acid, aluminum sulfate, or elemental sulfur.

[N.T.C.]

Bibliography: F. E. Bear, *Chemistry of the Soil*, 1955; C. A. Black, *Soil-Plant Relations*, 1957; M. L. Jackson and G. D. Sherman, Chemical weathering of minerals in soils, *Advances in Agron.*, 5:219–318, 1953; C. F. Marbut, *Atlas of American Agriculture*, vol. 3, 1935.

SOIL MANAGEMENT

Soil management may be defined as the preparation, manipulation, and treatment of soils for the production of crops, grasses, and trees. Good soil management involves practices which will maintain a high level of production on a sustained basis. Ideally, these practices should provide the crop with an adequate supply of air, water, and nutrients; maintain or improve the fertility of the soil for subsequent crops; and prevent the development of conditions which might be injurious to plants.

Several systems of land-use classification have been developed which help a farmer to know the kinds of soils he has on his farm and their suitability for various types of farming. One of these systems, developed by the U.S. Soil Conservation Service, involves land-use capability ratings.

Land capability survey maps. Land capability survey maps are worked out in conjunction with the soil survey and serve as a guide to the suitability of land for cultivation, grazing, forestry, wildlife, watersheds, or recreation, with primary consideration given to erosion control. There are eight capability classes which describe the characteristics of the land and the difficulty or risk involved in using it for one kind of crop production or another. These eight classes are sometimes distinguished on land capability survey maps by roman numerals as well as by standard colors. Four of the eight classes include land that is suited for regular cultivation with varying degree of erosion control measures and management practices required; three classes of land are not suited for cultivation but require permanent vegetation and impose severe limitations on land use; and one class includes lands suited only for wildlife, recreation, or watershed purposes. For a description of the land capability classes and the management practices recommended for each class, *see* SOIL CONSERVATION.

Cropping system. A cropping system refers to the kind and sequence of crops grown on a given area of soil over a period of time. It may be a regular rotation of different crops, in which the sequence of crops follows a definite order, or it may consist of a single crop grown year after year in the same location. Other cropping systems include different crops but have no definite or planned sequence.

Cropping systems that involve the systematic rotation of different crops generally include hay and pasture crops, small grains, and cultivated row crops. Legumes, such as alfalfa, clover, and vetch, are usually grown alone or mixed with grasses in the hay and pasture sequence in the rotation because they supply nitrogen and contribute to good soil tilth. The beneficial effect of legumes and grasses on tilth may be attributed to the fact that (1) the soil is not tilled while these crops are being grown, and (2) the organic matter returned to the soil by the extensive root systems and in the plowed-under top growth is particularly suited to the development of a stable, porous soil structure.

Small grains function somewhat like legumes and grasses in giving protection against soil erosion, but they add no nitrogen and remove moderate quantities of plant food from the soil. Since small grains do not provide maximum economic return from the high nitrogen residues left in the soil by legumes, and are likely to lodge owing to the stimulation of growth from these residues, they are not planted in the rotation following legumes. Small grains are generally planted either at the end of a rotation following row crops or as a nurse crop for legumes.

Row crops, such as corn, potatoes, cotton and sugar beets, are an excellent choice to follow legumes because they utilize the nitrogen supplied by the legumes and bring good cash returns. Since row crops in early stages of growth provide little protection against erosion and require considerable cultivation which breaks down soil structure, it has not been considered desirable to plant them continuously.

A cropping system that involves the growing of the same crop year after year generally depletes the soil and results in lower crop yields. This is particularly true if the crop is cultivated frequently and returns little crop residue to the soil. Weeds, diseases, and insects also become more of a management problem when the cropping system does not involve a rotation. Thus the farmer who intends to grow one crop year after year becomes completely dependent on disease-resistant varieties of plants, chemical insecticides and fungicides, soil fumigation, and other methods of controlling diseases, insects, and pests. Through the appropriate use of improved varieties, pesticides, and adequate amounts of fertilizers, farmers have succeeded in maintaining a high level of production on land repeatedly planted to the same crop. While such results are causing farmers to take another look at cropping systems that do not involve rotation, they are well aware that more intensive practices and costly supplements are required to maintain production.

Organic matter and tilth. The value of adding organic matter to the soil in the form of animal manures, green manures, and crop residues for producing favorable soil tilth has been known since ancient times. Only recently, however, has scientific research provided information that helps to explain the mechanisms for this effect.

Experiments reveal that during the decomposition of organic matter in the soil, microorganisms synthesize a variety of gumlike substances, at least partly polysaccharide in nature, which when dried with the soil bind the soil particles together into a porous, water-stable structure. While these binding substances are produced in relatively large quantities during stages of rapid organic-matter decomposition, they may in turn be decomposed by other organisms. Thus, to maintain a continuous supply of binding substances, organic matter must be added to the soil frequently.

In addition to the beneficial action of microbially synthesized binding substances, roots and fungal mycelia also contribute to the development of favorable soil tilth by molding the smaller gum-cemented granules into still larger aggregates. The aggregates adhering together form large pores that permit the rapid movement of air and water and form small pores that store water. Both conditions are essential features of good tilth.

Unfortunately not all the effects of organic matter on tilth are desirable. The growth of organisms in a fine-textured soil may interfere with the downward movement of water whenever the soil pores become clogged with microbial bodies. This condition is of particular significance where water is ponded to recharge underground water supplies or to leach excessive amounts of soluble salts from the soil.

Even the characteristic property of organic matter of promoting aggregate formation is not always desirable. Some surface mulches during decomposition induce the formation of a layer of small surface aggregates which are more susceptible to wind erosion than the fine soil particles initially present. In such cases, aggregation intensifies the hazard of severe wind erosion.

In spite of these negative aspects of the effect of organic matter on the physical properties of soil, the incorporation of organic matter with soil is the most suitable and practical way of developing and maintaining good tilth.

Soil conditioners and stabilizers. Soil conditioners and stabilizers include a wide variety of natural and synthetic compounds that, upon incorporation with the soil, improve its physical properties. The term soil amendment also is applied to these compounds but is a more general term since it includes any material, exclusive of fertilizers, that is worked into the soil to make it more productive, regardless of whether it benefits the physical, chemical, or microbiological properties of the soil.

Soluble salts of calcium such as calcium chloride and gypsum, or acid and acid-formers, including sulfur, sulfuric acid, iron sulfate, and aluminum sulfate, have been used as conditioners to improve the physical properties of soils that were made unfavorable by excessive quantities of sodium ions adsorbed on the soil colloids.

The tilth of dense clay soils, which are slow to take water and have a marked tendency to become cloddy, may be improved by the addition of gypsum and by-product lime from sugar-beet processing factories.

Limestone has improved the physical condition of acid soils, apparently by stimulating the activity of microorganisms to synthesize substances that bind soil particles into aggregates.

The discovery that soil microorganisms synthesize substances that improve soil structure stimulated the search for synthetic compounds that would be more effective than the natural products. While a wide variety of compounds have improved soil structure temporarily, three water-soluble, polymeric electrolytes of high molecular weight which are very resistant to microbial decomposition have been developed commercially for use in ameliorating poor soil structure. These are modified hydrolyzed polyacrilonitrile (HPAN), modified vinyl acetate maleic acid (VAMA), and a copolymer of isobutylene and maleic acid (IBMA).

Mixed with the soil in amounts ranging from 0.02 to 0.2% of soil by weight, these compounds are readily adsorbed by moist soils and tend to stabilize or fix the existing structure. They are therefore synthetic binding agents and should be added only to soils that have previously been worked into a desirable physical condition. These materials are not equally effective on all soils, and if improperly used they can be as effective in stabilizing a poor physical condition as in stabilizing a desirable one.

Fertility. Soil fertility may be defined as that quality of a soil which enables it to provide nutrient elements and compounds in adequate amounts and in proper balance for the growth of specified plants, when other growth factors such as light, moisture, temperature, and the physical condition of the soil are favorable.

Soil testing. Even though relatively fertile and of good physical condition, a soil may be lacking in one or more of 16 of its elements presently known to be essential to plant growth, or it may be strongly acidic, alkaline or salty, and thus unsuitable for plant growth. *See* PLANT, MINERALS ESSENTIAL TO. Fortunately, soil tests are available that indicate the existence of possible deficiencies or excesses in the soil. In most instances, these tests involve the use of various reagents for extracting from the soil the total or proportionate amount of the nutrient or compound in question. The amount of material extracted is then compared with values that have been correlated previously with crop response on the same or similar soil. No single test is reliable for all crops on all soils.

Control of pH. The availability of soil nutrients for plants is influenced greatly by the reaction of the soil. Soils may be classified as acid, neutral, or alkaline in reaction. The method commonly used in measuring and expressing degrees of acidity or alkalinity is in terms of pH. The pH value of soil may range from less than 4 to more than 8, the lower the value, the more acid the soil. Under most conditions lime is applied to acid soils to maintain their pH between 6.5 and 7.0. Under special conditions it may be desirable to maintain pH values either higher or lower than those stated. In any case the desirability of applying lime should be determined by the pH of the soil and the requirements of the plants to be grown.

It is occasionally necessary to make soils more acid. Materials commonly used to decrease pH are sulfur, sulfuric acid, iron sulfate, and aluminum sulfate.

Control of salinity. Restricted drainage caused by either slow permeability or high water table is the principal factor in the formation of saline soils. Such soils may be improved by establishing artificial drainage, if a high water table exists, and by subsequent leaching with irrigation water to remove excess soluble salts.

Soils can be leached by applying water to the surface and allowing it to pass downward through the root zone. Leaching is most efficient when it is possible to pond water over the entire surface.

The amount of water required to leach saline soils depends on the initial salinity level of the soil and the final salinity level desired. When water is ponded over the soil about 50% of the salt in the root zone can be removed by leaching with 6 in. of water for each foot of root zone; about 80% can be removed with 1 ft of water per foot of soil to be leached; and 90% can be removed with 2 ft of water per foot of soil to be leached.

Because all irrigation waters contain dissolved salts, nonsaline soils may become saline unless water is applied in addition to that required to replenish losses by plant transpiration and evaporation, to leach out the salt that has accumulated during previous irrigations and through the addition of fertilizer.

Regulating nutrient supplies. The nutrients supplied to crops can be regulated by modifying the availability of nutrients already present in the soil. This can be accomplished by changing soil reaction, turning under green manure crops, including legumes which add nitrogen, and adding fertilizers. *See* FERTILIZER.

By changing soil reaction through the addition of lime, acidulating agents such as sulfur, or residually acid fertilizers such as ammonium sulfate, solubility and availability of compounds of phosphorus, iron, manganese, copper, zinc, boron, and molybdenum can be increased or decreased. Phosphorus compounds are generally more available in the slightly acid to neutral pH range, whereas compounds of iron, manganese, zinc, and copper become more available as the acidity of the soil increases. The activity of microorganisms responsible for the transformation of nitrogen, sulfur and phosphorus compounds into forms available to plants also is influenced by soil reaction. A reaction which is too acid or too alkaline retards the activities of these organisms.

The decay of turned-under green manures and plant residues produces carbon dioxide, rendering soluble the nutrients from soil particles, and the nutrients which were absorbed from the soil during the growth of these crops are also made available. Although the turning under of a green manure affects the availability of nutrients, it does not add to the total nutrient supply unless the green manure is a legume which fixes atmospheric nitrogen.

The system of farming determines to a considerable extent the manner in which fertilizers are used to regulate nutrient supplies. Each system of farming depends upon the crop, soil, climate, kinds and rates of fertilizers applied, and available equipment, and for each system there are many ways of applying fertilizers.

Common methods include broadcasting, banding, deep placement, and foliar applications. Broadcasting fertilizer on the soil is usually less desirable than localized placement of the fertilizer in relation to the seed or plant. Banding fertilizers to the side of the rows in furrow bottoms or beds and drilling fertilizer with the seed give the best response from limited quantities of fertilizer. Deep placement of fertilizers is effective in arid regions where soils dry out to a considerable depth or where deep-rooted crops are grown. Foliar applications of fertilizers, particularly those containing micronutrients, circumvent soil interactions. Such interactions within the soil may render the applied fertilizer unavailable to the crop. Foliar applications also make it possible for the farmer to supply his crops directly with a number of essential plant nutrients at critical stages of growth. *See* PLANT, MINERAL NUTRITION OF; PLANT, WATER RELATIONS OF. [D.G.A.]

Bibliography: C. E. Millar, L. M. Turk, and H. D. Foth, *Fundamentals of Soil Science*, 3d ed., 1958; *Soil*, USDA Yearbook Agr., 1957; L. M. Thompson, *Soils and Soil Fertility*, 2d ed., 1957; E. L. Worthen and S. R. Aldrich, *Farm Soils: Their Fertilization and Management*, 5th ed., 1956.

SOIL EROSION

Soil erosion is that physical process by which soil material is weathered away and carried downgrade by water or moved about by wind. Two categories of erosion are recognized. The first, called geologic erosion, is a natural process that takes place independent of man's activities. This kind of erosion is always active, wearing away the surface features of the earth. The second kind, referred to as accelerated erosion, occurs when man disturbs the surface of the earth or quickens the pace of erosion in any way. It produces conditions that are abnormal and poses a problem for the future food supply of the world. To combat erosion successfully, it is important that man recognize the erosion processes and have a knowledge of the factors which affect erosion.

Types of erosion. Erosion by running water is usually recognized in one of three forms: sheet erosion, rill erosion, and gully erosion.

Sheet erosion. The removal of a thin layer of soil, more or less uniformly, from the entire surface of an area is known as sheet erosion. It usually occurs on plowed fields that have been recently prepared for seeding, but may take place after the crop is seeded. Generally only the finer soil particles are removed. Although the depth of soil lost is not great, the total loss of relatively rich topsoil from an entire field may be serious (Fig. 13). If continued for a period of years, the entire surface layer of soil may be removed. In many parts of the world only the surface layer is suited for cultivation.

Rill erosion. During heavy rains runoff water is concentrated in small streamlets or rivulets. As the volume or velocity of the water increases, it cuts narrow trenches called rills. Erosion of this type can remove large quantities of soil and reduce the soil fertility rapidly (Fig. 14). This type of erosion is particularly detrimental because all traces of the rills are removed after the land is tilled. The losses which occurred are often forgotten and adequate conservation measures are not taken to prevent further loss of soil.

Gully erosion. This type of erosion occurs where the concentrated runoff is sufficiently large to cut

Fig. 13. Sheet erosion showing how soil has been brought from the entire cultivated hillside. A large deposit has collected on the flat area at the bottom of the slope. (*USDA*)

Fig. 14. Rill erosion showing how water has followed the old corn rows. (*USDA*)

deep trenches, or where continued cutting in the same groove deepens the incision. Gullying often develops where there is a water overfall. The stream bed is cut back at the overfall and the gully lengthens headward or upslope. Once started gullying may proceed rapidly, particularly in soils that do not possess much binding material. Gully erosion requires intensive control measures (Fig. 15), such as terracing or the use of diversion ditches, check dams, sod-strip checks, and shrub checks.

Factors affecting erosion. The rate and extent of soil erosion depend upon such interrelated factors as type of soil, steepness of slope, climatic characteristics, and land use.

Type of soil. Soil types vary greatly in physical and chemical composition. The amounts of sand, silt, and clay constituents, colloidal material, and organic matter all have a bearing upon the ease with which particles or aggregates can be detached from the body of the soil. Such detachment is caused chiefly by the beating action of raindrops. The particles are then transported downgrade by moving water. Sandy or gravelly soils often have little colloidal material to bind particles together, and hence these materials are easily detached. However, because of their size, sand particles are more difficult to move than fine particles. For this reason sand particles are moved chiefly by rapidly flowing water on steep slopes or by streams at flood stage. Finer particles, such as silt, clay, and organic matter, can be carried by water moving at a slower rate. On gently sloping fields there is a tendency for more of the fine particles to be carried away, leaving the heavier sand particles behind. However, if rainfall is intense and the volume of runoff great, sand may be moved even on gentle slopes.

Slope. The relation of slope to the amount of erosion on different classes of soil is illustrated in Fig. 16. The amount of total runoff from rainfall increases only slightly with increase in the slope of the land above 1–2%, but the speed of the flowing water, or rate of runoff, may increase greatly. Since the capacity of moving water to transport soil particles increases in geometric ratio to the rate of flow, the amount of erosion increases greatly with increase in the slope of the land (Fig. 17).

Climate. In cold climates the frozen soil is not subject to erosion for several months of the year. However, if such areas receive heavy snow, serious erosion may take place when the snow melts. This

Fig. 15. A large gully in cultivated land. The gully is too large to cross with machinery. It could be repaired by plowing in and seeding to grass. (*USDA*)

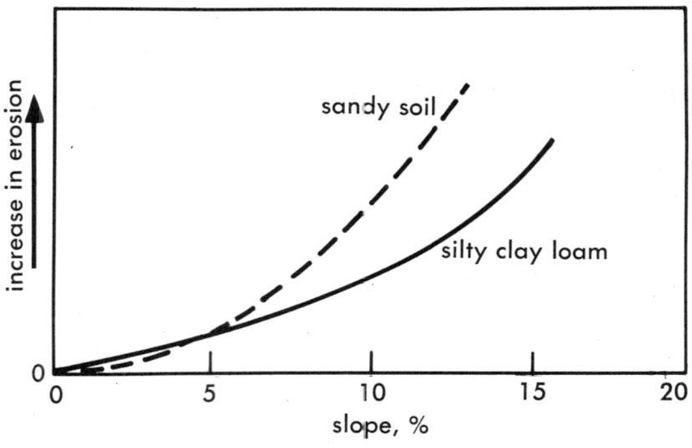

Fig. 16. Generalized diagram illustrating greater loss of fine-grained soil (silty clay loam) on gentle slopes (0–5%) and greater loss of sandy soil on steeper slopes.

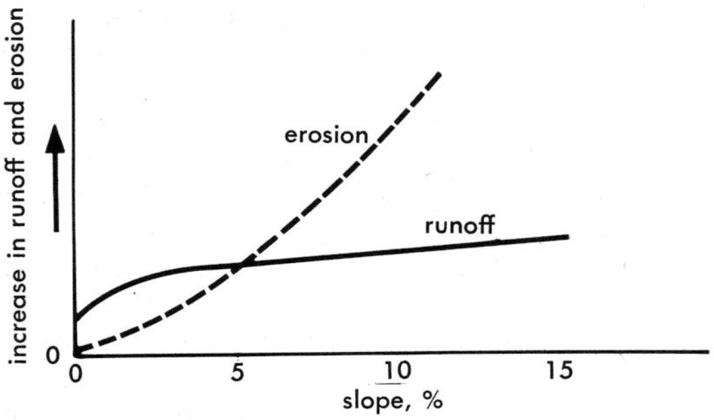

Fig. 17. Effect of slope on total amount of runoff and on rate of runoff and soil erosion.

is particularly true if the snow melts as the ground gradually thaws. As the water moves over the thin unfrozen layer of soil, it transports much of this soil material downgrade.

In warm climates soils are susceptible to erosion any time there are heavy rains. Such soils are particularly vulnerable to erosion if rains fall in winter and there is little vegetative cover.

The amount of rainfall is an important factor in determining the erosion that occurs in a given region. However, the character of the rainfall is usually a much more important factor than the total amount in determining the seriousness of erosion. A rain falling at the rate of 2 in./hour may cause 3–5 times as much erosion as a rainfall of 1 in./hour. Regions where most of the precipitation comes in the form of mist or gentle rain may undergo little erosion, even though the total rainfall may be high and other conditions conducive to erosion.

In some areas of dry climates strong winds cause soil movement and serious loss of soil. Wind erosion is more common on sandy soils, but it is by no means confined to them. Heavier soils, which have a fluffy physical condition produced by freezing and thawing or drying, may be moved in great quantities by the wind.

Land use. The type of crops and the system of management both influence the amount and type of erosion. Bare soils, clean uncultivated soils, or land in intertilled row crops permit the greatest amount of erosion. Crops that give complete ground cover throughout the year, such as grass or forests, are most effective in controlling erosion. Small-grain crops, or those that provide a fairly dense cover for only part of the year, are intermediate in their effect on erosion. Table 5 gives results of some of the earliest experiments in America on differences in land use and the effect on runoff and erosion. These results show that cultivated land, especially without a crop or protective cover of vegetation, is particularly vulnerable to erosion. In addition, excessive erosion usually occurs where cultivated crops like corn, cotton, and tobacco are grown on hilly or sloping land that is subjected to increased rates of runoff. In some areas where row crops have been grown continually the soil has been removed to the depth of the plow layer within a lifetime.

Pastures in humid areas usually have a tough continuous sod that prevents or greatly reduces sheet or surface erosion. Natural range cover, if in good condition, is usually effective in controlling erosion, but in areas of limited rainfall, where bunch grasses form most of the cover on range land, occasional heavy rain may cause severe erosion of the bare soil exposed between the bunches of grass. Forest lands, with their overhead canopies of trees and surface layers of decaying organic matter, have much greater water intake and much less surface runoff and erosion.

Table 5. Relative runoff and erosion from soil under different land uses, with mean rainfall of 35.87 in.*

Land use and treatment	Runoff, %	Tons soil per acre eroded annually	Years required to lose surface 7 in. of soil
Plowed 8 in. deep, no crop; fallowed to keep weeds down	28.4	35.7	28
Plowed 8 in., corn annually	27.4	17.8	56
Plowed 8 in., wheat annually	25.2	6.7	150
Rotation; corn, wheat, red clover	14.1	2.3	437
Bluegrass sod	11.6	0.3	3547

* Missouri Research Bull. 63, 1923.

Wind erosion. In the western half of the United States and in many other parts of the world, great quantities of soil are moved by the wind. This is particularly so in arid and semiarid areas. Sandy soils are more subject to wind erosion than silt loam or clay loam soils. The latter, however, are easily eroded when climatic conditions cause the soil to break into small aggregates, ranging from 0.4 to 0.8 mm in diameter. The coarse particles usually are moved relatively short distances, but the fine dust particles may be carried by strong winds for hundreds or even thousands of miles.

In some areas the coarse, or sand, particles are moved by the wind and deposited over extensive areas as dunes. The dunes move forward in the same direction as the prevailing winds, the particles being moved from the windward side of the dune to the lee side. If dunes become covered with grass or other vegetation, they cease to move. The sandhill region of Nebraska is a good example of such an area. *See* DUNE.

Control of soil erosion. The following are a few fundamental principles which will help control erosion and greatly reduce the damage done by soil erosion.

1. Keep land covered with a growing crop or grass as much of the time as possible. Cover increases intake of water and reduces runoff. The extent of erosion control will be roughly in proportion to the effective cover.

2. When there is no growing crop, retain a cover of stubble or crop residue on the land between

crops and until the next crop is well started. This can be done by using a system known as stubble-mulch farming. It utilizes the idea of preparing a seedbed for a new crop without burying the residue from the previous crop. Tillage tools that work beneath the surface and pulverize the soil without necessarily inverting it or burying the residue are used instead of moldboard plows. This system is best adapted to regions of low rainfall or warm climates.

3. Avoid letting water concentrate and run directly downhill. By doing this the soil is protected against water at its maximum cutting power. Construct terraces with gentle grades to carry the runoff water around the hill at slow speeds. These diversions should empty onto grassed waterways or on meadow land to prevent creation of gullies. *See* TERRACING (AGRICULTURAL).

4. Plant crops and till the soil along the contours.

5. Control wind erosion by keeping land covered with sod or planted crops as much of the time as possible. Maintain crop residue on the land between crops and while the next crop is getting started.

6. If wind erosion begins on a bare field or one where a crop is just getting started, the soil drifting may be stopped temporarily by cultivation. An implement with shovels that will throw up clods or chunks of soil to give a rough surface is usually effective. Often, only strips through the field need be so treated to stop erosion on the whole area.

7. Moving dunes may require artificial cover or mechanical obstructions on the windward side, followed by vegetative plantings, depending on climatic conditions. Along shorelines, beach grasses followed by woody plants and forests may be required.

For a discussion of the physical, economic, and social effects of soil erosion *see* SOIL CONSERVATION. [F.L.D.]

Soil (great soil groups)

A widely used category in the system of soil classification in the United States in the past has been the great soil group. Each great soil group consists of a large number of soils with several internal features in common. All soils of any one great soil group have the same numbers and kinds of definitive horizons in their profiles, although the horizons are not necessarily of the same thickness or expressed to the same degree. For all members of one great soil group, however, certain specific horizons must be recognizable.

In the United States the thousands of soil series, each of which may consist of more than one soil type, are classified into some 40 great soil groups. All of these groups occur in other parts of the world as well, along with some additional ones, perhaps 20 more in all. In parts of the world where the soils have been studied little, it is likely that some great soil groups are as yet unrecognized. For example, available information on soils in tropical regions is generally limited and, for some parts of the tropics, is completely lacking. Within the United States, several new groups have been defined in recent years.

Great soil groups are classified into broader classes called suborders and orders in two higher categories. All great soil groups (all soils) are included in three orders, known as zonal, intrazonal, and azonal. Zonal soils have profiles with evident to prominent horizons, which largely reflect factors such as climate and living organisms that operate more or less uniformly over a broad geographic region. Intrazonal soils also have profiles with evident to prominent horizons, but the latter reflect the dominating influence of some local factor of topography, parent material, or time. Azonal soils lack evident horizons in their profiles, largely because of youth, highly resistant parent materials,

Classification of soils

Order and suborder	Great soil group*
Zonal soils	
Soils of cold zones	Tundra soils Subarctic Brown Forest soils
Light-colored soils of arid regions	Desert soils Red Desert soils Sierozems Brown soils Reddish Brown soils
Dark-colored soils of semiarid, subhumid and humid grasslands	Chestnut soils Reddish Chestnut soils Chernozems Brunizems Reddish Prairie soils
Soils of the forest-grassland transition	Noncalcic Brown soils
Light-colored podzolized soils of timbered regions	Podzols Brown Podzolic soils Gray Wooded soils Sols Bruns Acides Gray-Brown Podzolic soils
Lateritic soils of forested warm-temperate and tropical regions	Red-Yellow Podzolic soils Reddish-Brown Lateritic soils Yellowish-Brown Lateritic soils Groups of Latosols
Intrazonal soils	
Halomorphic (saline and alkali) soils of imperfectly drained places	Solonchak (saline) soils Solonetz (alkali) soils Soloth soils
Hydromorphic soils of marshes, swamps, seep areas, and flats	Humic-Gley soils Alpine Meadow soils Bog soils Low-Humic Gley soils Planosols Ground-Water Podzols Ground-Water Laterite soils
Calcimorphic soils	Brown Forest soils Rendzinas Grumusols Calcisols
Dark soils on volcanic ash	Ando soils
Azonal soils	
(No suborders)	Lithosols Regosols Alluvial soils

* Each great soil group consists of several or many soil series.

or steep topography. Characteristics of azonal soils are mainly those of the parent materials.

Within each of the zonal and intrazonal orders are several suborders, defined largely on the basis of climate and natural vegetation. No suborders have been set apart in the azonal order.

Great soil groups, suborders, and orders, as they were defined in 1959, are listed in the preceding table. Of these three categories, that of the suborder has been used very little. Categories most used in the soil classification system in the United States have been the soil series, soil type, great soil group, and order.

Brief descriptions for major great soil groups are given in the following subsections in alphabetical sequence. Within these group descriptions mention may also be made of other great soil groups that are either less important or less well known. A revision of the basic or pedological system of soil classification in the United States is now being completed, which will effect important changes in the categories above the level of the soil series. The category of the great soil group, as used in the past, will most nearly correspond to the category of the suborder in the revised classification. A number of great soil groups will remain essentially intact in some category of the revised classification system, whereas others will no longer appear as separate entities. For a discussion of the revised system of soil classification, *see* SOIL.

Alluvial soils. An azonal group of soils, formed from geologically recent alluvium, Alluvial soils lack evident horizons in their profiles even though the surface layer may have gained some organic matter. The alluvium is stratified in most places. The soils range from wet to extremely dry and occur in all except very cold climates. Many of the soils are highly fertile, but some are not. A large proportion of the people in the world are dependent upon food production from Alluvial soils.

Ando soils. An intrazonal group of soils formed from volcanic ash, the Ando soils have thick, dark A horizons, high levels of acidity, and poorly crystalline clay minerals. These soils are widely distributed, occurring over a wide range of temperatures in humid climates, but apparently are restricted to regions of rather recent volcanic activity. They have been formed under both grass and forest vegetation. The name was coined from Japanese words meaning dark soils (Fig. 1).

Fig. 1. Ando soil profile showing thick, dark A horizon and general penetration of fine plant roots; numbers on scale indicate feet. (*Photo by R. W. Simonson, Soil Conservation Service, USDA*)

Bog soils. An intrazonal group, Bog soils consist of brown, dark-brown, or black peat or muck. These deposits consist of partly decayed remains of plants that have been preserved in wet places where they remain saturated with water. Some peats and mucks can be made very productive if they are drained, but many cannot. All Bog soils tend to shrink and to be oxidized rapidly if they are overdrained.

Brown soils. A zonal group of soils having brown A horizons of moderate thickness grading into lighter-colored B or C horizons, Brown soils commonly have an accumulation of calcium carbonate at depths of 1–3 ft. These soils were formed under short grasses, bunch grasses, and shrubs in temperate to cool, semiarid climates. Much used for wheat growing in North America, the soils have less dependable rainfall than do the darker-colored Chernozems and Chestnut soils. The related group of Reddish Brown soils differs from Brown soils in color, as indicated by the name, and occurs under warmer climates (Fig. 2).

Brunizems. A zonal group of soils, these soils have acid, thick, very dark brown to black A horizons rich in organic matter, brown B horizons that may or may not be mottled, and lighter-colored parent materials at depths of 2–5 ft. These soils were formed under tall grass vegetation in temperate, relatively humid climates. They are distinguished from Chernozems, which may be very similar in appearance, by the absence of a layer of accumulated calcium carbonate in the deeper profile. Brunizems are fertile and highly productive, comprising major soils in the Corn Belt of the

Fig. 2. Brown soil profile with A and B horizons of equal thickness in the top foot and highly calcareous C horizon; numbers on scale indicate feet. (*Photo by R. W. Simonson, Soil Conservation Service, USDA*)

United States. Called Prairie soils in most published literature, this group by no means extends over all of the grassland commonly known as prairie (Fig. 3).

Reddish Prairie soils are a closely related group, differing mainly in color and commonly occurring in warmer climates.

Calcisols. An intrazonal group, Calcisols have A horizons that are variable in thickness and color, prominent deeper horizons of calcium carbonate accumulation, and calcareous parent materials. These soils were formed from parent materials that were high to very high in carbonates. Recently proposed as a great soil group, Calcisols are associated with the zonal groups of Desert soils, Red Desert soils, Brown soils, Reddish Brown soils, Chestnut soils, Reddish Chestnut soils, and Chernozems. Surface horizons of the Calcisols tend to be like those of the associated zonal soils. The group has a wide geographic range in arid, semiarid, and subhumid climates.

Chernozems. These comprise a zonal group of soils having thick, black A horizons rich in organic matter, brown transitional B horizons which may also be higher in clay, light-colored C horizons, and carbonate accumulations at depths of 1½–4 ft. These soils were formed under tall grasses or a mixture of tall and short grasses in temperate to cool, subhumid climates. Chernozems are extremely fertile, but production from them is variable because of fluctuations in weather. These soils are prominent in the great grain-growing sections of North and South America and in the steppes of the Soviet Union and eastern Europe (Fig. 4).

Closely related great soil groups are the Chestnut soils and Reddish Chestnut soils. The first group has less dark A horizons, profiles that are not so deep, and carbonate accumulations nearer the surface. Chestnut soils are formed in regions between those of Chernozems and Brown soils, where the climate is intermediate in character. Reddish Chestnut soils differ in color, as their name indicates, and are formed in warm-temperate to warm, semiarid climates.

Desert soils. These comprise a zonal group of soils having light-colored A horizons low in organic matter, transitional B horizons that may be similar or darker in color and higher in clay, light-colored C horizons, and carbonate accumulations at depths of 1–3 ft. Desert soils have been formed under

Fig. 3. Brunizem profile showing thick, dark A horizon and filled former animal burrows in deeper horizons; numbers on scale indicate feet. (*Photo by R. W. Simonson, Soil Conservation Service, USDA*)

Fig. 4. Chernozem profile with thick, dark A horizon, dark B horizon of equal thickness, and C horizon marked with white spots of calcium carbonate; numbers on scale indicate feet. (*Photo by R. W. Simonson, Soil Conservation Service, USDA*)

scanty shrub vegetation in cool to temperate, arid climates. Profiles are generally neutral to mildly alkaline in reaction.

Closely related great soil groups are the Red Desert soils and Sierozems. The former have redder colors than the Desert soils and occur in warmer climates. Sierozems are similar in general appearance of the profile but have slightly thicker horizons with carbonate accumulations at greater depths and occur in slightly less arid climates. In many ways, Sierozems are intermediate between Desert soils and Brown soils (Fig. 5).

Gray-Brown Podzolic soils. A zonal group, these soils have thin surface layers, partly of leaf litter and partly of mild humus, thin dark A_1 horizons, lighter-colored leached A_2 horizons, brown B horizons that may or may not be mottled with more clay and a blocklike structure, and light-colored parent materials at depths of 2–5 ft. The entire profile is acid in reaction but less so than those of Podzols and Red-Yellow Podzolic soils. Gray-Brown Podzolic soils were formed under deciduous forest vegetation, for the most part, and under temperate, humid climates. Important in the northern half of the eastern United States and in western Europe, these soils are suitable for many crops and are responsive to good management, but they require fertilizers and other amendments for high yields.

Related great soil groups not described separately are the Gray Wooded soils, Sols Bruns Acides, and Brown Forest soils. The Gray Wooded soils may be thought of as analogs of the Gray-Brown Podzolic group, but are found in cooler regions where rainfall is also slightly lower. Gray Wooded soils have lighter colored A_2 horizons and more drab B horizons, as a rule. They also commonly have carbonate accumulations in the deeper profiles, which are absent from Gray-Brown Podzolic soils.

Sols Bruns Acides (brown acid soils) are more acid and have much less distinct B horizons than Gray-Brown Podzolic soils. Many occur in regions bordering Podzols and some are associated with the latter group. The Sol Brun Acide group has been recognized in the United States only within the last few years.

Fig. 5. Red Desert soil profile with pale silty A horizon, darker B horizon higher in clay, and calcareous C horizon; numbers on scale indicate inches; profile shown is 20 in. deep. (*Photo by R. W. Simonson, Soil Conservation Service, USDA*)

Fig. 6. Gray-Brown Podzolic soil profile with A horizon 12 in. thick over darker B horizon with blocklike structure; larger numbers on scale indicate feet. (*Photo by R. W. Simonson, Soil Conservation Service, USDA*)

Brown Forest soils are an intrazonal group having dark brown A horizons high in organic matter over calcareous, lighter-colored C horizons. These soils were formed under deciduous forest from parent materials rich in lime. Commonly slightly acid to mildly alkaline in reaction in the A horizon, the soils are calcareous or nearly so in the deeper profile and are high in exchangeable calcium. They are most commonly associated with Gray-Brown Podzolic soils (Fig. 6).

Grumusols. An intrazonal group, the Grumusols have profiles which are rather high in clay, relatively uniform in texture, and marked by signs of local movement due to shrinking and swelling as the soils wet and dry. Many soils in this group have thick dark A horizons over calcareous C horizons, but others are uniform in general appearance except for signs of churning. Grumusols were formed from parent materials that were high in clay and alkaline-earth elements or from rocks that provide abundant clay and alkaline earths upon weathering. Occurring chiefly in tropical and subtropical regions with alternating wet and dry seasons, the soils extend into cool regions as well. Soil movement due to shrinking and swelling of the mass as it wets and dries tends to produce a microrelief of small knolls or ridges and basins, often called gilgai. Among names used for this group are Regur in India, Black Cotton soils in parts of Africa, Black Earths in Australia, and, formerly, Rendzinas in the United States.

The name Rendzina was formerly used in the United States for soils now set apart as Grumusols and also for gray to black, shallow to very shallow soils formed from limestone. The name is now restricted to these latter soils which lack the capacity to shrink and swell with changing moisture contents.

Humic-Gley soils. An intrazonal group of poorly or very poorly drained soils having thick, black A horizons high in organic matter over gray or mottled B or C horizons, these soils were formed under

Fig. 7. Humic-Gley soil profile showing very thick dark A horizon and signs of mixing and movement by burrowing animals in deeper horizons; numbers on scale indicate feet. (*Photo by R. W. Simonson, Soil Conservation Service, USDA*)

herbaceous marsh or swamp forest vegetation in subhumid, cool to warm-temperate climates. Most of the soils are medium acid to mildly alkaline in reaction and rather fertile, but some are strongly acid and low in plant nutrients. Many of the soils are highly productive if drained and managed well, as indicated by their importance in the Corn Belt of the United States (Fig. 7).

Closely related is the group of Low Humic-Gley soils, largely restricted to humid cool to tropical climates. These soils have thin A horizons rather than thick ones but have very similar deeper profiles. On the whole, they are less desirable but some are highly productive under good management.

Wet soils having a mucky or peaty surface layer of appreciable thickness, formed under swamp forest, which would not qualify as Bog soils, were formerly placed in an intrazonal group known as Half-Bog soils. Under cultivation, these become indistinguishable from Humic-Gley soils.

An intrazonal group similar in profile appearance to Humic-Gley soils is that of Alpine Meadow soils, found in rather wet meadows at high altitudes near or above timber line. Such soils also have some affinities with those of the Tundra group, in that the deeper profile is cold or frozen much of the time.

Latosols. A broad zonal group of soils, Latosols are strongly weathered, high in sesquioxides, relatively low in silicates, porous throughout the profile, and marked by indistinct horizons. The soils have evident A horizons as a result of gains in organic matter, below which further distinctions are obscure. Latosols have been formed under forest vegetation in humid to subhumid, tropical climates. The soils are generally of low to very low fertility but are widely used for both subsistence and commercial crops. Yields are generally low under simple management but can be high on a number of Latosols under complex management, including fertilization (Fig. 8).

Fig. 8. Latosol profile showing lack of distinct horizons and the deep penetration by fine plant roots; numbers on scale indicate feet. (*Photo by R. W. Simonson, Soil Conservation Service, USDA*)

The term Latosol covers a broader range of soils than is normally included within a single great soil group. It may be that the broad group of soils now included in this class is more nearly a suborder than a great soil group. It seems clear that several great soil groups will need to be recognized for the soils now called Latosols, and several such groups have been proposed. As now used, Latosols include soils that some years ago were called Laterites, Reddish Brown Lateritic soils, and Yellowish Brown Lateritic soils. These last two groups are considered in a subsequent description of Red-Yellow Podzolic soils.

A related intrazonal group of Ground-Water Laterite soils occurs in the same general regions as the Latosols but was formed under conditions of impeded drainage. The profile commonly includes concretionary layers or completely cemented layers of laterite. Such soils are presently of little value. *See* LATERITE.

Lithosols. An azonal group of soils lacking evident genetically related horizons, Lithosols were formed in materials that are shallow or very shallow to bedrock and commonly stony. Lithosols commonly occur in hilly or mountainous regions, occupying steep slopes.

Noncalcic Brown soils. A zonal group, these soils have light to medium-colored A horizons low in organic matter, yellowish brown to red B horizons high in clay, and lighter-colored underlying parent materials. The soils are generally slightly acid to neutral in reaction, being less acid on the whole than Gray-Brown Podzolic soils. Characteristically, the A horizons become very hard or extremely hard when dry, although they are friable or very friable when moist. Some of the soils have hardpans in the deeper profile. These soils were formed under thin stands of deciduous trees mixed with brush and grass in temperate to warm-temperate, wet-dry, subhumid to semiarid climates. The climatic regime is similar to that of the Mediterranean region, and the Noncalcic Brown soils seem to be closely related or identical to a group known as Red and Yellow Mediterranean soils. The soils are of moderate fertility, on the whole, but production is variable because of uncertain rainfall. Under irrigation, many of the soils are highly productive.

Planosols. An intrazonal group, these soils have one or more horizons abruptly separated from and sharply contrasting to an adjacent horizon because of high clay content, cementation, or compactness. Some Planosols have B horizons very high in clay beneath A horizons that are low in clay, the two being separated by an abrupt boundary. Other Planosols have a fragipan, a compact, or brittle, seemingly cemented horizon, below a B horizon with some clay accumulation. Planosols have been formed under both grass and forest in temperate to subtropical, subhumid to humid climates. Because some horizon in the profile is slowly permeable to air, water, and roots, the soils are difficult to manage successfully for crop production and are better suited for shallow-rooted than for deep-rooted plants (Fig. 9).

Podzols. These comprise a zonal group of soils having a surface mat of leaf litter and acid humus;

Fig. 9. Planosol profile with pale A horizon about 18 in. thick resting on B horizon high in clay, a claypan; numbers on scale indicate feet. (*Photo by R. W. Simonson, Soil Conservation Service, USDA*)

Fig. 10. Podzol profile showing marked local variations in thickness of A and B horizons; scale marked to show 6-in. intervals. (*Photo by W. M. Johnson, Soil Conservation Service, USDA*)

a leached, acid, and light-colored A_2 horizon; a brown to black B horizon in which either organic matter or sesquioxides or both have accumulated; and a lighter-colored C horizon. Individual horizons are thin (a few inches) in many Podzol profiles, but they are thick in others. In some Podzols, the B horizon is cemented into a hard layer known as ortstein. Podzols have been formed for the most part under coniferous forest or heath vegetation, although some were formed under mixed forests of various kinds. Most Podzols are in regions of cool, humid climates; a few may be found in warm-temperate and tropical climates. These soils are all of low fertility but can be improved and made fairly productive with good management (Fig. 10).

Brown Podzolic soils are a closely related zonal group, like Podzols in many ways, but they lack a distinct A_2 horizon and generally have less distinct horizons.

A related intrazonal group is that of Ground-Water Podzols, formed under conditions of impeded drainage. These resemble Podzols in the sequence of horizons, but the B horizons are generally higher in organic matter and lower in iron oxides. Ground-Water Podzols have been formed mainly from sandy sediments under forest vegetation in humid, cool to tropical climates. Much improvement is needed for the successful growth of crop and pasture plants on such soils.

Red-Yellow Podzolic soils. These are a zonal group of soils having thin surface layers of litter and acid humus; thin organic-mineral A_1 horizons; thicker light-colored and leached A_2 horizons; thick red, yellowish red, or yellowish brown B horizons with some accumulation of clay and sesquioxides; and relatively siliceous C horizons. Coarse reticulate patterns of streaks are common in the deeper C horizons. Parent materials commonly contain appreciable amounts of quartz or its equivalent in the silt and sand sizes. Red-Yellow Podzolic soils have been formed under deciduous, coniferous, or mixed forests in warm-temperate, humid climates. They also extend into tropical regions where they were formed under broadleaf-evergreen, coniferous, or mixed forests. The soils are low in organic matter and plant nutrients but respond to good management and are used for a wide variety of crops. Large acreages are also in forest (Fig. 11).

Reddish Brown Lateritic soils and Yellowish Brown Lateritic soils comprise closely related groups having the same geographic distribution as Red-Yellow Podzolic soils but formed from less acid and less siliceous parent materials. Consequently, they lack the light-colored and leached

Fig. 11. Red-Yellow Podzolic soil profile with pale A horizon 14 in. thick over a darker B horizon higher in clay and iron oxides and of about equal thickness; numbers on scale indicate feet. (*Photo by R. W. Simonson, Soil Conservation Service, USDA*)

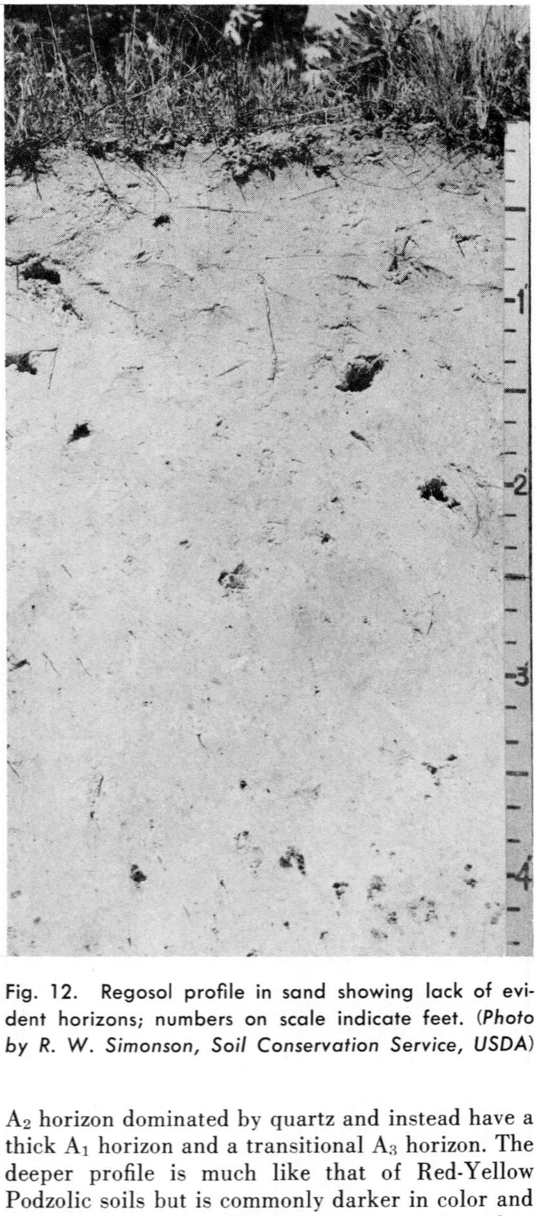

Fig. 12. Regosol profile in sand showing lack of evident horizons; numbers on scale indicate feet. (*Photo by R. W. Simonson, Soil Conservation Service, USDA*)

A_2 horizon dominated by quartz and instead have a thick A_1 horizon and a transitional A_3 horizon. The deeper profile is much like that of Red-Yellow Podzolic soils but is commonly darker in color and often higher in clay. These soils are more productive than Red-Yellow Podzolic soils for some crops and for pasture but are less desirable for other crops.

Regosols. An azonal group of soils, the Regosols lack evident genetically related horizons and were formed from deep unconsolidated regoliths, such as loess, sands, or glacial drift (Fig. 12). *See* LOESS; REGOLITH; TILL.

Solonchak. An intrazonal group of soils having high concentrations of soluble salts and lacking clearly differentiated horizons in the profile, these soils were formed under salt-tolerant grasses and shrubs in cool to tropical, arid to subhumid climates, commonly in places that receive seepage or runoff water. Some Solonchaks are used for crops under irrigation, following leaching and removal of the soluble salts.

Latosol profile under sugar cane, showing great uniformity of profile. Scale in feet.

Sierozem profile with brown A and B horizons of equal thickness over highly calcareous C horizon. Scale in feet.

Chernozem profile with thick dark A horizon, prismatic B horizon, and C horizon of calcareous loess. Scale in feet.

Photographs by W. H. Johnson, Soil Conservation Service, U.S.D.A.

Grumusol profile with wide cracks through A horizon, caused by shrinkage as soil dries. Scale in feet.

Red-Yellow Podzolic soil profile with thick, nearly white, and sandy A horizon over red B horizon that is much higher in clay content. Scale in feet.

Noncalcic Brown soil profile with hardpan below depth of 2 ft. Scale in feet.

Related intrazonal groups of soils are Solonetz and Soloth. If a Solonchak that is high in sodium salts undergoes partial leaching, it may become highly alkaline and gradually acquire a dark, hard, columnar B horizon marked by clay accumulation. This soil is a Solonetz, a member of another intrazonal group. Further leaching may largely remove sodium, slowly making the soil acid and breaking down the dark, hard B horizon, with formation of still another intrazonal soil, the Soloth. This last group is uncommon and of limited extent, whereas Solonetz soils comprise a widely distributed group in arid and semiarid regions. Some members of that group are used for crops under irrigation, although management is difficult and yields are often low.

Subarctic Brown Forest soils. These are a zonal group of soils having thin surface layers of leaf litter and humus, dark brown A horizons fairly high in organic matter, and deeper light-colored B or C horizons. These soils have been formed under forests of aspen, spruce, and birch in cold, subhumid climates, marginal to the zone of the tundra (Fig. 13).

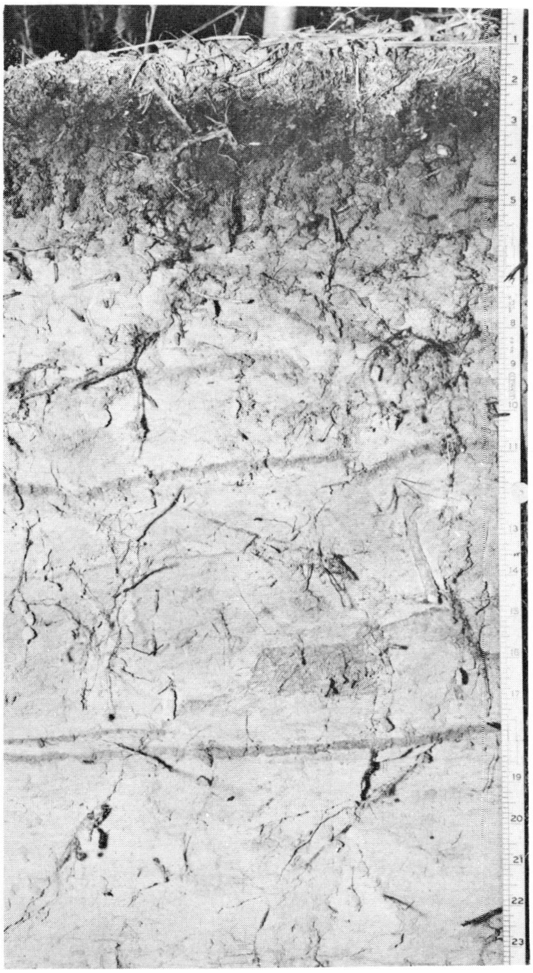

Fig. 13. Subarctic Brown Forest soil profile with forest litter on surface, dark A horizon 4 in. thick, and pale C horizon; deeper profile is marked by plant roots and traces of former roots; numbers on scale indicate inches. (*Photo by R. W. Simonson, Soil Conservation Service, USDA*)

Tundra soils. These have been defined in the past as a zonal group of soils having dark brown surface layers high in organic matter over grayish horizons which rest on an ever-frozen substratum. The soils have been formed under sedges, shrubs, and mosses in cold, semiarid to humid climates. Most soils in the tundra region seem to be poorly drained, resembling those of the Low Humic-Gley group in general morphology, except for the frozen substratum. Occasional sites are well drained and have profiles much like those of the Subarctic Brown Forest group, although the soils were formed under sedges, shrubs, and mosses rather than forest.

[J.G.S.; R.W.SI.]

Bibliography: *Soils and Men*, USDA Yearbook Agr., 1938; J. Thorp and G. D. Smith, Higher categories of soil classification: order, suborder, and great soil groups, *Soil Sci.*, 67(2):117–126, 1949.

Soil, zonal distribution

A world-wide consideration of zonal soil arrangement on the lands of the earth. Zonal soils are the well-drained soils developed on undulating to rolling topography. Local differences in parent material, vegetative cover and incorporation, age, and drainage characteristics produce variations in soil type which may be classified in great soil groups. Every soil type in any great soil group has the same number and kinds of definitive horizons within its vertical profile. The world contains perhaps 60 great soil groups. Some of these groups have dominant characteristics related to excess drainage or dominating parent material unrelated to any zone of the earth (intrazonal). Others are immature with poor horizon development also unrelated to earth zones (azonal). The qualities of the remainder of the great soil groups, the zonal soils, reflect the temperature, precipitation, and vegetative cover of broad areas; and within each of these zones certain combinations of great soil groups are characteristic. *See* SOIL (GREAT SOIL GROUPS).

Six broad classes of zonal soils may be recognized on a generalized map of the earth (see illustration). One consists of areas of high local relief such as mountains and rough hill lands. Within these areas there are relatively few zonal soils, and local patterns of soil arrangement are very complex. The remainder of the earth includes the other five categories of zonal soils: tundra, podzolic, latosolic, chernozemic, and desertic. Each category includes considerable variation and the zonal soils of each region are associated with soils other than the dominant variety. In some areas, azonal and intrazonal soil may predominate because of unusual local characteristics. Nevertheless, the broad zonal categories have much meaning to world patterns of farming and forestry.

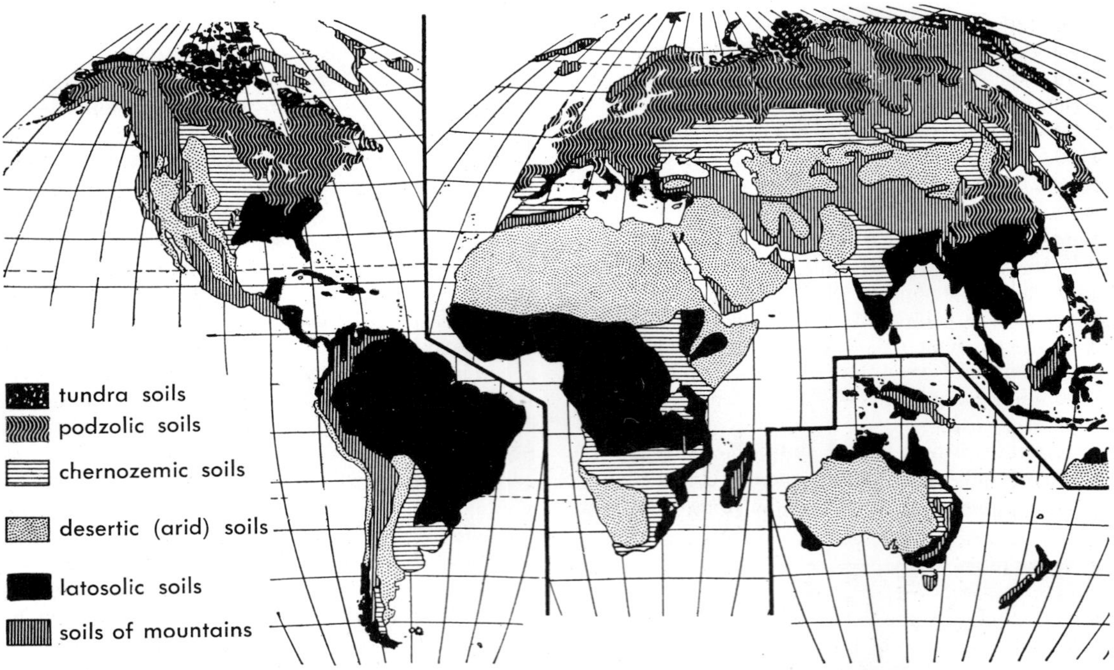

Map of the world, showing six broad soil zones. Important areas of organic soils, saline soils, and other intrazonals are omitted as well as important bodies of alluvial soils, along such great rivers as the Mississippi, Amazon, Nile, Niger, Ganges, Yangtze, and Yellow. (*Soil, USDA Yearbook of Agriculture, 1957*)

Tundra soils. These soils occur in the high latitudes where a cold climate and little vegetation are characteristic. The low temperatures restrict biologic and weathering activity; thus horizon differentiation is poor. Permafrost or permanently frozen subsoil is general; where drainage is relatively good, the soils have podzolic characteristics. *See* PERMAFROST.

Podzolic soils. Among the zonal soils, podzolic soils dominate in a broad area of humid subarctic climates in the Northern Hemisphere. They occur in a few small areas elsewhere, but the long, cold winters, short summers, and a forest vegetation of coniferous and mixed coniferous-deciduous broadleaf trees in the subarctic areas favor their development. Podzolic soils are grayish and strongly leached in the A horizon (upper part); they are commonly acid and low in bases, such as calcium, as well as in incorporated organic matter. For information on soil profiles and their horizons, *see* SOIL. The B horizon is usually strongly marked by a net receipt of materials (illuviated rather than eluviated) to an extent that chemical or mechanical cementation commonly produces a pan layer (such as clay pan or hard pan) which inhibits drainage and root penetration. Levels of fertility are low and their utilization for agriculture requires careful management.

Latosolic soils. Broad areas in central Africa, northern South America, Central America, southeast Asia and northern Australia are characterized by latosolic soils. Because of high temperatures and precipitation they are strongly weathered and leached to great depths. Horizon development is poor. They tend to have high concentrations of the oxides of iron and aluminum and are characteristically red or yellow. They are highly porous and less subject to erosion than middle-latitude soils. Extreme concentrations of iron oxide sometimes produce a pan layer called laterite. The upper eluviated portion of the deep profile is relatively infertile, and sustained cropping is not possible without considerable fertilization.

Chernozemic soils. Soils of this type are developed in areas of grass vegetation in the semiarid and subhumid regions of either the tropics or the middle latitudes. Their development for agricultural purposes is best in the middle latitudes in the eastern Great Plains of North America, the eastern plains of Argentina and adjacent areas, and a broad east-west belt in south-central U.S.S.R. from the Ukraine eastward. Chernozemic soils have dark-colored, thick A horizons, and are high in organic matter content owing to the automatic incorporation of the root masses of annual grasses. They have a neutral to basic reaction and are ordinarily very fertile. Moisture capacities are high.

The chernozemic soils of the middle latitudes are among the most productive soils of the world. They are easily worked and the grains produced from them have a high protein content, in contrast to the higher carbohydrate concentrations produced from podzols and latosols. Tropical and subtropical chernozems are neither so productive nor so easy to till.

Desertic soils. These have developed under mixed shrub and grass vegetation in arid climates in middle and low latitudes. Most of the desert areas contain relatively large regions of azonal and intrazonal soils. Desertic soils are slightly weathered and leached, and are low in organic matter and nitrogen. Profiles are faint and horizons are shallow. Nutrient levels, on the other hand, are commonly high, and with proper management and water provision they can be highly productive. *See* SOIL. [A.H.R.]

Bibliography: V. C. Finch, G. T. Trewartha, A. H. Robinson, and E. H. Hammond, *Elements of Geography*, 4th ed., 1957; *Soil*, USDA Yearbook of Agriculture, 1957.

Soil balance, microbial

The equilibrium between the diverse types of microorganisms in soil. Although the qualitative and quantitative composition of the soil microflora and microfauna fluctuates with temperature, moisture, and treatment (such as fertilization, cultivation, and cropping) of the soil, a balance exists which is characteristic of a given soil. The balance is determined chiefly by the available supply of nutrients required by groups of microorganisms of different nutritional needs. The numbers and types of the microorganisms also depend on the available nutrient supply. Associative and antagonistic effects exerted by certain organisms on others are factors in establishing the balance. The equilibrium is not easily upset. Natural soil resists attempts to change its balance when organisms are introduced.

Associative action. The process whereby one type of microorganism produces a substance required by another is widespread in soil. This action may be extended through successive groups to give a chain effect. Thus ammonia formed through decomposition of proteins by proteolytic microorganisms is used by nitrite-forming bacteria. Nitrite is required by nitrate-forming bacteria. Many cellulose-decomposing organisms utilize nitrate in hydrolyzing cellulose, and form glucose and organic acids which may be used by still other forms. Many soil bacteria synthesize vitamins needed by other organisms. Syntrophism is a form of associative action in which two organisms are mutually dependent, each producing a factor needed by the other.

Antagonisms between soil microbes. These are a factor in maintaining the equilibrium and are manifested in different ways. Many protozoa depend upon bacteria for food and ingest certain species in preference to others. Antagonism may rest on a competition for nutrients. It may also depend upon the production of substances inhibitory to other organisms, particularly antibiotics. Though synthesized only in small amounts, the antibiotics exert their effects in the microenvironments in which soil microbes are active. The advantage possessed by such microorganisms does not lead to their predominance, since capacity for antibiotic production is but one factor in the competition with other microbes. *See* RHIZOSPHERE; SOIL MICROBIOLOGY. [A.G.L.]

Soil conservation

The practice of arresting and minimizing artificially accelerated soil deterioration. Its importance has grown because cultivation of soils for agricultural production, deforestation and forest cutting, grazing of natural range, and other disturbances of the natural cover and position of the soil have increased greatly in the last 100 years in response to the growth in world population and man's technical capacity. Accelerated soil deterioration has been the consequence.

Geographic extent and intensity. Accelerated erosion has been known throughout history wherever men have tilled or grazed slopes or semiarid soils. There are many evidences of the physical effects of accelerated erosion in the eastern and central parts of the Mediterranean basin, in Mesopotamia, in China, and elsewhere. Wherever the balance of nature is a delicate one, as on steep slopes in regions of intense rainstorms, or in semiarid regions of high rainfall variability, grazing and cultivation eventually have had to contend with serious or disabling erosion. Irrigation works of the Tigris and Euphrates valleys are thought to have suffered from the sedimentation caused by quickened erosion on the range lands of upstream areas in ancient times. The hill sections of Palestine, Syria, southern Italy, and Greece experienced serious soil losses from grazing and other land use mismanagement many centuries ago. Accelerated water erosion on the hills of southern China and wind erosion in northwestern China also date far back into history. Exactly what effects these soil movements may have had on history has been a debated question, but their impact may have been serious on some cultures, such as those of the Syrian and Palestinian areas, and debilitating on others, as in the case of classical Rome and the China of several centuries ago.

The exact extent of accelerated soil erosion in the world today is not known, particularly as far as the rate of soil movement is concerned. However, it may be safely said that nearly every semiarid area with cultivation or long-continued grazing, every hill land with moderate to dense settlement in humid temperate and subtropical climates, and all cultivated or grazed hill lands in the Mediterranean climate areas suffer to some degree from such erosion. Thus recognized problems of erosion are found in such culturally diverse areas as southern China, the Indian plateau, south Australia, the South African native reserves, the U.S.S.R., Spain, southeastern and midwestern United States, and Central America.

Within the United States the most critical areas have been the hill lands of the Piedmont and the interior Southeast, the Great Plains, the Palouse area hills of the Pacific Northwest, southern California hills, and slope lands of the Midwest. The high-intensity rainstorms of the Southeast, and the cyclical droughts of the Plains have predisposed the two larger areas to erosion. The light-textured A horizon formed under the Plains grass cover was particularly susceptible to wind removal, while the high clay content of many southeastern soils predisposed them to water movement. These natural susceptibilities were repeatedly brought into play by agricultural systems which stressed corn and cotton in the Southeast, corn in the Midwest, and intensive grazing and small grains on the Plains, Palouse, and California. The open soil surface left in the traditional cotton, corn, and tobacco cultivation of the Southeast furnished almost ideal conditions for water erosion, and at the same time caused heavy nutrient depletion of soils thus cropped. The open fields of seasons between crops have also been susceptible to soil depletion. Open fields have been especially disastrous to maintenance of soil cover during the droughts of the Plains. Soil mismanagement thus has been a common practice in parts of the United States where the stability of soil cover hangs in delicate balance.

Types of soil deterioration. Soil may deteriorate either by physical movement of soil particles from a given site or by depletion of the water-soluble elements in the soil which contribute to the nourishment of crop plants, grasses, trees, and other economically usable vegetation. The physical movement generally is referred to as erosion. Wind, water, glacial ice, animals, and man's tools in use may be agents of erosion. For purposes of soil conservation, the two most important agents of erosion are wind and water, especially as their effects are intensified by the disturbance of natural cover or soil position. Water erosion always implies the movement of soil downgrade from its original site. Eroded sediments may be deposited relatively close to their original location, or they may be moved all the way to a final resting place on the ocean floor. Wind erosion, on the other hand, may move sediments in any direction, depositing them quite without regard to surface configuration. Both processes, along with erosion by glacial ice, are part of the normal physiographic (or geologic) processes which are continuously acting upon the surface of the earth. The action of both wind and water is vividly illustrated in the scenery of arid regions (Fig. 1). Soil conservation is not so much concerned with these normal processes as with the new force given to them by man's land use practices. *See* LAND USE PLANNING.

Depletion of soil nutrients obviously is a part of soil erosion. However, such depletion may take place in the absence of any noticeable amount of erosion. The disappearance of naturally stored nitrogen, potash, phosphate, and some trace elements from the soil also affects the usability of the soil for man's purposes (*see* PLANT, MINERAL NUTRITION OF; SOIL). The natural fertility of virgin soils always is depleted over time as cultivation continues, but the rate of depletion is highly dependent on management practices.

Accelerated erosion may be induced by any land use practice which denudes the soil surface of vegetative cover (Fig. 2). If the soil is to be moved by water, it must be on a slope. The cultivation of a corn or a cotton field is a clear example of such a practice. Corn and cotton are row crops; cultivation of any row crop on a slope without soil-conserving practices is an invitation to accelerated erosion. Cultivation of other crops, like the small grains, also may induce accelerated erosion, especially where fields are kept bare between crops to store moisture. Forest cutting, overgrazing, grading for highway use, urban land use, or preparation for other large scale engineering works also may speed the natural erosion of soil (Fig. 3).

Fig. 1. Erosion of sandstone caused by strong wind and occasional hard rain in an arid region. (*USDA*)

Fig. 2. Improper land use. Corn rows planted up and down the slope rather than on the contour. Note better growth of plants in bottom (deeper) soil in foreground as compared to stunted growth of plants on slope. (*USDA*)

Where and when the soil surface is denuded, the movement of soil particles may proceed through splash erosion, sheet erosion, rill erosion, gullying, and wind movement (Fig. 4). Splash erosion is the minute displacement of surface particles caused

Fig. 3. Rill erosion on highway fill. The slopes have been seeded (horizontal lines) with annual lespedeza to bind and stabilize soil. (*USDA*)

by the impact of falling rain. Sheet erosion is the gradual downslope migration of surface particles, partly with the aid of splash, but not in any defined rill or channel. Rills are tiny channels formed where small amounts of water concentrate in flow. Gullies are V-shaped or U-shaped channels of varying depths and sizes. A gully is formed where water concentrates in a rivulet or larger stream during periods of storm. It may be linear or dendritic (branched) in pattern, and with the right slope and soil conditions may reach depths of 50 ft or more. Gullying is the most serious form of water erosion because of the sharp physical change it causes in the contour of the land, and because of its nearly complete removal of the soil cover in all horizons. On the edges of the more permanent stream channels, bank erosion is another form of soil movement.

Causes of soil mismanagement. One of the chief causes of erosion-inducing agricultural practices in the United States has been ignorance of their consequences. The cultivation methods of the settlers of western European stock who set the pattern of land use in this country came from a physical environment which was far less susceptible to erosion than North America, because of the mild nature of rainstorms and the prevailing soil textures in Europe. Corn, cotton, and tobacco, moreover, were crops unfamiliar to European agriculture. In later years the plains environment, with its alternation of drought and plentiful moisture, was also an unfamiliar one to settlers from western Europe.

Conservational methods of land use were slow to develop and mismanagement was tolerated because of the abundance of land in the eighteenth and nineteenth centuries. One of the cheapest methods of obtaining soil nutrients for crops was to move on to another farm or to another region. Until the twentieth century, land in the United States was cheap, and for a period it could be obtained by merely giving assurance of settlement and cultivation. With low capital investments, many farmers had little stimulus to look upon their land as a vehicle for permanent production. Following the Civil War, tenant cultivators and sharecroppers presented another type of situation in the Southeast where stimulus toward conservational soil management was lacking. Management of millions of acres of Southeastern farm land was left in the hands of men who had no security in their occupancy, who often were illiterate, and whose terms of tenancy and meager training forced them to concentrate on corn, cotton, and tobacco as crops.

On the Plains and in other susceptible western areas, small grain monoculture, particularly of wheat, encouraged the exposure of the uncovered soil surface so much of the time that water and wind inevitably took their toll (Fig. 5). On range lands, the high percentage of public range (for whose management little individual responsibility could be felt), lack of knowledge as to the precipi-

Fig. 4. Two most serious types of erosion. (*a*) Sheet erosion as a result of downhill straight-row cultivation. Note onions washed completely out of ground. (*b*) Gully erosion destroying rich farm land and threatening highway. (*USDA*)

Fig. 5. Wind erosion. Accumulation of topsoil blown from bare field on right. (*USDA*)

tation cycle and range capacity, and the urge to maximize profits every year contributed to a slower, but equally sure denudation of cover.

Finally, the United States has experienced extensive erosion in mountain areas because of forest mismanagement (*see* FOREST CONSERVATION). Clearcutting of steep slopes, forest burning for grazing purposes, inadequate fire protection, and shifting cultivation of forest lands have allowed vast quantities of soil to wash out of the slope sites where they could have produced timber and other forest values indefinitely. In the United States the eastern Appalachian area and the southern part of California have suffered severely in this respect, but all hill or mountain forest areas, except the Pacific Northwest, have had such losses.

Economic and social consequences. Where the geographical incidence of soil erosion has been extensive, the damages have been of the deepest social consequence. Advanced stages of erosion may remove all soil and therefore all capacity for production. More frequently it removes the most productive layers of the soil—those having the highest capacity for retention of moisture, the highest soil nutrient content, and the most ready response to artificial fertilization. Where gullying or dune formation takes place, erosion may make cultivation physically difficult or impossible. Thus, depending on extent, accelerated erosion may affect productivity over a wide area. At its worst, it may cause the total disappearance of productivity, as on the now bare limestone slopes of many Mediterranean mountains. At the other extreme may be the slight depression of crop yields which may follow the progress of sheet erosion over short periods. In the case of forest soil losses, except where the entire soil cover disappears, the effects may not be felt for decades, corresponding to the growth cycle of given tree species. Agriculturally, however, losses are apt to be felt within a matter of a few years.

Moderate to slight erosion cannot be regarded as having serious social consequences, except over many decades. As an income depressant, however, it does prevent a community from reaching full productive potentiality. More severe erosion has led to very damaging social dislocation. For those who choose to remain in an eroding area or who do not have the capacity to move, or for whom migration may be politically impossible, the course of events is fateful. Declining income leads to less means to cope with farming problems, to poor nutrition and poor health, and finally to family existence at the subsistence level. Communities made up of a high proportion of such families do not have the capacity to support public services, even elementary education. Unless the cycle is broken by outside financial and technical assistance or by the discovery of other resources, the end is a subsistence community whose numbers decline as the capacity of the land is further reduced under the impact of subsistence cultivation. This has been illustrated in the hill and mountain lands of southeastern United States, in Italy, Greece, Palestine, China, and elsewhere for many millions of peasant people. Illiteracy, short life spans, nutritional and other disease prevalence, poor communications, and isolation from the rest of the world have been the marks of such communities. Where they are politically related to weak national governments, indefinite stagnation and decline may be forecast. Where they are part of a vigorous political system, their rehabilitation can be accomplished only through extensive investment contributed by the nation at large. In the absence of rehabilitation, these communities may constitute a continued financial drain on the nation for social services such as education, public health, roads, and other public needs.

Effects on other resources. Accelerated erosion may have consequences which reach far beyond the lands on which the erosion takes place and the community associated with them. During periods of heavy wind erosion, for example, the dustfall may be of economic importance over a wide area beyond that from which the soil cover has been removed. The most pervasive and widespread effects, however, are those associated with water erosion. Removal of upstream cover changes the regimen of streams below the eroding area. Low flows are likely to be lower and their period longer where upper watersheds are denuded than where normal vegetative cover exists. Whereas flood crests are not necessarily higher in eroding areas, damages may be heightened in the valleys below eroding watersheds because of the increased deposition of sediment of different sizes, the rapid lifting of channels above flood plains, and the choking of irrigation canals.

A long chain of other effects also ensues. Because of the extremes of low water in denuded areas during dry seasons, water transportation is made difficult or impossible without regulation, fish and wildlife support is endangered or disappears, the capacity of streams to carry sewage and other wastes safely may be seriously reduced, recreational values are destroyed, and run-of-the-river hydroelectric generation reaches a very low level. Artificial storage becomes necessary to derive the services from water which are economically possible and needed. But even the possibilities of storage eventually may disappear when erosion of upper watersheds continues. Reservoirs may be filled with the moving sediment and lose their capacity to reduce flood crests, store flood waters, and augment low flows. For this reason plans for permanent water regulation in a given river basin must always include watershed treatment where eroding lands are in evidence. *See* WATER CONSERVATION.

Conservation measures and technology. Measures of soil management designed to reduce the effects of accelerated erosion have been known in both the western world and in the Far East since long before the time of Christ. The value of forests for watershed protection was known in China at least 10 centuries ago. The most important of the ancient measures on agricultural lands was terrace construction, although actual physical restoration of soil to original sites also has been practiced. Terrace construction in the Mediterranean countries, in China, Japan, and the Philippines represents the most impressive remaking of the face of the earth before the days of modern earth-moving equipment (Fig. 6). Certain land management practices which were soil conserving have been a part of western European agriculture for centuries, principally those centering on livestock husbandry and crop rotation. Conservational management of the soil was known in colonial Virginia and by Thomas Jefferson and others during the early years of the United States. However, it is principally since 1920 that the technique of soil conservation has been developed for many types of environment in terms of an integrated approach. The measures include farm, range, and forest management practices, and the building of engineered structures on land and in stream channels.

Farm, range, and forest. A first and most important step in conservational management is the determination of land capability—the type of land use and economic production to which a plot is

Fig. 6. The Ifugao rice terraces, Philippines. (*Philippine Embassy, Washington, D.C.*)

suited by slope, soil type, drainage, precipitation, wind exposure, and other natural attributes. The objective of such determination is to achieve permanent productive use as nearly as possible. The United States Soil Conservation Service has developed one of the more easily understood and widely employed classifications for such determination (Fig. 7). In it eight classes of land are recognized within United States territory. Four classes represent land suited to cultivation, from the Class I flat or nearly flat land suited to unrestricted cultivation, to the steeper or eroded Class IV lands which can be cultivated only infrequently. Three additional classes are grazing or forestry land, with varying degrees of restriction on use. The eighth class is suited only to watershed, recreation, or wildlife support. The aim in the United States has been to map all lands from field study of their capabilities, and to adjust land use to the indicated capability as it becomes economically possible for the farm, range, or forest operator to put conservational use into force.

Once the capability of land has been determined, specific measures of management come into play. For Class I land few special practices are necessary. After the natural soil nutrient minerals begin to decline under cultivation, the addition of organic or inorganic fertilizers becomes necessary. The return of organic wastes, such as manure, to the soil is also required to maintain favorable texture and optimum moisture-holding capacity. Beyond these measures little need be added to the normal operations of cultivation.

On Class II, III, and IV lands, artificial fertilization will be required, but special measures of conservational management must be added. The physical conservation ideal is the maintenance of such land under cover for as much of the time as pos-

Fig. 7. Land capability classes. (*USDA*)

sible. This can be done where pasture and forage crops are suited to the farm economy. However, continuous cover often is neither economically desirable nor possible. Consequently, a variety of devices has been invented to minimize the erosional results from tillage and small grain or row crop growth. Where wind erosion is the danger, straw mulches, row or basin listing may be employed and alternating strips of grass and open-field crops planted. Fields in danger of water erosion are plowed on the contour (not up- and downslope). They may be terraced on the contour also and the terraces strip-cropped with alternating cover and grain or row crops. In the United States the bench terrace is now little used, its place having been taken by the broad-base or Mangum terrace, and variations on it. Design of conservational cultivation also includes provision for grass-covered waterways to collect drainage from terraces and carry it into stream courses without erosion. Where suitable conditions of slope and soil permeability are found, shallow retention structures may also be constructed to promote water infiltration. These are of special value where insufficient soil moisture is a problem at times. Additional moisture always encourages more vigorous cover growth.

The measures just described may be considered preventive. There are also measures of rehabilitation where fields already have suffered from erosion and offer possibilities of restoration. Grading with mechanical equipment and the construction of small check dams across former gullies are examples.

For the remaining four classes of land, whose principal uses depend on the continuous maintenance of cover, management is more important than physical conditioning. In some cases, however, water retention structures, check dams, and other physical devices for retarding erosion may be applied on forest and range lands. In the United States such structures are not often found in forest lands, although they have been commonly employed in Japanese forests. In forestry the conservational management objective is one of maximum production of wood and other services while maintaining continuous soil cover. The same is true for grass and other forage plants on managed grazing lands. For range lands, adjustment of use is particularly difficult because grazing must be tolerated only to the extent that the range plants still retain sufficient vitality to withstand a period of drought which may arrive at any time.

A last set of erosion-control measures is directed toward minimizing stream bank erosion which may be large over the length of a long stream. This may be done through revetments, retaining walls, and jetties, which slow down current undercutting banks and hold sand and silt in which soil-binding willows, kudzu, and other vegetation may become established. Sediment detention reservoirs also reduce the erosive power of the current, and catchment basins or flood control storage help reduce high flows (Fig. 8).

Conservation agencies and programs. Whereas excellent soil-conserving soil management was maintained for generations by some farmers and farm groups, as in Lancaster County, Pennsylvania, a major amount of the soil conservation activities in the United States is derived from Federal government assistance. The Soil Conservation Service of the USDA has been a focal agency in spreading knowledge of soil conservation in farm and range land management and aiding in its application. In practice, the local administration of a soil conserving program is within a Soil Conservation District, which usually is coincident with a county, and is organized under state law. The district is the liaison unit between the farmer and public assistance agencies at the state and Federal levels. It is managed by a board or committee, generally composed of five members, and usually elected by farmers within the district. Other local public bodies which may have soil conservation objectives include conservancy districts, wind-erosion districts, drainage or irrigation districts, Agricultural Stabilization and Conservation Service County Committees, and Farmers' Home Administration County Committees. In addition there are private groups with conservational interests, such as the farmers' cooperatives and national farm organizations like the Farm Bureau Federation.

The local districts may be aided technically and financially in their program. Much of the financial aid stems from Federal sources, and theoretically it is on a matching fund basis. In actual practice, however, a major part of the expenditures for special soil conserving programs is from Federal funds. Technical aid is provided throughout the nation by the Soil Conservation Service, and also by the U.S. Forest Service for its special fields of forestry and grazing-land management. Technical aid also has been provided by the Agricultural Extension Services and the Land Grant Colleges of the several states. The Tennessee Valley Authority has maintained a program of its own design, with the cooperation of the colleges and the Extension Services. The Soil and Moisture Conservation Operations Office of the Indian Service, U.S. Department of the Interior, likewise has conducted a program limited to specific Indian lands.

Financial assistance for soil conservation measures has been provided by the Federal government through the Soil Conservation Service, the Agricultural Stabilization and Conservation Service, the TVA, and the Farmers' Home Administration. Assistance has been particularly in the form of loans from the FHA, in low-cost fertilizer from the TVA, and as direct cash outlay from other agencies. Over the years, the program of the Agricultural Stabilization and Conservation Service has been the largest single source of financial aid for these purposes.

In addition to technical and financial aid, the farmers or other land operators of the United States are given valuable indirect assistance through the many research programs, basic and applied, which treat the fields related to soil conservation. The work of the Agricultural Research Service, of the Soil Conservation Service, of the Tennessee Valley Authority, and of the Land Grant Colleges has been especially helpful. Through these works new soil conserving plants, new fertilizers, improved means of physical control, and new methods of management have been developed. Through them soil conservation has not only become important but also an increasingly efficient public activity in the United States. *See* AGRICULTURE; FOREST AND FORESTRY; RANGE LAND CONSERVATION. [E.A.A.]

Bibliography: *See* CONSERVATION OF RESOURCES.

Fig. 8. Stream bank erosion control. Construction of a new conservation pool which will help reduce flooding, retard downstream erosion, and store water. (*USDA*)

Soil mechanics

The application of the laws of solid and fluid mechanics to soils and similar granular materials as a basis for design, construction, and maintenance of stable foundations and earth structures. Soil mechanics differs from other applications of mechanics to engineering, such as the design of concrete and steel structures, in that soil and similar materials have a much wider range of mechanical and other physical properties than concrete and steel. In addition, soil materials are usually present in layers and strata that vary widely in composition and physical character.

Soil mechanics as an applied science has a relatively small system of legitimate theory and a large and important methodology concerned with soil exploring, sampling, and testing. It also makes use of information from other fields such as geology and soil science.

Soil mechanics will always be an important part of soil engineering. However, the tools of soil engineering are no longer predominantly the principles of solid and fluid mechanics but to an ever-increasing extent those of other subdivisions of physics, such as thermodynamics, electricity and magnetism, acoustics, optics, and chemical, atomic, and nuclear physics. This is true also in the present development of soil exploration and testing techniques and in endeavors to improve soil properties.

Soil, in the engineering sense, comprises all accumulations of solid particles in the earth mantle that are loose enough to be moved with spade or shovel. Soils range from deep-lying geologic deposits to agricultural surface soils. Soil mechanics is also applicable to similar granular assemblies of artificial origin, such as fills of mineral waste materials.

SOIL TYPES AND COMPOSITION

Soils vary widely in composition and physical properties. At one extreme are the inert, granular, cohesionless sands and gravels; at the other are clay soils of great water affinity and well-developed cohesion in the moist and dry state. According to the ease or difficulty of working them, rather than their density, soils are called light when predominantly granular and noncohesive and heavy when predominantly clayey and cohesive.

Mineral origins of soils. The mineral composition of gravels, stones, and boulders is essentially that of the parent rock from which they have been derived, predominantly by mechanical weathering action.

Sands are largely quartzitic and siliceous in humid climates but may be any kind of mineral in dry climates or under special circumstances. The white sands of New Mexico consist of gypsum particles; coral and shell beach sands may have more than 90% of calcium carbonate particles; the black sand of Yellowstone Park, Wyo., and some of the blue and purple beaches of the Pacific consist of obsidian particles.

The silt particles resemble quite closely the minerals in the parent rock, with feldspars, micas, and quartz usually well represented. In certain tropical soils, however, the silt and even larger particles may have been formed by stable agglomeration of smaller-sized chemical rock decomposition products.

Clay particles are submicroscopic, plate-shaped crystalline minerals that have been divided into three main groups: the kaolinite, the hydrous mica or illite, and the montmorillonite group. The clays possess great affinity for water as a result of the large amount of surface per volume of particle and as a function of the number and kind of adsorbed cations that neutralize unbalanced electrical charges in the clay mineral structures.

The boulder, gravel, and sand fractions are called coarse-grained or granular; they may be considered as the bones, and the combined silt, clay, and water fractions as the meat of a soil. Depending on their size, composition, or granulometry, soils may have a continuous granular skeleton with the pores between the sand and gravel particles either empty or filled to various degrees with the silt-clay-water phase; they may possess a matrix of the latter in which the granular materials are discontinuously dispersed, or they may consist entirely of silt components, clay components, or both. The physical and mechanical properties of soils with sand and gravel skeletons are markedly different from those without. In practice, the limiting size between granular and nongranular (silt-clay) fractions is the opening of a 200-mesh sieve (74 microns, μ).

Types by deposition. Engineering soils include unconsolidated sediments transported to their present place by glaciers, water, and air; and soils formed at their present site by climatic and biologic forces from solid igneous, sedimentary, or metamorphous rock or from loose sediments transported by the above-named agents. The different agents have different carrying capacities and also affect the properties of their loads in different ways. It is, therefore, important to recognize and name such soils in accordance with the means of their transportation.

Glacial soils have been transported by glaciers whose action may be likened to giant bulldozers that push all sorts of materials ahead with droppings on the side and grinding underneath. Spring melting of the glaciers stops their forward movement and permits the settling out of finely ground material. Thus, glacial soils may vary in size composition from boulders to varved clays. Typical for these soils is a disordered landscape, often with inhibited drainage and development of bogs and peat.

Aeolian soils range from sand dunes to loess deposits whose particles are predominantly of silt size. The valley loesses (Missouri, Mississippi, Rhine and others) are formed from glacier-ground particles stirred up and dissipated by wind from the dry bottoms of glacier-draining rivers during the winter when the glacier supplies no water.

Fluvial soils are river deposits of relatively uniform particle size within the range from gravel to clay. The size itself depends upon the speed of water flow at the specific location.

Lacustrine soils are sediments formed on the bottom of lakes, and marine soils are sediments in ocean and other salt-water bodies. The salt content of the latter often flocculates the fine sediments and gives them a special type of structure. Because the carrying capacity of the respective agents varies with their speed, thus with weather and season, sedimentary deposits are usually stratified; that is, built up in layers of similarly sized particles.

Soils developed in place from either solid or loose rock parent material by the action of climate, plant growth, and animal life are usually shallow in temperate and cold regions. They are of importance mainly in the case of shallow foundations, as in highway and airport engineering. In tropical regions, however, they may extend to considerable depth and acquire importance for deep foundations. The science of pedology is concerned with their formation and characteristics.

Soil structure. Natural soil systems are characterized not only by the sizes and types of their component particles but also by the arrangement of these particles in relation to each other. For granular, noncohesive soils, the porosity or its supplement, the portion of the total volume filled by the solid particles, often suffices as an indicator of structure or packing. In natural clay and silt soils, typical secondary structures are formed in which disturbance can greatly alter their mechanical properties. The structure may be caused by and typical of flocculation, as in marine clays, or may have been developed by wetting and drying and freezing and thawing cycles. Since soils may be employed in engineering in the undisturbed, partly disturbed, or greatly disturbed condition, the existence of soil structure must be taken into account. For each specific purpose, the soil must be tested in a condition as close as possible to the one in which it is to be used.

Particle size and weight. The grain-size distribution of soils lacking or having negligible amounts of particles smaller than 74 μ is determined by dry sieving, and that of materials with all particles smaller than 74 μ by sedimentation methods. The latter are based on Stokes' law for the rate of fall in a viscous medium of lesser density:

$$v = \Delta G\, d^2/1800n \quad \text{cm/sec}$$

and

$$d = \sqrt{1800\, nv/\Delta G\, g}$$

in which ΔG = difference in specific gravity between particle and viscous medium, g = acceleration of gravity, d = diameter of particle in mm, n = viscosity of medium in poises, and v = rate of fall in cm/sec. Practical methods utilize the decrease of suspended particles with time at a definite distance from the surface of a soil-in-water suspension. This decrease is usually calculated from hydrometer measurements of the density of the suspension. For soils having both coarse and fine constituents, sieving (both dry and wet) is combined with sedimentation analysis.

At least four different sieve sizes are used, with openings of 4760, 2000, 420, and 74 μ, respectively. For specific purposes, other sieves are used to advantage. The results of mechanical analysis are best presented in the form of a grain-size accumulation curve (Fig. 1). The naming of soils in accordance with their texture is shown in Fig. 2.

Determination of the specific gravity of soil particles is necessary for the sedimentation analysis and also for calculating (from the weight per unit

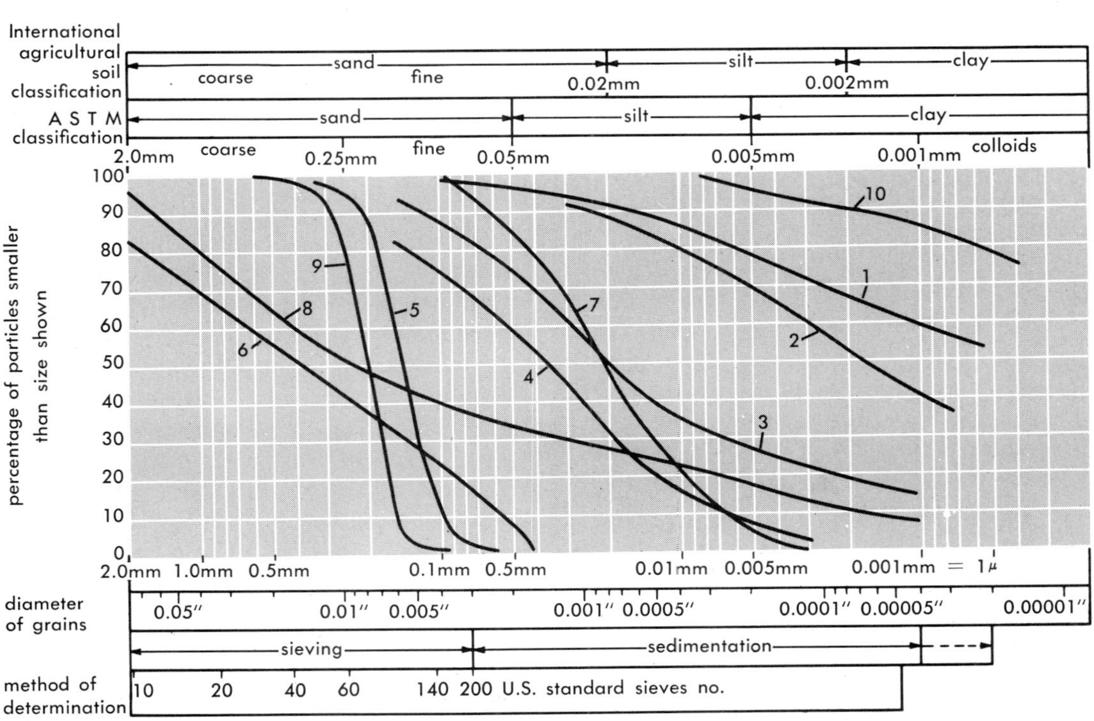

Fig. 1. Grain-size accumulation curves, plotted on semilogarithmic paper. Coarser fractions (*left*) measured by sieving; finer fractions (*right*) determined by sedimentation. Two of several systems of classification by grain size are shown at top of illustration. Curves 1 and 2, clay soils of the Nile Delta; 3 and 4, silts from the Nile Delta; 5, Port Said beach sand; 6, sand artificially graded for maximum density; 7, Vicksburg loess; 8, New Mexico adobe brick; 9, Daytona Beach sand; 10, Wyoming bentonite. (*From G. P. Tschebotarioff, Soil Mechanics, Foundations and Earth Structures, McGraw-Hill, 1951*)

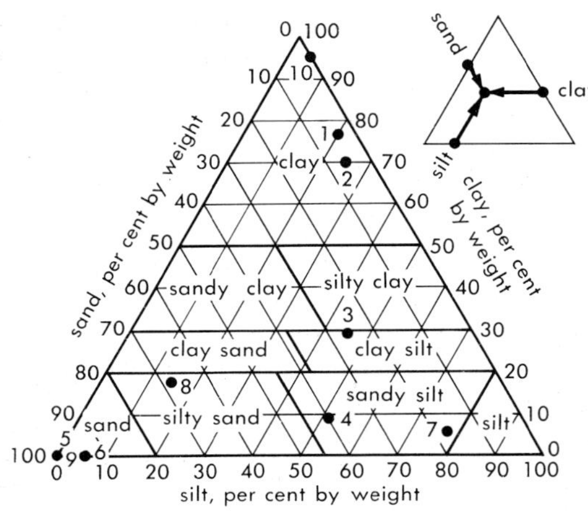

Fig. 2. Textural soil classification chart. Nomenclature of soil is determined by percentages of sand, clay, and silt contained. The 10 numbered soils refer to those in Fig. 1. (*From G. P. Tschebotarioff, Soil Mechanics, Foundations and Earth Structures, McGraw-Hill, 1951*)

volume and the moisture content) the actual volume relationships in the soil which are important for its mechanical strength properties. The ratio of the volume of water- and air-filled voids to the volume of solids in the sample is called the voids ratio e. Because of the variability of water content of soils, their unit weight is usually given in lb/ft^3 of dry soil.

Engineering soil classifications. Soils have been classified for engineering purposes by several public agencies and individuals. The most widely used system at the present time is the so-called unified classification of the U.S. Bureau of Reclamation. While the different systems vary in details and nomenclature, the physical classification principles are essentially the same. The first division is made between soils that contain a coarse granular skeleton of gravel, sand, or both and those without such a skeleton, the silt-clay soils. The coarse granular materials are then subdivided into gravel and sand soils, which are further differentiated by the degree of water affinity of silt-clay material found in the intergranular spaces. The silt clay soils which may or may not contain separate occluded coarse particles are subdivided into groups that differ in water affinity, plasticity, and swelling and shrinkage characteristics. A separate class of boulders and cobbles may be added to the coarse granular category and one of fibrous organic soil (peat) of great compressibility to the fine-grained category. The physical distinctions underlying engineering soil classifications entrain definite differences in other important physical properties, such as mechanical strength, elasticity, compressibility, shear and flow behavior, and permeability. The classification of a soil thus may serve to indicate its suitability for various engineering uses.

Soil-water relationships. Relationships between soil and water are of primary importance in soil mechanics. Their understanding and utilization presupposes a thorough knowledge of the properties of the water substance and of the physical and physicochemical characteristics of the surfaces of the soil minerals. Water is a peculiar substance. According to its molecular weight, it should be a gas at room temperature, but it is a liquid. Even as a liquid, it possesses structural properties commonly associated with solids and provable by means of x-rays. Its peculiarities, which are the consequences of the geometric-electric structure of the H_2O molecule, assert themselves also in its interactions with any type of surface.

Hygroscopic water. Dry soil minerals adsorb water from the atmosphere; the actual amounts depend on the physicochemical character of the surfaces and increase with rising relative humidity. The amount adsorbed is called hygroscopic water; it is usually measured at room temperature and expressed in per cent of the dry soil weight. Very small amounts of water may be in solid solution within the surface of the minerals. Additional increments build up water films whose consistency may range from solid near the particle, through plastic, to that of normal water at a certain distance from the particle surface. A plastic water film 10^{-6} cm thick may not even equalize the roughness of a sand or gravel particle, but if it is around a plate-shaped clay mineral 10^{-5} cm thick. the water film represents more than 20% of the total clay-water volume. Therefore, the smaller the soil minerals, the more important are their water affinity and its consequences, such as swelling and shrinkage.

Gravitational and capillary water. In addition to this physicochemically restrained water, there may be free water in the soil pores, which is called gravitational water if it moves freely under the force of gravity, or capillary water if it is controlled by the forces of capillarity. The height at which capillary water can exist above the groundwater level depends on the effective pore radius. The water affinity and the capillarity of clay and silt soils cause the entrance of water in either the liquid or vapor phase and also affect the ease with which water moves through a soil, or the difficulty of its removal.

SHEAR AND PLASTICITY

Shear in soils. Soils are composed of many separate particles of great range in size and shape. The particles may or may not be held together by water films or by a clay-water cement. Analogous systems are encountered in the molecular world. Systems composed of a single kind of molecule or atom have, at constant pressure, definite temperatures of transition from the solid to the liquid state. This transition generally involves an expansion, that is, an increase in the interparticle spacing, which represents the only real difference between the solid and the liquid phases at the melting point.

Introduction of molecules of different size and character into a pure substance lowers the melting point; also, the mixture will soften over a range of temperature instead of melting sharply at a single temperature. In multicomponent materials, such as asphalts and pitches, a wide softening and liquefaction range replaces a definite melting point. The same phenomenon occurs in macroparticle systems such as soils. Densely packed gravels and sands are macromeritic (large-particle) solids. If submitted to shear stresses, they must expand in the shear zone to voids ratios characteristic of the molten state, in order that shear may take place without breaking of the individual particles. At voids ratios above the critical (melting range) ratio, gravel and sand soils can be "liquefied" by vibration that reduces interparticle friction or, if sheared, they may collapse to the critical voids ratio.

In granular noncohesive soils, the shear resistance obeys the equation $S = N \tan \varphi$ in which S = shear resistance in psi, N = effective normal pressure on the shear plane in psi, and $\tan \varphi$ = coefficient of friction, or tangent of angle of friction related to angle of repose of a pile of the granular material. If the soil possesses a granular skeleton bonded together by moist or dry clay, the shear resistance can be approximated by $S = C + N \tan \varphi$ in which C denotes the cohesion of the system in psi. The numerical value of C depends on the amount, type, and water content of the clay binder, its physicochemical interaction with the coarse particles, and the history of the system. In soils without granular skeleton, S approaches C.

Plasticity. The property of plasticity is possessed by many crystalline solids within a certain temperature range below and adjoining the melting point.

Conditions for plasticity. Plasticity denotes the ability of a body to deform permanently without rupture under applied stresses. This presupposes that during deformation (1) no marked expansion takes place in the shearing zones which would alter the cohesive forces; (2) the particulate components (molecules, atoms, micro- and macroparticles) are in similar geometric arrangement after as before deformation; and (3) the rate of deformation and of breaking existing bonds does not exceed that of forming new bonds between atoms or molecules. Even in plastic bodies, very rapid deformation produces brittle fracture. In solid crystals, these requirements are best fulfilled if their building units are arranged with the highest degree of symmetry, as is the case with many metals. In macroparticle systems such as soils, the requirements for plasticity are essentially the same. Large masses of relatively uniform sands and gravels may deform plastically at a slow rate and without change in voids ratio.

Small soil masses, and especially laboratory samples, show plasticity only if they are moist and cohesive. If so, the reforming of bonds broken during deformation of the mass and the reestablishment of the original symmetry of the bonding elements pertain essentially to the molecules in the water films around the particles. These water films play the same role in soils (increasing the distances between gliding planes and decreasing cohesion) as elevated temperature does in crystalline solids. Because plasticity increases with increasing total area of gliding planes per unit volume, the greatest plasticity is possessed by moist systems of pure clays of smallest particle size and greatest force of interparticle attraction. Admixture of coarser-grained material to clay interferes with the normal development of gliding planes and "shortens" the soil to an extent that depends on the amount, type, and size distribution of the coarser components.

Consistency limits. The term consistency is preferentially employed for mechanical resistance properties in the twilight zone between true elastic solid and simple liquid behavior. Since several physical phenomena are usually involved in consistency, this property is normally defined by the use of specific apparatus and standardized procedure. Typical are the slump test for fresh concrete, the penetration test for asphalt, and the consistency limits tests for soils. The last indicate what water content (per cent of dry weight of soil) will bring soils to states of analogous consistency. The liquid limit is the water content at the transition between the plastic and the liquid state, the plastic limit the water content at the solid-plastic transition, and the shrinkage limit the moisture content that will just fill the soil pores in the solid state if that state is reached by the drying out of a soil paste. The difference between the liquid and plastic limits is the plasticity index. It indicates the moisture range in which a soil shows plastic behavior. These consistency indices are valuable tools if properly understood, but they do not as such make it possible to predict under what circumstances large soil systems will show rupture, plastic flow, or creep. This depends on complex factors of granulometry; amount, type, and water content of soil fines; and stress conditions. Analogous factors govern the behavior of all building materials.

STUDY OF SEEPAGE AND FROST

Seepage. Water movement in soils is normally caused by hydraulic pressure or tension gradients. This flow is ordinarily viscous or laminar rather than turbulent. Darcy's law is fundamental: $V = Akit$, where V = volume of water flowing in time t through soil with a cross section A and where pressure gradient $i = \Delta p/\Delta l$ (pressure drop per unit length of flow) and k = coefficient of permeability. The permeability coefficient varies from 100 cm/sec for clean gravel to 10^{-9} cm/sec for heavy clay.

Since water possesses an electric structure that interacts with the electrically charged soil minerals, it also moves in soil capillaries upon application of electric potentials. In this case, k of the Darcy equation is replaced by k_e, the electroosmotic transmission coefficient, and i by the electric potential gradient. Since the electric soil-mineral surface interaction structure is temperature-susceptible, water also moves in dense clay soils upon application of a thermal gradient with coefficient of thermoosmosis k_{th}. In unsaturated soils of high porosity, thermal gradients also cause water movement in the vapor phase.

Measurement of permeability. The coefficient of permeability may be determined either in the laboratory on disturbed or undisturbed samples by means of the constant or falling head permeameters, or in the field by pumping or injection tests. In the constant head permeameter, used for materials of high permeability, water maintained at a constant level flows through a soil sample. Permeability is computed from the rate of flow. In the falling head permeameter, used for materials of low permeability, the permeating water is supplied by means of a standpipe in which the hydraulic head falls from h_1 to h_2 during the time t while the water volume $a(h_1 - h_2)$ flows through a soil specimen of thickness L (Figs. 3 and 4). Then

$$k = 2.3 \frac{La}{At} \log \frac{h_1}{h_2}$$

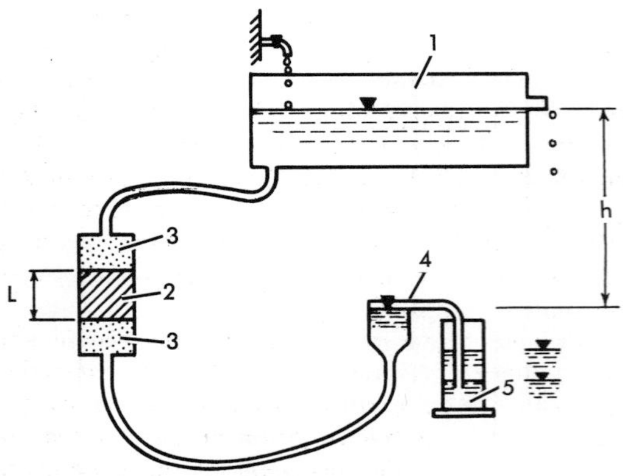

Fig. 3. Constant head permeameter. Water level (1) is kept constant. Water percolates through soil sample (2) of thickness L. Porous filters (3) hold soil in place. Tail water level is kept constant by overflow (4). Volume of discharge is measured in receiving vessel (5). (*From G. P. Tschebotarioff, Soil Mechanics, Foundations and Earth Structures, McGraw-Hill, 1951*)

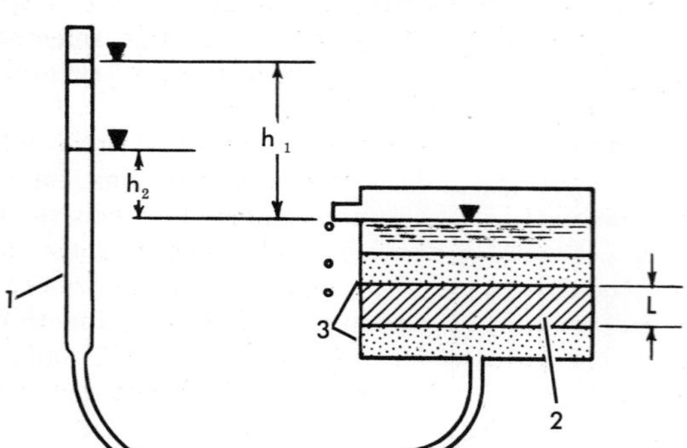

Fig. 4. Falling head permeameter. Water level h_1 in a thin glass tube (1) decreases to h_2 as water percolates through soil sample (2) of thickness L, restrained between porous filters (3). (*From G. P. Tschebotarioff, Soil Mechanics, Foundations and Earth Structures, McGraw-Hill, 1951*)

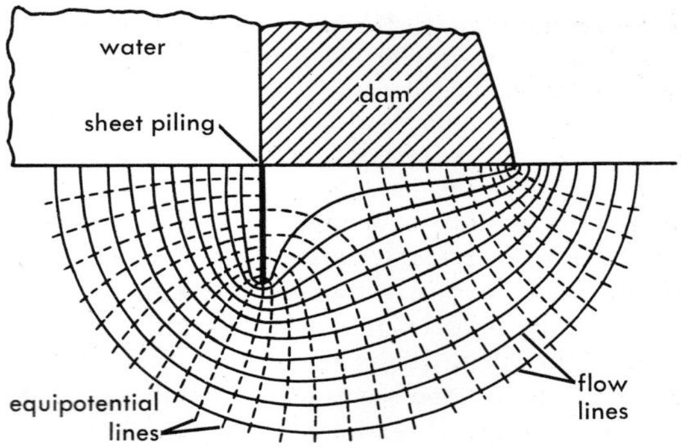

Fig. 5. Flow net under the cutoff wall of a dam. (*From D. P. Krynine, Soil Mechanics, 2d ed., McGraw-Hill, 1947*)

In calculating k from pumping and injection tests in the field, spherical symmetry is assumed in the pressure distribution around the well point or the injection nozzle. Moving water transmits momentum to the contacting soil particles. If in noncohesive soils the flow is upward and the hydrostatic uplift is sufficient to sustain the weight of the individual soil particles, a quick condition results. The minimum hydraulic gradient i_{cr} to cause quick condition can be estimated from

$$i_{cr} = \frac{G - 1}{1 - e}$$

where G = specific gravity and e = the voids ratio.

Flow nets. The study of seepage through natural and man-made soil structures is greatly facilitated by the use of flow nets. These are two nests of curves, one representing the flow lines and the other the equipotential lines. In accordance with the laws of laminar flow, flow lines, which follow the path of the water, may not intersect each other; and the equipotential lines, each connecting points of equal hydraulic head, must cross the flow lines at right angles. While an infinite number of flow lines exists, only a few are drawn and in such a manner that the quantities of flow between adjacent lines are equal. Then the equipotential lines are drawn so that they intersect the flow lines at right angles and form areas that resemble squares as closely as possible (Fig. 5).

A conventional flow net has the following properties: (1) all paths, each of which lies between two adjacent flow lines, carry the same seepage quantities; (2) potential differences are the same between all adjacent equipotential lines; (3) flow lines intersect equipotential lines at right angles; (4) all figures formed by adjacent pairs of lines resemble squares as closely as possible if the soil is homogeneous; and (5) at any point of the net, the spacing of equipotential lines is inversely proportional to the hydraulic gradient and the spacing of flow lines inversely proportional to the seepage velocity. For the drawing of flow nets, the boundary conditions must be known. Two-dimensional presentation is of course applicable only if the profile remains the same in the third dimension. From a satisfactory flow net, the seepage can be calculated using

$$Q = \frac{n_f}{n_d} kh_i$$

where Q = quantity of seepage in ft^3/sec per linear foot of the length of the structure, n_f = number of flow paths, and n_d = number of spaces between equipotential lines in the net, k = coefficient of permeability in ft/sec and h_i = head loss in ft between surfaces of the head and tail waters. Methods are available to correct for differences in permeability in different directions. Seepage in complicated structures is studied on models employing, instead of water, electric or other energy forms whose transmission obeys the same mathematical laws expressed in the Laplace equations for conjugate functions.

Frost action in soils. Frost action is of great engineering importance. The forceful expansion of confined water when it freezes can destroy porous building materials and loosen soil. Of greater importance is the accumulation of water in the form of ice lenses, with resulting frost heave in winter and morass formation during thawing.

A freezing front penetrating into moist soil starts only a limited number of crystallization centers. Water moves to these from the surrounding soil, especially from lower depths. As it freezes, it gives off about 80 cal/g of water. If the heat released by the freezing water just balances the heat lost by conduction to the earth surface, the freezing front becomes stationary and forms thick ice lenses, especially if the ground-water level or other water reservoir is close by. Without such close supply of free water, the frost may advance until a new moist layer is encountered. Here nuclei and ice formation begin anew.

Daily and other short-period temperature variations make the larger lenses grow at the expense of the smaller. In large areas of Siberia and also in Canada and Alaska, permanently frozen soil (permafrost) is encountered at depths that depend on the climate and the thermal conductivity of the surface soil. This frozen ground prevents the drainage of water from spring thawing. Structures founded on permanently frozen soil must be separated from it by insulating materials. Some of the permafrost is not in equilibrium with the present climate and if once melted would not be reformed. *See* PERMAFROST.

EXPLORATION, SAMPLING, AND TESTING

The extent to which field testing is necessary depends on the type, magnitude, and importance of the job. Before a sampling and testing program is started, all readily available information should be utilized. Such information may be in the form of construction experience records in the same general location; air photos on which recognizable erosion and vegetation features indicate the types of surface and subsurface soil to be encountered, as on a proposed highway route; geological maps that show the profiles of solid and loose rock and serve as reminders of troubles usually associated with certain rock types, as sink holes are with limestone; and pedologic maps, which are especially useful in highway soil work. Valuable information is often obtained from old maps which show soil and drainage patterns that have been covered up in urban or industrial developments.

Site investigations. These are made either to find out whether the soil in its natural site condition will support the planned structure or to establish the qualities of soils as construction materials for dams, embankments, subgrades, and other uses. For the first case, the natural site condition and the samples taken for laboratory testing must be disturbed as little as possible. For the second case, disturbed samples are useful, but they should not be permitted to dry out before being tested.

For exploration of relatively large sites or extended strips, such as airports and highways, electrical and mechanical energy transmission phenomena may be employed. The electric method furnishes information on the electric resistivity of the soil at various depths. Previous determination of relationships between soil type, water content, salinity, and electric resistivity permits the plotting of a soil profile and the detection of strata of specific materials.

Mechanical energy is employed in seismic and vibration tests. These tests utilize the reflection and refraction of earth waves at interfaces of different strata and the variation of their speeds of propagation with the character of the conducting medium. The seismic velocity of compression waves varies for different soils within a range of 500–8000 ft/sec and for solid rock from 6000 to more than 25,000 ft/sec. The velocity in water is about 4700 ft/sec. In refraction shooting, one usually explodes a blasting cap or a small charge of dynamite and records the first signal picked up by the seismographs located at three different distances. The methods employing excited vibration utilize either the difference in rate of propagation in different soils or the characteristic frequency of the soil-vibrator system.

Sampling. Soil sampling may be divided into shallow and deep, and into disturbed and undisturbed. Shallow disturbed samples of cohesive soils are usually obtained with a soil auger. Slightly disturbed samples of both cohesive and noncohesive soils may be obtained by carefully pushing thin-walled metal cylinders into the ground. The cylinder is subsequently capped on both ends to prevent moisture loss.

Deep samples can be taken by digging trenches and pits so that the actual profile is exposed. Also, machines are available for making open holes into the ground 10–36 in. wide and well over 10 ft deep. The most common method for taking deep samples is by driving a casing into the ground. The casing is at least 2½ in. wide. A record is kept of the hammer weight, height of drop, and the number of blows required to drive through each linear foot. While the boring proceeds, the casing is cleaned out with water conducted to the bottom of the casing by a pipe of about 1-in. diameter. Inspection of washings can give an idea of the character of the soil layer reached. If an undisturbed sample is to be taken at a certain depth, then the wash water is shut off above that depth and the wash pipe replaced by or changed into a pushing rod by which a thin-shelled sampler (Shelby tube) is pushed into the stratum to be sampled. The tubes with the samples are then extracted, sealed on both ends and sent to the laboratory for testing. A number of modifications of procedure and of type of sampler exists, but the essence of the method remains the same.

For important construction, the casing is driven to bedrock and through it for 5–10 ft in order to make sure that a boulder is not mistaken for a rock stratum. Because of the possibility of sinkholes, core drilling in limestone and dolomite should be as close to the actual construction site as possible.

Field tests. Static or dynamic tests at the site may be made for mechanical properties or other physical characteristics such as density, moisture content, permeability, and thermal resistivity. The in-place density and moisture content may be measured by digging a clean hole, measuring its volume, and weighing and determining the water content of the earth removed. Densities and moisture contents of surface soils and of deeper layers reached by bore holes may also be determined by nondestructive methods employing γ-rays for the former and neutrons for the latter. Thermal resistivity is usually determined with the thermal needle.

Common mechanical resistance tests include sounding, that is, pushing a steel rod or ¾–1 in.-wide pipe into the soil and noticing the resistance to driving as a function of depth; penetration tests, where the resistance to penetration by a cone of standardized form is measured at various levels reached by a bore hole; and vane shear tests, where a vane is pushed into a soil layer with subsequent application of horizontal shear forces. These tests possess only indicative value.

Load or bearing power tests are performed in open pits at foundation level. Round or square plates of practically undeformable material are employed for transmission of the applied load. The pressure bulb, or zone, created under the loaded plate bears a definite geometrical relationship to the shape and dimensions of the plate. Thus, the loading of small plates will not detect layers of low bearing power and large compressibility that may be reached by the pressure bulb of the actual foundations. Within limitations, load tests are valuable tools, especially for shallow foundations. To determine the effect of surface loading by a foundation on deeper soil layers, the stress acting on them is calculated from theory, and undisturbed samples taken from the location and depth concerned are submitted in the laboratory to the calculated stresses.

Stresses in soil masses may be due to their own weight or to outside forces. In either case, the geometry of the system is an important factor in the resulting stress distribution. The other decisive factor is the physical state of the soil mass. This physical state may range from an elastic solid to a macromeritic liquid like quicksand. Over the entire range, plasticity may be observable to various degrees. Every type of internal structure and response to stresses of different intensity, known from other construction materials, may be encountered in undisturbed and disturbed soils. This emphasizes the importance of obtaining representative soil samples and testing them carefully in the laboratory. Even then the results must be used with engineering judgment, since large masses of a material behave differently from small ones.

THEORETICAL STRESSES

Calculation of stresses would become too cumbersome if all pertinent soil factors were carried along. Theories are therefore based on idealized earth masses which for calculation of stress distribution are homogeneous, isotropic, elastic bodies and for nonelastic deformation or stability problems are idealized masses endowed with friction, cohesion, or both. The methods of stress analysis are the same as in statics and strength of materials. *See* STRUCTURAL ANALYSIS.

Pressure bulb. Stress distribution in a perfectly elastic continuum under a vertical point load at the horizontal boundary was first derived by J. Boussinesq. If the point of application of a force P serves as the origin of a coordinate system with vertical coordinates z and horizontal coordinates x and y, and $x^2 + y^2 = r^2$, then vertical pressure

$$\sigma_z = \frac{3P}{2\pi} z^3(x^2 + y^2 + z^2)^{-5/2}$$

By connecting points of equal vertical stress, a pressure bulb is obtained. Stress distribution in the elastic medium under a loaded area, rather than a point, is obtained by resolving the distributed force into a large number of point forces, and then employing the principle of superposition where stresses from different points overlap. Such solutions are available only for simple forms such as line, strip rectangle, square, and circle. However, the form of most foundations can be approximated by summation, subtraction, or both, of shapes for which solutions are available and the stresses calculated using the principle of superposition. For pressure bulbs under loaded areas of simple shape, see Fig. 6.

Accuracy of analysis. Despite the drastic simplifications in theoretical treatment, loading tests have shown that, except in the immediate vicinity of the loaded areas, the calculated data agree quite well with the experimental results.

Direct loading tests have shown differences in load acceptance by different soils. Contrary to theoretical assumptions, movement of soil material occurs close to the plate where stresses are inten-

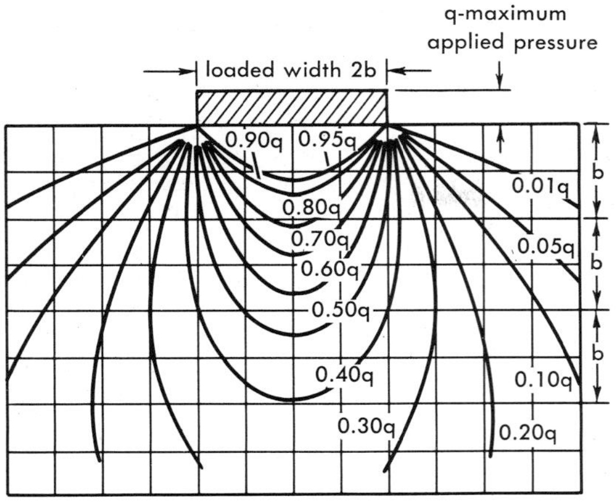

Fig. 6. Stress distribution for loaded areas of simple shape. Distribution pattern of vertical pressures through soil from a vertical load on the surface is called a pressure bulb, from the shape of equal pressure curves. (*From Road Research Laboratory, Dept. Sci. Ind. Research, London, Soil Mechanics for Road Engineers, 1954*)

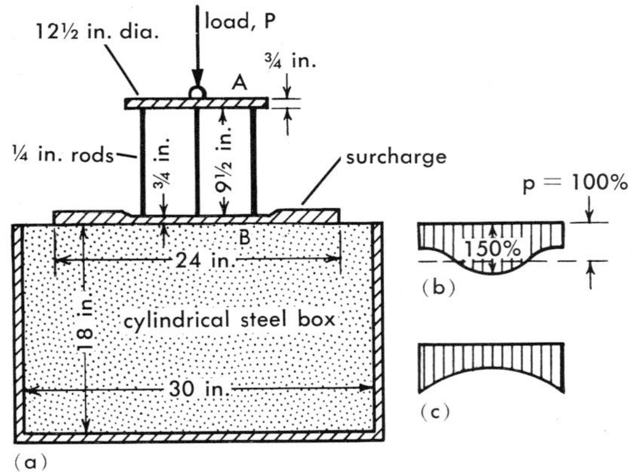

Fig. 7. Stress acceptance by sand and clay. (*a*) O. Faber's experiments in 1933 showed that under a load applied through a circular plate (*B*), stresses in sand would follow the pattern shown in (b), reaching a maximum at the center of the plate and that stresses in clay would follow the pattern shown in (c), reaching a maximum around the edges. Soil stresses were determined from contraction of the rods between the two disks (*A* and *B*) when load *P* was applied. (*From D. P. Krynine, Soil Mechanics, 2d ed., McGraw-Hill, 1947*)

sive. Coarse-grained soils in which shear resistance is directly proportional to normal stress resist movement most in the zone under the center of the plate and have lower resistance and more pronounced outward movement as the plate edges are approached. As a result, the center cone accepts most of the applied load, and the stress intensity under the center of the plate has been found in the case of sand to be 2–3 times the average stress under the plate. With plastic clay, where the shear resistance is pressure-independent, material moves from the central position outward until it encounters the resistance of the earth outside the plate edge. This equalizes the pressure under the plate and produces a maximum under the rim (Fig. 7).

CONSOLIDATION AND SETTLEMENT

The consequences of soil loading are affected by the presence of pore water. Normal pore water may be under simple hydraulic pressure γz if located at a distance z below the ground-water level and having a density γ. It may be stressless if located at the ground-water level or under tension if in capillaries above the latter. Sudden decrease in pore volume, occasioned by sudden loading, may result in a pressure increment which is called excess hydrostatic or neutral pressure. Presence of hydrostatic water decreases the weight of the solid soil particles by the weight of an equal volume of water. This must be considered in calculating the total effective stress due to the weight of overburden; it affects the friction part of shear resistance which is proportional to the normal pressure on the shear plane. Excess hydrostatic pressure as well as hydraulic uplift must be subtracted from the total interparticle pressure to obtain the effective pressure which governs frictional resistance. Excess hydrostatic pressure is dissipated by expulsion of water, at a rate determined by the hydraulic gradient and the permeability, with consolidation of the system until the total stress is carried by the skeleton of solid particles.

Consolidation theory. Consolidation has been studied theoretically only for systems of simple geometry, such as a soft clay stratum bounded by one or two pervious sand or gravel strata, under the assumptions that the soil is saturated, soil particles and water are incompressible, Darcy's law is valid and the coefficient of permeability remains constant during consolidation, the time lag of consolidation is due entirely to low permeability, the soil is laterally confined, the total and effective normal stresses are the same for any point on a horizontal plane and water flows out of the voids only in a vertical direction, and the change in effective pressure results in a corresponding change in voids ratio.

Loading of sand within its elastic range produces an immediate elastic deformation supplemented by some consolidation due to grain adjustment. Loading of a saturated clay produces an immediate, though small, elastic deformation and a slow deformation that can be called elastoid since, though water is expelled, no shear takes place in the system.

Settlement analysis. For actual structures, settlement analysis involves the following: (1) detection of a soil stratum of high compressibility by borings; (2) determination of pressure-voids ratio relationships and time factor in the laboratory on undisturbed soil samples from borings; (3) calculation of existing pressures due to natural overburden and of pressure increase by added foundation loads; (4) initial condition of consolidation; (5) theoretical curve of time factor versus consolidation; (6) predicted time-settlement curve assuming instantaneous loading and curve corrected for construction period; and (7) comparison with settlement experience of other buildings in location and follow-up by actual settlement observations on structure.

If plastic under the applied stresses, soil will flow sideways, resulting in secondary consolidation that may, like creep in metals, continue at a constant rate or may decrease with time. The rigidity of the foundation influences the stress acceptance of the contacting soil and thereby the consolidation behavior. Proper design of a foundation minimizes total settlement and eliminates as much objectionable differential settlement as possible.

STABILITY

Stability of an earth structure is primarily resistance to shear failure. Such resistance involves geometrical and soil-physical factors. Wide and deep foundations are more stable than narrow and shallow ones.

Bearing power. Several building codes contain so-called safe bearing values, which may have ranges of 25–40 tons/ft^2 for rock to 1 ton/ft^2 for soft clay. The minimum load causing failure is the ultimate bearing power which, divided by an arbitrary factor of safety, is called safe bearing power. More meaningful values are obtained from loading tests, shear tests, and unconfined and triaxial compression tests. Safe construction must be designed for the weakest possible condition of the soil on site. There exist a number of semiempirical laws with respect to foundation stability against shear failure.

Pile foundations are used for carrying building loads through soft compressible layers to layers strong enough to carry them. They may be end-bearing if resting on solid rock or floating if driven into sand or clay; if floating, the load is transmitted by skin and point resistance. Groups of floating piles develop, by superposition, pressure bulbs characteristic of the geometry of the group rather than of the individual pile. *See* PILE FOUNDATION.

Landslides. These may be due to slow soil flow such as creep, to fast liquid flow, to sinking into caves or mines, and to sliding. Often they occur along water-lubricated boundaries of inclined rock strata. Soil mechanics is particularly concerned with ensuring cuts, embankments, earth dams, and other earth structures against failure by sliding.

Stability analysis employs the concept of a sliding wedge, one surface of which is the surface of maximum shear stress in the soil. This surface is found from the geometry of the system. Next, the shear resistance on this surface is determined as a function of soil character and normal stresses. Specific calculation methods for plane and curved sliding surfaces have been developed. Since soil cohesion remains constant, while friction increases with pressure, slopes of cohesive soils possess critical heights, while slopes of purely frictional materials may be stable up to any height. This demonstrates the importance of proper choice of materials and densification to greatest possible

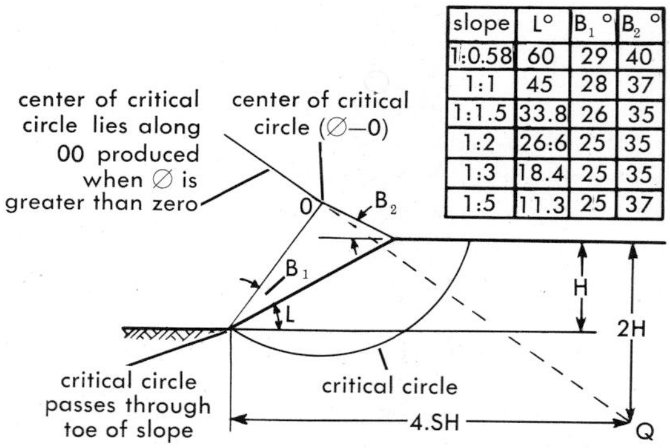

slope	L°	B₁°	B₂°
1:0.58	60	29	40
1:1	45	28	37
1:1.5	33.8	26	35
1:2	26.6	25	35
1:3	18.4	25	35
1:5	11.3	25	37

Fig. 8. Stability analysis of embankments. (*From Road Research Laboratory, Dept. Sci. Ind. Research, London, Soil Mechanics for Road Engineers, 1954*)

shear resistance of soil used for embankment and earth dam construction (Fig. 8).

Retaining walls. Stability analysis of retaining walls, that is, determination of the magnitude, distribution, and direction of soil pressure on such walls, is also based on the concept of a sliding wedge. One differentiates between neutral pressure (if no movement at all has taken place), active pressure (in the case of sufficient movement to mobilize the shear resistance of the sliding surface), and passive pressure (if the wall has pushed the soil back sufficiently to mobilize shear in a wedge which for geometric reasons involves a larger sliding surface than the active pressure). Special methodology for calculating earth pressures with varying simplifying assumptions is available. Further progress appears to depend more on a better understanding of soil-physical properties than on employing more complicated mathematics. *See* RETAINING WALL.

Conduits. Conduits are rigid or flexible pipes for the conduction of liquids, especially water, and are placed in ditches with backfilling or under embankments. Present knowledge and design formulas are based mainly on experimentation, and the closer the actual conditions resemble those of the experiment, the more reliable are the formulas. *See* COFFERDAM; SOIL; STRUCTURES (ENGINEERING). [H.F.WI.]

Bibliography: Am. Soc. Testing Materials, *Procedures for Testing Soils*, 1958; D. P. Krynine, *Soil Mechanics*, 2d ed., 1947; Natl. Research Council, *Water and Its Movement in Soils*, Highway Research Board Spec. Rep. 40, 1958; K. Terzaghi and R. B. Peck, *Soil Mechanics in Engineering Practice*, 1948; G. P. Tschebotarioff, *Soil Mechanics, Foundations and Earth Structures*, 1951; H. F. Winterkorn, Principles and practice of soil stabilization, *Colloid Chemistry*, 6:459–492, 1946.

Soil microbiology

A study of the microorganisms in soil, their functions, and the effect of their activities on the character of soil and the growth and health of plant life, particularly cultivated crops. It embraces the biology of microorganisms—their morphology, physiology, and taxonomy—as well as their biochemistry. It is related to soil chemistry and physics and, in its application, to agronomy, plant physiology, and plant pathology. The soil microorganisms are viruses, Myxomycetes (slime fungi), protozoa, algae, yeasts, fungi, actinomycetes, and bacteria. Soil is distinguished from subsoil chiefly by the presence of organic matter in the soil. The organic matter is composed of dead plant and animal tissues and the products of their decomposition by microorganisms. The microorganisms derive their energy by oxidizing plant and animal residues. Plants depend upon nutrients made available by microorganisms which form an essential link in the cycle of food in nature.

The soil is the greatest natural reservoir of microorganisms. These take part in many reactions in the soil. Some microorganisms are specific, taking part only in certain reactions (*Nitrosomonas*, which oxidize ammonia to nitrite). Some are more general in their metabolism (heterotrophic types, which depend on organic material for energy source) and take part in many reactions. One group of organisms may use the products of another, for example, during the course of decomposition of complex nitrogenous material to nitrate. In addition to forming simpler degradation products, microorganisms synthesize many complex substances. These are bound up in microbial protoplasm while excess amounts may be liberated to provide essential substances for other groups of microorganisms. In contrast to this associative action are antagonisms caused by competition for the same food or by the ability of many organisms to produce antibiotics.

The net result of such associations and antagonisms is the establishment of a microbial equilibrium, which varies with the geological origin and evolutionary history of the soil. Although the equilibrium is ever shifting under the influences of season, temperature, moisture, and state of cultivation, the micropopulation of a soil of definite type has a characteristic composition, consisting of organisms that have become adapted to the particular environment.

Microorganisms are not uniformly distributed throughout the soil, because soil is not homogeneous but comprises a variety of microenvironments. The organisms congregate in colloidal films about the surface of soil particles and are particularly abundant around fragments of decaying plant and animal debris.

The majority of soil microorganisms are most active at pH 6.0–6.8. Some sulfur-oxidizing bacteria tolerate a pH as low as 2.0, other organisms a pH as high as 10. A supply of free oxygen is important to most organisms. Almost all fungi, actinomycetes and protozoa as well as most bacteria, are aerobic. With abundant oxygen, decomposition proceeds rapidly to completion. However in close-textured and wet soils where anaerobic conditions (molecular oxygen not available) prevail, organic matter is decomposed slowly and incompletely. The number of microorganisms is highest in spring and fall. Proximity to growing plants markedly increases the number of soil organisms.

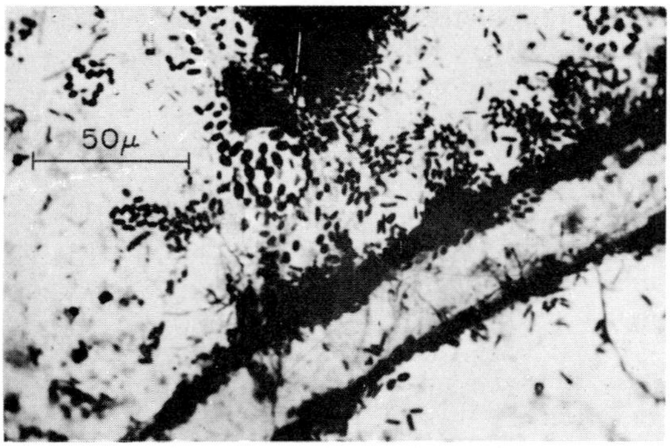

Bacteria attacking fragments of dead plant material in soil. (*T. Gibson, World Crops, vol. 3, no. 4, 1951*)

The soil micropopulation may be divided into two groups, autochthonous and zymogenous microorganisms. The autochthonous organisms, comprising the great majority, are the indigenous forms responsible for the processes occurring under normal conditions. Their character is relatively uniform in soil of definite type and they are little affected by amendments (changes in soil organic matter). In contrast, the zymogenic microorganisms, much less numerous and normally quiescent, increase temporarily to participate in primary decomposition processes upon addition of readily decomposable organic material.

Microbiological soil analysis. The enumeration of soil microorganisms is carried out by culture and microscopic methods. Culture methods depend upon the growth of organisms and permit their isolation for detailed study. In microscopic methods, counts are made from stained films, but since the organisms are killed no cultures can be made.

Culture methods. These include plate and elective culture procedures (*see* CULTURE, ELECTIVE). Plate cultures are used for counts of bacteria, actinomycetes and fungi, and are prepared by suspending a definite weight of soil in sterile water. Dilutions are made of the soil suspension with sterile water. Aliquots (representative portions) of these dilutions are placed in petri plates with an agar medium which will support the growth of the microorganisms in the soil sample. For counts of protozoa the soil aliquots are placed on the surface of agar plates, and after incubation wet mounts are examined microscopically. In order to group soil bacteria on the basis of taxonomy, nutritional requirements, and ability to synthesize substances such as vitamins or antibiotics, isolates (individual bacterial colonies) are taken nonselectively from plate cultures. The relative incidence in soil of a given category of bacteria can be estimated from a study of the individual isolates.

Elective culture methods are used to count microorganisms of special physiological groups, such as algae and the nitrifying, denitrifying, sulfur-oxidizing, and cellulose- or protein-decomposing bacteria. Portions of successive soil dilutions are added to liquid media selected to promote growth of the group in question. The presence of organisms is determined by visual or microscopic observation or by chemical tests for by-products. The use of replicates (that is, more than one sample of the particular soil) permits the calculation of the most probable numbers.

Microscopic methods. Such methods are based on the examination of a definite amount of soil spread over a definite area of a slide and stained. From the known area of the microscope field numbers of organisms may be estimated. As this procedure involves mixing the soil, the "contact slide" method is used to observe the localized distribution of the organisms. A glass slide is inserted in the soil and later removed, stained, and examined for organisms adhering to the surface.

Plating methods, which are more commonly used for counting soil microorganisms, give too low values, for there is no medium on which all will grow. Microscopic methods give far higher numbers than plate counts but suffer from the disadvantage that they do not distinguish readily between living and dead cells. The two methods are complementary. *See* HUMUS; MANURE; NITROGEN CYCLE; PATHOGEN (SOIL); RHIZOSPHERE; SOIL BALANCE, MICROBIAL; SOIL MICROORGANISMS; SOIL MINERAL (MICROBIAL UTILIZATION); SOIL PHOSPHORUS (MICROBIAL CYCLE); SOIL SULFUR (MICROBIAL CYCLE); *see also* ALGAE; BACTERIA, TAXONOMY OF; FUNGI; PROTOZOA; VIRUS; YEAST. [A.G.L.]

Bibliography: E. J. Russell, *Soil Conditions and Plant Growth*, 8th ed., 1950; S. A. Waksman, *Soil Microbiology*, 1952.

Soil microorganisms

Most microscopic organisms in the soil belong to the plant kingdom. Examples are bacteria, actinomycetes, fungi, and green and blue-green chlorophyll-containing algae. The animal kingdom is represented by the protozoa and the myxomycetes. The protozoa may include the soil euglenids, considered by some as organisms intermediate between algae and protozoa. The myxomycetes are often regarded as organisms bridging the gap between protozoa and fungi. Bacteria, actinomycetes, and fungi are the groups most active in decomposing organic residues and in rendering inorganic nutrients soluble. The final result of this activity is the liberation of such elements as carbon, nitrogen, phosphorus, potassium, and sulfur in forms available to plants.

Bacteria. The bacteria exceed all other soil microorganisms in numbers and in the variety of their activities. Numbers may surpass 100,000,000 per gram of soil by plate count or 1,000,000,000 per gram by microscopic count. The bacteria vary in size, shape, growth requirements, energy utilization, and function. Morphologically they are divided into straight or irregular rods, of both sporeforming and nonsporeforming types, thin flexible rods, cocci, vibrios, and spirilla. Short rods and cocci are most frequent, but many of the cocci are the coccoid stage of pleomorphic (varying in shape and size) rods or spores of actinomycetes.

Members of most taxonomic groups, with the exception of certain animal and human parasites, occur in soil. Some taxonomic groups are characteristic of soil alone. Although the identity of many bacteria engaged in specific processes, such as nitrification and nitrogen fixation, is known, a large proportion of the indigenous (autochthonous) organisms have not been classified. Of these, many have not been grown in any culture medium, a requisite for systematic study. Consequently taxonomic knowledge of the autochthonous microflora is imperfect.

On the basis of their nutrition, soil bacteria are divided into autotrophs and heterotrophs. Autotrophs are able to use carbon dioxide as the sole source of carbon for their body tissues; heterotrophs must obtain carbon from organic food substances.

Autotrophic bacteria. These comprise two groups (photosynthetic and chemosynthetic) according to their source of energy. The purple and the green sulfur bacteria are photosynthetic due to the presence of bacteriochlorophyll or chlorobium chlorophyll pigments. Like chlorophyll-containing algae and higher plants, they obtain energy from sunlight.

Other autotrophs are chemosynthetic, deriving energy from various oxidation reactions. Their requirements for food and energy are met by inorganic sources. These autotrophs carry out the process of nitrification in which two stages are distinguished, the oxidation of ammonia to nitrite, and that of nitrite to nitrate. Autotrophic sulfur bacteria derive energy from the oxidation of elemental sulfur, sulfides, sulfites, thiosulfates, and thiocyanates to sulfuric acid, which reacts with soil bases to form sulfates.

Heterotrophic bacteria. Heterotrophic bacteria, which comprise the great majority of soil bacteria, derive both food and energy from the decomposition of organic substances. They embrace a wide variety of morphological and taxonomic types including sporeformers or zymogenous forms, which are bacteria that develop in soil in response to the addition of certain substances like organic matter, or certain processes like aeration. Also included are the far more numerous nonsporeformers which comprise the great majority of the autochthonous microflora. The most abundant forms are short rods and pleomorphic rods.

The majority of the heterotrophs require combined nitrogen to build cell substance. The nitrogen-fixing bacteria utilize elemental nitrogen of the air. These bacteria include symbiotic organisms, such as species of the genus *Rhizobium* that live in symbiosis with leguminous plants, and nonsymbiotic organisms, such as species of the aerobic genera *Azotobacter*, *Beijerinckia*, and *Azotomonas*, and the anaerobic genus *Clostridium* (*see* AZOTOBACTERACEAE; BACILLACEAE; PSEUDOMONADACEAE). The indigenous heterotrophic bacteria may be classified according to their nutritional requirements. Though all require a source of energy, such as a simple sugar, the additional needs of some are satisfied by inorganic salts. Other soil bacteria require amino acids or more complex food sources, and some more exacting bacteria require factors present in soil extract. The proportion of each of the nutritional groups is fairly constant in soil of definite type. However, increased fertility, such as that resulting from fertilizer treatment, is reflected in an increase in the proportion of bacteria with complex requirements at the expense of those with simple nutritional needs. Although there is no precise correlation between nutritional requirements and morphological type or taxonomic grouping, *Pseudomonas* species are more abundant among the nutritional group with simpler needs, and the pleomorphic types, particularly *Arthrobacter* species, are relatively more numerous in the group with the most complex requirements.

As many as 25% of the indigenous soil bacteria capable of being isolated may require one or more vitamins for growth. The vitamins most essential are, in order of frequency, thiamine, biotin, and vitamin B_{12}. A smaller percentage require other B vitamins as well as the terregens factor, a substance found only in soil that promotes bacterial growth. *See* NITROGEN CYCLE; SOIL SULFUR (MICROBIAL CYCLE).

Actinomycetes. Next to the bacteria in numbers, the actinomycetes range from hundreds of thousands to several millions per gram of soil. They are more abundant in dry and warm soils than in wet and cold soils. With increasing depth of soil, their numbers are reduced proportionately less than those of bacteria. Actinomycetes are particularly abundant in grassland soil. The three genera of Actinomycetales occurring most commonly in soil are *Nocardia*, *Streptomyces*, and *Micromonospora*. Thermophilic forms, represented by the genus *Thermoactinomyces*, are active in rotting manure and also may be present, though inactive, in normal soil. *Streptomyces* species are the dominant types. Although they are largely saprophytes, a few species, such as those associated with potato scab, are parasitic. *See* ACTINOMYCETACEAE; STREPTOMYCETACEAE.

Less is known of the function of the actinomycetes in soil than of the bacteria. They are heterotrophic and are nutritionally an adaptable group, less demanding in growth requirements than many bacteria. They take part in the decomposition of a wide range of carbon and nitrogen compounds, including the more resistant celluloses and lignins, and are important in humus formation. Actinomycetes are responsible for the earthy or musty odor characteristic of soil rich in humus. *See* HUMUS.

As many as 60% of the actinomycetes isolated from soil by plating methods may show antagonism toward bacteria or fungi in artificial culture. The importance of this antibiotic-producing capacity under normal soil conditions is not known. However, although antibiotics can rarely be detected in soil and then only under abnormal conditions, they may be important in microenvironments where intense microbial activity takes place.

Fungi. Fungi are present in numbers ranging from several thousand to several hundred thousand per gram of soil. They occur extensively in the mycelial state, as well as in the form of spores. Since plate colonies may develop from fragments of mycelium or from spores, plate counts give only an approximation of the abundance of fungi in soil. As with bacteria, some fungi do not grow on plates; consequently plating methods give minimum counts. Most fungi require humid, aerobic conditions for growth and spore formation. They are most common near the surface of soil and are more abundant in lighter, well aerated soils than in heavier soils. Because the optimum pH range for fungi is 4.5–5.5, they are more prevalent in acid soils which are less favorable to bacteria and actinomycetes.

Ecologically, two broad groups of soil fungi may be recognized, the soil-inhabiting, and the root-inhabiting, fungi. The soil-inhabiting fungi are able to survive indefinitely as saprophytes and have a general distribution in soil. They include not only obligate saprophytes, but also some unspecialized parasites which are able to infect plant roots, but whose parasitism is only incidental to their saprophytic existence. Root-inhabiting fungi are specialized parasites that invade living root tissues. Their distribution in soil is localized and depends upon the presence of the host plant. Their activity diminishes following death of the plant and they persist in soil only as resting spores or sclerotia. Mycorhizal fungi are included among the root-inhabiting fungi.

Soil fungi are heterotrophic and have a wide variety of food requirements. All obtain their carbon entirely from simple carbohydrates, alcohols, or organic acids. But while some fungi can utilize inorganic nitrogen, others require more complex forms, or vitamins, chiefly thiamine and biotin.

Soil fungi do not comprise as many physiological groups as do the bacteria, but as a group they are more versatile in their ability to decompose a great variety of organic compounds. The saprophytes, the true soil inhabitants, may be divided into groups depending upon the nature of the substrate favoring their development. Two such groups are the sugar fungi and the cellulose-decomposing fungi. Other groups attack some of the most resistant substances, such as lignins, vegetable gums, and waxes. When plants die, or when fresh plant material is added to soil, the growth of fungi is greatly stimulated. Those able to attack the more soluble constituents, such as sugars and other simpler carbon compounds, develop rapidly. Chief among such forms are the Phycomycetes. As the special substrate is exhausted, other types flare up and attack progressively more resistant components of organic residues. Cellulose and hemicelluloses are decomposed by a variety of fungi including species of *Penicillium*, *Aspergillus*, *Sporotrichum*, and *Fusarium*. Fungi are the predominant lignin-decomposing organisms. Various simple fungi can attack the lignin of straw and leafy plant material although the higher Basidiomycetes are most active in decomposing lignin-rich residues.

Although polysaccharide-forming bacteria play a part, fungi are chiefly responsible for improving the physical structure of soil by exerting a binding effect on loose particles, thus forming water-stable aggregates. This binding effect is caused by the growth of mycelia which form fine networks that entangle the smaller particles. The soil-binding effect is favored by addition of fresh organic material whose decomposition products provide cementing substances. *See* FUNGI; PATHOGEN (SOIL); RHIZOSPHERE.

Yeasts. A group of simple fungi, yeasts occur in soil only to a limited extent, in the surface layers. In field soils their numbers are small, some samples being devoid of yeasts. They are found most frequently in the soils of orchards, vineyards and apiaries where special conditions, particularly the presence of sugars, favor growth of yeasts which invade the soil. Soil is not a favorable medium for their growth and yeasts do not play a significant part in soil processes. *See* CRYPTOCOCCALES; SACCHAROMYCETALES; SPOROBOLOMYCETALES.

Algae. These are widely distributed in soils, developing most abundantly in moist, fertile soils well supplied with nitrates and available phosphates. They contain chlorophyll and in the soil surface layers, where they are chiefly confined, function as green plants converting carbon dioxide and inorganic nitrogen into cell substance by means of energy derived from sunlight. Smaller numbers occur at lower depths where, in the absence of sunlight, they exist heterotrophically. The soil algae comprise the green algae (Chlorophyceae), the blue-green algae (Myxophyceae), and the diatoms (Bacillariaceae). In acid soils green algae predominate, while in neutral or alkaline soils the other groups are more prominent. Numbers vary widely, ranging from a few hundred to several hundred thousand per gram of soil.

As autotrophs, algae are of importance in adding to the organic matter of soils. They play a fundamental ecological role on barren and eroded lands by colonizing such areas and synthesizing protoplasm from inorganic substances. Several blue-green algae are able to fix atmospheric nitrogen and are of agricultural significance, particularly in rice culture. Under the water-logged conditions needed for this crop, blue-green algae develop abundantly and may increase the nitrogen supply by as much as 20 lb per acre. *See* ALGAE.

Protozoa. Occurring in all arable soils, protozoa are largely confined to the surface layers, although in drier, sandy soils they may penetrate more deeply. Numbers usually range from a few hundred in dry soil to several hundred thousand per gram in moist soils rich in organic matter. Most soil protozoa are flagellates and amebas; ciliates are less frequent although they are often found in wet soils and swamps. Protozoa are active in soil only when living in a water film. The majority are able to form cysts, and in this inactive state they can withstand desiccation.

Although a few flagellates, such as *Euglena*, have chlorophyll and are autotrophic, and others can live saprophytically by absorbing nutrients from solution; the majority of soil protozoa feed by ingesting solid particles, mainly bacteria. Not all bacteria are suitable as food. Amebas have decided preferences for certain bacterial species and will not ingest others, particularly pigmented bacteria. The formation of cysts by protozoa is favored by some bacterial types, not by others. Excystment of some amebas requires the presence of bacteria; others are independent of bacteria. Though protozoa are a factor in maintaining the microbial equilibrium in soil through their selective action on bacteria, their effect is limited and is not considered detrimental to the activities of the micropopulation as a whole. *See* PROTOZOA.

Myxomycetes. Myxomycetes, or slime fungi, form a minor group of soil microorganisms intermediate in character between the flagellated protozoa and the fungi. They possess a motile, flagellated stage in their life cycle, and later form large aggregates of cells, or coalesce into jellylike masses of naked protoplasm. These eventually form spores which give rise to flagellated forms. Like the protozoa, myxomycetes feed on bacteria. *See* ACRASIALES.

Viruses and phages. Although these ultramicroscopic organisms exist in soil, little is known of the part they play in soil processes. Viruses that attack plants and animals can in some cases be transmitted from the soil. Phages, which are active against bacteria and actinomycetes, limit susceptible microorganisms and thus affect the microbial balance. Those that attack the various species of symbiotic nitrogen-fixing bacteria may prevent effective inoculation of legumes, particularly in soils in which the same crop has been repeatedly grown. The deleterious effect of phage on the nitrogen-fixing bacteria was formerly ascribed to direct lytic action (dissolution of the bacterial cell). However, this is now considered to result from the development of phage-resistant mutants which are less effective in fixing nitrogen than the parent strains. *See* BACTERIA; BACTERIOPHAGE; SOIL MICROBIOLOGY; VIRUS. [A.G.L.]

Soil mineral (microbial utilization)

Microorganisms utilize soil minerals for their own growth, and also help make the essential nutrients available to higher plants. A soil is fertile mainly because it can supply growing plants with nutrients (calcium, magnesium, potassium, and phosphate) in forms which can be readily taken up by plants. The conversion of the nutrients in soil minerals into forms available to plants is of course partly carried out by the plant roots themselves, but experiments have begun to show that microorganisms play a part in this conversion.

Soil formation. New soil is perpetually being formed all over the world by the breaking up of rocks into fine particles and the dissolving out of minerals from them. This process, the so-called weathering of rocks, was at one time supposed to be caused entirely by physical and chemical agents—heat, frost, water, and the oxygen of the air. It is now known that exposed rocks are subject also to microbial attack. When the island of Krakatau was visited three years after the great volcanic explosion which blew off its top, microscopic blue-green algae were found growing on the bare rocks, the first living things to appear there. These blue-green algae are the most self-supporting of all forms of life, for they are photosynthetic, obtaining carbon from the carbon dioxide in the air, and also nitrogen-fixing, obtaining nitrogen from the atmosphere. In 1946, Russian microbiologists found that the blue-green algae on freshly exposed rocks are soon accompanied by bacteria, among which nitrogen-fixers and autotrophic nitrifiers are prominent. The nitrogen-fixers add to the nitrogen supply, and the nitrifiers, because they fix carbon dioxide, add to the carbon supply in the film of growth on the rock. The two groups pave the way for other bacteria and fungi, and the film of organic growth, which is continually dissolving mineral nutrients from the rock below it, increases until it can support the growth first of lichens, then of mosses, and finally of higher plants.

Buried rocks are also broken down to form new soil; if they are not buried too deeply, there is probably a succession of microbial growth on them. This has not been proved, because plant roots, at depths which they can reach, also break up rocks. Rock minerals are dissolved by carbon dioxide and by organic acids; both of these are given off by plants and by microorganisms. Since it is impossible to distinguish between them, in most cases it must be assumed that both plant roots and microorganisms are taking part in soil formation.

Utilization of minerals in formed soils. In soil which is already formed, microorganisms may release nutrients from the soil minerals and make them available to plants. In soil where a particular nutrient is scarce, they may render it unavailable by assimilating the nutrient as fast as they release it from the minerals; in this case they may be said to be competing successfully with the growing plants.

The mechanism of release of nutrients consists in dissolving out the nutrient from the soil mineral and converting it into a soluble salt or (in the case of heavy metals) into a complex with a chelating compound. Carbon dioxide is probably the most abundant dissolving agent, as all microorganisms and plant roots produce it in respiration. Organic acids such as lactic acid are also important dissolving agents which are secreted into the soil by plant roots and microorganisms, particularly fungi. Some fungi and bacteria from podzols (relatively infertile soils) are evidently adapted to dissolve minerals in a poor soil because in cultures of them more acid is formed in poor than in rich nutrient media. The gum, or slime, produced by some bacteria can take up phosphate, sulfate, and possibly potassium from culture media; if the same mechanism operates in soil, it should supply the bacteria with more of these nutrients.

Inorganic acids are produced by some autotrophic soil bacteria. For example, the nitrifiers *Nitrosomonas* and *Nitrobacter* produce nitric acid as the end result of their combined activity, and the sulfur-oxidizing *Thiobacillus* species produce sulfuric acid. These acids quickly combine with metals in the soil minerals to form nitrates and sulfates. The autotrophic bacteria are useful to plants not only because they supply nitrogen or sulfur in an assimilable form, but also because they make soluble salts of most of the metals essential for plant growth, such as potassium and magnesium. Hydrogen sulfide, produced during the bacterial decomposition of proteins, may render tertiary phosphates soluble. Even if it causes the precipitation of iron and other metals, the sulfides are subject to bacterial oxidation, with the formation of soluble sulfates.

Some of the bacteria that reduce ferric hydroxide convert the iron into a chelated form held in an organic complex, but the nature of these organic complex-forming substances is quite unknown.

Phosphates. The most definite evidence for the beneficial effect of soil microorganisms on plant growth has been obtained in work with phosphates in Holland, Russia, and Australia since 1950. Most of the phosphate in soil is combined with calcium or iron in insoluble compounds; many soil bacteria and fungi, when they are first isolated from soil, can dissolve these insoluble phosphates, though they lose this ability when kept in artificial culture. These phosphate-dissolvers are commoner in the rhizosphere (the zone immediately surrounding plant roots) than elsewhere, and it has been found that plants grow better and take up more phosphate from insoluble calcium phosphate in soil with rhizosphere organisms present than in sterile soil. Cultures of phosphate-dissolving microorganisms are added to compost heaps in Russia, and are said to improve the quality of the compost.

The mycorhiza fungi on the roots of pine trees are supposed to supply the trees with phosphate, but this is unlikely since trees without mycorhiza, such as oaks, also increase soluble phosphate in the soil around their roots. It is more probable that phosphate is brought up from the subsoil by the tree roots themselves.

Microorganisms may deprive plants of phosphate under certain conditions. All microorganisms need phosphate, and some soil species, the nitrogen-fixing *Azotobacter* spp. for instance, need quite large amounts. Fungi have a particularly high phosphate requirement, and if much organic matter with a low phosphate content is added to soil, the fungi that develop on it may fix so much phosphate in organic compounds that a temporary phosphate starvation is induced in plants growing in the same soil.

Potassium, calcium, and magnesium. Very little is definitely known about the effect of microorganisms on the supply of these three elements to plants. It may reasonably be assumed that all three are turned into soluble salts through the action of the sulfur-oxidizers and the nitrifying bacteria. Potassium is liberated in the breakdown of complex silicates by the so-called silicate-decomposing bacteria. It is quite probable that the breakdown of silicates is not a specific enzymatic process carried out by a special group of bacteria, but is caused rather by acids produced by many different soil bacteria and fungi.

Iron and manganese. Bacteria which reduce oxides of iron and manganese to soluble ferrous and manganous salts are very common in soil. The most efficient of them appear to be so because they make either organic acids or chelating substances. It is probable, but by no means proved, that these bacteria improve the supply of iron and manganese to plants. They also induce movement of both metals in the soil profile.

Microbial oxidation of iron and manganese compounds to insoluble oxides and hydroxides also takes place in soil. It has been claimed that the autotrophic iron bacteria, which are common in water and in bogs, also occur in drier soils, and that they make the iron concretions and hardpans in tropical lateritic soils, but there is no experimental evidence for this claim.

It is therefore probable, though not yet proved, that soil microorganisms can increase the supply of essential nutrients to plants. A great deal more experimental work will be necessary to decide how much microbes contribute to plant nutrition by modifying soil minerals. *See* NITROGEN CYCLE; RHIZOSPHERE; SOIL MICROBIOLOGY; SOIL MICROORGANISMS. [J.M.]

Soil phosphorus (microbial cycle)

This is essentially a phosphate cycle. Plants assimilate phosphorus as phosphate, the $H_2PO_4^-$ ion, and build it into organic compounds such as phytin, nucleic acids, and phospholipids. The microbial breakdown of dead plant tissues liberates the phosphate again, so that there is an alternation in the soil between organic and inorganic phosphate. Microorganisms themselves assimilate phosphate, and consequently much of the organic phosphate in soil is contained in microbial cells. Plants may be deprived of phosphate by the addition of phosphate-deficient organic wastes to soil, because the fungi which develop on the waste assimilate all the available phosphate.

There is also an alternation in soil between soluble and insoluble inorganic phosphates. Microorganisms can increase the phosphate supply to plants by dissolution of insoluble, tertiary phosphates through acid production; such microbes are particularly active in the rhizosphere. Others may precipitate soluble primary or secondary phosphates as tertiary salts as a result of the production of alkali. *See* SOIL MICROBIOLOGY; SOIL MINERAL (MICROBIAL UTILIZATION). [J.M.]

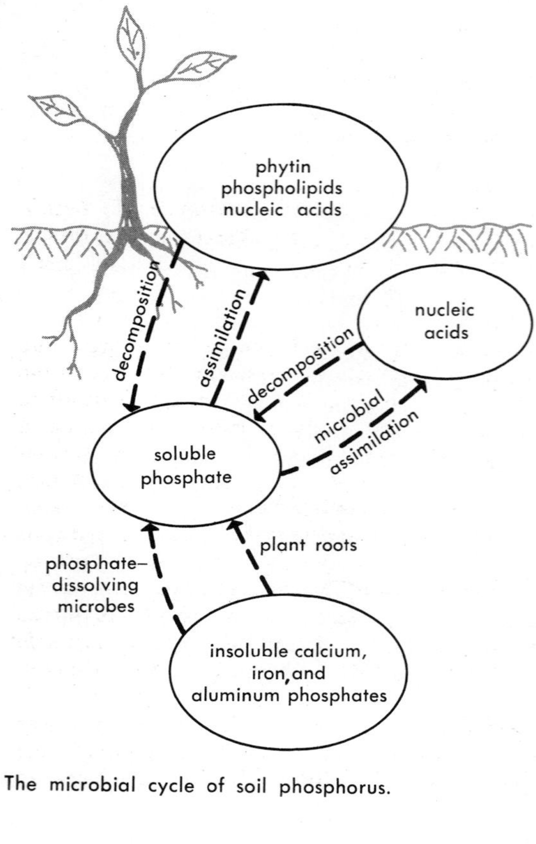

The microbial cycle of soil phosphorus.

Soil sterilization

As used in the field of weed control, and rather generally by agricultural chemists, the term soil sterilants refers to chemicals that render the soil unfit as a culture medium for the growth of higher plants. Plant pathologists, nematologists, and zoologists use the same term to connote a material used for complete sterilization, that is, the rendering of soil completely devoid of all life. This is often done with heat or by soil fumigation and is usually only temporary. *See* FUMIGANT; FUNGISTAT AND FUNGICIDE; NEMATOCIDE.

Soil sterilants, as used in weed control, may be classified into two general groups—temporary and permanent. However, these terms are only relative, and the persistence of toxicity from any one chemical depends upon many factors. Principal among these are rainfall, textural grade of the soil, organic matter in the soil, height of the water table, seasonal differences in temperature, differences among plant species, and volatility of the agent.

Prior to 1945, soil sterilants were used almost entirely on noncropped areas. The chemicals employed at that time were arsenic, borax, boron ores, and sodium chlorate; the latter two were often used in combination. Typical of noncropped areas are railway ballast, irrigation ditchbanks, parking areas, mill or factory yards, airport landing strips, and firebreaks. Such treatment is also used around signboards, telephone and power poles, oil storage tanks, and highway guardrails. Chemical treatments lessen the need for hand hoeing, weed knifing, burning, and other costly operations.

Inorganic compounds. Arsenic is the most permanent of the soil sterilants. It may be applied dry as white arsenic (arsenious oxide) or in solution as sodium arsenite. The latter is poisonous and also attractive to cattle so that it must be used with great care. Arsenic sterilization has been known to last for as long as 10 years.

Borax or boron ores, such as kramerite or colemanite, are less toxic to plants than arsenic and less persistent. However, they are not poisonous and they are easy to handle. These materials alone, and in combination with sodium chlorate, are used along thousands of miles of highways and railroads.

Sodium chlorate is only temporary in its effects in the soil; however, it is readily soluble in water, and it moves freely with moisture in the soil. Hence, it may be used to kill deep-rooted perennial weeds on both cropped and noncropped soils. Sodium chlorate is hazardous to use by itself because, when combined with organic material, such as straw, wood, or clothing, it forms a highly flammable mixture that ignites from friction and burns with an intense flame that cannot be smothered. Borax or sodium pentaborate reduces this flammability so that the mixture is safe to use. Also, the boron compounds are toxic and are less readily leached from soils. Borate-chlorate mixtures are used in very great quantities for soil sterilization in industrial areas all over the world.

Organic compounds. The organic herbicides introduced since 1945 are rapidly changing the picture in the field of chemical weed control. Many of these may be used in preemergence treatments to sterilize temporarily the soil in croplands, and some are highly selective, killing weeds without harming crops. Among the organic herbicides are the chlorophenoxy compounds, 2,4-dichlorophenoxyacetic acid or 2,4-D; the carbamates, isopropyl-*n*-(3-chlorophenyl) carbonate or CIPC; the substituted ureas, 3(3,4-dichlorophenyl)-1,1-dimethylurea; the symmetrical triazines; and many others.

Used at dosages of about 1 lb per acre, these newer materials may be used to control weeds on croplands; they are usually applied at seeding time or soon thereafter; this use is termed the preemergence method. The same materials used at rates of 10–40 lb or more per acre are used to bring about a more permanent type of soil sterility.

Because most of these materials are selective, they are often used in combination to control mixed weed populations. A few such combinations are 2,4-D plus monuron, monuron plus borax, monuron plus chlorate, monuron plus sodium trichloroacetate, 2,4,5-T plus dalapon. Eventually, chemicals should be available that will eliminate weeds in every situation in which they are pests. *See* AGRICULTURAL CHEMISTRY; AGRICULTURAL SCIENCE (PLANT); HERBICIDE. [A.S.C.]

Soil sulfur (microbial cycle)

Plants and microorganisms assimilate sulfur from soil sulfates, and convert it into organic sulfur compounds. Sulfates are formed in soil by oxidation of sulfides, which are derived from (1) the parent rock from which the soil is formed, (2) breakdown of organic sulfur compounds, (3) reduction of sulfates by anaerobic bacteria of the genus *Desulfovibrio* (*see* SPIRILLACEAE). Corrosion of water pipes and other buried iron structures is caused by species of *Desulfovibrio*. The oxidation of sulfides to sulfur in soil is probably not a microbial process but a purely chemical one. The sulfur is oxidized micro-

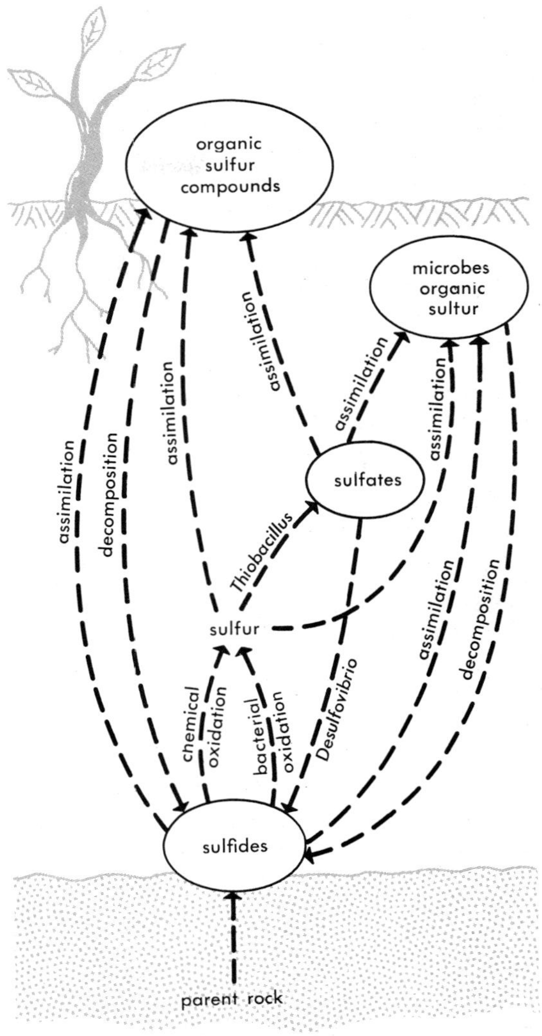

Microbial cycle of soil sulfur.

bially to sulfates, by the autotrophic *Thiobacilli*. *See* SOIL MINERAL (MICROBIAL UTILIZATION); THIOBACTERIACEAE. [J.M.]

Solar battery

A battery which converts the energy of light into electric energy. The most common solar cell is the silicon photovoltaic cell shown in the illustration.

Pure silicon is an *n*-type semiconductor. It has a high free-electron density, so that electric current is a flow of electrons. By exposure to boron vapor, the surface of the silicon becomes a *p*-type semiconductor. This type has a low electron density but a high hole density, so that current is a migration of holes, or positively charged sites, through the material. The region between the *p*- and *n*-type silicon is called the barrier region, and the whole structure is called a *p-n* junction. *See* SEMICONDUCTOR.

Exposure of the treated surface to light produces light absorption within a layer about 0.00001 cm thick. Each light photon absorbed displaces an

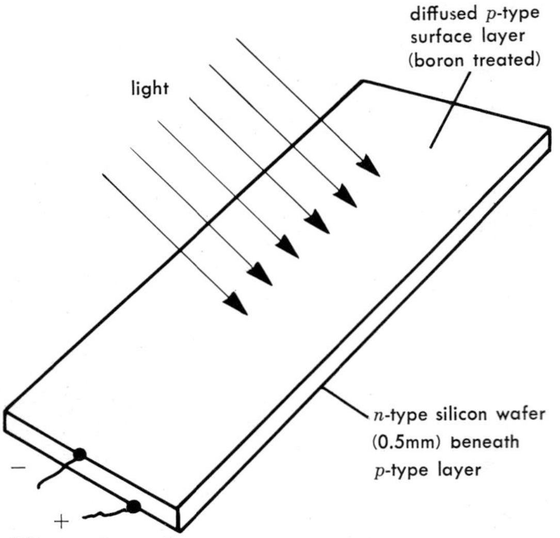

Silicon solar cell.

electron, producing both a free electron and an electron vacancy or hole. Since the original surface had a low electron density compared to the hole density, the effect of this photon absorption is to increase the electron density while increasing the hole density to a much lower extent.

A portion of the excess electrons will have sufficient energy to move through the barrier region into the *n*-type semiconductor region, where they are free to move into an external circuit and thus deliver power to a load.

The solar cell may be considered as a constant-current generator, the constant current being equal to that obtained on short circuit. This generator is shunted by the *p-n* junction, acting as a diode rectifier, and the load resistance. In addition, there are both shunt and series resistance elements present within the solar cell.

Solar battery installations have been made in high-altitude rockets. A bank of 8 cells, each 1 cm^2 in area, was connected in series and located behind a pyrex-glass window. A mounting was provided so that the cells were flush with the outer skin of the rocket. This was one battery. A total of five were used. Four were mounted 90° apart at the base of the nose cone. One was mounted directly above one of the lower four. The four batteries were connected in parallel to a load of 150 ohms. The single battery was also connected to a load of 150 ohms.

The output voltage of the four units in parallel varied from 2.7 to 1.0 volt, while the single-unit voltage ranged from 2.1 to 0.0. These variations were due to changes in the incident light as the rocket spun and changed its angle with respect to the sun. The peak outputs reported are equal to 4.5 ma at 0.338 volt/cm^2 (1.52 mw/cm^2) and 14.0 ma at 0.263 volt/cm^2 (3.68 mw/cm^2), respectively. These are equal to 1.41 watts/ft^2 and 3.42 watts/ft^2 respectively.

Other experimental installations have been made to determine the suitability of solar cells in applications such as furnishing power for transistorized telephone repeaters in remote areas.

Temperature has a considerable effect on the output of the silicon solar-cell battery. Contrary to the usual effect, the output goes up as the temperature decreases. The output is greater at −100°F than at room temperature. [S.EI.]

Solar constant

The rate at which energy is received from the Sun just outside Earth's atmosphere. At Earth's mean distance from the Sun, the solar constant is 1.36×10^6 ergs/(cm^2)(sec). Depending on the Sun's distance from the zenith, up to a third of this energy may be scattered in Earth's atmosphere. From the measured solar constant, the total radiation of the Sun is 3.86×10^{33} ergs/sec.

The energy emitted by the Sun has been thought to remain constant. However, recent observations indicate that the Sun increased 2% in brightness from 1954 to 1959. The increase was measured by comparing the blue portion of the spectrum of sunlight reflected by Uranus and Neptune with the direct light from 16 nearby stars. The increase in solar constant coincides with the peak of the 11-year sunspot cycle. Further observations, however, are necessary to establish a relationship. *See* SUN.

[J.W.E.]

Solar energy

The energy that can be obtained in the form of heat and power from the Sun's radiation. The total amount of energy showered upon the Earth from the Sun is about 10^{18} hp-hr per annum. To equal this rate, the Earth's total supply of combustibles would have to be burned up in about three days. However, not all this solar energy is available. About half of it is turned back into space by the atmosphere before it reaches the surface of the Earth. The world's vegetation utilizes about 6×10^{13} hp-hr per annum of solar energy but reflects back into space a thousand quanta for every quantum that it usefully employs. However, about 70% of incident sunshine is retained during the day. About 15% of retained solar energy is absorbed by bare earth. The other 85% is used to evaporate water from the hydrosphere and from vegetation, to raise the temperature of surface water, and to cause the growth of marine vegetation. The growth of marine vegetation has been estimated to use nearly 5×10^{14} hp-hr per annum, eight or nine times the rate of growth of terrestrial vegetation. Energy used for evaporation cannot be recovered as power because it is balanced by nocturnal radiation away from the Earth, with resulting condensation of water vapor. However, the energy required to lift water vapor above the Earth is partly recoverable in the form of water power.

In addition to water power, which can be and probably will be developed to the extent of about 0.001% of the sunshine absorbed by the Earth (5×10^{12} hp-hr per annum, 10 times as much water power as the world was recovering in 1957), solar energy can be utilized in several other more

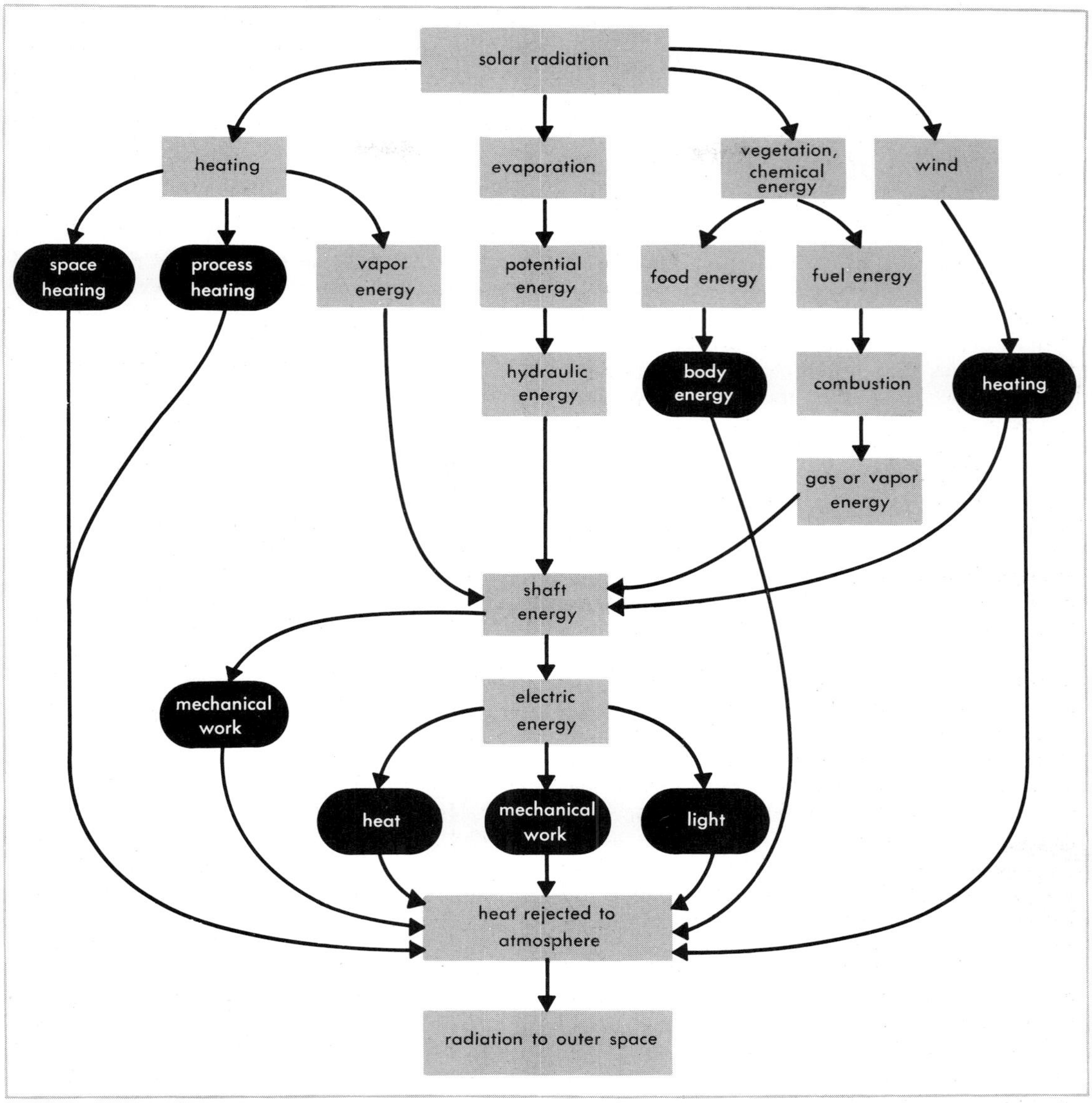

Utilization of solar energy. (*Power Magazine*)

important ways. A portion of the 3×10^{16} hp-hr absorbed by bare earth during the day undoubtedly can be recovered, and a portion of the 5×10^{14} hp-hr of solar energy now being utilized to form food through photochemical reactions undoubtedly could be diverted to the formation of fuels. This applies primarily to marine growth.

To get these figures in proper perspective, the 1957 total consumption of energy in the world was about 5×10^{13} hp-hr. To equal this amount from solar energy alone, it would be necessary to capture 0.2% of the sunlight absorbed by the bare earth of the entire world, or one-tenth of total marine vegetation, or smaller proportions of each to add to what may be secured by hydroelectric power. It is just conceivable that the world's future demand for energy might be met entirely from solar radiation. The potential is there, but many important inventions will have to be made.

Biological utilization. Photosynthesis leads to the average annual production on the Earth of around 100,000,000,000 tons of vegetation and a like amount of oxygen. For balanced production of wood, the energy requirement of the world (1957) could be satisfied with about 8,000,000 sq mi of forest, about one-half the total forested area of the world, but this assumes no other demands upon wood—no lumber or paper. As a matter of fact, in many parts of the world, wood has been cut at much too rapid a rate. The peak of production of wood in the United States was reached in 1907, and production has been declining irregularly ever since.

Wood is relatively expensive (and energy consuming) to transport as fuel. For widespread distribution of energy, vegetation must be converted to a liquid, that is, to alcohol by fermentation or to gasoline by the Fischer-Tropsch process. But the processing of all food crops in the United States would give less than half enough fuel to operate motor vehicles. The more expensive processing of all agricultural wastes would give only 30%. Costs would be unreasonably high, and more than three times as much energy can be obtained by burning vegetation directly as by burning liquid derivatives.

Work on the single-celled alga, *Chlorella pyrenoidosa,* has shown that it is possible to obtain a many-fold increase in the efficiency of utilization of sunlight through the agency of chlorophyll. Such marine vegetation as *Chlorella* could, of course, be used as fuel, but to produce enough fuel to operate motor vehicles alone would require an area for cultivation about the size of the state of Louisiana. Industrialization of photosynthesis is in its infancy and may have a brilliant future. The remarkably efficient job done by chlorophyll for the endothermic photochemical reaction of photosynthesis has been just about equaled by such synthetic dyes and inorganic chemicals as thionine and ceric perchlorate. Research on liquid and solid photogalvanic cells may some day lead to direct conversion of solar radiation to electric power.

Utilization as heat. The simplest utilization of quantum energy is, of course, in development of heat. In Arizona, it is estimated that about 440,000 hp-hr per annum per acre in the form of power might be developed from solar energy. This is more than seven times as much power as could be generated by burning the balanced production of terrestrial vegetation from a fertile acre, but to produce by optical means all the heat and power consumed in the United States in 1957 would require an area of 50,000 sq mi of arid country with maximum sunshine. The most efficient and economical method of optical capture may be the flat-plate collector (two or more panes of glass backed by a black sheet of copper), but much attention has been paid also to optical concentration of sunlight with mirrors, which is capable of attaining temperatures approaching 7000°C. The major problems in using solar heat for power (aside from cost) are the oscillation between night and day and the lack of known means for massive storage of energy.

For space heating, the major difficulties are that buildings are not all in sunny climates, and solar heating in most areas of congested living—city houses, apartments, hotels, and office buildings—would seem to be out of the question. Storage of solar heat has been reasonably solved with insulated masses of water or with chemicals that have high heats of fusion and that melt between 85 and 110°F. Some saving of fossil fuel consumption will undoubtedly be made by solar space heating, and some use of solar heat is being made now in southern states for heating of domestic water supplies.

Solar energy can be made to function as a useful auxiliary to the heat-pump. The coefficient of performance of the heat-pump is raised by supplying heat to the evaporator. The advantage of providing a source of heat at reasonably high temperature is that smaller heat-pump equipment is required, thus smaller interest on investment. Solar-heat collectors, under some circumstances, can accomplish this result, and devices for storage of heat can theoretically lower peak demand for electric power by collecting heat in mild weather to deliver in extremely cold weather. Solar-heated houses have been built in Dover, Mass., in Cambridge, Mass., and in Phoenix, Ariz. Under relatively unfavorable circumstances, it has been found possible to supply about 80% of needed comfort-heat with solar energy. In Arizona, the proportion is nearer 100%. However, until economic pressure exists, no house is likely to be built to employ all devices for optimum fuel economy. *See* ENERGY SOURCES; FUEL; HEAT PUMP; INSOLATION; PHOTOCHEMISTRY; PHOTOSYNTHESIS; SOLAR BATTERY; SOLAR ENGINE; SOLAR RADIATION; WATER POWER. [E.AY.]

Solar engine

An energy conversion device using solar radiation as primary source, to produce either thrust or auxiliary power. A measure of the radiative energy available to operate a solar engine is the solar constant which in space, at mean distance of Earth from Sun, has the value $S_♁ = 1.36$ erg/(cm²)(sec) = 1.36 kw/m² = 92 ft-lb/(ft²)(sec) ~ 1/30 cal/(cm²)(sec). Its variation with solar distance is shown in Fig. 1. The useful application of solar

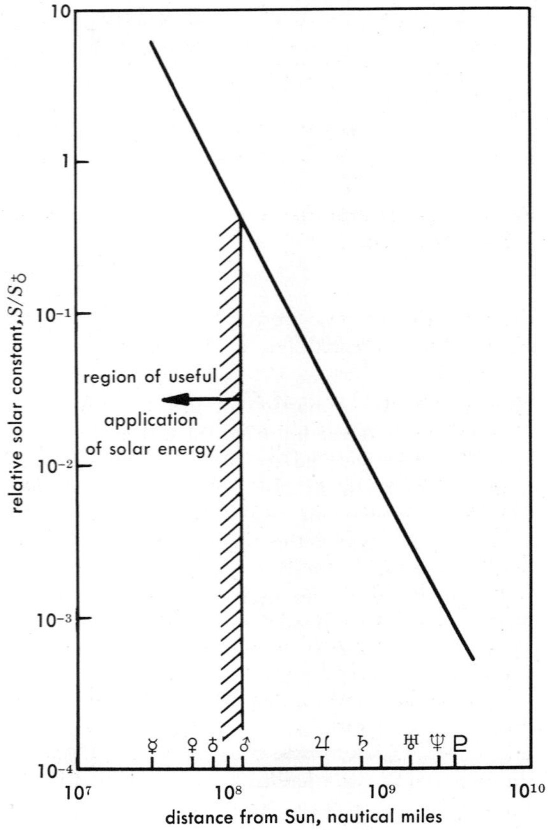

Fig. 1. Ratio of local solar constant to solar constant at Earth's distance.

energy apparently is restricted to the inner solar system.

Reflector boiler. One way to use solar energy in space is to collect it in a focal area, occupied by a boiler, to heat a working fluid which can be either expelled to produce thrust directly, called a solar heat exchanger drive (SHED), or used to drive a turbine-generator system producing electrical energy, called a solar turboelectric drive (STED).

At Earth's distance, large reflector areas are needed. Therefore, the collector must be of light weight. The lightest designs proposed so far involve transparent polyethylene spheres, stabilized by low pressure or electrostatically. Half the sphere is coated with a highly reflective material (Fig. 2).

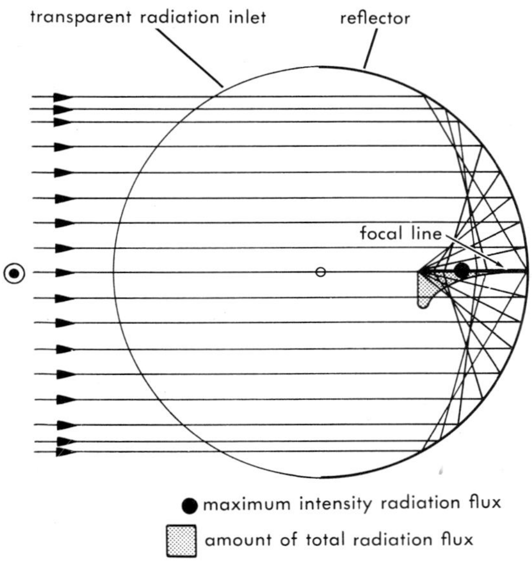

Fig. 2. Optical characteristics of collector sphere for solar-heated drive.

Electrostatically stabilized spheres would be less sensitive to meteoritic dust than gas-filled ones. A spherical collector focuses solar rays on an optical axis equal to half the radius (Fig. 2). This property allows more efficient heating with reduced radiation losses from the boiler surface than a parabolic reflector, which is valuable chiefly if high local temperatures are required. A representative SHED design is shown in Fig. 3. The best working fluid is hydrogen. The best duct material is tungsten, which allows the highest heating temperatures. Figure 4 shows the basic performance data for a SHED at Earth's distance. The STED is, in principle, similarly designed.

Photovoltaic cell. Solar radiation is converted directly into electricity by means of the photovoltaic effect, which produces an electric current if the junction of two dissimilar metals is illuminated (*see* SOLAR BATTERY). The most widely known photovoltaic cell consists of a silicon crystal whose radiation-exposed surface is treated by diffusing boron to the depth of about 1 micron. The crystal

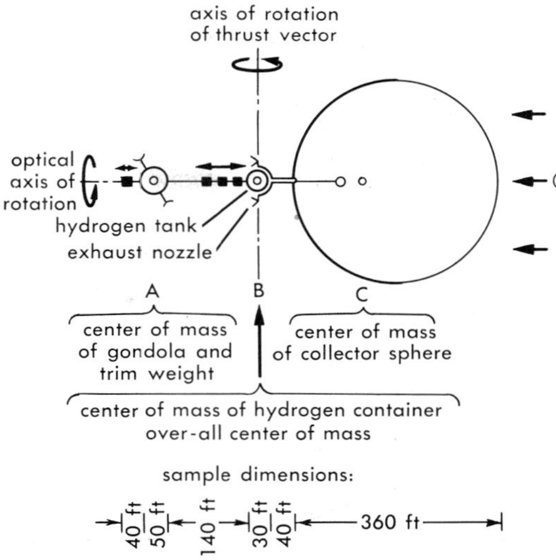

Fig. 3. Typical solar heat exchanger spacecraft. Hydrogen container and two exhaust nozzles at its poles are close to heater behind reflector. Gondola and trim weights are farther out to balance collector weight and keep over-all center of mass on thrust axis.

is about ½ millimeter thick. Absorption of light quanta causes an electric current (0.5–1 milliwatt). Photovoltaic cells are temperature sensitive (output decreases by 50% between 100 and 300°F) and extremely fragile. Their efficiency is 10–12%. Presently available cells with support structure weigh about 350 lb/kw, but may be reduced to half this value. At present, no practical method exists for using photovoltaic cells for large power outputs (kilowatts).

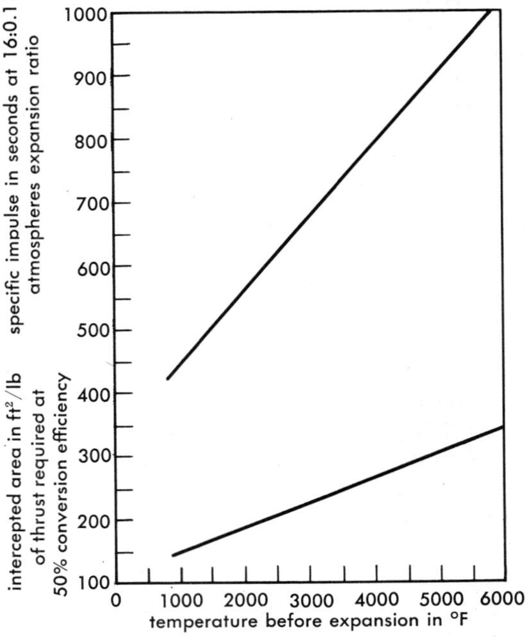

Fig. 4. Area requirement for radiation collectors at Earth's distance.

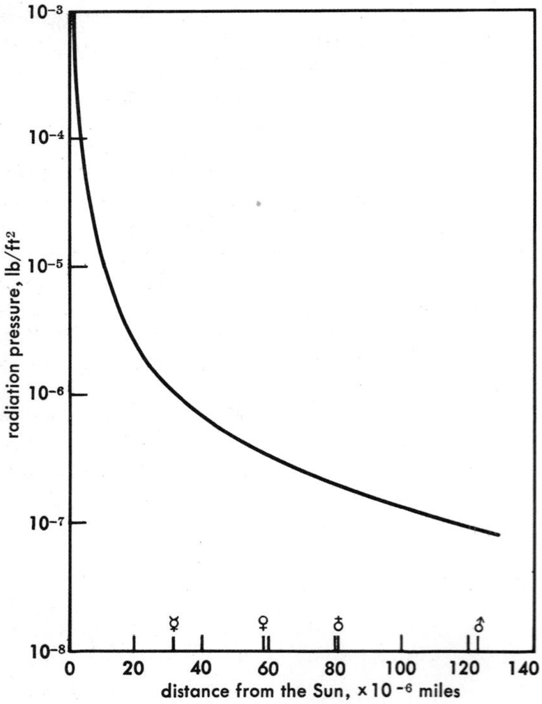

Fig. 5. Radiation pressure of the Sun (for normal incidence and perfect reflection).

Photon reaction. Radiation pressure could serve directly as a space drive. The solar radiation pressure is extremely small (Fig. 5). By properly inclining a reflective membrane in an orbit, one obtains a slight accelerating or retarding force. The membrane must be extremely light. A 0.1-mil thick reflective polyethylene foil weighs about 8.8×10^{-4} lb/ft^2. Assuming that each square foot is loaded on the average by 7.92×10^{-3} lb for accessories and payload, total weight equals 8.8×10^{-3} lb/ft^2. Under these conditions, at 45° inclination to the Sun's rays at Earth distance, the vehicle is accelerated in its orbit at $1.1 \times 10^{-5}g$ (flight time to Mars over 2 years), requiring a reflector surface of 12,700 ft^2/100 lb accessories and payload. Electric and magnetic forces in space, and gravitational torque, may prevent adequate control of such large, extremely thin foils. [K.A.E.]

Bibliography: K. A. Ehricke, *The Solar-Powered Heat Exchanger Drive*, Convair-Astronautics AZM-074, 1959; K. A. Ehricke, *The Solar-Powered Space Ship*, Am. Rocket Soc. Paper 310–356, 1956; P. Rappaport and J. J. Loferski, New solar converter materials, *Proc. 11th Annual Battery Res. and Dev. Conf.*, 1957.

Solar heating

Utilizing the energy of solar radiation to produce heat. The Sun's heat energy has been used to heat houses and domestic hot water, to cook food, to melt metals and refractory materials at temperatures approaching 7000°C, and to operate solar heat engines of various types. *See* SOLAR ENERGY.

House heating. Several solar-heated houses have been built. Such houses require a heat-collecting mechanism and a device in which heat can be stored during the day, to be withdrawn at night or on cloudy, overcast days. One of the first successful solar-heated houses was built in Cambridge, Mass., in 1948. Its heat collector consists of an expanse of glass-covered black metal surfaces 10 ft high and 72 ft long, so oriented as to receive maximum radiation from the Sun. The glass passes about 90% of the Sun's light and infrared radiation. The black surface under the glass absorbs this radiation and is warmed by it. Air circulated over the black surface is warmed and can then be used to heat the house. Part of the warmed air is also passed through the heat-storage device.

The heat-storage device consists of sealed cans containing approximately 21 tons of a chemical salt, which absorbs heat from the heated air passing around the cans. The chemical salt melts at approximately 90°F and, in melting, absorbs considerable heat (called the heat of fusion), which it will give up when it solidifies as the heat is withdrawn.

Other materials can be used to store heat. A house completed in 1958 in Lexington, Mass., uses water. In this house, the heat collector consists of 640 ft^2 of glass, two layers thick, over a thin aluminum sheet painted black. Water circulates through copper tubes attached to the aluminum sheet, and as it is heated, enters a 1500-gal storage tank which is the heat-storage device. Although this house has a small auxiliary oil furnace for emergency heating, the solar-heating system is designed to provide adequate heat for up to three dark, cloudy days in a row (provided the storage tank has been able to store sufficient heat).

A system similar to that used in the Lexington house has been successfully adopted in the southern and southwestern parts of the United States to heat domestic hot water; similar systems are also used to warm the water in swimming pools.

Solar cooking. Inexpensive solar cookers have been designed and are being supplied to peoples of depressed areas of the world where fuel supplies are scarce or relatively expensive. These cookers have reflectors to concentrate the Sun's rays on the container holding the food to be cooked.

Solar furnace. The solar furnace makes use of a parabolic reflector that focuses solar radiations in a small region. Materials held at the focal point will be heated to high temperatures. Since temperatures approaching 7000°C have been reached with solar furnaces, many experimental activities are possible. Solar furnaces are being adapted for vacuum melting of certain metals because this method avoids contamination of the metal from heating fuel.

Solar heat engines. Several solar heat engines have been built. The type uses flat or parabolic reflectors to concentrate heat in a steam boiler; the steam then drives a steam engine or turbine. Another depends upon the heat action on a series of thermocouples to produce electricity directly. As these engines need a constant source of sunlight, they are available only for intermittent operation. *See* SOLAR BATTERY; SOLAR ENGINE; SOLAR RADIATION. [W.H.C.]

Bibliography: American Society of Heating and Air-Conditioning Engineers, *Heating, Ventilating, Air-Conditioning Guide*, vol. 37, 1959.

Solar radiation

The electromagnetic radiation emitted by the Sun. Solar radiation represents nearly all solar energy. Its quantitative measurement is a very difficult problem of fundamental importance in determining the amount of energy generated in the Sun and its surface temperature. The best determination of the solar constant indicates a total energy output of 3.86×10^{33} ergs/sec and an effective surface temperature of 5780°K for the Sun. *See* SUN. [J.W.E.]

Solar system

The Sun and the bodies moving about it. The solar system consists of a star, the Sun, and associated bodies: planets, the satellites of the planets, asteroids, comets, and those meteor swarms that move about the Sun.

The planets move about the Sun in elliptic orbits of moderate eccentricities: 0.007 for Venus to 0.25 for Pluto. The orbits lie nearly in the same plane, Pluto having the greatest inclination. The asteroids have greater eccentricities and inclinations than the planets. *See* ASTEROID; CELESTIAL MECHANICS; COMET; METEOR; PLANET; SUN.

Various theories have been set forth from present evidence for the formation of the solar system. The planets and other bodies are revolving about the Sun with considerable momentum. The terrestrial planets, Mercury, Venus, Earth, and Mars, have significantly higher densities than the major planets. To account for these observations G. K. Kuiper assumes that as interstellar matter condensed to form the Sun, the matter broke up into eddies. The eddies contracted to form planets and satellites. H. C. Urey further postulates that the heat of compression so generated raised the lighter gases, chiefly hydrogen and helium, to their escape velocities at the surfaces of the terrestrial planets. The light gases were then swept away by the radiation pressure of the Sun. *See* COSMOGONY. [D.L.]

Soldering

The joining of metal by causing a lower-melting-point metal to wet or alloy with the joint surfaces and then freeze in place. A solder is defined as a joining material that melts below 427°C (800°F); brazing alloys melt above this temperature. Solders are used to establish reliable electrical connections, to make a liquid- or gastight joint, and to hold parts together physically. Most soldered joints will sustain loads of only 150–250 psi for long periods of time. The usual practice is to rivet, crimp, or otherwise support the load and to seal the space with solder.

Solder alloys. The most commonly used solders are alloys of tin and lead that melt below the melting point of tin. Antimony, bismuth, cadmium, silver, and arsenic are sometimes added to improve strength, wetting qualities, or grain size or to produce alloys having desired melting ranges.

The melting range and other features of all tin-lead compositions are shown in Fig. 1. An alloy of composition E, the eutectic, freezes uniquely at a single temperature, 183°C. All other compositions freeze over a temperature range in which liquid and solid coexist. Thus, an alloy near 38% tin gives a mushy mixture which can be manipulated with cloth pads to wipe joints, as in plumbing and in lead-cable splicing.

The microstructure of a 38% tin solder frozen in contact with copper is shown in Fig. 2. The tin-copper compound is exaggerated for illustration by prolonging the freezing time. The two-stage nature of freezing is apparent.

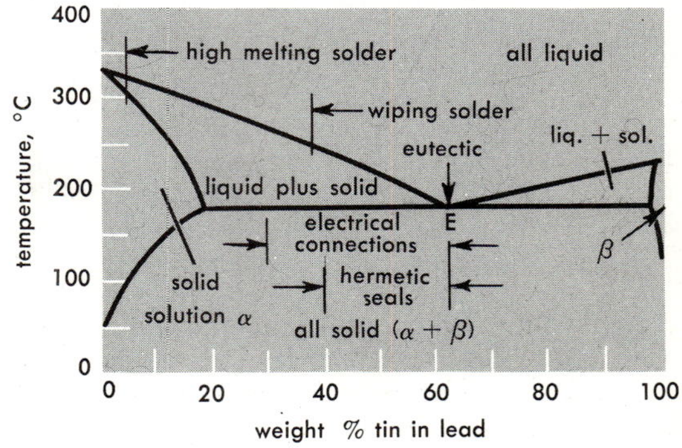

Fig. 1. Constitution diagram of tin-lead alloys.

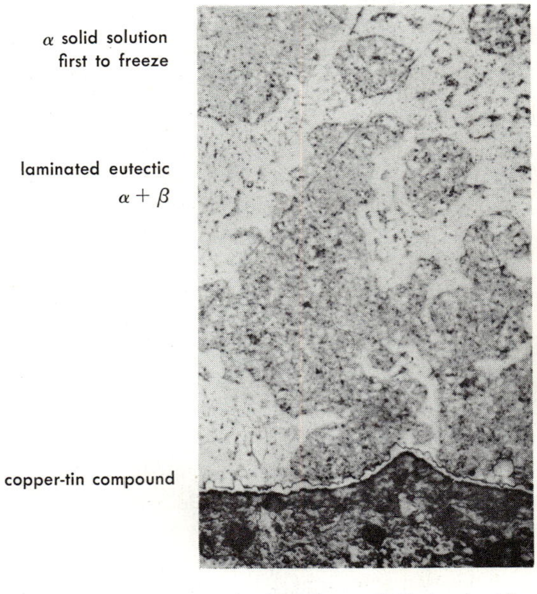

Fig. 2. Photomicrograph of 38% tin–62% lead solder frozen in contact with copper.

In order that surfaces will accept solder readily, they and the solder must be free from oxide or other obstructing films. When necessary, parts are cleaned chemically or by abrasion. Also, readily solderable coatings such as gold, silver, or tin may be applied. Even so, fluxes are usually used.

Fluxes. Fluxes range from very mild substances to those of extreme chemical activity. For centuries rosin, a pine product, has been known as an effective and practically harmless flux. It is used widely for electrical connections in which utmost reliability, freedom from corrosion, and absence of electrical leakage are essential. When less stringent requirements exist and when less carefully prepared surfaces are to be soldered, rosin is mixed with chemically active agents that aid materially in soldering. The activated rosin fluxes are offered by most leading solder manufacturers. The rosin type fluxes may be incorporated as the core of wire solders or dissolved in various solvents for direct application to joints prior to soldering.

Inorganic salts are widely used where stronger fluxes are needed. Zinc chloride and ammonium chloride, separately or in combination, are most common. They may also be obtained as so-called acid core solder wire or in petroleum jelly as paste flux. Many special purpose salt mixtures are on the market. All of the salt type fluxes leave residues after soldering that may be a corrosion hazard. Washing with ample water accompanied by brushing is generally wise.

Application. Under the usual favorable conditions, solders wet the joint surfaces well enough to be drawn into fine crevices or capillaries by surface tension. Joints such as lap seams or wires wound on electric terminals are designed with this in mind.

In applying solders, joints are heated by soldering irons, torches, induction heaters, furnaces, or by immersion in molten solder. In the first two instances, solder is fed to the joints by hand. In the third and fourth, preformed shapes are placed close to the joints before fluxing and heating.

Gold, copper, silver, tin, lead, and brass are examples of readily solderable materials; iron, nickel, and rhodium are moderately difficult. Stainless steel, nichrome, and germanium are typical materials requiring the strongest fluxes. Materials such as tungsten usually must be electroplated first with more solderable metals to permit solder attachments. *See* BRAZING; WELDING AND CUTTING OF METALS. [G.M.B.]

Bibliography: American Society for Metals, *Metals Handbook*, 1948.

Solenoid (electrical)

An electrically energized coil of insulated wire which produces a magnetic field within the coil. If the magnetic field produced by the coil is used to magnetize and thus attract a plunger or armature to a position within the coil, the device may be considered to be a special form of electromagnet and in this sense the words are synonymous. In a wider scientific sense the solenoid may be used to produce a uniform magnetic field for various investigations. So long as the length of the coil is much greater than its diameter (20 or more times), the magnetic field at the center of the coil is sensibly uniform, and the field intensity is almost exactly that given by the equation for a solenoid infinite in length.

Two identical solenoids of radii r units and spaced r units apart on the same axis constitute a Helmholtz coil. A uniform magnetic field is produced by such a device along the axis in the central space between the solenoids.

When used as an electromagnet of the plunger type, the solenoid usually has an iron or steel casing. The casing increases the mechanical force on the plunger and also serves to constrain the magnetic field. The addition of a butt or stop at one end of the solenoid greatly increases the force on the plunger when the distance between the plunger and the stop is small. The illustration shows a steel-clad solenoid with plunger and plunger stop. The relation of force versus distance with and without the stop is also shown.

The force for the solenoid rapidly increases as the plunger enters the coil due to the rapid rate of change of reluctance of the magnetic path. The force is then relatively constant and its maximum value is approximately

$$F_{\max} = \frac{KANI}{L} \quad \text{lb}$$

where I is the current in amperes, N is the number

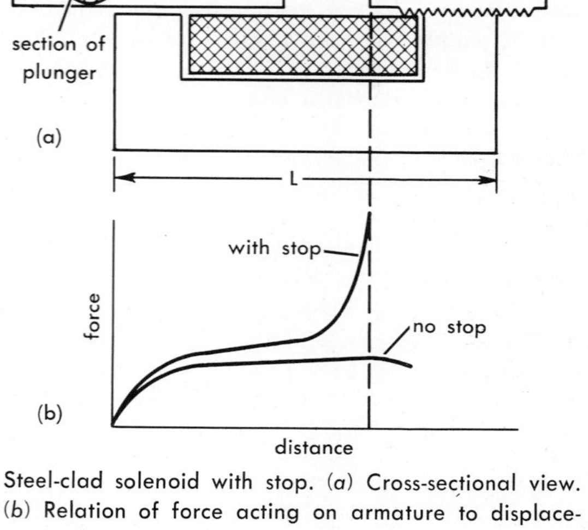

Steel-clad solenoid with stop. (*a*) Cross-sectional view. (*b*) Relation of force acting on armature to displacement of armature.

of turns, A is the cross-sectional area of the plunger in square inches, L is the solenoid length in inches and K is a parameter of the coil design and the material and dimensions of the plunger.

In principle a solenoid works with either ac or dc excitation. In the dc solenoid the flux is always at its maximum value. In an ac solenoid the force varies at twice the frequency of the supply voltage. The variation, which is caused by the variation of the magnetic flux, gives rise to excessive chattering or vibration unless a shading coil is imbedded in the face of the plunger stop. The shading coil acts to smooth the variation of flux and attractive force.

When dc is used, only the resistance of the coil wire limits the final value of current, while with ac excitation the inductance of the coil must also be considered. It is difficult to calculate the proper value of inductance, because it is a function of the position of the plunger. As the plunger moves into the solenoid, the inductance, or flux per ampere, becomes much greater because the reluctance of the flux path is much less. The current drawn by the coil for constant applied ac voltage becomes smaller as the plunger moves into the solenoid. It is common practice to laminate the plunger, stop, and casing in ac service to reduce the eddy current losses that would otherwise be incurred. *See* ELECTROMAGNET. [J.E.G.]

Bibliography: A. E. Knowlton (ed.), *Standard Handbook for Electrical Engineers*, 9th ed., 1957; E. Molloy, M. G. Say, R. C. Walker, *"Electrical Engineer" Reference Book*, 3d ed., 1948; H. Pender and W. A. Del Mar, *Electrical Engineers' Handbook*, vol. 1, 4th ed., 1949.

Solenoid (meteorology)

In meteorological usage, solenoids are hypothetical tubes formed in space by the intersection of the set of surfaces of constant atmospheric pressure (isobaric surfaces) and the set of surfaces of constant specific volume of air (isosteric surfaces). The isobaric and isosteric surfaces are such that the value of pressure and specific volume, respectively, change by one unit from one surface to the next. When the state of the atmosphere is barotropic, there are no solenoids. *See* BAROCLINIC FIELD; BAROTROPIC FIELD; ISOPYCNIC. [F.S.]

Solenopora

A genus of extinct calcareous red algae. The plants resembled the coralline algae in secreting calcium carbonate within and between the cell walls, but had much larger cells, commonly polygonal in outline, which formed rounded nodular or hemispherical masses in most cases. The genus appeared in the Late Cambrian and lasted until the Cretaceous; it flourished during the Ordovician, Silurian, Mississippian, and Jurassic. *See* ALGAE FOSSILS; CORALLINE ALGAE. [J.H.J.]

Bibliography: J. H. Johnson, Introduction to the study of rock building algae, *Quart. Colo. School Mines*, 49(2):46–51, 1954; J. Pia, Neue Arbeiten über fossile Solenoporaceae und Corallinaceae, *Neues Jahrb. Mineral. Geol.*, Referate 3:122, 1930; J. Pia, Sammelbericht über fossile Algen: Solenoporaceae 1930 bis 1938, *Neues Jahrb. Mineral. Geol.*, 3:731–760, 1939.

Solid (geometric)

A geometric solid, usually called a solid, is a set of points forming a finite portion of 3-dimensional space, continuously joined together, and separated from the rest of space by a set of points called its boundary. An interior point of the solid is a point I such that all points of space within a sufficiently small positive distance d from I are points of the solid. A boundary point is a point B such that, no matter how small a positive distance d is chosen, there are points of the solid and points not of the solid within distance d of B. A plane section of the solid consists of the points common to the solid and to a given plane that intersects the solid. A bounding section contains boundary points but no interior points of the solid. Any two bounding sections cut by parallel planes may be called bases of the solid. The distance between these planes is then called the altitude, and the section by a plane halfway between the bases is called a midsection of the solid. A solid is called convex if every line segment joining a pair of its points belongs to the solid. *See* POLYHEDRON; SURFACE AND SOLID OF REVOLUTION. [J.S.F.]

Solid solution

A homogeneous crystalline phase composed of several distinct chemical species. Thermodynamically and physically, a solid solution is completely analogous to the more common liquid solution except for the existence of a regular crystal lattice. The various substances are distributed essentially at random among the various lattice sites. For example, in alloys of silver with gold, in which solid solutions of all compositions from 100% silver to 100% gold are possible, a regular face-centered cubic lattice exists in which each atom is surrounded by 12 equidistant atoms, but no long-range order exists in the distribution of the two kinds of atoms.

A small degree of solid-solution formation must exist in all systems with two or more components, but this may be too small (frequently less than 0.001%) to be of any practical significance. Extensive solid-solution formation can occur only when the molecules or atoms of the two substances are very similar in size and shape. Solid solutions commonly are found in mixtures of monatomic substances (silver and gold, potassium and rubidium, or argon and krypton), but are less common for polyatomic substances because of shape differences. When size, shape, and energy factors are all compatible (for example nitrogen and carbon monoxide, *tert*-butyl chloride and carbon tetrachloride, sodium chloride and sodium bromide, fluorene and diphenylene oxide), the two polyatomic solids are miscible in all proportions.

When solid solutions are formed, the addition of a solute need not produce depression of the melting point. The usual methods of measuring mixed melting points to identify a substance are no longer applicable, and cryoscopic measurements of molecular weights become invalid.

In 1899, Bakhuis Roozeboom classified solid solutions into five types, depending upon the shape and character of the phase diagrams. For some typical phase diagrams, *see* EQUILIBRIUM, PHASE. *See* ALLOY STRUCTURES; CRYSTAL STRUCTURE. [R.L.S.]

Solid-state chemistry

The solid state of matter is characterized by a regular geometric arrangement of the atoms, ions, or molecules constituting each solid substance. The properties of solids are dependent on the nature of their ultimate structural units (atoms, ions, or molecules), the manner in which these are united, and the resulting electronic relationships. The properties relating to these factors are often altered as a consequence of deviations from the idealized geometric structure.

Simple ionic solids. Simple electrovalent salts have some of the most easily understood solid structures. These substances are composed of discrete monatomic cations and anions, arranged in a regular crystal lattice in a manner such that each cation is surrounded only by anions; conversely, anions are surrounded only by cations. The halides of the alkali metals provide familiar examples of such structures. In sodium chloride, each sodium ion (Na^+) is surrounded by 6 chloride ions (Cl^-). The sodium chloride structure belongs to the cubic system of crystals which is shared by all of the alkali metal halides except CsCl, CsBr, and CsI. Many other substances, such as the oxides and sulfides of the alkaline-earth metals, certain alloys, and some nitrides and carbides, also exhibit this arrangement of atoms. The cesium chloride structure is an example of a body-centered cubic structure.

The presence of positive and negative ions alternately arranged throughout a 3-dimensional lattice gives rise to the properties of simple electrovalent salts. The electrostatic attraction between these oppositely charged ions produces stable crystals from which the constituent ions may be separated only at the expense of considerable energy, thus giving rise to relatively high melting points and a fair degree of hardness. Because both the ions and their electrons are confined to definite regions in space, these purely electrovalent salts are, in the ideal case, nonconductors of electricity. Their brittleness may be explained in terms of the nonequivalence of the adjacent atom sites in the crystal lattice. A shearing motion would bring atoms of like charge into close proximity, resulting in repulsion and cleavage.

The complete determination of the crystal structures of electrovalent substances provides information regarding the distance between the centers of the neighboring positive and negative ions. If it is assumed that the positive and negative ions will be drawn together by the electrostatic force arising from their opposite charges until the repulsion caused by their extranuclear electrons causes them to come to rest, then this distance between centers may be regarded as the sum of the radii of the two ions. The crystal radii deduced by Linus Pauling are given in Table 1. Values for crystal radii may be calculated for species which probably do not exist in a purely ionic state. Because the measurement actually provides the distance between the centers of two atoms, their ionic nature is not directly confirmed.

Table 1. Crystal radii of a-subgroup ions, in angstroms

Li^+ 0.60	Be^{2+} 0.31	B^{3+} 0.20	C^{4+} 0.15	N^{3-} 1.71	O^{2-} 1.40	F^- 1.36
Na^+ 0.95	Mg^{2+} 0.65	Al^{3+} 0.50	Si^{4+} 0.41	P^{3-} 2.12	S^{2-} 1.84	Cl^- 1.81
K^+ 1.33	Ca^{2+} 0.99	Ga^{3+} 0.62	Ge^{4+} 0.53	As^{3-} 2.22	Se^{2-} 1.98	Br^- 1.95
Rb^+ 1.48	Sr^{2+} 1.13	In^{3+} 0.81	Sn^{4+} 0.71	Sb^{3-} 2.45	Te^{2-} 2.21	I^- 2.16
Cs^+ 1.69	Ba^{2+} 1.35	Tl^{3+} 0.95	Pb^{4+} 0.84			

In addition to those ionic solids which are composed of simple ions, many ionic crystals contain discrete complex ions. The salts of the familiar oxy acids, such as $NaNO_3$, K_2SO_4, and Na_3PO_4, contain discrete complex anions. Similarly, complex cations are NH_4^+, $N(CH_3)_4^+$, and $S(CH_3)_3^+$. The cationic and anionic metal complexes, such as $[Co(NH_3)_6]^{3+}$, $[Pt(NH_3)_4]^{2+}$, $[Ni(H_2O)_6]^{2+}$, $[Fe(CN)_6]^{4-}$, $[Co(NO_2)_6]^{3-}$, $[HgI_4]^{2-}$, $[PdCl_4]^{2-}$, and $[SiF_6]^{2-}$, also form ionic crystals. In most cases, these may be packed together with ions of opposite charge (either simple or complex) in much the same manner as is true of simple ions. For discussions of the structures of the individual complex ions, *see* CHEMICAL STRUCTURES; COORDINATION CHEMISTRY.

A particular property of crystals of these classes is their continuous nature. For example, the simple units (unit cells) of sodium chloride and cesium chloride repeat throughout the solid, in this way generating the macroscopic crystal and communicating to it some of the geometric properties of the unit cell. In addition to the ionic crystals just described, this continuous nature is shared by other types of substances. Among these are metals, alloys, solid inert gases, and a variety of materials involving continuous covalent bonding throughout the crystal lattice. Certain hydrogen-bonded materials also approximate this behavior. Ice is the classic example of this latter group of substances. *See* CRYSTAL STRUCTURE.

Van der Waals structures. The inert gases are composed of spherical monatomic molecules which crystallize at low temperature to form solids having the cubic close-packed structure. This structure is also described as face-centered cubic, for obvious reasons. As a consequence of the nature of the inert gas atoms, the only forces responsible for holding these solids are the weak attractions called van der Waals forces (*see* CHEMICAL BINDING). Similar structures are encountered among some relatively complex molecular substances in certain instances in which the molecules are free to rotate. Examples are found in solid HCl, N_2, and H_2. In substances such as these, the rotation may cease sharply at a characteristic temperature as the solid is cooled, thus producing a structure of lower symmetry.

Weak interactions are largely responsible for the aggregation of the molecules of many covalent substances in addition to the simple examples discussed. Although much variety may be found in the manner in which such molecular substances are arranged in the solid state, their most significant properties arise from their molecular structures, rather than their crystal structures. Low melting point, high volatility, softness, and very low electrical conductance are typical properties of these weakly bound solids.

Covalently bonded structures. Solid materials which are united throughout by covalent bonds exhibit properties quite different from those of materials involving discrete molecular units. The great strength of the covalent bond is manifested in the hardness, very high melting points, and extreme stabilities of these solids. Perhaps the most familiar example is the diamond, which is composed of tetrahedral carbon atoms joined by single covalent bonds to each of four other identical carbon atoms. The resulting crystal is composed of a single molecule because the 3-dimensional network of covalent bonds extends throughout. The diamond structure is shared by a number of other substances, such as silicon, germanium, carborundum (silicon carbide), and the cubic form of boron nitride. Zinc blende (ZnS) also possesses the diamond structure, as do the selenide and telluride of zinc. The same structure is encountered also among the sulfides, selenides, and tellurides of Be, Cd, and Hg, and among the phosphides and arsenides of aluminum and gallium. The wurtzite structure is closely related, involving tetrahedral bonding between alternate electropositive and electronegative atoms. In all cases, there is reason to believe that the bonds are not purely ionic but might well be relatively covalent in nature.

Numerous other 3-dimensional continuous structures are known to exist; however, none of these is related to the structure of a free element. Quartz, rutile, and corundum are examples.

Two-dimensional and one-dimensional infinite arrays are equally common among solid structures. Examples of the 2-dimensional structures will be discussed first. The carbon atoms in graphite (Fig. 1) are arranged in hexagonal polygons, extending in 2-dimensional array throughout the crystal, thus producing a layer structure. Within the layer, each carbon atom is covalently bound (by sp^2 hydridized orbitals) to three nearest neighbors, the fourth bond is a π bond; however, the resulting π-electron system extends throughout the layer. This fact results in mobility of electrons and high electrical conductance. The forces between the layers are relatively weak so that it is easy to cause slippage along these layers. This latter fact accounts for the lubricating properties of graphite. Hardness and high melting point arise from the same source as in the case of the 3-dimensional structures. Boron nitride, BN, also exists in a hexagonal structure of the graphite type. In BN, the boron and nitrogen atoms alternate throughout the layer structure. The mobility of the π electrons is not so great in this instance as in the case of graphite, possibly indicating the linking of boron and nitrogen atoms from different planes. MoS_2 provides an additional example of a layer structure which is of interest because the compound is known to be a solid lubricant.

One-dimensional structures are found in substances such as $PdCl_2$, SiS_2, SO_3, Se, and Te. Examples are given in Fig. 2.

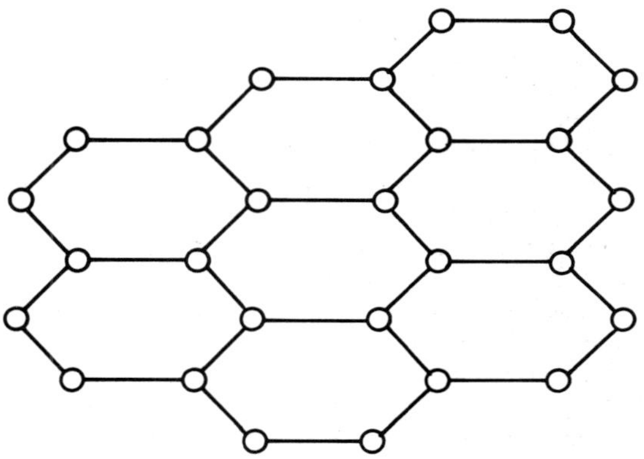

Fig. 1. Structure of graphite. Only a single layer of carbon atoms is shown.

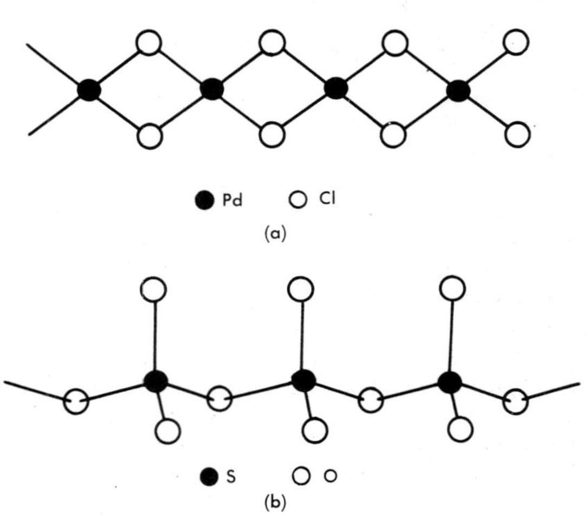

Fig. 2. One-dimensional continuous structures. (*a*) $PdCl_2$ structure. (*b*) β-SO_3 structure.

Infinitely extended complex ions. Ions of this type are closely related to covalently bonded compounds. They are more complex because their structures involve continuous covalent frameworks, which are ionic in nature, together with discrete ions of opposite charge. The silicates and aluminosilicates represent an extensive series of structures of this class. The zeolites provide extremely interesting examples of 3-dimensional anionic structures. Because of the open nature of the anionic framework, zeolites exhibit remarkable properties in addition to those resulting from their continuous structures. The cations in these crystals are relatively mobile and may be exchanged with other cations upon contact with the appropriate solutions. The large cavities in the crystals provide sites for very strong adsorption of polar molecules within the crystal. The size of the apertures to these cavities may be varied by structural modifications, thereby providing materials which can sort molecules on the basis of size (Fig. 3). These materials have been called molecular sieves. The ultramarines resemble the zeolites in structure; however, the feldspars, a third group of framework aluminosilicates, are relatively compact. Typical formulas are given in the following list.

Aluminosilicate	*Formula*
Feldspars	
Orthoclase	$KAlSi_3O_8$
Celsian	$BaAl_2Si_2O_8$
Albite	$NaAlSi_3O_8$
Zeolites	
Analcite	$Na(AlSi_2O_6)\cdot H_2O$
Pollucite	$Cs(AlSi_2O_6)\cdot xH_2O$
Thompsonite	$NaCa_2(Al_5Si_5O_{20})\cdot 6H_2O$
Ultramarines	
Ultramarine	$Na_8Al_6Si_6O_{24}\cdot S_2$
Sodalite	$Na_8Al_6Si_6O_{24}\cdot Cl_2$
Helvite	$Na_8Al_6Si_6O_{24}\cdot SO_4$

See MOLECULAR SIEVE; SILICATE MINERALS.

Two-dimensional anions are best exemplified by silicates, of which the micas and talc are the most familiar examples (Fig. 4). Similarly, pyroxenes and amphiboles involve infinite chain structures (Fig. 5). The structures of these materials are evident in the macroscopic properties of the crystals; that is, the layer structure of mica and the fibrous nature of asbestos.

The metallic state. A number of characteristic properties distinguish metallic structures from the types discussed above. These are (1) high thermal and electrical conductivities, (2) metallic luster, (3) malleability and ductility, and (4) close-packed arrangement of atoms with large number of nearest neighbors compared to the number of valence electrons. The characteristic structures of metals are cubic close-packed, hexagonal close-packed, and body-centered cubic. The metals of the actinide series are the only important exceptions to these structures. The number of nearest neighbor atoms to a given metal atom is 12 in the first two structures, whereas the third involves 8 nearest neighbors and 6 slightly more distant neighbors.

The simplest concept which provides a rational explanation of the typical metallic properties is essentially a free-electron model. According to this view, the valence electrons of a metal are substantially free to move about a lattice of positive ions, derived from the metal atoms by ionization. The expected mobility of electrons in metals is

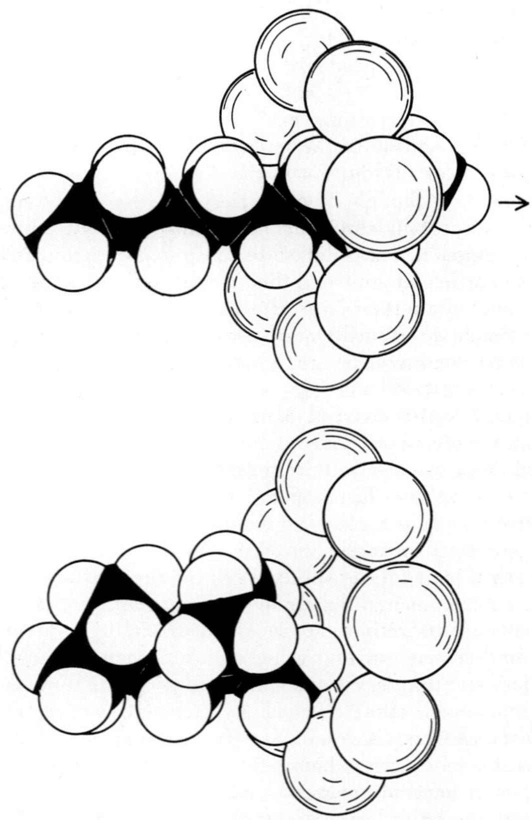

Fig. 3. Zeolite separates molecules on basis of size and shape. (*Linde Company*)

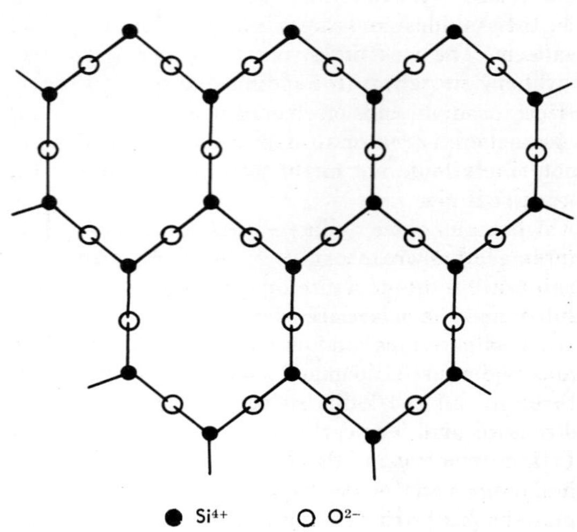

Fig. 4. Layer structure of mica and talc. Oxygen atoms above silicon atoms are omitted.

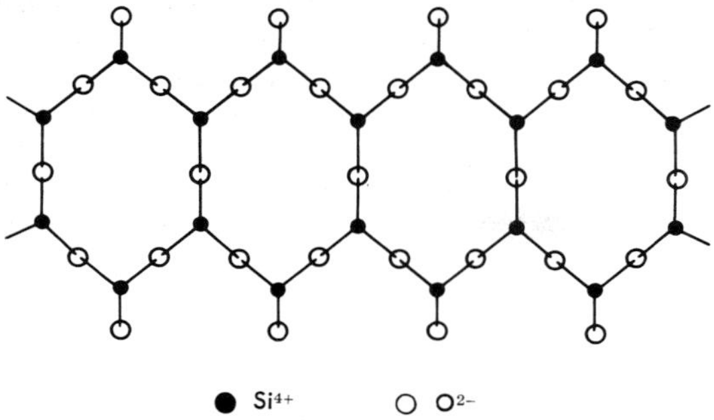

Fig. 5. Chain structure of anions in amphiboles. Oxygen atoms above silicon atoms are omitted.

thereby explained. The malleability and ductility of metals are associated with the equivalence of all the lattice points and the relatively slight directional property of the energy states confining the valence electrons. A more proper application of the same ideas may be made through the use of molecular orbital concepts. The crystal lattice of the metal is conceived to constitute a single molecule and molecular orbitals, extending over the entire crystal, are derived from the orbitals of the individual atoms. The number and energy spacing of the resulting molecular orbitals are related to the geometric arrangement of the metal atoms. In this way it is deduced that the valence electrons may be found in energy bands, composed of very closely spaced, discrete energy levels. It may also be shown that certain zones or bands of energy are forbidden; that is, electrons are not permitted to have energies in these regions. A characteristic number arising from these concepts is the number of electrons per atom which may be inserted into the lowest band of energies (or Brillouin zone). When this total number of electrons has been added, the band is filled and additional electrons would, of necessity, go into high energy states. The electrons enter the discrete energy levels within a band in accordance with the Pauli exclusion principle (that is, their spins are antiparallel if they occupy the same molecular orbital).

This view permits the systematization of the electrical conductivities of metallic conductors, semiconductors, and insulators. Paired electrons do not contribute to conduction. Therefore, it is only the unpaired electrons in incompletely filled zones which may conduct. The occupied Brillouin zones in insulators are completely filled. Conversely, a substance in which Brillouin zones are incompletely filled must be a metallic conductor. The high electric conductivities of such metals as copper, silver, and gold are readily explained in these terms. These three metals all exist in the cubic close-packed structure and each presumably contributes one valence electron to the lowest energy band. The maximum ratio of electrons to atoms in the cubic close-packed structure (first zone) is 2:1. Consequently, the zone is only half filled. In the case of insulators, conductivity is permitted only in the case that some of the electrons are promoted from the filled band to an empty band. The energy necessary to accomplish this promotion is dependent on the energy difference between the filled band and the lowest unoccupied energy band. This energy may be provided, depending on the individual example, by heat, by an electrostatic potential, or by the absorption of electromagnetic radiation (photoconduction). *See* BAND THEORY OF SOLIDS; BRILLOUIN ZONE.

Alloys and intermetallic compounds. In general, the melting together, or otherwise combining, of two metals may result in the formation of a new solid phase. Considering only binary systems (two metals), the new phases formed may be grouped roughly into the following three classifications: solid solutions; valency compounds; and electron compounds. This classification is not all-inclusive; however, it serves as an outline within which to systematize the predominant structural types. *See* ALLOY STRUCTURES.

Solid solutions. Solid solutions are characterized by random substitution of the solute atoms for those of the solvent metal with retention of the structure of the solvent metal. This represents the simplest possible manner of combining two metals. The range of composition over which such solid solutions may exist depends on the similarities of the two metals. It varies from zero to complete miscibility. Complete miscibility is observed only among very similar elements, such as K-Rb, Ag-Au, As-Sb, Mo-W, Ni-Pd, and Ni-Co. In these cases, the pairs of elements are closely related in atomic size, crystal structure, and electronegativity (attraction for electrons). Within pairs of similar metals, the range of composition is a function of the relative sizes of the atoms. This may be rationalized from Vegard's law which predicts that the unit cell dimensions in such a solid solution will vary linearly with the solute concentration. *See* SOLID SOLUTION.

In certain substitutional solutions of this general class, a tendency toward ordered rather than random array of the solute atoms is observed. This may be observed in solid solutions of copper and gold. These two metals form a continuous range of solid solutions, both metals exhibiting face-centered cubic structures. At low temperatures, the atoms of each metal tend to congregate in the same plane, producing layer structures.

Valency compounds. Valency compounds represent the other extreme in alloy structure because they exhibit stoichiometries which might be predicted on the basis of the usual valencies of the metals which are combined. Further, the structures of these substances are those expected of salts having similar formulas. Some structures of valency compounds are listed as follows: antifluorite structure, Mg_2Ge, Mg_2Sn, Mg_2Pb; anti-Mn_2O_3 structure, Mg_3As_2, Mg_3Sb_2; sodium chloride structure, MgSe.

Substances of this class occur among metals of substantial electronegativity difference. These systems usually exhibit very limited solid solutions, and the resulting intermetallic compounds are very poor electrical conductors (insulators or semiconductors). These properties do not necessarily require that these compounds be ionic. The application of the band theory of metals to the fluorite (CaF_2) structure reveals that the electron-atom ratio of the lowest energy band is 8/3. From the formulas of Mg_2Ge, Mg_2Sn, and Mg_2Pb, it is clear that this band should be filled and that these compounds should be insulators. *See* INTERMETALLIC COMPOUNDS.

Electron compounds. Certain binary pairs of metals, whose relative electronegativity differences are intermediate between those pairs forming the two extreme classes of systems already discussed, form the so-called electron compounds, or Hume-Rothery compounds. In these binary systems, the progressive change in composition is accompanied by a progression of phases, differing in crystal structure.

α-phase	→	β-phase	→
Face-centered cubic structure		Body-centered cubic structure	

γ-phase	→	ϵ-phase
Complex cubic structure		Hexagonal close-packed structure

The α-phase is that of the pure metal while the β-, γ-, and ϵ-phases are intermetallic compounds. As shown in Table 2, phases of widely varying stoichiometries may occur with these particular structures. Table 2 also reports the ratios of valence electrons to atoms for these compounds, and it is quite apparent that these ratios have characteristic values for the structures of the three intermetallic phases: β-phase, 3/2; γ-phase, 21/13; ϵ-phase, 7/4. In the cases involving Fe, Co, Ni, and Pt. these ratios are attained only by assuming that the metals named contribute no valence electrons. It should also be pointed out that Table 2 has been constructed from the best examples of this behavior. The band theory of metals provides an explanation of these compounds. This theory predicts that the lowest energy electron zone will be filled in the β-phase at a ratio of 1.480 electrons per atom and that the critical ratio will be 1.538 in the case of the γ-phase. From the point of view of the band theory, the atom ratios of the most stable β-, γ-, and ϵ-phases are not necessarily whole numbers. For example, in adding zinc atoms to copper, one is (according to this theory) merely increasing the number of electrons per atom by replacing monovalent atoms with divalent atoms. The initial structure, α-phase, will remain stable until the first Brillouin zone is about filled. The additional electrons must then go into a second zone of much higher energy. It is to be expected that a second structure having a greater electron to atom ratio in its lowest Brillouin zone might be more stable.

Interstitial compounds. The transition metals form a novel group of binary compounds with the lighter nonmetals, boron, carbon, nitrogen, and to a limited extent, oxygen. These substances exhibit metallic luster and conductance. Some of these compounds are among the hardest and most infusible substances known. The formulas, melting points, and relative hardness of several examples are shown in Table 3. In addition to the examples cited, many other transition metal carbides, nitrides, and oxides have similar properties; it is of special significance that almost all of these have the sodium chloride structure. These materials pos-

Table 3. Interstitial carbides and nitrides

Formula	Melting point, °K	Hardness*
TiC	3410	8–9
ZrC	3805	8–9
HfC	4160	
TaC	4150	
W_2C	3130	9–10
WC	3130	9
Mo_2C	2600	
MoC	2840	
TiN	3220	8–9
ZrN	3255	8
TaN	3360	

* Mohs scale, based on hardness of diamond as 10.

Table 2. Electron or Hume-Rothery compounds

β-Phase		γ-Phase		ϵ-Phase	
Composition	Electrons/atoms	Composition	Electrons/atoms	Composition	Electrons/atoms
CuZn	(1 + 2)/2	Cu_5Zn_8	(5 + 16)/13	$CuZn_3$	(1 + 6)/4
AgZn	(1 + 2)/2	Ag_5Zn_8	(5 + 16)/13	$AgZn_3$	(1 + 6)/4
AuZn	(1 + 2)/2	Au_5Zn_8	(5 + 16)/13	$AuZn_3$	(1 + 6)/4
Cu_5Sn	(5 + 4)/6	$Cu_{31}Sn_8$	(31 + 32)/39	Cu_3Sn	(3 + 4)/4
Cu_5Si	(5 + 4)/6	$Cu_{31}Si_8$	(31 + 32)/39	Cu_3Si	(3 + 4)/4
Ag_3Al	(3 + 3)/4			Ag_5Al_3	(5 + 9)/8
Cu_3Al	(3 + 3)/4	Cu_9Al_4	(9 + 12)/13		
		$Na_{31}Pb_8$	(31 + 32)/39		
CoAl	(0 + 3)/2				
$CoZn_3$	(0 + 6)/4	Co_5Zn_{21}	(0 + 42)/26		
FeAl	(0 + 3)/2				
NiAl	(0 + 3)/2				
		Ni_5Zn_{21}	(0 + 42)/26		
		Pt_5Zn_{21}	(0 + 42)/26		

sess properties falling into two of the classifications discussed above. Their extreme thermal stability, brittleness, hardness, and structure suggest a continuous covalent structure that is geometrically different from the diamond structure but of the same general type. On the other hand, the accompanying metallic conduction and luster resemble alloy systems. *See* CERMET.

Defect solid state. The considerations outlined above are conceived largely on the assumption that solids exist as perfect structures. This is not generally true, although in many respects the deviations from the ideal are of slight importance. The level of impurities in most chemical substances does not fall below a few parts per thousand, and rarely is it below a few parts per million. Also, many solid substances do not exhibit perfect integral ratios of atoms in their stoichiometries. Indeed, these nonstoichiometric compounds are sufficiently common to have received the general name Berthollide compounds. In addition to these factors, materials of ideal purity and stoichiometry would still exhibit imperfections in the solid state. These latter imperfections belong to two categories: crystal dislocations and lattice defects. The first category deals principally with nonideal structures at the bounding surfaces and edges of crystals or microcrystals, or with cooperative deviations from the ideal geometrical structures of crystals which extend over many atoms. This subject will not be considered further here.

The diffusion of ions through crystals, the semiconducting properties, and the optical properties of many solid substances cannot be explained on the basis of the perfectly ordered lattice envisioned in the discussions above. The explanations of such phenomena follow from the understanding that a number of atoms or ions may be displaced from their lattice sites even in ideally pure, stoichiometric crystals. Two classes of crystal defects of this type are recognized. In the Frenkel defects, an ion, usually a cation, has vacated its usual lattice site and occupies a void in the crystal; that is, it becomes an interstitial ion. In the Schottky defects, ions are removed from their normal lattice sites and relocated at the surface of the crystal. In Schottky disorder, cations and anions may be displaced in equal numbers. In either case, the motion of ions through the solid is permitted by their movement through the vacancies in the lattice. This explains the diffusion of ions in solids, solid-state reactions, and a number of other phenomena. *See* CRYSTAL DEFECTS.

The effects of small concentrations of impurities may be illustrated by their application in the conversion of germanium into a semiconducting material. As pointed out above, silicon and germanium crystallize in the diamond structure. Also, the valence electrons of the atoms, four in number, are all utilized in the formation of a continuous 3-dimensional network of covalent bonds. In consequence, these substances are very poor conductors when extremely pure. The addition of small amounts of arsenic or gallium to pure germanium results in enhanced electrical conductance. The arsenic or gallium atom enters the crystal structure in the same way as do the germanium atoms. In the case of arsenic, however, 5 valence electrons are available even though only 4 bonds are formed. The fifth valence electron of arsenic is relatively mobile, giving rise to electrical conduction. Because the electron is a negatively charged carrier of current, germanium "doped" with arsenic is called n-type germanium. In contrast to arsenic, gallium has only 3 electrons, yet the gallium atom occupies a site in which it should form 4 covalent bonds. In consequence of the resulting electron shortage of one, the material is a conductor. As a result of the electron shortage, the current carrier behaves as if it bore a positive charge, and gallium-doped germanium is regarded to be p-type. *See* SEMICONDUCTOR.

Deviations from ideal stoichiometry also give rise to defects in crystal structures. *See* NONSTOICHIOMETRIC COMPOUNDS; *see also* EQUILIBRIUM, PHASE. [D.H.B.]

Solid-state physics

The branch of physics centering about the physical properties of solid materials. Solid-state physics usually is concerned with the properties of crystalline materials only, that is, of materials in which the constituent atoms are arranged in a three-dimensional lattice periodic in three independent directions. Coverage may also extend to the properties of glasses or polymers, which normally have lower structural regularity than crystals. Glasses possess a three-dimensional network which is not periodic; polymers are constituted of parallel bundles of long molecules which may possess a crystalline structure under certain conditions.

As the name suggests, solids usually possess rigidity when subjected to stress. The strain accompanying stresses usually is reversible to a large extent, although there may be some creep, or viscous flow, under static loading.

Historical development. In a historical sense, the study of the physics of solids has extended over four phases of development: macroscopic properties, particularly macroscopic symmetry; lattice theory and lattice models; application of atomic theory to the properties of perfect crystals; and development of the theory of imperfections in crystals.

The first of these topics was of major interest from the dawn of the subject in the Renaissance to the end of the last century. Lattice theory was of relatively abstract or theoretical interest until the discovery just prior to World War I that x-rays are diffracted in a unique way by the atoms in crystal lattices; it then became a practical topic. Atomic theory was applied to the determination of the properties of crystals in a rudimentary form late in the last century, but did not reach its full flowering until after the development of modern quantum mechanics, or wave mechanics, in the period between 1925 and 1928. The systematic study of the imperfections in crystals began soon after 1925 but did not reach its peak until after 1945.

Classification of solids. There are two broad classes of solids when the substances are viewed from the standpoint of electrical and optical properties, namely, insulators and conductors. The common metals are ideal examples of conductors; organic crystals probably provide the best example of insulators. Actually, there is no sharp dividing line between the two types of materials, particularly at room temperature or higher temperatures, where substances which are insulating near the absolute zero may show intermediate electrical conductivity. Materials of this type normally are termed semiconductors and have been the object of much special study. In general, they may be regarded as insulators in which bound charges have been made mobile by temperature fluctuations.

There is another classification of solids which rests upon the analyses of chemical and structural properties. It contains four categories: metals, salts, valence crystals, and molecular crystals. The metals are very good electronic conductors of electricity and heat; moreover, they usually are quite ductile. The pure metals of commerce represent ideal examples. They may combine with one another to form alloys when the atoms have similar sizes. The compounds form over broad ranges of composition if the metallic elements in the alloys lie close to one another in the electromotive series. Otherwise, the alloys tend to resemble normal chemical compounds (for example, Mg_2Sn, an alloy of magnesium and tin). The salts, or ionic solids, are composed of neutral orderly arrangements of positive and negative ions combined in accordance with conventional rules of chemistry. They normally have a large static dielectric constant and exhibit electrolytic conductivity at sufficiently high temperatures. Substances such as sodium chloride and magnesium oxide are typical examples. Valence compounds are composed of atoms which are covalently bonded with one another in accordance with the rules of valence chemistry. Diamond, with its tetrahedral bonding, is the ideal example, although materials such as silicon carbide and boron carbide also are good examples. The crystals are usually very hard and strong. Molecular crystals correspond to lattice arrays of molecules such as hydrogen, nitrogen, carbon dioxide, methane, or more complex organic compounds. The crystalline polymers are in this class. These substances tend to have low melting points and are good electrical insulators. The constituent molecules retain much of their individuality in typical cases.

Any of the four crystal types may behave as semiconductors. This is true, for example, of alloys of metal such as Mg_2Sn, of ionic crystals such as zinc oxide or cuprous oxide, of valence crystals such as silicon and germanium, and of organic crystals such as phthalocyanine.

Areas of study. The goal of the physics of solids is to understand the various properties in terms of atomic and nuclear theory. The topics which have received principal attention up to the present time include the following: elastic and plastic behavior; cohesion and cohesive energies; electronic transport (electrical and thermal conductivity including superconductivity, Hall effect, and thermoelectricity); ionic transport (electrolytic conductivity, chemical and radioactive diffusion); specific heats; magnetic properties; dielectric properties; ferroelectric properties; optical properties (including photoconductivity and luminescence); and nuclear and electron resonance.

The preceding represent areas of basic scientific exploration. Certain large blocks of these areas have received special attention for practical or technological reasons. The principal breakdown into the technological fields is as follows: metallurgy; semiconductors (transistor technology is rapidly becoming a part of the field of electrical engineering); ceramics (this field of technology centers about the properties of salts and valence crystals); polymer chemistry; and magnetics. It appears that the subject of solid-state masers, which centers about materials exhibiting a particular type of electron resonance, is on the way to providing a major area of technology.

Band structure. It is possible to obtain an approximate description of the various solid types which is valid when the atoms are sufficiently in contact so that the valence electrons may be regarded as distributed throughout the solid. This condition is satisfied in the metals, salts, and valence crystals, although it may not be satisfied ideally in the molecular crystals. Under these circumstances, the energy levels of the electrons can be grouped into bands. The levels are occupied with electrons in accordance with the Pauli exclusion principle over an appreciable range of energy. In insulators, completely filled and completely empty bands are separated by so-called forbidden regions, whereas the occupied and unoccupied bands are contiguous in metals. The intrinsic semiconductors, which exhibit semiconductive behavior even when having ideal purity and composition, correspond to cases in which the filled and empty bands of levels are relatively close, so that the electrons may jump from one band to the other as a result of thermal excitation.

It is possible to use an alternate description in the case of insulators when dealing with the ground state of the entire crystal or the first few excited states. In this description, the electrons are regarded as localized about atoms or ions. This point of view has particular value when one is concerned with the first excited states of insulators which do not show electronic conductivity (the so-called exciton states).

Imperfections in crystals. Many of the properties of solids are affected in a radical way by the presence of imperfections, such as flaws in the ideal structure or foreign atoms. In certain cases, the imperfections are so important that the ideal lattice may be regarded primarily as a medium which supports the imperfections.

The imperfections may be of a surface or intergranular type. For example, they may be localized at the boundaries between the crystalline constituents in polycrystals in the second case, or they may be distributed throughout the volume of the crystals. Both types of imperfection are important in practical cases. Usually one deals with grain boundaries, adsorbed atoms, or structural flaws in the surface in the first two cases. The important volume imperfections which are common to large numbers of crystals are as follows: lattice vibrations (phonons), free electrons and holes in insulators, excitons in insulators, vacant lattice sites and interstitial atoms, impurity atoms, and dislocations. *See* CRYSTAL DEFECTS; *see also* ALLOY STRUCTURES; BAND THEORY OF SOLIDS; CERAMIC TECHNOLOGY; COHESION (PHYSICS); COLOR CENTERS; CONDUCTION (HEAT); CRYSTAL GROWTH; CRYSTAL OPTICS; CRYSTAL STRUCTURE; CRYSTALLOGRAPHY; DIFFUSION IN SOLIDS; ELASTICITY; ELECTRON DIFFRACTION; ELECTRON EMISSION; EXCITON; FERROELECTRICS; GLASS AND GLASS PRODUCTS; HALL EFFECT; HOLES IN SOLIDS; INSULATOR, ELECTRIC; IONIC CRYSTALS; LATTICE VIBRATIONS; LUMINESCENCE; MAGNETIC MATERIALS; MAGNETIC RESONANCE; MAGNETISM; MASER; METAL; NEUTRON DIFFRACTION; PHOTOCONDUCTIVITY; PLASTICITY; POLYMER; RESISTIVITY, ELECTRICAL; SEMICONDUCTOR; SINGLE CRYSTAL; SOLID-STATE CHEMISTRY; SPECIFIC HEAT OF SOLIDS; SUPERCONDUCTIVITY; THERMOELECTRICITY; X-RAY CRYSTALLOGRAPHY; X-RAY DIFFRACTION. [F.SE.]

Bibliography: A. J. Dekker, *Solid State Physics*, 1957; C. Kittel, *Introduction to Solid State Physics*, 2d ed., 1956; R. E. Peierls, *Quantum Theory of Solids*, 1955; F. Seitz, *Modern Theory of Solids*, 1940; F. Seitz and D. Turnbull (eds.), *Solid State Physics*, vols. 1–11, 1955–1960.

Solids pump

A device used to move solids upward through a chamber or conduit. It is able to overcome the large dynamic forces at the base of a solids bed and cause the entire bed to move upward counter to the force of gravity.

Solids pumps are used to cause motion of solids in process-type equipment in which treatment of solids under special conditions of temperature, oxidation, and reduction can be combined with upward motion and discharge of the spent solids overhead from the reacting vessel.

Solids pumps are inherently of the positive displacement type. One practical method uses a reciprocating piston mounted on a trunnion permitting it to swing into an inclined position for filling and then to swing back into vertical position for discharge. Figure 1 shows a mechanically driven solids pump in four positions through its cycle of operation. In position (*a*), it is filling with solids from the inlet hopper; in position (*b*), the piston is rotating on a trunnion toward its discharge position; in position (*c*), it is in discharge position and the piston is pushing the charge of solids upward; and in position (*d*), the piston is rotating back toward the original filling position. Figure 2 shows a large hydraulically operated pump used in units of capacity exceeding 1000 tons/day. Hydraulic activation permits very precise control of the feeder mechanism and good efficiency in operation.

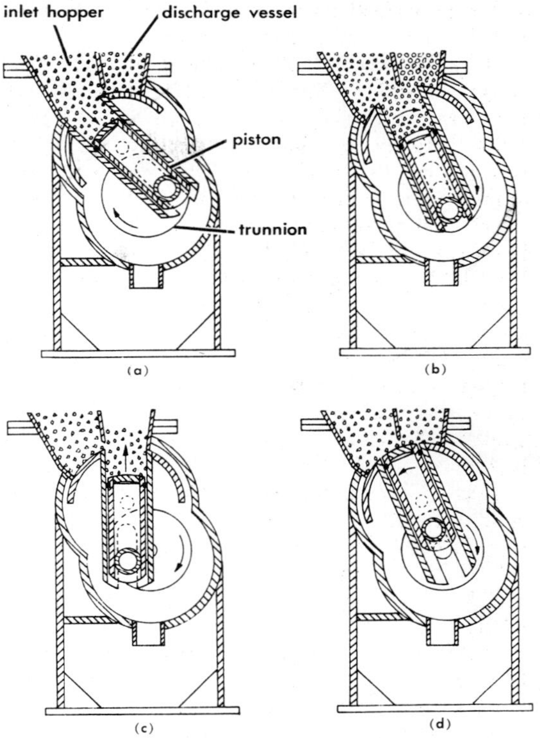

Fig. 1. (*a–d*) Mechanically driven solids pump.

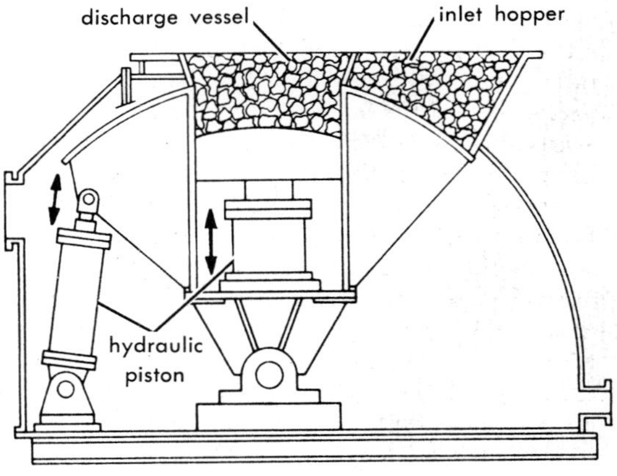

Fig. 2. Hydraulically operated solids pump.

The solids pump has found its principal application in the operation of oil-shale retorts. Here it is used to feed crushed shale into the bottom of a conical vessel and as the shale moves upward through this vessel, air is drawn downward countercurrent. At the top of the retort, the air burns the residual carbon on the shale ash. The hot flue gas so produced contacts shale in the midpoint of the reactor, educting the shale oil. These vapors together with the flue gas are cooled, and the oil condensed on the shale at the bottom of the retort. The oil flows out the bottom countercurrent to the upgoing bed of shale. *See* BULK-HANDLING MACHINES; UNIT OPERATIONS. [C.B.]

Solifluction

A variety of soil flow that is a characteristic form of subaerial denudation of land forms in polar and subpolar regions and in some high mountain areas of middle latitudes. It is a process of mass movement involving the flow-transfer of ill-sorted, incompletely consolidated superficial material which becomes saturated with water during the short summer season. The distortion is essentially downslope with the movement being intermediate between that of normal soil creep (slow distortion without obvious surface indication) and that of avalanche character, where much of the movement takes place on slip-plane boundaries within and below the moving mass. Solifluction attains optimum development in areas where abundant unstratified rock fragments occur in a matrix of fine wet soil subjected to periodic penetrations of frost. It is thus particularly common in regions of permafrost. Typically, during the warming season, ice melting occurs to soak the mantle and thoroughly lubricate it with water. A down-slope gravitational flow is thus induced, even on slopes with gradients as low as 5°. The debris usually moves imperceptibly, with the annual mass displacement in the order of only a few inches. Most of this movement takes place soon after the thawing of the frost-heaved debris mantle in the spring. The flow pattern is often manifested at the surface as a reticulated edge of a vast lobate sheet, or as a sequence of individual and multiple tongues of debris. By this process, gullies, niches, and other depressions, even broad valleys, become completely filled and covered with solifluction debris. The result is an increasingly subdued relief, tending to develop toward a smooth topographic configuration.

[M.M.M.]

Solion

A dynamic electronic circuit element that supplements vacuum tubes and transistors. The term solion is applied to a class of devices that use ions in solution instead of electrons as the charge carriers. A solion consists of two or more electrodes sealed in an electrolyte. At the electrode surfaces, conduction changes from ionic to electronic by means of electrochemical reactions. At the anode, ions lose electrons (or are oxidized) and at the cathode, they gain electrons (or are reduced). The reactions are reversible, the electrodes are not affected by the reactions, and the electrolyte contains both the oxidized and the reduced species of the ions. This is called a redox system.

Solions are low-frequency devices but where applicable they offer power reduction, circuit simplification, and increased reliability and ruggedness. In some cases, a single solion may replace a complete circuit assembly. The basic solion units are the diode, the integrator, the pressure transducer, and the electroosmotic driver. More complicated solions can be designed to perform various mathematical or process-control functions.

The solion diode uses platinum electrodes in an aqueous solution that may be iodine and potassium iodide. Forward-to-back current ratios of 500:1 are easily obtainable and are maintained at voltages below 0.9 volt. The diode exhibits a storage charge effect (hysteresis) at very low frequencies. The maximum voltage that may be applied is 0.9 volt. The most important use is as a low-frequency switch.

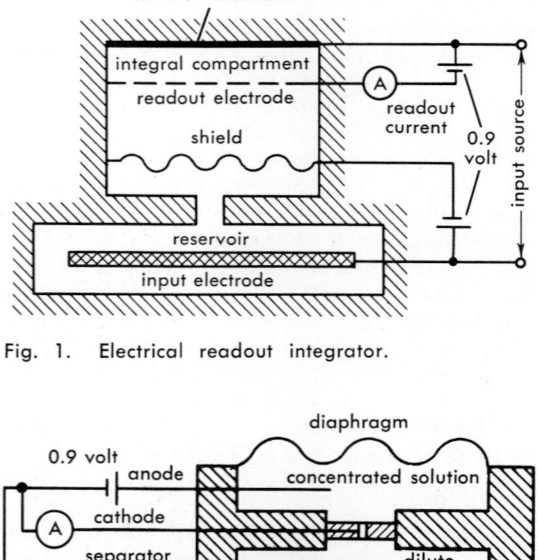

Fig. 1. Electrical readout integrator.

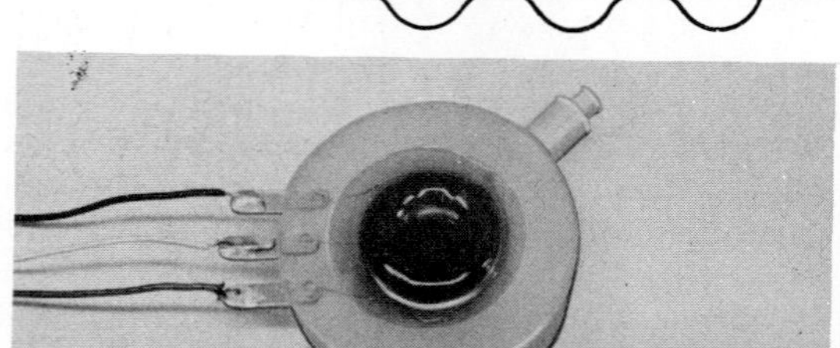

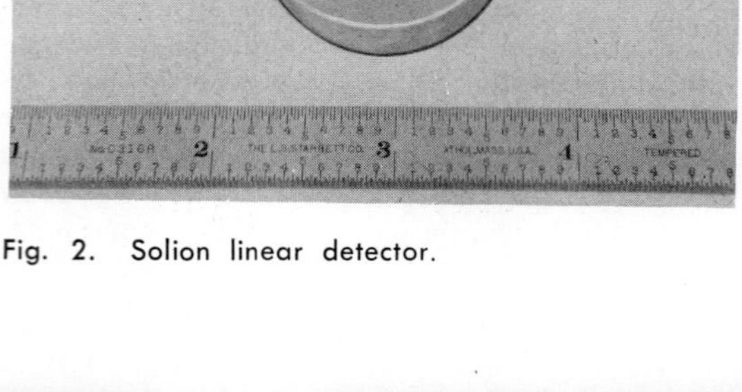

Fig. 2. Solion linear detector.

The solion electrical readout integrator uses the same electrochemical system as the diode (Fig. 1). Current flowing between the input and the common electrode oxidizes iodide to iodine in the integral compartment. The readout current through meter A is determined by the iodine concentration in the integral compartment and is proportional to the integral of the input current. The integrator is reversible and linear, and may be temperature-compensated externally. It is used as an integrator, a linear time-scale generator, an amplifier, and an adjustable constant-current element.

The solion pressure transducer (or linear detector) measures fluid flow through an orifice separating two electrolyte chambers (Fig. 2). A detecting cathode is located in the orifice. Pressure causes iodine to flow past the cathode where it is reduced to iodide. This results in a current increase in the cathode circuit. Pressure changes as small as 5 dynes/cm^2 can be detected. The transducer is used as a pressure-change detector, a vibration pickup, a hydrophone, and an accelerometer.

The solion electroosmotic driver, or micropump, converts voltage into fluid pressure. It uses a different electrochemical system, depolarizing electrodes, and operates through an effect known as the streaming potential; 10 volts can produce as much as 40 in. of water pressure. It is used as a pressure generator, an amplifier (with solion pressure transducer), and a part of mathematical function devices. *See* ELECTRODE POTENTIAL; ELECTROLYTIC CONDUCTANCE; OXIDATION-REDUCTION; STREAMING POTENTIAL. [D.B.C.]

Bibliography: R. M. Hurd and R. N. Lane, Principles of very low power electrochemical control devices, *Journal of the Electrochemical Society*, 104(12):727–730, 1957; R. N. Lane and D. B. Cameron, Current integration with solion liquid diodes, *Electronics*, 32(9):53–55, 1959.

Solo man

A relatively late but primitive form of fossil man from Java. The type is represented by 11 skulls and 2 tibias, recovered by C. ter Haar and others in 1931–1932 from the 20-meter terrace of the Solo River at Ngandong, central Java. They were associated with copious remains of the upper Pleistocene Ngandong fauna. Such accompanying artifacts as are known suggest a late date, corresponding to the Last Glacial phase of the Northern Hemisphere. Generally contemporary with Neanderthal man, and probably with Rhodesian man, the Solo form had a smaller brain size (about 1150 cm^3), heavy horizontal brow ridges, and a massive cranial base. All other facial and skeletal details are unknown, but the leg bones indicate no skeletal differences from *Homo sapiens* or other forms of Pleistocene man. The type has been named *Javanthropus soloensis* (Oppenoorth), *H. soloensis* (Oppenoorth), and *H. neanderthalensis soloensis* (von Koenigswald). *See* FOSSIL MAN. [W.W.H.]

Bibliography: F. Weidenreich and G. H. R. von Koenigswald, Morphology of Solo man, *Am. Museum Anthrop. Papers*, 43(3):205–290, 1951.

Solpugida

An order of nonvenomous, spiderlike, predatory arachnids found chiefly in arid and semiarid, tropical, and warm-temperate regions. They are also known as the Solifugae or sun spiders. The relatively large anterior appendages, or chelicerae, are used for holding and crushing prey. Each of the appendages of the second pair, the leglike and nonchelate palpi, is tactile and ends in a structure

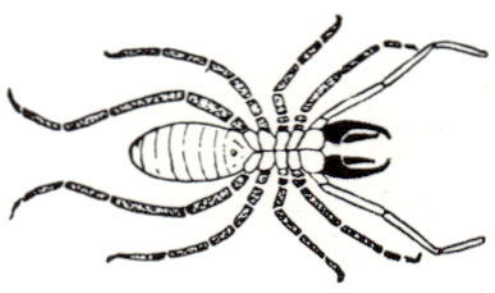

Solpugida, a sun spider. (*From T. I. Storer and R. L. Usinger, General Zoology, 3d ed., McGraw-Hill, 1957*)

that is said to be adhesive in function. The first pair of legs is tactile, while the other three pairs are ambulatory. Eggs are laid in subterranean burrows. The solpugids are extremely agile and usually stalk their prey during the night. A fossil form is known from Pennsylvanian time. *See* ARACHNIDA. [C.C.HO.]

Solstice

The two days during the year when the earth is so located in its orbit that the inclination (about 23½° or 23°.45) of the polar axis is toward the sun. This occurs on June 21, called the summer solstice, when the North Pole is tilted toward the sun; and on December 22, called the winter solstice, when the South Pole is tilted toward the sun. The adjectives, summer and winter, used above, refer to the Northern Hemisphere; seasons are reversed in the Southern Hemisphere.

At the time of the summer solstice the sun's rays are vertical overhead at the Tropic of Cancer, 23½° north. At the North Pole the sun will then circle 23½° above the horizon, and at the Arctic Circle, 66½° north, the noon sun will be 47° above

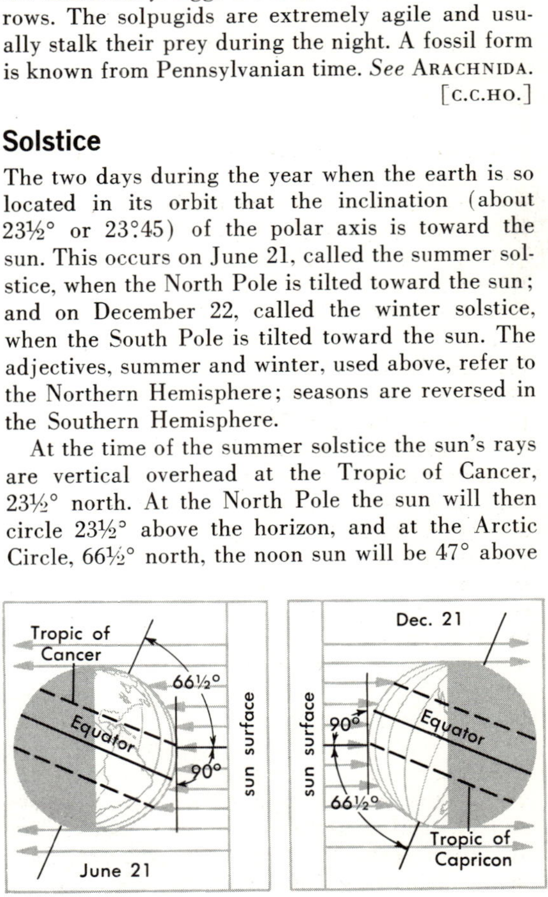

The earth at the time of the summer and winter solstices. The dates may vary because of the extra one-fourth day in the year.

the horizon and the setting sun will touch the horizon to the north. Thus, on this day every place north of the Arctic Circle will have 24 hours of sunlight and the length of day at all places north of the Equator will be more than 12 hours, increasing in length with increasing latitude.

Identical conditions are found in the Southern Hemisphere at the time of the Northern Hemisphere's winter solstice when the sun is vertical above the Tropic of Capricorn, 23½° south, and the South Pole is tilted toward the sun. [V.H.E.]

Solubility product constant

A special type of simplified equilibrium constant (symbol K_{sp}) defined for, and useful for, equilibria between solids (*s*) and their respective ions in solution, for example,

$$AgCl\ (s) \rightleftharpoons Ag^+ + Cl^- \quad (1)$$

For this relatively simple equilibrium:

$$[Ag^+]\,[Cl^-] \cong K_{sp} \quad (2)$$

and

$$(Ag^+)\,(Cl^-)/(AgCl) = K_{sp} \quad (3)$$

It can be demonstrated experimentally that a small increase in the molar concentration of chloride ion $[Cl^-]$ (produced, for example, by the introduction of NaCl, HCl, or other soluble chloride), causes a reduction in the concentration of silver present as Ag^+. Similarly, an increase in $[Ag^+]$ reduces $[Cl^-]$. The product of the two concentrations is approximately constant as indicated by Eq. (2) and equal to the K_{sp} of Eq. (3). Equation (3) is exact since the variables are activities instead of concentrations. In accordance with the choice of standard state usually made for a solid, the activity of solid AgCl is unity, hence

$$(Ag^+)\,(Cl^-) = K_{sp} = 1.8 \times 10^{-10}\ \text{mole}^2\ \text{liter}^{-2} \quad (4)$$

In practice, various complications arise: addition of too much of either ion produces more complicated ions and hence actually increases the apparent concentration of the other ion. Addition of a salt without a common ion (that is, a salt supplying neither Ag^+ nor Cl^-) either may react with Ag^+ or Cl^- or may merely increase the concentration of both ions by a lowering of the mean ionic activity coefficient. (Sodium nitrate at a concentration of 0.01 molar increases each concentration by about 10% and the product by 20%.)

It is usually assumed that an aqueous solution saturated with silver chloride contains only Ag^+, Cl^-, and the solvent. Some recent work indicates, however, that about 2.5% of the solute is present as undissociated AgCl. In practice, such effects are usually neglected. Equation (2) is especially useful in the explanation of analytical procedures in which it is desired to add a sufficient quantity of one ion to ensure (virtually) complete precipitation of the other.

An example of a salt of a different charge type is lead iodate $Pb(IO_3)_2$. The solubility equilibrium is represented by the equation:

$$Pb(IO_3)_2\ (s) \rightleftharpoons Pb^{2+} + 2IO_3^- \quad (5)$$

At about 25°C, the solubility of this salt is 0.024 g/liter. The mass of 1 gram-formula weight of the salt is 557 g. In the saturated solution, therefore, the concentrations are

$$[Pb^{2+}] = \frac{0.024}{557} = 4.3 \times 10^{-5}\ \text{mole/liter}$$
$$\text{and}\ [IO_3^-] = \frac{2(0.024)}{557} = 8.6 \times 10^{-5}\ \text{mole/liter} \quad (6)$$

Because the solid is in its standard state, its activity is unity. Hence

$$[Pb^{2+}][IO_3^-]^2 = K_{sp} = 3.2 \times 10^{-13}\ \text{mole}^3/\text{liter}^3 \quad (7)$$

It is important to note that the concentration, 8.6×10^{-5} mole/liter, is the concentration of IO_3^- ion in the solution; it is not twice the concentration of the IO_3^- ion, although it happens to be twice the concentration of Pb^{++}. It is also important to observe that the square of 8.6×10^{-5} mole/liter enters this product because the coefficient of IO_3^- in Eq. (5) is 2. *See* EQUILIBRIUM, IONIC; GRAVIMETRIC ANALYSIS; PRECIPITATION (CHEMISTRY). [T.F.Y.]

Solubilizing of samples

The process by which difficultly soluble samples are converted into different chemical compounds which are soluble. The sample may be heated in air to evolve volatile components or to oxidize a component to a volatile higher oxidation state with the formation of an acid-soluble form, as in the roasting of a sulfide to form the oxide and sulfur dioxide. Most frequently, the sample is treated with a solvent which reacts with one or more constituents of the sample.

Solvents. The choice of solvent is determined by the chemical reactions which are required. Reactions used include solvation, neutralization, complex formation, metathesis, displacement, oxidation-reduction, or combinations of these. Most water-soluble salts dissolve by solvation. Basic oxides such as ferric oxide dissolve in aqueous hydrochloric acid by neutralization followed by complex formation:

$$Fe_2O_3 + 8H^+ + 8Cl^- = 2FeCl_4^- + 2H^+ + 3H_2O$$

Amphoteric oxides dissolve in either acids or bases:

$$Al_2O_3 + 2Na^+ + 2OH^- = 2AlO_2^- + 2Na^+ + H_2O$$

Aqueous ammonia converts insoluble silver chloride into a soluble complex:

$$AgCl + 2NH_4OH = Ag(NH_3)_2^+ + Cl^- + 2H_2O$$

Carbonates are solubilized by treatment with hydrochloric acid to displace carbon dioxide and to form soluble chlorides:

$$CaCO_3 + 2H^+ + 2Cl^- = Ca^{++} + 2Cl^- + H_2O + CO_2$$

Many substances are converted to easily dissolved mixtures by metathesis. An example is the boiling of barium sulfate with aqueous sodium carbonate to form barium carbonate and sodium sulfate. The insoluble barium carbonate dissolves easily in hydrochloric acid.

Dissolution of metals and alloys. Metals above hydrogen in the electrochemical series will dissolve in a nonoxidizing acid by reduction of hydrogen ion. Other metals such as copper and lead require an oxidizing acid, usually nitric acid. The treatment of alloys is determined by the constituents present. All of the components of brass and bronze are usually dissolved by nitric acid except tin, arsenic, and antimony, which precipitate as hydrated oxides. Alloy steels are usually dissolved by combinations of hydrochloric, nitric, phosphoric, and hydrofluoric acids, depending on the elements present. Aluminum-base alloys are treated with sodium hydroxide solution and any residues are dissolved in acid.

Fusions. Many substances do not dissolve at temperatures obtainable in the presence of liquid water. In these cases, the treatment must be performed at elevated temperatures, usually 400–500°C. The material used as the solvent is called a flux, and the process of melting the mixture of dry, solid flux with the sample is called a fusion. Fusion is done in a crucible (usually platinum or nickel) which is not attacked by the flux or the sample constituents. The same types of reactions are used as with aqueous solvents. Sodium carbonate is used to attack acid materials such as silicates and for metathesis reactions with sulfates.

$$CaSiO_3 + Na_2CO_3 = CaCO_3 + Na_2SiO_3$$

Potassium bisulfate on fusion yields potassium pyrosulfate, an acid flux, which attacks basic oxides such as alumina and ferric oxide and metals such as chromium and tungsten. This flux must be used in porcelain crucibles. Oxidizing fluxes such as sodium peroxide and mixtures of sodium carbonate with potassium nitrate are used with sulfides, chromium, and tin ores and some silicon samples. These usually require nickel crucibles. Calcium carbonate plus ammonium chloride is used to free the alkali metals from silicates. See ANALYTICAL CHEMISTRY. [K.G.S.]

Bibliography: W. F. Hillebrand, G. E. F. Lundell, H. A. Bright, and J. I. Hoffman, *Applied Inorganic Analysis*, 2d ed., 1953.

Solution

A homogeneous mixture of two or more components whose properties vary continuously with varying proportions of the components. A liquid solution can be distinguished experimentally from a pure liquid by the fact that during transfers into other single phases at equilibrium (freezing and vaporizing at constant pressure) the temperature and other properties vary continuously, whereas those of a pure liquid remain constant. For an apparent exception *see* AZEOTROPIC MIXTURE. Gases, unless highly compressed, are mutually soluble in all proportions.

A solid solution is, similarly, a single phase whose composition and other properties vary continuously with changing composition of the liquid phase with which it is in equilibrium. *See* SOLID SOLUTION.

Types of intermolecular force. The extent to which substances can form solutions depends upon the kind and strength of the attractive forces between the several molecular species involved. It is necessary to consider the attractive forces exerted by molecules of the following types: (1) nonpolar molecules; (2) polar molecules, that is, those containing electric dipoles; (3) ions; and (4) metallic atoms.

London forces. The theory of attraction between nonpolar molecules, developed by F. London in 1930, is based upon the quantum-mechanical interaction between pairs of electron systems. For two molecules with electrons having frequencies ν_1 and ν_2, polarizabilities α_1 and α_2, and separated by the distance r between centers, the attraction potential is

$$\epsilon_{12} = -\frac{3\alpha_1\alpha_2 h}{2r^6} \cdot \frac{\nu_{0,1}\nu_{0,2}}{\nu_{0,1} + \nu_{0,2}} \tag{1}$$

where h is the Planck constant. For molecules of the same species, this reduces to

$$\epsilon_{11} = -\frac{3\alpha_1{}^2}{4r^6} h\nu_0 \tag{2}$$

The frequency ν_0 is that corresponding to $h\nu_0$, the zero-point energy, of the molecule in its unperturbed state. The perturbation by another molecule is related to its perturbation by light of varying frequencies, as seen in the variation of refractive index n with the frequency of light, that is, the dispersion. For this reason, London designated these forces as dispersion forces. It is equally appropriate to speak of London forces by analogy with the nearly equivalent term, van der Waals forces. In the case of gases, the dispersion n_ν is related to the frequency by

$$n_\nu - 1 = C/(\nu_0{}^2 - \nu^2)$$

where C is a constant. The polarizability α can be determined from the refractive index with the aid of the Lorentz-Lorenz formula. As a substitute for zero-point energy, London proposed the ionization potential.

The model upon which these relations are based is much simpler than the polyatomic molecules in most solutions of interest. In these, the potential field is not central and radial; the interaction is between the electrons in the peripheral bonds. A striking example is octamethylcyclotetrasiloxane, whose core of alternating silicon and oxygen atoms is so buried within the 8 methyl groups that it behaves toward other molecules essentially as an aliphatic hydrocarbon. The normal paraffins themselves are not symmetrical spherically. Moreover, the electrons in the molecules have many frequencies.

Although attempts have been made to extend London's basic concept to take account of such complexities, only the general implications of the concept as to the characteristics of the London forces are necessary here. These forces are (1) of very short range; (2) additive and nonspecific; (3) temperature independent; (4) operative between molecules of all types, whether nonpolar, polar, or metallic; (5) dependent in magnitude upon the number and "looseness" of the electrons; and (6) ordinarily less than average between molecules of different species. This last property can be seen by comparing ϵ_{12} with ϵ_{11} and ϵ_{22} as given by Eqs. (1) and (2). Eliminating the αs one obtains

$$\epsilon_{12} = \frac{(\nu_1\nu_2)^{1/2}}{(\nu_1 + \nu_2)/2} \cdot (\epsilon_{11}\epsilon_{22})^{1/2}$$

Most component pairs differ much less in ionization potential than in polarizability, and the factor representing frequencies in this expression is not far from unity. Therefore

$$\epsilon_{12} \approx (\epsilon_{11}\epsilon_{22})^{1/2} \quad (3)$$

This means that the interaction potential between unlike molecules is less than the arithmetic mean of the like potentials.

The pair potentials can be integrated over all the molecules in the pure components as well as the solution to obtain approximate attraction constants, as, corresponding to the attraction constant of the van der Waals equation. M. P. E. Berthelot discovered the relation $a_{12} = (a_{11}a_{22})^{1/2}$, which is a good approximation for attractive forces in gases. J. Hildebrand and H. M. Carter found the Berthelot relation to be valid within 1% for 7 liquid pairs. For example, $a = 31.21$ for liquid CCl_4 and 64.79 for $SnBr_4$. The calculated geometric mean is 46.46 and a_{12}, observed, is 46.86.

The consequence of this geometric mean relation is that the cohesion in a mixture of two liquids having different cohesion is less than their average. This results in expansion in volume, absorption of heat, and vapor pressures greater than additive upon mixing.

Dipole interaction. This is the attraction between molecules containing permanent electric dipoles; it includes both the London forces and an electrostatic interaction of the dipoles. The latter depends upon the dipole moments of the molecules; it is temperature dependent. because thermal agitation opposes the antiparallel orientation in which the interaction is greatest. Its magnitude depends also upon the geometry of the molecules, because it is related to the distance of approach of the dipoles, not the molecular centers; the dipoles of some molecules are buried more deeply than those of others. This is the case with chloroform, which has solvent properties similar to those of carbon tetrachloride, except in a few specific cases. J. G. Kirkwood has expressed the degree of interaction between the dipoles of pure liquids by a g-factor. For pyridine, the dipole moment μ is 2.20×10^{-18} debye units, the dielectric constant ϵ is 12.5, and the dipole interaction g is 0.9. For water, $\mu = 1.84 \times 10^{-18}$, $\epsilon = 78.5$, and $g = 2.7$; for ethyl alcohol, $\mu = 2.80 \times 10^{-18}$, $\epsilon = 24.6$, and $g = 3.0$.

It is the g-factor, not the dipole moment or the dielectric constant, that is most significant for understanding solubility relations. Furthermore, in the case of molecules having more than one polar bond, it is the separate polar bonds, not their vector sum of the over-all dipole moment, that determines solubility relations. The three isomeric dinitrobenzenes all affect the vapor pressure of benzene virtually to the same extent, even though their dipole moments are quite different.

The substances with the largest g-factors are those that form hydrogen bonds. These have exceptionally high boiling points and are poor solvents for nonpolar substances. These liquids resist penetration by nonpolar molecules. The best-known pairs of incompletely miscible liquids are composed of a nonpolar liquid and water.

Electron donor-acceptor interaction. In the modern theory of generalized acids and bases, initiated by G. N. Lewis, a base is a substance having electrons that may be "accepted" into the vacant orbitals of other molecules, termed acids. This acceptance of electrons takes place reversibly and with little or no activation energy. Typical bases or donors are: pyridine, acetone, ether, alkyl bromides, alkyl iodides, alkyl sulfides, iodide ion, thiocyanate ion, and aromatic hydrocarbons. Typical acids are: the pure and mixed halogens, sulfur dioxide and trioxide, boron trichloride and trifluoride, aluminum halides, and stannic chloride. R. S. Mulliken and his co-workers have pointed out the close relationship between base strength and ionization potential, and elaborated a theory of charge transfer complexes. H. A. Benesi and Hildebrand discovered the strong absorption in the ultraviolet characteristic of such complexes. They found that the basic strength increases in the order: benzene < toluene < xylene < mesitylene. R. L. Scott has determined that acid strength increases in the order:

$$Cl_2 < Br_2 < I_2 < BrI < BrCl < ICl$$

This type of interaction is specific and saturating, and it reduces the escaping tendencies of the components. It corresponds to $\epsilon_{12} > (\epsilon_{11}\epsilon_{22})^{1/2}$. In cases where it is weak it may reduce but not overcome the opposite effect of unequal London forces.

Ion-ion interaction. The ions in a solid or liquid salt attract and repel electrostatically according to Coulomb's law, but there is also a London force component, and large ions are polarized by smaller ones. This last effect is illustrated by solid silver bromide which is colored although its ions in aqueous solutions are colorless, and its crystals have the sodium chloride structure. The evidence is that in both the solid and the fused salt, the electron cloud of the bromide ion is distorted equally by each of its 6 neighboring silver ions. *See* FUSED-SALT PHASE EQUILIBRIA.

Ion-dipole interaction. In order to dissolve a solid salt, its lattice energy must be supplanted by the ion-ion action of another salt already in the liquid state, or by the predominantly electrostatic attraction of a polar solvent, or by the specific chemical interaction represented by complex ions.

Ideal solution. It is profitable to deal with actual solutions in terms of their departure from a simple idealized model—a mixture of components having the same attractive fields, which mix without change in volume or heat content. This is analogous to an ideal gas mixture, which is formed with no heat of mixing and in which the total pressure is the sum of the partial pressures. In such a solution, the escaping tendency of the individual molecules is the same, whether they are surrounded by similar or by different molecules. Therefore, the combined escaping tendency of all the molecules of species 1, f_1, is given by the equation

$$f_1 = f_1^0 x_1$$

where x_1 is the mole fraction of species 1, and f_1^0 is the escaping tendency of the molecules from the pure liquid. For a binary mixture, $x_1 + x_2 = 1$. If the gas imperfections of vapors are disregarded, vapor pressures may be substituted for fugacities to give Raoult's law (1886) in its usual form, $p_1 = p_1^0 x_1$. The total pressure is $P = p_1 + p_2 = p_1^0 x_1 + p_2^0 x_2$.

A more sophisticated derivation than the foregoing requires one to postulate molecules of the same size and shape. The gross structure of the solution containing n_1 plus n_2 molecules of two components is identical with those of the pure liquid components. The number of configurations of the components in the mixture within this structure is $(n_1 + n_2)!/n_1!n_2!$ and the configurational entropy of mixing is, by aid of Stirling's formula

$$\Delta S = k\left[n_1 \ln \frac{n_1 + n_2}{n_1} + n_2 \ln \frac{n_1 + n_2}{n_2}\right]$$

Substituting a number of moles of each N_1 and N_2 gives

$$\Delta S = R\left[N_1 \ln \frac{N_1 + N_2}{N_1} + N_2 \ln \frac{N_1 + N_2}{N_2}\right] \quad (4)$$

The partial derivative $(\partial \Delta S/\partial N_1)_{N_2}$ represents the partial molal entropy of transfer of component 1 from pure liquid into an ideal solution of mole fraction x_1,

$$\bar{s}_1 - s_1^0 = -R \ln x_1 \quad (5)$$

Because the model postulates no change in enthalpy,

$$T(\bar{s}_1 - s_1^0) = RT \ln (f_1/f_1^0)$$

and $f_1/f_1^0 = x_1$, which is Raoult's law, where f is fugacity.

But to arrive at this conclusion one must assume identical structures in the solution and the pure liquid components. This is very far from the case in solutions of high polymers in ordinary solvents, even though, as with polystyrene in benzene, the heat of mixing is practically zero.

Moderate difference in molal volume between components of high symmetry has little effect, as might be expected from the fact that the radius varies only with the cube root of the volume.

As a foundation for dealing with actual solutions in terms of the deviations of their properties from those of the model, it is necessary to derive other equivalent thermodynamic relationships.

Solubility of a crystalline solid. The fugacity of a crystalline substance, f^s, at temperature T is less than that of its supercooled liquid, f^0, to an extent depending upon its melting point, T_m, and heat of fusion, ΔH_F, as given by

$$\frac{d \ln (f^s/f^0)}{dT} = \frac{\Delta H_F}{RT^2}$$

If ΔH_F is assumed constant, this gives upon integration

$$\ln \frac{f^s}{f^0} = -\frac{\Delta H_F}{R}\left(\frac{1}{T} - \frac{1}{T_m}\right) \quad (6)$$

If the heat capacities of the solid and liquid forms are known, the variation of ΔH_F with temperature can be taken into account. This is hardly necessary for the present purpose, because the deviations from ideal solubility involve factors that are more uncertain than this.

Molecular weight measurements. If a solid dissolves to form an ideal solution, its heat of solution is the same as its heat of fusion, ΔH_F, and $f_1^s/f_1^0 = x_1$. Therefore

$$-\ln x_1 = \frac{\Delta H_F}{R}\left(\frac{T_m - T}{T_m T}\right) \quad (7)$$

This is the approximate equation for solubility of a solid that forms an ideal solution. It can be transformed into one much used for determining the molal weight of a solute by the depression of the freezing point of the solvent, here component 1. For $-\ln x_1$, one can write $\ln (1 + N_2/N_1)$. Expanding in powers of N_2/N_1 gives

$$\ln\left(1 + \frac{N_2}{N_1}\right) = \frac{N_2}{N_1}\left[1 - \frac{1}{2}\frac{N_2}{N_1} + \frac{1}{3}\left(\frac{N_2}{N_1}\right)^2 - \cdots\right]$$

When $N_2 \ll N_1$, the higher powers may be neglected to give

$$\frac{N_2}{N_1} \approx \frac{\Delta H_F}{RT_m^2}\Delta T \quad (8)$$

where $\Delta T = T_m - T$. By measuring ΔT for a known weight of solute in N_1 moles of solvent, the molal weight of the solute can be calculated. Because of the approximations made, and the fact that even good solvents for the solid in question are seldom ideal, the resulting molal weights are not very exact unless extrapolated to $x_2 = 0$ from a series of values.

The lowering of the vapor pressure of a solvent upon the addition of a nonvolatile solute (component 2) may be offset by raising the temperature to restore the pressure of the solvent. These changes are related as follows:

$$x_2 = \frac{\Delta H_{vap,1}}{RT_b^2} \qquad (9)$$

This relation is far less useful than that for the freezing point depression because the heat of vaporization is much greater than the heat of fusion; therefore the elevation of the boiling point is much smaller than the depression of the freezing point, and also is harder to determine.

Osmotic pressure. One mole of a solvent can be removed from a large quantity of a solution in which its mole fraction is x_1 in two reversible, and hence equivalent, ways. If it is distilled from the solution into pure liquid, the gain in (Gibbs) free energy is $\Delta F = RT \ln (f_1{}^0/f_1)$. If it is pressed out through a semipermeable membrane against the hydrostatic pressure difference, osmotic pressure ΔP, the gain in free energy is $\Delta P\bar{v}_1$, where $\bar{v}_1$ is the partial molal volume of the solvent. In an ideal solution, this is the molal volume. Equating the free energy of these processes gives

$$\Delta P\bar{v}_1 = RT \ln \frac{f_1{}^0}{f_1} = RT \ln \frac{N_1 + N_2}{N_1} \qquad (10)$$

Expanding as before and neglecting the higher powers gives $P\bar{v}_1 \approx (N_2/N_1)RT$, or $PV = RT$, where $V = N_2\bar{v}_1/N_1$, the volume containing 1 mole of solute.

This is the van't Hoff law for osmotic pressure, put forth in 1887. The theoretical basis of Raoult's law, discovered at almost the same time, was not yet appreciated. The formal correspondence between the van't Hoff law and the perfect gas law seemed to lend unique significance to osmotic pressure, and elevated the van't Hoff to the status of an ideal solution law. It is a limiting law only and not valid at high concentrations; it neglects the specific nature of the solvent. The solvent is regarded as furnishing space for a quasi-gas solute. Thus, the law cannot furnish evidence concerning molecular states in solutions of finite concentration.

The determination of osmotic pressure offers a valuable means for determining molal weights of high polymers in solution, where high weight concentrations correspond to mole fractions so low as to have only minute effects upon the vapor pressure and the freezing point of the solvent. For example, consider a solution of 0.001 mole of solute in 1 mole of benzene; $T_m = 279°$K and $\Delta H_F = 2370$ cal/mole. ΔT by Eq. (8) would be only 0.065° but ΔP, by Eq. (10), would be 194 mm. The latter is large enough to be measured with some precision.

Nonideal solutions. Unlikeness of the components of a binary mixture leads, as explained earlier, to fugacities in excess of ideal values. The excess is largest when the molecules of one species are surrounded mostly by those of the other, as illustrated in Fig. 1. They approach Raoult's law at the upper end and Henry's law, $p_1 = kx_1$ (or $f_1 = k_1x_1$) at the lower end, where k is an experimentally determined constant.

An important relation between the two components is given by the Gibbs-Duhem equation

$$\left(\frac{\partial \ln f_1}{\partial \ln x_1}\right)_T = \left(\frac{\partial \ln f_2}{\partial \ln x_2}\right)_T \qquad (11)$$

If Raoult's law holds in the limit when $x_1 = 1$, since $(\partial \ln f_1)/(\partial \ln x_1) = 1$, then also $(\partial \ln f_2)/(\partial \ln x_2) = 1$. Integrating gives $\ln f_2 = \ln x_2 + \ln k_2$ or $f_2 = kx_2$, where k is a constant of integration that cannot be evaluated unless Raoult's law holds for component 1 over the whole range. In that case, it also holds for component 2.

The activity in the case of nonelectrolytes is defined as $a_1 = f_1/f_1{}^0$, and so on. In an ideal solution. $a_1 = x_1$ and $a_2 = x_2$. The activity coefficient is $\gamma_1 = a_1/x_1$, and so on. Alternate, equivalent forms of the Gibbs-Duhem equation include

$$(\partial \ln a_1)/(\partial \ln x_1) = (\partial \ln a_2)/(\partial \ln x_2)$$

and $N_1 d\bar{F}_1 + N_2 d\bar{F}_2 = 0$, where $\bar{F}$ denotes partial molal free energies.

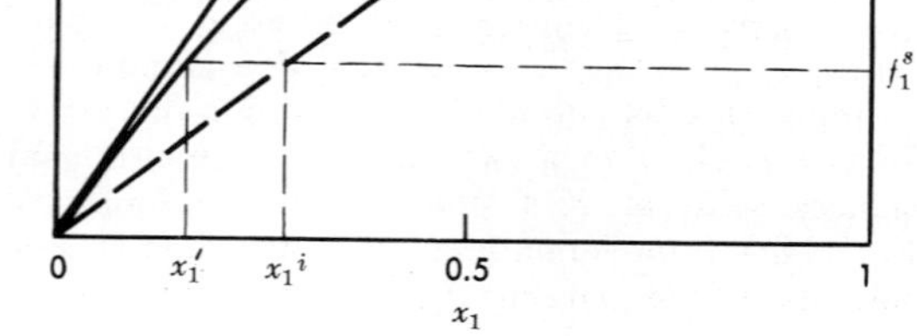

Fig. 1. Fugacity and mole fraction. Line A, ideal. Line B, typical nonideal. Line C, Henry's law.

If one component is a crystalline solid, its activity, a_s, is less than that of the liquid, which is 1, as given by Eq. (6); its maximum solubility would be $x_1{}^i$ in Fig. 1 if in an ideal solvent, and x'_1 if in a real solution.

Regular solutions. There are many mixtures of nonpolar components in which, except in the immediate neighborhood of a critical mixing temperature, thermal agitation suffices to neutralize tendencies to segregate and yields virtually complete randomness of mixing, with a close approach to ideal entropy of mixing, Eqs. (4) and (5).

The enthalpy of mixing can be calculated as the difference between the potential energy of the mixture and the sum of the potential energies of the liquid components. The potential energy of a mole of liquid may be related to the potential between a pair of molecules, $\epsilon(r)$. The lattice energy of a crystal is obtained by summation of $\epsilon(r)$ over all of the lattice distances; that of a liquid is obtained by integration over the continuous distribution function, $\rho(r)$. The expression for a pure liquid is

$$\Delta E_{\text{vap}} = -\frac{2\pi N_{Av}^2}{\mathrm{v}} \int \rho(r)\epsilon(r)r^2\,dr \qquad (12)$$

Here N_{Av} is the Avogadro number and v is the molal volume of the liquid. The corresponding expression for the potential of the mixture of N_1 and N_2 moles of the pure components involves their relative sizes. With certain simplifying assumptions, including the geometric mean for $\epsilon_{12}(r)$, Eq. (13*a*) was obtained for the energy of mixing N_1 and N_2 moles of two nonpolar liquids

$$\Delta E_M = \frac{N_1\mathrm{v}_1 N_2\mathrm{v}_2}{N_1\mathrm{v}_1 + N_2\mathrm{v}_2}\left[\left(\frac{\Delta E_1}{\mathrm{v}_1}\right)^{1/2} - \left(\frac{\Delta E_2}{\mathrm{v}_2}\right)^{1/2}\right]^2 \qquad (13a)$$

The corresponding partial molal energy of transferring pure liquid to solution, for component 2, is

$$\bar{\mathrm{E}} - \mathrm{E}_2^0 = \mathrm{v}_2\phi_1^2(\delta_1 - \delta_2)^2 \qquad (13b)$$

Here ϕ_1 denotes volume fraction, neglecting expansion on mixing, and the δ's are $(\Delta \mathrm{E}_{\text{vap}}/\mathrm{v})^{1/2}$, designated solubility parameters. Because energy and enthalpy are virtually identical for liquids, Eq. (13*b*) may be combined with the entropy of transfer as given by Eq. (5) to give the free energy of transfer,

$$\bar{\mathrm{F}}_2 - \mathrm{F}_2^0 = \bar{\mathrm{H}}_2 - \mathrm{H}_2^0 - T(\bar{\mathrm{S}}_2 - \mathrm{S}_2^0)$$

$$RT \ln a_2 = \mathrm{v}_2\phi_1^2(\delta_1 - \delta_2)^2 + RT \ln x_2 \qquad (14a)$$

or

$$RT \ln \gamma_2 = \mathrm{v}_2\phi_1^2(\delta_1 - \delta_2)^2 \qquad (14b)$$

Representative values of solubility parameters at 25°C are given in Table 1. Parameters for substances solid at 25°C have been calculated by Eq. (14*a*) from solubilities of these substances in solvents whose parameters are well determined.

Table 1. Solubility parameters and molal volumes, 25°C

Liquid	Formula	Molal volume, ml	Solubility parameter
Perfluoroheptane	C_7F_{16}	225	6.0
n-Heptane	*n*-C_7H_{16}	147	7.4
Silicone tetrachloride	$SiCl_4$	115	7.6
Cyclohexane	C_6H_{12}	109	8.2
Carbon tetrachloride	CCl_4	97	8.6
Chloroform	$CHCl_3$	81	9.0
Benzene	C_6H_6	89	9.2
Carbon disulfide	CS_2	60	10.0
Bromine	Br_2	52	11.5
Iodine	I_2	59	14.1

Table 2. Solubility of iodine at 25°C in mole per cent

Solvent	100 x_2	δ_2(calc.)
C_7F_{16}	0.018	14.2
$C_4Cl_2F_6$	0.124	14.4
$SiCl_4$	0.50	13.9
CCl_4	1.15	14.2
$CHCl_3$	2.28	14.3
CS_2	5.46	14.1
$CHBr_3$	6.27	14.4
Ideal	25.8	(14.1)

With paraffins, the solubility data give concordant δ-values a little greater than $(\Delta \mathrm{E}_{\text{vap}}/\mathrm{v})^{1/2}$.

Equation (14*a*) has found wide applicability for calculating solubilities from solubility parameters. Inaccuracies of the terms for enthalpy and entropy offset each other to give fairly reliable results for free energy. This is strikingly illustrated by comprehensive experimental data for iodine, where deviations from ideal become extremely large, and therefore provide a stringent test. The violet color of certain iodine solutions offers a simple means for distinguishing those in which London forces alone operate (physical solutions) from the red, brown, or yellow solutions which involve specific chemical interactions. A partial list of iodine solubilities in physical solutions is given in Table 2.

The values of δ_2 for iodine have been calculated from the solubilities and molal volumes. Their near constancy, despite the 350-fold range in x_2 indicates the degree of accuracy with which the reverse calculation of solubility could be carried out.

In Fig. 2 are plotted data for log x_2 vs log T. The solid lines represent violet solutions of iodine; the broken lines indicate solutions in which there is chemical interaction. The first lines constitute a family of curves indicating a common regularity, which led to the designation of these solutions

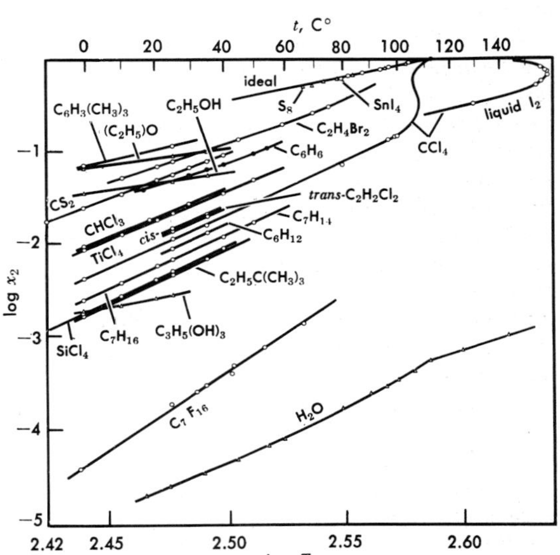

Fig. 2. Solubility of iodine in various solvents.

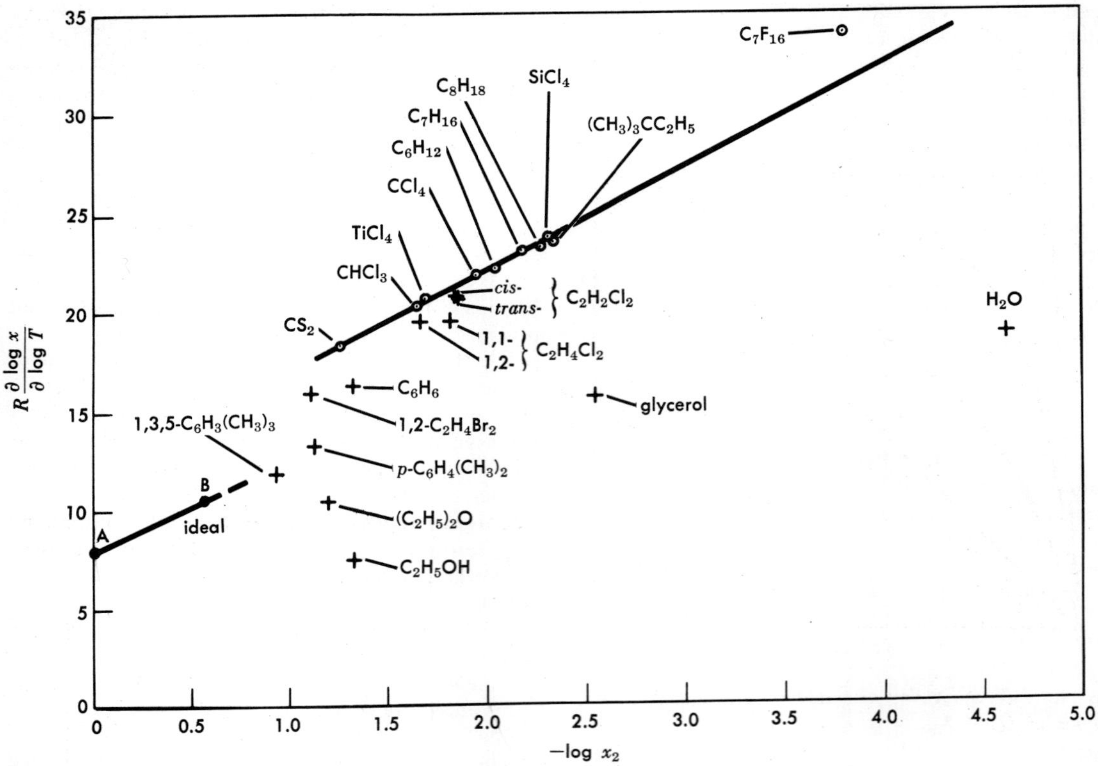

Fig. 3. Entropy of solution of iodine (25°C) vs. solubility.

as regular solutions. The same sort of regularity was evident with other solids, including sulfur, stannic iodide, and naphthalene.

A regularity in temperature coefficients is a matter of entropy. From the picture of a regular solution as one in which orienting and chemical effects are absent, and in which the distribution and orientations are random, just as in the ideal solution, it may be concluded that the probability is the same in the two solutions and, therefore, that the difference in entropy is zero. This conclusion cannot be expected to hold unless the random distribution of the molecules persists. Also, a small correction should be applied to take care of the change in entropy accompanying changes in volume, given by

$$(\partial S/\partial V)_T = (\partial P/\partial T)_V$$

A regular solution involves no entropy change when a small amount of one of its components is transferred to it from an ideal solution of the same composition and the total volume remains unchanged.

The slopes of the lines in Fig. 2, when multiplied by R, give the entropy of solution.

$$\bar{S}_2 - S_2^s = -\left(\frac{\partial(\bar{F}_2 - F_2^s)}{\partial T}\right)_x$$

$$= \left(\frac{\partial(\bar{F}_2 - F_2^s)}{\partial \ln x_2}\right)_T \left(\frac{\partial \ln x_2}{\partial T}\right)_{\Delta F}$$

$$\left(\frac{\partial(\bar{F}_2 - F_2^s)}{\partial \ln x_2}\right)_T = RT\left(\frac{\partial \ln a_2}{\partial \ln x_2}\right)_T$$

$$\bar{S}_2 - S_2^s = R\left(\frac{\partial \ln a_2}{\partial \ln x_2}\right)_T \left(\frac{\partial \ln x_2}{\partial \ln T}\right)_{\Delta F}$$

Most of these solutions are very dilute, hence a_2 approximates Henry's law and $\partial \ln a_2 / \partial \ln x_2 \approx 1$. The coefficients $(\partial \ln x_2 / \partial \ln T)_{\Delta F}$ are those of solutions in equilibrium with solid iodine, the slopes of the lines in Fig. 2. In Fig. 3 are plotted values of the entropy of solution against log x_2. The points for the violet solutions fall on a line whose slope corresponds to ideal entropy, with no divergence related to the great disparity between the molal volumes of the several solvents.

The points for nonviolet solutions all fall below the line, because solvation restricts the freedom of individual molecules, reduces entropy, and enhances solubility. The discrepancy increases in the order of increasing base strength: benzene $<$ p-xylene $<$ mesitylene. The colors change progressively from red to brown. The color of iodine in ethylene chloride is slightly red in harmony with its slight displacement. The entropy of solution in perfluoroheptane is enhanced by a very large expansion. The partial molal volume is 100 ml in perfluoroheptane, 65.6 in chloroform, 67.1 in silicon chloride, 62.7 in mesitylene, and 49.6 in ether.

Liquid-liquid miscibility. In Fig. 2 it can be seen that the line for CCl_4, although straight at lower temperatures, becomes S-shaped near the melting point, and is there cut by a liquid-liquid loop. If activity vs. mole fraction is calculated with increasing values of $(\delta_1 - \delta_2)^2$, one obtains curves

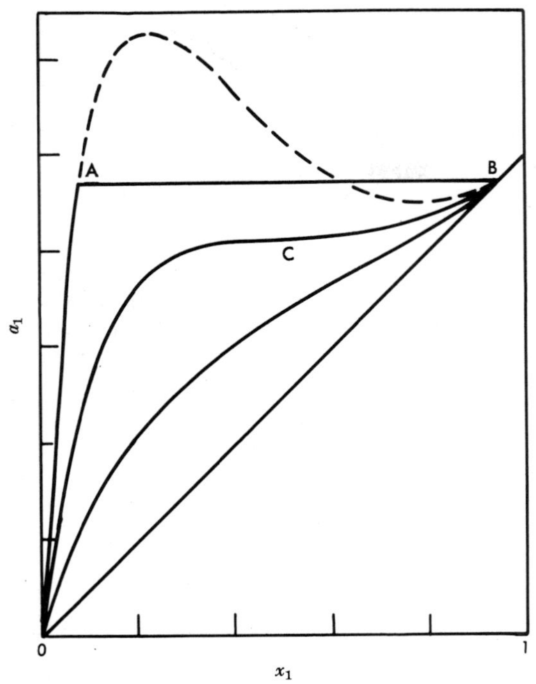

Fig. 4. Plot of activity vs. mole fraction for different values of $(\delta_1 - \delta_2)$.

that diverge progressively from the ideal line as shown in Fig. 4, and eventually become S-shaped, where there are three values of x_1 with the same value of a_1. The meaning of this is similar to that of the plot of the van der Waals equation; the two physically real roots, at A and B, represent the composition of two liquid phases in equilibrium. Upon raising the temperature, the enthalpy term in Eq. (14) diminishes and the roots coalesce in a critical point, C. This point is at $x_1 = x_2 = 0.5$ if $v_1 = v_2$, but is displaced when $v_1 \neq v_2$. By setting the first and second derivatives of $\ln a_1$ with respect to $\ln x_1$ both equal to zero, it is possible to solve the two equations to obtain expressions for both the critical temperature and composition. The former is sensitive, giving approximate values.

The curves in Fig. 5 for C_7F_{16}, $\delta = 6.0$, rise in order of increasing δ for the other components. The critical composition can, however, be calculated with remarkable accuracy. For the most unsymmetrical system in Fig. 5, that with $CHCl_3$, x_2 observed was 0.74; calculated, 0.78. For the pair $CHCl_3$ in $(C_3F_7COOCH_2)_4C$, where $v_1 = 82.6$ ml and $v_2 = 553$ ml at the critical temperature, 43.5°C, the mole fraction of the latter was 0.073 observed, 0.077 calculated.

Partition of a solute. Two liquid phases in equilibrium necessarily have very different cohesions and solvent powers. The difference becomes greater as their mutual solubility decreases. A third substance can therefore be expected to have very different activity coefficients in the two phases. Suppose that the solute has an a–x curve of type A, Fig. 6, in pure liquid A, and of type B in pure liquid B. If, for simplicity, it is assumed that A and B are not significantly miscible, the partition coefficient is x_A/x_B when the activity of the solute is a, and it is x'_A/x'_B when more is added and its activity is a'. It is obvious that this partition coefficient is not a constant. It is customary to refer all departures from constancy to some sort of chemical effect, association, dissociation, or solvation, in one of the solvents, and indeed type B is good evidence for solvation or ionization and type A may be caused, in part, by an association of solvent A; however, it may also be attributed to a purely physical difference in δ-values. The ratio, x_B/x_A, therefore, does not measure uniquely only what occurs in phase B. The interpretation is still further complicated by the fact that the two solvents are often somewhat soluble in each other, and this solubility is further affected by the concentration of the third component.

Vapor pressures of binary solutions. Equation (14) lends itself to evaluation of vapor pressure data for binary mixtures.

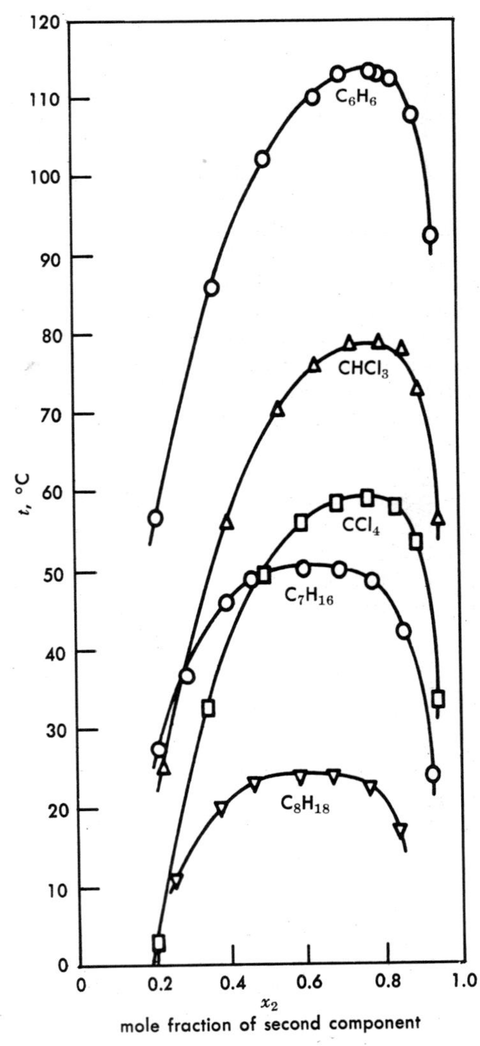

Fig. 5. Solubility of C_7F_{16}.

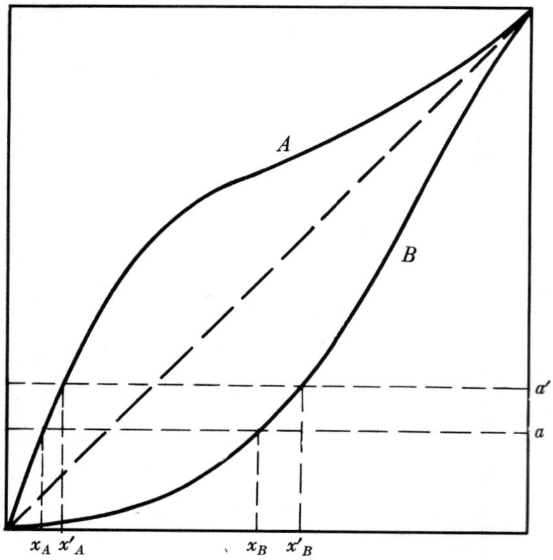

Fig. 6. Plot of activity vs. mole fraction for several types of liquid.

$$(\delta_1 - \delta_2)^2 = \frac{RT \ln \gamma_1}{\text{v}_1} \cdot \frac{1}{\phi_2^2} = \frac{RT \ln \gamma_2}{\text{v}_2} \cdot \frac{1}{\phi_1^2}$$

If one plots $(RT/\text{v}_1) \ln \gamma_1$ against ϕ_2^2 and $(RT/\text{v}_2) \ln \gamma_2$ against ϕ_1^2, using the same scale for abscissa, the experimental points for both components of a regular solution, if they are accurately determined, should fall upon the same straight line, whose intercept is $(\delta_1 - \delta_2)^2$ when $\phi^2 = 1$.

An example of activities less than ideal, corresponding to $\epsilon_{12} > (\epsilon_{11}\epsilon_{22})^{1/2}$, is the mixture of mercury with cadmium, shown in Fig. 7. Measured values of the vapor pressure of mercury over the amalgams were divided by the vapor pressure of pure mercury to get activities. The corresponding

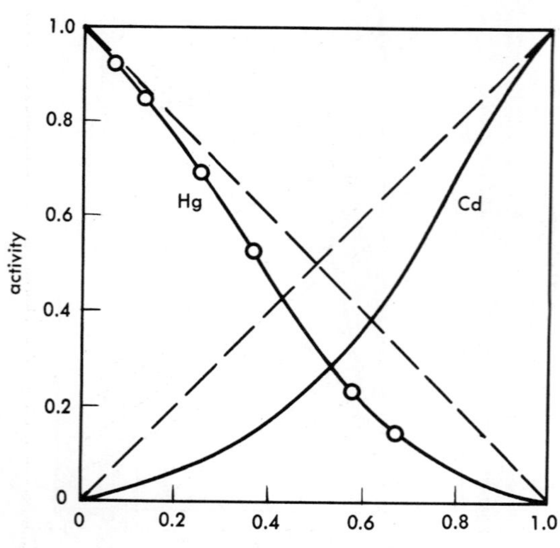

Fig. 7. Activities of cadmium and mercury in their amalgam, 322°C.

activities of cadmium were then calculated by means of the Gibbs-Duhem equation.

Azeotropic mixtures. If the total vapor pressure of a liquid mixture is higher in some intermediate region than the vapor pressure of either pure component, the composition of the vapor upon repeated distillation will approach that of the maximum of the total vapor pressure curve, not that of a pure component, and the pure components cannot be obtained by fractional distillation under these conditions (Fig. 8). Similarly, if the plot of total vapor pressure against composition passes through a minimum, the residual liquid, as distillation proceeds, will approach the composition of the minimum, not of the pure, less volatile component.

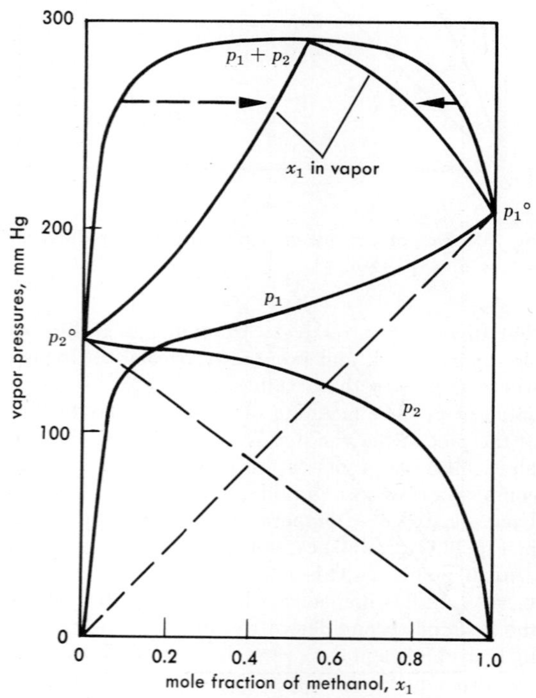

Fig. 8. Vapor pressures of mixtures of methanol with benzene, 35°C. (*Scatchard, Wood, and Muchel*)

Solubility of gases. A solution of a gas that can be liquefied can be treated as a liquid mixture. For example, chlorine has a vapor pressure of 3.66 atm at 0°C. Its ideal solubility at 1 atm and 0°C is accordingly $x_2 \approx p_2/p_2^0 = 0.273$. If pressures are corrected to fugacities, $x_2 = 0.286$. The measured solubility in CCl_4 is 0.298. In C_7F_{16}, whose δ value is 6.0, less than that of chlorine (about 9.8), the solubility is much less (0.164).

The value of p_2^0 becomes larger as the boiling point of the gas decreases, as seen in Table 3. A better measure is the force constant of the gas, ϵ/k, as determined from its *PVT* behavior. Figure 9 shows the linear relation between $\log x_2$ and ϵ/k for a series of gases in the same solvent. The relation between $\log x_2$ and δ_1 for different solvents is shown in Fig. 10. The change in solubility with temperature is to a considerable extent dependent upon the solubility itself. Figure 11 shows log x_2 vs. log T for argon in several solvents. The change in slope from one solvent to the next is mainly the result of added entropy of dilution, $R \ln (x_2/x'_2)$, where x'_2 is solubility in the poorer solvent.

The effect of pressure is given adequately by Henry's law for gases of low solubility.

The enormous drop in going from CS_2 to H_2O reflects the high cohesion of water caused by hydrogen bonding.

Water as a solvent. Water is a poor solvent for nonpolar substances, which cannot overcome the high cohesion or internal pressure that result from its small molal volume and strong hydrogen bonds.

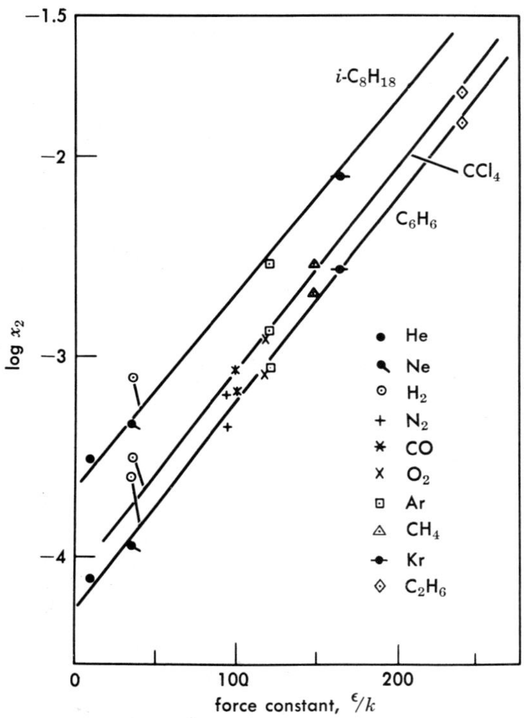

Fig. 9. Plot of mole fraction against force constant for several gases in the same solvent.

This is illustrated by its poor solvent power for the gases included in Table 3. It is a good solvent for SO_2, which hydrates to H_2SO_3 and for ammonia,

Table 3. Solubility of gases, $10^4 x_2$, at 25°C and 1 atm

Element:	He	H_2	N_2	O_2	Ar	Kr
ϵ/k =	10	37	95	118	121	165
Solvent						
C_7F_{16}		14.0	38.7	55.3		
n-C_7H_{16}	2.5	6.9			25.0	67.6
C_6H_{12}			7.6		14.9	46.7
CCl_4		3.2	6.4	12.0	13.4	
C_6H_6	0.78	2.6	4.5	8.2	8.9	27.3
CS_2		1.6	2.2	4.4	4.9	
H_2O	0.07	0.14	0.12	0.23	0.25	0.45

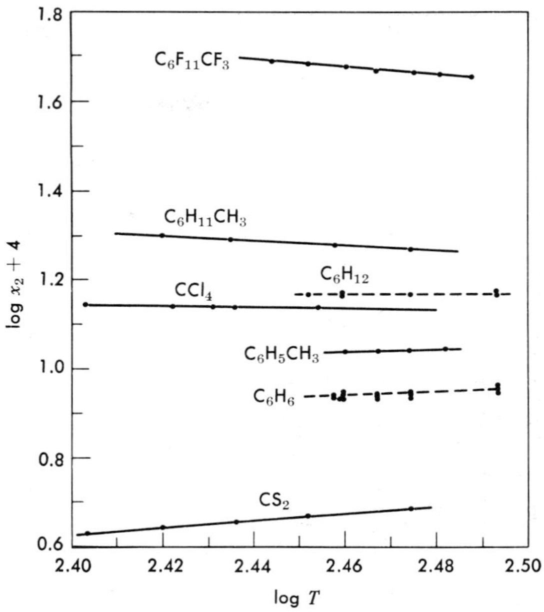

Fig. 10. Relation between x_2 and δ_1 for various solvents.

with which it forms hydrogen bonds. Its solvent powers for benzene and benzene derivatives are shown in Table 4. Benzene has by far the highest vapor pressure, but it nevertheless dissolves to the least extent. Nitrobenzene, with the highest dipole moment, and only a little handicapped by low

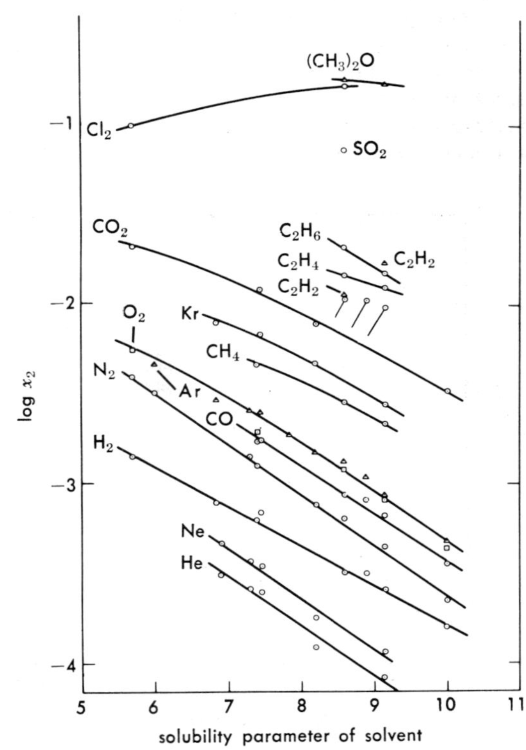

Fig. 11. Solubility of argon in various solvents.

Table 4. Solubility in water, 20°C

Solute	$100\ x_2$	Dipole moment	Boiling point
Benzene	0.013	0	80
Nitrobenzene	0.028	4.2	208
Aniline	0.70	1.5	184
Phenol	1.7	1.7	182

vapor pressure, is not nearly as soluble in water as aniline and phenol. Phenol has the highest solubility by virtue of being able to form the strongest hydrogen bonds.

Solution of electrolytes. In solutions of electrolytes, relations become simplest as dilution approaches infinity, where the ions can act independently. Here, their activities can be set at unity, $a_+ = 1$, $a_- = 1$, and the activity coefficient, $\gamma_\pm$, likewise becomes unity. Because the ionic activities cannot be measured separately, their mean activity, defined as $a_\pm = (a_+a_-)^{1/2}$ for a uni-univalent electrolyte, must be used. Departures occur as concentration increases. These were once thought to be caused by the formation of un-ionized molecules, but are now interpreted, at least in the case of electrolytes, as purely coulombic interactions, where the attractions between ions of opposite charge lead to activities less than unity. The theory was developed by P. Debye and E. Hückel (1923). The effect is measured by the ionic strength s defined by $s = \Sigma \frac{1}{2} mz^2$, where m is the molality, moles per 1000 g solvent, and z is the charge on each ion. The activity coefficient of an ion is in the limit $\log \gamma = -Az^2s^{1/2}$, where A is a function of the ionic charge, the dielectric constant of the solvent, the Boltzman constant, and the Avogadro number.

Table 5 gives activity coefficients at 25°C of typical electrolytes.

Table 5. Activity coefficients, 25°C

Solute	Molality		
	0.01	0.1	1.0
NaCl	0.90	0.78	0.66
NaI	0.91	0.83	0.80
KOH	0.90	0.80	0.76
$AgNO_3$	0.90	0.72	0.40
$ZnCl_2$	0.71	0.50	0.33
$MgSO_4$	0.40	0.18	0.06
$CdSO_4$	0.40	0.17	0.04

Ionic strength largely determines the extent to which an electrolyte can diminish the solubility of a nonelectrolyte in water, the salting-out effect.

[J.H.H.]

Bibliography: H. S. Harned and B. B. Owen, *The Physical Chemistry of Electrolytic Solutions*, 3d ed., 1958; J. H. Hildebrand and R. L. Scott, *Solubility of Nonelectrolytes*, 3d ed., 1950.

Solvation

The association or combination of a solute unit (ionic, molecular, or particulate) with solvent molecules. This association may involve chemical or physical forces, or both, and may vary in degree from a loose, indefinite complex to the formation of a distinct chemical compound. Such a compound contains a definite number of solvent molecules per solute molecule.

Solvation occurring in aqueous solutions is referred to as hydration. In aqueous ionic solutions, the highly polar water molecules become oriented about the ions, forming spheres of hydration. As a result, the mobility of an ion under an applied voltage gradient is decreased. The extent of hydration depends upon the size and charge of the ion.

In certain colloidal suspensions, solvation is, to a large extent, responsible for the stability of the sol. Particles in lyophilic sols strongly adsorb on their surfaces one or more layers of solvent molecules. This protective layer prevents the particles from colliding so closely as to adhere. In water starch and many proteins form suspensions which are highly solvated. Such systems exhibit a viscosity which is markedly higher than that of the pure solvent. *See* COLLOID; ELECTROLYTIC CONDUCTANCE; HYDRATION; SOLUTION. [F.J.J.]

Solvent

A liquid which dissolves other materials (solutes) to form single homogeneous phases (solutions). While water is the most widely used solvent, a common usage of the term excludes water or aqueous solutions. Occasionally, the term is used broadly to describe the major component in any liquid mixture. Solvents may be classified on the basis of physical properties, chemical properties, or industrial application. *See* SOLUTION; WATER.

Physical classification is usually with respect to volatility as measured by boiling point or evaporation rate. Chemical classification may be based on chemical composition or on general chemical properties. Most solvents are organic, although a few important ones are inorganic. Liquids made up of molecules with relatively symmetrical structure are nonpolar, while polar solvents are made up of molecules with permanent dipoles which, because of the attraction of unlike charges, tend to orient themselves into an ionic atmosphere around polar solutes. Since this promotes the separation of electrically charged ions, they are called ionizing solvents. Among polar solvents, distinctions may be made on the basis of acidity, although whether a solvent is acidic or basic depends on the solute used to establish this fact.

In a solution, the attraction between the unlike molecules of solute and solvent must be sufficiently great to prevent the separation into two phases. If solute and solvent are similar chemically, they usually have substantial mutual solubility, since there is no great difference between the forces attracting like molecules in a pure phase and the unlike molecules in the solution. In any liquid mixture, the forces between molecules may be due to chemical bond formation (of which hydrogen bonding is a special case), to the attraction of permanent dipoles, or to the more widely present van der Waals forces arising from electrostatic interactions of the electronic orbits of any adjacent molecules. *See* HYDROGEN BOND.

Chemical bonding may be rather permanent and the true solute a new compound rather than the undissolved solute. In such cases, the term reactive solvent, or chemical solvent, is often used. Because of the possibility of chemical binding of complementary, though widely differing materials, the general rule that like dissolves like often appears to be strongly modified. With the understanding that "likeness" refers to the intermolecular forces in the liquid state, the similarity rule is a useful one.

Solvents in industry. A variety of solvents is employed in large amounts in the coatings industry, in the form of paints, lacquers, and varnishes. The coating is usually a plastic mixture which may be applied to the surface in the liquid phase and which drys to form the final stable coating film. Usually, the solvent evaporates to leave the final coating, although a chemical reaction such as oxidation by the air to which the film is exposed is often necessary to complete the film formation. To some extent, use of solvents in adhesives is related to use in coatings, with the film formed between the two surfaces to be joined. The textile industry is another large consumer of solvents. Many synthetic fibers such as rayon are spun from solution. However, the solvent is usually recovered, and consumption is small compared to the volume actually employed. The application of pesticides in the field provides a growing market for solvents. A special group called plasticizers are fundamental raw materials in the molding of plastics.

Liquids with widely differing properties and no possibility of complementary behavior usually form two separate phases with limited mutual solubility. Since a pair of solutes would have differing solubilities in two immiscible solvents, the separation of such a pair may often be effected by taking advantage of these solubility differences. This process is called solvent extraction or liquid-liquid extraction and has been employed in such diverse fields as petroleum refining, production of penicillin, and the purification of irradiated uranium in the atomic energy industry, as well as in analytical chemistry. Solvent extraction often refers not only to liquid-liquid extraction, but also to the treatment of solids or gases with a solvent to remove a soluble product. This may be either a specific, unwanted impurity or a desirable by-product which may be recovered by evaporation of the solvent or by some alternative isolation procedure. Such uses of solvents are encountered in dry cleaning or in the extraction of fats and oils from natural products such as cottonseed and soybeans. *See* SOLVENT EXTRACTION.

For each of these industrial processes, the required solvent properties must include: the ability to dissolve the desired solute; chemical stability and compatibility with all materials contacted such as containers, processing equipment, and surfaces to be coated; and such physical characteristics in terms of boiling point, evaporation rate, surface tension, and viscosity, as are appropriate to the application. Ease of recovery may be important for uses such as cleaning, or in spinning fibers from solutions. In addition, criteria of hazard are often important and low flammability and low toxicity are widely sought.

Flammability may be measured in terms of the flash point, autoignition temperature, or for volatile solvents, in terms of the explosive limits when mixed with air. Toxicity in use is largely a function of the volatility; this governs the amount likely to be present. The toxicity of the resulting vapor and the ease with which excessive amounts can be detected are also important factors. A noxious material such as ammonia may be less dangerous than a solvent such as benzene which is not particularly objectionable at concentrations high enough to lead to severe chronic poisoning. For control of occupational hazards, maximum allowable concentrations in the air have been set by various authorities. Good industrial practice will usually require that average exposure be kept well below these maxima. *See* TOXICOLOGY.

Industrial solvents are usually either natural mixtures such as petroleum fractions, or synthetic blends such as an alcohol-ether mixture used to dissolve cellulose nitrate in the manufacture of celluloid. Even when only a single component is wanted, high purity is often not required and rather broad boiling ranges may be permissible for commercial grades. The data in Tables 1 and 2 refer to such grades and may differ noticeably from the properties found for reagent quality pure chemicals. Mixtures may be much more effective solvents than a single component, or good solvents may be mixed with cheaper, but less effective, diluents. In the latter case, solvent properties are often measured by tests which determine the amount of dilution that is permissible. A number of empirical tests have been employed to indicate solvent power. The aniline point, kauri-butanol value, and dilution ratio tests have been standardized and are used for petroleum hydrocarbon solvents. In the textile industry, ease of recovery is more important than original cost, and single, relatively pure solvents are used.

Economic criteria often decide the choice between a variety of acceptable solvent mixtures for industrial purposes. Since the petroleum hydrocarbons are rather cheap, commercial mixtures may employ small amounts of rather expensive solvents which permit dilution with larger amounts of these materials. Occasionally, solvents are used to form suspensions or emulsions which may be handled as a homogeneous liquid phase, though not a true solution.

Petroleum fractions are available as a large number of commercial mixtures. The most volatile fractions are called petroleum ether or ligroin (boiling range 35–60°C), naphtha fractions are less volatile (boiling up to 150°C), and kerosine fractions are

Table 1. Important solvents and their uses

Solvent	Solvent class	Typical use
Acetic acid	Miscellaneous	Textiles
Acetone	Ketones	Lacquers, textiles, adhesives, explosives, photographic films, and acetylene
Ammonia	Miscellaneous	Chemical synthesis
Benzene	Aromatic hydrocarbons	Coatings, artificial leather, rubber
n-Butyl acetate	Esters	Lacquers, paints
Carbon disulfide	Miscellaneous	Rayon
Carbon tetrachloride	Chlorinated hydrocarbons	Oil and fat extraction
Cyclohexanone	Ketones	Coatings
Dibutyl phthalate	Esters	Plasticizer
Diethylene glycol	Glycols	Cosmetics
Ethyl acetate	Esters	Lacquers
Ethyl alcohol	Alcohols	Shellac
Ethyl ether	Ethers	Celluloid
Hexane	Aliphatic hydrocarbons	Extraction of oils
Isopropyl alcohol	Alcohols	Medicinals
Methyl cellosolve	Ether alcohols	Lacquers
Methyl isobutyl ketone	Ketones	Solvent-extraction purification of irradiated uranium
Nitrobenzene	Nitrohydrocarbons	Shoe polish
Nitromethane	Nitrohydrocarbons	Chemical reaction media
Perchloroethylene	Chlorinated hydrocarbons	Dry cleaning
Sulfur dioxide	Miscellaneous	Oil refining
Toluene	Aromatic hydrocarbons	Lacquer diluent
Tributyl phosphate	Esters	Irradiated uranium purification, plasticizer
Trichloroethylene	Chlorinated hydrocarbons	Metal cleaning
Turpentine	Terpenes	Paint thinner

Table 2. Physical properties of important solvents

Solvent	Specific gravity* (at 20°C)	Boiling range* °C (at atm pressure)	Flash point†, °F	Maximum allowable vapor concentration‡, ppm
Acetic acid	1.049	116–119	135	20
Ammonia	0.65 (−10°C)	−33		100
Benzene	0.883	78–80	<5	100
n-Butyl acetate	0.880	115–130	72	200
Carbon disulfide	1.269	45–47	<0	10
Carbon tetrachloride	1.584	77–78	none	25–50
Dibutyl phthalate	1.048	181 (4 mm)	340	
Diethylene glycol	1.116	240–250	255	
Ethyl acetate	0.902	76–79	32	400
Ethyl alcohol	0.818	75–80	45	1000
Ethyl ether	0.717	34–35	−20	400
Hexane	0.687	66–69	<0	500
Isopropyl alcohol	0.790	81–83	59	400
Methyl cellosolve	0.965	122–126	105	25
Nitrobenzene	1.198	211	171	1
Nitromethane	1.139	101	112	200
Perchloroethylene	1.618	121–123	none	200
Sulfur dioxide	1.46 (−10°C)	−10		10
Toluene	0.867	109–111	45	200
Tributyl phosphate	0.973	289 (decomposes)	295	
Trichloroethylene	1.470	87–88	none	200
Turpentine	0.862	156–170	93	100

* These are approximate values as given for technical grade solvents and may differ somewhat from the literature value for the pure reagent.

† As measured by the standardized "Tag closed-cup" test.

‡ This value may vary depending on the authority setting the concentration limit.

still less volatile (boiling range up to 250°C). Both aliphatic- and aromatic-rich mixtures are available, the latter in general being slightly more expensive. The low price of these solvents leads to their widespread use as diluents for more effective solvents. Similar solvent fractions are obtained as by-products in the coking of coal.

The terpenes are obtained largely as by-products of pine-tree processing for lumber or paper. As a natural product, the composition of turpentine may vary rather widely, usually without affecting its use as a solvent.

Solvents in scientific research. Solvents play a major role in science as media for chemical reactions. While by far the greatest fraction of scientific investigations has involved aqueous solutions, there is growing interest in nonaqueous-solution chemistry. In this field, the solvent may serve as an inert diluent, it may participate directly, or it may modify one of the reactants so as to affect its manner of participation. When solvents are utilized as reaction media, their ability to promote ionization is often important. For this purpose, the molecules should preferably be small and electrically unsymmetrical so that they will tend to form an ionic atmosphere capable of separating cationic and anionic parts of the solute from each other. The dielectric constant measured for the bulk solvent is an important clue to this ability, but does not give a direct measure of the effectiveness of the solvent in promoting ionization.

The solvent plays a major role in acid-base chemistry. Some chemists prefer to restrict the term acid to substances which are capable of donating a proton and the term base to proton acceptors. Solvents such as water are amphoteric and may act either as acids or as bases, depending upon the solute.

Polar solvents with mobile hydrogen atoms may be arranged in a rough order of acidity from anhydrous sulfuric and hydrofluoric acids at one end to liquid ammonia and hydrazine at the other.

In any solvent to which an acid is added, the actual acidic species present is the protonated solvent. No matter what the intrinsic strength of the added acid, the solvent limits the effective strength. This is known as the leveling effect of solvents. Two acids of very different strength in one solvent may be essentially indistinguishable in another, so that an extended acidity scale must include measurements in a number of solvents.

The extension of the term acidity from proton donors to include electron pair acceptors was suggested by G. N. Lewis and has been widely applied to aprotic solvents. The electron pair is often associated with a single element anion such as the oxide, fluoride. or chloride ion. Acid-base systems may be established in a specific group of solvents involving such anion transfers. An acidity scale may be constructed for materials which contain no hydrogen and compared with the more familiar acidity scale. *See* ACID AND BASE.

Although solvents normally are employed in a relatively narrow temperature range, materials which are gaseous or solid at room temperature may be quite useful solvents at lower or higher temperatures, respectively. Liquid ammonia, propane, and sulfur dioxide are examples of the first. Fused salts as well as liquid metals are interesting examples of high-temperature solvents. They have not been extensively investigated or treated as solvents, although some important industrial processes employ such materials. For example, cryolite is used as a solvent for alumina in the electrolytic production of aluminum. [H.H.HY.]

Bibliography: L. F. Audrieth and J. Kleinberg, *Non-aqueous Solvents*, 1953; T. H. Durrans, *Solvents*, 7th ed., 1957; J. H. Hildebrand and R. L. Scott, *The Solubility of Nonelectrolytes*, 3d ed., 1950; I. Mellan, *Source Book of Industrial Solvents*, vols. 1–2, 1957.

Solvent extraction

A technique, also called liquid extraction, for separating the components of a liquid solution. This technique depends upon the selective dissolving of one or more constituents of the solution into a suitable immiscible liquid solvent. It is particularly useful industrially for separation of the constituents of a mixture according to chemical type, especially when methods that depend upon different physical properties, such as the separation by distillation of substances of different vapor pressures, either fail entirely or become too expensive.

Industrial plants using solvent extraction require equipment for carrying out the extraction itself (extractor) and for essentially complete recovery of the solvent for reuse, usually by distillation. *See* DISTILLATION; EVAPORATION; STRIPPING.

Applications. The petroleum refining industry is the largest user of extraction. In refining virtually all automobile lubricating oil, the undesirable constituents such as aromatic hydrocarbons, which have poor chemical and viscosity-temperature characteristics, are extracted from the more desirable paraffinic and naphthenic hydrocarbons. The principal solvents used are furfural, phenol, and a combination of phenol with propane and cresylic acid; nitrobenzene and 2,2′-dichloroethyl ether are used in minor amounts. Liquid propane is also used preferentially to extract the desirable constituents from unwanted asphaltic compounds.

By suitable catalytic treatment of lower boiling distillates, naphthas rich in aromatic hydrocarbons such as benzene, toluene, and the xylenes may be produced. The latter are separated from paraffinic hydrocarbons with such solvents as liquid sulfur dioxide, furfural, and ethylene glycol to produce high-purity aromatic hydrocarbons and high-octane gasoline.

Gasoline is "sweetened," or freed of its sulfur-containing compounds, by extraction with aqueous caustic solutions containing various naphthenic and aromatic acids, or methanol, to modify the solvent characteristics. Aqueous copper ammonium acetate is used to extract butadiene from other 4-carbon hydrocarbons in synthetic rubber production.

Vegetable oils are separated into relatively saturated and unsaturated glyceride esters with furfural or liquid propane as solvents. The former are edible products; the latter are drying oils used in paints. Fish oils are similarly treated and yield a high-vitamin fraction as well.

In by-product coke-oven plants, phenols and other tar acids are recovered from the ammoniacal liquors with benzene, tricresyl phosphate, butyl acetate, and other solvents in large installations. The pharmaceutical industry uses extraction to separate natural impurities or unwanted chemical by-products from products such as synthetic vitamins, penicillin, Aureomycin, antihistamines, reserpine, and a host of others.

All uranium for atomic energy purposes is freed of its impurities in aqueous solution by extraction into diethyl ether, tributyl phosphate, and other solvents. The reprocessing of atomic energy fuels for the recovery of plutonium, and the separation of many of the other fission products such as the rare-earth metals, utilize solvent extraction extensively. The otherwise hard-to-separate metal pairs, zirconium-hafnium and niobium-tantalum, are separated in quantity with comparative ease by these methods.

Equipment. Extractors bring about direct contact of the feed (solution to be separated) and extracting solvent in order to permit diffusional transfer of the constituents from the feed to the solvent. The rate of transfer depends upon the contact area of the two liquids and the degree of turbulence developed within them. The extractor disperses one of the liquids in the other to produce large surface area, and relative motion to produce turbulence. The extractor must also provide for the subsequent mechanical separation of the dispersion, based upon the different densities of the liquids, to permit withdrawal of the two effluent products, the extract (solvent containing the extracted constituents) and the raffinate (unextracted residue).

Mixer-settlers (Fig. 1) provide for these requirements in separate vessels. The feed and solvent flow continuously through the mixer, in which the rotating agitator disperses one of the liquids into small droplets immersed in the other. The size of this vessel must provide sufficient residence time for the liquids that the desired diffusional transfer occurs. The degree of agitation must be intense without,

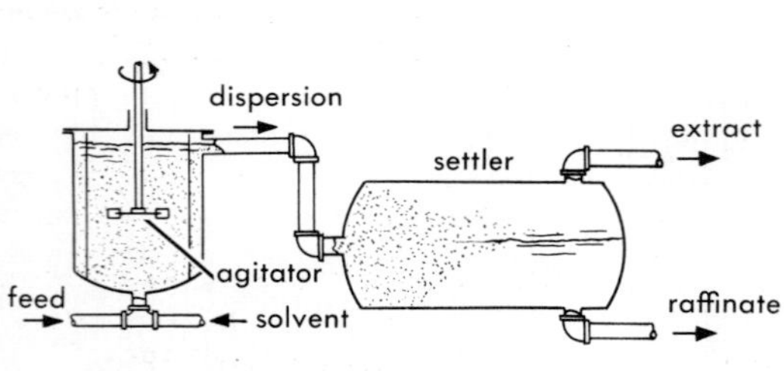

Fig. 1. Single-stage mixer-settler extractor.

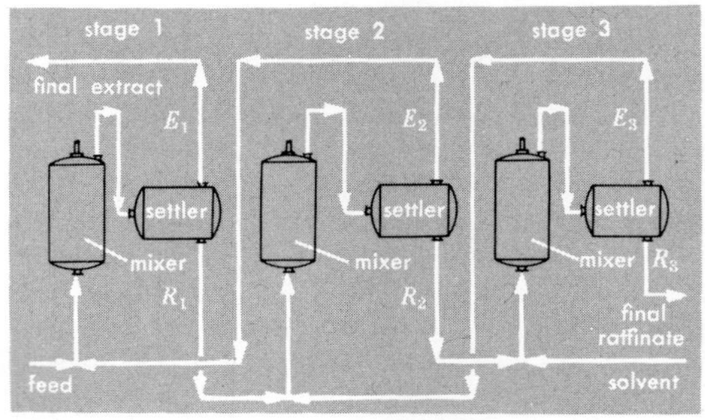

Fig. 2. Three-stage countercurrent mixer-settler extractor. *(From R. E. Treybal, Mass Transfer Operations, McGraw-Hill, 1955)*

however, producing so fine a dispersion that subsequent settling is difficult. The dispersion flows to the settler, most simply a drum, in which low velocity and lack of agitation promote gravity settling and coalescence of the drops to provide clear effluents.

Since in such single-stage apparatus the extractable substance approaches a concentration equilibrium in the effluents, nearly complete extraction requires a multiplicity of stages. An arrangement for countercurrent interstage flow of the liquids reduces the amount of solvent needed (Fig. 2). The compact modification of Fig. 3 has found particular favor in extraction of radioactive metals from aqueous solutions in processes associated with atomic energy operations.

To reduce the floor space and pump requirements for multistage extractors, a variety of vertical towers is also used. These involve countercurrent vertical flow, under gravity, of one of the liquids in dispersed form through a continuum of the other by virtue of the different liquid densities. A packed tower (Fig. 4*a*) is a cylindrical shell, the bulk of which is filled with manufactured packing, such as rings or saddles, randomly arranged (*see* GAS ABSORPTION OPERATIONS). The more dense liquid, introduced at the top, flows downward as a continuum. The less dense liquid enters at the bottom through small nozzles. The resulting small droplets rise through the heavy liquid, during which time ex-

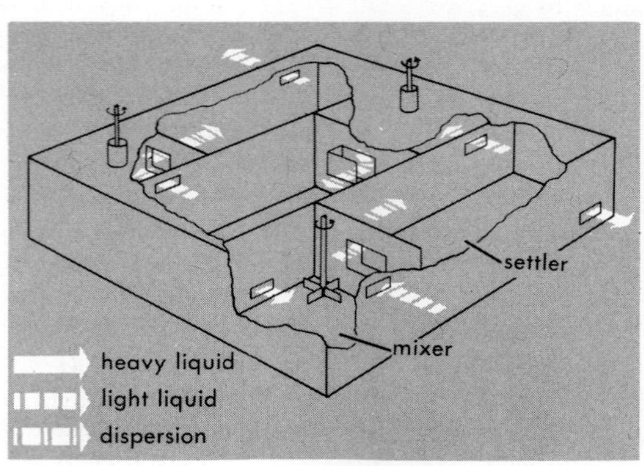

Fig. 3. Three-stage, box-type, mixer-settler extractor.

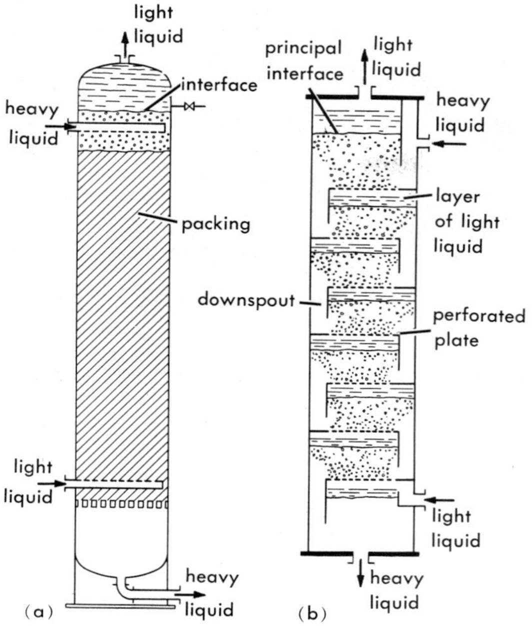

Fig. 4. (*a*) Packed-tower extractor. (*b*) Perforated-tray extractor. (*From R. E. Treybal, Mass Transfer Operations, McGraw-Hill, 1955*)

traction occurs, and then coalesce into a bulk and leave at the top. The packing serves to maintain the dispersion and provide moderate turbulence. The dispersed liquid may be either feed or solvent, light or heavy. If heavy, the droplets settle downward. Although the liquids are not repeatedly dispersed and settled as in the multistage mixer-settler, nevertheless multistage effects are obtained. Spray towers contain no packing and are not as effective.

In perforated-tray towers (Fig. 4*b*) the light liquid collects in a layer under each tray and is dispersed into droplets by the small perforations. The drops rise through the heavy liquid, which flows across each tray and through the downspouts. The frequent redispersion achieved makes these towers very effective. Alternatively, by turning the tower upside down, the heavy liquid may be dispersed.

Mechanical agitation, provided by rotating impellers as in the towers of Fig. 5*a*, *b*, and *c*, is used to obtain finer dispersions and increased turbulence. The pulsed tower (Fig. 5*d*) provides the mechanical agitation by rapid (20–100 cycles/min), small amplitude (0.25–2 in.), reciprocating motion superimposed upon the natural flow of liquids as they alternately pass through small perforations in the plates. This is particularly useful for handling radioactive liquids, since moving parts may be located in a place of safety.

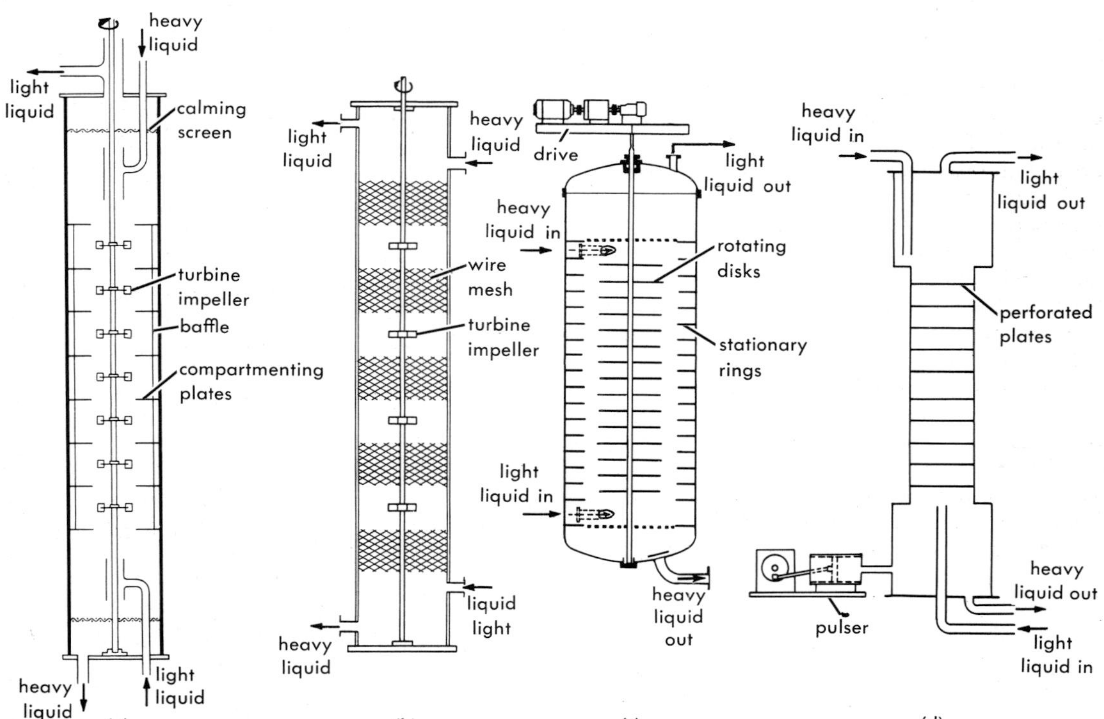

Fig. 5. (*a*) Oldshue-Rushton extractor; (*b*) Scheibel-York extractor (*from R. E. Treybal, Mass Transfer Operations, McGraw-Hill, 1955*); (c) Rotating-disk extractor (*from G. H. Reman and R. B. Olney, Chem. Eng. Progr., 51:141–146, 1955*); (*d*) Pulsed extractor (*from T. B. Drew and J. W. Hoopes, eds., Advances in Chemical Engineering, vol. 1, Academic Press, 1956*).

In all these designs, the tower diameter is governed by the quantity of liquids to be handled, the height by the number of stages of extraction required. Towers up to 15 ft in diameter and 125 ft tall have been built. Auxiliary equipment may include pumps for movement of the liquids, motor-drives for agitators, valves and flow meters for control of flow rates, and liquid-level control instruments.

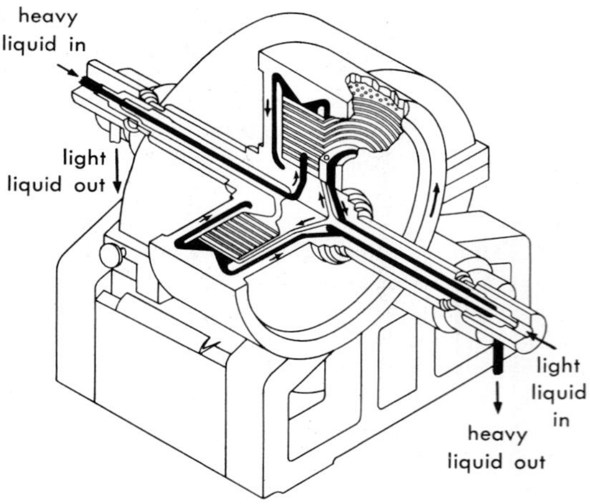

Fig. 6. Podbielniak centrifugal extractor. (*Podbielniak, Inc.*)

The centrifugal extractor (Fig. 6) consists of a series of perforated, concentric rings in a cylindrical drum, the whole rapidly rotated (2000–5000 rpm) on the horizontal shaft. Liquids enter and leave through the shaft; they flow radially and countercurrently in the rotating drum because the effects of density differences are increased by centrifugal force. The particular virtue of this machine is the low residence time of the liquids, which has made it especially useful in the extraction of antibiotic pharmaceuticals from fermentation broths. *See* COUNTERCURRENT MASS-TRANSFER OPERATION; EXTRACTION; MASS-TRANSFER OPERATION; MIXING; UNIT OPERATIONS. [R.E.TR.]

Bibliography: R. E. Treybal, *Liquid Extraction*, 1951.

Somásteroidea

A subclass of extinct Asterozoa in which the arm was devoid of an ambulacral groove. The arms were scarcely differentiated structurally from the rest of the body.

The skeleton was unique in having a pinnate structure in which successive rows of ossicles (virgalia) lay on either side of each radial row of ambulacral ossicles. Somasteroidea were probably sluggish sea stars which rested upon the sea floor and used centrally directed ciliary water currents to convey small food particles to the mouth. The food would be derived either from the plankton-shower or from the mud of the sea floor. They ranged the lower Paleozoic from the Lower Ordovician onward, and are believed to be ancestors of the other Asterozoa. *See* ASTEROIDEA; ASTEROZOA; OPHIUROIDEA. [H.B.F.]

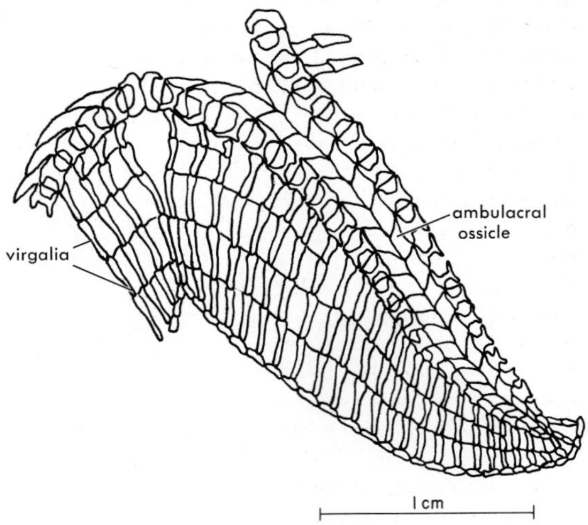

Chinianaster levyi, reconstruction of arm. (*After W. K. Spencer*)

Bibliography: W. K. Spencer, Early Palaeozoic starfish, *Phil. Trans., Roy. Soc. London*, B235:87–129, 1951.

Somatization

A type of neurosis. Perhaps the simplest and most diffuse form of neurotic symptomatology are the so-called neurasthenias in which the principal complaint is a general malaise involving a loss of energy and zest along with a vague sense of aches and pains and fatigue. The symptom pattern is generally taken to indicate a high price in energy expenditure in maintaining strict repression over unacceptable impulses along with a secondary gain in terms of care from others that exceeds the gain obtained from one's presymptomatic life. Neurasthenic symptoms tend to be centered about those activities in the person's life that have proved either notably unrewarding or internally stressful.

Hypochondriacal symptoms, in the form of exaggerated somatic complaints accompanied by a strong but ambivalent fear of falling ill, and with much concern about diet, elimination, and health, represent another form of somatization. A common origin of hypochondriacal symptoms is a family background where parents were excessively concerned with the health of children and in which sickness was rewarded with attention. Feelings of inadequacy on the part of the patient later in life may then be translated into reward-seeking somatic complaints.

Both hypochondriacal and neurasthenic symptoms are relatively nonspecific reactions in contrast to the conversion symptoms of the hysteric (*see* HYSTERIA). When one considers the more specific forms of psychosomatic complaints, such as essential hypertension, certain forms of ulcerative colitis, and gastric ulcer, however, there is a closer link to what has been described as conversion symptomatology. What characterizes the psychosomatic reaction in contrast to the other somatizing patterns is that there is an involvement of specific organs under the control of the autonomic nervous system. Although the origin of psychosomatic illness is presumed to be psychological in nature, the consequences may be irreversibly pathologic in the organic sense. There is much speculation and some research (for example by T. M. French and F. Alexander) on the relationship between psychodynamic patterns and specific organ involvements in psychosomatic disorders, usually stated in the form of the symbolic appropriateness of the organ system as an expression of a core problem in the person's life. However, the precise nature of "organ choice," as it is called, is poorly understood. It seems likely that the autonomic nervous system and its reactivity to anxiety and internal stress are closely related to the etiology of psychosomatic disturbances. Essential hypertension is a case in point. Where, for example, there is chronic unrecognized and unresolved hostility, to the point almost of inner rage, and other more appropriate means of expression are unavailable, a severe autonomic flare-up with dramatic elevation of blood pressure may occur and in turn result in physiological sequelae which are potentially irreversible and may threaten the patient's survival. *See* ABNORMAL BEHAVIOR; NEUROSIS. [J.S.B.; W.MS.]

Somesthesis

The general name for all systems of sensitivity present in the skin (cutaneous sensitivity), muscles and their attachments (kinesthesis), visceral organs (organic sensitivity), and nonauditory labyrinth of the ear (vestibular sensitivity). Somesthesis involves all three kinds of receptors in the common physiological classification of the senses based on source of stimulation: exteroceptors (cutaneous), interoceptors (organic), and proprioceptors (kinesthetic and vestibular).

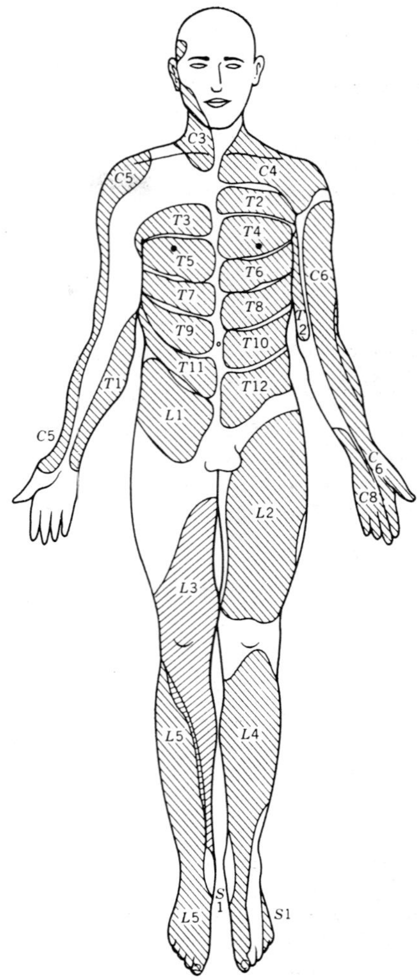

Fig. 1. Human dermatomes. The letters indicate the regions of the spinal cord represented: C, cervical; T, thoracic; L, lumbar. The numbers refer to particular segments of the spinal cord. (*T. Lewis, Pain, Macmillan, 1942*)

Nerves leading to and from the peripheral parts of the body are organized segmentally; there is an orderly spacing of vertebrae and spinal nerves. Each spinal nerve supplies a limited region of the skin and deep tissues lying beneath it. The skin region into which a given spinal nerve proliferates is called the dermatome of the nerve. Since peripheral branches of two or even three spinal nerves may overlap in a given skin region, mapping of the dermatomes becomes difficult. In animals, dermatomes may be outlined by cutting three spinal roots above and three below the one studied, thus isolating the spinal root of interest. This is the method of "remaining sensibility." In humans, dermatomes may be established by noting the distribution of skin eruptions in "shingles" (herpes zoster), a virus infection of spinal roots.

A nerve leading from a dermatome to the cord contains fibers promoting all somesthetic functions. Accordingly, severance of the nerve will produce anesthesia to pressure, cold, warm, and pain stimuli in the dermatome. After entrance to the spinal cord, however, there is a separation of paths into functional groups, rather than on a segmental anatomical basis. Fibers carrying impulses of pressure sensation pass into the cord by way of the dorsal roots of the spinal nerves, ascend the cord for a few spinal segments, and then terminate in the interior gray matter. Connecting fibers next do either of two things: (1) cross over to the opposite side of the cord to continue upward in the anterior spinothalamic tract, or (2) continue upward in the dorsal column on the same side, as far as the medulla. The net result is that such fibers, from a given dermatome, arrive at both sides of the medulla. Here the fibers that have not previously crossed to the opposite side now do so, joining those that crossed over at the spinal level.

Fibers for pain and temperature follow a course different from that for pressure. They invariably cross to the contralateral (opposite) side by way of collateral fibers immediately upon entering the cord. These ascend the cord in the lateral spinothalamic tract, pass through the medulla, to terminate in the thalamus, without further crossing or connecting with new fibers.

The over-all effect is that pressure and touch have dual representation throughout much of the length of the cord, while temperature and pain have single, and contralateral, representation.

From the thalamus, somesthetic fibers radiate upward to the cortex of the parietal lobe where they are projected on the postcentral gyrus, located just dorsal to the prominent fissure of Rolando. The "method of evoked potentials," a technique of recording electrical changes at the cortex induced by peripheral stimulation, may be used to demonstrate that the topographic relations of the body surface, apparently lost in transmission up the cord, are projected in an orderly fashion on the postcentral gyrus (*see* NEUROPHYSIOLOGY). Modal differentiation seems to have disappeared, that is, there are no "pressure areas," "cold areas," and so forth discoverable on the cortical surface.

Projection of body topography on the parietal cortex is not only surprisingly orderly, with each segment arranged in natural sequence, but there is also a "spare" projection region for somesthetic impulses in the cortex. A similar situation is seen in the second visual projection outside the primary area and the three, or possibly four, projections of the auditory apparatus on the cortex. Somatic area II, as it has been named, lies on the lateral surface of the cortex, between somatic area I and the primary auditory area of the temporal cortex. Careful mapping of somatic II, by the evoked potential method, reveals face, arm, and leg sections,

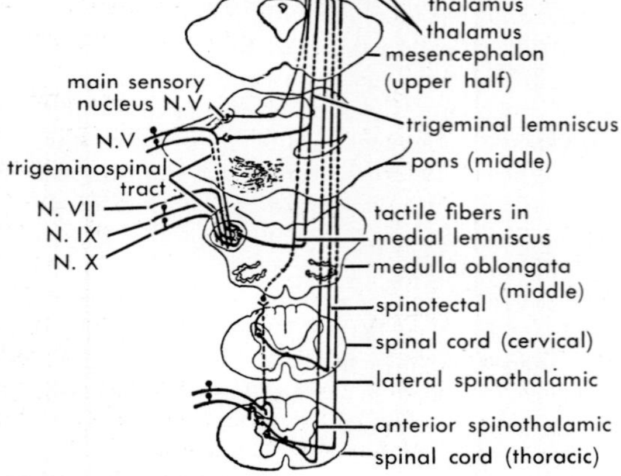

Fig. 2. Spinal pathways for somesthesis. (*S. W. Ranson, The Anatomy of the Nervous System, 7th ed., Saunders, 1943*)

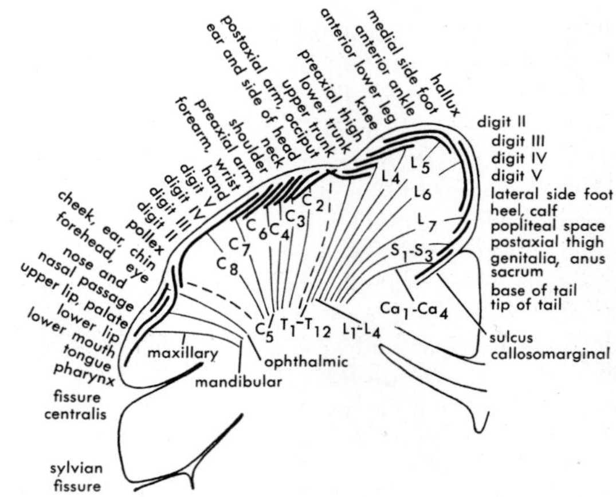

Fig. 3. Somesthetic projection on the parietal cortex. Areas: C, cervical; T, thoracic; L, leg; Ca, caudal; S, sacral. (*C. N. Woolsey, W. H. Marshall, and P. Bard, Bull. Johns Hopkins Hosp., 70:399–441, 1942*)

just as somatic I does. An important point of difference, however, is that somatic II is a bilateral projection area, receiving impulses from both sides of all regions of the body. Potentials evoked from the opposite side of the body are about twice as large as those coming from the same side. Somatic II also seems to be intimately linked to the motor mechanism since upon direct electrical stimulation, it will produce muscle twitches in the periphery, a phenomenon not obtainable in somatic area I. *See* CUTANEOUS SENSATION; ITCH; KINESTHETIC SENSATION; PAIN, CUTANEOUS; PAIN, DEEP; PARESTHESIA; STEREOGNOSIS; TEMPERATURE SENSES; TICKLE; TOUCH. [F.A.G.]

Bibliography: F. A. Geldard, *The Human Senses*, 1953; C. T. Morgan and E. Stellar, *Physiological Psychology*, 2d ed., 1950; R. S. Woodworth and H. Schlosberg, *Experimental Psychology*, rev. ed., 1954.

Sonar

A term that refers both to the application of underwater sound to the detection and location of objects in the sea, and to the apparatus used in such applications. The word is derived from "Sound Navigation And Ranging"; the British use the word asdic. Sonar methods are used widely in naval warfare, especially in the detection of submerged submarines. Since electromagnetic radiations, such as visible light or radar, do not penetrate the sea significantly, sonar is the most successful method of underwater detection. Except for underwater telephones, sonar is also the only means by which a fully submerged submarine can appraise itself of what is happening around it (*see* UNDERWATER TELEPHONE). Thus it is used by submarines to locate surface vessels and other submarines, and to navigate through a mine field or under the ice in the Arctic Ocean.

There are two general types of sonar methods, active and passive. In an active sonar system a pulse of sound is generated by the searcher and projected into the water. This sound is reflected back from the target and detected by the searcher as an echo. Since the speed of sound in sea water is known, the range and also the bearing of the target are found. This method is also called echo-ranging. In a passive system the searcher detects the noises emitted by the target. Unless more than one listening station is involved, passive sonar provides information only as to the existence of a noise source and its bearing from the searcher. Because the noise radiated from a ship is often characteristic of the type of ship, passive sonar, in contrast to active sonar, may help in classifying the target.

The particular sonar system used in a given military situation depends upon operational or tactical requirements. Since active sonar systems involve the deliberate emission of sound, which discloses the searcher to easy detection by other vessels, they are used sparingly by submarines, which rely on concealment for their safety. The active systems are therefore used mainly by surface antisubmarine vessels, such as destroyers, since they generally operate in an environment which is too noisy for successful passive detection of a quiet submarine. This interfering noise may be generated by the antisubmarine vessel itself moving at high speed or by other ships in the vicinity which it may be escorting. For a discussion of ship noise and other related information, *see* UNDERWATER SOUND.

There is a wide variety of sonar systems other than those carried by ships. Sonar may also be used by aircraft for submarine detection by dropping buoys (called sonobuoys) in the water, by mine sweepers in locating mines, or by acoustic torpedoes. There are also nonmilitary applications. Sonar may be used for detection of schools of fish or for the location of submerged wrecks. An echo sounder, or depth indicator, is essentially an active sonar system which sends sound pulses to the sea bottom. *See* ACOUSTIC TORPEDO; ANTISUBMARINE WARFARE; ECHO SOUNDER; SONOBUOY.

Active sonar. Active sonar systems consist of the following: one or more transducers to send and receive sound, electrical and electronic equipment for the generation and detection of the electrical impulses to and from the transducer, and a display or recorder system for the observation of the received signals (see illustration).

The transducer is located within a water-flooded, streamlined sonar dome in order to reduce noise generated by the motion of the transducer through the water. On a surface ship the sonar dome is usually mounted on the keel in a location that puts it as deep in the water as possible. In order to increase this depth, a towed sonar is sometimes used. Towed sonar also can be used from blimps or helicopters. The receiving transducer of an active sonar is directional in order to resolve the bearing of the target. *See* DIRECTIVITY; TRANSDUCER, UNDERWATER.

Sound pulsing. A narrow bandwidth pulse of electrical energy is applied to the transducer, which in turn generates a similar pulse of acoustic

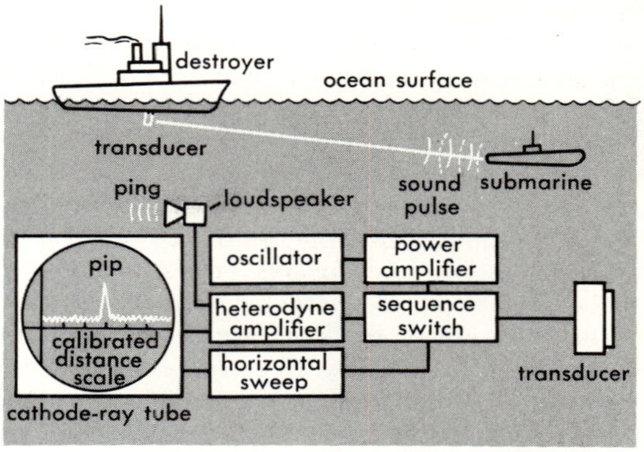

Active sonar system. (*From H. F. Olson, Acoustical Engineering, 3d ed., Van Nostrand, 1957*)

energy in the water. Targets in the beam reflect part of this energy back either to the same transducer or to another receiving transducer, which converts this to electrical energy. This signal is amplified and then displayed as a function of time after the original emission, thus indicating reflecting targets in the beam of the transducer. The sonar frequencies employed are generally in the range of 5,000–50,000 cycles per second. The pulse is repeated periodically at a rate which depends upon the maximum range obtainable.

Display systems. Various types of display or recording systems are used. Often both a visual and an aural presentation is made to the sonar operator. In an aural presentation the outgoing pulse (ping) is heard followed by the returning echoes which come back from along the sonar beam. (The frequencies which are presented are scaled down from the original ones.)

One type of visual indicator is a range recorder. In this a stylus moves across a chemically treated paper, which is darkened electrically by a received signal. The paper moves at a constant rate and the stylus is phased so that it starts each sweep as a sound pulse is emitted. The distance along the trace at which an echo appears is then a measure of the range to the target. Since successive echo pulses are displayed adjacent to one another, the motion of the target relative to the searching vessel may be determined.

An oscilloscope presentation is used with a scanning sonar. Here the emitted sound pulse is sent in all directions, and a directional receiving beam is rotated electrically in a horizontal plane. The location of the ship is at the center of the oscilloscope pattern, and the spot of the oscilloscope moves outward in a direction corresponding to the axis of the receiving transducer. The intensity of the illuminated spot varies with the received signal. This display, which gives both the range and direction of the various targets, is also used in radar and is called a plan-position indicator (PPI).

A cathode-ray tube may be operated in still another way so that the spot, initiated with the outgoing pulse, moves from left to right, the received signal deflecting the spot in the vertical direction.

This is a variable-displacement indicator and shows targets as a function of range for a fixed bearing. In radar such a display is called an A-Scan. *See* RADAR.

Performance factors. Many factors influence the performance of an active sonar system. Chief among these, particularly since it can be so variable, is the propagation loss in the water between the sending transducer and the target. This loss is determined by the frequency of the sound and by the thermal gradients in the ocean which refract the sound waves.

Another determining factor is the reflecting power of the target, or target strength. For an echo to be detected it must be of greater strength than other signals which may be received. These interfering sounds are generally due either to sonar self-noise or to reverberation.

Sonar self-noise is the noise which is generated by the motion of the ship carrying the sonar. Most of this sound is generated by cavitation and turbulence from regions very close to the transducer. Self-noise increases rapidly with the speed of the ship, especially above the speed where cavitation sets in. *See* CAVITATION.

Reverberation refers to the combination of all the echoes returned to an active sonar system from the ocean itself. This includes the surface and bottom, bubbles, suspended marine organisms, and inhomogeneities in the sea. A sonar operator hears reverberation as a quavering ring which sets in as soon as the outgoing sound pulse has been emitted. Since reverberation is the resultant of a large number of very weak scattered echoes, it is statistical in nature so that the individual echoes are not resolved. There are three kinds of reverberation: volume, surface, and bottom. Volume reverberation, which arises in the water itself, is evident as soon as the outgoing pulse leaves, and decays fairly rapidly thereafter. Surface reverberation appears as soon as the outgoing pulse encounters the surface of the sea. This type of reverberation increases markedly with increased sea state. (The term sea state refers to the conditions of the ocean surface; for details, *see* SEA STATE.) Bottom reverberation is due to irregularities in the ocean bottom and may be the most significant reverberation in shallow water.

Target strength is a term used with active sonar to give a quantitative measure of the echo returned from an object back along the direction from which the incident sound came. Since reflected or scattered sound intensities I_s are generally proportional to the incident intensity I_i, $I_s = kI_i/r^2$, where r is the distance from the target and k is a constant which measures the scattering or reflecting strength of the target. For most objects, k depends upon its orientation. The target strength T is defined as k measured in units of decibels (db); that is, $T = 10 \log k$. For a perfectly reflecting sphere of radius a (where a is larger than the wavelength), $T = 20 \log (a/2)$. Generally the yard is chosen as the unit of length in the definition of target strength. Since $\log 1 = 0$, a sphere of 2-yard radius has a target strength $T = 0$. Therefore, an object's target strength T is a comparison of its reflecting strength with that of a sphere of 2-yard radius. For example, an object with $T = -6$ db has the equivalent reflecting power of a perfectly reflecting sphere 1 yard in radius.

The sonar Doppler effect is the change in frequency of a sonar echo due to target motion. As in all wave motion, there is a frequency shift in a sound wave reflected from a moving target. Since the sound scatterers in the sea which are responsible for reverberation are relatively fixed, an echo from a moving target can often be detected in the presence of reverberation by having a different frequency. *See* DOPPLER EFFECT.

Passive sonar. This is an underwater acoustic system which gains detection by merely listening for noise radiated by a possible target. Passive systems have the distinct advantage of being undetectable themselves but have the disadvantage of having to take in all the sea and ship noise. Modern submarines carry passive sonar apparatus which is believed to be capable of detecting ships as far away as 100 miles under favorable conditions. A passive system consists of a highly directional and trainable transducer or array of transducers, electronic amplifiers, and a display system. The display is usually aural; consequently the frequencies employed generally lie in the audible region. These are also the most favorable frequencies in terms of the spectrum of noise radiated by ships. The noise emitted by a ship which is detected by passive sonar comes from the operation of machinery, primarily the propulsion machinery, and impulsive beats from the rotation of the ship's propellers.

The effectiveness of passive sonar depends upon the magnitude of the radiated noise, the propagation loss between the radiating ship and the sonar, and the background noise observed by the sonar. This limiting background noise may be self-noise or ambient sea noise. The former is noise generated either by machinery on the listening ship or by its motion through the water. If the ship carrying the sonar is in a quiet condition and is moving slowly, the self-noise may be reduced sufficiently so that ambient sea noise becomes the limiting factor. This is the general continuous spectrum of noise which is found naturally in the sea. Since ambient sea noise comes from all directions, its effect relative to a target may be decreased by improving the directivity of the sonar. *See* ULTRASONICS. [R.W.MO.]

Bibliography: J. W. Horton, *Fundamentals of Sonar*, 2d ed., 1959.

Sone

A unit of loudness. It has been found that, contrary to popular belief, the subjective loudness of a sound is not directly proportional to its sound pressure level in decibels (db). In one research study, listeners were asked to adjust the intensity of a 1000 cycles per second (cps) tone until it sounded

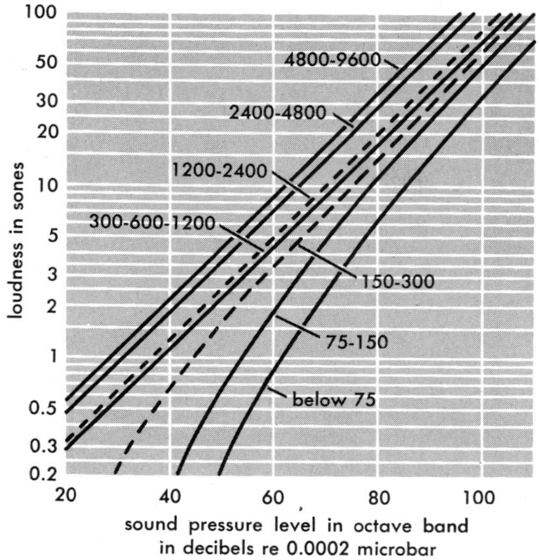

Loudness functions for octave bands in a diffuse sound field.

twice as loud as it did at a level of 40 db. The loudness of the 1000 cps tone at 40 db was arbitrarily assigned the value 1 sone. The sound pressure level of the tone that sounded twice as loud as the tone of 1 sone was assigned the value 2 sones. The listeners were then asked to double the loudness of the tone again, that is, to set its loudness at 4 sones, and then again and again to higher and higher levels.

The sone scale of loudness is based upon a number of studies, similar to that just described, in which subjects judged the loudness of pure tones and bands of noise. This sone scale can be used to evaluate the loudness of pure tones and of bands of noise. By definition, the loudness of a sound is its value in sones. The illustration shows the functions relating the loudness in sones to the sound pressure level in decibels of bands of noise 1 octave wide. *See* LOUDNESS; *see also* DECIBEL; HEARING; PHON; PSYCHOACOUSTICS; SOUND PRESSURE.

[K.D.K.]

Bibliography: S. S. Stevens, Calculation of the loudness of complex noise, *J. Acoust. Soc. Am.*, 28(5):807–832, 1956.

Sonic barrier

An aeronautical term coined during World War II to symbolize the technical difficulties for manned aircraft in accelerating through the speed of sound. In designing and testing subsonic airplanes to fly at increasingly higher speeds, aeronautical engineers found the following adverse effects as the speed of the airplane approached the speed of sound: a rapid rate of increase in drag, breakdown of lift, and loss of control and maneuverability of the airplane. Such experience suggested that these violent and uncontrollable aerodynamic phenomena might be inevitable in the transonic range of speeds (that is, in the immediate vicinity of the speed of sound). These adverse effects are collectively and loosely termed the sonic barrier. On the other hand, cannoneers had long known that ballistic artillery shells had no difficulty moving through the atmosphere at speeds considerably in excess of the speed of sound.

As a result of intensive research during and after World War II, the source of most of the difficulties has been traced to the formation of local shock waves, which in turn interact with the boundary layer and induce flow separation from the aerodynamic surfaces. By careful choice of geometrical shapes, such as use of thin wings and slim bodies, it has been found possible to design airplanes with relatively smooth flow characteristics through the transonic speed range. Although the increase in drag with flight speed is inevitable, it can be kept at a manageable level and be overcome by more powerful propulsion devices. *See* TRANSONIC FLIGHT.

Although flight at speeds in excess of the speed of sound has been achieved, the theoretical problem of prescribing the transonic flow field over objects of arbitrary geometric shape remains a formidable one. The difficulty is partly mathematical and partly the result of flow discontinuities, such as shock waves and flow separation, which make it difficult to specify the boundary conditions for the flow field, even though the shapes of the objects themselves are known. *See* AIRPLANE.

[S.C.L.]

Bibliography: T. Von Kármán, *Collected Works*, vol. 4, 1956.

Sonic boom

An explosivelike sound heard when an aircraft in the vicinity of an observer maneuvers at supersonic speed. The characteristics of a sonic boom or of sonic bangs vary according to the situation. Observers most often report hearing two bangs in quick succession. Sometimes they hear only one bang, but they occasionally report hearing three or more bangs from a single maneuver. Although some of the reported bangs may have been due to extraneous effects, such as sudden ignition of afterburners in some turbojet engines, actual sonic bangs are caused by the arrival at the observer of aircraft-induced shock waves.

Sonic boom can be sufficiently intense to damage property; loud bangs are certainly physiologically damaging. Thus the booms constitute a potential limitation to military and commercial flight at supersonic speed at low altitudes.

If the relative speed of approach of the aircraft toward the observer is always less than the speed of sound, no bang will be heard. If the relative speed of approach becomes sonic for a finite period, all disturbances generated during that period will reach the observer together with their combined local intensities and produce a single bang. If the relative speed accelerates through the speed of sound, there must exist two points along the flight path at which the relative speed of approach passes through the sonic value, if collision with the observer is to be avoided. In this case, two bangs will be heard. The bang originating from the point of relative sonic speed closer to the observer will be heard first and will also be the louder of the two. [S.C.L.]

Bibliography: T. Gold, The double bang of supersonic aircraft, *Nature*, 170(4332):808, 1952.

Sonics

A term used to describe the technology of sound, or elastic wave motion, as applied to problems of measurement, control, and processing. It is a branch of acoustics which relates primarily to processes and techniques, as distinguished from those studies which relate to man's hearing.

Ultrasonics is the term applied to sound whose frequency is in excess of approximately 20,000 cycles per second (cps). *See* ULTRASONICS.

Infrasonics is the term applied to sound whose frequency is less than approximately 15 cps. These sounds are also referred to as having subsonic frequencies. Many problems in vibration, oceanography, seismology, and the dynamic behavior of elastic materials are analyzed by treating the phenomenon being studied as sound waves of infrasonic frequency. *See* SOUND. [W.J.G.]

Sonobuoy

An acoustic listening buoy employed by aircraft for the detection of submerged submarines. A sonobuoy is dropped or parachuted from low altitude; after it contacts the water, a hydrophone (a transducer which receives underwater sound signals) is lowered and an antenna raised for radio transmission. The sonobuoy carries batteries which power a radio transmitter so that the sounds picked up by the hydrophone can be monitored by personnel in the aircraft. Several radio channels ordinarily are available so that the antisubmarine aircraft can lay sonobuoys in a pattern about the suspected position of the submarine. Since the sonobuoy detects the propeller or cavitation noises of the submarine, the range of detection depends upon the speed of the submarine and is relatively limited if the submarine is moving slowly.

Releasing a sonobuoy from a Navy ship. (*Official U.S. Navy photograph*)

The advantages of sonobuoys in antisubmarine warfare are considerable for they allow aircraft, which have a large speed and which are relatively invulnerable to attack by the submarine, to be used against submarines. If the submarine chooses to avoid detection by proceeding slowly, then time is allowed for the arrival of surface antisubmarine vessels, which can use active sonar detection methods. *See* ANTISUBMARINE WARFARE; SONAR; UNDERWATER SOUND. [R.W.MO.]

Sorghum

Sorghums include many widely cultivated grasses having a variety of common names. They were among the first wild plants to be domesticated by man, and were grown in Egypt prior to 2200 B.C. The cultivated sorghums in the United States are usually classified as a single species, *Sorghum vulgare*, although there are several varieties. The two major types are the grain, or nonsaccharine, sorghums cultivated primarily for grain and to a lesser extent for forage, and the sweet, or saccharine, sorghums used for forage and for making syrup.

Grain sorghum. This crop is grown primarily in those areas of the United States in which rainfall is below minimum or temperature is above optimum for corn. It is most extensively grown in the Great Plains area with the greatest acreages in Texas, Kansas, Oklahoma, New Mexico, Colorado, Nebraska, and the irrigated valleys in California. Grain sorghum is fed primarily to livestock and in feed value is nearly equal to corn. The average annual farm value for 1945–1954 was $175,674,800.

Origin and description. Sorghum is believed to have originated in Africa although it was early grown in China, India, and Near East countries. Grain sorghum was brought from Africa to the United States in 1850. The types first introduced were tall and not well adapted to mechanized harvest. For this reason grain sorghums did not expand rapidly in acreage until new varieties with shorter stems were developed that could be harvested with a combine. *See* AGRICULTURAL MACHINERY.

Grain sorghum, in the early stages of growth, closely resembles corn, but at later stages it becomes strikingly different. Sorghum plants may tiller (put forth new shoots) profusely, producing several headbearing culms from the lower nodes. *See* STEM (BOTANY). The drought tolerance of grain sorghum is due in part to its ability to produce new tillers from the crown of the plant when soil moisture again becomes favorable for growth. The inflorescence varies from a dense to a lax panicle, and spikelets produce perfect flowers that are subject to both self- and cross-fertilization. *See* FLOWER (BOTANY); REPRODUCTION, PLANT. Mature grain in different varieties varies widely in size and in color from white to buff, red, and brown. Color pigments are in the pericarp (outer covering) of the grain or in a layer of cells beneath the pericarp. *See* FRUIT (BOTANY). The endosperm (starchy material) of the seed of currently grown varieties is white, but a yellow endosperm type is now available. *See* SEED (BOTANY).

Grain sorghums are classified into types designated as milo, kafir, feterita, hegari, durra, shallu, and kaoliang. This classification is based on morphological rather than cytological differences, since all types (including also forage sorghums, broomcorn, and Sudan grass) have 10 pairs of chromosomes and freely intercross.

Varieties and hybrids. One of the first important advances in grain sorghum improvement was the development of dwarf and double-dwarf combine types, first released in 1941. By 1953, nearly 100% of the grain sorghum grown in the United States was of this type. Combine types usually grow to a height of only 2½–3 ft and are much more resistant to lodging than the earlier tall-growing varieties.

The second advance was the discovery of cytoplasmic male sterility in 1952. This type of sterility, as in corn, prevents formation of normal pollen and thus makes possible the production of commercial hybrid seed. Because flowers in grain sorghum contain both staminate (male) and pistillate (female) parts, the production of commercial quantities of hybrid seed was not possible without this sterility mechanism. In a period of only 5 years, grain sorghum varieties have been almost completely replaced by first generation hybrids adapted to a wide range in length of growing season from Texas to the Dakotas. The superiority in yield of many hybrids is as great as the superiority of corn hybrids over the older corn varieties. In the Corn Belt states, grain sorghum hybrids are generally equal to corn in yield under favorable moisture conditions, and superior to corn on droughty soils (Fig. 1).

Planting. Grain sorghums generally are planted in rows 38–42 in. apart with the same equipment used for corn. In some areas a grain drill is used by stopping holes in the drill box to provide a desired row width. Seed is drilled in the row at the rate of 4–6 seeds per foot of row, or about 5 lb per acre. Because it is susceptible to poor germination and low seedling survival under low soil temperature, grain sorghum is usually planted about 2 weeks later than corn. Depth of planting varies from 1 to 2 in. below the soil surface.

Cultivation. Seedling growth of sorghum is considerably slower than that of corn and consequently control of weeds becomes more difficult. The seed bed should be well prepared and early weed growth controlled prior to planting. A rotary hoe is effective in controlling weeds in the first cultivation. Subsequent cultivations are made as needed with the same equipment used for cultivating corn.

Fig. 1. (*a*) A typical head of a grain sorghum variety. (*b*) A hybrid.

Harvesting. Grain sorghums are always harvested with a grain combine (Fig. 2). The danger of high grain moisture is a serious problem in harvesting grain sorghum in years of excessive rainfall. Because sorghum grain packs tightly, owing to its high bushel weight, the grain should not contain above 12% moisture to insure safe storage without heating. In the southwestern states high moisture grain is not a problem because the crop is harvested earlier in the season when temperatures are

Fig. 2. Dwarf type of grain sorghum has made it possible to use combines for harvesting.

high and relative humidity is low. In states further north conditions are unfavorable for drying in the field and the use of grain driers may become a standard practice in these areas. [I.J.J.]

Sweet sorghum. Commonly known as sorgo, it was introduced into North America from China in 1850, although its ancestry traces back to Egypt. It is an annual, rather drought-resistant crop. The culms are from 2–15 ft tall and the hard cortical layer, or shell, encloses a sweet, juicy pith that is interspersed with vascular bundles (*see* CORTEX, PLANT; PITH; VASCULAR BUNDLES). At each node both a leaf and a lateral bud alternate on opposite sides; the internodes are alternately grooved on one side. Leaves are smooth with glossy or waxy surfaces and have margins with small, sharp, curved teeth. The leaves fold and roll up during drought. The inflorescence is a panicle of varying size having many primary branches with paired ellipsoidal spikelets containing two florets in each fertile sessile spikelet. The plant is self-pollinated.

Seed is planted in cultivated rows and fertilized similarly to corn. Maturity varies between 90–125 days. The juice contains about 12% sugar. The main sorghum-syrup producing area is in south central and southeastern United States (Fig. 3). Sweet sorghum and cane syrup production in these areas, primarily a farm industry, decreased from about 25,000,000 gal in 1939 to 7,000,000 gal in 1949, with sorghum-syrup production decreasing over 80%. [L.D.B.]

Sorghum diseases. In 1957 the total value of the United States sorghum grain, forage, silage, syrup,

Fig. 4. Head smut on *Leoti sorgo*. (*a*) Healthy head. (*b*) Smutted head.

Fig. 3. Sweet sorghum in Oklahoma. (*USDA*)

and broomcorn fiber crop was $586,045,000, and losses from diseases were estimated at $29,000,000. Diseases frequently are limiting factors in production and may be classified in four general categories: (1) those that rot the seed and kill seedlings; (2) those that attack the leaves, making the plants less valuable for forage; (3) those that attack and destroy the grain in the heads; and (4) those that attack the roots and stalks.

Fungi causing seed rotting and seedling diseases may be seed-borne or soil inhabiting and are most destructive when the soil is cold and wet after planting. Species of *Fusarium, Pythium, Helminthosporium,* and one of *Penicillium* are the most important fungi involved (*see* FUNGI). Damage may be considerably reduced by planting sound seed of recommended varieties treated with a good disinfectant in soil warm enough to insure prompt germination.

Leaf and sheath diseases caused by at least three species of bacteria and eight fungus species are generally favored by high temperature and humid conditions (*see* BACTERIA). Disease lesions occurring as discolored spots or streaks may coalesce to involve the entire leaf. Sanitation, seed treatment, and the use of resistant varieties are recommended control measures.

Three smuts—covered kernel, loose kernel, and head smut—are the principal diseases attacking the grain (Fig. 4). Kernel smuts are distinguished by the replacement of the grain with a dark sooty mass of spores, the covered type contained in a semipermanent membrane in contrast to the easily ruptured gall of the loose smut. Head smut destroys the entire head. Resistant varieties control kernel and head smuts to some extent, and seed treatment controls the kernel smuts.

Three diseases of the roots and stalks are of primary importance. Periconia root rot infecting the roots and crown caused extensive damage to milo and darso sorghums until it was controlled by resistant varieties. Charcoal rot, most evident as the plant approaches maturity under extreme conditions of heat or drought, causes shredding of the stalks and extensive lodging. The development of resistant varieties appears to offer the only effective method of control. [H.A.RO.]

Bibliography: *See* AGRICULTURAL SCIENCE (PLANT); PLANT DISEASE.

Sound

A mechanical disturbance in an elastic medium. Elastic media include solids, liquids, and gases. A mechanical disturbance is defined as any alteration in a property of a medium (such as pressure, particle displacement, or density) that can be detected by an instrument or a listener. Sound is a form of wave motion. Sound pulses propagated in solids, liquids, and gases are called acoustic waves, or sound waves.

The word sound is sometimes applied only to disturbances that produce an auditory sensation, a definition which is generally confined to nonscientific usage. For example, the scientist would say, "That sound is inaudible," while the layman would say, "There is no sound." Sound waves having frequencies above the audible or sonic range, that is, above about 20,000 cycles per second (cps), are termed ultrasonic waves; those with frequencies below the sonic range, that is, below about 16 cps, are called infrasonic waves. The spectrum of sound is depicted in Fig. 1.

This article discusses methods of detection of sound waves, radiation of sound, and propagation of sound waves, including propagation in the atmosphere. For related information, *see* WAVE MOTION and the articles listed therein; *see also* ABSORPTION (SOUND); ARCHITECTURAL ACOUSTICS; ECHO; HEARING; MUSICAL ACOUSTICS; NOISE, ACOUSTIC; NOISE CONTROL; NOISE MEASUREMENT; PSYCHOACOUSTICS; REFLECTION (SOUND); REVERBERATION; SOUND REPRODUCTION SYSTEMS, ELECTRICAL; STEREOPHONIC SOUND; ULTRASONICS; UNDERWATER SOUND; VIBRATION.

Detection of sound. When a mechanical disturbance occurs in a medium, it may be detected by observing a change in (1) position of the molecules, (2) pressure, (3) density, or (4) temperature.

A change in position of the molecules may be observed with a microscope focused on illuminated particles of the medium; in gases and liquids, from the forces produced on a small disk around which the disturbed medium flows; or, in solids, from the shaking of an electromechanical transducer attached to the solid (the shaking produces an observable voltage).

A change in pressure may be observed by a device containing a diaphragm that moves in response to pressure changes and that, in turn, actuates either (1) an electromechanical transducer (thereby producing a voltage); (2) a marker for scratching on a moving screen; or (3) a valve that regulates a flow of gas so as to produce a visible fluctuating flame. One or more surfaces of some electromechanical transducers may also serve as

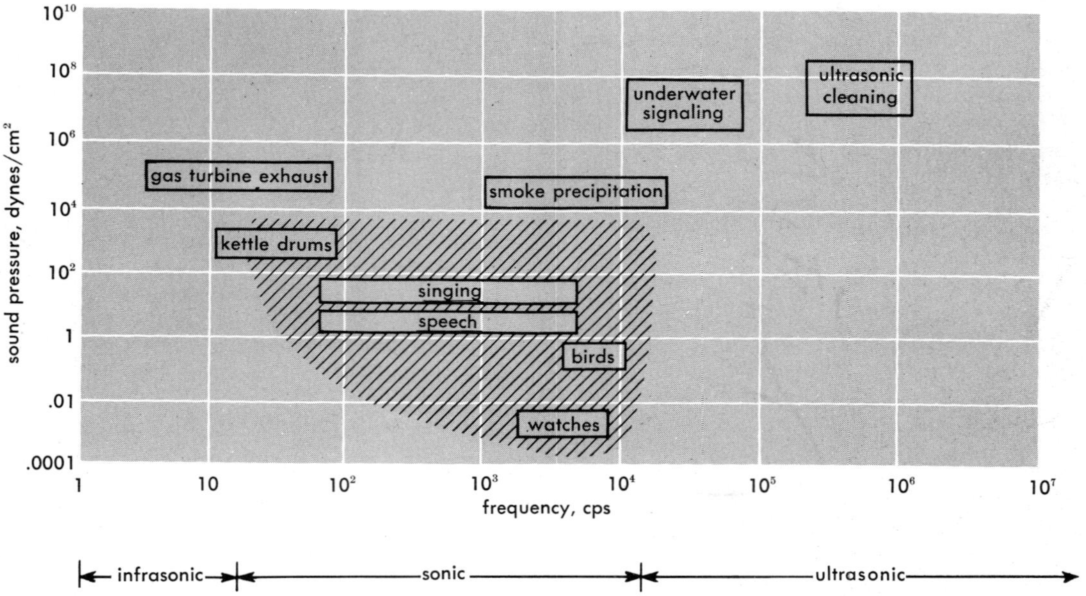

Fig. 1. Spectrum of sound. The intensity and frequency of several sonic phenomena are roughly located by the labeled boxes. The shaded area indicates the sound audible to the human ear. (*From G. E. Henry, Ultrasonics, Sci. American, 190(5):54–63, 1954*)

the diaphragm, for example, a surface of a piezoelectric crystal.

A change in density may be observed from the diffraction of a light beam passing through the medium.

A change in temperature may be observed by a sensitive thermocouple imbedded in the medium.

In practice, sounds are generally detected by electromechanical transducers or by listeners. In gases and liquids the transducers may be actuated by changes in pressure at a point, in which case they are called pressure microphones. Or, they may be actuated by changes in the difference between pressures at two adjacent points, in which case they are called pressure-gradient microphones. Because the average velocity of the particles (molecules) in longitudinal waves is proportional to this quantity, pressure-gradient microphones are often referred to as velocity microphones. *See* MICROPHONE.

In solids, it is generally easier to observe the motion of particles of the medium than the pressures in the medium. To do this, vibration pickups are used. These are attached to some portion of the medium and respond in proportion to the displacement, the velocity, or the acceleration of the medium at that point. Accordingly, they are called displacement, velocity, or acceleration pickups, respectively. *See* VIBRATION PICKUP.

Pressures may be measured in solids by one of two procedures. One may imbed in the solid a pressure-sensitive electromechanical transducer which will produce a voltage related to the pressures acting on its surfaces. Or, one may determine, by a separate experiment, the amount of internal force necessary to displace an observable part of the solid. Then by observation of the displacements of this part, the internal forces may be ascertained.

Definitions and formulas. To understand the subsequent discussion, one must establish a few basic items of terminology and their definitions.

Static pressure P_0 is the barometric pressure in the medium with no sound waves present.

Instantaneous sound pressure $p(x,t)$ is the incremental change from the static pressure at a given instant t and point x in the medium caused by the presence of a sound wave.

For a plane, outwardly traveling sound wave with a single frequency f, the instantaneous pressure is

$$p(x,t) = \sqrt{2}p_0 \cos [\omega(t - x/c)] \qquad (1)$$

where $\sqrt{2}p_0$ is the peak sound pressure, ω equals 2π times the frequency f, and c is the speed of sound in the medium.

For a spherical, outwardly traveling wave with a single frequency f,

$$p(r,t) = \frac{\sqrt{2}A_0}{r} \cos [\omega(t - r/c)] \qquad (2)$$

where $\sqrt{2}A_0$ is equal to the magnitude of the peak pressure at unit distance from the center of the spherical source, and r is the distance to the point of observation from the center of the source.

Effective (root-mean-square or rms) sound pressure $p(x)$ is the square root of the quantity obtained by squaring Eqs. (1) or (2) and averaging the squared result at a point x over a period $T = 1/f$. The effective sound pressure at any point x for the wave of Eq. (1) is therefore

$$p(x) = p_0 \qquad (3)$$

It is seen that for an outwardly traveling plane wave the effective sound pressure has the same value at all points.

The rms sound pressure for Eq. (2) is

$$p(r) = \frac{A_0}{r} \qquad (4)$$

It is seen that for an outwardly traveling spherical wave the effective sound pressure decreases linearly with distance r.

It is easy to show that if a sound wave is made up of a number of waves each of different frequency f_1, f_2, and so on, and with effective values p_1, p_2, and so on, the effective sound pressure for the total at a point x is

$$p(x) = \sqrt{p_1^2 + p_2^2 + p_3^2 + \cdots} \qquad (5)$$

Instantaneous particle velocity $u(x,t)$ is the velocity at point x of a given infinitesimal part of a medium at a given instant due to a sound wave. In other words, when a sound wave of frequency f passes through a medium, it causes the pressure to vary sinusoidally above and below barometric pressure and also causes the air molecules to vibrate back and forth sinusoidally in the direction of travel of the wave.

For a plane outwardly traveling sound wave,

$$u(x,t) = \frac{\sqrt{2}p_0}{\rho_0 c} \cos [\omega(t - x/c)] \qquad (6)$$

where ρ_0 is the ambient density of the medium and c is the speed of sound. The product $\rho_0 c$ is called the characteristic resistance (or characteristic impedance) of the medium and is the constant relating sound pressure and particle velocity in a plane outwardly traveling wave.

Effective (root-mean-square or rms) particle velocity $u(x)$ is found in the same manner as the effective sound pressure. The effective particle velocity at any point for the wave of Eq. (6) is

$$u(x) = \frac{p(x)}{\rho_0 c} = \frac{p_0}{\rho_0 c} \qquad (7)$$

For an outwardly traveling spherical wave the relation is given by

$$u(r) = \frac{p(r)}{\rho_0 c}\sqrt{\left(\frac{c}{\omega r}\right)^2 + 1} \qquad (8)$$

where $p(r)$ is given by Eq. (4). When the point of observation is far enough from the source,

$$u(r) = \frac{p(r)}{\rho_0 c} = \frac{A_0}{r\rho_0 c} \tag{9}$$

and the relation between $u(r)$ and $p(r)$ is the same as that between $u(x)$ and $p(x)$.

Sound intensity $I(x)$ is the acoustic power passing through a surface of unit area. For example, assume that a spherical sound source radiates a power W. To find the acoustic power per unit area of the sphere at radius r, one simply divides W by the area of the sphere, so that the sound intensity at a distance r from the center of the source is

$$I(r) = \frac{W}{4\pi r^2} \tag{10}$$

As can be seen from Eq. (4), the square of the rms sound pressure of a spherical wave also decreases as $(1/r^2)$, so that $I(r)$ must be proportional to $[p(r)]^2$. It is well known that the product of $p(x,t)$ and $u(x,t)$ averaged over a period $T = 1/f$ yields the intensity $I(x)$. Performing this operation on Eqs. (1) and (6) yields

$$I(x) = \overline{p(x,t) \times u(x,t)} = \frac{[p(x)]^2}{\rho_0 c} = \frac{p_0{}^2}{\rho_0 c} \tag{11}$$

Now, consider the sound intensity $I(r)$ for outwardly traveling spherical waves. By virtue of Eq. (9), which is valid for large values of r, and by virtue of the observation that $I(r)$ must always be proportional to $[p(r)]^2$, one is able to write

$$I(r) = \frac{[p(r)]^2}{\rho_0 c} = \frac{A_0}{r^2 \rho_0 c} \tag{12}$$

Therefore, the relation between $I(x)$ and $[p(x)]^2$ is the same as that between $I(r)$ and $[p(r)]^2$, namely $\rho_0 c$, independent of the value of x or r. It is noted, however, that $I(x)$ is independent of x, while $I(r)$ decreases with r^2, because $[p(r)]^2$ decreases with r^2.

Radiation of sound. Sound is radiated by different types of sound sources in different ways. The sound intensity produced at a point distant from a source is determined by the total power radiated by the source; by the type of source, which, in turn, determines its directional characteristics; by the location of a source relative to the ground or other sound reflecting on absorbing surfaces; by the nature of the medium through which the sound is traveling; and by the existence of sound-scattering, -reflecting, or -absorbing objects or barriers between the source and the point of observation.

Radiation from point sources. Spherical radiation: When sound is radiated through a lossless medium from a point source in free space, that is, a space in which the effects of boundaries are negligible, the magnitude of the sound pressure at a point removed from it decreases in inverse proportion to the distance from it (see Eq. 4).

In terms of the power radiated by the source (see Eqs. 10 and 12),

$$p(r) = \frac{1}{r}\sqrt{\frac{W\rho_0 c}{4\pi}} \tag{13}$$

Sectorial radiation: When sound is radiated into a portion of a sphere, for example, from a source mounted in a large plane surface (so-called half-space), more sound power will be radiated through unit area at any given distance from the source than would be radiated through a whole sphere, so that the intensity increases.

The directivity factor Q is equal to the factorial increase that would have to be made in the power radiated by a spherical source (with a base power W) to make it produce the same intensity as a source of the same power W radiating into a spherical sector of less than 360°. For example, if a source radiates acoustic power W into half-space, it will produce twice the intensity at a given distance as would a spherical source radiating the same power W. Hence, for half-space, $Q = 2$. In Table 1, the directivity factors for several common types of boundary near a source are shown.

Table 1. Relation between directivity factor Q and the size of space into which a spherical source radiates

Type of space	Description	Q
Full-space	Spherical source in free space	1
Half-space	Spherical source radiating to one side of an infinitely large, flat plane	2
Quarter-space	Spherical source radiating outward from the inside of the intersection of two infinitely large, flat planes	4
Eighth-space	Spherical source radiating outward from the inside of the intersection of three infinitely large, flat planes	8

Radiation from complex sources. A source may be directive of itself without the influence of bounding planes. In that case, the directivity factor is defined as

$$\frac{WQ}{4\pi r^2} = \frac{[p_D(r,\theta,\phi)]^2}{\rho_0 c} \tag{14}$$

where W is the acoustic power being radiated by a directional source that produces a sound pressure p_D at a distance r and directional angles θ and ϕ. As before, Q is the factor by which the power of a nondirectional (spherical) source must be increased in order to produce the same sound pressure p_D at a distance r.

Examples of the directivity pattern of three common sources are given in Figs. 2, 3, and 4.

The linear line array is common in underwater signaling and may be constructed from a number of identical sources closely spaced along a line, all radiating in phase.

The doublet sound source is made of two simple (spherically radiating) sound sources operating exactly out of phase. By out of phase is meant that one source is expanding while the other is contracting.

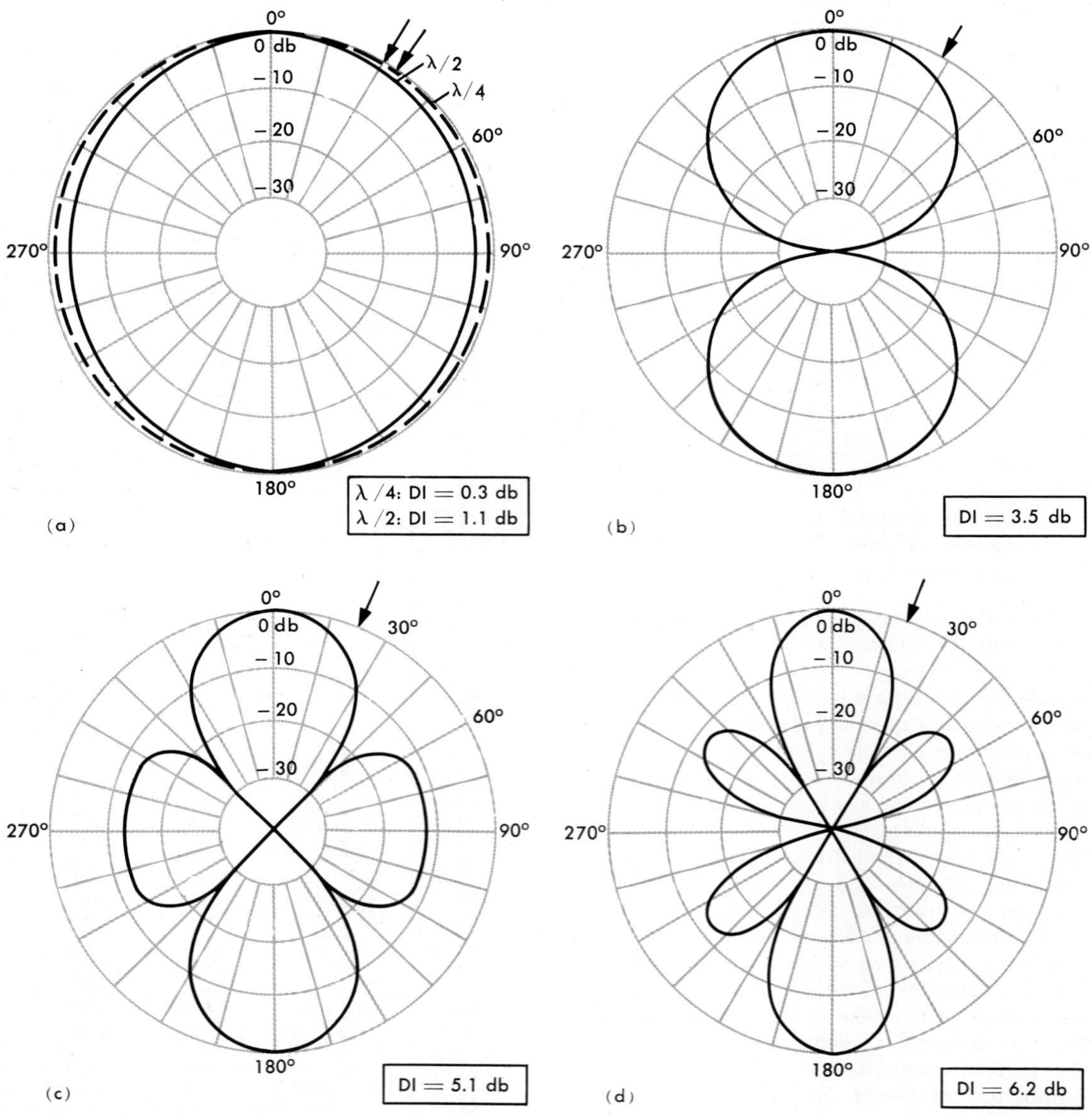

Fig. 2. Directivity patterns for a linear line array radiating uniformly along its length d. The boxes give the directivity index (DI = 10 $\log_{10} Q$) at $\theta = 0°$. One angle of zero directivity index is also indicated by the arrow for each length. (a) $d = \lambda/4$ indicated by broken line and $\lambda/2$ indicated by solid line. (b) $d = \lambda$. (c) $d = 3\lambda/2$. (d) $d = 2\lambda$.

If the distance b between the sources of a doublet is small compared to the wavelength λ of the sound being emitted, and if the distance r of the observer from the source satisfies the relation $r^2 \gg \lambda^2/36$, then the sound pressure at the observer's position is

$$pD = \frac{\rho_0 \omega^2 U_0 b}{4\pi r c} \cos\theta \tag{15}$$

where U_0 = the root-mean-square strength of each of the simple sources comprising the doublet (in cubic meters per second)

$\omega = 2\pi f$ = angular frequency

b = distance between simple sources

r = distance between observer and a point between the simple sources

c = speed of sound

θ = angle formed by a line drawn between the observer and the doublet and a line drawn through the center of the simple source comprising the doublet

A rigid circular piston in a so-called infinite baffle is similar to a circular loudspeaker in the wall of a room or in the side of a large box; none of the sound waves being radiated to the rear can come around to the front again.

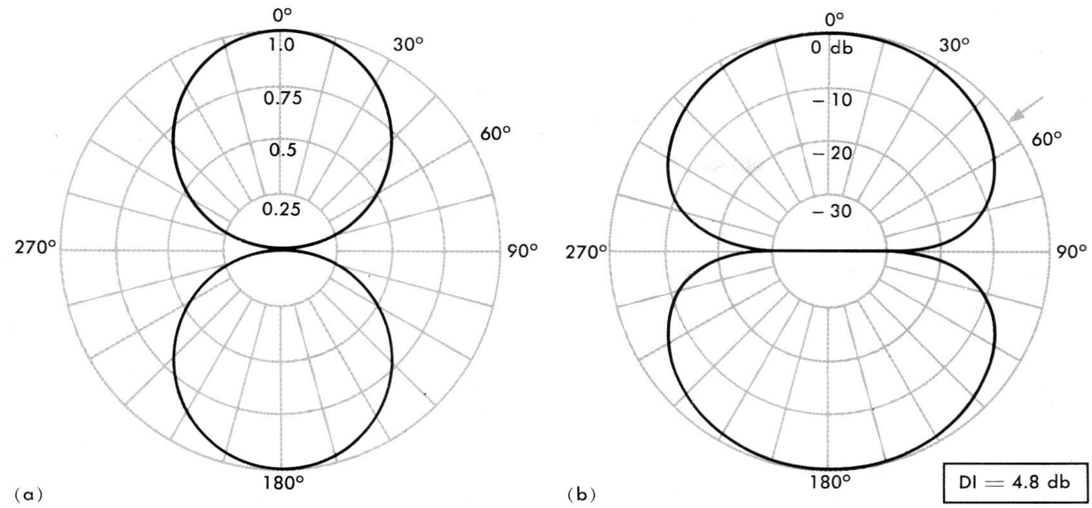

Fig. 3. Directivity patterns for a doublet sound source. (a) Sound-pressure ratio p/p_0 vs. θ. (b) 20 log p/p_0 vs. θ. The boxes give the directivity index (DI = 10 $\log_{10} Q$) at $\theta = 0°$. One angle of zero directivity index is also indicated by the arrow.

The directivity indexes for the principle axes of three kinds of radiators as a function of the ratio of the circumference $2\pi a$ of the radiator to the wavelength λ are given in Fig. 5. A circular piston in the end of a long tube is equivalent to a circular loudspeaker in a small box. An unbaffled piston is equivalent to a circular loudspeaker, both sides of which are free to radiate into the same medium.

One can obtain a variety of directivity patterns from a source composed of two small spherical radiators (two monopoles) by driving them with more than one frequency. *See* LOUDSPEAKER.

Most transducers are reversible. A familiar example is a megaphone, which, when used for speaking, increases the directivity of the voice. When used reciprocally at the ear it increases the directivity of the ear (*see* MEGAPHONE). Hence, all of the directivity patterns given in the previous paragraphs are also valid for receivers of the same construction.

Propagation of sound. Sound may be propagated in gases, liquids, and solids. Sound energy is lost during propagation due to viscous, heat, and molecular losses. In general, the speed (velocity) of sound in a medium is proportional to the square root of the absolute temperature. The velocity of propagation and losses in various media are discussed in the following paragraphs.

Gases. The accepted value of the velocity of sound in 17 different gases at 0°C is given in Table 2.

Two properties of a gaseous medium combine to attenuate a wave that is propagated in free space. The first property is molecular absorption and dispersion in polyatomic gases involving an exchange of translational and vibrational energy between colliding molecules. The second property is viscosity and heat conduction. The attenuation in the pressure amplitude of a plane progressive wave is expressed by $p(x) = p_0 e^{-\alpha x}$ where α is in nepers per unit distance.

Table 2. Velocity of sound in gases at 0°C*

Gas	Velocity, m/sec	Velocity, ft/sec
Air	331.45	1087.42
Ammonia, NH_3	415	1361
Argon, Ar	319	1046
Carbon monoxide, CO	337.1	1106
Carbon disulfide, CS_2	189	606
Chlorine, Cl_2	205.3	674
Ethylene, C_2H_4	314	1030
Helium, He	970	3182
Hydrogen, H_2	1269.5	4165
Illuminating gas	490.4	1609
Methane, CH_4	432	1417
Neon, Ne	435	1427
Nitric oxide, NO	325	1066
Nitrogen, N_2	337	1096
Nitrous oxide, N_2O	261.8	859
Oxygen, O_2	317.2	1041
Steam (100°C), H_2O	404.8	1328

* Compiled from *Handbook of Chemistry and Physics*, 37th ed., *International Critical Tables*, and *J. Acoust. Soc. Am.*

The problem of absorption and dispersion of sound by molecular collision has been treated analytically, and the attenuation due to molecular absorption α_m can readily be found for any ordinary set of conditions of temperature, humidity, and frequency.

The attenuation caused by heat conduction and viscosity of the air α_c is not known so accurately. The classical absorption due to these causes has been thoroughly described by Lord Rayleigh and was first derived by G. R. Kirchhoff and G. G. Stokes as the relation

$$\alpha_c = \frac{50\omega^2}{\rho_0 c^3}\left[\frac{4\eta}{3} + (\gamma - 1)\frac{\kappa}{C_p}\right] \quad \text{nepers/m} \quad (16)$$

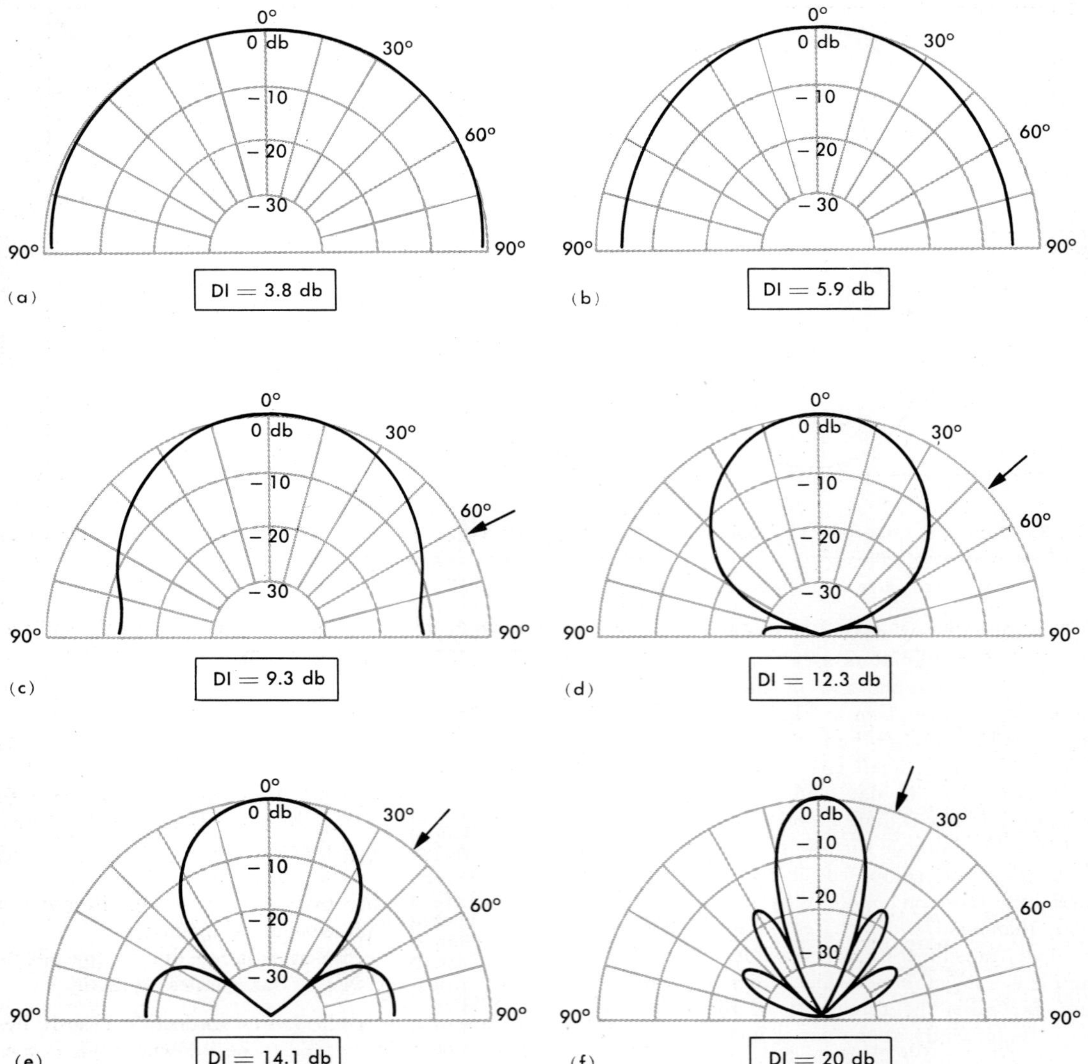

Fig. 4. Directivity patterns for a rigid circular piston in an infinite baffle as a function of $2\pi\ a/\lambda$, where a is the radius of the piston. The boxes give the directivity index at $\theta = 0°$. One angle of zero directivity index is also indicated by the arrow. (a) $2\pi a/\lambda = 1$. (b) $2\pi a/\lambda = 2$. (c) $2\pi a/\lambda = 3$. (d) $2\pi a/\lambda = 4$. (e) $2\pi a/\lambda = 5$. (f) $2\pi a/\lambda = 10$.

where $\omega/2\pi$ = frequency in cycles per second; ρ_0 = density in grams/cm^3; c = speed of sound in centimeters per second; η = coefficient of viscosity in poises; γ = ratio of specific heat at constant pressure to specific heat at constant volume; κ = coefficient of thermal conductivity in calories per second-degree-centimeter; and C_p is the specific heat at constant pressure in calories per gram-degree. To convert from nepers per meter to decibels per meter, multiply the former by 8.69.

It appears from measurements that the value of α_c as computed from Eq. (16) should be multiplied by a factor of about 1.3.

The total attenuation α_A due to both types of absorption is, therefore,

$$\alpha_A = \alpha_m + 8.69\alpha_c \quad \text{db/meter} \tag{17}$$

Liquids. The acoustical behavior of liquids is fundamentally identical to that of gases, but the great difference in the magnitudes of the basic properties (density and compressibility) gives rise to notable differences in the nature of practical sound fields in the two media. Up to 1000 megacycles, no measurable effect of frequency on velocity has been found. The measured value of the attenuation due to viscosity α_c varies in water inversely as the square of frequency. F. E. Fox and G. D. Rock have found that

$$\alpha_c/f^2 = 21.5 \times 10^{-17}\ \text{cm}^{-1} \tag{18}$$

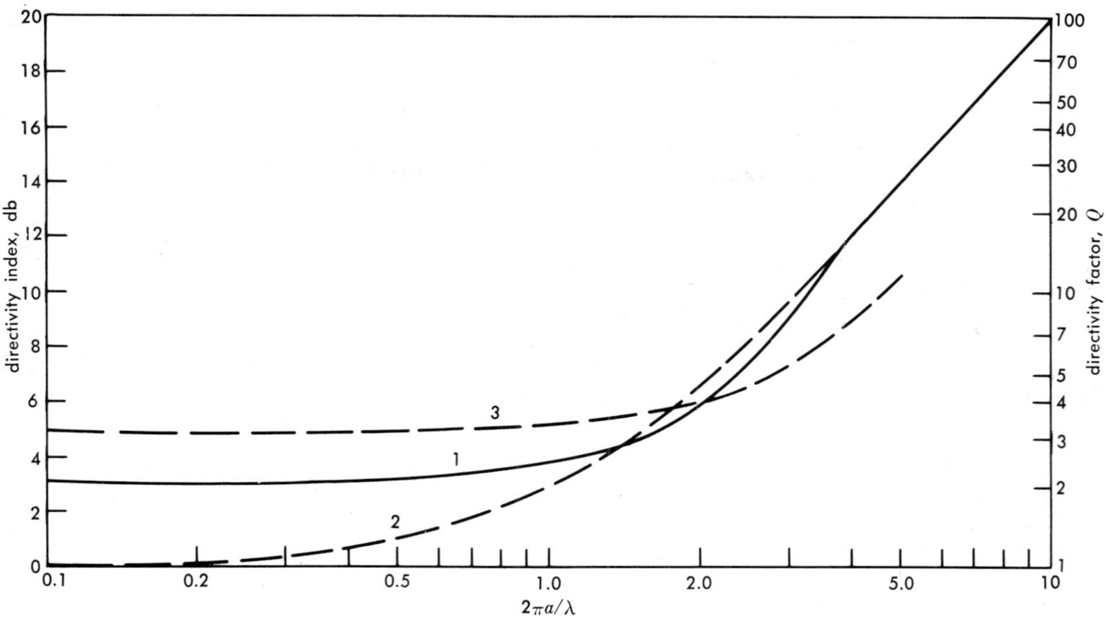

Fig. 5. Directivity indexes for the radiation from one side only of a piston in an infinite plane baffle (curve 1), piston in the end of a long tube (curve 2), and piston in free space without any baffle (curve 3).

Absorption in organic liquids shows no observable relation to the viscosity. The increments in sound velocity due to dissolved salts at the low concentrations found in sea water are observed to be proportional to the molar concentration for each salt. They are also additive for a number of salts.

The speed of sound in pure organic liquids covers little more than a 2:1 range; the lowest reported is for ethyl bromide (892 m/sec) and the highest is for glycerin (1986 m/sec). The absorption constant α_c/f^2 varies over a wide range, about 300:1. Numerically, the highest reported absorption constant for a simple liquid is about one-tenth that for dry air.

Solids. The velocities of propagation of sound in some solids are given in Table 3. The speed for a bar of zero diameter-to-wavelength ratio is the longitudinal bar speed and the speed for an infinite diameter-to-wavelength ratio is the plate (bulk) speed.

Atmospheric acoustics. Because of atmospheric conditions and the presence of obstacles, the sound pressure levels measured outdoors are generally lower, sometimes greatly so, than those predicted from spherical spreading alone.

Attenuation effects. The important factors that affect sound propagation along the ground are (1) sound absorption in the air; (2) presence of fog, rain, or snow; (3) presence of walls and trees; and (4) effect of wind and temperature gradients, atmospheric turbulence, and the acoustic effect of the presence of the ground.

The excess attenuation due to these effects can be added to Eq. (4) as

$$p(r,\theta) = \frac{A_0}{r}\sqrt{Q}\,\frac{1}{A_e} \tag{19}$$

where A_0 = constant depending on the strength of the source (dimensions of pressure times distance)
r = distance from the source
Q = directivity factor in the direction θ (dimensionless quantity)
A_e = excess attenuation due to the effects 1 to 4 (dimensionless quantity)

Attenuation in db due to absorption in still, homogeneous air is found to be $\alpha_A r$ where r is in meters.

No satisfactory data on attenuation due to fog, rain, or falling snow are available. It can only be said that the total is less than 5 db/1000 ft and is usually assumed to be negligible compared to the other attenuations; see Eq. (17).

Walls, buildings, and other large rigid barriers, if interposed between source and receiver, result in appreciable excess attenuation due primarily to the acoustic shielding effect of such structures. Since sound waves are diffracted around the obstacle, one would expect the excess attenuation to increase with frequency. This has, indeed, been confirmed both experimentally and theoretically.

Over open level ground, appreciable vertical temperature and wind gradients almost always exist, the former because of the heat exchange between the ground and the atmosphere, the latter because of the friction between the moving air and the ground. Because of these gradients, the speed of sound varies with height above the ground, and sound waves are refracted (bent upward or down-

Table 3. Velocity of sound in solids

Material	Longitudinal bar velocity, cm/sec $\times 10^5$	Plate (bulk) velocity, cm/sec $\times 10^5$
Aluminum	5.24	6.4
Brass	3.42	4.25
Copper	3.58	4.6
Gold	2.03	3.24
Iron	5.17	5.85
Lead	1.25	2.4
Nickel	4.76	5.6
Silver	2.64	3.60
Steel	5.05	6.1
Tin	2.73	3.32
Tungsten	4.31	5.46
Zinc	3.81	4.17
Cork	0.50	
Crystals		
Quartz, X-cut	5.44	5.72
ADP ($NH_4H_2PO_4$)		
45° Z-cut	3.28	4.92
Rochelle salt		
45° Y-cut	2.47	
Glass		
Heavy flint	3.49	3.76
Crown	5.30	5.66
Quartz	5.37	5.57
Granite	3.95	
Ivory	3.01	
Marble	3.81	
Slate	4.51	
Wood		
Elm	1.01	
Oak	4.1	

ward). Under such conditions, it is possible to have a shadow zone into which no direct sound can penetrate. These shadow zones are never sharp in the sense of light propagation because sound energy is diffracted into the shadow zone as well as scattered into it by turbulence.

A shadow zone is most commonly encountered upwind from a source, where the wind gradient bends the sound rays upward. Wind gradients near the ground are, on the average, always positive; that is, the windspeed increases with height. Downwind, the wind gradient bends the sound rays downward and no shadow zone is produced. Crosswind, there is a zone of transition. This asymmetric behavior is characteristic of wind-induced sound refraction. Temperature-induced sound refraction tends to be symmetrical about the source. A shadow may encircle a source completely in the presence of a strong negative temperature gradient and a low windspeed, such as may be expected on a calm sunny day. On the other hand, there will be no shadow at all within a mile or two of the source in the presence of a strong positive temperature gradient (large temperature inversion) and a low windspeed, such as may be expected on a clear calm night. *See* REFRACTION OF WAVES.

The measurement or estimate of the micrometeorological parameters which must be used in any computation is probably beyond the capabilities of the average person interested in noise control. On the other hand, useful estimates of at least the maximum of the excess attenuation to be expected from temperature and wind gradients over open level terrain can be had from considering experimental data.

Consider Fig. 6. Source and receiver are shown a distance r apart. The average direction from which the wind is blowing is indicated by a wind vane. The angle between the direction of the wind vane and the line connecting the source and receiver is designated ϕ. There will generally be a shadow zone (the shaded region) produced on the upwind side of the source because sound waves traveling upwind tend to be bent upward by the wind. Often the air near the ground is warmer than that farther up, so that there is a tendency for the sound waves to be diffracted upward on all sides of the sound source, in addition. However, any wind present tends to bend the sound waves downward in the downwind direction. At some critical angle ϕ_c the wind and temperature gradients cancel each other and the shadow zone vanishes. As a result the plane is divided into an upwind sector $2\phi_c$ and a downwind sector $360° - 2\phi_c$.

Experiments have shown that the excess attenuation is frequently radically different upwind and downwind, with a gradual transition at the boundaries ($\phi = \pm\ \phi_c$). On a sunny day, with moderate winds, the excess attenuation upwind, inside the shadow zone, is typically 20–30 db higher than that for the same distance downwind.

Extensive measurements in the frequency range of about 300–5000 cps have been made over open level terrain with sparse low ground cover (1 ft high), a source height of 12 ft, and a receiver height of 5 ft, using octave bands of random noise as the transmitted signal. Windspeeds encountered ranged from 2–3 mph to 10–15 mph. From these measurements, there are available empirical design curves with the aid of which the excess at-

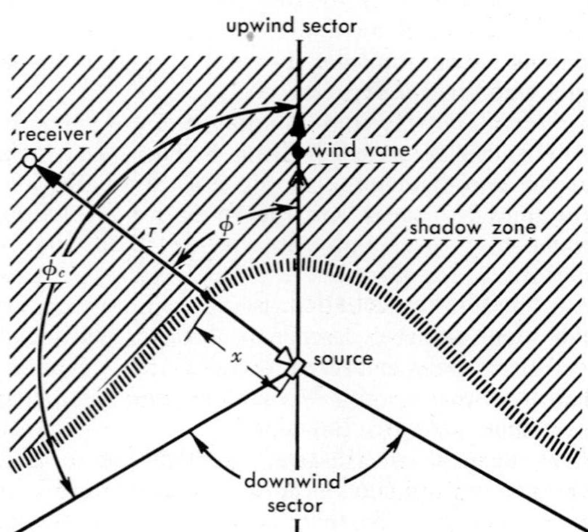

Fig. 6. Geometry of sound propagation over open level terrain (plan view). Average daytime conditions are shown. x = distance from source to shadow zone; ϕ = angle between wind and sound; ϕ_c = critical angle.

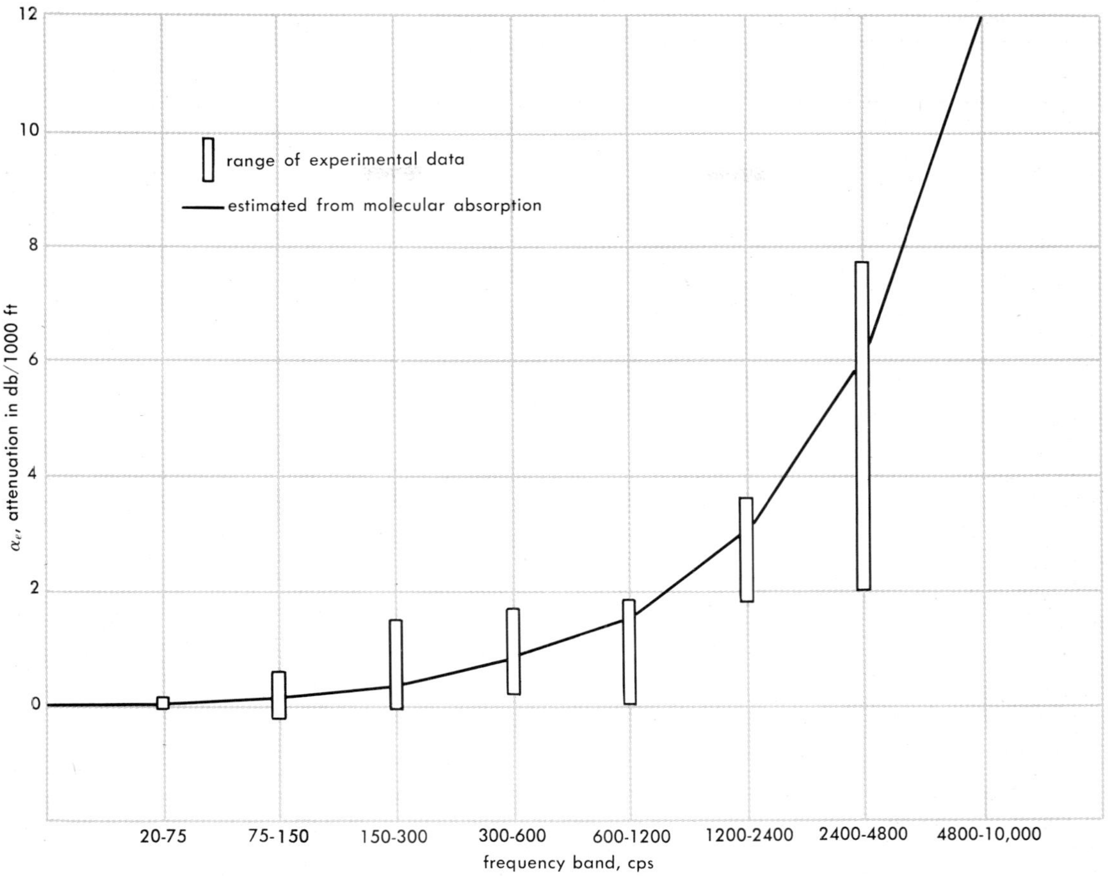

Fig. 7. Estimated excess attenuation coefficient α_e for air-to-ground propagation. The range of available experimental data is also shown.

tenuation can be estimated for any angle ϕ for distances r up to about 1 mile and the terrain conditions noted, provided the temperature and wind gradients are known from measurements made at approximately half the average source and receiver heights. Since the experiments did not include tests at the very low and very high audio frequencies, the design curves are subject to confirmation there.

Propagation from air to ground. This case is of considerable practical importance in estimating the sound-pressure level near the ground due to overhead aircraft. The excess attenuation of sound has been measured by several investigators using propeller aircraft, helicopters, and jet aircraft flying at moderate altitudes (½ mile or less) under various atmospheric conditions. It was assumed in every case that the excess attenuation measured for various distances to the airplane (slant range) can be represented in terms of an attenuation coefficient α_e which is independent of the slant range. Figure 7 shows, as a function of frequency, the attenuation coefficients obtained from the various studies. *See* AIRCRAFT NOISE.

The values of attenuation due to molecular absorption plotted in Fig. 7 account reasonably well for the results of the measurements of air-to-ground attenuation, assuming that precipitation was either absent or had no effect on the results.

As the aircraft passes overhead in level flight, the sound-pressure level at a fixed point on the ground will rise, go through a maximum, and fall again. The position of the aircraft for maximum sound-pressure level depends on its acoustic directivity pattern. As a first approximation, a 45° position can be assumed for jet aircraft. Consequently, this slant distance is the effective length of the transmission path and must be used to compute not only the excess attenuation but also the spherical divergence. For propeller aircraft the minimum distance to the flight path is relevant. [L.L.B.]

Bibliography: American Standards Association, *Acoustical Terminology*, 724.1, 1959; L. L. Beranek, *Acoustics*, 1954; L. L. Beranek, *Noise Reduction*, 1960; S. Fluegge (ed.), *Handbuch der Physik*, vol. 48, 1957; D. E. Gray (ed.), *American Institute of Physics Handbook*, 1957; L. E. Kinsler and A. R. Frey, *Fundamentals of Acoustics*, 1950; R. B. Lindsay, *Mechanical Radiation,* 1960; P. M. Morse, *Vibration and Sound,* 2d ed., 1948; Lord Rayleigh, *Theory of Sound,* reprint, 2 vols., 1945; R. W. B. Stephens and A. E. Bate, *Wave Motion and Sound,* 1950.

Sound intensity

A term which describes the rate of flow of sound energy. The sound intensity measured in a specified direction at a given point is the average rate at which sound energy is transmitted through a unit area perpendicular to the specified direction at the point considered. The unit of measure is the watt per square meter.

In a plane or spherical free-progressive sound wave, the intensity I in the direction of propagation is related to the root mean square, or effective sound pressure p by the equation $I = p^2/\rho_0 c$, where ρ_0 is the density of the fluid in which the wave is propagated and c is the speed of sound.

The amount of sound power radiated by an acoustical source is obtained by integrating the sound intensity over a surface which encloses the source. *See* SOUND; SOUND PRESSURE. [W.J.G.]

Sound pressure

The incremental variation in the static pressure of a medium when a sound wave is propagated through it. This incremental variation is also known as the excess pressure.

The unit of pressure commonly used in acoustics is the microbar.

$$1 \text{ microbar} = 0.1 \text{ newton/m}^2 = 1 \text{ dyne/cm}^2$$

which is approximately 10^{-6} times the normal atmospheric pressure.

Instantaneous sound pressure at a point is the incremental change from the static pressure of the medium at a given instant caused by the presence of a sound wave.

Peak sound pressure for any specified time interval is the maximum absolute value of the instantaneous sound pressure in that interval.

Effective sound pressure at a point is the root-mean-square (rms) value of the instantaneous sound pressure over a time interval at that point. The time interval should be an integral number of periods if the sound pressure is periodic. In the case of nonperiodic sound pressures, the interval should be long enough to make the value obtained essentially independent of small changes in the length of the interval. The effective sound pressure level is a highly useful quantity since the power transmitted by a sound wave is generally related to the mean square pressure in the wave. *See* SOUND INTENSITY.

Sound pressure level is the term applied to the specification of the effective sound pressure in decibels. It is defined as 20 times the logarithm to the base 10 of the ratio of the effective sound pressure of a sound to a reference effective sound pressure

$$SPL = 20 \log \frac{p}{p_{ref}}$$

In measurements dealing with hearing, and for sound-level and noise measurements in air and liquids, the reference level of 0.0002 microbar is generally employed. A reference level of 1 microbar has achieved wide usage in the calibration of microphones and loudspeakers and in underwater acoustics. It is essential to state the reference level employed whenever a sound pressure level is reported.

Pressure spectrum level is the specification in decibels, as a function of frequency, of the sound pressure level for a frequency bandwidth of 1 cps. *See* SOUND. [W.J.G.]

Sound recording

The process of recording sound signals so they may be reproduced at any subsequent time. The most common sound-reproducing systems are the disk phonograph, the sound motion picture, and the magnetic tape reproducer.

The disk phonograph, commercialized at the turn of the century, was the first sound-reproducing system employing a record. This device made it possible for all the people of the world to hear statesmen, orators, actors, orchestras, and bands previously heard only at first hand. The disk phonograph, used in every country, owes its popularity to the fact that the individual can select any type of information or entertainment and reproduce it whenever he wishes.

The addition of sound to motion pictures in the late 1920s made this type of entertainment complete. It was the first time that picture and sound were synchronized and reproduced simultaneously. The term optical recording is used to describe the recording of sound on photographic film.

The magnetic tape reproducer was commercially distributed shortly after World War II. One of its novel features is that a recorded program on the tape can be erased and a new program recorded. Prerecorded tapes containing programs similar to those on phonograph records have been commercialized on a wide scale. *See* DISK RECORDING; MAGNETIC RECORDING; OPTICAL RECORDING. [H.F.O.]

Sound reproduction systems, electrical

Systems for the transformation of sound waves into an electrical signal and the reconversion of that electrical signal back into sound waves. Figure 1 shows the boundaries, in both intensity and frequency, for most speech and music sound sources. At the bottom of the figure is a curve showing the softest sound of any frequency that the normal ear can hear, while at the top is a curve giving the level at which the sensation of feeling begins (*see* HEARING). Music requires a dynamic range of 70 decibels (db) from the softest sounds to the loudest sounds and a frequency range of approximately 40–15,000 cycles per second (cps). Conversational speech requires a dynamic range of 40 db and a frequency range of 100–8000 cps. In contrast to the ear's range of 120 db (a power ratio of 1,000,-000,000,000:1), 60 db (a power ratio of 1,000,-

000:1) is considered quite good even for a fine sound-reproducing system.

Figure 2 shows the frequency range required for the accurate reproduction of the various instruments of the symphony orchestra, speech, and a few noises, while Fig. 3 shows the frequency ranges usually found in different types of sound-reproducing system.

Articulation. The effectiveness of systems intended primarily for communication is indicated by the percentage of single-syllable nonsense words that can be correctly understood over the system. Figure 4 shows that the high-frequency response can be limited to 6000 cps and the low-frequency response limited to about 300 cps without affecting the syllable articulation score appreciably. Limiting the high-frequency response to 3000 cps, as in a standard telephone circuit, results in an articulation score of about 90%, which, since normal speech contains considerable redundancy, corresponds to a sentence articulation of virtually 100%.

Distortion. Nonlinearity in a reproducing system results in the introduction of unwanted tones. If harmonically related to the frequency of a single original tone, they are called harmonic distortion. If more than one original tone is being reproduced, sum and difference frequencies can result. This is called intermodulation. Figure 5 shows the percentage of harmonic distortion in reproduced speech or music that is just perceptible, tolerable, or objectionable.

Component matching. Distortion produced by driving an amplifier beyond its normal operating level (into the overload region) is likely to be much more annoying than the distortion that occurs in the normal operating range. Overload symptoms can also occur if the various elements of the system are not properly matched to work together. Load circuits in sound equipment are chosen to present an impedance to the source that will result in the desired frequency response, optimum signal-to-noise ratio, or in the maximum power being delivered to the load with a minimum of harmonic distortion. It is also necessary to keep the power or voltage handling capabilities of the components consistent at the junction points; otherwise a system, even though composed of excellent components, will show overload symptoms much of the time.

Transient distortion. It is desirable to avoid sharp resonances in the reproducing system since these resonances, once excited by the program material, can produce an output sound after the exciting sound has completely disappeared. This effect is known as transient distortion, and the verbal description of the sound of the phenomenon depends upon the frequency range in which it occurs. A resonant peak around 100 cps, such as that produced by an under-damped loudspeaker, tends to make the reproduced program material sound "boomy," while several sharp peaks in the region around 5000–8000 cps give the reproduction a

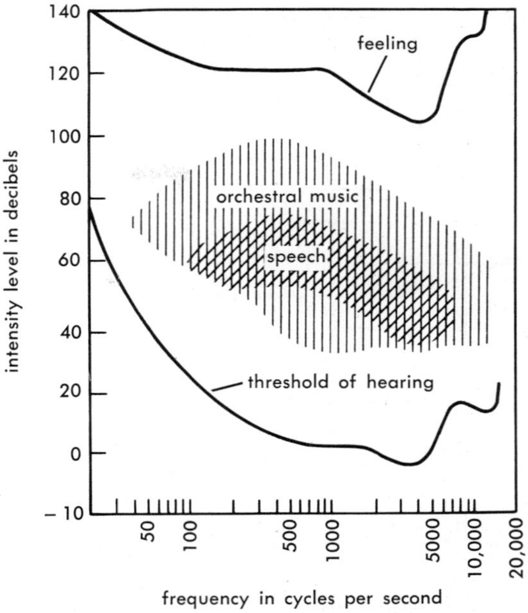

Fig. 1. Frequency and dynamic range of speech and music, the minimum audible sound contour and the contour corresponding to the threshold of feeling. *(From Bell Labs. Record, June, 314–315, 1934, and Robinson and Dadson, A redetermination of the equal-loudness relation for pure tones, Brit. J. Appl. Phys., vol. 7, May, 1956)*

"crackly" sound resembling that produced by crumpling a piece of cellophane.

The transient conditions that occur in some amplifiers during a momentary overload are also sometimes called transient distortion. The audible symptoms may include a complete blocking of the signal (giving a moment of silence), or the introduction of extraneous interrupted oscillations in the high-frequency region that sound much like mechanical parts rattling.

Input sources. A wide variety of input material makes up the source of sound signals for a sound-reproducing system, and the technical quality of the various inputs varies over extreme ranges. Studio equipment used for live pickup of the human voice, symphony orchestra, or chorus, whether for motion pictures, radio or television broadcast, recording, or sound reinforcement in an auditorium, is potentially good enough so that it should produce no limitations on the quality of the reproduced sound. Likewise the radio and television broadcast transmitters and the professional disk, film, or magnetic tape recording equipment, when operated properly, introduce virtually no quality limitation. The long telephone lines used to connect the network broadcast stations do introduce limitations in frequency range, as do AM radio receivers. Improperly controlled film-printing and tape-dubbing equipment, as well as poor-quality or poorly maintained and adjusted film, tape, and disk playback equipment, can result in serious

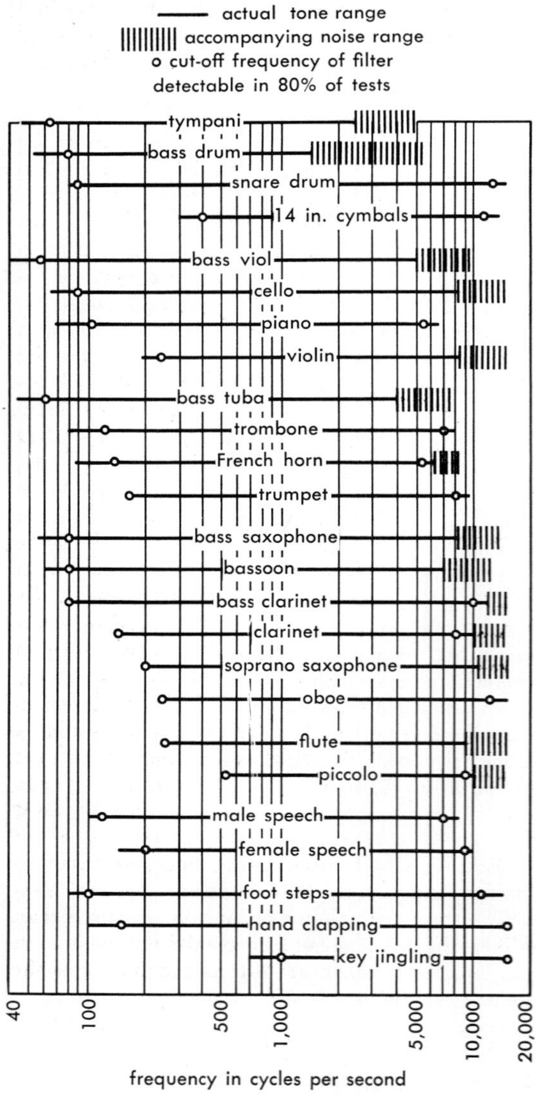

Fig. 2. Audible frequency ranges of music, speech and noise. (*From W. B. Snow, J. Acoust. Soc. Am., 3:155–166, 1931*)

noise, distortion, and irritating pitch changes in the reproduced sound. There is no inherent frequency limitation in FM receivers, but distortion is introduced when there is interference among various portions of the radio signal that have arrived at the receiver over different path lengths.

Home high-fidelity systems. With the advent of the long-playing high-fidelity microgroove record in 1948, the quality of recorded music became good enough to justify home sound-reproducing equipment covering a frequency range as wide as that of the best human ears and having distortion ratings below audibility. Varying individual taste, room acoustics, and recording techniques require the inclusion of elaborate tone-compensating circuits in the high-fidelity amplifiers. Some of these tone-compensating circuits, also called equalization circuits, are intended to provide the user with means for continuously adjusting the relative level of the high- and low-frequency response. Still other tone-shaping circuits provide automatic compensation for the lack of sensitivity of the ear to the extremely low and high frequencies, when the sound is reproduced at an intensity level lower than that of the original sound. This latter type of equalization is usually termed low-level equalization, or loudness equalization. Figure 6 shows a block diagram of typical elements in a home high-fidelity sound system. *See* AMPLIFIER; LOUDSPEAKER; RADIO RECEIVER; SOUND RECORDING; TELEVISION RECEIVER.

Presence. A high-quality system, in contrast to the performance of a limited-frequency-range system, gives the feeling of the actual presence of the performer. This feeling is enhanced by the introduction of a slight exaggeration of the response in the vicinity of 5000 cps. Equipment, especially home high-fidelity equipment, that has this subtle exaggeration is said to have presence. Since the air attenuates the high frequencies more than it does the low, a sound originating from a distance usually contains proportionately less energy in the high-frequency region than a sound originating near by. Conversely, it follows that the exaggera-

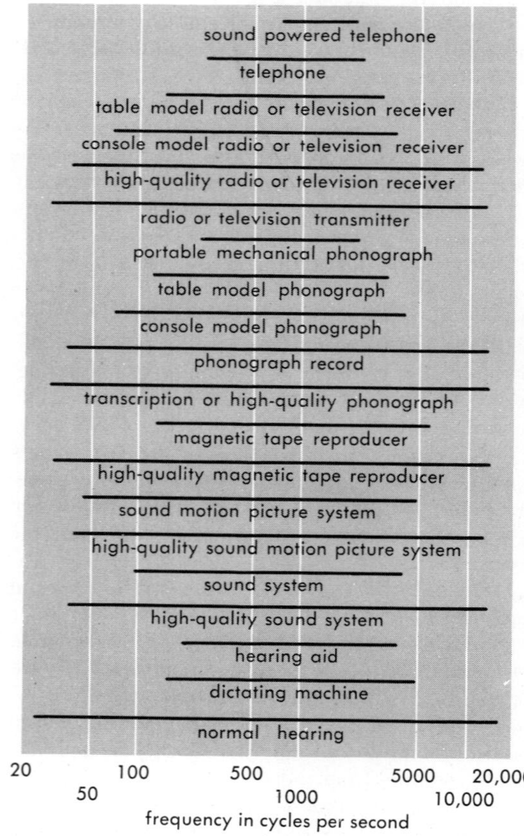

Fig. 3. Frequency ranges of sound-reproducing systems. (*From H. F. Olson, Acoustical Engineering, 3d ed., Van Nostrand, 1957*)

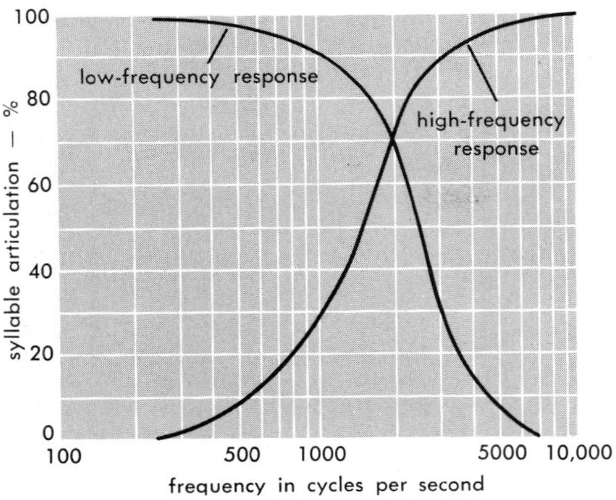

Fig. 4. Syllable articulation versus cut-off frequency of high-pass and low-pass filters. (*From N. R. French and J. C. Steinberg, J. Acoust. Soc. Am., 19:102, 1947*)

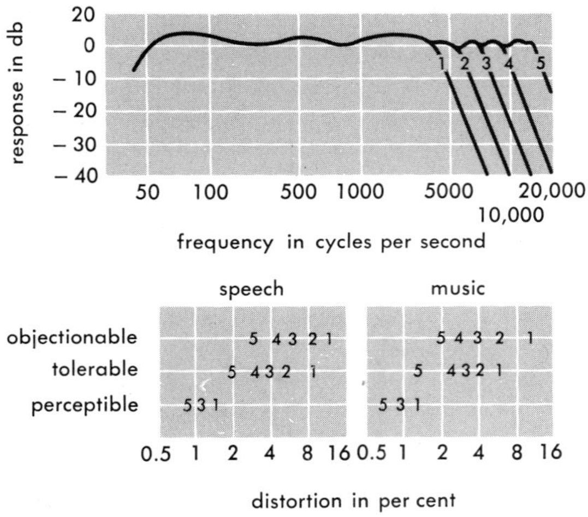

Fig. 5. Results of subjective tests of reproduced speech and music, depicting objectionable, tolerable, and perceptible nonlinear distortion for various high-frequency cut-offs. (*From H. F. Olson, Acoustical Engineering, 3d ed., Van Nostrand, 1957*)

tion of the high frequencies tends to make the sound source seem to come forward into the actual presence of the listener. The "presence peak" is usually introduced either in the loudspeaker or in the original recording, rather than by the alteration of the amplifier response.

The amount of power required for a sound-reproducing system is rather uncertain, since a sizable change in power makes little difference in loudness. It takes a power increase of approximately 10 times to make a sound seem twice as loud. Most home high-fidelity systems have 10–60 watts of electrical output power available.

Stereophonic and binaural sound. The reproduction of sound by a single-speaker system is likely to produce the impression that the sound is coming from a hole in the wall, with little feeling of the spatial extent or distribution of the original sound source. However, if two completely separate sound-reproducing systems are used, in which the two microphones are spaced by the distance between the ears on a person's head, the signal from the right-hand microphone can be reproduced by an earphone on the listener's right ear and the signal from the left-hand microphone reproduced at the left ear. Under these conditions the illusion of spatial orientation is completely convincing.

This type of two-channel system using earphones is usually called binaural. A good approximation of spatial distribution can be achieved by using a multiple-channel loudspeaker system, in which the individual microphones are connected through independent amplifiers to the corresponding loudspeakers. A sound-reproducing system of this latter type is called stereophonic. If it were economical, all of the directional and distance effects could be retained in the reproduced sound by using a screen of closely spaced microphones in front of the stage where an orchestra is playing. Then if the output of each microphone were amplified individually and reproduced by a loudspeaker located in a corresponding position in front of a second stage in another auditorium, a sound wave would leave the second stage and travel out over the audience in exactly the same way that the original wave did. A satisfactory and practical approximation of the sound originating from an orchestra on a stage can be achieved if only three channels are used instead of many. One center channel plus one on either side of the stage reproduces the effect of a live orchestra convincingly. A two-channel system, however, is much more common in home systems than a three-channel system. The results obtained from a two-channel system are not as convincing as the

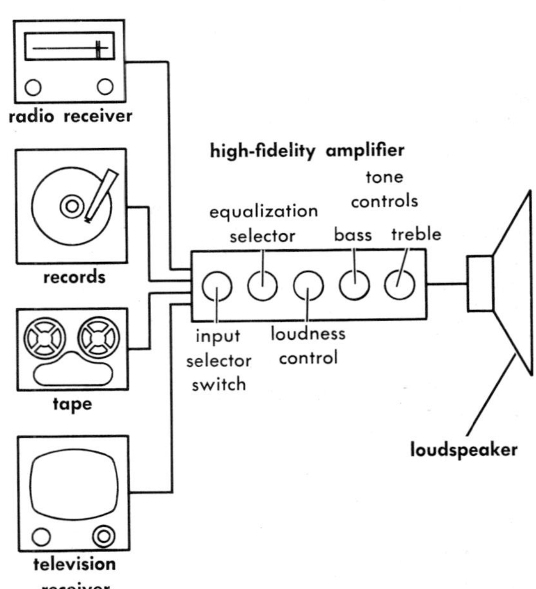

Fig. 6. Block diagram of typical elements in a home high-fidelity system.

results from a three-channel system, since there is likely to be difficulty from the lack of a real sound source in the center of the speaker arrangement. Because of this difficulty, the two speakers must be spaced at a narrower angular distance from the listener's position than the outside speakers in a three-channel stereo system. Nevertheless, with only two channels, the improvement in spatial effect is quite marked in comparison to the results obtained with a single-channel system. These sound-reproduction systems represent the ultimate application of all that has been learned in the investigation of the characteristics of human hearing, and the most realistic form of sound reproduction so far devised. *See* SOUND; *see also* PUBLIC ADDRESS SYSTEM. [F.H.SL.]

Bibliography: L. L. Beranek, *Acoustics*, 1954; H. F. Olson, *Acoustical Engineering*, 3d ed., 1957.

Sound track

A narrow band on a motion-picture film used for the sound record. In some cases, a number of such bands may be used, as in stereophonic sound recording. The two types of sound tracks in general use are variable area tracks and variable density tracks. For an extended discussion, *see* OPTICAL RECORDING. [K.W.P.]

Soundproofing

A term sometimes used to indicate the application of air-borne and solid-borne sound insulation constructions, as well as sound-absorptive constructions, in order to achieve a low noise level in an enclosure. For details, *see* NOISE CONTROL IN BUILDINGS. [C.M.H.]

Source flow

A source, in three-dimensional flow, is a point from which fluid issues at a uniform rate in all directions (Fig. 1).

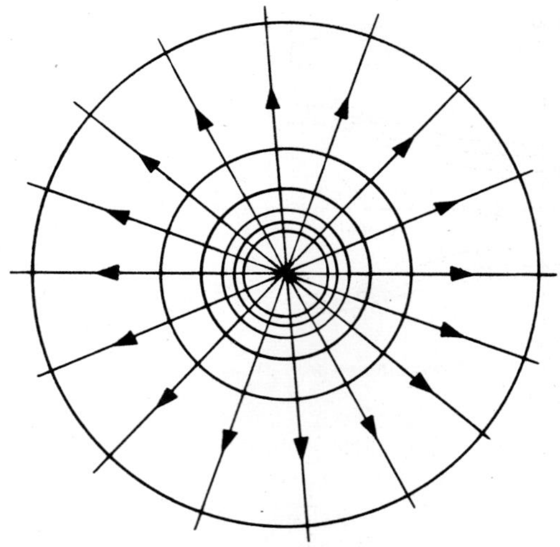

Fig. 1. Streamlines and equipotential lines for a three-dimensional source.

Characteristics of a source. The strength of a source is defined as the volume per unit time issuing from the point. Because the flow is outward and is uniform in all directions, velocity v_r at distance r from the source is the strength m divided by the area of sphere through the point with center at the source, or $v_r = m/4\pi r^2$. By defining velocity potential ϕ as a scalar function of space and time, whose negative derivative with respect to any direction is the velocity component in that direction

$$v_r = -\frac{\partial \phi}{\partial r} = \frac{m}{4\pi r^2}$$

As the velocity vector is entirely in the r direction $\phi = m/4\pi r$ by integration.

Streamlines are radial lines through the point of the source, and equipotential surfaces are given by letting ϕ = constant in the equation. A sink is a negative source, or a point into which fluid flows uniformly in all directions.

A source, in two-dimensional flow, is a line normal to the planes of flow, from which fluid is imagined to flow uniformly in all directions at right angles to the line (Fig. 2). The source appears as a point on the customary two-dimensional flow diagram. The total flow per unit time per unit length is the strength of the source. By calling the strength $2\pi\mu$, the velocity at distance r from the source is $2\pi\mu/2\pi r = \mu/r$. Then

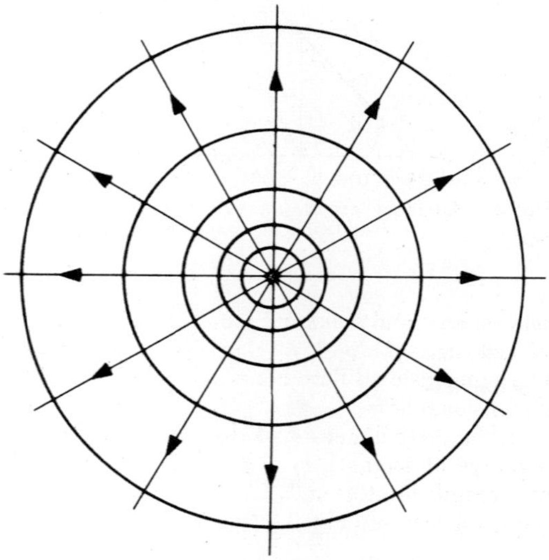

Fig. 2. Streamlines and equipotential lines for a two-dimensional source.

$$-\frac{\partial \phi}{\partial r} = \frac{\mu}{r} \qquad \frac{\partial \phi}{\partial \theta} = 0$$

in which θ is the angle in polar coordinates associated with r. By integrating, $\phi = -\mu \ln r$ in which ln is the natural logarithm. A negative source is a sink, into which fluid is imagined to flow uniformly from all directions at right angles to its line.

Application to flow about a body. Sources and sinks are used as flow elements in conjunction with doublets, vortices, and uniform flow to develop complex flow situations. The combination of a

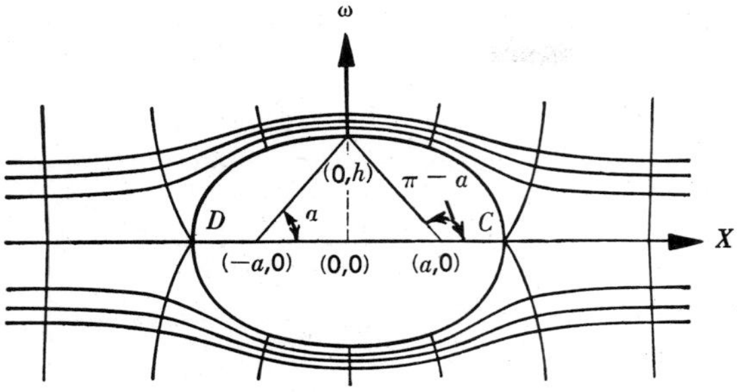

Fig. 3. Three-dimensional flow net about a Rankine body.

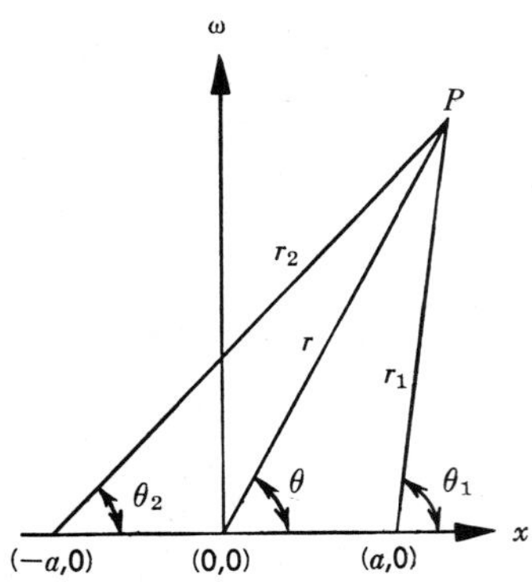

Fig. 4. Auxiliary coordinate systems used for Rankine body.

source, an equal sink, and a uniform flow, properly placed, results in flow about a closed body in three-dimensional flow and about a cylinder in two-dimensional flow.

In the three-dimensional case of a Rankine body, a source of strength m is located at $(a,0)$, a sink of strength m at $(-a,0)$, and a uniform flow U in the $-x$ direction (Fig. 3). In equation form

$$\phi = Ux + \frac{m}{4\pi}\left(\frac{1}{r_1} - \frac{1}{r_2}\right)$$

using auxiliary coordinates (Fig. 4). This flow case has axial symmetry; that is, all streamlines are in planes that pass through the x axis, and all such planes have identical streamlines. The length of the Rankine body is the distance between stagnation points D and C. With $(x_0,0)$ the coordinate of the upstream stagnation point C, the half length x_0 is

$$0 = \frac{x_0/a}{[(x_0/a)^2 - 1]^2} - \frac{\pi U a^2}{m}$$

which is most conveniently solved by trial. The value of (x_0/a) depends on the value of Ua^2/m. The half-breadth h, which occurs at the midsection of the body, is

$$\left(\frac{h}{a}\right)^2 \sqrt{\left(\frac{h}{a}\right)^2 + 1} = \frac{m}{\pi U a^2}$$

which also is solved most conveniently by trial.

By expressing dynamic pressure p as zero at a great distance from the body, where the velocity is U, from Bernoulli's equation

$$p = \frac{\rho}{2}(U^2 - q^2)$$

in which ρ is the mass density of fluid and q is the velocity at any point (x,ω) where the pressure is p. Speed q is determined from the components q_x and q_ω by

$$q = \sqrt{q_x^2 + q_\omega^2}$$

in which $\quad q_x = -\dfrac{\partial\phi}{\partial x} \qquad q_\omega = -\dfrac{\partial\phi}{\partial\omega}$

Frequently it is easier to compute the velocity due to each of the flow elements at a point, and then to add them vectorially to find the speed q at a point.

Other solutions to two- and three-dimensional flow are made up in an analogous manner. [V.L.S.]

Bibliography: V. L. Streeter, *Fluid Dynamics*, 1948.

Sourwood

A deciduous tree, *Oxydendrum arboreum*, of the heath family, indigenous to the southeastern section of the United States, and found from Pennsylvania to Florida and west to Indiana and Louisiana. Sourwood is hardy and often cultivated in the north. It is usually a small or medium-sized tree,

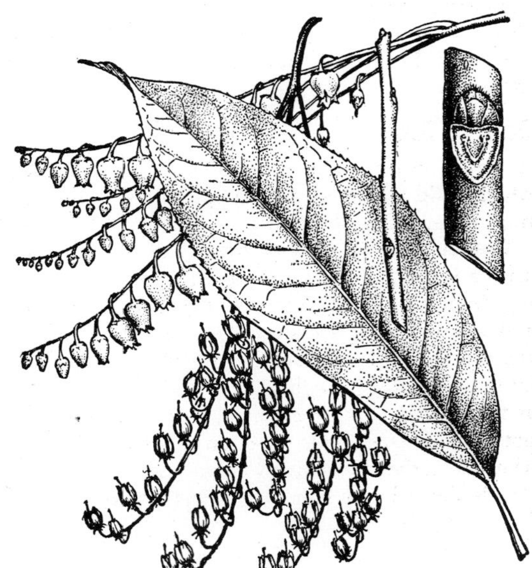

Sourwood, *Oxydendrum arboreum*. (From A. H. Graves, *Illustrated Guide to Trees and Shrubs*, rev. ed., Harper, 1956)

but it sometimes attains a height of 75 ft and a diameter of nearly 2 ft. The leaves are simple, finely toothed, long-pointed, 4–8 in. long, and have an acid taste which explains the popular name. See LEAF (BOTANY). They turn scarlet in the fall. The flowers are white and urn-shaped, and grow in long clusters. The dry fruit remains on the tree through the fall and winter. The wood is not used commercially. Sourwood is also known as sorrel tree, and it is widely planted as an ornamental. *See* FOREST AND FORESTRY; TREE. [A.H.G.]

South America

The southernmost of the New World or Western Hemisphere continents, three-fourths of which lies within the tropics. South America is approximately 4500 miles long and at its greatest expanse 3000 miles wide. Its total area is estimated to be about 7,000,000 square miles, more than two-thirds of which consists of plains. The land portions stand upon a continental platform somewhat larger than the land above sea level so that varying widths of shallow sea extend over the continental shelf to the steep scarp of the submarine continental slopes (*see* CONTINENT).

At least three-fourths of the South American coastline is composed of highland masses; consequently, the areas of maximum elevation on the continent generally lie not far from the ocean. This topography has an important influence on the continental climate and causes drainage of streams to be toward the interior more often than is true of any other continent. Moreover, there is a paucity of large independent rivers, for they have little chance of developing when the coasts throughout the well-watered parts are guarded by mountains and plateaus that overlook the seas, forcing the bulk of the drainage to find its ultimate outlet at a small number of well-marked gaps. Thus the Andes, the Brazilian and Guiana Highlands, and to some extent the Patagonian Plateau have determined the present form and configuration of the entire South American land mass.

Regional characteristics. On the western margin of the continent are the Andes, the highest and longest north-south mountain system on earth. Altitudes often exceed 20,000 ft and perpetual snow caps many of the peaks even within the tropics (Fig. 1). So high are the Andes in the northern half of the continent that few passes lie below 12,000 ft elevation. Both active and quiescent volcanoes are to be found.

Over most of their length the Andes are not a single range but two or three ranges (Fig. 2). Between the two ranges in the center and north lie a vast series of intermontane basins and plateaus. Ecuador contains a string of 10 such basins, and Bolivia has the Altiplano. This plateaulike basin, almost entirely surrounded by rugged and lofty peaks, is about 400 miles long and up to 200 miles wide. It is cold, high—about 12,000 ft—and mostly level. It differs from most interior basins of the world in that it possesses a large, fresh-water body, Lake Titicaca.

On the northeastern and eastern continental periphery lie the Guiana and the Brazilian Highlands—vast areas of hilly uplands and low mountains separated from each other by the great valley of the Amazon (Fig. 3). Together these two plateaus form the core of South America. Rocks of ancient igneous and metamorphic origin found there are but partially buried by younger sedimentary beds.

Plains. Plains under 1000 ft in elevation comprise more than 65% of South America's total area. Some 40% of the continent is less than 650 ft above sea level. The plains lie between the lofty Andean backbone on the west and the Guiana and the Brazilian Highlands on the east, and between the Rio Orinoco on the north and the Rio Colorado on the south—that is, from about 8°N to 40°S. In-

Fig. 1. View of the Andes Mountains in Chile. The Andes, which parallel the Pacific Coast, constitute one of the earth's most majestic land features. Only the Himalayas exceed them in elevation and mass. (*United Press-International*)

Fig. 2. Physical map of South America. (*Drawn by E. Raisz*)

Fig. 3. A representative view of the hilly uplands and low mountain country of the Guiana Highlands. One of five waterfalls on the Canaima River, Gran Sabana in southern Venezuela. (*Venezuelan Information Service*)

cluded are the Llanos, the Amazon Plain (see Fig. 6), the Chaco, and the Pampa. Some are flat; others, undulating. Some, such as the Llanos and Chaco, are alternately flooded and baked (Fig. 4); only 10% of the enormous Amazon Basin and almost none of the Pampa are exposed to flood. The northern plains consist of recent marine sediments mantled with alluvium carried down from the Andes and deposited by rivers. The southern plain, the Pampa, is composed of unconsolidated beds of fine sand, clay, silt, and windblown loess.

Fig. 4. Portion of the Orinoco Llanos (plain) in the state of Apure during rainy season. Thousands of square miles become inundated. (*Consejo de Bienestar Rural*)

Coastal zone character. Northward from Cape Horn to a little south of 41°S latitude, the western coastal zone consists of a broad chain of islands where a mountainous strip subsided and the ocean invaded its valleys. This is one of the world's finest examples of a fiorded coast. Nowhere along the Pacific Coast is there a true coastal plain. South of Arica the aspect is particularly forbidding, the bold, precipitous coast being broken by only a few deep gashes of streams, the majority of which carry no water for years at a time. Between Arica and Caldera, there are no natural harbors and almost no protected anchorages. The land has emerged in comparatively recent geologic time and on the narrow, discontinuous wave-cut terraces are perched the several towns that serve as ports for the mining industry. Western Ecuador consists of a broken, hilly belt and a low plain, the latter 50–100 miles wide and composed of alluvium washed down from the mountains. Along Colombia's Pacific coast, any semblance of a plain is snuffed out by the westernmost of the four north-south mountain ranges, the Serrania de Baudo which, although the lowest of the country's ranges, has steep slopes and sharply crested ridges.

The Caribbean coast of Colombia is a lowland formed largely of alluvium deposited by the Magdalena and Cauca Rivers and backed by mountains on three sides. In Venezuela the Central Highlands rise abruptly from the Caribbean, but there is lowland around Lake Maracaibo, west of Puerto Cabello, and from the mouth of the Rio Tuy to the port of Guanta. The littoral of the Guianas is a low, swampy alluvial plain 10–30 miles wide, although as much as 60 miles wide along the larger rivers. This coastal plain is being built up by sediments carried by the mighty Amazon to the Atlantic and then deflected westward by the equatorial current and cast upon the shore by the trade winds.

There is no broad coastal plain south of the Amazon and east of the Brazilian Highland to afford the easy access to the interior characteristic of North America's coast. The rise from the coastal strip to the interior is quite gradual in northeastern Brazil; but southward, between Bahia and Rio Grande do Sul, the steep Serra do Mar is a formidable obstacle to transportation. Along coastal Uruguay there is a transition between the hilly uplands and plateaus of Brazil and the flat Pampa of Argentina, whereas coastal Argentina as far south as the Rio Colorado, in Patagonia, is an almost featureless plain. In Patagonia steep cliffs rise from the water's edge. Behind these cliffs lies a succession of dry, flat-topped plateaus, surmounted occasionally by hilly land composed of resistant crystalline rocks. Separating southern Patagonia from Tierra del Fuego is the Strait of Magellan, which is 350 miles long and 2–20 miles wide. Threading through numberless islands, the Strait is lined on each side with fiords and bleak mountains.

Mountains. Because of the vast extent of the Andes, a greater proportion of South America than of any other continent lies above 10,000 ft. The young, rugged, folded Andean peaks stand in sharp contrast to the old, worn-down mountains of the eastern highlands (Fig. 1). Although the Andes appear to be continuous, most geologists believe that they consist of several structural units joined more or less closely. They are a single range in the south, two ranges in Bolivia, and dominantly three ranges in the north. They parallel the Pacific Coast, thus forming the continent's backbone. Topographically, they are similar in most parts.

Fig. 5. Uplifted beaches on desert coast, east side of San Juan Bay, Departmento de Ica, Peru. Sea cliff in left foreground is about 100 ft high. Each horizontal line represents a former beach. (*Photograph by F. Atchley*)

Throughout the Andean mountain mass there are high plateaus and deep longitudinal valleys, especially in the middle and northern parts. Some of these valleys are the result of folding and trough faulting much deepened by river erosion. Thus, in Colombia, the Magdalena and Cauca Rivers have appropriated and enlarged structural depressions.

Except in Bolivia where they attain their maximum width of 400 miles, the Andes are seldom more than 200 miles wide. They do not equal the Himalayas in height, but they offer at least 30 peaks above 20,000 ft. The average height of the Andes is estimated at about 13,000 ft. However, it is only north of latitude 35°S that the mountains exceed 10,000 ft in elevation.

Although in general the Andes resulted from folding and faulting, volcanic formations are common in southern Colombia, Ecuador, central and southern Peru, and western Bolivia. Possibly nowhere else on earth is there a more spectacular series of volcanic peaks than in western Bolivia and on each side of the structural depression in Ecuador.

Because of the great north-south extent of the Andes, the processes of their erosion and denudation have varied. Southward from about 40°S, and especially in the far south, the Andes were heavily glaciated during the Ice Age, and an extensive area north of the Strait of Magellan carries a broad mantle of permanent ice. Glaciers still descend to the heads of many fiords on the Pacific Coast or into lakes on the eastern side of the mountains.

Stratification of climate and vegetation with altitude can be observed in the Andes of the tropics. At the eastern base, the zone containing the hot, humid lowland and the foothills up to 3000 ft is known as the *tierra caliente*. In the *tierra templada*, the zone from 3000–7000 ft, the annual range in temperature does not exceed 5°F. From 7000–10,000 ft is the *tierra fria*, the zone most favorable for European settlement. In Bolivia and Peru the zone from 10,000–13,000 ft, though occasionally to 15,000 or 16,000 ft, is known as the *puna*. Here the hot days vary sharply and abruptly from the cold nights. Above the *puna*, from timberline to snowline, is the *paramo*, a desolate region of broadleaf herbs and grasses.

Gapways between continental barriers. The broadest lowland routes to the interior are gapways between the three great barriers separating coastal from interior South America: the Andes cordilleran belt, the Guiana Highlands, and the Brazilian Highlands.

The breach between the Andes and the Guiana Highlands is some 300 miles wide. The western part, near the Andes, is composed of rolling and irregular plains. Local gapways averaging 1000 ft

Principal Andean peaks*

Aconcagua, Argentina, 22,835 ft
Ampato, Peru, 21,702 ft
Caca Aca, Bolivia, 20,329 ft
Cachi, Argentina, 21,326 ft
Chimborazo, Ecuador, 20,577 ft
Cincel, Bolivia, 20,102 ft
Condoriri, Bolivia, 20,043 ft
Coropuna, Peru, 22,802 ft
Cuzco (Ausangate), Peru, 20,187 ft
Del Acay, Argentina, 20,801 ft
Dos Conos, Argentina, 22,507 ft
Falso Azufre, Argentina-Chile, 22,277 ft
Huascaran, Peru, 22,188 ft
Illampu, Bolivia, 21,276 ft
Illimani, Bolivia, 21,282 ft
Incahuasu, Argentina-Chile, 21,720 ft
Llullaillaco, Argentina-Chile, 22,015 ft
Mercedario, Argentina-Chile, 21,884 ft
Ojos del Salado, Argentina-Chile, 22,573 ft
Payachata, Bolivia, 20,768 ft
Pissis, Argentina, 22,245 ft
Porongos, Argentina-Chile, 20,512 ft
Pular, Chile, 20,342 ft
Sajama, Bolivia, 21,390 ft
Sarmiento, Chile, 20,670 ft
Socompa, Argentina-Chile, 19,787 ft
Tocorpuri, Bolivia-Chile, 22,163 ft
Tortolas, de las, Chile, 20,018 ft
Tres Cruces, Chile, 21,720 ft
Turpungato, Chile, 21,490 ft
Valadero, Argentina, 20,735 ft

* The elevations are approximate and some are controversial. Data based mostly on tables in *Goode's World Atlas*, Rand McNally & Co., 10th ed., 1957.

or less in elevation extend through the hill and low-mountain country of the eastern section. The climate of the gap is warm to hot at all times. Rainforest vegetation covers the southern part, and tropical savanna and semideciduous forest predominate in sections of the northern region. The gap is undeveloped and little traveled.

The low-relief plain between the hilly margins of the Brazilian Highlands and the Guiana Highlands is 150–200 miles wide, through it courses the Amazon River. The climate is monotonously warm to hot. Vegetation is evergreen tropical rainforest, or selva, along rivers and in low sections; semideciduous forest and areas of savanna are found on interstream uplands and the bordering hillier country. The gap is little developed except for the rivers, which are navigable for small ocean steamships far into the interior.

In the southeastern part of the continent a breach 300 or more miles wide extends between the Brazilian Highlands and the Andes. Savanna and semideciduous forest predominate in the northern portion. To the south there is a change from tropical to humid subtropical climate, and scrub forest gives way near the coastal zone to prairies and the cultivated lands of the Argentine Pampa. West of the Pampa the climate grows drier; vegetation becomes "monte" scrub and scrub as the Andes are approached.

Major rivers and river systems. There are three great river systems in South America and a number of important rivers which are not a part of these systems. First of the river systems in size is the Amazon which, with its many tributaries, drains a basin covering 2,700,000 square miles, or about 40% of the continent. Next is the system composed of the Paraguay, Paraná, and La Plata Rivers, the last being a huge estuary. The third river complex, located in southern Venezuela, is the Orinoco, which drains water from 365,000 square miles of land, emptying into the Atlantic Ocean along the northeast edge of the continent. The major independent rivers are the Sao Francisco, arising on the Brazilian plateau, and the Magdalena-Cauca, which empties into the Caribbean.

Amazon River system. The Amazon River system begins in the Andean highlands and stretches across the north central part of the continent, draining the interior lowlands. The Amazon itself is joined by some 500 tributaries descending the Andes and the Brazilian and Guiana Highlands.

After carving their way from the Andean highlands, several headwaters join near Iquitos, Peru, to form the *Amazonas Solimoes* (upper Amazon segment). The gradient is gentle, about 0.2 in. per mile, or about 35 ft for the last 2000 miles to the sea. Consequently the upper segment is in many parts a maze of channels and islands. For such a broad course a surprisingly small portion (about 10%) of the basin is chronically flooded, and high banks or even bluffs adjoin most of the upper and lower course. Flooded and marshy land widens in the lower course between Manaus, where the large

Fig. 6. A northeasterly oblique view over the confluence of the Rio Maranon and Rio Ucayali in Peru, where the real Amazon is born. This scene lies in a portion of one of the world's largest plains. (*Servicio Aerofotográfico Nacional*)

Negro River joins, and the junction with the Tapajos; but the basin plain narrows to 150 miles north-south between the hilly margins of the Guiana and the Brazilian Highlands. Below the confluence with the Xingu the alluvial and deltaic bottom lands widen; and the river and its tributaries, now a densely interwoven pattern, discharge huge volumes of water and silt, staining the Atlantic 200 miles from shore.

Because the highwater flows of Amazon tributaries occur at compensating times, the maximum flood height on the Amazon is only about 20 ft. From February to April the highwater stage moves down the Amazon from Peruvian headwaters to Manaus, reaching Belem in June. Because of the river's low gradient, the Atlantic tide reaches upstream about 600 miles to Obidos, the rise moving with a pronounced bore or *pororóca* (*see* TIDAL BORE).

The Amazon has been important as a way in and out of the broad tropical parts of the continent. It is navigable for ocean-going vessels of about 7000 tons and 14-ft draft. Navigation is difficult because the river shifts courses so rapidly that a permanent chart has little value.

Among the larger Amazon tributaries is the Madeira River, which extends 1200 miles to the northeast through the interior Brazilian plains. It is formed by the joining of the Mamore, Beni, and Madre de Dios Rivers and discharges into the Amazon about 90 miles east of Manaus.

The Maranon River is an Amazon headwater which originates within 85 miles of the Pacific near Cerro de Pasco. The river travels through a 5000-ft-deep canyon during its long northward course in the Andes, then swings east and joins the Ucayali to form the upper Amazon near Iquitos. The Maranon is 1000 miles long.

The Rio Negro is a major left-bank tributary of the Amazon. From its source in eastern Colombia to its junction with the Casiquiare on the Colombia-Venezuela border, the river is known as the Guainia. The Negro is 1400 miles long and attains a width of 20 miles above Manaus, but it is only 1½ miles wide at the mouth. Unlike the brown, muddy Amazon, the Rio Negro is inky in color. It is classed as a black-water river.

Paraguay–Paraná–La Plata system. From its headwaters in southwestern Brazil the Paraguay River courses southward 1300 miles, discharging into the Paraná at the southern edge of Paraguay. The western bank of the Paraguay is low, and during the rainy season the western side of the entire basin, thousands of square miles of low-lying country, becomes inundated.

The Paraná ranks among the world's major rivers. It is longer and carries more water than the Mississippi River in the United States. Its source in the Brazilian Highlands is 2450 miles from its outlet in the Rio de La Plata. The Paraná has cut a deep canyon in the flat-topped plateau of its upper course. As the river drops over the edge of the lava formation, famous waterfalls are created—La Guayra and Iguazu, the latter on the Iguazu, a tributary of the Paraná. However, from Corrientes, where the water of the Paraguay is added, the gradient is gentle. Since the Paraná rises in areas with characteristically heavy summer rainfall, the water volume fluctuates widely, the variation from high to low water reaching 15 ft in the lower course. At the mouth of the Paraná is an enormous delta, shared with the emptying Uruguay River. The delta consists of numerous low, flat islands which are submerged for weeks at a time.

From the head of the estuary of La Plata to the open Atlantic is some 200 miles—the largest indentation on the east coast of South America. At Buenos Aires the estuary is about 25 miles wide, at Montevideo 60 miles, and at its mouth 138 miles. The Paraná and the Uruguay transport huge quantities of silt into La Plata, some of which settles in navigation channels, necessitating costly dredging.

Navigation for vessels of 6- to 7-ft draft is possible as far as Corumbá, Brazil, some 1800 miles upstream from Buenos Aires via the Paraná and Paraguay Rivers. The remainder of the Paraná is little-used for navigation and its hydroelectric potential is as yet undeveloped because markets are too distant. The Paraguay, however, serves the interior and is considered the national life-line to the outside world for Paraguay.

Orinoco River system. The Orinoco of Venezuela, with its many tributaries, comprises the third largest river system in South America. The headwaters lie in southern Venezuela near the Brazilian border. Near Esmeraldas, in southern Venezuela, is a stretch of about 220 miles known as the Casiquiare Canal, joining the Orinoco with the Negro of the Amazon system.

The Orinoco is approximately 1600 miles long and is noted for its variability in volume from wet to dry season. It is a broad stream, being a mile wide even at Ciudad Bolivar—the "narrows." So heavy is the rainfall from June to October that the master stream and its tributaries are unable to handle all the water, and enormous areas become inundated. The river rises 39 ft at the Puerto Ordaz. From January to March the dry season holds sway. Not a drop of rain falls. The waters recede and only the larger rivers flow freely. The smaller ones are gradually converted into chains of pools and swamps along the valley bottoms.

Where the Orinoco flows into the Atlantic it forms a low, wide delta consisting of small islands and swamps and an intricate maze of distributaries.

Sao Francisco River. The Sao Francisco, regarded as South America's fourth largest river, is independent of the other river systems. Its source is in the hills of Minas Gerais, Brazil, and from there to its outlet near Maceio it is some 1800 miles. Together with a few tributaries the Sao Francisco drains a basin area of 300,000 square miles. After leaving the deep gorges the river has cut through the low mountains of southern Minas Gerais, it flows for a great distance between the ranges paralleling the coast. Here atop the surface of the plateau it appears to flow within a broad shallow valley bottom, or local peneplain, some 15–20 miles wide and 1200–1500 ft above sea level. After it turns in an eastward direction, the river passes through channels between granite canyon walls and plunges over an enormous cataract, the Paulo Afonso Falls, about 150 miles from the sea.

Above the falls the river is navigable almost to its source. The Sao Francisco, with a 72-mile section of railroad to circumvent the falls, rapids, and gorges, provides the chief access to a large part of eastern Brazil.

Magdalena-Cauca River. The Magdalena and Cauca Rivers arise in the cordilleran Andes about 1°N of the Equator and join some 600 miles farther north. About 200 miles beyond, the river empties its silt-choked waters into the Caribbean near Baranquilla.

In its upper course the Magdalena flows between the Oriental and Central cordilleras. Its middle course occupies a trough fault. Some geologists believe that the valley floor subsided in recent times, accounting for the widespread swamp areas. The river is interrupted by rapids from La Dorado to Hondo.

Over approximately 250 miles of its upper course the Cauca flows through a wide rift valley and over numerous rapids, then through a series of gorges cut into the rugged terrain where the two cordilleras are joined. Below this segment the Cauca, as well as the Magdalena, passes through poorly drained country with many shallow lakes, or overflow basins, along the river banks.

The Magdalena is of first importance to Colombia, for it provides the only surface link between Caribbean ports and the mountain-girt cities of the interior.

South American deserts. South America is unique in having a west-coast desert that extends almost to the Equator, an east-coast desert poleward from latitude 40° (the Patagonian), and a desert that probably receives less rain than any on the earth (the Atacama).

The Atacama-Peruvian Desert dominates the Pacific Coast for nearly 2000 miles, lying between the ocean and the higher slopes of the Andean ranges. In many places, particularly in Chile, the land rises abruptly from the sea to an elevation of 2000–3000 ft (Fig. 5). The driest stretch, one of the driest places in the world, is between Arica and Caldera. Above the seaward scarp is a low range of hills and a desert trough, 30–50 miles wide, from which nitrate deposits are extracted. Beyond are the Andes, whose western slopes are seamed by numerous dry ravines. In this 600-mile stretch the Rio Loa is the only stream flowing from the Andes toward the Pacific that reaches its goal.

The persistence of this desert through latitudes at which deserts are not found elsewhere is mainly the result of the presence of the northward-moving

Fig. 7. Port of Matarani on desert coast of southern Peru. This oblique aerial view looks coastwise southward over one of the world's driest and most barren coastal zones. (*Servicio Aerofotográfico Nacional*)

Humboldt Current. Cold water rises between the current and the shore, and prevailing winds blow the resulting cold air onto the warmer land with a drying effect.

The dryland climate of the Atacama-Peruvian Desert crosses the Andes (it is characteristic of the Bolivian Altiplano) and reappears in western Argentina. Here, between approximately 30 and 40°S, is one of Argentina's two deserts. It lies in a long band between mountain slopes and more humid land to the east. As it is primarily a region of interior drainage, there are many swamps, marshes, and salt flats. Although precipitation is highly variable, at least one-third of the region receives less than 10 in. of rainfall yearly.

South America's east-coast desert, the Patagonian, is found between the Rio Colorado and the Strait of Magellan. The desert is partly located in the rain shadow of the Andes, and the meager annual rainfall is ordinarily 5–6 in. Also contributing to the region's aridity and its extent to the Atlantic shores are the cold waters of the Falkland Current, moving northward off the eastern coast, and the ceaseless winds, which hasten evaporation. Wind velocities sometimes exceed 70 mph. What precipitation there is occurs mostly in winter. Winter temperatures, which are very cold, are the result of high latitude and high elevation.

Islands. Three major islands or island groups lie off the South American coast: Trinidad, in the north, contains a range of mountains exceeding at points 3000 ft elevation. The range is geologically an extension of Venezuela's Cumaná mountains. Also on the island is the largest asphalt lake known. *See* West Indies.

The Falklands stand on the shallow continental shelf off the rocky coast of Patagonia, near the southern tip of the continent. In a few places elevations exceed 2000 ft above sea level. The islands have a cool marine climate.

On the Equator and about 550 miles west of Ecuador are the Galapagos, a group of 13 volcanic islands. Those in the northwest have steep slopes, many of bare lava with well-preserved craters to attest to their extreme youth. The southeastern islands are older, with rounded heights, gentle slopes, and craters much altered by erosion. The climate of the Galapagos is arid. [C.L.W.]

Bibliography: F. A. Carlson, *Geography of Latin America*, 3d ed., 1951; P. E. James, *Latin America*, rev. ed., 1950.

South Pole

That end of the earth's rotational axis opposite the North Pole. It is the southernmost point on the earth and one of the two points through which all meridians pass (the North Pole being the other point). This is the geographic pole and not the same as the south magnetic pole, which is near the coast of Antarctica, south of Tasmania. The South Pole lies inland from the Ross Sea, within the land mass of Antarctica, at an elevation of about 9200 ft.

As the earth's axis is inclined 23½° from the vertical to the plane of the ecliptic, the South Pole experiences six months when the sun is continuously above the horizon, reaching a maximum altitude of 23½°. It also has six months when the sun remains continuously below the horizon, from March 22 to September 22. During the latter period, however, there is an interval of seven weeks at the beginning and at the end when twilight conditions preclude total darkness. Twilight prevails when the sun is between 0° and 18° below the horizon.

There is no natural way to determine local sun time because there is no noon position of the sun, and shadows point north at all times, there being no other direction from the South Pole. *See* GEOGRAPHY, MATHEMATICAL; NORTH POLE. [V.H.E.]

Southeast Asian waters

All the seas between Asia and Australia and the Pacific and the Indian Ocean. They form a geographical and oceanographical unit because of their special structure and position, and comprise an area of 8,940,000 km², or about 2.5% of the surface of all oceans.

Ocean floor. The ocean floor consists of two large shelves, and a number of deep-sea depressions. The Sunda Shelf is covered with littoral sediments and the submerged valleys of two large river systems are found on it. The Arafura Shelf connects New Guinea with Australia. The China Basin, whose maximal depth is 5016 meters (m), is connected with the Pacific Ocean over a sill about 2000 m in depth. The Sulu Basin (maximal depth 5580 m) has the highest sill (420 m). The Celebes Basin (maximal depth 6220 m) is connected with the Pacific Ocean over three sills of about 1400 m each. The Banda Basin is subdivided into several smaller basins, including the Weber Deep of 7440 m; its sill depth is 1880 m. In the Celebes, Banda, and Flores Basins volcanic ashes are found.

Surface circulation. The surface circulation is completely reversed twice a year (Figs. 1 and 2) by the changing monsoon winds.

Monsoon current. During the north monsoon, which is fully developed in February, the monsoon current is formed in the northern China Sea, flows along the coast of Vietnam into the Java Sea, and into the Flores Sea. Parts of its water masses return into the Pacific Ocean; parts sink in the Banda Sea or flow into the Indian Ocean.

During the south monsoon in August, the monsoon current is formed by water flowing into the Java Sea between the Molucca Islands and through

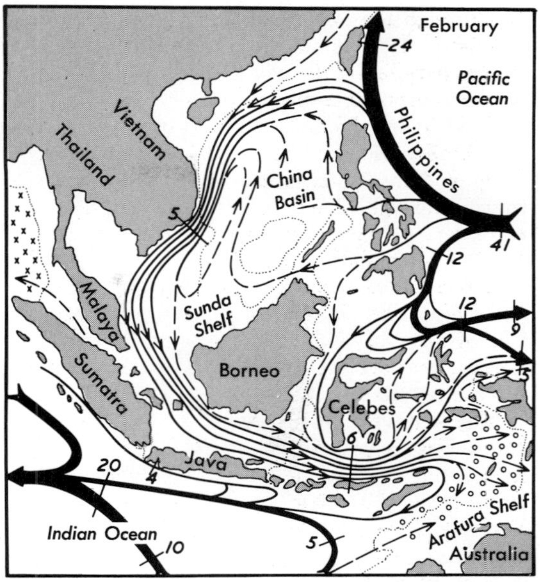

transports in million m³ sec — upwelling of water x x
sinking of water o o — 200 meters depth ……

Fig. 1. Water circulation in South China and Indonesian Seas, February.

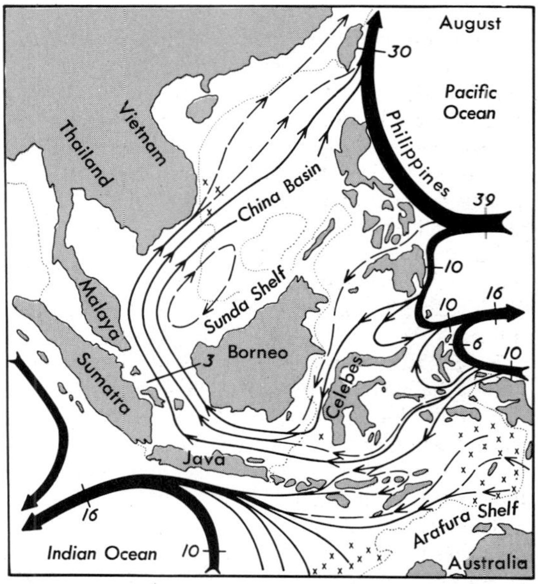

transports in million m³ sec — upwelling of water x x
200 meters depth ……

Fig. 2. Water circulation in South China and Indonesian Seas, August.

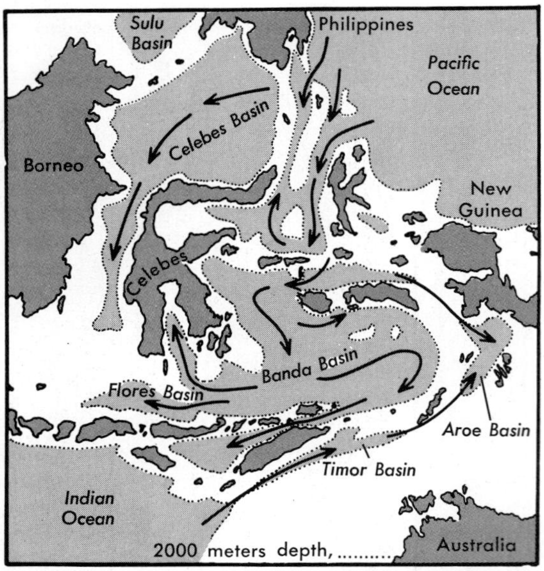

Fig. 3. Flow of bottom waters in the eastern part of the Indonesian Archipelago.

the Macassar Strait from the Pacific Ocean. It flows through the Java and the South China Seas and returns north of the Philippines into the Pacific Ocean. The water masses originating from the upwelling region in the Banda Sea flow chiefly into the Indian Ocean.

The transports of the monsoon current are 3,000,000 m^3/sec in August and 5,000,000 m^3/sec in February, but are small compared with those in the adjoining parts of the Pacific and Indian Oceans (Figs. 1 and 2).

Salinity and temperature. The monsoons cause a pronounced rainy and dry season over most parts of the region and consequently strong annual variations of the surface salinity, which normally increases during the dry season (north monsoon), when evaporation prevails, and decreases during the rainy season (south monsoon). Regions of permanent low salinities are in the Malacca Straits, the Gulf of Thailand, and the waters between Borneo and Sumatra, because of the discharge of the large rivers. The surface temperature is always high, between 26 and 29°C, and drops only in the northern parts of the South China Sea and in the Arafura Sea to about 24°C during the dry season.

Subsurface circulation. The subsurface circulation carries chiefly the outrunners of the intermediate waters of the Pacific Ocean into these seas. These waters are identified by salinity minima at depths of 300 and 1000 m. The general subsurface flow is from the Pacific to the Indian Ocean. The deep basins are supplied with Pacific Ocean deep water, that enters over the various sills (Fig. 3). Only the Timor and Aroe Basins are filled with Indian Ocean deep water.

Tides. The tides are mostly of the mixed type. Diurnal tides are found in the Java Sea, in the Gulf of Tonking, and in the Gulf of Thailand. Semidiurnal tides with high amplitudes occur in the Malacca Straits. Strong tidal currents occur in many of the small shallow passages. *See* INDIAN OCEAN; PACIFIC OCEAN. [K.K.W.]

Bibliography: *Oceanographic and meteorological observations in the China Seas and in the Western Part of the North Pacific Ocean*, Koninkl. Ned. Meteor. Inst. no. 115, 1935; G. Schott, *Geographie des Indischem und Stillen Ozeans*, 1935; *The Snellius Expedition 1929–30*, 6 vols., 1933–1938; K. Wyrtki, *Physical Oceanography of the Southeast Asian Waters*, 1960.

Sow bug

Any of several species of the order Isopoda, class Crustacea, phylum Arthropoda. Broadly speaking, all of the Isopoda may be called sow bugs, or pill bugs. More commonly the term sow bug is reserved for the terrestrial species. There are several terrestrial forms, all quite similar. There are a few fresh-water Isopoda and a large number of marine species.

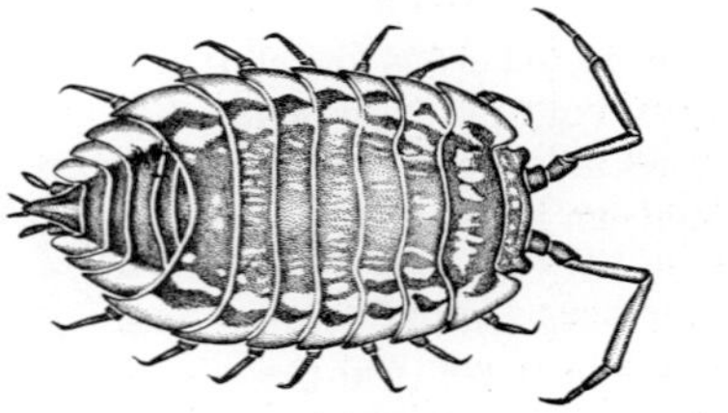

The sow bug, *Oniscus asellus;* length to 3/5 in. (*From E. L. Palmer, Fieldbook of Natural History, McGraw-Hill, 1949*)

Most of the sow bugs are of no great importance. They act as scavengers and thus aid in the breaking down of dead organic matter. They serve in a limited way as food for other animals. Some aquatic isopods eat living plants and animals. Several marine forms are parasitic on fishes and decapods, and one form is destructive to wood.

The Isopoda are flattened dorsoventrally; the head is a cephalothorax representing the true head and first thoracic segment; the remaining seven segments are alike and are expanded laterally into platelike extensions. The abdominal segments are short and fused. The legs of the terrestrial forms are short. The Isopoda are worldwide in their distribution.

There are two American terrestrial families, the Procellionidae, or sow bugs, which do not roll up in a ball; and the aptly named Armadillidiidae, or pill bugs, which roll into a ball in true armadillo fashion when disturbed. The former are flattened, with antennae about half the length of the body. The latter are convex, with the antennae rarely equal to more than one-third of the body length. Most land isopods are gray in color.

Most common of the American fresh-water Isopoda is the genus *Asellus*, with longer legs and antennae than the land forms; their color may be yellow, gray, blackish, reddish, or brown. Some white forms live in caves and subterranean waters. Their eyes are greatly reduced in size or entirely lacking.

Interesting marine forms include the salve bug, *Aega psora*, used as a salve by fishermen. It is parasitic on the cod, halibut, and other fishes from Europe across the North Atlantic and southward into the Gulf of Mexico.

The gribble, *Limnoria lignorum*, is another common marine form found on both sides of the Atlantic and on the Pacific coast of the United States. It burrows into submerged timbers and causes great damage to docks and other shoreline installations.

The sexes of the Isopoda are separate. Eggs are carried on the ventral surface of the female's thorax in a brood pouch. They hatch into forms resembling the adults, there being no metamorphosis. Their general anatomy is essentially like that of the crayfish. *See* Isopoda. [J.D.B.]

Soybean

This crop plant (*Glycine max*), an annual legume native to China and Manchuria, is now widely grown throughout the world as a source of oil and protein for human and animal consumption and for industrial usage. Originally brought to the United States in 1804 as ballast on a ship coming from China, soybeans were first grown as a curiosity, and then as a forage and soil building crop. Today soybeans are grown extensively in the Midwest and Midsouth areas and are the major source of vegetable oil in the United States. Illinois leads the nation with a production in 1958 of over 140,000,000 bushels (60 lb each), followed by Iowa, Minnesota, Indiana, Missouri, Arkansas, and Ohio. Basically the soybean production area of the United States corresponds to that of corn, the two crops requiring about the same soil, moisture, and climatic conditions. Machinery for seed-bed preparation, planting, and cultivation of the two crops is interchangeable (Fig. 1).

Economic importance. The United States is now the world's largest producer of soybeans. Of the 480,000,000 bushels (bu) produced in the United States in 1957, about 360,000,000 bu were crushed or processed domestically, 90,000,000 bu were exported, and 30,000,000 bu were used for seed, for livestock feed, and for other farm purposes. Soybeans rank second to corn in importance as a cash crop throughout the production area, bringing returns of over $1,000,000,000 to Midwest and Midsouth farmers. Since 1946 soybeans have grown in importance more rapidly than any other crop in this area.

In China, Manchuria, Korea, and Japan, soybeans are the major source of protein and fat for human consumption. A minor portion of the crop grown in the Orient is crushed and the extracted oil is used as a cooking or salad oil, the cake or flakes (residue) going into the production of food products. However, the major portion of the soybean production in these areas is used directly as food in the

Fig. 1. Cultivating a field of growing soybeans.

form of miso, tofu, shoyu, natto, kinako, and similar products. Some soybeans are also used directly for human consumption in the green or mature stages, but in most cases the protein is first prepared by a process of fermentation or water extraction. Soybeans are known as "meat of the fields" throughout the Orient where they take the place in the human diet which is commonly filled by milk, meat, and eggs in western civilizations. One of the most abundant of the vegetable proteins, soybean protein is also one of the best, its amino acids having a balance closely approaching that required for human nutrition. It is especially valuable as a supplement to wheat flour, being high in lysine, one of the amino acids that is deficient in wheat protein (*see* Wheat). Soy protein contains 16.6 times as much lysine as wheat flour. Furthermore, soy protein has 8.1 times as much isoleucine and leucine as wheat flour. Also, soy protein contains 7.1 times as much arginine, 5.3 times as much histidine, 8.5 times as much phenylalanine, 4.5 times as much tryptophan, and 12.9 times as much valine.

Several thousand varieties and selections of soybeans have been introduced into the United States from other areas of the world, and many of them were once grown commercially. Current production, however, is almost entirely devoted to varieties produced in the breeding program conducted by the United States Department of Agriculture and the cooperating states. Regional headquarters for these soybean breeding programs are maintained at Urbana, Illinois, and Stoneville, Mississippi.

Research on soybean utilization is centered at the Northern Utilization and Research Branch of the Agricultural Research Service in Peoria, Illinois. Extensive research programs are also being conducted by the companies engaged in soybean proc-

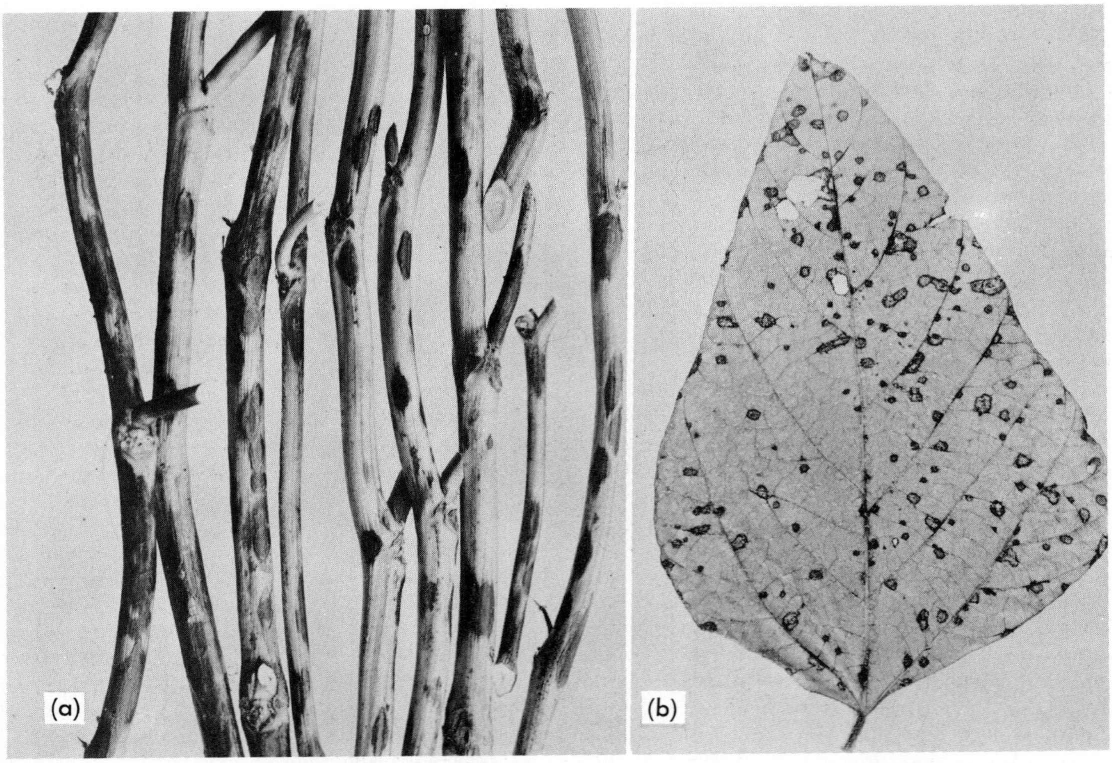

Fig. 2. Frogeye disease of soybeans. (*a*) On stems. (*b*) On leaf.

essing, in the refining of oil, and in the utilization of oil and protein in manufacturing processes.

Since 1946, the United States has become the world's largest producer and exporter of soybeans and their products. Japan is the largest customer for whole soybeans, but northern European countries also take heavy shipments. The largest export market for soybean oil is the Mediterranean area, where it is used as replacement for olive oil.

The United States now stands first in soybean production, with China and Manchuria in second place, followed by Brazil, Japan, Korea, Malaya, Indonesia, and the Union of South Africa.

Characteristics of the plant. Soybean seeds differ by varieties grown, and in relation to the weather conditions and the geographic region in which they are produced. Size varies from 1200 to 9000 seeds/lb, most species in current use producing between 1500–4000 seeds/lb. Yellow is the predominant color of the soybeans grown for crushing purposes, but varieties having green, black, brown, and varicolored seedcoats are also produced. Among the yellow-seed varieties, the color of the hilum (seed scar showing former point of attachment to pod) ranges from black to brown, gray, and colorless. *See* SEED (BOTANY). Oil content of the commercially grown varieties in the United States ranges from 15% to as high as 22%, and the protein content varies between 30 and 40%.

The soybean plant is usually erect and produces one, two, or three beans per pod. Flowers are self-fertilized and are white, pink, purple, and intermediate in color. *See* FLOWER (BOTANY). Pubescence (hairs) on the stems, leaves, and pods differs with variety, ranging from dark brown through varying degrees of grays (*see* EPIDERMIS, PLANT). Leaves are normally dark green.

Although the soybean plant is a legume and thus supports nitrogen fixation bacteria, the strains required are not native to soils outside the Orient and must be introduced to new areas by soil introductions or by bacterial cultures. Commercial inoculation cultures are available in most countries. *See* RHIZOSPHERE.

Measured from planting date to maturity, the growing season of different varieties of soybeans varies from about 70 days in the north to 180 days in the more tropical areas. Harvesting in the United States is normally done with a combine or harvester-thresher when the plant is mature, after the leaves have yellowed and dropped from the plant, and when seed moisture is below the 13% required for safe storage. *See* AGRICULTURAL MACHINERY.

Photoperiodism governs maturity of the soybean crop (*see* PHOTOPERIODISM IN PLANTS). The same variety may be planted on successive dates with all lots maturing at approximately the same time. Yield, however, will vary, depending on the growing season of the variety concerned. [G.M.ST.]

SOYBEAN DISEASES

Two widely prevalent bacterial diseases of soybeans are blight caused by *Pseudomonas glycinea* and pustule caused by *Xanthomonas phaseoli.* These are most conspicuous on the leaves where they result in dead areas. The bacteria causing both of these diseases are seed-borne and live through the winter in the dead, fallen leaves. *See* BACTERIA.

The more common leaf-spotting diseases are frogeye, caused by the parasitic fungus *Cercospora sojina;* brown spot caused by *Septoria glycines;* target spot by *Corynespora cassiicola;* and downy mildew by *Peronospora manshurica.* Heavily spotted leaves fall prematurely (Fig. 2). Pod and seed infections occur late in the growing season, and the parasites live from one season to the next in or on infected seed. *See* FUNGI.

Brown stem rot caused by *Cephalosporium gregatum* and stem canker by *Diaporthe phaseolorum* var. *caulivora* are fungus diseases that attack the stems and kill the plants prematurely. They are more common in the Midwest than in other regions of the United States. The brown stem rot fungus is soil-borne. The stem canker fungus is seed-borne and lives through the winter in diseased stems.

Fungus diseases that involve the roots and basal portion of the stem of the soybean plant include charcoal rot caused by *Macrophomina phaseoli,* sclerotial blight by *Sclerotium rolfsii,* stem rot by *Sclerotinia sclerotiorum,* Fusarium blight by *Fusarium oxysporum* f. *tracheiphilum* (Fig. 3), Pythium root rot by *Pythium* spp., Rhizoctonia root rot by *Rhizoctonia solani,* cotton root rot by *Phymatotrichum omnivorum,* and Phytophthora root rot by *Phytophthora* spp. These fungi all live from year to year in the soil.

Purple stain caused by *Cercospora kikuchii* is a discoloration of soybeans that is objectionable from the standpoint of the producer of pure seed. The fungus attacks the leaves, stems, and pods as well as the seeds.

Three virus diseases attack soybeans. These are soybean mosaic (*Soja* virus 1), yellow bean mosaic (*Phaseolus* virus 2), and bud blight (tobacco ring spot virus).

Fig. 3. Fusarium blight of soybeans. Two dead plants in the foreground. (*North Carolina Agricultural Experiment Station*)

Root knot caused by the nematode *Meloidogyne* spp. and yellow dwarf by *Heterodera glycines* both occur in the South and are most injurious on sandy soils. *Meloidogyne* produces knotlike swellings or galls on the roots, whereas *Heterodera* is identified by brown, egg-filled cysts on the roots.

Control of soybean diseases is obtained primarily by the use of disease-resistant varieties. Soybean varieties resistant to bacterial blight, bacterial pustule, frogeye, target spot, downy mildew, stem canker, purple seed stain, and root knot are known. Breeding for disease resistance continues. Crop rotation and sanitary measures also aid in controlling soybean diseases. Seed treatment has been recommended in some areas to guard against seed decay and damping-off of seedlings. [H.W.J.]

SOYBEAN PROCESSING

Soybean processing or milling is the source of a large number of food and industrial products which may be grouped in three classifications. Two are most important in volume. Over 70% of the soybeans raised in the United States are used in production of soybean oil meals, primarily for livestock feeds, and soybean oils for food and industrial uses. A third classification might encompass a long list of end products generally resulting from further refinements of soybean meals and oils—products of wide diversity in their nature and uses, but of comparatively limited volume individually.

There are more than 100 soybean mills in the United States. Production methods vary from plant to plant, but only two basic processing methods are in general use. One is the solvent extraction process, the other a continuous screw press process. A third process, using hydraulic presses, was little used for soybeans and may now be considered obsolete.

Solvent extraction process. The solvent extraction process uses hexane, a petroleum solvent, to leach oil from prepared soybean flakes. Almost all U.S. plants employing this process are of the continuous flow, rather than batch operating, type. Equipment designs vary considerably, but the process is similar.

Raw soybeans are first cleaned to remove foreign material, then cracked and heat treated to adjust temperature and moisture. The particles are then rolled into flakes, from which oil is removed by solvent extraction. The flakes are then desolventized.

If the flakes are to be used for livestock feed, the moisture is adjusted and the flakes heat treated or toasted to increase their feeding value. Toasted flakes are ground to size and marketed as soybean meal.

The solution of solvent and oil, called miscella, is distilled to remove the solvent. The solvent is reused, and the oil is sold as crude soybean oil or subjected to further refining or partial refining.

Solvent extraction gives the highest yield of oil, averaging about 11.1 lb of oil per bushel of soy-

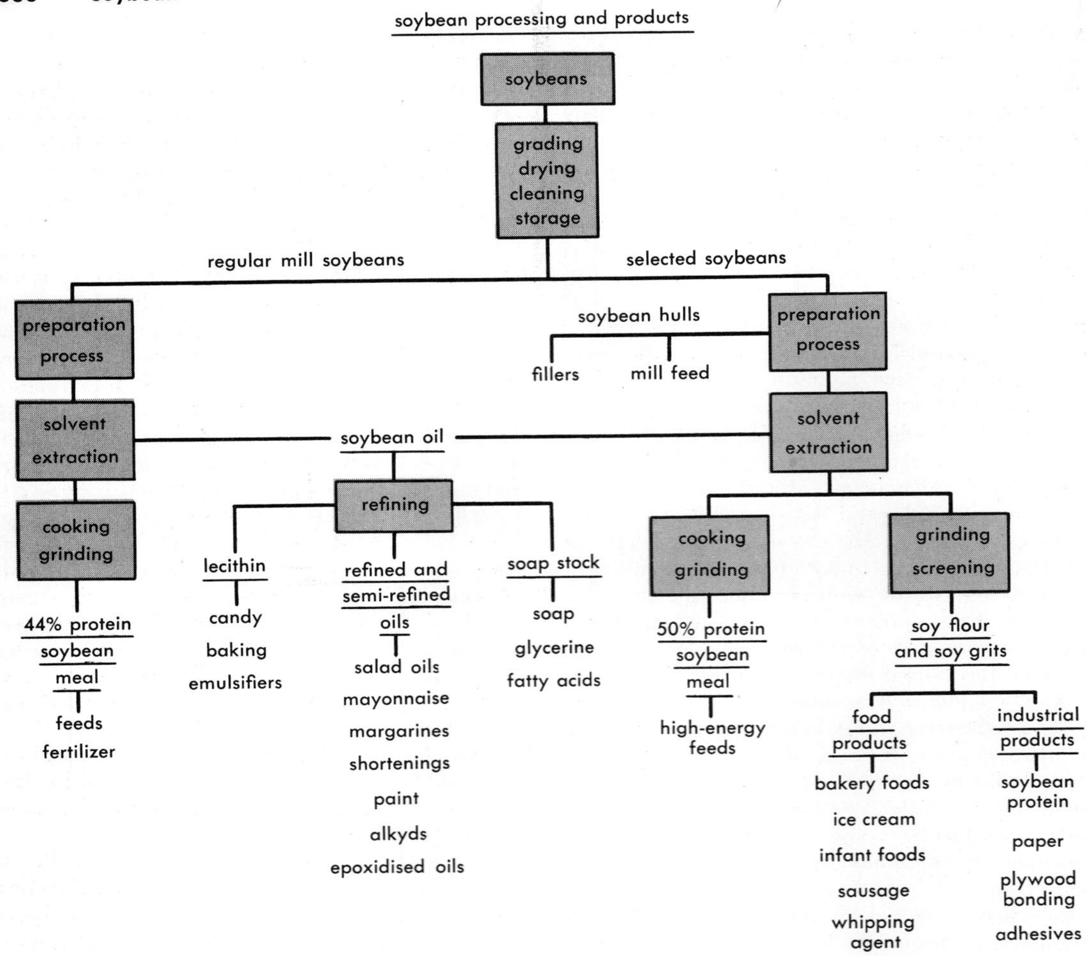

Fig. 4. Soybean processing and products.

beans, or about 2 lb more than the continuous screw press.

Soybean oil meal is a valuable ingredient in many livestock feed formulas, and it supplies a large part of all the supplementary protein used in livestock feeding in this country. Its crude protein content is usually about 44%. A 50% protein oil meal is produced by a modification in the milling process to remove hulls immediately after the cracking step. The hulls are also sold for cattle feed. *See* PROTEIN.

Continuous screw press process. Oil is squeezed out of cracked, dried soybeans in the continuous screw method. The resulting caked residue is milled to size for sale as soybean oil meal. Because the meal reaches a high temperature in passing through the screw press, no subsequent heat treatment or toasting to improve digestibility is necessary. Because it contains more oil, this meal is lower in protein content, containing about 41%.

The screw press process was almost universally used in the United States before World War II. Now the higher yield solvent extraction method has substantially replaced it.

Soybean milling is carried on in other countries with essentially the same methods. Solvent extraction was used earlier and spread more rapidly in Europe, where yield is of even more economic importance than in the United States.

Soy flour. Soy flour is a broad term for a number of soybean products, usually processed and milled flour-fine for specific food or industrial purposes. Oil content, flavor, color, urease content, water solubility of protein, fineness of grind and other variables are adjusted in processing to meet specific requirements.

These products are classified in three groups, according to fat content. The three groups—high or full fat, low fat, and extracted or defatted soy flour—are obtained by varying production methods.

High fat, or full fat, soy flour is produced by steaming or otherwise debittering selected raw beans, followed by drying, cracking, and careful removal of hulls by screen and air separation; then milling the debittered, cleaned bean meats to flour or whatever granulation is desired. These soy flours range from 18 to 20% fat content, and are principally used in foods.

Low fat soy flours are produced by debittering selected beans, carefully removing the hulls, and removing part of the oil, usually by means of continuous screw presses. The resulting cake is milled to required fineness. Fat content of these soy flours varies considerably, generally within the range of 4.5–9%.

Extracted, or defatted, soy flours are produced by essentially the same process as solvent extracted soybean oil meal, but with thorough removal of hulls. Heat treatment and steaming incident to extraction make preliminary debittering of the raw beans unnecessary and also produce some desirable variations in other properties of the product.

Extracted soy flours may be made into either high or low fat flours by adding soybean oil or other vegetable oil to the finished product. Substantial amounts of high and low fat soy flours are made in this way. Total soy flour production in the United States utilizes 1–2% of the soybeans grown.

Further modifications of these basic processes, sometimes by additional steps, yield other soy derivatives of economic importance, such as isolated protein, modified proteins, and special adhesives.

An even wider variety of soy derivatives has been produced in other countries, particularly in the Orient. Examples are substitute milk products, soybean curds, and soy sauce. All are to be found in U.S. production, but usually not in large volume. Some are produced by other than strictly milling processes.

Soybean oil. Soybean oil finds a ready market for both food and industrial uses in refined and semirefined forms.

Degummed or nonbreak oil is made by the addition of a small percentage of water to crude soybean oil and centrifugally separating the resulting precipitated gums. Degummed oil may either be sold for industrial uses or further refined for other uses. The separated gums are a source of lecithin.

Customary alkali refining used for other vegetable oils can be applied to either crude soybean oil or degummed oil. In this refining process, the oil is treated with enough sodium hydroxide to remove the free fatty acids, which in turn are precipitated and centrifugally separated from the oil in the form of soapstock. As the name suggests, this product finds an outlet in the soap industry or free fatty acid industry.

Subsequent bleaching and steam deodorization processes yield edible soybean oils. Large quantities of these are consumed in production of shortening, margarine, and salad dressings in the United States. Industrial uses of both refined and degummed oil include paints, blown and bodied oils, alkyd resins, epoxidized oils, and others. *See* FAT AND OIL, EDIBLE; MARGARINE.

Refining may also employ other chemicals, some of them acidic. For example, acetic anhydride and water are used in one of the most recent processes developed in the United States, which yields crude commercial lecithin and a refined oil. This process eliminates the need for alkali refining and can produce edible soybean oil with only the customary bleaching and deodorization treatment.

Refined soybean oil, after prolonged storage, especially under adverse conditions, may undergo flavor reversion, that is, develop an off-flavor described as "painty" or "fishy." This is seemingly not the same as rancidity which develops in all vegetable oils. There are several theories as to cause, such as unstable impurities not removed by refining, changes in some of the unsaturated acids in this particular oil, and others. The following procedures help to hold reversion in check: (1) careful refining and deodorizing, especially at higher than normal vacuums, (2) use of equipment made from noncorrosive metals such as nickel or stainless steel, (3) careful avoidance of contact with copper or copper-bearing metals, (4) uses of traces of metal sequestrants such as sorbitol, phytic acid, and other nontoxic sequestrants in the deodorizing step of the refining. Traces of dissolved metals promote onset of this reversion reaction. Hydrogenation, especially with properly selected catalysts, will reduce or eliminate reversion. Edible soybean oil finds its best acceptance in outlets where final consumption is not too long delayed, or in secondary products that themselves do not have a long shelf life such as margarine, certain salad dressings, and products which have pronounced masking flavors such as sardines or other fish products. Research is active on this problem in several laboratories.

Lecithin (soy). Crude, commercial soy lecithin is a mixture of phosphatides and oil obtained by drying the separated gums from the water washing and some of the nonalkaline refining methods.

This material consists mainly of phosphatides, not only of the recognized pure chemical called lecithin, but also cephalin, other fatlike phosphorous-containing compounds, and 30–35% unseparated soybean oil. *See* PHOSPHATIDE.

Crude lecithin may be treated with oxidizing agents, such as hydrogen peroxide, to produce the bleached grades of commercial lecithin. Fluid grades are produced by increasing the free fatty acid content of the mixture.

The gums may be further purified by washing with selected solvents, such as acetone, to remove soybean oil and other materials, leaving a product known as refined lecithin, pharmaceutical grade.

Lecithin is a natural bridge between water and oil, and is widely used in foods for its emulsifying and antioxidant properties, especially in the preparation of chocolate, margarine, and shortening compounds. It is also used industrially for its emulsifying and surfactant properties. *See* FOOD ENGINEERING. [R.E.G.]

Bibliography: K. S. Markley (ed.), *Soybeans and Soybean Products*, 2 vols., 1950–1951; American Soybean Association, *Soybean Blue Book*, 1958.

Space

Physically, space is that property of the universe associated with extension in three mutually perpendicular directions. Space may contain matter, but space exists apart from matter. Through usage, the term "space" has come to mean generally outer space or the region beyond Earth. Geophysically, space is that portion of the universe beyond the immediate influence of Earth and its atmosphere. From the point of view of flight, space is that region in which a vehicle cannot obtain oxygen for its engines or rely upon an atmospheric gas for support (either by buoyancy or by aerodynamic effects). Astronomically, space is a part of the space-time continuum by which all events are uniquely located (*see* SPACE-TIME). Biologically, one characteristic of space, in the flight sense, is that positional acceleration is comparable to gravitational acceleration.

Perceptually, space is sensed indirectly by the objects and events within it. Thus, a survey of space is more a survey of its contents. *See* GEOMETRY, EUCLIDEAN.

The space around the Earth and extending out perhaps 10 Earth radii (40,000 miles) has properties which differ from those of interplanetary space. The chief reasons are the existence of the gravitational field and of the magnetic field of Earth. These fields affect the densities and motions of neutral and charged particles in the vicinity of Earth. *See* AERONOMY; VAN ALLEN RADIATION.

The gravitational field allows Earth to retain its atmosphere but is not strong enough to prevent escape of the atmosphere completely. Above a level of about 500 km, where the atmosphere is so rarefied that collisions between atoms can be neglected, the atoms describe ballistic orbits and form the exosphere. Light atoms, especially hydrogen, but also helium, occasionally obtain enough energy so that they escape completely from Earth's gravitational field.

Earth's magnetic field deflects moving charged particles (*see* GEOMAGNETISM). Particles of the highest energy, namely cosmic rays, are deflected by the magnetic field so that only those having an energy of more than 15,000,000,000 electron volts can enter near the Equator. At the poles, however, cosmic rays of all energies may enter because there the lines of force of the magnetic field are vertical. As a result, cosmic ray intensity at the pole is 10–50 times higher than at the Equator (*see* COSMIC RAYS).

Solar corpuscular radiation, containing particles of much lower energies, is affected more strongly by Earth's magnetic field and is deviated into zones at high latitudes. Many of these solar particles are trapped in Earth's magnetic field and remain in it for long periods of time, moving back and forth along a line of force, but being reflected from each end. In this process particles may be accelerated to higher energies and then be energetic enough to produce luminous displays in the upper atmosphere (*see* AURORA). In their trapped condition these particles contribute to Earth's radiation belt, first observed in early Earth satellite experiments. Another source for the radiation belt may come from cosmic rays which plunge into Earth's atmosphere, there disintegrate atmospheric nuclei, and throw back into space debris which can be trapped there. *See* ASTRONOMY; INTERSTELLAR SPACE. [S.F.S.]

Space biology

An inclusive term for the various medical sciences and disciplines which are related to the rocket and man in space: physiology, biodynamics, biophysics, radiobiology, bacteriology, biochemistry, psychiatry, and medicine. *See* BACTERIOLOGY; BIOCHEMISTRY; BIOPHYSICS; EQUILIBRIUM, BIOLOGICAL; MEDICINE; PSYCHOLOGY, PHYSIOLOGICAL AND EXPERIMENTAL; RADIATION BIOLOGY.

To place man in space safely means to recognize and overcome various forces and deterrents, known and unknown, which will be imposed upon man by the space environment. Some of these deterrents, such as oxygen lack and low temperatures, are familiar problems and can be effectively controlled, while others, such as cosmic radiation and zero gravity, represent a major problem to space flight.

Oxygen problem. In limited space flight, the problem of no oxygen can be solved by transporting a supply of oxygen. The oxygen can then be administered in the proper amount either through an oxygen mask regulator system or by introducing the gas into the rocket cabin. For prolonged flight, however, such a method would be unsuitable because of the excessive weight. Therefore, some other method for supplying oxygen must be devised, for example, by the conversion of carbon dioxide to oxygen and starch, as in photosynthesis. The carbon dioxide, ammonia, and water vapor which normally accumulate from the respiratory processes can be chemically absorbed and possibly reused. *See* PHOTOSYNTHESIS.

Atmospheric pressure problem. In space flight, when one goes from the earth's surface into space, there is a rapid decrease in atmospheric pressure. In a man-carrying rocket or spaceship an atmospheric pressure similar to that of the earth's surface would have to be maintained. If not, with this fall in atmospheric pressure, the various gases which are normally dissolved in the blood and other tissue fluids will dissolve out and form bubbles. This condition is commonly called decompression illness or the bends and can be totally incapacitating. *See* DECOMPRESSION ILLNESS.

Temperature problem. The temperature problems in space flight are of two types: (1) the very high skin friction temperatures which can develop as the rocket leaves or reenters the earth's atmosphere, and (2) the low temperature normally expected in space and the stratosphere. Just as in the case of the meteor which burns on entering the earth's atmosphere, the rocket can develop such high friction temperatures that it could burn. By controlling the speed of the rocket and by using heat-resistant alloys, however, this problem can be overcome. As a matter of safety for the rocket occupants, all excessive temperatures must be guarded against and are best controlled through insulation and heat dissipators.

As one leaves the earth's surface, the outside temperature gradually decreases until at 11,000 meters, the temperature has fallen to −55°C. Except for certain unexplained temperature variations which are probably related to cosmic radiation, the −55°C temperature remains relatively constant throughout the stratosphere. Above the stratosphere, that is, in space or the thermosphere, there is no temperature since temperature refers to kinetic energy of molecules and atoms; the term absolute zero (−273°C) is commonly used in referring to space temperatures. However, heat energy may be derived from the sun and planets by radiation, including the earth's albedo. Through such radiation, very high temperatures can develop within a rocket, but these temperatures depend upon absorption or reflectance of the rocket's surface and, therefore, can be controlled. *See* ALBEDO; KINETIC THEORY OF MATTER.

Cosmic radiation problem. Cosmic rays are made up of high-energy primary particles which, on collision with atomic nuclei, produce secondary particles. It is now established that the earth is encircled by two zones of high-energy particles, one at approximately 2000 miles above the earth's surface and the other at 10,000 miles. Since these zones of high-energy particles are largely dependent upon the earth's magnetic field, it is likely that other planets will show a similar distribution. Not only the primary particles produce marked biological changes; such changes also result from the secondary and even the tertiary rays. For this reason, the space traveler must be shielded against radiation. Though the biological effects cannot yet be precisely described, it is well known that a high-energy particle can kill single cells and even unicellular organisms. Also, shattering of chromosomes, mutations, and hair follicle and skin effects have been recorded as the result of cosmic radiation. *See* MUTATION; RADIATION INJURY (BIOLOGY).

Gravity tolerance. The problem of acceleration-deceleration and gravity (*g*) tolerance in the human has been under extensive experimental investigation. The instrument used for such experimental work has been the human centrifuge. Data from these studies indicate that the human can best tolerate the various *g* forces imposed by forward acceleration when seated with legs flexed and the spine tilted 25° in the direction of the acceleration. For optimum tolerance in backward acceleration or deceleration, the subject is seated so that the spine is perpendicular to the direction of acceleration. However, a specially designed suit with restraining belts must be used, for only in this way can all parts of the subject be equally restrained in the seat as deceleration takes place. It seems likely that the occupants of space ships in the course of acceleration (exit of rocket) and deceleration (reentry of rocket) will occupy the face-forward position, as described above. Under these conditions, the *g* tolerance of the average human is well above that required to place a three-stage rocket into orbit.

In terms of the human, the earth's gravitational pull is commonly expressed as 1 *g*. If gravitational pull is less than 1, as may be observed in the outside loop of an airplane or in a roller coaster just as it starts downward, it is spoken of as sub *g* and is expressed as a fraction, that is, 0.1 *g*, 0.5 *g*, 0.9 *g*, 0.0 *g* or weightlessness, and negative *g*. The reverse of the above situation may be seen in a roller coaster just as it starts upward, or in an airplane as it pulls out of a dive. Here the *g* is in excess of 1 and is expressed as a multiple *g*, that is, 2.5 *g*, 3.0 *g*, 5.0 *g*, and so on. A subject's capacity to withstand *g* in excess of 1 is spoken of as his *g* tolerance. Most subjects have a *g* tolerance of 2.5–5.0 *g* and can withstand 8 *g* to 12 *g* for a very limited period.

Tolerances for forward acceleration are limited to about 5 sec at 12 *g*, 23 sec at 10 *g*, and 1 min at 8 *g*. Prolonged low acceleration (under 4 *g*) has a distinct advantage in that man could tolerate velocities in excess of 200,000 miles per hour but this is not possible with the propulsion systems

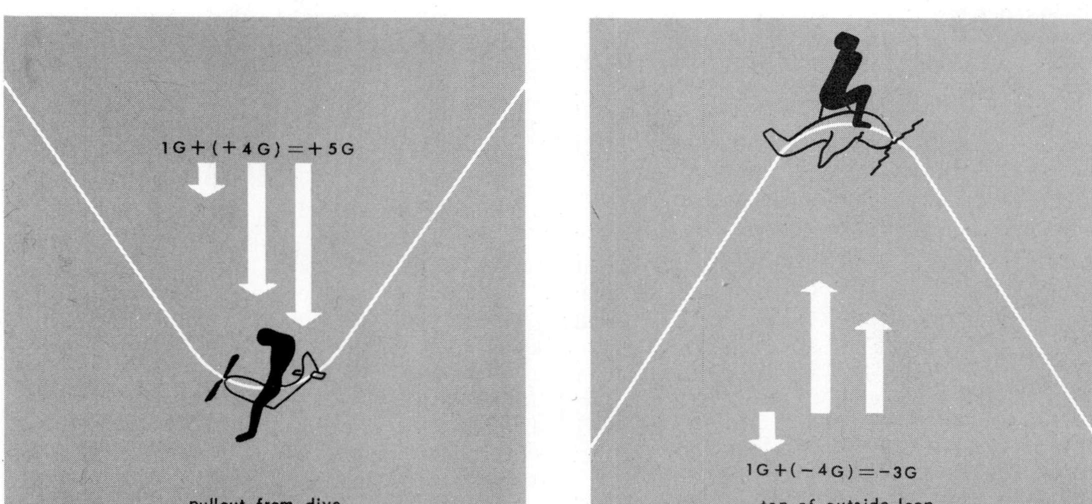

Relations illustrating positive and negative gravity forces.

that are currently used for rockets. Man's tolerance for reentry (deceleration) will depend entirely upon his ability to decelerate the rocket at a rate compatible with his *g* tolerance.

Once a man is in space and free of all gravitational pull, that is, at zero gravity, he will exist in a state of weightlessness. In such a state, man and objects tend to float (*see* WEIGHTLESSNESS).

In a space environment, man is largely deprived of his normal senses; he becomes isolated and detached. There is nothing to be seen and there is complete silence. Under a state of sensory deprivation, especially for prolonged periods, some become disoriented. *See* KINESTHETIC SENSATION.

Food and water problem. In space flight of limited duration, there is no real problem of food and water since these can be carried, but, in prolonged space flight where weight and time are critical, this problem becomes acute. Highly concentrated foods and vitamins may be the answer to the nutrient problem, but the water problem is somewhat more complicated. In all probability, the water normally lost by man through diffusion and renal clearance would have to be conserved, treated, and reused. [M.GO.]

Space charge

The net electric charge within a given volume. If both positive and negative electric charges are present, the space charge represents the excess of the total charge of positive sign diffused through the volume in question over the total charge of negative sign. Since electric field lines end on electric charge, the space-charge density ρ may also be defined in terms of the divergence of the electric field **E** or the Laplacian of the electric potential V by Poisson's equation

$$-\frac{4\pi\rho}{\epsilon} = -\text{div}\,\mathbf{E} = \nabla^2 V = \frac{\partial^2 V}{\partial x^2} + \frac{\partial^2 V}{\partial y^2} + \frac{\partial^2 V}{\partial z^2}$$

Here ϵ is the dielectric constant of the medium and x, y, and z are rectangular coordinates defining the position of a point in space. If, under the influence of an applied field, the charge carriers acquire a drift velocity v, the space charge becomes j/v, where j is the current density. For current carried by both positive and negative carriers, such as positive ions and electrons, the space charge density is given by

$$\rho = j_+/v_+ - j_-/v_-$$

Here the subscripts + and − indicate the current density and drift velocity for the positive and negative carriers, respectively. Thus a relatively small current of slow-moving positive ions can neutralize the space charge of a much larger current of high-velocity electrons. *See* LANGMUIR-CHILD LAW. [E.G.RA.]

Space flight

Travel beyond the naturally habitable region of the earth and its atmosphere. Space flight may be an exploratory trip well away from the earth and return, or it may be a more extended journey. This article surveys the interaction of engineering and biological requirements of manned space flight. The biological problems of exit and reentry are distinct from in-flight problems.

Exit and reentry. Tolerable acceleration (*g* loads), tolerable thermal loads, and emergency escape are of prime concern.

The solid curve in Fig. 1 shows the duration of acceleration required at various constant *g* values to reach an orbital velocity of 18,000 miles per hour. The hatched area shows the highest *g* tolerable in human centrifuge experiments. The best position is lying on the back in a seated position with the trunk flexed forward 25°, as illustrated. A peak tolerance of 22 *g* was demonstrated in 1958 on the U.S. Naval centrifuge at Johnsville, Pennsylvania, using a tight-fitting contoured couch in a similar position. Immersed in water and using suitable underwater breathing apparatus on the Aero Medical Laboratory centrifuge at Wright-Patterson AFB, subjects tolerated 12 *g* in 1958 for 4 min. The efficacy of a swinglike arrangement for protection from multidirectional *g* loads has been demonstrated by H. von Beckh of the Aero Medical Field Laboratory, Holloman Air Force Base, New Mexico.

Steep angles of reentry exceed the maximum tolerable accelerations for many seconds. Shallow angles of reentry by no-lift space craft produce lower peak *g*s for longer periods of time, the deceleration remaining within the limits of the curve. *See* REENTRY.

Ventilation. The body normally disposes of its metabolic heat by vasodilation (increased skin-surface blood flow) and by sweating. These mechanisms require skin ventilation with dry air cooler than skin temperature. The body can accept 21 Btu/ft^2 before performance is seriously impaired. Figure 2 illustrates improved tolerance using a U.S. Air Force Mk IV antiexposure suit assembly ventilated with 6 ft^3 of air per minute at 60°F.

Escape. Escape from a space vehicle approaching orbital speed during exit or reentry presents all the problems for which the primary vehicle is designed. For simple one-man vehicles such as Project Mercury, only a duplicate of the original capsule would suffice for emergency reentry. In such cases, the only protection is reliability. Escape throughout the exit phase can be provided as in Project Mercury by means of emergency rocket separation of the manned capsule portion from the booster rockets in the event of booster failure. Only on large multiman rockets would separate escape units, doubtless modeled after the Mercury capsule, be practical.

In-flight. Cabin design will be determined by in-flight biomedical factors, including the sealed-cabin atmosphere, psychic factors, nutrition and waste, weightlessness, radiation hazards, and emergency protection.

Psychic factors include isolation-confinement reactions and the work-rest cycle. Particularly on solo

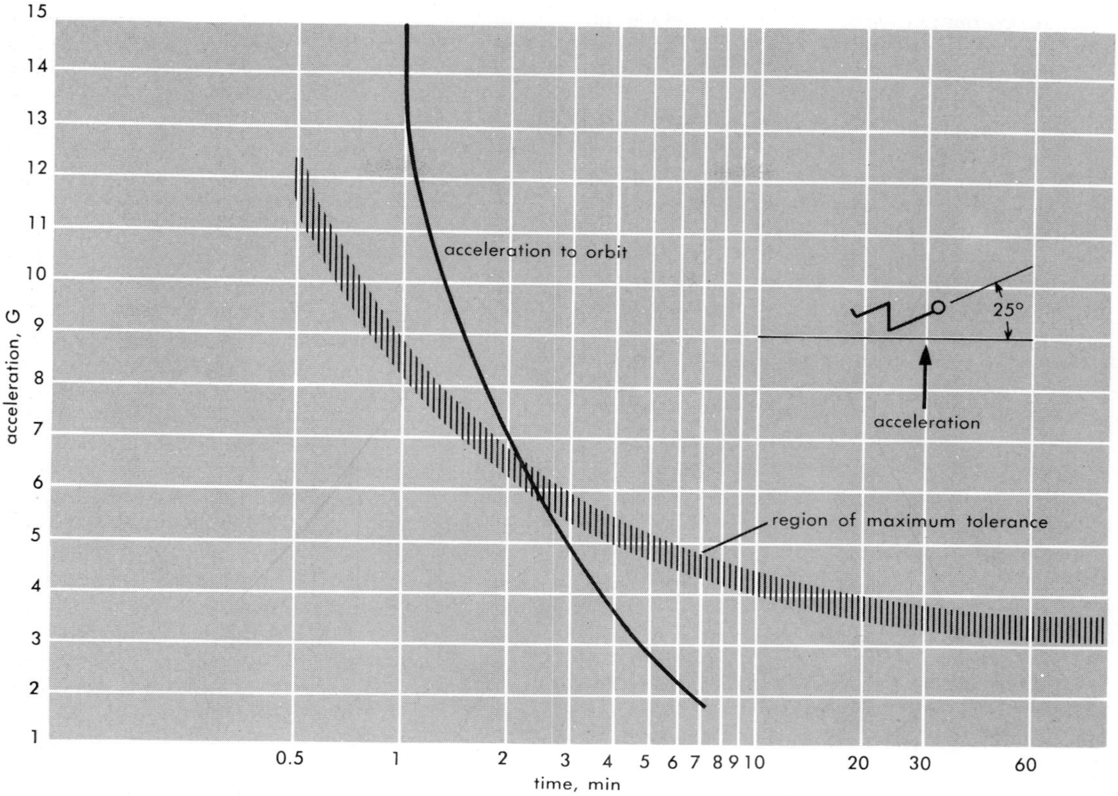

Fig. 1. Human tolerance to acceleration.

missions which explore the utmost range attainable, human reactions to isolation will be a major problem. The sense of belonging to the community of mankind versus a sense of identification with the emptiness and incomprehensible vastness of space will be influenced by communications available with other men, visibility outside the life compartment, its interior design, and the intensity of interest and the activity level realized from exploring the new region.

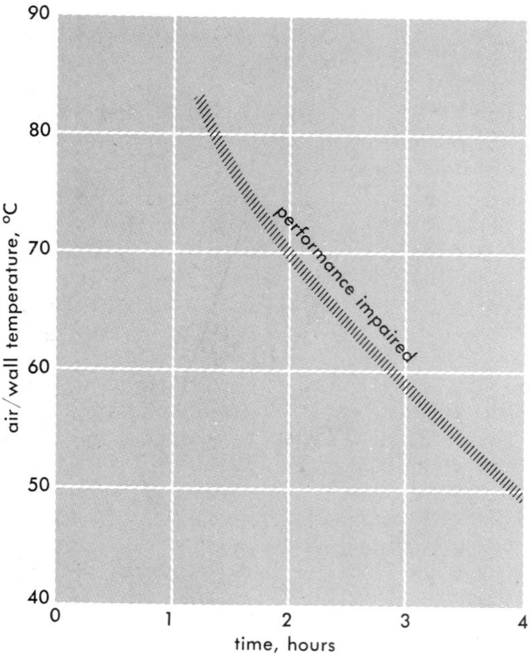

Fig. 2. Thermal tolerance with ventilated suit.

The severe confinement imposed by early exploratory solo missions, judging from Projects Manhigh and Mercury, will permit a man approximately 50 ft^3 of living space. The degree to which this is tolerable will depend largely on whether the man is motivated to prove he can "take it" or to study the unknown. Frequent opportunities for a change of activity, change of pace, and exercise become increasingly important as missions extend beyond 24 hours.

The work-rest cycle was studied by G. Hauty in 1959 with experiments at the School of Aviation Medicine in a space cabin simulator. Days divided evenly into three work and three rest periods permit higher total performance than the normal one work and one rest period per day.

Oxygen. The atmosphere in a sealed space cabin must provide a pressure of at least 150 mm Hg pO_2 (partial pressure of oxygen). This corresponds to the earth's atmosphere at 8000 ft or breathing 100% oxygen at 38,000 ft. The extra gas in the earth's atmosphere is 79% inert nitrogen.

There is no known requirement for nitrogen in inspired air. It can produce bends after loss of cabin pressure if the initial cabin (pressure) altitude is below 25,000 ft. Helium is theoretically preferable in this decompression situation and has no known toxicity but needs to be studied more fully. *See* DECOMPRESSION ILLNESS.

The percentage of inert gas in a sealed cabin atmosphere remains constant if pure oxygen is supplied to replace the oxygen metabolized. The carbon dioxide produced metabolically must be eliminated physically by an absorber or other means (Fig. 3). The selection of a cabin pressure between the surface atmospheric 15 lb/in.^{2} and the 38,000-ft pure oxygen 3 lb/in.^{2} will depend on the objectives and mission profile of the space flight.

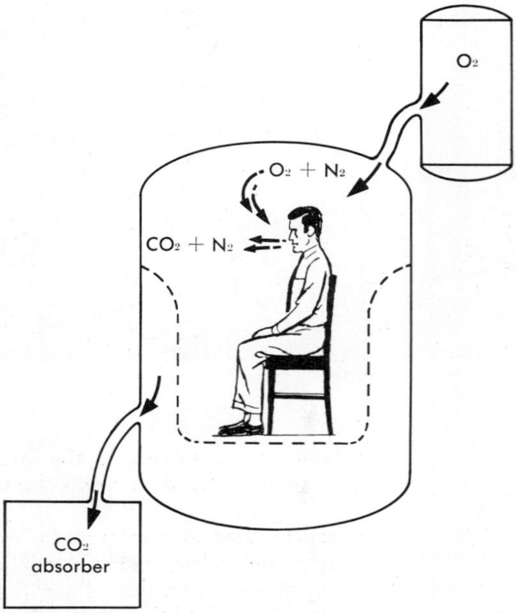

Fig. 3. Sealed-cabin atmosphere.

Nourishment. The caloric requirements for an astronaut will average close to 3000 calories/day. For a few days, this can be almost exclusively high-energy low-bulk carbohydrate foods. For longer missions, a completely balanced protein-fat-carbohydrate diet with necessary vitamins and minerals will be required. D. Keller of the School of Aviation Medicine has meticulously analyzed daily metabolic requirements including storage containers and necessary hardware. His values in the table do not include weight for temperature control of the cabin. *See* FOOD; NUTRITION.

It costs as much in weight to remove the metabolic by-products as it does to supply needs. Water is relatively simple to recycle; so doing saves 10 lb/day, not including the additional weight for the simple apparatus required.

Daily metabolic requirements, grams

Oxygen	2,636
Complete balanced diet	862
Drinking and wash water	2,967
CO_2 and H_2O absorption	6,953
Total	13,418 = 28 lb

Recycling of oxygen would reduce the daily weight requirement by more than one-third, not counting the weight of the more complicated processing equipment. The oxygen can be recovered from carbon dioxide by algae, bacteria, plants, catalysts, or a chemical system. All require considerable power. Experimental models of both water and oxygen recycling systems using mice operated successfully at the School of Aviation Medicine in 1958.

A completely recycled system including nutritional elements is more complicated and may well require animal intermediates, such as insects, to balance all portions of the cycle adequately.

Weightlessness. On short space flights of less than 24 hours, weightlessness is expected to present no serious problem. On longer flights, however, it will contribute to the type of psychic stress imposed by the hazards of the situation, isolation, and confinement. Many experimenters have found neuromuscular coordination only slightly impaired; corrections are quickly learned. By providing a space diet in toothpaste tubes and liquids in plastic squeeze bottles, feeding will be no problem. Visual clues will be sufficient to establish orientation with no difficulty. *See* WEIGHTLESSNESS.

Radiation. The risk from both the radiation of a nuclear rocket engine and ambient space radiations is significant. *See* RADIATION INJURY (BIOLOGY). Undoubtedly, the conservative dose rate of 0.3 roentgens/week must be exceeded. In high-risk space-flight missions, the absolute maximum acute dose would be about 200 rem (roentgen equivalent, physical) which produces apparently reversible radiation illness. The shield weight calculated for a 1,000,000-lb thrust nuclear rocket engine for a 3-man crew is 8000 lb at 100 ft permitting an exposure of 10 rem/5 min of reactor operation.

Space radiations are considered under three types: (1) local planetary, (2) solar, and (3) galactic (Fig. 4).

Of the local planetary radiation, the 0.5-Mev electrons and protons characteristic of the outer (soft) Van Allen radiation belt require only minimal shielding. The much higher energy (1 Bev and greater) protons of the middle (hard) Van Allen belt require prohibitive amounts of shielding to protect from the x-rays which they produce upon striking a space vehicle. The astronaut must either accept a calculated risk exposure during quick transit through this region or exit by a polar route avoiding this radiation zone which girdles the earth's equator. *See* VAN ALLEN RADIATION.

Giant solar flares occur without warning during active periods of the sunspot cycle. Still incompletely investigated, these are clearly a serious radiation hazard in space, producing 20–30 Bev protons. A single flare may continue to be a hazard for 12–18 hours.

Galactic cosmic radiation produces a unique, partly investigated pattern of radiation in tissue. Exposures of a few days in the region of the earth apparently represent no health hazard. Its effects are unknown if exposure time extends to 1 month

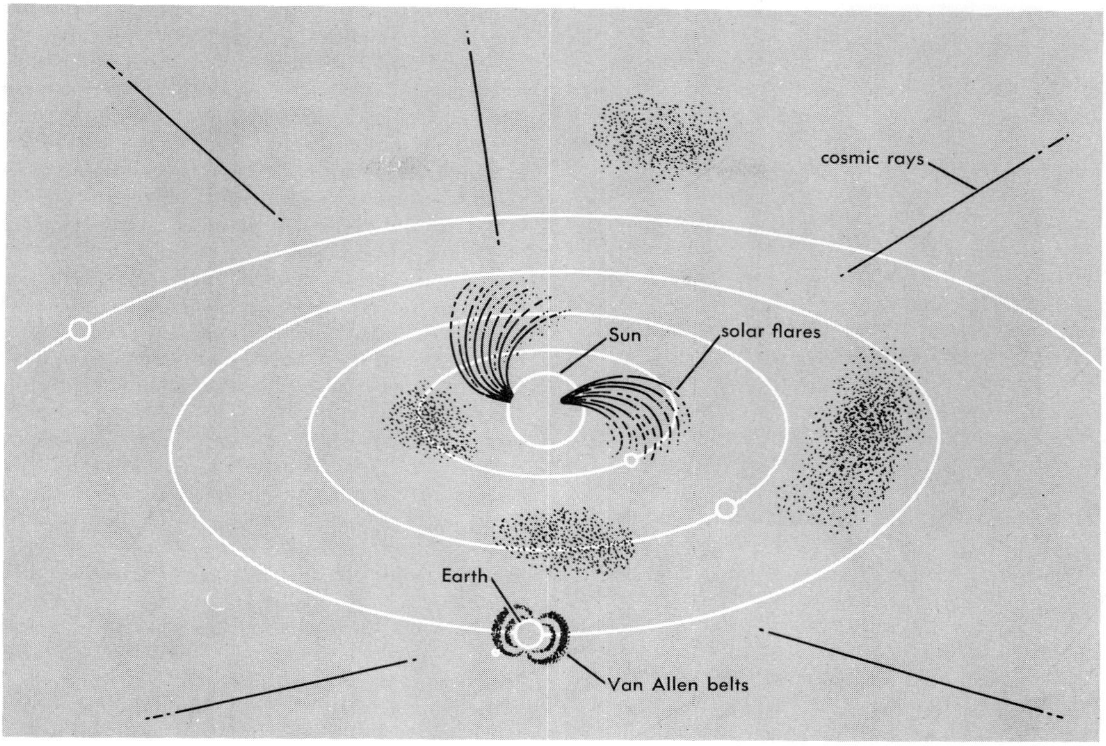

Fig. 4. Space radiations.

or more. Intensity of exposure is likely to increase significantly as astronauts explore remoter regions of the solar system (approaching the orbit of Mars), particularly during periods of low solar activity. *See* COSMIC RAYS.

Cabin puncture. The most serious emergency likely to occur is sudden loss of cabin pressure. The rapidity of such a decompression depends upon the volume of the cabin, the area of leak orifice, and the cabin pressure. The pressure in a 100-ft^3 cabin sustaining a 1-in. hole would leak from a pressure equivalent of 25,000 ft to 40,000 ft altitude in 10 sec. Estimates indicate that in space such a hole might be expected in a stainless-steel cabin wall 0.6 mm thick of 1000 ft^2 area approximately once per year.

When the cabin altitude (pressure) passes 40,000 ft the astronaut experiences fatal hypoxia, and as the pressure altitude passes 70,000 ft he goes on to ebullism with swelling of surface and internal tissues. One possible solution is an emergency chamber or bag into which the astronaut can seal himself within a few seconds and then, once inside, don an emergency pressure suit. *See* INTERPLANETARY PROPULSION; SPACE BIOLOGY. [D.G.S.]

Bibliography: C. C. Adams, *Space Flight*, 1958; H. Haber, *Man in Space*, 1953; W. Ley, *Rockets, Missiles and Space Travel*, rev. ed., 1957.

Space station

A laboratory in space where scientific and exploratory work can be done. Beginning with the International Geophysical Year (July, 1957–December, 1958), elementary space stations in the form of artificial satellites were placed in orbits about Earth.

Space stations can be used as platforms for astronomical telescopes and other scientific equipment designed to make measurements of the cosmos free from the blanketing and distorting effects of Earth's atmosphere.

Whether a space station should be manned or be operated by remote control will be determined by the relative advantages and the relative costs of the two types of operation (*see* SATELLITE, ARTIFICIAL). Building a space station capable of supporting a man for any extended period of time and placing it in orbit is a monumental undertaking.

Many designs for manned space stations have been formulated. Common to all of these are means to protect the occupant from environmental hazards such as high-energy radiation and meteor collision. The design must also be such that the atmosphere in the cabin will sustain life and that the temperature will stay within acceptable limits. One major problem is weightlessness. Material in a space station traveling in an orbit will be in a state of free fall and effectively weightless. Much is being learned about the physical and psychological effects on a man subjected to this condition for a long period. Gravity can be simulated, however, by building the station as a large wheel and causing it to spin on its axis; the resulting centrifugal effect then replaces gravity.

One means for establishing a manned station is to build up a massive Earth-circling space station

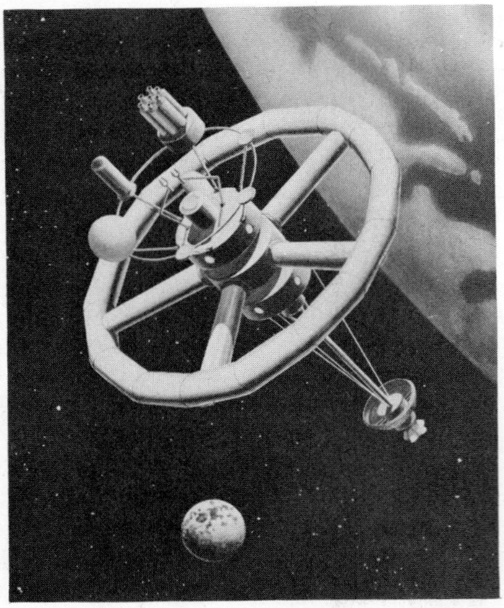

Proposed space station powered by nuclear rocket at shielded end of boom rotates on its axis to simulate gravity. (*Lockheed Aircraft* Co.)

by successively launching parts into satellite orbits and then assembling them. Another method that would achieve most of the objectives of an observatory in space would be to establish an observatory on the Moon. [J.P.H.]

Space technology

The systematic treatment of those arts and sciences concerned with exploration of outer space. Space technology encompasses the processes of going from Earth to space, the traveling and maneuvering in space, the gathering of knowledge of space, and the return of the gathered knowledge to Earth either by communication or by a returnable vehicle. For the present discussion, space is considered to be those regions of the universe sufficiently removed from either stars or planets so that local atmospheric envelopes have negligible effect.

During the initial and sometimes the final stages of space exploration, the atmosphere, the gaseous envelope surrounding the Earth, must be traversed. This involves the use of the specific technologies relating to vehicle design and construction, propulsion, aerodynamic stability and control, thermal effects due to aerodynamic heating, and the supporting technologies. For example, vehicle design and construction involves a knowledge of materials and their properties which are established through the technological aspects of physical metallurgy. Propulsion is dependent upon an understanding of the chemistry of combustion, as well as the thermodynamics and fluid mechanics of both liquids and gases, while aerodynamic stability and control not only require a knowledge of classical aerodynamics but may involve the interaction between aerodynamics and the structural deformations due to loads and temperature gradients which can be covered by the general term aerothermoelasticity. Electronics and communications are involved heavily in this phase of operation particularly if the vehicle is unmanned.

Beyond the Earth's atmosphere many of the same problems arise and new technologies must be brought into action. In interplanetary and space travel, celestial mechanics is required to use the gravitational fields in space. Whereas atmospheric propulsion can depend upon the oxygen of the air to oxidize the propellant, in outer space oxygen is no longer present and must be furnished or new propulsion schemes devised. Problems of radiation effects (which are not present on Earth because of the atmospheric blanket) must be taken into account. Some of these are damaging, and equipment and personnel must be shielded against them while others may be useful as possible propulsion or power sources. Because space contains electromagnetic radiations which are concentrated in the regions of the planets and stars and because in these same regions ionized gases are prevalent, a new science of magnetohydrodynamics is developing which in turn will lead to new technologies. Knowledge gained in the field of ionized gas flows will be useful in evaluating the possibilities of ion propulsion as a driver for space vehicles. *See* MAGNETOHYDRODYNAMICS.

Being outside the Earth's atmosphere brings new astronomical techniques into play, and the proper gathering, storing, and final analyzing of these data calls for advances in the fields of electronics and optics. Development of new techniques in the processing and handling of large quantities of data is required as the area of detailed investigation expands through our local planetary system and eventually to include far reaches of the universe.

The establishment of permanent satellite stations in itself requires the development of many technological areas. Lightweight construction to be assembled in space poses many new and difficult problems to the construction industry. Maintenance of these stations with a minimum of transported material from Earth calls for methods of growing and processing food and the obtaining and conservation of water and oxygen so as to permit life processes (human, animal, and plant) to continue with the least possible restrictions. The use of such permanent satellites as communications relay stations as well as space launching platforms brings many new facets into play in the field of communications, including low-weight power supplies as well as systems for the reception and transmission of signals.

Upon a vehicle's return to Earth, there are again aerodynamic problems to contend with. The rate of departure through the atmosphere can be regulated by the burning rate of the propellants whereas, on return, the major problem is to overcome the large energy stored in the returning vehicle in the form of high velocity. One solution is the use of retrorocket devices, but these employ fuel which involves weight so there is need for an energy absorbing system which is essentially weightless and which does not involve the generation of excessive amounts of heat.

Coupled with the technological problems of the actual space vehicle are the ground equipment problems for storage, launching, landing, and maintenance of these vehicles. Exotic fuels raise storage problems and the handling of these fuels in large quantities enters the field of economic logistics (*see* METAL-BASE FUEL). The bases themselves are complex installations covering nearly all phases of present knowledge; much of the control of the space vehicle may be from the ground during the early stages of flight and return to save weight of equivalent apparatus in the vehicle. *See* LAUNCHING PAD COMPLEX.

From a military standpoint, many advances in technological progress are essential. Control and guidance of ballistic missiles and space launched missiles represent only two of the many problems in the efficient utilization of space from a military standpoint. Camouflage and diversionary methods of attack must be developed concurrently with protection devices to prevent damage from enemy utilization of space as a military point of attack.

It is difficult to envision any technological or scientific advances in any area which are not of advantage in space exploration. Pure science in the fields of mathematics and physics enables the space research team to acquire new knowledge and to develop new technologies for practical application. The engineering sciences are called into play to develop vehicles and propulsion systems which will be light enough to carry reasonable payloads either scientific or military. Furthermore, the biological sciences require new techniques of determining the effects of space travel, weightlessness, and space radiations on the living cell so as to prepare the way for human space travel. Just as the industrial age brought into play all of the then current knowledge and developed new techniques of its own, so does the advent of the space age call upon the highest level of technological development in essentially all areas of knowledge. *See* ASTRONAUTICAL ENGINEERING. [E.E.SE.]

Space vehicle

A combination of a power plant, propellant, guidance and control systems, and a payload designed to operate outside of Earth's atmosphere. Space vehicles may be launched from and return to Earth, may be launched from Earth but not be expected to return, or may be assembled and launched from a satellite space station. Space vehicles are usually classified with regard to their mission or their flight path. The earliest space vehicle was the high altitude sounding rocket which was launched from Earth in essentially a vertical direction, passed through the atmosphere, penetrated a short distance into space, and returned to Earth. During its flight, information on space was collected and either transmitted to Earth by telemetry or returned to Earth by a parachute attached to a data capsule. The second most complex space vehicle is the Earth satellite.

Earth satellites. Some space vehicles are designed to revolve around Earth in circular or elliptical paths. For a vehicle to go into a circular satellite orbit around any celestial body, the vehicle must have a velocity $V_s = R\sqrt{g/(R+h)}$ where V_s is the required satellite velocity, R the radius of the central body, g its gravity constant, and h the altitude of the satellite. The time for one revolution is given by $t = 2\pi\,(R+h)/V_s$ sec. For an Earth satellite, typical values are given in the table. If, at any given height, the velocity and launching angle are not exact, the satellite will go into an elliptical rather than a circular orbit.

Typical orbital periods for Earth satellites at various heights

h, Earth radii	0.5	1	2	3	4	5	5.6
V_s, mi/sec	4.0	3.5	2.8	2.5	2.2	2.0	1.9
t, hours	2.6	4.0	7.3	11.3	15.8	20.7	24.0

Earth satellites can be used for the collection of data on conditions in outer space, as reconnaissance vehicles to determine happenings on Earth or in its atmosphere, as communication relay stations, or as a base for astronomical observations. If unmanned, the satellite can transmit its data to Earth stations by telemetry, or the satellite or some part of it can be brought back to Earth with its data records. Earth satellites may revolve about Earth either in planes through the polar axis (polar satellites), in the plane of the Earth's equator (equatorial satellites), or in planes which are inclined at any angle to Earth's axis. Polar satellites have the advantage that Earth revolves underneath the orbit so that eventually the satellite passes over all of Earth's surface. The equatorial satellite is of particular value in long-range communications systems. As indicated in the table, an Earth satellite at approximately 22,000 miles from the Earth (5.6 Earth radii) will have a period exactly equal to that of Earth and, if rotating in the same direction as Earth, would appear to be stationary. It therefore would make an ideal communications relay station because three such satellites equally spaced would cover essentially all of Earth's surface.

Large Earth satellites could also be adapted as space stations for the assembling, launching, and landing facilities for other space vehicles. In this manner, vehicles for interplanetary travel would not have to be designed to exit and return through Earth's atmosphere, thus avoiding many design problems. Such space stations would be manned and supplied by missiles from Earth. These large satellite platforms would also be ideal locations for astronomical observations because they would be completely outside of Earth's atmosphere.

Space probes. Vehicles designed to collect information on outer space and the bodies therein are space probes. The simplest probe is the vertical sounding probe previously mentioned. If the velocity of a vehicle from Earth is increased, a critical velocity will be reached which will enable the vehicle completely to escape Earth's gravitational field. The ideal escape velocity, unaffected by atmosphere or gravitational losses, is given by $V_e = R\sqrt{2g/(R+h)}$ where the symbols are as previously defined. The actual escape velocity from the surface of Earth is approximately 7 mi/sec. Once having escaped from Earth, the space vehicle can be directed toward the Moon, other planets, the Sun, or to regions beyond our solar system. The name of the vehicle designates its mission.

Lunar probes. Space vehicles designed to investigate the Moon and its surroundings may be directed to go closely by the Moon and return to Earth, to go past the Moon and enter a solar orbit, to go into a satellite orbit around the Moon, or to land on the Moon. In addition to the usual space information to be collected and returned to Earth, such as temperature, radiation strengths, and meteoric impact, television might be used to send back visual information on the lunar surface conditions. Unmanned vehicles will be used first, followed by manned probes with a much larger data collection ability.

Planetary or solar probes. As their name indicates, planetary or solar probes are space vehicles designed to investigate conditions near to and on either the planets or the Sun. Their possible orbits will be similar to those for the lunar probes except for the solar probes which will never attempt a close approach or a landing because of the high temperature of the Sun. The nearer planets will be investigated first and, as propulsion, guidance, and communications continue to improve, probes to the more distant planets may be expected to follow. It is not sufficient to be able just to reach one of the outer planets because such a feat would be valueless unless significant amounts of data could be collected and returned to Earth.

Cosmic space vehicles. The ultimate effort to leave our solar system and investigate some of the closer bodies beyond requires a truly cosmic vehicle, cosmic being used to differentiate such vehicles from those for use within the central portion of our solar system. (In Russian usage any space vehicle is termed cosmic.) To make significant trips even to the outer planets such as Neptune and Pluto, with mean radii from the Sun of approximately 27.9×10^8 and 36.7×10^8 miles, respectively, high velocities will be required if the journey is to be made in a reasonable proportion of a human lifetime. Because of the great distances to the nearest stars, velocities equaling a significant fraction of the velocity of light must be obtained if the time for the journey is not to be measured in generations.

Space structures. Vehicular structures suitable even for Earth satellites must be light in weight. To possess the lightest weight consistent with a given performance specification, the vehicle is designed and built in accordance with the most advanced knowledge available in structural mechanics, structural design and analysis, structural vibration, and structural materials. *See* SPACECRAFT STRUCTURE.

Structural mechanics. The behavior of a vehicle structure depends on the action of applied or induced forces. At any given flight stage, a space vehicle may be subjected to various combinations of forces such as thrust, aerodynamic lift and drag, gravitational fields of Earth or of other heavenly bodies, centrifugal forces, and control forces. In addition to these major forces, others which may be important are gyroscopic forces due to spin, magnetic forces due to magnetic fields in space, and light pressure from the Sun. If the forces are in equilibrium, the velocity of the vehicle will be constant, but if there are unbalanced forces, the vehicle will accelerate or decelerate, bringing into action mass inertia forces. For vehicles using liquid propellants, an internal force due to motion of the propellants and known as sloshing is important. The relationship of the forces to the gross vehicle motion follows the usual laws of mechanics, and the major task arises in attempting to distribute some of the more concentrated loads, such as the thrust, uniformly over the structure so that all the structural material carries its proportion of the load.

Structural design. The first concern in structural design is with the structural configuration, the second with the detailed stress analysis of all parts of the vehicle. Because so much energy is required to put a vehicle and its payload into space, a large percentage of the total structure consists of containers for the fuel, either liquid or solid. Even with nuclear or ion propulsion, a large proportion of the structural weight will be used by the power plant and a minor part for guidance and payload.

Whatever the propulsive technique, nearly all the structure will consist of a thin-walled container, subjected to internal pressure, and the problem is that of designing the lightest possible pressure vessel. To reduce the amount of useless weight during flight, empty fuel containers along with their engines and controls will be jettisoned (staging operation). Therefore, the basic structural configuration becomes a series of thin-walled pressure vessels, some of which will be separated during flight until only the payload compartment with its required power plant and control system remains.

The most serious single problem is that of trying to distribute large concentrated forces, such as thrust or control forces, into these thin shell structures. The stress analysis used on missiles and space vehicles differs from that in other structures primarily in that it must be more highly refined and more accurate so as to assure safety and reliability with no excess in weight. Any truss or beam-type structures are analyzed by standard methods of applied elasticity. The main difference between these designs and other vehicle designs lies in the emphasis on the analysis of thin-shell structures and the problems of strength and stability of this type of construction.

Thin-shell analysis. Thin structural materials (usually, but not necessarily, metals) are used to sustain loads under a wide range of stress conditions. A typical missile or space vehicle may consist of cylindrical or spherical pressure vessels connected by thin-walled truncated conical sections which are subjected to axial accelerations from thrust, and bending and shear from wind or control forces. During the storage and launching operations, the vehicle may be called upon to carry axial compressive loads without internal pressure.

The compressive load-carrying ability of an element of a circular cylinder of radius R and wall thickness t is given by $N_c = KEt^2/R$ lb/in., where E is the modulus of elasticity of the material and K is a factor depending upon ratios of radius to thickness and length to radius, and is always less than unity. Furthermore, for a thin-walled, cylindrical pressure vessel, the maximum tensile stress σ_m is given by $\sigma_m = pR/t$, where p is the internal pressure. Thus, if the vessel is designed for pressure and has an ultimate strength equal to σ_u psi, the thickness will be $t = pR/\sigma_u$ and the compressive load carrying ability will be $N_t = KEp^2R/\sigma_u^2$ lb/in. with no internal pressure acting.

For the usual structural materials, if the vessel is designed on the basis of a reasonable (2–4 atm) internal pressure, the compressive load-carrying ability is at most a few hundreds of pounds per inch, a value far insufficient to support the missile weight in the launcher. Two techniques are used to increase the compressive load-carrying ability of such structures. Stiffeners are added to the thin-skin wall, or internal pressure is applied to stabilize the structure. The use of internal pressure is usually the lighter technique, but it has the disadvantage that a constant pressure is required to prevent collapse of the thin-walled structure. The same analysis problems are present for conical and spherical shells.

For large internal pressures, such as are found in solid propellant combustion chambers, the shell walls must be thicker and compression buckling may not be a problem. In such structures, the major emphasis is on obtaining a smooth flow of the stress field from one part of the shell to another and to avoid sudden changes in shell thickness and shell curvature. Once these conditions have been taken care of, the remaining problem is to find the material which has the highest strength when subjected to the stress field of the particular pressure vessel.

Vibration. The vibrational modes of missiles and space vehicles determine the proper location of guidance and control systems. Such systems should be located at nodal points of the main structure so that only gross displacements and rotations of the structure as a whole will be measured and not local movements due to structural vibrations. For instrumental exploration of space, it is necessary that any perturbations of the vehicle be rapidly damped and that the system be as vibrationally stable as possible.

Structural materials. For portions of a vehicle subjected to compressive loads, the highest stiffness-weight ratio should be used; it may be necessary to use specialized techniques such as sandwich construction to raise the stiffness of the structure without increasing the weight. In tensile areas such as high pressure containers, the most suitable material is one which not only has a high strength, usually in a multiaxial stress field, but also one which has sufficient ductility so that it can accommodate local stress discontinuities without being subject to brittle failure. In the general structure of missiles and space vehicles, there will be requirements for materials having high strength-weight and high stiffness-weight ratios over wide ranges of temperatures and stress conditions.

[E.E.SE.]

Spacecraft structure

The supporting structure for systems capable of leaving the earth and its atmosphere, performing a useful mission in space, and generally returning to the surface of the earth. The technologies that enter into the design of spacecraft structures include aerodynamics, aerothermodynamics, heat transfer, structural mechanics, materials technology, and systems analysis. In applying these technologies to the structural design of spacecraft, the significant factors influencing the design are analyzed and trade studies are made to arrive at a structural design which fulfills system requirements.

The structural aspects of space flight can be divided into four broad regions or phases: (1) transportation, handling, and stowage; (2) boost; (3) orbiting or space flight; (4) reentry. Each phase has its own structural design criteria based upon the specific environment, the configuration existing during the phase, the mission objectives, and whether the craft is manned or unmanned. These design criteria require detailed consideration of heat, loads, shock, rigidity, radiation, and meteoroids. Statistical analyses are used to prevent designing for statistically improbable combinations of such factors as heat, load, and shock.

Transportation, handling, stowage. The growth factor for space vehicles, that is, the change in

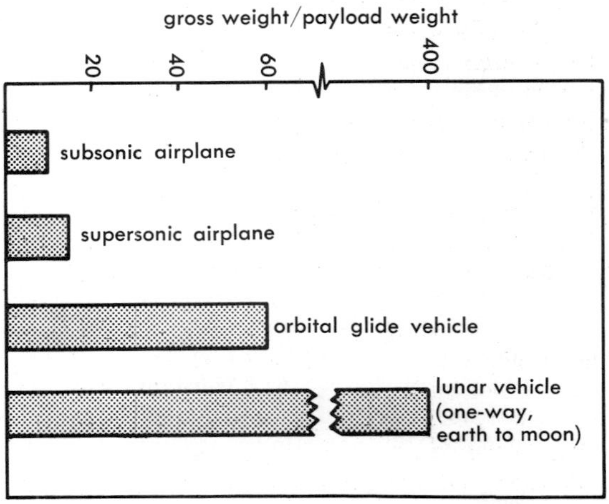

Fig. 1. Typical growth factors for air-borne and space vehicles.

take-off weight required for each additional pound placed in orbit or transported to the moon, is shown in Fig. 1. For comparison, growth factors for airplanes are also shown.

The resultant penalty for each unwarranted pound, for example, 400/1 for an earth-to-moon, one-way, three-stage trip, dictates that the transportation, handling, and stowage of the over-all system or its components must not impose a weight penalty on the flight article. The most severe environments encountered in transportation, handling, and stowage are shock and vibration. Special equipment is designed to reduce these loads to such a level that they do not govern the design of the flight article.

Boost phase. The purpose of the boost phase is to lift the vehicle above the sensible atmosphere and to accelerate the vehicle to the velocity required for its mission. For space missions, the required velocities range from 26,000 ft/sec for nearly circular orbits to 36,000 ft/sec for lunar or interplanetary missions. Achievement of these velocities requires boosters many times the size of the vehicle itself. Generally, this boosting is accomplished by a chemically powered rocket-propulsion system using liquid or solid-state propellants (*see* ROCKET ENGINE). Multiple stages are required to reach the velocities for space missions (*see* ROCKET STAGING). Launching may be accomplished vertically (from a surface stand or from underground silos) or horizontally from an airfield. Generally, vertical launching from a surface stand is preferable. *See* LAUNCHING PAD COMPLEX.

Vertical take-off requires a thrust or propulsive force that exceeds the weight of the complete flight system by approximately 30%.

The trajectories, that is, the variation of velocity with altitude and time, can be controlled to some extent by proper selection of the thrust-to-weight ratio and launch angle. The higher thrust-to-weight ratios are more efficient but also result in higher dynamic pressures q given by $q = \frac{1}{2}\rho V^2$ where q is dynamic pressure, lb/ft^2; ρ is density, slugs/ft^3; and V is velocity, ft/sec. The higher the dynamic pressure, the higher are the aerodynamic loads and heating.

Determination of boost phase loads requires consideration of ignition and release shocks, weight, thrust, acoustical or noise environment, gusts, winds, propellant temperature, boost trajectory, method of stabilization, vehicle geometry, exhaust radiation, and base heating.

The propulsive thrust force is reacted by inertia loading, and the local compressive force is dependent upon the acceleration and mass distribution of the vehicle. Usually, the thrust loads are critical at stage burn-out because, with constant thrust, the accelerations increase to a maximum as the propellant weight of the operating stage is reduced to zero.

Payload configuration. Probably the strongest influence on booster bending moments is the payload shape. Figure 2 indicates the relative bending moments imposed upon the same two-stage booster for a 50-ft/sec sharp-edge gust when the payload is a glide shape or a ballistic shape. The differences are due to the greater aerodynamic efficiency of the glide shape. It is possible, however, in certain cases, to use the control surfaces of the glider to reduce the bending moments drastically. The dashed line in Fig. 2 shows the bending moments for the glider when it is pivoted in pitch at its center of pressure and its elevons are actuated in a manner to nullify the glider aerodynamic load while it is present. Bending moments due to atmospheric disturbances can be substantially reduced through proper control-system design.

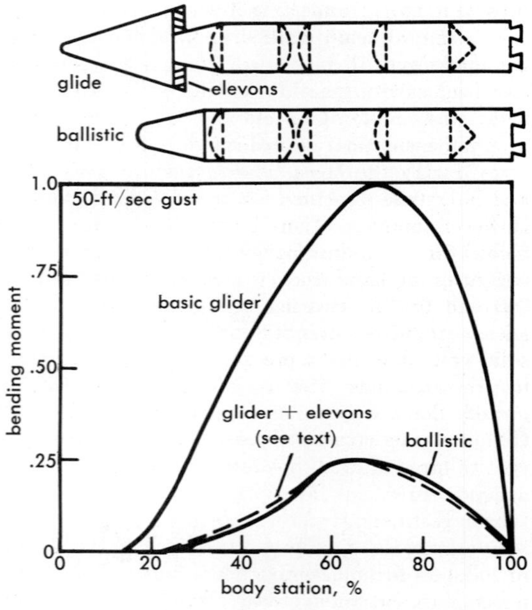

Fig. 2. Relative bending moment on same booster system when payload is a glide or ballistic shape.

Booster stiffness. The structural stiffness of the booster system must be carefully considered. The sensor location and the higher-order modes must be considered so that control-system coupling, which may result in over-all vehicle instability, is avoided. The liquid systems using fairly low pressurization as a means of stabilizing the thin shell are not as rigid as solid-propellant cases which must inherently withstand high internal pressures. The use of plastics for solid rocket cases reduces the stiffness due to their high strength-to-weight ratio and inherently low moduli of elasticity. This reduction in stiffness can be compensated for to a large extent by proper control-system design.

Sloshing in the case of liquid-propulsion systems is carefully evaluated because it affects the structural integrity of the thin shell and stability of the vehicle. The fuel-mass movements result in shifts of the center of gravity, causing pitch or yaw of the system. Control-system and aerodynamic coupling with either pitch or lateral sloshing can occur, unless proper baffling or damping is provided.

Fore and aft accelerations or surging of the liquid propellant mass due to thrust variations, such as at take-off or burn-out, result in radial tank expansion and fluid compression. Dynamic overloads can result if thrust variations at take-off, burn-out, and staging occur at rates or frequencies near the natural frequency of the tanks and their support structure.

High temperatures. The higher dynamic pressures result in higher heating rates and thus higher temperatures on the structural materials, because the temperature T_w of the boundary layer at the booster case surface in degrees Rankine is $T_w = T(1 + 0.2rM^2)$ where T is ambient temperature, °R; r is recovery factor (approximately 0.90 for turbulent flow); and M is Mach number.

The rate of temperature increase depends upon boundary-layer temperature, heat-transfer coefficient, emissivity, heat sinks or sources of the vehicle and contents, material conductivity, and whether insulation or cooling is used. Temperatures during the boost phase on the engine cases and interstage structure can be as high as 2000°R. However, many portions of the boost system are below this temperature with the first-stage temperatures of solid rocket cases ranging between 750 and 950°R. Heating of the base area of the missile requires attention to detail. Because the solid-propellant fuel temperature must be restricted to approximately 950°R, insulation may be required. Some designs use materials such as glass fiber for the structural case material and thus obtain the required insulation without paying a weight penalty.

The pertinent structural design parameters are temperature, load, stiffness, and time. In the design of boosters for spacecraft, these four are mutually dependent and interrelated. This relationship requires that many trade studies be made before the optimum approach can be selected.

The materials of interest for booster case and interstage design are alloys of aluminum, magnesium, titanium, steel, stainless steel, super alloys, and plastics. The primary properties of interest are those which contribute to stability of the shell against the high compressive stresses. The compressive buckling allowable is a function of the modulus of elasticity, and thickness. The modulus of elasticity at a given temperature is time dependent. Thus data used in spacecraft design are typically based upon an exposure time of 5–10 min, because the boost phase falls within this time span.

Pressure vessels. Another important material property, especially in pressure-vessel design, is notch sensitivity. Notch sensitivity, as used here, refers to the material's apparent brittleness under biaxial strain. This is usually obtained as the ratio of the failing stress of a notched specimen to that of an unnotched specimen. The notched specimen geometry is designed to produce some biaxial strain at the notch. A minimum value of this ratio is 1.00. This apparent brittleness contributed to premature failure of some early boosters.

In pressure-vessel design the variation of ultimate tensile strength with temperature is important. Again, the exposure to elevated temperature is short.

Tank structures for boosters differ in accordance with the type of propellant used. With solid propellants the tank is primarily a pressure vessel; it must contain the high pressures (500–1000 lb/in.^{2}) generated by combustion. Other loads are of secondary importance, and bursting strength of the tank is the significant factor. Care must be taken in design to avoid stress concentrations, particularly near joints, which might cause premature failure. Material selection must consider the sensitivity of strength to such concentrations, with the material in tension in two directions.

Tanks for liquid propellants, on the other hand, must contain relatively low pressures (10–100 lb/in.^{2}). However, flight loads in the form of thrust and bending moment become important in the design of the tank. One approach is to use the tank pressure to assist in carrying these loads, and to stabilize the light-gage unstiffened tank shell. The skin gage is chosen to be just adequate for containing the tank pressure stresses; the pressure is chosen on the basis of providing adequate stability under flight loads. Such pressures are generally in excess of those required to stabilize the tank to the point where the buckling strength predicted by linearized stability theory is developed. Alternative approaches are to provide stiffening in the form of stringers or frames to stabilize the tank, and lowering tank pressures only to that level required by the propulsion system. Care must be taken to prevent excessive stress concentrations from arising at the frames, when the tank is pressurized, or at the stringers because of thermal gradients.

Interstages between tanks are generally designed for flight loads; their construction is similar to airplane fuselages. They may be monocoque unstiffened shells or may have longitudinal (stringer) or circumferential (frame) stiffening. One problem is to provide a means for disconnecting stages as they are expended. This is generally accomplished by explosive actuated frangible bolts which are activated by command.

Multiple staging. Various arrangements of the multiple stages of a boost system are possible. By far the most generally used scheme is the tandem arrangement in which each succeeding stage is placed forward of the one it follows in operation (Fig. 3). This has the advantage of simplifying the separation of stages, but may result in an excessively long vehicle when a large number of stages is involved. This may lead in turn to large flight loads on the tank structure and to aeroelastic instabilities.

An alternative is a clustered arrangement of stages. Although a clustered system reduces vehicle length and flight loads, it usually involves a structural penalty in the fittings and tie structure required to transmit loads and thrust from the outer stages through the cluster.

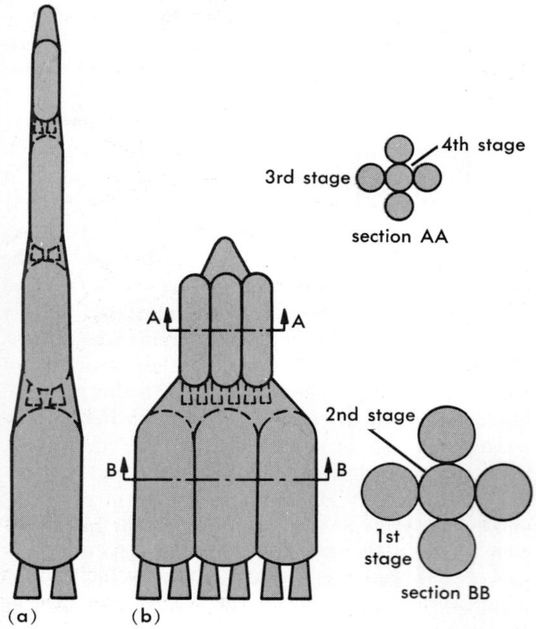

Fig. 3. Booster systems. (*a*) Tandem. (*b*) Clustered.

Launching of the vehicle and booster system may be accomplished in a variety of ways. The most common is launching from an aboveground site with the vehicle rising vertically. This requires consideration of the effects of ground winds on the vehicle prior to and during the time just after launching. Special launching structures are required to withstand the impingement of high-temperature gases from the rocket exhaust during launching.

An alternative is launching from an underground silo. This generally imposes extremely high noise levels (150–170 db) on the booster and vehicle structure, and also may subject it to extremely high pressure pulses of short duration. Launching from an airfield runway using a horizontal take-off requires wings for lift on the booster system and is not generally employed.

Space phase. Structural design of vehicles for space operation is still in its infancy because there are large unknowns in the environment which the structure must sustain, and differing requirements for various space missions. In general, vehicles which will reenter the earth's (or another planet's) atmosphere are designed to meet the reentry requirements. These are much more severe, and if modifications are required to provide capability for the space portion of the mission, such modifications are minor.

The primary function of the structure in a space-mission vehicle is to provide an enclosure containing an environment satisfactorily conditioned for operation of the crew and equipment. This environment generally consists of an artificial atmosphere tailored to payload requirements (such as oxygen for manned vehicles), and protection against extremes in temperature and the influx of radiation and particles present in the space environment.

The most significant structural loads in space will probably arise from the pressure in the enclosed artificial atmosphere. Efficient tension structure, generally of the shell type, is used to contain this. The principal problem is one of insuring pressure tightness so that leakage rates approach zero to a far greater extent than in the past. The maximum accelerative loads experienced will be those occurring during boost phase. Acceleration loads would not be a problem for vehicles initially assembled in orbit.

Temperature extremes in the structure and the enclosed environment are controlled by treatment of the exterior surface to maintain a balance between absorbed and emitted thermal radiation. This is readily accomplished by surface coatings. Because incident solar energy varies inversely with the square of distance from the sun, means of adjusting surface conditions will be required for interplanetary missions. Heat generated by internal equipment or other source must be considered in the heat balance equations.

A principal problem in the design of space vehicles arises from the presence of meteoric particles in space. These particles may have extremely high velocities relative to the space vehicle (up to 225,000 ft/sec). Probability of collision with larger particles is extremely low, but small-particle collisions will be frequent (Fig. 4). Hence protective structure of some type may be required to prevent rupturing of pressurized compartments, damage to equipment, or erosion of surfaces needed for thermal protection in the space or reentry phase by collision with such particles.

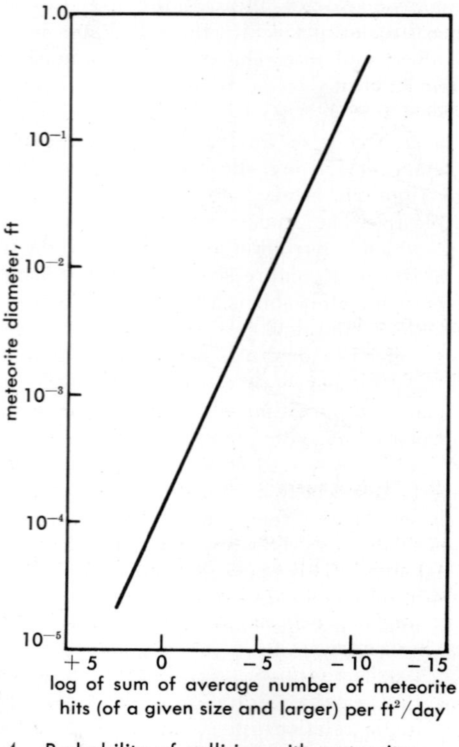

Fig. 4. Probability of collision with meteorites.

Radiation shielding may be required for some vehicles, particularly those operating for extended times within the earth's magnetically trapped radiation belts or during times of high sunspot activity (*see* VAN ALLEN RADIATION). The shielding may or may not be an integral part of the structure. Effects of radiation on the structure are not expected to be severe because dosages required to produce structural damage are generally extremely high.

Reentry phase. Although the atmospheric layer of the earth is relatively thin, it is responsible for the reduction of vehicle velocity and the resulting deceleration loads as well as for the severe heating experienced by reentering vehicles. A body entering the earth's atmosphere possesses a large amount of energy. This energy must be dissipated in a manner which allows the reentering vehicle to survive. Most of the vehicle's original energy can be transformed into thermal energy in the air surrounding the vehicle, and only part of the original energy is retained in the vehicle as heat. The fraction that appears as heat in the vehicle depends upon the characteristics of the flow around the vehicle. In turn, the flow around the vehicle is a function of its geometry, attitude, velocity, and altitude (*see* NOSE CONE). Several different approaches are used, such as the zero-lift, blunt, high-drag body, as exemplified by the intercontinental ballistic missiles and the man in space programs, and high lift-to-drag ratio glide vehicles. The latter approach is used in the Dyna-Soar and intercontinental glide missile programs.

A third approach is to combine the rather compact ballistic shape with some of the lift capability of the glide vehicles; Fig. 5 shows typical examples of the three shapes.

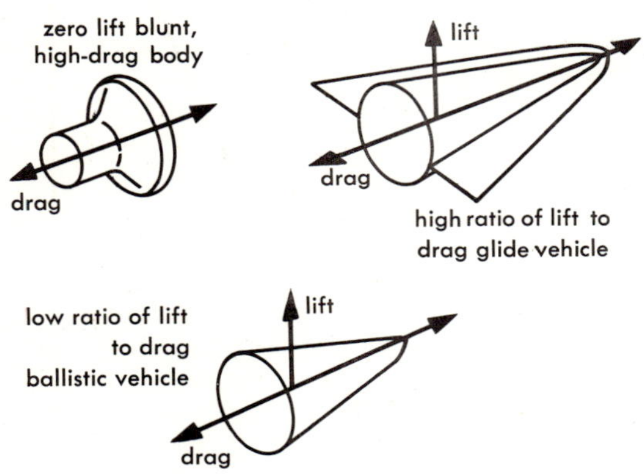

Fig. 5. Typical reentry shapes.

Figure 6 illustrates the basic difference in reentry paths for lifting and nonlifting vehicles. The ballistic missile enters the atmosphere at a steeper entry angle γ between the vehicle path and the local horizontal.

Ballistic zero-lift reentry. Figures 7, 8, and 9 show the velocity, altitude, deceleration, and reentry angle relationships. The ballistic coefficient W/C_DA, where W is vehicle weight, C_D is drag

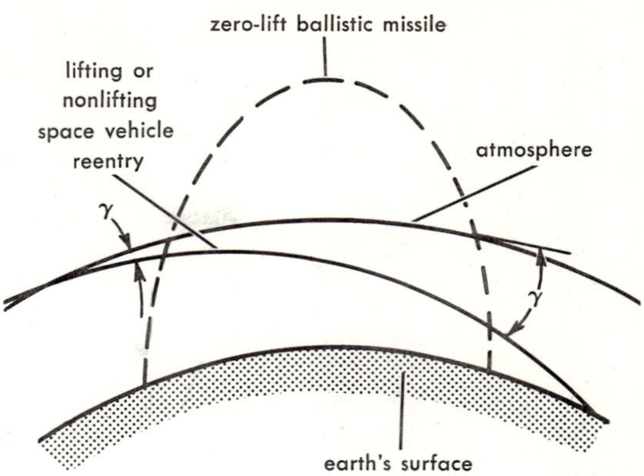

Fig. 6. Comparison of space reentry and ballistic missile trajectories.

coefficient, and A is frontal area, is a measure of the slenderness of the entering vehicle. The higher the ballistic coefficient, the higher the velocity at a given altitude and hence the higher the heating rate (Fig. 10).

The maximum deceleration is independent of drag characteristics and depends only on entry angle, the entry velocity, and the atmospheric characteristics. The drag characteristics determine the altitude at which the maximum deceleration occurs.

Heating rate $\dot{q}$ is a function of geometry, attitude, velocity, and altitude. Ballistic entry produces higher peak heating rates but less total heat is transferred to the entry body because its descent is so rapid.

Initial entry velocity and entry angle at the top of the effective atmosphere determine the maximum load factor and are significant parameters in determining heat rates and total heat input during entry. The initial velocity, which is a function of the vehicle's mission, can be controlled with retro rockets—but with a significant weight penalty.

The heating severity can be controlled to some extent by selection of the entry angle. Blunt noses can divert a large portion of the kinetic energy of the nose cone of a ballistic missile into the air flowing behind the shock wave. This does not eliminate the need for some type of nose-cone protection.

For ballistic missiles, or for any vehicle exposed to high heating rates which exceed the material capability for a relatively short time (up to 5–10 min), several protection schemes are possible.

Ablation is the absorption of thermal energy by melting, vaporization, or sublimation of the surface material to protect the vehicle and payload.

Heat sink is the nonmelting shielding material whose thermal capacity is used to provide protection to the interior structure and equipment.

Cooling protects the shell of the vehicle by using a fluid to absorb the thermal energy by temperature rise or by phase change of the material.

Transpiration cooling operates by diffusing a gas or vapor through a porous skin or opening into the boundary layer to carry heat away and insulate the vehicle.

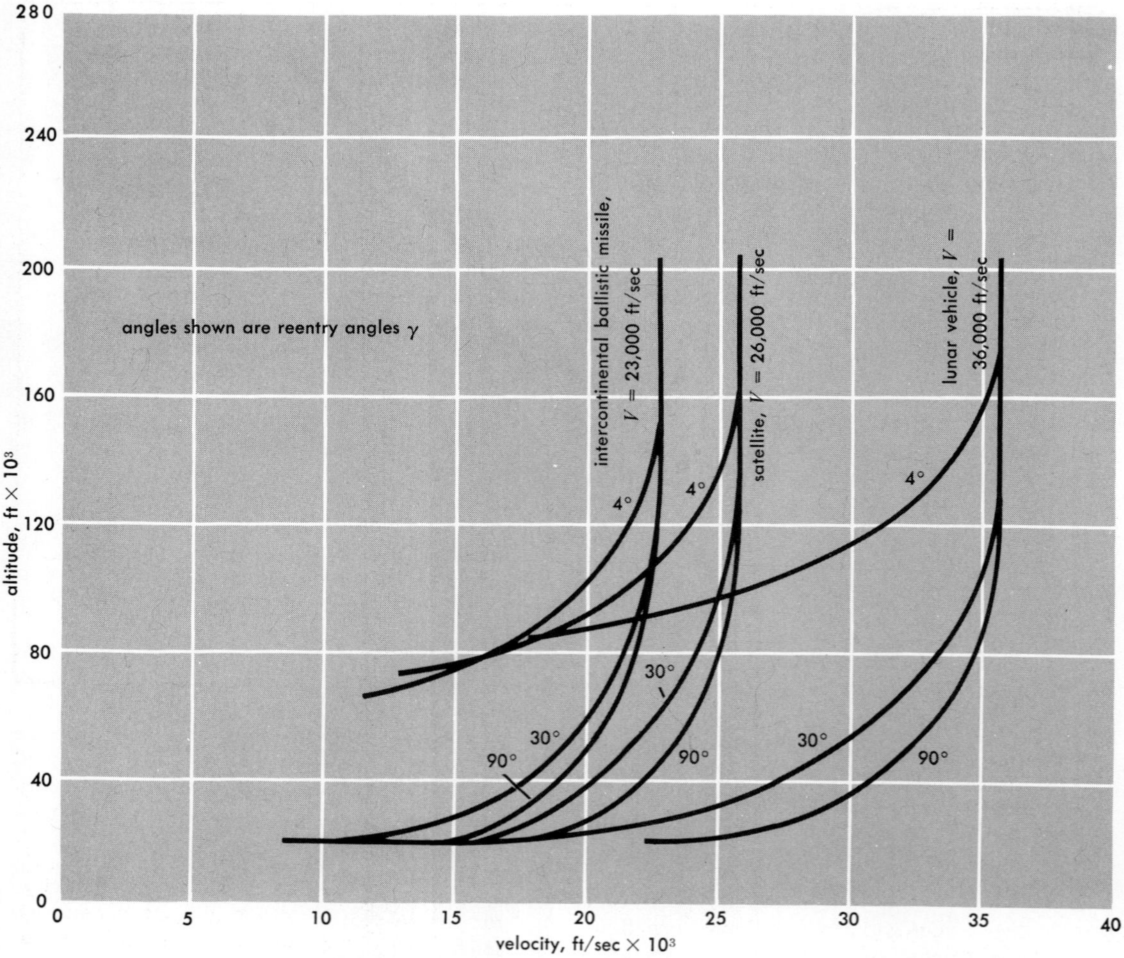

Fig. 7. Typical reentry trajectories for ballistic space vehicles.

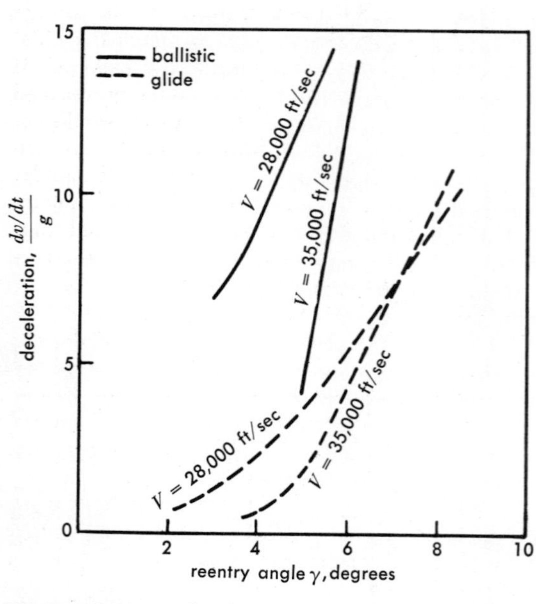

Fig. 8. Maximum deceleration vs. initial reentry angle for glide and ballistic vehicles.

Combinations of these are possible. Several trades can be made. The heating rate $\dot{q}$ at the stagnation point varies inversely with the square root of the diameter, thus $\dot{q} \sim 1/\sqrt{D}$. Therefore, blunting the nose cone, that is, increasing the diameter, will reduce the heating rate. The total energy that must be absorbed per unit time is the product of heating rate $\dot{q}$ times surface area S. When cooling or ablation schemes are used, the total weight, that is, weight of ablation material or weight of the entire cooling system, must be minimized. In these systems the weight is proportional to total heat input Q. The optimization tends toward smaller nose-cone diameters because the rate of heat input varies inversely as the square root and the area varies directly as the square of the diameter. Therefore $Q = \dot{q}S = (1/\sqrt{D})D^2K = KD^{3/2}$. In the case of a cooling system, the weight of any pipes and pumps must also be considered.

Ablation. The normal application of ablation is to limit the temperature of the structural shell or internal components of the payload. This temperature limitation is the principal criterion. The most desirable material, therefore, is one which has a

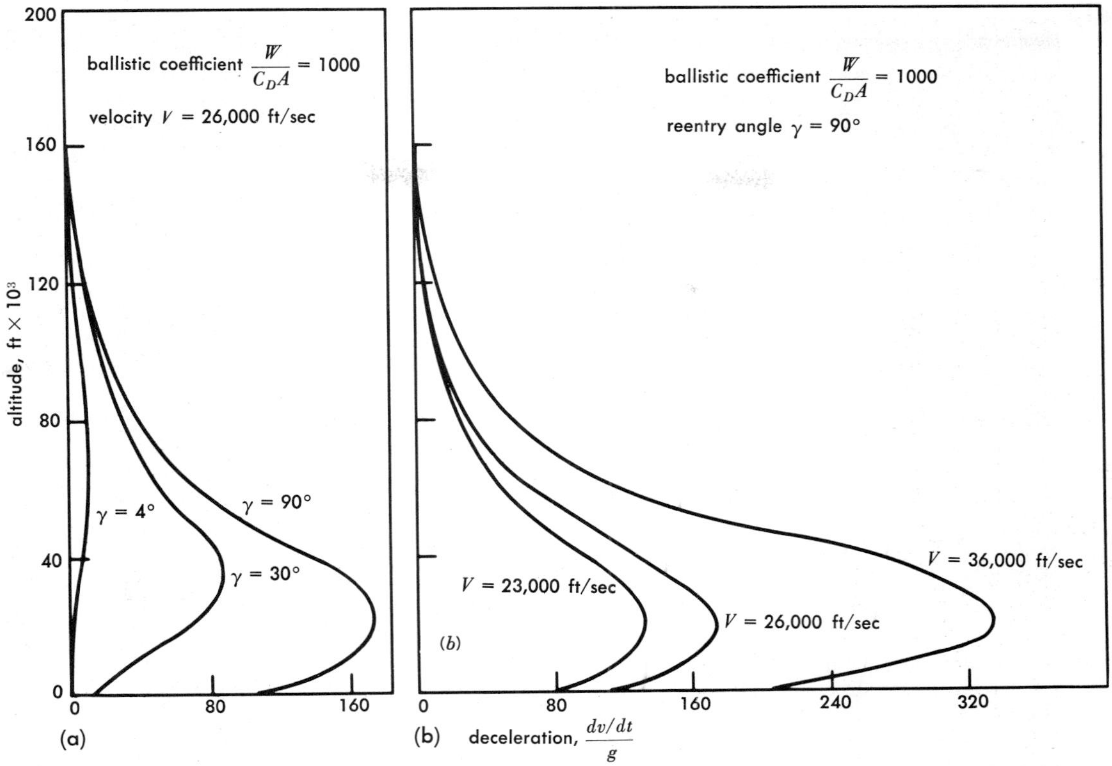

Fig. 9. Deceleration vs. altitude. (a) For constant reentry velocity of 26,000 ft/sec; (b) for constant reentry angle of 90°.

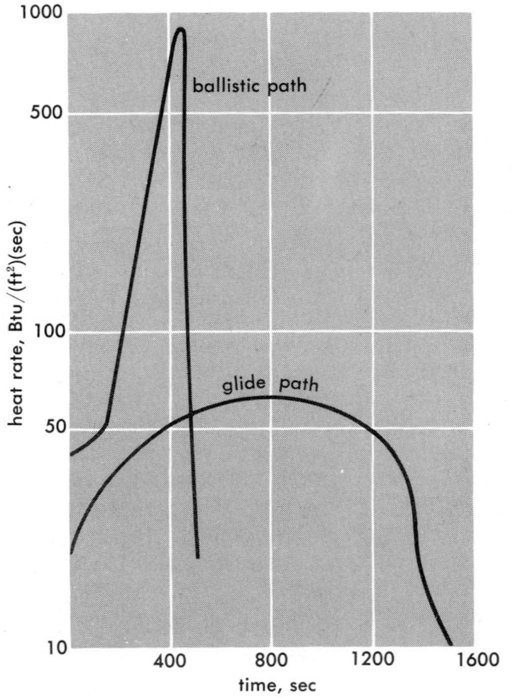

Fig. 10. Heating rate at the stagnation point of an hemispherical body vs. time for glide and ballistic entry paths.

high heat of ablation and low thermal conductivity. Material properties, however, do not necessarily possess these characteristics simultaneously, to the same degree. Therefore, specific investigations or trade studies must be made to determine the minimum weight of ablation material required to maintain the structural temperature within design limits. Care must be exercised to match the ablation material to the environment. For example, if the maximum temperature that would be reached without ablation is below 3000°F, quartz would not be considered because it does not ablate appreciably below 3000°F. Many materials will rapidly deteriorate without ablating if exposed to temperature below their ablation temperatures and thus will not provide the needed protection for the structure. The table lists typical ablation materials.

Heat sink. The heat-sink method is used for low-ballistic-coefficient entering bodies. The basic ma-

Typical ablation materials and properties

Material	Ablation temperature, °F	Thermal conductivity, Btu/(in.)(sec)(°F)	Specific heat, Btu/(lb)(°F)	Density, lb/in.³
Quartz	4400	0.1292×10^{-4}	0.25	0.08391
Acrylic	250*	0.4033×10^{-3}	0.38	0.04282
Teflon	1000*	2.896×10^{-6}	0.25	0.07523
Refrasil + 41% phenolic resin	2000*	3.475×10^{-6}	0.23	0.06366

* Estimated.

terial properties of importance are the thermal conductivity, specific heat, and melting temperature. For high ballistic coefficients, that is, $W/C_DA \geqq 150$, the heat-sink material melts, because it cannot dissipate the heat rapidly enough either by external radiation or internal conduction.

Cooling and transpiration. Cooling the vehicle by circulation of water, lithium, or other mediums, or by transpiration of gas or vapor, is limited by system complexity, reliability, and adverse weight penalties.

Thermal protective systems are time limited because the heat-absorption capacity of the protection system is continuously expended during the process. These systems, therefore, apply primarily to short durations (5–10 min).

Lifting reentry. In manned applications, vehicles employing aerodynamic lift during reentry have several advantages over zero-lift ballistic bodies. First, the use of lift allows a more gradual descent, thus reducing the deceleration forces on both vehicle and occupants (Fig. 8). Second, its ability to glide and maneuver within the atmosphere gives it greater accuracy in either hitting a target or landing at a predetermined spot. Third, it can accommodate greater errors of guidance systems because for a given deceleration it can tolerate a greater range of entry angles. Fourth, greater temperature control is afforded because aerodynamic lift may be varied to control altitude with velocity.

Figure 11 shows the equilibrium flight paths for a design acceleration of unity and for several values of W/SC_L where W is vehicle weight, S is lift area, and C_L is lift coefficient. Parameter W/SC_L may be interpreted as a vehicle with fixed wing loading W/S and varying lift C_L.

The total time required for a lifting vehicle to descend through the atmosphere depends on the drag characteristics of the vehicle. With high drag

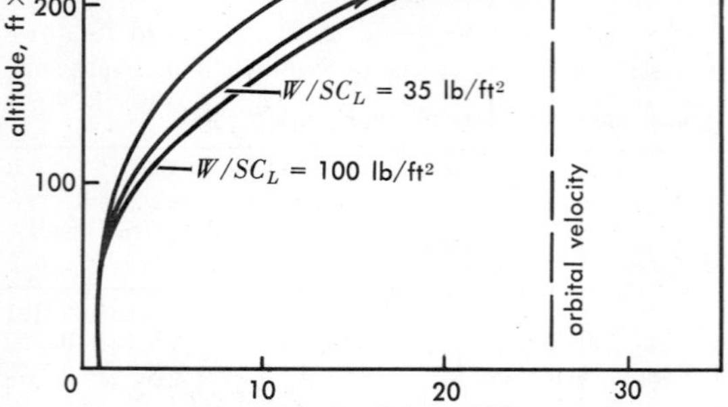

Fig. 11. Typical reentry trajectories for glide space vehicles.

coefficients the time may be as low as 30 min and with low drag, approximately 1–2 hours. In the case of grazing reentries, several orbital loops may be required to bleed off the energy, and reentry duration can be measured in days. The relatively long times required for descent make the use of time-dependent ablation or cooling impractical.

Heat dissipation. The rate of heating along a glide path is sufficiently low to permit use of thin-skin radiation-cooled structure except perhaps for local hot spots near the nose and leading edges.

The structural shell without insulation can reject heat by radiation, its ability to do so being a function of its emissivity ϵ. At equilibrium, the heat input is balanced by the heat radiated away.

An emissivity of 0.90 is readily available. A typical radiation equilibrium temperature for a typical point on a flat plate is plotted in Fig. 11. Although the temperature at points A, B, and C is the same, the loading condition is not, a portion of the weight being carried by the centrifugal force which is velocity dependent. The structural load at A is greater than that at B or C. The designer can, within limits, by proper selection of vehicle geometry, weight, and aerodynamic lift, control the phasing of the temperature and loads.

The designer can further reduce the structural temperatures by using insulation (with a cover of externally high emissivity) over the hottest area of the vehicle and reradiating the heat to a cooler area of the vehicle. For example, the lower surface of a glide vehicle which is at an angle of attack is hotter than the upper surface.

The designer has two options. First, the structural margin of safety can be increased by lowering the structural temperature with the above technique; second, the reentry corridor can be extended in terms of altitude by allowing the insulated vehicle to drop more deeply into the atmosphere until the original design temperature is attained.

Thermal stress. One of the most severe structural problems that can occur is the thermal stresses, which arise as a result of temperature gradients. These temperature gradients result from such causes as heat sinks, transient heating, and manned-compartment cooling. Thermal stress f_{th} is a function of the coefficient of expansion α, the modulus of elasticity E, and the temperature gradient ΔT. That is, $f_{th} = K\Delta T$, within the proportional limit, where K is a function of geometry, material, and restraints.

Thermal stresses will be produced in sandwich or conventional sheet-stringer construction if the thermal deformations are externally constrained. Thermal stresses will also be produced in the absence of external constraints, solely because of the incompatible deformations of the different parts of the body, as for example, a rectangular, anisotropic, sandwich panel with a temperature gradient through its thickness. The thermal stresses result in an effective reduction in the material allowable and leave little strength to carry the airload.

Figure 12 illustrates the thermal stresses for two different materials used in brazed honeycomb con-

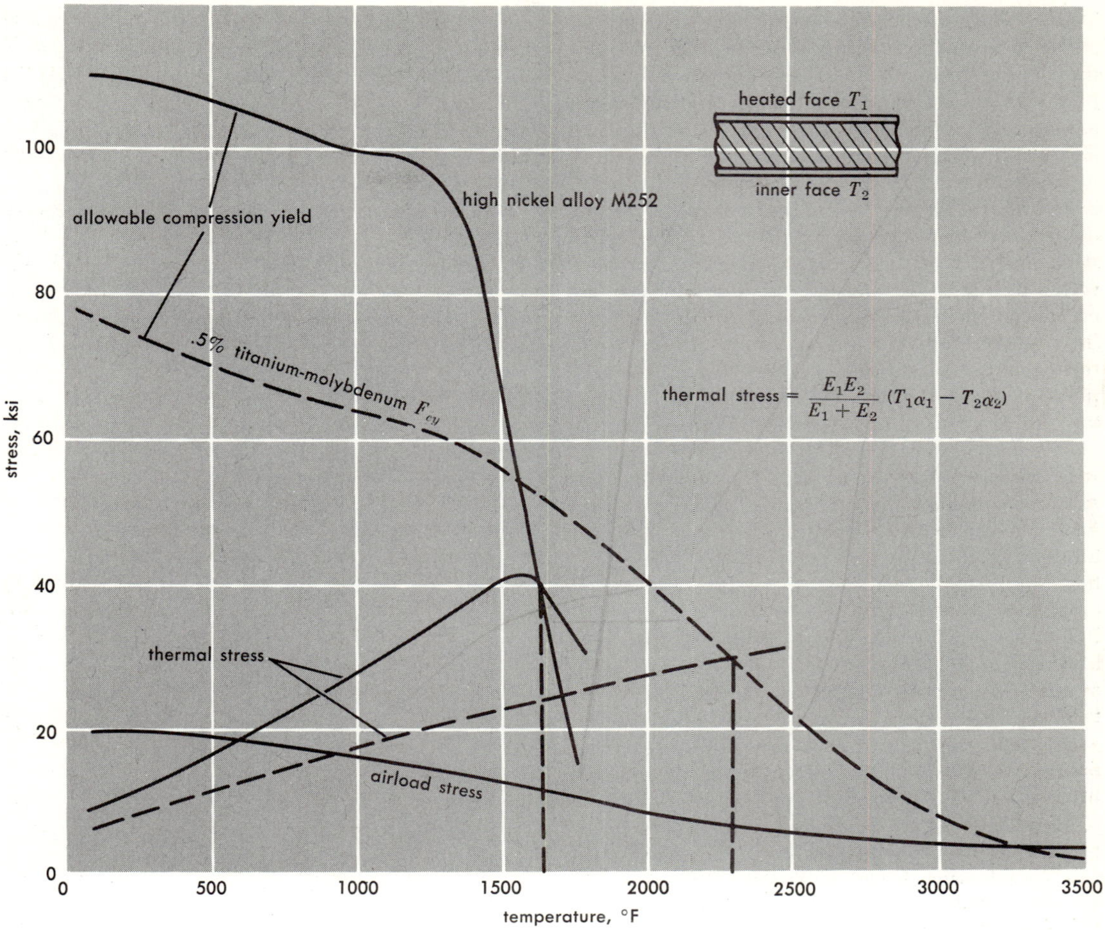

Fig. 12. Typical allowable, thermal, and airload stresses vs. temperature.

struction. It also gives the material compression-yield allowables vs. temperature and a typical airload stress vs. temperature.

The elastic thermal stresses are computed for the typical temperature gradients associated with the maximum temperatures. If the vehicle is intended for repeated use, the compressive yield point cannot be exceeded. As an example, sandwich construction of high-nickel alloy M-252 cannot be used above 1660°F because the thermal stress equals the compressive yield stress at that temperature. The airload stresses should be added to the thermal stresses, thus reducing even further the allowable temperature for material and type of construction under these conditions. The same curves are drawn for a titanium-molybdenum alloy. Its more favorable (lower) coefficient of thermal expansion and its higher allowable stress in the higher-temperature region result in a better sandwich. However, there are many practical problems to be solved in the brazing of molybdenum alloys.

If the vehicle were designed for a single shot, the thermal stresses could exceed the yield point considerably without seriously jeopardizing the mission. However, tremendous benefits accrue if thermal stresses are eliminated.

It is possible, with judicious design, to eliminate or drastically reduce the thermal stresses. One solution is to use corrugated panels (Fig. 13). The direction of the corrugations should be normal to the lines of constant temperature. For example, in Fig. 13, lines AB and CD represent lines of constant but not necessarily equal temperatures. The cover sheet should be as thin as possible or eliminated if not required for aerodynamic reasons. The thermal expansions or strains are accommodated by buckling of the cover skin at low stress.

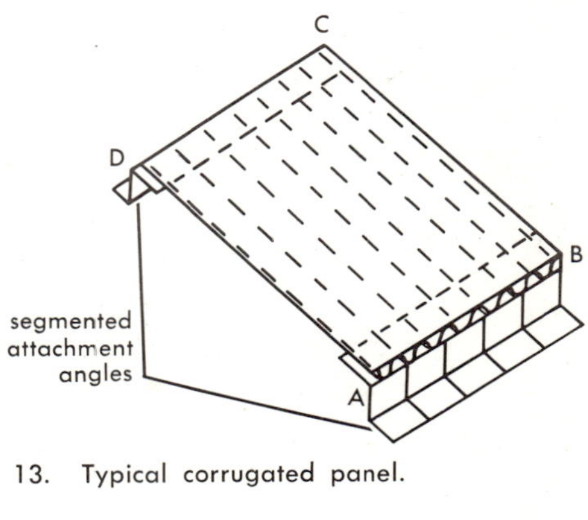

Fig. 13. Typical corrugated panel.

The airloads normal to the panel are carried by simple beam action to edges AB and CD and into the internal framework by short attachments. In-plane shear is transmitted to the internal framework along all four sides by these short attachments. There, the panel is simply supported with ability to rotate at the edges.

The internal framework in turn must be statically determinate in respect to the thermal expansion of its members. A statically determinate truss or space frame meets this requirement. No axial loads are created in a statically determinate pinned truss irrespective of the temperature distribution throughout the truss. Other approaches are possible. The elimination or reduction in thermal stresses results in large dividends in either margins of safety or mission flexibility.

High-temperature materials. The extreme thermal environment has required materials with higher temperature capability than the standard aircraft steels and aluminum alloys. These materials are the super alloys, refractory alloys, and ceramics.

The super alloys which contain high percentages of either nickel or cobalt have a maximum useful temperature range, under low stresses, of 1600–2000°F. These alloys are readily available in thin gages and can be formed, riveted, brazed, and fusion- or spot-welded with essentially standard procedures. The surface of these alloys is rapidly oxidized above 1300°F. However, the oxide forms an adherent coating that protects the metal from further oxidation and provides a high surface emissivity (0.8–0.9+).

The refractory alloys (columbium, molybdenum, tantalum, and tungsten) have higher densities than the super alloys but retain their useful strength from 3300°F (columbium) to over 4000°F (tungsten). All refractory metals are oxidized catastrophically at high temperatures and surface protection is required. Coatings have been developed which protect molybdenum for approximately 300 hours at 3000°F and columbium for several hours at 2600°F. Columbium alloys are available in thin gages and can be formed by standard procedures. The quality of thin-gage molybdenum alloys is not consistent and requires hot forming because of their low ductility at room temperature. Manufacture of tungsten alloy sheet is in the development stage. Tantalum is similar to columbium except that it has a much higher density. The refractory alloys can be joined by riveting, but fusion-welding requires such special procedures as inert atmosphere. Spot-welding is extremely difficult.

Conventional ceramics retain fairly high strengths up to 4000°F, and the more exotic ceramics are potentially usable to over 8000°F. Their high melting points and relative inertness to oxidizing atmospheres make them attractive materials for leading edges and nose cones of glide vehicles. However, the lack of ductility presents serious design problems.

Cost, on the basis of dollars per pound, is not necessarily the criterion for material selection. Availability, strength, stiffness, weight, producibility, and cost of fabrication must also be considered. The final material selected will be the one which results in the least expensive end item and best meets the design requirements. *See* SATELLITE, ARTIFICIAL; INTERPLANETARY PROPULSION; SPACE FLIGHT. [E.G.C.]

Bibliography: R. A. Anderson and W. A. Brooks, Jr., *Effectiveness of Radiation as a Structural Cooling Technique for Hypersonic Vehicles*, Inst. Aeronaut. Sciences Preprint 59–65; E. G. Czarnecki and M. T. Braun, Structural aspects of earth glide reentry vehicles, *Proc. Am. Astrounautical Soc.*, vol. 3, 1956; P. E. Grafton and E. F. Styer, *Booster Case Design for Hypersonic Vehicles*, Inst. Aeronaut. Sciences Preprint 835.

Space-time

A term used to denote the geometry of the physical universe as suggested by the theory of relativity. It is also called space-time continuum. Whereas in Newtonian physics space and time had been considered quite separate entities, Albert Einstein and H. Minkowski showed that they are actually intimately intertwined. Isaac Newton's ideas on space and time may be summarized as follows:

1. Given two events, each of which is clearly localized in space and lasts only for an instant in time, such as two strokes of lightning striking small targets, all observers will agree as to which of the two events took place earlier in time, or whether they were actually simultaneous.

2. If the events were not simultaneous, the interval of time between them is an absolute entity, agreed on by all competent observers.

3. The spatial distance between the two events is an absolute entity, agreed on by all competent observers.

Of these three, the first assumption, concerning the concept of simultaneity of distant events, is the crucial one, the other two depending on it. Simultaneity, however, can be given an unambiguous meaning only if there is available some instantaneous method of signaling over finite distances. Actually, according to the theory of relativity, the greatest speed of transmission of intelligence of any kind is the speed of light, c, equaling about 3×10^{10} cm/sec (*see* RELATIVITY). Moreover, any signal traveling precisely at the speed c appears to travel at that same speed to all conceivable observers, regardless of their own states of motion. This is the only reasonable interpretation of the results of the Michelson-Morley experiment and the effect of aberration (*see* ABERRATION OF LIGHT; LIGHT). Accordingly, the question of whether two given events are simultaneous or not can be decided only with the help of signals that at best have traveled from the sites of these events to the station of the observer at the speed of light.

Under these circumstances, Einstein showed that in general two observers, each using the same techniques of observation but being in motion relative to each other, will disagree concerning the simultaneity of distant events. But if they do disagree, they are also unable to compare unequivocally the rates of clocks moving in different ways, or the lengths of scales and measuring rods. Instead, clock rates and scale lengths of different observers and different frames of reference must be established so as to assure the principal observed fact. Each observer, using his own clocks and scales, must measure the same speed of propagation of light. This requirement leads to a set of relationships known as the Lorentz transformations (*see* LORENTZ TRANSFORMATIONS).

In accordance with the Lorentz transformations, both the time interval and the spatial distance between two events are relative quantities, depending on the state of motion of the observer who carries out the measurements. There is, however, a new absolute quantity that takes the place of the two former quantities. It is known as the invariant, or proper, space-time interval τ and is defined as follows:

$$\tau^2 = T^2 - \frac{1}{c^2} R^2 \qquad (1)$$

In this equation T is the ordinary time interval, R the distance between the two events, and c the speed of light in empty space. Whereas T and R are different for different observers, τ has the same value. In the event that Eq. (1) would render τ imaginary, its place may be taken by σ, defined thus:

$$\sigma^2 = R^2 - c^2T^2 \qquad (2)$$

If both τ and σ are zero, then a light signal leaving the location of one event while it is taking place will reach the location of the other event precisely at the instant the signal from the latter is coming forth.

The existence of a single invariant interval led the mathematician Minkowski to conceive of the totality of space and time as a single four-dimensional continuum, which is often referred to as the Minkowski universe. In this universe, the history of a single space point in the course of time must be considered as a curve (or line), whereas an event, limited both in space and time, represents a point. So that these geometric concepts in the Minkowski universe may be distinguished from their analogs in ordinary three-dimensional space, they are referred to as world curves (world lines) and world points, respectively.

Minkowski geometry. The geometry of the Minkowski universe in some respects resembles the geometry of ordinary (euclidean) space but differs from it in others. The Minkowski universe has four dimensions instead of the three dimensions of ordinary space; that is to say, for a complete identification a world point requires four pieces of data, for instance three space coordinates and a time reading. But there are in the Minkowski universe world points, world lines, two-dimensional surfaces (including planes), three-dimensional surfaces (often called hypersurfaces), and four-dimensional domains. A hypersurface may, for instance, be a spatial domain (volume) at one instant in time, or it may be a two-dimensional (ordinary) surface for an extended period of time. Thus one may form all the geometric figures that are also possible in a four-dimensional euclidean space.

The indefinite metric. In a euclidean space there are the cartesian coordinate systems, those rectilinear systems of coordinates that are mutually perpendicular and whose coordinate values correspond to real lengths. The distance S between two points whose coordinate differences are X, Y, and Z, respectively, is given by

$$S^2 = X^2 + Y^2 + Z^2 \qquad (3)$$

The coordinate transformations that lead from one cartesian coordinate system to another are called orthogonal coordinate transformations. Formally, they are the coordinate transformations that preserve the precise form of Eq. (3). Likewise, the Lorentz transformations preserve the precise form of Eqs. (1) and (2), respectively. A coordinate system in which these two equations hold is called a Lorentzian frame of reference.

Whereas the form on the right of Eq. (3) is positive definite, that is, always greater than or equal to zero, the right-hand sides of Eqs. (1) and (2) are indefinite, that is, they may be positive or negative. This fact represents the single but all-important difference between a four-dimensional euclidean space and a Minkowski universe. The forms (1), (2), and (3) are called metrics. Hence the Minkowski universe is said to possess an indefinite metric.

As a result of the indefinite character of the metric, a triangle in the Minkowski universe may possess one side that is longer than the sum of the two others; conversely, one of the three sides may have the length zero. Depending on whether τ or σ is real, or both vanish, an interval is classified as timelike, spacelike, or a null-interval.

Improper Lorentz transformations. In the Minkowski universe there are three different types of Lorentz transformations involving some kind of reflection, in addition to the more usual Lorentz transformations (called proper Lorentz transformations). The first type of improper Lorentz transformation changes the sign of all three spatial coordinates but leaves the sense of the time axis unchanged. This transformation changes a right-handed screw into a left-handed screw; it is also called a parity transformation. The second type of improper Lorentz transformation interchanges the future with the past but leaves the space coordinates unchanged. This transformation is called time reversal. The third improper Lorentz transformation reflects both the space and the time coordinates; it bears no special name of its own. The original arguments which led to the formulation of the special theory of relativity all support the proposition that the laws of nature are invariant

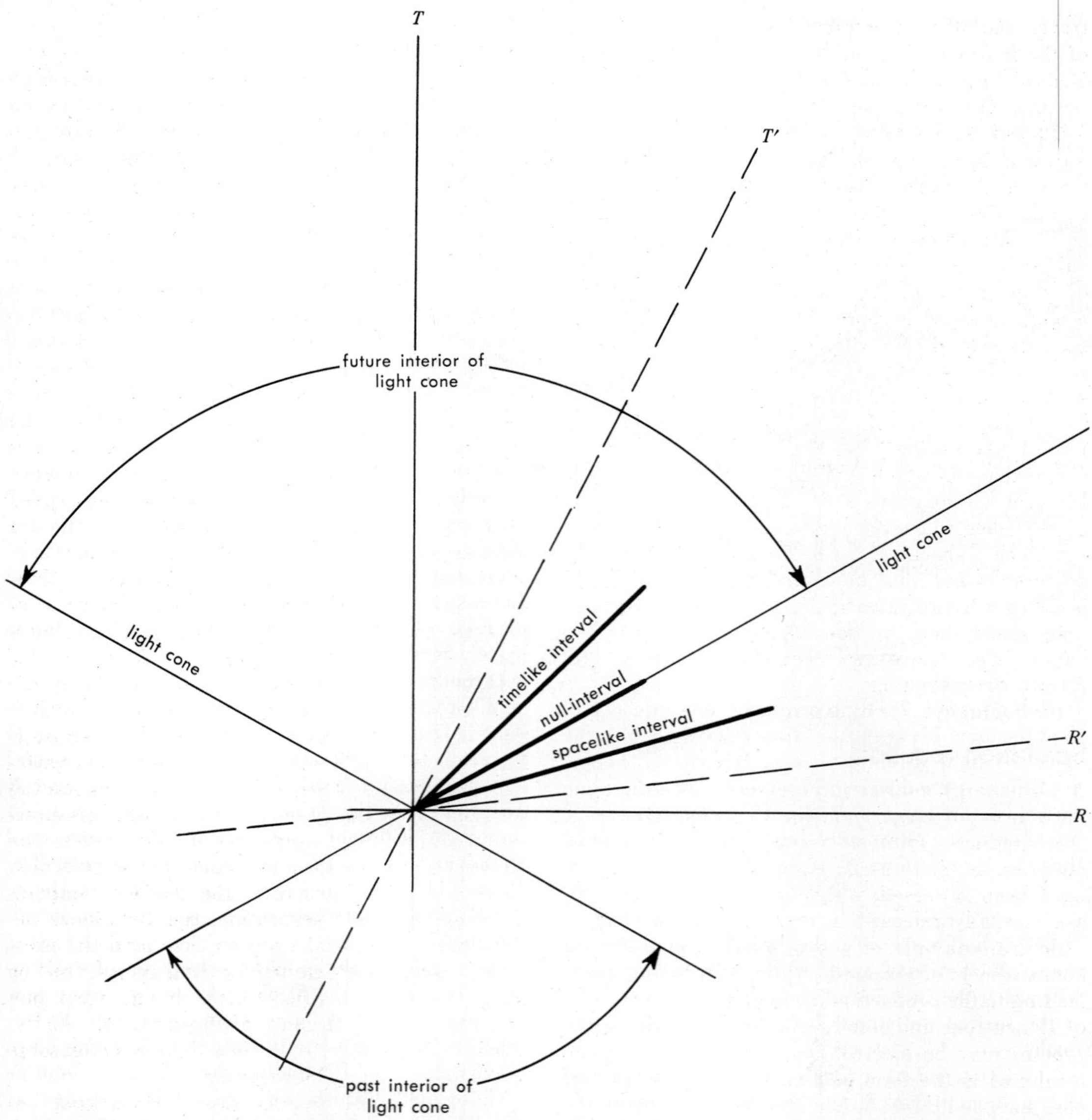

Two-dimensional sketch showing the three possible classes of intervals. The null-directions actually form a three-dimensional surface having the shape of a double cone, called the light cone. An alternative Lorentz frame is indicated by the dashed lines.

under proper Lorentz transformations. They are inconclusive as to whether the laws of nature should also be invariant under the improper Lorentz transformations. The laws of mechanics and of electrodynamics have this property; the second law of thermodynamics distinguishes between past and future in a nonsymmetric manner, but it is usually assumed that this feature of thermodynamics is to be explained by its statistical nature and that it has no bearing on the properties of the underlying basic dynamics. *See* SYMMETRY LAWS (PHYSICS).

Until recent times, it had therefore been taken for granted that the basic laws of nature should make no distinction between right-handed and left-handed screws, and that they should not discriminate between future and past. Certain difficulties in the interpretation of the decay of *K*-mesons led T. D. Lee and C. N. Yang in 1956 to suspect that these assumptions might not be tenable, and they suggested some experiments on meson decay and on radioactive β-decay, which showed that in these particle transformations nature certainly discriminates between left-handed and right-handed screws. *See* PARITY (QUANTUM MECHANICS). It appears there are two kinds of neutrinos (electrically neutral particles which travel at the speed of light and are endowed with an intrinsic spin equal to that of electrons). One kind of neutrino always spins clockwise when viewed in the direction of its travel, the other counterclockwise. The exact role of the improper Lorentz transformations in nature is now under intensive theoretical as well as experimental investigation.

Curved space-time. Whereas the Minkowski universe is the appropriate geometric model for the special theory of relativity, the general theory of relativity makes use of a further generalization. In the Minkowski universe a particle that is not subject to external forces and which therefore travels along a straight line (in ordinary space) and at a uniform speed is represented by a straight world line. In general relativity, an external gravitational force is indistinguishable from an inertial force, which in special relativity would arise if a non-Lorentzian frame of reference were to be employed. Accordingly one requires a four-dimensional space in which it is impossible to distinguish between a Lorentzian and a non-Lorentzian frame of reference whenever a gravitational field is present. Such a space cannot be flat, as the Minkowski universe is, but must be *curved*. Such an aggregate is called a Riemannian space. In a general Riemannian space there are no straight lines but only curves. *See* GEOMETRY, RIEMANNIAN; *see also* FRAME OF REFERENCE. [P.G.B.]

Bibliography: *See* RELATIVITY.

Spallation reaction

A high-energy nuclear interaction which results in the release of large numbers of nucleons as reaction products. With sufficiently high bombarding energies, as many as 20 or more ejected particles have been observed, although only 3–7 nucleons are normally released in the reactions of highest yield. An example of a spallation reaction is the bombardment of arsenic-75 with 180-Mev deuterons, leading to the production of iron-59 and the release of 11 neutrons and eight protons. The neutrons and protons may be ejected as individual nucleons or combined in the form of α-particles or heavier nuclei, such as lithium-8. *See* NUCLEAR REACTION. [W.W.BU.]

Sparganosis

An infection by the metacestode called the sparganum, or plerocercoid, of certain species of the genus *Spirometra*. The adult stage normally occurs in dogs and cats. The genus has a life cycle much like *Dibothriocephalus latus*, and the plerocercoid commonly develops in the musculature of frogs, snakes, or aquatic mammals. The whitish, unsegmented plerocercoid mav be more than a foot long. If a host containing plerocercoids is eaten by an animal other than a suitable definitive host, the worms can reinvade the musculature or body cavities of the new host and remain as plerocercoids. In the Orient freshly opened frogs are sometimes used as poultices on sores, particularly around the eyes, and if plerocercoids are present, they may invade the human tissues. Human sparganosis is rare in North America. *See* PSEUDOPHYLLIDEA. [R.S.FR.]

Spark, electric

A transient form of gaseous conduction. This type of discharge is difficult to define, and in fact no universally accepted definition exists. It can perhaps best be thought of as the transition between two more or less stable forms of gaseous conduction. For example, the transitional breakdown which occurs in the transition from a glow to an arc discharge may be thought of as a spark.

Electric sparks play an important part in many physical effects. Usually these are harmful and undesirable effects, ranging from the gradual destruction of contacts in a conventional electrical switch to the large-scale havoc resulting from lightning discharges. Sometimes, however, the spark may be very useful. Examples are its function in the ignition system of an automobile, its use as an intense short-duration illumination source in high-speed photography, and its use as a source of excitation in spectroscopy. In the second case the spark may actually perform the function of the camera shutter, because its extinction renders the camera insensitive. *See* SPECTROSCOPY; STROBOSCOPIC PHOTOGRAPHY.

Mechanisms. This phenomenon is probably the most complicated of all forms of gaseous conduction. It is exceedingly difficult to study, because it is a transient and because there are so many variables in the system. Some of these variables are the components of the gaseous medium, the gas pressure, the chemical form of the electrodes, the physical shape of the electrodes, the microscopic physical surface structure, the surface temperature, the electrode separation, the functional dependence of potential drop on time, and the presence or absence of external ionizing agents. One or more of these conditions may change from one spark to the next. Because of the great complexity, it will be impossible to do more than touch on some of the main features in this article.

The dependence of breakdown, or sparking, potential on pressure, p, and electrode separation, d, may be considered first. It was shown experimentally by F. Paschen and theoretically by J. S. Townsend that the sparking potential is a function of the product pd and not of p or d separately (Fig. 1). Further, there is a value of pd for which the sparking potential is a minimum. Thus, if it is desired to prevent sparking between two electrodes, the region may be either evacuated or raised to a high pressure. The latter method is used in the case of accelerators of the electrostatic generator variety. Here the entire apparatus is placed in a pressurized tank.

Qualitatively, one of the aspects of a spark is that the entire path between electrodes is ionized. It is the photon emission from recombination and decay of excited states which gives rise to the light from the spark. Further, if the spark leads to a stable conduction state, the cathode must be capable of supplying the needed secondary electrons, and

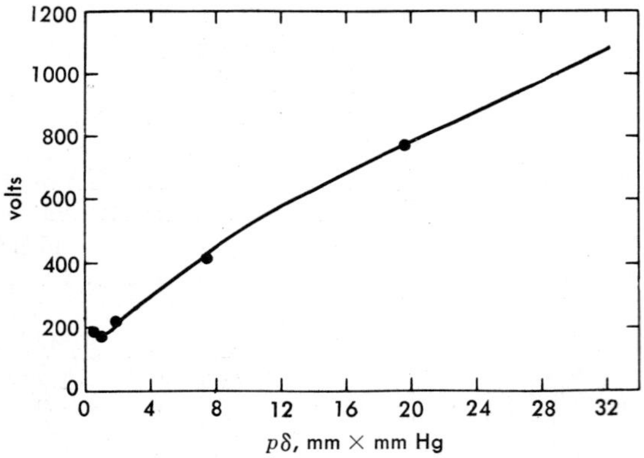

Fig. 1. Dependence of sparking potential on pd for a sodium cathode in hydrogen gas. (*From L. B. Loeb and J. M. Meek, The Mechanism of the Electric Spark, Stanford Univ. Press, 1941*)

the conduction state produced must permit the discharge of the interelectrode capacitance at the very minimum. *See* ARC DISCHARGE; ELECTRICAL CONDUCTION IN GASES; GLOW DISCHARGE.

In a consideration of the mechanism involved in the spark, the time required for the breakdown of the gas in a gap is an important element. L. B. Loeb has pointed out that this time is often less than that required for an electron to traverse the gap completely. This implies that there must be some means of ionization present other than electron impact and that the velocity of propagation of this ionizing agent or mechanism must be much greater than the electron velocity. It seems definitely established that this additional method must be photoionization. In the intense electric field which is necessary for the spark, the initial electron will produce a heavy avalanche of cumulative ionization. Light resulting from the decay processes will produce ionization throughout the gas and electrons at the surfaces by the photoelectric effect (Fig. 2). The electrons resulting from this will in turn produce further avalanches through the entire region, so that in a time of the order of 10^{-8} sec the entire path becomes conducting. If the pressure is approximately atmospheric, the spark will be confined to a relatively narrow region, so that the conducting path, while not straight, will be a well-defined line. If the external circuit can supply the necessary current, the spark will result in an arc discharge. At lower pressure the path becomes more diffuse, and the discharge takes on either a glow or arc characteristic.

Figure 2 shows A, the electron multiplication of electrons by the cumulative ionization of a single electron liberated from the cathode by a photon; B, a secondary electron emitted from the cathode by a positively charged ion; C, the development and structure of an avalanche, with positively charged ions behind electrons at the tip; D, the avalanche has crossed the gap, spreading by diffusion; and F, the later history of an avalanche when electrons

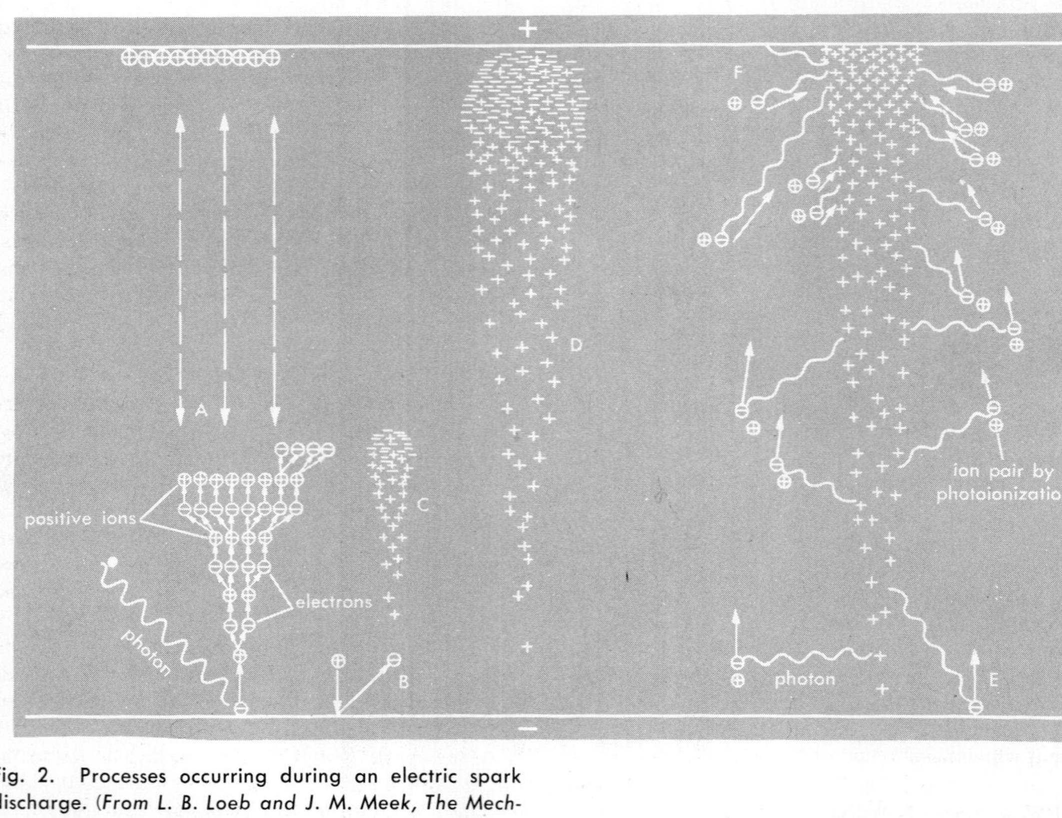

Fig. 2. Processes occurring during an electric spark discharge. (*From L. B. Loeb and J. M. Meek, The Mechanism of the Electric Spark, Stanford Univ. Press, 1941*)

have disappeared into the anode. A positive space-charge boss appears on the cathode at F. Ion pairs out from the trail indicate the appearance of photoelectric ion pairs in the gas produced by photons from the avalanche. E shows a photoelectron from the surface of the cathode produced by the avalanche.

Theory. Mathematically, the theory of Townsend predicts that the current in a self-sustained discharge of the glow variety will follow the equation:

$$I = I_0 \frac{e^{\alpha x}}{1 - \gamma e^{\alpha x}}$$

In this equation, I is the current with a given plate separation x, I_0 is the current when x approaches zero, and α and γ are constants associated with the Townsend coefficients. This equation represents the case where the electrode separation is varied while the ratio of electric field to pressure is held constant. The condition for a spark is that the denominator approach zero, or:

$$\gamma = e^{-\alpha x}$$

It has been indicated by Loeb that this criterion must be handled carefully. Townsend's equation really represents a steady-state situation and it is here being used to explain a transient effect. If the processes which are involved are examined more carefully, it appears that there should also be a dependence on I_0 as well. [G.H.MI.]

Bibliography: L. B. Loeb, *Fundamental Processes of Electrical Discharge in Gases*, 1939; L. B. Loeb, Statistical factors in spark discharge mechanisms, *Revs. Modern Phys.*, 20:151–160, 1948; L. B. Loeb and J. M. Meek, *The Mechanism of the Electric Spark*, 1941.

Spark counter

A particle detector which uses the ionization produced in a gas by high-speed charged particles to trigger a spark between two electrodes (*see* PARTICLE DETECTOR). Spark counters react in a very short time (about 10^{-9} sec) to the particle, and thus can be used for fast timing. The spark is visible and can be photographed.

The principal components of a spark counter are two plane-parallel metallic electrodes, with a gas between the electrodes consisting of a mixture of argon and an organic gas such as xylene (see illustration). A potential difference of about 2000 volts is placed across the electrodes, which are spaced about 2 mm apart. The function of the xylene gas is to aid in quenching the discharge which occurs between the plates when ions are produced in the gas by a charged particle passing through. Although the response time of a spark counter is very fast, the counting rate is very low, since additional quenching must be provided by an electronic circuit which has a recovery time of 0.25 sec. [W.B.F.]

Bibliography: J. W. Keuffel, *Parallel-plate counters, Rev. Sci. Instr.*, 20:202—208, 1949.

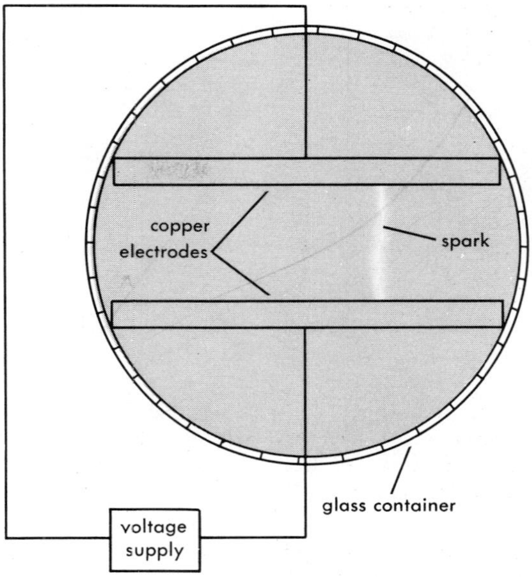

Spark counter. The spark may be observed visually or photographed. The gas is a mixture of argon and xylene.

Spark gap

The region between two electrodes in which a disruptive electrical spark may take place. The gap should be taken to mean the electrodes as well as the intervening space.

Such devices may have many uses. The ignition system in a gasoline engine furnishes a very important example. Another important case is the use of a spark gap as a protective device in electrical equipment. Here, surges in potential may be made to break down such a gap so that expensive equipment will not be damaged.

The dependence of spark-gap operation on such factors as pressure, length, electrode characteristics, and time dependence of potential is quite complicated. *See* BREAKDOWN POTENTIAL; SPARK, ELECTRIC. [G.H.MI.]

Spark plug

A device that screws into the cylinder of an internal-combustion engine to provide a pair of electrodes between which an electrical discharge is passed to ignite the combustible mixture. The plug consists of an outer steel casing that is electrically grounded to the engine and a ceramic insulator that is sealed into the casing and through which a central electrode passes. The high-tension current jumps the gap between this electrode and a similar one fixed to the outer casing. The electrodes are made of alloys that resist electrical and chemical erosion. The plug parts exposed to the combustion gases are designed to operate at temperatures hot enough to prevent electrically conducting deposits, but cool enough to avoid ignition of the mixture before the spark occurs. (Illustration on following page.) *See* IGNITION SYSTEM. [A.R.R.]

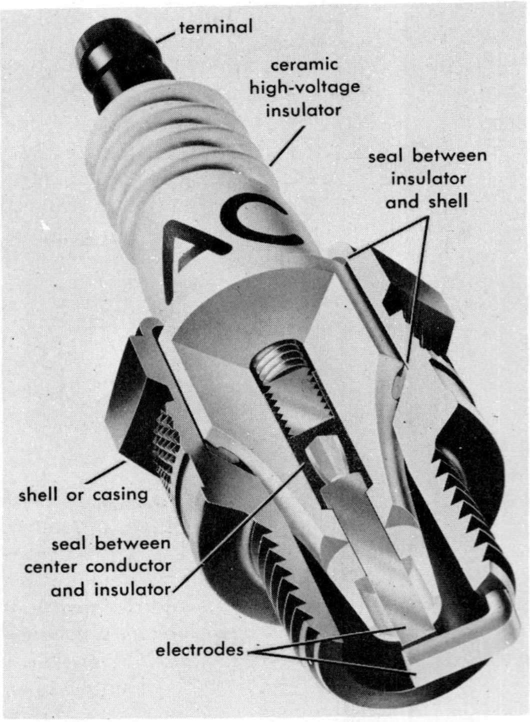

Construction of typical spark plug. (*AC Spark Plug Division, General Motors Corp.*)

Sparrow

Any of a number of perching birds, of the order Passeriformes, and divided variously into families by different authors. The sparrows and their relatives, the finches and grosbeaks, are worldwide in their distribution. Those species that are streaked brown birds are usually called sparrows. In the eastern United States, 60 species are listed as sparrows; all but 2 of these are native American sparrows of the family Fringillidae. The English sparrow, *Passer domesticus*, and its close relative the European tree sparrow, *P. montanus*, belong to the Old World sparrows or weaver finches of the family Ploceidae. The latter is limited in this country to a small area in and near St. Louis, Missouri. The English sparrow is well known as one of the most successful of all introduced birds, being rivaled only by the starling in this respect, and equally regarded as a pest.

The native sparrows vary in size from the small chipping, field, and grasshopper sparrows, scarcely 5 in. long, to the relatively large and showy Harris's sparrow, 7–8 in. long. Many nest on the ground; others nest in bushes and trees. Most sparrows are migratory; some northern nesting species winter in the central states, and many winter in the southern states. Most sparrows are seed eaters as adults, but the young are fed a diet of insects. The utilization of seeds rather than insects as adult food is one of the principal reasons sparrows may winter in relatively cold sections. The song of many species is feeble and unmusical, even insectlike, but some, like the song sparrow and white-throated sparrow, have melodious and pleasing songs. *See* PASSERIFORMES; STARLING. [J.D.B.]

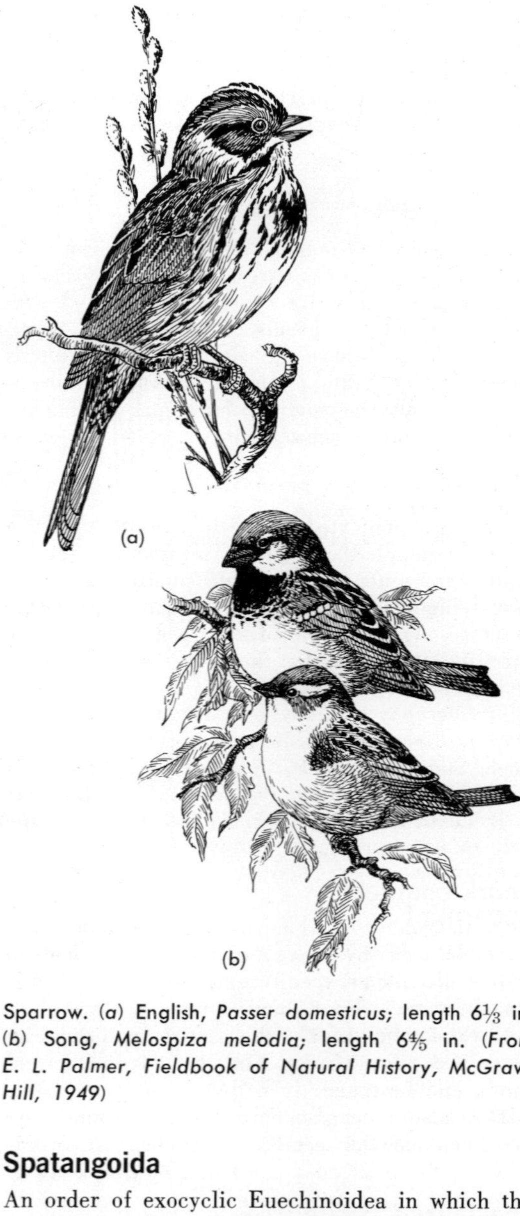

Sparrow. (*a*) English, *Passer domesticus;* length $6\frac{1}{3}$ in. (*b*) Song, *Melospiza melodia;* length $6\frac{4}{5}$ in. (*From E. L. Palmer, Fieldbook of Natural History, McGraw-Hill, 1949*)

Spatangoida

An order of exocyclic Euechinoidea in which the posterior ambulacral plates form a shield-shaped area behind the mouth. The plates are arranged in two similar parallel longitudinal series. The structure is termed an amphisternous plastron. Teeth are lacking; phyllodes develop, but without bourrelets. Thus there is no floscelle (*see* CASSIDULOIDA). The apical system is compact. The families are defined mainly by reference to the fascioles, which are ribbonlike bands of minute, close-set uniform, ciliated spinules on various parts of the test (see illustration). According to their position, they are defined as anal, subanal, inner (surrounding the

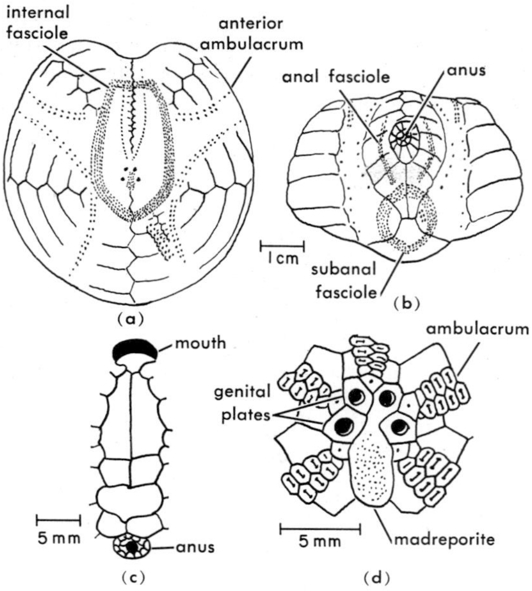

Diagnostic characters of spatangoids. (*a*) *Echinocardium cordatum,* aboral view. (*b*) Posterior aspect. (*c*) Amphisternous plastron. (*d*) Apical system.

apical system), peripetalous (surrounding the petals), and marginal. The earliest forms were the Toxasteridae of the lower Cretaceous, which lacked fascioles and petals. The surviving Palaeopneustidae are deep-sea forms with an oval test, long spines, and weakly developed fascioles and petals. The Spatangoida are more specialized, heart-shaped forms which live in mud or sand. The order reached its maximum in the mid-Tertiary, but is still richly represented today in most seas. *See* ECHINOIDEA; EUECHINOIDEA. [H.B.F.]

Spearmint

This plant, *Mentha spicata,* a common mint now widely distributed in all parts of the world, has long been regarded a prized aromatic. The leaves, both fresh and dried, are used as tea, for mint sauce and mint jelly, and to flavor sauces, soups, stews, and beverages such as mint juleps. Spearmint is also a popular flavoring in chewing gum. *See* TUBIFLORALES; *see also* SPICE AND FLAVORING. [P.D.S.]

Specialized tissue

There exists a common, fundamental chemical ground plan, of composition as well as of metabolism, to which all animals conform. The composition of cells, the energy-yielding mechanisms, the synthetic sequences, and the elaboration of enzymes are remarkably similar. With evolution, tissue differentiation becomes more and more prominent, each tissue specializing more or less completely in some particular form, or forms, of activity. This differentiation does not fabricate new metabolic devices, but rather numerous secondary, specific, and adaptational variations, some of addition, often a perfection of a fundamental device, and others of omission. In the sections that follow, the special partitions of metabolic duties and adaptations to them by muscle, nerve, connective tissue, bone, tooth, blood, eye, liver, and skin will be surveyed. Some tissues are perfected in such functions as motility, transmission of impulses for the control and coordination of other tissues, the processing of metabolites for other tissues, the storage of nutrients for future use, the segregation of tissues into discrete organs, the protection of tissues against physical and chemical injury, the transportation of nutrients and waste products to their respective destination, and vision. For a further discussion of these tissues *see* BLOOD; BONE; EYE; INTEGUMENT; LIVER; MUSCULAR SYSTEM; NERVOUS SYSTEM; SKELETAL SYSTEM; TOOTH.

Muscle. This is a tissue that occurs in three types, striated (skeletal), smooth (involuntary), and cardiac (heart), all concerned with contraction and relaxation. Their composition and metabolism are not greatly different. The structural unit of striated muscle, the type for which most information is available, is the long, threadlike myofibril; it takes several of these to make up a muscle fiber, and many fibers to make up a whole muscle. Muscle consists chiefly of the protein actomyosin, formed by the association of two proteins, actin and myosin. The breakdown of adenosinetriphosphate (ATP) is the earliest detectable reaction in the contractile cycle, and hence is the most immediate known source of free energy for conversion into the mechanical energy of contraction. Myosin catalyzes this breakdown. The continuous regeneration of ATP is accomplished by the hydrolysis of creatine phosphate and by other means. A most important property of muscle is its ability to do work anaerobically and to accumulate an oxygen deficit in which lactic acid is generated from glycogen at a rate in excess of its utilization. It is oxidized later in a recovery period. Besides glucose, the muscles use fatty acids as an energy source, at a faster rate during exercise. The mechanical efficiency of work is somewhat higher on carbohydrate than on fat diets. *See* ADENOSINEDIPHOSPHATE (ADP); ADENOSINETRIPHOSPHATE (ATP); CARBOHYDRATE; GLYCOGEN; MUSCULAR SYSTEM.

Nerve tissue. Nerve tissue is distinguished chemically from other tissues by the high proportion of lipid material and, in the latter, the very low proportion of neutral fat. For example, the brain, accounting for some 85% of the total nerve tissue in man, contains 54% of lipid on the dry basis. The brain also differs from other organs in the body in its exclusive reliance upon the metabolism of carbohydrate for its activity. The rate at which oxidation occurs in proportion to the weight of brain is many times that of other tissues under basal conditions. *See* CARBOHYDRATE METABOLISM; LIPID.

The nerves transmit impulses from the brain to the tissues and vice versa. These impulses are associated with biochemical and electrical phenomena, but their nature is not fully understood. See BIOPOTENTIALS AND ELECTROPHYSIOLOGY.

Brain metabolism. The activity of the brain is closely correlated with the biochemical events occurring within it. Its temperature is higher by about 0.5°C in the waking condition than that of its arterial blood supply; its oxygen uptake parallels its functional state. While the energy cost of mental activity is so low as to be difficult to dissociate from the incidental concomitant muscular hypertonicity, the continuing activity of the midbrain and the medulla in controlling the functions of the viscera, the circulation, and respiration represent a metabolic load equivalent to one-fourth of the total oxygen used by the body under resting conditions. The activity of the brain is supported by the energy-rich phosphate bonds in such organic compounds as adenosinetriphosphate and creatine phosphate. The known involvement of certain vitamins in these activities is indicated by the neurological symptoms associated with the dietary deficiencies of thiamine, pyridoxine, and nicotinic acid. See BRAIN; NIACIN; THIAMINE; VITAMIN B_6.

The nerve impulse. The passage of an impulse along a nerve fiber, or from one fiber to another, or from a terminal fiber to a receptor of some kind in the target organ is associated with many different phenomena, electrical and chemical, changes in membrane permeability and polarization, and in ion movement. The difficulty in deciding which of these phenomena are primarily responsible for the impulse and its conduction and transmission is enhanced by the speed of these events; the velocity of the impulse is of the order of 120 m/sec, and their frequency may well exceed 300 per sec. The energy released per impulse is small, of the order of 10^{-11} cal/cm^2. Whether the action potential accompanying a nerve impulse is the cause or the result of the impulse is unknown. The theory of chemical transmission has assumed a dominant role in this confusing situation, with primary emphasis upon the changes undergone by acetylcholine during nerve activity: free acetylcholine is necessary for generating the electric potentials that propagate the impulse in conducting tissue; the ion movements across nerve membranes appear to be associated with the rapid intracellular hydrolysis of acetylcholine; the transmission of the impulse across nerve junctions, the synapses, is presumably due in parasympathetic nerves to the liberation of acetylcholine, while the recovery processes in the nerve, restoring it to its original condition, are associated with the synthesis of acetylcholine. The transmitter of nerve impulses across the synapses of sympathetic nerves or to a receptor in the target organ has been named sympathin. Its chemical structure is not known. See ACETYLCHOLINE; SYMPATHETIC NERVOUS SYSTEM.

Glutamic acid and its derivatives. Among the free amino acids of brain tissue, the molar concentration of glutamic acid, the monoamide, and the semidecarboxylated product, γ-aminobutyric acid (GABA), account for some 70% of the total. For this reason, its therapeutic effect in mental deficiency and in epileptic disorders of the petit mal type has been studied extensively. Among a wide range of animals tested, GABA and the enzyme responsible for its production, L-glutamic acid decarboxylase, are found almost exclusively in the brain. See AMINO ACIDS; ENZYME; SEROTONIN.

Connective tissue. Connective tissue binds together and is the support of the various structures of the body. In modified form it occurs in cartilage, bones, and teeth. Considered as a highly integrated organ, it is the largest in quantity in the mammalian body. Its maintenance and function depend on cellular metabolism. It is characterized by the nature and dominant role of its intercellular substance, consisting of an amorphous matrix in which are embedded fibers of various types. It functions in the synthesis and storage of fat in adipose tissue and the storage of minerals in bone, in the healing of wounds, and in antibody formation. See ANTIBODY; BONE; CARTILAGE; TOOTH.

Adipose tissue. This tissue has a special structure and a special type of cell capable of synthesizing fat from carbohydrate at a rate far superior to that of the liver cell. The fat stored in this tissue may amount to 50% of the body weight of the animal.

Collagen. Collagen is the dominant protein in the white fibers of connective tissue, which are so prominent in tendon. It changes to gelatin in boiling water. Its content of the amino acid hydroxyproline, 13.5 ± 0.24%, is so characteristic that this has become the basis of a method for its quantitative analysis in animal tissues. Ascorbic acid (vitamin C) is essential for its formation, but it is so inert metabolically that deficiency of this vitamin in mature guinea pigs has little effect on the collagen content of their tissues. Reticulin fibers differ in structure from collagen fibers, but reticulin is similar to collagen in its chemical nature. See ASCORBIC ACID; PROTEIN.

Elastin. This is the characteristic protein of the yellow fibers in the connective tissue, found especially in the aortic arch, the chorda tendineae of auricular valves of the heart, and the ligamentum nuchae. It differs greatly from collagen in its amino acid composition, containing only 1.5–2.3% of hydroxyproline, depending on its source.

Mucoproteins. Mucoproteins are components of connective tissue fibers and greatly influence their physicochemical behavior. They contain in their molecules, as a prosthetic group, a mucopolysaccharide such as chondroitin sulfate (in cartilage) or heparin (in many tissues).

Mucopolysaccharides. These are structural carbohydrates found in the ground substance of connective tissue and to some extent in the fibers, generally being loosely bound to protein. In these compounds, amino sugars and sugar acids, such as uronic acids, are the principal units of structure. The groups of mucopolysaccharides and the units to which they are reduced by chemical procedures or enzymatic methods follow: hyaluronic acids—glucuronic acid, glucosamine, and acetic acid; chondroitin sulfates—glucuronic acid, galactosamine, acetic acid, and sulfuric acid; heparins (potent blood anti-coagulants)—glucuronic acid, glucosamine, acetic acid, and sulfuric acid. *See* HYALURONIC ACID.

Bone metabolism. The metabolism of bone can be broken down into matrix and ground substance formation, mineralization, growth remodeling, Haversian remodeling, and the regulation of the ionic composition of the body fluids.

Matrix formation. This is concerned with the production of collagen fibers in an organic crystalline form capable of refracting x-rays. The osteoblasts, or bone-making cells, are presumably responsible for this activity, but collagen formation cannot proceed in the absence of vitamin C.

Ground substance. This substance fills all of the spaces between the collagen fibers and the crystals of bone mineral. It consists of protein and carbohydrates, some of which are polymers of glucuronic acid and hexosamines in both sulfated and unsulfated forms.

Bone mineral. Bone mineral corresponds most closely to hydroxyapatite, $3Ca_3(PO_4)_2{\cdot}Ca(OH)_2$. The mineral also contains considerable amounts of carbonate and citrate, probably concentrated on the surfaces of the apatite crystals, together with small amounts of Na^+, K^+, Mg^{++}, Cl^-, and F^- as contaminants. The tremendous surface area of the minute mineral crystals (200–300 m^2/g), as well as the hydrated ions bound to the surfaces, contribute to the reactivity of the crystal complex. As the crystals increase in numbers with age, they displace the water essential for their reactivity.

Growth remodeling. This accompanies increase in bone length and diameter, and consists of resorption and redeposition of mineral and matrix with a continuance of high reactivity.

Haversian remodeling. Haversian remodeling relates to the bone units, the Haversian systems (osteons), characteristic arrangements of collagen fibers, matrix and ground substance, mineral deposits, and vascular elements. The remodeling of these systems insures a continuing fresh supply of reactive bone. Otherwise, the continuing acquisition of minerals with age would lead to a metabolically inert bone, incompatible with the life of the organism.

Regulatory agents. The maintenance of the calcium content of the blood plasma at the physiologic level of about 10 mg/100 ml may occur both by passive ion transfer and by an active mechanism depending upon calcium citrate transfer under the control of the parathyroid hormone and vitamin D. The maintenance of a constant plasma concentration of inorganic phosphate is less well understood. The involvement of vitamins A and C, and the ratio of dietary calcium (Ca) to potassium (P) in bone growth and maintenance is also well established. *See* CALCIUM METABOLISM; HORMONE; PHOSPHATE METABOLISM.

Tooth metabolism. Tooth metabolism involves three different mineralized structures: enamel, dentine, and cementum, enclosing a pulp chamber containing vascularized connective tissue and cells. Polysaccharide reactions are found in each, suggesting the presence of mucoprotein. The fibrous protein of the dentine and cementum is largely collagen, while that of enamel contains keratin. The mineral in all three structures is apatite. A two-way permeability of dentine and enamel to labeled ions has been demonstrated by isotope-tracer techniques. However, except for the shedding of deciduous teeth, it has never been demonstrated that any portion of the fully formed tooth is physiologically removed by cellular resorption. Cementum and dentine are subject only to acquisition of minerals; the enamel is completely protected from remodeling. The participation of vitamins A, C, and D in tooth metabolism is revealed in the corresponding avitaminoses. *See* POLYSACCHARIDE; PROTEIN, FIBROUS; VITAMIN A; VITAMIN D.

Blood. Blood is a continuously circulating tissue that is in dynamic equilibrium with all other tissues via the capillary bed that penetrates each and the lymph that bathes them. It consists of cells, the erythrocytes and leukocytes, and platelets, making up 40–45% of the volume, and of plasma. The blood volume of mammals amounts to 7–10% of the body weight. *See* BLOOD.

Formed elements. The red cells (erythrocytes) carry the hemoglobin concerned with oxygen and carbon dioxide transport. The white cells (leukocytes) are nucleated. They are concerned with the defense of the body against bacterial and viral infection and with the removal of cellular debris. The platelets are colorless oval disks containing proteins and phospholipids, much of which is cephalin. They are concerned with blood coagulation. In the adult, the erythrocytes are formed in the red bone marrow, the leukocytes in bone marrow but also in the spleen and lymph nodes, and the platelets in the giant cells (megakaryocytes) of bone marrow, spleen, and lungs. *See* HEMOGLOBIN.

Destruction of the red cells takes place in the spleen, bone marrow and other similar (reticulo-endothelial) tissues. The leukocytes are destroyed by the macrophage cells of the spleen, bone marrow, and the Kupffer cells of the liver. When blood is shed, the platelets disintegrate and liberate thromboplastic substances necessary for coagulation.

Plasma proteins. These are classified as albumins or globulins, depending upon their solubilities and salting out characteristics. Each of these groups may be separated into several (albumin) or many (globulin) components. Many of the globulins are mixtures of lipoproteins and lipid-free proteins. The immune bodies or antibodies of blood belong to the γ-globulin fraction. Fibrinogen is a globulin possessing the unique property of conversion into fibrin in blood coagulation. *See* ALBUMIN; GLOBULIN.

The liver forms all of the fibrinogen, essentially all of the albumin, and the larger proportion of the globulin fractions of plasma. Plasma proteins may serve for the formation and maintenance of tissue proteins. After protein depletion, or fasting, the reserve proteins of tissues can be used for plasma protein formation. *See* FIBRINOGEN.

Transport mechanisms. These mechanisms in internal respiration relate to the transport of oxygen from lungs to tissues and of carbon dioxide (CO_2) from tissues to lungs. The transport of oxygen depends upon the reversible reaction between the hemoglobin contained in the red cells and the oxygen dissolved in the blood plasma. While the mass action principle governs both the loading and unloading phases of this reaction, the variations in the partial pressure of CO_2 in the plasma increase the affinity of hemoglobin for oxygen in the pulmonary capillaries, and decrease it in the capillary bed of the tissues. The transport of CO_2 is effected mainly as bicarbonate dissolved in the plasma and in combination with hemoglobin as carbaminohemoglobin.

The transport of minerals in blood, with the exception of sodium and potassium, is generally effected by loosely bound combinations with plasma proteins. The combination of ferric iron with a specific plasma globulin has been established; the unloading of the iron is probably preceded by reduction of the iron to the ferrous state.

The lipoproteins of blood provide a mechanism for the transport of lipids in an aqueous environment. Clinical interest in the plasma lipoproteins derives from the finding that they may reflect predisposition toward the occurrence of atherosclerosis, the characteristic lesion of which is a fatty deposition in the arterial walls.

The blood plasma is an important transport agent for hormones, some of which (thyroxin, estriol) are known to be carried in combination with proteins. *See* IRON METABOLISM.

Eye metabolism. The metabolism and structure of the eye are concerned with vision, its efficient performance, and its stabilization. The eye's supply of nutritive material and its removal of metabolites and waste products must be by diffusion, since no blood vessels are found in the cornea, aqueous or vitreous humor, or the lens. The osmotic pressure of the eye and its various compartments must be very delicately controlled to maintain their proper relative hydration and a fairly constant intraocular pressure of 20–25 mm Hg. The transparency of the cornea is dependent upon its hydration. For a constant degree of refraction, the size and shape of the eyeball must be maintained. *See* EYE.

The cornea is the window of the eye; the iris is the shutter regulating the amount of light admitted; the lens focuses the image upon the retina, the light-sensitive film in the fundus of the eye which absorbs the light and transforms it into another form of energy, presumably chemical, which is carried by the optic nerve to the brain.

Photochemistry of vision relates to the chemical changes occurring in the two light receptors of the retina, the rods, responsible for vision in dim light, located in the peripheral area of the retina, and the cones, responsible for color vision in bright light, concentrated in the central (macular) area. The substances absorbing light are pigments, rhodopsin (visual purple) in the rods and iodopsin in the cones; both are proteins, resolvable by light into retinene (the aldehyde of vitamin A_1) and a protein (opsin) which is different in the two pigments. Each pigment undergoes a cycle of degradation when exposed to light and regeneration in the dark.

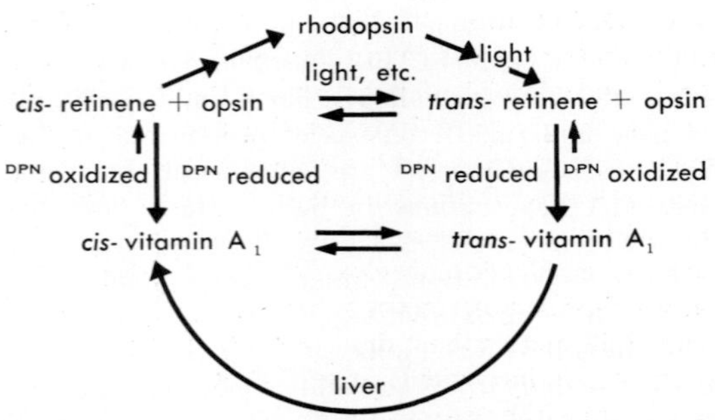

Transformations of the carotenoids in the visual cycle. (From A. White, P. Handler, E. L. Smith, and D. Stetten, Jr., Principles of Biochemistry, 2d ed., McGraw-Hill, 1959)

The visual cycle for rhodopsin involves the cis-trans isomerism of retinene: light isomerizes the *cis*-retinene to the all-trans configuration while it is still attached to opsin. This probably is the reaction (endergonic in nature) responsible for visual excitation. The reversible oxidation-reduction of vitamin A_1 and its aldehyde involves a dehydrogenase and the diphosphopyridine nucleotide (DPN), containing the vitamin nicotinamide. In mammals, the isomerization of vitamin A_1 possibly occurs in the liver, the cis form being supplied continuously to the retina from the circulation. In continuous light, the system attains a steady state in which the continuous restitution of rhodopsin allows vision to persist indefinitely at a lower level of acuity (light adaptation). *See* DIPHOSPHOPYRIDINE NUCLEOTIDE (DPN); VISION.

Liver metabolism. Liver metabolism has bearings on the processes of digestion and on the disposition of the end products of digestion. In the formation of bile, it aids in the digestion of lipids by promoting their emulsification in the intestine through the action of the bile salts in lowering surface tension. The bile also carries the bile pigments, cholesterol, and various other components into the duodenum. The liver is the site of most of those reactions that involve alteration of foreign compounds or intestinal putrefaction products that cannot be metabolized readily. These detoxication reactions involve acetylation, methylation, mercapturic acid and hippuric acid syntheses, oxidation, and glucuronide and ethereal sulfate synthesis. *See* LIPID METABOLISM.

In protein metabolism, the liver aids in maintaining the amino acid concentration of the blood within narrow limits; it synthesizes all of the fibrinogen and albumins of the blood plasma and most of the globulins; it is the chief site of removal of amino groups from amino acids by oxidative deamination or transamination, and the principal site of urea formation; it synthesizes and stores the reserve or dispensable protein. *See* PROTEIN METABOLISM.

In carbohydrate metabolism, it maintains the glucose concentration of the blood within narrow limits under normal conditions, mainly by the synthesis and storage of glycogen, or as occasion demands, by glycogenolysis; it converts nonglucose sugars into glucose. Here, also, segments of the carbon skeletons of the amino acids metabolized in the body are converted into intermediates that may be employed for glucose and glycogen synthesis. *See* CARBOHYDRATE METABOLISM.

In lipid metabolism, the liver is important in the synthesis of fatty acids from 2-carbon fragments, which are either stored in the liver or released to the circulation for oxidation or storage elsewhere. The liver modifies dietary fatty acids by shortening or lengthening their carbon chains, or by saturating or desaturating them to make them conform more nearly to the fatty acids characteristic of the species. The liver synthesizes cholesterol from acetic acid, with acetyl coenzyme A (CoA) and acetoacetyl CoA as the reactive intermediates. *See* STEROID.

The liver is preeminent in its ability to store the lipid-soluble vitamins. It is a notable storage depot for iron as ferritin.

Skin metabolism. This relates to a large extent to the production and maintenance of the epidermis which is undergoing continual disintegration and replacement. The outer layers of the epidermis consist of flattened cornified cells, the characteristic protein of which is keratin, a tough insoluble albuminoid. The desquamation of these cells sets the pace for the formation, differentiation, and keratinization of a new layer of cells, above the basal epidermal layer, and its extrusion to the surface. This renewal process in the human may take weeks for completion. Beneath the protective epidermis is the dermis, consisting of connective tissue with its collagen fibers and the mucopolysaccharide matrix. The basal layers of the epidermis contain melanoblasts, the cells in which the dark pigment, melanin, is elaborated by oxidation from the amino acid tyrosine, a reaction catalyzed by the copper-containing enzyme, tyrosinase. The color of skin depends upon the concentration of melanin or of its state of oxidation. Ultraviolet irradiation of the skin causes its darkening because of the oxidation of tyrosine, the formation of vitamin D by the alteration of 7-dehydrocholesterol, and the transformation of porphyrins resulting in skin photosensitivity. [H.H.MI.]

Bibliography: E. C. Albritton (ed.), *Standard Values in Blood*, 1952; R. W. Brauer (ed.), *Liver Function*, Am. Inst. Biol. Sci. Publ. 4, 1958; H. M. Leicester, *Biochemistry of the Teeth*, 1949; R. W. Miner (ed.), Recent Advances in the Study of the Structure, Composition, and Growth of Mineralized Tissue, *Ann. N.Y. Acad. Sci.*, vol. 60, 1955; W. F. Neuman and M. W. Neuman, *The Chemical Dynamics of Bone Mineral*, 1958; S. Rothman, *Physiology and Biochemistry of the Skin*, 1954; D. M. Surgenor, *Blood: Some Functional Considerations*, in D. E. Green (ed.), *Currents in Biochemical Research*, 1956; R. E. Tunbridge (ed.), *Connective Tissue*, 1957.

Speciation

The term applied to the evolution of species as distinct from the evolution of genera and other more inclusive taxonomic categories. Speciation is an aspect of evolution and, as such, the modern period of its study began with the publication in 1858 of Charles Darwin's and Alfred Russel Wallace's theory of evolution by natural selection and in 1859 of Darwin's *Origin of Species*.

There was no problem of speciation during the period when it was believed that species were the product of special creation and, once created, were stable. This concept of the fixity of species was held almost universally before the middle of the nineteenth century but subsequently to an ever-decreasing extent.

Darwin believed that there are two main components to the evolutionary process. First, within the population of a given species there is variation. That is, the individuals of the population differ from one another in structure, physiology, and habits. Second, natural selection acts upon this variation by eliminating the less fit. Thus, if any individual of a population enjoys an advantage over another in ability to find a mate, secure food, escape from predators, or survive the rigors of the climate, it would have a better chance of leaving offspring. Over the course of time its characteristics would become more common in the population. Darwin believed that natural selection acted by preserving these differences of an adaptive nature but, since the differences were usually slight or even imperceptible, evolution took a very long time. It is still believed today that the dominant pattern of speciation is the slow accumulation of minute adaptive differences. There are other known mechanisms, especially in plants, but these are of lesser importance.

Since Darwin's time knowledge of the origin and nature of variability has increased tremendously (*see* GENETICS; MUTATION). The variability that is of importance in evolution originates as a consequence of gene mutation and other changes in the chromosomes. Under natural conditions mutations are spontaneous in the sense that the stimulus for them to occur is not known. Mutations are also random in the sense that no specific relation exists between the environment and the type of mutation that occurs. The basis of all evolutionary change, therefore, is an indeterminate event. If the new mutant form, that arises by chance, happens to be better adapted than the other members of the population, it will have a superior chance of surviving and leaving offspring.

The species concept. In sexually reproducing organisms a species is a group of individuals that interbreed if given the opportunity. They share a common genetic heritage. Generally the individuals that comprise the species are similar to one another in habits, structure, and physiology. Under natural conditions the individuals of one species usually do not interbreed with individuals of other species. This is nearly always true of animals, and to a lesser degree of plants, even though interbreeding between species may occur under artificial conditions. Under natural conditions, therefore, species are characterized by intragroup matings and intergroup isolation. In organisms that do not reproduce sexually it is not possible to give an entirely satisfactory definition of a species. In these cases morphological, physiological, or biochemical criteria are used in defining species.

Types of speciation. It is necessary to distinguish two general sorts of speciation. The origin of new species can occur either through phyletic evolution or splitting. In phyletic evolution one species evolves until, after a long time, the descendants are so different from the ancestral type that one would say the two are different species. At any one time, of course, there is but a single species and a distinction cannot be made as to where one species ended and the other began. In splitting, a single species evolves into two or more species. The events involved are the same as in phyletic evolution except that some extrinsic factor serves to split the single population into two or more groups. As a consequence there are two or more slowly evolving phyletic lines.

Geographic speciation. A form of splitting known as geographic speciation is a common pattern of speciation in sexually reproducing animals, especially in the vertebrates (fish, amphibians, reptiles, birds, and mammals). In fact, geographic speciation is the only sort that is well documented in animals, though other patterns are theoretically possible.

A hypothetical island or continent is used to illustrate this type of speciation (see illustration). On this land mass there is a temperature gradient running in the north-south direction and a moisture gradient in the east-west direction. The land mass is divided into three zones by barriers, such as high mountain ranges, that produce an isolated zone in the southwest (A), northwest (B), and southeast (C). The southwest is warm and moist, the northwest, cold and dry, and the southeast warm and dry.

A species of terrestrial vertebrate arrives in zone A and becomes established (Time 1 in illustration). During this period genetic changes will be occurring and selection of favorable variations will result in an increasingly better-adapted population. In all probability the population in zone A will now differ somewhat from the population from which it was derived.

If a few individuals are able to pass the barriers and enter zones B and C, they will encounter two new sorts of environments for which they are not well adapted (Time 2 in illustration). As a result the course of evolution in zones B and C will be different from that in zone A. In zone B any variants that are better adapted to the cool dry environment would have a better chance of surviving and leaving offspring. In zone C, on the other hand, the course of evolution would be toward the perfecting of a population adapted to a warm and dry environment.

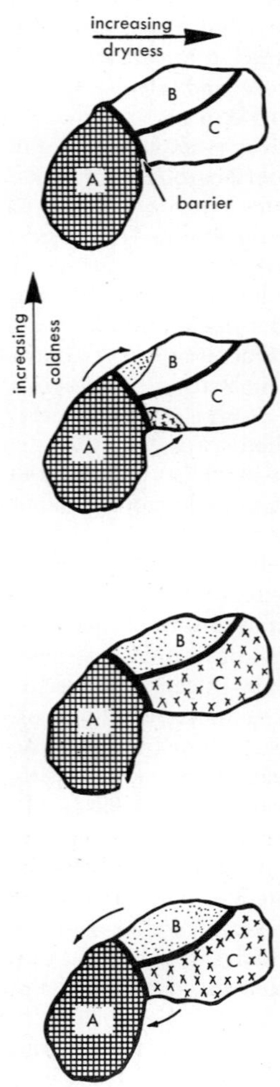

Time 1. Single Species. The population is restricted to Zone A.

Time 2. Single Species. Some individuals have migrated to Zone B and others to Zone C. Selection will promote the development of a population adapted to a cold, dry environment in B and to a warm, dry environment in C.

Time 3. Three Subspecies. Evolution through the selection of spontaneous mutations in the three isolated populations has reached the point where each zone has its own adapted population. These populations, though slightly different from one another, could still interchange genes, consequently they are subspecies.

Time 4. Three Species. The three isolated populations have diverged to the point where isolating mechanisms prevent the interchange of genes. If individuals from either Zone B or Zone C migrate back to Zone A, they can occupy this zone together with the Zone A species without the danger of losing their specific identity.

A model for geographic speciation. (From J. A. Moore, *Principles of Zoology*, Oxford Univ. Press, 1957)

With the passage of time, and to the extent that the barriers between zones A, B, and C are effective, there will be three different phyletic evolutions (Time 3 in illustration). The barriers are of the utmost importance; without them there would be a continuously interbreeding population, and adaptation to the local environments would be slow. The population would become more variable but it would not form different species; phyletic evolution could occur but there would be no splitting.

As each of the three populations continued to evolve in its own region, it would be difficult to designate exactly a time when three species exist in place of one. After some divergence had occurred each population might be recognized as a distinct subspecies or geographic race. If the divergence became great, each might be tentatively regarded as a distinct species. A decision is commonly difficult because the best way of telling whether or not two or more closely related forms are in fact different species is to observe their behavior under natural conditions and in the same region. If such populations do not interbreed they are different species; if they interbreed and eventually merge they form, of course, a single species. On the hypothetical land mass each of the populations remains in its own region, and the critical test of whether or not they are different species, therefore, cannot be applied. The practical taxonomist would call them different species if they had diverged to approximately the same degree as other "good" species in the same genus or family living in the same territory. It would also be possible to obtain some data that might suggest an answer by attempting to cross the different forms. If they could not be crossed this would be sufficient evidence that the forms had reached the species level of divergence. If they could be crossed and produced normal offspring, no conclusion could be reached, since it is well known that many different forms can be crossed under artificial conditions, yet under natural conditions they behave as distinct species. If some of the individuals of zones B and C migrate back to zone A and thereafter do not interbreed with the endemic population of zone A (Time 4 in illustration), their distinctness as a species would be demonstrated.

Isolating mechanisms. The pattern of geographic speciation requires a regional isolation for the beginning of divergence. Within each region evolutionary divergence occurs, and after the passage of a long period of time the single original species will have split into two or more. The new species are recognized by their possession of mechanisms that allow them to maintain their genetic integrity. Such mechanisms are known as isolating mechanisms, since they serve to isolate the individuals of one species from all other species.

There are many sorts of isolating mechanisms. Some isolating mechanisms serve to keep individuals apart at the time of breeding. Thus different species may breed at different times (seasonal isolation) or in different places (habitat isolation). Even if two species breed at the same time and at the same place they may be kept from crossing by sexual isolation. This involves a selection by males or females of their mates from individuals of their own group and the rejection of individuals belonging to other groups. *See* BREEDING (ANIMAL).

Seasonal, habitat, and sexual isolation operate by keeping the males and females from meeting and mating. It is probable that these three isolating mechanisms are the ones that usually prevent species from crossing in nature. There are additional ones, however, that are characteristic of those species that have diverged to such an extent that their sex cells cannot combine and produce normal offspring. Thus, isolation may occur through a failure of the sperm to fertilize the ovum, or by the death of the hybrid at some stage of its development. Finally, even if hybrids are produced and reach the adult stage they may be sterile (as with the mule) and hence gene exchange between the two populations is impossible. *See* POLYPLOIDY.

There is a tremendous body of evidence, assembled during the past century, to support the notion that geographic speciation is the common pattern for the formation of new species in the vertebrates. All gradations exist between populations showing slight geographic variation to those split into well-marked species. Geographic speciation is especially noteworthy in those populations occupying large land masses divided by effective barriers (as in the western part of the United States) or islands (as in the southwest Pacific).

Although geographic speciation is best known in the vertebrates it is undoubtedly common in other groups of animals. Most other groups are less well known but, when their species have been carefully studied, it is often apparent that geographic speciation has been operating. In many groups of parasitic animals, however, it is clear that speciation is associated with the acquisition of new host organisms. Thus the roundworm, *Ascaris*, has one species in man and another in swine. There is no need to assume that geographic speciation was responsible for this distinction. It is probable that the isolation of the two groups in different host species was sufficient to allow their independent evolution.

Hybridization and speciation. It is a common belief that the ability, or lack of ability, of two forms to produce hybrids is a sure indication of whether or not they are different species. This is not so. Among the ducks, for example, there are many species that can be crossed under artificial conditions. Numerous intrageneric crosses have been made between various species of frogs and toads and in nearly half of the cases the hybrids develop at least to the juvenile stage and could probably reach the adult stage. Intrageneric crosses of fish species frequently result in normal offspring and there are a few cases of crosses of species belonging to different genera. Among the mammals species as different as the lion and tiger or cattle and American bison can be crossed.

In spite of the widespread ability to cross under artificial conditions, hybridization is rare under natural conditions.

There is one special sort of speciation, however, in which hybridization plays an important role. Up to now, this is known to occur only in plants. It consists of hybridization of two different species followed by the doubling of the chromosomes of the hybrid. In order to understand this type of speciation it is first necessary to explain one sort of hybrid sterility.

It is a common occurrence for the hybrids between two species to be sterile. Sterility may be present even though in other respects the hybrid individual is normal and vigorous. A familiar example is the mule, a sterile hybrid resulting from the cross of a female horse and a male ass.

The sterility of hybrids is usually a matter of their inability to produce normal ova or sperm. This, in turn, is often the result of their having inherited chromosomes from the two parental species that are so different that sperm or ovum meiosis is impossible (*see* MEIOSIS).

In a normal individual (not a hybrid) the chromosomes in the cells from which ova or sperms will be formed are present in homologous pairs. Thus, if there is a total of eight chromosomes, these would consist of four pairs. We could designate these chromosomes as AA, BB, CC, and DD. One of the essential features of normal ovum or sperm formation is the pairing of homologous chromosomes. Thus the two A chromosomes will pair with each other, the two Bs and so on. This occurs at the beginning of the process, known as meiosis, which will result in the number of chromosomes being halved. At the end of meiosis each ovum or sperm will have one chromosome of every kind, normally one each of A, B, C, and D.

Suppose that a hybrid is formed of two parents, one having chromosomes AA, BB, CC, DD and the other having EE, FF, GG, HH. It will receive chromosomes A, B, C, D from one parent and E, F, G, H from the other. These two groups of different chromosomes might permit normal development of the hybrid individual but there would be difficulty in meiosis. It would be impossible for pairing to occur because no two chromosomes in the set, A B C D E F G H, are alike. As a consequence meiosis could not occur and no ova or sperms would be formed. These events are a frequent cause of hybrid sterility.

There are a few instances where hybrids have circumvented these difficulties by doubling their chromosomes. For example, if a hybrid with chromosomes A B C D E F G H should have an abnormal mitosis (cell division) in which the chromosomes would duplicate but the cell would fail to divide, then this cell would have twice as many chromosomes as before. Of greater importance, the chromosomes would be in homologous pairs, since the set would now be AA BB CC DD EE FF GG HH. If, in the course of development, this cell with the double set of chromosomes gave rise to a part of the plant where the flowers formed, normal meiosis could occur and the plant could produce normal sex cells. Each sex cell would contain the chromosomes ABCDEFGH. The fusion of two such sex cells would produce a new plant with the chromosomal constitution AA BB CC DD EE FF GG HH.

The original parental plants, each with eight chromosomes in the premeiotic cells, have a diploid number of chromosomes. The hybrid when first formed also had eight chromosomes and was likewise diploid. When these doubled in the atypical mitosis to result in 16 chromosomes, a tetraploid arises. It could not cross normally with either parental species and hence it would constitute a new species. In plants, therefore, hybridization followed by a doubling of the chromosomes can lead to the formation of new species in a very brief period of time.

There are several well-analyzed instances of the formation of new species by this method. *Primula kewensis* originated in this manner from *P. floribunda* and *P. verticillata*. *P. floribunda* and *P. verticillata* each have 18 chromosomes and their sex cells would have nine. When these are crossed a hybrid with 18 chromosomes is formed. This individual is sterile because the eight *floribunda* chromosomes will not pair normally at meiosis with the eight *verticillata* chromosomes. *P. kewensis* was formed when one of the hybrids doubled its chromosomes to produce a complement of 32. When this occurred every type of chromosome was represented twice and meiosis could occur normally.

There is a similar origin for a grass that occurs along the southern coast of England, *Spartina townsendii*. It has 126 chromosomes and was produced by the crossing of *S. stricta* (sex cells with 28 chromosomes) and *S. alterniflora* (sex cells with 35 chromosomes) to form a hybrid with 63. The hybrid doubled this number to produce 126.

Polyploid series in plants. There are other types of speciation in plants, and possibly in animals, that are associated with changes in chromosome number. Thus, in some closely related species the chromosomes are in multiples of some basic number. There are three species of wheat (*Triticum*) with chromosome numbers of 14, 28, and 42. Here the basic haploid number is 7 and the other species have doubled (14) and tripled (21) this number. There are many genera of plants in which the chromosomes of different species exist in polyploid series of this kind.

Introgressive hybridization. Numerous examples, among both plants and animals, suggest that the process of geographic speciation is, to some extent, reversible. These are situations where two or more "species" that originally occupied different areas have spread and invaded one another's territory. In the zones of overlap there is considerable hybridization with some gene exchange between the different species. It seems probable that this introgressive hybridization can be explained by assuming that the extrinsic isolating mechanisms that originally kept the groups apart have broken down. Furthermore, this occurred before the isolated populations had evolved effective intrinsic isolating mechanisms. As a result interbreeding occurs and there is a partial merging of the partially differentiated populations.

Introgressive hybridization is frequently associated with changes in the environment as a result of man's activities. Original habitats are destroyed, new ones are created, and barriers to dispersal are erased.

There is considerable introgression among the species of oak trees in the eastern United States. Among animals, introgression is common in the toads of the genus *Bufo*. The American toad and Fowler's toad, for example, have lost some of their distinctiveness in areas where they occur together. In these regions the individuals vary all the way from typical examples of American toad, through numerous intermediate stages, to typical examples of Fowler's toad.

Conceivably introgressive hybridization could result in the fusion of two originally distinct populations into one.

Ecotypes. Another pattern of evolution that has been studied in plants is the formation of races and species in relation to characteristic and restricted environments. The term ecotype was introduced by G. Turesson, who observed that, in many plant species, well-marked varieties exist and each variety is restricted to one habitat. In southern Sweden *Hieracium umbellatum* consisted of five ecotypes: a shifting-dune ecotype; a stationary-dune ecotype; two different ecotypes on the sea cliffs; and a woodland ecotype. These five ecotypes differed in many respects, such as structure of the leaves, flowers, and stems; time of flowering; and ability to regenerate. These ecotypes maintained their distinctive features when grown together in an experimental garden.

Numerous studies of plants living in the varied environments of the western United States have revealed the presence of ecotypes. In California the yarrow, *Achillea*, occupies a variety of habitats, from sea level to an altitude of 11,000 ft, and differing considerably in rainfall and temperature. Every type of habitat except the true desert has a race of *Achillea*. A study of the local populations in a 200 mile east-west transect showed that eleven distinct ecotypes are present. This represents the number of differently adapted populations necessary for this one species to occupy all the habitats in this transect that encompasses sea cliffs, coastal mountains, valleys, and high mountains. There is some interbreeding in the areas where the ecotypes come in contact with one another.

The factors involved in the evolution of ecotypes are the same as in geographic speciation. Genes that confer an advantage upon the individuals of the populations are increased in frequency by natural selection. The noteworthy aspect of ecotypes is that they can be formed in rather restricted habitats. They are examples of geographic speciation where the areas involved may be small in size and consist of a fairly uniform and restricted habitat.

[J.A.MO.]

Bibliography: T. Dobzhansky, *Evolution, Genetics, and Man*, 1955; T. Dobzhansky, *Genetics and the Origin of Species*, 3d ed., 1951; G. G. Simpson, *The Major Features of Evolution*, 1953.

Species concept

The idea of the species as a biological unit or entity. The term species has its roots in the typological philosophy of the Greek logicians, but the origin of the concept in biology is generally credited to the English naturalist, John Ray (1628–1705). Its firm establishment in biological science resulted from the influence of the Swedish naturalist Linnaeus (1707–1778), who not only applied the concept uniformly to the animal and plant kingdoms but at the same time provided the useful binomial system of nomenclature by which species are designated (*see* PLANT CLASSIFICATION; ZOOLOGICAL NOMENCLATURE). The idea of species is fundamental to all biological research and is especially important in those fields of biological science which are comparative. The species problem is at the core of animal and plant systematics.

Historical aspects. Linnaeus, at least in his earlier writings, referred to species as having been separately "created" and their number unchanged "from the beginning." Under the prevailing doctrine of separate creation each species was assumed to be descended from an originally created pair. Linnaeus conceived of the species as a class of similar individuals which, with due consideration to obvious differences resulting from age and sex, not only exhibited a constancy of form and structure permitting of precise and objective definition, but had a reality in nature. Individual variations were known to him but they were classed mostly as freaks or varieties ("varietas"). This concept of the constancy and morphological distinctness of species was the natural result of studying local faunas and floras where the lines of demarcation (gaps) between species are greatest. As a result, the "species" of such a concept are essentially nondimensional (E. Mayr).

Darwin, a traveling naturalist and student of domesticated plants and animals, was impressed with the variation exhibited by species and the arbitrary nature of their limits. From these studies emerged his concept of evolution expressed in *The Origin of Species through Natural Selection* (1859), which gradually led to a concept of species having dimensions in time and space (multidimensional species). The rediscovery of Mendel's laws of inheritance, and developments in the field of genetics, provided an explanation for individual variation and resulted ultimately in the concept of species as populations or groups of populations which actually or potentially interbreed, but are reproductively isolated from other such populations; that is, individuals of different species are unable, or rarely able, to interbreed and produce fertile offspring. This genetical concept, often called the biological species concept, which has been emphasized particularly in the writings of Ernst Mayr, is accepted in principle by most biologists concerned with living groups of sexually reproducing plants and animals, but it can only be applied indirectly to fossil forms and scarcely at all to groups of organisms in which reproduction is asexual or parthenogenetic. Furthermore, since speciation is a continuing process, the degree of differentiation of related species in nature will vary greatly. Nevertheless, although biologists have been unable to agree on a species definition applicable to all kinds of organisms, most do accept and utilize a species concept which is interpreted with remarkable uniformity and agreement in actual taxonomic practice. In part this no doubt results from the fact that although morphological characters still provide the basis for practical species recognition in most groups of organisms, physiological, genetic, ethological, and ecological criteria are being utilized to an increasing degree to test the validity of preliminary conclusions reached on morphological grounds.

Population aspects. Emphasis on population aspects of the species problem has focused attention on variations within the species and their significance in taxonomy (*see* ANIMAL EVOLUTION; ANIMAL SYSTEMATICS). The most significant of such categories of variation is the geographically defined infraspecific population (deme) or group of local populations, the subspecies. Subspecies differ taxonomically and genetically as well as geographically. Since the taxonomic characters which separate them are usually quantitative, opinions vary as to how different two or more populations should be before they should be recognized formally in zoological nomenclature. Some systematists regard them as namable if 75% of the individuals of one or two subspecies can be distinguished from all those of another; others regard them as namable if 75% of the individuals in the two populations can be correctly assigned by means of taxonomic characters without reference to locality data. However, neither of these applications of the so-called 75% rule has as yet become generally accepted taxonomic procedure.

A species made up of recognizably distinct populations, especially subspecies, is referred to as polytypic; one that is relatively uniform throughout its range, as monotypic. However, the taxonomic characters of a variable species may exhibit a character gradient (cline) and change gradually through a series of continuous or adjacent infraspecific populations. Two or more species which occupy identical or broadly overlapping geographical areas are termed sympatric; those which occupy mutually exclusive, but usually adjacent, geographical areas are allopatric. Species which occur together in time, more particularly at the same geological level, are designated as synchronic; those which are separated in time are allochronic. A variant which occurs at random in the species and which comprises only certain individuals of a population is termed a variety. Since these are arbitrarily selected individuals and the term does not apply to populations, it has little taxonomic significance though Latin or latinized names are sometimes applied to these and to seasonal and other polymorphic forms. [E.G.LI.]

Bibliography: J. Huxley, *The New Systematics*, 1940; E. Mayr, *Systematics and the Origin of Species*, 1942; E. Mayr, E. G. Linsley, and R. L. Usinger, *Methods and Principles of Systematic Zoology*, 1953; *The Species Problem*, AAAS Publ. 50, 1957; P. C. Sylvester-Bradley (ed.), *The Species Concept in Paleontology*, 1956.

Species population

A group of similar organisms residing in a defined space at a certain time. Although species have geographic ranges, their individuals typically are not scattered over the entire area, but occur in groups, the species populations. These follow more or less discontinuous spatial patterns. The size of such groups and the number of populations into which a species may be divided vary. In the almost extinct whooping crane (*Grus americana*) the entire species forms a single population. For a common bird like the English sparrow (*Passer domesticus*) many groupings may be defined, whose size depends upon convenience for some particular study. All the sparrows of a single farm, or of Great Britain or even America may constitute a species population. Often a population is not completely separable from neighboring groups.

One can describe populations as static units at some instant of time, but they can be explained only in developmental terms. The component living individuals, which are born, respond to their environment, and ultimately die, confer on the population certain statistical attributes. These attributes provide the basis for the group concept. Birth, death, immigration, and emigration rates determine density, age distribution, and sex ratio, and are related to dispersion and genetic constitution of the population. *See* POPULATION DYNAMICS.

Population density. This is a familiar concept in human populations, where it is customarily expressed as the number of people per unit area. For other organisms different scales are more suitable; bacteria are measured in numbers per cubic centimeter. Often the biomass, the amount of living material, is more instructive than are numbers alone. Numbers of fish in a lake are complemented by information about their size. Among many populations, whose individuals move about, density must be estimated by elaborate sampling techniques. The development of better methods of estimation is itself a significant part of population research.

Dispersion. Dispersion modifies the interpretation of density. Organisms within a population are often not randomly distributed, but occur in clusters. Such clumping or infradispersion may be historically caused—heavy seeds of a plant fall close to the parent. Local differences in habitat may be responsible—in dry weather earthworms concentrate in moist spots. Animals are often social—conspicuous examples occur in the family groups of some warm-blooded vertebrates, but less apparent yet significant sociality is a widespread phenomenon. In contrast to clumping, which is common, uniform spacing or superdispersion is rare. Except in artificial examples such as a corn field, carefully planted with stalks equidistant, superdispersion is caused by intolerance between individuals, as in the territories established by nesting birds. *See* POPULATION DISPERSION; SOCIAL ANIMALS.

Age distribution and sex ratio. In some organisms, as in annual plants, all members of the population are equally old. More commonly many age groups exist simultaneously whose populational effects are not the same. Age distribution affects subsequent population growth through birth and death rates and is in turn affected by these. A constant proportion of organisms of any age group is attained by any population with constant age-specific birth, death, emigration and immigration rates. Age distributions often fluctuate as a result of variations in these rates. In certain species similar changes in sex ratio are common. Characteristically a growing population has a high proportion of young individuals, with age structure shifting more toward older organisms when population growth rate declines or even becomes negative. Age distribution in some, but far from all populations, can be determined as a result of some discontinuous growth process. Annual rings of trees, fish scales, and certain bones yield information as to the age of individuals.

Genetic constitution. Except for clones of lower organisms—populations derived from a single ancestor by repeated asexual reproduction—members of a population differ in hereditary makeup. The heredity of a group is definable as an array of gene frequencies for various inherited characteristics. This array is constantly and sometimes rapidly modified by mutation, selection, and immigration. Genetic change is a major source of adjustment of population to environment.

Birth, death, immigration, and emigration rates. Crude rates state the number of individuals born, dying, and moving in or out per unit of population for some time interval. Although they describe what has happened to the group, these rates have little predictive power, and reveal almost nothing about the age structure of the remainder. Age-specific rates which treat births, deaths, and movements as functions of age, are the major immediate determinants for the population. Knowledge of the causes for their change is therefore essential for predicting population growth. Variation of these rates within and between species is enormous. The only generalization possible is that high birth rates normally imply high death rates and vice versa. Many components of environment influence the rates. One major factor may be the population itself, which is part of its own environment.

Knowledge of the properties of populations results in an understanding of their dynamics. During their existence, populations require a constant input of energy and materials. The energy is in part transformed into metabolic heat and in part is stored as new protoplasm and body food reserves. Some remains in dead bodies. Usually little of the energy used by animals, and none of that used by plants, becomes available twice to the same population. The materials required, however, often recirculate through the population any number of times. A population newly released into an adequate environment grows in characteristic fashion, and modifies its environment. Given a continuing source of energy and raw materials and a sufficiently stable environment, it may reach a more or less steady state. Otherwise a population peak and decline ensue. This may have cyclic characteristics. Particular causes for growth, stability, decline, and extinction must be sought in complex interactions between the physical and biological environment, and birth, death, immigration, and emigration rates modified by genetic change. Species populations often are systems with considerable powers of self-regulation. As a result they may persist as recognizable entities for indefinite periods. *See* ECOLOGY; EVOLUTION, ORGANIC; GENE; MUTATION; SPECIATION; SPECIES CONCEPT. [P.W.F.]

Bibliography: W. C. Allee et al., *Principles of Animal Ecology*, 1949; H. G. Andrewartha and L. C. Birch, *The Distribution and Abundance of Animals*, 1954.

Specific charge

The ratio of charge to mass, e/m, of a particle. The acceleration of a particle in electromagnetic fields is proportional to its specific charge. Specific charge can be determined by measuring the velocity v which the particle acquires in falling through an electric potential V ($v = \sqrt{2eV/m}$); by measuring the frequency of revolution in a magnetic field H (the so-called cyclotron frequency $\omega = eH/mc$, where c is the velocity of light); or by observing the orbit of the particles in combined electric and magnetic fields. In the instrument known as the mass spectrograph, the fields are arranged so that particles of differing velocities but of the same e/m are focused at a point. *See* MASS SPECTROSCOPE; *see also* ELECTRON MOTION IN VACUUM; ELEMENTARY PARTICLE. [C.J.G.]

Specific fuel consumption

Weight flow rate of fuel required to produce a unit of power or of thrust, often abbreviated as SFC. In reciprocating piston engines, SFC is usually expressed as pounds of fuel per hour required to deliver one shaft horsepower. For ram and turbojet engines it is given as pounds per hour for one pound of thrust. For turboprops, where both shaft and jet power are taken from the engine, SFC is given as fuel rate for equivalent shaft horsepower (*see* TURBOJET).

SFC varies inversely with the fuel's heat of combustion. It can be calculated from

$$\mathrm{SFC} = \frac{2545}{\text{heat of combustion}} \times \frac{100}{\text{cycle efficiency}}$$

where SFC is in pounds per horsepower-hour, heat of combustion is in Btu/lb, and cycle efficiency is in per cent. Alternatively, it can be computed from

$$\mathrm{SFC} = \frac{4.63}{\text{heat of combustion}} \times \text{flight speed} \times \frac{100}{\text{cycle efficiency}}$$

where SFC is in pounds per hour per pound of thrust and flight speed is in feet per second.

Cycle efficiency increases with increasing compression ratio for piston, turbojet, and turboprop engines, and with increasing flight speed in ramjets. Efficiency also increases with decreasing fuel-air ratio, with, however, a loss in power.

With hydrocarbon fuels, minimum SFCs of about 0.4 lb/hp-hr are obtained with reciprocating and turboprop engines and about 0.8 lb/(hr) (/lb of thrust) for turbojet engines. Ramjet SFCs range from about 10–3 lb/(hr) (/lb of thrust). [R.R.H.]

Specific gravity

The specific gravity of a material is defined as the ratio of its density and the density of some standard material, such as water at a specified temperature, for example, 60°F, or (for gases) air at standard conditions of temperature and pressure. Specific gravity is a convenient concept because it is usually easier to measure than density, and its value is the same in all systems of units. *See* DENSITY. [L.N.]

Specific heat

The ratio of the amount of heat required to raise unit mass of a material one degree in temperature to the amount of heat required to raise the same mass of a reference substance one degree in temperature. Both measurements are made at a reference temperature and in nearly all cases at either constant volume or constant pressure. Water is usually the reference substance. Because the heat capacity of water is nearly unity, the value of specific heat for a material is nearly equal to its heat capacity. Specific heat, as defined here, is a ratio without units, although it is often defined differently (*see* SPECIFIC HEAT OF SOLIDS). For clarity it is recommended that thermodynamic discussion be carried out in terms of heat capacity instead of specific heat. Also, it is desirable to define heat units in electrical terms. *See* BRITISH THERMAL UNIT (BTU); CALORIE; HEAT CAPACITY; THERMODYNAMICS (CHEMICAL). [H.C.W.]

Specific heat of solids

When 1 gram (g) of a material absorbs an amount of heat ΔQ and this causes the temperature of the material to increase an amount ΔT, then the ratio $s = \Delta Q/\Delta T$ is often called the specific heat of the material, although other definitions are also used. The heat capacity C of a body of mass M is the product $C = Ms$. The atomic and molecular heats are the heat capacities of a gram-atomic weight and a gram-molecular weight of material, respectively.

The measured heat capacity of solids is usually made at some constant pressure P, such as atmospheric pressure, and is represented by the symbol C_P. The theoretical heat capacity is most often calculated for constant volume V, and is denoted by C_V. The difference $C_P - C_V$ is essentially the heat per degree required to expand the solid against its internal elastic forces. The difference is

$$C_P - C_V = \alpha_V{}^2 VT/\chi \qquad (1)$$

Here α_V is the temperature coefficient of volume expansion (at constant pressure), V the volume, T the temperature in °K, and χ the isothermal compressibility. The quantities represented by the symbols C_P and C_V are often referred to loosely as specific heats, although they are really heat capacities. *See* HEAT CAPACITY.

Dulong-Petit law. P. Dulong and A. Petit observed in 1819 that although the specific heats of the solid elements at room temperature differ widely from one another, the atomic heats are nearly all the same, the values being about 6.3 cal/°C. A theoretical explanation was given by F. Richarz in 1893. It is an extension of the theory of the specific heat of an ideal gas. According to the kinetic theory of gases, the thermal energy of an ideal monatomic gas is the same as its kinetic energy (*see* KINETIC THEORY OF MATTER). From this, it was deduced that the atomic heat of such a gas is $3R/2$, where R, the gas constant, is about 2.0 cal/°C. The thermal energy of a solid, however, is the energy of the harmonic motion of the atoms, and this, on the average, is half kinetic and half potential. Richarz then supposed that $3R/2$ is the atomic heat arising from the mean kinetic energy, and $3R/2$ that arising from the mean potential energy, yielding a total atomic heat of $3R$ or 6.0 cal/°C.

The Dulong-Petit law is quite accurate at room temperature. To find s for many solid elements, one need only substitute the atomic weight A from a periodic table into the formula $s = 6/A$. However, it was noticed, even in the nineteenth century, that there are important exceptions to the law, notably diamond, germanium, and silicon, whose atomic heats at room temperature are considerably smaller than $3R$. Furthermore, many solids showed a decrease in C_V as the temperature was lowered to that of liquid nitrogen, which is 77°K or −196°C.

Einstein theory. The quantum hypothesis which M. Planck introduced into the theory of black-body radiation in 1900 did not become a general principle until Albert Einstein applied it with success to the photoelectric effect in 1905 and to the theory of specific heats in 1907. In his theory of specific heats, Einstein sought to show that the observed failure of the classical theory, which gives $C_V = 3R$ for the atomic heat, could be explained in terms of the quantum hypothesis.

A so-called Planck oscillator can absorb or emit radiation only in integral amounts $nh\nu$, where n is an integer, h is Planck's constant, and ν is the natural frequency of the oscillator. The temperature is introduced by considering the mean value of the energy $\bar{\epsilon}$ of such an oscillator, using the classical Boltzmann statistics. The result is

$$\bar{\epsilon} = h\nu/[\exp(h\nu/kT) - 1] \tag{2}$$

where k is the Boltzmann constant and T is the absolute temperature. *See* BOLTZMANN STATISTICS; HEAT RADIATION; QUANTUM MECHANICS.

Einstein's theory assumes that each atom of the solid oscillates with the same frequency ν_E and that this is the frequency observed in infrared absorption studies in crystals. Each atom vibrates in three dimensions and therefore has the mean energy $3\bar{\epsilon}$. The energy E of the solid is $3N\bar{\epsilon}$, if it contains Avogadro's number of atoms N. The quantum hypothesis then leads to

$$E = 3Nh\nu_E/[\exp(h\nu_E/kT) - 1] \tag{3}$$

The frequency ν_E is called the Einstein frequency. *See* ABSORPTION (ELECTROMAGNETIC RADIATION); INFRARED SPECTROSCOPY.

A parameter called the Einstein characteristic temperature Θ_E is defined by equating one quantum of energy $h\nu_E$ to the classical energy kT of an

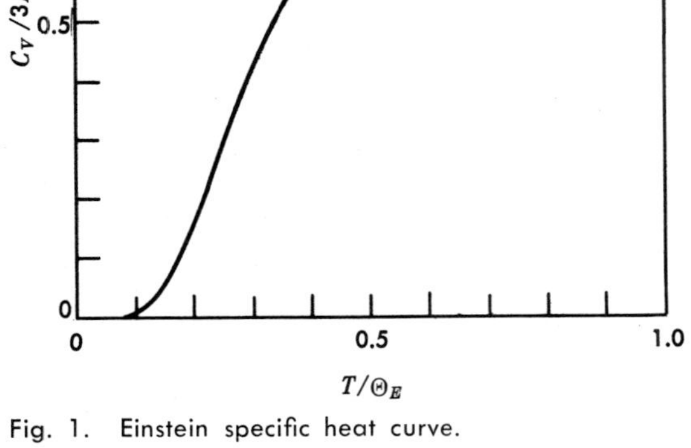

Fig. 1. Einstein specific heat curve.

oscillator and denoting the particular value of T obtained in this manner by Θ_E. According to Einstein, the thermal energy Q of the solid is just the energy E of vibration, so that $C_V = dQ/dT = dE/dT$. This yields the Einstein formula of specific (atomic) heats,

$$C_V = 3Ry^2e^y/(e^y - 1)^2 \tag{4}$$

Here $y = h\nu_E/kT = \Theta_E/T$ and $Nk = R$, the gas constant.

A plot of $C_V/3R$ versus T/Θ_E is shown in Fig. 1. At $T = \Theta_E$, the value of $C_V/3R$ is 0.92, which means that at this temperature, C_V has 92% of the Dulong-Petit value. Above this temperature, C_V approaches $3R$ with increasing temperature. Below this temperature, C_V decreases to zero, practically vanishing at $T < 0.1\Theta_E$. Einstein's theory thus concludes that C_V is temperature-dependent. Furthermore, the observation that $C_V/3R = 0.31$ for diamond at $T = 331$°K is explained by stating that diamond has a value of Θ_E equal to about 1300°K, which corresponds to an infrared wavelength of 11 microns.

The prediction contained in the theory that C_V practically vanishes below $T = 0.1\Theta_E$ stimulated W. Nernst and his assistants to make experimental investigations of C_V down to 16°K. It was found that C_V is still appreciable at $T < 0.1\Theta_E$ for all substances examined; therefore Einstein's theory fails at these low temperatures. However, it appeared from the data that C_V approaches zero at 0°K, in keeping with deductions from Nernst's heat theorem.

Debye theory. The next advance in the theory of specific heats began with the suggestion of E. Madelung and W. Sutherland that the Einstein frequency is equivalent not only to the infrared absorption frequency of the crystal but also to the frequency of the shortest sound wave (or elastic wave) which can propagate through the crystal. This wave travels with the velocity of sound and has a wavelength of about twice the interatomic distance. Since sound waves of longer wavelength can also propagate through the crystal, Madelung made the further suggestion that a whole spectrum of acoustical frequencies should be used in computing C_V rather than just the single frequency ν_E.

In 1912 two theories of the specific heats of solids appeared incorporating these ideas, one by P. Debye and the other by M. Born and T. von Kármán. Both theories use an acoustical spectrum containing so many frequencies that the spectra can be treated as continuous for purposes of computation. The number of waves (or modes) with frequencies between ν and $\nu + d\nu$ in the solid is thus represented by $g(\nu)\,d\nu$. The energy associated with each of these waves is that of a Planck oscillator, so that one obtains for the total energy E the expression

$$E = \int_0^\infty \frac{g(\nu)h\nu\,d\nu}{e^{h\nu/kT-1}} \tag{5}$$

The two theories differ in the manner of estimating $g(\nu)$. Only the simpler Debye theory is discussed in this article.

In order to estimate $g(\nu)$, Debye made two assumptions. One is that the solid is a continuous medium. With this idea, $g(\nu)$ is computed in a manner analogous to that employed in the theory of black-body radiation with the result

$$g(\nu) = 4\pi V \left(\frac{1}{U_l^3} + \frac{2}{U_t^3}\right) \nu^2 \quad (6)$$

The symbol U_l represents the velocity of longitudinal sound waves and U_t that of transverse waves. The volume of the solid is V. The second assumption is that the total number of waves is equal to $3N$, where N is the number of atoms in the crystal. This assumption implies that the solid is not really continuous after all and that the shortest permissible wavelengths are those of about two interatomic distances. The restriction is expressed mathematically by

$$\int_0^{\nu_D} g(\nu)\, d\nu = 3N \quad (7)$$

which serves to define a Debye frequency ν_D. The Debye frequency is the maximum allowable frequency. Thus for $\nu > \nu_D$, $g(\nu)$ is zero and the value of the integral above this limiting frequency is zero. This allows the upper limit in Eq. (5) to be replaced by ν_D.

Debye temperature. It is customary to replace the Debye frequency ν_D by the Debye characteristic temperature Θ, defined by the relation $k\Theta = h\nu_D$. From this, and from Eqs. (5), (6), and (7), the energy E becomes

$$E = 9R \frac{T^4}{\Theta^3} \int_0^{\Theta/T} \frac{z^3\, dz}{e^z - 1} \quad (8)$$

where $z = h\nu/kT$.

Equation (8) can be integrated and then C_V deduced from $C_V = dE/dT$. The result is the infinite series

$$C_V/3R = \frac{4\pi^4}{5}\left(\frac{T}{\Theta}\right)^3 - \frac{3\Theta/T}{e^{\Theta/T} - 1} + 12 \log (1 - e^{-\Theta/T}) - 36 \frac{T}{\Theta} \sum_{n=1}^{\infty} \left\{\left[1 + \frac{2T}{n\Theta} + \frac{2}{n^2}\left(\frac{T}{\Theta}\right)^2\right] \frac{e^{-n\Theta/T}}{n^2}\right\} \quad (9)$$

As T approaches infinity, the Dulong-Petit value of C_V is obtained. To see this, note that z approaches zero in this limit and the integrand in Eq. (8) reduces to z^2. Integration then leads to $E = 3RT$, from which $C_V = 3R$. On the other hand, as T approaches 0°K, the Debye T^3 law results:

$$C_V = \frac{12\pi^4 R}{5}\left(\frac{T}{\Theta}\right)^3 \quad (10)$$

This law is contained in the first term of Eq. (9). The other terms in Eq. (9) contribute less then 1% to C_V at temperatures below $T = \Theta/12$. Figure 2 shows a plot of $C_V/3R$ versus T/Θ as given by the Debye theory. The accompanying table lists the values of Θ required to fit the Debye formula for C_V to the experimental data of solid elements in the region near where C_V is about half the Dulong-Petit value. The corresponding values of Θ_E determined in this manner are smaller and are approximately $3\Theta/4$.

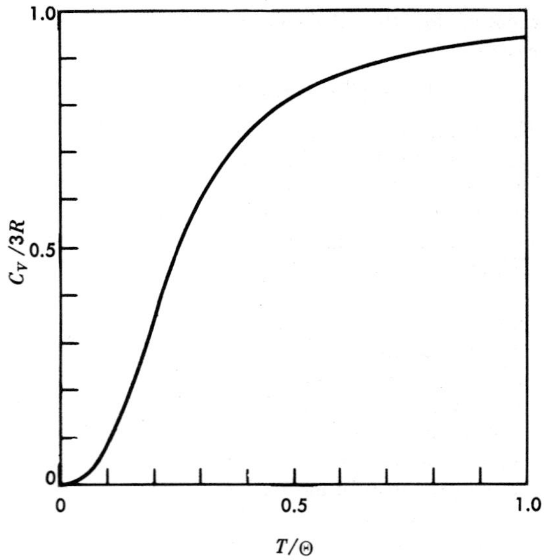

Fig. 2. Debye specific heat curve.

Debye characteristic temperature of solid elements, °K

Element	Θ	Element	Θ	Element	Θ
Ar	85	Ga	240	Pd	275
Ag	215	Ge	360	Pr	74
Al	394	Gd	152	Pt	230
As	285	Hg	100	Sb	200
Au	170	In	129	Si	625
B	1250	K	100	Sn, gray	260
Be	1000	Li	400	Sn, white	170
Bi	120	La	132	Ta	225
C, diamond	1860	Mg	318	Th	100
Ca	230	Mn	400	Ti	380
Cd	120	Mo	380	Tl	96
Co	385	Na	150	V	390
Cr	460	Ne	63	W	310
Cu	315	Ni	375	Zn	234
Fe	420	Pb	88	Zr	250

The courses of the two curves shown in Figs. 1 and 2 are quite similar for T above about 0.2Θ. The critical test distinguishing between the two theories must therefore be made at temperatures below about 0.1Θ, where the Debye T^3 law should hold. The T^3 law was first verified by A. Eucken and F. Schwers in 1913 by measuring the heat capacity of a number of insulators. It failed for metals. The reason for this failure is now understood, for A. Sommerfeld's theory of metals (1928) shows that the conduction electrons can make an important contribution to the heat capacity (*see* FREE-ELECTRON THEORY OF METALS). According to Sommerfeld, there must be a linear term in the temperature included in the expression for C_V in order to account for the electron contribution. Thus one writes

$$C_V = \gamma T + (12\pi^4 R/5)(T/\Theta)^3 \qquad (11)$$

The coefficient γ in the electron term is sometimes called the Sommerfeld gamma. In order to analyze low-temperature C_V data for metals, C_V/T is plotted versus T^2. According to Eq. (11), this should give a straight line of slope $12\pi^4 R/5\Theta^3$ and of intercept γ on the C_V/T axis.

Deviations. As experimental measurements became more precise, it was noticed that the data could not be fitted to a Debye curve. One sensitive test is to calculate the Debye Θ for each experimental value of C_V and T, after correcting for the electron contribution. If the data satisfy the Debye formula, then Θ should be independent of the temperature. In most cases, the data do not satisfy this criterion.

A particularly marked deviation from Debye's theory occurs for cadmium. Figure 3 shows a plot of Θ versus T for this element. The data were treated in the following manner. First, C_V was calculated from the measured C_P data, using essentially Eq. (1). Then the 12 items of data below about 3°K were plotted on the basis of Eq. (11) and γ and Θ determined from the straight line graph. Next, all C_V data were corrected for the electron term and then the Θ for each point computed from tables based on Eqs. (8) or (9). The result is plotted in Fig. 3.

Agreement with Debye theory in the case of cadmium exists only below about 3°K, the only part of the curve where Θ is substantially constant. Thus the T^3 law holds (to better than 1%) only below about $T = \Theta/50$, instead of $\Theta/12$ as required by the Debye theory. For most of the solids examined, this limitation of the range of the validity of the T^3 law to the region $T < \Theta/50$ seems to occur. M. Blackman explained this in 1935 on the basis of the lattice dynamics of Born and von Kármán. He gave the estimate $T < \Theta/50$ as the "true" range of the T^3 law for most solids, an estimate which he made when the experimental evidence was still rather meager.

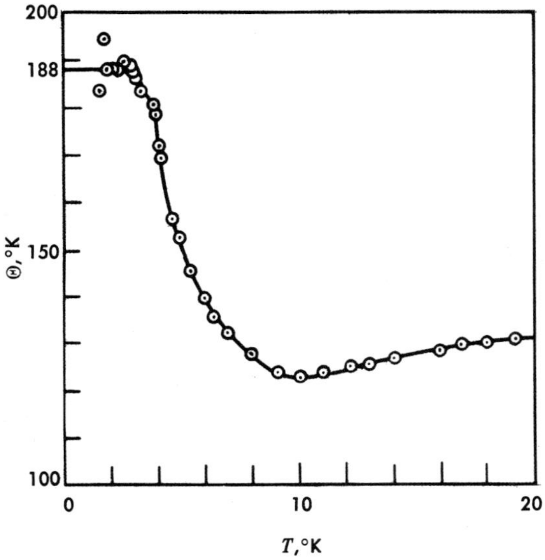

Fig. 3. Plot of Θ versus T for cadmium. (*After P. L. Smith and N. M. Wolcott, Phil. Mag., ser. 8, 1:854–865, 1956*)

Experimental verification. An important independent check of the theory of specific heats can be made, based on Eq. (6). The velocity of sound is measured for single crystals at temperatures in the "true" T^3 region of specific heats. Since the velocity of sound depends on the direction of propagation through the crystal (an effect known as anisotropy of the velocity of sound), the inverse cube of the velocity must be averaged over all directions. This is done theoretically from velocity of sound measurements made in several appropriate directions in a single crystal. When this is done, Θ at 0°K can be calculated. The comparison of the values calculated in this manner with those obtained from low temperature specific heat data has been made for copper, silver, and gold. The velocity of sound values of Θ for these elements are respectively 345, 226, and 162°, which compare well with the corresponding specific heat values 345, 226, and 165°K. *See* CONDUCTION (HEAT); LATTICE VIBRATIONS; ULTRASONICS. [J.D.L.]

Bibliography: S. Fluegge (ed.), *Handbuch der Physik*, vol. 7, pt. 1, 1955; F. Seitz and D. Turnbull (eds.), *Solid State Physics*, vol. 2, 1956.

Specific impulse

The quantity used to define the amount of thrust potentially available from rocket propellant combinations; also called specific thrust (*see* THRUST). Specific impulse is the thrust that theoretically can be obtained when unit weight of the propellant reacts in unit time; that is, lb thrust per lb per sec of propellant flow. This ratio for specific impulse has the dimension of time and can be expressed as seconds. For example, a propellant having an impulse of 300 sec will theoretically deliver 300 lb of thrust when the propellant is consumed at the rate of 1 lb/sec. Alternatively, 1 lb of propellant will deliver 1 lb of thrust for 300 sec.

Specific impulse can be calculated from

$$I_{sp} = 9.80\sqrt{\frac{T_c}{M}}\sqrt{\frac{k}{k-1}}\sqrt{1-\left(\frac{p_e}{p_c}\right)^{(k-1)/k}}$$

where I_{sp} is specific impulse; T_c, combustion temperature, °R; M, molecular weight of combustion products; k, specific heat ratio of combustion products, C_p/C_v; p_e/p_c, ratio of external to chamber pressures. The first square root term shows the importance of attaining high combustion temperatures and low molecular-weight products. The next term varies but little, since k ranges only between 1.2 and 1.3 for most propellants. The last term shows that the impulse of a given propellant varies with the operating conditions of the thrust cham-

Specific impulses of typical propellant systems

Propellant system	Chamber pressure, psi	Specific impulse, sec
Liquid oxygen–ammonia	300	255
Liquid oxygen–kerosene	300	248
Liquid oxygen–liquid hydrogen	340	335
Hydrogen peroxide–hydrazine	500	252
Red fuming nitric acid–aniline	300	221
Nitrogen tetroxide–hydrazine	300	249
Fluorine–hydrazine	300	299
Various castable composite solids	1000	165–240

ber. Highest impulse is obtained when the chamber discharges into a vacuum so that this term becomes unity.

At the high temperatures in rocket chambers, the products include not only those calculated for usual chemical reactions but also many additional components which result from dissociation. For example, the stoichiometric combustion of hydrocarbons with oxygen yields not only carbon dioxide and water but also carbon monoxide, hydrogen, oxygen, the hydroxyl radical OH, and atoms of hydrogen and oxygen. Temperature, molecular weight, and specific heat ratio are all influenced by the various equilibria and resulting compositions. Therefore the calculation of specific impulses is a complex procedure and is greatly facilitated by the use of computing machines.

High specific impulse is the quality most desired in rocket propellants. Values for typical propellant systems exhausting to 1 atmosphere pressure are presented in the table. *See* ROCKET ENGINE.

[R.R.H.]

Spectrography

The use of photography to record the electromagnetic spectrum displayed in a spectroscope. The technique is used mainly in atomic and molecular physics, in analysis of the chemical composition of materials, and in astronomical photography.

The sources of radiation for spectrography are incandescent or electrically excited, and the spectrum contains lines characteristic of the elements and gives definite evidence of their presence. Continuous spectra are also emitted by incandescent sources, but are little used except in absorption measurements. Band spectra are characteristic of molecules. The position and intensity of the lines and bands in the spectrum is a measure of the nature and amount of the elements present.

A great variety of photographic plates and films is made, having a range of speeds, contrast, resolving power, and spectral sensitivity, permitting spectrography from very short wavelength ultraviolet to the infrared at about 13,000 A. *See* ASTRONOMICAL PHOTOGRAPHY; PHOTOGRAPHY; SPECTROCHEMICAL ANALYSIS; SPECTROSCOPY.

[W.C.]

Spectrochemical analysis

A technique used in qualitative or quantitative chemical elemental analysis. It is performed by observing the spectrum of the light emitted by the incandescent vapor of the material analyzed. The technique is used chiefly to detect and determine metals (qualitative and quantitative analysis, respectively) although it can also be applied to many metalloids and to a few nonmetals. Under optimum conditions, absolute detection sensitivities for individual elements range from 10^{-6} g to below 10^{-9} g.

The steps in emission spectrochemical analysis are (1) vaporization of sample, (2) excitation of vapor to luminescence, (3) resolution of the resultant radiation into a spectrum, and (4) observation of spectra.

Vaporization. Nonconducting solids, sometimes diluted with graphite or other specially chosen matrix, often are placed in a cavity in the end of a pure graphite rod. This serves as the anode of a dc arc, the heat of which volatilizes the sample. Finely powdered samples of the same type also may be mixed with pure flaky graphite, briqueted, and (partially) volatilized by means of a spark discharge. Metal powders or chips may be treated by either of these methods. Metal specimens having a flat surface are usually subjected to a spark discharge, with a graphite rod or a piece of the sample material as counterelectrode. Solutions may be (1) dried on a graphite or copper electrode and volatilized by arc or spark, (2) continuously introduced into a spark discharge as a thin film on a graphite electrode (see Fig. 1), (3) sprayed into a spark gap, or (4) sprayed or aspirated into the gas supply of a flame (*see* FLAME PHOTOMETRY).

Excitation. In most spectrochemical light sources, the same electrical discharge is used to produce the sample vapor and to excite it to luminescence.

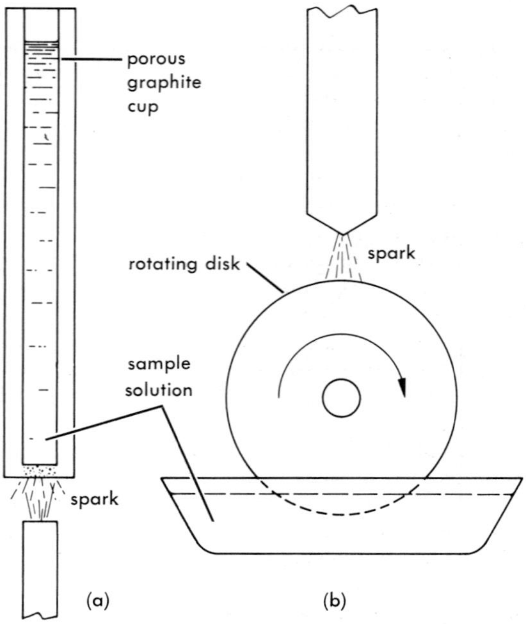

Fig. 1. Excitation of liquids by high-voltage spark. (*a*) Porous cup technique. (*b*) Rotating disk technique.

When the atoms or molecules of sample vapor enter the analytical gap, they are bombarded by highly energetic electrons and other particles. Thus, their outer electrons are raised to higher energy levels; if the excitation is sufficiently intense, one or more electrons may even be removed. The ions so produced can be excited further by continued bombardment. The higher the temperature of the particle's immediate environment, the higher the state of excitation the particle can attain. Thus, various individual particles of a given element may exist at different discrete energy levels, depending on their past and present environments. Each of these excited states has a characteristic normal lifetime, which ends with the spontaneous descent of the particle to a lower level of excitation. This descent is accompanied by the emission of a quantum of radiation of energy content equivalent to the difference in energy between the upper and lower levels. The energy content of the quantum uniquely determines the wavelength of the emitted radiation. (Occasionally, an excited particle loses energy by collision before this transition can occur, and no radiation is emitted.) Since particles in each of the higher levels may descend to any one of a number of lower levels, a considerable number of spectral lines may be produced by a single element. Each state of ionization of an element or compound has its own characteristic set of energy levels and, thus, its own spectrum. The lines of a given element that are actually emitted during a given discharge depend on the temperatures existing in that discharge. The relative intensities of these lines depend principally on the relative numbers of atoms existing at these respective temperatures and on the relative inherent probabilities of the energy transitions involved.

In the core of a continuous low-voltage arc, the temperature is 3000–7800°C, varying directly with the ionization potential of the components of the vapor. The resultant spectra are due principally to neutral and singly ionized atoms and diatomic molecules.

In the oscillatory 15,000- to 35,000-volt sparks used in spectrochemical analysis, core temperatures depend principally on instantaneous current. Peak currents up to several hundred amperes may occur, producing core temperatures up to 40,000°C. At such times, the core may contain doubly or triply ionized species. At lower currents, ionization in the core is less severe. Peripheral regions of the discharge, being cooler, contain only neutral and singly ionized species.

Resolution. To permit analysis, the component wavelengths must be separated and arranged in order of wavelength. This is done with a spectroscope, spectrograph, or spectrometer, using visual, photographic, or photoelectric detection, respectively. For general analytical use, the instrument's dispersion should be 10 or less angstroms per millimeter (A/mm) on the photographic plate. Most prism instruments have adequate dispersion in the ultraviolet but inadequate dispersion in the red and infrared. Their wavelength scale is thus nonuniform, making wavelength interpolation somewhat tedious. Grating instruments have an almost uniform wavelength scale and can be used in higher orders to obtain high dispersion. Difficulties due to the resultant overlapping of orders can usually be overcome by appropriate choice of photographic emulsion, filters, or supplemental dispersion in the vertical plane. *See* DIFFRACTION GRATING; SPECTROSCOPY.

Observation of spectra. Visual observation is used for instruction and for on-the-spot sorting of metals.

Photographic observation is used for most qualitative and nonroutine quantitative analysis. Plates or 35-mm film may be employed. Spectrograms usually are viewed with a projection comparator, which projects sample and standard plates on a split-field, ground-glass screen, permitting any spectrum on one plate to be brought adjacent to and in register with any spectrum on the other. Many comparators also incorporate a microphotometer. Any emulsion is suitable for the 2300- to 4300-A range; at higher wavelengths, special sensitization is needed. Emulsion sensitivity varies directly with grain size; thus high sensitivity is obtained at the cost of resolving power, and vice versa. Sensitivity and contrast vary with wavelength (and age) for all emulsions. Aging changes can be retarded by storage at 0°F.

Different production batches of a given emulsion are seldom identical; if quantitative work is to be done, a 6–12 months' supply of plates or film should be bought from a single batch and stored at 0°F. *See* PHOTOGRAPHIC MATERIALS.

Measurement of line intensity ratios requires calibration of the emulsion, that is, construction of a curve showing blackening of the emulsion as a function of the intensity of incident light of the wavelength used. Lines whose intensities are to be compared photographically should be as close as possible in wavelength.

Densitometers or microphotometers measure spectral-line blackening by scanning the appropriate section of an illuminated spectrum with a fine receiving slit or by projecting a fine illuminated line onto and through the spectrogram and onto a photocell.

The shape of the calibration curve obtained with spectrographic equipment depends not only on emulsion characteristics, but also on instrumental characteristics such as slit widths and aperture ratios. Calibration curves must be made therefore with the same spectrograph and densitometer which will be used for producing the sample spectra to be measured. Correction of line intensity for contribution by the underlying spectral background is desirable, especially if the ratio of line to background intensities is less than 5:1.

Photoelectric observation is used when the analysis is highly repetitive. A receiving slit and photomultiplier cell are placed in the spectrometer at the position of each line of interest. The output of each cell may be used to charge a capacitor; at the end of the exposure, the voltage on each capacitor is proportional to the intensity-time integral for the appropriate spectral line. Alternatively, this integral may be measured by counting the number of times that the photomultiplier output is able to charge a much smaller capacitor during the exposure. Intensity ratios (actually energy ratios) of lines may be measured (1) by comparing intensity-time integrals (ITI) after a fixed exposure time, (2) by measuring the ITI for all subject lines when a single comparison line has reached a set ITI, or (3) operating as in (2), but using undispersed light as the comparison "line." Photomultiplier installations have an immense advantage over photographic instruments in speed and some advantage in accuracy. They are used wherever large numbers of similar analyses must be performed rapidly. They are, however, relatively costly, inflexible, and unsuited for exploratory work.

Qualitative analysis. For this purpose, the sample is vaporized and excited by one of the above techniques, and its spectrum is photographed in the 2200- to 4300-A region. This permits detection of all metals except small amounts of the alkali metals. The spectrum of an iron arc is photographed adjacent to the sample spectrum. A master plate (or film) is prepared containing an iron spectrum adjacent to a master spectrum in which the positions of the strongest lines of all detectable elements are marked. The two plates are placed in a split-field projection comparator and the iron spectra aligned with each other. If a line occurs in the sample spectrum at the position indicated for a given element in the master plate, the element is usually present (Fig. 2). Since lines of two different elements may happen to occur at the same wavelength, the actual presence of an element should be confirmed by looking for several of its spectral lines.

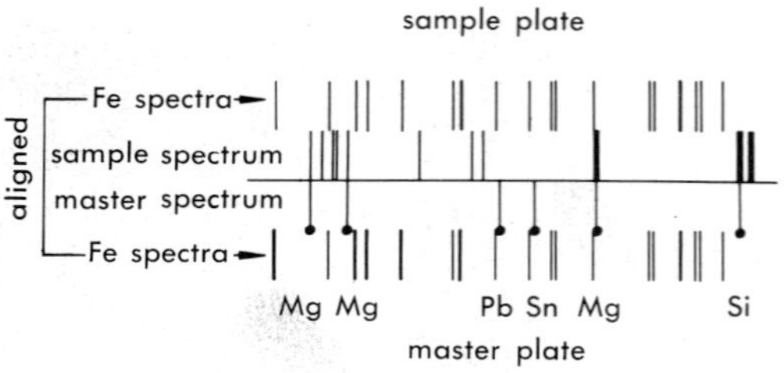

Fig. 2. Identification of elements with aid of master plate (schematic). The sample shown contains magnesium and silicon, but no lead or tin.

Some elements, for example, uranium, have a large number of closely spaced lines, making it difficult to detect other elements in their presence. This difficulty can sometimes be circumvented by preliminary chemical removal of the offending element or by preferential distillation of the impurities from the electrode crater. Interference is also encountered from spectral bands produced by cyanogen molecules formed when carbon vapor from the electrodes reacts with atmospheric nitrogen, by hydroxide radicals (from water vapor), and by silicon monoxide molecules (from silicates).

The alkali metals have their most sensitive lines in the visible and infrared regions. In many cases, a separate exposure is necessary for their detection.

Quantitative analysis. The intensity of a spectral line emitted by one component of a sample is affected by the concentration of that component in the sample, by the composition of the sample as a whole, and by the type of excitation used. In performing a quantitative spectrochemical analysis, it is necessary therefore to prepare for comparison a set of synthetic standards which resembles the sample chemically and physically, since either the excitation level, the intensity level, or both, may fluctuate during a given exposure. To overcome this variation, sample and standards are treated with equal concentrations of some element not previously present, called an internal standard. The standard samples are exposed spectrographically, and the ratio of the ITI of the subject element line to the ITI of the internal standard line is plotted against the concentration of the subject element. The sample is exposed by the same technique, and the same ratio observed and converted to a concentration figure.

In the metal industries, where graded metallic standards are available, much chemical analysis is done spectrochemically. Samples are sparked essentially as received, and intensity ratios registered photoelectrically for 30 or more elements simultaneously. Computing equipment can be used to convert this information to chemical concentrations by comparison with stored information from previously sparked standards. Analyses can be available within a few minutes after receipt of the sample, thus permitting the composition of furnace heats to be adjusted while the charge is still molten.

Since, in this technique, only a very small quantity of matter is vaporized by the spark, both standards and samples must be quite homogeneous to prevent sampling errors. If proper homogeneity cannot be achieved, samples of adequate size can be dissolved and compared with synthetic solutions.

Minerals usually are analyzed by arc methods. As successively higher-boiling components distill into the arc, the temperature and excitation conditions change continuously. The time at which a given minor component distills into the arc depends on the original state of combination of that element in the sample and on chemical reactions which it may undergo in the hot carbon crater. As a result, it is very difficult to synthesize standards for mineral analysis from laboratory reagents; it is much preferable to prepare these standards from analyzed mineral samples or mixtures of the type to be analyzed.

When standards that are chemically and physically similar to the sample are not available, it is often possible to dissolve the sample, to prepare comparable standard solutions, and to process samples and standards similarly from that point on. The use of solutions also permits one to eliminate interfering elements and to concentrate trace elements, thus effectively increasing the sensitivity of the method. *See* ATOMIC STRUCTURE AND SPECTRA; MOLECULAR STRUCTURE AND SPECTRA; X-RAY FLUORESCENCE ANALYSIS. [C.F.]

Bibliography: L. H. Ahrens, *Spectrochemical Analysis*, 1950; American Society for Testing Materials, *Methods for Emission Spectrochemical Analysis*, 1957; N. H. Nachtrieb, *Principles and Practice of Spectrochemical Analysis*, 1950.

Spectrohelioscope

An instrument for the monochromatic visual observation of the Sun. A telescope projects an image of the Sun on the first slit of a powerful spectroscope (Fig. 1). The resulting spectrum is imaged in the plane of a second slit which permits only a

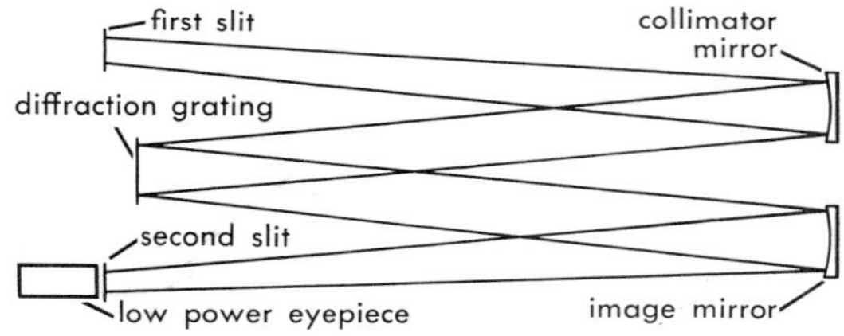

Fig. 1. The Hale spectrohelioscope.

single line element of the spectrum to emerge from the instrument. The emergent line element is a monochromatic image of that part of the Sun that falls on the first slit. The widths of the slits are generally chosen to isolate a spectral interval ½ angstrom (A) or less in width. When the two slits are vibrated synchronously at high frequency, persistence of vision permits monochromatic observation of an area of the solar surface. The slits may also be moved at a slow rate and the image recorded photographically (Fig. 2). This modification of the spectrohelioscope is one of the simple forms of the spectroheliograph. [R.R.M.]

Spectrophotometric analysis

A method of chemical analysis based on the absorption or attenuation by matter of electromagnetic radiation of a specified wavelength or frequency. The region of the electromagnetic spectrum most useful for chemical analysis is that between 2000 angstroms (A) and 300 microns (μ). Since the sample being analyzed absorbs the radiation, spectrophotometric analysis is sometimes referred to as absorptimetric analysis.

The instruments used in this work are referred to as spectrophotometers. A simple spectrophotometer consists of a source of radiation, such as a light bulb; a monochromator containing a prism or grating which disperses the light so that only a limited wavelength, or frequency, range is allowed to irradiate the sample; the sample itself; and a detector, such as a photocell, which measures the amount of light transmitted by the sample. (See Fig. 1.)

The Bouguer-Lambert-Beer law. By using a spectrophotometer, the intensity of the light transmitted through an absorbing substance may be compared with the light intensity when no such substance is in the light beam. Two fundamental laws govern the intensity of the light transmitted by an absorbing material. The first law, called the Bouguer-Lambert law, states that

$$\log (I_0/I) = Kb$$

where I_0 is the intensity of the light beam with no sample present, I is the intensity of the light beam after passing through the sample, K is a constant depending on the sample and wavelength of the light, and b is the thickness of the absorbing solution. The second law, called Beer's law, states that

$$\log (I_0/I) = K'c$$

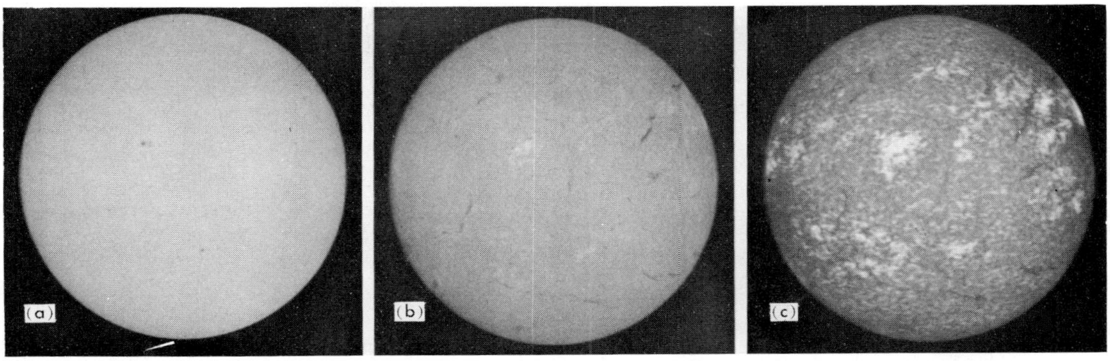

Fig. 2. Spectroheliograms of the Sun (1958 June 20^{d} 12^{h} 10^{m} UT) formed of line elements 0.3 A wide. (*a*) At 6600 A. (*b*) At 6563 A (Hα). (*c*) At 3933 A (K).

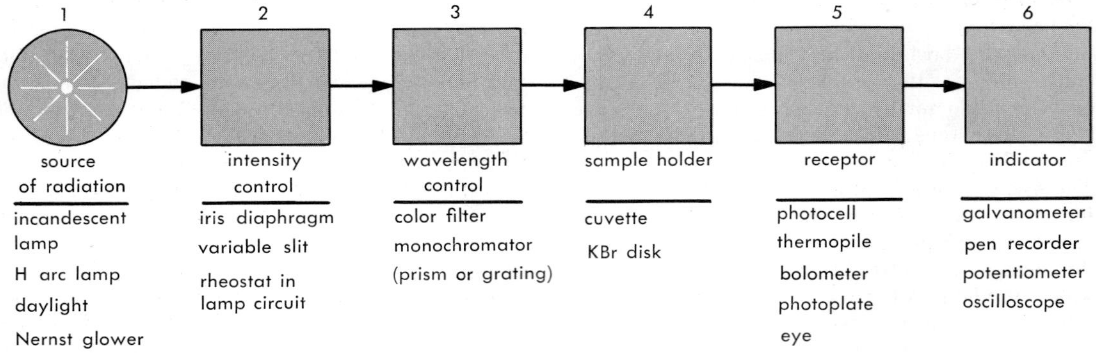

Fig. 1. Block diagram of generalized spectrophotometer.

where I and I_0 are as above, K' is a constant depending on the sample and wavelength of the light, and c is the concentration of absorbing material in the sample. Usually these two laws are combined in the form

$$\log (I_0 / I) = abc$$

where I_0, I, b, and c are as described above, and a is a constant called the absorptivity or extinction coefficient.

Two other terms are commonly used in spectrophotometric analysis. These are transmittance T and absorbance A

$$T = I/I_0$$
$$A = \log (1/T) = \log (I_0 / I) = abc$$

The absorbance, A, is directly proportional to the length of the light path through the sample, and to the concentration of the absorbing material. It is the term most used in quantitative spectrophotometric work.

The Bouguer-Lambert-Beer law is strictly obeyed only when monochromatic radiation, that is, radiation of a single wavelength or frequency, is used. The monochromators of most commercial spectrophotometers produce radiation which is close enough to monochromatic so that the deviations from the law from this source are minor, except in the infrared region. There are, however, occasional deviations from this law for both chemical and instrumental reasons.

In most quantitative analytical work, a calibration or standard curve is prepared by measuring the absorption of known amounts of the absorbing material at the wavelength at which it strongly absorbs. Such a calibration curve is shown in Fig. 2 for the absorbing material whose absorption spectrum is shown in Fig. 3. The absorbance of the sample is read directly from the measuring circuit of the spectrophotometer. Calibration curves are usually linear, as in Fig. 2. Occasionally, instrumental or chemical factors lead to nonlinear curves.

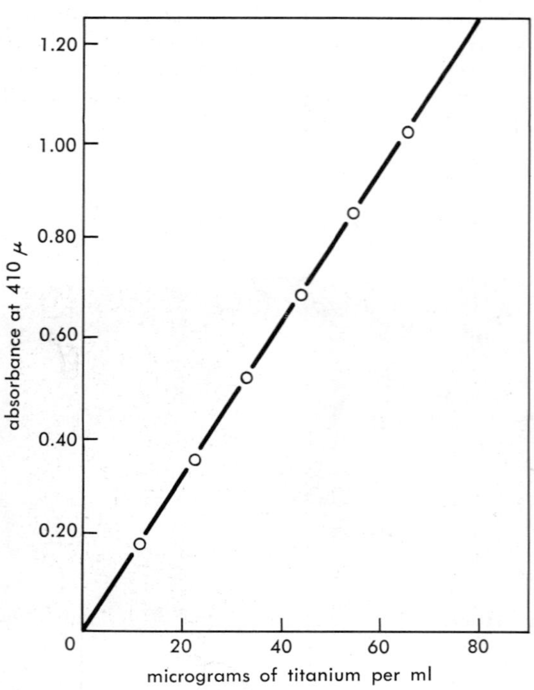

Fig. 2. Calibration curve for the determination of titanium by its color formed with hydrogen peroxide.

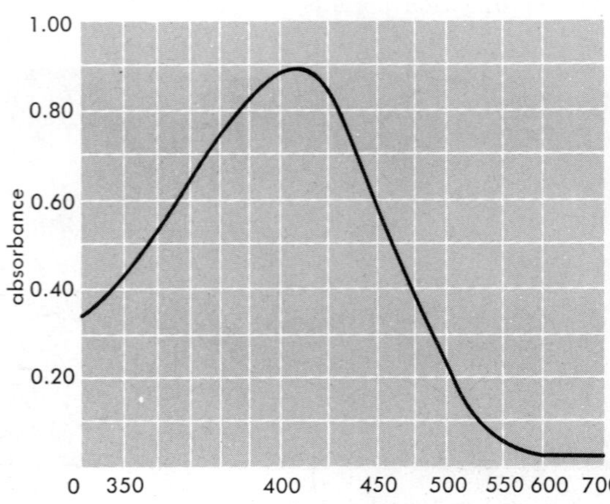

Fig. 3. Absorption spectrum of the peroxytitanate complex in the region from 340 to 700 mμ.

Absorption spectra. When the transmittance or the absorbance of a sample is measured and plotted as a function of wavelength, an absorption spectrum (Fig. 3) is obtained. This spectrum indicates that the sample transmits the least light at 410 millimicrons (mμ) and transmits the most around 700 mμ. Because of instrumental sensitivity limits, absorption spectra are obtainable only on samples which are relatively transparent to the radiation being used.

Reflectance spectra. In opaque samples, such as solids or highly absorbing solutions, the radiation reflected from the surface of the sample may be measured and compared with the radiation reflected from a nonabsorbing or white sample. If this reflectance intensity is plotted as a function of wavelength, it gives a reflectance spectrum. Reflectance spectra are used most often in matching colors of dyed fabrics or painted surfaces; they are used occasionally in qualitative analysis but seldom in quantitative analysis.

Chromogen. A molecule which absorbs radiation in a particular spectral region, usually in the visible or ultraviolet, is called a chromogen.

Chromophore. Groups of atoms within a molecule which are responsible for the absorption of light in the visible or ultraviolet regions are called chromophores. These chromophores are usually resonating structures which absorb at the same wavelength, or frequency, regardless of the molecule to which they are attached. Examples of such groups are the phenyl group, C_6H_5, which absorbs at 2700 A and the azo group N═N, which absorbs at 3700 A.

Auxochrome. Substituent groups which affect the wavelength of the spectral regions of strong absorption of chromophores are called auxochromes. Auxochromes cause two types of wavelength shifts. A shift to longer wavelength, or lower frequency, is called a bathochromic shift. Conversely, a shift to shorter wavelength is called a hypsochromic shift.

Infrared spectrophotometry. The interaction with matter of electromagnetic radiation of wavelength between 1 and 300 μ (frequencies of 10,000 and 33 cm^{-1}) induces either rotational or vibrational energy level transitions, or both, within the molecules involved. This region from 1 to 300 μ is usually referred to as the infrared. The frequencies of infrared radiation absorbed by a molecule are determined by its rotational energy levels and by the force constants of the bonds in the molecule. Since these energy levels and force constants are usually unique for each molecule, so also the infrared spectrum of each molecule is usually unique. The qualitative analytical use of the infrared region is based on this fact. Because of their individuality, infrared spectra of organic compounds are considered equivalent to, or superior to, the preparation of chemical derivatives for the identification of species in organic chemistry. The infrared portion of the spectrum is often called the fingerprint region. (See Figs. 4 and 5.)

For radiation sources, infrared instruments usually use a hot filament called a Nernst glower, or a hot carborundum rod called a Globar. Various inorganic prisms are used in the monochromators for different regions, for example, rock salt (sodium chloride) from 2 to 15 μ, potassium bromide from 15 to 27 μ, or cesium bromide from 12 to 40 μ. Gratings are also used in monochromators, either alone or in conjunction with prisms. A wide variety of detectors are used; examples are thermocouples and thermistors. These detectors must be very sensitive, as the amount of energy they must detect is quite small. Infrared cells, or sample containers, are prepared from materials transparent in the region of interest, and usually are made of rock salt, potassium bromide, or some other inorganic salt.

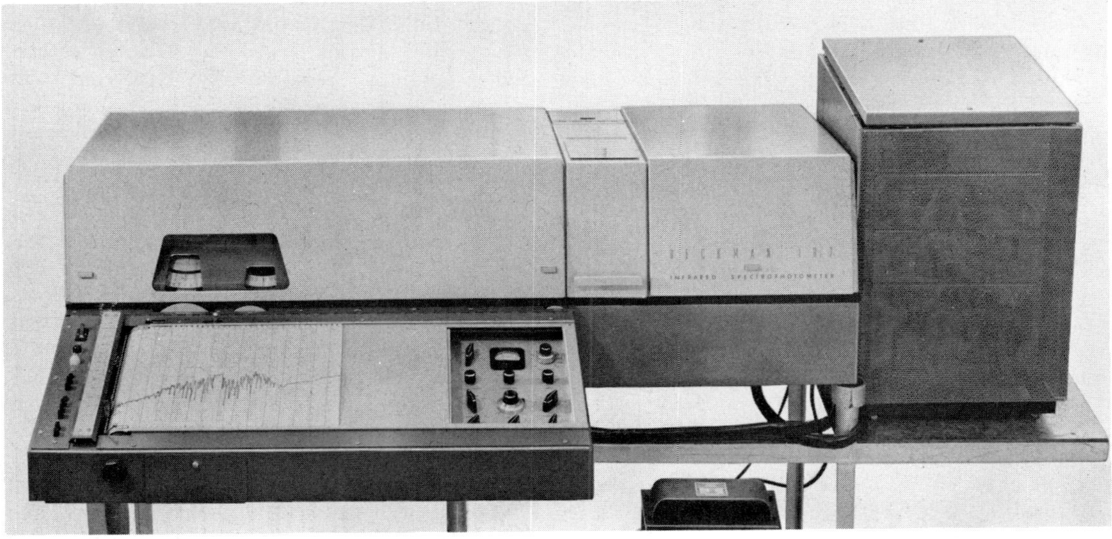

Fig. 4. A high resolution infrared spectrophotometer. (*Beckman, Scientific Instruments Division*)

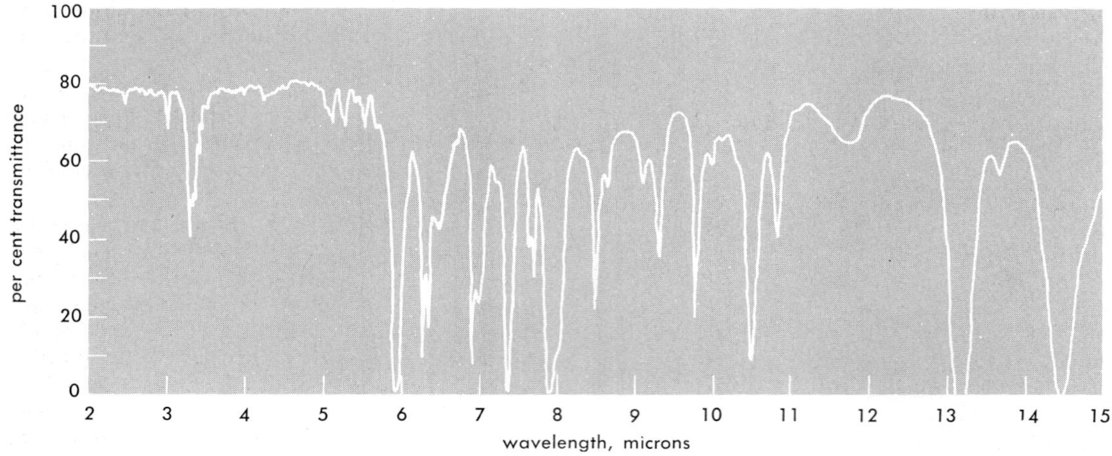

Fig. 5. Infrared spectrum of acetophenone.

Although almost every compound has a unique infrared spectrum, various groupings within a molecule have well-defined regions of absorption. For example, hydroxyl groups in alcohols absorb strongly at 2.8, 7.3, and about 8.5 μ, ester carbonyls absorb from 5.7 to 5.8 μ, and free amino groups absorb at 3.0 and from 6.1 to 6.4 μ. A typical infrared spectrum, that of acetophenone ($C_6H_5COCH_3$), is shown in Fig. 5. The band at 3.3 μ is due to the methyl group (CH_3), that at 5.9 μ to the conjugated carbonyl group (CO), and those at 13.2 and 14.4 μ are due to monosubstituted benzene (C_6H_5).

Quantitative analysis is based on the Bouguer-Lambert-Beer law as applied to a specific absorption band, usually unique to the compound being determined, as shown in Figs. 2 and 3. Occasionally, adequate quantitative results can be obtained using data for a similar compound containing the same functional group as in the material measured. The accuracy and precision of most infrared analyses is ± 3–5% of the amount of material present.

Infrared spectrophotometry can be applied to gaseous, liquid, or solid samples. For gaseous samples, cells (sample containers) from 1 cm up to 50 m long are used in order to get enough molecules in the light path to measure. In the long cells, the length is obtained by using mirrors to make the light traverse the cell numerous times before it is measured. Liquid samples are handled in cells whose thicknesses vary from 0.1 mm to 1 cm. The most common solvents for liquids are carbon tetrachloride and carbon disulfide, since these solvents have few absorption bands in the infrared region. Water is a very poor solvent because it absorbs strongly in this spectral region. Solid samples are analyzed by (1) preparing a thin film which may be either a free film or a film cast on a salt plate; (2) preparing a paste or mull by grinding up the solid with a viscous material such as mineral oil (Nujol), which has few infrared bands; or (3) pressing a disk of an intimate mixture of the solid with potassium bromide, KBr. This latter, the so-called KBr disk method, appears to be the best for quantitative infrared work with solids.

The infrared region is used primarily for analyses of organic compounds because they are readily soluble in a desirable solvent and because they have unique and complex spectra. However, work has been done on the infrared spectra of inorganic compounds in the forms of KBr disks or Nujol mulls.

Near-infrared spectrophotometry. This designates work carried out between 0.78 and 3 μ. The instruments in use in this region have quartz prisms in their monochromators and lead sulfide photoconductor cells as detectors. The absorption bands in this region are mainly overtones (harmonics) of bands in the infrared region. These bands are quite sharp and are of great value in quantitative analysis for various functional groups, particularly those containing hydrogen atoms; examples are terminal methylene groups (CH_2), hydroxyl groups, and amines. The cells used in this region are usually made of quartz, and hence are more durable than infrared cells. The near-infrared spectra give some of the same information as those in the infrared region, but occasionally more specifically, more inexpensively, or more rapidly. The accuracy and precision of ± 1–3% of the amount present is also somewhat better than in infrared spectrophotometry.

Visible spectrophotometry. The visible region of the spectrum covers the narrow range from about 380 to 780 mμ. The spectrophotometers for this region use tungsten lamps as light sources, glass or quartz prisms or gratings in the monochromators, and photomultiplier cells as detectors. Within this narrow portion of the electromagnetic spectrum, a majority of the spectrophotometric analyses are made. The absorption of light in this region is caused by the excitation of the outer electrons of the molecule by the impinging light beam. Figure 3 shows a typical visible absorption spectrum. The substance, peroxytitanate ion, absorbs light in the region below 500 mμ, that is, it absorbs violet, blue, and green light and transmits red, orange, and yellow. For analytical work, the wavelength of maximum absorption is usually used, in this case 410 mμ. The calibration curve in Fig. 2 was made by plotting the absorption at 410 mμ versus the concentration of the absorbing material in the solution. Unknown samples are then analyzed by measuring the absorbance of the solution after appropriate reagents have been added. The amount of material in question, in this case titanium, is obtained from the calibration curve.

Although few materials, in particular inorganic ions, have visible colors, there are spectrophotometric methods utilizing visible colors for most of them. The method used above for titanium is typical. Hydrogen peroxide, when added to a colorless titanium solution, forms the highly colored peroxytitanium complex. Similar reactions are those of thiocyanate ion with ferric iron and of ammonia with copper. In recent years, organic reagents have been prepared which form intense colors with different metal ions. Many of these reagents are so specific that they form colors with but one or two inorganic ions. Examples of these are *o*-phenanthroline which reacts with ferrous ion and 2,2′-biquinoline with copper.

Visible spectrophotometry is used extensively because there are methods available for determining a wide variety of materials, especially inorganic cations, with great sensitivity and selectivity. The equipment is relatively inexpensive ($300–1000).

Visible spectrophotometry is usually carried out with liquid samples and is usually used for quantitative, rather than qualitative purposes. Visible reflectance spectra of solid samples are run where the matching of colors is important or in cases where the samples are opaque. Gaseous samples seldom are run in the visible region since they seldom absorb visible light intensely.

Ultraviolet spectrophotometry. The spectral region from 2000 to 4000 A, called the near ultraviolet, is commonly used in chemical analysis. The absorption of ultraviolet radiation by a molecule is usually the result of exciting the outer, or valence, electrons of the molecule in question. The more easily the electrons are excited, the longer the wavelength of the absorption peak.

Ultraviolet spectrophotometers usually have a hydrogen lamp as a radiation source; a quartz prism, or a grating in the monochromator; and a photomultiplier tube as a detector. Quartz or silica cells of 1-mm to 10-cm length are commonly used for the samples.

Simple inorganic ions and their complexes as well as organic molecules can be detected and determined in this region. Useful solvents are water, saturated hydrocarbons, aliphatic alcohols, and ethers. Organic compounds which absorb ultraviolet radiation have at least one unsaturated linkage, such as C=C, C=O, N=N, or S=O, which acts as a chromophore. The wavelength of the absorption peak increases with the degree of unsaturation within the chromophore.

Inorganic groups which absorb in the ultraviolet region owe their activity to numerous valence electrons, as in the complex $FeCl_4^-$, or to electrons in a single atom, possible hydrated, as in the rare-earth elements.

Quantitative work in the ultraviolet region is common, and most substances obey Bouguer-Lambert-Beer law over a wide range. In the field of organic analysis, the ultraviolet region is most applicable to aromatic compounds. The spectra of some compounds, such as phenols, may be greatly enhanced by using basic solutions of the samples. Since most compounds which absorb in the region have intense bands, it is possible to analyze either dilute solutions or extremely small samples. For example, the spectrum in Fig. 6 is that of 0.01% acetophenone in isooctane in a 1-cm cell. By using longer cells, it is possible to detect 0.00001% acetophenone in isooctane. Many substances absorb much more intensely than acetophenone.

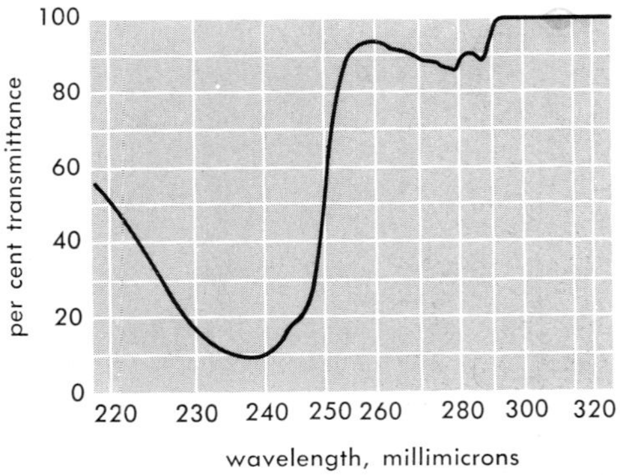

Fig. 6. Ultraviolet absorption spectrum of acetophenone.

Similarly, ultraviolet spectrophotometry is especially useful for the determination of inorganic ions as simple complexes, such as $FeCl_4^-$, $PtCl_6^{--}$, or I_3^-. Accuracy and precision are usually about ±1–2% of the amount of material being determined.

As indicated above, samples are usually liquids although gases may also be analyzed. Because of relatively greater scattering of short-wavelength radiation, transmission measurements on turbid samples are difficult to interpret, and opaque samples are seldom run by reflectance.

Filter photometry. In filter photometry, the monochromator of the spectrophotometer is replaced by a filter. This filter passes a band of light of a much wider range of wavelengths than those passed by even the poorest monochromator. For the example in Fig. 3, a filter which transmits blue light would be used in order to obtain maximum sensitivity. The filter chosen is usually of the color complementary to that of the solution, that is, the filter is chosen so as to transmit best the light which the sample absorbs most. In the visible region, colored glass or gelatin films containing dyes have been most widely used. Interference filters, based upon selective transmission of radiation through very thin metallic films between two glass plates, usually yield more nearly monochromatic bands and higher transmittance. They are used when their somewhat greater costs can be justified. Absorbing liquids and gases have been used as filters but are generally more cumbersome.

Filter photometers are generally much less expensive than spectrophotometers. Because they do not use monochromatic light, the calibration curves obtained often do not obey the Bouguer-Lambert-Beer law. However, by careful use of calibration curves, filter photometers can give sufficiently accurate and precise results for a wide variety of applications.

Although filter photometers are most often used in the visible region, some filters for the infrared and ultraviolet regions are available. *See* ANALYTICAL CHEMISTRY; MOLECULAR STRUCTURE AND SPECTRA; OPTICAL METHODS OF CHEMICAL ANALYSIS. [R.F.G.]

Bibliography: G. R. Harrison, R. C. Lord, and J. R. Loofbourow, *Practical Spectroscopy*, 1948; M. G. Mellon, *Analytical Absorption Spectroscopy*, 1950; W. West (ed.), *Chemical Applications of Spectroscopy*, Technique of Organic Chemistry, vol. 9, 1956.

Spectroscopy

Spectroscopy pertains to the production and investigation of spectra, the phenomena observed when all the electromagnetic radiations characteristic of a particular source of radiant energy are separated into an array of constituent colors, wavelengths, or frequencies (*see* ELECTROMAGNETIC RADIATION; SPECTRUM). The wavelengths are separated by refraction in a transparent prism, by diffraction from a ruled grating, or by diffraction in crystalline solids. Instruments designed for this purpose are called spectroscopes, spectrographs, spectrometers, or spectrophotometers.

The significance and importance of spectroscopy dates from 1860 when G. Kirchhoff and R. Bunsen, by systematically comparing the sun's spectrum with flame or spark spectra of salts and metals, made the first chemical analysis of the Sun's atmosphere, and thus laid the foundation for spectrochemical analysis and astrophysics. Most of the progress in the description, interpretation, and applications of spectra dates from 1910 when the first international standards of wavelength were adopted. These and later standards made it possible to measure with unprecedented accuracy the wavelengths occurring in any spectrum whatsoever. In this manner, spectroscopy has gathered wavelength data for several million lines observed in atomic and molecular spectra extending from the extreme ultraviolet to the far infrared and embracing more than 60 octaves, as compared with the single visible octave known in 1800.

Spectroscopic units. In stating the wavelengths of spectral lines, various units of length are commonly employed in different parts of the spectrum, namely,

$$1 \text{ micron } (\mu) = 10^{-4} \text{ cm}$$
$$1 \text{ millimicron } (m\mu) = 10^{-7} \text{ cm}$$
$$1 \text{ angstrom (A)} = 10^{-8} \text{ cm}$$
$$1 \text{ X-unit, or Siegbahn unit (XU)} \cong 10^{-11} \text{ cm}$$

For more exact definitions, *see* ANGSTROM; MICRON; X-UNIT.

Classification by wavelength. Because so small a portion of the entire gamut of electromagnetic radiation is visible, spectroscopy has been subdivided into several ranges of wavelength, according to the method of either producing or detecting the radiations. Thus, there is x-ray spectroscopy, ultraviolet spectroscopy, visual spectroscopy, infrared spectroscopy, microwave spectroscopy, and radiofrequency spectroscopy. No sharp division exists between these successive ranges, and there is usually a considerable overlap.

X-ray spectroscopy. This covers 17 octaves, extending from about 0.006 to over 1000 A, and the methods of detection include photography, fluorescence, and ionization. *See* X-RAY FLUORESCENCE ANALYSIS.

Ultraviolet spectroscopy. This extends from about 6 to 3800 A, and the methods of detection are photographic, photoelectric, and radiometric.

Visual spectroscopy. Because of certain properties of the human eye, visual spectroscopy is limited to a relatively small portion of the electromagnetic spectrum. As officially defined by the Commission Internationale de l'Éclairage (International Commission on Illumination), the average human eye responds to light waves between about 3800 A (violet) and 7800 A (red) in length, with a maximum response at 5550 A in the yellow-green region (Fig. 1). The most common visible spectrum presented by nature is the rainbow. In addition to visual perception, this octave of the

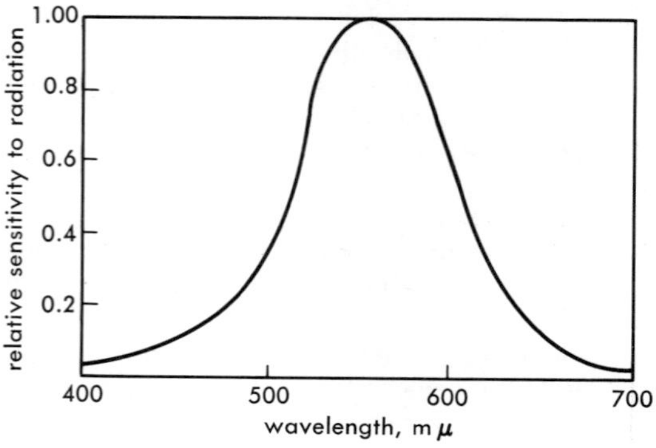

Fig. 1. Spectral sensitivity of the human eye. (*From G. R. Harrison, R. C. Lord, and J. R. Loofbourow, Practical Spectroscopy, Prentice-Hall, 1948*)

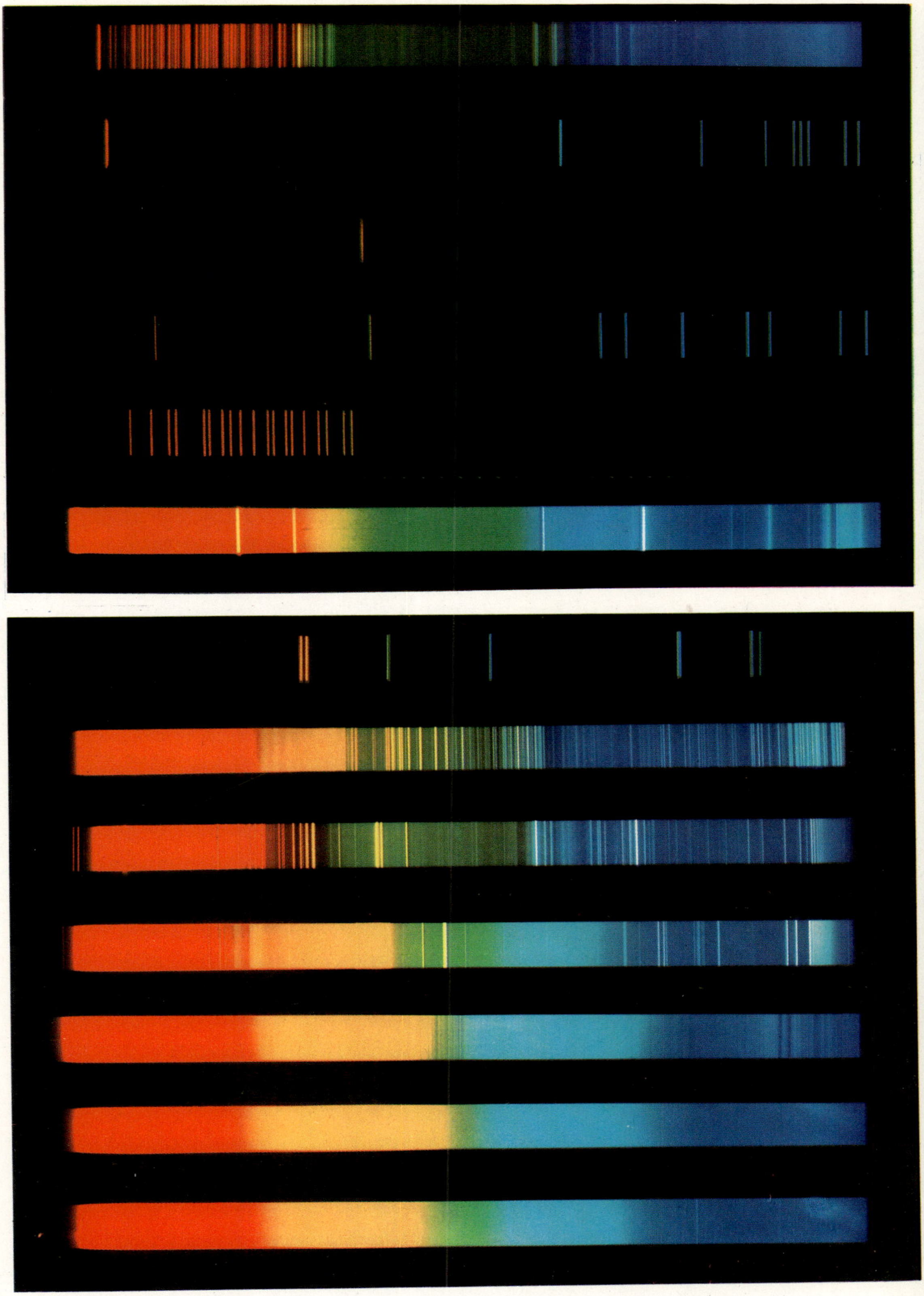

Typical spectra obtained in the visible region. (*a*) Molecular hydrogen. (*b*) Atomic hydrogen. (*c*) Sodium-vapor lamp (D lines). (*d*) Helium. (*e*) Neon. (*f*) Lithium. (*g*) Mercury. (*h*) Iron. (*i*) Barium. (*j*) Calcium. (*k*) Fraunhofer absorption lines. (*l*) Tungsten filament lamp. (*m*) Fluorescent lamp. (*Bausch and Lomb*)

spectrum is readily detected by photography, phototubes, and radiometers.

Infrared spectroscopy. This extends from the long-wave (7800-A) limit of visibility to about 1 mm or 10^7 A. Photographic detection stops a little beyond 10,000 A; photoconductive detectors such as lead sulfide or lead telluride are generally used between 10,000 and 50,000 A, but radiometers, that is, bolometers or thermopiles, are used to detect waves of greater length, up to 1 mm. *See* INFRARED SPECTROSCOPY; RADIOMETRY.

Microwave spectroscopy. This includes waves from about 1 mm to about 30 cm in length, or frequencies from 3×10^{11} sec^{-1} to 10^9 sec^{-1}. Because neither the ordinary methods for infrared spectroscopy nor the conventional radio circuits were capable of generating or detecting microwaves, this gap remained uninvestigated until radar was developed during World War II. Now microwaves are efficiently generated by cavity resonators driven by velocity-modulated electron beams, transmitted by hollow wave guides, and detected by crystal rectifiers. *See* MICROWAVE SPECTROSCOPY.

Radio-frequency spectroscopy. This is concerned with wavelengths ranging from about 30 cm to many kilometers, generated in more or less conventional radio circuits such as vacuum-tube oscillators. Resonant receiving circuits serve to detect these waves. Radio waves and microwaves are produced by the ordered, synchronized motions of free electrons moving under the control of a resonant circuit or cavity. All electrons in a radio-frequency source oscillate with the same phase and frequency. The output frequency can be varied or modulated to scan a spectral range, so that no dispersive instrument is needed in radio-frequency spectroscopy. The most fruitful application of the method of molecular or atomic beams has been as a spectroscopic device in this range of frequencies. This method is usually characterized by the production of an extremely well defined beam of neutral atoms or molecules which is then subjected to further study by the application of one or more influences which modify the subsequent trajectory of the particles in the beam. *See* MOLECULAR BEAMS; RADIO-FREQUENCY SPECTROSCOPY.

Interpretation of spectra. After spectra have been produced, observed, and recorded in detail, certain measurements must be made before they can be interpreted and applied for specific purposes. The measurements in line spectra consist of wavelength or frequency determinations corresponding to each radiation or line, relative or absolute intensities, line widths, shapes, hyperfine structures, and isotope effects. The interpretation of such measurements leads either (1) to chemical identifications and quantitative determinations (spectrochemical analysis), (2) to deductions as to the structure of atoms and molecules, or (3) when hyperfine structures are resolved, to evaluation of intrinsic properties of atomic nuclei. *See* ATOMIC STRUCTURE AND SPECTRA; HYPERFINE STRUCTURE; ISOTOPE SHIFT; MOLECULAR STRUCTURE AND SPECTRA.

Spectrochemical analysis. This refers to the chemical identification and quantitative analysis of matter by means of spectra that are uniquely characteristic of atoms and molecules. The basic principle of spectrochemical analysis—that emission and absorption spectra are characteristic of the atoms and molecules that produce them—was first clearly stated by G. Kirchhoff and R. Bunsen in 1860; it led almost immediately to the discovery of four new chemical elements (cesium, rubidium, thallium, and indium) and a preliminary chemical analysis of the sun's atmosphere. In 1874, N. Lockyer proposed a second principle in his statement that "while the qualitative spectrum analysis depends upon the positions of the lines, the quantitative analysis depends . . . upon their brightness and number as compared with the number visible in the spectrum of pure vapor." It is observed that when one element is progressively diluted in another, the spectrum of the diluted element becomes weaker until the strongest line vanishes when the concentration falls below the limit of spectroscopic detection. *See* SPECTROCHEMICAL ANALYSIS.

Procedures similar to those in spectrochemical analysis are used in x-ray spectroscopy, except that Geiger counters or scintillation counters receive the selected radiations, which range in wavelength from about 0.1 to 10 A.

Astronomical spectroscopy. This is a major branch of astrophysics in which light from celestial objects is dispersed into spectra for the purpose of obtaining information concerning the sun, planets, stars, nebulae, comets, and meteors. The light from these natural sources is spread out into spectra either by placing a prism in front of an astronomical telescope or by attaching a spectrograph at the observing end of a telescope. The types of information obtained from astronomical spectra include chemical composition, temperature, pressure, density, magnetic fields, electric forces, and radial velocity (or motion in the line of sight) which often give additional information about stellar rotation, convection, and turbulence, and serve to identify so-called spectroscopic binaries. The original visible range of wavelengths (3800–7800 A) of astronomical spectroscopy was extended to 3000–10,000 A by photography. The infrared solar spectrum has been explored with photodetectors and radiometers, and still longer waves in the radio-frequency range are received by radio telescopes. Wavelengths shorter than about 3000 A from celestial sources cannot be detected on the earth's surface because they are absorbed by oxygen and ozone in the terrestrial atmosphere, but since 1946, rockets have carried spectrographs sufficiently high to photograph the sun's spectrum in the extreme ultraviolet to about 900 A. *See* ASTRONOMICAL SPECTROSCOPY; RADIO ASTRONOMY.

Raman spectroscopy. This records the scattered radiations resulting from the illumination of transparent matter with an intense beam of light containing approximately monochromatic radiation. When bright beams of monochromatic light pass through clear gases, liquids, or solids, a small fraction of the incident energy is absorbed by molecules and reradiated as light of (usually) greater wavelength. The differences between the frequencies of the incident light and the scattered rays correspond to vibration or rotation frequencies of the molecules normally observed in infrared spectra. Thus, Raman spectroscopy supplements infrared spectroscopy by determining the fundamental frequencies of molecules from measurements in the visible and ultraviolet (between 6700 and 2500 A). Because of the very low intensity of Raman lines compared with the incident radiation, intense line sources and efficient spectrographs have been developed for Raman spectroscopy. For an extended discussion, *see* RAMAN EFFECT.

Light sources. The sources of light employed for spectroscopic studies may be divided into two groups, depending on whether they emit continuous or discontinuous spectra. Hot, incandescent solids, such as a Welsbach gas mantle, tungsten-lamp filament, Nernst glower, or Globar lamp, always emit broad continuous spectra with a maximum intensity at a wavelength that varies approximately inversely as the absolute temperature. These spectra are dependent only upon temperature and emissivity of the radiator; their principal use in experimental spectroscopy is to provide continuous backgrounds for the observation of absorption spectra of gases, vapors, solutions, or solids in the near ultraviolet, visible, and infrared ranges. High-voltage discharges in hydrogen, or between metal electrodes under water, are also suitable sources of continua for the study of absorption spectra in the visible and ultraviolet. For the study of absorption in the vacuum ultraviolet, electrical discharges in noble gases at high pressures produce the continuous backgrounds.

Light sources that produce discontinuous spectra are flames, furnaces, and electrical discharges in arcs and sparks at atmospheric pressure or in lamps containing gases or metal vapors at reduced pressure. In these sources, bright lines with different wavelengths are emitted by atoms, ions, or molecules excited to radiate uniquely characteristic spectra. The individual particles are excited by absorbing energy either from collisions with other atoms or electrically charged particles, or from incident radiation.

Flames. It has long been customary in elementary chemistry classes to demonstrate spectrochemical identification by dipping a platinum wire into salt solutions, inserting it in a bunsen burner, and observing the spectrum of the colored flame with a bunsen spectroscope or amici prism. Because the temperature of this flame is near 2000°C, the kinetic energy of atomic collisions excites only the stronger lines of simple spectra. Flame photometry has become a popular method for determining the alkali or alkaline-earth content of solutions sprayed into the flame. In the hotter oxyacetylene flame, atomic spectra are more efficiently excited, and at least 34 chemical elements can be thus recognized and measured. *See* FLAME PHOTOMETRY.

Electric furnaces. Metallic spectra in which the excitation is pure thermal energy, as in flames, are more effectively produced in an evacuated furnace in the form of a carbon tube heated by forcing large electric currents through it. The temperature may be controlled between 1500°C, where spectra begin to appear, and 3600°C, where the carbon tube tends to fail. Small samples of metal are placed in a porcelain boat in the carbon tube and the luminous vapor is imaged on the slit of a spectrograph. If a carbon plug is inserted in one end of the tube, it emits a continuous background for the study of absorption spectra of metal vapors.

Electric sparks. The high-voltage condensed spark is the most energetic source for generating atomic and ionic spectra. The spark consists of discharges from one or more electrical condensers connected, in parallel with the spark gap, to the secondary terminals of an alternating-current transformer that elevates the primary (110 or 220) voltage to 11,000 or more volts (Fig. 2). The condensers are charged on every half cycle to the breakdown potential of the spark gap. An oscillating current then flows in the spark circuit with an initial value of $I = V\sqrt{C/L}$, where V is the condenser voltage, C the capacitance in farads, and L the circuit inductance in henrys. This initial current may be many hundreds of amperes, resulting in high effective temperatures. When operated in the highest attainable vacuum, such sparks produce multiple ionization of atoms, even to the point of temporarily removing all their electrons, and the corresponding temperature is estimated at several million degrees.

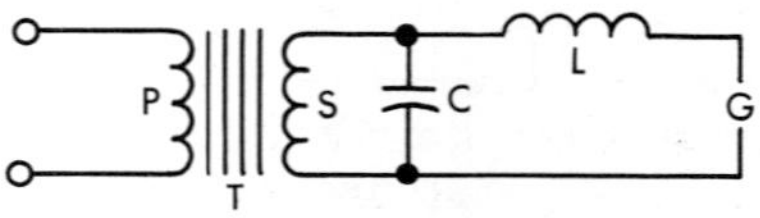

Fig. 2. Electrical circuit for operating a spark. T, high-voltage step-up transformer; P, primary; S, secondary; C, condenser; L, self-inductance (used if it is desired to suppress air lines); G, spark gap. (*From G. R. Harrison, R. C. Lord, and J. R. Loofbourow, Practical Spectroscopy, Prentice-Hall, 1948*)

Geissler tubes. In 1858, H. Geissler prepared an efficient light source by passing electrical discharges through gases at reduced pressure contained in a small-bore glass tube connecting two larger-bore tubes provided with internal electrodes (Fig. 3). Because Geissler tubes operate at ordinary temperatures and low gas or vapor pressures, they emit sharp lines with high intensity, and therefore have been used to study the spectra of all natural gases and of metals with low boiling

Fig. 3. Geissler tube. (*From G. R. Harrison, R. C. Lord, and J. R. Loofbourow, Practical Spectroscopy, Prentice-Hall, 1948*)

points. Since 1930, modified forms of Geissler tubes containing helium, neon, argon, or mercury have become familiar to everyone as luminous signs or fluorescent tube lamps.

Hollow-cathode discharge. Another type of light source, designed especially for high-resolution spectroscopy, consists of a hollow cylindrical metal cathode and an anode enclosed in a glass chamber containing a small amount of pure noble gas (Fig. 4). When a potential difference of about 1000 volts is applied, some of the cathode material, or any other metal within it, is vaporized by bombardment of the noble gas ions and excited by collisions with electrons. Since the cathode radiates at low pressure and may be cooled with liquid air, this lamp yields atomic spectrum lines of the great sharpness desired for the investigation of isotope shifts and hyperfine structures.

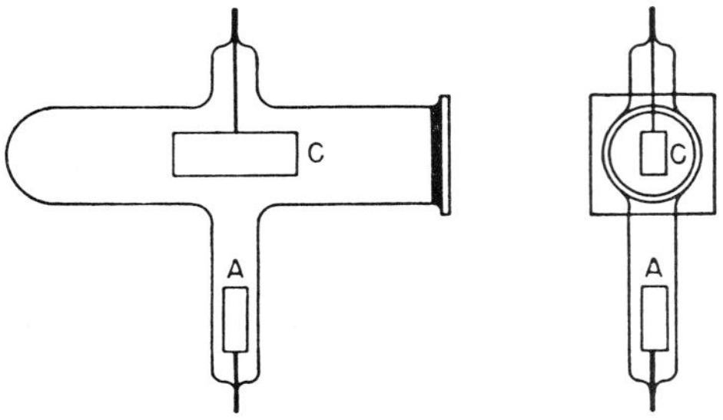

Fig. 4. Hollow-cathode discharge tube. A, anode; C, cathode. (*From G. R. Harrison, R. C. Lord, and J. R. Loofbourow, Practical Spectroscopy, Prentice-Hall, 1948*)

Electrodeless discharges. A simple and extremely useful light source has been developed since 1946 when radar and microwave generators became generally available. It consists of an electrodeless glass or quartz tube containing a minute amount of volatile metal or compound plus a trace of argon gas excited in alternating electromagnetic fields of ultra-high frequency, 200–3000 megacycles per second. Because the pressure is low and the temperature is moderate, these lamps produce intense spectra of sharp lines, and because they operate with microgram samples, they are extremely useful for investigating spectra of artificial and highly radioactive elements.

Instrumentation. The important instruments that are used in spectroscopy include spectroscopes, spectrometers, spectrographs, interferometers, and spectrophotometers.

Spectroscope. This is an optical instrument that separates composite light into its components, thus producing a spectrum for visual observation. A complete spectroscope, consisting of slit, collimator lens, prism, and telescope, was first assembled by R. Bunsen in 1859 (Fig. 5). When a light source illuminates the slit, the transmitted rays are collimated by a lens to fill a glass prism which refracts and disperses the rays by virtue of different refractive indices for different colors which are then focused by a telescope to form monochromatic images of the slit in a spectral array from red to violet with increasing refraction.

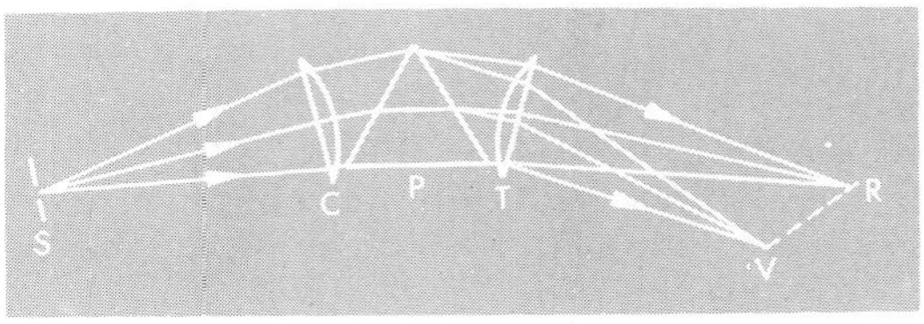

Fig. 5. Prism spectroscope or spectrograph. S, slit; C, collimator lens; P, 60° prism; T, telescope; RV, red to violet spectrum. For producing ultraviolet spectra, the glass items in C, P, and T must be replaced by equal amounts of left- and right-handed crystal quartz.

The simplest spectroscope (which can be pocket-sized) consists of a cemented train of light crown and dense flint glass prisms (Fig. 6). Invented in 1860 by G. B. Amici, this direct-vision spectroscope consists of two prisms of crown glass, having a low ratio of dispersion to index of refraction, cemented to a prism of flint glass with a high ratio of dispersion to refractive index. The prism angles are chosen so that the central spectral ray is undeviated upon emergence, but all other rays are refracted and dispersed. The dispersion and resolving power can be increased by assembling trains of five or seven prisms. *See* DISPERSION (RADIATION); PRISM, OPTICAL; REFRACTION OF WAVES; RESOLVING POWER (OPTICS).

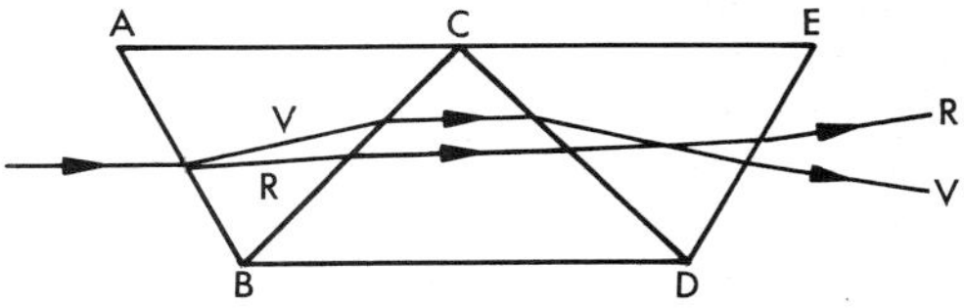

Fig. 6. Direct-vision spectroscope. ABC and DEF, crown glass prisms; BCD, flint glass prism; RV, red to violet spectrum.

The most commonly used visual spectroscope is probably the constant (right-angle) deviation type (Fig. 7) introduced in 1904. Provided with calibrated wavelength scales coiled on a drum connected with a rotating prism of tetragonal form, the best of these instruments enable one to measure with an accuracy of 1–2 A from 3900 to 8000 A.

Spectrometer. This is a spectroscope provided with scales for the measurement of wavelengths or

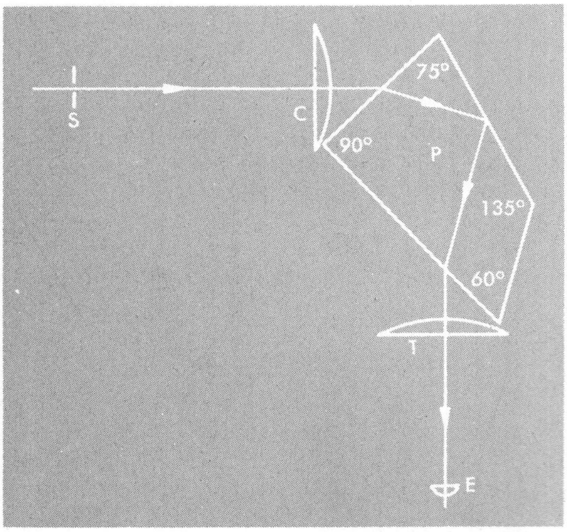

Fig. 7. Constant-deviation spectroscope. S, slit; C, collimator lens; P, glass prism; T, telescope lens; E, eyepiece.

for the measurement of indices of refraction of transparent prism materials. The constant-deviation spectroscope with calibrated drum (Fig. 7) is the most convenient spectrometer for measuring wavelengths. The bunsen spectroscope (Fig. 5) is converted to a spectrometer if circular scales are attached to measure the angles of incidence and refraction referred to normals to the prism faces. Then the index of refraction of the prism for a wavelength is the ratio of the sine of the angle of incidence to the sine of the angle of refraction for that wavelength.

Spectrograph. This is a spectroscope provided with a photographic camera or other device for making a record of the spectrum, called a spectrogram. In principle, the spectroscope becomes a spectrograph when a photographic plate or film is substituted for the eyepiece. However, another substitution is required in glass prism spectrographs if the ultraviolet is to be recorded; glass is opaque to radiation of wavelength below about 3500 A and must be replaced by transparent crystal optics such as natural quartz, which transmits to about 1800 A. Two types of quartz spectrographs are in common use for recording ultraviolet and visible spectra. Because of double refraction in crystal quartz, it is necessary to use equal amounts of so-called left- and right-handed quartz to avoid double images of the slit (*see* BIREFRINGENCE; CRYSTAL OPTICS; OPTICAL MATERIALS). One type of quartz spectrograph is made of two 30° prisms, left- and right-handed, as well as left- and right-handed lenses, so that the rotation occurring in one-half the optical path is exactly compensated by the reverse rotation in the other (Fig. 5). Prism spectrographs employing the principle of autocollimation accomplish this by the use of a Littrow 30° quartz prism with a rear reflecting surface (Fig. 8) that reverses the path of the light through prism and lens, thus compensating for rotation of polarization in one direction by equal rotation in the opposite direction. A spectrograph with optics of lithium fluoride can be used to record spectra to 1100 A, but all crystalline materials are opaque to shorter waves until the x-ray range is reached, where atoms of crystal planes act like diffraction gratings to produce spectra.

The most powerful and useful spectrographs employ ruled diffraction gratings. The grating spectrographs now available surpass prism spectrographs in dispersing and resolving powers. Furthermore, these powers remain practically constant throughout a grating spectrum, whereas in prisms they vary considerably with the wavelength. Modern diffraction gratings are reflection gratings; when combined with concave mirrors, they can form spectrographs whose freedom from all absorbing material makes them uniquely useful in the spectral extremes (extreme ultraviolet and far infrared) where all prisms are opaque. *See* DIFFRACTION; DIFFRACTION GRATING.

Interferometer. This is an optical device that divides a beam of light into two or more parts which travel different paths and then recombine to form interference fringes. Since an optical path is the product of the geometric path and the refractive index, an interferometer measures difference of geometric path when the two beams travel in the

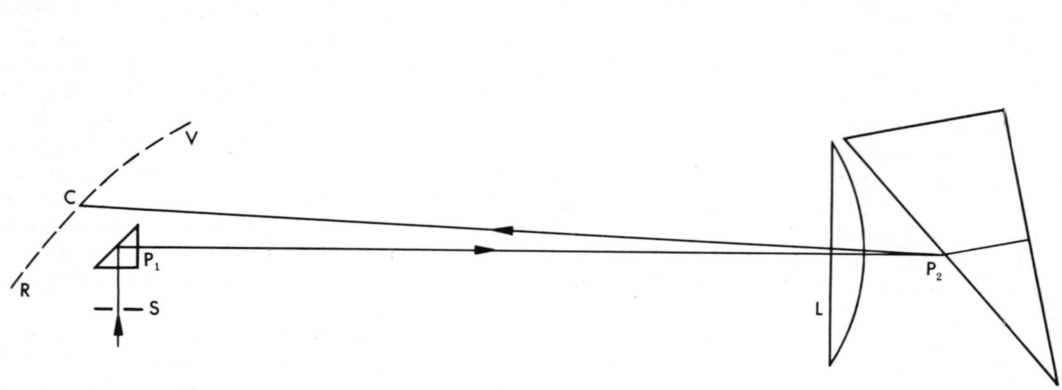

Fig. 8. Littrow quartz spectrograph. S, slit; P_1, totally reflecting quartz prism; L, autocollimating quartz lens; P_2, Littrow quartz prism; C, camera; RV, red to violet spectrum.

same medium, or the difference of refractive index when the geometric paths are equal. Interferometers may be used to measure a length, a difference in optical path or wavelength, or a refractive index. Because interference fringes enable one to measure distances which are small fractions of the length of a light wave, as well as determine without error any number of integral waves, interferometers are used for all precision measurements in optics and spectroscopy. Interferometers that produce interference fringes with multiple beams possess the maximum practical resolving power; they may resolve spectral lines that differ in wavelength by less than 10^{-6} of a wave, such minute differences frequently being encountered in investigations of isotopic effects on spectra. *See* INTERFERENCE OF WAVES; INTERFEROMETRY.

Spectrophotometer. This is an optical instrument devised for the measurement of radiant energy distribution or radiant flux as a function of wavelength. It is essentially a spectroscope provided with calibrated spectral energy detectors (visual, photographic, photoelectric, or radiometric) and is used for measuring the relative absorbance, transmittance, or reflectance at different wavelengths characterizing any materials. *See* SPECTROPHOTOMETRIC ANALYSIS.

Identification of spectrum lines. Spectrum lines are identified as to their chemical origin first and foremost by their wavelengths, the most accurately measured property of the lines. Several hundred wavelengths characteristic of a half-dozen chemical elements (cadmium, iron, thorium, neon, argon, krypton, and mercury) have been determined to eight figures to serve as standards (*see* WAVELENGTH STANDARDS). Relative to these standards, many thousands of wavelengths from atoms, ions, and molecules have been measured to seven figures, and hundreds of thousands to six figures. Operating with wavelengths between 2000 and 9000 A, spectroscopists readily identify and determine more than 70 chemical elements in the analysis of complex mixtures. If the atomic and ionic spectra of these 70 elements were fully recorded, they would be represented by approximately 500,000 lines, and, if evenly distributed, their average separation would be only 0.014 A. However, the density of lines per angstrom is much greater in the ultraviolet than in the infrared and visible regions, so that it would seem that six-figure accuracy in wavelength measurements is insufficient for unambiguous identification of spectral lines. Fortunately, this problem is simplified by the fact that elements diluted in mixtures exhibit simplified spectra as a function of dilution, until in the limit a single line remains to represent a spectroscopic trace. Furthermore, no samples for spectrochemical analysis (excepting synthetic mixtures) ordinarily contain more than 12–24 chemical elements. It therefore appears that six-figure accuracy in wavelength is usually ample for positive identification of a chemical element.

A second property of spectral lines that is useful in confirming spectral identification of chemical elements is the intensity ratio of two or more lines of each element. For example, if the line 5895.92 A has been identified as sodium, the line 5889.95 A must also be present because it is twice as intense. Although line intensities cannot be measured with the same accuracy as wavelengths, approximate relative intensities of many thousands of spectral lines are known, and should be used to confirm the spectroscopic identification of a chemical element whenever its abundance exceeds a spectroscopic trace. [W.F.M.]

Bibliography: E. U. Condon and G. H. Shortley, *The Theory of Atomic Spectra*, reprint, 1951; G. R. Harrison, R. C. Lord, and J. R. Loofbourow, *Practical Spectroscopy*, 1948; G. Herzberg, *Atomic Spectra and Atomic Structure*, 1944; G. Herzberg, *Molecular Spectra and Molecular Structure*, 2 vols., 2d ed., 1950; R. A. Sawyer, *Experimental Spectroscopy*, 2d ed., 1951; H. E. White, *Introduction to Atomic Spectra*, 1934.

Spectroscopy of combustion

An experimental technique for obtaining data from flames without interfering with the combustion process. The spectrum of light emitted or absorbed in a flame is a physical property of the materials present in the flame. *See* SPECTROSCOPY.

Flame spectra are used in interpreting combustion mechanisms and in determining flame temperatures. Various spectrographic techniques are used depending on the type of measurement desired. The method of line reversal, in which the radiation intensity from thermally excited metal atoms is compared to a black body lamp filament of controllable brightness, is one technique that gives a measure of temperature; it can be handled by simple equipment with only filters to isolate the necessary radiation. An example is the addition of small amounts of sodium salts to gaseous or liquid fuels; the technique is known as the sodium D line reversal method.

Band spectra from reactive combustion intermediates are usually studied with a spectroscope of either the prism or grating type; for extremely fine resolution, an interferometer is sometimes used. The dispersed radiation is detected either photographically or by photomultiplier tubes. Photography is of value in recording spectra over a range of wavelengths simultaneously. If only specific lines are of interest, two or more photomultiplier tubes can be used to record relative line brightnesses. A continuous scan of a spectrum can be achieved by moving one phototube and slit along the focus of the spectral radiation and displaying the output on an oscilloscope.

The band spectra from flames has been associated with the quantized energy changes in molecules due to rotation and vibration. Each spectral line in a band spectrum represents a discrete energy level. Under equilibrium temperature conditions, there is ideally a Maxwell-Boltzman distribution of molecules in different energy states. According to the kinetic theory of gases, this is a dynamic equilibrium with molecules having their energy distributed in a specific way over these energy states. The radiation of light is a measure of the number of molecules in the process of changing energy levels. From these measurements, on band spectra, flame temperatures and reaction intermediates can be determined. *See* BURNING VELOCITY MEASUREMENT; COMBUSTION; KINETIC THEORY OF MATTER. [R.S.S.]

Bibliography: A. G. Gaydon, *Spectroscopy and Combustion Theory*, 2d ed., 1948.

Spectrum

The term spectrum is applied to any class of similar entities or properties strictly arrayed in order of increasing or decreasing magnitude. In general, a spectrum is a display or plot of intensity of radiation (particles, photons, or acoustic radiation) as a function of mass, momentum, wavelength, frequency, or some other related quantity. For example, a β-ray spectrum represents the distribution in energy or momentum of negative electrons emitted spontaneously by certain radioactive nuclides, and when radionuclides emit α-particles, they produce an α-particle spectrum of one or more characteristic energies. A mass spectrum is produced when charged particles (ionized atoms or molecules) are passed through a mass spectrograph in which electric and magnetic fields deflect the particles according to their charge-to-mass ratios (*see* MASS SPECTROSCOPE). The distribution of the energy of sound waves over a given range of frequencies is also called a spectrum (*see* SOUND).

In the domain of electromagnetic radiation, a spectrum is a series of radiant energies arranged in order of wavelength or of frequency. The entire range of frequencies is subdivided into wide intervals in which the waves have some common characteristic of generation or detection, such as the radio-frequency spectrum, infrared spectrum, visible spectrum, ultraviolet spectrum, x-ray spectrum, and so on (*see* ELECTROMAGNETIC RADIATION). Spectra are also classified according to their origin or mechanism of excitation as emission, absorption, continuous, line, and band spectra.

An emission spectrum is produced whenever the radiations from an excited light source are dispersed. Excitation of emission spectra may be by thermal energy, by impacting electrons and ions, or by absorption of photons. Depending upon the nature of the light source, an emission spectrum

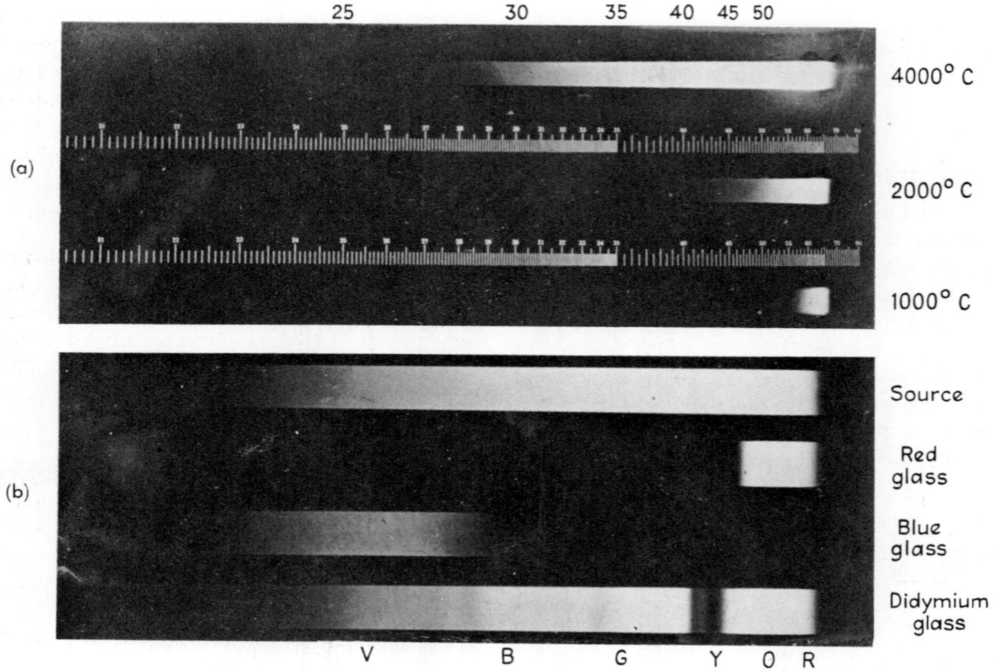

Photographs of continuous spectra of a black body. (*a*) Continuous emission spectra of a solid at the three temperatures indicated, taken with a quartz spectrograph. The spectra for 1000 and 2000°C were obtained from a tungsten filament. That for 4000°C is from the positive pole of a carbon arc. The wavelength scale is marked in hundreds of angstroms. (*b*) Continuous absorption spectra. The upper spectrum is that of the source alone, extending roughly from 4000 to 6500 A. The others show the effect on this spectrum of interposing three kinds of colored glass. (*From F. A. Jenkins and H. E. White, Fundamentals of Optics, 3d ed., McGraw-Hill, 1957*)

may be a continuous or a discontinuous spectrum, and in the latter case, it may show a line spectrum, a band spectrum, or both.

An absorption spectrum is produced against a background of continuous radiation by interposing matter that reduces the intensity of radiation at certain wavelengths or spectral regions. The energies removed from the continuous spectrum by the interposed absorbing medium are precisely those that would be emitted by the medium if properly excited. This reciprocity of absorption and emission is known as Kirchhoff's principle; it explains, for example, the absorption spectrum of the sun, in which thousands of lines of gaseous elements appear dark against the continuous-spectrum background.

A continuous spectrum contains an unbroken sequence of waves or frequencies over a long range (see illustration). All incandescent solids, liquids, and compressed gases emit continuous spectra, for example, an incandescent lamp filament or a hot furnace. In general, continuous spectra are produced by high temperatures, and under specified conditions the distribution of energy as a function of temperature and wavelength is expressed by Planck's law. *See* PLANCK'S RADIATION LAW; *see also* HEAT RADIATION.

Line spectra are discontinuous spectra characteristic of excited atoms and ions, whereas band spectra are characteristic of molecular gases or chemical compounds. *See* BAND SPECTRUM; LINE SPECTRUM; *see also* ATOMIC STRUCTURE AND SPECTRA; MOLECULAR STRUCTURE AND SPECTRA; SPECTROSCOPY. [W.F.M.]

Spectrum analyzer

A device which sweeps over a portion of the radio-frequency spectrum, responds to signals whose frequencies lie within the swept band, and displays them in relative magnitude and frequency on a cathode-ray-tube screen.

In essence, it is a superheterodyne receiver having a local oscillator whose frequency is varied cyclically, usually at the power-line frequency. The block diagram of a typical spectrum analyzer is shown in the illustration.

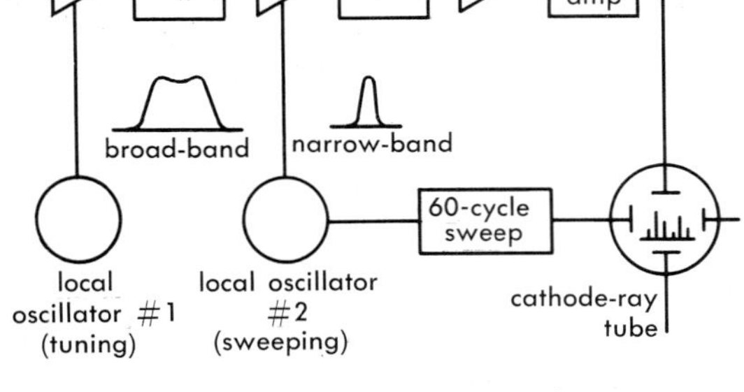

Block diagram of typical spectrum analyzer.

Signals whose frequencies lie within a range equal to the bandwidth of broad-band i-f (intermediate-frequency) amplifier 1, can be heterodyned by local oscillator 1 and converted to within the pass-band of that i-f amplifier (*see* RADIO RECEIVER). The resulting spectrum is scanned by local oscillator 2 and, in a second conversion, swept back and forth across narrow-band i-f amplifier 2. This amplifier is tuned to a frequency outside the pass-band of i-f amplifier 1. Whenever a signal lying within the spectrum is swept across the pass-band of i-f amplifier 2, a burst of energy passes through that amplifier to the detector and produces a pulse that is applied to the vertical-deflection plates of a cathode-ray tube. Since the heterodyne process maintains a linear relation between input and output, the magnitude of the pulse is proportional to the strength of the input signal. The horizontal deflection of the cathode-ray spot is obtained from the same source as the sweep signal applied to local oscillator 2, thereby providing a display corresponding to frequency.

Spectrum analyzers are used specifically to study the spectra of pulsed transmitters, such as radar, to make sure that they are operating properly, without spurious emissions. Spectrum analyzers are often built into test equipment and military identification receivers. As an aid in making accurate frequency measurements they are used as comparison devices to indicate when coincidence occurs between a known and an unknown signal or to locate an unknown frequency with respect to a "picket-fence" spectrum generated from a standard frequency. *See* ELECTRIC POWER MEASUREMENT; FREQUENCY MEASUREMENT. For discussion of sound spectrum analyzers see NOISE MEASUREMENT. [D.B.S.]

Speech

A set of audible sounds produced by disturbing the air through the integrated movements of certain groups of anatomical structures. Humans attach symbolic values to these sounds for communication. There are many approaches to the study of speech.

SPEECH PRODUCTION

The physiology of speech production may be described in terms of respiration, phonation, and articulation. These interacting processes are activated, coordinated, and monitored by acoustical and kinesthetic feedback through the nervous system.

Respiration. Most of the speech sounds of the major languages of the world are formed during exhalation. Consequently, during speech the period of exhalation is generally much longer than that of inhalation (*see* RESPIRATION). In providing a subglottal air supply, the respiratory muscles coordinate with the laryngeal and supralaryngeal muscles to produce suitable driving pressures for the formation of many speech sounds (Fig. 1). The

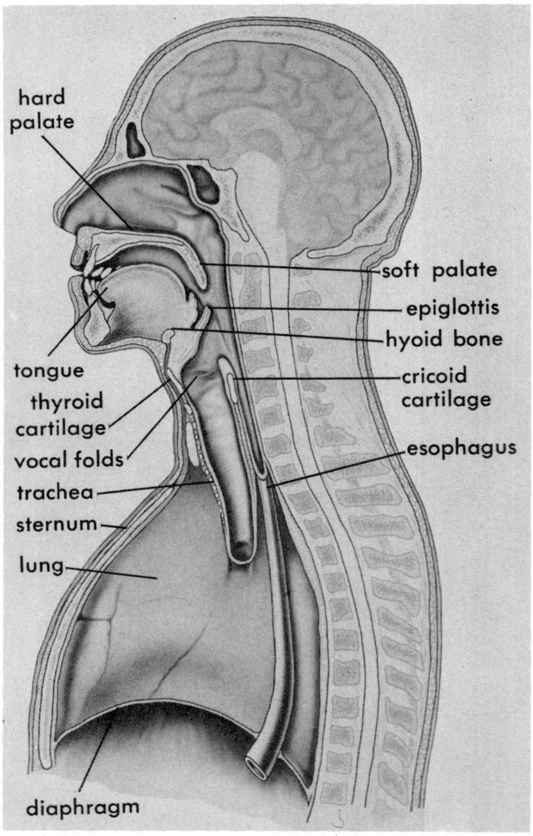

Fig. 1. Sagittal section of the head and thorax. (*G. E. Peterson*)

aerodynamics of the breath stream influence the rate and mode of the vibration of the vocal folds. This involves interactions between the pressures initiated by thoracic movements and the position and tension of the vocal folds.

The pressure pattern of the sublaryngeal air is closely related to the loudness of the voice and appears to be correlated with the perception of stress. For example, the word "permit" may be either a noun or a verb, depending on the placement of stress. Various attempts have been made to correlate units such as the syllable and the phrase to the contractions of specific respiratory muscles. However, the relation between the thoracic movements and the grouping of speech sounds is poorly understood.

Experimental studies on respiration for speech production employ techniques such as x-ray photography to investigate anatomical movements, and electromyography to record action potentials at various points of the thorax.

Phonation. The phonatory and articulatory mechanisms of speech may be regarded as an acoustical system whose properties are comparable to those of a tube of varying cross-sectional dimensions. At the lower end of the tube, or the vocal tract, is the larynx. It is situated directly above the trachea and is composed of a group of cartilages, tissues, and muscles. The upper end of the vocal tract may terminate at the lips, at the nose, or both.

Mechanism of phonation. The larynx is the primary mechanism for phonation, that is, the generation of the glottal tone. The vocal folds consist of connective tissue and muscular fibers which attach anteriorly to the thyroid cartilage and posteriorly to the vocal processes of the arytenoid cartilages. The vibrating edge of the vocal folds measures about 23–27 millimeters in men, and considerably less in women. The aperture between the vocal folds is called the glottis (Fig. 2).

During phonation the thyroid cartilage remains relatively immobile. The tension and position of the vocal folds are adjusted by the intrinsic laryngeal muscles, primarily through movement of the two arytenoid cartilages. By contraction of groups of muscles, the arytenoid cartilages may be pulled either anterior-posteriorly or laterally in opposite directions. They may also be rotated by the cricoarytenoid muscles.

Air pressure. When the vocal folds are brought together and there is a balanced air pressure to drive them, they vibrate laterally in opposite directions. During phonation, the vocal folds do not

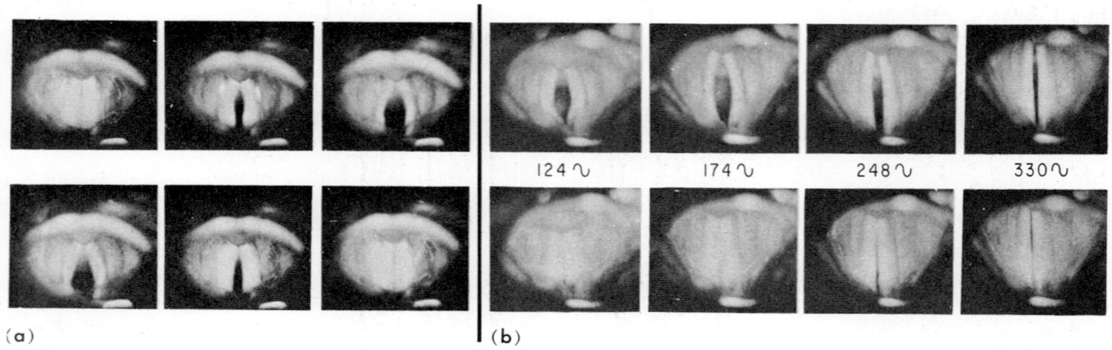

Fig. 2. (*a*) Six successive positions of the vocal folds in one vibration cycle at 120 cps. (*b*) Relation between pitch and vocal fold tension. (*D. W. Farnsworth*)

transmit the major portion of the energy to the air. They control the energy by regulating the frequency and amount of air passing through the glottis. Their rate and mode of opening and closing are dependent upon the position and tension of the folds and the pressure and velocity of air flow. The tones are produced by the recurrent puffs of air passing through the glottis and striking into the supralaryngeal cavities.

As the air passes through the glottis with increasing velocity, the pressure perpendicular to the direction of air flow is reduced. This allows the tension of the folds to draw them together, either partly or completely, until sufficient pressure is built up to drive them apart again. It is generally assumed that the volume flow is linearly related to the area of the glottis, and that the glottal source has a wave form that is approximately triangular.

Voiced sound. Speech sounds produced during phonation are called voiced. Almost all of the vowel sounds of the major languages and some of the consonants are voiced. In English, voiced consonants can be illustrated by the following words, bathe, vary, dog, jelly. The speech sounds produced when the vocal folds are apart and are not vibrating are called unvoiced; examples are the consonants in the words hat, cap, sash, faith. During whispering, all the sounds are unvoiced.

Frequency and pitch. The rate of vibration of the vocal folds is the fundamental frequency of the voice and correlates well with the perception of pitch. The frequency increases when the vocal folds are made taut (Fig. 2*b*). Relative differences in the fundamental frequency of the voice are utilized in almost all languages to signal some aspects of linguistic information. Many languages use them to distinguish meanings between words. In Mandarin Chinese, for example, the utterance "ma" can mean "mother" or "horse," depending upon the variations in the fundamental frequencies. The mode of laryngeal vibration can be varied to change the acoustical properties of glottal tone. For example, if the vocal folds remain sufficiently apart during the vibration the voice is perceived to be breathy.

Study methods. The properties of the movements of the vocal folds have been studied by placing microphones at different points in relation to the larynx. Experiments have also been conducted with electrodes inserted directly into the laryngeal structures of hemilaryngectomized subjects and simultaneously recording their speech. Much information about the glottal vibration has been gathered through direct observation of the vocal folds by means of x-ray photography, the stroboscope, and high-speed motion picture photography. The relation between the volume flow, the mode of glottal vibration, and the spectrum of the glottal tone has been studied by these techniques. Several electronic models have been constructed to simulate the glottal tone, based on the collated evidence from the acoustics and physiology of the larynx.

Articulation. The activity of the structures above and including the larynx in forming speech sounds is called articulation. It involves some muscles of the pharynx, palate, tongue, and face and of mastication (Fig. 1).

The primary types of speech sounds of the major languages may be classified as vowels, plosives, and fricatives. They may be described in terms of degree and place of constriction along the vocal tract.

Vowels. These are characterized by a relatively open vocal tract with associated cavity resonance and a minimum of friction. The shape and position of the tongue, lips, and velum are the major factors which control the shape of the supralaryngeal cavities. These cavities determine the phonetic quality of the vowel by their resonances and antiresonances of the glottal source.

The nasal cavities can be coupled onto the resonance system of the vocal tract by lowering the velum and permitting air flow through the nose. Vowels produced with the addition of nasal resonances are called nasalized vowels. Nasalization may be used to distinguish meanings of words which are made up of otherwise identical sounds such as the French *bas* and *banc.* If the oral passage is completely constricted and air flows only through the nose, the sounds which result are nasal consonants as in "meaning."

Vowels are grossly described as rounded when produced with the contraction of the muscles of the lips. This contraction frequently results in a quasicircular aperture of the mouth, as in "who." They are called unrounded when the corners of the lips are retracted, as in "he."

Vowels are referred to as high, mid, and low, depending on the elevation of the tongue for their production as in "feed, fed, fad," respectively. Those produced with the front part of the tongue humped up are called front vowels as in "beet." Back vowels are produced when the body of the tongue is drawn back and the size of the pharynx is reduced as in "bought." Vowels produced with the tongue in the intermediate range are called central as in "but."

Plosives. These are characterized by the complete interception of air flow at one or more places along the vocal tract. The pressure which is built up behind the intercepting mechanism may not be immediately released or may be released through the oral or nasal orifice. The places of constriction and the manner of the release are the primary determinants of the phonetic properties of the plosives. Words like car, papa, and technique are plosives.

When the interception is brief and the constriction is not necessarily complete, the sound is classified as a flap. By tensing the articulatory mechanism in proper relation to the air flow, it is possible to set the mechanism into a vibratory movement which quasi-periodically intercepts the air flow. These sounds are called trills and can be executed with the velum, the tongue, and the lips.

Fricatives. These are produced by partial constriction along the vocal tract which results in turbulence. Their properties are determined by the place or places of constriction and the shape of the modifying cavities. The fricatives in English can be illustrated by the words vase, this, faith, hash.

Acoustical analysis. The muscular activities of speech production influence each other both simultaneously and with respect to time. It is frequently difficult to segment a sequence of sounds because the physiological activities which produce them form a continuum. Consequently the physical aspects of a sound type are determined to some extent by neighboring sounds. Each physiological parameter, for example lip-rounding and fundamental frequency of the glottal tone, may be varied continuously. Within its physiological limits, the human speech mechanism is able to produce an infinite number of different sounds. The physiological and acoustical criteria in a classification of these sounds are therefore dependent on external conditions such as the threshold of the analyzing instruments.

There is a complex, many-to-one relation between the physiology of production and the resultant acoustical waves. Acoustically, speech sounds may be regarded as the simultaneous and sequential combinations of pulse, periodic, and aperiodic forms of energy interrupted by silence of varying duration. These energy patterns are labeled as A, B, C, and D respectively in the sound spectrogram of Fig. 3*a*. The pitch of the utterance is illustrated in the sound spectrogram of Fig. 3*b*. The sound spectrograph is the major instrument in research on the acoustical aspect of speech. Essentially, it makes a Fourier type of analysis on the acoustical wave. The results are then translated into graphic form, usually with frequency along the ordinate and time along the abscissa.

Neurology. The ability to produce meaningful speech is dependent in part upon the association areas of the brain. It is through them that the stimuli which enter the brain are interrelated. These areas are connected to motor areas of the brain which send fibers to the motor nuclei of the cranial nerves and hence to the muscles. Three neural pathways are directly concerned with speech production, the pyramidal tract, extrapyramidal, and cerebellar motor paths. It is the combined control of these pathways upon nerves arising in the medulla and ending in the muscles of the tongue,

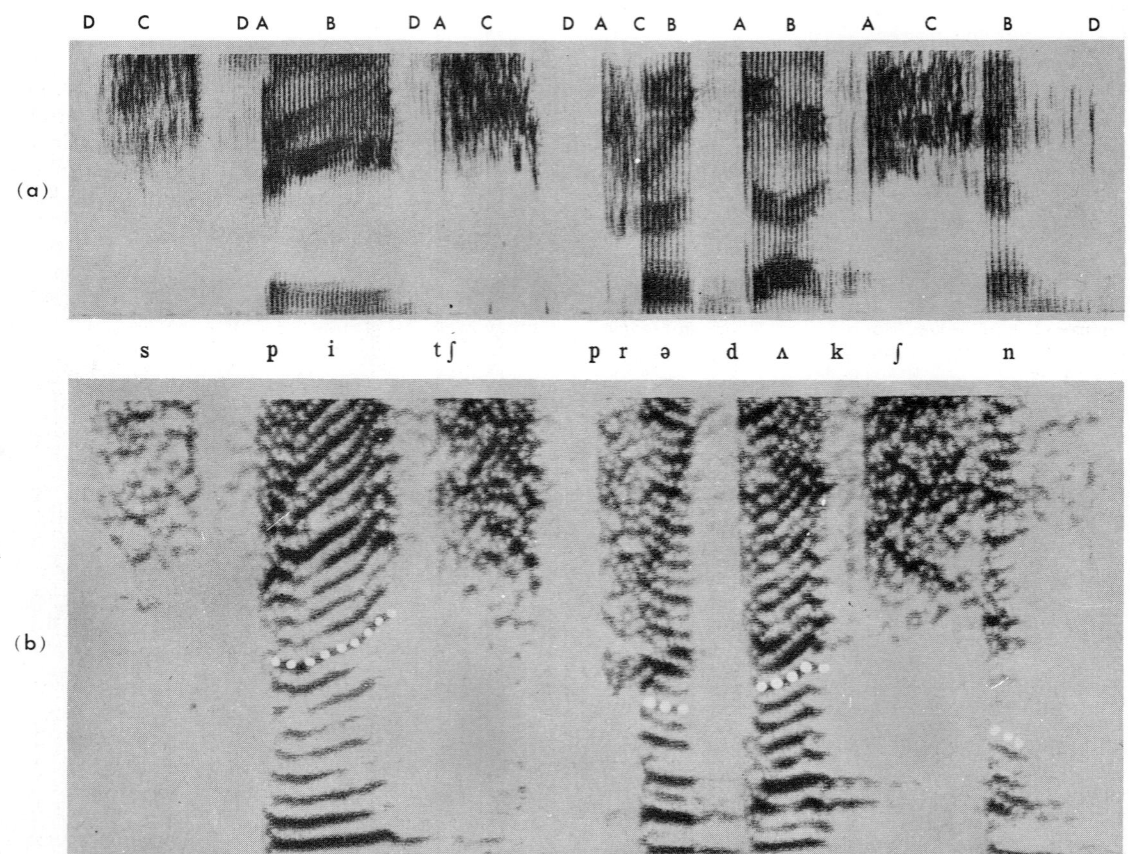

Fig. 3. Sound spectrograms of the utterance, "speech production." (*a*) Energy patterns: A, pulse; B, periodic; C, aperiodic; and D, silence. (*b*) The dotted lines represent the pitch pattern of the utterance.

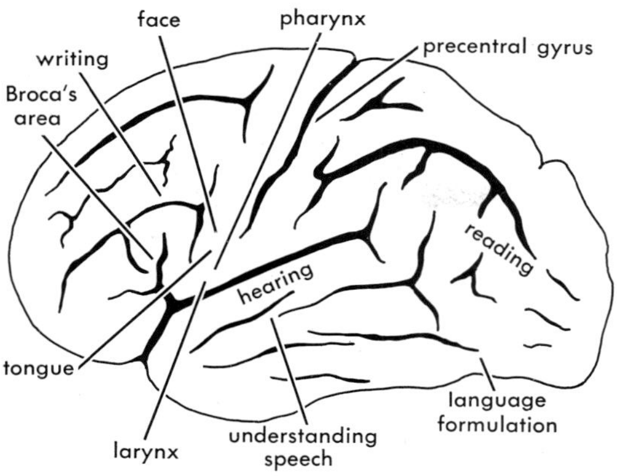

Fig. 4. Cerebral cortex showing the areas related to the speech movements and supposed association areas.

lips, and larynx which permits the production of speech. *See* NERVOUS SYSTEM.

The part of the pyramidal tract which is most important for speech has its origin in the lower portion of the precentral gyrus of the cerebral cortex. In close proximity to the precentral gyrus on the left side of the brain is Broca's area. This area is one of several which are believed to activate the fibers of the precentral gyrus concerned with movements necessary for speech production (Fig. 4). Areas of the extrapyramidal tract which contribute to speech are the caudate nucleus, globus pallidus, and the thalamus. Some of the fibers from these centers also go to the same motor nerves as those of the pyramidal tract. The chief function of these paths in regard to speech is their regulatory and refining actions. It is through the influence of the extrapyramidal system that the regulation and inhibition of opposing sets of muscles are controlled. The cerebellar motor paths also contribute to the coordination and tonus of muscle structures requiring movements of paired muscles. In addition, the cerebellar motor paths are important in coordinating breathing for speech production.

Six of the 12 cranial nerves send motor fibers to the muscles involved in the production of speech. These nerves are the trigeminal, facial, glossopharyngeal, vagus, spinal accessory, and the hypoglossal. They represent the link between the neural activity which begins in the cerebral cortex and the coordinated muscular movements which produce speech.

The relations between the neurology, physiology, and acoustics of speech are extremely complex and poorly understood. Within normal anatomical variation, the skill in producing particular speech sounds is almost entirely determined by the early linguistic environment of the speaker. Through the interactions between the linguistic environment and the speech of the individual, both the sounds and their symbolic values change from community to community and through time. *See* PSYCHOACOUSTICS; PHONETICS.

DEVELOPMENT

In the early stages of speech development the child's vocalizations are quite random. The control and voluntary production of speech are dependent upon physical maturation and learning. The relation between age and development of speech behavior is shown in Fig. 5.

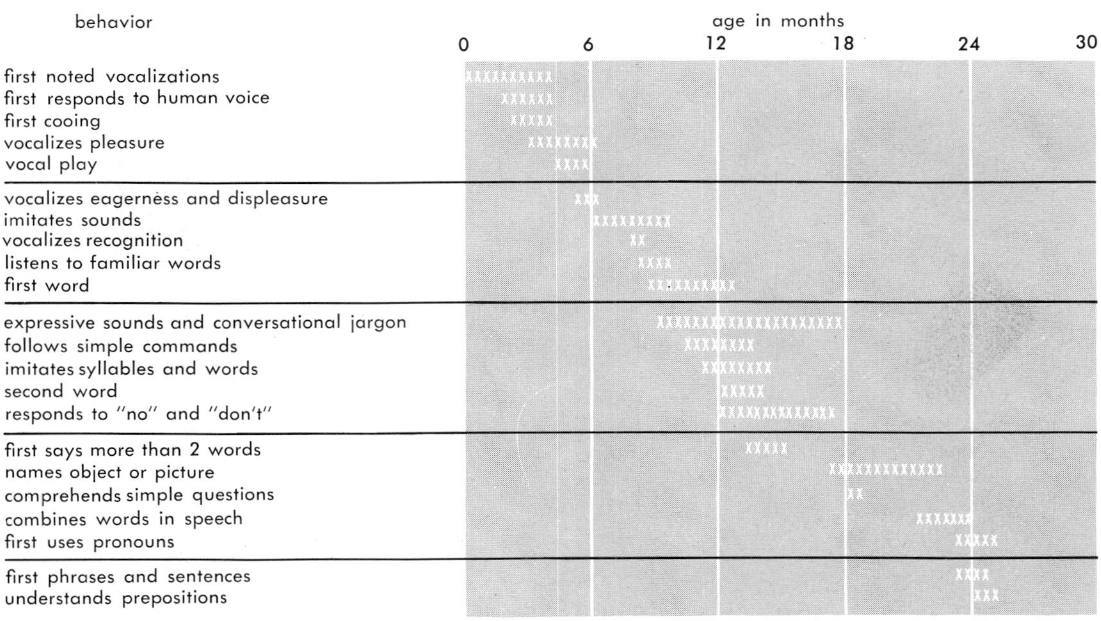

Fig. 5. Composite table showing age in months at which selected items are reported in eight major studies of infant development. (*After McCarthy, 1946, from G. A. Miller, Language and Communication, McGraw-Hill, 1951*)

Developmental stages. It is possible to describe the development of speech in five stages. In the first stage the child makes cries in response to stimuli. These responses are not voluntary but are part of the total bodily expression. The second stage begins between the sixth and seventh week. The child is now aware of the sounds he is making, and appears to enjoy this activity. During the third stage the child begins to repeat sounds that he hears himself make. This is the first time that the child begins to link speech production to hearing. Sometime during the ninth or tenth month the child enters the fourth stage when he begins to imitate without comprehension the sounds that others make. The last stage begins between the twelfth and eighteenth month. It is at this time that the child intentionally employs conventional sound patterns in a meaningful way. The exact time at which each stage may occur varies greatly from child to child.

Although most children begin to use speech meaningfully by the eighteenth month, they are not able to articulate all the necessary sounds. The first sounds the child uses are mostly front vowels. The back vowels become more frequent as the child develops. The ability to produce vowels correctly seems to be almost completely learned by 30 months. The earliest consonants the child uses are those formed with both lips. By 54 months, the child can produce all the necessary consonants. The approximate age level at which each of 23 consonant sounds of English is mastered by American children is shown below.

Average age, months	*Sounds*
3½	b(*baby*), p(*papa*), m(*mama*), w(*wet*), h(*he*)
4½	d(*dada*), t(*two*), n(*nose*), g(*go*), k(*cat*), ng(*sing*), y(*yet*)
5½	f(*fun*)
6½	v(*very*), th(*that*), z(*azure*), sh(*shoe*), l(*lie*)
7½	s(*see*), z(*zoo*), r(*rain*), th(*thin*), wh(*who*)

DISORDERS

Speech disorders may be classified according to their causes or symptoms. The major causes are organic, imitative-environmental, and psychogenic. Organic disorders may result from disease, impairment, or absence of the organs of speech. Imitative disorders occur when the child imitates defective speech. A speech disorder has a psychogenic origin when there is a psychological basis for its presence. A classification includes disorders of articulation, rhythm, voice, and symbolization.

Articulation. Disorders of articulation may be so severe that the resultant speech is unintelligible. Specific sounds or groups of sounds may be omitted, added, substituted, or distorted. The following are some examples of this type of disorder. Lalling, misarticulated r, l, t, and d sounds, may be caused by poor control of the tongue tip. Lisping, misarticulated sibilant sounds, particularly s and z, are often substituted by the th sound. Delayed speech, the absence of many consonants, and poor intelligibility, is often caused by slow physical or psychological maturation. Dysarthria, generalized sound substitutions and distortions, is caused by lesions in the peripheral or central nervous system.

Rhythm. Disorders of rhythm are characterized by disruptions of the normal rate of speech. Two common disorders of rhythm are stuttering or stammering and cluttering. Stuttering usually begins between 3 and 4 years of age, and occurs most frequently in males. Many theories have been advanced to explain the cause of the disorder but none have been completely accepted. Primary stuttering is the repetition of words, phrases, syllables, or the initial sounds of words, without apparent awareness by the speaker. In secondary stuttering the repetitions become prolonged fixations of the speech musculature, accompanied by physical and psychological tension, and often contortions of the face and body. In cluttering, words are slurred and syllables are omitted because of improper phrasing and excessive speed of utterance.

The following examples involve lesion of the nervous system. The cerebral palsied often show disorders of rhythm. One type is spastic speech. This is an inability to make smooth transitions from sound to sound and insufficient breath control to produce polysyllabic words or phrases. Another is athetotic speech, which involves a general jerkiness in speech production which interferes with the normal rate of speech. Similar rhythm disorders accompany Parkinson's disease, multiple sclerosis, and cerebellar tumors. *See* NERVOUS SYSTEM DISORDERS.

Voice. Voice disorders are usually described as defects of pitch, loudness, and voice quality. Improper use of the voice may cause an injury to the vocal folds and intensify an existing voice disorder. Some examples of these disorders are presented below.

Disorders of pitch are characterized as too high, too low, monotonous, and repeated pitch patterns. A high-pitched voice is most often caused by psychological tension. The muscles of the larynx are contracted so that the pitch is raised beyond its normal range. It may be caused also by a small larynx where the vocal folds are short and thin or by the inability to perceive changes in pitch. Monotonous voices and repeated pitch patterns are usually also the result of a failure to perceive the pitch variations.

Voice quality disorders are usually described as hoarseness, nasality, denasality, and similar conditions. Hoarseness may be caused by pharyngeal or laryngeal pathologies. Cleft palate speech is a striking example of nasality. It is the result of a failure of the bilateral structures of the palate to unite during fetal life. Nasal speech may also follow a paralysis of the palatal muscles (Fig. 6).

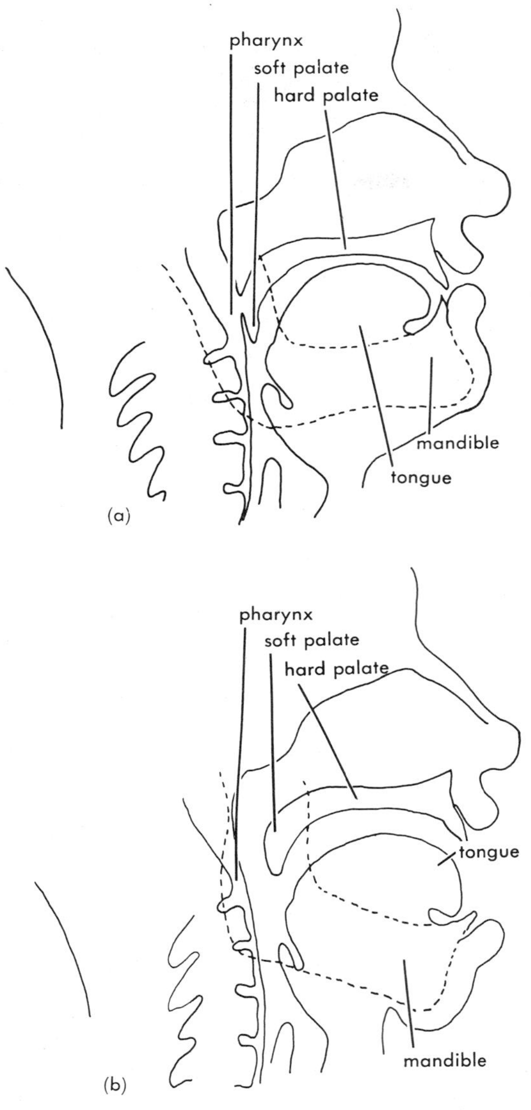

Fig. 6. (*a*) Schematic diagram showing normal nasopharyngeal closure. (*b*) Schematic diagram showing failure to achieve normal nasopharyngeal closure because of a palatal cleft, paralysis, or injury. (*Based on x-rays from the collection of H. H. Bloomer*)

The nasal speech results because the palate is unable to aid in achieving nasopharyngeal closure. *See* PALATE.

Aphonia, the absence of voice, does not fall into the above categories. It may result from paralysis, inflammation of the vocal folds, or surgical removal of the larynx. Hysterical aphonia is the result of a marked emotional disturbance. *See* PHOBIC REACTION.

Symbolization. Symbolization disorders involve an impairment of language formulation and expression. They may occur without a concomitant impairment of speech production. Common types of these disorders are aphasia and delayed speech.

Aphasia is the inability to use or comprehend the symbolic value of language as a result of a brain lesion. The specific types of aphasia usually fall into one of four groups: expressive, receptive, amnesic, or a combination of these three. Expressive aphasia is the inability to express ideas in speech or writing when there is no significant muscular impairment. Receptive aphasia is the inability to comprehend spoken or written symbols when there is no significant impairment of the sensory organs. Amnesic aphasia is the inability to evoke the appropriate names for objects, conditions, or relations. In adults the disorder is usually preceded by a cerebral vascular accident, brain tumor, or injury to the brain. In children these disorders are usually the result of a failure of the development of the brain or of brain damage incurred before, during, or after birth. Many neurologists have postulated centers in the brain which are said to be related to various types of aphasic disorders. Aphasic arrest of speech has been demonstrated by electrical stimulation applied to certain areas of the brain (Fig. 7).

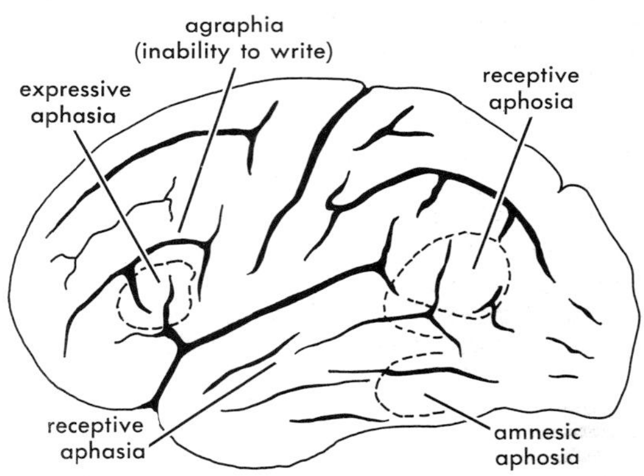

Fig. 7. Areas of the cerebral cortex which when damaged supposedly result in aphasia. The areas within the dotted lines are areas which when stimulated electrically result in aphasic arrest.

Delayed speech may also be considered as a symbolization disorder. The symptoms range from complete absence of vocalization to vocalizations which have no communicative value. Speech is considered delayed when it fails to develop by the second year, and is judged to be retarded if there are significant deviations from the norm for the age and mental abilities of the child. The causes of delayed speech are quite varied. Some of the most common causes are deafness, impaired hearing, severe illness during childhood, or emotional disturbances.

The patterns of speech development and disorders show great variability among individuals. Research in these areas requires the integrated knowledge of experts in the many fields dealing with language and speech. [R.S.T.; W.S.Y.W.]

Bibliography: G. Fant, *Acoustic Theory of Speech Production*, 1959; G. W. Gray and C. M. Wise, *The Bases of Speech*, 3d ed., 1959; L. Kaiser (ed.), *Manual of Phonetics*, 1957; L. E. Travis (ed.), *Handbook of Speech Pathology*, 1957.

Speed

The time rate of change of position of a body without regard to direction. It is the numerical magnitude only of a velocity and hence is a scalar quantity. Linear speed is commonly measured in such units as meters per second, miles per hour, or feet per second. It is the most frequently mentioned attribute of motion.

Average linear speed is the ratio of the length of the path s traversed by a body to the elapsed time t during which the body moved through that path:

$$\text{Speed (average)} = \frac{s_f - s_0}{t_f - t_0} = \frac{s}{t}$$

where s_0 and t_0 are the initial position and time, respectively, and s_f and t_f are the final position and time.

Instantaneous speed is the limiting value of the foregoing ratio as the length of path is made infinitesimally small and as the elasped time approaches zero.

$$\text{Speed (instantaneous)} = \lim_{t \to 0} \frac{s_f - s_0}{t} = \frac{ds}{dt}$$

See VELOCITY. [R.D.RU.]

Speed regulation

The change in speed of direct-current motors from no load to full load, expressed as a percentage of full-load speed with the motor at rated temperature.

$$\%\ \text{speed regulation} = \frac{\text{no-load speed} - \text{full-load speed}}{\text{full-load speed}} \times 100$$

[A.F.P.]

Speedometer

An instrument that indicates the speed of travel of a vehicle. Speedometers were originally driven from a gear attached to the front wheel meshing with a pinion to which was attached a flexible cable to the speedometer head. Almost all speedometers now are driven from a helical gear at the rear of the transmission, which meshes with a helical pinion at right angles to it (Fig. 1). This pinion drives a flexible multistranded cable approximately 0.130 in. in diameter, running in a lightly lubricated casing to the speedometer head, 1000 rpm corresponding to 60 mph.

The speedometer head carries a face on which is a dial registering speed measured in miles per hour in the United States and its possessions, the United Kingdom, India, and Japan; and in kilometers per hour in the rest of the world. The face also carries an odometer, which reads the total miles the car has been driven, and some units carry a second trip-

Fig. 1. Speedometer driving-gear assembly in transmission with nylon pinion. (*Buick*)

measuring odometer, which can be returned manually to zero.

Speedometers generally have carried circular dials with pointers to indicate speed. A new type carrying the figures on a horizontal scale (Fig. 2) has become popular in recent years. A cylinder carrying a helical indicating line, identified by one color on one side of it and another color on the other side, and which is visible in a slot as a pointer above or below the figures, serves to indicate speed on the horizontal scale located above or below the slot.

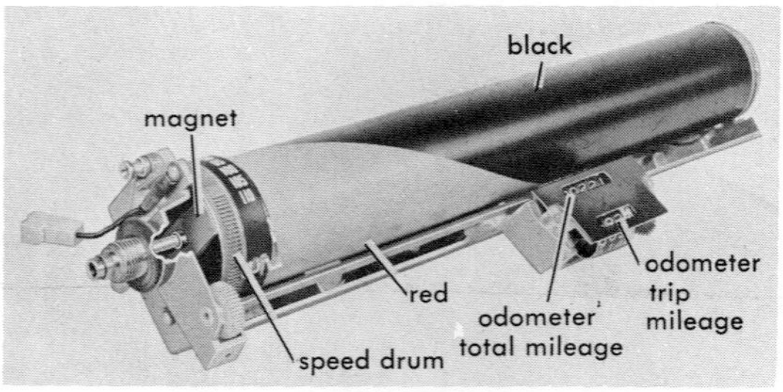

Fig. 2. Mechanism for speedometer drum to indicate speed. (*Buick*)

The speedometer pointer, or cylinder, is operated by a magnet driven by the cable. In Fig. 3, the magnet is located inside a metal cup which is attached to a shaft carrying the pointer. A spring on the shaft tends to hold the pointer at zero mph. However, as the magnet rotates, it exerts a magnetic drag on the metal cup, tending to turn it and the pointer. The faster the magnet rotates, the greater is the movement of the pointer so that it indicates the higher speed. *See* TACHOMETER.

The odometer mechanism is simply a gear train of proper reduction to show miles and tenths of miles on the final gears, carrying small drums on the face of which are numbers 0–9. In 1957 a de-

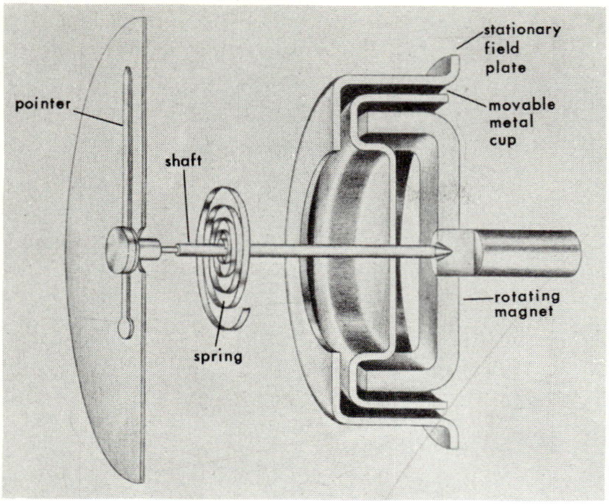

Fig. 3. Mechanism of circular-dial speedometer (*AC Spark Plug Div.*)

vice introduced by Buick employed a warning buzzer which operated when a predetermined speed was attained. The action of the buzzer was controlled by the driver by the rotation of a knob to any desired speed shown on an opening in the speedometer face. Governor devices have been used on trucks for years. [F.R.M.]

Spelaeogriphacea

A peracaridan order of the higher Crustacea, Malacostraca, erected in 1957. The only known species, *Spelaeogriphus lepidops* (see illustration), inhabits a pool in a cave on Table Mountain, South Africa. It is a small, blind, transparent, shrimplike animal, 6–9 mm in length. The short shell or carapace coalesces dorsally with the first thoracic somite. Behind the carapace the body is fully segmented and the abdomen, comprising six somites and the telson, exceeds half the total length. The lash or flagellum of the second antenna is almost as long as the body, and the antennal scale is minute. Most unusual, however, is the presence of three pairs of oval, vesicular gills attached to thoracic limbs 5–7.

A small ocular scale arises from each side of the triangular rostrum. The first pair of thoracic limbs is modified as mouthparts, the maxillipeds, which are isopodan in form. At the base of each maxilliped is a large cup-shaped respiratory organ, recalling the spoon-shaped organ of *Apseudes* (Tanaidacea). The other seven pairs are slender walking legs. The exopodites of the first three pairs are lashlike whereas those of the next three pairs are the gills already mentioned. The first four pairs of abdominal appendages are biramous swimming paddles; the fifth pair is vestigial; the sixth pair forms, with the telson, a conspicuous tail fan.

The few large eggs are incubated in a brood pouch composed of five pairs of overlapping plates, the oostegites.

The affinities of Spelaeogriphacea are probably with the Tanaidacea and the Isopoda. *See* PERACARIDA. [I.GO.]

Bibliography: I. Gordon, On *Spelaeogriphus*, a new cavernicolous crustacean from South Africa, *Bull. Brit. Museum Natl. Hist., Zoology*, 5(2):31–47, 1957.

Sperm cell

The male gametes, or spermatozoa, capable of uniting with an egg in the process of fertilization (*see* REPRODUCTION, ANIMAL). In most animals, they are elongate cells with a thin, cylindrical, motile tail attached to a somewhat thicker head. In vertebrates, the tail is generally 50–150 μ long by 0.5 μ wide, while the head is about 2×4 μ, but there are marked divergences from this, as in the 2 mm-long

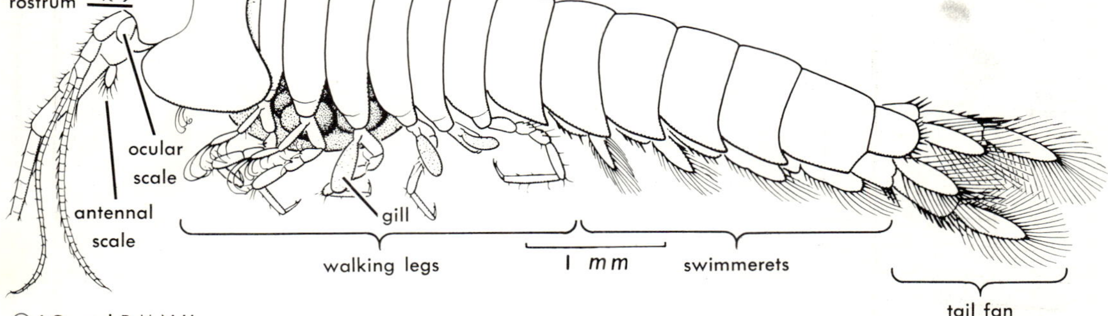

Spelaeogriphus lepidops Gordon, ovigerous female in dorsolateral aspect. (*British Museum of Natural History*)

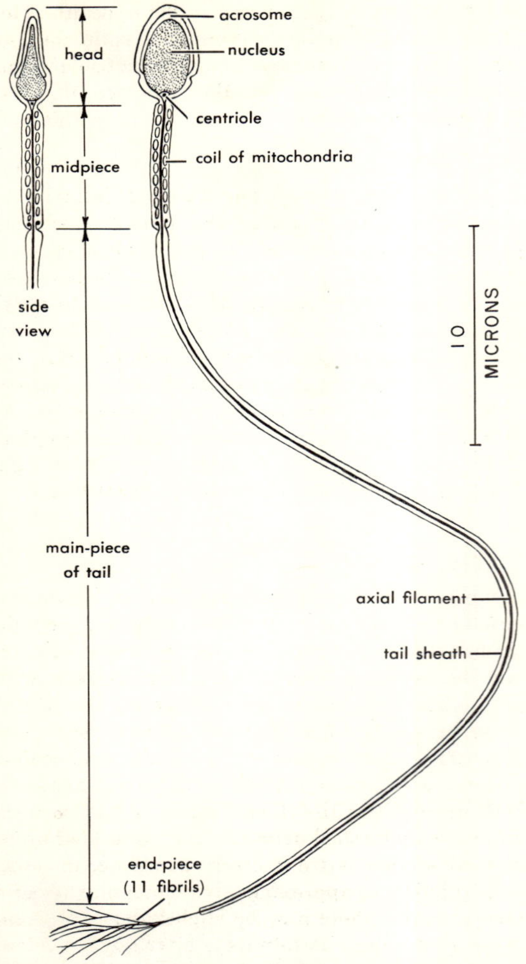

Diagram of human spermatozoon based on electron micrographs.

sperm of the anuran *Discoglossus*. The head consists of an acrosome and a nucleus. In some animals it also includes the mitochondrial midpiece, but in others, for instance mammals, the latter comprises the proximal portion, about one-tenth of the length of the tail (*see* SPERMATOGENESIS).

The nucleus is mainly in the form of the condensed chromosomes, of which at least 50% is the chemical substance desoxyribonucleic acid (DNA) and the rest mainly protein. The acrosome is anterior to the nucleus and varies in appearance in different species. It is cap-shaped in humans, sickle-shaped in rodents, and sharply conical in chickens. In mammalian sperm, the acrosome most probably is the source of the enzyme hyaluronidase, which is effective in dissolving the gelatinous material, hyaluronic acid, between the cumulus cells that surround the freshly ovulated egg. The midpiece is composed largely of mitochondria. These cytoplasmic bodies are now known to carry many of the important oxidative enzymes of cells. Very likely, then, they represent the initial site of energy-supplying reactions for motility of the tail. The tail has a central core, or axial filament, which is typically made up of 2 central fibrils surrounded by 8 double and 1 triple peripheral fibrils. This 2 + 9 fibrilla structure has been found, by electron microscopy, not only to be typical of sperm tail filaments but of cilia and flagella of other kinds of cells of all animals and plants so far examined.

Among the nematodes, myriapods, cladocerans, and decapod crustaceans there occur nonflagellate, slow-moving, or immotile, sperm. Their basic plan of construction is, however, similar to that of the flagellate type. [A.TY.]

Spermatogenesis

Sperm formation comprises the process by which certain cells, spermatogonia, of the testis undergo meiosis and transform into spermatozoa (*see* GAMETOGENESIS). The postmeiotic transformation of spermatids into spermatozoa is termed spermiogenesis.

In their early stages, cells undergoing spermatogenesis resemble those in oogenesis, and at the time of synapsis of the chromosomes, primary spermatocytes and primary oocytes are both about 2–4 times the size of the terminal gonia. The oocytes continue to enlarge until final egg size is attained, but the spermatocytes proceed with meiosis wherein, by two cell divisions accompanied by a single splitting of the chromosomes, the chromosome number is reduced from a double, or diploid, set to a single, or

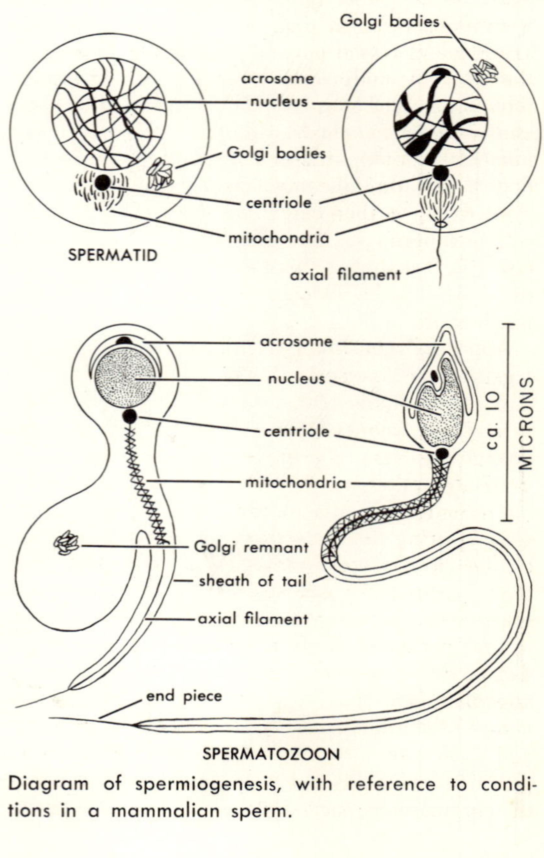

Diagram of spermiogenesis, with reference to conditions in a mammalian sperm.

haploid, set. Each of the four spermatids resulting from these divisions transforms into a functional spermatozoon, in contrast to the formation of a single egg cell and three nonfunctional polar body cells from the corresponding two meiotic divisions of the primary oocyte. Experiments with mice show that about 30 days is required for the transition of terminal spermatogonium to ripe sperm, about half this time being taken to reach the spermatid stage.

Spermiogenesis. Spermiogenesis comprises certain nuclear and cytoplasmic transformations that are essentially alike in practically all animals. The vesicular nucleus of the spermatid becomes progressively smaller, as its chromosomes condense, until in the mature sperm its volume is about equal to that of a tightly packed haploid set of metaphase chromosomes (*see* MITOSIS). As this occurs, the spermatid also loses almost all of its cytoplasm. At the same time certain cytoplasmic organelles, namely, the Golgi bodies, the mitochondria, and the centriole, participate in the formation of the acrosome, midpiece and tail filament, respectively, of the spermatozoon. *See* CELL (BIOLOGICAL). In this process the Golgi bodies move from the region of the centriole to the opposite side of the nucleus, where they form one or more vesicular structures, called the acroblast, in association with which the acrosome for the tip of the sperm is formed. The remnants of the Golgi bodies then move back and are discarded along with the surplus cytoplasm. The mitochondria assemble in filamentous form near the centriole. In most vertebrates they then wind themselves in one or more spiral filaments about the proximal portion of the axial filament of the tail. This midpiece is of different length in different species, being roughly one-tenth that of the entire tail in mammals. From the centriole, or an immediate product thereof, a filament grows out, at first penetrating the membrane of the cell for a short distance, then extending itself along with the cell membrane so that the latter forms a sheath about it. It is during this process of elongation that the bulk of the cytoplasm of the spermatid is pinched off.

Abnormal spermatogenesis. Abnormal spermatogenesis occurs sporadically in many groups of animals. In snails of the genus *Viviparus*, in which it is rather frequent, its study has led to an interesting discovery concerning the relation of the centriole to the structure known as the centromere, the spindle-fiber attachment body, of the chromosome. During the abnormal spermatogenesis many of the chromosomes may become detached from their centromeres and degenerate. The separated centromeres, however, assemble next to the centriole in the spermatid. Each now behaves as a centriole and produces an axial filament for the sperm tail. It appears, then, that centriole and centromere are bodies of similar nature. Both are self-reproducing and both can give rise to filaments in spermiogenesis. Very likely, too, the centromere, as part of the chromosome, normally functions in cell division to form filaments that extend to the opposite poles of the mitotic spindle and pull the split chromosomes apart. At the same time the centriole forms other filaments for the spindle itself. *See* MEIOSIS; SEX DETERMINATION. [A.TY.]

Sperrylite

A mineral with composition $PtAs_2$ (platinum diarsenide), crystallizing in the isometric system. Crystals are usually cubes or cubo-octahedrons; there is an indistinct cubic cleavage. Its hardness is 6–7, and its specific gravity 10.59. The luster is metallic, and the color tin-white. Sperrylite is a rare mineral, found originally in the Sudbury district of Ontario, Canada, where, associated with the copper-nickel ores, it is mined as an ore of platinum. It has also been found in the Bushveld Igneous Complex, Transvaal, South Africa, and in gravels of the Timpton River, eastern Siberia. *See* PLATINUM. [C.S.HU.]

Sphaeractinoidea

Limestones formed in the warm-water Mediterranean (Tethyan) seas of Mesozoic times contain abundant nodular, or lamellose, or ramose calcareous fossils, often encrusting, or intergrown with each other, and from a few millimeters to a meter or so across. On their surface and when sectioned they are seen to consist of fine rods (trabeculae, pillars) and plates (lamellae, tabulae, laminae), both generally 0.01–0.1 mm thick, linked to form a fine evenly-meshed network (reticulum) which in vertical section is transversely lamellate, or vertically tubular, or approximately rectangular. Within this reticulum there may be slightly wider vertical tubes (autotubes, astrotubes), often associated as astrosystems with radially arranged and branching surface grooves or lateral canals or corridors (astrorhizae, astrocorridors). Such fossils are, by

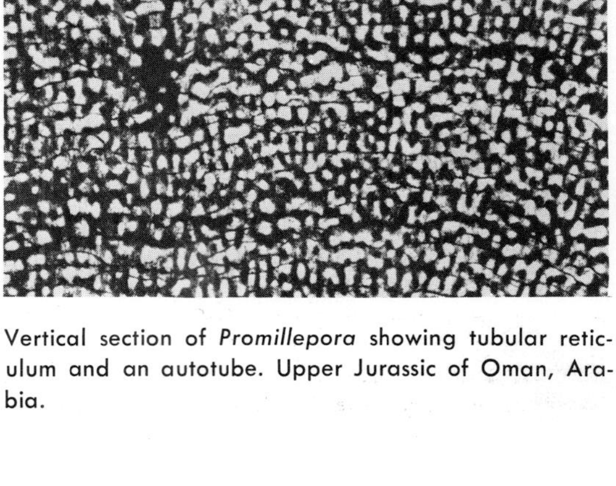

Vertical section of *Promillepora* showing tubular reticulum and an autotube. Upper Jurassic of Oman, Arabia.

analogy with recent forms, the basal exoskeletons (coenostea) of colonial polypoid hydrozoans. They ranged from the Permian to the Cretaceous, and were most abundant in relatively quiet offshore shallow waters during the Late Jurassic.

The Sphaeractinoidea are similar both to the Palaeozoic Stromatoporoidea, from which they differ in that their skeletal tissue (sclerenchyme) is fibrous, and to the Hydroida and Hydrocorallina, mainly Tertiary and Recent, with which they cannot be grouped since they lack evidence of the specialized polyps characteristic of these orders. Though formerly allocated to the Stromatoporoidea, they are now usually grouped in the Sphaeractinoidea (O. Kühn, 1927), an order probably polyphyletic, variously developed from the Stromatoporoidea, and possibly giving rise to various forms in the Hydroida and Hydrocorallina.

Of the major groups within the Sphaeractinoidea some are distinguished by the orientation of the sclerenchymal fibers which may be at an upward acute angle to the axis or plane from which they orginate (clinogonal, Milleporellacea) or similarly at right angles (orthogonal) and either bilaterally (Actinostromariacea) or unilaterally (Burgundiidae) developed. Another group, the Spongiomorphidae, are doubtfully characterized by astrosystems with pseudosepta. *See* HYDROIDA; HYDROZOA; STROMATOPOROIDEA. [R.G.S.H.]

Bibliography: O. Kühn, Hydrozoa, *Handbuch der Paläozoologie*, pt. 5, vol. 2A, 1939.

Sphaerioidaceae

A family of fungi of the order Sphaeropsidales containing many plant pathogens. This family is also called Sphaeropsidaceae or Phomaceae. There are

Important genera of the Sphaerioidaceae

Spore group	Genus	Genus description	Disease	Caused by
Hyalosporae 1-celled bright hyaline spores	*Phoma* 200 spp.	Plant pathogens; pycnidia on plant stems only	Dry rot of turnip; canker of cabbage	*P. lingam*
			Blackleg of beet seedlings	*P. betae* (Fig. 2*a*)
			Scab disease of celery	*P. apiicola* (Fig. 2*b*)
			Dry rot of carrot	*P. rostrupii* (Fig. 2*c*)
	Phyllosticta 500 spp.	Plant pathogens; pycnidia on leaves only	Apple blotch	*P. solitaria*
			Leaf spot of beet	*P. betae* (Fig. 2*d*)
	Macrophomina	Root parasite; pycnidia on plant stems, conidia longer than 15 μ		
	Pyrenochaete	Plant pathogens; pycnidia with stiff bristles	Pink root rot of onion	*P. terrestris* (Fig. 2*e*)
	Phomopsis[a] 100 spp.	Plant pathogen; pycnidial wall thick; 2 types of spores	Canker of Douglas-fir	*P. pseudotsugae* (Fig. 2*f*)
	Cytospora[b] 100 spp.	Pathogen; pycnidia in a stroma; irregular, incompletely separate cavities; conidia elongate curved	Canker on twigs of *Prunus*	*C. leucostoma* (Fig. 2*g*)
Phaesporae 1-celled dark spores	*Coniothyrium* 100 spp.	Pathogen; conidia small, varying from flask-shaped to elliptical	Canker of rose	*C. wernsdorffiae* (Fig. 2*h*)
	Sphaeropsis 30 spp.	Pathogen; conidia large, typically 1-celled (sometimes 2-celled)	Black rot and canker of apple	*S. malorum*[c] (Fig. 2*i*)
Hyalodidymae 2-celled hyaline spores	*Ascochyta* 400 spp.	Pathogen; pycnidial wall thin	Leaf and pod spot of pea	*A. pisi*
			Black stem of alfalfa	*A. imperfecta*
Phaeodidymae 2-celled dark spores	*Diplodia*	Pathogen; conidia 15 μ or more long	Dry rot of corn	*D. zeae*
Phaeophragmiae 2-celled spores with cross septa	*Hendersonia*	Saprophyte		
Hyaloscolecosporae Long, hyaline threadlike spores	*Septoria*[d] 1000 spp.	Pathogen; pycnidial wall thin	Leaf spot of celery	*S. apii* (Fig. 2*j, k, l*)
			Leaf blight of tomato	*S. lycopersici*

[a] Several species are stages of *Diaporthe* (Ascomycete) spores. [b] Several species are stages of *Valsa*.
[c] Conidial stage of *Physalospora obtusa*. [d] Some species are stages of *Mycosphaerella* or *Leptosphaeria*.

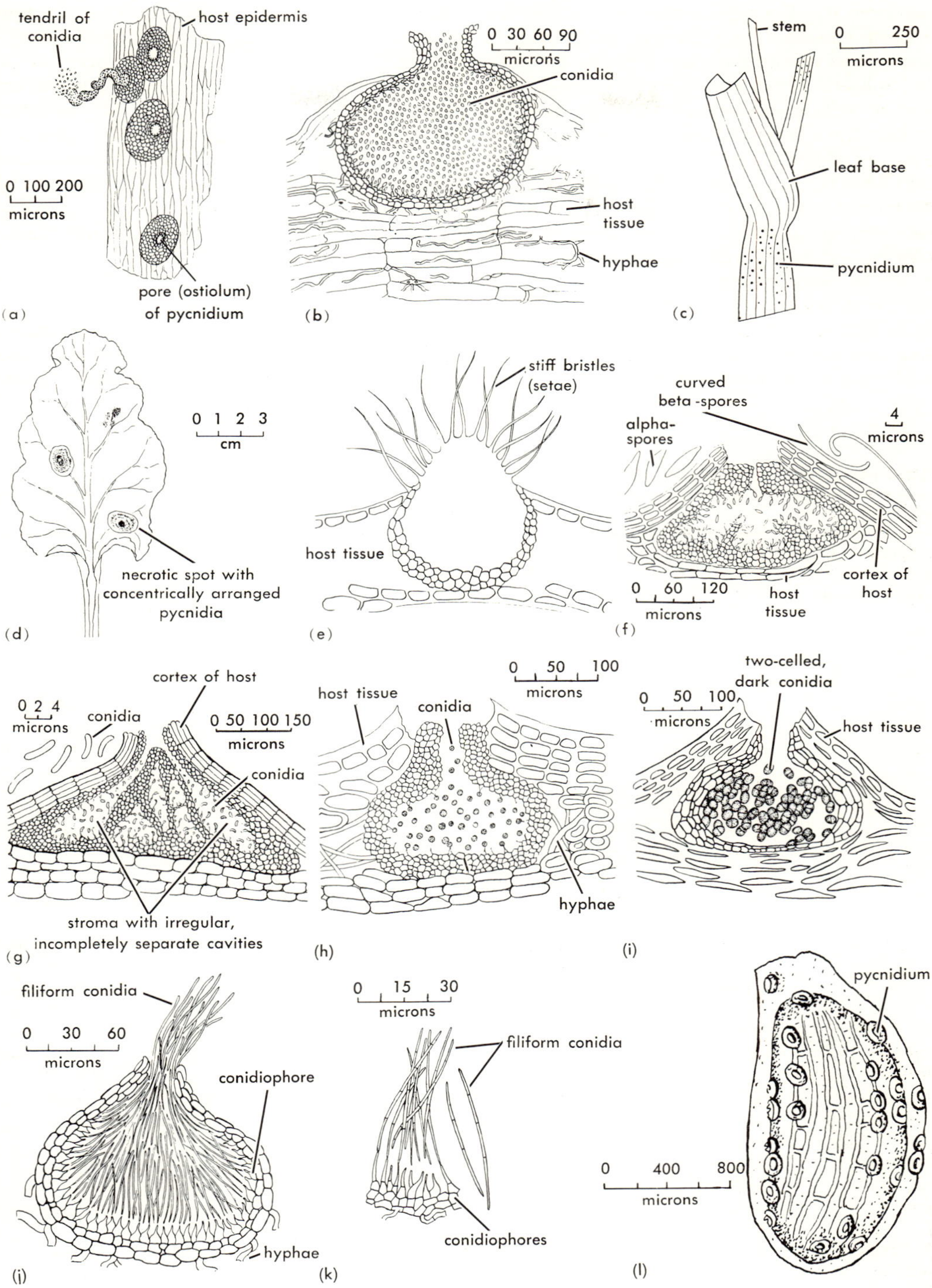

Fig. 1. (*a*) Four pycnidia of *Phoma betae* (*after A. B. Frank, 1892*). (*b*) Pycnidium and conidia of *Phoma apiicola* (*after H. Klebahn, 1910*). (*c*) Stem of carrot with pycnidia of *Phoma rostrupii*, (*d*) Beet leaf showing spots due to *Phyllosticta betae*. (*e*) Pycnidium of *Pyrenochaete ilicis*. (*f*) Pycnidium, α- and β-spores of *Phomopsis conorum* (*after G. G. Hahn, 1930*). (*g*) Pycnidium and conidia of *Cytospora leucostoma* (*after R. Alderhold, 1903*). (*h*) Pycnidium and conidia of *Coniothyrium hellebori* (*after Griffon and Maublanc, 1911*). (*i*) Pycnidium and conidia of *Sphaeropsis malorum* (*after L. R. Hesler, 1916*). (*j*) Pycnidium and conidia of *Septoria apii* (*after H. Klebahn, 1910*). (*k*) Piece of pycnidial wall with conidiophores and conidia of *Septoria apii* (*after H. Klebahn, 1910*). (*l*) Seed of celery with pycnidia of *Septoria apii* (*after H. Klebahn, 1910*).

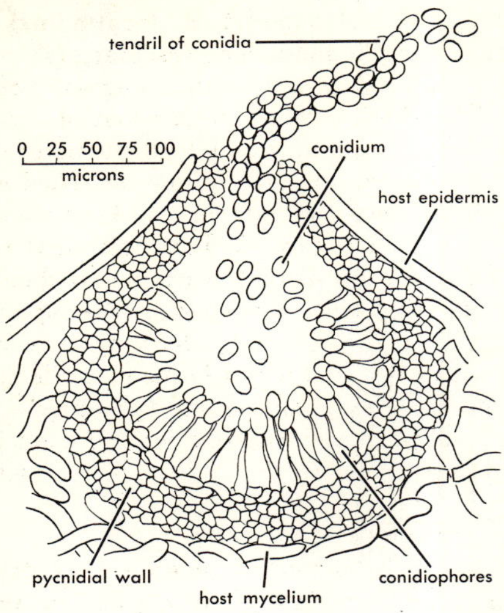

Fig. 2. Pycnidium of *Phoma* with a tendril of conidia. (*After Quaintance and Shear, 1907*)

300 genera and 5000 species recognized by some authorities.

The pycnidia, or fruit bodies, containing the asexual spores (conidia) are either black or dark colored. They are flask-, cone-, or lens-shaped structures with thin walls and a round, relatively small pore. The conidia are discharged from the pycnidia in tendrils or cirrhi.

The genera are usually arranged into spore groups, depending on the number of cells, shape of spore cluster, and whether the spores are bright or dark (*see* FUNGI IMPERFECTI). The important genera in the spore groups of the Sphaerioidaceae are listed in the table. *See* ASCOMYCETES; PLANT DISEASE; SPHAEROPSIDALES. [N.F.B.]

Sphaeropsidales

An order of fungi, also known as Phomales or Phyllostictales, of the class Fungi Imperfecti. Some members of this group are saprophytes and others are parasites causing diseases of plants. There are 520 genera with 5000–6000 species known.

The asexual spores (pycnospores or conidia) are formed in globose or flasklike fruit bodies called pycnidia. The pycnidia, found with or without openings (ostioles), may be separate or joined by vegetative hyphae (stromatic tissue). Conidiophores (hyphae bearing the conidia) are short or absent. The conidia are always slime spores, that is, they become separated with the slime from the cells producing them. They are rarely staurospores (star-shaped) or helicospores (corkscrew-shaped) and never catenulate (spores in chains).

This order is divided into 4 families: Sphaerioidaceae, Zythiaceae, Leptostromataceae, and Discellaceae (Excipulaceae). Sphaeropsidales are sometimes united with Melanconiales into one group, Coelomycetes. *See* DISCELLACEAE; FUNGI IMPERFECTI; LEPTOSTROMATACEAE; MELANCONIALES; PLANT DISEASE; SPHAERIOIDACEAE; ZYTHIACEAE. [N.F.B.]

Sphalerite

A mineral, also called blende, with composition ZnS (zinc sulfide); the major ore of zinc. It crystallizes in the isometric system, and crystals show the tetrahedron, dodecahedron, and cube, but are frequently complex and malformed. The mineral is most commonly in coarse to fine, granular, cleavable masses. The luster is resinous to submetallic; the color, white when pure, is commonly yellow, brown, or black, darkening with increased percentage of iron. There is perfect dodecahedral cleavage; the hardness is 3½, and specific gravity is 4.0.

Pure sphalerite contains 67% zinc and 33% sulfur, but iron is usually present, substituting in the structure for zinc, and may amount to 36%. The amount of iron in sphalerite varies directly with temperature at the time of crystallization. Cadmium and manganese may be present in small amounts.

Sphalerite is a common and widely distributed mineral associated with galena, pyrite, marcasite, chalcopyrite, calcite, and dolomite. It occurs both in veins and in replacement deposits in limestones. As the chief ore mineral of zinc, sphalerite is mined on every continent and in many countries. The United States is the largest producer, followed by Canada, Mexico, U.S.S.R., Australia, Peru, Belgian Congo, and Poland. *See* WURTZITE; ZINC. [C.S.HU.]

Sphene

A nesosilicate mineral, composition $CaTiSiO_5$, also known as titanite. It is monoclinic and usually occurs as well-formed crystals with characteristic wedge shape. There is prismatic cleavage and frequently a well-developed parting. Hardness is 5–5½ on Mohs scale; specific gravity is 3.4–3.5. The luster is resinous to adamantine and the color brown, green, yellow, gray, or black. Sphene is a common accessory mineral in igneous rocks, particularly syenites and nepheline syenites; it is also present in gneisses, schists, and crystalline limestones. Fine crystals are found at Binnental and St. Gothard, Switzerland; Arendal, Norway; and in Ontario and Quebec, Canada. Sphene, associated with nepheline and apatite, occurs in huge

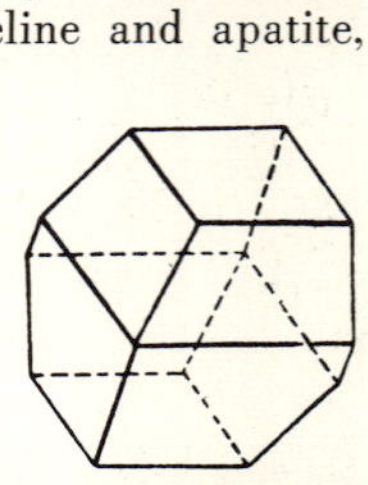

Wedge-shaped crystal of mineral sphene, or titanite. (*From C. S. Hurlbut, Jr., Dana's Manual of Mineralogy, 16th ed., Wiley, 1952*)

masses on the Kola Peninsula, U.S.S.R., where it is mined as a source of titanium. *See* SILICATE MINERALS. [C.S.HU.]

Sphenisciformes

The order of penguins, the most completely aquatic of living birds, contains the single family Spheniscidae. The 16 living species breed on coasts and islands of the Southern Hemisphere. The range of the Galapagos penguin (*Spheniscus mendiculus*) reaches the equator. Only two penguins, the Adelie (*Pygoscelis adeliae*) and the emperor (*Aptenodytes forsteri*), actually inhabit the Antarctic continent. The latter is the largest living species, attaining a weight of over 90 lb, but one fossil species may have weighed over 200 lb. In spite of the highly modified paddlelike wings, erect posture, scalelike feathers, and other peculiarities, penguins undoubtedly represent merely a highly specialized offshoot of a flying ancestral type, possibly related to the petrels (*Procellariiformes*). Their segregation as a distinct superorder, Impennes, of the Neognathae, advocated by some authors, does not seem warranted.

Behavior patterns of penguins are complex and varied. Surface-nesting species tend to be communal in their social relationships, while the hole-nesters usually have more strongly-developed family bonds. *See* AVES. [K.C.P.]

Sphenophyllales

An extinct group of articulate land plants, common during Late Pennsylvanian and Early Permian times. They are typified by *Sphenophyllum*, a small, branching plant, probably of trailing habit. The long, jointed stems rarely exceeded 1 cm in diameter, and had superposed, longitudinal, superficial ribs between nodes. The vascular system contained a solid xylem core with triangular primary wood. The leaves were wedge-shaped, usually shorter than 2 cm, and had toothed, notched, or rounded distal margins. They were attached at the nodes by their narrow ends, in whorls of usually 6 or 9 leaves each, rarely 18. Long, terminal cones, usually called Bowmanites when found detached, contained sporangia and spores. The sporangia terminated slender stalks, forming concentric whorls that alternated with whorls of sterile bracts. Most species were homosporous (produced spores of a single type). *See* PALEOBOTANY; PLANT KINGDOM; STELE. [S.H.M.]

Reconstruction of *Sphenophyllum cuneifolium*, showing whorled leaves and two terminal cones. (*From G. M. Smith, Cryptogamic Botany, vol. 2, 2d ed., McGraw-Hill, 1955*)

Sphenopsida

A subphylum of the plant phylum Tracheophyta containing the horsetails. It is a group of ancient origin being abundant in the Paleozoic era and reaching its highest development in the Carboniferous period (*see* GEOLOGY; PALEONTOLOGY). However, there seems to be little doubt that the one living genus, *Equisetum*, is a direct descendant of some of the fossil forms. The most conspicuous characters of this group are the whorled arrangement of the stem branches and appendages and the jointed conspicuous nodes on the hollow, ribbed stems.

The sporophyte (spore-producing generation) is the dominant phase of the life cycle and it produces the sporangiophores in compact distinct strobili, or cones. Each sporangiophore bears several sporangia (spore sacs) on the under surface of a terminal shieldlike structure. The spores produced are alike (homosporous).

The gametophyte (gamete-producing generation) is dioecious (male and female sex organs on different plants). *See* TRACHEOPHYTA. [P.A.V.]

Bibliography: *See* EQUISETALES.

Sphere

Both in euclidean solid geometry and in common usage the word sphere denotes a solid of revolution obtained by revolving a semicircle of radius r about its diameter. Its total volume is $V = (4/3)\pi r^3$.

However, in analytic geometry, and more generally in modern mathematics, the word sphere denotes a spherical surface that bounds a solid sphere. In this sense a sphere is the locus of all points P in 3-dimensional space whose distance from a fixed point O (called the center) is equal to a given number. The word radius may refer either to one of the segments OP, or to their common length r. A plane that intersects a sphere in just one point is called a tangent plane and is perpendicular to the radius drawn from the center of the sphere to that point. A plane that intersects a sphere in more than one point intersects it in a circle. The circle is called a great circle or a small circle of the sphere according to whether the plane does or does not pass through the center of the sphere. If two

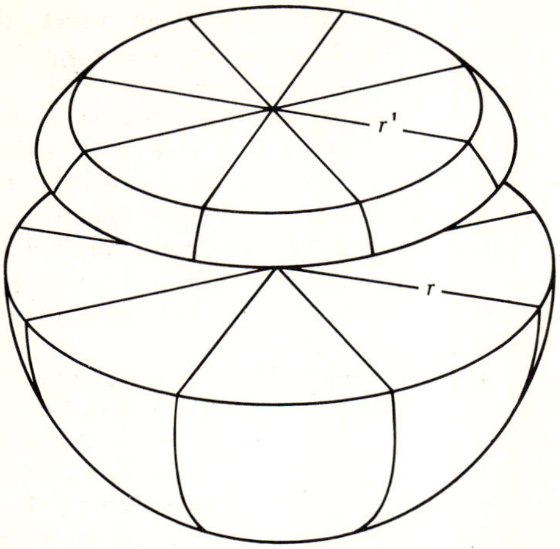

Segment of a sphere formed by intersection of two parallel planes with sphere.

parallel planes intersect a sphere, the spherical surface between them is called a zone, and its area is $2\pi rh$, where the height h of the zone is the distance between the parallel planes. The volume bounded by the zone and the two base planes is called a spherical segment, and is measured by the formula $V = (\pi h/6)(3r_1^2 + 3r_2^2 + h^2)$, where r_1 and r_2 are the radii of the two base circles. It is also measured by the prismoidal formula $V = (h/6)(B_1 + 4M + B_2)$, where B_1, B_2 denote the areas of the two bases, and M the area of the midsection halfway between the bases.

Any great circle of a sphere divides it into two hemispheres. A second great circle cuts a hemisphere into two lunes, each having vertices at the two points where the great circles intersect and each having an area proportional to the angles at the vertices between the tangents to the great circle boundaries. A third great circle cuts each lune into two spherical triangles. In general a spherical triangle is the figure formed by connecting any three points A, B, C on the surface of a sphere (but not on the same great circle) by great-circle arcs called sides. The sides BC, CA, AB are measured by the angles a, b, c that they subtend at the center of the sphere. If the angles between the pairs of sides at the vertices A, B, C respectively are denoted by α, β, and γ respectively, and if $2\sigma = \alpha + \beta + \gamma - 180°$ denotes the excess of the sum of these angles over two right angles, then the area of a spherical triangle of excess 2σ on a sphere of radius r is $(\sigma/90°)\pi r^2$ (or $2\sigma r^2$ if σ is measured in radians). Important relations between the sides and angles of a spherical triangle are

Law of sines: $$\frac{\sin \alpha}{\sin a} = \frac{\sin \beta}{\sin b} = \frac{\sin \gamma}{\sin c}$$

Law of cosines:

$$\cos a = \cos b \cos c + \sin a \sin b \cos \alpha$$

See SURFACE AND SOLID OF REVOLUTION; TRIGONOMETRY, SPHERICAL. [J.S.F.]

Spherical and aspheric surfaces, optical

Many common optical instruments contain optical materials having spherical surfaces with a wide range of curvatures. Such surfaces are capable of forming real images. A surface that is not a sphere is called an aspheric surface. Examples of such aspheric surfaces with symmetry of rotation that are occasionally used in optical instruments are conic sections (ellipsoids, hyperboloids, and paraboloids). Cylindrical and toroidal lenses are called anamorphotic systems.

Since it is much easier to grind and polish spherical surfaces than aspheric ones, most optical systems consist of lenses that are bounded by two spherical surfaces. These surfaces are arranged so that their centers lie on a line known as the axis of the system. The point of intersection of the surface with the axis is called the vertex. A plane surface is generally considered as a special case of a spherical surface with center at infinity and which can therefore be said to have an infinite radius.

Aspheric surfaces. The rays normal to a given wave surface can be refracted or reflected by a suitable aspheric surface so that they are normal to any desired wave surface. This is most often done for the aperture rays to improve the aperture errors and is successful where the field to be imaged is very small. Thus, large telescopes are figured by hand, and corrector plates are added to an optical system. If the unfigured system is good, the figuring may not destroy the other corrections. This is the secret of the Schmidt camera, which is the best-known application of aspheric lenses (*see* SCHMIDT CAMERA). Aspheric surfaces can also be used to correct the principal rays (the rays through the center of the diaphragm).

Aspheric surfaces are difficult to manufacture. When many identical aspheric elements are to be made, special methods of grinding and polishing are used. For instance, templates are used for spectacle lenses and for paraboloidal and elliptical mirrors. Where extreme correction is desired, as, for instance, in astronomical telescopes, retouching by hand is the proper procedure. For condensers and for other lenses where extreme accuracy is not needed, a molding process has been developed. Large paraboloidal mirrors for searchlights and ellipsoidal mirrors for arc lamps used to project motion pictures are made by a "dropping" process. A sheet of plate glass is laid on a suitable concave mold and heated until it softens and can be sucked into the mold.

Lenses with more than one aspheric surface have been proposed and designed by M. Linnemann to correct aperture and first-order asymmetry errors. There are a few patents in which two or more aspheric surfaces are successfully used to balance errors, but a complete theory for doing this has not yet been developed.

Ray-tracing formulas. To obtain ray-tracing formulas that are valid for all surfaces, it is convenient to place the coordinate origin at the vertex of the refracting (reflecting) surface, though for a refracting sphere alone, formulas with the origin at the center are somewhat simpler. If the z axis is in the direction of the system axis and the x and y axes are normal to it, the equation of the surface is then

$$\bar{z} = 0 \quad \text{(for the plane)}$$
$$\bar{z} = \rho\bar{u} \quad \text{(for the sphere of curvature } \rho)$$
$$\bar{z} = \tfrac{1}{2}\rho\bar{u} + \tfrac{1}{2}A_2\bar{u}^2 + \tfrac{1}{6}A_3\bar{u}^3 + \cdots \quad \text{(for an aspheric surface)}$$

where
$$\bar{u} = \tfrac{1}{2}(\bar{x}^2 + \bar{y}^2 + \bar{z}^2) \tag{1}$$

The object point and the object ray may be given by the coordinates x, y of the intersection point with the plane at the vertex and the optical direction cosines ξ, η, ζ of the ray. The procedure is to find the value of a parameter λ such that

$$u = \tfrac{1}{2}[(x + \lambda\xi)^2 + (y + \lambda\eta)^2 + (\lambda\zeta)^2] \tag{2}$$

and
$$\bar{z} = \lambda\zeta$$

fulfill Eq. (1). This can be achieved by an iteration process starting from $\lambda = 0$. With λ found, the coordinates of the point of intersection

$$\begin{aligned} \bar{x} &= x + \lambda\zeta \\ \bar{y} &= y + \lambda\eta \\ \bar{z} &= \lambda\zeta \end{aligned} \tag{3}$$

are computed.

Equation (1) gives, for the direction $\mathbf{O}$ of the unit vector along the surface normal (coordinates $\mathbf{O}_1$, $\mathbf{O}_2$, $\mathbf{O}_3$),

$$\begin{aligned} O_1 &= \bar{z}_u\bar{x}/[1 + \bar{z}_u{}^2(2\bar{u} - \bar{z})]^{1/2} \\ O_2 &= \bar{z}_u\bar{y}/[1 + \bar{z}_u{}^2(2\bar{u} - \bar{z})]^{1/2} \\ O_3 &= [1 + \bar{z}_u{}^2(2\bar{u} - \bar{z})]^{-1/2} \end{aligned}$$

where

$$\begin{aligned} \bar{z}_u &= \tfrac{1}{2}\rho && \text{(for the sphere)} \\ &= \tfrac{1}{2}\rho + A_2\bar{u} + \tfrac{1}{2}A_3\bar{u}^2 && \text{(for an aspheric surface)} \\ &= 0 && \text{(for the plane)} \end{aligned}$$

The directions of the normal and of the entering ray being known, the refraction law enables the refracted ray to be computed. The corresponding formulas are

$$\begin{aligned} \xi' - \xi &= \Gamma O_1 \\ \eta' - \eta &= \Gamma O_2 \\ \zeta' - \zeta &= \Gamma O_3 \end{aligned} \tag{4}$$

with
$$\Gamma = n' \cos i' - n \cos i$$

where n and n' are the refractive indices of the media, separated by the refracting surface, and i and i' are the angles formed by the incident and refracted rays with the surface normal and

$$\begin{aligned} \cos i &= \xi O_1 + \eta O_2 + \zeta O_3 \\ \cos i' &= +[n'^2 - n^2 + \cos^2 i]^{1/2} \end{aligned}$$

Refracting sphere. Rays through the center of a sphere are unrefracted. The center of a sphere is sharply imaged, and since the sine condition is fulfilled, a surface element through the center is imaged without errors of asymmetry.

The points having a center distance $c = n'r/n$ are sharply imaged upon points with center distance $c' = nr/n'$. The spheres on which these points respectively lie are called aplanatic spheres. Either the object or the image is virtual. The magnification with which these two spheres are imaged upon each other is $m = n^2/n'^2$. Again the sine condition is fulfilled, and thus first-order asymmetry errors are corrected. Lenses consisting of a refracting centered sphere or a refracting aplanatic sphere, or both, are often added to a given system to achieve a desired effect without destroying corrections previously achieved.

Cartesian surfaces. The centered sphere and the aplanatic sphere are special cases of surfaces which image the rays coming from an object point so that they all converge to another point. The general surface of this kind is determined by the fact that on all rays the light path from the object point to the surface and thence to the image point is constant.

Such a surface is in general of the fourth order. For an infinitely distant object, it is a hyperboloid. For $C = 0$, it becomes the aplanatic sphere.

The conic sections are refracting cartesian surfaces. The refracting ellipsoid images the rays from one geometrical focus to the other; the paraboloid images the geometrical focus sharply at infinity, or conversely, a set of parallel rays sharply at the focus; the hyperboloid images the rays from a real (or virtual) point sharply at a virtual (or real) point. Both points are situated at the geometrical foci. The only cartesian surface that is free from asymmetry errors is the aplanatic sphere. *See* OPTICS, GEOMETRICAL. [M.H.]

Bibliography: C. Carathéodory, *Hamburger Math. Einzelschriften*, 1940; A. Gullstrand, *K. Svenska Akademy Handlingar*, 1919; M. Linnemann, *Diss. Göttingen*, 1905; F. Twyman, *Prism and Lens Making*, 1952; A. Warmisham, British Patent 548730, 1942.

Spherical harmonics

A spherical harmonic or solid spherical harmonic of degree n is a homogeneous function, $R_n(x,y,z)$, of degree n which satisfies Laplace's equation

$$\Delta R \equiv \frac{\partial^2 R}{\partial x^2} + \frac{\partial^2 R}{\partial y^2} + \frac{\partial^2 R}{\partial z^2} = 0$$

where n is any number. $(x^2 + y^2 + z^2)^{(-n-1)/2} R_n(x,y,z)$ is a spherical harmonic of degree $-n - 1$. There are analogous definitions for spaces of any number of dimensions. In the present article, n will be a nonnegative integer and R_n, a polynomial in x, y, z (polynomial spherical harmonic). In terms of spherical coordinates r, θ, ϕ, $R_n(x,y,z) = r^n S_n(\theta,\phi)$ where S_n, a polynomial in $\cos\theta$, $\sin\theta$, $\cos\phi$, $\sin\phi$, is a spherical surface harmonic of degree n. There are $2n + 1$ linearly independent spherical surface harmonics of degree n; any spherical surface harmonic of degree n is a linear combination of these, and conversely any linear combination of spherical surface harmonics of degree n is again a spherical surface harmonic of degree n.

Applications. Spherical harmonics occur in potential theory. They occur in connection with Laplace's equation not only in spherical coordinates but also in spheroidal coordinates (spheroidal harmonics) and confocal coordinates (ellipsoidal surface harmonics). In the latter case their occurrence is due to the circumstance that the natural affine mapping of an ellipsoid onto a sphere carries the partial differential equation of ellipsoidal surface harmonics into the partial differential equation satisfied by spherical surface harmonics. In spherical coordinates, spherical surface harmonics occur in connection with Laplace's and Poisson's equations, the wave equation, the Schrödinger equation and generally in connection with partial differential equations of the form $\Delta U + f(r)U = 0$. In the latter case one has special solutions of the form $F(r)S_n(\theta,\phi)$, where F satisfies the ordinary differential equation

$$\frac{d^2F}{dr^2} + \frac{2}{r}\frac{dF}{dr} + \left[f(r) - \frac{n(n+1)}{r^2}\right]F = 0$$

In geometry, spherical surface harmonics are used in the theory of surfaces. In mathematical physics, spherical harmonics appear in the theories of gravitation, electricity and magnetism, hydrodynamics, and in other fields.

Spherical harmonics of degree n. The equation

$$\int_{-\pi}^{\pi} (x \cos u + y \sin u + iz)^n f(u)\, du$$

with $f(u)$ an integrable function, is a polynomial spherical harmonic of degree n, and every such spherical harmonic can be so represented. The representation is not unique. If c_n is a constant, h_1, h_2, . . . , h_n are n directions (not necessarily distinct), and $\partial/\partial h$ denotes directional differentiation in the direction h, then

$$c_n r^{2n+1} \frac{\partial^n}{\partial h_1 \cdots \partial h_n} \frac{1}{r}$$

is a polynomial spherical harmonic of degree n, and every such spherical harmonic can be so represented. The representation is unique. In a zonal spherical harmonic the n directions coincide, and in a sectorial spherical harmonic they are in a plane at angles of π/n. If $n - m$ directions coincide in the axis and the remaining directions are in the plane perpendicular to the axis at angles of π/m, one has a tesseral spherical harmonic of degree n and order m.

Explicit forms. For spherical harmonics whose axis is the z axis, ($m = 0, 1, 2, \ldots, n$), the equation

$$S_n^{\pm m}(\theta,\phi) = \frac{(-1)^{n-m} r^{n+1}}{(n-m)!} \frac{\partial^{n-m}}{\partial z^{n-m}} \left(\frac{\partial}{\partial x} \pm i \frac{\partial}{\partial y}\right)^m \frac{1}{r}$$
$$= P_n^m (\cos \theta)\, e^{\pm im\phi}$$

represents a linearly independent system of spherical surface harmonics of degree n. $S_n^m \pm S_n^{-m}$ is a zonal, sectorial, tesseral spherical surface harmonic of degree n and order m according as $m = 0$, $m = n$, $1 \leqq m \leqq n - 1$. The $P_n^m(w)$ are associated Legendre functions which satisfy the associated Legendre equation

$$(1 - w^2)\frac{d^2P}{dw^2} - 2w\frac{dP}{dw} + \left[n(n+1) - \frac{m^2}{1-w^2}\right]P = 0$$

$P_n^0 = P_n$ is the Legendre polynomial of degree n.

Properties of spherical harmonics. A function is said to be harmonic in a region if it is a twice continuously differentiable solution of Laplace's equation there, and if, in addition, it vanishes at infinity in case the point at infinity is an interior point of the region. Every function harmonic inside a sphere about the origin can be expanded in a series $\sum_{n=0}^{\infty} r^n S_n(\theta,\phi)$ convergent inside that sphere. Every function harmonic outside a sphere about the origin can be expanded in a series $\sum_{n=0}^{\infty} r^{-n-1} S_n(\theta,\phi)$ convergent outside that sphere. The reciprocal distance of the points (x,y,z) and $(0,0,a)$ is harmonic in the regions $r < a$ and $r > a$ and possesses in these regions the expansions

$$\frac{1}{\sqrt{a^2 - 2ar\cos\theta + r^2}} = \begin{cases} \sum_{n=0}^{\infty} \frac{r^n}{a^{n+1}} P_n(\cos\theta) & r < a \\ \sum_{n=0}^{\infty} \frac{a^n}{r^{n+1}} P_n(\cos\theta) & r > a \end{cases}$$

θ, ϕ determine a point on the unit sphere. The scalar product of two functions, f and g, on the unit sphere is suitably defined as

$$(f,g) = \int_0^{\pi} \int_{-\pi}^{\pi} f(\theta,\phi)\overline{g(\theta,\phi)} \sin\theta\, d\theta\, d\phi$$

where $\bar{g}$ is the complex conjugate of g. If $(f,g) = 0$, f and g are orthogonal. Spherical surface harmonics are functions on the unit sphere. Any two spherical surface harmonics of different degrees are orthogonal. The spherical surface harmonics

$$S_n^m(\theta,\phi)\quad m = -n, -n+1, \ldots, n \qquad n = 0, 1, \ldots$$

form an orthogonal system; that is, $(S_n^m, S_{n'}^{m'}) = 0$ unless $m = m'$ and $n = n'$. This orthogonal system is complete; that is, a continuous function which is orthogonal to all the S_n^m vanishes identically. With an integrable function f is associated the Laplace expansion

$$\sum_{n=0}^{\infty} \sum_{m=-n}^{n} C_{mn} S_n^m \quad \text{where} \quad C_{mn} = \frac{(f,S_n^m)}{(S_n^m,S_n^m)}$$

Under suitable conditions, the Laplace expansion will converge to f. For instance, if f is continuous and continuously differentiable on the unit sphere, then the Laplace expansion converges to f uniformly.

Let (θ_0,ϕ_0) be a fixed point, and let $\cos\gamma = \cos\theta\, \cos\theta_0 + \sin\theta\, \sin\theta_0\, \cos(\phi - \phi_0)$ be the spherical distance of (θ,ϕ) and (θ_0,ϕ_0). The Laplace expansion

$$P_n(\cos\gamma) = P_n(\cos\theta)P_n(\cos\theta_0) + 2\sum_{m=-n}^{n}\frac{(n-m)!}{(n+m)!} \cdot P_n{}^m(\cos\theta)P_n{}^m(\cos\theta_0)\cos m(\phi-\phi_0)$$

is the addition theorem of Legendre polynomials and expresses the change to a new axis through the point (θ_0,ϕ_0). Other spherical surface harmonics have corresponding addition theorems.

Let $\cos\gamma$ be the spherical distance of (θ,ϕ) and (θ_0,ϕ_0), and let $K(w)$ be a continuous function for $-1 \leqq w \leqq 1$. Then for any spherical surface harmonic S_n of degree n,

$$\int_0^\pi \int_{-\pi}^{\pi} K(\cos\gamma)S_n(\theta,\phi)\sin\theta\,d\theta\,d\phi = \lambda_n S_n(\theta_0,\phi_0)$$

where $\lambda_n = 2\pi \int_{-1}^{1} K(w)P_n(w)\,dw$

See DIFFERENTIAL EQUATION; POTENTIALS (MATHEMATICS). [A.ER.]

Bibliography: P. Appell and J. Kampé de Fériet, *Fonctions hypergéometriques et hypersphériques, Polynomes d'Hermite*, 1926; A. Erdélyi et al., *Higher Transcendental Functions*, 3 vols., 1953–1955; E. W. Hobson, *The Theory of Spherical and Ellipsoidal Harmonics*, 1931; J. Lense, *Kugelfunktionen*, 1950; T. M. MacRobert, *Spherical Harmonics*, 2d ed., 1947; G. Prasad, *A Treatise on Spherical Harmonics and the Functions of Bessel and Lamé*, 2 vols., 1930–1932; E. T. Whittaker and G. N. Watson, *A Course of Modern Analysis*, 4th ed., 1940.

Spheroid

A term sometimes used loosely for any surface that is almost spherical in shape, but denoting specifically a surface of revolution formed by revolving an ellipse about one of its axes. A spheroid is oblate or prolate according to whether the ellipse generating it is rotated about its minor or major axis, respectively (see illustration). If $y^2/a^2 + z^2/b^2 = 1$, $a > b$, is the equation of an ellipse in the yz plane, the equation of the oblate spheroid obtained by revolving the ellipse about the z axis is $x^2/a^2 + y^2/a^2 + z^2/b^2 = 1$; the equation of the prolate spheroid obtained by revolving about the y axis is $x^2/b^2 + y^2/a^2 + z^2/b^2 = 1$. The volume of the prolate spheroid is $(4/3)\pi ab^2$, and that of the oblate spheroid $(4/3)\pi a^2 b$. The surface of the prolate spheroid is $2\pi b^2 + 2\pi(ab/e)$ arcsin e, where eccentricity e is $[a^2 - b^2]^{1/2}/a$; that of the oblate spheroid is $2\pi a^2 + (\pi b^2/e)\ \ln[(1+e)/(1-e)]$. Gravitational attraction of spheroids, a problem of fundamental importance in astronomy, was investigated by Isaac Newton, A. M. Legendre (who was lead by his investigations to the class P_n of polynomials known by his name), J. L. Lagrange, and Karl F. Gauss. *See* ELLIPSOID AND SPHEROID; SURFACE AND SOLID OF REVOLUTION. [L.M.BL.]

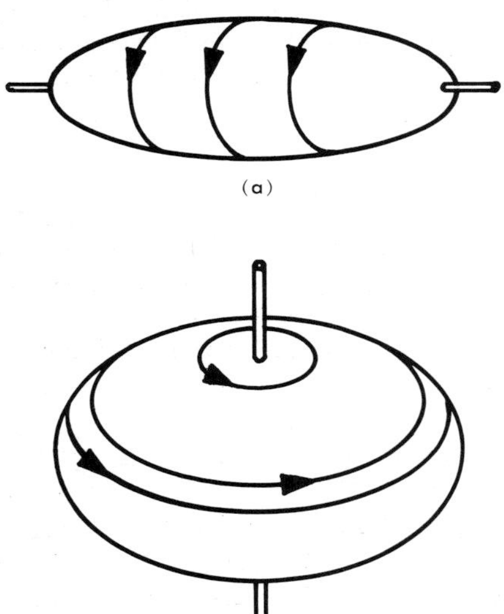

Spheroids. (*a*) Prolate spheroid. (*b*) Oblate spheroid.

Sphingolipid

A complex lipid which contains the amino alcohols sphingosine (I), dihydrosphingosine or phytosphingosine (II).

$$CH_3(CH_2)_{12}CH{=}CH{-}\underset{OH}{CH}{-}\underset{NH_2}{CH}{-}CH_2OH \quad \text{(I)}$$

$$CH_3(CH_2)_{13}\underset{OH}{CH}{-}\underset{OH}{CH}{-}\underset{NH_2}{CH}{-}CH_2OH \quad \text{(II)}$$

Ceramides are fatty acid amides of sphingosines. Ceramides of sphingosine have been isolated from animal tissues and are also the basic components of cerebrosides (glycosides of ceramides) and sphingomyelins (phosphoryl choline esters of ceramides). A hexasaccharide phosphate ester of the ceramides of phytosphingosine (phytoglycolipid) is present in plant seeds. Since the fatty acids of the sphingolipids are amide bound, these compounds can be readily separated from those lipids in which the acids are ester bound by hydrolytic decomposition of the latter with dilute alkali. *See* GLYCOLIPID; LIPID; PHOSPHATIDE. [H.E.C.; R.H.G.]

Spica

Alpha Virginis is one of the brightest and nearest of the hot, main-sequence stars, spectral type B1. Located at a distance of 75 parsecs, Spica is 1500 times brighter than the Sun and has a temperature of nearly 20,000°K. Spica is a spectroscopic binary of 4-day period consisting of two nearly identical stars. Rapidly rotating, and relatively close, they are nearly unstable and are connected by streams of matter. The relative intensities of the spectral lines of the two components vary around the orbit because of these circumstellar streams.

Maps of the intensity in the far ultraviolet (near 1300 angstroms) obtained by Naval Research Laboratory rockets show that Spica is surrounded by a large nebulosity of unknown nature and origin. *See* STAR. [J.L.GR.]

Spice and flavoring

Substances added to food to enhance savoriness. Spices are aromatic vegetable materials used for food seasoning. Flavorings are compounded blends of materials used to produce particular flavors or simulate other known flavors.

A spice may be derived from any part of the plant: leaf (bay), flower bud (cloves), fruit (pimento), bark (cassia), rhizome (turmeric), root (horseradish), or seed (anise). Spices may be used whole as are caraway and poppy seed, or ground as is cinnamon bark. With the exception of red peppers and a few others, spices used in the United States are imported, since most grow in tropical or subtropical climates. In the food industry spices are used in many physical states. Besides in the whole and ground conditions, spices are procured as mixed materials ready for use. There has been a growing use of spice extracts, or oleoresins—either as such, or spread on the surface of sugar, salt, or dextrose to make so-called "soluble spices." These have the advantage of showing no specking in light-colored foods. Premixed oleoresins are available for use in such products as canned foods and pickles.

The oleoresins are prepared by solvent extraction of the spice and may be separated into two fractions by steam distillation. The volatile fraction which comes over in steam distillation is called the essential oil; the oleoresin portion remaining behind is called the fixed oil. Spices vary markedly in their proportion of volatile to fixed oils as well as in their total oleoresin content.

In the preparation of oleoresins and essential oils it is usual commercial practice in the United States to remove the active flavoring constituents by solvent percolation. The essential oil is then removed from the oleoresin by steam distillation. It is not uncommon, however, to remove the oil from the flavor-bearing substance by direct steam distillation. In solvent extraction for oleoresin the spice material is usually ground to increase extraction efficiency. Particle size is extremely important as in all solvent extraction procedures. The time of extraction and solvent used are determined by the material being extracted. Hexane, perchlorethylene, trichlorethylene, acetone, and ethyl and propyl alcohols are used as solvents. Each spice presents a peculiar problem; for example, cassia extract polymerizes readily and if not treated promptly will "set up" in the receiver so that it is virtually impossible to remove.

Spices. The economically important spices are shown in the table.

Economically important spices

Spice	Plant and source	Plant part used	Principal use
Allspice* (*Pimenta*)	Evergreen of myrtle family from Jamaica and Guatemala	Dried fruit	Pickles, roast meat, catsup
Anise*	Parsley family from Mediterranean	Seed	Baked goods, Anisette, a liqueur
Basil	Mint family grown in United States and many other parts of world	Leaves and tender stems	Sauces and soups especially tomato based
Bay	Evergreen of laurel family from Mediterranean, Turkey, Greece, and Portugal	Leaf	Pickles, stews, soups and sauces
Caraway*	Biennial of parsley family from north central Europe and south England	Seed	Bread and baked goods, cheese and sauces, in Kummel, a liqueur
Cardamon*	Ginger family from Guatemala and India	Seed	Baked goods and in coffee blends, curry powders
Cayenne	Capsicum family grown over most of the world (name used for those high in pungency)	Pod or fruit	Many foods, pickles, sauces, meats, curry powder
Celery*	Parsley family principally from France and India	Seed	Sauces, salads, pickles, and soups
Cinnamon*	Evergreen member of laurel family from Ceylon	Bark	Baked goods, pickles, candy
Cloves*	Evergreen tree of the myrtle family from Madagascar and Zanzibar	Unopened bud flower	Pork products, pickles, stews, meats and gravies
Coriander*	Plant of parsley family from Yugoslavia and Morocco	Dried fruit	Frankfurters, bologna, baked goods
Cumin*	Annual plant of parsley family from Iran and Morocco	Dried fruit	Chili powder
Dill*	Member of parsley family from India and domestic sources	Seed and leaves	In pickles, soups and sauces

Economically important spices (cont.)

Spice	Plant and source	Plant part used	Principal use
Fennel*	Perennial of parsley family from India and Rumania	Seed	Baked goods, salad dressings, meat products
Garlic*	Member of lily family from U.S. and Mediterranean	Bulb	Baked meats, sauces, dressings, soups and so on
Ginger*	Tuberous perennial from Jamaica, India, and Africa	Rhizome	Soft drinks, baked goods, pickles, puddings
Mace*	Tropical tree similar to rhododendron from Indonesia and West Indies	Aril covering nutmeg	Baked goods and processed meats
Marjoram	Perennial of mint family from France, Peru, and Chile	Leaves	Soups, stews, and sauces
Mint*	Perennial herb of many varieties grown in U.S.	Leaves	Confections, sauces, jellies, and gums
Mustard*	Annual of mustard family from U.S. in Montana and Calif.	Seeds	Sauces, baked meats, and processed meats and gravies
Nutmeg*	Tropical tree similar to rhododendron from Indonesia or West Indies	Seed	Baked goods, processed meats, fruit sauces
Oregano	Perennial of mint family from U.S., Mexico, Italy, and France	Leaves	Sauces, stews especially those tomato based
Paprika	Member of capsicum family from Spain, U.S., and Hungary	Pod or fruit	For red colored dressings, sauces, meats, and condiments as catsup
Pepper* black white	Perennial climbing vine from India, Borneo, Malaya, and Indonesia	Berry (black); berry with cortex removed (white)	Endless use in virtually all food products, and table use
Poppy seed	Annual of poppy family from Poland, Argentina, Iran, and Turkey	Seed	Toppings for baked goods and confections
Rosemary	Evergreen of mint family from France, Spain, and Portugal	Leaf	Meats, sauces and gravies
Saffron*	Plant of crocus family from Spain	Stigma of flower	Coloring rice and other specialties
Sage*	Member of mint family from Dalmatia, Greece, and Yugoslavia	Leaf	Sausage, poultry and poultry stuffing
Savory	Member of mint family from France and Spain	Leaf	Meats, stuffings, salads, and sauces
Sesame	From sesame plant in U.S., Nicaragua, Salvador, Egypt, Brazil	Seed	Baked goods and confections
Thyme	Perennial of mint family from France and U.S.	Leaf	Sauces for shellfish and with fresh tomato
Turmeric*	Member of ginger family from India, Haiti, Jamaica, and Peru	Rhizome	Yellow color in pickles, sauces, and fish

* See individual articles for further information.

Flavoring. A great deal has been written on the use of various flavoring materials, but much of the field remains in the art state and reliance is placed on experienced personnel with knowledge of the materials available and of the nature of the finished products desired.

Flavor ester mixtures are flavoring materials which closely resemble the flavor and odor of natural herbs, nuts, fruits, and seeds. A solution of flavor ester mixtures in alcohol is known as a flavor extract; a flavor ester mixture in a solvent other than alcohol is termed a "flavor."

Blended flavorings are mixtures of natural and synthetic compounds. These may be naturally occurring substances, such as lemon oil; substances like eugenol, isolated from naturally occurring flavoring materials; substances like vanillin, prepared synthetically but also occurring in natural material; or synthetic substances, such as methyl anthranilite used in grape essence. Methyl anthranilite is also used in banana, currant, and melon flavors. Compounding these materials calls for careful balancing of the ingredients. *See* CITRUS FLAVORING; FOOD ENGINEERING; FOOD MANUFAC-

TURING; SALT (FOOD); VANILLA EXTRACT; *see also* ESSENTIAL OILS; ESTER. [R.E.M.]

Bibliography: E. Guenther, *The Essential Oils*, vols. 1–6, 1948–1952; M. B. Jacobs, *Synthetic Food Adjuncts*, 1947; L. W. Jones (ed.), *A Treasury of Spices*, 1956.

Spider

Any member of the order Araneae, class Arachnida, phylum Arthropoda. Spiders are worldwide in their distribution, and are among the most common land animals. There are about 20,000 species. Most of them are terrestrial.

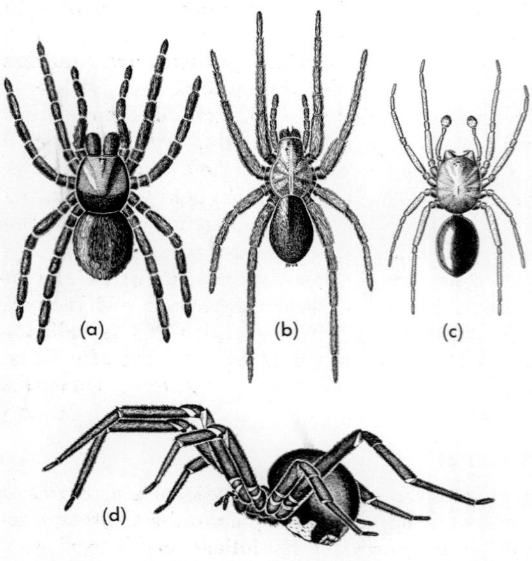

Spider. (*a*) Tarantula, *Aphonopelma (Eurypelma) californica*; length to 2 in. (*b*) Wolf, *Lycosa helluo*; length to ½ in. (*c*) Balloon, *Erigone autumnalis*; length to 1/20 in. (*d*) Black widow, *Latrodectus mactans*; length to ¼ in. (From E. L. Palmer, *Fieldbook of Natural History*, McGraw-Hill, 1949)

Spiders are predatory, carnivorous animals, subsisting almost entirely upon insects which they catch in a variety of ways. They are useful animals in their predatory activities upon harmful insects, although they do catch and eat useful insects as well.

A very few spiders have bites that are dangerous to man although any bite is a potential source of inflammation. The only really dangerous spiders are the four members of the genus *Latrodectus*. Most common of these is *L. mactans*, the black widow spider.

Morphology. As Arthropoda, spiders have a chitinous exoskeleton, although it is thin and weak, making them soft-bodied animals. They also possess jointed appendages, but in common with other Arachnida there are no locomotory appendages on the abdomen. Their eyes are simple; the cuticle is often equipped with sensory hairs or scales; they lack gills; and they are mostly oviparous. There is no metamorphosis.

Spiders are distinguished from other Arachnida by lacking visible segmentation on both the cephalothorax and abdomen, which are joined by a narrow waist. They produce silk from spinnerets on the abdomen. Eggs are frequently laid in cocoons. There are other technical distinctions, including a pair of poison claws on the short chelicerae.

Although all spiders spin silk, only a few spin the elaborate webs usually associated with spiders. Best known of the web spinners are the orb weavers. Spider web silk is sometimes used as cross hairs in delicate optical instruments. Many species lie in hiding for their prey, while others stalk their victims. They may either jump on their prey, like the jumping spiders, or run it down like those of the genus *Lycosa*.

Common species. Among the more interesting spiders in the United States are the trap-door spiders. *Bothriocyrtum californicum*, of the Southwest. These spiders live in burrows, covered by a silken trap door, which they can hold closed from inside. They make quick, short dashes from their burrow to capture food.

The commonly feared tarantulas of the Southwest, *Aphonopelma (Eurypelma) californica*, are of little danger to man, and their bite is no worse than the sting of a wasp. Their venom, however, quickly overpowers most invertebrates, and sometimes even small mice, snakes, and lizards. The black widow spider carries a very toxic venom, and numerous deaths have been caused by its bite. This species ranges from Canada southward into South America and occurs in the Hawaiian Islands. It is most common in the southwestern United States, but is also found in the Middle West and East. Only the females bite. They are plump, black spiders, with a body about ¼ in. long, marked with a red hourglass on the under side of the abdomen. A similar, harmless species has red markings on the upper side of the abdomen. Like many other spiders, the males are about half as large as the females, and are eaten by the female after mating.

Among the few aquatic spiders are the interesting water spiders of the genus *Dolomedes*. These animals build a silken nursery under water into which they laboriously carry bubbles of air and where they deposit their eggs. They also construct similar underwater rafts from which they make forays to catch aquatic insects and planktonic animals for food.

Ballooning, the technique of spinning a long strand of silk and then riding this strand on the wind, is employed by the young of many species of spiders as a method of dispersal. Individuals may be carried great distances in this manner. *See* ARANEAE. [J.D.B.]

Spilite

An aphanitic (microscopically crystalline) to very fine-grained igneous rock, with more or less altered appearance, somewhat resembling basalt but composed of albite or oligoclase, chlorite, epidote, calcite, and actinolite.

In spite of the highly sodic plagioclase, spilites are generally classed with basalts because of the low silica content (about 50%). They also retain many textural and structural features characteristic of basalt.

Under the microscope, laths of albite or oligoclase usually appear clouded or closely associated with abundant epidote, calcite, and chlorite. Actinolite and additional chlorite and epidote appear to have formed from augite, and small relic grains of augite may survive. Olivine is uncommon and has usually been changed to serpentine. Certain textures of basaltic rocks may still survive; others may be completely destroyed.

Spilites form small intrusive masses (dikes and sills) as well as lava flows. Intrusive spilites appear to grade into diabase; flows grade into basalts. Spilites are most commonly and abundantly associated with stratified rocks of geosynclines and may have had a submarine origin (*see* GEOSYNCLINE). Pillow structure, in which the rock appears composed of closely packed, pillow-shaped masses up to a few feet across, is typical and more perfectly developed than in other rock types. Vesicules, commonly filled with various minerals, may give the rock an amygdaloidal structure (*see* AMYGDULE).

Spilites are generally believed to represent rocks of basaltic composition in which calcic plagioclase (labradorite) has been converted largely to albite. Albitization may have been accomplished during the late stages of crystallization of the basaltic lava or shortly thereafter. Sodium from the sea water may have gradually replaced calcium in the plagioclase to form albite, and some of the displaced calcium may have gone to form epidote and calcite. Another source for requisite sodium may have been emanation from deeper masses of molten rock.

A less popular theory supposes spilite to form by direct crystallization of spilitic magma (rock melt), but the origin of such a magma poses something of a problem. Many so-called spilites may be of metamorphic and metasomatic origin; they may have formed by reconstitution and partial replacement of normal basaltic rocks. *See* BASALT; IGNEOUS ROCKS; METAMORPHISM; METASOMATISM; PETROGRAPHIC PROVINCE. [C.A.CA.]

Spin (quantum mechanics)

The intrinsic angular momentum of a particle. It is that part of the angular momentum of a particle which exists even when the particle is at rest, as distinguished from the orbital angular momentum (*see* ANGULAR MOMENTUM). The total angular momentum of a particle is the sum of its spin and its orbital angular momentum resulting from its translational motion. The general properties of angular momentum in quantum mechanics imply that spin is quantized in half integral multiples of $\hbar$ ($\hbar = h/2\pi$, where h is Planck's constant); orbital angular momentum is restricted to half *even* integral multiples of $\hbar$. A particle is said to have spin 3/2, meaning that its spin angular momentum is $\frac{3}{2}\hbar$.

A nucleus, atom, or molecule in a particular energy level, or a particular elementary particle, has a definite spin; for instance, a deuteron has spin 1, a $_3\mathrm{Li}^5$ nucleus in its ground state has spin 3/2, and an electron has spin 1/2. The spin is an intrinsic or internal characteristic of a particle, along with its mass, charge, and isotopic spin. *See* ISOTOPIC SPIN; SYMMETRY LAWS (PHYSICS).

A particle of spin s has $2s + 1$ spin states, since according to quantum mechanics the projection of an angular momentum of magnitude j along an axis can have the $2j + 1$ integrally spaced values $j, j - 1 \cdots -j + 1, -j$. These spin states represent an internal degree of freedom of a particle, in addition to its external freedom of motion in 3-dimensional space.

In field theory, in which particles are regarded as quanta of a field, the spin of the particle is determined by the tensor character of the field. For instance, the quanta of a scalar field have spin 0 and the quanta of a vector field have spin 1. A celebrated theorem of quantum field theory, proved first by W. Pauli, states a connection between spin and statistics: a particle with half even integral spin obeys Bose-Einstein statistics and is called a boson; a particle with half odd integral spin obeys Fermi-Dirac statistics and is called a fermion. *See* QUANTUM STATISTICS; *see also* ELECTRON SPIN; QUANTUM FIELD THEORY; QUANTUM MECHANICS. [C.J.G.]

Spinach

A cool-season annual of Asiatic origin, *Spinacia oleracea*, belonging to the plant order Centrospermales. It is grown for its foliage and served as a cooked vegetable or as a salad (*see* ANNUAL PLANTS). New Zealand spinach, *Tetrogonia expansa*, and Mountain spinach, *Atriplex hortense*, are also called spinach but are less commonly grown. Spinach plants are usually dioecious. *See* FLOWER (BOTANY).

Propagation is by seed, commonly planted in rows that are 10–20 in. apart. *See* SEED (BOTANY). Cool weather favors maximum production. High temperatures and long daylight periods encourage seedstalk formation and reduce vegetative growth

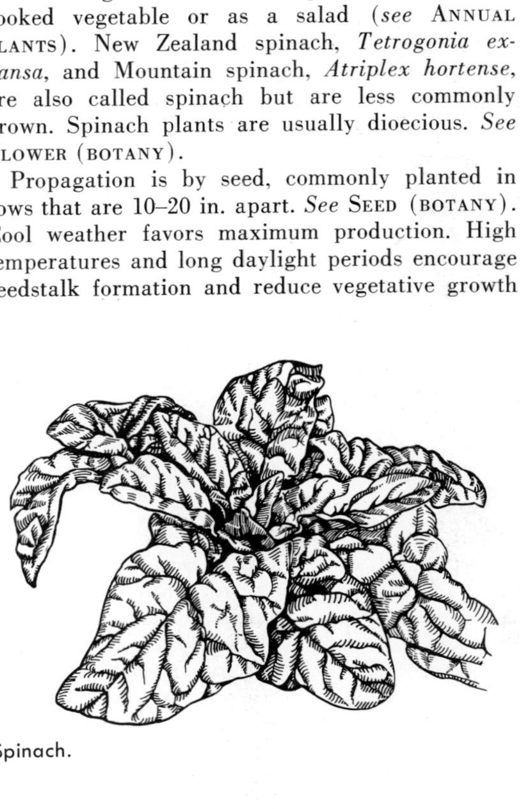

Spinach.

(*see* PHOTOPERIODISM). Fall-seeded spinach is overwintered and harvested in the spring in areas where the weather is mild or the crop is protected by snow cover. *See* VEGETABLE GROWING.

Varieties are classified according to (1) seed (smooth or prickly), (2) leaves (smooth or savoyed), and (3) seeding tendencies (early or late, or long standing). Popular varieties for fresh market are Long Standing Bloomsdale and America; for canning, Hollandia and Viroflay. Certain hybrid varieties now under investigation look promising.

Harvesting by hand or machine begins when the plants have reached full size and before seedstalks form, usually 40–50 days after planting.

In the United States, the average annual value of spinach for fresh market is approximately $11,000,000; Texas and Pennsylvania are the important producing states. The nation's average annual farm value of spinach for canning is about $5,000,000, with California and Oklahoma the principal producing states. *See* CENTROSPERMALES. [H.J.C.]

Spinal cord

The lower extension of the central nervous system, located in the spinal canal of the vertebral column. It contains major sensory and motor fibers and nerve cells in its white (outer) and gray (inner) portions, respectively. Sensory elements are posterior; motor elements are anterior. Functionally and anatomically, the cord is bilaterally symmetrical, and serves each half of the body with 31 spinal nerves (in man) that emerge from between successive vertebrae. These contain representative motor and sensory fibers for each body segment. The former primarily supply muscles and glands whereas the latter carry sensations of pain, pressure, and touch from peripheral sense organs. In the adult, the cord extends from the skull to the second lumbar vertebra. *See* CENTRAL NERVOUS SYSTEM.

All vertebrates display a similar basic pattern with variations typical of specific classes and orders. [E.G.ST.]

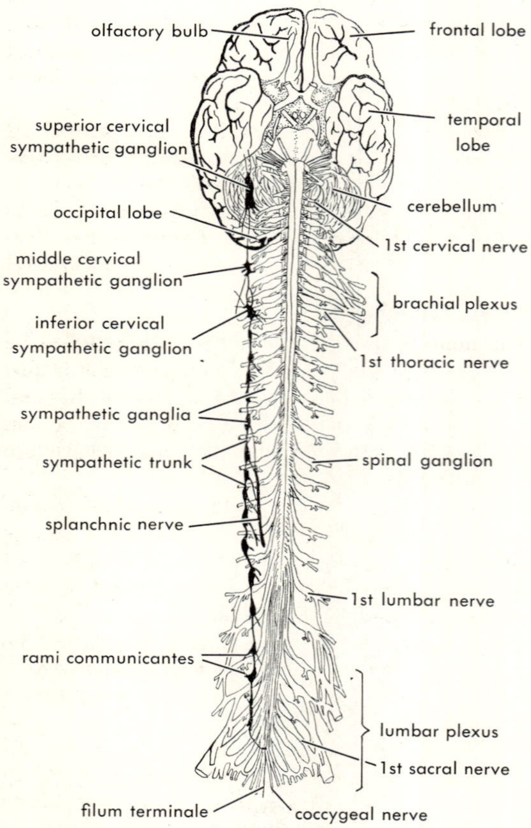

Brain and spinal cord (ventral aspect). Nerve roots and chief autonomic ganglia are shown. (*After H. Morris and A. Thomson, H. V. Neal and H. W. Rand, Comparative Anatomy, Blakiston–McGraw-Hill, 1936*)

Spinal cord disorders

In addition to those disorders common to the brain, the spinal cord is subject to certain lesions because of its position or structure. A few of the more important are mentioned.

Injury. Spinal cord injury results from dislocation, fracture, or compression in many cases, but a special form, called spinal shock, may result from a severe blow without actual distortion of adjacent tissue. In this case, there is a temporary paralysis which gradually clears in a length of time related to the severity of the paralysis. In direct damage, the cord may be slightly, partially, or completely damaged at one or more levels. Typical motor and sensory losses follow, with a poor prognosis for recovery if the nerve tissue is severely injured. Occasionally, similar damage may result from a hemorrhage produced by injury.

A fairly common type of potential cord injury is seen in a number of cases of slipped disks, in which the inner, soft part of the vertebral column extrudes into the spinal canal. If this compresses the cord, functional loss of temporary or permanent degree can follow; more often, pressure is exerted on spinal roots so that pain, numbness, and some type of muscle weakness intervene.

Tumors. Spinal cord tumors are not infrequent and most of these are of two types, the metastatic, from a primary source elsewhere in the body, and the tumors of the meninges or connective tissue related to the cord. The latter include neurofibromas, meningiomas, and gliomas, which occur most often. The signs and symptoms and the extent of damage relate largely to the physical compression of the cord at a particular level. *See* TUMOR.

Congenital defects. A few of the more common congenital defects involving the cord include an unclosed neural canal, or spina bifida, and reduplicated or otherwise malformed cords, such as those caught in an external sac of other tissues, the meningomyelocoele. *See* TERATOGENESIS.

Inflammation. Inflammations may result from known or unknown agents and in meningitis may involve primarily the coverings; in myelitis, the cord itself. The meningococcus, pneumococcus, streptococcus, tubercle bacillus, and other microorganisms frequently cause meningitis. The most widely known cause of myelitis, of course, is the poliomyelitis virus group, although other agents will also produce inflammations, such as typhus, Rocky Mountain spotted fever, and *Treponema pallidum*, the cause of syphilis. Malaria, amebiasis, trichinosis, elephantiasis, toxoplasmosis, and blastomycosis may display their effects on the spinal cord. *See* AMEBIASIS; BLASTOMYCOSIS; FILARIASIS; TOXOPLASMOSIS; TRICHINOSIS.

Vascular diseases. Not uncommonly motor or sensory deficits in the spinal cord may result from specific vascular diseases and from such nonspecific conditions as injury, aneurysms, or congenital defects, and rarely, from blood vessel tumors.

Nerve tract degeneration. An ill-defined group of disorders characterized by degeneration of nerve tracts or myelin sheaths of the cord is found more often than one would suspect. In these, some unknown or poorly understood mechanism causes the deterioration of cells and fibers so that function is altered, then lost, and the nervous tissue is either replaced by a scar or a softening or cystlike area remains. Multiple sclerosis, combined degeneration associated with pernicious anemia, Parkinson's disease, postinfectious encephalomyelitis, and syringomyelia are examples. *See* MULTIPLE SCLEROSIS; PARKINSON'S DISEASE.

Drugs. Finally, the effect of drugs or chemicals and metabolic disorders is often seen as cord depression or stimulation with variable prognoses and permanence of effects. These include alcoholism, carbon monoxide poisoning, lead intoxication, vitamin deficiency, and certain familial diseases. *See* SPINAL CORD. [E.G.ST.]

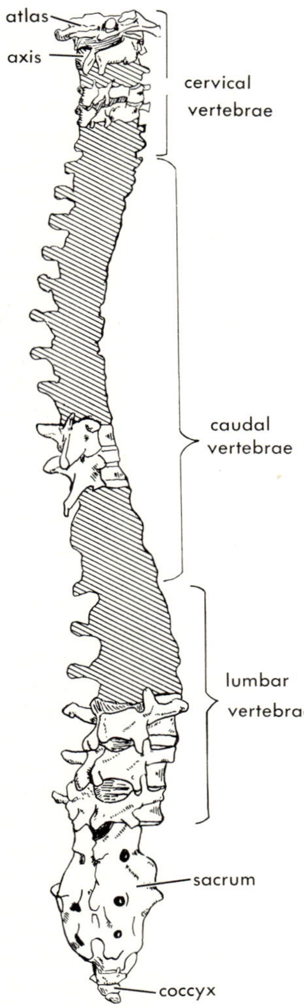

Three-quarter back view of the vertebral column. (From W. T. Foster, *Anatomy*, Foster Art Service)

Spine

The backbone, or vertebral column, composed of 33 vertebrae in man. These include 7 cervical (neck), 12 thoracic, 5 lumbar, 5 sacral, and usually, 4 coccygeal vertebrae. The cervical segments curve forward, while the thoracic segments curve backward and also articulate with the 12 pairs of ribs. The heavy lumbar vertebrae of the loin are supported by the fused sacral bones which form the rear wall of the pelvis. The coccyx consists of a small, curved group that forms a semiflexible "tail" at the base of the spine. Despite common general features, the vertebrae vary in shape and size. Their thick anterior bodies form the vertebral column and are separated by cartilaginous intervertebral disks.

The spine is the characteristic structure of all vertebrates, but wide variations in numbers of vertebrae and development of particular regions are common. Most classes, however, present a fairly typical pattern for that group of animals. *See* VERTEBRA. [E.G.ST.]

Spinel (chemical)

The mineral $MgAl_2O_4$, which is considered to be a mixed oxide, $MgO \cdot Al_2O_3$. The term spinel is also used to refer to other mixed oxides of bi- and trivalent metals, $M^{II}O \cdot M_2^{III}O_3$. M^{II} may be Zn, Mn, Fe, Ni, Co, or Cd and M^{III} may be Al, Co, Fe, Cr, or Ga.

Spinel, $MgAl_2O_4$, is found in various colors including red and black. Some of the red variety called ruby spinels, are of gemstone quality. Spinel has a specific gravity of 3.6 and hardness of 8 on the Mohs scale. Synthetic spinels are manufactured as gemstones, bearings for instruments, and refractory materials. *See* SPINEL (MINERAL). [E.E.WR.]

Spinel (mineral)

A mineral with composition essentially $MgAl_2O_4$ (magnesium and aluminum oxides) crystallizing in the isometric system, usually in octahedrons. It also is found in twinned octahedrons (spinel twins). The hardness is 8 and the specific gravity varies from 3.5 to 4.1, depending on the composition. Its luster is vitreous, and the color may be

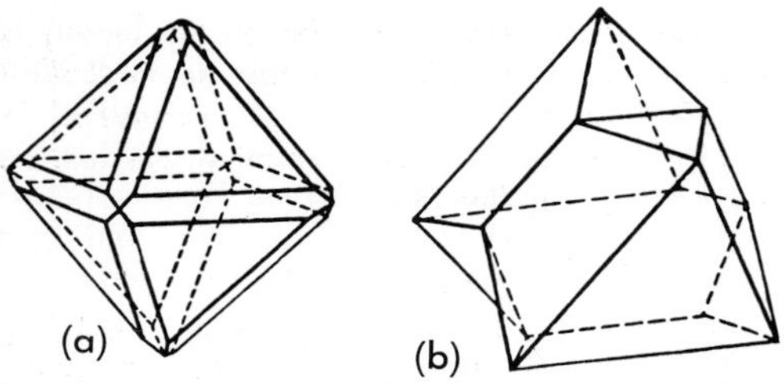

Crystals of spinel. (*a*) Dodecahedron with small truncations. (*b*) Twinned octahedrons or spinel twins. (*From C. S. Hurlbut, Jr., Dana's Manual of Mineralogy, 16th ed., Wiley, 1952*)

white, red, purple, blue, green, brown, or black.

Pure spinel, $MgAl_2O_4$, is rare, and there is usually considerable substitution of ferrous iron and manganese for magnesium and of ferric iron and chromium for aluminum. Ruby spinel is the red, nearly pure magnesium variety; pleonaste is the dark green to black iron spinel; and picotite is the yellow to greenish-brown chrome spinel.

Spinel occurs as an accessory mineral in some dark igneous rocks but is more common as a metamorphic mineral in crystalline limestones and gneisses. It also forms as a contact metamorphic mineral associated with phlogopite, pyrrhotite, chondrodite, and graphite. Transparent and fine colored spinel, most of which is found in rolled pebbles in stream gravels in Ceylon, Siam, Burma, and Madagascar, is used as a gem stone. Gem spinel is manufactured in the same manner as corundum by the Verneuil process. *See* GEM; GEM, MANUFACTURED. [C.S.HU.]

Spinning (textile)

The making of thread or yarn for weaving and sewing from any of several fibrous materials by drawing the fibers out of the carded (cleaned and untangled) mass of raw material and attenuating and twisting them for strength, evenness, and fineness. Spinning with a single spindle and wheel was essentially unchanged until the eighteenth century, when Richard Arkwright developed the water-twist or jack frame (1769) for continuous spinning; James Hargreaves, the jenny (1767) for operating many spindles at once; and Samuel Crompton, the mule (1779) for fine spinning. After the introduction of ring spinning and the traveler by an American, John Thorp, in the 1840s, and of electric power operation a few decades later, the development of modern high-speed automatic spinning machines was swift and dramatic. Since 1950, centrifugal pot spinning has sped up the process greatly by eliminating the need for spindles, rings, travelers, and bobbins in many applications. *See* TEXTILE. [G.CO.]

Spinning of metal

A production technique for shaping and finishing metal. In the spinning of metal, a sheet is rotated and worked by a round-ended tool, controlled manually or mechanically. The sheet is formed over a mandrel.

Spinning operations are usually carried out on a special rigid lathe fitted only with a driving headstock, tail spindle, and tool rest. Surface speeds of 500–5000 ft/min are used, depending on material and diameter. The work is rubbed with soap, lard, or a similar lubricant during working. The operation can be set up quickly, and thus is desirable for short runs or for experimental units subject to change. Spinning may serve to smooth wrinkles in drawn parts, provide a fine finish, or complete a forming operation as in curling an edge of a deep-drawn part. Other operations include smoothing, necking, bulging, burnishing, beading, and trimming. Spun products range from precision reflectors and nose cones to kitchen utensils. Such materials as steel, aluminum, copper, and their softer alloys are spun in thicknesses up to ⅛ in. It may be necessary to anneal the metal during the spinning. *See* SHEET METAL FORMING. [R.L.F.]

Bibliography: ASTE Handbook Committee, *Tool Engineers Handbook*, 1949.

Spinor

A complex vector whose transformations, in a two-dimensional space, are intimately connected with three-dimensional rotations in physical space. More precisely, to every three-dimensional rotation R there corresponds a spinor transformation S, such that if $R_1R_2 = R_3$, $S_1S_2 = S_3$; that is spinors yield a two-dimensional representation of the group of three-dimensional rotations.

Spinor representations are two-valued, namely after rotation through an angle 2π about the z axis, (which in physical space is equivalent to no rotation at all) a spinor is transformed into its negative. As a result physically measurable quantities (which cannot change if there has been no rotation) never transform like spinors or like odd powers of spinors, but may transform like even powers of spinors. For instance the two-component wave function of a particle of spin 1/2 (in units of $h/2\pi$) transforms like a spinor and is not directly observable; its square, which represents the probability density, is. In fact from even powers of spinors expressions can be constructed which transform like any of the directly observable scalars, vectors, tensors, etc. of classical physics. [E.G.]

Bibliography: W. T. Payne, Elementary spinor Theory, *Am. J. Physics*, vol. 20(5):253–262, 1952.

Spinulosa

An order of Asteroidea in which pedicellariae rarely occur and are never of the crossed type. The marginal plates bounding the arms and disk are small and inconspicuous. Papulae are usually present on both the upper and lower surfaces. On the upper side the spines usually occur in groups and are not surrounded by wreaths of pedicellariae (see illustration). The tube-feet have suckers and

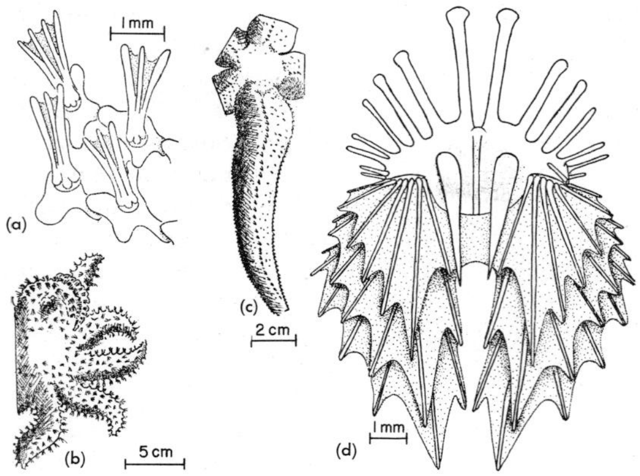

Spinulosa. (*a*) Clusters of spinules on the loosely arranged plates of the aboral surface in *Peribolaster lictor*. (*b*) *Crossaster japonicus*. (*c*) *Echinaster farquhari*. (*d*) Webbed adambulacral spine fans and oral spines of *Pteraster bathami*.

usually lie in two longitudinal rows along the ambulacral groove. The order includes 11 families in existing seas at all depths except the ultra-abyssal. The best-known groups are the sun stars (Solasteridae), the starlets (Asterinidae), and the deep-water Pterasteridae. The latter have webbed spine fans and carry the brood in a pouch on the upper surface of the body. *See* ASTEROIDEA. [H.B.F.]